Building Up Efficient and Sustainable Transport Infrastructure 2017 (BESTInfra 2017)

IOP Conference Series: Materials Science and Engineering
Volume 236

Prague, Czech Republic
21 – 22 September 2017

ISBN: 978-1-5108-4810-8
ISSN: 1757-8981

Printed from e-media with permission by:

Curran Associates, Inc.
57 Morehouse Lane
Red Hook, NY 12571

Some format issues inherent in the e-media version may also appear in this print version.

Printed by Curran Associates, Inc. (2017)

For permission requests, please contact the Institute of Physics
at the address below.

Institute of Physics
Dirac House, Temple Back
Bristol BS1 6BE UK

Phone: 44 1 17 929 7481
Fax: 44 1 17 920 0979

techtracking@iop.org

Additional copies of this publication are available from:

Curran Associates, Inc.
57 Morehouse Lane
Red Hook, NY 12571 USA
Phone: 845-758-0400
Fax: 845-758-2633
Email: curran@proceedings.com
Web: www.proceedings.com

Table of contents

Volume 236

Building up Efficient and Sustainable Transport Infrastructure 2017 (BESTInfra2017)

21–22 September 2017, Prague, Czech Republic

Accepted papers received: 4 September 2017
Published online: 15 September 2017

Railways

Tunnels

Use of Secondary and Recycled Materials

Safety of Transport Infrastructure

Diagnostics of Transport Infrastructure

Management of Transport Infrastructure

IOP Conf. Series: Materials Science and Engineering **236** (2017) 011001 doi:10.1088/1757-899X/236/1/011001

Preface

The world is continuously on the move, the mobility is understood as one of key freedoms and as a crucial condition for the economic growth and social development. It is hard to imagine a sector more important for the everyday life of all the people than transport infrastructure. In the EU's Europe 2020 growth strategy, under the flagship initiative "Resource Efficient Europe", the need to "develop smart, upgraded and fully interconnected transport infrastructures" is defined as one of the main tasks for the member states within the following years.

The transport infrastructure structures are designed for many decades ahead, therefore our today's decisions and outputs will affect several future generations. At the same time, infrastructures embody huge volumes of non-renewable resources and the efficient use and management of these resources is therefore more than important. Research and innovations play a significant role in the responsible attitude to planning. We have to search for new technical and management approaches that will allow for savings of raw materials and energy, as well as for solutions that are smart, durable, friendly to the environment and cost-efficient in long-term prospects.

The intention of the Building up Efficient and Sustainable Transport Infrastructure 2017 (BESTInfra 2017) international conference is to enable the transfer of the latest ideas regarding the transport infrastructure design, construction, operation and maintenance between the industrial, public and academic sector, between various specializations and across countries. By bringing together experts from various fields, the organizers would like to contribute to the implementation of the development of the transport infrastructure sector.

It is a great pleasure for me to present to you the Proceedings of BESTInfra 2017. You can find more information about the conference at www.bestinfra.cz.

Alena Kohoutková

Head of the Centre for Effective and Sustainable Transport Infrastructure
Dean of the Faculty of Civil Engineering, Czech Technical University in Prague

IOP Conf. Series: Materials Science and Engineering **236** (2017) 011001 doi:10.1088/1757-899X/236/1/011001

Organizing Institutions

The conference was organized by the Centre for Effective and Sustainable Transport Infrastructure (CESTI) and the Faculty of Civil Engineering, Czech Technical University in Prague. CESTI is a research center focused on technical innovations of transport infrastructure. CESTI joins important technical universities, construction companies, designers, consultants and laboratories involved in the field of transport infrastructure in the Czech Republic.

Centre for Effective and Sustainable Transport Infrastructure (CESTI)	www.cesti.cz
Czech Technical University in Prague, Faculty of Civil Engineering	www.fsv.cvut.cz
Brno University of Technology, Faculty of Civil Engineering	www.fce.vutbr.cz/en
Technical University of Ostrava, Faculty of Civil Engineering	www.fast.vsb.cz/en
Transport Research Centre	www.cdv.cz/en
Skanska a.s.	www.skanska.cz/en
Metrostav a.s.	en.metrostav.cz/en
DT - Výhybkárna a strojírna, a.s.	www.dtvm.cz/dtvs/en
VPÚ DECO PRAHA a.s.	www.vpupraha.cz

IOP Conf. Series: Materials Science and Engineering **236** (2017) 011001 doi:10.1088/1757-899X/236/1/011001

EUROVIA CS, a.s.	www.eurovia.cz/en/home
SMP CZ, a.s.	www.smp.cz
HOCHTIEF CZ a.s.	www.hochtief.cz/en
TOTAL ČESKÁ REPUBLIKA s.r.o.	www.total.cz
KOLEJCONSULT & servis, spol. s r.o.	www.kcas.cz
3G Consulting Engineers s.r.o.	www.3-g.cz/en
Pontex, spol. s r.o. (Pontex Consulting Engineers, Ltd.)	www.pontex.cz/profile
GEOSTAR, spol. s r.o.	www.geostar.cz/en
CONSULTEST s.r.o.	www.consultest.cz
Valbek, spol. s r.o.	www.valbek.eu/en
VIAKONTROL, spol.s r.o.	www.viakontrol.cz

Supporting Institutions

The conference was organized under the auspices of Minister of Transport of the Czech Republic Mr. Dan Ťok, COST TU 1406, Plataforma Tecnológica Ferroviaria Española (PTFE) and Technology Agency of the Czech Republic (TA ČR).

Ministry of Transport	Ministry of Transport of the Czech Republic	www.mdcr.cz
TU1406 COST ACTION	COST TU 1406: Quality specifications for roadway bridges, standardization at a European level (BridgeSpec)	www.tu1406.eu
PTFE Plataforma Tecnológica Ferroviaria Española	Plataforma Tecnológica Ferroviaria Española (PTFE)	www.ptferroviaria.es/en
T A Č R Technology Agency of the Czech Republic	Technology Agency of the Czech Republic (TA ČR)	www.tacr.cz/en

Media and Other Partners

BETON	Časopis BETON – technologie, konstrukce, sanace (BETON TKS)	www.betontks.cz
ČESKÁ TUNELÁŘSKÁ ASOCIACE ITA AITES CZECH TUNNELLING ASSOCIATION	Czech Tunnelling Association ITA-AITES	www.ita-aites.cz/en
ATENA Reinforced Concrete Modeling	Červenka Consulting s.r.o.	www.cervenka.cz
cpp conference partners prague	Conference Partners Prague s.r.o.	www.conferencepartners.cz/en/

IOP Conf. Series: Materials Science and Engineering **236** (2017) 011001 doi:10.1088/1757-899X/236/1/011001

Scientific Committee

Dr. Angelos Amditis (Greece)
Dr. Petr Bílý (Czech Republic)
Prof. Zdeněk Bittnar (Czech Republic)
Dr. Daniel Castro-Fresno (Spain)
Dr. David Fletcher (Great Britain)
Dr. Ing. Ulf Gerber (Germany)
Dr. Thierry Goger (France)
Dipl.-Ing. Dr. Andreas Goricki (Austria)
Dr. Umberto Guida (Italy)
Prof. Meor Othman Hamzah (Malaysia)
Dr. Leoš Horníček (Czech Republic)
Prof. Libor Ižvolt (Slovakia)
Dr. Piotr Jaskula (Poland)
Dr. Noemi Jiménez-Redondo (Spain)
Dr. Rafiq Kakar (Pakistan)
Prof. Alena Kohoutková (Czech Republic)
Prof. Stjepan Lakusic (Croatia)
Mgr. Roman Ličbinský (Czech Republic)
Dr. Evangelos Manthos (Greece)
Prof. Jose C. Matos (Portugal)
Dr. Ing. Konrad Mollenhauer (Germany)
Ing. Carlo Polidori (Italy)
Dr. Otto Plášek (Czech Republic)
Dr. Jan Pruška (Czech Republic)
Dr. Tomáš Rotter (Czech Republic)
Dr. Pavel Ryjáček (Czech Republic)
Dr. Dharamveer Singh (India)
Prof. Alfred Strauss (Austria)
Dr. Josef Stryk (Czech Republic)
Dr. Vít Šmilauer (Czech Republic)
Dr. Jan Valentin (Czech Republic)
Dr. Wim Van den Bergh (Belgium)
Dr. Christiane Weise (Germany)
Dr. Martins Zaumanis (Latvia)

Organizing Committee

Dr. Petr Bílý (Czech Republic)
Prof. Zdeněk Bittnar (Czech Republic)
Prof. Alena Kohoutková (Czech Republic)
Dr. Marcela Pavlíková (Czech Republic)
Dr. Otto Plášek (Czech Republic)
Dr. Jan Pruška (Czech Republic)
Dr. Pavel Ryjáček (Czech Republic)
Alexandra Sternberg (Czech Republic)
Dr. Jan Valentin (Czech Republic)
Martin Zein (Czech Republic)

IOP Conf. Series: Materials Science and Engineering **236** (2017) 011001 doi:10.1088/1757-899X/236/1/011001

Conference Paper Review Process

All papers published in this volume of *IOP Conference Series: Materials Science and Engineering* have been peer reviewed through processes administered by the Editors. Reviews were conducted by expert referees listed below to the professional and scientific standards expected of a proceedings journal published by IOP Publishing. Conference organizers would like to thank all the reviewers for their time and effort.

Dr. Tomáš Apeltauer (Czech Republic)
Dr. Fátima Batista (Portugal)
Dr. Petr Bílý (Czech Republic)
Prof. Zdeněk Bittnar (Czech Republic)
Ing. Rut Bízková (Czech Republic)
Dr. Silvia Capayová (Slovakia)
Dr. Andrew Dawson (Great Britain)
Dr. Daniela Ďurčanská (Slovakia)
Ing. Tomáš Ebermann (Czech Republic)
Dr. Jana Frankovská (Slovakia)
Dr. Ing. Ulf Gerber (Germany)
Prof. Meor Othman Hamzah (Malaysia)
Prof. Matouš Hilar (Czech Republic)
Dr. Leoš Horníček (Czech Republic)
Dr. Jan Hradil (Czech Republic)
Dr. Vladislav Hrdoušek (Czech Republic)
Dr. Eduard Hromada (Czech Republic)
Dr. Eva Hrubešová (Czech Republic)
Dr. Petr Hýzl (Czech Republic)
Dr. Piotr Jaskula (Poland)
Dr. Noemi Jiménez-Redondo (Spain)
Ing. Milan Kalný (Czech Republic)
Dr. Rafiq Kakar (Pakistan)
Prof. Jozef Komačka (Slovakia)
Dr. Vítězslav Křivánek (Czech Republic)
Prof. Stjepan Lakusic (Croatia)
Mgr. Roman Ličbinský (Czech Republic)
Ing. Martin Lipl (Czech Republic)
Dr. Evangelos Manthos (Greece)
Dr. Soňa Masarovičová (Slovakia)
Prof. Jose C. Matos (Portugal)
Dr. Radek Matula (Czech Republic)
Dr. Ciaran McNally (Ireland)
Dr. Lumír Míča (Czech Republic)
Dr. Monika Mitew-Czajewska (Poland)
Dr. Ing. Konrad Mollenhauer (Germany)
Dr. Petr Mondschein (Czech Republic)
Ing. Jakub Motl (Czech Republic)
Dr. Leoš Pelikán (Czech Republic)
Dr. Otto Plášek (Czech Republic)
Ing. Barnabas Polák (Czech Republic)
Ing. Carlo Polidori (Italy)
Dr. Jan Pruška (Czech Republic)
Dr. Tomáš Rotter (Czech Republic)
Dr. Alexandr Rozsypal (Czech Republic)
Dr. Pavel Ryjáček (Czech Republic)
Dr. Dharamveer Singh (India)
Prof. Jaroslav Smutný (Czech Republic)
Prof. Alfred Strauss (Austria)
Dr. Josef Stryk (Czech Republic)
Prof. Jiří Šejnoha (Czech Republic)
Dr. Vít Šmilauer (Czech Republic)
Ing. Pavel Šourek (Czech Republic)
Dr. Radek Štefan (Czech Republic)
Dr. Jan Tichý (Czech Republic)
Dr. Jan Valentin (Czech Republic)
Dr. Wim Van den bergh (Belgium)
Dr. Michal Varaus (Czech Republic)
Prof. Jan L. Vítek (Czech Republic)
Dr. Lukáš Vráblík (Czech Republic)
Dr. Ing. Christiane Weise (Germany)
Dr. Martins Zaumanis (Latvia)
Ing. Tomáš Zítko (Czech Republic)

Peer review statement

All papers published in this volume of *IOP Conference Series: Materials Science and Engineering* have been peer reviewed through processes administered by the proceedings Editors. Reviews were conducted by expert referees to the professional and scientific standards expected of a proceedings journal published by IOP Publishing.

IOP Conf. Series: Materials Science and Engineering **236** (2017) 012001 doi:10.1088/1757-899X/236/1/012001

Towards improved correlations between bitumen properties and rutting resistance of bituminous mixtures – FunDBitS literature review

Fátima A Batista[1], Bernhard Hofko[2], Joëlle De Visscher[3], Tine Tanghe[3] and Margarida Sá da Costa[1]

[1]Laboratório Nacional de Engenharia Civil, Av. do Brasil, 1750-434 Lisboa, Portugal
[2]Technische Universitaet Wien, Karlsplatz 13, Wien 1040, Austria
[3]Belgian Road Research Centre, Fokkersdreef 21, 1933 Sterrebeek, Belgium

E-mail: xana@lnec.pt

Abstract. Bitumen is the most common binder used in the production of bituminous mixtures for road paving, having a fundamental influence on the performance of the pavement. However, it is not yet fully understood which bitumen properties have a significant influence on the behaviour of bituminous mixtures. In this context, the FunDBitS project was developed. Its main objective was to prepare recommendations on the properties of the bitumen to be specified in order to obtain suitable bituminous mixtures, namely in what concerns to its resistance to permanent deformation (rutting), stiffness, low temperature cracking, fatigue cracking and binder / aggregate interaction. In this framework, the research studies that have become internationally available since the BiTVal project were reviewed in order to assess performance-based bitumen characteristics, which may be introduced into bitumen specification standards. This paper specifically presents the main conclusions regarding the properties of bitumen related to the behaviour to permanent deformation of bituminous mixtures. It was concluded that the most promising test is the "non-recoverable creep compliance" (J_{nr}) from the Multiple Stress Creep and Recovery (MSCR) test method, although Zero/Low Shear Viscosity (ZSV/LSV) by creep or oscillation test method can also give good correlations with permanent deformation of bituminous mixtures.

1. Introduction

Bitumen is the most common binder used in the production of bituminous mixtures for road paving, having a fundamental influence on the performance of the pavement. However, it is not yet fully understood which bitumen properties have a significant influence on the behaviour of bituminous mixtures. In order to improve this understanding, it was recently developed the FunDBitS – "Functional Durability Related Bitumen Specification" project (funded by CEDR), that intended to forward the BiTVal – "Bitumen Test Validation" project (set up by FEHRL) carried out in the first decade of 2000 [1].

Within the scope of FunDBitS project, the data that has become internationally available since the BiTVal project was reviewed in order to assess performance-based bitumen characteristics which may be introduced into bitumen European specifications EN 12591 (paving grade bitumens), EN 14023 (polymer modified bitumens) and EN 13924 (hard paving grade bitumens). Through the literature review, possible correlations between the bitumen and asphalt properties were identified [2]. For this, the following five main asphalt properties were considered: permanent deformation (rutting); stiffness; low temperature cracking; fatigue cracking and binder/aggregate interaction.

This paper deals specifically with the properties of bitumen related to the behaviour to permanent deformation of bituminous mixtures. Therefore, it will address the main conclusions and recommendations related with that asphalt performance characteristics, resulting from FunDBitS project [3, 4].

2. Test methods to address bitumen properties and asphalt behaviour in terms of permanent deformation (rutting)

Permanent deformation or rutting is the result of visco-plastic deformations in the asphalt layer caused by the repeated passage of heavy vehicles, particularly under low speed traffic and high temperature conditions.

Due to the viscoelastic nature of bituminous bound materials, the permanent deformation mode of deterioration strongly depends on the temperature. For this reason, most test methods to assess both bitumen properties and asphalt behaviour in terms of permanent deformation are performed at elevated temperatures. At an early stage of FunDBitS project [2], the following relevant bitumen properties for elevated service temperatures were identified: viscosity; softening point; elastic and recovery properties; complex modulus and phase angle; performance grading. Furthermore, a review on the available test methods to measure the referred bitumen properties was carried out, as follows:

- Viscosity: Capillary Viscometer Test; Coaxial Cylinder Viscosity (CCV) Test; Cone and Plate Viscosity Test; Creep Zero/Low Shear Viscosity (ZSV/LSV) Test and Oscillation Zero/Low Shear Viscosity (ZSV/LSV) Test.
- Softening point: Ring and Ball (R&B) Test method
- Elastic and recovery properties: Multiple Stress Creep and Recovery (MSCR) Test; Elastic Recovery Test.
- Complex modulus and phase angle: Dynamic Shear Rheometer (DSR) Test.
- Performance Grade (PG) classification: Superpave high temperature parameter

As regard the relevant asphalt tests on permanent deformation, they are as follows:

- in Europe: Wheel Tracking test, WTT (EN 12697-22); Cyclic compression test, CCT (EN 12697-25)
- other used tests: SUPERPAVE shear tester, SST; Simple Performance Tests, SPT (e.g. Dynamic modulus test; Flow Number; Flow Time)

In the following sections, an analysis of the possible relationship between bitumen relevant properties and the permanent behaviour of bituminous mixtures is presented.

3. Evaluation of the relationship between bitumen properties and behaviour of permanent deformation of bituminous mixtures

3.1. Bitumen viscosity

In respect to possible correlations between bitumen viscosity properties and bituminous mix permanent deformation only a limited number of relevant studies was found, particularly on Zero/Low Shear Viscosity (ZSV/LSV) either by creep and oscillation tests, according to what is synthesized in Table 1.

The data presented in Table 1. shows that in many cases the correlation between different test methods to address the resistance of mixtures to permanent deformation and viscosity tests on binder is above 0.8 and therefore good. However, in four out of 17 studied possible correlations, the coefficient of determination is below this value. These negative cases happened when the correlation analysis included samples with unmodified and polymer modified bitumen. Nevertheless, some good correlations are also obtained when testing both unmodified and modified bitumen binders.

Table 1. Correlations found between binder viscosity and bituminous mixes permanent deformation.

Paper	Correlated data	Type of correlation	Data sets	Constant *a*	Constant *b*	R^2	*Comment*
[8] (Robertus *et al.*, 2012)	WTT rut rate (mm/s) *vs.* ZSV (kPa*s)	Power (y= a.x^b)	14	N/A	N/A	0.93	U / UA
			30	N/A	N/A	0.49	U & PMB
[9] (Morea, 2012)	WT rut rate (µm/min) vs. LSV (Pa*s)	y=a+b/x	29	1.06	2287.9	0.87	U & PMB / UA & A
				2.04	4494.5	0.86	
[10] (Gungor & Sağlik, 2012)	Axial deformation (mm) TCCT *vs.* ZSV (Pa*s)	Power (y= a.x^b)	12	2.7886	-0.3368	0.92	U & PMB / A (TCCT@40°C)
			12	4.0315	-0.3606	0.84	U & PMB / A (TCCT@50°C)
[12] (Guericke & Schlame, 2008)	HWT rut depth (mm) *vs.* ZSV (Pa*s)	Power (y= a.x^b)	25	N/A	N/A	081	U & PMB / A (WTT at +40°C)
			6	N/A	N/A	0.91	U & PMB / A (WTT at +50°C)
			11	N/A	N/A	0.91	U & PMB / A (WTT at +60°C)
[16] (De Visscher & Vanelstraete, 2009)	PRD (%) *vs.* EVT1 (°C)	Linear (y= a.x + b)	6	-0.61	42.69	0.86	U
			11	-0.60	42.12	0.83	U & PMB
	PRD (%) *vs.* EVT2 (°C)	Linear (y= a.x + b)	6	-0.61	42.71	0.85	U
			11	-0.23	22.4	0.49	U & PMB
	Creep rate (µm/m/n) *vs.* EVT1	Linear (y= a.x + b)	6	-0.01	0.97	0.83	U
			11	-0.01	1.06	0.77	U & PMB
	Creep rate (µm/m/n) *vs.* EVT2	Linear (y= a.x + b)	6	-0.01	0.97	0.83	U
			11	-0.01	0.71	0.76	U & PMB

LEGEND: U – Unmodified bitumen; PMB – Polymer modified bitumen; UA – Unaged bitumen; A – Aged bitumen; PRD – Percentage of rut deformation in wheel tracking tests (WTT); ZSV – Zero shear viscosity; EVT1 – Equiviscous temperature 1; EVT2 – Equiviscous temperature 2; HWT –Hamburg wheel tracking test; TCCT – Triaxial cyclic compression test..

The following conclusions from the presented studies can be drawn:

- Studies on creep Zero/Low Shear Viscosity (ZSV/LSV) show that correlation between binder viscosity and wheel tracking parameter is good only when non modified binders are used. However, when the axial strain from Triaxial Cyclic Compression Tests (TCCT) is linked to ZSV/LSV, a good correlation can be achieved even when modified binders are taken into consideration:
- Studies on oscillation Low Shear Viscosity reveal that good correlations could be achieved when: the binder viscosity is linked with the rut depth (WT) and the equiviscous temperature 1 (EVT1, temperature at which the viscosity measured at very low shear rate is 2000 Pa.s) is linked to the proportional rut depth (WT) and the creep rate (TCCT).

3.2. Bitumen softening point

As regards the softening point, the Ring and Ball (R&B) test is considered as a traditional test method with a large background in data. However, it is of general understanding that this test is not suitable for modified bituminous binders. Correlation analysis between bitumen softening point and bituminous mix testing was found in seven of the reviewed papers (Table 2.), six of them correlating R&B results with wheel tracking test results and one with triaxial cyclic compression test results. It was concluded that some studies find a reasonable correlation between R&B and wheel tracking results, even when polymer modified binders are used for testing. However, these studies are generally limited to samples from the same binder source, i.e. one unique unmodified bitumen-base binder was used to produce the other polymer modified samples. In fact, poor correlations are found when a mix of unmodified and modified binders that are not from the same base bitumen is used.

Table 2. Correlations found between binder R&B and bituminous mixes permanent deformation.

Paper	Correlated data	Type of correlation	Data sets	Constant *a*	Constant *b*	R^2	*Comment*
[6] (Eckmann *et al.*, 2012)	RD (mm) *vs.* R&B (°C)	Linear (y= a + b x)	4	-0.2323	18.695	0.82	U (20/30 pen grade) + PMB
			4	-0.1221	11.642	0.99	U (35/50 pen grade) + PMB
[8] (Robertus *et al.*, 2012)	WTS (mm/s) *vs.* R&B (°C)	Linear (y= a + b x)	19	N/A	N/A	0.68	U + PMB
			7	N/A	N/A	0.95	U
[12] (Guericke & Schlampe, 2008)	RD (mm) *vs.* R&B (°C)	Linear (y= a + b x)	11	0.5893	46.523	0.84	U + PMB
[15] (Dreessen & Pascal, 2009)	R&B (°C) *vs.* RD (mm)	Logarithmic (y= a ln x + b)	13	-22.635	91.949	0.60	U + PMB
[18] (Tusar *et al.*, 2009)	RD (mm) *vs.* R&B (°C)	Logarithmic (y= a ln x + b)	7	-2.702	12.916	0.91	U + PMB
[20] (Renken, 2012)	RD (mm) *vs.* R&B (°C)	No correlation found	N/A	N/A	N/A	N/A	N/A
[21] (Reyes-Lizcano *et al.*, 2009)	Accumulated axial strain (%) *vs.* R&B (°C)	Linear (y= a + b x)	9	-0.0188	1.9494	0.54	PMB

LEGEND: U – Unmodified bitumen; PMB – Polymer modified bitumen; R&B – Ring and ball softening point; RD – Rut depth in wheel tracking tests (WTT); WTS – Wheel tracking slope in wheel tracking tests (WTT).

3.3. Bitumen elastic and recovery properties

With regard to the elastic and recovery properties, no relevant papers were found from the literature database on the Elastic Recovery test. On the other hand, a significant number of papers on Multiple Stress Creep and Recovery (MSCR) had been identified from the literature database [1], but only a limited number of them discuss correlations with asphalt permanent deformation (Table 3.). In general, these studies report quite good correlations (often at higher stress levels) for both paving grade and polymer modified binders. It is worth noticing that in the studies where negative cases occur ($R^2 < 0.7$), also quite good correlations where obtained using the same set of binders but varying the parameters that are correlated, the MSCR test conditions or the ageing of the tested binders. Most studies showed a very good correlation between non-recoverable creep compliance (J_{nr}) and bituminous mixture testing to address permanent deformation behaviour (especially for wheel tracking tests) and this for all types of binder. In fact, only one study (addressed in papers [7], [15] & [17]) obtained a lower correlation ($R^2 = 0.44$), but this was improved by using higher stress levels ($R^2 = 0.77$). The other studies found good correlations even at 'normal' stress levels (like $\tau = 3.2$ kPa, the stress level prescribed in the new EN 16659). It seems that better correlations are obtained when binders are tested after short-term ageing (e.g. after RTFOT) and for higher stress levels (≥ 3.2 MPa) used in the MSCR tests. This is thus recommended to be further investigated in the future.

Table 3. Correlations found between MSCR test results (J_{nr} or % recovery) and bituminous mixes permanent deformation.

Paper	Correlated data	Type of correlation	Data sets	Constant *a*	Constant *b*	R^2	*Comment*
[5] (Dueñas *et al.*,2012)	RT (WTT) *vs.* J_{nr} (T=60 °C; τ=3,2 kPa)	Linear (y= a + b x)	4	N/A	N/A	0.87	U + PMB + Crumb Rubber Modified
	Compliance in CCT *vs.* J_{nr} (T=60 °C; τ=3,2 kPa)	Linear (y= a + b x)	4	N/A	N/A	0.69	U + PMB + Crumb Rubber Modified
	% recovery (CCT) *vs.* %R MSCR (T=60 °C; τ=3,2 kPa)	Linear (y= a + b x)	4	N/A	N/A	0.96	U + PMB + Crumb Rubber Modified
[7] (Dreessen & Gallet, 2012), [15] (Dreessen & Pascal, 2009) & [17] (Dreessen *et al.*, 2009)	RD (WTT – Large SD) *vs.* J_{nr} (T=60 °C; τ=3,2 kPa)	Linear (y= a + b x)	15	N/A	N/A	0.44	7 U + 6 PMB + 2 special A (RTFOT)
	RD (WTT – Large SD) *vs.* J_{nr} (T=60 °C; τ=25,6 kPa)		15	N/A	N/A	0.77	7 U + 6 PMB + 2 special A (RTFOT)
[8] (Robertus *et al.*, 2012)	WTS (WTT – Small SD) *vs.* J_{nr} (T=45 & 60 °C; τ=1 kPa)	log y=a*log x	20	N/A	N/A	0.79	5 U + 11 PMB + 2 wax modified + 2 special UA
						0.90	5 U + 11 PMB + 2 wax modified + 2 special A (RTFOT)
[13] (D'Angelo *et al.*, 2007)	ALF of FHWA *vs.* J_{nr} (T=64 °C; τ=25,6 kPa)	Linear (y= a + bx)	6	-0.112	0.493	0.81	U + PMB
	Field rutting (after 6 yrs) *vs.* J_{nr} (T=64 °C; τ=0,8 kPa)	Linear (y= a + bx)	7	0	0.0116	0.77	U + PMB
[19] (Laukkanen *et al.*, 2014)	WTT large size device *vs.* J_{nr} (τ=3,2 kPa)	Linear (ax+b) rut rate	9	N/A	N/A	0.98	U & PMB / UA
		Parameter b from the power law: linear	9	5.94	0.84	0.98	U & PMB / UA
[22] (Tabatabaee & Tabatabaee, 2010)	Unconfined cyclic creep test *vs.* J_{nr} (τ=3,2 kPa)	Not clear (Linear ?)	6	N/A	N/A	0.83	U & CMB / A
							U.D.C. at 40°C

LEGEND: U – Unmodified bitumen; PMB – Polymer modified bitumen; UA – Unaged bitumen; A – Aged bitumen; CCT – Cyclic Compression Test; WTT- Wheel Tracking Test; T – test temperature; τ – stress level.

3.4. Bitumen complex modulus (G) and phase angle (δ)*

Concerning bitumen binders complex shear modulus (G*) and phase angle (δ) in sinusoidal oscillatory shear mode (using a dynamic shear rheometer), only six papers from the database were found correlating bitumen rheological properties with the asphalt resistance to permanent deformation (Table 4.). The following parameters are generally used in the studies: bitumen complex shear modulus (G*), G*/sin δ characteristic and the unified evaluation index (R_J) (inversely proportional to compliance).

Table 4. Correlations found between binder complex modulus and phase angle and bituminous mixes permanent deformation.

Paper	Correlated data	Type of correlation	Data sets	Constant *a*	Constant *b*	R^2	*Comment*
[8] (Robertus *et al.*, 2012)	WT Rut Rate *vs.* G* (kPa) @ 1,59Hz & 45°C / 60°C	Power ($y= a.x^b$)	14	N/A	N/A	0.94	U / UA
			30	N/A	N/A	<0.70 *(1)*	U & PMB / UA
[10] (Gungor & Sağlik, 2012)	TCCT Deformation (mm) *vs.* G*/sinδ (kPa) @ 1,59Hz & 40°C / 50°C	Power ($y= a.x^b$)	12	14.22	-0.73	0.37	U & PMB / A (TCCT@40°C)
			12	4.69	-0.55	0.40	U & PMB / A (TCCT@50°C)
[11] (Beckedahl *et al.*, 2008)	WT Rut Depth (mm) *vs.* G*/sinδ (kPa) @ 1,59Hz & 60°C *(2) (3)*	Power ($y=a.x^b$)	*3 (3)*	39.39	-0.29	0.93	U & PMB / UA
			3 (4)	20.53	-0.22	0.95	U & PMB / UA
[12] (Guericke & Schlame, 2008)	HWT Rut Depth (mm) *vs.* $T_{(G^*/\sin\delta=2.2kPa)}$ (°C) @ 60°C	Linear	11	N/A	N/A	0.77	U & PMB / A
[14] (Tan *et al.*, 2014)	WT dynamic stability (time/mm) *vs.* R_J @ 1,59Hz & 60°C	Linear (y= a + bx)	7	788.07	323.02	0.99	U & PMB / UA
		Grey relational analysis	5	N/A	N/A	[0.89] *(4)*	U / UA
			2	N/A	N/A	[0.90] *(4)*	PMB / UA
	WT dynamic stability (time/mm) *vs.* G*/sinδ (KPa) @ 1,59Hz & 60°C	Linear (y= a + bx)	7	-298.59	925.37	0.99	U & PMB / UA
		Grey relational analysis	5	N/A	N/A	[0.89] *(4)*	U / UA
			2	N/A	N/A	[0.59] *(4)*	PMB / UA
[15] (Dreessen & Pascal, 2009)	FWT rutting (%) *vs.* G*/sinδ (KPa) @ 1,59Hz & 60°C	Logarithmic (y= a lnx + b)	15	-30.55	59.29	0.27	U & PMB / A

LEGEND: U – Unmodified bitumen; PMB – Polymer modified bitumen; UA – Unaged bitumen; A – Aged bitumen; FWT – French Wheel Tracking Test; HWT - Hamburg Wheel tracking Test; TCCT – Triaxial Cyclic Compression Test; WT – Wheel tracking Test.

NOTES: (1) Paper [8] refers $R^2 < 0,7$ for the correlation of G* and G*/sinδ with WT Rut Rate for both unaged and aged polymer modified binders, but the type of correlation is not clearly identified. However, the paper refers that for rheological complex binders (i.e. most PMBs) there is clearly no correlation between rut rate and G*; (2) Rut depth of Stone Mastic Asphalt (SMA); (3) Ruth depth of Asphalt Binder mixture (ABM); (4) Coefficient from the Grey relational analysis.

The data presented in Table 4. shows that there are a significant number of studies were no correlation or very weak correlation ($R^2 < 0.7$) is obtained when relating G*/sin δ with the asphalt resistance to permanent deformation. In fact, only two papers from the six considered, found a good correlation between the referred properties. In these two cases, a limited number of binders was addressed, since in both of the studies only two modified bituminous binders were considered.

Therefore, most of the papers that determine G*/sin δ, find that it is not suitable to evaluate the asphalt resistance to permanent deformation, when analysing an ensemble of unmodified and polymer modified bituminous binders. Nevertheless, better correlations are achieved for lower frequencies.

3.5. Bitumen performance grade

No relevant papers presenting a clear relationship between the performance grade of the used/recovered binder and the permanent deformation behaviour of bituminous mixtures were found in the database. Nevertheless, it is of common understanding that the high temperature of PG grade is not suitable for ranking the bituminous mixture behaviour to permanent deformation, mainly when modified bitumen binders are used. Therefore, the present PG specification requirements based on the parameter G*/sin δ (at the high temperature grade and a frequency of 10 rad/s / 1.59 Hz) are not suitable for assessing rutting resistance. Nevertheless, new specification requirements to address rutting (e.g. MSCR test parameters) are being considered in the USA having in view to provide better parameters related to asphalt performance characteristics at elevated service temperatures.

4. Conclusions

Among the addressed bitumen properties and its relation with the permanent deformation of bituminous mixtures, the most promising tests providing better correlations are the Zero/Low Shear Viscosity (ZSV/LSV) by creep or oscillation test method [CEN/TS 15324 (Oscillation LSV); CEN/TS 15325 (Creep ZSV)] and the non-recoverable creep compliance (J_{nr}) from MSCR tests [EN 16659].

Comparing both type of tests (Creep or oscillation Zero/Low Shear Viscosity and MSCR tests), it seems that the MSCR test method is more promising in a near future, given that, at the present, it seems to be an easier test method for the laboratories to implement, there is a European standard specifying the test and it is a method preferred in other countries as well, such as USA. From a technical point of view, it is believed that the MSCR test is better than the LSV/ZSV tests, because the MSCR test also run at high stress levels which are more representative for the stresses induced by rutting and it is a repeated test which better simulates the loading due to traffic. Nevertheless, the test conditions (e.g. applied stress level) need to be further investigated in the future in order to get even better results and correlations for all types of binders.

Acknowledgements
The research presented in this paper was carried out as part of the CEDR Transnational Road Research Programme Call 2013. The funding for the research was provided by the national road administrations of Denmark, Germany, Ireland, the Netherlands, the United Kingdom and Slovenia.

References

[1] FERHL 2006 BiTVal - Analysis of Available Data for Validation of Bitumen Tests *Report on Phase 1 of the BiTVal Project* ed. Cliff Nicholls

[2] Nicholls C, Valentin J, Soukupova L, Mollenhauer K, Vansteenkiste S, De Visscher J, Tušar M, Bueche N, Bressi S, Karcher C, Batista F, Sá da Costa M, Malkoc G, Hofko B and Soenen H 2015 Identified correlations between bitumen and asphalt properties *Interim report – Deliverable 1 of FunDBitS project*

[3] Batista F, Sá da Costa M, Hofko B, De Visscher J and Tanghe T 2016 Correlations between bitumen and asphalt properties – Permanent deformation (rutting) *Interim report – Deliverable 2.b of FunDBitS project*

[4] Batista F, Sá da Costa M, Hofko B, De Visscher J and Tanghe T 2016 Proposal for specification requirements - Permanent Deformation (rutting) *Deliverable 3.b of FunDBitS project*

[5] Dueñas A P, Lepe A P, Martinez E M and Ibañez V C 2012 Relationships between zero shear viscosity, low shear viscosity and MSCRT tests and EN 12697-22 rutting test *Proc 5th Eurasphalt & Eurobitume Congress (Istanbul)*

[6] Eckmann B, Mazé M, Largeaud S and Dumont S F 2012 The contribution of cross-linked polymer modified binders to asphalt performance *Proc. 5th Eurasphalt & Eurobitume Congress (Istanbul)*

[7] Dreessen S and Gallet T 2012 MSCRT: Performance related test method for rutting prediction of asphalt mixtures from binder rheological characteristics *Proc. 5th Eurasphalt & Eurobitume Congress (Istanbul)*

[8] Robertus C, van Rooijen R and Thimm L 2012 A comparison of binder tests that relate to asphalt mixture deformation *Proc. 5th Eurasphalt & Eurobitume Congress (Istanbul)*

[9] Morea F 2012 Performance of asphalt mixtures at different temperatures and load, their relation with the asphalt low shear viscosity (LSV) *Proc. 5th Eurasphalt & Eurobitume Congress (Istanbul)*

[10] Gungor A G and Sağlik A 2012 Evaluation of rutting performance of neat and modified binders using zero shear viscosity *Proc. 5th Eurasphalt & Eurobitume Congress (Istanbul)*

[11] Beckedahl H J, Sivapatham P and Neutag L 2008 Impacts of the compaction degree of asphalt mixes on the asphalt pavement performance *Proc. 4th Eurasphalt & Eurobitume Congress (Copenhagen)*

[12] Guericke R and Schlame K 2008 A new "softening-point" based on asphalt pavement performance figures *Proc. 4th Eurasphalt & Eurobitume Congress (Copenhagen)*

[13] D'Angelo J, Kluttz R, Dongré R, Stephens K and Zanzotto L 2007 Revision of the Superpave high temperature binder specification: The Multiple Stress Creep Recovery test *Journal of the Association of Asphalt Paving Technologists* **76** pp 123-62

[14] Tan Y, L Shan L and Li X 2014 Unified evaluation index for high- and low-temperature performance of asphalt binder *Proc. 93rd TRB Annual Meeting (Washington, D.C.)*

[15] Dreessen S and Pascal J P 2009 Seeking for a relevant binder test method for rutting prediction *Proc. 2nd Int. Conf. on Environmentally Friendly Roads ENVIROAD (Warsaw)*

[16] De Visscher J and Vanelstraete A 2009 Equiviscous temperature based on low shear viscosity: Evaluation as binder indicator for rutting and critical discussion of the test procedure *Proc. 7th Int. RILEM Symposium ATCBM (Rhodes)*

[17] Dreessen S, Planche J P and Gardel V 2009 A new performance related test method for rutting prediction: MSCRT *Proc. 7th Int. RILEM Symposium ATCBM (Rhodes)*

[18] Tusar M, Turk M R, Bankowski W, Wiman LG and Kalman B 2009 Evaluation of modified bitumen, high modulus asphalt concrete and steel mesh as materials for road upgrading *Proc. 7th Int. RILEM Symposium ATCBM (Rhodes)*

[19] Laukkanen O-V, Soenen H, Pellinen T, Heyrman S and Lemoine G 2015 Creep-recovery behavior of bituminous binders and its relation to asphalt mixture rutting *Materials and Structures* **48** pp 4039-53

[20] Renken P 2012 Walzasphalte mit viskositätsabsenkenden Additiven-Entwicklung und Optimierung der Erst- und Kontrollprüfungsverfahren und Bestimmung der Einflüsse auf die per-for-mance-orientierten Asphalteigenschaften *AiF research project No. 15589 N (Technische Universität Braunschweig)*

[21] Reyes F A, Lizarazo C M, Figueroa A S, Candia M, Flintsch G and Katicha S 2009 Dynamic Characterization of Hot-Mix Asphalt Mixtures Using Modified and Conventional Asphalts in Colombia *Proc. 88th TRB Annual Meeting (Washington, D.C.)*

[22] Tabatabaee N and Tabatabaee H 2010. Multiple Stress Creep and Recovery and Time Sweep Fatigue Tests: Crumb Rubber Modified Binder and Mixture Performance *Transportation Research Record* **2180** pp 67-74.

Field investigation of low-temperature cracking and stiffness moduli on selected roads with conventional and high modulus asphalt concrete

Józef Judycki[1], Mariusz Jaczewski[1], Dawid Ryś[1], Marek Pszczola[1], Piotr Jaskula[1] and Adam Glinicki[2]

[1]Gdansk University of Technology, Department of Highway Engineering, 11/12 Narutowicza str., 80-233 Gdansk, Poland
[2]General Directorate for National Roads and Motorways, Bialystok Division, 2 Zwycięstwa Str, 15-703 Bialystok, Poland

E-mail: mariusz.jaczewski@wilis.pg.gda.pl

Abstract. High Modulus Asphalt Concrete (HMAC) was introduced in Poland as a one of the solutions to the problem of rutting, type of deterioration common in the 1990s. After first encouraging trials in 2002 HMAC was widely used for heavily loaded national roads and motorways. However some concerns were raised about low-temperature cracking of HMAC. This was the main reason of the studies presented in this article were started. The article presents the comparison of performance of pavements constructed in typical contract conditions with the road bases made of HMAC and conventional asphalt concrete (AC). The field investigation was focused on the number of low-temperature cracks, bearing capacity (based on FWD test) of road sections localized in coldest region of Poland. Also load transfer efficiency of selected low-temperature cracks was assessed. FWD test confirmed lower deflections of pavements with HMAC and two times higher stiffness modulus of asphalt courses in comparison to pavements constructed with conventional AC mixtures. Relation of stiffness of asphalt layers and amount of low-temperature cracks showed that the higher stiffness modulus of asphalt layers could lead to increase of the number of low-temperature cracks. FWD test results showed that the load transfer efficiency of low-temperature cracks on pavements with HMAC presents very low values, very close to lack of load transfer. It was surprising as section with HMAC road base were aged from 2 to 5 years and presented very good bearing capacity.

1. Introduction

1.1. High Modulus Asphalt Concrete in Polish conditions

High Modulus Asphalt Concrete (HMAC) is type of asphalt mixture developed in France [1]. HMAC mixture has several advantages such as higher rutting resistance and improved fatigue life in comparison to typical AC mixtures. HMAC wasintroduced in Poland as a solution to the problem of rutting, type of deterioration common in the 1990s. After first encouraging trials in Poland in 2002, HMAC was widely used for heavily loaded national roads and motorways, especially after encouraging trials and laboratory tests conducted worldwide [1–7]. However between Polish experts some concerns were raised about susceptibility to low-temperature cracking of HMAC. Raised concerns were confirmed after winter of 2012 when very high amount of thermal cracks appeared on constructed and existing sections of roads on which HMAC was used in base and binder courses. This issue was described in details in [8–10]. In this article field test results from the coldest region of Poland, where the highest amount of low-temperature cracks appeared, are presented.

High Modulus Asphalt Concrete (HMAC) that was applied in Poland include some modifications in comparison to its French predecessor. It is more dense mixture (from 2% to 4% voids) and has slightly lesser amount of bitumen (usually around 5%, m/m). High value of stiffness modulus (at least 14 000 MPa in 4PB test at 10°C and 10 Hz) is obtained by using neat 20/30 bitumen. The fatigue characteristics remained unchanged and ε_6 is equal to 130 μstrains.

1.2. Aim and scope of the paper

The main aim of this study was to evaluate the performance of road sections constructed in the coldest area of Poland with the usage of HMAC mixtures in comparison to the typical Polish pavement structure constructed using conventional Asphalt Concrete (AC) mixtures. For this purpose three aspects were evaluated – low-temperature cracking intensity, bearing capacity of the pavement and load transfer of the thermal cracks on selected test sections.

2. Field investigation

2.1. Characterization of the test sections

Field investigation of selected road sections constructed using HMAC and AC mixtures was one of the major part of the wider study [11,12] prepared for the GDDKiA. Totally in the years of 2012-2014 80 road sections located in Poland were surveyed for thermal cracking and rutting of the pavement. 33 of them were constructed using HMAC base and binder course and 47 using conventional AC mixtures. All of selected sections were new pavement structures, aged 1-12 years, in which the lower base course was made using crushed stone. The localisation of all tested sections is presented on the Figure 1. Sections analysed in this paper are marked in circle.

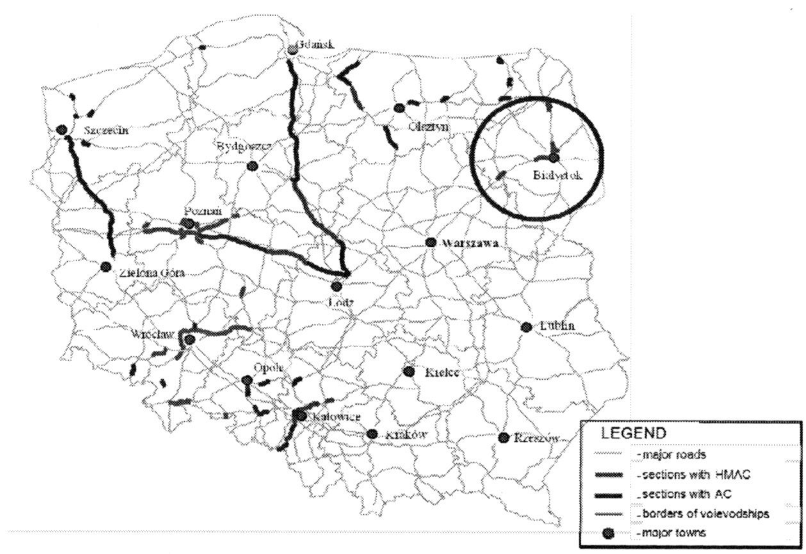

Figure 1. Localisation of tested road sections.

FWD tests were conducted on 3 sections localized along the national road no 8. Due to very heavy transit traffic all of them were dual carriageways. Pavement structures are the same in both cases:

sections with AC and sections with HMAC. The typical construction of the upper layers of road section with HMAC comprised of 27 cm of asphalt layers and 22 cm of crushed stone base course. The construction of section with conventional AC comprised of 28 cm of asphalt layers and 20 cm of crushed stone base course. On all analyzed sections all HMAC layers were produced using hard grade 20/30 neat bitumen. Conventional AC mixture was produced using 35/50 neat bitumen. Below the crushed stone base layer in all sections improved subgrade was constructed using cement bound soils. Detailed description of selected test sections along with the amount of low-temperature cracks is given in Table 1.

2.2. Low-temperature cracks investigation

For each selected road section detailed pavement distress survey was conducted according to the Polish SOSN [13] methodology supplemented by FHWA distress manual [14]. The majority of distresses comprised of transverse cracks (Figure 2 and Figure 3), which in most cases originated from the effect of low-temperatures. Mechanism of development of low-temperature cracks is described in details in [15,16]. Nevertheless other distresses were also observed, but their impact on pavement performance was minor or even negligible. What was surprising on very short sections small ruts in the right wheel path were observed, even on sections constructed using HMAC mixtures.

Figure 2. Not repaired low-temperature crack.

Figure 3. Repaired low- temperature crack.

Table 1. Description of analysed road test sections.

No.	*Description of road section (year of construction)*	*Pavement construction* [a)]	*FWD test section* [b)]	*No. of low-temperature cracks per km*
1	S8 Expressway, Jeżewo – Białystok (2012)	4 cm SMA 23 cm HMAC 23 cm CS	1) 623+000 – 624+000 L 2) 620+000 – 621+000 R	5,3 1,7
2	S8 Expressway, Zambrów Bypass (2012)	4 cm SMA 23 cm HMAC 23 cm CS	3) 3+000 – 4+000 L 4) 1+500 – 2+500 R	0,7 2,3
3	DK8, Białystok – Katrynka (2009)	4 cm SMA 23 cm HMAC 22 cm CS	5) 649+000 – 650+000 R 6) 648+000 – 649+000 L 7) 649+000 – 650+000 L	8,7
4	One of Polish Motorway section (2008/2010)	4 cm SMA 24 cm AC 20 cm CS	8) 25+000 – 29+000 L 9) 83+000 – 87+000 L	1,7 1,2

Remarks: a) SMA – stone matrix asphalt, HMAC – high modulus asphalt concrete, AC – asphalt concrete, CS – crushed stone
b) L – left carriageway, R – right carriageway

2.3. Analysis of the conducted FWD tests

FWD tests were conducted in November 2013 and in October 2014. The average temperatures of asphaltic layers ranged from 8 up to 21°C. Deflection of the pavements with HMAC mixtures were measured on slow traffic line every 25 meters on the selected sections of 1000 meter length. In the case of the motorway structure with conventional AC, the measurements were made every 100 meters on the selected sections of 4000 meter length. Obtained during FWD test pavement deflections were later standardized to the load of 50 kN and the temperature of +20°C.

2.3.1. Load transfer efficiency determination. During determination of deflection using FWD test equipment on two road sections additional test was made, to determine the load transfer efficiency of the low-temperature cracks. Test was conducted on two heaviest cracked sections. The load transfer efficiency factor was calculated using equation (1) given after Polish regulations [17].

$$k = \frac{y_2}{y_1} \qquad (1)$$

where: k – load efficiency factor, y_1 – deflection of the loaded edge, y_2 –. deflection of the unloaded edge.

According to [17] the value of $k < 0,7$ indicates lack of load transfer efficiency, $0,7 < k < 0,95$ indicates partial load transfer efficiency and k > indicates full load transfer efficiency.

2.3.2. Stiffness moduli back-calculation. For every tested road section on the basis of deflection bowls stiffness moduli were calculated using ELMOD6 program using backcalculation methodology described in details in COST 336 Report [18]. The pavement was modelled as three-layered (asphalt layers, crushed stone base and improved subgrade) half-space, characterized by E – elastic modulus, ν - Poisson's ratio and h – thickness of the layer. The thicknesses of the pavement layers were assumed as stated in the Table 1. Poisson's ratios were assumed as either 0,3 for asphalt layers and crushed stone base or 0,35 for improved subgrade. Calculated stiffness moduli were standardized to the temperature of +10°C using equation 2 [17].

$$E_{10} = E_T \times (0.77 + 0.023 \times T) \qquad (2)$$

where: E_{10} – standardized stiffness modulus (+10°C), T – temperature of the test.

3. Obtained results and discussion

Test results of FWD tests are presented in Tables 2, 3 and 4, respectively standardized deflections, load transfer efficiency factors and standardized stiffness moduli. For each parameters mean values and coefficient of variations COV were calculated.

Table 2. Standardized deflection from FWD test results (T = +20°C, 50 kN load).

Property	*Deflections on sections[μm]*								
	S8 Expressway, Jeżewo – Białystok		S8 Expressway, Zambrów Bypass		DK8, Białystok – Katrynka			One of Polish Motorway section	
no of test section:	1	*2*	3	*4*	5	*6*	*7*	*8*	*9*
Mean ($U_{śr}$)	95	94	108	94	95	103	105	139	161
80% percentile	103	106	114	107	103	118	111	155	174
Reliable (U_m)	115	121	137	121	118	130	131	187	195
CoV, %	10	14	13	14	12	13	12	17	10
No. of points	34	34	34	34	34	34	34	40	40

Remarks: a) $Um = U_{śr} + 2\sigma$, where σ - standard deviation
b) CoV – coefficient of variation

All nine tested road sections, both with HMAC courses (after 2 and 5 years of service) and AC courses (after 2 and 6 years of service) showed very high homogeneity of measured deflections along the whole length of the test sections. The coefficients of variation of values of deflections on all tested sections are lower than 20%, what according to COST 336 [18] classifies all testes road sections as very homogeneous.

Table 3. Load transfer efficiency factors "k".

Property	Load transfer efficiency factor "k" [-] No. of the test section, (acc. to table 1)		
	2	*6*	*7*
Mean	0.66	0.73	0.76
CoV, %	29	13	14
No. of points	90	10	30

The load transfer efficiency factors were evaluated only for sections with HMAC layers (see Table 3). On the DK8 national road (no. 6 and 7 test sections; 5 years of service) the results of "k" factor indicate partial load transfer efficiency, but the value is in the lower part of the range, just above the limit of lack of load transfer. On the S8 road (no. 2 test section; 2 years of service) the "k" factor indicates the lack of load transfer. Obtained results are surprising, as all test sections are relatively new and "k" factor should show very high values. The probable reason, responsible for reduction of the "k" values can be higher values of linear coefficient of thermal contraction, due to the usage of higher amount of very hard grade neat bitumen. The lack of load transfer between of low temperature cracks can be big issue in the later service of the road section, as this is the weakest point of the entire pavement, which can cause faster deterioration of the whole pavement [19,20].

Table 4. Standardized back-calculated stiffness moduli (T = +10°C, 50 kN load).

Property	*Stiffness modulus [MPa]*								
	S8 Expressway, Jeżewo – Białystok		S8 Expressway, Zambrów Bypass		DK8, Białystok – Katrynka			One of Polish Motorway section	
no of test section:	1	*2*	3	*4*	5	*6*	*7*	*8*	*9*
Mean	19736	17351	21718	20017	20766	18333	19008	9073	11718
20% percentile	15520	13564	18327	14904	16188	14684	14916	6473	10112
80% percentile	23399	21062	25514	23293	24813	20785	22662	11400	13398
CoV, %	25%	26%	21%	28%	28%	26%	24%	34%	22%
No. of points	34	34	34	34	34	34	34	40	40

20% percentile value in the case of improved subgrade and crushed stone base changed in small range and were similar for every section. Values of elastic modulus ranged from 398 up to 482 MPa in the case of crushed stone base and from 164 up to 259 MPa in the case of improved subgrade.

As for the assessment of the asphaltic layers, apart from mean values also 20% and 80% percentiles were calculated to describe the variability of the obtained stiffness moduli. The results are presented in the Table 4 and in Figure 4. As can be seen, the obtained results for the road sections with HMAC layers differ very much from the values obtained for sections with conventional AC layers. The obtained values for all sections with HMAC layers are two times higher in comparison to the section with AC layers. Additionally in the Figures 5 and 6 authors presented respectively the relation of thermal cracking intensity vs obtained values of stiffness moduli and the relation of thermal cracking intensity vs years of service. Different colors were used to distinguish between different road test sections.

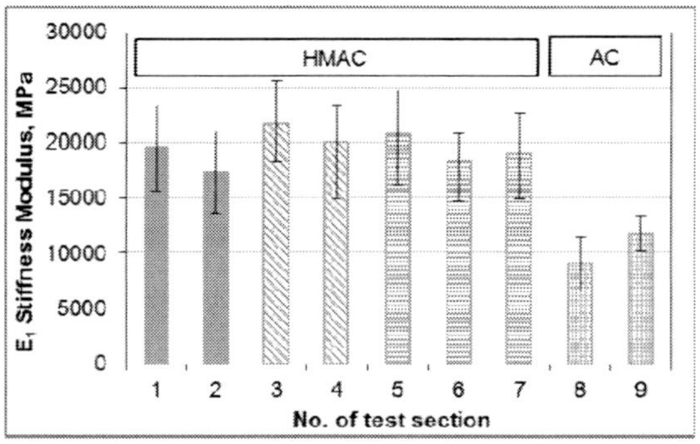

Figure 4. Asphalt layers stiffness moduli obtained from field FWD test (+10°C).

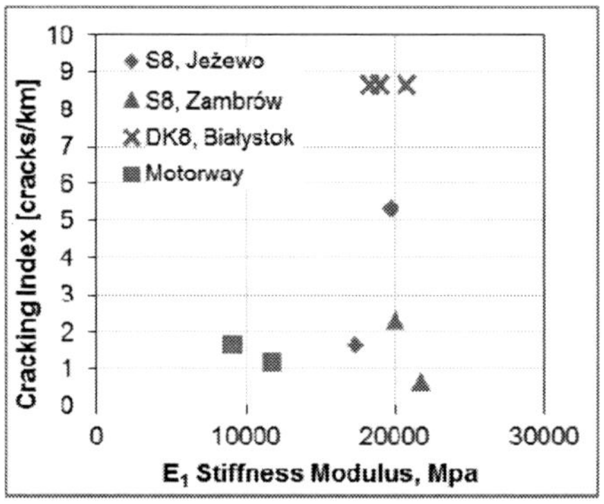

Figure 5. Relation of low-temperature cracking intensity vs FWD E_1 stiffness moduli (mean).

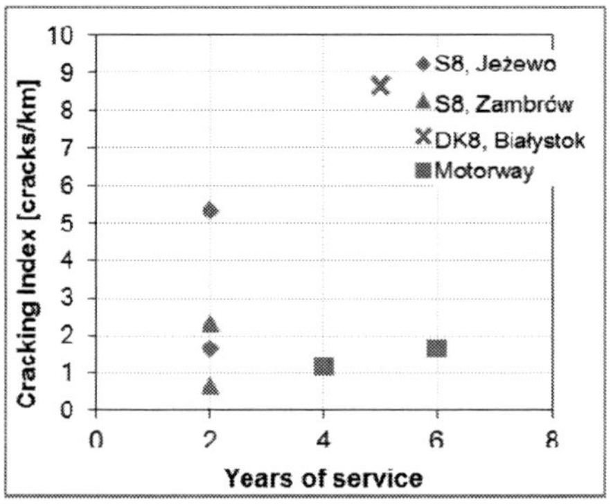

Figure 6. Relation of low-temperature cracking intensity vs years of service.

While there is no clear correlation between stiffness modulus and cracking intensity (R^2 ranges from about 0.1 to 0.2 depending on the used function) some tendencies are visible. The stiffer the asphalt layers of the pavement, the higher number of observed low-temperature cracks. Among the factors which could disturb the correlations probably the most important ones are: small number of test sections, different age of the analyzed pavements (ranged from 2 to 6 years). The last of the factor is quite important as not all winters in the recent years were very harsh with very low temperatures.

More disturbing is relation presented in the Figure 6. Test section with conventional AC layers (represented by gray squares) presents one of the lowest number of low-temperature cracks, despite that this test sections were among the oldest. On the other side, sections with HMAC layers after 2 years of service presents very high number of cracks (range of 0.7 up to 5.3), what are unusually for such new pavements. As it was stated in the introduction, presented results are only a part of the bigger road survey (presented in details in [10]), in which HMAC sections also presented increased cracking intensity in comparison to test sections with conventional AC mixtures.

4. Conclusions

On the basis of both field survey and FWD test conducted on the selected road sections it can be stated:

- Test sections constructed using HMAC layers in general presents higher number of low-temperature cracks in comparison to sections constructed with conventional AC layers. In average, the cracking intensity on test sections with HMAC layers is two times higher.
- The FWD test results showed, that all tested sections were constructed according to best practices and are homogeneous in the case of pavement deflections. The bearing capacity of all tested sections is very high.
- The stiffness moduli determined for the asphalt layers is almost two times higher for test sections with HMAC layers in comparison to test sections with conventional AC layers. The elastic moduli of crushed stone base and improved subgrade is almost the same for the all tested layers and their values are around 450 MPa and 200 MPa respectively.
- While there are no strong correlations between E_1 stiffness moduli and cracking intensity, some trends between the cracking index and stiffness moduli of the asphalt layers are visible. The number of low-temperature cracks increases with the increase of the value of the stiffness moduli.
- The cracking intensity of new (2 years of service) road sections constructed with HMAC base is disturbing. The cracking index ranges from 0.7 up to 5.3 cracks per kilometre, what is almost two times higher value than obtained for sections constructed with conventional AC layers, despite their longer time of service.
- The load transfer efficiency on the sections with HMAC layers is poor. Two out of three tested sections presents only partial load transfer, while the last section presents lack of load transfer. This situation can be big issue in the later service of the road section, as this is the weakest spot of the entire pavement, from which the deterioration of the pavement can be accelerated.

Acknowledgements

It is gratefully acknowledge that the field investigation were performed in a research project sponsored by the Polish General Directorate for National Roads and Motorways. Authors would like also to thank the Bialystok Division of General Directorate for National Roads and Motorways for conducting the FWD tests.

References

[1] Corté J F 2001 Development and Uses of Hard-Grade Asphalt and of High-Modulus Asphalt Mixes in France *Perpetual Bitum. Pavements, Transportation Research Circular* pp 12–31

[2] De Backer C, Glorie L and Reynaert R 2008 Test Section in High-Modulus Asphalt: A Comparative Experiment with Ten Variants *Proc. 4th Eurasphalt Eurobitume Congr. (Copenhagen)*

[3] De Visscher A, Vansteenkiste J and Vanelstraete S 2008 Test Section in High-Modulus Asphalt: Mix Design and Laboratory Performance Testing *Proc. 4th Eurasphalt Eurobitume Congr. (Copenhagen)*

[4] Rohde L, Ceratti J A P, Nunez W P and Vitorello T 2008 Using APT and Laboratory Testing to Evaluate the Performance of High Modulus Asphalt Concrete for Base Courses in Brazil *Third Int. Conf. Accel. Pavement Testing. APT 2008 (Madrid)*

[5] Wielinski J C and Huber G A 2011 Evaluation of French High Modulus Asphalt (EME) in Pavement Structural Design (MEPDG) *Journal of the Association of Asphalt Paving Technologists* **80** pp 697–718

[6] Carbonneau X, Michaut J P, Andersen T, Thorup C and Ladehoff L 2008 High Modulus GAB II: A Danish Experiment *Proc. 4th Eurasphalt Eurobitume Congr. (Copenhagen)*

[7] Lee H J, Lee J H and Park H M 2007 Performance evaluation of high modulus asphalt mixtures for long life asphalt pavements *Construction and Building Materials* **21** pp 1079–87

[8] Judycki J, Jaskuła P, Dołżycki B, Pszczoła M, Jaczewski M, Ryś D and Stienss M 2015 Investigation of low-temperature cracking in newly constructed high-modulus asphalt concrete base course of a motorway pavement *Road Materials and Pavement Design* **16** (sup 1) pp 362–88

[9] Judycki J, Jaskuła P, Dołżycki B, Pszczoła M, Jaczewski M, Ryś D and Stienss M 2016 The Impact of Homogeneity of High Modulus Asphalt Concrete Layer on Low-Temperature Crackin, *8th RILEM Int. Conf. Mech. Crack. Debonding Pavements (Nantes)* pp 319–26

[10] Ryś D, Judycki J, Pszczoła M, Jaczewski M and Mejłun Ł 2017 Comparison of low-temperature cracks intensity on pavements with high modulus asphalt concrete and conventional asphalt concrete bases *Construction and Building Materials* **147** pp 478–87

[11] Judycki J, Pszczoła M, Jaczewski M, Mejłun Ł and Ryś D 2014 *Research on the Effect of Application of High Modulus Asphalt Concrete (HMAC) in Pavement Structures on Low - Temperature Cracking and Decrease of Permanent Deformations (in Polish)* (Gdansk University of Technology)

[12] Jaczewski M 2016 *The influence of high modulus asphalt concrete on low temperature cracking of pavements - Ph.D. Thesis* (Gdansk University of Technology)

[13] General Directorata for National Roads and Motorways (GDDKiA) Warsaw 2002 Evaluatuion of pavement condition system *(SOSN 2002)*

[14] Miller J S and Bellinger W B 2003 Distress Identification Manual for the Long-Term Pavement Performance Program FHWA-RD-03-031 (Federal Highway Administration)

[15] Judycki J 2016 A new viscoelastic method of calculation of low-temperature thermal stresses in asphalt layers of pavements *International Journal of Pavement Engineering* (in press)

[16] Zofka A and Braham A 2009 Comparison of Low-Temperature Field Performance and Laboratory Testing of 10 Test Sections in the Midwestern United States Transportation Research Record *Journal of the Transportation Research Board* **2127** pp 107–114

[17] GDDP 2001 *Catalogue of reinforcement and reconstruction of flexible and semi-rigid pavements*

[18] Beuving E 2005 Falling Weight Deflectometer *COST 336 Final report of the action*

[19] Mackiewicz P 2013 Fatigue Tests of Bituminous Mixtures with Inclusion of Initial Cracks *Journal of Materials in Civil Engineering* **25** pp 140–147

[20] Mrugała J and Chomicz-Kowalska A 2017 Influence of the Production Process on the Selected Properties of Asphalt Concrete *Procedia Engineering* **172** pp 754–59

Effect of wax crystallization on complex modulus of modified bitumen after varied temperature conditioning rates

Diana Simnofske[1] and Konrad Mollenhauer[1]

[1]University of Kassel, Mönchebergstraße 7, 34125 Kassel, Germany

E-mail: d.simnofske@uni-kassel.de

Abstract. Most of European roads are paved with asphalt materials. Mechanical properties as well as durability depend on bituminous binder properties. To influence viscous binder properties wax additives are applied in asphalt mixture for reducing temperature during production process. The crystallization of wax additives results in a rapid viscosity changes within a small temperature span. This allows the reduction of asphalt mix temperatures as well as affects the complex modulus within the performance temperature range. In order to evaluate the effect of wax crystallization in bituminous binders, three binders of different viscosity are modified with 0%, 1.5% and 3% Fisher-Tropsch wax. For the rheological characterization complex shear modulus and phase angle are measured by variation of the cooling rate after sample trimming. Furthermore physical properties were determined by softening point ring and ball again with varied cooling of the bitumen sample after specimen preparation.

1. Introduction

The majority of European road network is paved with asphalt materials. Mechanical properties as well as durability depend on bituminous binder properties. To influence viscous binder properties wax additives are used in asphalt mixture for reducing temperature during production process. Wax additives show crystallization effects resulting in a rapid viscosity change within a specific temperature range. Above the crystallization temperature, binder viscosity is reduced significantly allowing lower mixing and compaction temperatures. Below the wax crystallization temperature, the viscosity of the modified binder is increased which results in improved rutting resistance.

The wax crystallization takes place during cooling in a temperature range between 80 °C and 100 °C for Fischer-Tropsch (FT) wax [1,5]. Within this temperature range, the time available for crystallization affects the size and type of crystallization. FT-wax consists of long hydrocarbon chains up to C_{100} forming a lattice microcrystalline structure and crystallize as small needles in a small melting area [2]. The process of needle crystallization process of paraffin wax was studied by Rhodes et al. 1927 in detail [6]. Several fractions of paraffin wax from crude petroleum were analyzed by microscopy with a detailed view to crystallization process within fast and slow cooling rates. In dependence of the cooling speed the wax is solidifying in plates, generally. Rhodes et al. observed during rapid cooling rate that needles are developed by curling of the edges of the plates. The developing from wax crystal plates to needles can be seen in Figure 1.

Figure 1. Converting plate crystals to needles during crystallization process with rapid cooling. [6]

In contrast to hypothesis in [1], Rhodes et al. described that during slow cooling over a period of several hours paraffin fractions solidify in plates with less potential to form well defined needles (Figure 2). The process of needle crystallization is very slow resulting in a loose structure with less defined needles.

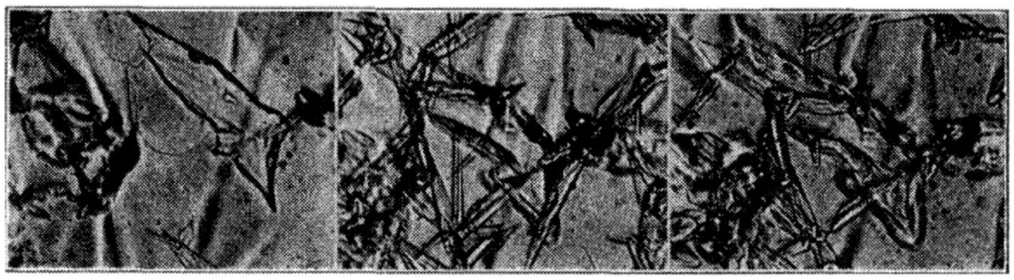

Figure 2. Converting plate crystals to needles during crystallization process with slow cooling. [6]

In a previous publication [1] the effect of FT-wax modified binders by using different ageing procedures was investigated. On the one hand standard RTFOT ageing procedure was carried out to simulate ageing process of hot mix asphalt, on the other hand modified RTFOT ageing procedure at 143 °C was carried out to simulate warm mix ageing conditions. The complex modulus results showed an influence in the ageing stage (indicated as binder viscosity) on the viscosity changing effect provoked by wax modification. In aged binders less effects of the waxes were identified. The neat binder viscosity was estimated as responsible factor for these test result: Within (aged) binder of high viscosity, the wax crystallization is hindered by high bitumen viscosity which results in smaller wax crystals and therefore lower effects of the modification on the complex modulus.

2. Experimental

In order to validate this theory, a new experiment was designed, which results are discussed in this distribution. In order to validate the results discussed in [1], varied binders were tested in complex modulus with and without wax modification. In addition, the cooling rate between specimen trimming and complex modulus test temperature was varied. According to the drafted model [1], a fast cooling rate would result in a weak wax crystal network whereas slow cooling rates creates a strong crystal wax network. In order to evaluate the effect of wax crystallization in dependency of the binder viscosity, three penetration grade binders (20/30, 30/45 and 50/70) were modified by 0%, 1.5% and 3% FT-wax.

The complex modulus was assessed in DSR tests according to EN 14770 in temperature-sweep tests between 20 °C and 120 °C. After trimming of the specimens (diameter 25 mm, gap 1 mm) at a temperature of 80 °C above the softening point the temperature was increased to 120 °C. Then two different temperature conditioning procedures (TCP) were applied:

- TCP I included a slow cooling rate of 10 K/h from 120°C to 20°C simulating cooling conditions of asphalt layers in field.

- TCP II included a fast cooling rate of 1.800 K/h from 120°C to 20°C as usually applied within DSR tests.

The temperature sweep tests were started at 20 °C and complex modulus was measured for 20, 30, 40… up to 120 °C with a frequency applied of 1.59 Hz.

The DSR tests were complemented by softening point ring and ball according to EN 1427. Again, in addition to standard procedure with non-controlled cooling a second set of samples were prepared which were cooled by applying a constant temperature rate of -10 °K/h within a temperature control chamber.

3. Results of DSR temperature sweep tests

The complex shear moduli G* and the phase angles δ versus the test temperature are plotted for assessed temperatures above 30 °C in Figure 3 to Figure 8. The results obtained at 20 °C showed some scattering and were not plotted for improving the readability of the figures. G* and δ measured on pure FT-wax samples after slow and fast cooling are included in the figures.

In Table 1 and Table 2 the shear moduli G* and phase angles δ are given for the test temperatures 20°C to 120°C with a temperature difference of 20 °C.

For the unmodified binders, the difference between complex modulus measured after slow and fast cooling are within ± 6 % and therefore neglectable. With increasing wax content, the effect of cooling rate is increasing. For the binders with 1.5 % added wax, the complex modulus measured after fast cooling rate (TCP II) is up to 80 % higher compared to the values measured at slow cooling rate (TCP I). With 3 % wax addition, this difference increases up to 250 %. The difference between fast and slow cooling rate is further affected by the viscosity of the neat bitumen. With decreasing viscosity (20/30 - 30/45 – 50/70), the differences between slow and fast cooling rate grow stronger.

The addition of wax results in a decrease of phase angle for temperatures below 110 °C indicating the stiffening effect of the crystalized wax within the bitumen. Regarding the effect of cooling rate, δ of the neat binders are only little higher after fast cooling compared to slow cooling. At test temperatures below 60 °C the binder samples with 1.5 % wax addition indicate higher phase angles (and therefore more viscous properties) when cooled down faster compared to slow cooling. At higher temperature, the fast cooling results in lower values of δ. The binders with 3 % wax modification indicate systematically lower phase angles for TCP II (fast cooling) compared to TCP I (slow cooling). As also observed for the complex modulus, this difference is highest for the bitumen 50/70.

Table 1. Results for complex shear moduli G* [Pa].

binder	T [°C]	Complex shear modulus G* [Pa] TCP I (slow) 0% wax	1.5% wax	3.0% wax	TCP II (fast) 0% wax	1.5% wax	3.0% wax	Difference between fast and slow cooling G^*_{fast}/G^*_{slow} [-] 0% wax	1.5% wax	3.0% wax
20/30	20	1497829	5256462	5230966	851142	3963015	3468841	0.57	0.75	0.66
	40	631283	737190	854239	591865	847774	1153666	0.94	1.15	1.35
	60	33859	45098	59128	32631	56307	98081	0.96	1.25	1.66
	80	2544	3331	4409	2533	4764	9969	1.00	1.43	2.26
	100	298	364	399	303	417	722	1.02	1.15	1.81
	120	54	43	38	56	44	45	1.04	1.02	1.18
30/45	20	847636	3457321	3997503	913065	3966638	2632852	1.08	1.15	0.66
	40	225851	334451	467444	213491	400207	710068	0.95	1.20	1.52
	60	12120	20465	31857	11830	27232	64480	0.98	1.33	2.02
	80	1076	1576	2726	1069	2821	6162	0.99	1.79	2.26
	100	154	200	219	156	245	408	1.01	1.23	1.86
	120	34	29	27	35	30	28	1.03	1.03	1.04
50/70	20	2238714	3102375	2943019	1857368	3145357	2798119	0.83	1.01	0.95
	40	116705	219415	276953	112073	245390	495529	0.96	1.12	1.79
	60	5881	14436	17556	5785	15416	42808	0.98	1.07	2.44
	80	549	1174	1358	552	1499	3411	1.01	1.28	2.51
	100	87	111	101	89	144	247	1.02	1.30	2.45
	120	21	18	16	22	19	17	1.05	1.06	1.06

Table 2. Results for phase angle δ [°].

binder	T [°C]	Phase angle δ [°] TCP I (slow) 0% wax	1.5% wax	3.0% wax	TCP II (fast) 0% wax	1.5% wax	3.0% wax	Difference between fast and slow cooling $\delta_{fast} - \delta_{slow}$ [°] 0% wax	1.5% wax	3.0% wax
20/30	20	48	36	33	59	34	20	11	-2	-13
	40	60	56	53	61	57	52	1	1	-1
	60	72	67	63	73	69	61	1	2	-2
	80	82	77	69	83	75	61	1	-2	-8
	100	88	83	78	88	83	69	0	0	-9
	120	90	89	90	90	90	87	0	1	-3
30/45	20	54	40	39	52	51	38	-2	11	-1
	40	66	61	57	67	62	55	1	1	-2
	60	77	69	65	78	71	58	1	2	-7
	80	85	80	65	85	72	57	0	-8	-8
	100	89	85	80	90	83	69	1	-2	-11
	120	90	90	90	90	90	86	0	0	-4
50/70	20	54	45	47	54	43	33	0	-2	-14
	40	69	61	60	71	64	56	2	3	-4
	60	80	67	68	81	70	58	1	3	-10
	80	87	76	71	87	71	59	0	-5	-12
	100	90	85	86	90	82	69	0	-3	-17
	120	90	90	90	90	90	90	0	0	0

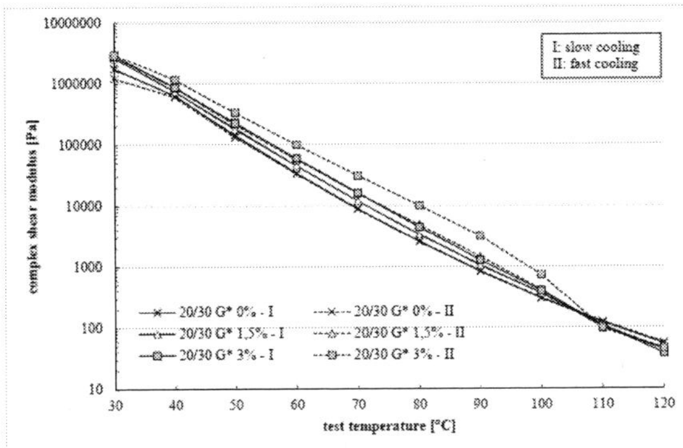

Figure 3. Shear moduli G* for 20/30.

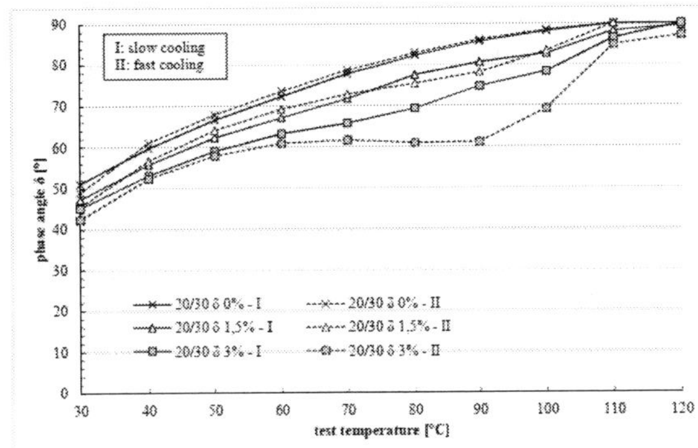

Figure 4. Phase angle δ for 20/30.

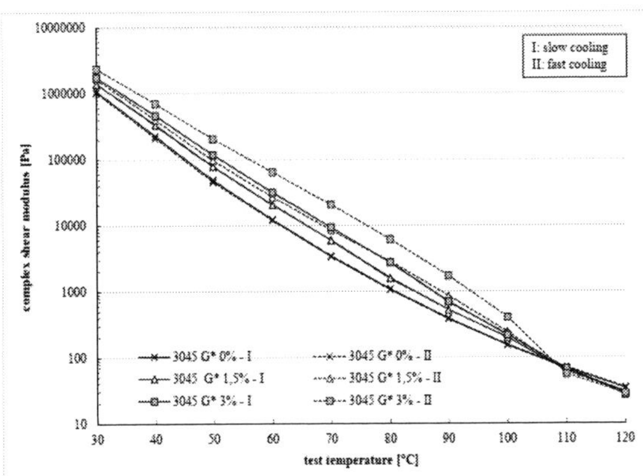

Figure 5. Shear moduli G* for 30/45.

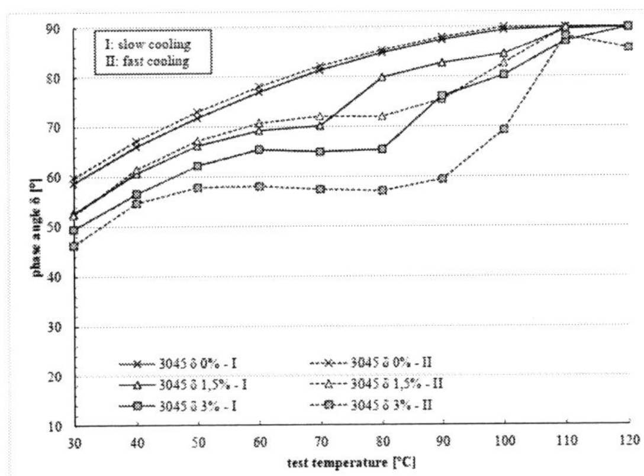

Figure 6. Phase angle δ for 30/45.

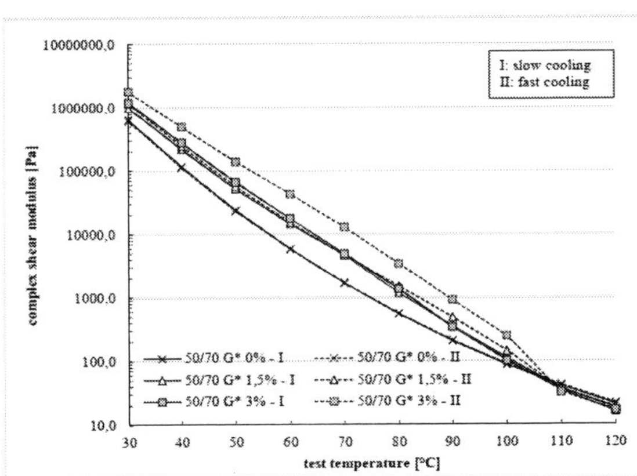

Figure 7. Shear moduli G* for 50/70.

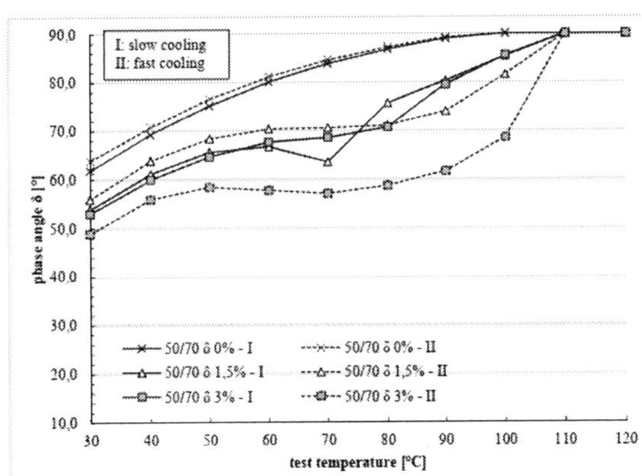

Figure 8. Phase angle δ for 50/70

In order to check for bias effects caused by binder sample flowing from the 1 mm gap during slow cooling time, softening point Ring and Ball was measured. The standard procedure for test sample preparation is uncontrolled cooling at room temperature which is significantly faster compared to site condition. For applying the slow cooling rate, the samples representing TCP I were cooled within a temperature-control cabinet. The resulting softening points are summarized in Table 3.

Table 3. Results for softening point ring and ball dependent of binder wax content.

	20/30		30/45		50/70	
Wax content	Normal cooling TCP II	Slow cooling TCP I	Normal cooling TCP II	Slow cooling TCP I	Normal cooling TCP II	Slow cooling TCP I
0%	66,0	-	58,5	-	54,5	-
1.5%	78,5	(77,0)	68,0	75,0	60,5	68,5
3.0%	86,0	81,0	81,5	78,5	81,0	72,5

4. Discussion

As known from [3, 4] wax modification results in an increase of binder stiffness. Softening point ring and balls as well as rheological properties in a temperature range from 30°C to 100°C show increased complex modulus and an increase in elasticity which can be observed in a decrease of phase angle.

Figure 9 and Figure 10 show the measured phase angles after TCP I (slow cooling) and TCP II (fast cooling) for all binders with 0%, 1.5% and 3.0% wax modification. The phase angles obtained are spreat wider after fast cooling (TCP II) indicating more distinct differences between the samples.

For the unmodified binders the phase angle increases continuously with increasing test temperature. For wax modified binders, the phase angle curve can be divided into three phases: Below the temperature of around 50°C, an increasing temperature results in increasing phase angle. Here the complex modulus of the bitumen is higher compared to the modulus of the wax itself and the phase angle is controlled by the properties of the neat bitumen. With further increasing temperatures up to approx. 70°C the temperature increase results in a phase angle decrease or a constant phase angle. This can be explained by the decreasing bitumen viscosity which results in increasing proportion of the wax properties within the overall sample properties resulting in increasing elasticity. With further increase of the temperature up to 110°C the wax network indicates decreasing stiffness at temperatures near its melting point which results again in an increase of phase angle.

Comparing phase angle development of unmodified and wax modified binders for TCP I in Figure 9, no characteristic phase progress produced by wax content in binder 20/30 can be recognized as described before. The results of δ are similar to unmodified binders. For binder 30/45 and 50/70 distinct wax characteristics can be observed.

Figure 10 shows results of δ for TCP II with wax characteristic for all modified binders. It should be noted that between test temperature 60 °C and 90 °C the influence of wax crystallization effect is less for binder 20/30 which results in higher phase angles with less elastic binder properties compared to 30/45 and 50/70. As a result, the influence in wax content is significantly stronger after TCP II compared to TCP I and binder viscosity influence wax ability forming crystal network.

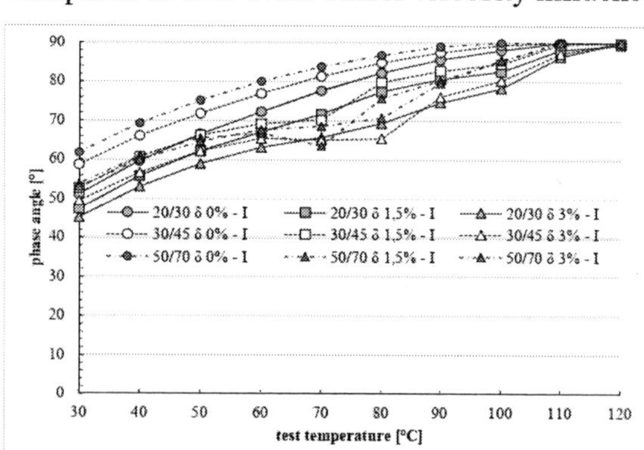

Figure 9. Phase angle in TCP I (slow).

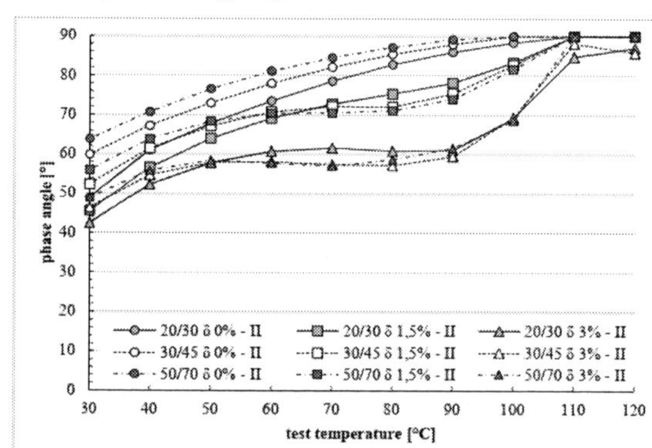

Figure 10. Phase angle in TCP II (fast).

After evaluation of the test results between TCP I and TCP II and with regard to the crystalled wax structure within all binders the hypothesis described in [1] has to be withdrawn. In [1] it was discussed that the sizes of the emerging crystals depend on the cooling rate and the viscosity of the material. In the results obtained it can be observed that a fast cooling rate results in more elastic properties than a slow cooling rate. That means that the wax crystal network within the bitumen resulting from fast cooling indicate higher stiffening effects compared to slow cooling rate.

With regard to the presented results of dynamic shear rheometer this structure is more sensitive to shear force as occurring within DSR tests and the measured results are similar to non wax modified binder sample.

The observations made within the DSR tests are confirmed by the softening points ring and ball. For the wax modified binders, the slow cooling results in differing values of $T_{R\&B}$. The difference between slowly and rapidly cooled samples is strongest for samples with 3 % wax addition.

5. Conclusions

The following conclusions can be drawn from the results of the presented investigations

- The cooling rate applied after casting (and trimming) specimen doesn't affect the properties of the tested neat bitumen.
- The cooling rate affects the wax crystallization in FT-wax modified binders significantly. This results in systematically varying complex modulus, phase angle and also softening point. These differences has to be taken into account when comparing results of binder tests (usually measured on rapidly cooled samples) with laboratory asphalt mix properties in field conditions, for which slower cooling rates apply.
- The viscosity of the binder affects the wax crystallization. The differences between slow and fast cooled samples are highest for the 50/70 and lowest for the 20/30.
- In order to avoid wrong interpretation of bitumen tests, the cooling rate has to be taken into account for wax modified binders.

References

[1] Mollenhauer K, Simnofske D, Oelkers C and Butz T 2016 Benefits of F-T wax based warm asphalt mixes for short-term binder aging and pavement durability *Proc. 6th Eurasphalt & Eurobitume Congress (Prague)*

[2] Gandhi T 2008 *Effects of warm asphalt additives on asphalt binder and mixture properties* (Clemson University)

[3] Simnofske D, Mollenhauer K, Oelkers C and Butz T 2015 Aging characteristics of warm mix asphalt binders modified with F-T wax – Benefits for efficient application and adequate performance *Proceedings 11th Conference on Asphalt Pavements for southern Africa - CAPSA 2015 (Sun City, South Africa)*

[4] Edwards Y 2005 Influence of Waxes on Bitumen and Asphalt Concrete Mixtures Performance *Road Materials and Pavement Design* **10**(2) pp 313-35

[5] Alisov A and Wistuba M 2016 Bestimmung des Ersarrungsbeginns von Wachsen in Bitumen mittels Dynamischen Scherrheometer *Straße und Autobahn* **11**

[6] Rhodes F H, Mason C W and Sutton W R 1927 Crystallization of Paraffin Wax *Journal of Industrial and engineering chemistry* **19**(8) pp 935–938

Performance properties of asphalt mixes for rich bottom layers (RBL)

Petr Bureš[1], Jiří Fiedler[1], Jiří Kašpar[1], Michal Sýkora[1] and Petr Hýzl[2]

[1]EUROVIA Services s.r.o., Národní 138/10, 110 00 Praha-Nové Město, Czech Republic
[2]Vysoké učení technické v Brně, Antonínská 548/1, 601 90 Brno, Czech Republic

E-mail: petr.bures@eurovia.cz

Abstract. The binder content of asphalt mixes has an important influence on the performance properties. Higher binder content improves fatigue resistance. That is why the concept of RBL was developed in USA and applied for "perpetual pavements". However excessive binder content could lead to the decrease of the mix stiffness and to permanent deformations of asphalt pavement during hot summer. The advantages and limitations of RBL concept have been studied in research project CESTI. Fatigue tests of mixes with road bitumen and polymer modified bitumen and RBL were realised. Deformation behaviour of these mixes was also evaluated. The experience from the test section with RBL laid in 2015 will be presented. The results corresponded to expectations. However, low void content was obtained on one subsection. In spite of it, there were no permanent deformations during summer 2016. The analysis of methods for the prediction of the permanent deformation was also undertaken in research project CESTI. Some information about the results of these analysis related to the use of RBL will be also briefly mentioned.

1. Introduction

There is a tendency to increase a pavement life for roads with high traffic load. These pavements are often called long life pavements in Europe or "perpetual pavements" in USA. Pavement distresses due to the fatigue of asphalt mix are reduced or avoided. Higher fatigue resistance of the mix can be achieved either by the use of modified bitumen or by the increase of its binder content. The base course mixes with the higher binder content have been used recently in USA and Canada. They are called "Rich Bottom Layer" (RBL) or "Fatigue Resistant Layer" (FRL). The term RBL is used in this paper.

The RBL mixes have been tested in the Czech Republic in the framework of the research project CESTI carried out by Technical Universities of Prague (ČVUT) and Brno (VUT), Eurovia and other road companies. Stiffness and fatigue properties of some standard mixes and RBL mixes tested in the CESTI research are presented in this paper. Some test results on these mixes with road bitumen have been published in [1] and tests on mixes with polymer modified bitumen in [2] (both in the Czech language). The impact of measured fatigue test results on the pavement design was discussed in [3] written in the Hungarian language. This paper presents tests mentioned in [1], [2], [3] and other fatigue tests carried out at VUT laboratory in 2016 and their statistical evaluation. It discusses also some aspects not mentioned in quoted publications. As the fatigue behavior of asphalt mixes has a big impact on the pavement design, the research in CESTI was focused on fatigue properties. Thus fatigue behavior is mainly presented also in this text.

2. Composition and volumetric properties of tested mixes

Mixes for the base course with two binders were tested. One binder was road bitumen 50/70 another was PMB 25/55-60. RBL mixes had binder content 0,5 % higher than normal AC. All mixes were compacted with 2 x 50 blows of Marshall hammer. Mix with the road bitumen was compacted at 150°C and mix with PMB at 160°C. Aggregates from the quarry Litice (spilite) were used. Limestone filler was supplied by the lime company Hasit from Velké Hydčice. Gradation of the mixture is on Figure 1. Volumetric properties of these mixes are in Table 1.

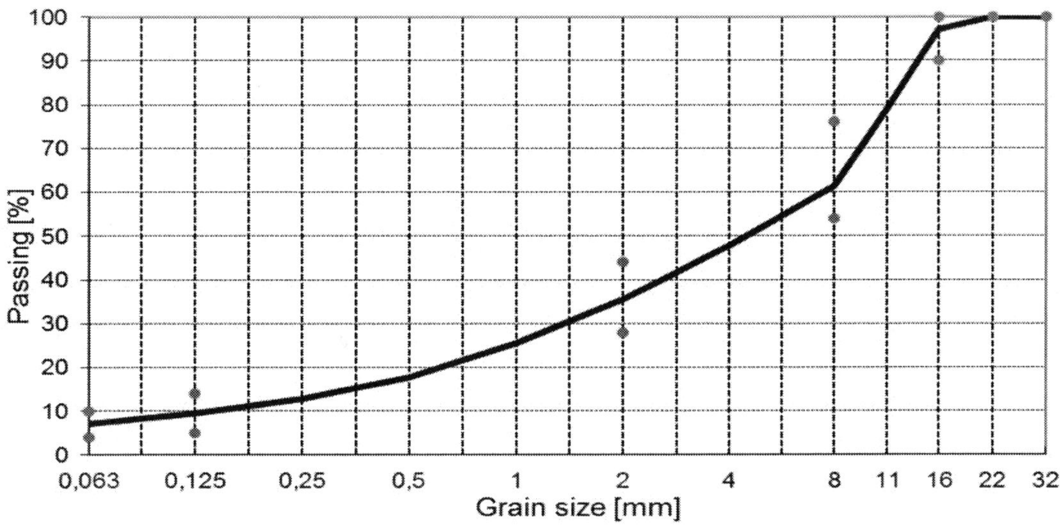

Figure 1. Gradation of the mixture.

Table 1. Volumetric properties of tested mixes.

Mix type			AC 16 base	AC 16 RBL	AC 16 base	AC RBL
Binder type			50/70		PMB 25/55-60	
Binder content	B_{mass}	%	4.100	4.600	4.100	4.600
Binder content	B_{vol}	%	9.700	11.100	9.700	11.000
Density	ρ_{bssd}	Mg/m^3	2.505	2.533	2.496	2.527
Maximum density	ρ_{max}	Mg/m^3	2.659	2.648	2.659	2.648
Void content	V_m	%	5.800	4.300	6.100	4.600
Bitumen filled voids	VFB	%	62.600	72.000	61.300	70.600

3. Stiffness properties

Indirect tensile stiffness tests on the cylindrical specimen after European standard [4] for the temperature range 0 to 30 °C were carried out in the road laboratory at the Technical University in Prague (ČVUT). Results are on Figure 2. Each point represents an average stiffness for 6 specimens. Both mixes with harder polymer modified bitumen had distinctly higher stiffness than mixes with road bitumen. However, there was a small difference in the stiffness (≈ 10 %) of the AC mixes and RBL mixes with the same binder.

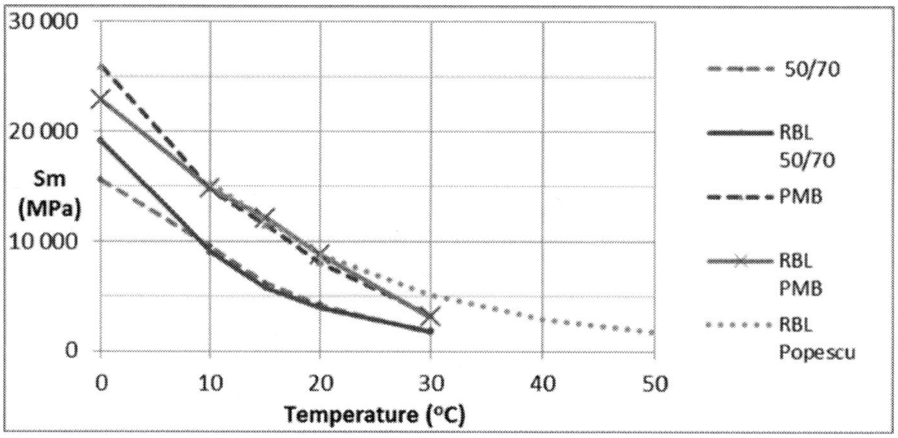

Figure 2. Stiffness IT-CY after EN [4].

Further increase in the binder content of RBL mixes would probably lead to the decrease of stiffness. Then the difference between AC and RBL will be important. This is illustrated by the test results on the other mix for the base course with the PMB 25/55-55 carried out also at ČVUT presented in [5]. The stiffness of the mix with the binder content B_{mass} 4.8, 5.1 and 5.4 % decreased with the increase of the binder content. The general relationship between binder content, void content and stiffness for the mix with road bitumen 50/70 was well illustrated in older French paper [6]. The Czech pavement design method, as well as the French one, uses the concept of equivalent temperature usually fixed at 15 °C. Thus the indirect tensile stiffness tests are usually carried out only for medium temperatures. It assures that the test conditions allow the evaluation of the stiffness modulus from the formulae based on elastic theory.

In order to get an approximate idea about the possible decrease of the stiffness of RBL for higher temperatures, the results of the test on RBL mix for the temperature 10 – 50 °C given in [7] were added to the graph in the Figure 2, as that mix had similar stiffness as our mix with PMB for the temperature range 10 – 20 °C.

The repeated loading at higher temperatures and for longer loading time could lead in some cases to distinctly higher permanent deformation for RBL mixes. This was demonstrated by the results of the shear test with the constant height (RSST-CH) used in USA at 50 °C in [7]. RBL mix with binder content B_{mass} = 5.2 % and void content m = 3 % had about 50 % higher permanent deformation after 5000 cycles than AC with B_{mass} = 4.7 % and m = 6.0 %. The impact of higher binder content on permanent deformations in the triaxial cyclic compression test (TCCT) after European standard EN 12697-25 [8] is shown for example in [9]. Creep rate f_c for 2 mixes with B_{mass} = 4.8 % and one mix with B_{mass} = 4.9 % was higher than for the same mix with B_{mass} = 4.5 %. Even steeper decrease of stiffness was observed for tested mixes for binder content ≥ 5 %. According to [9], the optimum binder content derived from the TCCT test is about 0.3 to 0.1 % lower than the optimum binder content found by the traditional Marshal mix design method

Repeated triaxial load tests are normally not realized in our country. Some repeated load tests on different mixes are now under way at the VUT Brno in the framework of CESTI research. The results will be presented when more experience with this test will be obtained.

4. Fatigue properties

4.1. Fatigue properties of mixes with road bitumen 50/70

Fatigue tests according to European standard EN 12697-24 [10] on trapezoidal specimens (2 PB) were carried out with the frequency 25 Hz at 10 °C in the laboratory of VUT Brno in 2015 on specimens with bitumen 50/70 fabricated in the laboratory of Eurovia. The fatigue properties of RBL were much better than for AC base, as it is illustrated in Table 2.

Table 2. Fatigue parameters for mixes with bitumen 50/70 tested in 2015.

		AC base 50/70	RBL 50/70
Binder content	(%)	4.1	4.6
Slope B = - 1/b	(-)	4.7	5.0
ε_6	(μs)	81.5	107.9
$\gamma_{úp} = \varepsilon_{6,50} / \varepsilon_{6,5}$ after Czech TP 170	(-)	1.47	1.22
$\Delta\varepsilon_6$ after EN12697-24	(μs)	14.3	8.5
$\varepsilon_{6min} = \varepsilon_6 - \Delta\varepsilon_6$	(μs)	67.2	99.4
Standard error $SN_{N/\varepsilon}$		0.481	0.267
Coefficient of determination R^2		0.796	0.918
Number of specimens		16	17

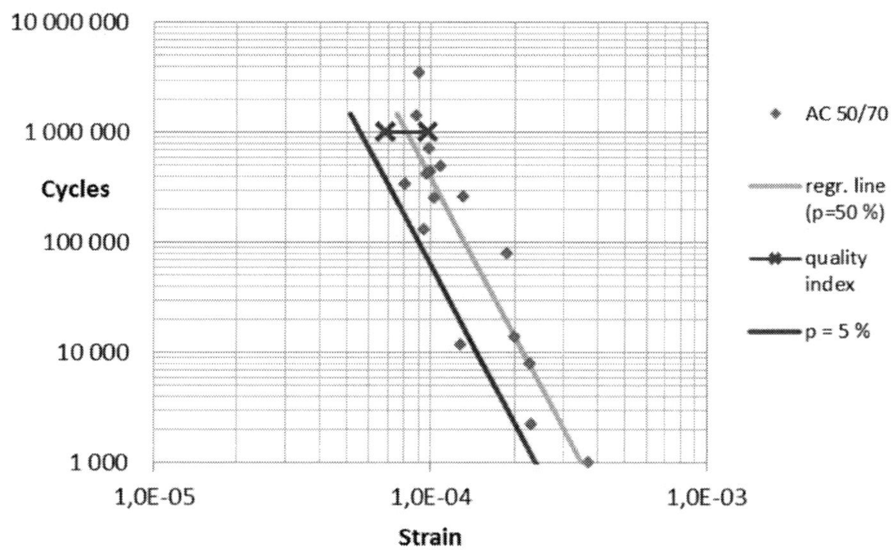

Figure 3. Shift of the fatigue line after TP 170 for *AC base* with 50/70.

Parameters ε_6 , $\Delta\varepsilon_6$, b in Table 2 are defined in EN 12697-24 [10]. The slope of the fatigue line B and the coefficient $\gamma_{úp}$ are used in the Czech pavement design method after TP 170 [11] and [12]. The coefficient $\gamma_{úp}$ represents the shift of $\varepsilon_{6,50}$ on the fatigue line obtained from the regression analysis of the test results (which corresponds to the probability p = 50 %) to the value $\varepsilon_{6,5}$ (lower values than $\varepsilon_{6,5}$ have the probability of p = 5 %). It is represented by the red line on Figure 3. Thus the shift is calculated in a similar way as in the French pavement design method in [13]. This shift in both methods corresponds to the prediction interval for the regression line in the terminology of statistical analysis. The strains $\varepsilon_6 \pm \Delta\varepsilon_6$ are also shown in Figure 3 (They are marked with the crosses).

Table 2 shows that RBL mix has a distinctly higher value of ε_6 than *AC base*. The standard error for RBL is lower and coefficient of determination is higher. That is why the "quality index" $\Delta\varepsilon_6$ for

RBL is lower (the term used in EN 12697-24 is somewhat misleading, as lower "quality index" for RBL corresponds to a better homogeneity of measured mix fatigue resistance).

However, even for RBL calculated value of $\gamma_{üp}$ was higher than preliminary design value in TP 170. Higher values of $\gamma_{üp}$ than in TP 170 were obtained also for the other asphalt mixes tested during the last few years at VUT. It indicates that the value of $\gamma_{üp}$ recommended in TP 170 (published in 2004) which was based on a few tests realized about 15 years ago, was too optimistic for all mixes. The dispersion is often larger, especially for mixes with binder content close to minimum binder content given in asphalt mix specifications. This is important observation as these mixes are frequently used due to economic reasons. However, there are some pessimistic assumptions in the pavement design method after TP 170 which might compensate optimistic value of $\gamma_{üp}$. The explication of these problems is out of the scope of this paper. These discrepancies can be adjusted during the next novelization of TP 170 which is planned in the future.

The first test section where these AC and RBL were laid was realized in autumn 2015. Some samples of the mixes produced in the mixing plant were taken. Test specimens were prepared at Eurovia laboratory. The fatigue tests were realized in 2016 at the laboratory of VUT on the same testing equipment as in 2015. Fatigue parameters of RBL were worse than measured in 2015. The reason for this difference might be due to the natural dispersion and due to the fact that the mixes tested in 2016 had to be reheated in the laboratory before the fabrication of the specimens. It can be supposed that the short term aging of the binder was more important than in 2015. The binder extracted from the specimen after fatigue test in 2016 had the penetration at 25 °C only pen=26. The softening point *R&B* was 59,6 °C (fresh binder 50/70 has *R&B* = 46 – 54 °C). Values for extracted binder correspond to parameters of 50/70 bitumen after 3 times RTFOT test. This confirms that aging of the road bitumen was important on this job site.

It is worth to mention that the differences between laboratory prepared specimens, samples of mixtures produced in the mixing plants and samples taken from pavements were observed in a huge German research project finished in 2015. Comprehensive evaluation and analysis of the data from 21 road sections showed large spans of the results that can be explained only in part by material properties or by the installation quality, as stated in [14].

4.2. *Fatigue properties of mixes with PMB 25/55-60*

Fatigue parameters for mixes with PMB were also better for RBL mix than for AC. Results are in Table 3. Due to various reasons, only 14 specimens of RBL with PMB were tested in 2016, even if EN 12697-24 demands at least 18 specimens. The value of $\varepsilon_{6min} = \varepsilon_6 - \Delta\varepsilon_6$ for 2016 RBL mix was only slightly lower than $\varepsilon_{6max} = \varepsilon_6 + \Delta\varepsilon_6$ for 2015 RBL mix. However, confidence limits for two regression lines for the probability of 95 % distinctly overlapped, especially for very low and very high strains levels (The graph is not presented here). That is why fatigue parameters for all 32 specimens were also calculated. They are presented in table 3 as well as the design parameters after TP 170 for *AC base* and on the Figure 4.

Measured values of B and ε_6 of all mixes in table 3 are distinctly higher than the design values in TP 170. It is well known that fatigue parameters of mixes with PMB are normally better than for the same mixes with road bitumen of similar stiffness. However that difference is considered only for high modulus mixes (HMAC) in the Amendment of TP 170 [12]. HMAC have $\varepsilon_6 = 125\,\mu s$ for mixes with road bitumen and $\varepsilon_6 = 135\,\mu s$ for mixes with PMB in [12]. The difference between the design value of ε_6 for the mixes with road bitumen and PMB is not considered for ordinary mixes for the base layer in [12]. It is supposed that this difference will be considered in the future version of the design guide.

Table 3. Fatigue parameters for mixes with PMB.

	ACP PMB 2015	RBL PMB 2015	RBL PMB 2016	RBL 2015+2016	AC base TP 170
B	5.7	6.0	6.2	6.1	5.0
ε_6 (10^{-6})	112.6	119.1	129.5	123.6	100
$\gamma_{úp}$	1.31	1.15	1.16	1.17	1.15
$\Delta\varepsilon_6$ (10^{-6})	11.4	6.4	6.7	4.9	
$\varepsilon_{6min} = \varepsilon_6 - \Delta\varepsilon_6$	101.2	112.7	122.8	118.7	
$SN_{N/\varepsilon}$	0.399	0.227	0.244	0.25	
R^2	0.878	0.885	0.926	0.89	
Number of spec.	20	18	14	32	

Results for 32 specimens of RBL with PMB, confidence and prediction limits are on the Figure 4.

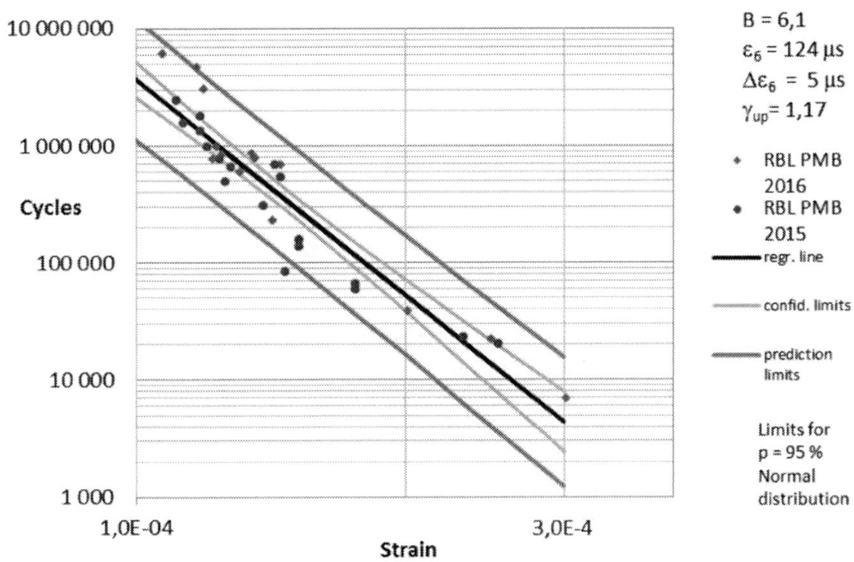

Figure 4. Statistical evaluation of the results of fatigue tests for RBL with PMB.

The application of laboratory measured values of fatigue parameters for the pavement design is allowed in TP 170 with some restrictions which lead to similar limitations as in the French pavement design method in [13]. Different way of the limitation on the application of the laboratory fatigue tests is considered in the new Austrian pavement design method which will be published this year. Differences in these 3 approaches are out of the scope of this text. Paper devoted to this is under preparation for the Czech conference Asphalt pavements AV 17 in November 2017.

5. Information about the test section realized in 2015

The first test section (composed of 4 sub-sections) was laid on one road near the town Pilsen in autumn 2015 during its reconstruction. Pavement design was carried out according to TP 170. Parameters used for RBL were based on detailed review of foreign literature in the framework of CESTI project, as own fatigue tests had not been finished at that time. The same slope of the fatigue line for AC and RBL B=5,0 was used. The same difference between ε_6 for AC and RBL (10 μs) was used for both binder types. This corresponds well to the difference between ε_6 for AC and RBL for 32 specimens tested in 2015 and 2016 as presented in the Table 3.

Fabrication and placement of RBL mixture was realized generally without problem, in spite of some complications related to the fact that it was a fragmented reconstruction of the main road in urban area without allowance for any interruption of the traffic (This led to rest periods between the realization of individual sub-sections, partial closure of one direction, etc.) Void content measured on cores and by nuclear asphalt density gauge was generally in the range 3 to 6 %. Unfortunately low void content 2 % was measured in one relatively small part of RBL section both by Eurovia laboratory and independent laboratory engaged by the investor.

Some risk of permanent deformation during the short period of very hot summer days could not be excluded in this subsection, especially as the trucks sometimes stop before crossings or due to the traffic jam. After an intensive review of the foreign literature on permanent deformations of RBL, some preliminary evaluation and discussion with partners on the project, it was decided to let this subsection in place and observe the behavior.

No permanent deformation of the pavement surface has been observed in 2016, in spite of quite warm summer. According to the results obtained at the state meteorological station in the region, there were 4 subsequent tropical days in August 2016 with air temperature over 30 °C and maximum air temperature 32,4 °C. Authors used the database from the continuous measurements of air and pavement temperature on one PPP project in Slovakia some years ago to estimate the surface temperatures on test section with RBL in 20016. The surface temperatures were probably in the range 50 – 57 °C for a couple of hours for some days. It has to be kept in mind that such analogy is approximate as the temperature on the surface of the pavement depends not only on air temperature, but also on the wind conditions, etc. Anyway, it was higher than the temperature of 50 °C which is specified for rutting test in the framework of initial type test according to the present Czech AC specifications.

Distinctly higher temperatures were observed in that region in extremely hot summer 2015 (before the reconstruction of the road). There were 17 tropical days during summer with maximum temperature 38 °C. The maximum pavement surface temperature could be under these conditions around 60 °C. The suitable temperature for rutting test in the Czech conditions could be 55 °C in our opinion, but European standard for rutting test [15] specifies only test temperature 50 or 60 °C.

6. Evaluation of shear stress and elastic and permanent deformation for the design axle load

French pavement design program Alize was used to calculate elastic shear stress and strains for various temperature distributions in the pavement. This program is very user-friendly and permits to present horizontal and vertical stress and strains distribution in various points in a graphical form. The surface temperature and the distribution of the temperature in the pavement have a big impact on calculated elastic strains. The temperature distribution given in the German pavement design method [16] was used for a couple of selected surface pavements temperatures. Permanent deformations of the pavement were calculated from elastic strain by the method proposed in [17]. As we had no own test results from repeated cyclic tests in 2016, we made only preliminary calculations for a pavement similar to the test section with parameters given in [17].

7. Conclusions

Tests of performance properties of AC and RBL showed the differences in their behaviour. The stiffness of RBL was similar to AC, but the fatigue properties of RBL were much better. Valuable experience was obtained from the first test section. Some care had to be taken to avoid a too low void content of RBL in the pavement. It is planned that calculations of permanent deformations by different methods given in the literature will be carried out. Some cyclic loading tests are now under way.

The review of the literature and preliminary calculations showed that any forecast by the methods presented in the literature is very approximate due to the sensitivity of the results to the temperatures in asphalt layers and difficulties of its forecast. However, the back-analysis of permanent deformations observed on real pavements during hot summer period by these methods is possible. Based on laboratory cyclic loading tests results of used materials and temperature measurements obtained from meteorological stations the back-analysis could lead to the detection of the probable source of deformations. It could be useful in cases that the causes of observed damage are not evident from classic empirical tests and/or judgment based on experience.

Acknowledgements
This paper was prepared thanks to the support of the Technology Agency of the Czech Republic (TAČR), project "Centre for Effective and Sustainable Transport Infrastructure (CESTI)" (no. TE01020168).

References

[1] Fiedler J, Bureš P and Hýzl P 2015 Influence of bitumen content on the performance properties of asphalt mixes *Proc. XX. Conf. Ivana Poliačka (Jasná)*
[2] Fiedler J and Bureš P 2015 Possibilities for extending asphalt pavement life (In Czech) *Proc. Conf. on Asphalt pavements AV 15 (České Budějovice)*
[3] Fiedler J 2016 Influence of bitumen content on the mix performance and pavement design (In Hungarian) *XVII Conf. Hungarian Asphalt Pavement Association (HAPA)* later published in *Aszfalt* **1** pp 73-77
[4] European Committee for Standardization (CEN) 2012 Bituminous mixtures; Test methods for hot mix asphalt. Stiffness. *(EN 12697-26)*
[5] Mondschein P, Valentin J, Vacková P and Dočkalová A 2015 Asphalt pavements with HMAC and RBL mixes (in Czech) *Proc. XX. Conf. Ivana Poliačka (Jasná)*
[6] Travers F 2005 Effets de facteurs de formulation sur la fatigue et le module *Proc. Seminaire LCPC* http://media.lcpc.fr/ext/pdf/sem/2005_ch15_25.pdf
[7] Popescu L, Leahy R and Monismith C, 2010 Analysis and Design of Airfield Pavements Using Laboratory Tests and Mechanistic-empirical Methodology *FAA Worldwide Airport Technology Transfer Conference (New Jersey)*
[8] European Committee for Standardization (CEN) 2012 Bituminous mixtures; Test methods for hot mix asphalt. Cyclic compression test. *(EN 12697-25)*
[9] Blab R 2015 Performance-based asphalt mix and pavement design *Romanian Journal of Transport Infrastructure* **1**(2)
[10] European Committee for Standardization (CEN) 2012 Bituminous mixtures; Test methods for hot mix asphalt. Resistance to fatigue. *(EN 12697-24)*
[11] Ministry of Transport of the Czech Republic 2004 Technické podmínky - Navrhování vozovek pozemních komunikací /Pavement design method/ *(TP 170)* http://www.pjpk.cz/data/USR_001_2_8_TP/TP_170.pdf
[12] Ministry of Transport of the Czech Republic 2010 Technické podmínky - Navrhování vozovek pozemních komunikací – Dodatek TP /Pavement design method – Amendment/ *(TP 170)*
[13] French Committee for Standardization (AFNOR) 2011 Dimensionnement des structures de chaussées routières - Application aux chaussées neuves *(NF P 98-086)*
[14] Wörner T, Hase M and Roos R 2016 Performance-relevante Asphalteigenschaften als Grundlage für neue Vertragsbedingungen *Strasse und Autobahn* **1**
[15] European Committee for Standardization (CEN) 2012 Bituminous mixtures; Test methods for hot mix asphalt. Wheel tracking. *(EN 12697-22)*
[16] Wellner F et al 2009 *Vorstellung der Richtlinien für die rechnerische Dimensionierung des Oberbaus von Verkehrsflächen mit Asphaltdeckschicht* (RDO Asphalt 09)
[17] Dragon I and Reinhart U 2016 Improvements regarding the procedure of rutting prediction of asphalt pavements *E&E congress (Praha)* p 416

Effect of interlayer bonding quality of asphalt layers on pavement performance

Piotr Jaskula[1] and Dawid Rys[1]

[1]Gdansk University of Technology, Faculty of Civil and Environmental Engineering, Gdansk, Narutowicza 11/12, 80-233, Gdansk, Poland

E-mail: piotr.jaskula@wilis.pg.gda.pl

Abstract. The quality of interlayer bonding at the interfaces between the asphalt layers in flexible pavements affects the overall pavement performance. Lack or partial lack of interlayer bonding between asphalt layers can cause pavement's premature failures such as rutting, slippage of the wearing course, cracking or simply a reduction in the calculated fatigue life of the pavement structure. This paper shows the case studies of investigation of actual or potential premature failure of newly reconstructed and constructed pavements where low quality of interlayer bonding has a dominant meaning. In situ and laboratory tests were performed and followed by analytical calculation of pavement structure where thicknesses of layers and maximum shear strengths obtained from the tests were used. During the investigation it was found out that a low quality of tack coat as well as the same aggregate gradation in the bonded asphalt mixtures were the main reasons behind the weak quality of interlayer bonding. Partial interlayer bonding has a strong influence on reduction of calculated fatigue life of pavement. The summary of the paper includes recommendations on how to avoid the low quality of interlayer bonding of asphalt layers.

1. Introduction

1.1. Background

The problem of lack or partial lack of interlayer bonding quality between asphalt layers is a common reason of dispute between contractor and investor. Poor quality of interlayer bonding causes pavement's premature failures such as rutting, slippage of the wearing course, cracking or simply a reduction in the calculated fatigue life of the pavement structure[1]–[12]. The laboratory test of interlayer shear strength performed in a Leutner device [13] is the standard procedure in Poland for controlling the quality of interlayer bonding. Polish requirements [14] state minimum levels of shear strength for specimens drilled out from pavement. The problem described in this paper concerns three cases of pavements where shear strength obtained for specimens did not fulfill the Polish requirements. The article is focused on the problem of weak interlayer bonding between the asphalt base and the binder course or between two lifts of the asphalt base course. A detailed analysis of fatigue life is presented for one selected case.

The interlayer bonding strength depends on several factors including quantity of tack coat, type of bitumen emulsion used for the tack coat, difference between asphalt mix grading in respective layers, poor compaction of the upper layer etc. [7], [15]–[20]. In some cases when interlayer bonding is limited or does not pass the requirements but there is some degree of bonding still, the interlayer bonding can be expected to increase due to aging, traffic load and simply time [21]–[23].

1.2. Objectives

The primary objectives of this study were as follows:

- to reveal how limited interlayer bonding can impact the decreases in calculated fatigue life of pavement structure,
- to show the investigation of three cases of newly constructed pavements where the quality of interlayer bonding did not fulfill Polish requirements,
- to present some recommendation on how to avoid low quality of interlayer bonding of asphalt layers.

2. Case studies

Three cases of different road sections were used in the analysis:

- Road 1 is a newly constructed expressway, the highest (express) technical standard (R1S), for heavy traffic and represents the highest construction standard and investor control.
- Road 2 is a newly constructed national road, the high (highway) technical standard (R2H), for heavy traffic and represents the highest construction standard and investor control.
- Road 3 is a reconstructed city street, the medium (local) technical standard (R3L), for low medium traffic and represents the medium construction standard and investor control.

In roads R1S and R2H the problem of interlayer bonding occurred between two lifts of asphalt base course made from asphalt concrete AC22P with neat bitumen 35/50. The tack coat was applied properly with the usage of bitumen emulsion C60B3ZM (bitumen emulsion dedicated for tack coat of asphalt layers acc. to PN-EN13808) in an amount of bitumen within the range of 0.3–0.5 kg/m^2. In terms of interlayer bonding, the construction process followed good practices. Nevertheless, bonding strength measured in the Leutner test was under the required value of 0.6 MPa at the temperature of +20°C but higher than 0 MPa (see Figure 1). The reason of low interlayer bonding strength was the non-homogenous surface of the bottom base course due to segregation of mixture (see Figure 3) and the same gradation of aggregate used for two lifts of the base course. In consequence the interlocking between two lifts of the asphalt base course was relatively weak.

In the case of road R3L the problem of interlayer bonding occurred between an asphalt base course made from AC22P and a binder course made from AC16W, both with neat bitumen 35/50. In this case shear strengths between two layers were much lower than the required 0.7 MPa at the temperature of +20°C. Additionally, 30% of cored specimens disintegrated during coring, that is the core of the binder course and the base course separated spontaneously, therefore those specimens were marked with 0 MPa of interlayer shear strength (see Figure 2). Unlike in the case of roads R1S and R2H, the reason of weak bonding quality in road R3L was not gradation of aggregate in the asphalt mix, but rather poor quality of the tack coat sprayed in combination with improper compaction of the upper layer – the binder course. Despite the usage of the proper bitumen emulsion C60B3ZM in amount of 0.3–0.5 kg/m^2, the sprayed surface was not homogenous, which is visible in the Figure 4. Another problem identified in road R3L was the insufficient thickness of the asphalt base course, which additionally and ultimately contributed to the decrease in fatigue life of the pavement structure.

In all the aforementioned cases, the unsatisfactory bonding strength was the reason of dispute between the contractor and the investor. On the one hand, lower quality of interlayer bonding contributes to a decrease in fatigue life but, on the other, milling of layers and their subsequent reconstruction causes considerable losses for the contractor. The presented analyses were performed in order to assess whether the existing pavement structure can bear the designed traffic or whether its layers need to be re-built.

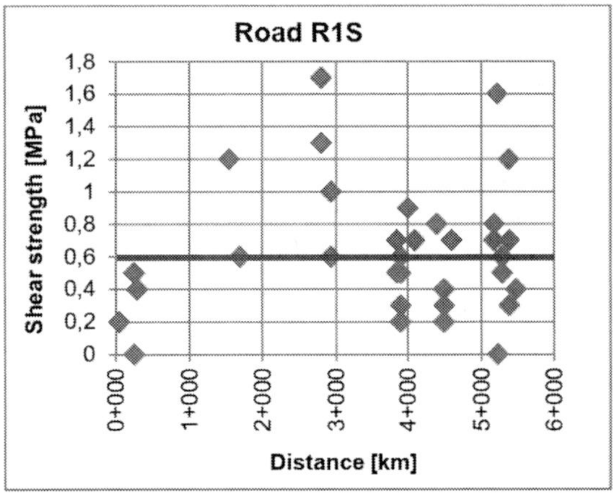

Figure 1. Results of shear strength obtained from the Leutner test between two lifts of the asphalt base course AC22P. Road R1S.

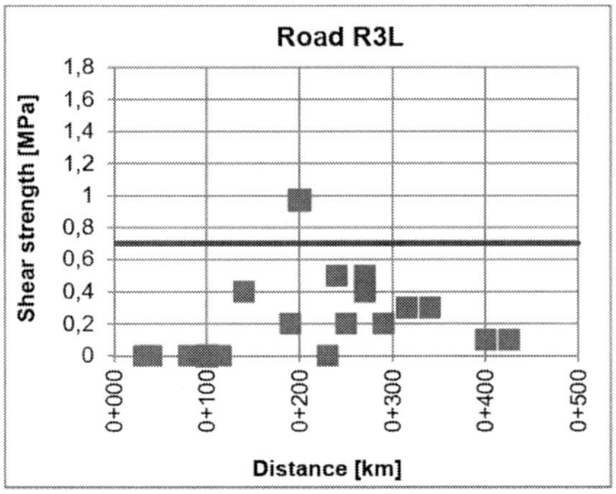

Figure 2. Results of shear strength obtained from the Leutner test between the asphalt base course AC22P and the binder course AC16W. Road R3L.

Figure 3. Surface of the lower asphalt base course AC22P. Non-homogenous surface. Road R1S.

Figure 4. Poor quality and lack of tack coat on the asphalt base course. Road R3L.

3. Mechanistic-empirical analysis of pavement performance

3.1. Pavement structure model

For the sake of analysis the pavement structure is modeled as a multilayer elastic half-space. The scheme of the model is presented in Figure 5. It is assumed that each layer is homogenous, isotropic and elastic. The calculations of stresses, strains and deflections inducted in pavement by a single wheel load were performed in BISAR® software [24].

The load is a vertical force $Q_V = 50$ kN, applied at a circular contact area and uniform contact pressure q=850 kPa. Two cases of load were considered: a load moving with constant speed and with horizontal force $Q_H = 0$ kN; and a load moving with deceleration and horizontal force $Q_H = 0.6$ kN $Q_V = 30$ kN. The additional horizontal force was considered in the analysis of shear stresses at the interlayer surface.

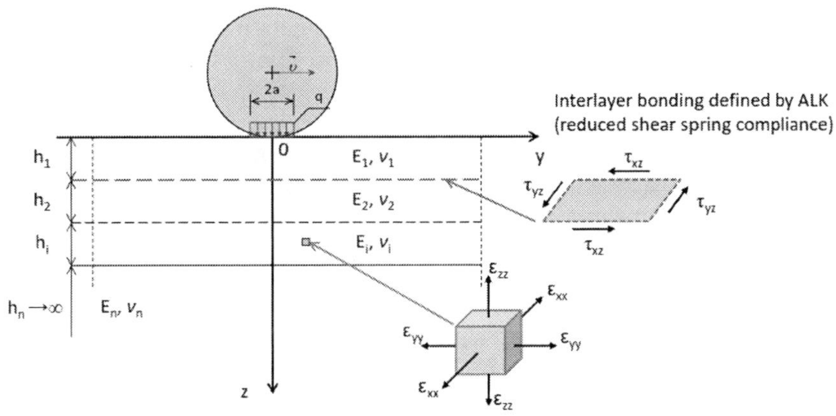

Figure 5. Model of pavement structure assumed for analysis.

Material properties of respective asphalt layers were assumed for two temperatures:

- +20°C for analysis of shear stress (similar to results obtained from the Leutner test),
- +13°C for analysis of fatigue life (the equivalent temperature for Poland [25]).

The detailed mechanical properties of each layer of pavement structure are given in Table 1.

Table 1. Properties of pavement layers.

Layer and material	Thick-ness (cm)	Binder content (%)		Air voids (%)	Stiffness/Elastic modulus E in temperature		Poisson ratio ν in temperature	
		m-m	v-v		+13°C (MPa)	+20°C (MPa)	+13°C (-)	+20°C (-)
1. Wearing course (SMA11 PMB 45/80-55)	4	6.4	16	3.0	7330	3630	0.3	0.38
2. Binder course (AC16W 35/50)	8	4.2	10.5	6.0	10300	6450	0.3	0.38
3. Asphalt base (AC22P 35/50)	Varies from 6 to 12	4.0	10.0	7.0	9820	6190	0.3	0.38
4. Subbase (Crushed stone)	20	-	-	-	400	400	0.3	0.3
5. Subgrade (Improved subgrade)	-	-	-	-	120	120	0.35	0.35

BISAR® software enables modeling of interlayer friction by reduced shear spring compliance ALK [24]. Full bonding between layers occurs when the shear strength of interlayer bonding is close to or a little lower than the shear strength of the used asphalt mixtures, that is when the results of the Leutner test meet the requirements. Full slip between two asphalt layers practically does not occur because of roughness of their surfaces. The modeling of interface parameter for limited bonding requires the performance of a series of calculations with different values of parameters ALK in the range from 0 to 200 m [26], [27], that is, respectively, full bond and full slip (non-bound).

3.2. Effect of reduced interlayer bonding on shear stresses

First stage of mechanistic analysis considered shear stresses on the interface between the asphalt base and the binder course. Calculations were performed for temperature of +20°C, same as in the Leutner test. The results of calculations are given in Figure 6. It can be concluded that for full (100%) bonding (parameter ALK = 0 m) shear stresses obtained from calculations are two times lower than the required 0.7 MPa from the Leutner test. It is also visible in Figure 6, that at low values of bonding, below 30%, the shear stresses are close to 0 MPa. Thus it was assumed that for places where specimens cored out from pavement had disintegrated, the bonding can be assumed as 30% due exclusively to friction at the layer interface. It is visible from comparison of Figures 2 and 6 that shear strength for the majority of specimens is lower than stresses obtained from mechanistic calculations. It means that two layers in pavement will displace horizontally in relation to each other, but interlayer bonding will bear some part of the shear stress, lower than its strength. It is also possible that when the maximum shear stresses caused by the traffic exceed its strength, the two layers will slip. Slip of layers – especially in the case of the wearing course – results in high displacement between them and causes characteristic semi-circular cracks, leading to loss of pavement roughness [4]. The risk of slip is especially high for sections with frequent acceleration and deceleration of vehicles, e.g. within the zones of crossroads. However, layer slipping is more often observed for cases of poor bonding quality between the wearing course and the binder course and it is rather not expected to occur between the binder course and the base or between two lifts of the asphalt base course.

It should be noted that the direct comparison of shear strength derived from the Leutner test with the stresses from mechanistic calculations is a considerable simplification due to different shape of the load. There is still a gap of knowledge pertaining to relation between the results of laboratory tests and the mechanistic calculations. Nevertheless, it is certain that at the level of stress between 0 and 0.3 MPa the interlayer bonding takes on values lower than 100% but the full slip between layers does not occur, and it is expected that interlayer bonding is greater than 30%.

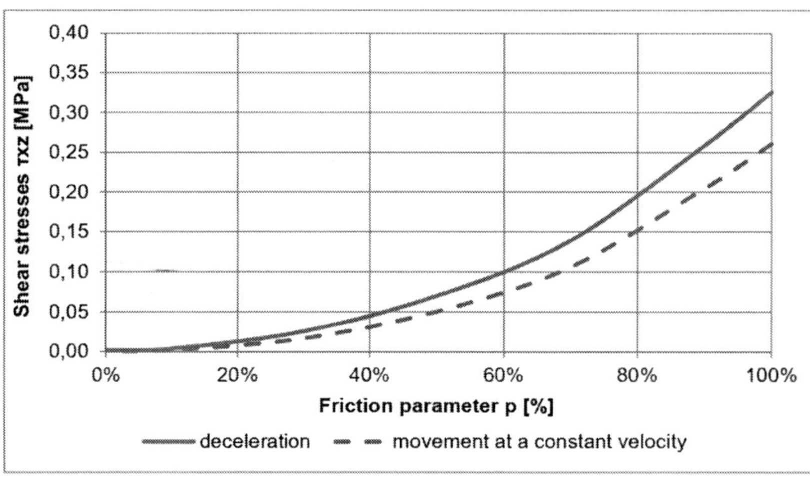

Figure 6. Effect of reduced bonding on shear stresses at the layer interface level.

3.3. *Effect of reduced interlayer bonding on fatigue life of pavement structure*

The analyses of fatigue life were performed for properties of asphalt mixtures obtained for equivalent temperature of +13°C, used for design of flexible pavements in Poland [25]. Calculations were performed for different levels of interlayer bonding from 0% (full slip) to 100% (full bonding) and for 4 thicknesses of asphalt layers from 18 cm to 24 cm. Two criteria were considered: bottom-up fatigue cracking of asphalt mixtures acc. to AASHTO 2004 [28] and the Asphalt Institute criterion of subgrade permanent deformation [29]. The fatigue life of pavement structure is represented by the minimum value obtained from the criteria of fatigue asphalt cracking and subgrade permanent deformation. The results of calculations are given in Table 2 for thickness of asphalt layers equal to 24 cm, and in Figure 7 for 4 thicknesses from 18 to 24 cm.

Reduced bonding (<100%) between asphalt base and binder course causes the increase in strains at critical points in pavement structure. For the AASHTO 2004 bottom-up fatigue cracking criterion the critical point is determined directly under wheel load at the level where the horizontal strains ε_{xx} in asphalt layers reach their maximum. For full bonding of asphalt layers the critical point (maximum ε_{xx}) occurs at the bottom of all asphalt layers in the asphalt base course. With the deterioration of quality of interlayer bonding between the base and the binder course, strains at the bottom of the asphalt base and strains at the bottom of the binder course increase simultaneously. For a limited level of bonding (parameter 0<ALK<200 m) strains at the bottom of the binder course are higher than at the bottom of the asphalt base and, therefore, fatigue cracks will be initiated in the binder course (see Table 2). However, for the considered case of pavement structure in R3L the fatigue life obtained from permanent deformation criterion is lower than fatigue life obtained for bottom-up cracking initiated in the binder course.

The reduction of interlayer bonding also results in the increase in vertical strains ε_{zz} at the top of subgrade. For full interlayer bonding and relatively thick pavements the criterion of permanent deformation provides higher fatigue life than the criteria of fatigue cracking and the pavement reaches the critical level of fatigue cracks first. It was revealed in the analysis that for low values of bonding the criterion of permanent deformation provides lower fatigue life. Finally, the minimum values of fatigue life were obtained for the criterion of fatigue cracking of the asphalt base and the binder course, as well as for the criterion of permanent deformation of subgrade. The minimum fatigue life values are highlighted in Table 2 and the critical criteria are marked in graphs in Figure 7. Figure 7 also shows the trend of impact of bonding level and asphalt layers thickness on the type of critical fatigue criteria.

Table 2. Fatigue life in millions of 100 kN ESAL's calculated for respective fatigue criteria. Road R3L with total asphalt layers thickness 24 cm.

Interlayer bonding (%)	Fatigue life (millions of 100 kN ESAL's)		
	Bottom-up cracking initiated in		Permanent deformation of subgrade
	asphalt base	binder course	
0%	8.00	5.91	**4.13**
10%	7.96	6.21	**4.28**
30%	8.17	10.41	**6.46**
50%	**9.77**	27.93	12.71
70%	**13.48**	130.89	26.03
100%	**28.72**	>>200	66.82

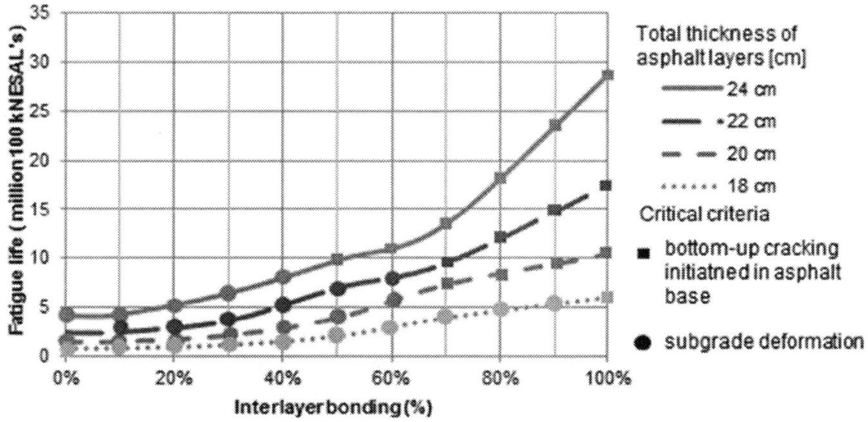

Figure 7. Effect of reduced interlayer bonding and total thickness of asphalt layers on fatigue life of pavement structure.

It is visible in the Figure 7 that a slight decrease in interlayer bonding causes a significant decrease in fatigue life, and this decrease is higher for thicker pavements. For example, with total asphalt layers thickness of 24 cm, slight decrease in interlayer bonding from 100% to 70% causes significant decrease in fatigue life by 50%. Further decrease to 30%, which represents zero shear strength, causes decrease in fatigue life by almost 85%. It can be concluded from Table 2 and Figure 7 that lack of interlayer bonding cannot be accepted by the investor and rebuilding of pavement layers is necessary. Partial lack of interlayer bonding, shown by shear strength in the Leutner test below required values, but above the maximum shear stress actually present in the pavement, can be accepted, provided that the quality of the asphalt mix and layer thicknesses fulfill all the other requirements.

4. Summary and recommendations

As shown by means of mechanistic-empirical analysis, lack or partial lack of interlayer bonding between asphalt layers can cause reduction in fatigue life of pavement structure.

Case studies included analyses of three road sections. Despite proper technical specification and technology of construction the shear strength of interlayer bonding did not fulfill the requirements. Potential reasons of unsatisfactory interlayer shear strength obtained in the Leutner test are as follows:

- Lack or low quality of an interlayer tack coat.
- Too low or too high quantity of bitumen emulsion for spraying a tack coat.
- Non-homogenous surface of the lower layer and weak interlocking due to insufficient compaction of the upper layer or the same gradation of bonded layers.

These reasons – in case of an otherwise good practice – would imply the main recommendations for contractors willing to reduce the risk of limited or zero interlayer bonding between asphalt layers.

Reduction of interlayer bonding between the asphalt base and the binder course or two lifts of the asphalt base course will result in:

- Premature initiation of fatigue cracks both on the bottom of asphalt layers in the base course and in the binder course, which accelerates the distress of pavement.
- Much faster occurrence of critical level of asphalt fatigue cracks for thicker pavements.
- Faster occurrence of permanent deformation rather than fatigue cracks for thinner pavements.
- For very low quality of interlayer bonding critical level of permanent deformation of thick asphalt pavement can occur before fatigue cracks.

References

[1] Uzan J, Livneh M, and Eshed Y 1978 Investigation of adhesion properties between asphaltic-concrete layers *Journal of AAPT* **47** pp 495–521

[2] Judycki J 2003 Sczepność między warstwami asfaltowymi nawierzchni *Drogownictwo* **9** pp 275–279

[3] Jaskula P 2006 Ocena sczepności między warstwami asfaltowymi nawierzchni *Drogownictwo* **12** pp. 412–414

[4] Romanoschi S A and Metcalf J B 2001 Effects of Interface Condition and Horizontal Wheel Loads on the Life of Flexible Pavement Structures *Transp. Res. Rec. J.* **1** pp 123–131

[5] Romanoschi S A and Metcalf J B 2002 The characterization of pavement layer interfaces *Ninth International Conference on Asphalt Pavements* pp 1–18

[6] Zofka A, Maliszewski M, Bernier A, Josen R, Vaitkus A and Kleiziene R 2015 Advanced shear tester for evaluation of asphalt concrete under constant normal stiffness conditions *Road Materials and Pavement Design* **16** (supl) pp 187-210

[7] Al-Qadi I L, Hasiba K I, Cortina A S, Ozer H, Leng Z, Parish D C and Worsfold S J 2012 *Best practices for implementation of tack coat: Part 1 Laboratory study* (Springfield)

[8] Ozer H, Al-Qadi I L, Wang H and Leng Z 2012 Characterisation of interface bonding between

hot-mix asphalt overlay and concrete pavements: modelling and in-situ response to accelerated loading *Int. J. Pavement Eng.* **13**(2) pp 181–196

[9] Raab C and Partl M N 1998 Shear strength properties between asphalt pavement layers *Arch. Civ. Eng.* **3** pp 353–365

[10] Raab C and Partl M N 2004 Effect of tack coats on interlayer shear bond of pavements *8th Conference On Asphalt Pavements For Southern Africa* p 9

[11] Górszczyk J and Malicki K 2012 Influence of the tack coat material on interlayer bonding properties in asphalt layers system *Proceedings of the 4th European Pavement and Asset Management Conference EPAM*

[12] Judycki J, Jaskula P, Dolzycki B, Pszczola M, Jaczewski M, Rys D and Stienss M 2015 Investigation of low-temperature cracking in newly constructed high-modulus asphalt concrete base course of a motorway pavement *Road Mater. Pavement Des.* **16**(1) pp 362–88

[13] Leutner R 1979 Untersuchung des Schichtenverbundes Oberbau *Bitumen* **3** pp 84–91

[14] Jaskula P 2014 *Instrukcja laboratoryjnego badania sczepności międzywarstwowej warstw asfaltowych wg metody Leutnera i wymagania techniczne sczepności* (Politechnika Gdańska Poland) p 12

[15] West R C, Zhang J and Moore J 2005 *Evaluation of bond strength between pavement layers* (Auburn, Alabama)

[16] Jaskula P 2014 Badania i analizy sczepności międzywarstwowej w nawierzchniach asfaltowych *Bud. i Archit.* **13**(4) pp 117–125

[17] Jaskula P 2014 Influence of compaction effectiveness on interlayer bonding of asphalt layers *The 9th International Conference Environmental Engineering* p 7

[18] Vaitkus A, Čygas D, Laurinavičius A, Vorobjovas V and Kleizienė R 2012 Research of asphalt layer bonding in Lithuanian pavement structures *Građevinar* **64**(11) pp 915–20

[19] White G 2016 State of the art: interface shear resistance of asphalt surface layers *Int. J. Pavement Eng.* **8436** pp 1–15

[20] Szydło A and Malicki K 2016 Analysis of the Correlation between the Static and Fatigue Test Results of the Interlayer Bondings of Asphalt Layers *Achives Civ. Eng.* **62**(1) pp 83–98

[21] Raab C and Partl M N 2009 Long term performance of asphalt pavement interlayer bonding *EMPA Act.* **08/09**(3) p 1

[22] Raab C, Grenfell J, Abd El Halim A O and Partl M N 2015 The influence of age on interlayer shear properties *Int. J. Pavement Eng.* **16**(6) pp 559–69

[23] Partl M N, Bahia H U, Canestrari F, de la Roche C, Di Benedetto H, Piber H and Sybilski D 2013 Advances in Interlaboratory Testing and Evaluation of Bituminous Materials, State-of-the-Art *Report of the RILEM Technical Committetee 206-ATB* (Springer)

[24] Shell Bitumen Business Group 1998 *BISAR 3.0 User Manual*

[25] Judycki J, Jaskuła P, Pszczoła M, Ryś D, Jaczewski M, Alenowicz J, Dołżycki B and Stienss M 2014 *Analizy i projektowanie konstrukcji nawierzchni podatnych i półsztywnych* (Wydawnictwa Komunikacji i Łączności: Warsaw, Poland)

[26] Judycki J, Jaskula P, Fronc T and Sierosławski T 2006 *Badania i ocena wpływu sczepności międzywarstwowej na trwałość konstrukcji nawierzchni asfaltowej, etap II* (Politechnika Gdańska, Poland)

[27] Stöckert U 2002 *Ein Beitrag zur Festlegung von Grenzwerten für den Schichtenverbund im Asphaltstraßenbau* (Technische Universitat Darmstadt)

[28] NCHRP Project 1-37A 2004 *Guide for Mechanistic-Empirical Design of New and Rehabilitated Pavement Structures, Final Report, Part 3 – Design and Analysis* Champaign

[29] Research and Development of the Asphalt Institute's 1982 *Thickness Design Manual (MS -1): Ninth edition* RR-82-2

[30] Iwanski M, Buczynski P, Mazurek G 2016 Optimization of the road binder used in the base layer in the road construction *Construction and Building Materials* **125** pp 1044–54

The influence of mixture composition, adhesion promotor and compaction degree on the groove stability of grooved Marshall asphalt

Cedric Vuye[1], Karolien Couscheir[1], Leen Lauriks[1], Wim Van den bergh[1] and Philippe Van Bouwel[2]

[1]Faculty of Applied Engineering, EMIB research group, University of Antwerp, 171 Groenenborgerlaan, Antwerp, 2020, Belgium
[2]Infrastructure & Real Estate, Brussels Airport Company, 1C Luchthaven Brussel Nationaal, Zaventem, 1930, Belgium

E-mail: cedric.vuye@uantwerpen.be

Abstract. After the first rehabilitation of runway 07R/25L in 2015, runway 01/19 was reconstructed in the summer of 2016, as part of a cycle where all runway pavements at Brussels airport are completely renovated each thirty years. The top layer is a Marshall asphalt with a polymer modified bitumen. To optimize the water drainage the central part of the runway (47 m wide) is grooved instead of applying an anti-skid layer. In this paper the focus is on the durability of the grooved top layer. Two different Marshall asphalt mixtures with a different maximum granulate size (10 mm or 14 mm) are compared, both in the laboratory and in a full-scale trial. In the laboratory the resistance against rutting and raveling are investigated for both mixtures with and without adhesion promotor, which did not show a positive effect. In the full-scale trial the compactability and impact of both a longer curing period and a variation in the degree of compaction on the groove stability is investigated for both mixtures using a heavy truck. No visual differences could be found except in areas which were undercompacted and showed more damage to the grooves.

1. Introduction

Grooving of the surface course for runways is used often as an alternative for so-called anti-skid layers. Both layers are used to improve the wet skidding resistance to prevent hydroplaning of the airplanes while landing or taking off. According to [1] there are four important factors that control the performance of runway surface grooves:

- Groove geometry (width and depth)
- Asphalt mixture properties
- Curing time before grooving
- Aircraft loading

As the groove geometry is fixed in the technical specifications and aircraft loading cannot be changed, this research focuses on the asphalt mixture properties and the curing time before grooving. A curing time of at least 30 days is still recommended by the Federal Aviation Administration (FAA)

[2, item P-621] and in Japan even up to two months [3], but for a commercial airport it is very difficult/expensive to wait such a long time. Grooving is performed mostly within one week of the laying of the asphalt. As shown in [3] the use of a polymer modified binder (PMB) can reduce the curing period and improve the groove stability drastically. This was both shown in laboratory test results as in a full-scale test, but it still remains unclear how long the curing period should remain. This was the inspiration for our full-scale tests.

The structure of the paper is the following: section two gives a description of the rehabilitation project for runway (RWY) 01/19 at Brussels airport and the different test tracks which were used for this research. In section three an overview is given of the laboratory results, including the resistance against rutting and raveling of the proposed mixtures with and without adhesion promotor (again to increase the groove stability). Next the results of a groove durability test on the full-scale test tracks are discussed in section four. The paper is finalized with some conclusions in section five.

2. Project description

In this section an overview is given of the actual runway and the full-scale test tracks that were constructed before the actual reconstruction to investigate the differences between the two Marshall asphalt types that were proposed by the contractor for the top layers. The focus of this research was on the top layers of the runway so no information will be provided about the foundation or base layers.

2.1. Runway 01/19

All runways at Brussels airport are being completely renovated as part of a 30 year cycle. Runway 07R/25L was rehabilitated as the first one in 2015, followed by runway 01/19 in the summer of 2016. The rehabilitation of the last runway 07L/25R is scheduled for 2018. An overview of the different runways and taxiways is shown in Figure 1.

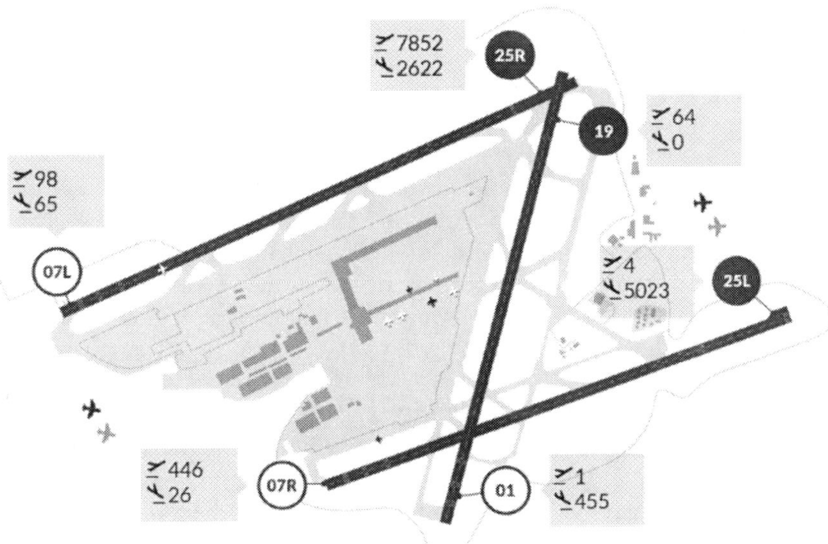

Figure 1. Overview of the different runways at Brussels airport.
[http://www.batc.be/en/statistics-runway-in-use]

The complete rehabilitation of runway 01/19 included:

- 239000 m² of asphalt on the runway and taxiways
- 62000 m² of asphalt on the shoulders (the area immediately beyond the lateral edges of the runway for aircraft running of the pavement)
- 373 kilometer of renewed cabling
- 5.6 kilometer of renewed runway drainage channels
- 3.2 kilometer of replaced sewage pipes
- replacement of 1850 lights by LED-lights

The current project had a total cost of approx. 21 million euro of which the asphalt pavements had a cost of approx. 5 million euro.

Originally the surface course of the runway consisted of a tar containing antiskid surface layer, but because of its high initial cost; environmental burden and its very strict installation instructions, it was decided to replace this antiskid surface layer by a grooved Marshall asphalt in the central part of the runway (47 m wide), as described in [4]. In [4] the binder type and content are described as between 5.0 and 7.0 % of a B70/100 or B100/150 paving grade bitumen (for easier compaction) and the aggregate size should be 0/14 mm when the thickness is between 40 and 50 mm for the surface course. A grading of 0/10 is used for regulating courses. To improve the groove stability and several mechanical properties such as fatigue and the resistance to rutting it was decided to use a polymer modified bitumen.

After the rehabilitation of RWY 07R/25L (2015) using a Marshall asphalt 0/14 (MA 0/14) some areas of the runway showed a surface texture which was too open, and could therefore lead to an early degradation, especially after the surface course was grooved. It was therefore decided to optimize the gradation of the mixture to obtain a more dense asphalt by including more fine material, while still maintaining the grading limits from [4]. As an alternative for this optimized MA 0/14 a Marshall asphalt 0/10 (MA 0/10) was proposed by the contractor Aswebo,

In order to be able to select the best mixture for RWY 01/19 an extensive laboratory study (see section 3) and tests on full-scale test tracks were performed (see section 2.2 and section 4).

2.2. Description of the full-scale test tracks

All test tracks have a length of 50 m and a width of 7.5 m as shown in Figure 2. The first goal of these test tracks was to assess the compactability of both asphalt mixtures in situ and to increase the thickness of the MA 0/14 from 4.2 cm (thickness of the surface course of RWY 07R/25L) to 5.0 cm (predetermined thickness for RWY 01/19) while still maintaining a good compaction degree. The second goal was to assess the influence on the groove stability due to a change in:

- Asphalt mixture grading: MA 0/10 vs. MA 0/14
- Compaction degree: normal compaction degree and instructions to roller operators to under- or overcompact
- Time of grooving: 3 days vs. 5 weeks after installation of the top layers

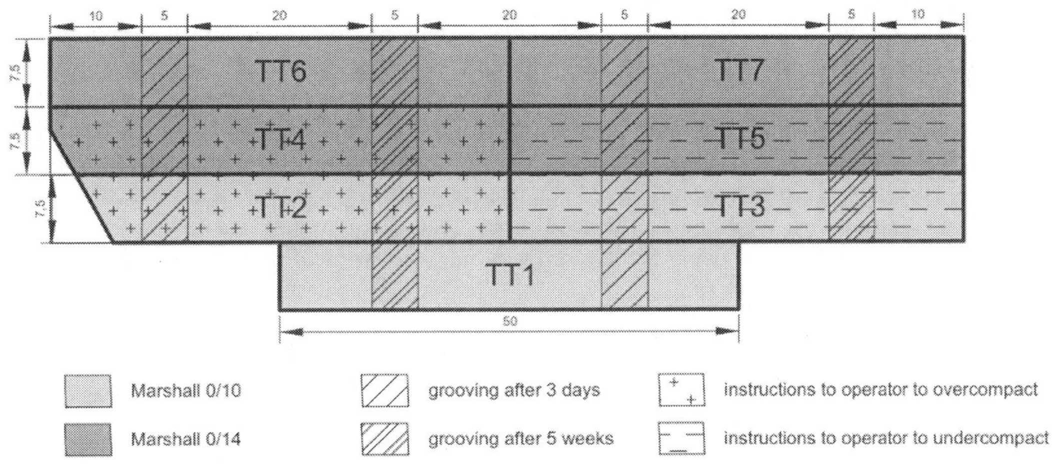

Figure 2. Overview of the full-scale test tracks.

The properties of all the different test tracks (TT) can be found in Table 1.

Table 1. Overview of the different test tracks

Test track ID	Grading	Compaction	Thickness [cm]
TT1	0/10	normal	4.2
TT2	0/10	instructions to overcompact	4.2
TT3	0/10	instructions to undercompact	4.2
TT4	0/14	instructions to overcompact	5.0
TT5	0/14	instructions to undercompact	5.0
TT6	0/14	normal	5.0
TT7	0/14	normal	4.2

The grooving dimensions are described in the tendering documents and are similar to the FAA grooving dimensions [2]. Transverse grooves saw-cut in the surface course must form a 6.0 ± 1.5 mm wide by 6.0 ± 1.5 mm deep by 38 ± 3 mm center-to-center configuration, as shown in Figure 3. To evaluate the possible effect of a longer curing period before grooving, as described in [3], two grooving areas of each 5 m wide were saw-cut in the asphalt, respectively one and three weeks after installation of the test tracks.

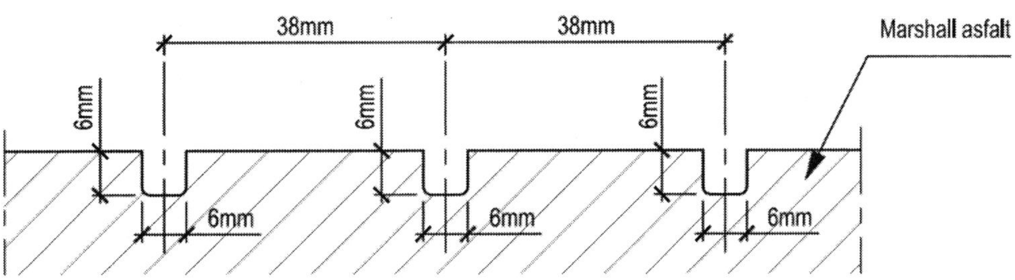

Figure 3. Detail of the grooving dimensions.

3. Laboratory results

The main goals of the laboratory tests were to determine the influence of the mixture grading (MA 0/10 vs. MA 0/14) and to evaluate the potential beneficial effect of an adhesion promotor (Wetfix AP17® by AkzoNobel at 0.3 % of the mass percentage of the binder). The purpose of an adhesion promotor is to enhance the bond between the binder and the aggregates which should lead to a better groove stability and resistance to raveling. If the adhesion promotor showed a beneficial impact on the laboratory results it would have been used as well in the full-scale test tracks.

The asphalt mixtures were mixed in the laboratory according to standard [5] and asphalt plates were compacted according to [6] using heavy compaction. Two types of plates were manufactured per mixture:

- Two plates of 50 x 18 x 5 cm which were not grooved and used for rutting tests
- Two plates of 60 x 40 x 5 cm which were then cut in two plates of 26 x 26 x 5 cm and grooved to be used for the raveling tests (4 tests per mixture), see Figure 4 (a)

The actual grooves were measured before performing the raveling tests. As mentioned in section 2.2 both the groove width and depth should be 6 ± 1.5 mm. The groove width ranges from 6.1 to 6.9 mm but the groove depth ranges from 4.5 up to 10.0 mm as the grooves were saw-cut manually. As all mixtures show a similar range of groove dimensions it was concluded that this was not an issue for comparing the different raveling results.

(a)

(b)

Figure 4. (a) Grooved plate of 26 x 26 x 5 cm used for raveling tests (b) Darmstadt Scuffing Device @ BRRC, adapted from [7].

3.1. *Resistance to rutting*

Two test samples of the same mixture are tested simultaneously with a load of 5000 N, a temperature of 50 °C and at a frequency of 1 Hz, according to [8]. The rutting is determined after 30000 cycles on both plates, averaged and expressed as a percentage of the sample height, as shown in Table 2.

The MA 0/14 outperforms the MA 0/10 in the rutting tests, although rutting isn't expected to be an issue on the runway itself. The influence of the adhesion promotor is unclear in these results as in one case the rutting is lowered and in the other case the rutting depth is higher.

Table 2. Results for the wheel tracking tests.

Mixture type	Average rutting depth [mm]	**Average rutting depth [%]**
MA 0/14	2.40	**4.90**
MA 0/14*	1.20	**2.30**
MA 0/10	2.55	**5.50**
MA 0/10*	2.90	**6.65**

* with adhesion promotor

3.2. Resistance to raveling

Four test samples of the same mixture are tested consecutively according to [9] with a tyre pressure of 3 bar, a vertical load of 1000 N, a test temperature of 25 ± 2 °C and a total of 10 cycles. The tests were performed by the Belgian Road Research Centre (BRRC) using a Darmstadt Scuffing Device (DSD), see Figure 4 (b).

The resistance against raveling is determined as the average weight loss in g/m² after 10 cycles, as shown in Table 3.

Table 3. Results for the raveling tests (measurements by BRRC).

Mixture type	Average weight loss [g/plate]	**Average weight loss [g/m²]**
MA 0/14	0.85	**21.4 ± 6.0**
MA 0/14*	2.67	**67.3 ± 11.5**
MA 0/10	0.97	**23.3 ± 4.8**
MA 0/10*	2.90	**51.0 ± 4.8**

* with adhesion promotor

The MA 0/14 and the MA 0/10 perform equally in the raveling tests. The influence of the adhesion promotor is surprisingly enough negative as the average weight loss increases. It was therefore decided not to use the adhesion promotor on the full-scale test tracks and on the actual runway.

More research is needed to verify if there might be a potential compatibility issue between the used PMB and the adhesion promotor.

4. Durability test on the test tracks

During the installation of the test tracks different measurements were performed:

- Determination of the compaction degree using different methods, for more details see [10]:
  - drilled cores
  - a non-nuclear density gauge (TransTech PQI-380)
  - a gamma density measurement device (Troxler 3450), performed by the Belgian Road Research Centre
- IR thermography (FLIR T640)
- Taking of samples to be analysed in the laboratory for mixture grading and bitumen content

Five weeks after the installation of the test tracks a heavy loaded truck with 5 axles (43 ton total weight including 30 ton ballast) was used to investigate the groove durability. The truck makes 100 straight passes on each TT at a speed of 20 km/h followed by 2 full-stop emergency brakes at the location of the grooves. A speed of 45-50 km/h was obtained before executing a full stop in under two seconds. This was chosen to simulate an airplane performing an emergency landing and to check how much the grooves would be damaged by this manoeuver. Another series of 150 straight passes was performed three days later. The straight passes simulate an airplane taxying and can be considered as pre-loading which is considered as beneficial [11].

Pictures and measurements with a caliper were taken before and after each of the different series of passes. After the brake tests with a heavy loaded truck a significant amount of damage to the grooves became apparent. No visual differences could be found between the two types of Marshall asphalt or between the two different grooved sections (different curing time), but more damage was apparent in the sections which were undercompacted. This shows the importance for the groove stability to achieve the correct compaction degree. Figure 5 shows an example of the grooves before and after the brake test and the loose granulates which were recovered from the test tracks. As the loose granulates are completely covered with bitumen it is clear that the damage occurs in the mortar matrix and is not an adhesion problem.

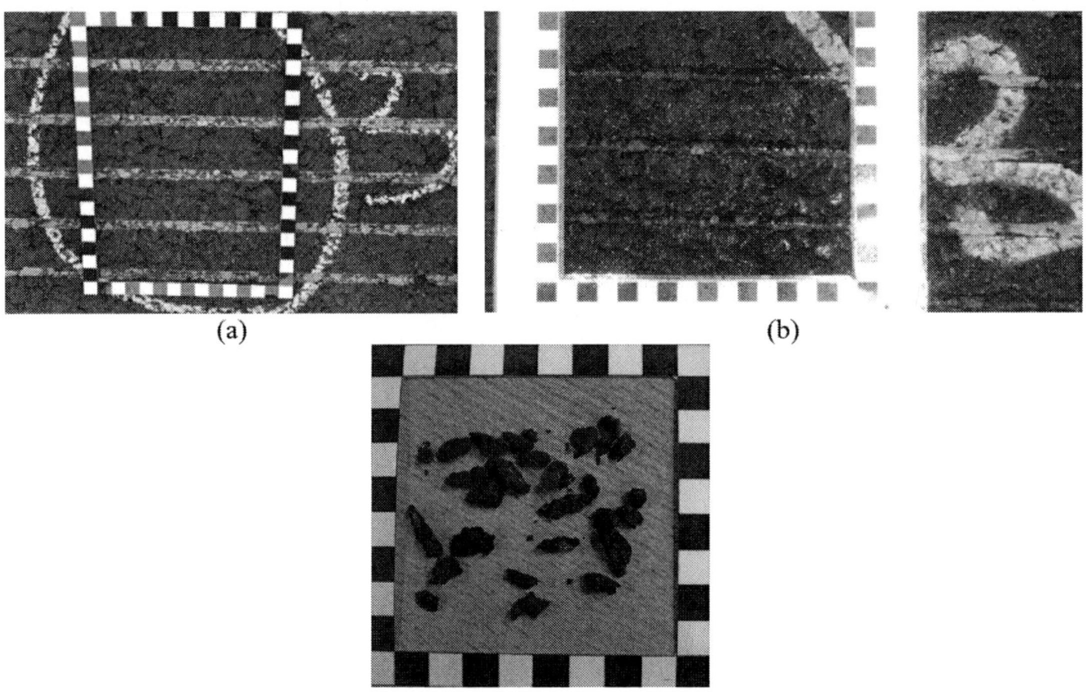

(a) (b) (c)

Figure 5. (a) Measurement location before testing in an "undercompacted" MA 0/14 TT
(b) Same location after testing (c) Example of the loose granulates recuperated after testing.

5. Conclusions

In order to obtain a more closed surface of the top layer of the grooved runway a Marshall asphalt 0/10, prescribed in [4] for regulating courses, was proposed by the contractor as an alternative for a MA 0/14. In both laboratory and a full-scale test these two mixtures were compared. Furthermore the effect of an adhesion promotor was investigated in the laboratory study. The following conclusions can be made:

- The MA 0/14 outperforms the MA 0/10 in the rutting tests, although rutting isn't expected to be an issue on the runway itself. The influence of the adhesion promotor is unclear in these results as in one case the rutting is lowered and in the other case the rutting depth is higher.

- The MA 0/14 and the MA 0/10 perform equally in the raveling tests. The influence of the adhesion promotor is surprisingly enough negative as more ravelling occurs, but more research is needed to verify these results and check the compatibility between the used PMB and the adhesion promotor.
- After the brake tests with a heavy loaded truck a significant amount of damage to the grooves became apparent. No visual differences could be found between the two types of Marshall asphalt or between the two different grooved sections (different curing time), but more damage was apparent in the sections which were undercompacted. This shows the importance for the groove stability to achieve the correct compaction degree.

Acknowledgements
The authors would like to thank the personnel of the Infrastructure & Real Estate department of Brussels Airport Company and the contractor Aswebo. The authors would also like to acknowledge the support of the Belgian Road Research Centre who performed the measurements with the gamma density measurement device and the experiments regarding the resistance against raveling. Without their support this research would not have been possible.

References
[1] Mejías-Santiago M, Doyle J D and Rushing J F 2015 Performance of Grooved Warm-Mix Asphalt Pavement Surfaces Under Heavy Aircraft Load and High Tire Pressure *Transportation Research Record* **2501** pp 40-5
[2] U.S. Department of Transportation, Federal Aviation Administration 2014 Advisory Circular AC150/5370-10G https://www.faa.gov/regulations_policies/advisory_circulars/
[3] Kawamura N, Kawana F and Maekawa R 2012 Stability of Runway Grooving under Repeated Large Aircraft Loads *Engineering Journal* **16**(4) pp 127-35
[4] Ministry of Defence, Defence Estates 2009 Specification 13 Marshall Asphalt for Airfields 2009 https://www.gov.uk/government/publications/specifications
[5] European Committee for Standardization (CEN) 2007 Bituminous mixtures - Test methods for hot mix asphalt - Part 35: Laboratory mixing (*EN 12697-35:2004+A1:2007*)
[6] European Committee for Standardization (CEN) 2007 Bituminous mixtures - Test methods for hot mix asphalt - Part 33: Specimen prepared by roller compactor (*EN 12697-33:2003+A1:2007*)
[7] De Visscher J and Vanelstraete A 2017 Ravelling by traffic: Performance testing and field validation *International Journal of Pavement Research and Technology* (available online)
[8] European Committee for Standardization (CEN) 2007 Bituminous mixtures - Test methods for hot mix asphalt - Part 22: Wheel tracking *(EN 12697-22:2003+A1:2007*)
[9] European Committee for Standardization (CEN) 2014 Bituminous mixtures – Test methods for hot mix asphalt – Part 50: Resistance to scu*ffing (prTS12697-50*)
[10] Van den bergh W, Vuye C, Kara P, Couscheir K, Blom J and Van Bouwel P 2017 The use of a non-nuclear density gauge for monitoring the compaction process of asphalt pavement *Proc. BESTInfra 2017 (Prague)*
[11] Wu S, Hao P and Hachiya Y 2002 Laboratory evaluation on durability of grooving for airport runways *Journal of pavement Engineering JSCE* **7** pp 30p1–30p7

IOP Conf. Series: Materials Science and Engineering **236** (2017) 012007 doi:10.1088/1757-899X/236/1/012007

The influence of nanoclay on the durability properties of asphalt mixtures for top and base layers

Johan Blom[1], Bram De Kinder[1], Jannes Meeusen[1] and Wim Van den bergh[1]

[1]Faculty of Applied Engineering, EMIB-research group, University of Antwerp, 171 Groenenborgerlaan, Antwerp, 2020, Belgium

E-mail: johan.blom@uantwerpen.be

Abstract. To avoid traffic congestion, due to road works, a continuous research into asphalt pavement and especially its durability is of great importance. This research focuses on improving the mechanical performance and the durability of asphalt mixtures by nanoclay modified bitumen. This promising technique of introducing nanoclays or nano particles into bitumen could offer an significant improvement on the fatigue properties and rutting performance and thus the durability of the asphalt top layer.

1. Introduction

For the heavely used "congested" Belgian road network, a continuous research into asphalt pavement and especially its durability is of great importance to avoid traffic congestion, due to road works.

This research focuses on improving the mechanical performance and the durability of asphalt mixtures by nanoclay modified bitumen.

A specific asphalt mixture type for base layers in Flanders, AC 14 base 35/50 (EN 13108-1), using a relatively hard bitumen was modified. Both physical and rheological properties of the modified binder were evaluated. This included the determination of the penetration and softening point and the characterization of dynamic rheological parameters, by using a Dynamic Shear Rheometer (DSR).

The prefix 'nano' indicates the occurrence of a material that measures less than 100 nm, in at least one dimension. From literature, general trends on organic montmorillonite nanoclay (OMMT) modified bitumen can be derived. It is indicated that the penetration decreases and the softening point increases by adding nanoclays[1, 2, 3, 4, 5, 6]. An explanation can be found by the reinforcement effect of the silicate layers on the bitumen [7]. Based on tests with a DSR, the bitumen seems to become more elastic and stiffer [2, 8, 9, 10, 11]. This effect is more pronounced at high temperatures and/or low frequencies. The viscosity increases also after the OMMT modification [2, 6, 8, 9, 12]. The indirect tensile strength does not change at 5°C and 25°C but does improve at 40°C [1, 2, 5, 12]. In addition, nanoclay can form a barrier in bitumen through which penetration of oxygen and water is prevented [2]. The effects of the modification by nanoclay are more pronounced by increasing the amount or when using soft bitumen. Nevertheless, there is still a need for sufficient research on the use of nano-modified bitumen in asphalt. Especially as there are almost no research publications regarding the application of nanoclay in harder bitumen, a type which is preferably used in Belgian mixtures to provide rutting, though cracking must be avoided. This exploratory study will evaluate physical and rheological properties of a harder bitumen modified with nanoclay. Additionally, the modified mixture was also processed into an asphalt mixture of which the changes in mechanical and physical properties were identified.

2. Experimental

Because adequate research on the use of nanoclay in hard bitumen is lacking, the relevance of the subject is tested in this study. The study is carried out on an base layer because the literature shows that nano-modified bitumen can have a positive influence on durability properties.

2.1. Materials

The bitumen B35/50 used in this study is an oil derivative from Dunkirk provided by Total. The binder has a softening point of 52,2°C and a penetration of 41.0 dmm. The binder is used in an AC base 35/50 mixture. The sandstone in the asphalt mixture was provided by Bremanger and delivered by Antwerp Stone Terminal (AST). The mixture also contains sand from the Lower-Zeescheldt, of the firm Delahaye – Lauwers. The filler in the asphalt mixture is from Sibelco, a Maastricht based company (Figure 1).

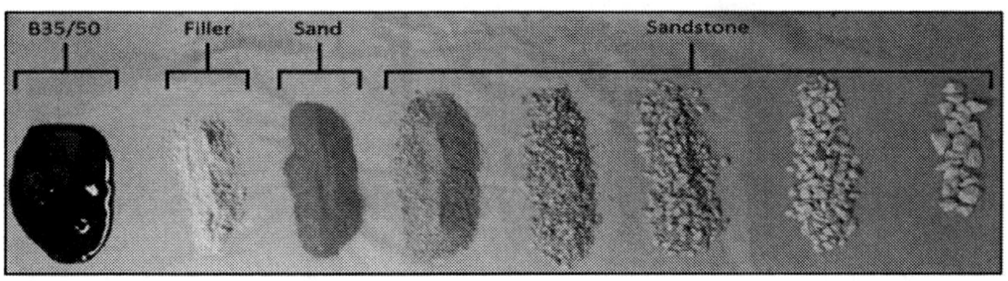

Figure 1. Materials for the manufacturing the bituminous mixture AC 14 base 35/50.

The nanoclay used to modify the bitumen is cloisite-15 from BYK Chemie GmbH. The product consists of bentonite treated with an alkyl group to form an OMMT. The chemical name of this surfactant is: bis(hydrogenated tallow alkyl)dimethyl. It stems from hydrogenated tallow [13, 14, 15]. The nanoclay has an interlayer spacing of approximately 35 Angström. For this study cloisite-15 was chosen to modify the bitumen because of its organic nature and larger interlayer spacing. Additionally, a modification with 3 and 5% nanoclay was chosen as higher percentages could lead to unstable and very viscous bitumen [1]. The auto-ignition temperature of cloisite-15 is 190°C. For this reason, a temperature of 175°C was selected as maximum temperature for the bitumen as well as the granules. To modify the bitumen with nanoclay, a high shear mixer was selected. Using this mixer an intercalated or an exfoliated structure can be obtained [2, 16, 17, 18]. The layer separations in the nanoclay result in higher surface work which intensives the interaction with the asphalt binder (Figure 2).

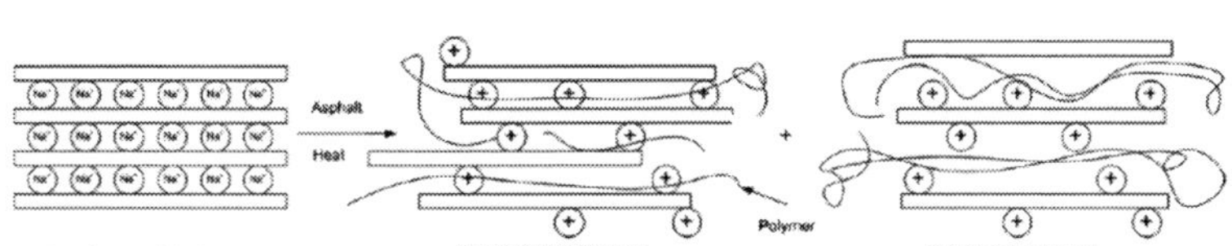

Figure 2. Blended with non-modified and polymer-modified nanoclay. [19]

2.2. Processing/mixing

The modification process was performed on binder batches of approximately 600g. The nanoclay was pre-sieved with a sieve of 0.5mm and dried for 50 min. at 150°C. The bitumen was heated in sealed cans for one hour in a ventilated oven at 150°C. The mixing process can be divided in two stages. At first the nanoclay is pre-mixed with the bitumen. An IKA®RW 16 basic mixer was used with a speed of (400±100) rpm for 10 min. To pull apart the silicate layers a high shear mixer "Silverson L5" was used during the second mixing stage. A detail of the Square Hole High Shear Screen stator is shown in Figure 3. The mixing process was executed at a speed of (4500±100) rpm [2]. In both stages the bitumen was heated on a heating plate at 145°C (Figure 3).

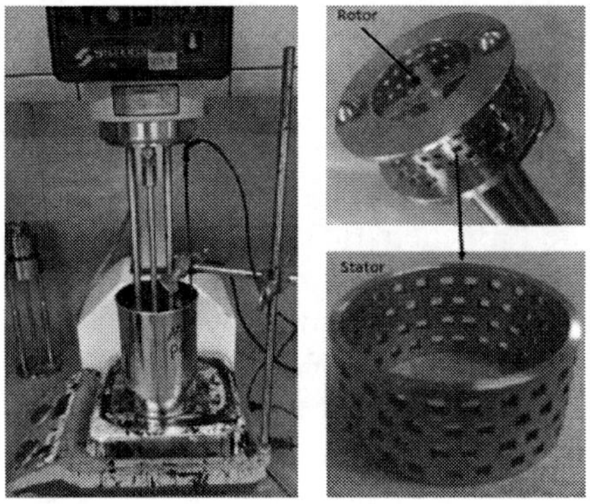

Figure 3. Experimental set-up of the mixing process.

2.3. *Experimental setups*

2.3.1. Penetration. The penetration is determined in accordance with standard NBN EN 1426 with an Anton Paar automatic penetrometer type PNR12. Immediately after the entire modification process the samples were moulded. Both cooling-down to room temperature as conditioning took 90 min.

2.3.2. Softening point. The softening point is determined in compliance with standard NBN EN 1427 using an automatic appliance by Herzog, type HRB 754. The samples of this test were moulded immediately after the pre-mixing and mixing processes as well. Cooling-down to room temperature took 90 min.; conditioning 15 min.

2.3.3. Fundamental rheological parameters. The complex modulus and phase angle were determined in conformity with standard NBN EN 14770. An Anton Paar DSR type MCR 102 was used. Frequency sweeps were performed at 45°C and ended at -5°C whit temperature steps of 10°C. All samples were moulded immediately after the modification process. Afterwards they were stored in a cold store (5±1 °C) for 20 hours before the test commenced.

2.3.4. Attenuated Total Reflectance – Fourier Transform Infrared spectroscopy (ATR - FTIR). The ATR-FTIR-tests were performed using an Thermo Scientific infrared spectrometer, type Nicolet iS10. The absorption diagram was compiled using OMNIC software based on one reflection of an infrared radiation beam. Before the testing commenced, an environmental scan was carried out. All samples have been examined with 32 consecutive scans with a 4cm-1 resolution [19].

2.3.5. Storage stability. The storage stability is evaluated in accordance with standard NBN EN 13399. The stability is appraised in line with testing the softening point (NBN EN 1427).

2.3.6. ITSR. ITSR-tests are carried out in accordance with standard NBN EN 12697-12. The samples were created manually in keeping with NBN EN 12697 31 and compacted with 25 gyrates in conformity with the Belgian 'Standaardbestek 250' (Version 3.1). The gyrator used is from ELE International, type ELE-Servopac. The measurements (diameter and height) and bulk density of the samples were established in accordance with standard NBN EN 12697-29 and NBN EN 12697-6 (procedure B). The samples were cut perpendicularly on the longitudinal axis having an average height of 41mm and an average diameter of 100mm. The samples were divided into a conditioned and a non-conditioned series (NBN EN 12697-12) to ensure the same average bulk density was reached. To determine the ITSR-level, the samples were tested in compliance with standard NBN EN 12697-23 on their indirect tensile strength (ITS) at a temperature of 20°C. For this a Controls appliance model 82 – P0375 was used. The Uniframe program was used to process the load and deformation data.

2.3.7. Wheel tracking. The creating of the plate samples was in accordance with standard NBN EN 12697-35. A Baustoff-Prüfsysteme Wennigsen GmbH, type GZM-30+ mixing drum was used. The pre-mixing process took 5 minutes. The granulates, filler and bitumen were pre-heated until they reached a pre-mixing temperature of 170°C. A 175°C pre-mixing temperature was determined for the 5% nano-modified mixture because of the increased viscosity. The plates were compacted with a BBPAC mlpc® roller compactor in accordance with standard NBN EN 12697-33 using the high compaction method. Each sample was tested on rutting in accordance with standard NBN EN 12697-22 one week after manufacturing. The standard recommends to test tracking after 30 000 cycles but in this study it was also tested after 40 000 and 50 000 cycles The experimental setup is shown in Figure 4.

Figure 4. Manufacturing and compaction of the asphalt.

3. Results and discussion

3.1. Empirical rheological parameters

The tests show that with an increasing amount of nanoclay the penetration remains consistent while the softening point increases (Table 1).

Table 1. Results of the rheological, penetration & softening point.

	Amount of cloisite-15 (%)		
	0	3	5
Penetration (dmm)			
Before modification	41		
After mixing process	41	42	43
Softening point (°C)			
Before modification	52.2		
After pre - mixing process	52.2	53.3	53.8
After mixing process	52.2	54.0	57.8

The latter is only significant in bitumen with 5% cloisite-15 which can be attributed to the reinforcement effect of the silicate layers on the bitumen. This result could indicate an increased rutting resistance. Subsequently, the penetration and softening point do not change during the modification process which could indicate that the bitumen does not age. Furthermore, the softening point seems to increase after the mixing process in comparison to the pre-mixing process. This result can indicate that an intercalated, possibly exfoliated, structure was obtained.

3.2. Fundamental rheological parameters
The master curves show that a modification with 3% cloisite-15 does not change the fundamental rheological parameters. The with 5% cloisite-15 modified bitumen displays a different behaviour. Namely, for frequencies lower than 0.1Hz, the complex modulus increases. For frequencies, higher than 60Hz the complex modulus decreases (Figure 5a). The phase angle reduces for frequencies lower than 0.1Hz and increases for frequencies higher than 40Hz. The black diagram (Figure 5b) shows that when the amount of nanoclay increases, the bitumen becomes more elastic at a given stiffness. This effect is more pronounced when the complex modulus is lower than 10 GPa.

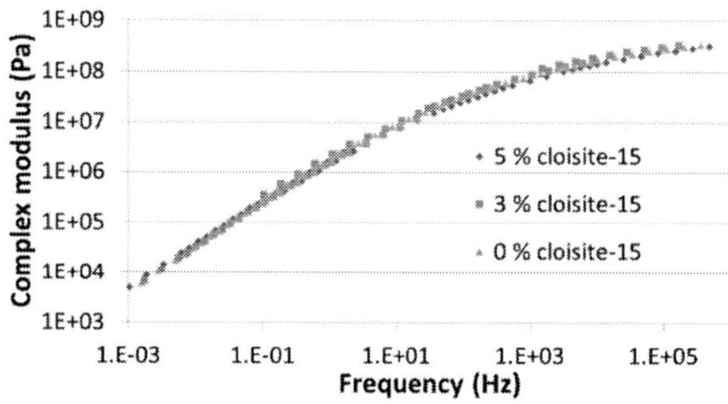

Figure 5a. Master curves (25°C).

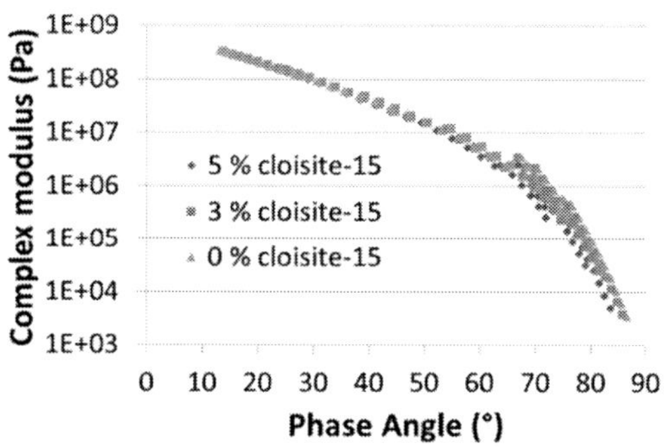

Figure 5b. Black curves.

3.3. ATR - FTIR
The FTIR analysis shows that in modified bitumen (3 and 5%) a greater absorption is perceived around a wave length of 1040cm^{-1} compared to the reference bitumen. This result is attributed to the presence of nanoclay and is related to the amount and the mixing process (complete modification). This could indicate that an intercalated, possibly exfoliated, structure was obtained. An explanation is that after the mixing process the silicate layers remain clustered as a result of which the concentration of cloisite-15 in the bitumen is overall relatively low. Subsequently, the results show that based on the carbonyl bonds (C=0; peak value at 1700cm-1) the bitumen does not age during the modification process.

3.4. Storage stability
After testing the storage stability, the mixtures seemed stable with respect to the sedimentation of silicate layers, because the softening point of the top and bottom of the samples remained almost identical after storage at high temperature. A broader interpretation of the aspect stability could mean that the silicate layers could possibly rearrange themselves when stored at high temperatures resulting in a decrease of interlayer spacing. This will partially undo the positive influence of cloisite-15. This conclusion is reached since the softening point of the modified bitumen (3 and 5%), after storage at high temperature, is reduced to the same value as after the pre-mixing process (Table 2).

Table 2. Results of the storage stability test.

Storage stability	Amount of cloisite-15 (%)		
	0	3	5
softening point (°C)			
Upper part	52.3	53.7	53.4
Lower part	52.4	53.3	53.4

3.5. ITS- and ITSR values

All asphalt mixtures in this work show a resistance against water sensitivity of approximately 70%, no significant changes occur. The ITS curves however, demonstrate that the modified asphalt mixtures (3 and 5%) allow for a greater deformation before the peak load is reached. However, these peak loads are lower compared to the reference mixture. This applies to both the conditioned as the non-conditioned series. Analysis of the fracture surfaces shows that in the samples containing 5% cloisite-15 the fracture pattern is very different compared to other series.

A possible explanation is a lesser internal adhesion between the raw materials which would cause the sample rather to crumble at the edges than a diametric fracture to be formed. The aforementioned ITS curves do not take the volume of the samples into account, therefore the fracture and failure energy was analysed. The fracture energy is virtually the same in all non-conditioned samples. Since the volume of these samples is almost the same and a higher peak load is registered in the reference mixture, the modified samples deform more when the maximum load is reached. The fracture energy in conditioned samples decreases, but because of the high standard deviation, it is difficult to draw a conclusion. The failure energy of the non-conditioned samples remains approximately the same. It follows that the modified samples allow for a slightly larger deformation before failure. For conditioned samples the failure energy increases slightly. It is presumed that the conditioned samples allow for a greater deformation before failure. Based on the standard deviation, this conclusion needs to be taken with caution. The results presented above follow the hypothesis that the failure mechanism of modified samples may differ due to reduced adhesion. Subsequently it is stated that the modified samples can deform more before failure because of the reinforcement effect of nanoclay. The mixture bonds better in its entirety even though the bitumen's adhesion is reduced.

3.6. Wheel tracking

The tested AC 14 base 35/50 mixture for base layers already has an excellent rutting resistance, therefore only marginal changes were detected in the modified mixtures. The with 5% nanoclay modified mixture is the only one that shows visible improvement.

From the beginning the rut depth is less in comparison to both other mixtures. The rut depth at 30 000 cycles is 2.4% which in absolute numbers is a small change. This compares to an improvement in rutting resistance of on average 15%, which is relatively speaking a strong improvement. Even after 50 000 cycles the improvement remains visible in mixtures with 5% nanoclay. There is still some space for

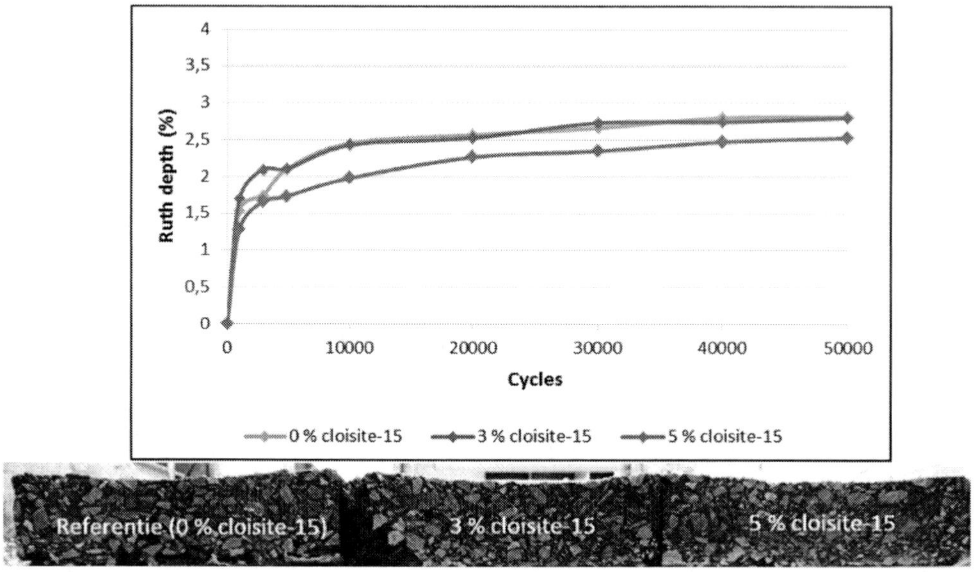

Figure 6. Deformed asphalt slabs after wheel tracking experiment.

4. Conclusions

In generall, nanoclay seems to have a slightly positive influence on durability properties in asphalt for base layers. Essential to this is the intercalated structure that might be obtained which was proved by the FTIR. The softening point increases when the amount of added nanoclay is increased, which can be attributed to the reinforcement effect of nanoclay. The penetration however remained consistent when nanoclay was added. A possible explanation for this latter, is that the bitumen is already so hard that the added nanoclay has no effect on the penetration value. In bitumen modified with 5% nanoclay, a slight increase of stiffness and elasticity takes place at low frequencies. It is not a significant increase according to which a modification of a relatively hard bitumen with nanoclay may not be efficient. Cloisite-15 seems to have good storage stability in the bitumen. In the broader interpretation of this notion, the silicate layers seem to have the ability to rearrange themselves at high temperatures so that the reinforcement function of nanoclay is nullified. It appears that cloisite-15 modified bitumen should be processed into asphalt in a relatively short period of time. The water sensitivity remains consistent whereas the rutting resistance increases slightly when bitumen is modified with 5% nanoclay. There is a limited influence, which is why the modification of existing asphalt mixtures for base layers in higher construction class is less recommended. The ITS value of nanoclay decreases, which can be explained by a reduced adhesion in modified asphalt mixtures. It is therefore important to determine the optimum binder content in order to obtain an equal bonding characteristic in modified asphalt mixtures. Nanoclay does seem to provide a better reinforced asphalt structure. For fatigue properties in asphalt mixtures, this can be positive.

Overall, based on this study, two research paths are recommended. Firstly, it is recommended to make a comparison between an existing asphalt mixture and this asphalt mixture with a softer bitumen that is modified with nanoclay. Secondly, it is recommended to conduct research into the use of nanoclay in top layers. Mixtures for top layers more often contain a softer bitumen which can lead to the positive effects of the nanoclay being more pronounced. Furthermore, the aging resistance can increase because of the barrier properties. In addition, this research shows that rutting resistance can increase through modification, which is of great importance for the durability properties of top layers.

Acknowledgements

The authors would like to thank the "Energy & Materials in Infrastructure & Buildings" (EMIB) research group, University of Antwerp, and Hilde Soenen of Nynas to make this research possible.

References

[1] Ghile D B 2006 *Effects of Nanoclay Modification on Rheology of Bitumen and on Performance of Asphalt Mixtures* (Technische Universiteit Delft)

[2] Liu G 2011 *Characterization and Identification of Bituminous Materials Modified with Montmorillonite Nanoclay* (Technische Universiteit Delft)

[3] Agubra V A, Owuor P S and Hosur M V 2013 Influence of Nanoclay Dispersion Methods on the Mechanical Behavior of E-Glass/Epoxy Nanocomposites *Nanomaterials* **3** pp 550-63

[4] Pei J, Wen Y, Li Y, Zhang Z, Shi X, Zhang J, Li R, Cao L and Du Q 2014 Organic Montmorillonite Modified Asphalt Materials: Preparation and Characterization *Journal of Testing and Evaluation* **42**

[5] Jahromi S G and Khodaii A 2009 Identification Effect of Nanoclay on Engineering Properties of Asphalt Mixtures *Amirkabir Int. Journ. of Modeling, Identific., Simulat. And Control* **1** pp 49-57

[6] Jianying Y, Xuan Z, Shaopeng W, Lin W and Gang L 2007 Preparation and properties of montmorillonite modified asphalts *Materials Science and Engineering* **1-2** pp 233-8

[7] Muniandy R, Mahdi L M J, Yunus R B, Hasham S and Aburkaba E 2013 *Effect of Organic Montmorillonite Nanoclay Concentration on the Physical and Rheological Properties of Asphalt Binder* (University Putra Malaysia)

[8] Mahdi L. M. J, Muniandy R, Yunus R B, Hasham S and Aburkaba E 2013 Effect of Short Term Aging on Organic Montmorillonite Nanoclay Modified Asphalt *Indian Journal of Science and Technology* **6**

[9] Jamshidi A, Rosli M, Hasan M, Yao H, You Z and Hamzah M O 2015 Characterization of the rate of change of rheological properties of nano-modified asphalt *Construction and Buiding Materials* **98** (Elsevier) pp 437-46

[10] Jahromi S G and Khodaii A 2009 *Effects of nanoclay on rheological properties of bitumen binder* (Amirkabir University of Technology Tehran Iran)

[11] Muniandy R, Mahdi L M J, Yunus R B, Hasham S and Aburkaba E 2013 *Effect of Organic Montmorillonite Nanoclay Concentration on the Physical and Rheological Properties of Asphalt Binder* (University Putra Malaysia)

[12] Jahromi S G, Andalibizade B and Vossough S 2010 Engineering properties of nanoclay modified asphalt concrete mixtures *The Arabian Journal for Science and Engineering* **35**

[13] Ash M and Ash I 2004 *Handbook of preservatives* (Synapse Information Resources New York USA)

[14] Thomas S, Durand D, Chassenieux C and Jyotishkumar P 2015 *Handbook of Biopolymer-Based Materials* (John Wiley nad sons)

[15] Koo J 2016 *Fundamentals, Properties, and Applications of Polymer Nanocomposites* (Cambridge University Press)

[16] You Z, Millis-Beale J, Foley J M, Roy S, Odegard G M, Dai Q and Goh S W 2010 Nanoclay-modified asphalt materials: Preparation and characterization *Cosntruction and building materials* **25** (Elsevier) pp 1072-78

[17] Pei J, Wen Y, Li Y, Zhang Z, Shi X, Zhang J, Li R, Cao L and Du Q 2014 Organic Montmorillonite Modified Asphalt Materials: Preparation and Characterization *Journal of Testing and Evaluation* **1** (ASTM) pp 118-25

[18] Yao H, You Z, Li L, Shi X, Goh S W, Mills-Beale,J and Wingard D 2012 Performance of asphalt binder blended with non-modified and polymer-modified nanoclay *Construction and building materials* **35** (Elsevier) pp159-70

[19] Usuki A, Tukigase A and Kato M 2002 Preparation and properties of EPDM-clay hybrids *Polymer* **43** (Elsevier) pp 2185-89

Evaluation of elevated temperature properties of asphalt cement modified with aluminum oxide and calcium carbonate nanoparticles

Shaban Ismael Albrka Ali[1,2], Amiruddin Ismail[2], Ramez A AlMansob[3] and Dhawo Ibrahim Alhmali[2]

[1]Civil Engineering Department, Near East University, Near East Boulevard, 99138 Nicosia, North Cyprus, Mersin 10, Turkey
[2]Department of Civil and Structural Engineering, Universiti Kebangsaan Malaysia, Pekan Bangi, 43600 Bangi, Selangor, Malaysia
[3]Faculty of Engineering, Technology and Built Environment, UCSI University, 56000 Cheras, Kuala Lumpur, Malaysia

E-mail: shabanismael.albrka@neu.edu.tr, shabarofking10@gmail.com

Abstract. Higher temperature properties of the asphalt cement have been characterized before and after modification using dynamic shear rheometer (DSR) and viscosity testing. In this study, calcium carbonate nanoparticles ($CaCO_3$) and aluminum oxide nanoparticles (Al_2O_3) have been added to the base asphalt cement with concentrations of 3, 5 and 7%.wt by the weight of the asphalt cement. The increase of $CaCO_3$ and Al_2O_3 content has significant effect on the properties of asphalt cement. The viscosity of the modified asphalt cement increased up to 90 and 108% respectively compared to the base asphalt cement. In addition, the results showed that both modifiers have great storage stability and compatibility at elevated temperature. The evaluation of the rheological properties of asphalt cements revealed that the stiffness of the modified samples improved with additional increase of the modifier concentration of up to 5%, which indicates better resistance to rutting parameter. The enhancement was up to 388.89% for Al_2O_3 and 74.07% for $CaCO_3$. As a result, the usage of $CaCO_3$ and Al_2O_3 nanoparticles can be considered as appropriate alternative materials to modify asphalt cement.

1. Introduction

Asphalt cement has been used for many years in the construction of highways and road networks. Therefore, it is able to resist the distresses caused by traffic loads and change in temperatures [1]. However, due to the limitation of temperature susceptibility, the low, intermediate and high temperature and temperature performance of asphalt need to be enhanced. As a result, modification of asphalt cement is crucial to improve the performance of pavement [2-5]. There are numerous types of asphalt cement modifiers, such as polymers, nanomaterials and rubbers. Currently, nanomaterials have been rapidly incorporated into the field of asphalt cement with various types of nanomaterials being used as modifiers of asphalt cement and asphalt mixes. It was found that nanomaterials are able to improve the performance of asphalt cement and asphalt mixes. A study conducted by Wang et al. (2017) showed that the performance investigation demonstrated that multi-walled carbon nanotubes

(CNTs) performed a positive role to enhance the properties of modified asphalt cements. In addition, the anti-aging and workability properties of modified asphalt cements were significantly improved [7]. Moreover, Ezzat et al. (2016) performed a research focusing on the influence of using nanosilica and nanoclay as modifiers of asphalt cement. From the results, it was noted that the penetration value decreased while the softening point and viscosity increased. The addition of increasing amount of both modifiers led to the increase in the high-performance grade of the modified asphalt cements and the rutting parameter [8]. The main aim of this study is to investigate and compare the influence of Al_2O_3 and $CaCO_3$ on the properties of base asphalt cement at high temperatures. In addition, the study compares the Superpave rutting specification ($G^*/\sin\delta$) with Aroon Shenoy parameter ($G^*/(1-(1/\tan\delta\sin\delta))$).

2. Experimental design

2.1. Materials

The base asphalt cement used in this study was 60/70 penetration grade, supplied by the factory at Port Klang Malaysia. The nanomaterials used were aluminum oxide nanoparticles (Al_2O_3) and calcium carbonate nanoparticles ($CaCO_3$) in white powder form supplied by the factory of the Shijiazhuang Chanchiang Corporation company in China. The physical properties of the base asphalt cement and nanoparticles are shown in Table 1.

Table 1. The physical properties of the asphalt cement and nanoparticles.

Material	Properties	Test Method	Value
	Specific Gravity	ASTM D70	1.30
	Penetration @ 25 °C	ASTM D5	70
Asphalt cement	Softening point (°C)	ASTM D36	47.0
60/70	Viscosity @ 135 °C (Pa.s)	ASTM D4402	0.5
	Ductility (cm) @ 25 °C	ASTM D113	≥100
$CaCO_3$	Density	-	2.70
	Size	nm	40
	Form	-	Powder
	Density	-	
Al_2O_3	Size	nm	13
	Form	-	Powder

2.2. *Preparation of Al_2O_3 $CaCO_3$ Modified Asphalt Cement*

The base asphalt cement was heated until it became fluid and stirred for about 10 min at the temperature of 170 °C. Then, three different percentages of both nanoparticles (Al_2O_3 and $CaCO_3$) at 3, 5 and 7% by the weight of the asphalt cement (%.wt) were added regularly into the base asphalt cement under the high shear mixture speed of 5000 rpm for 90 min until it realized a homogenous asphalt cement blend for each percentage.

2.3. Testing procedures

This section describes the tests conducted in the laboratory to evaluate the influence of both modifiers on the base asphalt cement.

2.3.1. Viscosity. The Brookfield Rotational Viscometer was used for the assessment of the viscosity of the base asphalt cement, Al_2O_3 and $CaCO_3$ modified asphalt cement samples. Three readings were recorded for each test, and the average reading was accepted as the test result. The test temperatures were 135 °C and 165 °C respectively.

2.3.2. Storage Stability Test. The base asphalt cement and the modified asphalt cement were poured into an aluminum foil tube with the height of 16 cm and width of 3 cm. The foil tubes were closed and stored vertically at a temperature of 163 ± 5 °C in an oven for 48 hours. Therefore, the samples were cooled and divided horizontally into three sections. The storage stability samples were taken from the top and bottom sections and the softening points of the sections were determined. If the difference between the top and the bottom sections was less than 2.5 °C, then the sample was considered to have great high-temperature storage stability. If the softening points differed by more than 2.5 °C, then the sample was considered to be unstable [8, 9].

2.4. Dynamic Shear Rheometer (DSR)
Dynamic Shear Rheometer (DSR) is used to determine the rheological properties of asphalt cements, including complex shear modulus (G*) and phase angle (δ) at various temperatures and frequencies. These parameters are able to describe the viscoelastic behavior of asphalt cement. The G* (complex shear modulus) and δ (phase angle) of asphalt cement are extremely dependent on the test temperature and frequency of loading. G* (complex shear modulus) is a measure of the total resistance of a material to deformation when exposed to a sinusoidal shear stress load. G* consists of both elastic and viscous components. The δ (phase angle) is an indicator of the relative amounts of viscoelastic mechanisms. In this study, the DSR test investigated the high temperature properties of Al_2O_3 and $CaCO_3$ nanoparticles modified asphalt cement using a frequency sweep test. The frequency sweep applied was 0.159 to 15 Hz and the temperature was from 45 °C to 75 °C. The plate used in the test was a 25 mm diameter spindle with a gap of 1 mm.

3. Results and discussion

3.1. Rotational Viscosity
Figure 1 shows the results of the viscosity of the asphalt cement samples which were measured using the Brookfield viscometer at 135 °C and 165 °C. It was noted that the increase of both nanoparticles in the base asphalt cement led to the increase of viscosity. Moreover, the base asphalt cement had the lowest viscosity among the blends, while 7% Al_2O_3 had the highest viscosity. Figure 1 also indicates that the viscosity of the base asphalt cement and the modified asphalt cements within the specifications of the Superpave™ Standard required limit at 135 °C was less than 3 Pa.s. In addition, asphalt cements modified with Al_2O_3 nanoparticles showed higher viscosity value than modified samples with $CaCO_3$ nanoparticles. This indicates that Al_2O_3 nanoparticles show better resistance to rutting parameter. The highest viscosity leads to the highest mixing and compaction temperatures and the lowest value leads to lowest mixing and compaction temperatures.

3.2. Storage stability of $CaCO_3$ and Al_2O_3 nanoparticles modified asphalt cement
The difference between the top and the bottom sections of the softening point of the base and modified asphalt cement indicates its storage stability. The less the difference between the sections indicates better storage stability for asphalt cements [10, 11]. Figure 2 shows the results of storage stability of the base asphalt cement, and $CaCO_3$ and Al_2O_3 nanoparticles modified asphalt cement. It was found that the differences of the softening points for modified asphalt cements were 1 °C and less. Therefore, the entire asphalt cements passed the requirement value of storage stability as the differences were less than 2.5 °C for all asphalt cement samples. This indicates that $CaCO_3$ and Al_2O_3 nanoparticles modified asphalt cements are stable and are able to be stored at high temperatures.

IOP Conf. Series: Materials Science and Engineering **236** (2017) 012008 doi:10.1088/1757-899X/236/1/012008

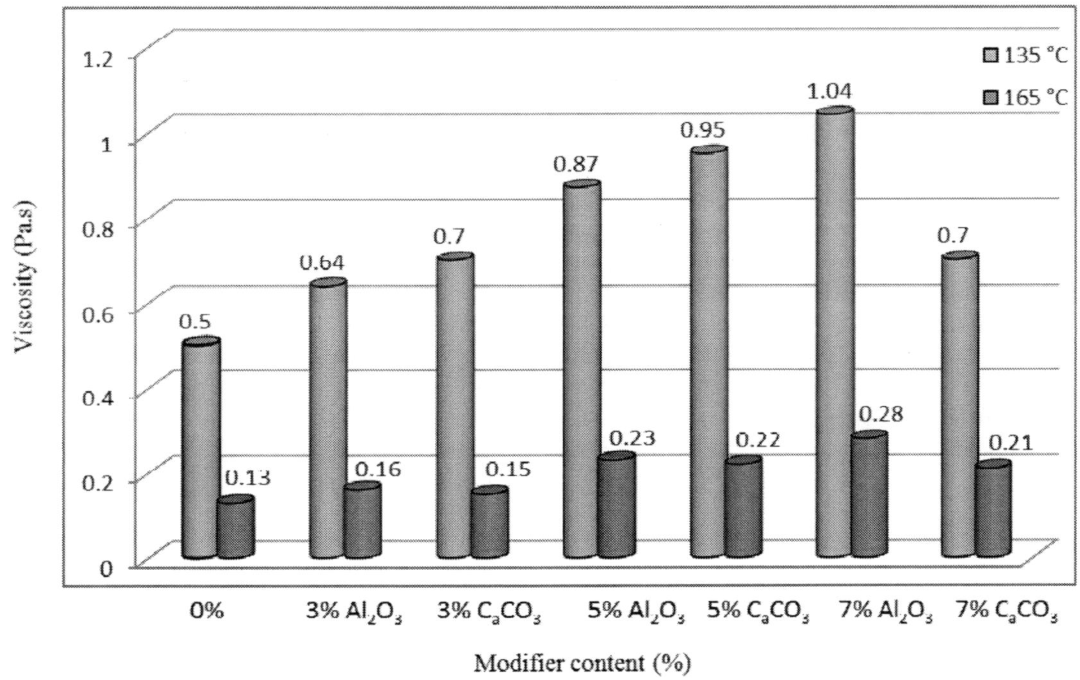

Figure 1. Viscosity of base and modified asphalt cement at 135 °C and 165 °C.

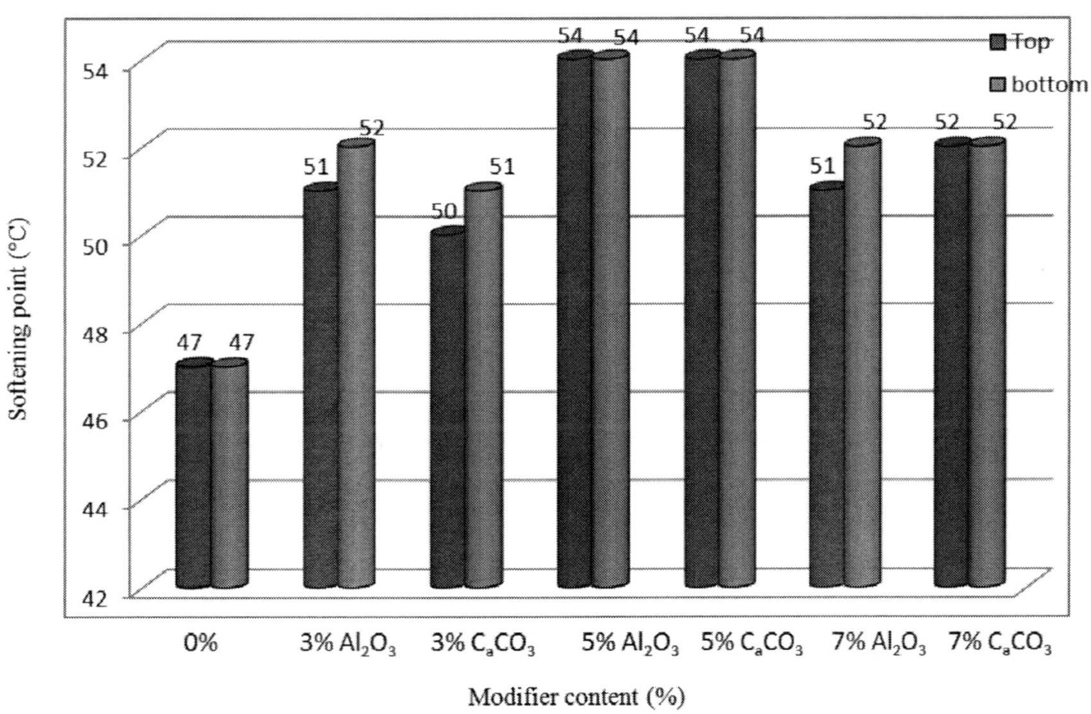

Figure 2. Storage stability of the base and modified asphalt cements.

3.3. The Dynamic Shear Rheometer (DSR)

3.3.1. Isochronal Plots. According to the compatibility between asphalt cement and the modifier, the modification of asphalt cement may result in two types of blends. The first type is a heterogeneous blend, in which the modifier and asphalt cement are incompatible and lead to two separate phases. The second type is a homogeneous blend, in which the modifier and asphalt cement are quite compatible [11]. Figure 3 shows the isochronal plots of the modified asphalt cements. It was observed that the increase of modifiers content caused an increase in the stiffness of the modified asphalt cements, regardless of the temperatures. Moreover, it was noted that the samples modified with Al_2O_3 nanoparticles had higher complex shear modulus value compared to those modified with $CaCO_3$ nanoparticles. This suggests that the Al_2O_3 nanoparticles have higher resistance to permanent deformation. Both modifiers at 7% showed different behaviour as the complex shear modulus was reduced. This may be due to the agglomeration among the nanoparticles with additional increase of modifier in asphalt cement blends.

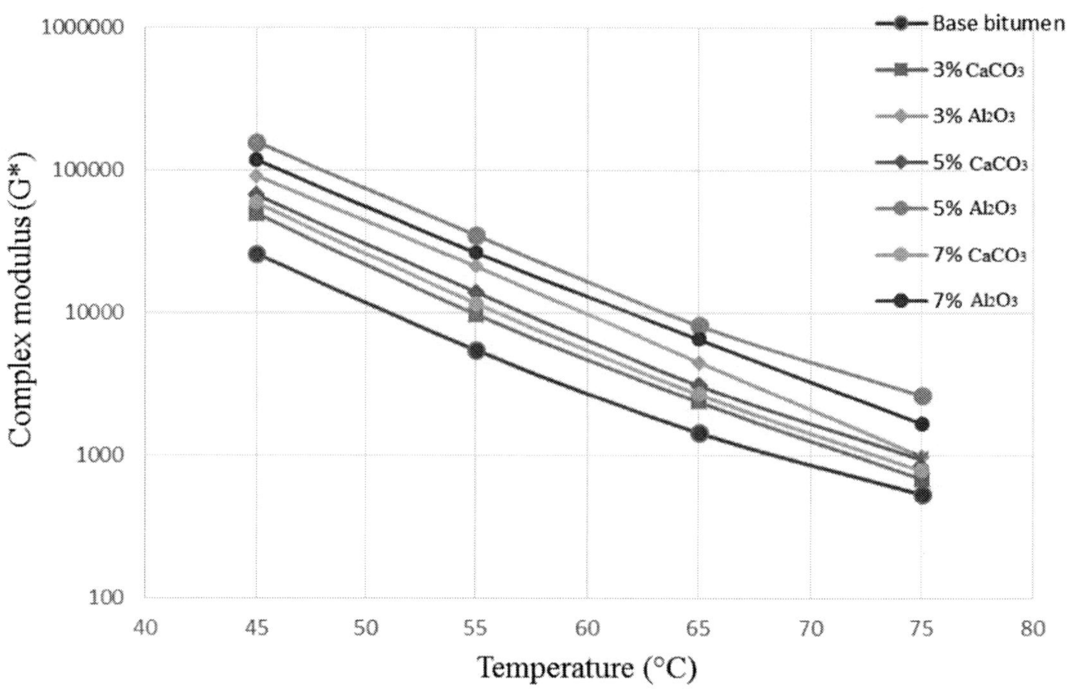

Figure 3. Isochronal plots of base and modified asphalt cements.

3.3.2 Rutting Parameter. The Superpave method requires G*/sinδ = 1 kPa as the minimum of SHRP rutting parameter for unaged samples of asphalt cement. Figure 4 shows that the lowest rutting value was displayed by the base asphalt cement, while the modified asphalt cement with Al_2O_3 nanoparticles displayed the highest value among the blends. The higher G*/sinδ depicts asphalt with better rutting resistance. Therefore, these results prove that Al_2O_3 nanoparticles modified asphalt cement performs much better than $CaCO_3$ nanoparticles modified samples in rutting performance. However, the 7% samples showed different behavior as the value of G*/sinδ decreased slightly. Overall, it can be observed that all asphalt cements had G*/sinδ values that were larger than 1.0 kPa at 65 °C. The complex shear modulus (G*) and phase angle (δ) at high elevated temperatures were proven as indicators of the rutting susceptibility of the asphalt (G*/sinδ) and the limitation was identified to be G*/sinδ = 1 kPa as the minimum value according to Superpave specification. Aroon Shenoy suggested the term G*/(1-(1/tanδ sinδ)) as a refinement of the Superpave specification parameter G*/sinδ. Figure 4 shows the evaluation of the rutting parameter of the base and modified asphalt cements with nanoparticles using the Superpave specification (SPS) and the Aroon Shenoy parameter (ASP) methods. From the results, it was observed that there were no significant differences between the two techniques. This indicates that both methods are able to evaluate and determine the rutting parameters of modified asphalt cements.

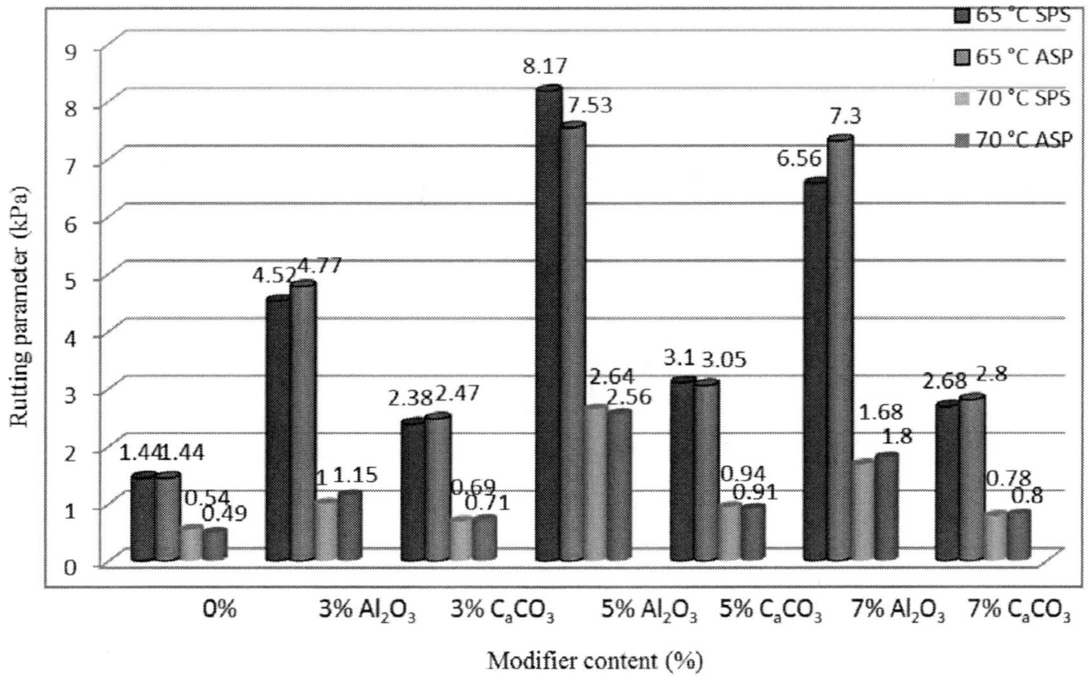

Figure 4. Evaluation of rutting depth of base and modified asphalt cements using SPS and ASP.

4. Conclusions

In this study, the effects of introducing Al_2O_3 nanoparticles and $CaCO_3$ nanoparticles modified asphalt cements were investigated and the high temperatures rheological properties of the modified asphalt cements with both modifiers were also evaluated. The results indicated that the modified asphalt cements showed greater stiffness at elevated service temperatures. It was also found that the rutting influence of the modified asphalt cements with Al_2O_3 and $CaCO_3$ nanoparticles was less than the base asphalt cement, and the improvement of the modified samples was up to 4.6 and 1.4 times respectively compared to the base asphalt cement. This indicates higher performance of modified samples at high temperatures. Moreover, the investigation approved that both Superpave and Aroon Shenoy parameters are able to evaluate the permanent deformation of modified asphalt cement. The evaluation of rheological examinations indicated that the concentrations of 5% Al_2O_3 and 5% $CaCO_3$ can be considered as the best performance condition of the asphalt cement for both modifiers and was selected as the optimum content of modifiers.

Acknowledgments

The authors would like to thank the Sustainable Urban Transport Research Centre (SUTRA) at the Faculty of Engineering and Built Environment of Universiti Kebangsaan Malaysia (UKM) for providing research facilities. The authors also acknowledge the Ministry of Science, Technology and Innovation (MOSTI) of Malaysia for the research funding provided through projects 03-01-02-

SF0999SF. Moreover, the authors would also like to acknowledge the Civil Engineering Department, Near East University for their collaboration and support.

References

[1] Albrka S I, Ismail A, Yahia H A and Ladin M A L 2013 Application of Transyt-7f on signalized road junction networks in Shah Alam and Petaling Jaya *Jurnal Teknologi* **69**(2) pp 59-64

[2] Fang C, Yu R., Liu S and Li Y 2013 Nanomaterials applied in asphalt modification: A review *Journal of Materials Science & Technology* **29**(7) pp 589-94

[3] Yildirim Y 2007 Polymer modified asphalt binders *Construction and Building Materials* **21**(1) pp 66-72

[4] Peters S, Rushing T, Landis E and Cummins T 2010 Nanocellulose and microcellulose fibers for concrete *Transportation Research Record* **2142**(1) pp 25-28

[5] Alhamali D I, Wu J L, Quan H, Norhidayah A, Yusoff N I M and Shaban I A 2016 Physical and rheological characteristics of polymer modified bitumen with nanosilica particles *Arabian Journal for Science and Engineering* **41**(4) pp 1521-30

[6] Wang P, Dong Z J, Tan Y Q and Liu Z Y 2017 Effect of multi-walled carbon nanotubes on the performance of Styrene–Butadiene–Styrene copolymer modified asphalt *Materials and Structures* **50**(1) pp 1-11

[7] Ezzat H, El-Badawy S, Gabr A, Zaki E S I and Breakah T 2016 Evaluation of Asphalt Binders Modified with Nanoclay and Nanosilica *Procedia Engineering* **143** pp 1260-67

[8] Zhang F, Yu J and Han J 2011 Effects of thermal oxidative ageing on dynamic viscosity, TG/DTG, DTA and FTIR of SBS-and SBS/sulfur-modified asphalts *Construction and Building Materials* **25**(1) pp 129-37

[9] Al-Mansob R A, Ismail A, Alduri A N, Azhari C H, Karim M R and Yusoff N I M 2014 Physical and rheological properties of epoxidized natural rubber modified bitumens *Construction and Building Materials* **63** pp 242-48

[10] Yu R., Fang C, Liu P, Liu X, and Li Y 2015 Storage stability and rheological properties of asphalt modified with waste packaging polyethylene and organic montmorillonite *Applied Clay Science* **104** pp 1-7

[11] Zhang F, Yu J and Wu S 2010 Effect of ageing on rheological properties of storage-stable SBS/sulfur-modified asphalts *Journal of Hazard Materials* **182**(1) pp 507-17

[12] Socal da Silva L, de Camargo Forte M M, de Alencastro Vignol L and Cardozo N S M 2004 Study of rheological properties of pure and polymer-modified Brazilian asphalt binders *Journal of Materials Science* **39**(2) pp 539-46

IOP Conf. Series: Materials Science and Engineering **236** (2017) 012009 doi:10.1088/1757-899X/236/1/012009

Bitumen performance and chemistry in solvent refined bitumen blends

Glynn Holleran[1], Irina Holleran[1] and Douglas J Wilson[1]

[1]The University of Auckland, 20 Symonds Street, Auckland, 1010, New Zealand

E-mail: ghol310@aucklanduni.ac.nz

Abstract. In years gone past most oil companies in Australia and New Zealand (NZ) developed experts that bridged the divide between refining and paving. This was supported by laboratories in Australia and sometimes Asia. This is no longer the case and many refineries have ceased bitumen production or closed. With the market moving towards imports and control to supply companies disconnects on bitumen passing a national specification and performance on the road. This reduces both durability and increases costs. This has been addressed by development in NZ of a performance specification for hot mix asphalt binders (including modified) and work being done on sealing grades. This paper discusses the development of the HMA specification with respect to crude sources and the development of methodologies to assess imported binders for suitability in all applications including emulsion. The conclusion is that bitumen quality may be maintained by use of these methodologies that include, chromatographic analysis, measurement of thermodynamic internal stability (Heithaus), aging, and Dynamic Shear Rheometry testing and mix performance testing in the laboratory. This forms a regime capable of use in any context and this leads to better durability of surfaces and extended service life.

1. Introduction

Bitumen in New Zealand (NZ) has two sources, the first is the New Zealand Refining Company (NZRC) and the second is imports, these may come from any part of the world. The penetration specification of New Zealand NZTA M-1 [1] is an empirical consistency specification originally based on Safaniya (Saudi heavy crude) based partially blown bitumen and then adjusted in 2006. This was in an attempt to widen the allowed bitumen imports. The result was achieved by relaxing the viscosity specification slightly but without reference to performance. The result was thermally susceptible bitumen being introduced that created issues of flushing, rutting, tenderness and shoving in early life. In NZ the primary premature failure mode of new surfaces, both chip seal and hot mixed asphalt (HMA) are related to higher temperatures and heavy traffic. To address long term deformation, a new specification requirement based on wheel tracking was introduced in 2006. The industry tolerated this issue until it became obvious that some complaint penetration bitumen asphalt grades rutted prematurely in service and could not pass the basic requirements of the specifications and produce a compliant design for heavily trafficked roads. It was also clear that early tenderness made compliance with field volumetric specifications difficult. This paper discusses the steps taken to ensure that correct mixes could be designed and applied for durable surfacing of NZ roads.

2. Background and literature review

The relationship between bitumen chemistry and performance has been studied extensively for many decades. The major collection of this work was during the SHRP program in USA in the 1990's. The effect of crude source was described by Harnsberger [2] who reported crudes as an important contributor to field performance even within a given Performance Grading (PG grading).

In earlier work [3] the relationship between bitumen composition and HMA properties was considered using Dynamic Shear Rheometry, and testing laboratory mixes using wheel tracking and conventional asphalt testing. This showed that bitumens made with lighter crudes produced higher phase angles in DSR testing. In turn, mixes were produced that were more shear and temperature susceptible in laboratory testing leading to flushing and tenderness in application. Premature failure under traffic via shoving and rutting followed.

The conclusion, that simply blending to penetration was insufficient to ensure HMA performance in heavy traffic and high temperature conditions has led to a new specification NZTA M-1A [4,5] that uses creep compliance and other rheological parameters to relate to performance of HMA , being based on NZTA M-10 2014 [6] performance related variables such as deformation (wheel tracking) fatigue (4 point bending) plus the Asphalt Performance Mix tester (AMPT) [7] overlay, dynamic modulus and push pull fatigue testing.

Modern refineries are collections of processing units balanced by refinery economics to produce the most advantageous product mix. Bitumen is rarely a main aim except in specialty refiners such as Nynas in Europe. In NZ the refinery has clients who provide crudes for processing into products for downstream marketing. The main processing routes for the NZRC refinery are shown in Figure 1. In this process bitumen is only shipped in small parcel sizes although demand is continuous. Added complications are crude shipment, refinery scheduling, tank segregation, blending and ship loading.

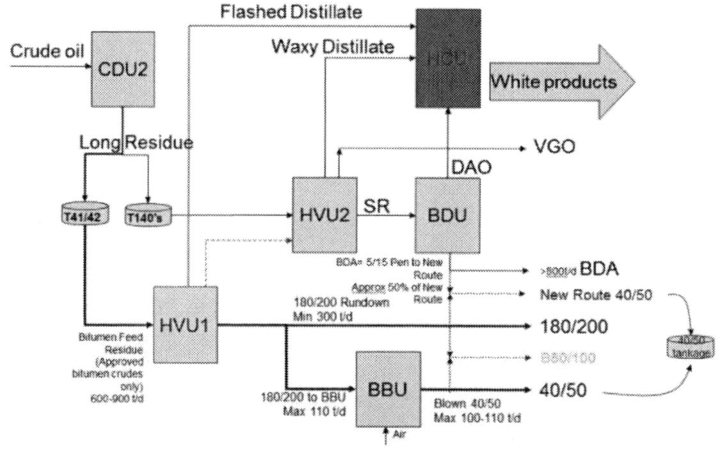

Figure 1. Refinery process NZRC New Zealand.

In NZRC only the soft grade (180/200 penetration) is made from certified heavy bitumen crudes - Ratawi and Arab Heavy. In some instances it is of higher penetration than specification and Butane pitch (BDA), a residue of general purpose crude processing that is solvent refined, is used in small percentages. This of course means the 180/200 is no longer straight run and can have variable rheology. The general purpose crudes used are light to extra light in nature. Classification uses the American petroleum institute consistency rating (API) of crudes.This measures degrees API which is an inverse function of its specific gravity relative to that of water. That is the greater the API, the lighter the crude.

Heavy is 29 and less, Medium is 29-32, Light is 32-34, Extra Light is 36-41 and Superlight is 49-52. The harder grade is made from 180/200 blown to 40/50 and then mixed with a blend of BDA (50%) and 180/200 – so called New Route bitumen 40/50 (NR). This BDA and its variability in composition is responsible for higher levels of thermal susceptibility in blends. Thus a high performing bitumen composition has a low tolerance for BDA addition. Solvent modified materials such as BDA have been shown in earlier work to create issues for emulsion stability [3] due to the chemical composition being rich in aromatic oils. For the same reasons BDA blends can produce hot mixed asphalts that do not perform satisfactorily in the field. The way in which it is blended affects the total composition of the final blended binder. Despite its low penetration the BDA has lower elasticity (as measured by phase angle using AASHTO T315 and Dynamic shear rheometry [8]) at any given temperature which leads to greater thermal and shear susceptibility in the final bitumen.

3. Research motivation and objectives

The issue that arises is one of specification and screening, it also touches on areas of making a given bitumen work in the required application. This research is an ongoing direction for the authors that includes aspects of renewable resources, durability in field situations and modification. The paper reports this development in the context of development of an assessment framework based on chemical knowledge and understanding of mix performance.

4. Test methods and material preparation

Bitumen is a complex material chemically. Being drawn from naturally occurring crudes it will vary both in component percentages based on SARA (Saturates, Aromatics, Resins and Asphaltenes) fractions but also in the chemistry of those fractions. This is due to differing levels of polar associations and the level and type of hetero atoms. Sulphur in the form of sulphoxides and other compounds is particularly influential on mechanical properties [9].

This paper uses Dynamic Shear Rheometry (DSR) AASHTO T315 testing [8], Multiple Stress Creep Recovery Test (MSCR) testing AASHTO TP70 [10] to look at differences in base binder rheology with differing BDA levels and composition. AASHTO T315 employs a fixed strain level and frequency based on field shear levels, and varying temperature to measure phase angle and complex modulus. MSCR uses an oscillatory test method to determine creep compliance. Mix performance is characterised by wheel tracking as per Austroads AGPT-T231 [11]. Fractionation by Iatroscan is carried out using an established methodology [3]. Work reported by the authors in this paper discussed the background of this in detail including the use of TLC- FID Iatroscan methodology. Heithaus testing was carried out using ASTM 6703 [12] and the Western Research Institute approach of Pauli [13]. The Heithaus parameters are from ASTM D6703. The most useful parameter is "P" the bitumen state of peptisation-internal stability. This is a combination of the ability of associated species (asphaltenes) to be stable in the dispersed medium (p_a) and the ability of the dispersing medium to disperse the associated species (p_o).

Moisture sensitivity, resilient modulus, beam fatigue testing and overlay testing were also measured on the HMA samples but this is not reported in this paper. However the results indicated that the effect of softening by BDA does not reduce fatigue life, overlay crack resistance, or increase moisture sensitivity, but resilient modulus is reduced. The bitumen samples used are shown in Table 1. The API ratings shown are: XL - extra light, L- light, M - medium, H - heavy.

Table 1. Information on crude type used in BDA.

Sample	Crude type	% BDA in 40/50
1	XL	2.5
2	L	11.5
3	L	17
4	L	20
5	L	25
6	L	26.5
7	L	34.5
8	XL	38.5
9	M	27.5
10	M	33.5
11	M	20.5
12	H	0

5. Test results, analysis, and discussion

5.1 Effect on structure of BDA

The effects of BDA on binder rheology are shown in Figures 2-5. Figure 2 shows the phase angle relationship for % BDA. This clearly shows that the higher percentage BDA has higher phase angles and hence is less structured (Note that some lines have been removed for clarity). Even small amounts of BDA have a significant effect on the structure. Figure 3 compares like BDA percentages but with different crude bases in the BDA. The comparison shows that phase angle is higher for 40/50 made with the lighter crude based BDA.

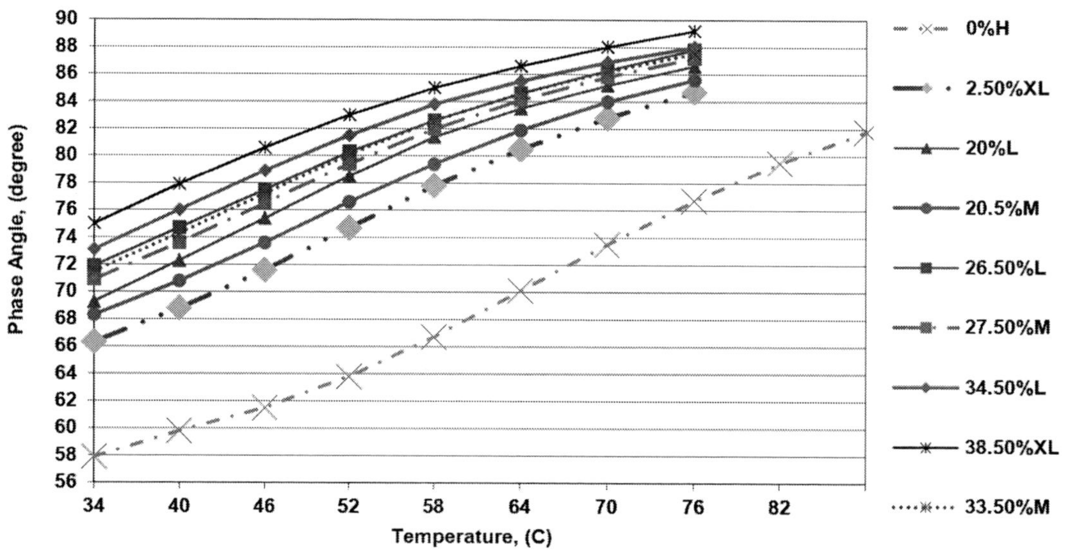

Figure 2. Phase angle relationship for BDA addition.

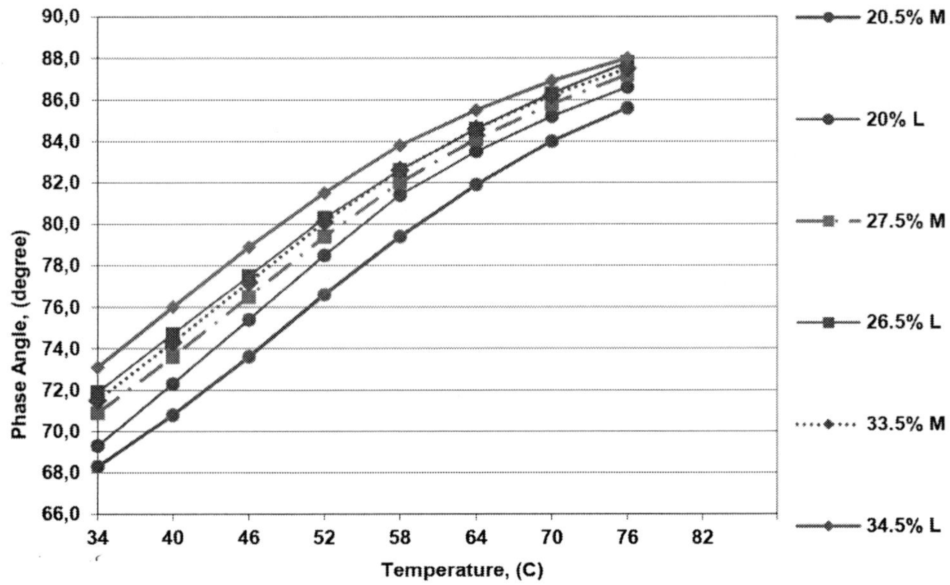

Figure 3. Effect of crude base on phase angle.

5.2 Effect on Stiffness of BDA

The results shown in figure 4 and 5 for Complex modulus (G*) follows the same trend and that higher BDA levels result in a decrease in elevated service temperature stiffness and higher low service temperature stiffness compared to well performed bitumen . Higher viscous modulus (G") and lower elastic modulus (G') indicate that the bitumen will be less deformation resistant and more prone to crack.

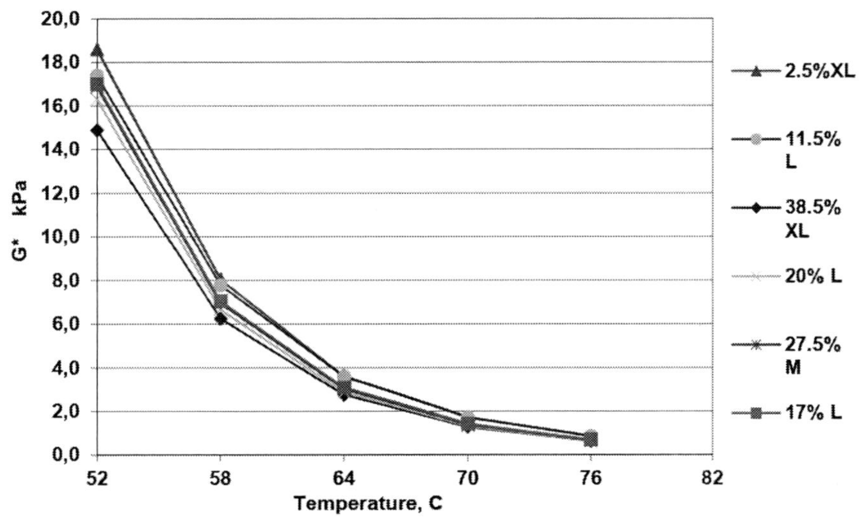

Figure 4. Complex modulus for different crudes.

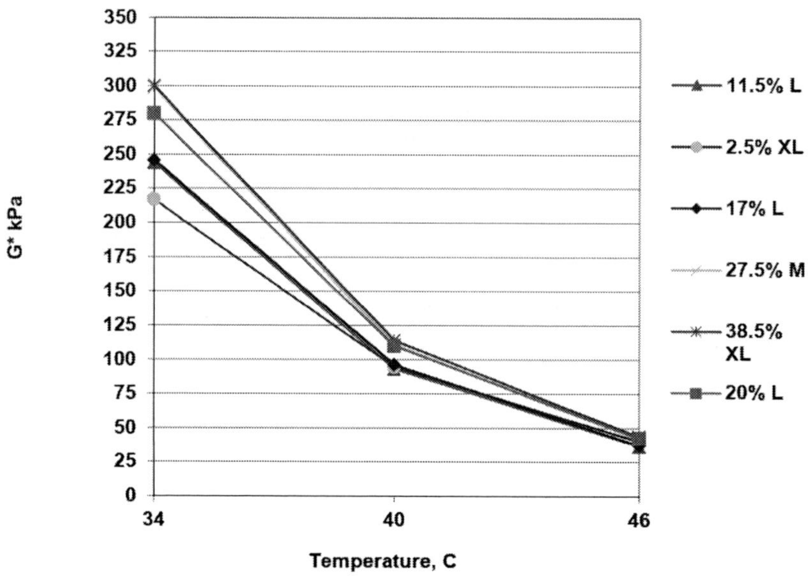

Figure 5. Complex modulus for different BDA from different crudes.

5.3 Composition changes

The Iatroscan measures SARA fractions and in this analysis the total structuring phase is defined as resinsA plus resins B and the total dispersing phase as saturates plus aromatics. Figure 6 indicates, that as BDA increases the associated or structuring phase is reduced as the dispersing phase increases this can lead to instability in the bitumen matrix.

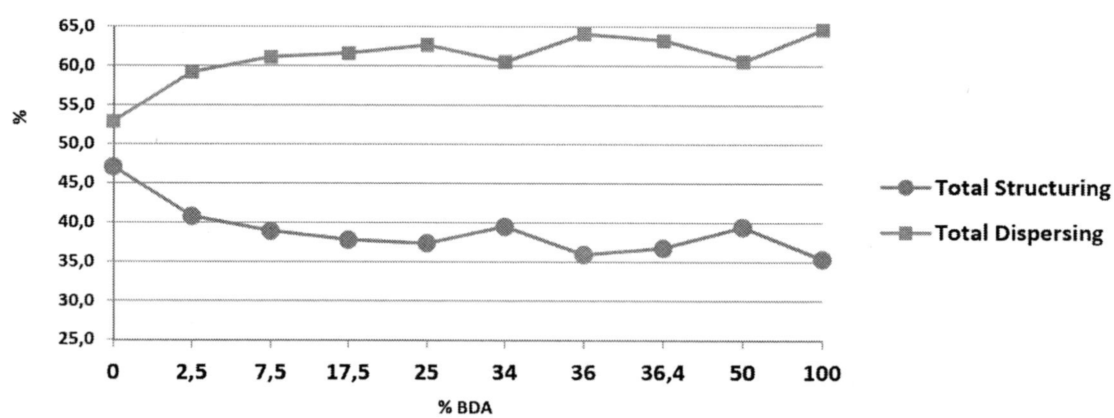

Figure 6. Iatroscan results.

This is also reflected in the Heithaus tests as per ASTM D6703 (using Western Research procedure) parameters. The total stability factor P is shown in Figure 7.

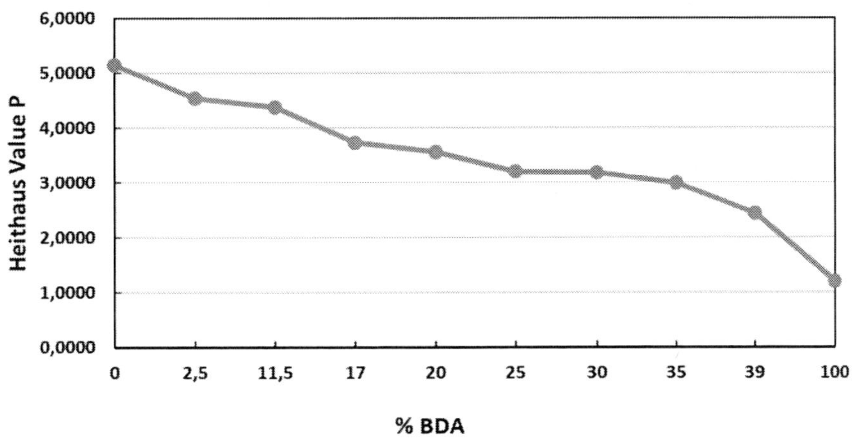

Figure 7. Thermodynamic stability by Heithaus test ASTM D6703.

A "P" value of around 3 is the normal point of internal instability of a binder that has been found ,in field materials ,to be an issue , with respect to deformation and cracking resistance ,when blended to a penetration grade of 60/70. When made into blends to a penetration grade of 60/70, the grade of binder tested on MSCR showed that with same base straight run 180/200 the PG grade went from PG64H to PG64S at around 30% BDA in the 40/50 blend. Wheel tracking results are reported in Figure 8 and show that, increasing BDA content resulted in increased deformation. It has been established that this correlates with field observations of increased propensity to flush, rut and exhibit tenderness in laying [3] (Note that this refers to 60/70 binders blended to NZTA M-1 grade using the same 180/200 straight run heavy crude base bitumen and the 40/50 with varying BDA).

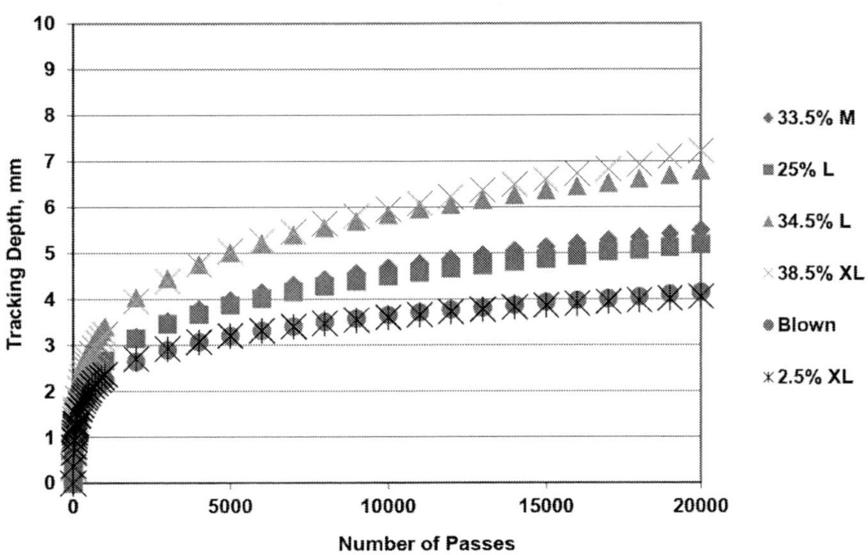

Figure 8. Wheel tracking results.

6. Conclusions and recommendations

1. The composition of bitumen determines its rheology.
2. Solvent refined materials such as BDA while being able to be used in blends to penetration grades can result in compositions that are unbalanced and hence created rheological properties unfavorable to performance.
3. The effects are crude dependent, lighter crudes give poorer results.
4. Performance specifications have been shown to address these issues.

References

[1] New Zealand Transport Agency (NZTA) 2011 Specification For roading bitumens *(NZTA M-1 2011)*

[2] Harnsberger P, Farrar M,ShinChe H and Roberston R 2014 Comparative field performance of asphalts from multiple crude oil sources *Transportation Research Record*

[3] Holleran G and Holleran I 2010 Bitumen chemistry using cheaper sources - An improved method of measurement by TLC-FID and the characterisation of bitumen by rheological and compositional means *24th Australian road research board conference (Melbourne)*

[4] New Zealand Transport Agency (NZTA) 2016 Performance specification for asphalt binders *(NZTA M-1A 2016)*

[5] Holleran G, Holleran I,Donbavand J,Vercoe J, D'Angelo A, Bearsley S and Stevens A 2015 Bitumen in New Zealand – performance based asphalt binder specification *14th Australian asphalt pavement association international flexible pavements conference (Gold Coast Queensland)*

[6] New Zealand Transport Agency (NZTA) 2014 Specification for dense graded and stone mastic asphalts *(NZTA M-10 2014)*

[7] Holleran G, Holleran I and Jayalath C 2013 Overlay testing and surface mixes Australian asphalt pavement association *15th International Conference on Flexible Pavements (Brisbane)*

[8] American Association of State Highway Technical Officiers (AASHTO) 2012 Standard Method of Test for Determining the Rheological Properties of Asphalt Binder Using a Dynamic Shear Rheometer *(AASHTO T315 2012)*

[9] Green J ,Yu S, Pearson C and Reynolds J 1993 *Analysis of Sulfur Compound Types in Asphalt* (Strategic Highways Research Program Publication SHRP-A-667 National Academy Of Sciences)

[10] American Association of State highway Technical Officiers (AASHTO) 2012 Standard Method of Test for Multiple Stress Creep Recovery (MSCR) Test of Asphalt Binder Using a Dynamic Shear Rheometer *(AASHTO TP70 2012)*

[11] Australian Association Of State Road Authorities (Austroads) 2006 Deformation Resistance of Asphalt Mixtures by the Wheel Tracking Test *(AGPT-T231 2006)*

[12] American Society for Test Methods 2014 Standard Test Method for Automated Heithaus Titrimetry *(ASTM D6703 2014)*

[13] Pauli A 1996 *Asphalt Compatibility by using the Austomated Hiethaus Titration Test* https://www.researchgate.net/publication/255254498

Optimizing the durability of the coarse fraction of porous asphalt RAP for effective recycling

Irina Holleran[1], Douglas J Wilson[1], Philippa Black[1], Glynn Holleran[1] and Lubinda F Walubita[2]

[1]The University of Auckland, 20 Symonds Street, Auckland, 1010, New Zealand
[2]Texas A&M Transportation Institute, College Station, Texas, 77843, USA

E-mail: imot002@aucklanduni.ac.nz

Abstract. Porous asphalt (PA) durability depends not only on the binder used to manufacture the mix, but also on the aggregates chosen, particularly the coarse fraction component. Aggregates for PA should be of the highest quality and highly durable to withstand the effects of weather and traffic. To recycle PA into a new PA mix, without compromising the long-term performance, the durability of the recovered aggregates from PA-derived reclaimed asphalt pavement (RAP) should be assessed alongside the aged binder properties. In this study, the Micro-Deval (MD) Abrasion test, combined with water absorption, was found to be a good predictor of asphalt mix performance for PA. Minerology of the aggregates is an important factor when setting limits for MD loss. New Zealand (NZ) aggregates are significantly younger in geological terms, and chemically and physically less stable compared to the aggregates used in many other countries. This is especially true for greywacke, the most used aggregate in NZ for road construction. If the MD limits reported in some literature are applied to NZ PA-derived RAP aggregates, poor performing material can be erroneously incorporated in asphalt mixes. Findings from this study contributes in understanding how PA-derived RAP can be recycled into new value PA mixes.

1. Introduction

Porous asphalt (PA) is a stone-on-stone mix designed to drain water from road surfacing and attenuate noise. It provides significant safety and environmental benefits; and, is primarily utilized in warmer climates especially where frequent heavy rainfalls are present. In New Zealand (NZ), the Auckland motorway network is the greatest user of PA mixes. The PA life span is reported to be 10 years compared to 7 years nationwide for conventional asphalt mixes [1]. All PA mixes used on this network are polymer-modified. However, unmodified PA mixes are still sometimes used outside the Auckland region in NZ.

Currently, in accordance with NZ specifications, no reclaimed asphalt pavement (RAP) of any type can be utilized in PA mixes. In contrast, 30% of RAP in dense-graded (DG) pavement surfacing mixes and up to 50% in pavement base mixes are routinely used. It is logical to utilize PA-derived RAP in high value mixes such as PA because the aggregates required for PA manufacture have significantly more stringent control compared to aggregates used for conventional DG mixes. Furthermore, PA-derived RAP millings can contain valuable aggregate resources, such as melter slag (a byproduct from steel making), that has historically performed well in relation to skid resistance. Demand for this aggregate outstrips the supply to the extent that New Zealand Transport Agency (NZTA) has recently decided to take control of the supply chain to ensure greater optimal use of this material in critical areas to maximize network safety outcomes.

To be able to recycle PA-derived aggregates into new PA mixes, RAP aggregate durability needs to be understood and measured, especially for the coarse fraction component. Resistance of PA to withstand traffic forces relies on the aggregate interlock structure provided by the coarse aggregates. Airey et al. [2] found that gap-graded mixes develop aggregate contact stresses in the coarse fraction at the time of mix compaction, which can result in fracture planes in coarse aggregate stones and aggregate degradation compared to DG mixes, where stresses are more evenly spread through the interaction of the coarse and fine aggregate structure. In PA surfacing, the durability of coarse aggregates is likely to be affected by wet and dry cycles. The PA mix and layer structure will operate in higher moisture environments due to the higher relative air voids in comparison to DG mixes. It is likely that the moisture level increases within the PA mat once the PA mix starts to trap finer particle size material fractions (both detritus and abraded materials) that over time reduces the capacity of the PA to drain water efficiently from its surface. This is especially the case in countries like NZ, where the PA is not cleaned (vacuumed), and frequent and heavy rainfall events are relatively common.

It has been reported in the literature that the Micro-Deval (MD) abrasion loss is currently the best indicator for predicting the PA aggregate performance, and if combined with other tests such as absorption, will increase the chance of better enabling aggregate performance prediction and therefore selection of appropriate aggregates. A considerable amount of research work has been carried out in various countries on setting the MD limits to separate good and fair performing materials from poor performers. Maximum MD limits vary from country to country, ranging from about 5% to 30%. The original work on MD was performed using virgin materials, not recycled aggregates. This paper looks at MD and abrasion testing as indicative tests for PA-derived RAP assessment for the suitability of this PA-derived RAP to be incorporated into new PA mixes containing RAP.

2. Background and literature review

MD is a wet abrasion test, which better reflects the real field conditions compared to the commonly used Los Angeles (LA) Abrasion test performed in the dry aggregate state. Under field conditions, especially in PA surfacing in countries where rainfall is frequent, the PA mat rarely remains dry by design. Some aggregates can have reduction in strength or slake when subjected to wet environments [3]. Bartley et al. [4] suggested that the durability of NZ aggregates should always be performed under wet conditions due to the unique mineralogy of these aggregates, which potentially contain highly absorbing clay minerals and zeolites that results in high water absorption and swelling.

Despite the majority of research work showing that LA Abrasion is not a good test for predicting field performance, many countries continue to use this test since ASTM adopted it in 1937 [5]. This is still the case for NZ, where LA Abrasion is required by NZTA specifications for the screening and selection of aggregates for PA manufacture; although it is reported only as an item with no limits currently set. MD testing is not required for any aggregates, virgin, or RAP. Very limited data on MD testing on NZ virgin aggregates is available from aggregate suppliers, and currently no MD limits, even tentative, have been proposed. There is no published data on MD testing of RAP aggregates from NZ. Using the MD limits from other countries may be an issue as they significantly vary based upon aggregate geological source properties, and these limits were mostly established based on research on virgin aggregates.

It is reported that the 18% limit in MD loss is suitable to differentiate good and fair virgin aggregates from poor field performers [6, 7]. Williamson [7] reported that even though the 18% MD limit is a good indicator of aggregate performance, it is possibly too stringent for Wisconsin aggregates; a maximum of 25-30% MD loss limit was suggested instead. The limit of 15% in MD loss together with a maximum of 2.2% water absorption were investigated as the indicators to predict virgin aggregate performance in PA mixes [5]. The Missouri Department of Transportation requires MD testing to be done on RAP aggregates; RAP materials with a loss of above 20% are prohibited for use in asphalt mixes [8]. The Ontario Ministry of Transportation specification presented by Lane et al.

[9] takes into account the minerology of the aggregate, as well as traffic loading and type of asphalt course. For the more heavily trafficked motorway asphalt surfacing mixes, the MD abrasion loss limit is reported to be 5% for igneous and metamorphic gravels; 10% for traprock, diabase, and andesite; and 15% for dolomitic sandstone, granitic meta-arkose, and gneiss. This is a more tailored specification, which is designed to accommodate real field conditions and parent rock mineralogy, compared to a generalized MD abrasion loss limit. Canada also has a long history of using MD testing compared to most countries, and their specification is most likely a reflection of the substantial database collected on MD, mineralogy of the parent aggregate, and field performance.

The combination of MD with other tests can increase the accuracy of field performance prediction [10]. For instance, aggregate absorption combined with MD gave a 93% success prediction rate compared to 82% with MD alone for PA performance [5]. If field observation data is collected for each PA site and linked to the actual aggregate MD loss and absorption, and the aggregate minerology is considered, this information can be used as a baseline to set the MD and absorption limits for NZ aggregates.

3. Research motivation and objectives

Due to the projected maintenance work on Auckland motorways in the next 6 years, more PA-derived millings will be available, and thus, more PA mix will be paved. If these millings can be recycled back to new quality PA mixes, it will optimize the re-use of valuable resources, aggregate and binder, in this high value application. Currently, these millings are primarily used in low level applications such as filling trenches, drainage layers, or as dust mitigation layers on unsealed roads, or places in landfills. The objective of this research paper was to assess the suitability of MD and Abrasion tests to predict the performance of coarse aggregate fraction from PA-derived RAP in PA surfacing.

4. Materials used in the study

Seventeen samples (minimum 150 kg each) of PA-derived millings were collected over a period of 13 months from various AMA sites ranging in age from 4 to 12 years. Information for each site is shown in Table 1.

Table 1. Site information used for sampling of RAP.

Site ID	Age (years)	Type of milling	PA mat condition
PAV108	8	Full mill at 25mm	Ravelling, highly aged and "hungry" surface
PAV112	7	Full mill at 25mm	No site visit
PAV117	8	Key millings	No site visit
PAV118	7	Full mill at 25mm	Cracking and ravelling. Base failure
PAV119	12	Key millings	Highly aged and ravelled
PAV123	7	Full mill at 25mm	Aggregate disintegration and ravelling
PAV125	5	Full mill at 35mm	Cracking with no ravelling. Base failure
PAV128	8	Key millings	Cracking and ravelling. Base failure
PAV129	6	Full mill at 35mm	Cracking with no ravelling. Base failure
PAV131	6	Full mill at 25mm	Cracking with no ravelling. Base failure
PAV133	4	Full mill at 25mm	Cracking with no ravelling. Base failure
SUR215	7	Key millings	Cracking and ravelling. Base failure
SUR216	8	Full mill at 25mm	Cracking with no ravelling. Base failure
SUR220	7	Full mill at 25mm	Severe scabbing and cracking. Base failure
SUR232	5	Full mill at 25mm	Cracking and scabbing. Base failure
SUR529	6	Full mill at 25mm	Cracking with no ravelling. Base failure
SUR530	7	Full mill at 25mm	Cracking and ravelling. Base failure

Observations on PA mat conditions were recorded. Overwhelmingly, the PA sites visited had issues with the base layers. This was especially true on sites where foamed stabilized bases were used. Three sites, which did not have any visible signs of base failure were severely ravelled (PAV108, PAV119, and PAV 123). There were no records available on aggregate source or binder type used on the majority of the sites. All sites except one were below the expected reported PA life span on Auckland motorways of 10 years. Most sites were between 7 and 8 years old. The way the PA sites were milled is divided into three categories, namely: 1) full milling, where milling was done at 25mm depth; 2) full milling at 35mm, this was done on some sites exhibiting severe issues with base failures, the milling depth of 35mm means some aggregate materials underneath the PA layer were present in the PA millings; and 3) key milling, where only profiling using the miller was carried out to link new areas to old areas.

5. Test methods and material preparation

Aggregates were recovered from PA-derived RAP by using 1-Bromopropane based solvent. Two aggregate fractions, passing the 9.5mm sieve and retained on the 6.7mm sieve, and passing the 6.7mm sieve and retained on the 4.75mm sieve, were prepared and tested for specific gravity (SG) and water absorption in accordance with ASTM C127 [11]. For the MD abrasion testing, the 750g of 9.5-6.7mm plus 750g of 6.7-4.75mm fraction were prepared and tested in accordance with ASTM D6928 [12]. The soaking time for MD was kept constant at 1.5 hours during the test. Each test (MD, SG and absorption) was performed three times. Thin sections of aggregates were prepared for petrographic examination and used to determine the minerology of PA-derived aggregates.

6. Test results, analysis, and discussion

Test results on RAP aggregates are shown in Table 2. The 9.5-6.7mm and 6.7-4.75mm fractions were tested separately for SG and absorption because some of the RAP aggregates were a mixture of different aggregate source. Average SG and absorption values shown in Table 2 were calculated from the 9.5-6.7mm and 6.7-4.75mm fractions tested. The SG value shown is oven dried (OD), as per the ASTM C127 definition [11]. The MD loss value is an average of three test results performed on the 9.5-6.7-4.75 fraction. The first named aggregate in the final Minerology column represents the majority of the aggregate present in the 9.5-6.7-4.75 fraction used for MD testing.

Table 2. Test results on PA-derived coarse RAP aggregates.

Site ID	Average SG (OD)	Absorption 9.5-6.7 mm (%)	Absorption 6.7-4.75 mm (%)	Average absorption (%)	MD loss (%)	Minerology
PAV108	2.593	1.45	1.65	1.55	15.4	greywacke
PAV112	2.781	0.44	0.56	0.50	4.7	greywacke, slag
PAV117	3.713	0.35	0.55	0.45	4.8	slag
PAV118	2.784	0.98	1.24	1.11	8.2	basalt, greywacke
PAV119	2.836	1.19	1.22	1.20	6.4	basalt
PAV123	2.618	1.58	1.46	1.52	11.1	andesite, greywacke, chert
PAV125	2.692	0.68	0.88	0.78	6.2	greywacke, diabase
PAV128	2.722	0.70	0.70	0.70	5.8	greywacke, slag
PAV129	2.698	0.80	0.83	0.82	6.2	greywacke
PAV131	3.050	0.86	0.87	0.87	8.1	slag, greywacke
PAV133	2.683	0.56	0.61	0.58	4.7	greywacke
SUR215	2.752	0.61	0.69	0.65	4.7	greywacke
SUR216	2.696	0.57	0.65	0.61	6.2	greywacke
SUR220	2.717	0.50	0.52	0.51	5.7	greywacke
SUR232	2.686	0.62	0.78	0.70	4.8	greywacke
SUR529	2.670	0.64	0.76	0.70	5.4	greywacke
SUR530	2.678	0.72	0.84	0.78	5.3	greywacke

Most of the aggregates that could be identified in the PA-derived coarse RAP aggregate were from a single source. However, there were 6 sites out of 17 that had a mixture of aggregates. In NZ, coarse aggregates for PA manufacture can be from different sources as long as 85% of all the coarse fraction to be made with the chosen aggregate has the required polished stone value for skid resistance performance [13]. This means that 15% of the coarse aggregates can come from other aggregate sources.

Test results from MD and absorption testing were plotted in Figure 1. A significant positive correlation (r=0.873) was found between absorption and MD loss in the tested RAP aggregates. Higher absorption of RAP aggregates resulted in higher MD loss. This relationship is consistent with and reinforces the findings from previous studies by Rismantojo [14], Khandal & Parker [15], and Jayawickrama et al. [10] that were performed on virgin coarse aggregates. Higher porosity aggregates usually have a tendency to be weaker [16]. Absorption of the 6.7-4.75mm fraction tested was generally higher compared to the 9.5-6.7mm fraction possibly due to bigger surface area in the finer fraction. This may mean that finer aggregates of the coarse fraction may have more effect on MD abrasion loss compared to the coarser size aggregate if MD loss is a function of water absorption.

Based on the findings from the NCHRP study by Khandal & Parker [15], MD testing was recommended for prediction of field failures such as pot-holing, pop-outs, and ravelling in hot-mix asphalt (HMA). The triangular dots in Figure 1 represent the PA sites PA108 (8 years old), PAV119 (12 years old), and PAV123 (7 years old), which had significant ravelling with no base issues recorded. The rest of the PA sites visited (circular dots) had base failures (although, it is unknown if the sites PAV112 and PAV117 had base failures as no site visit was possible), so it is hard to say if this ravelling is a function of the aggregate break down (or binder aging), or the flexing of the structurally compromised bases during traffic loading, resulting in aggregate loss in the PA surfacing.

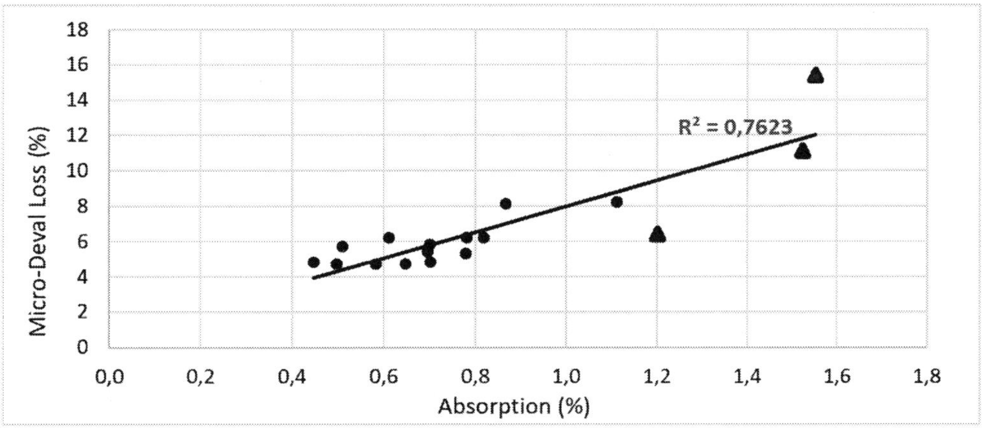

Figure 1. Relationship between Micro-Deval abrasion loss and water absorption in PA-derived RAP coarse aggregates.

All RAP aggregates tested had MD abrasion losses below the: 1) 18% limit reported by Wu, Parker, & Kandhal [6] and Williamson [7] on virgin aggregates; 2) 20% limit for RAP aggregates [8]; and 3) 15% limit for PA virgin aggregates [5] – except for one site, PAV108, which had MD loss of 15.4%. Furthermore, water absorption of all RAP aggregates was well below 2.2% – the maximum limit, which was used together with MD loss limit of 15% to predict PA. If these reported limits were to be implemented in NZ, PA-derived aggregates that had already visually disintegrated in the field prior to milling could end up being used in new PA mixes, and thus, highly likely compromise performance of the new PA mix. This highlights the importance of studying local materials and setting the limits for MD and absorption at a local level rather than relying on other countries' specifications.

It is likely that the age of aggregates in terms of geological terms will play a role in how aggregates perform in the field and therefore local geology properties would be influential.

Over 70% of RAP coarse aggregates from a single aggregate source were identified as greywacke. Most of the greywackes quarried in NZ are Permian-Mesozoic in age, weakly metamorphosed (prehnite-pumpellyite or, less commonly, zeolite facies), may be either volcaniclastic or lithic arkosic in composition, and on a macro-scale consist of interbedded sandstones, siltstones and mudstones. Even within one quarry, the properties of such aggregates may be different due to the heterogeneous nature of the parent rock [4]. In NZ, greywacke is the major source of aggregates for road construction.

Based on the limits set for each mineralogical group by the Ontario Ministry of Transportation [9], greywacke RAP aggregates fall into meta-arkose category with a maximum of 15% MD abrasion loss allowed for premium asphalt wearing courses for heavily traffic motorways. That is all of the PA-derived coarse aggregates from greywacke source tested in this study meet this selection criterion. The highest MD loss for this RAP aggregate was 15.4%. However, the greywacke from the PAV108 site gave very different MD, SG, and absorption results compared to the rest of the greywacke aggregates that were tested. The MD, SG, and absorption and thin section petrographic examination showed that aggregates from the PAV108 site came primarily from the Moutohora quarry, with a small amount of aggregate chips coming from the Hunua Winstones quarry. They are both greywacke aggregates, but are very different in properties. Greywacke aggregates from the Hunua Winstones quarry are still routinely used for asphalt manufacture. In contrast, the Moutohora quarry aggregates are no longer utilized in any asphalt mixes in NZ due to their propensity to erode under traffic. Bartley et al. [4] reported that in the Moutohora aggregates, the sand grains are easily dislodged from the highly porous matrix, which indicates a higher propensity to disintegration of the aggregate under traffic. This possibly explains why ravelling and "hungry" PA surfaces were observed on the PAV108 site, and in agreement with the high MD loss and the high absorption values for the RAP aggregates recovered from this site.

To understand how greywacke RAP aggregates may change in the PA mat with time, the results from the selected sites (PAV125, PAV129, PAV133, SUR216, SUR220, SUR232, SUR529, SUR530) were compared with the virgin greywacke aggregates from two quarries, which are routinely used in PA manufacture for Auckland motorways. The test results are graphed in Figure 2.

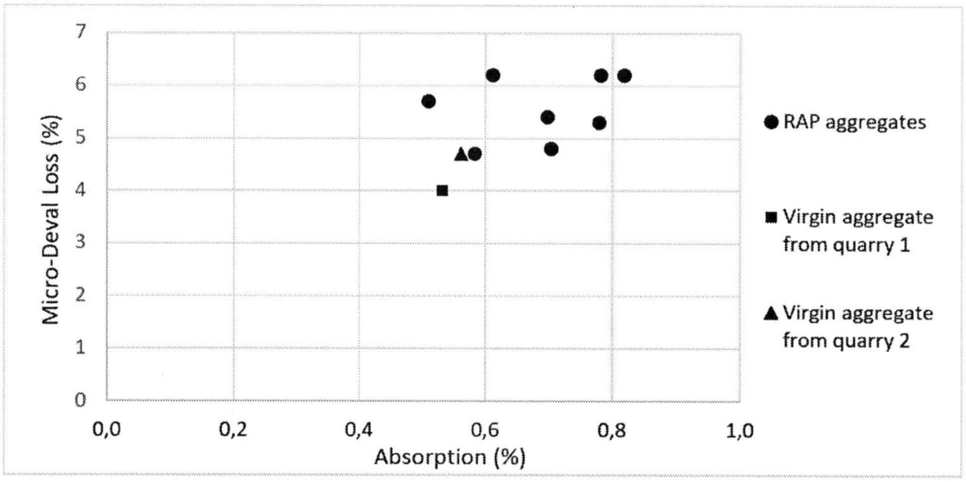

Figure 2. Micro-Deval abrasion loss and water absorption in PA- derived RAP coarse aggregates and virgin aggregates from greywacke aggregate source.

Each virgin greywacke aggregate was tested three times using the same testing conditions as for the RAP aggregates. The virgin aggregates' MD abrasion loss and water absorption were mostly lower compared to the RAP aggregates. This may indicate that greywacke aggregates go through chemical and (or) mechanical structural breakdown in a PA mat which is associated with traffic, water, and environment. Furthermore, entrapped pollutants in PA may also contribute to the mechanical breakdown. PA can accumulate various types of pollutants compared to DG surfacing [17]. This property of PA is beneficial to the environment, as it improves storm-water quality, but it may be a contributor to the aggregate degradation in a PA mat. Hajj et al. [18] reported that assessment of RAP aggregate abrasion is questionable as these aggregates were tested prior to use in hot mix asphalt. But if the aggregates are changing, physically or chemically, during service in PA surfacing, the aggregate durability should be taken into account, especially for RAP aggregates which are to be used in high stress demand conditions such as PA surfacing.

Aggregate disintegration was observed on the 7-year old PAV123 site. The MD loss testing on the RAP aggregates from this site gave 11.1% abrasion loss and an absorption of 1.52%, this was the second highest MD loss and absorption in RAP aggregates tested. RAP aggregate from this site were a mixture of predominately andesite with some inclusion of a small aggregate particle sized greywacke, and chert. Chert is a red color aggregate usually used for delineation of the road shoulder. It is not used in PA production. This means that during the milling process some chert material was incorporated in the millings. Chert is rich in silica and iron and it is a very tough material. Based on the petrographic examination, greywacke in this RAP sample matched the greywacke from the Hunua Winstone Quarry, a good performing aggregate currently utilized in many asphalt mixes. The andesite highly likely came from the Bay of Plenty region. After several years of utilizing andesite aggregates in HMA surfacing mixes, it was found that this aggregate could disintegrate under traffic loading. It is currently not used in major NZTA work in NZ for this reason. Again, high MD loss together with high absorption test results in the laboratory were able to reflect poor in-field aggregate performance in PA on the PAV123 site.

The PAV119 site, which had ravelling with no base failures observed, was 12 years old. This site was likely to have failed due to binder aging and subsequent, structural breakdown. RAP aggregates recovered from this site were of volcanic origin, basalt – a highly durable rock. MD abrasion loss and absorption results were relatively low, 6.4% and 1.20%, respectively, compared to the RAP aggregates from the PAV108 and PAV123 sites.

NZ aggregates are mineralogically different and geologically younger than the aggregates used in Europe and North America, where methodologies for assessment of aggregate engineering properties were originally developed. This is especially true for greywacke, the main aggregate used for road construction in NZ. It is less chemically and physically stable compared to sandstone aggregates, which are commonly used in countries like Australia, Europe, North America, and South Africa [4]. Limits for MD loss and water absorption for NZ aggregates need to be established that take into account the NZ aggregate minerology to be able to adequately predict aggregate field performance in asphalt mixes including PA. The term "greywacke" is too broad for setting the limits based on minerology.

7. Conclusions and recommendations

The following conclusions and recommendations can be drawn from this research study:

- Minerology plays an important role when setting limits for MD abrasion loss on PA-derived coarse aggregates
- Adopting limits for MD and water absorption without taking minerology into consideration can result in poor quality RAP aggregates to be erroneously utilized in asphalt mix manufacture.
- MD together with water absorption tests are very useful for the assessment of RAP aggregates for recyclability in new asphalt mixes.

- A strong relationship was found between MD and absorption in RAP coarse aggregates. Aggregates with higher absorption gave higher losses in the MD abrasion test. This finding was consistent with research reported on virgin aggregates.
- Greywacke virgin aggregates had generally lower MD and absorption compared to PA-derived greywacke RAP aggregates. This may indicate chemical and or physical changes in the aggregate during service of the PA surfacing.
- MD and absorption limits for RAP aggregates need to be developed for NZ based on the local aggregate mineralogy, and real field conditions that take into account the type of asphalt mix and traffic volume. The term" greywacke" is too broad for use as a mineralogical group for setting limits for MD and absorption, respectively as there is significant variance in properties and performance by greywacke source across NZ.

References

[1] Chappell B and Bennett D 2015 Asphalt Life Analysis *16th NZ Transport Agency & NZIHT Annual Conference (Bay of Islands, New Zealand)*

[2] Airey G D, Hunter A E and Collop A C 2008 The effect of asphalt mixture gradation and compaction energy on aggregate degradation *Construction and Building Materials* **22**(5) pp 972-80

[3] Rogers C A and Senior S A 1994 Recent Developments in Physical Testing of Aggregate to Ensure Durable Concrete *Proc. Engineering Foundation Conf. on Advances in Cement and Concrete (Durham)* pp 338-57

[4] Bartley F G, Peploe R J and Black P M 2010 *Abrasion resistance of aggregates in asphalt* (NZ Transport Agency research report 433)

[5] Fowler D W, Allen J J, Lange A and Range P H 2006 *The prediction of coarse aggregate performance by Micro-Deval and soundness related aggregate tests* (Research Report ICAR 507-1F)

[6] Wu Y, Parker F and Kandhal P S 1998 *Aggregate toughness/abrasion resistance and durability/soundness tests related to asphalt concrete performance in pavements* (NCAT Report 98-4)

[7] Williamson G S 2005 *Investigation of testing methods to determine long-term durability of Wisconsin aggregate resources including natural materials, industrial by-products, and recycled/reclaimed materials* (Virginia Polytechnic Institute and State University, Faculty of Engineering)

[8] Hajj E Y, Sebaaly P E and Kandiah P 2008 Use of Reclaimed Asphalt Pavements (RAP) in Airfield HMA Pavements *AAPTP Project 05-06: Final Report Appendices* (University of Nevada, Reno)

[9] Lane B C, Rogers C A and Senior S A 2000 The Micro-Deval Test for Aggregates in Asphalt Pavements *8th Annual Symposium of the International Center for Aggregate Research (Denver)*

[10] Jayawickrama P W, Hossain M S and Phillips F 2007 Micro-deval test for the assessment of bituminous aggregate durability *Journal of ASTM International* **4**(1) pp 1-13

[11] ASTM 2015 Standard Test Method for Relative Density (Specific Gravity) and Absorption of Coarse Aggregate *(ASTM C127-15)*

[12] ASTM 2010 Standard Test Method for Resistance of Coarse Aggregate to Degradation by Abrasion in the Micro-Deval Apparatus *(ASTM D6928-10)*

[13] New Zeeland Transport Agency 2005 Performance Based Specification For Hotmix Asphalt Wearing Course Surfacing *(TNZ P/23:2005)*

[14] Rismantojo E 2002 *Permanent deformation and moisture Susceptibility Related Aggregate Tests for use in Hot-Mix Asphalt Pavements* (Purdue University, Faculty of Engineering)

[15] Kandhal P S and Parker F 1998 Aggregate tests related to asphalt concrete performance in pavements *NCHRP Report 405*

[16] West T R 1995 *Geology Applied to Engineering* (Long Grove, Illinois: Waveland Press Inc.)
[17] Holleran I, Wilson D J, Holleran G, Walubita L F and James B 2016 Porous Asphalt - More Than Just Safety *IPENZ Transportation Group Conference (Auckland, New Zealand)*
[18] Hajj E Y, Sebaaly P E and Kandiah P 2008 Use of recliamed asphalt pavements (RAP) in airfields HMA pavements *AAPTP Project 05-06* (University of Nevada Reno)

Test of cold asphalt storability based on alternative approaches

Zora Abaffyová[1] and Jozef Komačka[2]

[1]CESTY NITRA a.s. Groupe COLAS, Murgašova 6, 949 78 Nitra, Slovakia
[2]Žilinská univerzita v Žiline, Univerzitná 8215/1, 010 26 Žilina, Slovakia

E-mail: jozef.komacka@fstav.uniza.sk

Abstract. Cold asphalt products for potholes repairs should be workable (soft enough) for long time to ensure their applicability. Storability is assessed indirectly using various tests of workability. Therefore, simple test methods (self-compaction and disintegration test) was developed and verified to investigate changes of storability of this group of cold asphalts. Self-compaction of the tested mixture in the upturned Abram`s cone for the cement concrete slump test and in the mould for the California Bearing Ratio test was assessed in first stage. After that the video record of disintegration test was taken. During this test, the mould was lifted up and the mixture fell off the mould (Abram`s cone) or disintegrate (CBR mould). The drop of surface after 10 min self-compaction and netto time related to falling out or disintegration of the mixture were used to evaluate the mixture from storability point of view. It was found out the self-compaction test has not a potential to reveal and prove changes of mixture properties. Based on the disintegration test results it can be stated this test at 5 °C using the upturned Abram`s cone could be a suitable approach to determine qualitative changes of a cold mixture from storability point of view.

1. Introduction

Various cold asphalt mixtures are used in asphalt road construction and maintenance. Depending on intended use, different requirements are defined for individual cold asphalt applications. Mostly they relate to a mixture composition and performance parameters in operation. However, other characteristics are also important in the case of some cold asphalt products. Particularly, cold asphalts used for pothole repairs should fulfill a lot of requirements from applicability and performance point of view. The exhaustive list of requirements was prepared and many of laboratory tests were performed in the project POTHOLE (7 FP EU, ERA–NET ROAD call). Based on the test results it was stated [1] it seems useful to determine some requirements for components of cold asphalts and parameters of a final mixture.

From applicability point of view it is important a cold mixture for pothole repair should be workable (soft enough) for long time regardless a storage method (plastic bags, buckets or roofed free air piles). A binder is a main factor that influences consistency (or stiffness) of a cold mixture. Basically, consistency changes in two ways: curing (a binder get stiff) and disintegration (loss of binder adhesion). It is suitable to know what stage of changes a cold mixture is in. Any test methods do not exist to evaluate storability of a cold mixture. Therefore, various tests of workability ([2], [3]) or the compaction test [4] are used to assess storability indirectly. These methods use obviously compaction energy and it can be biased when storability is appraised. Unsatisfactory workability according to a standardized method does not mean automatically negative evaluation of storability. A reason is the comparable level of workability can be achieved using various intensity of compaction energy. On the other hand, a good result of workability can be obtained in the situation when a binder get lose its adhesiveness. In this case the results of workability test are biased and unsatisfactory state of a cold mixture can be proven using tests of stiffness.

Therefore, simple test methods were proposed and verified to investigate changes of storability of cold asphalts for pothole repairs. These procedures do not need any compaction energy.

2. Experimental methods

The main idea was to find a procedure that would be able to reveal whether a cold mixture is "alive" or not. It was supposed a loose cold mixture in a good state should have a potential to self-compact. On the other hand, if a soft cold mixture is "alive" it should be able to disintegrate, when the mould is taken away. These assumptions are close to the requirements on cement concrete. Therefore, the cement concrete slump test was taken as an example and self-compaction of a cold mixture was investigated using the Abram`s cone. To eliminate possible influence of the shape of the Abram`s cone the cylindrical mould for the California Bearing Ratio (CBR) test was also used.

Some tentative tests were carried out to reveal possible rigours. Based on obtained experience the optimal procedure was established as follows.

2.1. Self-compaction test

- A mould was positioned on a metal plate placed on a laboratory scale (the Abram`s cone was positioned his narrower side on the metal plate, i.e. upturned in comparison to the cement concrete slump test)
- A cold mixture was continuously placed into the mould without any compaction till a small pile above the upper edge of the mould was originated
- Immediately, the excessed material was removed by a board and aligned to the upper edge of the mould
- Weight of the mixture in the mould was determined
- After 10 min the level of self-compaction was determined based on the distance between the surface of the mixture and the upper edge of the mould; this distance was determined as the average of five values measured at points evenly distributed on the inner area of the mould (to make the surface of the mixture even, a thin plate (approx. 2 mm thick) with the diameter 10 mm less than the diameter of the mould was placed on the surface of the mixture in the mould)

2.2. Disintegration test

- The disintegration test was performed immediately after the self-compaction test, i.e. approx. 15 min after aligning the mixture to the upper edge of the mould; based on experience in the tentative tests a video record of the test was taken
- The Abram`s cone was lifted up about 300 mm and it remained at this position till the mixture fell off the mould
- The CBR mould was taken away (lifted up)
- Time intervals related to beginning and end of falling out the mixture (Abram`s cone) or disintegration (CBR mould) were determined from the video record

The self-compaction values and time intervals were used for evaluation of the mixture from storability point of view.

3. Test results and discussion

One of cold mixtures for pothole repairs commonly available on the market was used for testing according to the mentioned procedures. The cold mixture consists of gravel 0/4 mm (36 %), crushed aggregate 4/8 mm (56.4 %) and a slow setting cationic bitumen emulsion C 65 BF 8 (7.6 %) prepared exactly for the used aggregate.

The cold mixture for the tests was taken from a roofed free air pile at a plant in monthly intervals since the cold asphalt production in November. Last sample was taken in March when stripping got started.

As cold asphalt mixtures for pothole repairs can be used at various temperatures, the tests were performed at the laboratory temperature of 20 °C ± 2 °C with the cold mixture temperature of 5 °C ± 1 °C and 20 °C ± 2 °C to cover a probable range of application temperatures. The cold mixture was conditioned to the test temperature during 24 h. The mould was also conditioned to the laboratory temperature of 20 °C ± 2 °C.

The following test outputs were evaluated for both test temperatures:

- Weight of the mixture necessary to fill in the mould
- The distance between the surface of the mixture and the upper edge of the mould (a drop of the surface) after 10 min of self-compaction
- Time since the mould was lifted up to the moment when the mixture started falling out the mould (Abram`s cone) or to disintegrate (CBR mould)
- Netto time of falling out the mixture from the mould or disintegation

The test results of self-compaction tests are shown on the next figures. There are no noticeable differences and no clear trend as for weight of the mixture at the temperature of 20 °C. May be a slight trend could be supposed for weight of the mixture at the temperature of 5 °C. There is a relatively strong decrease between November and February (about 400 g for both moulds) with the leap in March to the value higher than in November. The single one statement valid for all periods is weight of the mixture at 5 °C is lower when compared to 20 °C. It is reasonable and with high probability relate to the binder viscosity and workability of the mixture.

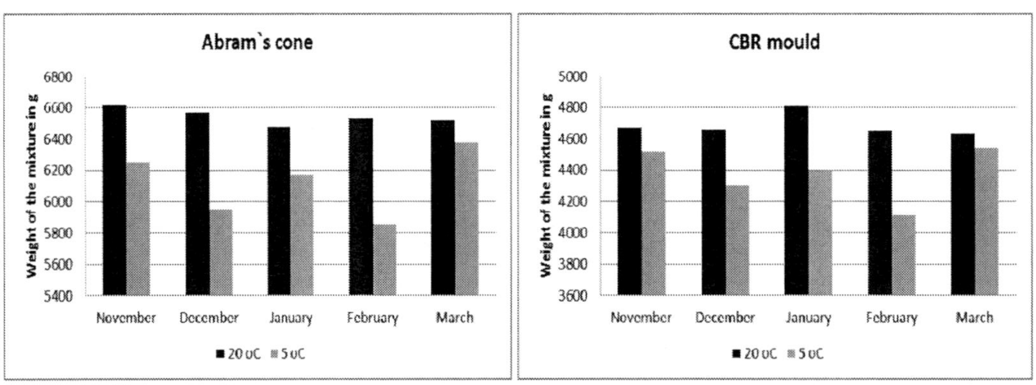

Figure 1. Weight of the mixture necessary to fill in the mould.

The drop of the mixture surface in the mould (self-compaction) after 10 min (Figure 2) demonstrates differences between 5 °C and 20 °C.

The self-compaction at the temperature of 20 °C is the highest in November. When the results for the Abram`s cone are evaluated, it is evident weight of the mixture from December to March was a little bit lower than in November. As the volume of the mould was the same, it means, air void content of the mixture in the mould in these months was higher than in November. If the mixture would be still "alive" at the same level as in November, an anticipation of the higher drop of the surface is legitimate. However, the results are in contradiction with the expectation. It can be perceived as some proof of change (worsening) of the mixture properties. Worsening of the mixture properties can be also deducted from the CBR mould results. Weight of mixture was approximately the same (except of January). It means, air void content of the mixture in the mould was also approximately the same. Therefore, the values of the self-compaction should be similar to the value in November. This did not become, although the differences are not significant. It seems the drop of surface after 10 min of self-compaction at the temperature of 20 °C could indicate a change of mixture properties (regardless the used mould) but there is not any clear difference in consecutive month that could be used as an unequivocal indication that a mixture is not in good state.

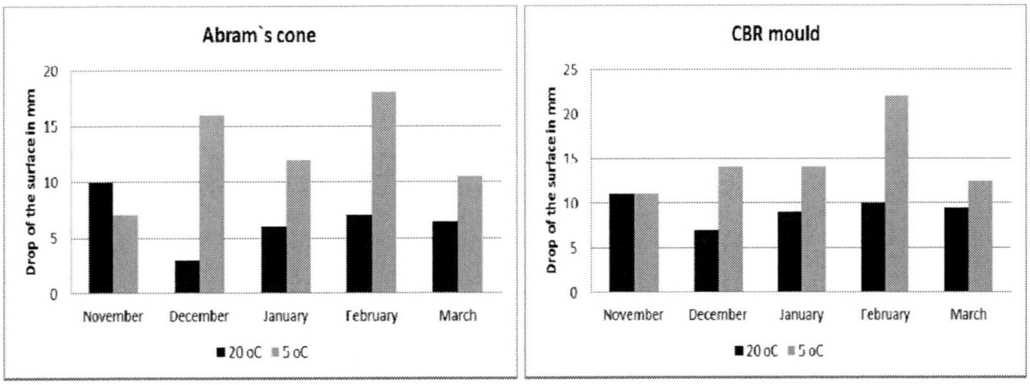

Figure 2. Drop of the mixture surface after 10 min self-compaction.

The values of the drop of the surface at the temperature of 5 °C have an inverse course when compared to the values at the temperature of 20 °C. The lowest value is in November, the next are higher. No trend can be observed. However, when weight of the mixture in the mould is taken into account, it seems the drop of the surface correspond to air void content of the mixture in the mould due to different weight of the mixture. Only exception is in March when weight of the mixture was higher than in November but the drop of the surface was also higher. It could be also perceived as an indication of a change of the mixture properties but it is not any indisputable proof.

The results of the disintegration tests (Figure 3, Figure 4) show starting and netto time determined from the video records. It is obvious the both time are higher for the temperature of 5 °C. It is in accordance with the assumption that the binder should have different viscosity that subsequently influences the mixture consistency. It was also supposed if a binder in a mixture lost its adhesive properties starting and netto time should be shorter. However, this expectation became real only in the case of netto time for Abram`s cone and the temperature of 5 °C. There is a clear difference among the value in March and the previous values. A reason can be netto time is sufficiently long in this case (unlike the others situations) and the differences are more visible.

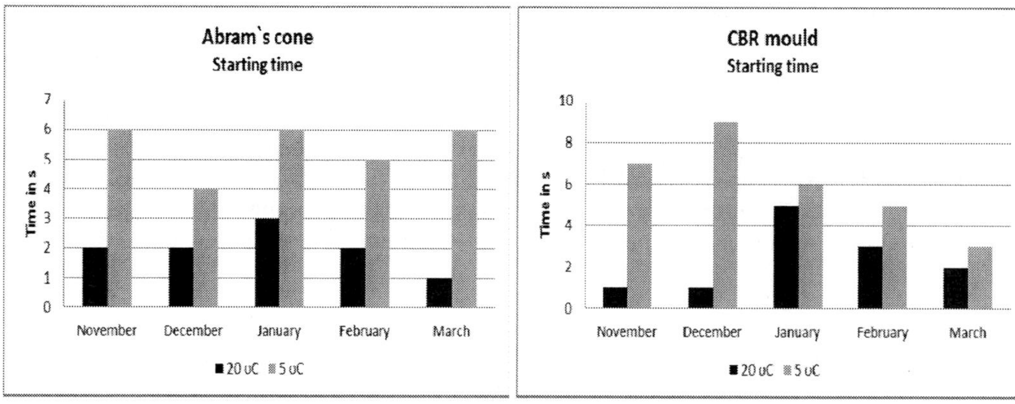

Figure 3. Starting time to falling out or disintegration.

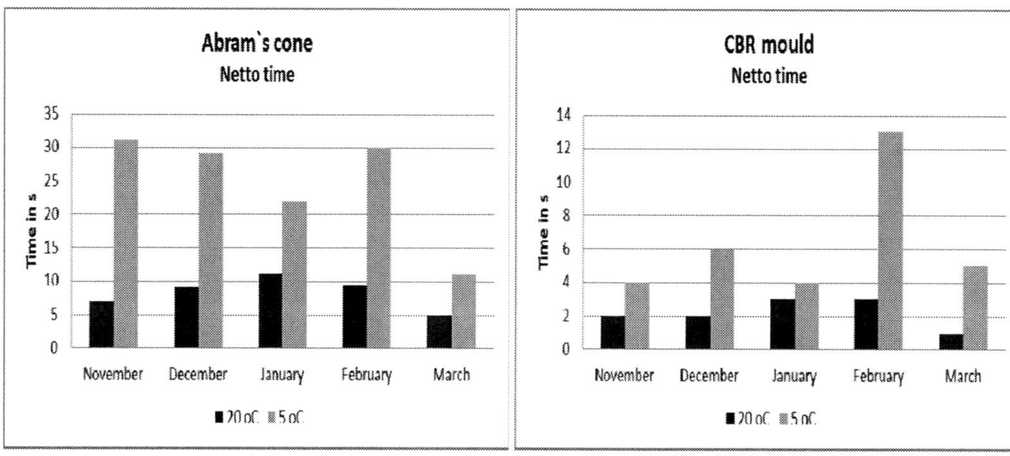

Figure 4. Netto time of falling out or disintegration.

Taking into account the outputs of the disintegration test, probably only the test at the temperature of 5 °C using Abram`s cone has a potential to reveal and prove changes of a mixture from storability point of view due to worsening a binder parameters (mainly adhesion to aggregate).

4. Conclusions

The alternative self-compaction and disintegration tests were performed five times in the monthly intervals on one cold mixture for pothole repairs using upturned Abram`s cone and CBR mould. All tests were carried out at laboratory temperature of 20 °C ± 2 °C with the cold mixture temperature of 5 °C ± 1 °C and 20 °C ± 2 °C. The test outputs taken into evaluation were the drop of the surface (self-compaction test) and starting and netto time of the mixture disintegration (disintegration test).

No clear trend, as for weight of the mixture in the mould, was founded out. The single one statement is weight of the mixture in the mould at 5 °C is lower when compared to 20 °C.

The results of the self-compaction test have showed the drop of surface after 10 min of self-compaction at the temperature of 20 °C could indicate a change of mixture properties (regardless the used mould). However, there is not any clear difference in the consecutive month that could be used as an unequivocal indication that a mixture changed its properties. The same statement has to be accepted based on the self-compaction test results at 5 °C. It seems, the alternative self-compaction test has not a potential (or it is not suitable) to prove changes of a mixture properties.

Starting and netto time determined from the video records of the disintegration test was higher for the temperature of 5 °C. The expectation of shorter netto time in subsequently month due to the loss of adhesive properties of the binder was confirmed only in the case of netto time for the Abram`s cone and the temperature of 5 °C. It is probably only the test that has a potential to reveal and prove changes of a mixture from storability point of view.

Acknowledgements
The authors are grateful for the financial support from the research project VEGA 1/0300/17 Research of performance related and rheological properties of bituminous binders and a good cooperation between University of Žilina and CESTY NITRA a.s. Groupe COLAS.

References

[1] Nicholls C, Kubanek K, Karcher C, Hartman A, Adesiyun A, Ipavec A, Komačka J and Nielsen E 2016 Durable pothole repairs *Materials and Infrastructures 1* ed J M Torrenti, F La Torre (London: Elsevier) pp 317-33
[2] Chatterjee S, Smit A F, White R P and Prozzi J A 2006 Ddevelopment of a slump-based workability test for bituminous maintenance mixtures *85th Transportation Research Board Annual Meeting* CD-ROM
[3] American Society for Testing and Materials (ASTM) 2015 Standard Test Method for Determining the Workability of Asphalt Cold Mix Patching Material *(ASTM D6704-08)* (ASTM International West, Conshohocken, PA)
[4] European Committee for Standardization (CEN) 2001 Bituminous mixtures. Test methods for hot mix asphalt. Compactibility. *(EN 12697-10)*

Analysis of the binder yield energy test as an indicator of fatigue behaviour of asphalt mixes

Johan O'Connell[1], Georges A J Mturi[1], Julius Komba[1] and Louw Du Plessis[1]
[1]CSIR Built Environment, Meiring Naude Road, Pretoria 0184, South Africa

E-mail: joconnell@csir.co.za

Abstract. Empirical binder testing has increasingly failed to predict pavement performance in South Africa, with fatigue cracking being one of the major forms of premature pavement distress. In response, it has become a national aspiration to incorporate a performance related fatigue test into the binder specifications for South Africa. The Binder Yield Energy Test (BYET) was the first in a series of tests analysed for its potential to predict the fatigue performance of the binder. The test is performed with the dynamic shear rheometer, giving two key parameters, namely, yield energy and shear strain at maximum shear stress (γτmax). The objective of the investigation was to perform a rudimentary evaluation of the BYET; followed by a more in-depth investigation should the initial BYET results prove promising. The paper discusses the results generated from the BYET under eight different conditions, using six different binders. The results are then correlated with four point bending beam fatigue test results obtained from asphalt mix samples that were manufactured from the same binders. Final results indicate that the BYET is not ideal as an indicator of fatigue performance.

1. Introduction

Bahia *et al* [1] investigated how a simple monotonic shear test, that they named the Binder Yield Energy Test (BYET), gave two parameters, yield energy and shear strain at maximum shear stress (γτmax), that showed good correlation with the time sweep results published by Martono and Bahia [2]. Good correlation between yield energy and crack length was also found for one project carried out by using an accelerated loading facility (ALF) [1]. The BYET parameters and correlations are illustrated in Figures 1 and 2.

However, other studies have shown poor correlation with cracking [3]. They reported challenges with this test, which include:

- Double-peak behaviour with certain modified binders, which complicate modelling for continuum damage analysis to estimate the fatigue life of the binder,
- Modifiers showing a more pronounced contribution to fatigue resistance at higher strains (i.e. post maximum stress behaviour).

One of the major forms of premature pavement distress in South Africa is fatigue cracking and it was decided to perform a rudimentary evaluation of the BYET as a possible improved indicator of fatigue performance. If the BYET results showed it could potentially serve as an indicator of fatigue performance, the BYET would be analysed further for incorporation into binder specifications used for asphalt mix design.

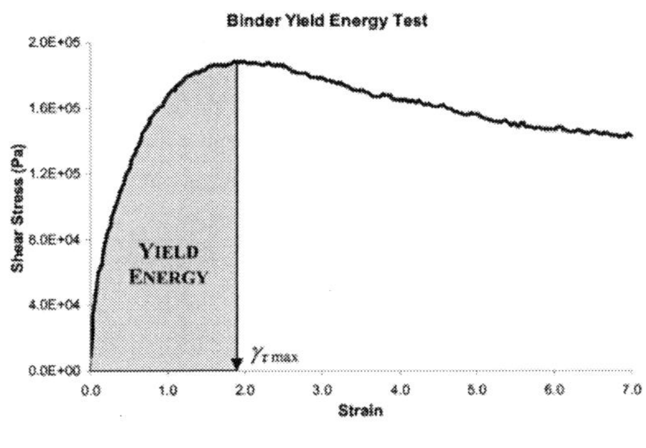

Figure 1. Illustration of the test [1].

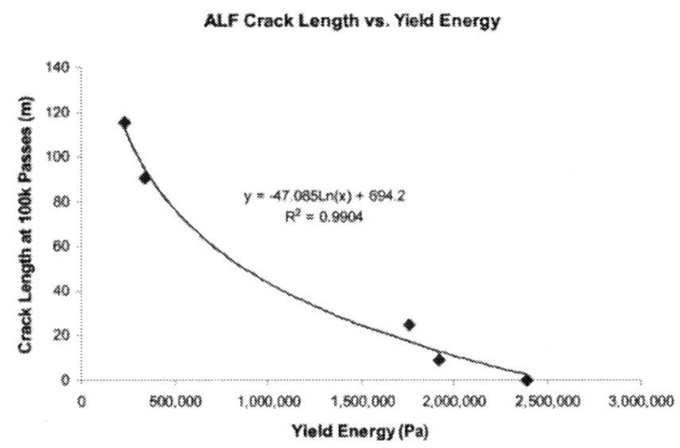

Figure 2. Correlation with ALF data [1].

2. Experimental

2.1. BYET evaluation of six binders

The parameters adopted for the BYET investigation were a combination of the 2009 and 2013 versions of the test method as shown in Table 1. The binders were evaluated after short-term rolling thin film oven test (RTFOT) ageing as opposed to long-term pressure ageing vessel (PAV) ageing, given that the asphalt fatigue specimens had only undergone short-term ageing. The following six binders were evaluated:

- Experimental 50/70 penetration grade binder created in a laboratory.
- 50/70 penetration grade bitumen (produced March 2010)
- 50/70 penetration grade bitumen (produced June 2012)
- EVA-modified binder, locally classified as A-P1 [4]
- Terpolymer-modified binder (Elvaloy), locally classified as A-E1 [4]
- SBS-modified binder, locally classified as A-E2 [4]

Table 1. BYET parameters.

Parameter	2009 Method	2013 Method	Adopted Conditions
Strain Rate (%/s)	1%/s	2.3%/s	1%/s and 2.3%/s
Temperature (°C)	Intermediate Temperature	Any Suitable Temperature	5, 10, 20, 28

2.2. Four point beam fatigue testing of asphalt mixes

The four point beam fatigue test is widely accepted to relate laboratory-derived fatigue life models to observed field performance where bottom-up cracking is the failure criterion [6]. Our methodology included an evaluation of the correlation of the BYET results with the beam fatigue test, which was interpreted as an accelerated representation of fatigue cracking in the field.

The binders that were selected for the BYET were used to manufacture medium continuously-graded hot mix asphalt. It was the goal to keep the asphalt mix properties the same for all the mixes in order that the influence of the binder on the four point beam testing results could be ascertained. When comparing these asphalt mixes, they had:

- Similar gradings (Table 2, Figure 3)
- Binder contents varying between 4.6 and 5.0%
- Void contents compacted to between 6 and 8%

The fatigue testing was conducted using an IPC standalone four-point bending testing setup. The tests were performed according to the CSIR test protocol [5] under controlled-strain loading conditions. The conventional failure criterion defined as the number load cycles to reach 50% reduction in the initial stiffness (N50%) was adopted for testing. A test temperature of 10°C was selected and strain amplitude levels ranged from 200 to 600 microstrain, representing a minimum number of 6 evaluations per mix. Results were plotted on a logarithmic scale for all mixes and the best-fit power graphs were used to calculate the number of repetitions to failure at 300 microstrain, which was the lowest strain amplitude that all the mixes had in common. The calculated value was used for correlation with the BYET results

Table 2. Grading and binder contents of asphalt mixes.

	AE2: SBS	50/70: 2010	50/70: 2012	AP1: EVA	AE1: Elvaloy	Exp. 50/70
Sieve Size			% Passing			
13.2	100	100	100	100	100	100
9.5	95	97	98	91	95	98
6.7	80	75	79	75	80	79
4.75	62	59	64	58	62	64
2.36	40	42	44	43	40	44
1.18	28	30	30	30	28	30
0.6	21	21	20	22	21	20
0.3	16	14	14	15	16	14
0.15	10	9	9	10	10	9
0.075	5.5	5.8	6.6	6.9	5.5	6.6
Binder Content	4.7	5.0	4.7	4.8	4.7	4.7

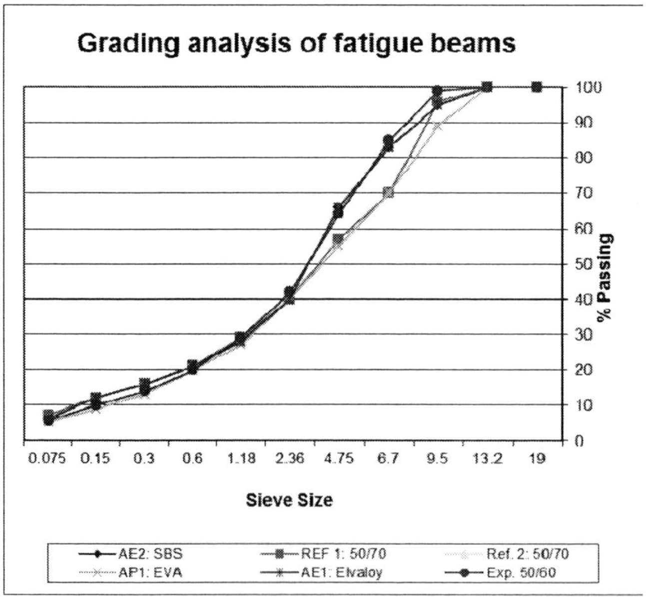

Figure 3. Grading of asphalt mixes.

3. Results

3.1. BYET evaluation of six binders

The repeatability of the BYET was tentatively evaluated by randomly re-testing some of the samples at various temperatures. The results are shown Figure 4. The repeatability was evaluated qualitatively, and a quantitative analysis of the effect of the differences in results was deferred to a later date, should the method prove to have potential.

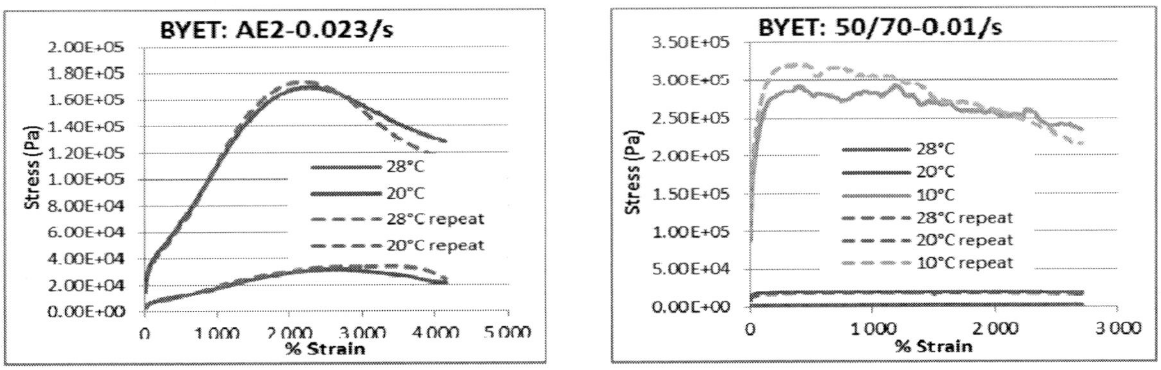

Figure 4. Repeatability of the BYET.

Examples of the BYET results are illustrated in Figures 5 to 7. Each figure shows the results obtained at the four temperatures investigated. In each figure, the test results from the lowest temperatures are removed consecutively as the y-axis is expanded in order to amplify the results obtained for the higher temperatures. Only three examples are illustrated due to space considerations.

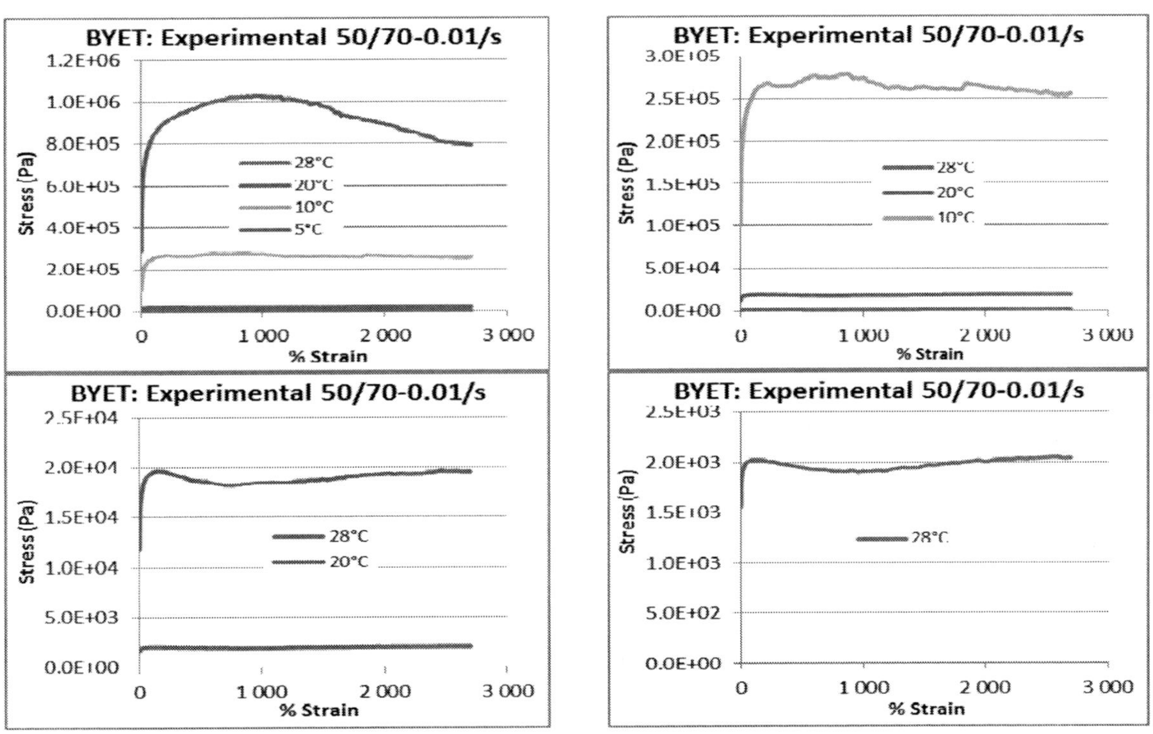

Figure 5. BYET test results for Experimental 50/70 penetration grade binder at 1%/s.

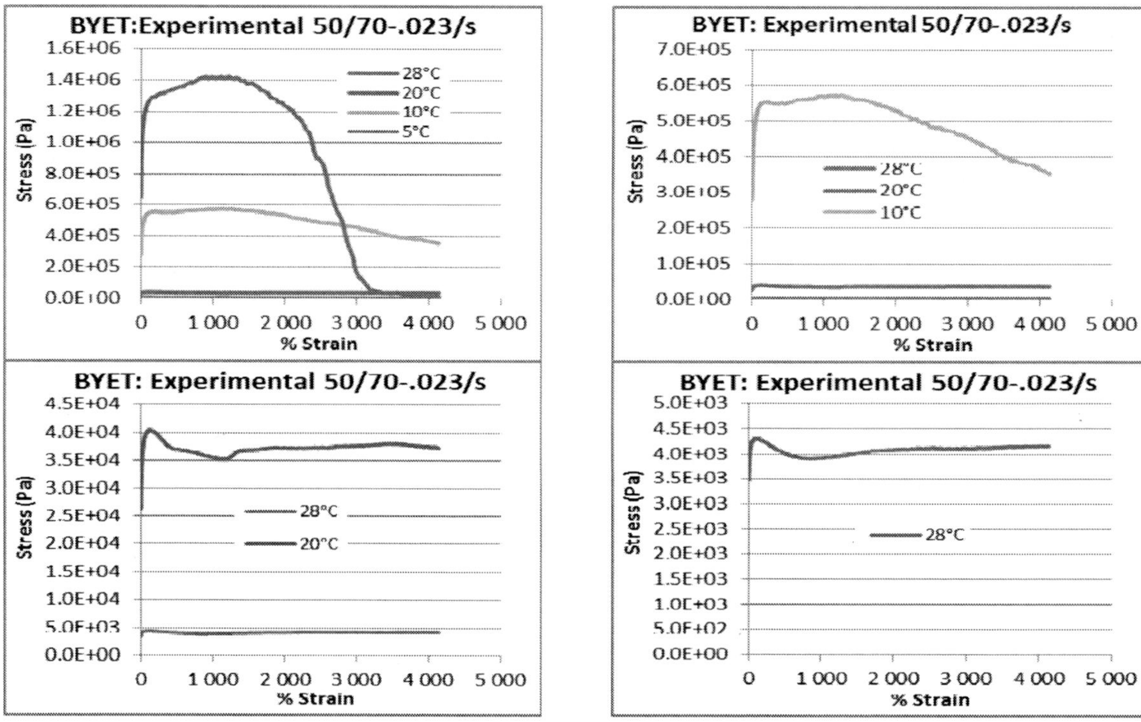

Figure 6. BYET test results for Experimental 50/70 penetration grade binder at 0.023%/s.

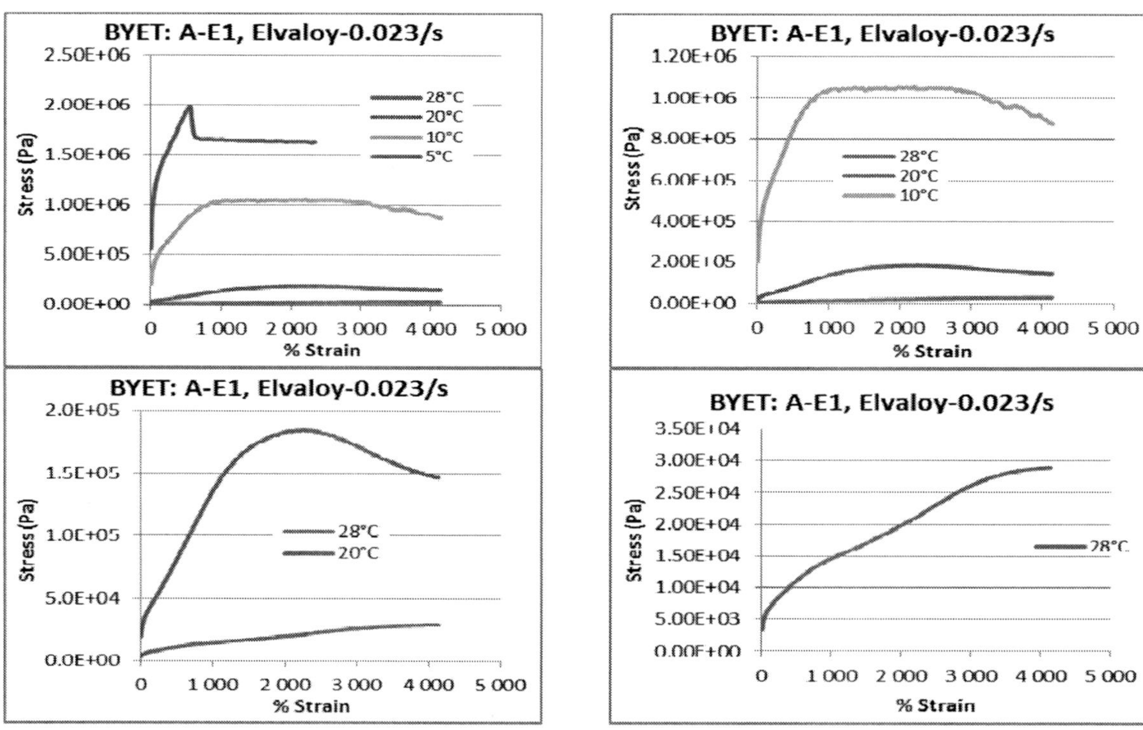

Figure 7. BYET test results for A-E1 (Elvaloy) binder at 0.023%/s.

An appraisal of the results led the authors to the following conclusions:

- The information at 5 °C contains anomalies (as illustrated in Figure 7), probably as a result of sample slippage caused by the large strains at low temperature. Therefore results at this temperature were abandoned with regards to further analysis.
- The maximum stress at 28 °C was often not attained by the end of the BYET (as illustrated in Figures 5 and 7). Similarly, the results at this temperature were also abandoned with regards to further analysis.
- At the temperatures of 10 °C and 20 °C, the graphs generated by a strain rate of 2.3%/s gave better defined maxima compared to the graphs generated at a strain rate of 1%/s.

The BYET results for 10°C and 20°C are summarised in Table 3 and 4, using the Binder Yield Energy, calculated up to $\gamma_{\tau max}$ (at maximum stress), and $\gamma_{\tau max}$, along with the predicted fatigue ranking suggested by the results.

Table 3. Yield energy calculated from the BYET, followed by the predicted fatigue ranking.

Binder	20 °C, 1%	20 °C, 2.3%	10 °C, 1%	10 °C, 2.3%
Exp. 50/70	2.89E+04, **6**	4.97E+04, **6**	2.30E+06, **5**	6.97E+06, **4**
50/70 (2010)	1.33E+05, **5**	1.34E+05, **4**	3.23E+06, **4**	1.27E+06, **6**
50/70 (2012)	1.84E+05, **4**	1.22E+05, **5**	1.09E+06, **6**	1.48E+06, **5**
EVA-modified	1.68E+06, **1**	3.92E+05, **3**	4.71E+06, **3**	9.68E+06, **2**
Elvaloy	1.57E+06, **3**	2.95E+06, **1**	6.73E+06, **2**	2.05E+07, **1**
SBS-modified	1.61E+06, **2**	2.63E+06, **2**	7.01E+06, **1**	8.21E+06, **3**

Table 4. $\gamma_{\tau max}$ calculated from the BYET, followed by the predicted fatigue ranking.

Binder	20 °C, 1%	20 °C, 2.3%	10 °C, 1%	10 °C, 2.3%
Exp. 50/70	155, **6**	132, **6**	871, **4**	1 259, **3**
50/70 (2010)	681, **5**	302, **4**	1 177, **2**	247, **6**
50/70 (2012)	909, **4**	256, **5**	347, **6**	279, **5**
EVA-modified	2 227, **2**	353, **3**	858, **5**	1 150, **4**
Elvaloy	2 193, **3**	2 256, **2**	1 168, **3**	2 200, **1**
SBS-modified	2 468, **1**	2 300, **1**	1 545, **1**	1 325, **2**

3.2. Four point beam fatigue testing of asphalt mixes

The four point beam fatigue results are given in Table 5. The calculated number of repetitions to failure at 300 microstrain is reported, followed by the ranking of four point beam fatigue performance.

Table 5. Asphalt mix fatigue evaluation.

Asphalt mix binder	No. of repetitions to failure at 300μm	Ranking
Exp. 50/70	103 177	6
50/70 (2010)	124 788	5
50/70 (2012)	144 813	4
EVA-modified	375 620	3
Elvaloy	507 650	2
SBS-modified	3 743 218	1

3.3. Correlation of binder and asphalt mix results

The BYET, using four different test conditions and using two different parameters, produce eight different sets of predicted fatigue rankings, of which not one corresponds to the four point beam fatigue results. The fatigue prediction rankings calculated with $\gamma_{\tau max}$ was poor, sometimes ranking unmodified binders above polymer-modified binders, which is contrary to four point beam results.

Technically, a correlation exercise should not be carried out when the ranking of the binder testing cannot be matched the ranking of the asphalt mix testing. However, for the sake of completeness, a limited correlation was done between the beam fatigue results and the yield energy. The results are presented in Figure 8.

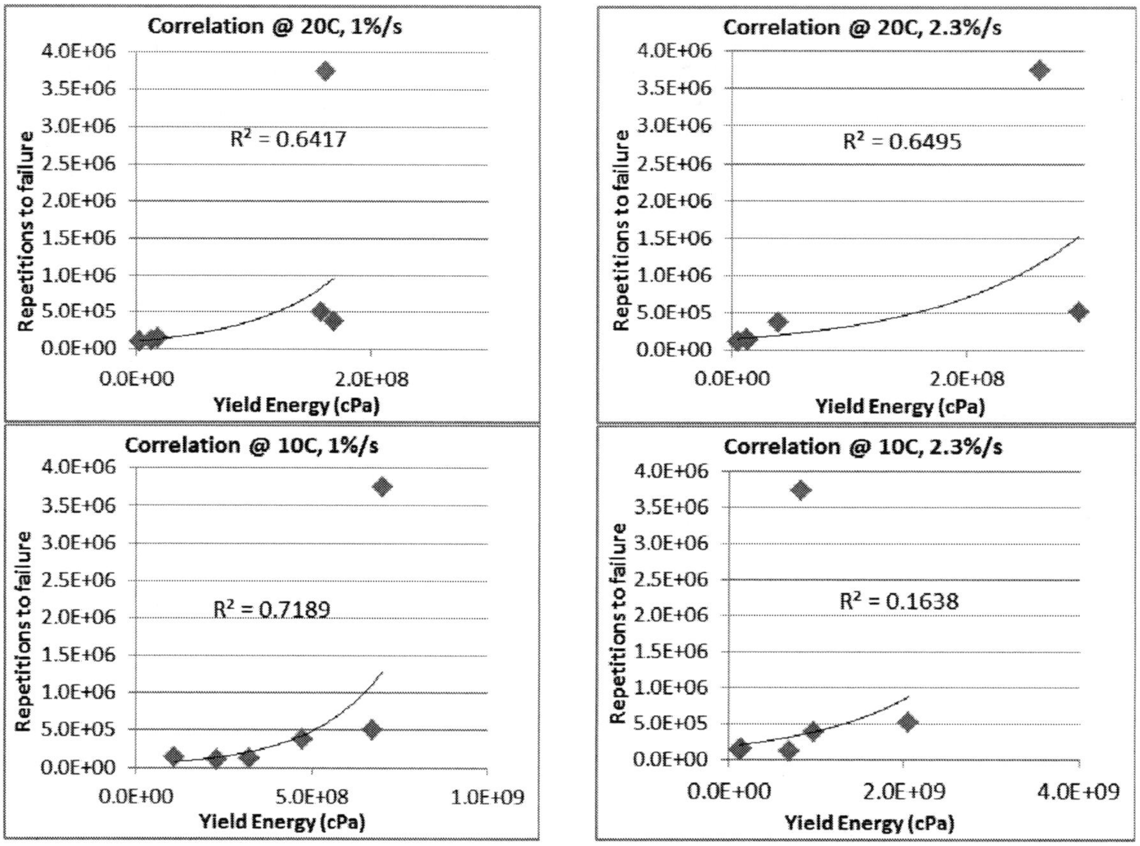

Figure 8. Correlation between beam fatigue results and yield energy.

4. Discussion and Conclusion

The BYET fatigue prediction rankings changed significantly as the temperature changed from 10°C to 20°C or as the rate of strain changed from 0.01/s to 0.023/s. The fact that slight changes in test conditions gave rise to inconsistency in the results gives rise to doubt whether the test relates well to the fundamental fatigue properties of the binder. None of the binder rankings attained from the BYET corresponded to the beam fatigue rankings. Furthermore, the double peak behaviour made it difficult to calculate BYET parameters consistently.

Based on these preliminary investigations, the BYET was not recommended for further investigation as a possible fatigue parameter for the Performance Graded binder specification in South Africa. The double peak behaviour resulted in changes in the fatigue rankings of binders with small changes in test parameters, i.e. poor repeatability. This, in turn, leads to poor correlation with asphalt mix fatigue behaviour.

Acknowledgements
We acknowledge the Southern African Bitumen Association (SABITA) for their funding of this work.

References

[1] Bahia H, Wen H and Johanson C M 2010 Developments in intermediate temperature binder fatigue specifications *Transportation Research Circular Number* **E-C147** pp 25-33
[2] Mortono W, Bahia HU and D'Angelo J 2009 Effect of testing geometry on measuring fatigue of asphalt binders and mastics *J Mat Civ Eng* ***19*** pp 746-752
[3] Johnson C M, Bahia H U and Coenen A 2009 Comparison of bitumen fatigue testing procedures measured in shear and correlations with four-point bending mixture fatigue *Paper presented at the 2nd Workshop on Four Point Bending (Guimaraes, Portugal)*
[4] Asphalt Academy TG1 2007 *Technical Guidelines; The use of modified bituminous binders in road construction*
[5] Wirczak M, Mamlouk M, Souliman M and Zeiada W 2013 Validating an Endurance Limit for Hot-Mix Asphalt (HMA) *NCHRP Project 9-44A, The national Cooperative Highway Research Program of the Transportation Research Board*
[6] Anochie-Boateng J, Denneman E, O'Connell J and Ventura D 2010 Test Protocols for determining properties of asphalt materials for SAPDM CSIR *Technical Report No: CSIR/BE/IE/IR/2010/0001/B, CSIR* (Pretoria)

Analyzing the stripping potential of warm mix asphalt using imaging technique

Muhammad Rafiq Kakar[1,2], Meor Othman Hamzah[3] and Jan Valentin[4]

[1]EMPA, Swiss Federal Laboratories for Material Science and Technology, Ueberlandstr. 129, Duebendorf, CH-8600, Switzerland

[2]Highway Research and Training Center (HRTC), National Highway Authority of Pakistan, NHA HQ 27, Mauve Area, G-9/1, Islamabad, 44000, Pakistan

[3]School of Civil Engineering, Universiti Sains Malaysia, Nibong Tebal, Penang, 14300, Malaysia

[4]Department of Road Structures, Faculty of Civil Engineering, Czech Technical University, Prague, Czech Republic

E-mail: cemeor@yahoo.com

Abstract. In asphalt mixtures, stripping occurs when the bond between the asphalt and the aggregate is broken due to the intrusion of water within the asphalt aggregate interface. Warm mix asphalt (WMA) is a technology that allows significant reduction in mixing and compaction temperatures of conventional hot mix asphalt. However, WMA is susceptible to moisture damage due to its lower production temperature. This can cause adhesive failure, hence stripping of asphalt binder from the aggregates. In this study, direct tensile strength (DTS) and indirect tensile strength (ITS) tests were applied to fracture the mixture specimen. Imaging technique was applied on the fractured faces of asphalt mixture to quantify the adhesive failure susceptibility due to the destructive effects of moisture. The results showed that adhesive failure increased with the number of freeze and thaw cycles and mixtures prepared with PG-76 binder exhibited lower adhesive failure compared to PG-64 binder. From fractured ITS samples, most of broken aggregates were found located in the vicinity where the indirect tensile load was applied. On the other hand, high adhesive failure was obtained at the center portion where maximum tensile stresses were developed. The image analysis method employed in this work has proven to be very effective to analyze the deterioration of asphalt mixtures subjected to moisture conditioning.

1. Introduction

Moisture is a major concern in asphalt pavements and it can exert adverse effects on the performance of hot mix asphalt (HMA) [1,2]. In the case of WMA, mixing at low temperature might result in insufficient drying of the aggregates, which in turn adversely affects the adhesion between aggregates and asphalt binder [3,4]. According to Lottman [5], two typical mechanisms of moisture damage are:

i. Loss of adhesion due to moisture at the asphalt aggregate interface, stripping away the asphalt film
ii. Loss of cohesion due to softening of the mastics.

The adhesion between asphalt and aggregate can be classified based on chemical reaction, surface energy, molecular orientation, and mechanical adhesion [6]. These can be influenced by factors such as surface tension of the asphalt binder and aggregate, chemical composition of asphalt and aggregate, asphalt viscosity, surface texture of the aggregate, aggregate porosity, aggregate cleanliness and aggregate moisture content and asphalt binder temperature at the time of mixing [7]. Lytton et al. [8] measured the thickness of asphalt film during which the failure from adhesion to cohesion takes place. They identified that thin films asphalt mixtures fail in tension (adhesive bond failure), while failures within the mastic in thick films asphalt mixtures are due to cohesive failure. Another study by Kim et al. [9] showed that moisture damage is strongly influenced due to the cohesive and adhesive bond failure. Stripping in asphalt mixtures can occurs when the bond between asphalt binder and aggregate is broken due to the intrusion of water within the binder aggregate interface. Moisture can disrupt the bonds between asphalt aggregate by means of molecular diffusion through asphalt film and has the potential to extend towards the aggregate surface [10]. Stripping mechanisms can occur in the form of detachment, displacement, spontaneous emulsification, pore pressure, hydraulic scour, pH instability, and the effects of the environment on bitumen aggregate material systems [11,12]. Stripping can also increase due to variations in temperature, freeze thaw cycles, and wetting-drying cycles. Stripping of asphalt mixtures is a phenomenon resulting from moisture damage and is more likely to take place in the context of WMA. Furthermore, the water nature to which the mix is exposed (salt content, pH) also affects the stripping of asphalt mixture [13]. Fillers such as hydrated lime, Ordinary Portland Cement, and many other liquid anti stripping agents have been used in HMA to mitigate stripping of asphalt binder from aggregate [14].

Presently, the water sensitivity of asphalt pavements are studied using relatively simple tests that can be divided into two groups. Firstly, tests conducted on loose mix to determine aggregate coating properties during water immersion or in boiling water (qualitative measurement). Secondly, tests conducted on compacted mixes to evaluate the retained strength or stiffness or more commonly, the mechanical properties (quantitative measurement). Unfortunately, these tests were not very effective as they evaluate the adhesive failure via visual inspection [15]. The use of mechanical test to measure strength followed by imaging technique to quantify the percent amount of adhesive failure is more convenient to achieve both quantitative and qualitative measurements. Such a technique can enable testing of parameters using only one material in addition to saving time and cost involved in the preparation and testing of loose mix. Compared to visual inspection the use of imaging technique is more precise, as the human vision is relatively poor at differentiating between the perceived brightness and color features of the failure plan [16]. Image analysis software has been widely used worldwide in a variety of applications. Salem et al. [17] described that imaging methods have been well-recognized to be very effective in analyzing deterioration status of concrete structures. This paper evaluates the adhesion behavior of asphalt mixtures related to its mechanical strength both in direct tension and indirect tension when subjected to moisture conditioning. In asphalt technology, moisture susceptibility testing of loose and compacted asphalt mixtures are carried out in a qualitative and quantitative manner. To address the above stated parameters, a novel method using image analysis technique was adopted which fractured the specimen in direct tensile and indirect tensile followed by examining the fractured surface. With this method, a more precise quantification is expected compared to the conventional method of visual inspection.

2. Materials and method

Two conventional asphalt binders, PG-64 and PG-76 (polymer modified binder), were used in this study. The asphalt binders were transported from Shell Bitumen Singapore to the laboratory in sealed containers to minimize the effects of oxidation and premature binder aging. Surfactant based WMA additive results in a significant improvement on the production conditions of sustainable asphalt mixtures. Cecabase RT 975 additive is consumed in a very limited amount of 0.2 to 0.5% by weight of the binder [18]. This additive, which is liquid at room temperature, can be easily mixed with the hot asphalt binder before the asphalt mixture production.

2.1 Filler

Mineral fillers are fine particles of mineral material that exhibit specific characteristics, helpful in the design and construction of asphalt pavements. The typical proportions are 94 to 96% aggregate comprising 6% of mineral filler, 4 to 6% asphalt binder by weight. To investigate the effects of filler as anti-stripping agent in asphalt mixture, ordinary Portland cement (OPC) and pavement modifier (PMD) were used. PMD is a greyish-black powder mineral filler developed in Malaysia used as anti-stripping agent. The addition of approximately 5% of PMD by aggregate weight, act as mineral filler in hot mix asphalt surface course mixes.

2.2 Specimen preparation

All specimens were prepared according to the Asphalt Institute manual (MS-2) (Asphalt Institute, 2001). The hot binder and aggregate were mixed together at 160°C, 180°C and compacted at 150°C, 170°C for PG-64 and PG-76 binders, respectively. PG-64 and PG-76 WMA mixtures were prepared at mixing temperatures shown in Table 1 and relatively 10°C lower temperatures were used for compaction of each WMA mixtures. The specimens tested in this study were 100 mm diameter and 120 mm in height for direct tension test (DTT) and 100 mm diameter and 63.5 mm height for indirect tensile strength (ITS). The samples were compacted to the desired heights by using the Servopac Gyratory Compactor. The air voids of the specimens were kept at 7 ± 1% and the specimen heights at 120 ± 0.5 mm. Since Cecabase RT 975 content has no effect on the optimum binder contents. In addition, the supplier recommends 0.3% Cecabase RT 975 and this value was adopted to test the DTT and ITS specimens. In Table 5, the specimens are designated according to their type (H for HMA and W for WMA), filler type (O for OPC and P for PMD) and their mixing temperature for example (180 for 180°C). Therefore, specimen HP160 denotes a HMA incorporating PMD filler mixed at 160°C.

The optimum binder content (OBC) utilized in the preparation of both DTT and ITS specimens are 4.5 and 5.2%, respectively. For mix design, the target air voids of all the specimens were 4 ± 0.1%. This showed that OBC for the WMA was found to be independent of the amount of additive used.

Table 1. Mixture designation (DTT and ITS).

Mixture Type	Filler Type	Mixing Temperature (°C)		Designation	
		PG-64	PG-76	PG-64	PG-76
HMA	OPC	160	180	HO160	HO180
HMA	PMD	160	180	HP160	HP180
WMA	OPC	140	170	WO140	WO170
WMA	OPC	130	160	WO130	WO160
WMA	OPC	120	150	WO120	WO150
WMA	PMD	140	170	WP140	WP170
WMA	PMD	130	160	WP130	WP160
WMA	PMD	120	150	WP120	WP150

The DTT specimen end faces were cleaned and dried out at room temperature. To avoid non-uniform stress concentration due to eccentricity, special care was taken while gluing samples to ensure flat end faces. The specimens were glued with specially designed tensile test molds which utilized a low temperature high adhesive strength epoxy resin. The Araldite high performance epoxy adhesive was used to glue the specimen with the capping mold. After gluing, the specimens were kept at room temperature for 24 h to allow strong bond between the molding cap and specimen to develop [15].

2.3 Laboratory moisture conditioning

The moisture conditioning was performed to evaluate the effects of accelerated water conditioning employing the freezing-thawing cycle on compacted asphalt mixtures as described in the previous study by Hamzah et al. [15]. The conditioning of all the compacted specimens was done according to

ASTM D4867 (2006) procedures with the only modification of using distilled water with addition of Na_2CO_3 at 6.62 gm. Water with Na_2CO_3 was used to increase the pH value so as to enhance the stripping rate or damage inside asphalt specimens [19]. The specimens were immersed in the solution and vacuumed for 15 min to achieve saturation levels between 55% and 80%. For the next step, these specimens were exposed to freezing condition at -18 ± 3 °C for 16 h and thawing at 60°C for 24 h as one cycle. Three sets of specimens; unconditioned dry, conditioned 1 freeze thaw cycle and 3 freeze thaw cycles were prepared and set aside. The conditioned DTT specimens were left for 24 h before gluing them to perform the test. Finally, the effects of conditioning on HMA and WMA stripping were evaluated using imaging technique on the fractured faces of DTT and ITS specimens [15].

2.4 Imaging technique

Images of DTT and ITS fractured specimens were obtained using a high-resolution optical device and were then processed by means of Environment for Visualizing Images (ENVI) software (Exelis Visual Information Solutions). Optical imaging is a simple and economic way as it can acquire information (images) of a component without direct contact (non-destructive type). It is also an area scanning method, capable of instantaneous recording of a two-dimensional image of an object. Once the live images are taken, appropriate image processing reveals potential characteristics and changes in image under investigation [15]. The Region of Interest (ROI) tool given by ENVI was used to classify the area marked as brown and white tinctures on the specimen images. The selected brown and white regions were identified as stripping/failure due to adhesion and failure within the aggregates/broken aggregates, respectively. More details on the process of imaging technique can be found in Hamzah et al. [15].

3. Results and discussion

In this paper, for the purpose of comparing results of ITS test with DTT, the results of DTT were obtained from the authors already published work elsewhere Hamzah et al. [15]. The following sections will describe the results of percent adhesive failure, tensile strength (direct and indirect) broken aggregates, and statistical analysis and correlations obtained from the results of DTT and ITS tests.

3.1 Effects of moisture conditioning on adhesive failure of DTT samples

The image analysis provides measurements of failure plan in terms of area in square meters. However, for analysis of the results, failures caused by adhesion and broken aggregates are expressed in percentage, while strength values obtained from DTT are expressed in Mega Pascal.

The adhesive failure results of both PG-64 and PG-76 mixtures at different conditioning cycles are respectively presented in Figures 1(a) and (b). The results show that the number of freeze-thaw cycles influences the percent amount of adhesive failure showing that with the increase in F-T cycles the percent adhesive failure increases. It was also observed that the percent adhesive failure due to moisture conditioning (F-T cycles) was higher in PG-64 compared to PG-76 mixtures. The PG-64 and PG-76 WMA mixtures show high resistance against moisture damage at 1 F-T cycle compared to HMA. However, this trend was reversed after 3 F-T cycles. Mix WO130 and WO120 exhibit the highest percentage adhesive failure after 1 and 3 F-T cycles. This might be due to the influence of lower mixing temperature, where WMA's are thought to be more prone to adhesive failure at higher F-T cycles.

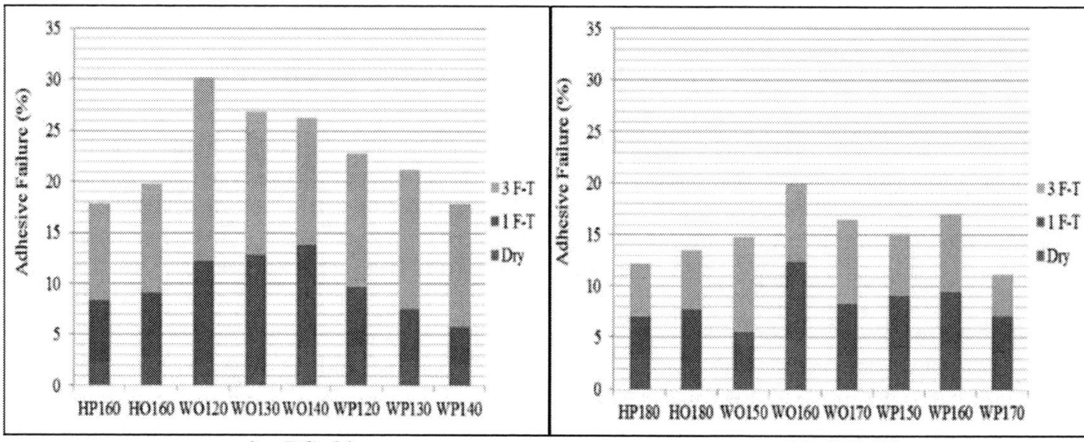

a) PG-64 b) PG-76

Figure 1. Percent adhesive failure on fractured surfaces of DTT specimens.

3.2 Failures due to broken aggregates on DTT samples

Figures 2(a) and (b) shows the percent area of failure due to broken aggregates quantified using imaging technique. Interestingly, these broken aggregates appeared on the fractured surface of every specimen after the test. These broken aggregates are thought to be due to the influence of compaction, as it is believed that some micro cracks are developed in aggregates during compaction. Therefore, when the specimen was subjected to tensile load, these aggregates can be easily split and appeared on the fractured plan. The results illustrates that the percent broken aggregates do no exhibit a specific trend and are not dependent on the moisture conditioning (F-T cycles).

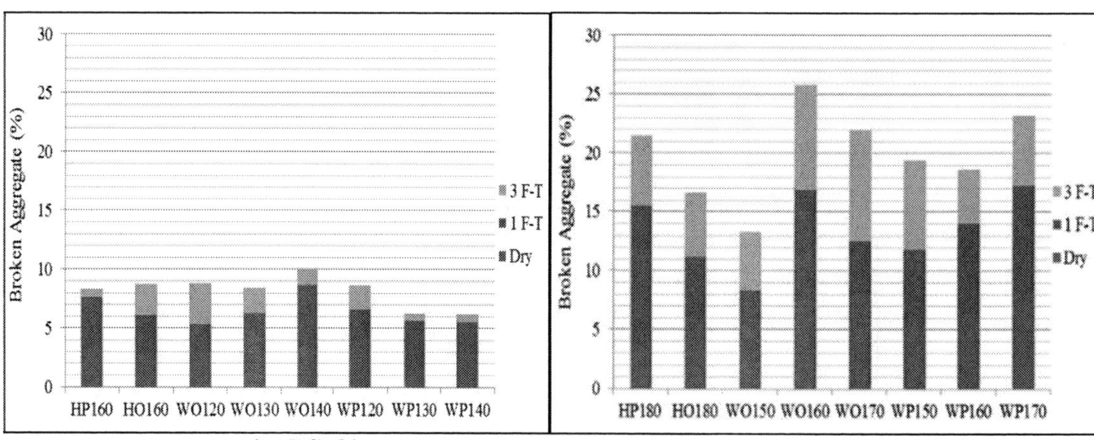

a) PG-64 b) PG-76

Figure 2. Percent broken aggregate on fractured surfaces of DTT specimens.

3.3 Effects of moisture conditioning on direct tensile strength

The direct tensile strength (DTS) results of both PG-64 and PG-76 mixtures at different F-T cycles are presented in Figures 3(a) and (b). The result illustrates that the mixtures DTS decreases with the increase in F-T cycles. This reduction in DTS is more noticeable after 3 F-T cycles. The reduction in DTS after 1 and 3 F-T cycles for HMA modified with PMD filler remain higher however, mixture WP150 exhibits the lowest DTS after 3 F-T cycles. Overall the DTS of PG-76 WMA remains higher when compared to PG-64. Considering the DTS results of PG-76 mixtures, it is evident that the OPC as a filler improved the DTS of HMA and WMA when compared with PMD filler after F-T cycles.

More detailed discussion on the results of percent adhesive failure, broken aggregates and direct tensile strength can be found elsewhere Hamzah et al. [15].

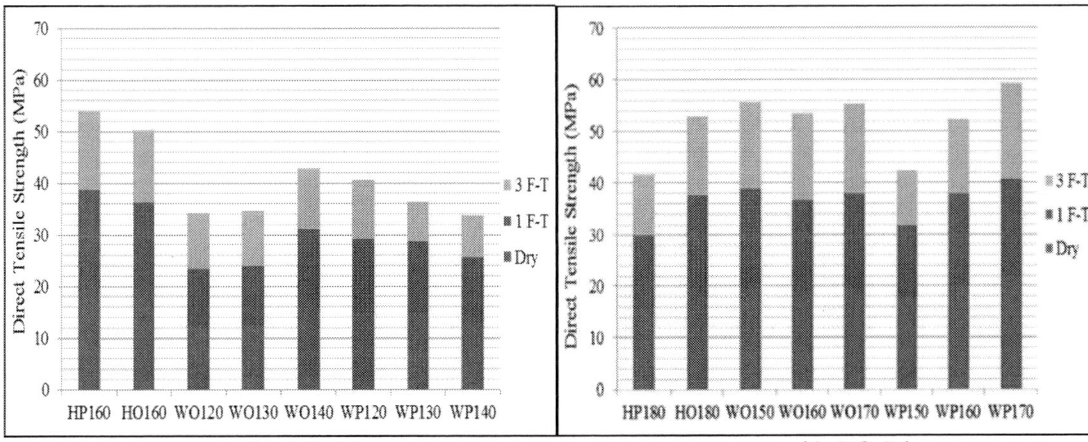

a) PG-64 b) PG-76

Figure 3. Direct tensile strength of fractured surfaces.

3.4 Statistical analysis on the results of percent adhesive failure after DTT

A two-way ANOVA at 95% confidence level ($\alpha = 0.05$) was conducted to analyze the effects of percent adhesive failure on asphalt mixtures. Tables 2 to 6 represent the statistical analysis on the limited results obtained. Table 2 shows that the freeze-thaw (F-T) conditioning on PG-64 mixtures significantly affects the percent adhesive failure, while mixing temperature and type of anti-stripping agent (ASA) have no significant effect on the percent adhesive failure. Similarly, Table 3 shows that the F-T conditioning on PG-76 mixture significantly affects the percent adhesive failure, while mixing temperature and anti-stripping agent have no significant effect on the percent adhesive failure.

A paired t-test was run to analyze the mean difference between percent adhesive failures of mixtures prepared with PG-64 and PG-76 binders. Tables 4 to 6 show a significant difference in the mean values of the percent adhesive failure between the mixtures in dry, 1 F-T and 3 F-T conditions. The mean values of the percent adhesive failure in PG-76 mixtures were lower than that of PG-64 mixtures after 1 F-T and 3 F-T cycles. Therefore, mixtures prepared with PG-76 binders are more resistant to moisture damage compared to PG-64. However, in dry conditions, the mean value of the percent adhesive failure for PG-64 remained lower than PG-76 mixtures.

Table 2. Two-way ANOVA on effects of percent adhesive failure (PG 64 Mixture).

Source	DF	SS	MS	F	*p*-value	Significant
F-T condition	2	435.2517	217.6259	50.66292	< 0.05	Yes
Temperature and ASA	7	47.86365	6.837664	1.591797	0.217265	No
Error	14	60.1379	4.295564			
Total	23	543.2533				

Table 3. Two-way ANOVA on effects of percent adhesive failure (PG 76 Mixture).

Source	DF	SS	MS	F	*p*-value	Significant
F-T condition	2	41.55308	20.77654	15.12766	< 0.05	Yes
Temperature and ASA	7	18.8185	2.688357	1.957427	0.134913	No
Error	14	19.22779	1.373414			
Total	23	79.59936				

Table 4. t-Test: Paired two sample for means (DTT, Dry).

Dry Mixture	PG-64	PG-76
Mean	2.51	6.7125
Variance	0.796542857	3.032764286
Observations	8	8
Hypothesized Mean Difference	0	
DF	7	
P(T<=t) two-tail	< 0.05	

Table 5. t-Test: Paired two sample for means (DTT, 1 F-T).

1 F-T Mixture	PG-64	PG-76
Mean	7.4075	4.85625
Variance	8.181335714	1.300055357
Observations	8	8
Hypothesized Mean Difference	0	
DF	7	
P(T<=t) two-tail	< 0.05	

Table 6. t-Test: Paired two sample for means (DTT, 3 F-T).

3 F-T Mixture	PG-64	PG-76
Mean	12.935	3.5025
Variance	6.450914286	1.102364286
Observations	8	8
Hypothesized Mean Difference	0	
DF	7	
P(T<=t) two-tail	< 0.05	

3.5 Indirect tensile strength test

The indirect tensile strength test was used as a standard moisture test to evaluate moisture susceptibility of mixtures according to ASTM D4867 [20]. The moisture conditioning procedure for indirect tensile strength test samples was similar to DTT specimens, as described in Section 2.3. Figures 4 and 5 shows the original and transformed ITS test specimens, respectively. These images were captured for the adhesive failure evaluation using imaging technique. It can be observed in Figures 4(a) and 4(c) that most of the broken aggregates were found in the vicinity of ITS load application. These observations are more visible in the transformed images obtained as shown in Figures 5(a) and (c).

a) Dry (b) 1 F-T (c) 3 F-T

Figure 4. ITS original images of WO150 fractured surfaces.

a) Dry (b) 1 F-T (c) 3 F-T

Figure 5. ITS transformed images of WO150 fractured surfaces.

3.6 Effects of moisture conditioning on adhesive failure using ITS

Figure 6 represents the percent adhesive failure quantified using image analysis on the fractured faces of mixtures after being subjected to indirect tensile strength test. The results, shown in Figure 6(a), illustrate that the percent adhesive failure of PG-64 mixture increases with F-T cycles. Mixture HP160 and mix WO120, respectively exhibits the lowest and highest percent adhesive failure. The percent adhesive failure of WMA is higher than HMA. The results further show that the overall percent adhesive failure quantified in ITS specimens are lower compared to the percentage failure obtained in mixtures tested using direct tensile.

Figure 6(b) shows the percent adhesive failure of PG-76 mixtures. The percent adhesive failure increases when the mixtures are exposed to F-T moisture conditioning. The PG-76 mixtures follow similar trends compared to mixtures prepared with PG-64 binder. Similarly, mixture HP160 exhibits the lowest percent adhesive failure, while the highest percent adhesive failure is observed in mixture WO120. The overall percent adhesive failure of PG-64 mixtures is higher than mixtures prepared with PG-76 binders. These results show similar trends compared to DTT results.

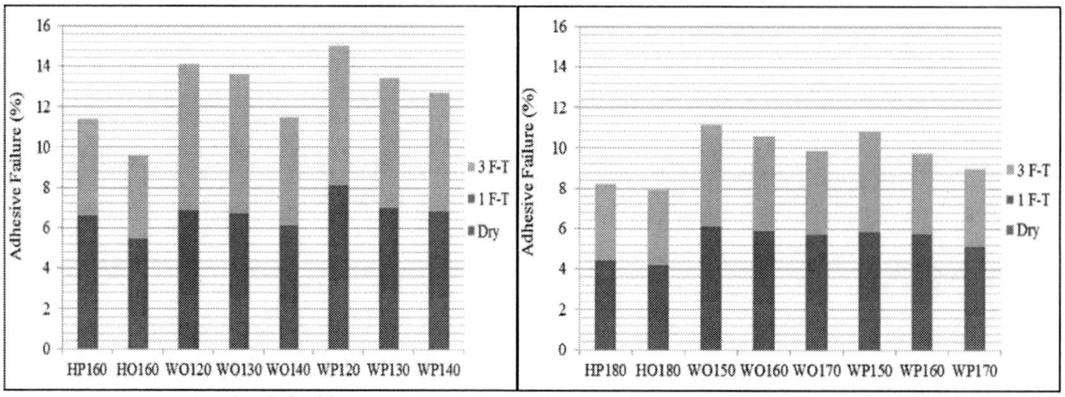

a) PG-64 b) PG-76

Figure 6. Percent adhesive failure on fractured surfaces of ITS specimens.

3.7 Failures due to broken aggregate using ITS test

Figures 7(a) and (b) presents the percent of broken aggregates on the fractured faces of mixtures tested using indirect tension tests. Similar to the results of DTT specimens, the broken aggregate on the fractured faces were assumed to be caused by fractures developed during the compaction process or due to the indirect tensile force applied on the specimen. The percent broken aggregate failure shows no trends when mixtures were moisture conditioned. However, the comparative result of PG-76 and PG-64 mixtures show the higher failures are observed in the former.

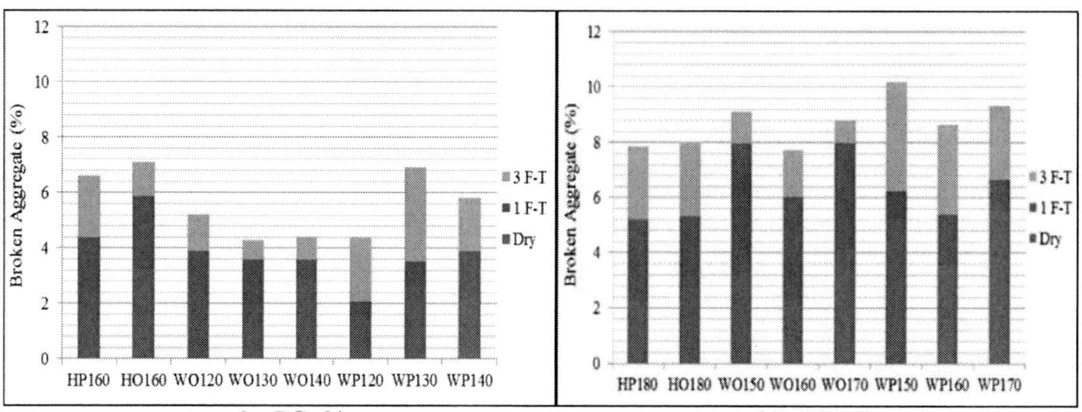

a) PG-64 b) PG-76

Figure 7. Percent broken aggregate on fractured surfaces of ITS specimens.

3.8 Effects of moisture conditioning on ITS

The indirect tensile strength results of PG-64 and PG-76 mixtures are shown in Figures 8(a) and (b). The results show that the ITS decreases when the samples are subjected to F-T moisture conditioning. HP160 mixture exhibits the highest strength in dry conditions, while the lowest strength is observed for mixture WO120 after being subjected to 3 F-T cycles. Similar trends are observed for mixtures prepared with PG-76 binder. The overall strength of PG-76 mixtures is slightly higher than PG-64 mixtures. This trend is similar with specimens tested for direct tensile strength.

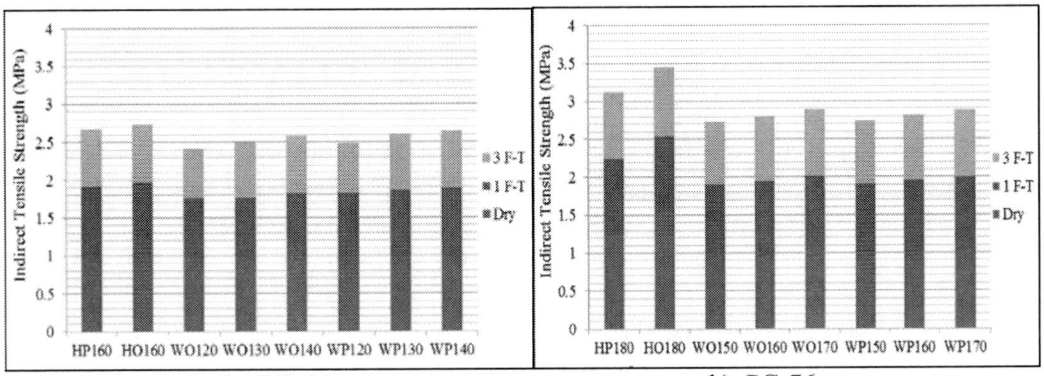

a) PG-64 b) PG-76

Figure 8. Indirect tensile strength of fractured surfaces.

3.9 Statistical analysis on indirect tensile test results

Tables 7 to 11 show the statistical analysis on the results of the indirect tensile test. A two-way ANOVA at 95% confidence level ($\alpha = 0.05$) was carried out to analyze the effects of the percent adhesive failure. Table 7 shows that F-T conditioning, mixing temperature and ASA on PG-64 mixtures significantly affect the percent adhesive failure. Similarly, Table 8 shows that the F-T conditioning, mixing temperature and ASA on PG-76 mixtures significantly affects the percent adhesive failure.

A paired t-test was carried out to analyze the mean difference between the percent adhesive failures of mixtures tested for ITS. Tables 9 to 11 show a significant difference in the mean values between the mixtures subjected to dry, 1 F-T and 3 F-T conditions. The mean percent adhesive failures of PG-76 mixtures are lower than PG-64 mixtures after dry, 1 F-T and 3 F-T cycles. The results also show that PG-76 mixtures are more resistant to moisture damage compared to PG-64 mixtures.

Table 7. Two-way ANOVA on effect of percent adhesive failure (PG-64 Mixture).

Source	DF	SS	MS	F	*p*-value	Significant
F-T condition	2	44.99065	22.49533	79.03236	< 0.05	Yes
Temperature and ASA	7	7.179337	1.02562	3.603288	< 0.05	Yes
Error	14	3.984881	0.284634			
Total	23	56.15487				

Table 8. Two-way ANOVA on effect of percent adhesive failure (PG-76 Mixture).

Source	DF	SS	MS	F	*p*-value	Significant
F-T condition	2	20.26369	10.13184	207.6009	< 0.05	Yes
Temperature and ASA	7	3.348172	0.47831	9.80055	< 0.05	Yes
Error	14	0.683262	0.048804			
Total	23	24.29512				

Table 9. t-Test: Paired two sample for means (ITS, Dry).

Dry Mixture	PG-64	PG-76
Mean	2.596	2.037
Variance	0.28133442	0.158715889
Observations	8	8
Hypothesized Mean Difference	0	
DF	7	
P(T<=t) two-tail	< 0.05	

Table 10. t-Test: Paired two sample for means (ITS, 1 F-T).

1 F-T Mixture	PG-64	PG-76
Mean	4.1259	3.3464
Variance	0.088146462	0.115922518
Observations	8	8
Hypothesized Mean Difference	0	
DF	7	
P(T<=t) two-tail	< 0.05	

Table 11. t-Test: Paired two sample for means (ITS, 3 F-T).

3 F-T Mixture	**PG-64**	**PG-76**
Mean	5.9460	4.2775
Variance	1.225407447	0.301280745
Observations	8	8
Hypothesized Mean Difference	0	
DF	7	
P(T<=t) two-tail	< 0.05	

4. Conclusions

The image analysis method employed in this work has proven to be very effective to analyze the deterioration of asphalt mixtures subjected to moisture conditioning. Optical imaging is a simple and economic way to acquire information in the form of images without direct contact (non-destructive type). It is also an area scanning method, capable of instantaneous recording of two-dimensional images of an object, which is very applicable. In this research, a novel method using imaging technique to evaluate moisture damage responsible for stripping in HMA and WMA in terms of adhesive failure quantification. The percent failure due to adhesion showed that PG-76 mixtures exhibited higher resistance to moisture damage compared to PG-64 mixtures. In addition, the adhesive failure increased when the mixtures were subjected to F-T cycles. Because the percent adhesive failure in PG-76 mixture was lower therefore, it facilitated the failure along the cracks that were already induced during compaction. However, PG-64 mixtures exhibited a higher percent adhesive failure, which resulted in lower aggregate breakage. Therefore, the failures due to broken aggregates in the PG-76 mixtures were higher than PG-64 mixtures. The DTS results of PG-64 and PG-76 mixtures showed that a higher percentage failure due to adhesion corresponded to lower DTS of HMA and WMA after F-T cycles. From fractured ITS samples, most of broken aggregates were found located in the vicinity where the indirect tensile load was applied. On the other hand, high adhesive failure was obtained at the center portion where maximum tensile stresses were developed. The ITS test results showed similar trends as observed for DTT specimens. However, the overall percent adhesive failures were lower compared to DTT results, which showed the effectiveness of the DTT method.

Acknowledgments
The authors would like to acknowledge the Malaysian Ministry of Higher Education through the Fundamental Research Grant Scheme (FRGS grant number 203/PAWAM/6071277) that enabled this paper to be written.

References

[1] Obaidat A F 1999 *Evaluation of Bituminous Mixtures Stripping Using Conventional and Image Processing Techniques* (Civil Engineering Department, Jordan University of Science and Technology, Irbid, Jordan)
[2] Abo-Qudais S and Al-Shweily H 2007 Effect of aggregate properties on asphalt mixtures stripping and creep behavior *Construstion and Building Materials* **21**(9) pp 1886-98
[3] Kakar M R, Hamzah M O, Akhtar M N and Woodward D 2016 Surface free energy and moisture susceptibility evaluation of asphalt binders modified with surfactant-based chemical additive *Journal of Cleaner Production* **112** pp 2342-53
[4] Mogawer W, Austerman A and Bahia H 2011 Evaluating the effect of warm-mix asphalt technologies on moisture characteristics of asphalt binders and mixtures *Transport Research Records* **2209** pp 52-60
[5] Lottman R P 2001 Predicting Moisture Induced Damage to Bitumen Concrete *Transport Research Board NCHRP Report no 192*
[6] Hamzah M O, Kakar M R and Hainin M R 2015 An overview of moisture damage in asphalt mixtures *Jurnal Teknologi (Sciences & Engineering)* **73**(4) pp 125-31
[7] Terrel R L and Shute W J 1989 *Summary Report on Water Sensitivity* (National Research Council, Washington D.C)
[8] Lytton R L, Uzan J, Fermando E G, Roque R, Hiltuene D and Stoffels S M 1993 *Development and Validation of Performance Prediction Models and Specifications for Asphalt Binders and Paving Mixes* (National Research Council, Washington, D.C.)
[9] Kim Y R, Little D N and Lytton R L 2003 Fatigue and healing characterization of asphalt mixtures *Journal of Materials in Civil Engineering* **15**(1) pp 75-83
[10] Jahromi S G 2009 Estimation of resistance to moisture destruction in asphalt mixtures *Construstion and Building Materials* **23**(6) pp 2324-31
[11] Taylor M A and Khosla N P 1983 Stripping of asphalt pavements: state of the art. *Transport Research Records* **911** pp 150-8
[12] Kakar M R, Hamzah M O and Valentin J 2015 A review on moisture damages of hot and warm mix asphalt and related investigations *Journal of Cleaner Production* **99** pp 39-58
[13] Roberts F L 1991 *Hot Mix Asphalt Materials, Mixture Design and Construction* (NAPA Educational Foundation)
[14] Eriksen K 1993 *Microscopical Analysis of Asphalt-aggregate Mixtures Related to Pavement Performance* (Report No 245 Danish Road Institute Roskilde, Denmark)
[15] Hamzah M O, Kakar M R, Quadri S A and Valentin J 2014 Quantification of moisture sensitivity of warm mix asphalt using image analysis technique *Journal of Cleaner Production* **68** pp 200-8
[16] Kabir S 2010 Imaging-based detection of AAR induced map-crack damage in concrete structure NDT & E Interantional **43**(6) pp 461-9
[17] Salem A, Kabir S and Musbah A 2011 Optical image analysis based concrete damage detection. *Proc. Progress in Electromagnetic Res Symposium (Marrakesh)* pp 1-4
[18] Oliveira J R, Silva H M, Abreu L P and Fernandes S R 2013 Use of a warm mix asphalt

additive to reduce the production temperatures and to improve the performance of asphalt rubber mixtures *Journal of Cleaner Production* **41** 15-22
[19] Hicks R G, Santucci L and Aschenbrener T 2003 Introduction and seminar objectives *Proc Nat. Seminar on Moisture Sensitivity of Asphalt Pavements (San Diego)* pp 3-20
[20] ASTM 2006 Standard Test Method for Effect of Moisture on Asphalt Concrete Paving Mixtures *(ASTM D4867)*

IOP Conf. Series: Materials Science and Engineering **236** (2017) 012014 doi:10.1088/1757-899X/236/1/012014

The use of a non-nuclear density gauge for monitoring the compaction process of asphalt pavement

Wim Van den bergh[1], Cedric Vuye[1], Patricia Kara[1], Karolien Couscheir[1], Johan Blom[1] and Philippe Van Bouwel[2]

[1]University of Antwerp, Faculty of Applied Engineering, EMIB research group, 171 Groenenborgerlaan, Antwerp 2020, Belgium
[2]Infrastructure & Real Estate, Brussels Airport Company, 1C Luchthaven Brussel Nationaal, Zaventem 1930, Belgium

E-mail: wim.vandenbergh@uantwerpen.be

Abstract. The mechanical performance of an asphalt pavement affects its durability – thus carbon footprint. Many parameters contribute to the success of a durable asphalt mix, e.g. material selection, an accurate mix and even the road design in which the asphalt mix quality is quantified. The quality of the asphalt mix, by its mechanical properties, is also related to the compaction degree. However, and specifically for high volume rates, the laying process at the construction site needs an effective method to monitor and adjust immediately the compaction quality before cooling and without damaging the layer, which is now absent. In this paper the use of a non-nuclear density gauge (PQI – Pavement Quality Indicator) is evaluated, based on a site at Brussels Airport. Considering the outcome of the present research, this PQI is advised as a unique tool for continuous density measurements and allow immediate adjustments during compaction, and decreases the number of core drilling for quality control, and as a posteriori asphalt pavement density test where cores are prohibited. The use of PQI could be recommended to be a part of the standard quality control process in the Flemish region.

1. Introduction

Asphalt pavement material is a part of the high-carbon emission model [1]. The International Energy Agency (IEA) reported that CO_2 emissions from the transport industry account for approximately 27.8% of global emissions [2]. The quality of the pavement surface affects the use of energy and safety. The mechanical performance of an asphalt pavement affects its durability – thus even enlarges carbon footprint. Hence, increasing the quality of an asphalt pavement leads to beneficial improvements towards a reduction of the carbon footprint. In order to avoid a high carbon footprint, as mentioned in [1], the use of asphalt should be optimized by quality and performance increase and recycling technologies. Many parameters contribute to the success of a durable asphalt mix, e.g. material selection, an accurate mix and even the road design in which the asphalt mix quality is quantified. The quality of the asphalt mix, by its mechanical properties, is also related to the compaction degree. Asphalt technologists agree that the density of an asphalt mix is one of the most prominent measures for road quality [3-5]. Reaching a certain desired density optimizes the desired mix characteristics, like: stiffness, fatigue characteristics, resistance against permanent deformation and moisture [4]. Poor density can lead to settlements in a later stage with a shorter life span of the pavement as a result, because rutting will occur in the wheel tracks. Therefore, good compaction is very important. Currently, this problem occurs more frequently: there is less time available to construct the road, many new asphalt mixes are introduced, often work is planned outside the ideal paving season and as a result, work is undertaken in less than ideal weather conditions and circumstances. So, reaching the desired quality during the compaction process is under threat, which often makes the result more and more uncertain [6]. In Belgium a project called 'intelligent compactors' was worked out using IT systems on different types of compaction rollers. No real practical procedure is worked out, so far, but the level of compacting is agreed to be crucial for the service life and performance. Other more practical methods are searched for quality control.

Key problem within the asphalt paving process is the lack of knowledge about the ongoing process of compaction: determining the level of compaction and if an additional compaction is needed; specifically for high volume rates and avoiding a high number of core drillings for controlling the compaction process where cores are prohibited, or repaving is not permitted due to limited closure of the pavement, for example, at the airport. Airport pavements are constructed to provide adequate support for the loads imposed by airplanes and to produce a firm, stable, smooth, all-year, all-weather surface free of debris or other particles such as loose stones from raveling that may be blown or picked up by propeller wash or jet blast. To produce such pavements requires a coordination of many factors of design, construction, and inspection to assure the best possible combination of available materials and monitoring the compaction of the pavement. Considering this, core drilling is strictly prohibited at those pavements as it weakens the paved layer locally, thus contributes to a higher risk for damaging the layer, resulting in a decreased service life or water infiltration problems, e.g. on bridges. Therefore, the laying process at the construction site needs an effective monitoring method with accurate measurements of density in real-time, the ability to adjust immediately the compaction quality before cooling and without damaging the layer for quality control, quality assurance and performance prediction of the pavement layers.

2. Background

In fact, there has always been a need for a quick, accurate and non-destructive instrument for measuring in-place density in the paving industry. In the late fifties and early sixties, several manufacturers began experimenting with the use of low level radiation as a source of energy for measuring the in-place density of soil, aggregates and asphalt concrete. Over the ensuing years the paving industry has realized the importance of timely compaction results. The first generation of nuclear gauges were quite heavy and required a scaler attached by a long umbilical cord to the density device. The nuclear source shielding for operator protection was not yet refined and required the operator to read the results from a scaler at some distance from the gauge. Over the years nuclear gauge manufacturers have refined instrument accuracy and operator safety, however, owning and operating nuclear gauges still requires a State Radioactive Materials license. While nuclear density gauges have served paving industry well, the need for an instrument without the difficulties of a radioactive materials license, special handling and a more operator-friendly device was needed [7].

Within the last decade, non-nuclear technology has been developed for the purpose of measuring the density of in-place HMA (hot-mix asphalt) materials. These devices operate based on the principles of a constant voltage, radio frequency and electrical impedance for a current passing through the HMA material, adjusted for moisture variations and mix types [8]. PQI non-nuclear gauge offers: the ability to start the reading of the measurements straight ahead after the equipment is switched on, to take numerous density readings in a very short period of time, possibility to re-test results, low standard deviation between tests and less surface texture effect in comparison to nuclear gauge for example [7]. The PQI gauge is easy to use in practice, it can be stored and transported anywhere and can be purchased without a Radioactive Materials license. Unlike the nuclear gauges, it does not require extensive and periodic calibrations either by the manufacturer.

However, in order to standardize this technology, its accuracy must be equal or even better compared to the nuclear gauge and core measurement methods [9]. However, the readings provided by the PQI gauge are not absolute. They are relative measurements based on a given known value. Thus, it is necessary to 'calibrate' the device (or adjust the reference point) with a pavement section of known density and made from of the exact same material as the one where density measurements are desired [10]. A number of studies have been performed to determine the ability of the non-nuclear devices to accurately and precisely measure in-place density of asphalt pavements starting from the eighties. Two early studies [11-12] used both nuclear gauge and core methods to determine density. Analyses indicated no statistically significant differences in densities among the measurements, but the study did identify that the measured means were statistically different and that density variability was dependent on the type of mix. In the study of Rogge & Jackson [13] a large amount of data was collected, where it was reported that neither nuclear nor non-nuclear densities correlated well with core densities, and neither method was determined to be adequate for controlling field compaction. In all recent studies the outcome of the research is indicated by the significant factors affecting the pavement density after cold rolling compaction like: gauge type, proper gauge calibration, the influence of weather conditions on measurements, human factor such as correct placement of the gauge, the contact zone between HMA and gauge (evenness of the surface), humidity of the surface, pollution of the road (dust, leaves, organics), temperature of the road, mix type, thickness of the surface layers, etc.

In quality control it is not sufficient to only check the apparent density, but also the compaction temperature and the representative (mechanical) properties should be taken into account [6]. In the asphalt paving industry, both researchers and practitioners, postulate that the temperature of the asphalt mix during compaction is important for the final quality of the pavement. There are many parameters that influence the temperature of the asphalt mix. These include the ambient temperature, the temperature of the underlying surface, the compaction process and roller regimes, layer thickness, wind speed and rain. This makes it difficult for operators to predict the material temperature and adjust their actions to this information and to compact in the 'ideal compaction window'. The compaction window is based on viscosity-properties and the density that can be reached at this viscosity. To determine the compaction window based on mechanical properties, laboratory experiments and a field study are usually conducted. Compaction outside the compaction window can reduce the cracking toughness by 35% and increase the crack propagation rate by 40%, despite the target density being reached [6]. It is advisable to determine the compaction window based on temperature and the resulting mechanical properties instead of bitumen viscosity and the density that can be reached at this viscosity.

Also, a new trend is noticed in road construction engineering: the implementation of internet of things. In the ROAD_IT project (University of Antwerp & Belgian Road Research Centre) an IT-architecture is demonstrated to evaluate the digital data transfer of different types of new technologies during asphalt production and the laying process. Density measurements during compaction, between e.g. an infrared line scanner on the finisher and the roller compactors, can improve the compaction homogeneity by adjusting the compaction pattern of the rollers. Moreover, when the gauge is equipped with a GPS system, the compaction degree can be set for each measured point and stored for later use. This work method will decrease a posteriori tests, which may damage the pavement.

3. Data collection

In this paper the method using a non-nuclear density gauge is evaluated for in-place density measurements, based on a work at Brussels Airport during the rehabilitation project of runway 01/19 and at the area of the Port of Antwerp. The density tests were performed using drilled cores, a non-nuclear gauge TransTech PQI-380 and a nuclear density device Troxler 3450.

A comprehensive test program was set up to evaluate how the TransTech PQI380 equipment works in terms of accuracy and precision. As it is known, a critical step in using impedance gauges effectively is to calibrate them in a manner that will increase the accuracy of results. Density measurements are relative measures of compaction, and can thus be adjusted mathematically in order to more accurately represent the "true" density of the pavement. Although an "absolutely true" measure of pavement density cannot be reasonably achieved, the most accurate measures are typically believed to result from the bulk specific gravity measurement of a pavement core drilled from the compacted mat. Therefore, an alternative measure of density is believed to be accurate if it can produce results similar to those generated by core densities [14]. The use of the non-nuclear density gauge is an alternative method avoiding a high number of core drillings for controlling the compaction process performing a continuous measurement of the compaction quality where cores are prohibited (at the edges of the pavement, airport runaway pavement, bridges, parking lots on the roof, etc.). In U.S. this device complies with ASTM (D7113-05) and AASHTO (T 343-12) standards.

Data for this study were collected from a certain selection of points (in total of 168 points) along and across seven newly paved test tracks; two grading HMA mix types (Marshall-designed asphalt; MA 0/10 and MA 0/14) are used with different compaction ratios asked to the roller operator (normally, less and extra compacted) at Antwerp Euro Terminal (see Figure 1). Each test track was constructed by two parallel finishers (width of each finisher - 3.75m). This way, a comparison can be made between different finishers, compaction grade, the side effect and differences in mix types. Drilled cores were collected and tested. Also others tests like grooving were performed [15], but they are out of the scope of this contribution. In this study, only a comparison is made between drilled cores and PQI-measurements, compared to the optimal compaction, derived from the technical justification note (TJN).

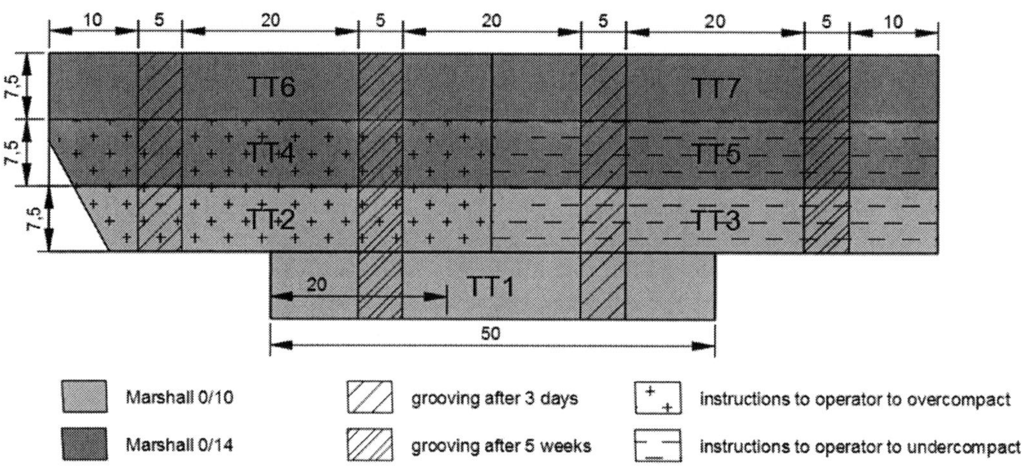

Figure 1. Overview of the full-scale test tracks (TT).

To estimate the density at each test track at different locations six measuring points were performed immediately after total rolling process. During the first test, the gauge was outlined with chalk and numbered for an easy detection. Between tests, the gauge was lifted from the pavement and then placed again at the same location. All the measurements were carried out with offset value 0. The offset value was subsequently determined on the basis of drilled cores. The nuclear density measurements were performed by Belgian Road Research Center at locations determined by University of Antwerp and carried out with the Troxler 3450 gamma density meter. A total of 35 measurements were carried out, five measurements per test section. The results of the non-nuclear density meter and the density determined by cores were compared. Temperature measurements during the day were: in the morning ±5°C and in the afternoon ±12°C. There was no wind and it was dry and sunny. The surface temperature of HMA was taken during the first test using a ScanTemp440 infrared thermometer. Surface temperature during the measurements were indicated in the readings of PQI-measurements.

4. Data analysis

Surface temperature measurements showed that the temperature of HMA surface varied between 179 and 147 °C with an average of 163 °C. In order to observe the effect of how temperature influences the density of asphalt, the non-nuclear gauge was repositioned twice in the test tracks 1&3 with time difference of 3-4 hours. In theory, as materials cool, they tend to shrink or become denser [16]. This process can be observed from the field data in figure 2 when the asphalt temperature decreased from 38°C to 31 °C and the measured density increased (in particular it can be seen in figure 2(right)).

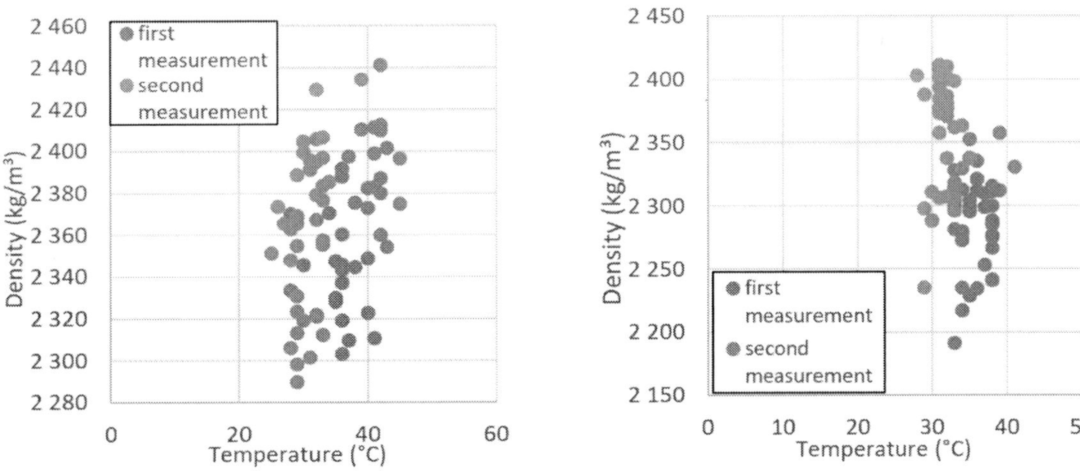

Figure 2. Densities compared to PQI – test track 1 (left) &3 (right).

The measurements were performed on the test tracks (see figure 1) at several positions, approximately each 10 m along the test track; 6 measurements per width are taken, taken into account that each test track was constructed by two parallel finishers. In that way it was possible to evaluate how the compaction evolves from left to right side (including middle part) of each lane and along the test track, and if a difference in compaction between two finishers at equal conditions is apparent. In this paper only the comparison with nuclear measurements and core results are reported. The average compaction from those measurements is summarized in table 1. During the measurements it has been noticed that there might be a slight difference in the results if the gauge is not firmly connected to the surface in each measurement (one of the reasons could be because of the unevenness of the surface). Also it was observed that the middle part of the lane was always better compacted than the sides. The compaction degree (%TJN) - is expressed as: PQI-measurement of the density after last rolling or density from cores, divided by the optimal density derived from the technical justification note (TJN) approved by the Flemish Road Agency. For example, if the PQI-measurement equals the optimal density a value of 100% is obtained. The difference between normally compacted test track, less compacted and extra compacted is the number of rolling movements on the location, without knowing if this difference does affect the compaction degree.

The primary objective of this study was to establish a correlation, using the non-nuclear method, nuclear method and drilling cores, between asphalt density measured immediately after finishing cold rolling. As it can be seen from the figures 3&4, the results obtained from PQI-measurements differ for 1-3% from drilled cores results, similar as for the nuclear measurements. In these figures also the deviation with the theoretical value from the technical justification note is observed. Points at the end of test tracks such as B5 and B10 show high deviations from measuring points in the middle of test tracks. This can be seen as a clear signal of less compaction at the start and end of the tracks. Both figures demonstrate also a trend for the different compaction conditions - however outliners are registered (e.g. B23). The results of PQI, the nuclear measurements and the cores are consistent. As observed in table 1, there is no large difference in compaction in spite that the roller operation had clear instructions to over- or undercompact. This demonstrates again that a tool like the PQI is necessary to control the compaction. In a subsequent project for Brussels Airport, the PQI is used just after first compaction movement until the correct compaction is obtained. Considering this, the PQI-380 non-nuclear gauge could be advised as a unique tool for continuous or a posteriori asphalt pavement density measurements and in time to be a part of the standard quality control process in the Flemish region.

Table 1. PQI-measurements compaction of test tracks pavements (B - are measuring points)

HMA mix type	Test track	Compaction degree PQI (%TJN):	Compaction degree Core (%TJN):	Test track target condition
	1	99	98	normally compacted (B1-B5)
MA 0/10	2	100	100	extra compacted (B6-B10)
	3	97	97	less compacted (B11-B15)
	4	101	100	extra compacted (B16-B20)
MA 0/14	5	98	98	less compacted (B21-B25)
	6	100	99	normally compacted (B26-B30)
	7	98	98	normally compacted (B31-B35)

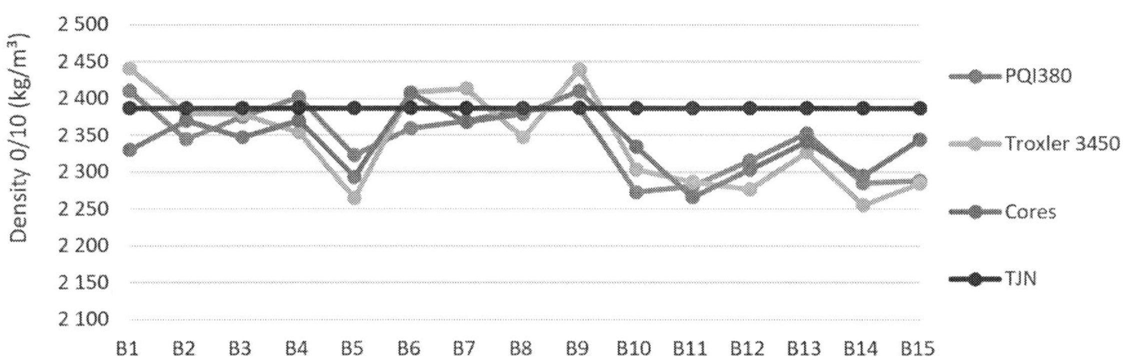

Figure 3. Density result comparison for HMA 0/10.

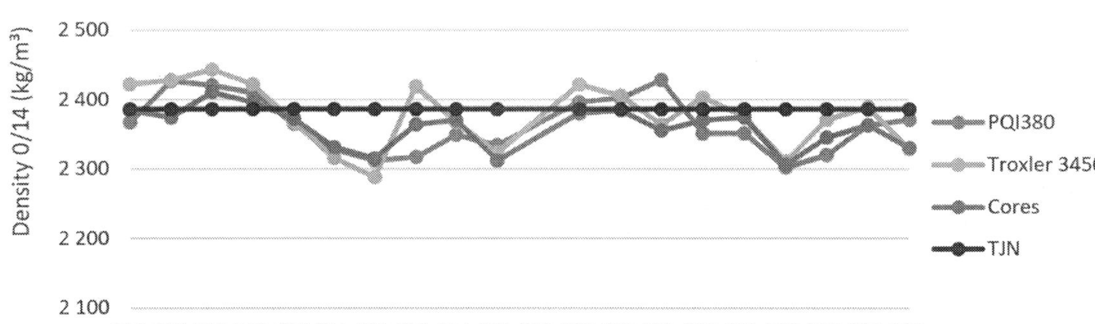

Figure 4. Density result comparison for HMA 0/14.

It is commonly known that compaction is always better in the middle of the lane (width 3.5m), in order to see the compaction homogeneity of asphalt lanes PQI-measurements were taken during another project in the Port of Antwerp (PSA Terminal). The results are reflected in the table 2. It can be seen that the middle part of the lane is better compacted than the sides of the lane, also there is no clear distinction between the right zone that is applied against another lane and the left zone which is free.

Table 2. PQI-measurements compaction of test track, expressed as [PQI density/TJN*100%] (L - left, ML – middle left, M - middle, MR – middle right, R - right).

			Location 1		
	L	ML	M	MR	R
SVM (%)	97.8	99.8	100.3	99.9	100.3
SVM (kg/m^3)	2352	2399	2410	2401	2411
T (°C)	62	70	71	72	76
			Location 2		
SVM (%)	96.7	98.1	99.6	99.2	98.5
SVM (kg/m^3)	2352	2355	2393	2385	2368
T (°C)	73	72	71	70	73
			Location 3		
SVM (%)	99.5	100.7	98.8	98.6	98.0
SVM (kg/m^3)	2392	2421	2376	2369	2355
T (°C)	74	73	71	71	73

5. Conclusions

In the present study results from different projects, where a non-nuclear gauge TransTech PQI-380 was used by University of Antwerp, are summarized. The purpose of the present investigation was to clarify whether a PQI non-nuclear density gauge can provide precise measurements in comparison with for example cores or a nuclear gauge. This device is an easy-to-use tool for site supervisors during the compaction process to determine whether the pavement is compacted properly or not and which locations require more compaction. After more validation, this device could be complying with the local standards and selected as one of the instruments for permanent quality control e.g. in a tender, as an effective tool for density measurements of pavements, and in cases where cores are prohibited.

The important results of this study can be concluded as follows:

1. During the rolling process immediate evaluation of the compaction process is important to increase the efficiency by permanent control during the time that the roller passes; local adjustments can be accomplished by PQI-measurements during the compaction process and not after reporting the cores results for example;
2. It can be a low cost time efficient measurement after a one time investment;
3. Also a lower number of cores is required which induces less damaged newly constructed roads or not damaged at all in the case of strict regulations like at the airports;
4. After the rolling process it is possible to determine the homogeneity of a lane over the relative measurements of the density at a constant temperature and to determine the absolute values of the density at various locations by correlating them with (single) core samples at known locations;
5. The process control provides a tool to obtain a higher quality level and durability effect by an increased service life and avoids repaving; resulting in decreased costs and carbon-footprint;
6. This device can be used as an intermediate control between new technologies such as an infrared-line scanner behind the finisher and so-called 'intelligent compactors';
7. Differences between start and end compaction process can be observed in the test tracks. It would be useful to register these bad points for later evaluation e.g. cracks, rutting and ravelling;
8. As disadvantages could be mentioned: training for an effective usage of the PQI non-nuclear gauge is required, the offset absolute value must be adapted to the local situation, optimal

compaction or current mix data are required since actual measurement of the density of an unknown mix in place without initial values is not possible; measured density values are temperature dependent. However, all of these disadvantages can be anticipated by a manual, training and knowing the limits of the method.

Acknowledgements
The authors would like to thank the personnel of the Infrastructure & Real Estate department of Brussels Airport Company and the contractor NV Aswebo. The authors would also like to acknowledge the support of the Belgian Road Research Centre who performed the measurements with the gamma density measurement device. Without their support this research would not have been possible.

References

[1] Peng B, Cai C, Yin G, Li W and Zhan Y 2015 Evaluation system for CO_2 emission of hot asphalt mixture *Journal of traffic and transportation engineering* **2**(2) pp 116-124
[2] International Energy Agency 2016 *Key world energy statistics*
[3] Huerne H L 2004 *Compaction of Asphalt Road Pavements Using Finite Elements and Critical State Theory* (PhD thesis University of Twente Enschede, The Netherlands) p 319
[4] Decker D S 2006 State-of-the-Practice for Cold-Weather Compaction of Hot-Mix Asphalt Pavements *Factors Affecting Compaction of Asphalt Pavements* (Transportation Research E-Circular Issue No. E-C105 (TRB Washingron DC USA) pp 27-33
[5] Miller S R, Hartmann T and Dorée AG 2011 Measuring and visualizing Hot Mix Asphalt Concrete Paving Operations *Automation in construction* **20** pp 474-81
[6] Bijleveld F, Miller S, De Bondt A and Doree A 2012 Too hot to handle, too cold to control – influence of compaction temperature on the mechanical properties of asphalt *Proc. 5th Eurasphalt & Eurobitume Congress (Istanbul, Turkey)* pp A5EE-231
[7] Sully-Miller Contraction CO 2000 *A summary of operational differences between nuclear and non-nuclear density measuring instruments* p 5
[8] Von Quintus H L, Minchin Jr R E, Nazarian S, Maser K R and Prowell B 2009 NDT Technology for quality assurance of HMA pavement construction *National Cooperative Highway Research Program Report 626* (TRB Washingron DC USA) p 112
[9] Zhuang Z 2011 *Effectiveness study of Non-Nuclear Gauge for Hot Mix Asphalt (HMA) Pavement Construction* (University of Nebraska-Lincoln, USA)
[10] Romero P 2002 *Evaluation of non-nuclear gauges to measure density of hot-mix asphalt pavements Final Report* (Department of Civil and Environmental Engineering, University of Utah USA)
[11] Sanders S R, Rath D and Parker F 1994 Comparison and core pavement density measurement *Journal of Transportation Engineering* **120**(6) pp 953-66
[12] Hanna A S, Russell J S, Noyce D A and Schmitt R L 1995 *Asphaltic pavement density measurement comparison using coring and nuclear methods - International Technical Peport* (University of Winsconsin-Madison USA)
[13] Rogge D F and Jackson M A 1999 Compaction and Measurement of Field Density for Oregon Open-Graded (F-Mix) Asphalt Pavement *Final Report SPR 386* (US: Department of Civil, Construction, and Environmental Engineering, Oregon State University)
[14] Williams S G 2008 Non-nuclear methods for HMA density measurements *MBTC 2075 Final Report* (University of Arkansas USA) p 77
[15] Vuye C, Couscheir K, Lauriks L, Van den bergh W and Van Bouwel Ph 2017 The influence of mixture composition, adhesion promotor and compaction degree on the groove stability of grooved Marshall asphalt *Proc. BESTInfra 2017 (Prague)*
[16] Hanna A S, Lee M-J, Schoenfelder S M and Lee C 2008 Temperature-density relationship of hot-mix asphalt pavement after completed rolling *Canadian Journal of Civil Engineering* **35** pp 1468-75

Evaluation of permanent deformation and durability of epoxidized natural rubber modified asphalt mix

Ramez A Al-Mansob[1], Amiruddin Ismail[2], Riza Atiq O K Rahmat[2], Muhamad Nazri Borhan[2], Jamal M A Alsharef[3], Shaban Ismael Albrka[4] and Mohamed Rehan Karim[5]

[1]Dept. of Civil Engineering, Faculty of Engineering, Technology & Built Environment, UCSI University, Malaysia
[2]Sustainable Urban Transport Research Centre (SUTRA), Dept. of Civil & Structural Engineering, Universiti Kebangsaan Malaysia, Malaysia
[3]Dept. of Civil & Structural Engineering, Universiti Kebangsaan Malaysia, Malaysia
[4]Civil Engineering Department, Near East University, Turkey
[5]Dept. of Civil Engineering, Universiti Malaya, Malaysia

E-mail: aismail@ukm.edu.my

Abstract. The road distresses have caused too much in maintenance cost. However, better understandings of the behaviours and properties of asphalt, couples with greater development in technology, have allowed paving technologists to examine the benefits of introducing additives and modifiers. As a result, modifiers such as polymers are the most popular modifiers used to improve the performance of asphalt mix. This study was conducted to investigate the use of epoxidized natural rubber (ENR) to be mixed with asphalt mix. Tests were conducted to investigate the performance characteristics of ENR-asphalt mixes, where the mixes were prepared according to the wet process. Mechanical testing on the ENR-asphalt mixes have demonstrated that the asphalt mix permanent deformation performance at high temperature was found to be improved compared to the base mixes. However, the durability studies have indicated that ENR-asphalt mixes are slightly susceptible with the presence of moisture. The durability of the ENR-asphalt mixes were found to be enhanced in term of permanent deformation at high and intermediate temperatures compared to the base asphalt mixes. As conclusion, asphalt pavement performance can be enhanced by using ENR as modifier to face the major road distresses.

1. Introduction

During last decades, a lot of research works has been carried out to modify asphalt and asphalt mix [1-6]. The potential value of polymer modified asphalt (PMA) is that polymers can significantly improve the performance of hot mix of asphalt (HMA) and substantially increase the service life of highway surfaces [8]. Specifically, the addition of polymers significantly improves various asphalt properties, such as elasticity, cohesion, stiffness and adhesive, resulting in a substantial improvement in the performance and quality of asphalt pavements with the HMA being more stable at warmer temperatures and more flexible at colder temperatures. In addition to rutting resistance, a premium polymer can provide a degree of flexibility or elasticity to an HMA, thereby improving the fatigue and thermal cracking characteristics of the asphalt mix.

One of the most well-known, widely used category of polymers is thermoplastic elastomers (TE). TE are polymers with thermoplastic and elastomeric properties [8]. TE polymers derive their strength and elasticity from a physical cross-linking of molecules within a three-dimensional network. Styrene-butadiene-styrene (SBS) is the most commonly known TE that can improve the rheological properties of asphalt [9]. Furthermore, SBS increases binder elasticity at high temperatures and improves flexibility at low temperatures. This improvement led to increase the resistance to asphalt rutting at high temperatures, and decrease cracking at low temperatures [10]. Styrene-butadiene-rubber (SBR) is another example of an elastomer polymer and it increases the ductility of asphalt pavement and consequently becomes more flexible and crack-resistant at low temperatures [11]. In addition, other types of additives have also been used as modifiers, i.e., as TE, such as styrene isoprene styrene (SIS), styrene ethylene butadiene styrene (SEBS), ethylene-propylene-diene-terpolymer (EPDM), isobutene-isoprene-copolymer (IIP), crumb-rubber, polybutadiene (PBD), polyisoprene and natural rubber (NR) [7, 12]. However, the major types of styrenic block copolymers such as SBS, SIS and hydrogenated styrenic block copolymers (HSBC) have the best modifying potential when blended with asphalt [14-16].

The potential application of epoxidized natural rubber (ENR) as a modifier was realized in the 1980s [15]. ENR is a chemically modified natural rubber created by reacting natural rubber with proxy formic acid [15]. This material has good properties, offering high strength because of its ability to bear strain crystallization, along with increased glass transition temperature. These properties facilitate increased oil resistance, enhanced adhesion properties, damping and reduced gas permeation [16, 17]. According a previous study, ENR is able to increase the viscosity, stiffness and decreased the temperature susceptibility for the binder [18, 19]. The best results were recorded for polymer-modified asphalt containing 6% of ENR [20]. This study was conducted to present a laboratory evaluation of performance of unaged, short and long term aged of base asphalt mix (HMA-0) and asphalt mix modified with 6% of ENR (as a percentage of asphalt by weight) (HMA-6) in terms of dynamic creep, rutting and to evaluate the susceptibility to moisture.

2. Experimental works

2.1 Materials

The asphalt used in this study is 80/100 penetration grade, supplied by the asphalt factory at Port Klang, Malaysia. The ENR was obtained from the Malaysian Rubber Board under the trade name of ENR 50, with 53% epoxidization and passed through 2.36 mm mesh sieve (before shearing). The physical properties of the asphalt and the ENR used are shown in Table 1. The aggregate used for preparing the mix samples was obtained from Negeri Roadstone Sdn. Bhd. Malaysia.

Table 1. Materials properties of asphalt and ENR.

Material	Properties	Value	Specification
Asphalt 80/100	Specific Gravity	1.03	ASTM D70
	Penetration @ 25 °C	82	ASTM D5
	Softening point (°C)	45.7	ASTM D36
	Viscosity @ 80 °C (Pa.s)	12.6	ASTM D4402
	Ductility (cm) @ 25 °C and 5 cm/min	>100	ASTM D113
ENR	Size (before shearing)	2.36 mm	-
	Specific Gravity	0.94	-

ENR-modified asphalt was produced by mixing 6% of ENR (by weight of asphalt) with base asphalt. ENR-modified asphalt was prepared using high shear mixer at 160˚C (±1 ˚C) under 4000 rpm of speed for 62 min [20]. ENR was added to the asphalt after the temperature stabilized at 160 ˚C.

The asphalt mix design is based on the Superpave volumetric mix design. The design equivalent single axle loads (ESALs) for this study are assumed to be less than 107. This makes the design in the category of 3x106 to 107 ESALs (80 KN/ESAL), or traffic level 4 [21]. Traffic levels are used to determine design requirements such as number of design gyrations for compaction, aggregate physical property requirements, and mix volumetric requirements. The traffic level also determines the level of mix design required. For traffic levels of less than 107 ESALs, a level 2 mix design is recommended [21]. The mix in this study will have a nominal maximum particle size aggregate of 19.0 mm. however, six stockpiles of aggregate consisting of three coarse aggregates and three fine aggregates.

2.2 Asphalt Mix Performance Tests

2.2.1 Dynamic creep test. The dynamic creep test is a testing which also can be implemented using UTM to determine the permanent deformations of asphalt mix in the laboratory. In this study, the dynamic creep test under haversine load pulse was conducted for the base asphalt mixes and ENR-modified mixes in accordance with protocol developed by NCHRP 9-19 Superpave Models [22]. Each category of ENR mixes consists of three replicate samples compacted at 4% air voids with a diameter of 100 mm and 150 mm height were conditioned at provided chamber at a temperature of 40°C for at least four hours before testing is initiated. During initial stage of testing, sample was pre-loaded with conditioning stress at 10 kPa for 120 seconds to ensure the platen is loaded flat on the sample. Then, sample is applied with a haversine wave load cycle which consists of 100 kPa stress pulse with 100 ms pulse width followed by a 900 ms rest period. The test was terminated when the accumulative stain reached 10000 micro-strains or until 10000 cycles whichever come first.

2.2.2 Rutting test. This test simulates wheel passes within a pavement. The procedure is described in BS-598-110 [25]. A laboratory scale of this test was conducted using the Wessex wheel tracking device. Wheel tracking test samples were produced for each mix having the respective design aggregate and OBC obtained earlier. In this study, one mould was used to hold the rut sample in the wheel tracking machine. The height of the mould follows exactly the original slab mould of the Wessex wheel tracking device. Approximately 3,700 gm of mix was compacted to 7±0.5% air voids with a size of 300×300 mm, and a final height of 65±1 mm. The samples were left to cool in room temperature for 24 hours after compaction before calculating the air voids content to meet test requirements. The rut test was conducted at 50°C. Before testing, the samples were conditioned for at least four hours at test temperature. The test was conducted only in dry conditions. The samples were subjected to a simulated trafficking with a simple harmonic motion by applying 520 N load for one hour.

2.2.3 Moisture Susceptibility. It is very important to evaluate the moisture sensitivity of the design mix. This step is accomplished by performing AASHTO Method on the design aggregate structure at the design asphalt binder content [24]. Samples, unmodified and ENR-modified asphalt, with 100 mm in diameter and 63.5± 2.5 mm in height are compacted to approximately 7 % air voids. One subset of three samples is the control subset. The other subset of three samples is the conditioned subset. The conditioned subset is subjected to partial vacuum saturation followed by an optional freeze cycle, followed by a 24 hour thaw cycle at 60°C. All samples kept at a temperature of 25°C for a period of 2 hours without soaking, and tested to determine their indirect tensile strengths. The moisture sensitivity is determined as a ratio of the tensile strengths of the conditioned subset divided by the tensile strengths of the dry (control) subset. The minimum criterion is 80 % retained tensile strength [24].

3. Results and Discussion

3.1 The design of asphalt mixers

Initially, the bulk specific gravity (Gsb) test of coarse and fine aggregate was calculated for all sieve sizes and shown in Table 2. The apparent specific gravity (Gsa) and effective specific gravity (Gse) were also determined from these tests. Further analysis on the aggregates was conducted as a critical requirement in the Superpave system to analyze the consensus and source aggregate properties in order to achieve a high performance of a HMA pavement. From the results, aggregate from the quarry complied with the specifications set by the standard criteria. Table 2 shows the results of the aggregate property tests.

Table 2. Aggregate property test results.

Aggregate properties	Result	Criteria	Standard
Gsb of coarse aggregate	2.58	-	ASTM C 127
Gsb of fine aggregate	2.61	-	ASTM C 128
Flakiness (%)	6.00	<20	BS 812 : section 105.1 : 1989
Fine Aggregate Angularity (%)	51.5	>45	AASHTO T33
Elongation (%)	16.0	<20	BS 812
Sand Equivalent Test	48.5	>45	AASHTO 176
Los Angeles Test (%)	32.13	<45	ASTM C: 131-81
Soundness Test (%)	6.1	<12	ASTM C88
Deleterious Materials (%)	0.5	0.2 to 10	ASTM C142

In general, aggregate fulfilled the Superpave mix design requirements and is suitable for use as pavement material. Table 3 shows the aggregate gradation used in this study. Aggregate gradation was prepared using the 19.0 mm nominal maximum size to comply with the Superpave gradation limits.

Table 3. Aggregate Gradation.

Sieve size (mm)	Upper Limit specs.	Lower Limit specs.	Pass (%)	Retained (%)
25	100	100	100	0
19	100	90	100	0
12.5	90	-	82	18
9.5	-	-	76	6
4.75	-	-	51	25
2.36	49	23	29	22
1.18	-	-	21	8
0.6	-	-	16	5
0.3	-	-	11	5
0.15	-	-	7	4
0.075	8	2	4	3
Pan	-	-	-	4

The material characterisations, based on standard and Superpave tests are presented in Table 4 [25, 26]. These results compiled with the specifications set by the standard criteria, with considering the binders to be PG 76.

Table 4. Binder properties.

	Binder description (and criteria)	ENR content (by % weight of asphalt)	
		HMA-0	HMA-6
Original binder	ENR content (by the weight of asphalt)	0%	6%
	Penetration at 25°C, 1/10mm	82	64
	Softening point,°C	45.70	53.80
	Viscosity @ 135°C (Maximum 3 Pa.s), Pa.s	0.244	0.606
	Flash point (Minimum 230°C),°C	275	329
	Dynamic Shear at 10 rad/s (G*/sinδ), (Minimum 1 kPa), kPa	0.177	2.103
RTFOT	RTFOT Weight Loss (Maximum 1%), %	0.36	0.29
	Dynamic Shear at 10 rad/s (G*/sinδ), (Minimum 2.20 kPa), kPa	0.969	3.826
PAV	Dynamic Shear at 10 rad/s (G*.sinδ), (Maximum 5000 kPa), kPa	3962	2414

Asphalt mixes were successfully developed in accordance with the procedures described in Superpave mix design standard [21]. In this study, mixes with five percentages (4, 4.5, 5, 5.5, 6%) of asphalt content were prepared (with three replications for every sample) for each unmodified and modified HMA. Specimens were prepared with blended mineral aggregates at an increment of 0.5% of binder (4.0, 4.5, 5.0, 5.5, 6.0%) by weight of aggregate for base and modified HMA.

The optimum binder content (OBC) to obtain acceptable volumetric properties when compared to the established mix criteria is based on the Superpave gyratory compactor (SGC) specimens with 4% air voids. The volumetric properties consists of OBC, effective binder content (Pbe), voids in mineral aggregate (VMA), voids filled with asphalt (VFA), ratio of dust to effective binder content (P0.075/Pbe). The major components in determining stability and durability of Superpave HMA mix consists of VMA, VFA, air voids and dust proportion. Table 5 summarises the volumetric properties of design mixes corresponding to OBC of the mix along with mix design criteria. For all mixes, results showed that the mixes properties satisfy all the criteria set by the Superpave system.

Table 5. Volumetric Properties of all mixes.

Mix Properties	HMA-0	HMA-6	Criteria
OBC (%)	4.86	5.08	-
Air Voids (%)	4.00	4.29	-
VMA (%)	14.44	17.44	≥13
VFA (%)	74.14	75.00	65-75

3.2 Dynamic creep test

This test is used to determine the resistance of HMA mixes to plastic deformation. The dynamic creep curve consists of three parts: primary, secondary and tertiary. The accumulated strain is recorded at each load cycle, and in this test, termination occurred when the numbers of cycles reaches 10000 cycles. As the loading period is required to terminate at 10000 cycles, tests showed that not all specimens failed before reaching the maximum number of cycles. The tests were conducted on the unaged and aged mixes. Under the 10000 load cycles, all axial strains exhibit curve relationship with load cycles in the strain versus load cycles plot. In this test, mixes where subjected to short and long term oven ageing to evaluate the influence of the added ENR on the asphalt mix, and the results were compared before and after short term ageing (STA) and long term ageing (LTA) ageing.

Figure 1 shows the results of comparison between the accumulated strain of unaged HMA-0 and HMA-6. The dynamic creep for the HMA-0 mix is extremely more than for HMA-6; this behaviour can be attributed to the influence of the ENR on the binder, and thus, on the mix. The mechanical properties of the HMA-6, at high temperatures, rely on the properties of the binder, especially for the permanent deformation. It is expected that the modified mix resistant to permanent deformation will increase. Therefore, adding ENR to asphalt mixes significantly decreased its susceptibility to permanent deformation.

The behaviour of a polymer on asphalt mix can be compared with a previous research with using various percentages of styrene butadiene styrene (SBS) as modifier [27]. Khodaii and Mehrara [27] found that adding 5% SBS to the asphalt increased the resistance of the asphalt mix to permanent deformation.

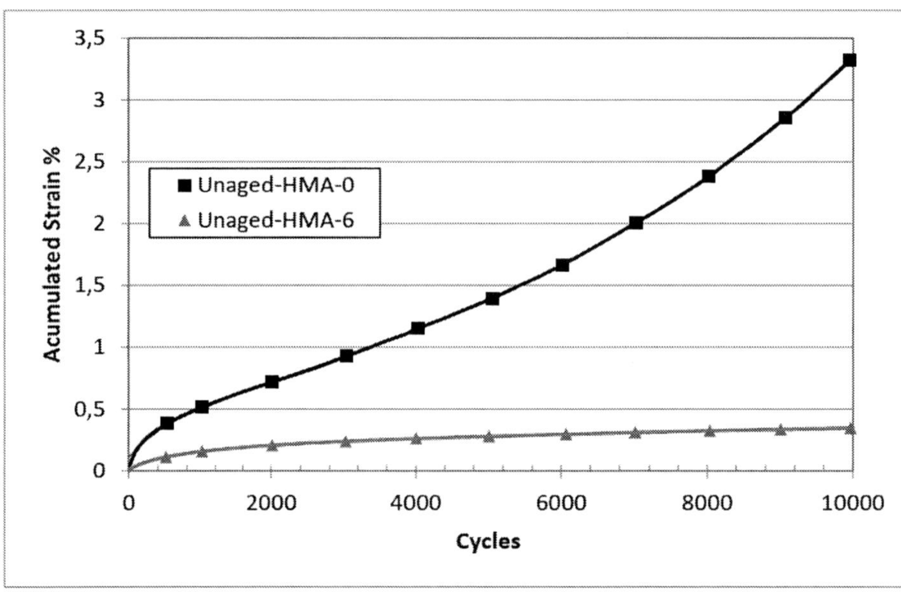

Figure 1. Dynamic Creep Curves for unaged mixes.

Figures 2 and 3 show the permanent deformation results for the mixes before and after STA and LTA. As expected, the results of both HMA-0 and HMA-6 mixes show the same behaviour of accumulated strain. The HMA-0 and HMA-6 after STA and LTA ageing produced less accumulated strain compared with the unaged mixes. However, the effect of STA and LTA on the permanent deformation performance of the HMA-0 mixes is high compared with HMA-6, indicating that ageing has less influence on the mixes with polymer. Table 6 shows the ageing index of the ultimate strains for the dynamic creep test. Compared with HMA-0, the accumulated strain for HMA-6 decreased more than three times after STA and six times after LTA.

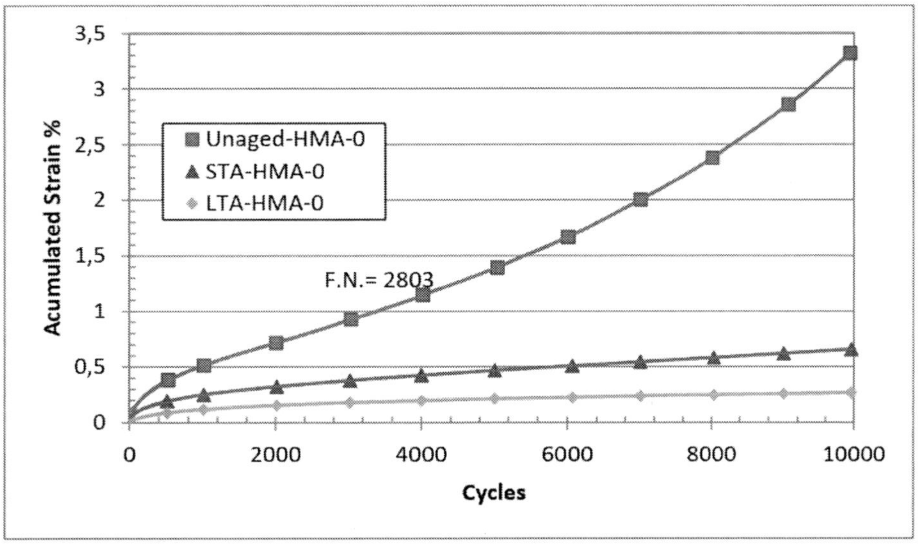

Figure 2. Dynamic Creep Curves for HMA-0 before and after oven ageing.

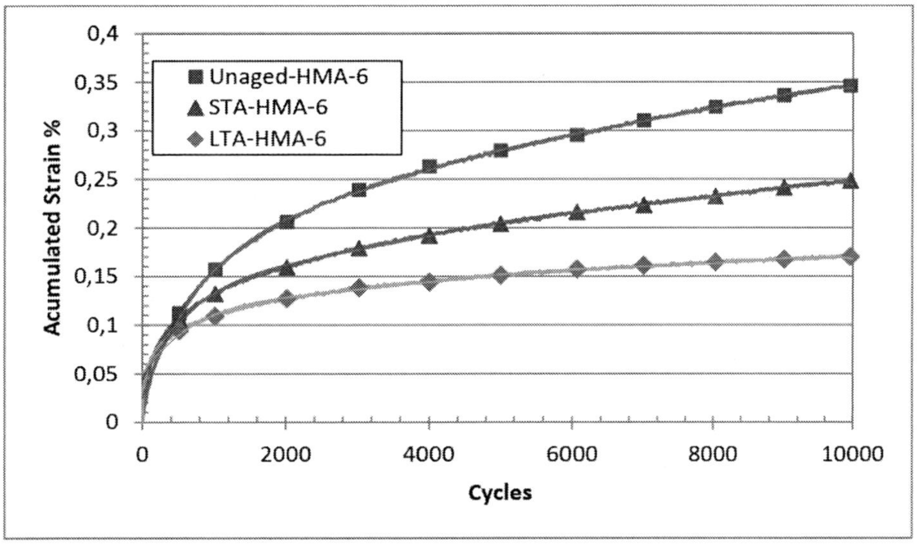

Figure 3. Dynamic Creep Curves for HMA-6 before and after oven ageing.

Table 6. Ageing index for the dynamic creep test results.

Mix	Ultimate Strain %	Ageing index %
Unaged-HMA-0	3.334	1.00
STA-HMA-0	0.657	5.07
LTA-HMA-0	0.267	12.48
Unaged-HMA-6	0.348	1.00
STA-HMA-6	0.248	1.40
LTA-HMA-6	0.170	2.05

IOP Conf. Series: Materials Science and Engineering **236** (2017) 012015 doi:10.1088/1757-899X/236/1/012015

3.3 Rutting test

The rutting test is a simulative test, also known as torture test conducted using the dry wheel tracking device. The test temperature of 50°C was chosen to simulate extreme environmental conditions for the HMA mixes. The wheel passes over HMA specimens and terminates after the test duration reaches 45 minutes. The result of rutting deformation is visible from the wheel tracking test conducted on the specimen as shown in Figure 4. In general, HMA-6 mixes exhibited better rutting resistance compared to HMA-0 mixes as shown in Figure 4. The maximum rutting values for HMA-6 mixes ranged from 2.05 mm to 3.24 mm, while HMA-0 mixes exhibited high rutting, with the maximum rutting values ranging from 3.07 mm to 5.71 mm. This obviously indicates the high resistance of HMA-6 mixes to rutting compared to HMA- 0.

On the other hand, it was found that the STA and LTA caused a significant impact on the rutting resistance for both HMA-0 and HMA-6, which is expected. STA and LTA decreased the rutting susceptibility for both mixes, HMA-0 and HMA-6. However, for HMA-6 showed more resistance to ageing impact. Table 7 shows the ageing indexes for all mixes. The process of oxidation in binders (ageing), form polar compounds that tend to increase the amount of asphaltenes. These asphaltenes contribute to a solid structure of asphalt binder that leads to increased binder stiffness and viscosity. However, in term of post curing, there is no visible swelling in the modified HMA, this is because of the existing of the epoxide in ENR which prevents the swelling of the polymer or the ENR modified asphalt. In addition, the ageing of asphalt mix was associated with the degradation of the ENR polymer and oxidation of asphalt, leading to a decrease in polymer molecular weight and an increase in the content of polar oxygen-containing molecules and asphalt molecular weight. Therefore, the general trend of ENR modified asphalt was to became more resistance to ageing effect because the asphalt content in HMA-0 is less than in HMA-6 [28-31].

An improvement in the performance of the asphalt mix has been observed on addition of SBS [32]. Improved resistance to rutting of the pavement was observed with SBS modification, compared with the base asphalt mix. On the other hand, an eight years old SBS modified asphalt pavement was found to have experienced intensive oxidative age hardening. Polymer modified asphalt mixes were found to be stiffer, thus, increased rutting resistance of the HMA [33].

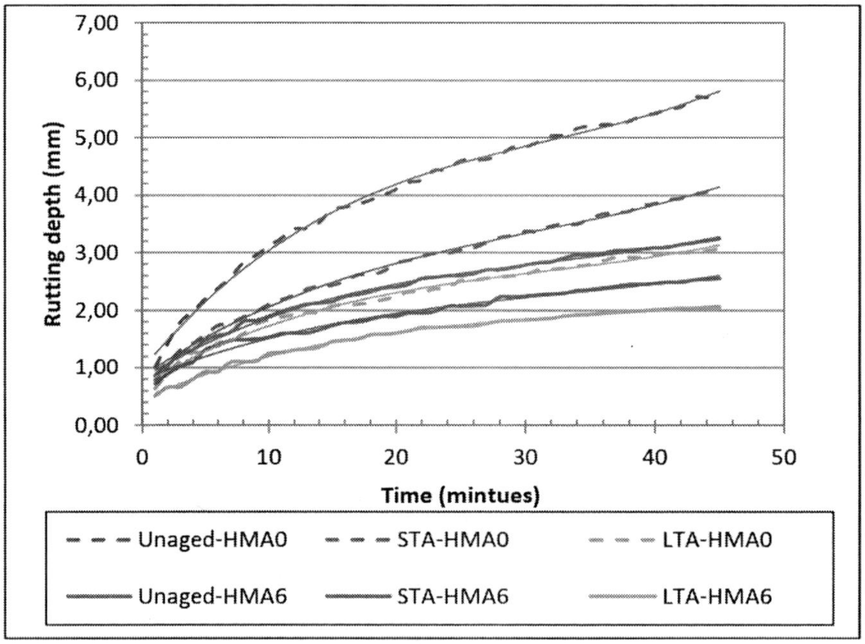

Figure 4. Rutting Test Results for unaged and aged mixes.

Table 7. Ageing index for the rutting test results.

Mix	Rutting depth (mm)	Ageing index %
Unaged-HMA0	5.71	1.00
STA-HMA0	4.07	1.40
LTA-HMA0	3.07	1.86
Unaged-HMA6	3.24	1.00
STA-HMA6	2.55	1.27
LTA-HMA6	2.05	1.58

3.4 Moisture Susceptibility

Indirect tensile strength (ITS) results of each mix under the dry and conditioned case were averaged based on results of three specimens [24]. The results for ITS test, are listed in Table 8. According to the AASHTO standard, all mixes tested were prepared to have air voids content in the range of 7±0.5%. The trend showed that ITS of all specimens decreased from the dry to conditioned specimens implying the presence of deterioration in the mixes which affects the strength of the HMA mix.

The tensile strength ratio (TSR) result is an indication of the HMA mix susceptibility to moisture damage. HMA-0 and HMA-6 mixes met the required minimum 80% TSR value as specified in AASHTO T283. The indirect tensile strength test showed that HMA-6 mixes exhibit higher tensile strength values than HMA-0 before and after conditioning, this showed that HMA-0 is more likely to be susceptible to hydrolysis. Khudyakova, Strizhev and Mashkova [34] investigated the chemical nature of the hydrolysis of asphalt and concluded that asphalt–mineral adhesiveness is primarily determined by the chemical composition of the dispersion medium (i.e. resins, aromatics and saturates), and by the concentration of the polar compounds capable of forming bonds with acid-based minerals which are not destroyed by water. In a previous study, it was found that asphalt mix modified with SBS (HMA-SBS) reduced the susceptibility of HMA to moisture damage [29]. It was found that the SBS-HMA asphalt samples achieved a higher TSR value than the unmodified HMA, 88.9% and 76.7% respectively.

Table 8. Moisture susceptibility test results.

Mix	ITS- Dry (Kpa)	Air void (%)	ITS-Conditioned (Kpa)	Air void (%)	TSR (%)
HMA-0	374.65	6.60	357.00	6.70	0.86
	397.72	7.03	318.18	6.99	
	401.47	6.85	335.86	7.29	
HMA-6	424.94	6.48	386.56	7.12	0.89
	419.63	7.02	363.56	7.17	
	443.53	6.98	394.04	7.04	

4. Conclusions

The mechanical performance of HMA-6 mixes for unaged, STA and LTA ageing were discussed in this study and compared to the performance of HMA-0 mixes. The mechanical performances which were investigated consisted of dynamic creep, rutting, and moisture susceptibility. In addition, the mixes were subjected to short term ageing and long term ageing. Permanent deformation, at 40 and

50°C, of ENR-HMA is dominated by the ENR content. Where, ENR-HMA produced less deformation comparing with base-HMA. This is due to the presence of highly elastic rubber in HMAs at high temperature. The effect of the ageing on the permanent deformation performance of the base-HMA is extremely high comparing with ENR-HMA, indicating that ageing has less effect on HMA with the existence of the ENR. Moisture damage has reduced after modifying the mixes with ENR, indicating that ENR-HMA is less susceptible to moisture damage comparing with the base-HMA. This improvement is not significant, but the results still within the standard requirements.

Acknowledgments
The authors would like to thank The National University of Malaysia (UKM), Malaysia for providing research funding through the project Dana Impak Perdana DIP-2015-019 and the Sustainable Urban Transport Research Centre (SUTRA) for providing research facilities. We also thank the faculty at the University of Malaya (UM) for their experimental collaboration.

References
[1] Brûlé B, Brion Y V and Tanguy A 1988 Paving asphalt polymer blends: relationships between composition structure and properties (with discussion) *Association of Asphalt Paving Technologists Proc* **57** pp 41-64
[2] Collins J H, Bouldin M, Gelles R and Berker A 1991 Improved performance of paving asphalts by polymer modification (with discussion) *Journal of the Association of Asphalt Paving Technologists* **60** pp 43-79
[3] Goodrich J L 1991 Asphaltic binder rheology asphalt concrete rheology and asphalt concrete mix properties (with discussion) *Journal of the Association of Asphalt Paving Technologists* **60** pp 80-120
[4] Khadivar A and Kavussi A 2013 Rheological characteristics of SBR and NR polymer modified bitumen emulsions at average pavement temperatures *Construction and Building Materials* **47** pp1099-1105
[5] Veeraragavan A 2011 Dynamic mechanical characterization of asphalt concrete mixes with modified asphalt binders *Materials Science and Engineering* **A 528**(21) pp 6445-54
[6] Arabani M and Mirabdolazimi S M 2011 Experimental investigation of the fatigue behaviour of asphalt concrete mixtures containing waste iron powder *Materials Science and Engineering* **A 528**(10) pp 3866-70
[7] Read J and Whiteoak D 2003 *The shell bitumen handbook* (Thomas Telford)
[8] Ferry J D 1980 *Viscoelastic properties of polymers* (John Wiley and Sons New York)
[9] Airey G D 2003 Rheological properties of styrene butadiene styrene polymer modified road bitumens *Fuel* **82**(14) pp1709-19
[10] Lu X and Isacsson U 1997 Rheological characterization of styrene-butadiene-styrene copolymer modified bitumens *Construction and Building Materials* **11**(1) pp 23-32
[11] Becker Y, Mendez M P and Rodriguez Y 2001 Polymer modified asphalt *Vision tecnologica* **39-50**
[12] Mo L, Shu D, Li X, Huurman M and Wu S 2012 Experimental investigation of bituminous plug expansion joint materials containing high content of crumb rubber powder and granules *Materials & Design* **37** pp 137-43
[13] Bull A L and Vonk W C 1984 Thermoplastic rubber/bitumen blends for roof and road Shell *Chemical technical manual TR 8* pp15 (London)
[14] Jun L, Yuxia Z and Yuzhen Z 2008 The research of GMA-g-LDPE modified Qinhuangdao bitumen *Construction and Building Materials* **22**(6) pp1067-73
[15] Moreno F, Sol M, Martín J, Pérez M and Rubio M C 2013 The effect of crumb rubber modifier on the resistance of asphalt mixes to plastic deformation *Materials & Design* **47** pp 274-80
[16] Gelling I R 1991 Epoxidised natural rubber *Journal of Natural Rubber Research* **6(3)** pp184-205

[17] Ismail H and Hairunezam H M 2001 The effect of a compatibilizer on curing characteristics mechanical properties and oil resistance of styrene butadiene rubber/epoxidized natural rubber blends *European Polymer Journal* **37**(1) pp 39-44
[18] Al-Mansob R A, Ismail A, Alduri A N, Azhari C H, Karim M R and Yusoff N I M 2014 Physical and rheological properties of epoxidized natural rubber modified bitumens *Construction and Building Materials* **63** pp 242-48
[19] Ismail A, Al-Mansob R A, Yusoff N I B M and Karim M R 2012 Effect of toluene as disperser on the bitumen modified with epoxidized natural rubber with aging simulation *Australian Journal of Basic and Applied Sciences* **6**(12) pp 97-101
[20] Al-Mansob R A, Ismail A, Yusoff N I M, Albrka S I, Azhari C H and Karim M R 2016 Rheological characteristics of unaged and aged epoxidised natural rubber modified asphalt *Construction and Building Materials* **102** pp 190-9
[21] Cominsky R J, Huber G A, Kennedy T W and Anderson M 1994 The superpave mix design manual for new construction and overlays *Washington DC Strategic Highway Research Program No. SHRP-A-407*
[22] Witczak M W 2005 Simple performance tests: *Summary of recommended methods and database* (Transportation Research Board) **46** *pp 15*
[23] British Standard Institute 1998 Sampling and Examination of Bituminous Mixtures for Roads and Other Paved Areas–Part 110 Methods of Test for the Determination of Wheel Tracking Rate and Depth *(BS 598-110-1998)*
[24] Canadian Council of Independent Laboratories 2003 Resistance of Compacted Asphalt Mixtures to Moisture-Induced Damage *(AASHTO T283-03)*
[25] Christensen D W and Anderson D A 1992 Interpretation of dynamic mechanical test data for paving grade asphalt cements (with discussion) *Journal of the Association of Asphalt Paving Technologists* **61** pp 67-116
[26] Asphalt Institute 2007 *The asphalt handbook* (Asphalt Institute USA)
[27] Khodaii A and Mehrara A 2009 Evaluation of permanent deformation of unmodified and SBS modified asphalt mixtures using dynamic creep test *Construction and Building Materials* **23**(7) pp 2586-92
[28] Al-Azri N, Jung S, Lunsford K, Ferry A, Bullin J, Davison R and Glover C 2006 Binder oxidative aging in Texas pavements: hardening rates, hardening susceptibilities, and impact of pavement depth. Transportation Research Record *Journal of the Transportation Research Board* **1962** pp 12-20
[29] Bennert T and Martin J-V 2010 Polyphosphoric Acid in Combination with Styrene-Butadiene-Styrene Block Copolymer Laboratory Mixture Evaluation *Proc. Asphalt Paving Technology (Sacramento CA)* pp 773-792
[30] Farrar M J, Harnsberger P M, Thomas K P and Wiser W 2006 Evaluation of oxidation in asphalt pavement test sections after four years of service *Int. Conf. on Perpetual Pavement (Columbus)* (Ohio University)
[31] Li X, Williams R C, Marasteanu M O, Clyne T R and Johnson E 2009 Investigation of in-place asphalt film thickness and performance of hot-mix asphalt mixtures *Journal of Materials in Civil Engineering* **21**(6) pp 262-70
[32] Sargand S M and Kim S S 2001 Performance evaluation of polymer modified and unmodified Superpave mixes *2nd Int. Symp. on Maintenance and Rehabilitation of Pavements and Technological Control (Auburn, Alabama)* No. 01-147
[33] Mohammad L N, Negulescu I I, Wu Z, Daranga C, Daly W H and Abadie C 2003 Investigation of the use of recycled polymer modified asphalt binder in asphalt concrete pavements (with discussion and closure) *Journal of the Association of Asphalt Paving Technologists* **72** pp 551-94
[34] Khudyakova TS, Strizhev E F and Mashkova I 1989 The Adhesion Capacity of Bitumen as a Function of Its Chemical Composition and Structure *Zhurnal Prikladnoi Khimii* **62**(8) pp 1849-1953

IOP Conf. Series: Materials Science and Engineering **236** (2017) 012016 doi:10.1088/1757-899X/236/1/012016

Influence of selected test parameters on measured values during the MSCR test

Lucie Benešová[1] and Jan Valentin[1]

[1]CTU in Prague, Faculty of Civil Engineering, Department of Road Structures, Thákurova 7, Prague, 166 29, Czech Republic

E-mail: lucie.benesova@fsv.cvut.cz

Abstract. One of today's most commonly used test on a Dynamic Shear Rheometer (DSR) is the Multiple Stress Creep Recovery (MSCR) test. The test is described in the standard EN 16659, which is valid in the Czech Republic since October 2016. The principle of the test is based on repeated loading and recovering of a bitumen sample, according to which it is possible to determine the percentage of elastic recovery (R) and non-recoverable creep compliance (J_{nr}) of the bituminous binder. This method has been recently promoted as the most suitable test for assessing the resistance of bituminous binders to permanent deformation. The test is performed at higher temperatures and is particularly suitable for modified bituminous binders. The paper deals with the comparison of the different input parameters set on the DSR device - different levels of stress, temperature of test, the geometry of the measuring device and also a comparison of the results for a different number of loading cycles. The research study was focused mainly on modified bituminous binders, but to compare the MSCR test it is performed even with conventional paving grade binders.

1. Introduction

The durability of asphalt mixtures is largely dependent on the quality and performance behavior of the bituminous binder. Basic empirical tests can characterize or classify bituminous binders, but in order to more accurately describe their behavior, functional tests are needed, for example by using Dynamic Shear Rheometer (DSR).

In recent years there has been an increasing demand on road traffic, which lead to the necessity for developing bituminous binders with improved performance behavior relative to conventional paving grade bitumen. This causes the progress and use of a wide range of bitumen modifiers which improve the performance of the basic bitumen and thus the asphalt mixture [1, 2, 3, and 4]. The most widely used technical standards for bituminous binders in the Czech Republic are based on the behavior of neat binders that cannot properly characterize potentials of modified binders.

One of the fundamental failures of asphalt pavements at higher operating temperatures is the formation of ruts. In the past rutting prediction of asphalt mixtures was based on $G^*/\sin\delta$ criteria. This parameter works reliably for conventional speed and moderate traffic volume, but for pavements with slow speed loading and/or high traffic volume it has been shown that it does not correspond enough to permanent deformations [4, 5, and 6]. Values G^* and δ are measured in the linear viscoelastic region, but permanent deformations are not a linear viscoelastic phenomenon and therefore the characteristic $G^*/\sin\delta$ does not sufficiently correlate with permanent deformations [7]. From the perspective of permanent deformation of bituminous binders there is a need to pay attention on the area of nonlinear behavior of bituminous binders, which include J_{nr} parameter obtained during the MSCR test. According to the authors [8, 9] parameter J_{nr} has a significantly better correlation with rutting measurements on asphalt mixtures than conventional oscillatory frequency sweep test.

2. Multiple Stress Creep and Recovery test (MSCR)

Multiple Stress Creep and Recovery test is a test of repeated loading and relaxation/recovery, using a device called Dynamic Shear Rheometer (DSR). The test is used for measuring the accumulated strain in the bitumen within the prescribed stress level. The MSCR test uses a constant stress creep of 1 second duration followed by a zero stress recovery of 9 seconds duration. This cycle is according to present standards in Europe or the U.S. repeated 10 times for each stress level. Figure 1 presents the typical creep and recovery cycle in MSCR test.

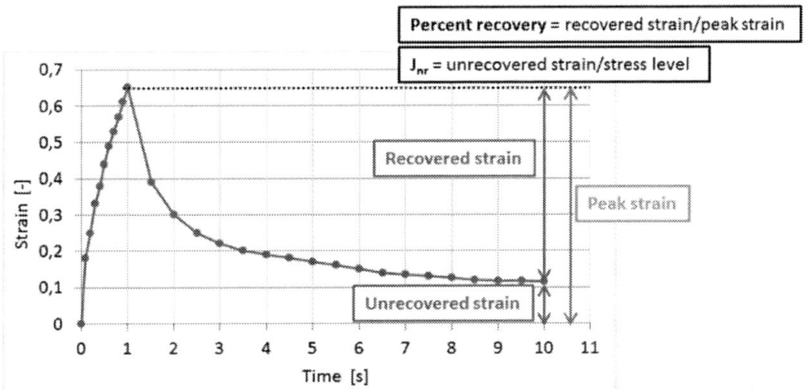

Figure 1. Typical creep and recovery cycle.

Two most important parameters of bituminous binder calculated from the measured strain at individual stress levels are non-recoverable creep compliance (J_{nr}) and percent recovery (R). J_{nr} is a measure of the amount of residual strain left in the specimen after repeated creep and recovery, relative to the amount of stress applied. Parameter J_{nr} is currently perceived as the best rheological parameter for the assessment of the susceptibility of bituminous binder to resist permanent deformation in asphalt mixture. The lower the J_{nr} is, the greater proportion of the stress the binder is able to absorb and the less susceptible to plastic deformation is. But there are any limiting values neither in the Czech technical standards nor in the European standards. Besides the mentioned characteristics, stress sensitivity parameters R_{diff} and $J_{nr,diff}$ are calculated. Golalipour [10] indicated that determine the stress level is not a simple task and no one level can be clearly defined as most appropriate of what a binder is exposed to in a real mixture. Therefore it seems most reliable to use the sensitivity parameter ($J_{nr,diff}$). The $J_{nr,diff}$ is important mainly when pavement is exposed to unexpected traffic loads or extremely high temperatures. In case the $J_{nr,diff}$ value is greater than 75 %, the bituminous binder is classified as a very stress sensitive material, which is not recommended for use in the construction of a pavement structure, [11].

MSCR is a test that has received a lot of attention and holds great promise for predicting rutting resistance of asphalt mixtures, but there are still many problems in its implementation, evaluation and interpretation of results. Some of these concerns are related to the number of cycles, the value of stress level or the temperature of testing. Conventional test set is based on the standard EN 16659 [12] and indicates the stress levels (0.1 and 3.2 kPa), number of loading/unloading cycles at each stress level (10) and the geometry of the device (25 mm parallel plate with a 1 mm gap). The test temperature is not exactly defined, which remains still a questionable task for the future.

Some research works showed [8, 13, and 14] that in order to study the behavior of the binder in the nonlinear range, it is necessary to apply a higher stress level than 3.2 kPa, depending on the type of bitumen. Moreover, according to studies [13, 14] the higher stress level is used, the better correlation coefficient for resistance to permanent deformation of mixtures (French rutting test) is received.

3. Testing procedures

This paper compares different groups of bituminous binders due to the widest possible coverage of the different behavior of bituminous binders. Some variants are conventional paving grade bituminous binders, but in most cases modified binders (crumb rubber – labeled CR, polymer modified binders – PMB or low viscosity additives) are tested and analyzed. Attention has been given to assessing the impact on the selection of the applied stress level and temperature of testing. Furthermore, two different testing geometries were used on DSR device and finally calculations were compared using the average of 10 cycles with values calculated from the last 5 cycles.

3.1. Effect of selected temperature and stress level

The attention in this chapter is pay to the use of different test temperatures and stress levels. In the technical standard EN 16659 [12] test temperature is not prescribed, but should be selected suitably in the range 50-80 °C (the temperature corresponding to the normal service temperature of asphalt mixture). The exact prescription of one temperature is even not planned for the revised EN 14023 standard for PMB binders. Different polymer modified binders (PMB) and one neat bitumen were compared. PMB are used in highly loaded pavement structures and for this reason test was measured at normal test temperature (60 °C) and elevated test temperature (70 °C). On the contrary neat bitumen 50/70 was measured at normal test temperature (60 °C) and at a reduced test temperature (50 °C). Another parameter that can be determined during the MSCR test is using different stress levels. The standard EN 16659 prescribes stress level of 0.1 kPa and 3.2 kPa. But as mentioned above, values obtained at higher stress level correlate better with permanent deformations. In addition, the application of a higher stress can cause significant increase of J_{nr}. For this reason, the standard stress levels were completed with 8.0 kPa stress.

The following Table 1 presents the calculated values of elastic recovery R, non-recoverable creep compliance J_{nr} and stress sensitivity characteristics R_{diff} and $J_{nr,diff}$.

Table 1. Values of elastic recovery, non-recoverable creep compliance and sensitivity parameters.

Binder type	0.1 kPa		3.2 kPa		8.0 kPa		0.1-3.2 kPa		3.2-8.0 kPa		0.1-8.0 kPa	
	R [%]	J_{nr} [1/kPa]	R [%]	J_{nr} [1/kPa]	R [%]	J_{nr} [1/kPa]	R_{diff} [%]	$J_{nr,diff}$ [%]	R_{diff} [%]	$J_{nr,diff}$ [%]	R_{diff} [%]	$J_{nr,diff}$ [%]
50/70_50 °C	2.5	1.59	1.2	1.09	0.5	1.08	53.7	-31.2	60.1	-0.9	81.5	-31.8
50/70_60 °C	0.5	4.93	0.1	4.41	0.0	4.57	87.3	-10.7	60.3	3.7	95	-7.4
45/80-65_60 °C	83.1	0.12	85.5	0.09	87.2	0.07	-2.9	-23.3	-2.0	-17.1	-5	-36.4
45/80-65_70 °C	89.8	0.19	85.6	0.20	57.4	0.64	0.1	-15.8	32.9	213.3	33	163.8
25/55-80_60 °C	69.4	0.15	66.6	0.16	66.4	0.17	4.1	3.8	0.3	5.1	4.3	9.1
25/55-80_70 °C	70.4	0.42	58.9	0.52	52.8	0.63	16.3	25.4	10.4	20.9	25	51.7
25/55-60_60 °C	76.8	0.13	38.5	0.49	29.7	0.57	49.9	267.0	22.9	14.7	61.4	320.8
25/55-60_70 °C	55.1	1.06	24.7	1.85	13.5	2.38	55.2	73.6	45.3	28.8	75.5	123.6

As seen in Table 1 it is obvious that the PMBs have high value of elastic recovery and very low non-recoverable creep compliance. As the temperature increases the shear strain increases and that is associated with an increase of J_{nr} value. Increased susceptibility to permanent deformation is logically at higher temperature (higher J_{nr} value). This is connected with the reduction of elastic recovery, but surprisingly, slight increase of R was observed in two PMBs at a stress level of 0.1 kPa. This increase is probably caused due to the variability of the measured values (detailed in section 3.3).

The unmodified binder 50/70 has high stress sensitivity R_{diff} (Figure 2.), which is caused by very low elastic recovery values (Table 1.). This neat bitumen also experienced the most pronounced reduction in elastic recovery at a higher test temperature. The most significant J_{nr} change between test temperatures for PMB is in the binder 25/55-60. It could be said that this binder seems no longer to be in the linear viscoelastic range at 70 °C. Figure 2 presents that this binder has a high sensitivity R_{diff} (at both temperatures) and Figure 3 shows an extremely high sensitivity $J_{nr,diff}$ at 60 °C. An extreme increase is due to the fact that at stress 3.2 kPa, there was a significant increase of J_{nr} (compared with a stress 0.1 kPa), which leads to a high parameter $J_{nr,diff}$. This binder is the only one that has failed the US condition $J_{nr,diff} < 75$ % for stresses 0.1 kPa and 3.2 kPa. The same binder has a lower $J_{nr,diff}$ (0.1-3.2 kPa) at temperature of 70 °C because it has a relatively high J_{nr} value at stress 0.1 kPa, (the binder is in nonlinear region at higher temperature).

By contrast, the PMB 45/80-65 changed the measured values significantly between stress 3.2 kPa and 8.0 kPa at 70 °C (Figure 2.). And thanks to it, this level can be described as marginal behavior in the nonlinear region for binder 45/80-65. Other binders can be described as very low stress sensitive. Generally, the increasing temperature increases deformation, reduces elastic recovery and increases non-recoverable creep compliance. Difference is more pronounced with higher stress level.

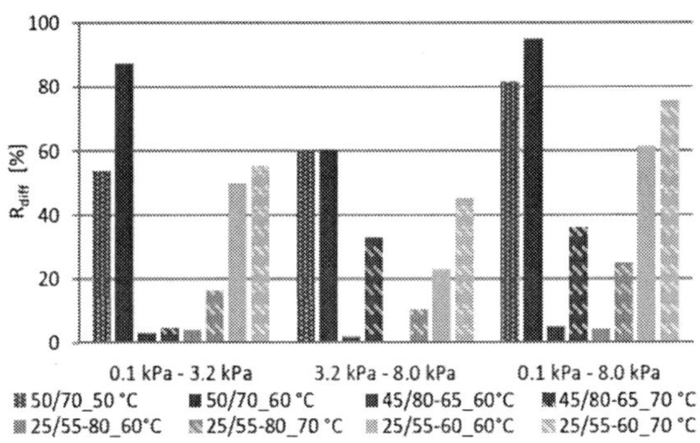

Figure 2. Stress sensitivity parameter R_{diff}.

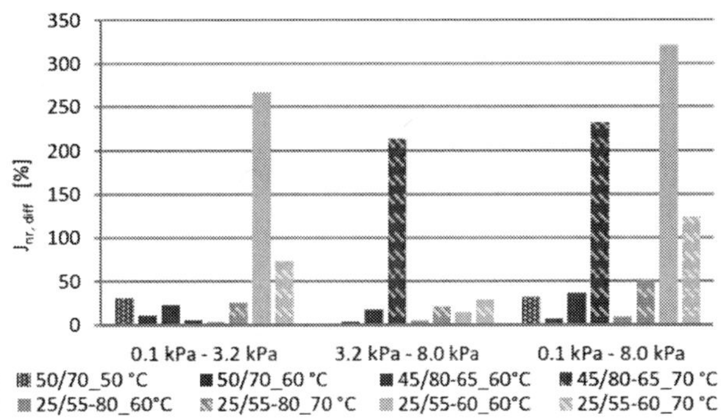

Figure 3. Stress sensitivity parameter $J_{nr,diff}$.

Higher stress level 8.0 kPa was also conducted for the second group of bituminous binders containing variants with the basic paving grade bitumen 20/30. Bitumen 20/30 was modified by low-viscosity waxes in an amount of 2-4 M% or by combination of crumb rubber CR (15 M%) with low viscosity waxes. Higher stress level was measured only for the selected variants.

Table 2. Values of elastic recovery, non-recoverable creep compliance and sensitivity parameters.

	0.1 kPa		3.2 kPa		8.0 kPa		0.1-3.2 kPa		3.2-8.0 kPa		0.1-8.0 kPa	
Binder type	R [%]	J_{nr} [1/kPa]	R [%]	J_{nr} [1/kPa]	R [%]	J_{nr} [1/kPa]	R_{diff} [%]	$J_{nr,diff}$ [%]	R_{diff} [%]	$J_{nr,diff}$ [%]	R_{diff} [%]	$J_{nr,diff}$ [%]
20/30	9.4	0.425	6.7	0.402	-	-	28.6	-5.3	-	-	-	-
2% CM	15.4	0.209	8.9	0.296	4.0	0.385	42.6	41.7	55.4	29.9	74.4	84.0
3% CM	18.1	0.222	14.1	0.236	-	-	22.0	6.4	-	-	-	-
3% BM	11.1	0.656	5.4	0.644	-	-	51.2	-1.7	-	-	-	-
4% BM	14.6	0.462	7.3	0.495	2.3	0.590	50.0	7.0	68.4	19.2	84.2	27.5
15% CR	73.5	0.015	67.9	0.018	-	-	7.6	20.9	-	-	-	-
15% CR+2% CM	81.7	0.006	74.6	0.008	73.3	0.009	8.7	31.8	1.8	6.3	10.3	40.2
15% CR+2% BM	78.6	0.007	75.7	0.008	73.3	0.009	3.7	12.5	3.2	11.4	6.8	25.3
15% CR+2% E10	72.7	0.011	70.6	0.012	67.6	0.013	2.9	1.8	4.2	9.1	7.0	11.1

Table 2 shows the fundamental values calculated from MSCR test. Crumb rubber modification resulted in a significant increase in the percent recovery and decrease in J_{nr} in comparison with bitumen 20/30. High J_{nr} values are expected to mean low resistance to permanent deformation. Application of low-viscosity additives does not significantly influence R or J_{nr} values. R values are very low and vice versa J_{nr} relatively high, when low-viscosity waxes are used with the bitumen 20/30, which leads logically to higher stress sensitivity characteristics R_{diff} (Figure 4.). On contrary, CR binders have high percent recovery, which helps to have low R_{diff} (below 10 %) for all stress levels. This indicates good elastic properties even at a stress 8.0 kPa. When comparing the stress sensitivity parameter $J_{nr,diff}$, it is obvious that the US limit for stress level 0.1 kPa and 3.2 kPa ($J_{nr,diff}$ < 75 %) would be fulfilled by all tested binders. But it is clear that a binder that has 2% CM is the most stress sensitive and especially $J_{nr,diff}$ is in this case high comparing the lowest (0.1 kPa) and the highest (8.0 kPa) stress level.

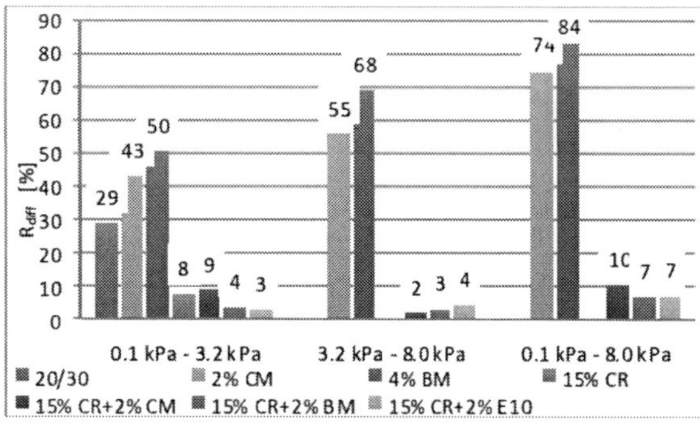

Figure 4. Stress sensitivity parameter R_{diff}.

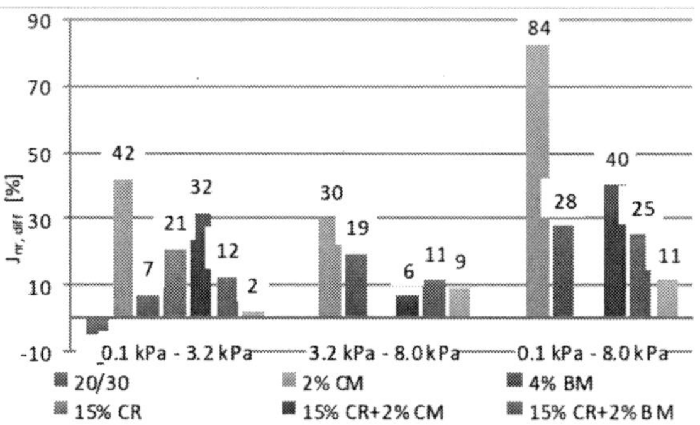

Figure 5. Stress sensitivity parameter $J_{nr,diff}$.

3.2. Effect of selected testing geometry used for Dynamic Shear Rheometer

In general, for more than two decades two testing geometries are used with the DSR for bitumen testing, namely 8 mm diameter plate-plate with a 2 mm gap between the lower and upper plate (labelled PP08) or 25 mm diameter plate-plate and 1 mm gap (PP25). Selecting geometry of the testing device is based on the operating conditions. Geometry PP08 is generally used at low and medium temperatures (-5-30 °C) or for aged binders (frequency sweep test). Geometry PP25 is normally selected for measurements at high temperatures (30-90 °C) and is therefore selected as a base for performing MSCR test. Different geometries were compared for PMBs and neat 50/70 after ageing, and in addition for the binder with a crumb rubber (CR).

Table 3. Values of elastic recovery, non-recoverable creep compliance and sensitivity parameters.

Binder type	PP25				PP08			
	0.1 kPa		3.2 kPa		0.1 kPa		3.2 kPa	
	R [%]	J_{nr} [1/kPa]	R [%]	J_{nr} [1/kPa]	R [%]	J_{nr} [1/kPa]	R [%]	J_{nr} [1/kPa]
50/70+10% CR	35.5	0.50	7.8	0.81	33.7	0.42	7.5	0.70
50/70_R	5.6	1.59	1.1	1.57	8.1	0.54	5.1	0.58
25/55-60 _R	73.9	0.16	73.2	0.15	73.9	0.07	74.8	0.06
40/100-65_R	84.8	0.09	88.9	0.04	84.8	0.03	89.2	0.02
50/70_R+P	17.0	0.28	14.0	0.23	25.2	0.06	23.9	0.06
25/55-60 _R+P	64.5	0.07	64.2	0.06	69.0	0.02	65.1	0.03
40/100-65_R+P	88.6	0.02	88.2	0.02	91.7	0.01	90.1	0.01
50/70_3R	16.9	0.32	13.0	0.28	21.5	0.11	22.2	0.08
25/55-60 _3R	70.0	0.07	65.6	0.09	67.0	0.05	66.9	0.04
40/100-65_3R	92.1	0.02	91.8	0.02	82.7	0.02	90.4	0.01

Table 3 presents a relatively low difference of the percent recovery between the two geometries. The percentage difference is up to 10 % with the exception of all variants 50/70, which have very low values of R and for this reason the percentage change is significant. The percentage difference between the two geometries of J_{nr} value is shown in Figure 6. The J_{nr} value is always lower in the case of PP08 and even more than about half in most variants (with the exception of variant with CR). Overall, this implies that the selected testing geometry has a major influence on the J_{nr} values.

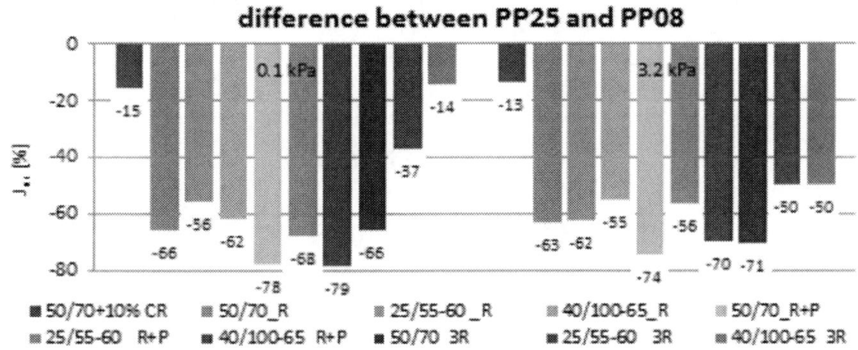

Figure 6. Percentage difference between two geometries.

3.3. *Effect of number of cycles for calculating characteristics of the MSCR test*

Taking the average of the 10 cycles to calculate characteristics, as required in the MSCR standard procedure, could be misleading because the response is significantly changing with number of cycles. The following figures show the percentage difference for the calculation from the last 5 cycles compared to the calculation from all 10 cycles for each group described in the previous chapters.

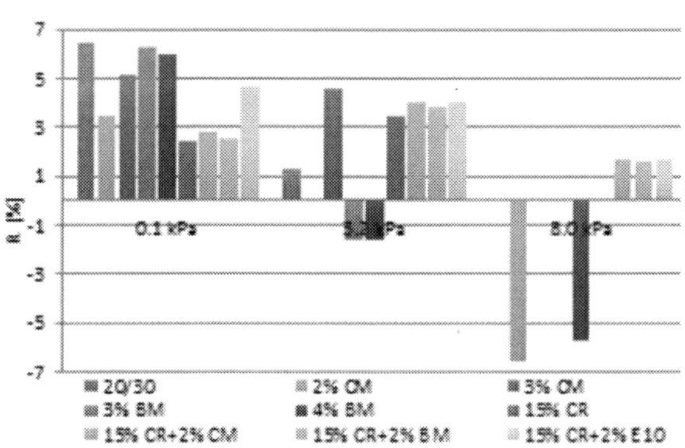

Figure 7. Percent change between two methods of calculating R.

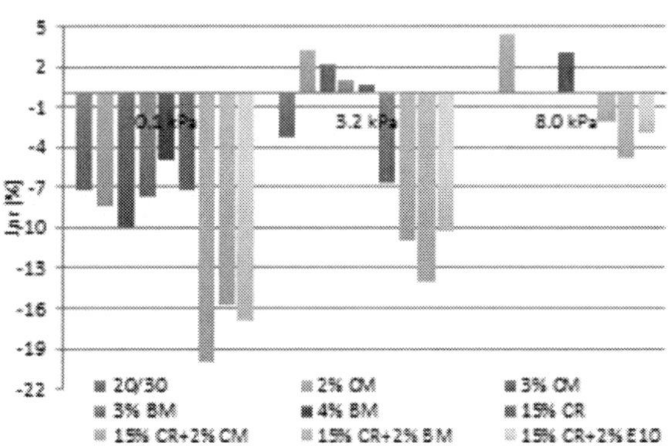

Figure 8. Percent change between two methods of calculating J_{nr}.

Figure 7 presents the percent change of percent recovery between the values calculated from all 10 cycles, and only the last 5 cycles. The calculated differences are not significantly high at all stress levels (less than 7 %). Figure 8 shows the percentage difference of J_{nr} between two methods of calculating. The maximum difference is about 20 % at stress level of 0.1 kPa, which is quite significant change that may be a source of variability of the results. This change is lower at higher stress and it is about 5 % when the stress level is 8.0 kPa, which can be considered negligible. Higher difference of J_{nr} is evident for binders with CR.

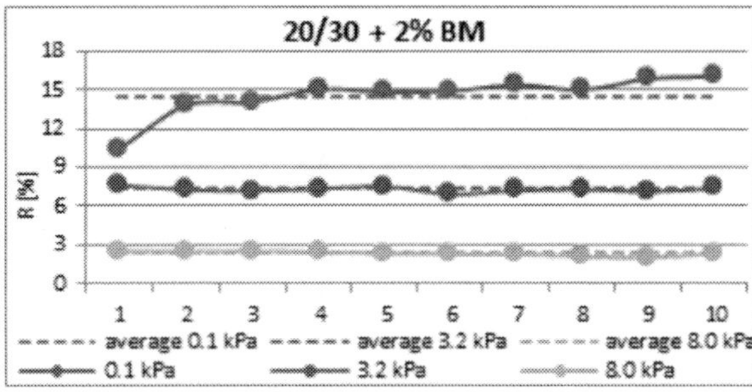

Figure 9. R values at each cycle.

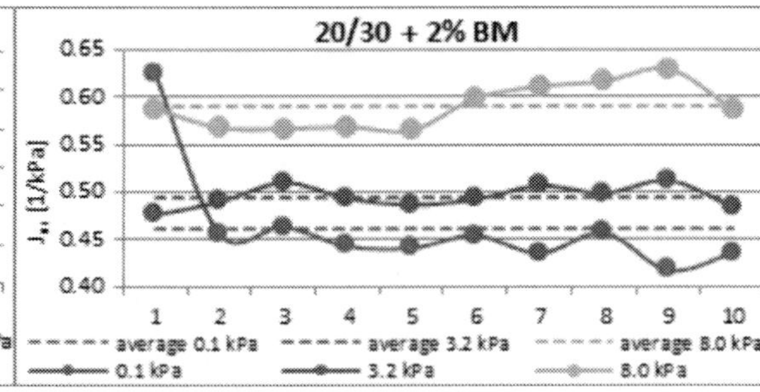

Figure 10. J_{nr} values at each cycle.

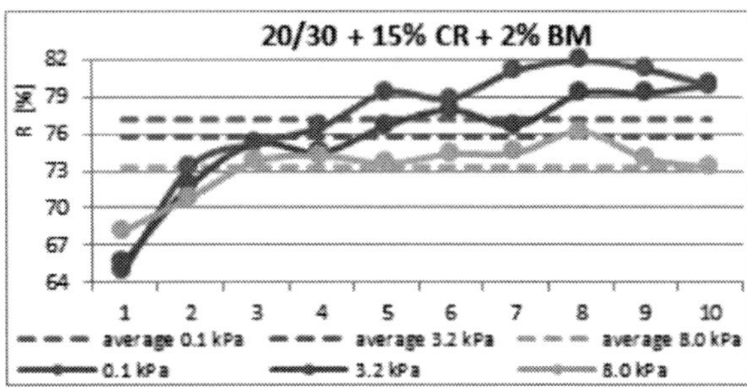

Figure 11. R values at each cycle.

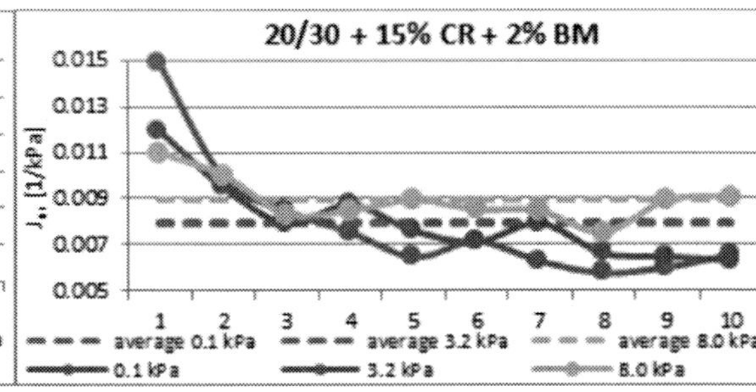

Figure 12. J_{nr} values at each cycle.

Figures 9-12 show that applying higher stress level leads to reduction of result variability in the test. The dotted lines in the figures show the average values calculated from all 10 cycles. Individual values have larger fluctuations at a low stress level, while a higher stress level contributes to a better reproducibility of results. Modification with low-viscosity waxes (Figure 9 and 10) have only a first calculated value significantly different from other values at the stress level 0.1 kPa. Conversely in case of modification with CR (Figure 11 and 12), there is not only the first value significantly different for all stress levels. For these binders (modified by CR with max. 15 M% content) would be therefore certainly appropriate to increase the number of cycles to obtain steady values (especially at lower stress level). This might by as hypothesis valid only for CRmB binders where the content of used crumb rubber would not exceed 15 M% and the maximum particle size would not exceed 1.0 mm.

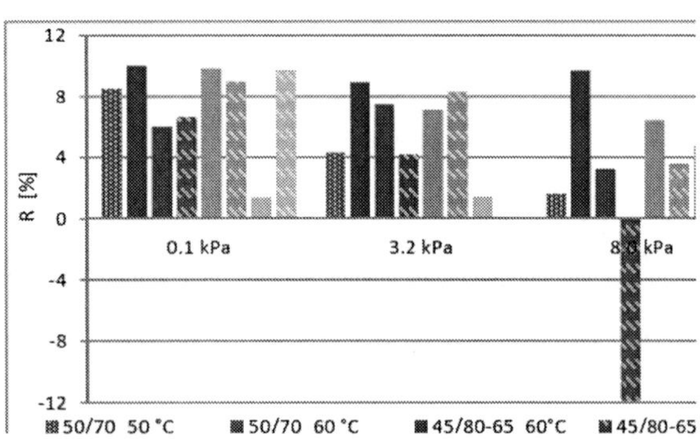

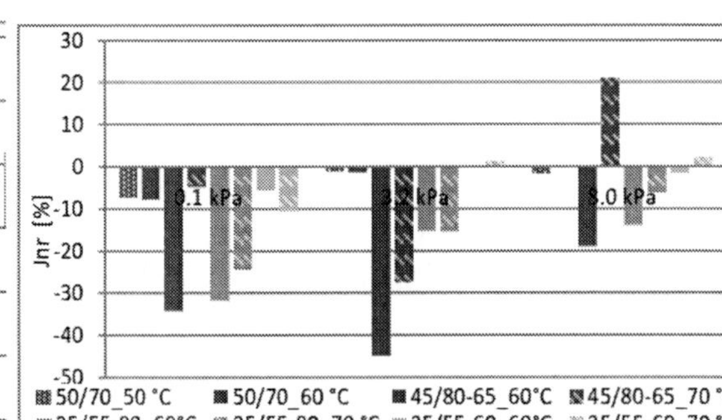

Figure 13. Percent change between two methods of calculating R.

Figure 14. Percent change between two methods of calculating J_{nr}.

Figures 13 and 14 present the difference of the values calculated as the average of all 10 cycles in comparison with the average of last 5 cycles. The difference in percent recovery ranges up to 12 %, but for certain binders are differences in the calculation of J_{nr} considerably higher. All PMB binders have similar trend which is seen in Figure 16 and 17, wherein the R values in the first 3-4 cycles gradually grew up (J_{nr} values decrease) and in the following cycles are stabilized. This pattern is evident among stress levels 0.1 kPa and 3.2 kPa. Increase of percentage change looks like a very low because of the relatively high values of R. Conversely, the variants that have very low J_{nr} have percent change significant between values calculated of 10 cycles and 5 cycles. Important change occurred in PMB 45/80-65 at 70 °C (Figure 18), J_{nr} value is increasing in each cycle at stress 8.0 kPa, which is the opposite trend than at other stress levels. This likely indicates the fatigue of the material which occurs in the nonlinear viscoelastic region. The closest values of all cycles are evident for neat binders (represented by Figure 15), only for the lowest stress level 0.1 kPa are apparent differences, but very low.

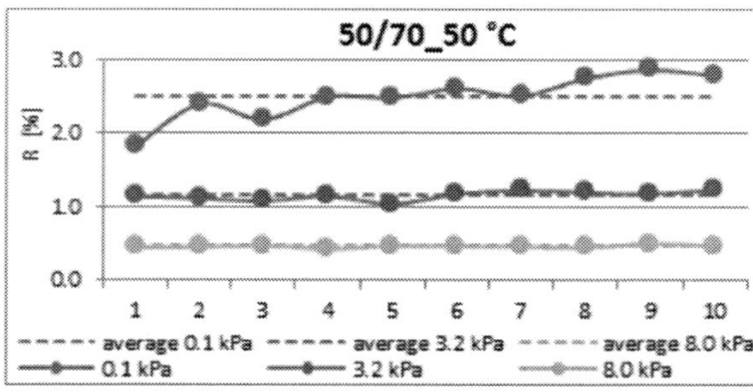

Figure 15. R values at each cycle.

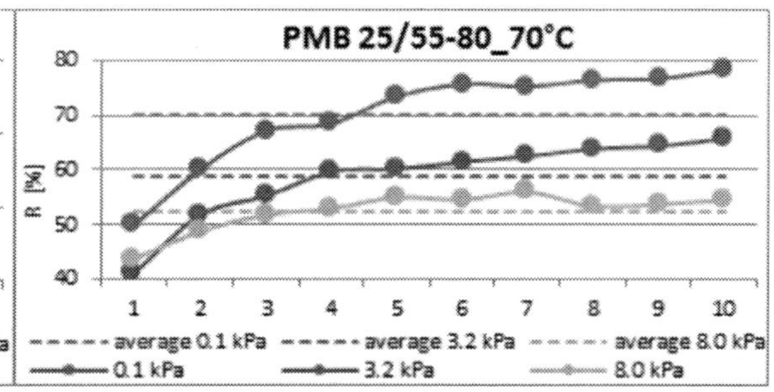

Figure 16. J_{nr} values at each cycle.

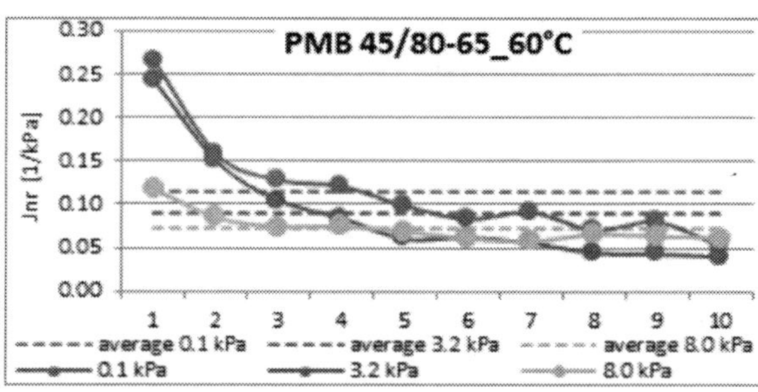

Figure 17. R values at each cycle.

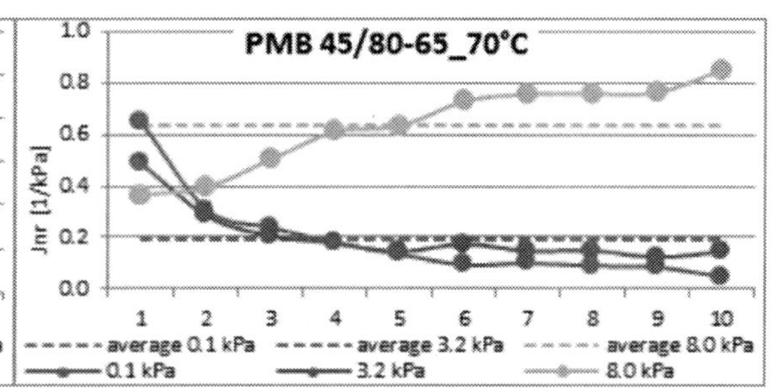

Figure 18. J_{nr} values at each cycle.

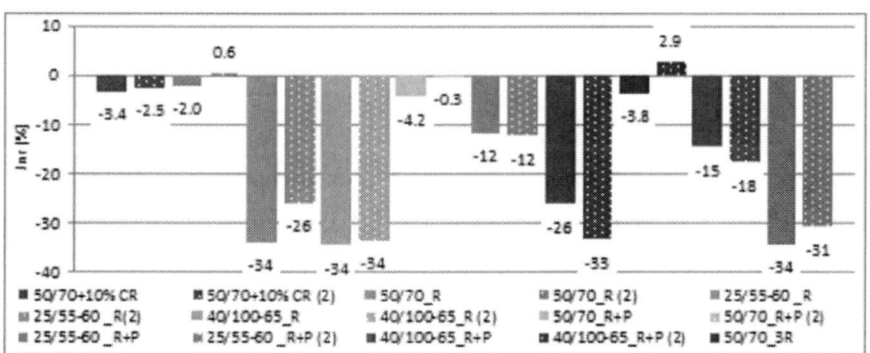

Figure 19. Percent change between two methods of calculating J_{nr} at stress level 3.2 kPa.

Figure 19 shows the last group of bituminous binders with different geometry of the test device. Variants with geometry PP08 are labeled (2).While the change in the calculation of R, considering the last 5 cycles, is less than 10 % for all variants of this group (excluding 50/70 + 10% CR variant, which has a 15 % difference), the change of J_{nr} value is around 30 % for most PMBs, which generally corresponds to the previous group with PMB binders. Unmodified binders have the smallest difference of J_{nr}, but this is again mainly related to high J_{nr} value thanks to which, the percentage difference is very small between the J_{nr} value calculated from all 10 cycles and the last 5 cycles.

4. Conclusions

Currently, the MSCR test is promoted as the most appropriate method for assessing the resistance of bituminous binders to permanent deformation. The advantage of MSCR test compared to Frequency Sweep test is applying a larger load and deformation which reflects better the real situation in the pavement. Parameter G */sin δ characterizes bituminous binder in its linear region (lower stress/strain) but permanent deformation occurs in nonlinear region and therefore it is necessary to describe the characteristics in nonlinear field with different parameter, e.g. J_{nr}.

This paper analyzes the results of MSCR test comparing the influence of temperature, stress level, type of geometry on measuring device and number of cycles for calculation individual values.

Performed measurements indicate that the method of averaging the response for 10 cycles at each stress level can lead to misleading values due to the changes during the cycles. Since the greatest differences are during the initial cycles, it would be appropriate to calculate non-recoverable creep compliance and percent recovery from the second half of the cycles at each stress level.

Furthermore research shows that it would be preferable (especially for PMB) to increase the number of cycles since the 10 cycles for most of PMBs is according to our findings not sufficient to achieve a stable secondary creep. An alternative option is to select a higher stress level, which leads to faster stabilization of test results.

In case the strain is very small (on the border of the measuring device range) it is not possible to calculate sufficiently the accuracy of determined values. This problem occurs especially at lower test temperatures or at low stress level (0.1 kPa) when the divergence between cycles is high. Additionally, correlation with asphalt mixtures (resistance to permanent deformation) is lower when the stress level is 0.1 kPa [15]. For these reasons, it is preferable to use a higher stress level, e.g. 8.0 kPa.

Selected testing geometry used for Dynamic Shear Rheometer primarily affects the J_{nr} value. Classical geometry PP25 is simpler for handling in DSR, as well as sample preparation (pouring) is easier. But in case of modification by crumb rubber with particle size of 1 mm or larger, it is necessary to use different geometry with a larger gap (PP08) or to analyze the possibility to use the standard geometry PP25 with a larger gap. The results show that the use of different geometry produce varying results and therefore it is always necessary to specify the type of used measured geometry.

Performing the test at a higher temperature leads to a higher strain, thereby reducing the risk of inaccurate measurements due to the extent of the device. Temperature 60 °C would be an appropriate temperature conducting the MSCR test assuming that the test resistance to permanent deformation for asphalt mixtures would be performed at the same temperature of 60 °C (currently, the test is carried out at 50 °C), in order to better compare the measured values.

Acknowledgements

This paper was prepared under the funding scheme of Competence centre program by the Technology Agency of the Czech Republic within the project Centre for effective and sustainable transport infrastructure (CESTI), project no. TE01020168.

References

[1] Neutag L, Zielke U, Beckedahl H J and Johannsen K 2009 Crumb rubber modified binders for flexible pavements *Environmentally friendly roads - ENVIROAD 2009*

[2] Chen C, Podolsky J H, Hernandez N, Hohmann A, Williams R Ch and Cochran E W 2016 Use of bioadvantaged materials for use in bituminous modification *Transport Research Arena*

[3] Valentin J, Mondschein P and Miláčková K 2013 Selected performance-based characteristics of viscosity improved bituminous binders *Proc. 9th Int. Conf. on the Bearing Capacity of Roads, Railways and Airfields* pp 593-605

[4] Bahia H U, Hanson D I, Zeng M, Zhai H, Khatri M A and Anderson M R 2001 *NCHRP Characterization of Modified Asphalt Binders in Superpave Mix Design: Report 459 Prepared for the National Cooperative Highway Research Program* (Transportation Research Board, National Research Council, Washington DC)

[5] Laukkanen O, Soenen H, Pellinen T, Heyrman S and Lemoine G 2014 Creep-recovery behavior of bituminous binders and its relation to asphalt mixture rutting *Materials and Structures* **12** pp 4039-53

[6] Stuart K, Mogawer W and Romero P 2000 Evaluation of the Superpave asphalt binder specification for high-temperature pavement performance *Journal of the Asphalt Paving Technologists* pp 148–176

[7] D'Angelo R, Dongre R and Reinke G 2006 Evaluation of Repeated Creep and Recovery Test Method as an Alternative to SHRP+ Requirements for Polymer Modified Asphalt Binders *Proc. Fifty-First Annual Conference of the Canadian Technical Asphalt Association (Charlottentown Canada)*

[8] D'Angelo J, Kluttz R, Dongré R, Stephens K and Zanzotto L 2007 Revision of the 361 Superpave High Temperature Binder Specification: The Multiple Stress Creep 362 Recovery Test *Journal of the Association of Asphalt Paving Technologists* **76** pp. 123-162.

[9] Domingos M D I and Faxina A L 2014 Creep-recovery behavior of modified asphalt binders with similar high-temperature performance grades *Transportation Research Board (TRB) 93rd Annual Meeting*

[10] Golalipour A 2011 *Modification of Multiple Stress Creep and Recovery Test Procedure and Usage in Specification* (University of Wisconsin, Madison)

[11] Standard by American Association of State and Highway Transportation Officials 2014 *Standard Specification for Performance-Graded Asphalt Binder Using Multiple Stress Creep Recovery (MSCR) Test (AASHTO M 332-14)*

[12] Úřad pro technickou normalizaci, metrologii a státní zkušebnictví (ÚNMZ) 2016 Zkušební metody pro asfalty a asfaltové výrobky; Asfalty a asfaltová pojiva - Zkouška MSCR /Multiple Stress Creep and Recovery Test/ *(ČSN EN 16659)*

[13] Wasage T L J, Stastna J and Zanzotto L 2011 Rheological Analysis of Multiple Stress Creep and Recovery (MSCR) test *International Journal of Pavement Engineering* **6**

[14] Dreessen A, Planche J P and Gandel V 2009 A new performance related test method for rutting prediction *Advanced Testing and Characterization of Bituminous Materials – Chap.93* (CRC Press)

[15] Dreessen S and Planche J-P 2009 Seeking for a relevant binder test method for rutting prediction *Environmentally friendly roads - ENVIROAD 2009*

IOP Conf. Series: Materials Science and Engineering **236** (2017) 012017 doi:10.1088/1757-899X/236/1/012017

Viscoelastic behaviour of cold recycled asphalt mixes

Zuzana Cizkova[1] and Jan Suda[1]

[1]CTU in Prague, Faculty of Civil Engineering, Department of Road Structures, Thákurova 7, 166 29, Prague 6, Czech Republic

E-mail: cizkova.zu@gmail.com

Abstract. Behaviour of cold recycled mixes depends strongly on both the bituminous binder content (bituminous emulsion or foamed bitumen) and the hydraulic binder content (usually cement). In the case of cold recycled mixes rich in bitumen and with low hydraulic binder content, behaviour is close to the viscoelastic behaviour of traditional hot mix asphalt. With decreasing bituminous binder content together with increasing hydraulic binder content, mixes are characteristic with brittle behaviour, typical for concrete pavements or hydraulically bound layers. The behaviour of cold recycled mixes with low content of both types of binders is similar to behaviour of unbound materials. This paper is dedicated to analysing of the viscoelastic behaviour of the cold recycled mixes. Therefore, the tested mixes contained higher amount of the bituminous binder (both foamed bitumen and bituminous emulsion). The best way to characterize any viscoelastic material in a wide range of temperatures and frequencies is through the master curves. This paper includes interesting findings concerning the dependency of both parts of the complex modulus (elastic and viscous) on the testing frequency (which simulates the speed of heavy traffic passing) and on the testing temperature (which simulates the changing climate conditions a real pavement is subjected to).

1. Introduction

As the stress-strain ratio of a linear elastic material is expressed by Young's modulus of elasticity, so the stress-strain ratio of a linear viscoelastic material is expressed by the complex modulus. The complex modulus is determined at steady harmonic varying oscillation with regard to the time displacement of the stress and strain. The strain is phase-delayed due to the viscoelastic behaviour of the asphalt mixtures. The designation "complex" comes from the fact that complex numbers are used for this characteristic in the analysis of results. The basic formula for calculating of the complex modulus is given by the equation (1):

$$E^*(i\omega) = \frac{\sigma^*}{\varepsilon^*} = \frac{\sigma_0 \cdot \sin(\omega t)}{\varepsilon_0 \cdot \sin(\omega t - \phi)} = \frac{\sigma_0 \cdot e^{i\omega t}}{\varepsilon_0 \cdot e^{i(\omega t - \phi)}} = \frac{\sigma_0}{\varepsilon_0} \cdot e^{i\phi} \quad (1)$$

where

σ_0= maximum stress (stress amplitude), (MPa),
ε_0 = maximum axial strain (amplitude of the cyclic component), (MPa),
ω = angular speed (rad/sec), $\omega = 2\pi f$, where f is the frequency of loading
i = the imaginary component of the complex modulus, $i^2 = -1$,
ϕ = phase angle (rad).

The relationship of the sinusoidal load and the strain response, including the phase shift and the stress and strain amplitudes is shown in Figure 1. The phase angle indicates the time delay between the peaks of the sinusoidal load and the peaks of the response of a linear viscoelastic material. In other words it determines the shift between stress application and the induced strain. In viscoelastic materials the phase angle reaches values between 0° to 90°, in a completely viscous material it equals to 90°. In completely flexible materials, the phase angle is 0 degrees, and the complex modulus of a resilient material is reduced only to the Young's modulus.

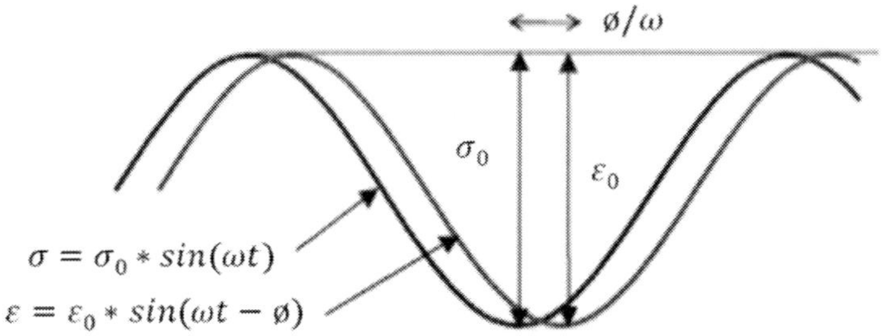

Figure 1. Sinusoidal stress and strain waveform in time.

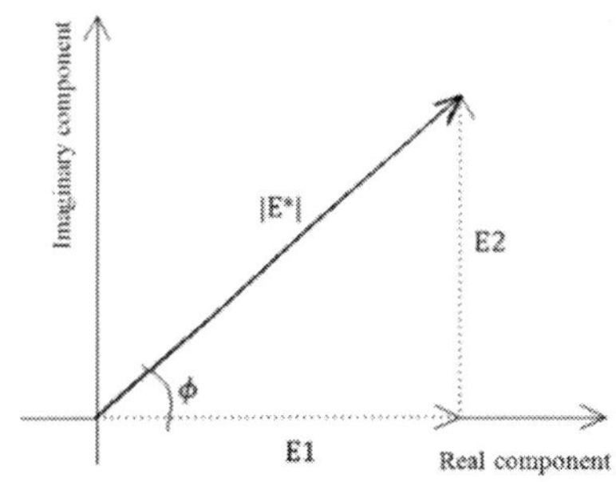

Figure 2. Graphic representation of the complex modulus components including the phase angle ø.

The term dynamic modulus is then used for the absolute value of the complex modulus calculated as the proportion of the maximum stress and the reversible axial deformation of the material, which is subjected to the repeated sinusoidal axial load pressure in the range at least from 0 to 25 kN at a specified temperature and load frequency. [1]

According to [2], at any combination of temperature and time, the linear viscoelastic behaviour of any material can be characterized by two properties. These are the total resistance to deformation and the relative distribution of this resistance between the elastic and viscous component. This pair of properties can be expressed in two ways according to [3]. The first method consists in the separation of the complex modulus into its real (elastic) part E1 or E', also called storage modulus, and the imaginary (viscous) part E2 or E", known as the loss modulus. The relationship of these two components is expressed by equations from (2) to (4):

$$E^* = E_1 + iE_2 = \left|E^*\right| \cdot (\cos(\phi) + i \cdot \sin(\phi)) \tag{2}$$

$$E_1 = \left|E^*\right| \cdot \cos(\phi) \tag{3}$$

$$E_2 = \left|E^*\right| \cdot \sin(\phi) \tag{4}$$

The second and more common way of expressing the total resistance to deformation and the distribution of this resistance between elastic and viscous component is described in (5) and (6). It is also evident from Figure 2.

$$\left|E^*\right| = \sqrt{E_1^2 + E_2^2} \tag{5}$$

$$\phi = arctg\left(\frac{E_2}{E_1}\right) \tag{6}$$

The elasticity of the material corresponds to its ability to store the deformation energy, while tg(ø) is the ratio of energy lost (converted into heat) and energy stored under cyclic deformation. The relative distribution of the elastic and viscous component is then the function of the material composition, the temperature and the loading time.

During the stiffness test, the force required for deforming the specimen and the phase delay between the signal of force and the signal of deviation are measured as the function of time. Based on these values, the value of complex modulus is set using the equations (7), (8), (9), (10), (11) and (12). These equations are based on the modified first-order approximation of the exact solution of the deformation due to pure bending.

$$E_1 = \gamma\left(\frac{F}{Z}\cos(\phi) + \mu\omega^2\right) \tag{7}$$

$$E_2 = \gamma\left(\frac{F}{Z}\sin(\phi)\right) \tag{8}$$

where

F = force applied to the beam (N),
Z = beam deformation as measured (m),
$ø$ = phase angle (rad),
ω = angular speed (rad/sec),
γ = shape factor (1/m),
μ = weight factor, for classical test conditions the weight factor is negligible.

The γ and μ factors are determined as follows:

$$\gamma = \frac{L^2 A}{bh^3}\left(\frac{3}{4} - \frac{A^2}{L^2}\right) \tag{9}$$

$$\mu = R\{x\}\left(\frac{M}{\pi^4} + \frac{m_1}{R(A)}\right) \tag{10}$$

$$R\{x\} = \frac{12L}{A}\left(\frac{1}{3\frac{x}{L} - 3\frac{x^2}{L^2} - \frac{A^2}{L^2}}\right) \tag{11}$$

$$A = \frac{L - l}{2} \tag{12}$$

where

L = effective length of the beam (m),
h = height of the beam (m),
b = width of the beam (m),
A = distance between the external clamp and the next internal clamp (m),
l = distance of internal clamps (m),
M = weight of the beam (kg),
m_1 = weight of the movable part (kg) in location A (m),
m_2 = weight of the movable part (kg) in location y (m),
x = location of the probe for the beam deformation measurement (m). [3]

2. Measurements of complex modulus by the 4PB-PR method

Complex modulus was measured by the four-point bending test method (4PB-PR) according to [3]. Test specimen in the shape of beams with the dimensions of 50 x 50 x 400 mm were produced by cutting slabs (305 x 50 x 400 mm), which were produced by using the segmental compactor in accordance with [4]. The test specimens were cured for at least 28 days at laboratory temperature and relative humidity of 40-70%. The test was conducted in the controlled strain mode with the desired deformation of 50 microstrain. The measurements were performed at the temperatures of 0 °C, 10 °C, 20 °C and 30 °C. Lower or higher test temperatures are not necessary, because the use of the tested type of cold-recycled mixes is expected to the base layers, where the temperature variations usually don't exceed that spectrum. For each tested temperature the measurements were performed at 9 chosen test frequencies (50, 30, 20, 10, 8, 5, 2, 1 and 0.5 Hz).

Composition of the tested mixes is presented in Table 1. For production of the tested mixes two kinds of bituminous binder were used – firstly the cationic slow-breaking bituminous emulsion C60B7, commonly used in the Czech Republic and secondly the foamed bitumen, which was produced by using the laboratory equipment Wirtgen WLB10S from common road bitumen 70/100. When preparing foamed bitumen, 3.8% of water were added into the hot asphalt binder (quantity determined in accordance with the procedure recommended by Wirtgen Manual [5]).

Table 1. Composition of the tested mixtures.

	Mix A	**Mix B**	**Mix C**	**Mix D**
RAP 0/22 Středokluky	91.0 %	90.5 %	94.0 %	93.5 %
Water	2.5 %	2.0 %	2.5 %	2.0 %
Bituminous emulsion	**3.5 %**	-	**3.5 %**	-
Foamed bitumen	-	**4.5 %**	-	**4.5 %**
Cement	**3.0 %**	**3.0 %**	-	-
	Mix E	**Mix F**	**Mix G**	**Mix H**
RAP 0/22 Středokluky	93.0 %	89.0 %	91.5 %	-
RAP 0/11 Středokluky	-	-	-	**92.0 %**
Water	2.5 %	3.5 %	3.5 %	3.5 %
Bituminous emulsion	-	-	**5.0 %**	-
Foamed bitumen	**4.5 %**	**4.5 %**	-	**4.5 %**
Pulverized concrete	-	**3.0 %**	-	-

Each measurement was always carried out on five specimens. The higher number of specimens was necessary because of the difficulties arising from the application of the test designed for hot mix asphalt to the cold recycled mixes. Nevertheless, the cold recycled mixes are characterized by high heterogeneity, bigger grain size, higher voids content and lower cohesion. The difficulties with the application of the 4PB-PR test on the cold recycled mixes were worsened also by the fact that it is a test highly prone to imperfections. As a result, even the production of the test specimens was already demanding, because the edges were often breaking off (see Figures 3 and 4).

The above mentioned reasons resulted in destruction of several specimens during the test itself, although the measurements of the complex modulus should be a strictly non-destructive test and the maximum strain occurring during the test is therefore limited to a much lower value than the one that should lead to the specimen breakdown. The specimen damage occurred in all cases at the test temperature of 30 °C. Possible difficulties at the highest temperature were assumed, based on the previous experience with the application of this test on cold recycled mixes. Therefore, the measurement at 30 °C was carried out as the last one after all the other test temperatures, and thus at least the number of test specimens for other temperatures was not reduced. Another consequence was the need not to count the value measured on some specimens to the average values, which were used for subsequent calculations. If the value measured on a specimen differed too much from the average, it was not further used. Therefore, the resulting values were always the average of 3-5 measurements.

Figure 3. Test specimens of mix E after measurements of the complex modulus by the 4PB-PR method.

Figure 4. Test specimens of mix F after measurements of the complex modulus by the 4PB-PR method.

3. Discussion of results

The main advantage of the complex or dynamic modulus measurements consists in the fact that based on the values of the dynamic modulus set at various temperatures and frequencies it is possible to construct the so-called master curve. Master curve is currently the most comprehensive way to analyse the deformation behaviour of a bituminous mix under different conditions. Master curves for all tested mixtures were created by horizontal shifting of the experimental data in the program IRIS Rheo Hub. As the reference temperature, to which all the results of conducted measurements are related, was chosen 20 °C.

Figure 5 depicts the master curves for both components of the complex modulus of mixtures A-D and Figure 6 shows the phase angle master curves of these mixtures. For better comparison of the behaviour of the tested mixtures in a wide range of temperature and frequency with the behaviour typical for hot mix asphalt, these graphs also include values obtained on a conventional asphalt concrete ACO 11+.

Figures 5 and 6 show clearly that the influence of contained bituminous or hydraulic binder in a cold recycled mix has lower influence on the behavior of the two components of the complex modulus, than the fact that it is a cold recycled mix. In other words, all four cold recycled mixtures have relatively similar shape of master curves but they are all significantly different from the shape of the complex modulus master curves of hot mix asphalt. All cold recycled mixes have significantly lower viscous component of the complex modulus and thus also the phase angle. Their behaviour is therefore somewhere between the behaviour of elastic materials and the viscoelastic behaviour of conventional hot mix asphalt. The effect of bituminous binder type, contained in the cold recycled mixes, is almost negligible. The effect of contained hydraulic binder is, however, apparent from the graph. Mixtures A and B with 3 % of cement reach much higher values of the elastic component in the low frequencies and high temperatures. In the entire range of test temperatures and frequencies, mixtures A and B show lower values of the phase angle. Included cement causes a shift of the behaviour closer to the behaviour of an elastic material.

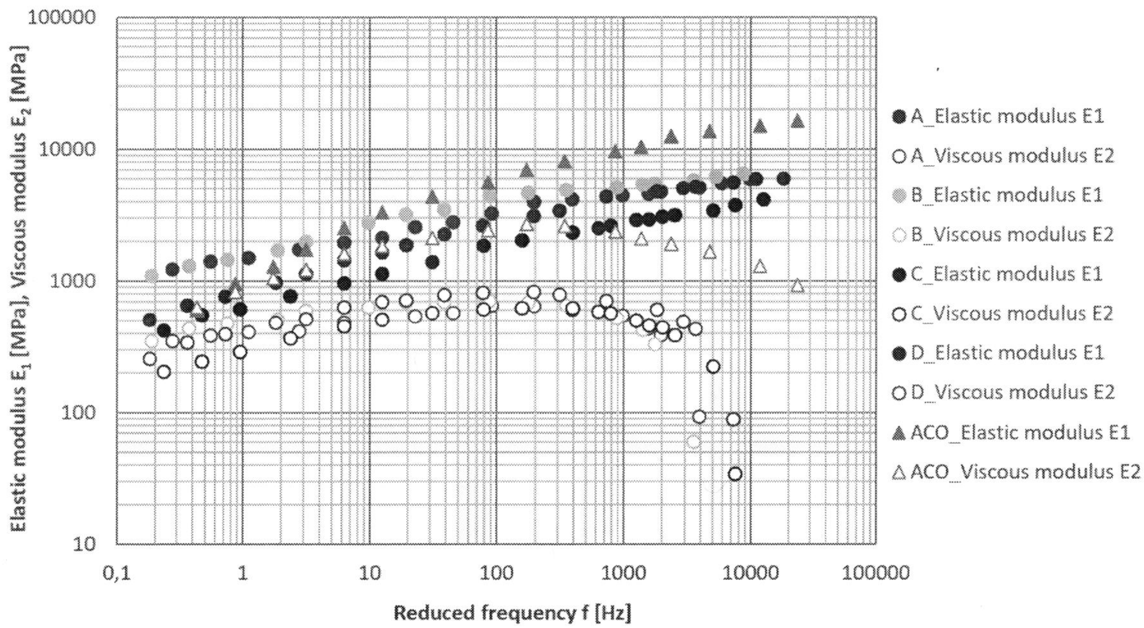

Figure 5. Master curves for the elastic and viscous component of the complex modulus of tested mixes A, B, C a D.

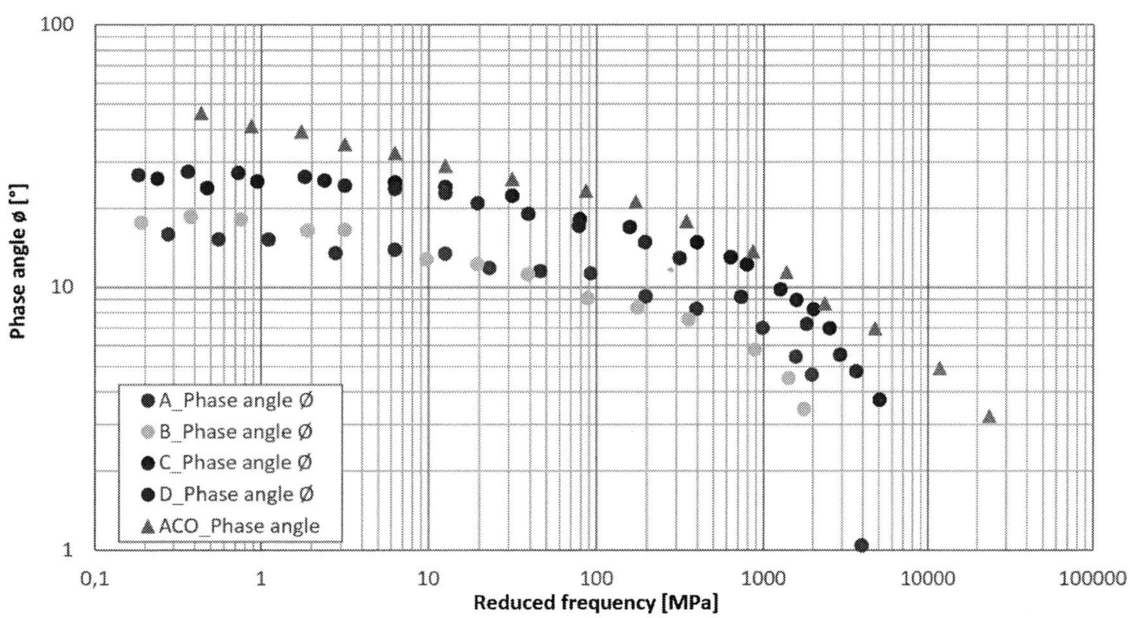

Figure 6. Master curves for the phase angle of tested mixes A, B, C and D.

Overall, it is possible to describe the behaviour of cold recycled mixes under different conditions and the differences compared to the reference hot asphalt mix as follows. In the high test temperatures and low frequencies, both components of complex modulus are low for cold recycled mixes, which do not contain cement (mixture C and D), and the reference mix ACO 11+. Although in the case of mixes A and B the included cement results in higher values of the elastic component in the high test temperatures and low frequencies, they are still much lower than the values in the rest of the test spectrum. Significant variation between the complex modulus components of different mixes under these conditions is interesting as well – in the mix ACO 11+ elastic modulus is almost equal to the viscous modulus and the phase angle therefore reaches around 45 °. In contrast, the phase angle of cold recycled mixes achieves approximately half of the ACO values, resulting in less viscous behaviour of the mix. This is a positive result. The mix will be less susceptible to permanent deformation under slow transport loading and exposure to high summer temperatures.

With increasing frequency and decreasing temperature, both complex modulus components of all five mixes are rising, however, the growth rate differs significantly. ACO 11+ shows more rapid increase in the elastic component and rapid reduction in the phase angle. On the other hand, the E_1 and E_2 master curves of cold recycled mixes are almost parallel and the phase angle of these mixtures remains almost unchanged at high temperatures and low frequencies. At medium temperatures and frequencies, the behavior of all five mixes is similar – both components of complex modulus are rising, phase angle decreases. Although the mix ACO 11+ generally achieves higher values of complex modulus at medium temperatures and frequencies, both of its components are in similar proportion as for the mix C and D. Hence, the phase angle of these three mixtures is also similar. Therefore, at the medium temperatures and frequencies, the behaviour of cold recycled mixes without added cement is the most similar to the behaviour of classic hot asphalt mixtures, which can be considered as an expected phenomenon. At the low temperatures and high frequencies there is a significant decrease in the loss modulus and the phase angle. This indicates that mixes lose their viscous behaviour and the material becomes completely elastic. The phase angle of the cold recycled mixes decreases faster than the phase angle of the mix ACO 11+. To conclude, in the low temperatures and high frequencies the cold recycled mixes show inferior qualities than the mix ACO 11+, therefore it can be assumed that they will be more susceptible to cracking.

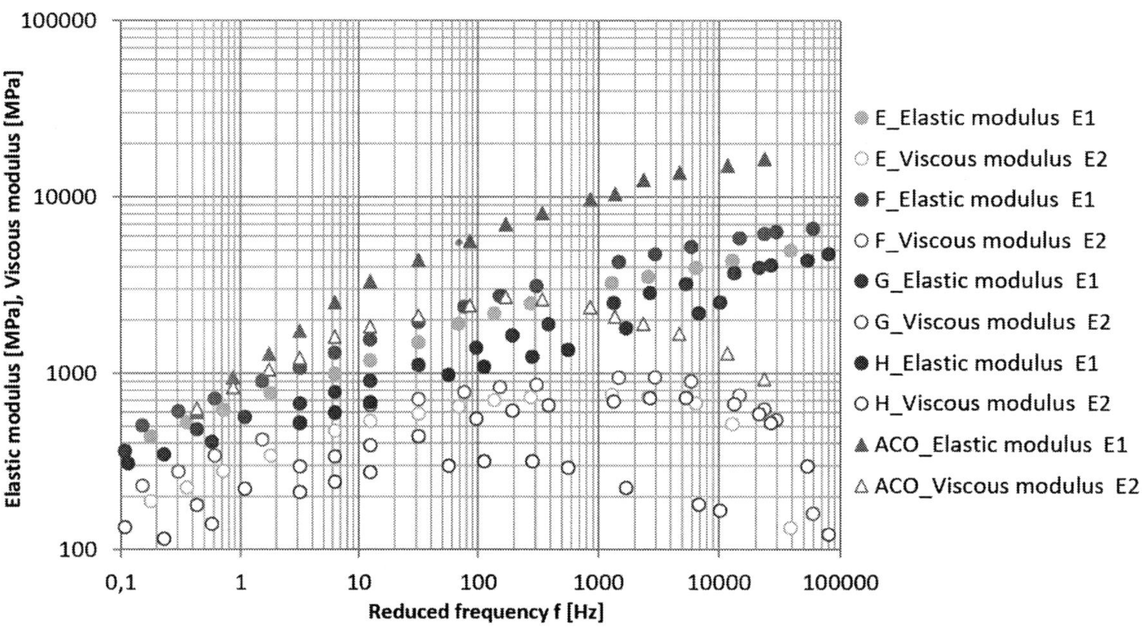

Figure 7. Master curves for the elastic and viscous component of the complex modulus of tested mixes E, F, G a H.

Figures 7 and 8 show master curves for mixes E-H, again compared to the reference mixture ACO 11+. The graphs clearly show that the addition of 3 % of pulverized concrete has no significant effect on the development of the measured characteristics as well. Further it was confirmed that the shapes of master curves are very similar, whether the mix comprises bituminous emulsion or foamed bitumen, and that the different binder dosage also doesn't bring any essential variance. From the group of four tested cold recycled mixes, the most different was the mix H, which contained RAP 0/11. The most distinct behaviour has again the reference mix ACO 11+.

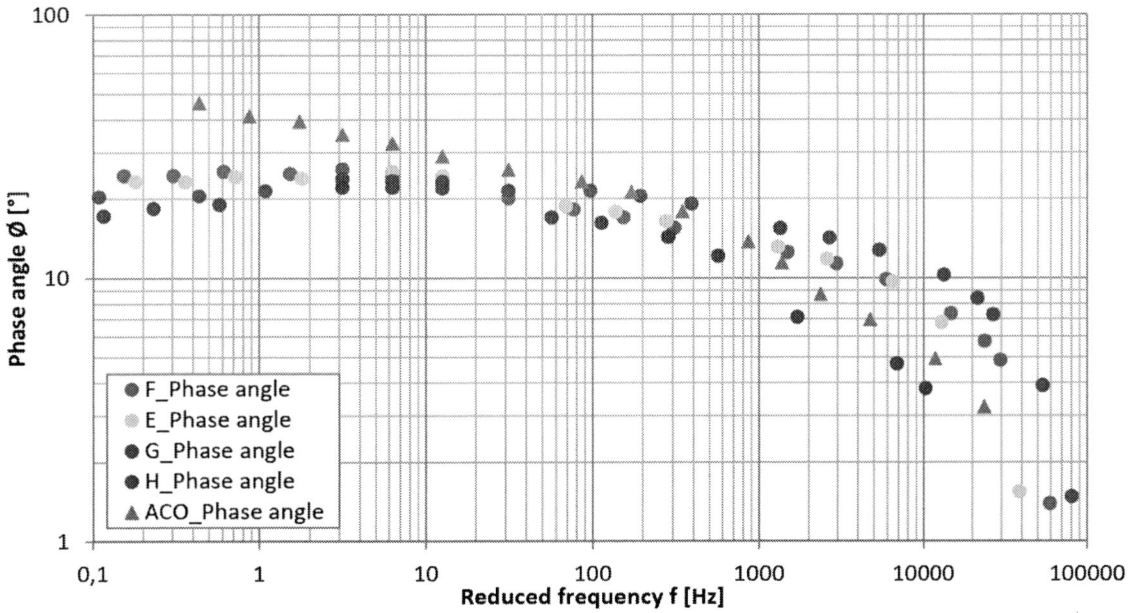

Figure 8. Master curves for the phase angle of tested mixes E, F, G and H.

In the entire range of test temperatures and frequencies, both complex modulus components of mix F achieve higher values compared to mix E. Included pulverized concrete has according to the results similar effect as added cement, i.e. it increases the mixture stiffness and it increases the elastic component of the complex modulus. From the perspective of the entire spectrum of the viscoelastic behaviour of cold recycled mixes, mix F is closer to the purely elastic behaviour typical e.g. for concrete. On the other hand mix E behaves more viscous (achieves higher values of the phase angle), and therefore its behaviour is closer the behaviour of hot asphalt mixtures.

The graphs 7 and 8 show also that in the entire frequency spectrum, both complex modulus components of mix G are lower than for mixes E and F, and mix H achieves substantially lowest values of both complex modulus components. Besides that, the progress of elastic and viscous modulus of all tested cold recycled mixes is quite similar. As in mixtures A-D, with the increasing frequency and the decreasing temperature the real component of the master curve increases and its derivative gradually decreases till it almost reaches the zero value. This phenomenon correlates well with the transition of the material properties into the elastic area and reaching the equilibrium modulus. It was also confirmed that the cold recycled mixes and the reference mix ACO 11+ achieve similar phase angle values at middle frequencies and temperatures. The decline in the phase angle values in mixes E, F and G occur even at higher frequencies than in the mix ACO 11+, suggesting lower susceptibility of cold recycled mixes to cracking at low temperatures. The difference, however, is not significant and it does not correspond with the trend identified in mixes A-D.

Mix H shows the worst qualities also in terms of the phase angle at high frequencies and low temperatures – as the only one of the tested mixes, mix H shows inferior qualities than the reference mixture ACO 11+ (a decrease in the phase angle occurs already at much lower frequency values than in other cold recycled mixes). The behaviour of mix H is in this perspective closer to the behaviour of the reference mix ACO 11+, which is more temperature-dependent and frequency-dependent.

4. Conclusion

The main findings resulting from the investigation of the complex dynamic modulus and the subsequent master curves construction for the cold recycled mixes can be summarized in the following points:

- Cold recycled mixes are thermo-mechanically sensitive. However, the dependence on the test temperature and frequency is lower than in conventional hot asphalt mixes. This fact is evident from the flatter phase angle curve and the flatter elastic component curve of cold recycled mixes.
- At the lowest temperatures and highest frequencies, this material tends to behave almost elastically.
- The behaviour of cold recycled mixes is closest to hot asphalt mixtures behaviour at the medium temperatures and frequencies (similar value of the phase angle).
- At the highest temperatures and lowest frequencies, cold recycled mixes achieve roughly half lower phase angle values than the conventional hot asphalt mixes. The behaviour of cold recycled mixes is therefore closer to the behaviour of an elastic material. As a result the cold recycled mixes should be less susceptible to permanent deformation.
- Mixtures with 3 % of cement reach significantly higher values of the elastic complex modulus component at low frequencies and high temperatures. In the entire range of test temperatures and frequencies, they show lower phase angle than the mixes with no cement. Added cement causes a shift of the behaviour closer to the behaviour of an elastic material.

Acknowledgement

This paper was prepared under the funding of project TA04031256, supported by the Technology Agency of the Czech Republic, program Alfa.

References

[1] Valentin J and Suda J 2015 Complex dynamic modulus and master curves of cold recycling mixes *CoRePaSol Project - Deliverable D2.1*

[2] Kim Y R 2009 *Modeling of asphalt concrete* (American Society of Civil Engineers)

[3] Úřad pro normalizaci, metrologii a státní zlušebnictví - ÚNMZ 2012 Asfaltové směsi – Zkušební metody pro asfaltové směsi za horka – Část 26: Tuhost /Bituminous Mixtures - Test Methods for Hot Mix Asphalt - Part 26: Stiffness/ (*ČSN EN 12697-26)*

[4] Úřad pro normalizaci, metrologii a státní zlušebnictví - ÚNMZ 2007 Zkušební metody pro asfaltové směsi za horka – Část 33: Příprava zkušebních těles zhutňovačem desek /Bituminous mixtures - Test methods for hot mix asphalt - Part 33: Specimen prepared by roller compactor/ (*ČSN EN 12697-33+A1)*

[5] Wirtgen GmbH 2012 *Wirtgen Cold Recycling Technology* (Manual, first edition, Windhagen, Germany)

Comparison of influence of ageing on low-temperature characteristics of asphalt mixtures

Pavla Vacková[1], Jan Valentin[1] and Lucie Benešová[1]

[1]Czech technical university in Prague, Thakurova 1, Praha 6, 160 00, Czech Republic

E-mail: pavla.vackova.1@fsv.cvut.cz

Abstract. Ability of relaxation of asphalt mixtures and thus its resilience to climate change and traffic load is decreasing by influence of aging – in this case aging of bituminous binder. Binder exposed to climate and UV ages and becomes more fragile and susceptible to damage. The results of the research presented in this paper are aimed to finding a correlation between low-temperature properties of referential and aged asphalt mixture specimens and characteristics (not low-temperature) of bituminous binders. In this research there were used conventional road binders, commonly used modified binders and binders additionally modified in the laboratory. The low-temperature characteristics were determined by strength flexural test, commonly used in the Czech Republic for High Modulus Asphalt Mixtures (TP 151), and semi-cylindrical bending test (EN 12697-44). Both of the tests were extended by specimens exposed to artificial long-term aging (EN 12697-52) – storing at 85 ° C for 5 days. The results were compared with characteristics of binders for finding a suitable correlation between characteristics of binders and asphalt mixtures.

1. Introduction

Asphalt mixture is a complex material used in asphalt pavement layers. The pavement is exposed to traffic load, weather conditions, UV radiation, climate changes etc. All these factors can cause ageing of asphalt mixtures which can lead to micro cracks and surface damages [1], which is significant source of premature degradation especially at low temperatures.

The asphalt mixtures are composed of aggregate, filler, bituminous binder and air voids. Also some additive can be provided for artificially improvement of asphalt mixture's properties. Asphalt mixture behaviour depends not only on properties of particular components, but also on air voids content, therefore compaction rate on the pavement.

Ageing in asphalt mixtures occurs mainly because of ageing (oxidation) of bituminous binder. The influence of ageing of aggregate was not proven yet, all the researches are focused solely on binder ageing. The binder oxidases due to its exposing to air especially at high temperature. The binder becomes harder (its penetration lowers), but more brittle – it can crack more easily, especially at low temperatures, and low temperature cracking can occur. The influence of binder ageing has been found to vary due to crude oil source, used additives, characteristics of asphalt mixture and characteristics of climate [2].

Resistance to fatigue is one the most significant characteristics of asphalt mixture. Unfortunately the fatigue test is very time and cost consuming. Further very important characteristics in countries like the Czech Republic is the material behaviour at low temperature range. Low temperature causes shrinkage of asphalt layers and therefore the tensile stresses are growing. The tensile stress increases with decrease of the temperature. When the stress reaches the asphalt mixtures tensile strength, the crack arise. The low temperature cracks occurs from the bottom of the layer and expands to the top. The climate changes and traffic load repetitions causes the propagation to the pavement surface. When the cracks occur on the surface, the water can infiltrate the pavement and the bearing capacity of pavement is significantly lowered [3].

Low temperature characteristics of asphalt mixtures can be determined by a lot of different laboratory tests. The tests differ in shape of test specimens, testing temperatures, testing devices etc. The method are for example [3] - [8]:

- Single-Edge Notched Beam (SENB) under Three or Four-Point Bending;
- Unnotched Beam Three-Point Bending;
- Disc-Shaped Compact Tension test (DCT);
- Indirect Tensile Test (IDT);
- Semi-Circular Bending Test (SCB);
- etc.

For this research two testing methods were chosen – Semi-Circular Bending Test (SCB) and Three-Point Bending of unnotched beams (3PB). The SCB test is worldwide popular, so many researchers use it in their work and therefore there are many comparable data available. The test is easy for required equipment and the specimens can be also easily prepared from core sample. For appropriate results it is very important to pay an increase attention to specimen preparation. The preparation is easy on its self, but it has to be done with precise. Compacted Marshall's specimens or core samples are cut to height of 50 mm, then halved and notched in the middle of bottom side. All the cutting has to be done exactly parallelly and centrally. The improper cutting can negatively influence the test results.

The three point bending test of unnotched specimens was chosen because of its common use in the Czech Republic for High Modulus Asphalt Mixtures (HMAC). The test is also easy for required equipment. The beam specimens are cut from the slabs used for rutting test. As in previous test the precise cutting work is required for adequate data.

2. Test methods of asphalt mixtures

2.1. Resistance to crack propagation

The resistance to crack propagation (semi-cylindrical bending test) was determined in compliance with the requirements of EN 12697-44 [14]. The essence of the test is a three-point bending using semi-cylindrical specimens with a defined notch in the middle of bottom specimen face. The test procedure was performed at a test temperature of 0 °C applying a standardized loading rate of 5 mm/min.

The test specimens were manufactured according to EN 12697-30 (Specimen preparation by impact compactor) using 2x50 blows of the Marshall hammer (50 blows from each side of the specimen) and then cut on laboratory saw.

The test specimens were divided into two sets. First set ("virgin") was left at laboratory conditions and the second set ("aged") was exposed to artificially ageing procedure specified in prEN 12697-52 [13]. The simulation of ageing was realized by conditioning the specimens in laboratory oven at temperature of 85 °C for 5 days. After the ageing procedure the Marshall's specimens were cut into the test specimens.

The significant resulting parameter is a fracture toughness [$N*mm^{-3/2}$]. The fracture toughness is defined as force multiplied by correction factor comprising the dimensions of the specimen. For fracture toughness and maximum stress the ageing indexes were calculated as a ratio of resulting value of aged specimen to virgin (unaged) specimen's value. The ageing index shows the thermo-oxidation degradation potential of the bituminous binder in an asphalt mixture.

2.2. Three-point bending (Tensile bending strength)

Tensile bending strength (simple destructive 3-point beam test) was tested in observance to the method given by the Technical specifications of the Ministry of Transport of the Czech Republic - TP 151 [9]. It is necessary to point out, that this is a specific test used mainly in the Czech Republic and required presently only for HMAC mixtures. The test was used in this research for its long history and sufficient results with HMAC and its practicability and repeatability which can be provided by a regular road laboratory.

Tensile bending strength is tested on beams, which can be cut from slabs used for wheel tracking test. The slab can be used for cutting the test specimens only in the case, when it is not too deformed from wheel tracking test. If the track would be too deep or the slab would be anyhow deformed from the test, it is necessary to prepare new slab specially for this bending beam test. The part of the slab under the wheel track is thrown away and only unloaded part are used as test specimens.

TP 151 [9] specifies testing temperature in a range from -20 °C to 0 °C. For sake of this research, temperature of 0 °C was chosen. TP 151 [9] specifies two testing velocities of loading (loading rates): 50 mm/min and 1.25 mm/min. The slower speed was chosen to allow certain comparability with the results of the SCB test.

The increase of the force and deformation during the test was recorded and data points were used for drawing a stress-strain diagram. The diagram was used additionally to determine the fracture energy, which can be defined as energy expended for beam to break. Fracture energy is calculated as the area under the curve of the force-deformation diagram. The data recorder used in the research records 50 data points per second.

The significant resulting parameter of three point beam bending test is flexural strength and flexural modulus of elasticity, both in [MPa]. Again the ageing indexes were calculated.

3. Test methods of bituminous binders

The results of test performed on asphalt mixtures were compared with results of bituminous binder tests for possible identification of any correlation between these parameters. On the bituminous binders the following test set was applied:

- Ring and ball softening point test (EN 1427);
- Needle penetration test at 25 °C (EN 1426);
- Complex shear modulus G* and phase angle δ determination at 40 °C for selected frequencies within the 0.1-10 Hz range and under shear stress of 2000 Pa (controlled stress mode) using dynamic shear rheometer HAAKE MARS (ČSN EN 14770);
- Multiple stress creep recovery test (MSCR) according to US methodology, AASHTO 70-09 and in compliance with prEN 16659 at 60 °C.

The bituminous binders were additionally artificially aged by 2 different methods in the laboratory. These methods were chosen because of the available laboratory equipment. The ageing methods were:

- TFOT (Thin-Film Oven Test) – where the binder is exposed to temperature of 163 °C for 5 hours (EN 12607-2);
- PAV (Pressurized Ageing Vessel) – where the binder is under pressure exposed to the temperature of 100 °C for 20 hours (EN 14769). Ageing was performed on a bituminous binder sample which had been subjected to short-term ageing first.

4. Test variants of asphalt mixtures

For this research 2 different asphalt mixtures were chosen. The first one was a superior stone mastic asphalt mixture with maximum particle size of 11 mm (SMA 11S) with 5.8 M% of bituminous binders. And the second one was high-stiffness modulus asphalt mixture with maximum particle size of 22 mm (HMAC 22S) with 4.8 M% of bituminous binders. Both mixtures were optimized within research studies which are presented in [11, 12].

The variants cover common unmodified binder with gradation 50/70, polymer modified binder PMB 40/80-85 and four binders modified in laboratory by "VP" additive. The "VP" additive is alfa-polyolefin based synthetic wax. The additive should improve adhesion of bitumen to aggregate and increase the strength characteristics. The additive was dosed to the bitumen in 1 M%, 5 M% and 6 M%. The modified binder was subsequently used for mixture manufacture. In one of the variants "VP" was combined with 3 M% of SBS modifier. Detailed data about bitumen characteristics and advanced functional tests can be found e.g. in [10]. The test variants with used binders with air void contents of mixtures (determined in compliance with EN 12697-8) are summarized in Table 1.

Table 1. Summary of test variants of SMA 11S and HMAC 22S mixtures and their air voids content.

Mix		Binder	Binder content (%)	Air voids content (%)
SMA	**50/70**	50/70	5.8%	3.4%
	5V	50/70 + 5 % VP		3.7%
	6V	50/70 + 6 % VP		4.1%
	SBS+VP	50/70 + 3 % SBS a 1% VP		3.0%
HMAC	**5V**	50/70 + 5 % VP	4.8%	2.6%
	6V	50/70 + 6 % VP		3.3%
	PMB	PMB 40/80-85		2.6%
	PMB+VP	PMB 40/80-85 + 1% VP		4.2%

5. Results of the test methods

Due to the exposure of the asphalt specimens to high temperatures and air flow (simulation by artificial laboratory ageing at temperature of 85 °C for 5 days), the bituminous binder in asphalt mixture oxidized and changed the properties of the mixture. The binder penetration lower and therefore the strength properties of asphalt mixture grows to a certain level. If the temperature of mixture decreases due to weather conditions the binder becomes more brittle. At low temperatures and under repeating traffic load the freeze cracks can occur and it causes the degradation of the pavement. This risk of freeze crack occur increases with ageing of the binder.

5.1. Resistance to crack propagation

As written above the resistance to crack propagation was determined according to EN 12697-44 [14] on notched semi-cylindrical specimens. In the results of this test there is visible a little unusual phenomenon. The expected trend of low temperature test parameter measured at 0 °C is that aged specimens will reach higher value. The aged binder increases to certain point the strength characteristics of the mixture. Nevertheless the fracture toughness of aged specimen of SMA mixtures in most cases decreases in comparison to virgin specimen. This phenomenon can be followed for surface layer mixtures through many recurrences taken in CTU Road laboratory during last 2 years. The fracture toughness of surface layers mixtures reaches usually lower values for aged specimens. These unusual results are described for example in [13].

The HMAC mixtures evince usual predictable trend, where fracture toughness is increased for aged specimens.

For all mixtures the ageing index was calculated. The closer the ageing index (labelled as "IA" in figures) to 1.0 is, the less sensitive the mixture is to bitumen ageing effects. The most stable results were achieved by SMA mixture with 50/70 and 6% of VP and HMAC mixture with 5 % of VP. Like a bad variant seems to be combination of 1 M% of VP with polymer modified binder. That causes high increase in fracture toughness and a high ageing index.

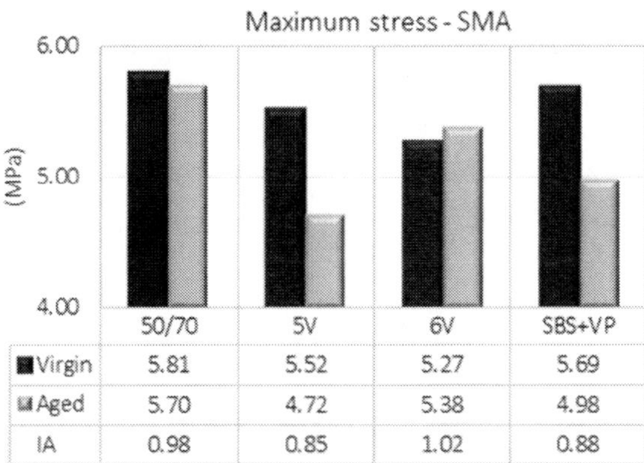

	50/70	5V	6V	SBS+VP
Virgin	5.81	5.52	5.27	5.69
Aged	5.70	4.72	5.38	4.98
IA	0.98	0.85	1.02	0.88

Figure 1. Maximum stress – SMA mixtures.

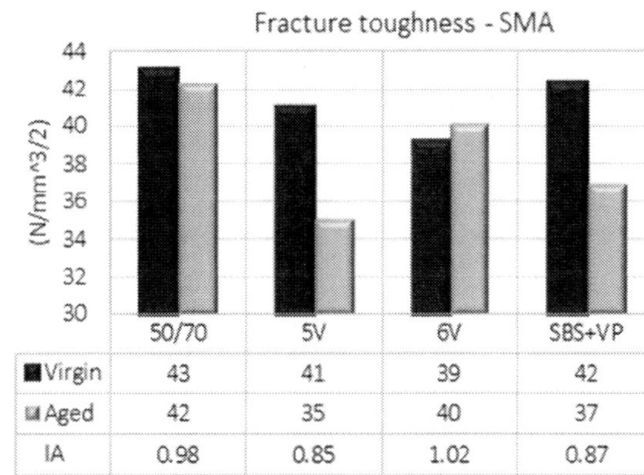

	50/70	5V	6V	SBS+VP
Virgin	43	41	39	42
Aged	42	35	40	37
IA	0.98	0.85	1.02	0.87

Figure 2. Fracture toughness – SMA mixtures.

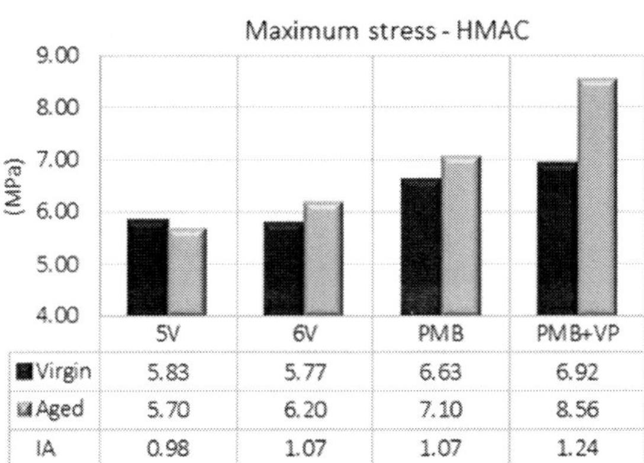

	5V	6V	PMB	PMB+VP
Virgin	5.83	5.77	6.63	6.92
Aged	5.70	6.20	7.10	8.56
IA	0.98	1.07	1.07	1.24

Figure 3. Maximum stress – HMAC mixtures.

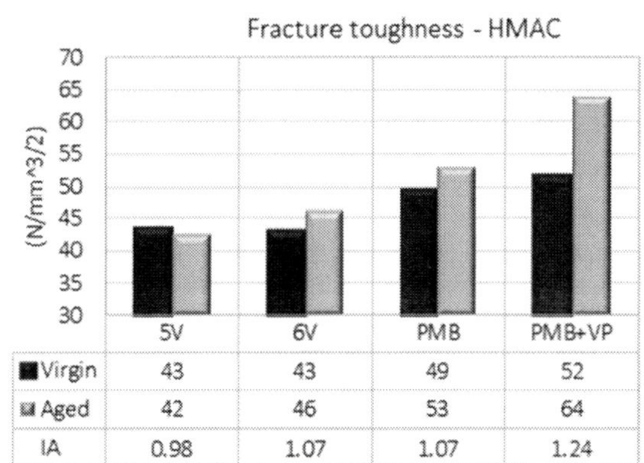

	5V	6V	PMB	PMB+VP
Virgin	43	43	49	52
Aged	42	46	53	64
IA	0.98	1.07	1.07	1.24

Figure 4. Fracture toughness – HMAC mixtures.

5.2. Three-point bending (Tensile bending strength)

The tensile bending strength was determined in compliance with the Czech Technical specifications TP 151 [9] on unnotched beam specimens.

The common trend of flexural strength is that flexural strength of aged specimen is higher than flexural strength of virgin specimen. Again it applies the closer the ageing index (IA) is to 1.0, the less sensitive is the mixture to ageing.

The most stable values were reached by the SMA with 50/70 and HMAC with PMB. It appears as the VP additive enables the binder to age (oxidize) more easily and so the mixture becomes volatile over the life time. The same trend was visible even in the previous test, but the influence was not that noticeable.

Quite surprisingly the SMA mixture with 50/70 reached again very high values in comparison to modified mixtures. The SMA mixture with 50/70 reached the highest values in both of the low temperature tests. It would be expectable, that modified binder would achieve better low temperature properties.

Among the HMAC mixture the highest results were achieved by the asphalt mixture with application of 6 % VP. For this mixture the influence of aged binder is the least visible (the ageing index is only 1.08).

Additionally to common test parameter the fracture energy was calculated. The fracture energy was determined as an area under the stress-strain diagram. The fracture energy can be defined as the energy needed for the beam to expend its bearing capacity and collapse. This is another parameter, which can and should be determined during the test. Focusing only on maximum values of force and deformation can be misleading. For example in the case of HMAC mixture with PMB, the flexural strength and the corresponding modulus of elasticity are very similar, but the virgin specimens needed almost 30 % more of fracture energy to collapse. So with certain point of vigilance it is possible to say, it would take more repetitions of loading for this virgin mixtures to collapse.

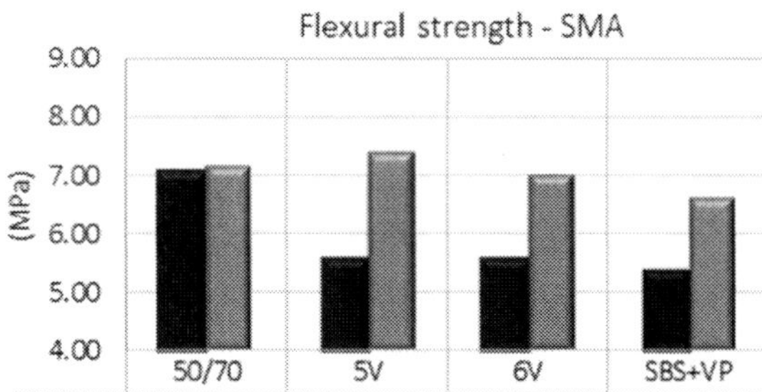

	50/70	5V	6V	SBS+VP
Virgin	7.05	5.55	5.56	5.36
Aged	7.18	7.42	7.02	6.64
IA	1.02	1.34	1.26	1.24

Figure 5. Flexural strength – SMA mixtures.

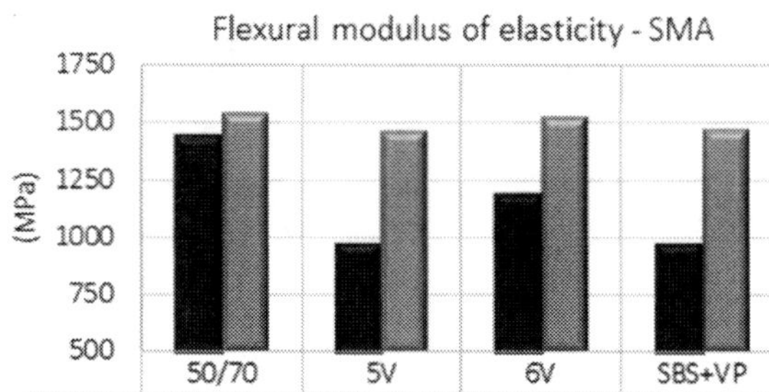

	50/70	5V	6V	SBS+VP
Virgin	1440	971	1190	969
Aged	1538	1466	1523	1472
IA	1.02	1.34	1.26	1.24

Figure 6. Flexural modulus of elasticity.

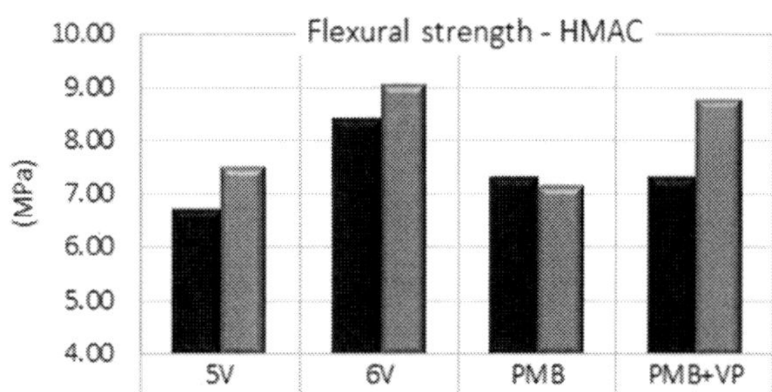

	5V	6V	PMB	PMB+VP
Virgin	6.71	8.37	7.29	7.30
Aged	7.50	9.06	7.17	8.78
IA	1.12	1.08	0.98	1.20

Figure 7. Flexural strength – HMAC mixtures.

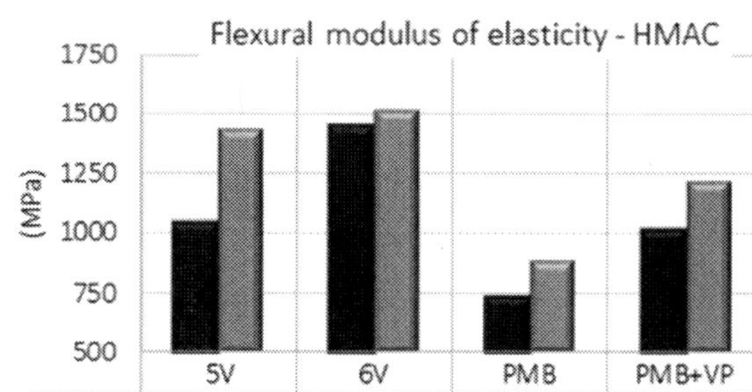

	5V	6V	PMB	PMB+VP
Virgin	1046	1452	738	1013
Aged	1442	1520	887	1222
IA	1.12	1.08	0.98	1.20

Figure 8. Flexural modulus of elasticity – HMAC mixtures.

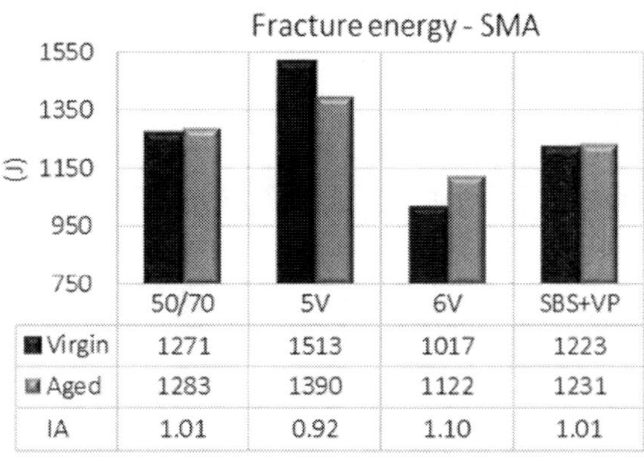

Figure 9. Fracture energy – SMA mixtures.

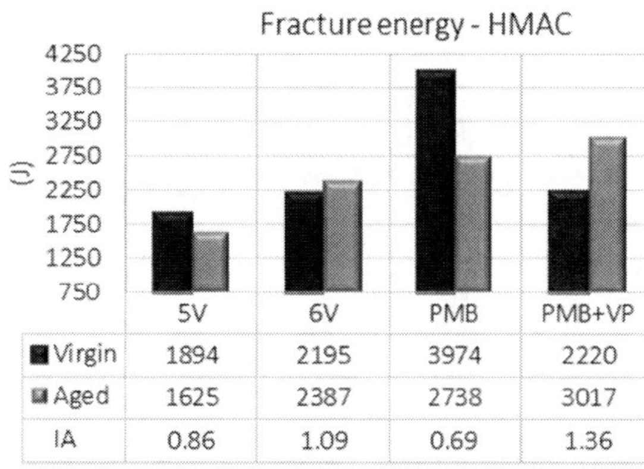

Figure 10. Fracture energy – HMAC mixtures.

5.3. Properties of bituminous binder

The bituminous binder used for manufacturing of the asphalt mixtures were tested virgin, then were exposed to laboratory ageing by TFOT method and then the TFOT sample were additionally subjected to PAV ageing method. The results of performed tests are summarized in Table 2.

Table 2. Selected performance characteristics of used bituminous binders.

	Penetra-tion (x 0.1 mm)	Pen. index (-)	Softening point (°C)			Complex shear modulus at 40 °C, G*/sin(δ) (kPa)			Complex shear modulus at 40 °C, G*x sin(δ) (kPa)			Elastic recovery 3.2 kPa (%)
	Virgin	*Virgin*	*Virgin*	*TFOT*	*TFOT +PAV*	*Virgin*	*TFOT*	*TFOT +PAV*	*Virgin*	*TFOT*	*TFOT +PAV*	*Virgin*
50/70	64	-1	49	56	65	50.4	156.6	721.0	48.6	125.2	389.6	0.4
PMB	54	5.2	85	90	92	68.7	114.7	312.4	53.9	86.7	207.3	88.5
5V	42	1.7	66	73	83	175.9	531.2	1708.2	140.2	310.3	645.4	3.3
6V	39	2.5	72	75	83	190.7	735.1	1308.5	152.4	391.6	507.6	2.9
SBS+VP	32	1	74	75	82	226.6	486.7	632.2	163.5	279.8	331.8	28.8
PMB+VP	46	5.1	88	92	92	88.2	158.6	358.6	68.6	114.9	229.7	90.4

6. Correlation between low temperature properties of asphalt mixtures and the bitumen characteristics

The correlation indexes were calculated always as R-squared value for linear trend line. For preserving the paper extent the data were shown in tables not in more comprehensible figure. The Figure 11 and 12 (shown below) are used for demonstration of calculations and easier understanding the results.

The first correlation data were the comaprison of the properties of virgin mixtures and virgin binders. The fracture toughness and flexural strenght were compared with penetration, softening point and penetration index, complex shear modulus measured at 40 °C and elastic recovery measured by MSCR at 60 °C and 3.2 kPa.

Very good correlation appears between fracture tougness and penetration index. It applies for both types of asphalt mixtures. The higher the penetration index is, the higher the fracture tougheness is.

The correlation index R^2 for HMAC mixtures seems to be higher, which can be most likely caused by usage of only modified binders. In case of SMA mixtures there was one non modified binder (50/70) which behaves differently from the others. Very good correlation index can be also found for elastic recovery measured by MSRC. If the binder 50/70 would be eliminated, the correlation index R^2 would reach 0.671 (instead of 0.085). Also the correlation index for G*sin(δ) would increase to 0.374.

Table 3. Correlation between fracture toughness of virgin asphalt mixtures and virgin binders.

Mix	Binder	Fracture toughness (N/mm$^{3/2}$)	Penetration (x 0,1 mm)	Softening point (°C)	Penetration index (-)	Complex shear modulus at 40 °C G*/sin(δ) (kPa)	Complex shear modulus at 40 °C G*x sin(δ) (kPa)	Elastic recovery 3.2 kPa (%)
		Virgin	*Virgin*	*Virgin*	*Virgin*	*Virgin*	*Virgin*	*Virgin*
SMA	50/70	43	64	50	-1	50.4	48.6	0.36
	5V	41	42	66	1.7	175.9	140.2	3.33
	6V	39	39	72	2.5	190.7	152.4	2.89
	SBS+VP	42	32	74	1	226.6	163.5	28.82
	Correlation index R^2 =		**0.235**	**0.343*)**	**0.814**	**0.243**	**0.341**	**0.085**
HMAC	5V	43	42	66	1.7	175.9	140.2	3.33
	6V	43	39	72	2.5	190.7	152.4	2.89
	PMB	49	54	86	5.2	68.7	53.9	88.46
	PMB+VP	52	46	88	5.1	88.2	68.6	90.44
	Correlation index R^2 =		**0.496**	**0.937*)**	**0.918**	**0.888**	**0.893**	**0.967**

*) shown in Figure 11

The correlation between flexural strength and the bitumen characteristics is not that noticeable as for the previous test. There are good correlations for SMA mixtures, but they are not proofed in any possibility with HMAC mixtures. For SMA mixtures the comparison with complex shear modulus works very well, however there is a zero correlation for HMAC mixtures. In case of fracture toughness the correlation index R^2 with complex shear modulus reached almost 0.9 for HMAC and 0.24, resp. 0.34 for SMA mixtures.

Table 4. Correlation between flexural strength of virgin asphalt mixtures and virgin binders.

Mix	Binder	Flexural strength (MPa)	Penetration (x 0,1 mm)	Softening point (°C)	Penetration index (-)	Complex shear modulus at 40 °C G*/sin(δ) (kPa)	Complex shear modulus at 40 °C G*x sin(δ) (kPa)	Elastic recovery 3.2 kPa (%)
		Virgin	*Virgin*	*Virgin*	*Virgin*	*Virgin*	*Virgin*	*Virgin*
SMA	50/70	7.05	64	49	-1	50.4	48.6	0.36
	5V	5.55	42	66	1.7	175.9	140.2	3.33
	6V	5.56	38.8	72	2.5	190.7	152.4	2.89
	SBS+VP	5.36	32.1	74	1	226.6	163.5	28.82
	Correlation index R^2 =		0.961	0.942	0.748	0.969	0.988	0.276
HMAC	5V	6.71	42	66	1.7	175.9	140.2	3.33
	6V	8.37	38.8	72	2.5	190.7	152.4	2.89
	PMB	7.29	54.3	85	5.2	68.7	53.9	88.46
	PMB+VP	7.30	45.6	88	5.1	88.2	68.6	90.44
	Correlation index R^2 =		0.133	0.001	0.001	0.091	0.092	0.044

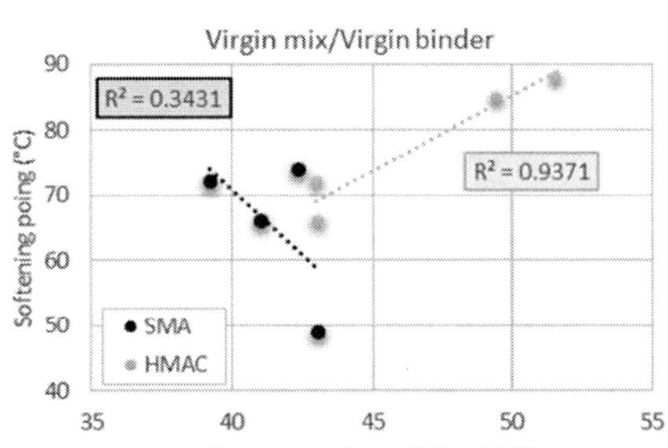

Figure 11. Comparison of properties of virgin mixtures and virgin binders.

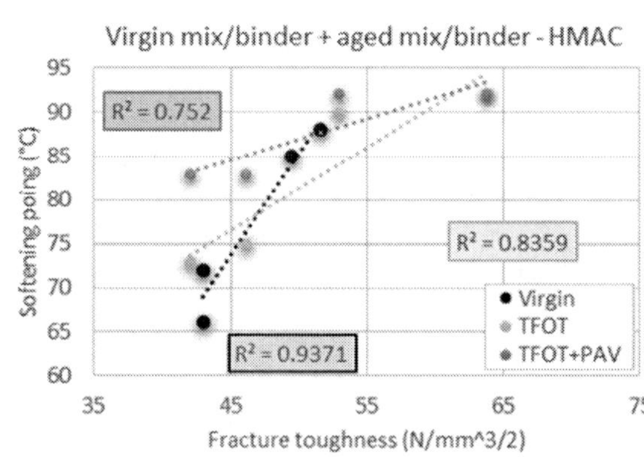

Figure 12. Comparison of properties of virgin mixtures and virgin binders and aged mixtures and aged binders.

Due to the higher correlation indexes for further comparison presentation only the fracture toughness was chosen. The fracture toughness of virgin specimens was compared with virgin binder properties and the aged specimens were compared with binder performed to two different ageing methods.

The correlation indexes R^2 for HMAC mixtures demonstrates quite interesting trend, the indexes are in all cases higher than 0.6. More than half of the indexes almost reaches 0.9. The highest indexes are achieved for parameter G*/sin(δ) of complex shear modulus and combination of TFOT + PAV ageing methods. From point of view of correlation between these tests, the combination of TFOT + PAV seems to evince results which correlate more with asphalt mixture specimens' ageing (5 day at 85 °C).

For SMA mixtures TFOT + PAV combination also evinces higher correlation, but the correlation indexes on its own are too low to be meaningful.

Table 5. Correlation between fracture toughness of virgin asphalt mixtures and virgin binders and aged asphalt mixtures and aged binders.

Mix	Binder	Fracture toughness		Softening point (°C)			*Com. shear modulus at 40°C,* G*/sin(δ) (kPa)			*Com. shear modulus at 40 °C,* G*x sin(δ) (kPa)		
		Virgin	*5d@85 °C*	*Virgin*	*TFOT*	*TFOT +PAV*	*Virgin*	*TFOT*	*TFOT+ PAV*	*Virgin*	*TFOT*	*TFOT+ PAV*
SMA	50/70	43	42	49	56	65	50	157	721	49	125	390
	5V	41	35	66	73	83	176	531	1708	140	310	645
	6V	39	40	72	75	83	191	735	1309	152	392	508
	SBS+VP	42	37	74	75	82	227	487	632	164	280	332
	Correlation index R^2 =			**0.343**	**0.480**	**0.554**	**0.243**	**0.172**	**0.243**	**0.172**	**0.250**	**0.341**
HMAC	5V	43	42	66	73	83	176	531	1708	140	310	645
	6V	43	46	72	75	83	191	735	1309	152	392	508
	PMB	49	53	85	90	92	69	115	312	54	87	207
	PMB+VP	52	64	88	92	92	88	159	359	69	115	230
	Correlation index R^2 =			**0.937*)**	**0.836*)**	**0.752*)**	**0.888**	**0.569**	**0.888**	**0.569**	**0.759**	**0.893**

*) shown in Figure 12

7. Conclusion

Within the scope of the presented research study, two different asphalt mixtures were tested in range of low temperature. The low temperature properties were tested by two different methods – semi-cylindrical bending test and unnotched beam three point bending. The parameters were compared with properties determined on bituminous binder. The assignment on binder was softening point, penetration, complex shear modulus at 40 °C and elastic recovery at 60 °C and after stressing by 0.1 kPa and 3.2 kPa (MSCR test).

The set of test specimens was enlarged by the set of specimens aged exposed to artificially ageing procedure (5 days in 85 °C). The set of binders was enlarged by the set of binders exposed to artificially ageing by the TFOT method (5 hours at 163 °C) and by the combination of TFOT and PAV method (20 hours under the pressure at 100 °C).

The parameters were compared in intension to found some correlation between the properties of asphalt mixtures and input bituminous binders.

Even though the Czech Republic belongs to countries with cold winters and snow season the low temperature properties of asphalt mixtures are not ordinarily required. The only exception are HMAC mixtures, where the Czech Technical specifications TP 151 [9] demands the minimum value determined by unnotched beam three point bending test.

The correlation indexes determinate for HMAC shows certain connection between the properties of asphalt mixtures and binders. The higher the softening point is, the better low temperature characteristics are achieved. The smaller the complex shear modulus is, the better low temperature characteristics can be achieved. There is also a correlation between the aged binder by combination of TFTO and PAV method.

For point of view of time consumption the determination of binder properties is easier and faster than that for asphalt mixtures. The comprehensive binder properties can show a lot of final mixtures properties without time and monetary consuming tests. The asphalt mixture tests are always taken on larger amount of the tested composite material. The binder can be tested in amount of a few grams. It is very important to include more extensive sets of binder testing (functional based) into the standard and standardized test methods.

It is necessary to point out that the compassion was taken only for a narrow set of asphalt mixtures and only with 4 different types of binders for each mixture. It would be important to enlarge the set of different kind of mixtures (ACs) and bigger variety of assessed binders.

Acknowledgements

The research work presented in is paper was funded within the activities of the research centre of competence CESTI under the project no. TE01020168 of the Technology Agency of the Czech Republic

References

[1] Guowei Z, Xinhua Y, Anyi Y and Fan B 2014 Simulation of damage evolution and crack propagation in three-point bending pre-cracked asphalt mixture beam *Construction and Building Materials* **55** pp 323-32

[2] Ziari H, Behbahani H, Izadi A and Nasr D 2013 Long term performance of warm mixture asphalt versus hot mixture asphalt *Journal of Central South University* **20**(1) pp 256-66

[3] Hadi S and Iman A 2016 Investigating the effects of aging and loading rate on low-temperature cracking resistance of core-based asphalt samples using semi-circular bending test *Construction and Building Materials* **126** pp 82-690

[4] Wagoner M, Buttlar W and Paulino G 2005 Development of a Single-Edge Notched Beam Test for Asphalt Concrete Mixtures *Journal of Testing and Evaluation* **33**(6) pp 452-60

[5] Vacková P and Valentin J 2017 Comparison of flexural strength and crack propagation test characteristics of SMA mixtures including the ageing impact *10th Int. Conf. on the Bearing Capacity of Roads, Railways and Airfields (Athens)*

[6] Behnia B, Buttlar W and Reis H 2016 Cooling cycle effects on low temperature cracking characteristics of asphalt concrete mixture *Materials and Structures* **47** pp 1359-71

[7] Molenaar J M M, Liu X and Molenaar A A A 2003 Resistance to crack-growth and fracture of asphalt mixture *Sixth International RILEM Symposium on Performance Testing and Evaluation of Bituminous Materials (Zurich)* pp 618-25

[8] Ameri M, Mansourian A, Pirmohammad S, Aliha M R M and Ayatollahi M R 2012 Mixed mode fracture resistance of asphalt concrete mixtures *Engineering Fracture Mechanics* **93** pp 153-167

[9] Ministry of Transport of the Czech Republic - Department of Road Infrastructure 2010 Technické podmínky: Asfaltové směsi s vysokým modulem tuhosti (VMT) /Technical specifications for high-stiffness modulus asphalt mixtures (HMCA)/ *(TP 151)*

[10] Valentin J and Soukupová L 2016 *Evaluation of the impact of VP on bitumen 50/70 using selected performance-related characteristics* (CTU in Prague)

[11] Valentin J et al 2016 *Asphalt mixtures of SMA type utilizing chemical additive VP* (CTU in Prague)

[12] Valentin J et al 2016 *Asphalt mixtures of HMAC type with use of organic additive VP* (CTU in Prague)

[13] Kotousova A 2016 *Laboratorní simulace stárnutí asfaltové směsi a vliv na vybrané funkční charakteristiky* (in Czech) */Laboratory simulation of asphalt mix ageing and its impact on selected functional characteristics/* (CTU in Prague)

[14] European Committee for Standardization (CEN) Bituminous mixtures — Test methods — Part 52: Conditioning to address oxidative ageing, standard in development (in preparation) *(prEN 12697-52)*
[15] European Committee for Standardization (CEN) 2010 Bituminous mixtures – Test methods for hot mix asphalt – Part 44: Crack propagation by semi-circular bending test *(EN 12697-44)*

Effect of new type of synthetic waxes on reduced production and compaction temperature of asphalt mixture with reclaimed asphalt

Tereza Valentová[1], Lucie Benešová[1], Jan Mastný[1] and Jan Valentin[1]

[1] Faculty of Civil Engineering, Czech Technical University in Prague, Thakurova 7, Prague 6, 166 29, Czech Republic

E-mail: tereza.valentova@fsv.cvut.cz

Abstract. Lower mixing and paving temperatures of asphalt mixtures, which are an important issue in recent years, with respect to increased energy demand of civil engineering structures during their processing, allow reduction of this demand and result in minimized greenhouse gas production. In present time, there are many possibilities how to achieve reduction of production temperature during the mixing and paving of an asphalt mixture. The existing solutions distinguish in target operating temperature behaviour which has to be achieved in terms of good workability. This paper is focused on technical solutions based on use of new types of selected synthetic and bio-based waxes. In case of bio-based additive sugar cane wax was used, which is free of paraffins and is reclaimed as waste product during processing of sugar cane. The used waxes are added to bituminous binder in form of free-flowing granules or fine-grained powder. Synthetic waxes are represented by new series of Fischer-Tropsch wax in form of fine granules as well as by polyethylene waxes in form of fine-grained powder or granules. Those waxes were used to modify a standard paving grade bitumen dosed into asphalt mixture of AC_{surf} type containing up to 30 % of reclaimed asphalt (RA).

1. Introduction

First technologies of warm asphalt mixtures (WMA) were available and further tested already about 20 years ago in Europe. It was developed in response to the demand of the road industry to decrease the energy consumptions and reduce the release of greenhouse gases as well as other harmful gases, which cause serious air pollution. Warm mix asphalt is a kind of new pavement material, generally, they are produced and compacted at lower temperatures compared to hot-mix asphalt (HMA). The range of mixing temperature for WMA is 100 °C to 140 °C and this type of mixtures is placed on the temperature scale between hot mix asphalt and cold mix asphalt [1, 2]. The main mechanism of this technology involve the use of organic or synthetic waxes (F-T wax, polyethylene wax, amide wax, Montan wax, fatty acid amides etc.), chemical additives (include combination of emulsification agents, surfactants, polymers, adhesion promoters), foaming techniques (which are divided into water-based and water containing), that helps to reduce the viscosity of bitumen at lower temperature and improve coating, mixture workability, and compaction. In such a way, good workability and reasonable long-term performance of the asphalt mixture should be achieved [3].

Lowering asphalt production emissions in the plant and compaction emissions in the field are one of the most important benefits of utilization of WMA. The properties of bitumen are improved by means of organic and synthetic WMA additives and the viscous behavior of used binders is changed, which leads to decrease mixing and compaction temperature. [2, 4, 5] The lower production temperature also reduces the ageing of the bitumen during the production stage, which results in an improved thermal and fatigue cracking resistance. [2] For the maximization of the mixing temperature reduction, the best solution for the production of WMA mixtures with the incorporation of synthetic waxes is by using softer base bituminous binders. [6] Consequences shown in this respect by the research [6] was that, the addition of synthetic waxes does not significantly change the water sensitivity of the studied WMA mixtures. The use of these additives increase the stiffness modulus and the rut resistance of the WMA in comparison with the HMA mixture produced with the same bitumen, allowing the use of a softer bitumen to obtain a WMA mixture with a similar performance.

Further, it should be stated that WMA additive enables the utilization of reusing more reclaimed asphalt (RA) in the pavement structure. Variations in RA properties complicate the characterization of the rheological properties of binder, which consists of added virgin bitumen and reclaimed asphalt binder. [4] Based on the study [7], which presents impact of selected types of WMA technologies involved also some kind of organic wax combined in an asphalt mixture with 35 % of reclaimed asphalt, which was applied in real production conditions as well. The results of this study showed that WMA produced with a high RA content is feasible, contributing to the reduction in energy consumption and in costs. The level of mechanical performance of WMA was in this case satisfactory and similar to the values obtained for the HMA used as reference. Nevertheless, the warm-mix with surfactant additive revealed high sensitivity to temperature. WMA with RA can be seen as a suitable alternative to conventional HMA for base and binder layers of pavements. [7]

2. Basic definition of the asphalt mixture variants for surface courses under assessment

2.1. Design of mixture ACs_{urf} 11+ with 30 M% reclaimed material

The reference variant for the starting design of a mixture with 30 M% RA was the AC_{surf} 11+ mixture of a known and optimised grading curve using aggregate from the Měrunice quarry. The mixture was designed with 5.8 M% bituminous binder. A typical paving grade bitumen 50/70 was chosen as the basic binder for the reference mixture. This reference variant served as the base for asphalt mixture comprising 30 M% sorted reclaimed asphalt of 0/11 mm gradation, sourced from the Měcholupy mixing plant deposit. This is material obtained from various pavements which is subsequently crushed and sorted, i.e. no exact origin and homogeneous composition can be determined for the material. Using asphalt extraction lab procedure, the quantity of bituminous binder in the reclaimed material was determined (5.9 M%) and a grading analysis was performed.

Table 1. Mix design and particular mix components of assessed asphalt mixtures.

Labelling of the asphalt mixture	AC_{surf} 11+		AC_{surf} 11+ 30 M% RA	
	-			
Bitumen	50/70		50/70	
Bitumen content [%]	5.80 %		3.60 % (5.40 %)	
Mix design	Content [%]	In the mix [%]	Content [%]	In the mix [%]
Měrunice 8/11 mm	21	19.8	23	22.2
Měrunice 4/8 mm	21	19.8	15	14.4
Měrunice 0/4 mm	52	49.0	26	25.0
RA 0/11 mm	0	0.0	30	28.9
Filler	6	5.7	6	5.8
Bitumen		5.8		3.6

The input characteristics were then used to design mixture AC_{surf} 11+ with 30 M% RA; this allowed determining the quantity of the bituminous binder added. The mixture needed optimisation where the original reference asphalt mixture best corresponds with the variant with a total of 5.4 M% binder, which is the equivalent of 3.6 M% of virgin bitumen 50/70 addition (note: the quantity of binder addition was modified depending on the analysis of the reclaimed material concerned). The mix design and representation of the individual components are indicated in Table 1.

2.2. Definition of low temperature mixtures with selected types of natural and synthetic waxes on test

There are a number of possibilities nowadays of reducing production temperature during the preparation as well as processing of asphalt mixtures using so-called warm mix asphalts. However, they principally differ by their behaviour in the working temperature range which we intend to achieve to ensure sufficient processing. The paper concentrated on the technology of applying a new generation of additives based on selected types of both synthetic and natural (bio-based) waxes. These are added directly to the bituminous binder. The main advantages of using synthetic and natural waxes in asphalt mixtures, or bituminous binders to be precise, are lower viscosity and improved workability under reduced working temperatures while the required asphalt mixture characteristics are maintained. In the case of natural waxes, a product based on sugar cane extract with no paraffin content was used – this is waste from sugar cane processing (hereinafter referred to as "BIT"). The synthetic wax group supplied representatives of a new generation of polyethylene waxes (E-wax). Another option from the synthetic wax range was the RH wax of Chinese origin. The additives were applied to modify the basic binder, paving grade 50/70, used to prepare asphalt mixture type AC_{surf} 11+ which allowed lowering the working temperature to 125-130 °C. Simultaneously, the variants of binders modified by the selected types of synthetic and natural waxes were applied to mixture AC_{surf} 11+ with 30 M% reclaimed asphalt. In the case of the variants with RA, two types of rejuvenating additives were used to ensure activation of the degraded binder in the reclaimed asphalt. The degraded binder had been exposed to oxidative ageing. In case we wish to re-activate the aged binder, a rejuvenator needs to be added to the reclaimed material. We chose a Czech-made rejuvenator, RE182 – refined rape seed oil-based additive, and S-RP1000, a rejuvenator based in tall oil from lignin depolymerisation in pinewood processing. The selected rejuvenators were added to heated reclaimed materials and mixed together; only afterwards was this material combined with new aggregate and the bituminous binder.

3. Selected test methods to examine the cold asphalt mixture behaviours

3.1. Basic characteristics of cold mixture variants

The major asphalt mixture characteristics include the voids content (for mixtures AC_{surf} 11+ the test specimens are compacted by 2x50 blows in the Marshall compactor) and maximum bulk density under CSN EN 12697-8. The individual parameters are listed in Table 2. In compliance with the requirements of the relevant product standard, the voids content in AC_{surf}11+ asphalt mixtures should fall within the range of 2.5-4.5 %. The test specimen preparation temperature was identical for all versions, 150 °C; contrastingly, the compaction temperature could be reduced by 20-25 °C with respect to the application of the individual additives.

Table 2. Determination of fundamental empirical characteristics of assessed asphalt mixtures.

Asphalt mixture	Mix type acronym	Bitumen	Bitumen content [%]	Temperature for production / compaction	Bulk density [g.cm^{-3}]	Maximum density [g.cm^{-3}]	Voids content [%]
	Ref	50/70		150/150 °C	2.587	2.692	3.91
	3E	50/70 + 3% E-wax		150/125 °C	2.629	2.694	2.44
AC_{surf} 11+	R	50/70 + 1.5% RH + 1.5% E-wax	5.8	150/125 °C	2.648	2.713	2.41
	2E	50/70 + 2% E-wax		150/125 °C	2.613	2.722	4.00
	BIT	50/70 + 3% BIT		150/125 °C	2.583	2.739	5.72
	RefR	50/70	3.4	150/150 °C	2.606	2.671	2.45
	3ER	50/70 + 3% E-wax		150/130 °C	2.629	2.677	1.80
AC_{surf} 11+ 30 % RA RE182	RR	50/70 + 1.5% E-wax+ 1.5% RH	4.0	150/130 °C	2.546	2.632	3.26
	2ER	50/70 + 2 % E-wax		150/130 °C	2.513	2.605	3.54
	BR	50/70 + 3 % BIT		150/130 °C	2.560	2.646	3.25
	RefS	50/70		150/150 °C	2.609	2.681	2.69
	3ES	50/70 + 3 % E-wax		150/130 °C	2.597	2.665	2.55
AC_{surf} 11+ 30 % RA S-RP1000	RS	50/70 + 1.5 % E-wax + 1.5 % RH	3.6	150/130 °C	2.587	2.671	3.16
	2ES	50/70 + 2 % E-wax		150/130 °C	2.569	2.665	3.58
	BS	50/70 + 3 % BIT		150/130 °C	2.565	2.673	4.02

A superior graphic representation of the voids content obtained is given by Figure 1 which also indicates the limits for minimum and maximum void content in AC_{surf} 11+ type asphalt mixtures for wearing courses. The reference asphalt mixture as well as the version with 2% E-wax with no reclaimed asphalt met the requirements for the range. This means that in this case, a lower temperature still allows achieving results identical to those of the reference mixture compacted at 150 °C. In the case of the variant with natural BIT wax with no reclaimed asphalt, the voids content of the mixture 5.72 % exceeded the maximum limit for the variable. A high void content might result in poor resistance to water in the wearing course; an asphalt layer of a high voids content is more water-permeable and the connection between bituminous binder and the aggregate deteriorates more easily. With respect to comparing the voids content of asphalt mixtures with RA content, the values are lower than those for mixtures without RA. This is not the case of variant R whose voids content was higher while the standard requirements were met. With a higher dose of E-wax (variant 3E) with rejuvenator RE182, the voids content dropped to 1.88 % to the minimum limit; this suggests that the compaction temperature could be lowered even further to achieve the required optimum.

3.2. Determining resistance to water and frost (ITSR)

The test specimen resistance to water test was conducted in compliance with CSN EN 12697-12 technical standard, applying the modified method under AASHTO T-283 which considers one freeze cycle on top of the standard requirements. The test delivers the ratio of indirect tensile strength of a group of saturated specimens to the indirect tensile stress of a dry specimen group (for ITSR-CSN values see Figure 2, ITSR-AASHTO see Figure 3). The required indirect tensile strength ratio according to CSN EN 13108-1 is currently set to at least 70 % for AC_{surf} 11+ mixtures. Within the framework of harmonised standard review in relation to the wearing course asphalt mixture requirements, this ITSR ratio requirement for this type of mixture is expected to become more stringent, to a minimum of 80 %. However, all of the variants on test exceeded even the 80% limit which suggests that the selected variants do not tend to lose indirect tensile strength due to water. The same may be noted for ITSR evaluation under the modified AASHTO test method which currently has no specific requirements in place and is not required by the European standards.

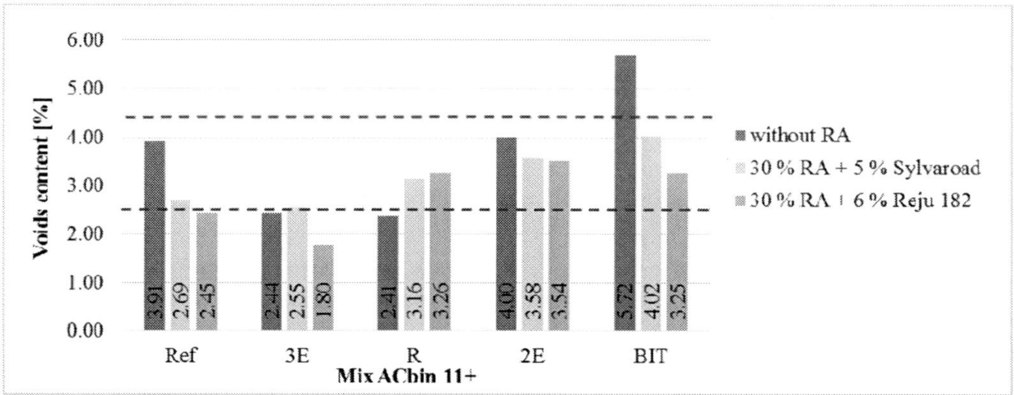

Figure 1. Comparison of voids content for assessed asphalt mixtures and the requirements according to ČSN EN 13108-1 (NA).

From the point of view of assessing the resistance to negative effects of water and frost test on selected mixtures, the generally best results were achieved by AC_{surf} 11+ mixture with 30 M% reclaimed asphalt and rejuvenator RE182 in all variants, except for BIT mixtures according to the Czech technical standard. In the case of AC_{surf} 11+ mixtures with 30 M% RA, to which rejuvenator S-RP1000 was applied, the ITSR values are generally similar to the mixtures without RA.

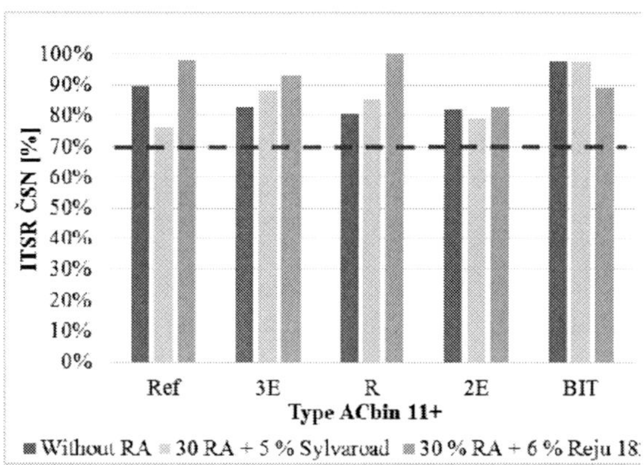

Figure 2. Comparison of water susceptibility test results of assessed asphalt mixtures with the requirements of CSN EN 13108-1 (NA).

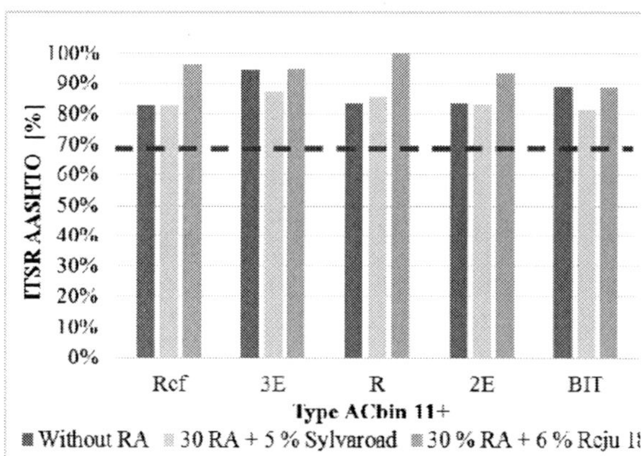

Figure 3. Comparison of water susceptibility test results of assessed asphalt mixtures according to AASHTO.

3.3. Stiffness modulus test

The next comparison of AC_{surf} 11+ mixtures without and with reclaimed asphalt was related to the stiffness modulus determination according to CSN EN 12697-26 by the IT-CY non-destructive test of indirect tensile stress on cylindrical specimens. The stiffness values of individual mixtures were determined under three different temperatures (0 °C, 15 °C a 27 °C) and are listed in Table 2. If we focus on the stiffness modules measurements, the essential ones, according to the usual application of the characteristic in the Czech pavement design manual, are those taken under the temperature of

15 °C – the determining temperature from the perspective of pavement structure design under Czech technical conditions TP 170.

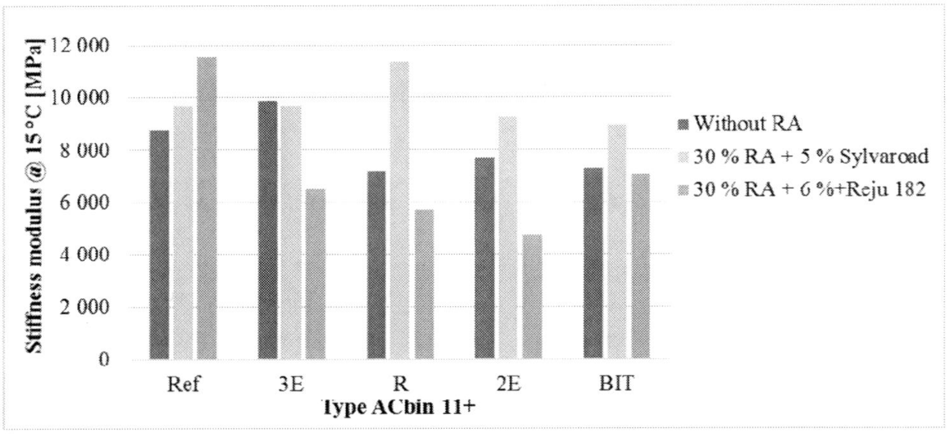

Figure 4. Comparison of stiffness @ 15 °C for tested asphalt mixtures.

The stiffness results measured on unaged test specimens at 15 °C (see Figure 4) clearly show that the lowest values were taken for the mixture with 30 M% reclaimed asphalt with RE182, except for the reference mixture variant which recorded its highest stiffness modulus at 15 °C. In the case of the highest scores for the characteristic, we can conclude that aside from the aforementioned reference mixture and 3E variant, the highest stiffness values were achieved by the mixture variants with reclaimed asphalt which included rejuvenator S-RP1000. The stiffness values taken are rather high (the minimum stiffness modulus value for HMAC mixtures with required high stiffness modulus is set by the standard to 9,000 MPa, this limit would be met even by the asphalt mixture for wearing courses), which, from the perspective of long-term behaviour in the pavement structure, could potentially have negative impact due to reduced life under cold temperatures (increased cracking risk).

Another characteristic assessed in the case of the stiffness modules measured at varying temperatures could be the indicator of thermal susceptibility of the asphalt mixture (Table 3). The lower the S_0/S_{27} stiffness ratio, the less susceptible the mixture is to temperature changes. In the case of mixture variants with no reclaimed asphalt, the lowest susceptibility was measured for the reference mixture and, contrastingly, the highest susceptibility was taken for the variant with a combination of two waxes. Focusing on thermal susceptibility in variants with RE182, the lowest susceptibility was noted for the reference asphalt mixture while a double value with the greatest thermal susceptibility and, therefore, the highest difference between the stiffness modules measured at 0 °C and at 27 °C, was recorded for mixture 2ER, which contained a lesser quantity of E-wax. For mixtures with rejuvenator S-RP1000, the RS variant had the lowest thermal susceptibility while 2ES had the highest thermal susceptibility indicator.

A part of the measurements taken within the experimental study was also dedicated to a laboratory simulation of long-term asphalt mixture ageing which was applied to the test specimens within the framework of the stiffness modulus test. The ageing simulation utilised one of the possible methods viable for standard road construction laboratories at the moment. A simple method where the cylindrical test specimens are put in a thermal chamber with forced air circulation at 85 °C for 5 days (120 hours) according to the method as stipulated in the prEN 12697-52 standard. The comparison of the stiffness modulus values measured in the test specimens exposed to ageing (see Figure 5) suggests that, with respect to the incomplete data set, the highest stiffness modulus value was achieved by the variants with S-RP1000. The asphalt mixtures with RE182 recorded a drop in comparison to the variants with no RA. However, the results have not been verified for all of the variant designs.

Table 3. Determination of stiffness modulus of assessed asphalt mixtures.

Asphalt mixture	Mix type acronym	Bitumen	Bitumen content [%]	Stiffness modulus [MPa] @ T= 0 °C	15 °C	27 °C	Thermal suscepti-bility S_0/S_{27} [-]	Voids content [%]
AC_{surf} 11+	Ref	50/70	5.8	23 627	8 755	3 138	7.53	3.91
	3E	50/70 + 3% E-wax		23 735	9 858	2 820	8.42	2.44
	R	50/70 + 1.5% RH + 1.5% E-wax		22 092	7 192	1 961	11.26	2.41
	2E	50/70 + 2% E-wax		20 840	7 647	2 257	9.23	4.00
	BIT	50/70 + 3% BIT		20 347	7 290	2 245	9.06	5.72
AC_{surf} 11+ 30 % RA RE182	RefR	50/70	3.4	22 770	11 525	4 270	5.33	2.45
	3ER	50/70 + 3% E-wax		19 662	6 511	2 371	8.29	1.80
	RR	50/70 + 1.5% E-wax+ 1.5% RH	4.0	17 629	5 697	1 855	9.50	3.26
	2ER	50/70 + 2 % E-wax		16 461	4 745	1 565	10.52	3.54
	BR	50/70 + 3 % BIT		18 560	7 043	2 226	8.34	3.25
AC_{surf} 11 + 30 % RA S-RP1000	RefS	50/70	3.6	24 895	9 676	3 337	7.46	2.69
	3ES	50/70 + 3 % E-wax		23 611	9 696	3 161	7.47	2.55
	RS	50/70 + 1.5 % E-wax + 1.5 % RH		22 770	11 358	3 945	5.77	3.16
	2ES	50/70 + 2 % E-wax		22 629	9 257	2 793	8.10	3.58
	BS	50/70 + 3 % BIT		20 861	8 956	3 031	6.88	4.02

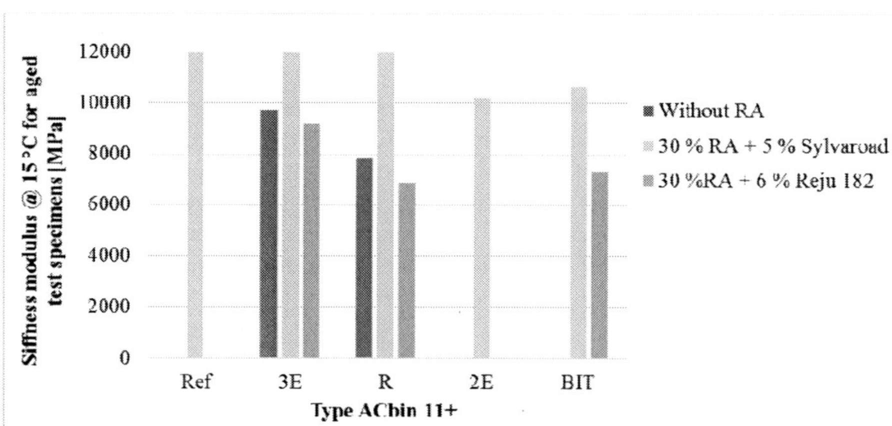

Figure 5. Comparison of stiffness @ 15 °C for aged tested asphalt mixtures.

3.4. Comparison of the results from the perspective of resistance to crack propagation

The comparison of the results of resistance to crack propagation by the bending test on semi-cylindrical test specimens in compliance with CSN EN 12697-44 is indicated in Figure 6. The only difference to the mentioned standard is, that 100 mm diameter test specimens were used. With respect to the assessment of low-temperature asphalt mixture behaviour in the low temperature range, the critical fracture toughness values for mixtures with reclaimed asphalt and RE182 record almost the same course as in the case of mixtures with S-RP1000. Only the variant with a combination of waxes RH + E-wax had a lower result (mixture with RE182). The values of mixture resistance to crack propagation in relation to both rejuvenator types show no influence of the higher stiffness values as measured which would induce a lower resistance in the low temperature range; this might be obvious from the results taken after the long-term ageing simulation.

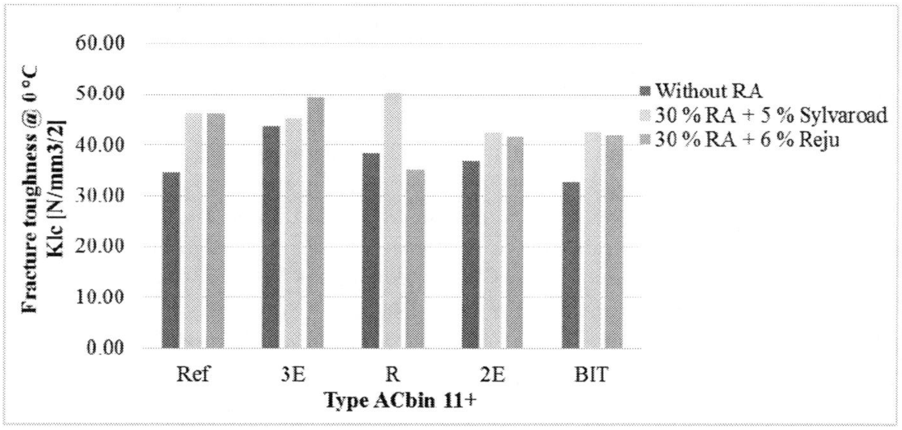

Figure 6. Comparison of fracture toughness @ 0 °C for tested asphalt mixtures.

The critical fracture toughness values were also measured for aged test specimens. The test, aiming to determine the asphalt mixture resistance to crack propagation by bending on semi-cylindrical specimens, was conducted with the Marshall test specimens compacted by 2x50 blows in the compactor and subjected ty an ageing cycle. The laboratory simulation of long-term ageing was performed according to the method described in prEN 12697-52. For the comparison of mixture resistance to crack propagation prior to and after the ageing simulation, it is obvious that AC_{surf} 11+ mixtures without reclaimed asphalt record increased critical fracture toughness values due to the ageing. The same trend can be seen in AC_{surf} 11+ mixtures with reclaimed asphalt and RE182. The opposite trend is visible in the case of variants with S-RP1000 where the critical fracture toughness value fell as a consequence of ageing.

4. Conclusion

The experimental tests taken and the comparison of the individual low-temperature mixture variants including the determination of selected functional asphalt mixture characteristics depending on a lower compaction temperature (i.e. asphalt mixture production temperature) yield the following findings. The application of a new generation of synthetic/natural waxes should primarily achieve a reduction in the production and paving temperature for the asphalt mixes; this has been verified as viable based on the results as presented. With respect to the parameters examined, the application of selected types of natural and synthetic waxes in asphalt mixtures for wearing courses using reclaimed material allows reducing the compaction temperature in the laboratory while maintaining the quality of common hot asphalt mixtures. This, as a consequence, facilitates a more favourable recycling of reclaimed material which would not have to be heated to extremely high temperatures; that, in turn, would prevent excess degrading thereof. Simultaneously, this opens the option of a more effective use of rejuvenators. Nonetheless, some variants of the asphalt mixture designs with various waxes and reclaimed asphalt, the voids contents were lower than the minimum value required by the relevant product standard; this points out the potential for an even bigger reduction of the compaction temperature, or for a reduction of the additive dose. At the same time, no significant deterioration in the functional characteristics examined was observed for any of the assessed mixture variants.

Acknowledgements
This paper was prepared under the support and financing scheme of research project No. TA04031255 funded by Technology Agency of the Czech Republic.

References

[1] EAPA 2010 *The use of warm mix asphalt – EAPA position paper* (European Asphalt Pavement Association Brussels)

[2] Rubio M C, Martinez G, Baena L and Moreno F 2012 Warm mix asphalt: an overview *Journal Cleaner Production* **24** pp 76–84 http://dx.doi.org/10.1016/j.jclepro.2011.11.053

[3] Su et al 2009 Laboratory evaluation of WMA mixture for use in airport pavement rehabilitation *Construction and Building Materials* **23**(7) pp 2709-14

[4] Oner J, Topal A, Sengoz B and Dokandari P A 2016 Permanent deformation characteristics of warm mix asphalt *6th Eurasphalt & Eurobitume Congress* (Prague)

[5] Topala A, Sengoza B, Kokb B V, Yilmazb M, Dokandaric P A, Onerc J and Kayac D Evaluation of mixture characteristics of warm mix asphalt involving natural and synthetic zeolite additives *Construction and Building Materials* **57** pp 38–44

[6] Silva et al 2010 Optimization of warm mix asphalts using different blends of binders and synthetic paraffin wax contents *Construction and Building Materials* **24**(9) pp 1621-31

[7] Martinho F C G, Picado-Santos L G and Capitão S D Mechanical properties of warm-mix asphalt concrete containing different additives and recycled asphalt as constituents applied in real production conditions *Construction and Building Materials* **131** pp

[8] Valentin J et al 2015 Odborná zpráva o postupu prací a dosažených výsledcích za rok 2015 *Research report project WMARec*

[9] Valentin J et al 2016 Odborná zpráva o postupu prací a dosažených výsledcích za rok 2016 - *Research report project WMARec*

[10] Český normalizační institut (ČNI) 2008 Asfaltové směsi - Specifikace pro materiály - Část 1: Asfaltový beton /Bituminous Mixtures - Material Specifications - Part 1: Asphalt Concrete/ *(ČSN EN 13108-1)*

[11] Český normalizační institut (ČNI) 2005 Asfaltové směsi - Zkušební metody pro asfaltové směsi za horka - Část 12: Stanovení odolnosti zkušebního tělesa vůči vodě /Bituminous Mixtures - Test Methods for Hot Mix Asphalt - Part 12: Determination of the Water Sensitivity of Bituminous Specimens/ *(ČSN EN 12697-12)*

[12] American Association of State and Highway Transportation Officials 2007 Standard Method of Test for: Resistance of Compacted Asphalt Mixtures to Moisture-Induced Damage *(AASHTO DESIGNATION: T 283-03)*

[13] Český normalizační institut (ČNI) 2005 Asfaltové směsi - Zkušební metody pro asfaltové směsi za horka - Část 23: Stanovení pevnosti v příčném tahu /Bituminous Mixtures - Test Methods for Hot Mix Asphalt - *Part 23:* Determination of the Indirect Tensile Strength of Bituminous Specimens/ *(ČSN EN 12697-23)*

[14] Úřad pro normalizaci, metrologii a státní zkušebnictví (ÚNMZ) 2011 Asfaltové směsi - Zkušební metody pro asfaltové směsi za horka - Část 44: Šíření trhliny zkouškou ohybem na půlválcovém zkušebním tělese /Bituminous Mixtures - Test Methods for Hot Mix Asphalt – Part 44: Crack Propagation by Semi-Circular Bending *(ČSN EN 12697-44)*

[15] Úřad pro normalizaci, metrologii a státní zkušebnictví (ÚNMZ) 2012 Asfaltové směsi - Zkušební metody pro asfaltové směsi za horka - Část 26: Tuhost /Bituminous Mixtures – Test Method for Hot Mix Asphalt - Part 26: Stiffness/ *(CSN EN 12697-26)*

IOP Conf. Series: Materials Science and Engineering **236** (2017) 012020 doi:10.1088/1757-899X/236/1/012020

Practical experiences with new types of highly modified asphalt binders

Petr Špaček[1], Zdeněk Hegr[1] and Jan Beneš[2]

[1]Skanska a.s., Pavelkova 1133/6, 779 00 Olomouc, Czech Republic
[2]Total ČR s.r.o., Pobřežní 620/3, 186 00 Praha 8, Czech Republic

E-mail: petr.spacek@skanska.cz

Abstract. As a result of steadily increasing traffic load on the roads in the Czech Republic, we should be focused on the innovative technical solutions, which will lead to extending the life time of asphalt pavements. One of these ways could be the future use of bitumen with a higher degree of polymer modification. This paper discusses experience with comparison of new highly polymer modified asphalt binder type with conventional polymer modified asphalt binder and unmodified binder with penetration grade 50/70. There are compared the results of various types laboratory tests of asphalt binders, as well as the results of asphalt mixtures laboratory tests. The paper also mentions the experience with workability and compactability of asphalt mixture with highly polymer modified asphalt binder during the realization of the experimental reference road section by the Skanska company in the Czech Republic.

1. Introduction

In these days, there are a lot of challenges in the field of asphalt technologies like increasing traffic load and higher requirements on durability of asphalt pavements. One of the options to face these challenges could be usage of highly polymer modified asphalt binders [19], [20], [21], [22].

For practical evaluation of asphalt mixtures with highly polymer modified asphalt binder just in real pavement there was carried out experimental road section by Skanska company in 2016. Before the realization of experimental road section there were performed laboratory comparisons of highly polymer modified asphalt binder with conventional PmB and paving grade bitumen 50/70. There were also carried out various laboratory tests of asphalt mixture for friction courses ACO 11 S with all above mentioned types of asphalt binders.

The experimental road section made of friction course ACO 11 S with highly polymer modified asphalt binder by TOTAL company (PmB 25/55-75 Styrelf) was carried out by Sknaska company in October 2016. This experimental road section is placed in logistical complex, where high point loads and high radial forces from heavy trucks and forklifts are expected.

This research study consists of three basic parts:

1) Testing of three different asphalt binders properties (paving grade bitumen 50/70, conventional PmB 25/55-65, highly polymer modified PmB 25/55-75).
2) Testing of asphalt mixture properties with three above mentioned compared types of asphalt binders.
3) Experimental road section paving and evaluation.

2. Input materials

Into the asphalt mixture for experimental road section, there were used below mentioned input materials:

- highly polymer modified asphalt binder STYRELF PmB 25/55-75 from TOTAL;
- mineral aggregate from Zarubka quarry fraction 0/4, 4/8, 8/11;
- limestone mineral filer from Prachovice.

The geological type of used mineral aggregate from Zarubka quarry is Granodiorite. All used mineral aggregate was in accordance with technical requirements according to technical standard ČSN EN 13108-1 [12] for mineral aggregate into the friction courses.

3. Asphalt binder STYRELF PmB 25/55-75 production

The highly polymer modified asphalt binder was produced in two steps. The first phase consisted of the blending of a granulated linear SBS polymer into a purely distilled bitumen with the help of an IKA shear mill in a TOTAL BITUMEN DEUTSCHLAND refinery. This concentrate has been chemically cross-linked and additivated with a scavenger to reduce the hydrogen sulfide concentrations in the gas phase of the dead space of the tanks, in which the bitumen is transported and stored.

This concentrate was shipped into the bitumen depot belonging to TOTAL ČESKÁ REPUBLIKA s.r.o. in Kouřim where it was blended with other bitumen through the INDAG DLM/S in-line mixer to produce a highly polymer modified asphalt binder with a penetration range of 25 – 55 and a softening point higher than 75 °C.

The bitumen was produced directly into the tanker truck and has not been intermediately stored. During the production process another additive was added to the binder in order to improve the adhesion, reduce the ageing sensitivity and thus increase the lifespan of the pavement. Sampling has been done during the loading – from the main stream pipe with a Strahman piston sampling valve – according to technical standard ČSN EN 58 – Sampling bituminous binders.

No complications were observed during the loading process – the temperature during production was approx. 176 °C and the loading time was close to 35 minutes.

4. Asphalt binder laboratory testing results

The asphalt mixture ACO 11 S used for the experimental road section was produced with a highly polymer modified asphalt binder PmB 25/55-75. To further analyze the binder properties a series of laboratory tests was performed. In order to have a comparison to these results, other conventional bitumen have been tested as well – the set consists of paving grade bitumen Azalt 50/70 and polymer modified bitumen Styrelf PmB 25/55-65. All bitumen were supplied by TOTAL. The set of tests was repeated for all binder also after various ageing methods:

1) RTFOT (Short term ageing – Rolling thin film oven test) [10]
2) RTFOT + PAV (Pressure ageing vessel – Long term ageing test) [10], [11]
3) 3 x RTFOT

The results of all tests performed on the binders are presented in Tables 1, 2, 3 and 4.

Table 1. Selected performance properties of the un-aged binders.

	Property	Unit	50/70	PmB 25/55-65	PmB 25/55-75
Virgin binder	Penetration [4]	[0.1mm]	63	35	42
	Softening point [5]	[°C]	47.7	69.6	77
	Fraass breaking point [8]	[°C]	-10	-13	-18
	Flash point [9]	[°C]	317	336	341
	Elastic recovery @ 25°C [6]	[%]	-	70	88
	Force ductility @ 10°C [7]	[J/cm^2]	-	4	5.7

Table 2. Selected performance properties of the RFFOT aged binders.

	Property	Unit	50/70	PmB 25/55-65	PmB 25/55-75
RFOT Aged binder [10]	Retained penetration [4]	[%]	75	89	93
	Increase in softening point [5]	[°C]	5	8.4	8.5
	Change of mass	[%]	0	0.1	0.1
	Penetration [4]	[0,1mm]	47	31	39
	Softening point [5]	[°C]	52.4	78	85.5
	Elastic recovery @ 25°C [6]	[°C]	-	64	83
	Force ductility @ 10°C [7]	[J/cm^2]	-	1	1.3
	Recovery MSCR 0.1kPa	[%]	3.19	78.56	90.98
	Non-recoverable creep compliance MSCR 0.1 kPa	[1/kPa]	2.11	0.03	0.02
	Recovery MSCR 3.2kPa	[%]	1.23	77.92	90.91
	Non-recoverable creep compliance MSCR 3.2 kPa	[1/kPa]	2.14	0.03	0.02
	R_{dif}	[%]	61.446	0.812	0.073
	$J_{nr\text{-}diff}$	[%]	1.557	4.422	1.429

A clear difference between the properties of the paving grade bitumen and the polymer modified bitumen can be seen from the results. The modified asphalt binders outperform the paving grade bitumen in all followed parameters.

Based on the fact that the real deal in binder performance is its behaviour in the asphalt mixture, it can be said – based on the differences in the empirical properties, but mainly the functional MSCR (Multiple Stress Creep Recovery) and the DSR (Dynamic Shear Rheometer) phase angle differences – that both polymer modified asphalt binders show a greater resistance to rutting in comparison with the paving grade bitumen.

The properties of all three binders were in accordance with the technical standard ČSN EN 12591 [2] for paving grade bitumen and technical standard ČSN EN 14023 [3] for polymer modified bitumen, although it should be fairly concluded that the demands of these technical standards are in the Czech Republic quite low. An interesting difference can be seen in the adhesion of the binders where both modified asphalt binders showed better results than the paving grade bitumen. Thanks to this fact and in combination with a relatively high deformation energy – reaching almost 200 % of the technical standard demand for the gradation PmB 25/55 – 75 – good adhesion and cohesion properties of the highly polymer modified binder can be expected in the asphalt mixture.

If we compare only the modified asphalt binders which were tested, a difference in the softening point can be seen, which was due to the nature of the test only limited informative potential about the binder performance. Another difference is the Fraass breaking point temperature, which would point to a higher modification of the PmB 25/55-75 binder.

Table 3. Selected performance properties of the RTFOT + PAV aged binders.

	Property	Unit	50/70	PmB 25/55-65	PmB 25/55-75
RFOT [10] + PAV [11] aged binder	Penetration [4]	[0,1mm]	20	22	29
	Softening point [5]	[°C]	65.8	86.5	89
	Elastic recovery @ 25°C [6]	[°C]	-	50	77
	Force ductility @ 10°C [7]	[J/cm^2]	-	-	2.5
	Recovery MSCR 0.1kPa	[%]	21.72	82.64	91.4
	Non-recoverable creep compliance MSCR 0.1 kPa	[1/kPa]	0.17	0.01	0.01
	Recovery MSCR 3.2kPa	[%]	19.75	82.85	91.37
	Non-recoverable creep compliance MSCR 3.2 kPa	[1/kPa]	0.17	0.01	0.01
	R_{dif}	[%]	9.063	-0.256	0.031
	$J_{nr\text{-}diff}$	[%]	1.557	-2.271	-1.858

Table 4. Selected performance properties of the RTFOT + PAV aged binders.

	Property	Unit	50/70	PmB 25/55-65	PmB 25/55-75
3 x RTFOT aged binder	Penetration [4]	[0,1mm]	30	26	25
	Softening point [5]	[°C]	58.8	87	91.5
	Change of mass	[%]	0.1	0.1	0.1
	Elastic recovery @ 25°C [6]	[°C]	-	59	72
	Force ductility @ 10°C [7]	[J/cm^2]	-	1.4	2.2
	Recovery MSCR 0.1kPa	[%]	10.68	83	92.62
	Non-recoverable creep compliance MSCR 0.1 kPa	[1/kPa]	0.62	0.01	0.01
	Recovery MSCR 3.2kPa	[%]	7.45	83.67	92.68
	Non-recoverable creep compliance MSCR 3.2 kPa	[1/kPa]	0.63	0.01	0.01
	R_{dif}	[%]	30.181	-0.817	-0.068
	$J_{nr\text{-}diff}$	[%]	1.804	-4.093	0.253

Difference can be seen in the force ductility and elastic recovery on the virgin binder and the recovery of the MSCR test on the RTFOT aged binder – which shows the higher level of polymer modification of the PmB 25/55-75 binder. These results show the higher resistance of the PmB 25/55-75 binder against fatigue cracking and low temperature cracking. Due to the fact, that the asphalt mixture has to withstand mainly local static load and stresses caused by forklifts, a better resistance against rutting can be expected in the long term [1].

From the results it can be concluded that the paving grade bitumen 50/70 and the polymer modified asphalt binders react differently to different ageing methods and thus it is difficult to draw clear conclusions from the above mentioned test results. It is recommended to pay attention to this fact in the future research due to the increase number of binder properties that have to be measured after RTFOT + PAV according to new asphalt binder technical standards.

If we focus on the ageing test methods used in the EN technical standards for polymer modified bitumen – RTFOT + PAV – we can see that the paving grade bitumen undergoes a greater ageing change. To further conclude – if we would follow only the results of the binder testing in this paper – the higher polymer modification improves the ageing resistance and also the lifespan of the pavement. These results are in line with the currently longest ageing research on asphalt binder ageing in the pavement – which has been done as part of the LAVOC (Le Laboratoire des voies de circulation) study [19] in the years 1988 – 2007 – and that shows certain limits in the ageing of highly polymer modified asphalt binders, after which there are no further changes in the colloidal composition of the bitumen and thus no further performance changes.

During the elastic recovery and force ductility test of the PmB 25/55-65, the test sample broke before reaching the test length, while the samples of the highly polymer modified asphalt binder PmB 25/55-75 stayed intact.

Differences can be seen in the recovery of the binders, while the DSR phase angle show almost the same values.

All results point of the fact, that the PmB 25/55-75 binder has a greater resistance against ageing in comparison with standard polymer modified asphalt binder PmB 25/55-65 and paving grade bitumen 50/70 [1].

5. Production of asphalt mixture ACO 11 S PmB 25/55-75

The asphalt mixture was produced in the asphalt mixing plant Teltomat V in Polička - Modřec. The temperature of the asphalt mixture during production ranged from 175 to 185°C. Subsequently, the mixture was dispatched from the container of hot asphalt mixture to trucks and delivered to about 25 km distant construction site, where the realization of the experimental road section took place. Due to the transport distance and routing mainly on the first class road, the transport time from mixing plant to the place of laying asphalt mixtures does not exceed 45 minutes.

To verify physical-mechanical and functional properties of the asphalt mixtures sampled from the production in the mixing plant, a number of control laboratory tests was performed. The tests were performed both for the mixture ACO 11 S with highly polymer modified asphalt binder PmB 25/55-75, and for a conventional mix ACO 11 S with unmodified road bitumen 50/70 as well as for the mixture ACO 11 S with conventional polymer modified asphalt binder PmB 25/55-65. Afterward, properties of these mixtures were compared and the results are given in the following chapter [1].

6. Results of laboratory tests of asphalt mixtures

6.1. Comparison of basic asphalt mixtures properties

The properties of the mixture ACO 11 S PmB 25/55-75, which was used for realization of the experimental road section, were compared with the properties of conventional mixtures ACO 11 S 50/70 and mixtures ACO 11 S PmB 25/55-65. All these mixtures have the same gradation curve, only difference is the asphalt binder and the temperature at which the mixtures were compacted. A mixture with modified asphalt binder requires higher temperature during compaction, specifically 155°C compared to 150°C for a mixture with conventional paving grade bitumen 50/70. This results in achieving different values of bulk density and consequently porosity. Specific values measured in control tests are shown in Table 5 [1].

Table 5. Comparison of the basic properties of asphalt mixtures ACO 11 S with different types of bituminous binders.

Mixture:	ACO 11 S PmB 25/55-75	ACO 11 S PmB 25/55-65	ACO 11 S 50/70
Maximal specific gravity ρ_{mv} ($Mg.m^{-3}$) [14]	2.465	2.452	2.444
Bulk specific gravity ρ_{bssd} ($Mg.m^{-3}$) [13]	2.393	2.382	2.370
Air voids content Vm (% of volume) [15]	2.9	2.9	3.0

6.2. Determination of asphalt mixture resistance against the water damage

A set of six specimens was prepared in the Marshall compactor for the asphalt mixure ACO 11 S PmB 25/55-75, a comparative mixture ACO 11 S 50/70 and mixture ACO 11 S PmB 25/55-65. Compacting was done by 2 x 25 hits of Marshall compactor at 155°C for the mixtures with the polymer modified asphalt binder and 2 x 25 hits of Marshall compactor at 150°C for the conventional mix ACO 11 S with unmodified paving grade binder 50/70. Resistance to water was subsequently tested according to the standard ČSN EN 12 697 – 12 on these compacted test specimens. Results of the tests are shown in Table 6 [1].

Table 6. The test results ITSR [16].

Mixture:	ACO 11 S PmB 25/55-75	ACO 11 S PmB 25/55-65	ACO 11 S 50/70
Ø indirect tensile strength (wet group) [kPa]	885	1010	830
Ø indirect tensile strength (dry group) [kPa]	1040	1050	920
ITSR [%]	85.1	96.2	90.2

The lowest value of ITSR was measured for the specimens prepared from a mixture of ACO 11 S PmB 25/55-75. However, the indirect tensile strength of the group of dry specimens is comparable with the mixture ACO 11 S PmB 25/55-65. The results show that although ITSR value is higher for the mixture ACO 11 S 50/70 than the mixture ACO 11 S PmB 25/55-75, indirect tensile strength of the mixture ACO 11 S PmB 25/55-75 has significantly higher values in both dry and wet groups of specimens. The values of ITSR of all tested mixtures fulfilled the requirement of ČSN EN 13 108-1 - not less than 80%. The surprising lowest value of ITSR, determined for the mixture ACO 11 S PmB 25/55-75, could be partly influenced by production inaccuracies in the asphalt mixing plant. For a more precise determination of the ITSR parameter, it would be appropriate to repeat the test on laboratory prepared mixtures.

6.3. Wheel tracking test

The mixture ACO 11 S PmB 25/55-75 was also tested to wheel tracking (resistance to permanent deformation). Comparative mixture ACO 11 S 50/70 and the mixture ACO 11 S PmB 25/55-65 also underwent this test. Test plates with thickness of 40 mm, compacted at 99.0 to 100.0%, were laboratory prepared from each mixture. Bulk density of the specimen prepared within the control analysis of mixtures was used as a reference value. The compaction temperature of the test specimens was 155°C for mixtures with polymer modified asphalt binder and 150°C for mixture with conventional road bitumen 50/70. Testing temperature was 50°C and the results of the test are shown in Table 7 [1].

Table 7. Results of rutting test (50 °C) [17].

Mixture:	Rutting PRD_{AIR} [%]	Rutting WTS_{AIR} [mm/10^3cyklů]
ACO 11 S 50/70	2.6	0.021
ACO 11 S PmB 25/55-65	2.4	0.010
ACO 11 S PmB 25/55-75	2.3	0.023

The results of the test determining resistance of asphalt mixtures to permanent deformation performed at 50°C do not indicate any large differences. In regard to the experience that the surface temperature of surface course may reach temperature exceeding 50°C on hot summer days, the test of resistance to permanent deformation was also performed at 60°C and the results of this test are shown in Table 8.

Table 8. Results of rutting test (60 °C) [17].

Mixture:	Rutting PRDAIR [%]	Rutting WTSAIR [mm/10^3cyklů]
ACO 11 S 50/70	5.6	0.105
ACO 11 S PmB 25/55-65	3.5	0.051
ACO 11 S PmB 25/55-75	3.0	0.036

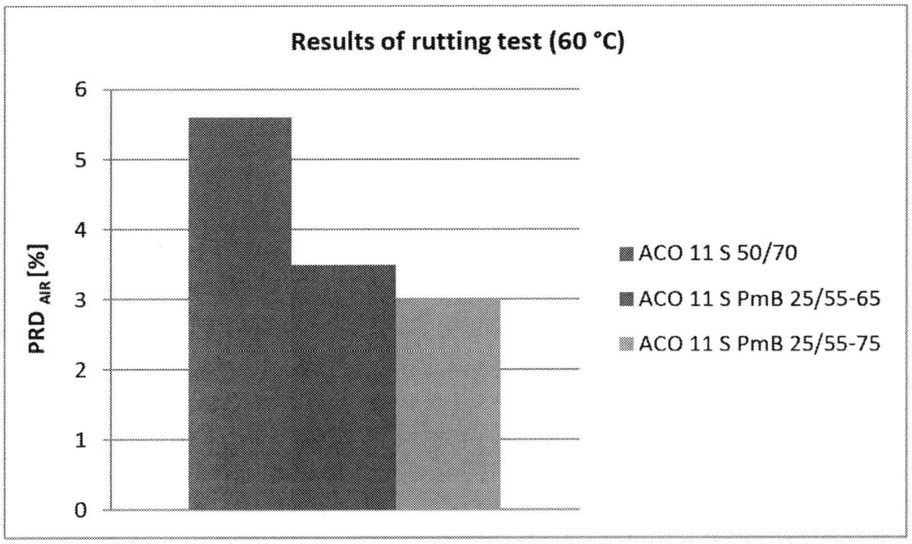

Figure 1. Results of rutting test – Proportional Rut Depth (60 °C) [17].

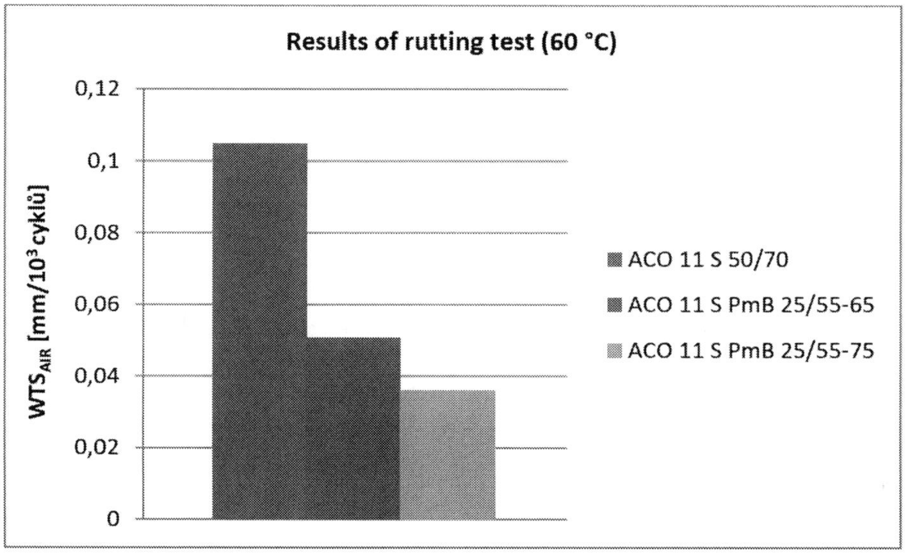

Figure 2. Results of rutting test – Wheel Tracking Slope (60 °C) [17].

Results of tests of asphalt mixture resistance to permanent deformation performed at 60°C clearly show how the type of asphalt binder influences the resistance against the permanent deformation. This is significant especially for surface courses which are directly confronted with high temperatures in summer.

6.4. Low-temperature properties and crack formation by using uniaxial tensile test

Finally, mixtures ACO 11 S PmB 25/55-75 and comparative mixture ACO 11 S PmB 25/55-65 were underwent to low-temperature properties test and crack formation by using uniaxial tensile test. Three specimens of dimensions 50 x 50 x 200 mm for each of the mixtures were prepared and tested for this test. The test results are summarized in Table 9 [1].

Table 9. Results of low temperature properties tests [18].

Mixture:	ACO 11 S PmB 25/55-65			ACO 11 S PmB 25/55-75		
sample	**1**	**2**	**3**	**1**	**2**	**3**
The temperature in the chamber at failure [°C]	-21.7	-23.8	-24.1	-25.0	-25.6	-25.4
The temperature of the sample at failure [°C]	-17.3	-20.3	-20.5	-21.5	-21.3	-21.4
The average temperature of the samples at failure [°C]		-19.3			-21.4	

It is apparent from the test results of low-temperature properties of asphalt mixtures ACO 11 S PmB 25/55-65 and ACO 11 S PmB 25/55-75 that the asphalt mixture with highly polymer modified asphalt binder is less susceptible to the formation of low temperature cracks. Generally, it can be concluded from the laboratory tests mentioned above that the mixture with highly polymer modified asphalt binder achieves better parameters in tests running at high temperatures, as well as low temperatures.

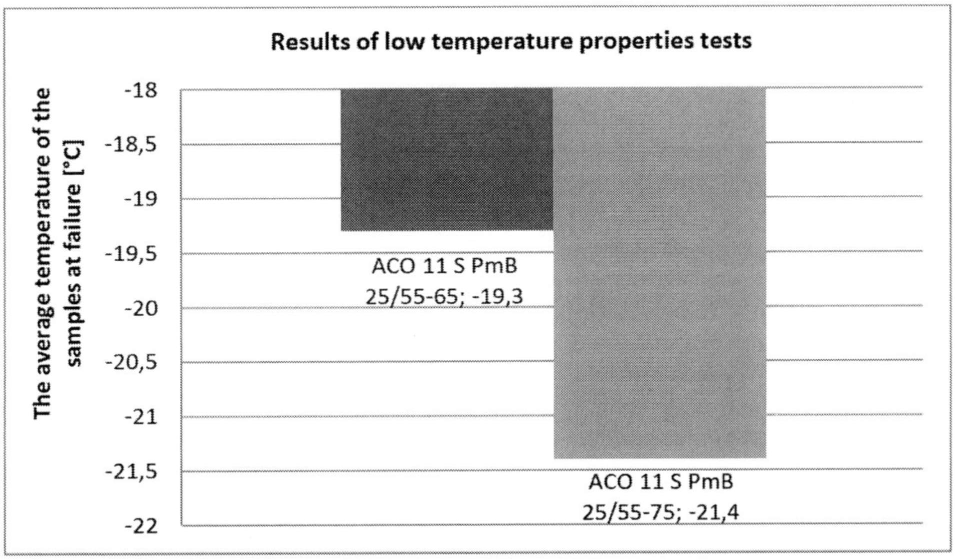

Figure 3 Results of low temperature properties tests [18].

7. Paving of asphalt mixture ACO 11 S PmB 25/55-75

Paving and compacting of asphalt mixture ACO 11 S PmB 25/55-75 was performed by the following machinery in the thickness of 40 mm:

- Asphalt Paver 1803-3i Vögele
- Compaction roller HAM HD 90
- Compaction roller HAM HD 12

During the paving, workability and compactability of the mixtures were mainly monitored. During the machine paving, no differences were observed compared to the machine paving of conventional asphalt mixture ACO 11 S with paving grade bitumen 50/70, or the polymer modified asphalt binder PmB 25/55-65. However, noticeable difference was observed during hand paving in places where machine paving was not possible. Already at temperatures around 160°C, the mixture exhibited a high degree of consistency and it was difficult to work with. The degree of compaction of the asphalt layer was checked by laboratory staff of company Skanska Asfalt s.r.o. using nondestructive nuclear gauge Troxler 4640B. The measurement was made after each pass of roller [1].

Table 10. The results measure the degree of compaction by radiosonde Troxler.

	Bulk specific gravity	Rate of compaction [%]	Air voids content in layer [%]
1. compaction cycle	2.271	94.9	7.9
2. compaction cycle	2.329	97.3	5.6
3. compaction cycle	2.378	99.4	3.6

7.1. Workability and compactability of asphalt mixtures

Correct paving and compaction is crucial for proper function of heavily loaded surface course. During compacting of asphalt layer ACO 11 S PmB 25/55-75, a compaction experiment was performed, in which compaction of layer immediately after laying by paver and then between passes of the rollers was measured by nondestructive nuclear gauge Troxler 4640 B. Individual measurements and reached values of bulk density and degree of compaction are shown in Table 10 [1].

8. Conclusion

The target of above described research task carried out in research project TE01020168 CESTI (Center for Effective and Sustainable Traffic Infrastructure) was practical verification of asphalt mixture with polymer modified asphalt binder which has higher degree of modification than conventional used PmBs in the Czech Republic. At first there were performed laboratory evaluations of highly polymer modified asphalt binder PmB 25/55-75. which were compared with results of laboratory tests carried out on conventional PmB 25/55-65 and paving grade bitumen 50/70. Next step for highly polymer modified asphalt binders possible benefits quantification was comparison of this binder with conventional PmB 25/55-65 and paving grade bitumen 50/70 directly in the asphalt mixture. The final step was realization of experimental road section with usage of highly polymer modified asphalt binder from TOTAL company (Styrelf PmB 25/55-75) into the heavy traffic load friction course.

With respect to results of performed asphalt binder laboratory tests there is possible to expect better functional properties and better durability for asphalt mixtures with highly polymer modified asphalt binder. Highly polymer modified asphalt binder PmB 25/55-75 shows much higher resistance of the binder against the thermo-oxidative aging. better elasticity and cohesion. compare with conventional PmB 25/55-65 and paving grade bitumen 50/70. On the bases of carried out tests there can be expected higher resistance against the fatigue and low temperature cracking. stripping and revealing and better resistance against the rutting in case of highly polymer modified asphalt binder usage in to the asphalt mixture.

By the experimental road section realization there was demonstrated that asphalt mixtures with highly polymer modified asphalt binder PmB 25/55-75 are possible to produce in standard way without any adjustment of asphalt mixing plant. The workability of the mixture on the construction site was good during the machinery paving. but little bit worse during the hand paving in conventional temperature. The compactability was very similar like in case of conventional asphalt mixtures with PmB. which was demonstrated by nondestructive density test by nuclear surface gauge TROXLER. The experimental road section will be observed in the future to evaluate field performance of the asphalt layer with highly polymer modified asphalt binder in longer time period.

9. Acknowledgement

The article has been prepared with the support of project TE01020168 Centre for Effective and Sustainable Traffic Infrastructure (CESTI) co-financed by Technology Agency of the Czech Republic.

10. References

[1] Špaček P, Hegr Z and Beneš J 2016 *Zpráva o realizaci pokusného úseku s použitím vysoce polymerem modifikovaného pojiva PmB 25/55-75 – Report CESTI 1.1.6 (in Czech) /Report about realization of experimental road section with usage of highly polymer modified bitumen PmB 25/55-75 - Research report CESTI 1.1.6* p35 p
[2] European Committee for Standardisation (CEN) 2009 Bitumen and bituminous binders – Specifications for paving grade bitumen *(EN 12591)*
[3] European Committee for Standardisation (CEN) 2017 Bitumen and bituminous binders – Specification framework for polymer modified bitumen *(EN 14023)*
[4] Úřad pro normalizaci, metrologii a státní zkušebnictví (ÚNMZ) 2015 Asfalty a asfaltová pojiva – Stanovení penetrace jehlou /Bitumen and bituminous binders – Determination of needle penetration/ *(ČSN EN 1426)*
[5] Úřad pro normalizaci, metrologii a státní zkušebnictví (ÚNMZ) 2015 Asfalty a asfaltová pojiva – Stanovení bodu měknutí – Metoda kroužek a kulička /Bitumen and bituminous binders – Determination of softening point – Ring and Ball method/ *(ČSN EN 1427)*
[6] Úřad pro normalizaci, metrologii a státní zkušebnictví (ÚNMZ) 2010 Asfalty a asfaltová pojiva – Stanovení vratné ductility modifikovaných asfaltů /Bitumen and bituminous binders – Determination of the elastic recovery of modified bitumen/ *(ČSN EN 13398)*

[7] European Committee for Standardisation (CEN) 2003 Bitumen and bituminous binders – Determination of the tensile properties of modified bitumen by the force ductility method *(EN 13589)*

[8] European Committee for Standardisation (CEN) 2007 Bitumen and bituminous binders – Determination of the Fraass breaking point *(EN 12593)*

[9] International Organization for Standardization (ISO) 2000 Determination of flash and fire point – Cleveland open cup method *(ISO 2592)*

[10] European Committee for Standardisation (CEN) 2007 Bitumen and bituminous binders - Determination of the resistance to hardening under influence of heat and air – Part 1: RTFOT method *(EN 12607-1)*

[11] Úřad pro normalizaci, metrologii a státní zkušebnictví (ÚNMZ) 2012 Asfalty a asfaltová pojiva – Urychlené dlouhodobé stárnutí v tlakové nádobě (PAV) /Bitumen and bituminous binders – Accelerated long ageing conditioning by Preassure Ageing Vessel (PAV)/ *(ČSN EN 14769)*

[12] Český normalizační institut (ČNI) 2008 Asfaltové směsi - Specifikace pro materiály - Část 1: Asfaltový beton /Bituminous mixtures – Material specifications – Part 1: Asphalt Concrete/ *(ČSN EN 13 108-1)*

[13] Český normalizační institut (ČNI) 2005 Zkušební metody pro asfaltové směsi za horka – Část 6: Stanovení objemové hmotnosti asfaltového zkušebního tělesa /Bituminous mixtures – Test methods for hot mix asphalt – Part 6: Determination of bulk density of bituminous specimens/ *(ČSN EN 12 697-6)*

[14] Český normalizační institut (ČNI) 2005 Zkušební metody pro asfaltové směsi za horka – Část 5: Stanovení maximální objemové hmotnosti /Bituminous mixtures – Test methods for hot mix asphalt – Part 5: Determination of the maximum density/ *(ČSN EN 12 697-5)*

[15] Český normalizační institut (ČNI) 2005 Zkušební metody pro asfaltové směsi za horka – Část 8: Stanovení mezerovitosti asfaltových směsí /Bituminous mixtures Test methods for hot mix asphalt – Part 8: Determination of void characteristic of bituminous specimens/ *(ČSN EN 12 697-8)*

[16] Český normalizační institut (ČNI) Zkušební metody pro asfaltové směsi za horka – Část 12: Stanovení odolnosti zkušebního tělesa vůči vodě /Bituminous mixtures – Test methods for hot mix asphalt – Part 12: Determination of the water sensitivity of bituminous specimens/ *(ČSN EN 12 697-*12*)*

[17] Český normalizační institut (ČNI) 2005 Zkušební metody pro asfaltové směsi za horka – Část 22: Zkouška pojíždění kolem /Bituminous mixtures – Test methods for hot mix asphalt – Part 22: Wheel tracking/ *(ČSN EN 12 697-22)*

[18] Český normalizační institut (ČNI) 2005 Zkušební metody pro asfaltové směsi za horka – Část 46: Nízkoteplotní vlastnosti a tvorba trhlin pomocí jednoosé zkoušky tahem /Bituminous mixtures – Test methods for hot mix asphalt – Part 46: Low temperature cracking and properties by uniaxial tension tests/ *(ČSN EN 12 697-46)*

[19] Dreessen S, Gallet T, Dumont A G and Pittet M 2012M Durability study: Field experience of long-term evolution of SBS polymer modified bitumen *5th Euroasphalt & Eurobitume Congress (Istanbul)*

[20] Kluttz R Q, Dongré R, Powell R B, Willis J R and Timm D H 2016 Long term performance of a highly modified asphalt pavement and application to perpetual pavement design 6th *Euroasphalt & Eurobitume Congress (Prague)*

[21] Eckmann B, Mazé M, Largeaud S and Dumont S F 201The contribution of cross-linked polymer modified binders to asphalt performance *5th Euroasphalt & Eurobitume Congress (Istanbul)*

[22] Saba R G, Uthus N and Aurstad J 2012 Long-term performance of asphalt surfacing containing polymer modified binders 5th *Euroasphalt & Eurobitume Congress (Istanbul)*

Asphalt mixtures with a high amount of RAP – case study

Tomas Koudelka[1] **and Michal Varaus**[1]

[1]Brno University of Technology, Veveří 331/95, Brno 602 00, Czech Republic

E-mail: koudelka.t@fce.vutbr.cz

Abstract. A case study of one trial section in the Pilsen region is presented. The pavement in the section was newly constructed in 2015 using one type of an asphalt concrete mixtures with varying RAP content. The constructed surface course comprises of 0% to 50% RAP. In order to restore the aged binder properties and to avoid the embrittlement of the produced mixtures, a rubber-based modifier/rejuvenator was employed. For technological reasons during manufacturing processes, which engage a parallel drying drum, a crude oil-based rejuvenator was also added. This article contains the preliminary data from an on-going project focused on monitoring the properties of bituminous binders contained in asphalt mixtures. The actual bituminous binders were extracted straight after production, after 6 months and after 12 months. The binder characteristics are evaluated using empirical testing as well as functional tests. Low temperature properties are measured by a Bending Beam Rheometer (BBR). The preliminary results show, that the bituminous binders properties change significantly in a relatively short period of time. The progress in binder' characteristics is contradictory to up-to date knowledge. The probability that the phenomenon of diffusion between aged binder and rejuvenator agents occurs exists. Moreover, the data might indicate that the process of rejuvenator evaporation takes place.

1. Introduction

The technology for recycling of asphalt mixtures has been known for several decades [1], but only the development of mixing technologies, production processes and, ultimately, environmental awareness, has led to its massive expansion [2].

In 2015 the Czech Republic produced about 8 million tonnes of virgin asphalt mixtures, whilst 2 million tonnes of asphalt mixtures were available for recycling purposes. It should be added that despite greater awareness of recycling technology in the Czech Republic, only 16% of the amount of available material is recycled in the production of hot asphalt mixtures. When compared with other EU member states, the Czech Republic lags behind in this respect, even when compared with countries that have a similar or even lower amounts of reclaimed asphalt pavement (RAP) material available: Austria 45%, Denmark 52%, Belgium 64%, and for example Finland 100% [3].

If milled material or reprocessed asphalt mixtures from flexible pavements meet the classification criteria according to Czech technical condition regarding recycling material classification (TP 210) and European standards (EN 13108-8), it may be classified as RAP that can be reused in the production of hot asphalt mixtures in the asphalt plant. At present in the Czech Republic, the addition of RAP material is only allowed in mixtures of type asphalt concrete, and the maximum allowance is determined in the National Annex to EN 13108-1 Table NA-E-4.5 Asphalt concrete wearing courses can contain up to 25% RAP material. For the production of hot asphalt mixtures with a higher content of RAP (more than 25%), the asphalt plant should be equipped with appropriate technologies that preheat RAP material or at least allow its gradual dosing into the pugmill [4].

One of the main parameters for testing the durability of asphalt mixtures is the use of fatigue tests. Furthermore, individual functional parameters are tested, such as stiffness, or low temperature properties. When adding a higher amount of RAP material, there is often concern that the durability and functional parameters of such mixtures are worse than in case of conventional mixtures. Asphalt mixture after the addition of RAP material becomes more rigid due to the aged binder which it contains [5]; the higher level of rigidity then causes greater resistance to permanent deformation [6]. On the other hand, such mixtures may encounter the previously mentioned disorders associated with temperature and fatigue failures [7, 8].

To prevent the above mentioned problems in mixtures with high levels of RAP material, it is important to use one of the methods of "rejuvenating" the aged binder. The normal method of rejuvenation of aged binder is to use softer-grade binder, or to use special additives. Rejuvenation of aged binder is the key process during production, affecting both functional parameters of asphalt mixtures and binders as well as mixture's lifespan [9, 10].

Aside from the effects of rejuvenating agents on the rheological properties of bituminous binder, it is also necessary to examine the durability of such "rejuvenation" in respect to the volatility of some of the added agents as well as the thermo-oxidative instability. Laboratory conditions, in which these tests are usually carried out, don't always accurately simulate the real situation in road construction. In 2015, for the reasons mentioned above, under the new technologies of the State Transport Infrastructure Fund, the construction of a test area was begun, with the application of higher levels of RAP material, which goes beyond the permissible limits stipulated in EN 13108-1 Asphalt concrete. This step will not only test the effects of higher dosing of RAP material, but will also explore the possibility of monitoring the long-term development and verify the influence of rejuvenation agents on the rheological properties of the bituminous binder.

2. Interaction between rejuvenating agents and aged asphalt binder

A crucial step during asphalt recycling is the process of mixing [11]. After manufacture, the consistency of the rejuvenated binder should be uniform and equal to virgin binder which would be chosen for the new construction without the use of RAP. However, blending aged binder and other constituents (virgin binder + rejuvenator) is a complex process which does not have to occur necessarily already in the course of the mixing period but continues until the equilibrium in consistency is reached over the majority of the asphalt film thickness [12]. During manufacture, the constituents are being continuously mixed; the rejuvenator is simultaneously attracted to virgin and aged binder. Both materials start to soften albeit in different rates. Also the procedure used to determine the optimal amount of rejuvenator and the way of dosing during recycling operation are focal points in mixture design [13]. Figure 1 illustrates possible options acquired after mixture manufacture which might be acquired through mixture manufacture which were proposed by Kriz [14].

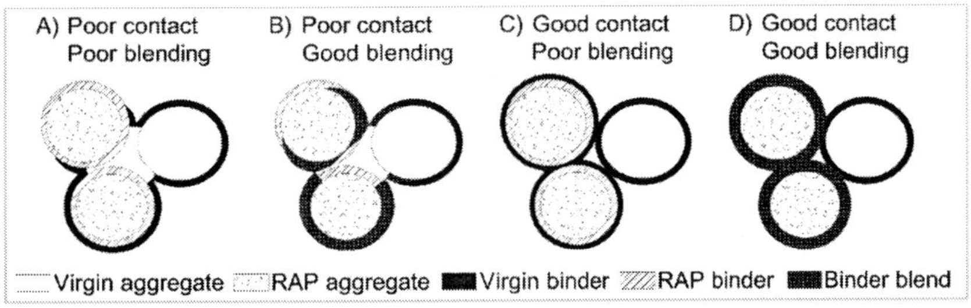

Figure 1. RAP-virgin binder contact and blending due to diffusion. Schematic representation of possible options after manufacture [14].

When it comes to a poor blending occurs (A), binder properties vary within the mixture. Although a rejuvenator might diffuse into the aged binder in the subsequent period following mixing process, binder homogeneity remains low leaving the mixture prone to failures (B). Rejuvenator added into the new fresh bitumen can also form a thin uniform layer that surrounds aged binder. Nevertheless, the blending process might not accomplish during manufacture period because the diffusion process is too slow which in turn causes the mixture being too stiff after production (C). Only when a sufficient degree of blending is reached, the functional properties are predictable (D) [14]. Furthermore, Kriz concluded that the rate of diffusion is greatly influenced by the binder film thickness and temperature.

From the above it is seen that the diffusion and subsequent material (rejuvenator) transport procedure are important processes in obtaining long-term performance of recycled asphalt pavements.

The process of diffusion might be in lab conditions simulated by the method of step extraction. Multi-stage extraction method was utilized in the work of Nourdelin and Wood in 1987 [15], lately supported by the work of Huang [16]. Both studies agreed that in most cases a multilayered system occurs after manufacture period instead of having well blended material with uniform properties. Comprehensive study conducted by Navaro [17] concluded that the degree of interaction between constituents is a function of both parameters, i.e. temperature (aggregate and RAP) and mixing time. It is therefore a major issue, yet not fully solved, to predict a degree of blending and particularly interaction between individual constituents within rejuvenated binder after mixture manufacture at a plant.

The reason for the actual situation is that there is an overabundance of products with different origin (petroleum, bio-based, artificial) and characteristics (viscosity, chemical composition, flash point, etc.) available on the market and no one really knows how they in fact affect individual performance characteristics and how they impact aged binder especially from a long-term perspective. In spite of this, all producers claim that the products significantly improve binder performance [18, 19 and 20].

It is reasonable to assume that the compatibility among binder constituents is the key parameter which governs rejuvenated binder‘s long term performance. Incompatible oils or rejuvenator overdose can result in a migration of light oils which results in bleeding or flushing and also getting the base binder harder [21]. The migration of light substances of which are some rejuvenators unquestionable composed off is promoted by heat [22].

3. Objectives

The main objective of this paper is to present and compare test results of bituminous binders which have been obtained through extraction and subsequent distillation of asphalt mixtures with different amounts of RAP.

Part of the aim is then the description of the effect of rejuvenation / modification additives on the properties of the bituminous binder over time.

4. Trial section description

The test section is located in a rural area between the villages of Mrtník and Kaznějov in the Plzeň region, on road number II / 204.

The total length of the test section is 2.765 km, which is divided into four subsections. The average daily traffic volume on this stretch of road is 1306 vehicles in 24 hours (RMD traffic census 2010) and the traffic load on the road is as follows: 466 heavy vehicles, 823 cars and 17 motorcycles. There is a kaolin quarry nearby and it can therefore be assumed that the road will be greatly burdened by heavy freight transportation.

5. Materials

5.1. Road construction

During reconstruction an experimental type of asphalt concrete mixture was used, designed as asphalt concrete for surface courses (ACO 11+), with a graduated dosing of RAP material. Optimum amounts of bituminous binder and grading curve for the individual mixtures were chosen to meet the requirements stated in EN 13108-1. The composition of each mixture is described in Table 1. These in a plant produced mixtures were designed and manufactured so as to hold all possible variables during the production process constant (binder content, discharge mixture temperature, virgin bitumen type) except for the amount of RAP.

Table 1 Individual wearing course mixture designs

Section	Designation	Bitumen binder	Modifier/rejuvenator (A)	Rejuvenator (B)	RAP content	Binder content
1	1.1	50/70	YES	NO[1]	0%	5,5%
2	2.1	50/70	YES	YES	30%	5,4%
3	3.1	50/70	YES	YES	50%	5,5%
4	4.1	PMB 45/80-65	NO	YES	30%	5,4%

Table 1 also indicates the content of the bituminous binder, which is closely related to the thickness of the asphalt film.

5.2. Rejuvenating agents

Rejuvenation / modification agent (A) is composed primarily of rubber granules with a maximum size of <1 mm, whose properties are modified by the addition of Fischer Tropsch wax and oil, which is obtained during the distillation process of petroleum products (additive B). Additive A was dosed in the amount of 13% of the weight of bituminous binder (virgin + aged) in the resulting mixture according to the manufacturer's recommendations. Since this additive contains mainly rubber granules, the manufacturer states that the addition of this agent during the manufacturing process will result in modification of the bituminous binder and the resulting mixture should thus behave the same way as after the use of modified binder. Characteristics of this modifier were the ground for also choosing the polymer modified binder (PMB) in the project with the aim to compare these two.

Due to technological reasons during the manufacturing process of a mixture in the mixing plant using a parallel drum, petroleum based oil (ingredient B) was also added to the mixtures containing RAP material, in the amount of 0.2% of the weight of RAP material.

5.3. Reclaimed asphalt pavement material

During production process each mixture containing RAP material was made using mixed RAP categorized as RA 0/11. Before manufacturing the material was rescreened and homogenized using a granulator, which is specially designed to avoid excessive formation of fine particles which would preclude adding larger amounts of RAP without further screening process. The proper screening process ensures the RAP is of excellent homogeneity. Figure 2 shows the granulator along with a detail of the shaft in the milling chamber. In the background of the figure 2 is an asphalt plant equipped with a parallel drying drum, which was used during the production of the mixtures.

[1] Rejuvenator B was not purposely dosed because the mixture contains no RAP material. However, rejuvenator B is a compositional part of modifier/rejuvenator A

Figure 2. Reclaimed Asphalt Pavement granulator (left), a detail of the crushing chamber with the milling shaft (right), an asphalt plant with the fitted parallel drying drum in the background.

5.4. Mixture production
Reclaimed asphalt pavement material was heated prior to mixing to 130°C. The temperature of mixture discharged from the pugmill was set to 170°C for mixtures with paving bitumen and 180°C for polymer modified bitumen.

5.5. Material collection
All asphalt mixtures were collected at the plant straight after production. Subsequently, the cores were sampled at the construction site after 6 and 12 months. The cores after 12 months were drilled right next to the cores which were taken after 6 months in order to have comparable materials.

6. Methods
The binder was extracted according to EN 12697-1 using a centrifuge and recovered by using a rotary evaporator according to EN 12697-3. Solvent used during distillation was trichloroethylene (TCE). Empirical tests which were carried out were: Penetration test determined according to EN 1426 at 25°C and Softening point test determined according to EN 1427. Low temperature behavior was tested by means of Bending Beam Rheometer according to EN 14771.

7. Results and discussion

7.1. Empirical testing
Bituminous binder at the present trial section was extracted from the mixtures after production and after 6 and 12 months from the date of construction. Figure 3 shows the test results of needle penetration and softening point graphically. Table 1 also indicates the content of the bituminous binder, which is closely related to the thickness of the asphalt film. The thickness of the asphalt film is important to the age-resistance of the bituminous binder. If the binder content in the mixture is lower, the thickness of the bituminous film would be also lower. This would result in a lower resistance of the bituminous binder especially against oxidative aging and individual characteristics would be not comparable.

Penetration values measured on extracted binders from mixtures with paving grade bitumen after manufacture (1.1, 2.1 and 3.1) produced a very small variance, which implies that the penetration is not adversely affected by the presence of RAP material, even in the case when a mixture contains 50% of RAP material. The almost equal penetration values (binder 1.1, 2.1, 3.1) after manufacture mean that aged binder was soften already during manufacture.

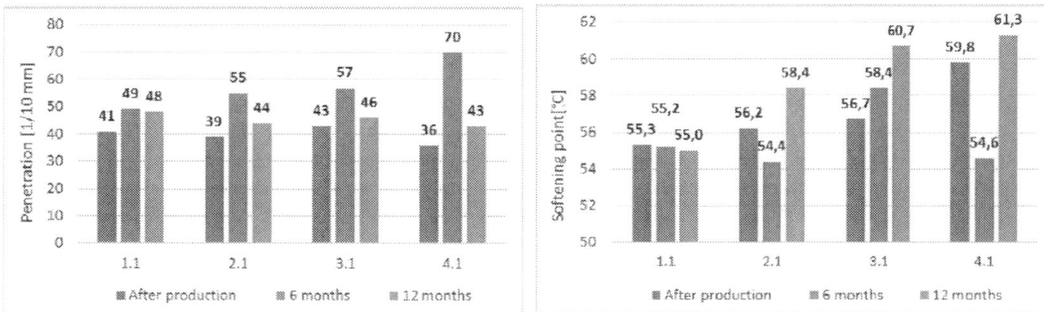

Figure 3. Empirical test results in the given period of time – Penetration (left), Softening point (right).

Likewise, the values of softening points in mixtures with paving bitumen after manufacture fluctuated within a very narrow interval of 56 ± 1 ° C. The results of empirical tests also show that after 6 months the penetration of all binders increased, including modified binder. The value of penetration increased by a value of 20% (mixture 1.1) to a value of up to 94% (mixture 4.1). This suggests that during production of asphalt mixtures with RAP at plants there does not have to occur a complete blending process among individual constitutes as described in chapter 2. This would mean that even after laydown operations, there is still an interaction between the rejuvenating agent and the aged binder – the rejuvenation agent diffuses further into the aged binder, thus changing its properties and "soften'' it. This development in penetration values is contrary to the generally known fact, which says that binder becomes stiffer with age and that over time the penetration value decreases and the softening point increases [23]. Mixtures which contain rejuvenation agents can therefore display completely the opposite behavior.

After 12 months, "softening process" is no longer present, but instead we see a noticeable decline in the values of penetration compared with the values after 6 months. In the case of a mixture free of RAP material (1.1) there is the smallest drop, only about 2%. For other mixtures which contain RAP material, the decrease in penetration value is more pronounced: mixture 2.1 shows a drop of 20%, mixture 3.1 of 19% and mixture 4.1 of 39%. Penetration values after 12 months tend to return to the original values after manufacture.

Looking at Figure 3, it can be seen that mixture 4.1, which contains the modified binder PMB 45/80-65, shows the biggest changes in empirical values. In this case, rejuvenation additive B (oil) was only dosed in a minimal quantity; 0.2% of the weight of the RAP material in the mixture. When comparing mixtures 2.1 and 4.1, which used different binders but the same content of RAP material, it is possible to state that even small amounts of rejuvenation petroleum based oil change the behavior of the modified binder substantially more than in the case of paving bitumen. This behavior may be caused by a reaction between rejuvenation agent and modifiers contained in PMB. The presence of oil might degrade the actual modifier. However, this findings would need to be further proved.

7.2. Functional testing – Bending Beam Rheometer (BBR)

Table 3 compares the various parameters that are measured by BBR. For comparison purposes temperature of -16 ° C was employed.

Table 2. Comparison of individual parameters obtained during low temperature measurement by BBR

Designation	Creep Stiffness S_m [-16°C]			m-value [-16°C]		
	After production	6 months	12 months	After production	6 months	12 months
	[MPa]	[MPa]	[MPa]	[-]	[-]	[-]
1.1	119.0	80.5	88.5	0.333	0.333	0.341
2.1	108.3	79.1	105.4	0.326	0.346	0.325
3.1	111.7	78.9	102.8	0.323	0.361	0.35
4.1	136.5	52.8	131.6	0.344	0.429	0.382

The presence of RAP material in mixtures 2.1 (30%) and 3.1 (50%) does not adversely affect the creep stiffness of the bituminous binder compared to the reference binder from mixture 1.1 after production. After 6 months there was a decrease in creep stiffness of 27% in the case of binder 2.1,

30% for binder 3.1 and of 61% in the case of binder 4.1. Even at low temperatures, the properties of modified binders are affected more significantly by rejuvenating agents than in case of paving bitumens. Binders extracted from mixtures with RAP material after 6 months have also improved their ability to relax, which is evidenced by the increase of m-value and the consequent decrease of the critical temperature. On the other hand, it is seen that after 12 months, creep stiffness increases compared to the situation after 6 months, by 10% for mixtures 1.1 and as much as 149% for mixture 4.1.

Figure 4 graphically illustrates the resulting critical temperature, which has been calculated based on measurements at three different temperatures (-16 ° C, -22 ° C and -28 ° C). The critical temperature is always the greater of the two values.

It is clear that rejuvenating agents have a major impact on low temperature properties. The most striking changes can be observed at critical temperatures, calculated on the basis of m-values. In a time period of 6 months there was a change in the temperature of binder mixture 2.1, which at production had a critical temperature of 20.3 ° C, of about 13.3 ° C and then 12 months after laying the value is the same as after manufacturing.

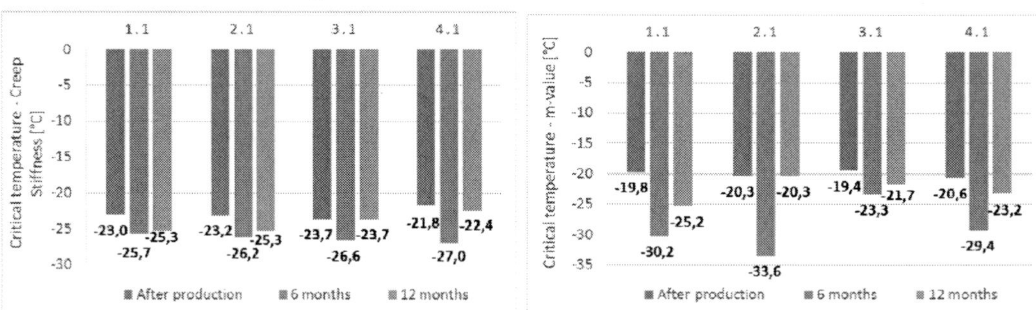

Figure 4. Critical temperatures based on creep stiffness (left) and on m-value (right).

8. Summary of results

The aim of the tests was to compare the rheological behavior of bituminous binders extracted from asphalt mixtures with different amount of RAP material. The properties of individual binders were tested immediately after production and then after 6 and 12 months from the date of laydown operations. Partial task was to determine the effect of RAP material on properties of the bituminous binder and to what extent these properties are affected by the presence of rejuvenating agents.

The following observation were made during the first 12 months of testing :

- The needle penetration test found that all the binders displayed increased penetration 6 months after laying by 20% to 41% for mixtures with paving bitumen. The most striking changes were observed for the mixture with the modified binder which showed an increase of 94%. At the same time, after 6 months a decrease in softening point occurred. After 12 months the "softening" process ceased, a characteristic which was observed in all mixtures regardless of RAP material presence. Penetration decreased after 12 months in comparison to the situation after 6 months whereas the softening point increased.
- With regards to low temperature properties creep stiffness decreased in all bituminous binders after 6 months as well as the critical temperature, Tc. Reduction of creep stiffness at low temperatures corresponds with the results of empirical tests and tests on DSR - the 6-month-old binder is "softer". When comparing individual binders, mixture 4.1, which contained the modified binder, reached the lowest critical temperature both after production and after 6 months. After 12 months, it can be seen that it loses its original effect of "rejuvenation"

(softening) of the binder. The resulting critical temperature values were closer to their initial state after manufacture.

9. Comments and discussion

The first round of extraction (results after manufacture) was performed on mixtures taken at the plant, whereas the extraction after 6 and 12 months was conducted on cores drilled at the site. This means that the first extraction could have been theoretically conducted on a material which lies in the same section where the cores were taken but on a different spot, i.e. the mixture is of the same designation but from a different manufacturing batch. Therefore, the dosing of the rejuvenators could vary slightly because of the dosing device deviation as well as of the properties of RAP binder due to possible inhomogeneity. Nevertheless, it is seen that the softening effect after 6 months occurred for all mixtures (even in the case of mixtures from binder courses – not mentioned in the article) and therefore the results are assumed to be correct. Thus the authors anticipate that the process of rejuvenator diffusion took place after laydown operation.

The weak point of the study is the process of distillation carried out according to EN 12697-3. During the extraction process all binder constituents are mixed together and homogenized. Supposing that the same amount of rejuvenator is present after manufacture as well as after 6 months, the binder characteristics should be equal, provided the homogenization occurs during distillation. The authors are not familiar with any study which would be focused on these points at issues (rejuvenated binder distillation) and thus cannot confirm or deny the above mentioned statement. However, the authors admit that the acknowledgement of the presence of the diffusion processes should not be based solely on the data provided in the article. Diffusion process might have been detected and confirmed by multi-stage extraction which was not carried out.

Another ground for binder softening after 6 months might be an incomplete washing process during extraction. Such error would leave harder binder stuck to aggregate. The authors find this option as highly improbable as this would have to be case for all mixtures. Moreover, aggregate was checked upon washing stage for traces of binder.

The cores studied after 6 and 12 months were drilled exactly at the same spot which in any case means that the tested materials are from the same batch, with the same amount of rejuvenator, rejuvenator distribution and with the same RAP material. Here is therefore no space for any misconduct in terms of material substitution. In this case, the authors suppose that the decrease in penetration after 12 months compared to 6 months might be explained by the phenomena of rejuvenators' migration. Higher ambient temperatures during summer season promoted oily substances to migrate. The trend of penetration decrease manifested itself in all mixtures.

The obtained data provide new insights into the behavior and development of the properties of the bituminous binder contained in mixtures with the addition of a higher amount of RAP material. The results are particularly unique, because they confirm that there is an interaction between the rejuvenating agent and aged bituminous binder contained in RAP material after manufacturing and laying period.

10. Perspective

Further research should be undertaken to confirm, whether the process of diffusion occurs and what rate of rejuvenator transport might be expected. The study focuses on interaction between rejuvenator and asphalt binder during the process of extraction and especially distillation according to EN 12697-3 is a must. As a further step, another trial section is planned in the Czech Republic using various kinds of rejuvenators. Validation of the diffusion process by multi-staged extraction should take place.

Acknowledgements

This paper has been worked out under the project No. LO1408 "AdMaS UP - Advanced Materials, Structures and Technologies", supported by Ministry of Education, Youth and Sports under the "National Sustainability Programme I" and under the Competence Centre TE01010168 "CESTI -

Centre for Effective and Sustainable Transport Infrastructure" supported by the Technology Agency of Czech Republic TAČR and project FAST – S – 17 – 4691 "The assessment of bituminous binder properties over time".

References

[1] Gomes V, Nolfo D M, Vlachos P and Bocci M 2016 Two case studies with high levels of RA enabled by a rejuvenating agent *Proc Int. Conf. 6th Euroasphalt and Eurobitumene Congress*
[2] Wayman M, Andersson-Sköld Y, Bergman R, Huag Y, Parry T, Raaberg J and Enell A 2012 Re-road – End-of-life strategies of asphalt pavements – Life cycle assessment of reclaimed asphal, *Final report of SCP7-GA-2008-218747* pp 1-94
[3] European Asphalt Pavement Association 2016 *Asphalt in Figures* (EAPA Brussel)
[4] Varaus M, Hýzl P, Mondschein P, Valentin J, Pazyna R, Svoboda P and Matoušek D 2013 Recycling of asphalt mixtures in parallel drying drum with the use of rejuvenators and additives for the reduction of manufacturing temperature *Proc. Int. Conf. Asfaltové vozovky: article 3.10*
[5] Al-Qadi L I, Elseifi M and Carpenter H S 2007 Reclaimed Asphalt Pavement – A literature Review *Reserach report FHWA-ICT-07-001* (Centre for Transporatation Ilinois)
[6] Patrick J E, Reilly S J and Cook G K 2006 Trials of recycled asphalt and rubber materials in hot mix asphalt for New Zealand roads *Wellington – Land Transport New Zealand: reserach report 309*
[7] Shah A, McDaniel R S, Huber G A and Gallivan V L 2007 *Investigation of low and high temperature properties of plant-produced rap mixtures* (Georgetown Pike – Federal Highway Administration)
[8] Zaumanis M, Mallick R and Poulikakos L 2014 Influence of six rejuvenators on the performance properties of reclaimed asphalt pavement (rap) binder and 100% recycled asphalt mixtures *Construction and Building Materials* **71** pp 538-50
[9] Radenberg M 2016 Einsatz von Rejuvenatoren bei der Wiederverwendung von Asphalt *Straße und Autobahn* **7** pp 19-25
[10] Zaumanis M and Mallick R 2015 Review of very high-content reclaimed asphalt use in plant produced pavements: state of the art *International Journal of Pavement Engineering* **16** pp 39-55
[11] Karlsson R and Isacsson U 2002 Bitumen rejuvenator diffusion as influenced by ageing *Road Materials and Pavement Design* **3**(2) pp 167-82
[12] Carpenter S H and Wolosick J R 1980 Modifier influence in the characterization of hot-mix recycled materia, *Transportation Reserach Record* **777** pp 15-22
[13] Varaus M and Koudelka T 2016 Performance properties of asphalt mixtures containing reclaimed asphalt pavement *Proc. Int. Conf. of XXIth Seminar of Ivan Poliacek* pp 184-190
[14] Kriz P, Grant L D, Veloza B A, Gale M J, Blahey A G, Brownie J H, Hirts R D and Maccarone S 2014 Blending and diffusion of reclaimed asphalt pavement and virgin materials *Road Materials and Pavement Design* **15**(S1) pp 78-122
[15] Noureldin A S and Wood L E 1987 Rejuvenator diffusion in binder film for hot/mic recycled asphalt pavement *6th annual meeting of the Transporatation Board* pp 51-61
[16] Huang B, Guogiang L, Vukosavljevic D, Shu X and Egan B 2005 Laboratory investigation of Mixing Hot-Mix Asphalt with Reclaimed Asphalt Pavement *Transporatation Reserach Record Journal* **1929** pp 37-45
[17] Drouandaine I, Navaro J, Bruneau D, Poteau B, Colin J and Dony A 2012 Rap binder in reclaimed asphalt concrete through mixing conditions: observation, measurement and mechanical consideration of the blending *Proc. 6th Euroasphalt and Eurobitumene Congress*
[18] Bailey H and Zoorob S E 2012 The use of vegetable oil as a rejuvenator for asphalt mixtures *Proc. Int. Conf. 6th Euroasphalt and Eurobitumene Congress*
[19] Oldham J D and Fini E H 2015 Application of a bio-binder as a rejuvenator for wet processed

asphalt shingles in pavement construction *Construction and building materials* **86** pp 75-84

[20] Somé C, Pavoine A, Cailleux E, Andrieux L, DeMarco L, Philippe D S and Stephan B 2016 Rheological behaviour of vegetable oil-modified asphaltite binders and mixes *Proc. Int. Conf. 6th Euroasphalt and Eurobitumene Congress*

[21] Zaumanis M, Mallick R B and Frank R 2103 Use of rejuvenators for production of sustainable high content rap hot mix asphalt *Proc. Int. Conf. The XXVIII International Baltic Road Conference*

[22] Cong P, Hao H, Yhang Y, Luo W and Yao D 2016 Investigation of diffusion of rejuvenator in aged asphalt *International Journal of Pavement Reserach and Technology* **9** pp 280-8

[23] Read J and Whiteoak D 2003 *Shell Bitumen Handbook 5th. edition* (London: Thomas Telford Publishing)

Epoxy asphalt concrete is a perspective material for the construction of roads

Valerii Vyrozhemskyi[1], Ivan Kopynets[1], Sergii Kischynskyi[1] and Nataliia Bidnenko[1]

[1] M.P. Shulgin State Road Research Institute State Enterprise – DerzhdorNDI SE, Peremohy Ave. 57, Kyiv 03113, Ukraine

E-mail: orgviazhuchi@ukr.net

Abstract. An effective way to increase the durability of asphalt concrete pavements that are subject to high traffic loads and adverse weather and climatic factors is the use of polymer additives which drastically improve the rheological and physical-mechanical properties of bitumen. The use of thermosetting polymers including epoxy resins for asphalt and bitumen modification is seen as a perspective solution for this issue. Conducted at DerzhdorNDI SE studies have proved high riding qualities of asphalt pavements that contain epoxy resins. When replacing 20-35% of bitumen with epoxy component, a significant improvement in strength characteristics of asphalt pavement is noted, especially at elevated temperatures. Specific feature of epoxy asphalt concrete is its ability to gain strength over a long-term operation. Thus, despite the increased cost of epoxy asphalt concrete, long service life of pavements on its basis (up to 30 years as predicted) ensures a high profitability of using this material, especially on the roads with heavy traffic and severe traffic conditions.

1. Introduction

Today in Ukraine, actual service life of asphalt pavements is 6-12 years instead of the design 15-20 years. Premature repairs and rehabilitation of pavements require millions of additional costs that could be directed to the construction of new and reconstruction of existing roads. One of the main reasons of rapid deterioration of pavements in conditions of heavy traffic is bitumen properties. Of all the road construction materials bitumen is the most sensitive to traffic loads and impact of weather and climate factors. Being a thermoplastic material, bitumen is softened at high summer temperatures and it becomes fragile at low temperatures in winter. Along with low adhesion of bitumen with stone materials this leads to the formation of ruts, cracks, potholes and pits on the pavement.

Due to the above, a targeted change of bitumen properties is the most effective way to increase efficiency and durability of road pavements.

Modification of bitumen and asphalt concrete by different additives, particularly by polymers, is widespread.

Today, great interest in the use of thermosetting polymers especially epoxy resins as modifiers have shown the countries of the Western Europe, the USA, China and Australia [1-5].

Epoxy resins are oligomers containing epoxy groups and can form cross-linked polymers under the hardeners impact. The most common epoxy resins are products of epichlorohydrin polycondensation with phenols, and with bisphenol A are the most widespread.

Materials having various properties from rubbery to rigid and solid that exceed steel by strength characteristics can be obtained by combining different resins and hardeners. Today, epoxy resin which properties exceed all other materials properties based on synthetic resins is used in all branches of industry.

Epoxy glue is considered as the most all-purpose and reliable glue. Thanks to its strength and good adhesion, this glue bonds equally well both shoes and the parts of the planes. Epoxy resin is used as an impregnating material for glass thread and fabrics to manufacture glass fiber plastics which are used in shipbuilding, construction, aircraft construction, manufacturing of the machines parts and others. Nowadays epoxy resins are used in road construction for preparation of asphalt mixtures. Different types of asphalt containing epoxy resins are called epoxy asphalt concrete.

2. Materials

The following materials are used for research of asphalt concrete based on epoxy resin:

- Bitumen of BND 60/90 grade and production of PJSC "Ukrtatnafta" that meets the requirements of DSTU 4044 [6] (Table 1);
- Crushed stone of 5 - 10 mm fractions from Malyn granite quarry that meets the requirements of DSTU B V.2.7-75 [7];
- Sand from crushing screenings that meets the requirements of DSTU B V.2.7-76 [8];
- Filler produced by PJSC "Asphalt Plant" that meets the requirements of DSTU B V.2.7-121 [9];
- Epoxy resin of CHS-EPOXY 525 grade produced by Spolchemie (Czech Republic);
- Hardener of TELALIT 410 grade produced by Spolchemie (Czech Republic).

Table 1. Bitumen properties used for the research purposes.

Indicator	Test results
Penetration, 0,1 mm	61
Softening point, °C	47.5
Brittleness temperature, °C	55

Grading of fine grain asphalt concrete of B type was designed according to the requirements of DSTU B V.2.7-119 [10] "Asphalt mixtures and road and airfield asphalt. Specifications" (Table 2).

The content of the epoxy component was 25% by weight of bitumen.

Table 2. Grading of asphalt concrete.

Materials	Content by weight, % of mineral grains smaller than this size, mm								
	15	10	5.0	2.5	1.25	0.63	0.315	0.14	0.071
Crushed stone of 5 - 10 mm fractions	100	88	3	0	0	0	0	0	0
Sand from crushing screenings	100	99	89	68	57	41	28	17	9
Filler	100	100	100	100	100	99	98	94	81
Designed grading (full passages)	100	100	60.42	46.8	40.2	30.54	22.68	15.84	10.26
Requirements to grading (full passages) DSTU B V.2.7-119 B type	100	100-90	65-55	53-43	43-33	33-23	25-16	18-11	14-8

3. Technology of asphalt mixtures and asphalt concrete preparation

Preparation of the original asphalt concrete was performed using traditional technology.

The procedure of epoxy asphalt concrete preparation included the following operations:

- batching and mixing of crushed stone, sand and filler according to the designed grading;
- heating of aggregate to the required temperature;
- introduction of the hardener into bitumen heated to the required temperature followed by mixing during (5 - 10) minutes;
- introduction of bitumen together with the hardener into a mixture of aggregate followed by mixing;
- introduction of epoxy resin into the mixture followed by mixing.

Temperature of the components heating at asphalt mixes preparation shown in Table 3.

Table 3. Temperature of the components heating at asphalt mixes preparation.

Mixture components	Temperature of the components heating, °C	
	Original asphalt concrete	Epoxy asphalt concrete
Bitumen	150	130
Aggregate	170	140
Epoxy resin	-	80
Hardener	-	80

Preparation and testing of asphalt concrete and epoxy asphalt concrete samples were performed according to the standard DSTU B V.2.7-89 (GOST 12801) [11].

Given that the actual content of crushed stone aggregate in mixtures was more than 35%, the samples obtained compacted mixture under the pressure of 30 MPa for 3 minutes. The temperature of original asphalt concrete samples at the beginning of compaction was 160 °C, and at the end of compaction - 150 °C. For epoxy asphalt concrete it was 130 °C and 100 °C, respectively. The temperature of asphalt concrete preparation was taken in accordance with section 2 of DSTU B V.2.7-89 (GOST 12801) [11]. Due to the plasticization of bitumen as a result of the epoxy components introduction, as well as to reduce the speed of hardening of epoxy components, the temperature of the epoxy asphalt concrete preparation was reduced.

To study changes in the properties of epoxy asphalt concrete dependent on the structure formation stage of the material under the impact of thermal-oxidative processes in the epoxy components, epoxy asphalt concrete samples were cured for 1, 14 and 28 days at 20 °C.

4. Test metods

For prepared asphalt concrete the following was determined: porosity of the mineral skeleton, residual porosity, water saturation, swelling, limit compressive strength, rut resistance.

4.1. Determination of the mineral skeleton porosity

Porosity of the mineral skeleton was determined in accordance with section 11 of DSTU B V.2.7-89 (GOST 12801) [11]. The main point of the method is to determine the volume of pores contained in the mineral skeleton of asphalt concrete.

4.2. Determination of residual porosity

The residual porosity was determined in accordance with section 12 of DSTU B V.2.7-89 (GOST 12801) [11]. The main point of the method for determining the residual porosity is to determine the volume of pores contained in asphalt concrete.

4.3. Determination of water saturation
Water saturation was determined in accordance with section 13 of DSTU B V.2.7-89 (GOST 12801) [11]. The main point of the method is to determine the amount of water absorbed by the sample at the given saturation mode.

4.4. Determination of swelling
Swelling was determined in accordance with section 14 DSTU B V.2.7-89 (GOST 12801) [11]. The main point of the method is to determine the increment in the volume of the sample after its saturation with water.

4.5. Determination of compression strength limit
The compression strength limit was determined in accordance with section 15 DSTU B V.2.7-89 (GOST 12801) [11]. The main point of the method is to determine the limit stress which leads to the destruction of the sample under the given conditions of the test.

4.6. Determination of rut resistance
The main point of the method is to determine the total value of residual deformations at the cyclic loading of asphalt concrete.

Test procedure is shown in Figure 1.

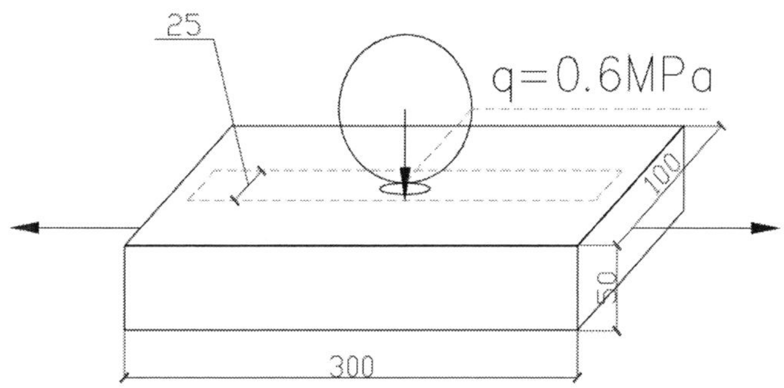

Figure 1. Rutting test procedure.

The tests were conducted at a temperature of 45 °C for the original asphalt concrete after 1 day of curing, for epoxy asphalt concrete – after 14, 28, 36 and 42 days of curing. The samples were cured at 20 °C. The sample size was 300 × 100 × 50 mm, wheel diameter - 120 ± 5 mm, track width - 25 ± 5 mm, track length - 250 mm. Test loading was 0.6 MPa. The results were recorded after 1; 2.5; 5; 7.5; 20; 25 days of curing and after 30.000 wheel passages.

5. Test results and their analysis
The results of asphalt concrete tests are shown in Table 4.

According to Table 4, the values of porosity, water saturation, swelling and strength at temperatures of 0 °C of epoxy asphalt concrete were the same as for original asphalt concrete and within the requirements range of DSTU B V.2.7-119 [10].

Compressive strength of a newly compacted epoxy asphalt concrete sample of CHS-EPOXY 525 at 20 °C (R_{20}) and 50 °C (R_{50}) is slightly lower than of the original asphalt concrete (0.5 MPa and 0.2 MPa, respectively). Strength of epoxy asphalt concrete samples increases with increasing the curing period. Particularly intense increase in strength takes place from 7 to 14 days of curing. After 28 days the strength of epoxy asphalt at 20 °C reaches 5.4 MPa versus 3.3 MPa of the original asphalt concrete and at 50 °C it reaches 2.9 MPa versus 1.4 MPa. Thus, the strength at 20 °C is increased by 64%, and at 50 °C - in 2 times.

Table 4. Test results of asphalt concrete.

Indicator		Test results: Original asphalt concrete	Test results: Epoxy asphalt concrete	Requirements of DSTU B V.2.7-119
Porosity of the mineral skeleton, % by volume		15	15	from 15 to 18
Residual porosity, % by volume		2	2	from 2 to 4
Water saturation, % by volume		1.0	1.2	from 1.0 to 2.5
Swelling, % by volume		0.2	0.2	0.5, not more
Limit compressive strength, MPa, at temperature:				
0 °C	1 day	8.0	–	12, not more
	14 days	8.0	–	-
	28 days	8.0	9.0	-
20 °C	1 day	3.3	2.8	2.5, not less
	7 days	3.3	3.0	-
	14 days	3.3	5.0	-
	28 days	3.3	5.4	-
50 °C	1 day	1.4	1.2	1.2, not less
	7 days	1.4	1.6	-
	14 days	1.4	2.6	-
	28 days	1.4	2.9	-

An important property of asphalt concrete is its ability to retain strength at higher temperatures which can be characterized by the coefficient of temperature sensitivity. The higher the ratio of the limit strength of asphalt concrete at a temperature of 50 ° C to the limit strength of asphalt concrete at the temperature of 20 ° C, the less the properties of asphalt concrete change under increasing temperature.The coefficient of temperature sensitivity is calculated by the formula:

$$R_T = \frac{R_{50}}{R_{20}} \tag{1}$$

Where R_{50} is limit compressive strength at 50 °C, MPa and R_{20} is limit compressive strength at 20 °C, MPa. The coefficient of temperature sensitivity for the original asphalt concrete is 0.42, and for epoxy asphalt concrete it is 0.54.

Given the fact that the introduction of epoxy component increases the strength and heat resistance of asphalt concrete, its impact on rutting resistance of pavements at high summer temperatures was investigated. The results of testing rutting intensity depending on the number of passages of a wheel are shown in Table 5.

According to test data, after 14 days of epoxy asphalt concrete samples curing rutting intensity was almost the same as of the original asphalt concrete (after 30.000 passages the rut depth on both pavements was 4.6 mm).

After curing for more than 28 days, the intensity of rutting on epoxy asphalt pavements was significantly reduced. If the rut depth on the original asphalt concrete after 1000 wheel passages was 1.7 mm and it was 4.6 mm after 30.000 of wheel passages – then on epoxy asphalt pavements after 42 days of curing it was 0.3 and 1.7 mm respectively. Thus, as the structure of epoxy asphalt concrete is formed, rutting resistance of pavements increases. After 42 days of curing it grows almost in 3 times.

Table 5. Depth of rut in asphalt concrete

Number of cycles, wheel passages	Intensity of rutting at the temperature 45 °C, mm				
	For original asphalt concrete	For epoxy asphalt concrete			
	1 day	14 days	28 days	36 days	42 days
1000	1.7	1.3	1.1	0.9	0.3
2500	1.9	1.9	1.5	1.2	0.6
5000	2.6	2.5	2.1	1.6	0.8
7500	2.9	2.9	2.3	1.8	1.0
20000	3.7	3.8	3.2	2.5	1.4
25000	4.3	4.4	3.4	2.7	1.6
30000	4.6	4.6	3.6	2.9	1.7

6. Conclusions

When introducing epoxy component to asphalt concrete, the trend of improving its strength characteristics at elevated temperatures can be clearly observed.

Epoxy asphalt concrete is characterized by high rutting and shear resistance. When applying EPOXY 525, the rut depth on epoxy asphalt after 42 days of curing was 3 times lower than that on the original asphalt concrete.

Low temperature characteristics of asphalt concrete can deteriorate as a result of hardening of epoxy components, so the further studies should be aimed at establishing the crack resistance properties of epoxy asphalt concrete.

References

[1] Kang Y, Song M, Pu L and Liu T 2015 Rheological behaviors of epoxy asphalt binder in comparison of base asphalt binder and SBS modified asphalt binder *Construction and Building Materials* **76** pp 343-50

[2] Yu J, Cong P and Wu S 2009 Laboratory Investigation of the Properties of Asphalt Modified with Epoxy Resin *Journal of Applied Polymer Science* **113** pp 3557-63

[3] Lu Q and Bors J 2015 Alternate uses of epoxy asphalt on bridge decks and roadways *Construction and Building Materials* **78** pp 18-25

[4] Xiao Y, van de Ven M F C, Molenaar AAA, Su Z and Chang K 2013 Design approach for epoxy modified bitumen to be used in antiskid surfaces on asphalt pavement *Construction and Building Materials* **41** pp 516-25

[5] Wei J and Zhang Y 2012 Study on the Curing Process of Epoxy Asphalt *Journal of Testing and Evaluation* **40** pp 1-8

[6] DSTU 4044-2001 Oil road viscous bitumen. Specifications.

[7] DSTU B V.2.7-75-98 Dense natural crushed stone and gravel for building materials, products, structures and works. Specifications.

[8] DSTU B V.2.7-76-98 Sand for building rock of crushing screenings from ore mining and processing enterprises of Ukraine. Specifications.

[9] DSTU B V.2.7-121-2003 Filler for asphalt mixes. Specifications.

[10] DSTU B V.2.7-119 2011 Asphalt mixtures and road and airfield asphalt. Specifications.

[11] DSTU B V.2.7-89-99 (GOST 12801-98) Materials based on organic binder for road and airfield construction. Test methods.

High temperature impact on fatigue life of asphalt mixture in Slovakia

Ján Mandula[1] and Tomáš Olexa[1]

[1]Institute of Structural Engineering, Faculty of Civil Engineering, Technical University of Košice, Vysokoškolská 4, 042 00 Košice, Slovakia

E-mail: jan.mandula@tuke.sk

Abstract. Temperature dependence of materials bonded with bitumen is a well-known fact. The impact of temperature changes the behaviour of asphalt mixtures from elastic to viscous state, and it also influences the complex modulus, phase angle and other properties of asphalt mixtures. This study observed the summer temperature influence on fatigue behaviour of an asphalt mixture for the surface course of roads in conditions of Slovakia. Measurements were made using the four-point bending method on the asphalt mixture with maximum grain size of 11 mm bonded with polymer modified bitumen. Summer conditions were represented by environmental temperature of 27 °C according to the Slovakian pavement design method. Ordinary temperatures for fatigue measurements are 10 °C, 15 °C and 20 °C according to European standards for asphalt mixture testing. Structural changes in the material were observed by dissipation energy calculations for each loading cycle. The aim of the study was to find out if the influence of high environmental temperature is positive or negative for the lifespan of asphalt mixtures.

1. Introduction

Asphalt mixtures consist of the optimum combination of two basic ingredients, the aggregate (various fractions) and asphalt binder (bitumen), and other optional additives. The term bituminous materials is generally used to denote substances in which bitumen is present or from which it can be derived. Bitumen is defined as an amorphous, black or dark-colored, (solid, semi-solid, or viscous) substance, composed principally of high molecular weight hydrocarbons, and soluble in carbon disulfide [1]. In road surfaces the combination is designed to meet diverse and often conflicting performance parameters, for example resistance to fatigue, deformation, cracking and moisture damage, durability, skid resistance, workability and economy. The mix designer generally manipulates three variables, namely type of aggregates, asphalt binder, and the ratio of asphalt binder to aggregates, and thus seeks to achieve the above-mentioned performance requirements. The components of an asphalt mixture play an important role in the mixture's behavior during the service period, so from the beginning of the design process it is very important to know their influence. Nowadays we face important changes in the way bitumen and aggregates are tested, and the development of new evaluation techniques for asphalt mixture components is essential for improving our understanding of the chemical composition and behavior of different binders and aggregates in order to develop performance-based specifications.

The asphalt mixture as the basic construction material in roads is described with specific deformation parameters. These are negatively influenced by temperature, and this is one of the problems which causes short pavement lifespan. Increasing temperature affects the behavior of bitumen-bonded materials from elastic to viscous state. Due to this the asphalt mixture is considered as a visco-elastic material. When a visco-elastic asphalt mixture is subjected to sinusoidal varying stress, a steady state will eventually be reached in which the resulting strain is also sinusoidal, having the same angular frequency but retarded in phase by an angle (φ). This is analogous to the delayed strain observed in creep experiments [2], and it means that the material structure is linear only in specific temperature and loading conditions. The influence of high temperatures on bitumen-bonded materials is a current research topic, and plenty of quality publications have focused on this problem. For example the study by Tao Ma [3] focused on the relation between air void content and high temperature creep behavior, using discrete element modeling. Other studies have focused on the temperature influence on permanent deformation of asphalt layers [4,5] and a few of them have also observed the high temperature fatigue performance of these materials [6]. The measurements applied in this study represent the fatigue behavior of a specific asphalt mixture and its structural changes at a environmental temperature of 27 °C, which is considered as a non-standard temperature because ordinary temperatures for fatigue measurements are 10 °C, 15 °C and 20 °C according to European standards for asphalt mixture testing.

2. Conditions of measurement

The material chosen for the study was an asphalt mixture which is one of the most frequently-used road construction materials for surface and binder courses of roads in Slovakia. Asphalt concrete (AC 11) with continuous gradation curve with maximum grain size of 11 mm was used, and the chosen binder was polymer-modified bitumen with penetration grade 25/55 - 65 with adhesive additive Wetfix BE. The mixture and binder were also tested for empirical properties according to the type testing standards. The results of the measurements can be seen in Table 1.

Table 1. Empirical parameters of the studied binder and mixture.

	Parameter	Unit	Value	Limit
BINDER	Penetration	mm/10	25	25~ 55
	Softening point	°C	67.7	65 >
MIXTURE	Binder content	%	5.8	5.4 >
	Air void content	%	3.4	2.5 ~ 4
	Void content filled by bitumen	%	80.1	70 >
	Indirect tensile strength ratio	%	98	80 >
	Wheel tracking slope	mm/10^3cycles	0.04	< 0.07
	Proportional rut depth	%	3.7	< 5.0

The results show that this mixture is good enough for use in pavement construction and on heavy-loaded road sections as well. Negative influence of moisture on this asphalt mixture was also observed, because it can cause structural damage to the mixture [7], which was measured in this study using the indirect tensile strength ratio. Differences between the results and the limit values were minimal. These results support the assumption that the fatigue parameters of the mixture should reach appropriate values also at unconventional temperatures.

In the study the four-point bending method with a prismatic beam specimen was used for measurement of fatigue according to the European standards. The specimen dimensions were 400 x 50 x 50 mm and the loading scheme of the equipment was a beam supported at four points with outer clamps free to rotate. The beam was loaded at the two inner points and it was fixed at the two outer points in a vertical direction. Load was applied with sinusoidal shape using the inner clamps. Observation of the asphalt mix fatigue lifespan was done at constant frequency 10 Hz. In this study the fatigue parameters of the asphalt mixture were observed at high temperature as a simulation of summer temperature conditions in Slovakia. The chosen temperature 27 °C is a typical summer environmental temperature in Slovak region, and at this temperature the asphalt mixture behavior should tend to viscous state. The four-point bending apparatus used can be seen in Figure 1.

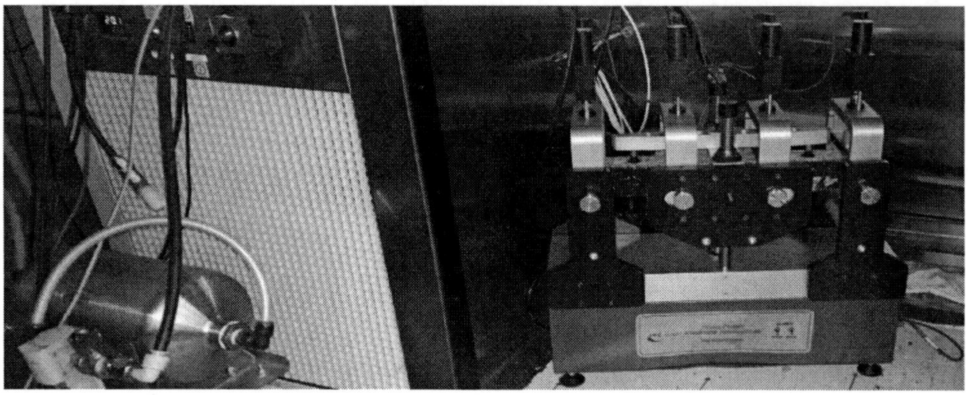

Figure 1. Four-point bending apparatus for fatigue measurement in a climatic chamber.

3. Dissipated energy and asphalt mixture testing

Asphalt mixture dissipates energy through mechanical work (loading and relaxation). Usually in an elastic material the energy is stored in the system when the load is applied. All the energy is recovered when the load is removed, in this case the unloading and the loading curves coincide. Visco-elastic materials are characterized by a hysteresis loop because the unloaded material traces a different path to that when it is loaded (phase lag is recorded between the applied stress and the measured strain), in this case the energy is dissipated in the form of mechanical work, heat generation, or damage. The area of the hysteresis loop represents the dissipated energy in the load cycle, and the following equation can be used to calculate its value in visco-elastic material [8].

$$w_i = \pi\,\sigma_i\,\varepsilon_i\,sin(\varphi_i) \qquad (1)$$

Where: w_i is the dissipated energy for *i-th* cycle (kJ/m^3); π is the mathematical constant; σ_i is the applied stress (MPa); ε_i is the strain level (μm/m); φ is the phase leg (degrees).

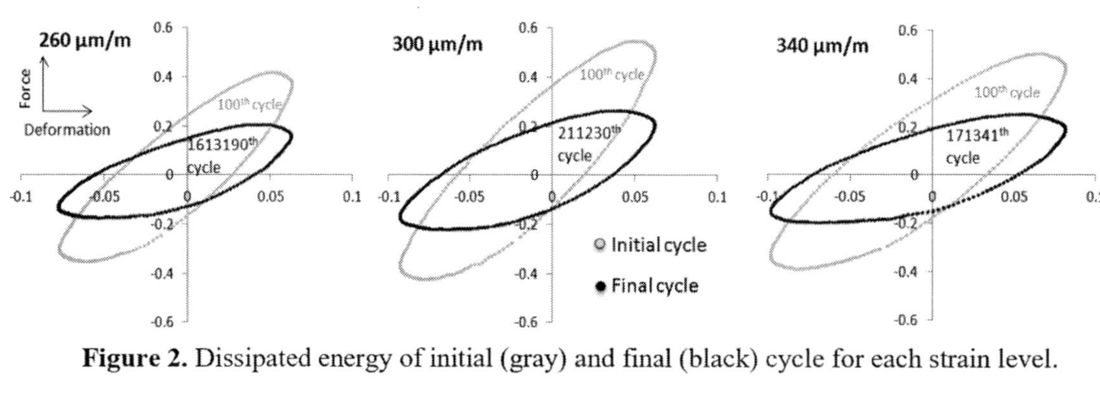

Figure 2. Dissipated energy of initial (gray) and final (black) cycle for each strain level.

Three different strain levels are illustrated in Figure 2. Each strain level shows the dissipated energy loops for the initial and final loading cycles. The value of dissipated energy in constant strain mode decreasing during test duration and also the value of initial dissipated energy are directly proportional to the strain level. The progress of dissipated energy and cumulative dissipated energy at constant temperature 27 °C have different shapes (logarithmic and linear) and can be seen in Figure 3.

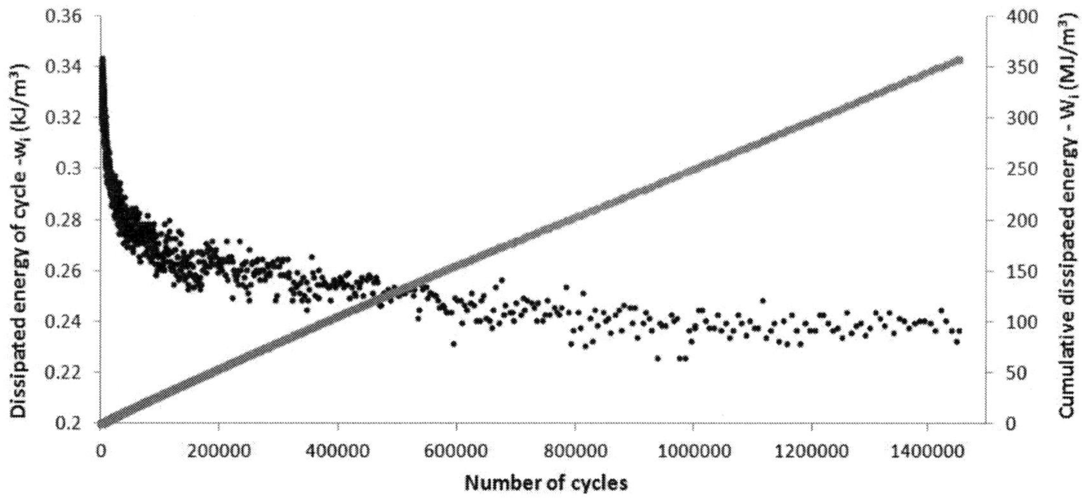

Figure 3. Progress of dissipated energy (black) and cumulative dissipated energy (gray).

According to the classical analysis, fatigue life is conventionally defined as the number of cycles at which the stiffness modulus has decreased to half of its initial value. However, dissipated energy methods also started being considered in order to determine the number of cycles to failure. The energy ratio (ER_{HP}) method considered here was introduced by Hopman and Pronk in 1989. Failure point according to the energy ratio concept is the cycle number where cracks are considered to initiate. Failure point is defined as the point at which the slope of the energy ratio versus the number of cycles deviates from a linear line.[8] The Hopman and Pronk method was chosen for this study because of its simplicity and its clearly noticeable failure point. The energy ratio calculation is shown below (2).

$$ER_{HP} = n\left(\frac{w_0}{w_i}\right) \tag{2}$$

Where: w_o is the initial dissipated energy (MJ/m^3); w_i is the *i-th* cycle dissipated energy (MJ/m^3); n is the load cycle value.

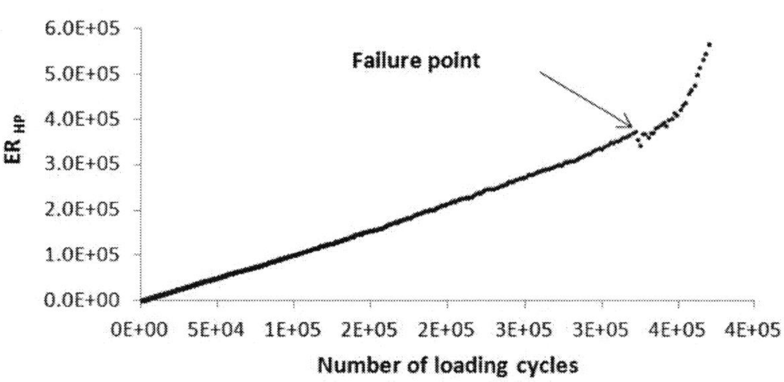

Figure 4. Progress of the energy ratio according to Hopman and Pronk during long-term dynamic testing, and fatigue failure point according to this method (conditions: 10 °C and 10 Hz).

4. Results and evaluation of fatigue measurements

The durability of the asphalt mixture was measured by means of fatigue testing. During the testing the phase angle, complex modulus and the dissipated energy were calculated for every loading cycle. The presented results were obtained at constant temperature 27 °C and three different constant strain levels 260 μm/m, 300 μm/m and 340 μm/m. Frequency of loading 10 Hz in line with the standards, and the shape of dynamic loading was sinusoidal. The fatigue failure criterion was set in the control software to the standard AASHTO criteria where the test is stopped when the complex modulus goes down below 50 % of the initial complex modulus. The dissipated energy failure criterion was calculated after the measurements.

Before fatigue measurements were made stiffness modulus measurements for scale of temperatures (0 °C; 11 °C; 20 °C and 27 °C) because of creation of master curve for studied mixture which can be seen in Figure 5. For calculation of main curve was chosen polynomial function and applied was Arrhenius equation.

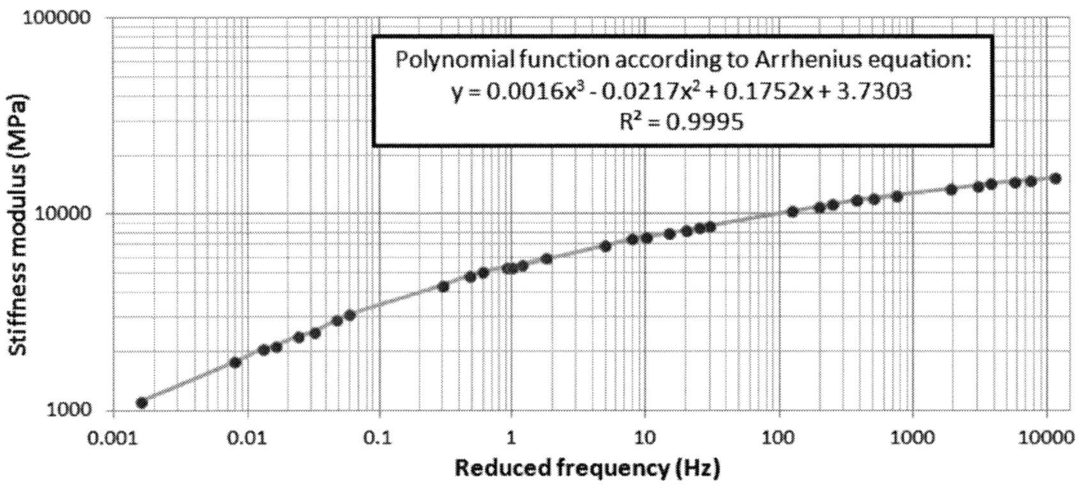

Figure 5. Master curve of studied mixture (Arrhenius equation).

The progress of the complex modulus during fatigue testing of the asphalt mixture should have a characteristic curve which can be seen in Figure 6. This progress has three specific phases: during the first the reduction of the complex modulus is steeper until the second phase, during which the complex modulus value is almost constant, or its reduction is much slower. In the final third phase the complex modulus reduction is again steeper until the fatigue failure point [9].

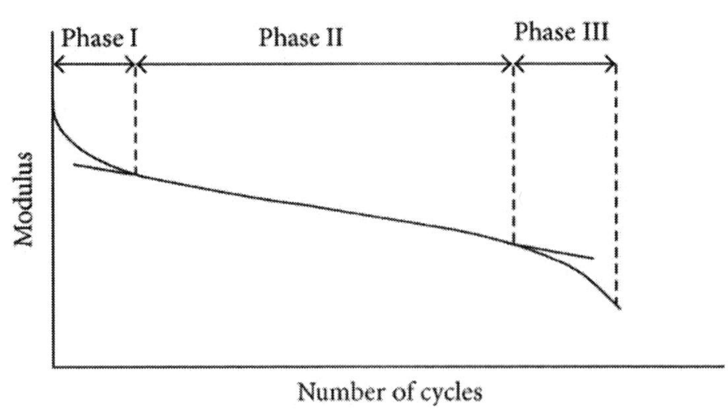

Figure 6. Progress of the complex modulus (stiffness) during the fatigue testing of asphalt mixture with standard temperature and frequency conditions [9].

The studied temperature influenced the shape of the complex modulus progress during testing, an example of which is shown in Figure 7. As can be seen, the third phase in this case was missing, and phase one was approx. 10 % of the total lifespan of the sample. In this phase the complex modulus was reduced to approx. 65 % of its initial value. Phase two represented 90 % of the lifespan of the sample, and the reduction of the complex modulus was much slower than in phase one. Progress of the complex modulus in these conditions can be described with a semi-logarithmic mathematical function.

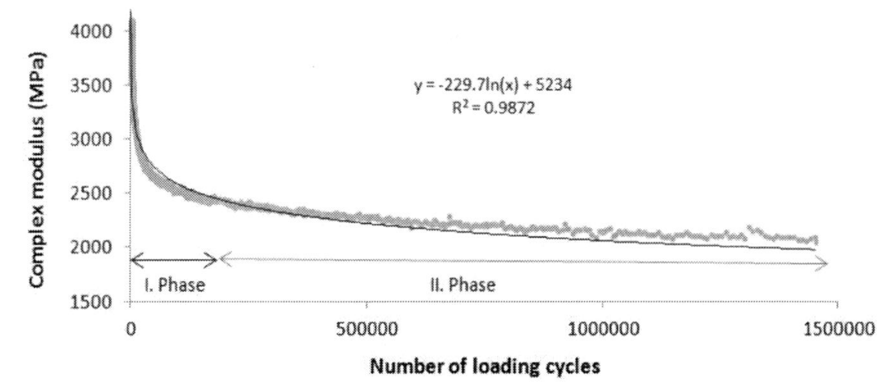

Figure 7. Progress of the complex modulus (gray), strain level 260 μm/m.

The shape of the complex modulus reduction was the same for each tested sample, and the proportion of the phases was also similar. Figures 3 and 7 show the same trend in the complex modulus and the dissipated energy, and this observation falsely suggests that the failure point should be the same for both methods. Evaluation of the fatigue measurements clearly shows that this prediction was not confirmed. An example of the energy ratio progress according to the dissipated energy method and the phase angle can be seen in Figure 8. In the plot of the energy ratio and number of cycles there is no deviation from the straight line, which means that there was no structural damage according to the Hopman and Pronk method. The failure point according to the dissipated energy method was not reached in every tested sample for each of the strain levels.

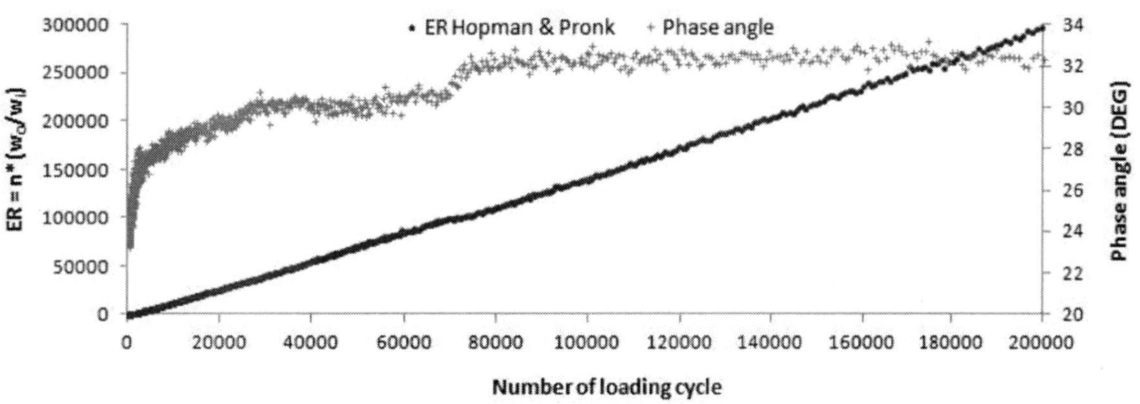

Figure 8. Progress of the energy ratio (black) and the phase angle (gray) during the fatigue testing, strain level 300 μm/m.

An interesting fact is that the progress of the phase angle was similar to the complex modulus progress, but in the opposite direction. The phase angle too achieved an almost constant value after the first third of the sample's lifespan. After a few tested samples there was an attempt to set a lower percentage for the fatigue failure point according to the AASHTO method, but then there were unwanted problems with the control software used. Because of these facts the evaluation of the fatigue testing measurements was finally based only on the AASTHTO criteria. The Wöhler curve and correlation coefficients can be seen in Figure 9.

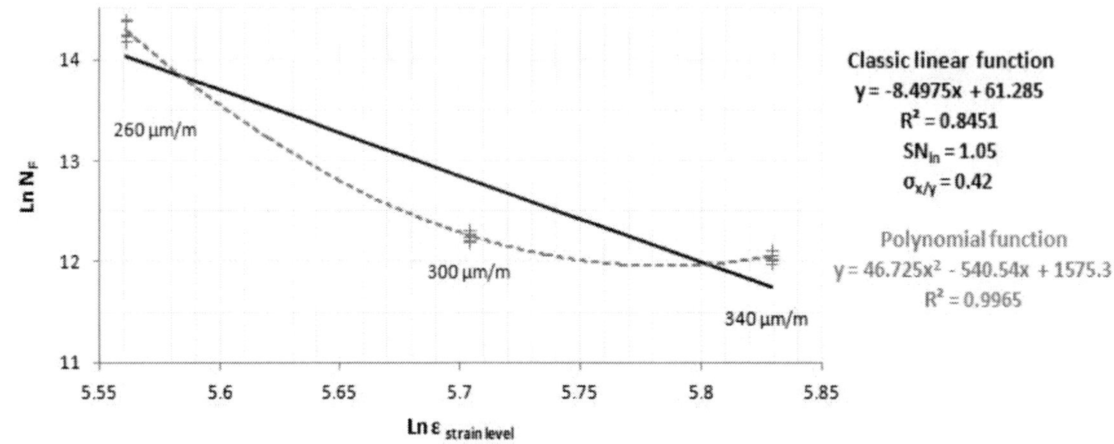

Figure 9. Fatigue curve according to current European standards (black) and estimation of polynomial function (gray).

Evaluation of the results shows that the bi-logarithmic relation between strain level and number of cycles until sample failure is not quite linear. The standard correlation coefficient should be within the interval of 0.9 - 1.0. As can be seen in Figure 8, the correlation coefficient of the polynomial function is higher, which means that the studied relation is closer to the polynomial than the linear function. Based on this curve it might be presumed that by reducing the strain level, the lifespan of the sample could become endless [10].

5. Conclusion

This study observed the summer temperature influence on fatigue behavior of an asphalt mixture for the surface course of roads in conditions of Slovakia. For fatigue evaluation were applied standard AASHTO method and Hopman and Pronk method.

Based on these results, the fatigue failure point according to the Hopman and Pronk method was not significantly clear for each of the strain levels. Energy ratio progress according to Hopman and Pronk was applied as main indicator of structural changes but results indicate that there were no structural changes such as micro-cracks, during environmental temperature 27 °C. Observation of the dissipated energy proved that the behavior of the material was more viscous than elastic, which is also documented by the phase angle observation.

Only the AASHTO method based on complex modulus reduction was utilized for this temperature. The lifespan of the sample did not have the third final phase, so the failure point was at the end of the second phase. Despite the fact that 50 % of the initial complex modulus was reached at the end of second phase, during the first phase there was reduction of the initial complex modulus to 65 %, which is also significant. From the failure point of view the fatigue lifespan of the asphalt mixture was positively influenced by increasing temperature. On the other hand, if we consider the fact that only 10 % of the sample lifespan had the complex modulus higher than 65 % of its initial value, the temperature influence appears more negative than positive.

In summary, the failure criteria used for fatigue testing measurements at higher temperatures should be based on the complex modulus, and moreover the reduction in bearing capacity of material during the first phase of the sample lifespan has to be considered for the better design of pavement structure.

Acknowledgements
This work was supported by the Slovak Research and Development Agency under the contract No. APVV-15-0777.

The contribution was incurred within Centre of the cooperation, supported by the Agency for research and development under contract no. SUSPP-0013-09 by businesses subjects Inžinierske stavby and EUROVIA SK.

The research was performed within project NFP 26220220051, Development of Progressive Technologies for Utilization of Selected Waste Materials in Road Construction Engineering, supported by the European Union Structural Funds.

References

[1] Goetz W H and Wood L E 1960 *Bituminous Materials and Mixtures* (Highway Engineering Handbook) ed K B Woods (New York: McGraw-Hill)

[2] Maröti G 2013 Closed form solution for bending oscillations of beams *Pollack Periodica* **8**(3) (Pécs: University of Pécs) pp 111–8

[3] Ma T, Zhang D, Zhang Y and Huang X 2015 Effect of air voids on the high-temperature creep behavior of asphalt mixture based on three-dimensional discrete element modeling *Materials and design* **89** ed A M Korsunsky (Amsterdam: Elsevier) pp 304-13

[4] Al-Mosawe H 2016 *Prediction of permanent deformation in asphalt mixtures* (University of Nottingham, Department of Civil Engineering)

[5] Remišová E and Zatkalíková V 2016 Evaluation of bituminous binder in relation to resistance to permanent deformation *Proc. XXV Polish-Russian-Slovak Seminar "Theoretical Foundation of Civil Engineering"(Žilina)* (Elsevier) pp 584-589

[6] Chen J, Wang H, Li M and Li L 2016 Evaluation of pavement responses and performance with thermal modified asphalt mixture *Materials and Design* **111** ed A M Korsunsky (Amsterdam: Elsevier) pp 88-97

[7] Kakara M R, Hamzaha M O and Valentin J 2015 A review on moisture damages of hot and warm mix asphalt and related investigations *Journal of Cleaner Production* **99** pp 39–58

[8] Maggiore C, Airey G and Marsac P 2014 A dissipated energy comparison to evaluate fatigue resistance using 2-point bending *Journal of Traffic and Transportation Engineering* **1**(1) (Nottingham: University of Nottingham) pp 49–54

[9] Mashaan N, Hassan Ali A and Rehan Karim M 2014 A review on using crumb rubber in reinforcement of asphalt pavement *The scientific world Journal* **2014** ed Flores T H D and Ma Q (Hindawi Publishing Corporation) pp 1-21

[10] Nagórska M, Blazejowski K and Nagorski R 2015 Selected aspects of design lifespan of flexible pavement with anti-fatigue course *Archives of Civil Engineering* **61** (Warszaw: Warszaw University of Technology) (De Gruyter) pp 311-327

Asphalt mix reinforced with vegetable fibers

Peter Gallo[1]

[1] CTU in Prague, Faculty of Civil Engineering, Thákurova 7, 166 29 Praha 6, Czechia

E-mail: peter.gallo@fsv.cvut.cz

Abstract. The use of a larger share of renewable materials in road construction is a trend that in the long term cannot be avoided. In some cases, due to this pressure, new innovative opportunities are generated. This article attempts to outline and bring one of such opportunity. The article describes selection and the use of special natural fibers from renewable natural resources adapted for use in various types of asphalt mixtures to improve the range of properties. Experimental results showed an improvement in stiffness modulus, indirect tensile strength (ITS) and good resistance to permanent deformation of blends containing vegetable fibers. This is a new topic in the road construction. But the results have so far proven that the used type of fibers can be a perspective way, as simple and in line with the policy of sustainable development, to improve the properties (reinforce) of the asphalt mixtures.

1. Introduction

This article is devoted to analyse the use of rather new, potentially perspective type of asphalt mix reinforcement. Its main purpose is to provide an introduction to the theme of the use of natural fibers in asphalt mixtures with a focus on fibers obtained from plants, which are increasingly starting to be used also in other parts of construction industry. In this industry, natural materials are currently mainly used for example as an insulation of buildings. However, natural fibers have greater potential than just work as simple insulators. In the automotive industry, they are used as reinforcement of plastic parts to form a composite structure with greater strength and lower weight than conventional plastic parts. In construction, the natural fibers as a reinforcing material begins to enforce slowly, but can be expected to increase their usage. Currently, for example, there are experiments with natural fibers working as a reinforcing additive in concrete, with potential application in the concrete pavement. However, the situation is different for the use in asphalt mixtures. The only natural material which is commonly added to asphalt mixtures, particularly mixtures of the SMA type, is cellulose or fibres based on basalt. Though, its primary function is not a reinforcement of the asphalt mixture.

The application of natural fibers for reinforcing asphalt pavement is still just a vision for greater use of renewable natural materials in construction industry. The high volume of traffic, the requirements for increased durability, sustainable development and lower cost of asphalt mixtures are a motor of discovery and use of new, perhaps unusual solutions.

2. Background

The use of fibers in asphalt mixes dates back many decades. Or longer: Button and Epps (1981) maintain that the earliest use of fibers in asphalt was the use of straw in ancient Egyptian building specifications. In the United States, asbestos fibers were used as early as the 1920s (Serfass and Samanos 1996), and this usage continued until the 1960s, when health and environmental concerns put an end to it (Busching et al. 1970). Cotton fibers were used in the 1930s (Busching et al. 1970), but they tended to degrade over time (Freeman et al. 1989). Since then many types of fibers have been used in various applications and different parts of the world. Fibers were reportedly used in asphalt mixtures for decades to provide the following benefits (Busching et al. 1970; Peltonen 1991):

- Increased tensile strength resulting in increased resistance to cracking,
- Reduced severity of cracking when it did occur,
- Increased fatigue resistance,
- Increased rutting resistance as a result of lateral restraint within the mixture,
- Increased abrasion resistance,
- Higher asphalt contents leading to increased durability, and
- Potential lower life cycle costs arising from longer service life. [1]

Early applications were in dense-graded asphalt mixtures. Beginning in 1991, the first SMA mixtures were placed in the United States after more than 30 years of successful use in Europe (Cooley and Brown 2001). These mixes were designed in Europe mainly to resist studded tire wear but were found to be highly resistant to permanent deformation as well. Usage in the United States increased rapidly: by 1997 more than 140 SMA projects across the country were evaluated by the National Center for Asphalt Technology (NCAT) (Cooley and Brown 2001). These mixes generally used cellulose or mineral fibers to help hold the asphalt binder in the gap-graded aggregate structure; that is, to prevent draindown of the binder. [1]

Fibers are also used in open-graded friction courses (OGFCs) or porous asphalt mixes to prevent draindown. These mixes have open-graded aggregate structures and high air voids to create stone-on-stone contact to resist rutting, reduce noise (McGhee et al. 2013), reduce splash and spray, and improve friction (Watson et al. 1998). [1]

In summary, there are two main uses for fibers:

- to prevent draindown in gap- and open-graded mixes, and
- to strengthen dense-graded asphalt mixes to resist rutting and cracking. [1]

Generally, it can be said, that for the first main usage are suitable cellulose (or natural) fibers. For the second main use are, with their greater elastic modulus and tensile strength, more suitable synthetic fibers such as aramid, fiberglass, polyester or similar. Only special type of synthetic fibers with irregular cross section are suitable to prevent draindown of bituminous binder. But there are few issues with synthetic fibers. The main issues are an environmental issue and economical aspect of usage of synthetic materials in construction industry. Others are e.g. polypropylene fibers which have low melting point requiring precise control of production temperatures, fiberglass fibers are brittle and may break during mixing and compaction, aramid fibers are very expensive. To secure the sustainable development policy we should looking for optimal solutions for common problems.

3. Vegetable fibers as asphalt mix reinforcement

Main goal of the research described in this article was to find an affordable and environment friendly material, which would simultaneously cover draindown and reinforce asphalt mixtures. After few thoughts it was clear, that the only material suitable to fulfill every request will be a natural fiber. Then another question arose, which one specifically? After few months of tests with many materials obtained from natural resources, the most promising candidates appeared. The most suitable fibers for use in asphalt mix seems to be hemp and flax fibers, processed into a yarn form.

Figure 1. Samples before thermal durability test.

Figure 2. Samples after thermal durability test.

3.1. Choice of fibers

Various materials were available at the beginning of the testing process. Waste materials, such as straw or raw sheep wool, hemp, flax and cotton fibers in both raw and yarn form (cut into pieces 40 – 60 mm long) and cellulose based fibers e.g. fine chipboard particles or recycled comminuted paper. Overall number of tested samples was around 22.

The amount of samples was too high, so their amount was decreased for the final testing in asphalt mix. Therefore, a special three steps selection process was designed.

First step was to find out a thermal durability of input samples, which were exposed to temperatures 150 °C, 180 °C and 220 °C for 120, 60 and 5 minutes. Weight decrease and color change of samples were monitored. Unsatisfactory samples were excluded from next steps.

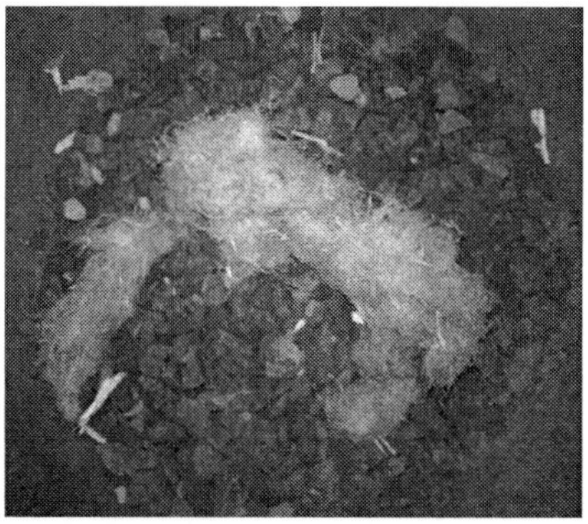

Figure 3. Tufts of raw hemp fibers mixed with gravel.

Second step was about mixing samples with gravel for verification of a workability of samples. Samples with the most even dispersion in gravel were selected for further testing.

In the third and the last step, tensile strength was measured for all assessed samples. Only yarn samples were tested in this step. This process decreased the number of samples into final three, Flax 500 tex, Hemp 100x3 tex and Hemp 120 tex yarns. These samples were excellent in every testing step except sample of Hemp 120 tex, which had the lowest tensile strength but was chosen because of comparison reasons.

Natural fibers twisted into yarns have few advantages. First, twisted fibers interact with each other. Therefore, a yarn has greater tensile strength than single fibers. Second, there's a workability factor.

Raw, single fibers tend to make tufts more easily when they are blended into asphalt mix and are hard to be cut into single pieces with exact length. Yarns prove great thermal durability too. This is major factor in case of use of vegetable fibers in hot asphalt mixes.

3.2. Yarns made of vegetable fibers

Yarns are available in many variations. Hemp and flax yarns are basically made of bast fibers, with a variety of thicknesses or better said different fineness, as it is an established term in textile industry. Eventual tensile strength of yarn depends on many factors. From conditions under which plants are grown through process of obtaining fibers from plants and it ends with spinning method, yarn fineness and cross-section. The very type of processing (spinning) fibers into yarns is crucial.

There are basically two main spinning methods:

- Ring spinning, which is the most common spinning method
- Open-end (or rotor) spinning

Ring spinning produces uniform and high strength yarns. Rotor spun yarns are more even, somewhat weaker and have a harsher feel than ring spun yarns. They are approximately 15-20 % weaker as the yarn is coarser because of loose outer layers. A further disadvantage of the loose outer layers is their sensitivity to axial rubbing. Since these open layers are not firmly secured in the core, they tend to accumulate in small knots during passage of the yarn over edges, guide elements, etc.[2,3]

Cross-section can consist of twisted fibers which create single yarn (Figure 2, 3), or can be more complex and consist of two or more single yarns/hanks twisted together (Figure 4). Yarn fineness is measured in "tex" unit; in other words, it can be defined as a unit of measure for the linear mass density of yarns.

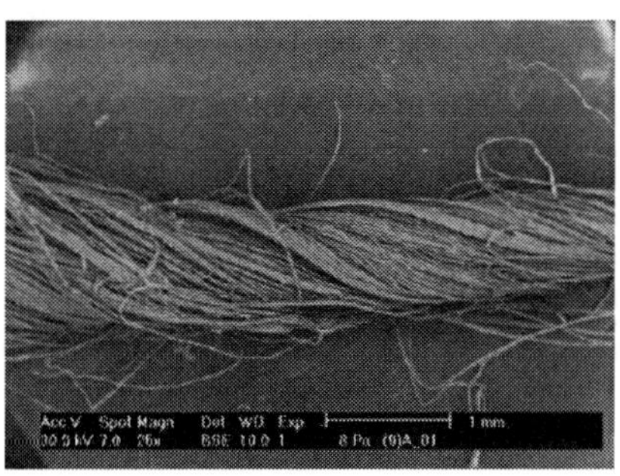

Figure 4. Ring spun flax yarn, 500 tex.

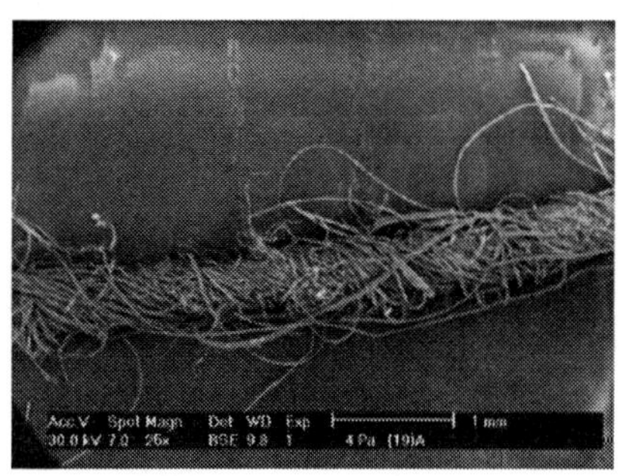

Figure 5. Rotor spun hemp yarn, 120 tex.

Figure 6. Plied cotton yarn made of ten single hanks, 50x10 tex.

4. Results from application of hemp and flax yarns in asphalt mixtures

For complex research of yarn addition influence on asphalt mixture mechanical properties two types of asphalt mixtures – AC_{bin} and SMA – were chose. Though, results from SMA mixture are not included in this article. Before mixing into a mixture, yarns were cut into 40-60 mm long elements. Yarn content in the mixture depends on the mixture type. Asphalt mixtures with yarns were compared with a reference mixture with identical mix design as well as with another reference where polymer modified bitumen was used. Binder in the asphalt mixtures with yarns and in the control mixture had a penetration grade 50/70. All tests and samples were designed and mixed in accordance with Czech standards, ČSN EN 13108-1 or ČSN EN 13108-5 depending on the assessed type of asphalt mixture.

4.1. Results of testing AC_{bin} 16 S mixtures

In this type of asphalt mixture 3 types of yarns – hemp 120 tex, hemp 100x3 tex and flax 500 tex – were tested. Yarn content in the mixture was 0.2 M%.

Table 1. Results of AC_{bin} 16 S tested mixtures – densities and voids content.

Asphalt mixture	Aspahlt content (%)	Bulk density (g.cm^{-3})	Max. bulk density (g.cm^{-3})	Voids content (%)
ACL 16 S 50/70 ref.	4.4	2.404	2.486	3.29
ACL 16 S 50/70 + Hemp 120	4.5	2.401	2.544	5.61
ACL 16 S 50/70 + Hemp 100x3	4.4	2.399	2.512	4.51
ACL 16 S 50/70 + Flax 500	4.4	2.424	2.522	3.90
ACL 16 S PmB 25/55-55	4.1	2.433	2.519	3.41

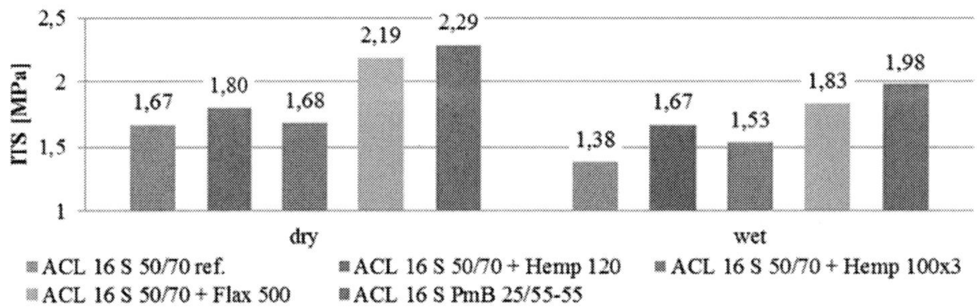

Figure 7. ITS results of asphalt mixtures AC_{bin} 16S with natural fibres.

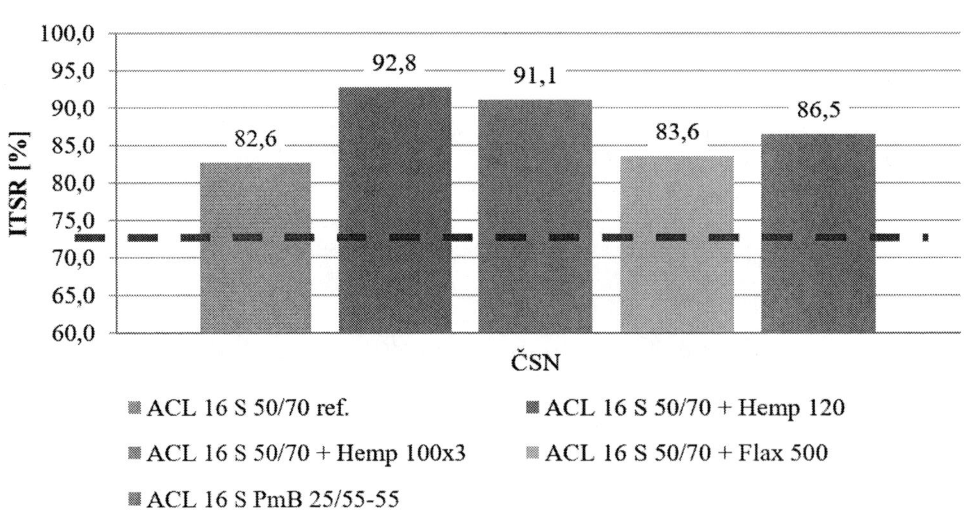

Figure 8. ITSR results of asphalt mixtures AC_{bin} 16S with natural fibres.

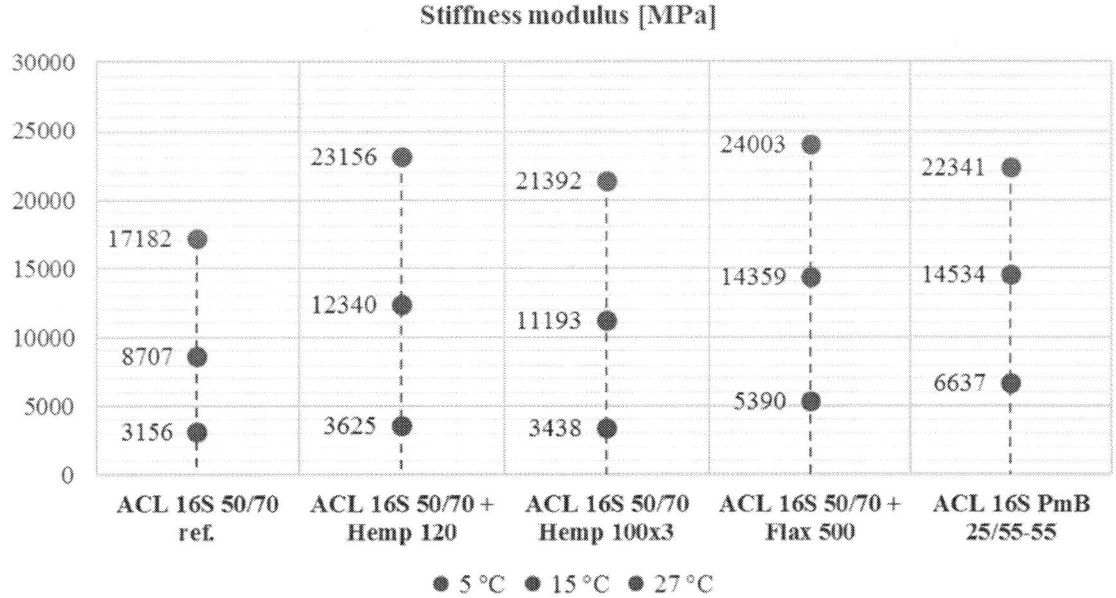

Figure 9. Stiffness modulus results of reinforced AC_{bin} 16S mixtures.

Results of performed tests show some potential of use of vegetable yarns as asphalt mix reinforcement. Asphalt mixtures with vegetable yarns show better tensile strength and rigidity than the reference mixture (Figure 7 and Figure 9). Especially the asphalt mixture with flax yarn reached promising values and results were close to the mixture with polymer modified bitumen. All mixtures fulfilled ITSR test requirement according to the Czech standards, which is at least 80 %. That is a proof of good water resistance of blends with vegetable fibers, although are water absorbent. However, when mixed in asphalt mixture, vegetable fibers soak up with bitumen binder, which protects them against water. That is also a key element in SMA type mixtures, when vegetable fibers hold bitumen binder and decrease binder pourability. Blends with yarns also shown good resistance to permanent deformation (Table 2.). Asphalt mixture laboratory slabs with 6 cm thickness were used for resistance to permanent deformation test. Test was conducted in air-bath at 50 °C in a small testing device according to EN 12697-22. Key material parameters were the proportional rut depth (PRD_{AIR}) and wheel tracking slope (WTS_{AIR}).

Table 2. AC_{bin} tested mixtures – densities and resistance to permanent deformation.

Asphalt mixture	Thickness (mm)	Bulk density ($g.cm^{-3}$)	Degree of compaction (%)	Rut after 10 000 cycles (mm)	WTS_{AIR} [$mm/10^3 cycles$]	PRD_{AIR} [%]
ACL 16 S 50/70 ref.	60.31	2.354	98.2	2.26	0.101	2.9
ACL 16 S 50/70 + Hemp 100x3	60.29	2.365	98.6	1.54	0.038	2.2
ACL 16 S 50/70 + Flax 500	60.13	2.386	98.4	2.01	0.039	3.0
ACL 16 S 50/70 + Hemp 120	60.22	2.372	98.5	1.52	0.033	2.1

5. Conclusion

A great potential of use of materials from renewable energy sources in asphalt pavements has been shown in this article. Except cellulose fibers, these materials are not common in road construction. Perhaps this is due to prejudice, distrust or underestimation of natural materials. It is understandable, that using vegetable fibers in an environment of relatively high temperatures, which undoubtedly hot asphalt mixtures are, may create some concerns. Only empirical concerns are not sufficient justification for ignoring any attempt to research this field. And the results of this article only confirm that. Asphalt mixtures with vegetable fibers achieved better results than the reference mixture; showing better tensile strength, stiffness and resistance to permanent deformation.

Issue of using the natural materials in construction industry is therefore a challenge for the future and the possibility of such simple, relatively inexpensive and significantly improve serviceability and the real life of asphalt pavements.

Acknowledgements

This paper was supported by the project No. TE01020168 within the funding scheme of the Technology Agency of Czech Republic.

References

[1] McDaniel R 2015 *Fiber additives in asphalt mixtures* (Washington DC: Transportation Research Board of the National Academies)

[2] Open-end spun yarns [Online] *http://www.rieter.com* Retrieved from http://www.rieter.com/en/rikipedia/articles/technology-ofshort-staple-spinning/yarn-formation/assembly-of-fibers-to-make-up-a-yarn/the-positions-of-the-fibers-in-the-yarn-structure/open-end-spun-yarns/

[3] Ring vs. Open-end Spinning [Online] *http://textilecentre.blogspot.com* Retrieved from http://textilecentre.blogspot.com/2014/01/ring-vs-open-end-spinning.html

IOP Conf. Series: Materials Science and Engineering **236** (2017) 012025 doi:10.1088/1757-899X/236/1/012025

Microtexture diagnostics of asphalt pavement surfaces

Zuzana Florková[1] and Ľubomír Pepucha[1]

[1]Research centre University of Zilina, Univerzitna 8215/1, Zilina SK-010 26, Slovak Republic

E-mail: zuzana.florkova@rc.uniza.sk

Abstract. The microtexture of asphalt pavement surface is an essential parameter from the traffic safety point of view and it closely relates to a geometrical, petrological and physical properties of aggregate particle used in asphalt pavement. Microtexture has a significant influence for assurance basic friction values between tire and pavement in relation to a skid resistance properties. Therefore, the microtexture detecting methods are necessary. The British pendulum tester measurements have been carried out on selected sections of roads with different asphalt surfaces. Individual grains of aggregates were taken from the surface of each section from the sliding path and also from the core sample after the extraction. The laboratory profilometry measurements have been practiced on these aggregate samples and subsequently the surface microtexture was investigated based on commonly used texture characteristics and the filtration approach was applied in calculation process. The results have shown the degradation of microtexture values occurs due to polishing of aggregate under loading from traffic in relation to the type of used aggregate. Some correlation between BPN values and texture characteristics was found.

1. Introduction

In relation to road safety, the roughness of pavement surface is the most important variable parameter which is characterizing quality of asphalt pavement surface from skid resistance properties point of view. Roughness of pavement surface can be evaluated from physical aspect as a skid resistance or in terms of geometrical aspect as a pavement surface texture. The texture is a function of the pavement surface which assure the basic interaction between tire and pavement. It is defined as a morphological configurations of material particles on the pavement surface. It is a combination of irregularities with different wavelengths and amplitudes within defined range. According to amplitudes and wavelengths of irregularities is texture dividing into three categories: microtexture, macrotexture and megatexture [1].

The Figure 1 describes a particular components of pavement texture and their influence on the fundamental effects which are closely relates to tire and pavement interaction. It is obviously from Figure 1, that friction level is affected by macro and microtexture values. Therefore the sufficient values of these texture components are important for assurance of friction between tire and pavement.

Macrotexture is the degree of roughness defined by the morphological configuration of aggregate particles on pavement surface (wavelengths in range 0.5 to 50 mm) and is mainly responsible for reducing the potential for separation of tire and pavement surface due to hydroplaning and for inducing friction caused by hysteresis for vehicles traveling at high speeds.

Microtexture is the degree of roughness imparted by aggregate particles [2]. According to [3] is microtexture defined as a configuration of particular peaks on the surface of aggregate particle (wavelengths in range 0.5 mm to 1 μm). The main function of asphalt pavement microtexture is to assecurate basic friction level on microscopic scale (adhesion) and primarily affects the frictional properties of a pavement surface at lower speeds. Its importance also consist in disconnecting of residual water film on the pavement surface. It leads to close contact between tire and pavement surface [4]. The importance of microtexture from the skid resistance point of view is closely presented in [2]. It stands to reason, that type of used aggregates by its petrological and physical properties and their ability to be resisting to the climatic effects and the mechanical abrasion due to traffic loading predetermine the level of asphalt pavement microtexture.

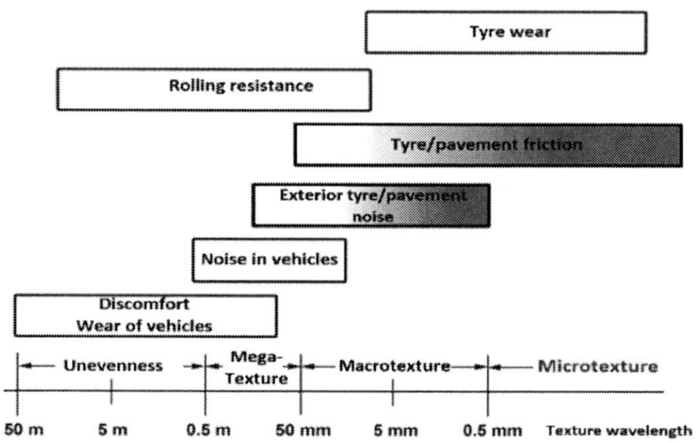

Figure 1. Influence of particular components of pavement texture on tyre/road interaction (a blue shade means a favourable effect of texture over this range, while a white area means an unfavourable effect). [1]

Over the years, a lot of studies have been practised to correlate the different friction and texture measurement methods. The determined correlations are significant in specifying how macro and microtexture influence friction performance between tire and pavement. From aspect of microtexture there are same gaps, because of microtexture wavelength range in micron scale. It is complicated not only because of the accuracy of measurement but also because of their evaluation and then possibility to direct microtexture measurement in the field is still problematic. Currently, there is no way to capture of entire microtexture range on pavements in the service because of laser scanning resolution and optical illusion [5]. Account on that, the indirect measurement techniques related to low slip speed friction such as the pendulum tester (BPT) [6] and dynamic friction tester (DFT) [7] are used as an alternate. Hence, the unavailability of the measuring device which allows direct measurements in real conditions on the pavement surfaces limits research in this field on laboratory conditions.

2. Experimental measurements

2.1. Microtexture measurements on the asphalt pavement surfaces

The measurements of friction coefficient by British pendulum tester was carried out in the track on the six selected pavement sections with different asphalt surfaces, because of measured BPN (British pendulum number) values obtained by this device are worldwide used as an indirect representation of microtexture level of pavement surface. It is a standard method, determines skid resistance of pavement surface by measurement of the energy loss when a rubber slider edge is swings across a test surface.

Measurements was practised according to standard EN 13036-4 [6] and carried out on the ten measurement points within each road section in the wheel track on the pavement surface. The mean of these ten *BPN* values gives a representative value of pavement microtexture for each pavement section. The results of *BPN* measurements are presented in the Table 1.

2.2. Laboratory determination of microtexture of aggregates taken form asphalt pavement surfaces
After the BPN measurements, the individual grains of aggregates were taken from the surface of each section from the sliding path and also from the core sample after the extraction. The laboratory contact profilometry measurements have been practiced on these aggregate samples.

Profilometry contact method is commonly used in mechanical engineering as a surface roughness measurement method. Contact profilometer ZEISS Countourecord 1700 allows roughness measurement with a periodic transformation of information about profile during mechanical movements of measurement detector on the investigated surface. In this method, by direct contact of sensor with investigated surface the vertical traverse of measurement detector is transformed on the electrical signal. Electrical signal is subsequent converted on the digital (numerical) signal which is represents changes of vertical interval (z-direction) in relation with the x-direction position. The obtained output in the form of profile, is then expressed by x and z coordinates (Figure 3) and represents real contours of the aggregate surface. The measurement accuracy in the longitudinal direction was 1 μm.

2.2.1. Filtration of profile data. However, in addition to microtexture information, the obtained output in the form of profile of investigated particle captures the shape of aggregate particle and the wavelengths and amplitudes that are out of the microtexture range (higher than 0.5 mm). It can leads to influence of results in evaluation process. Therefore, the filtration approach on the principles signal processing was applied in this research. Data filtration can be used to remove components which are irrelevant in regard to required evaluation within microtexture range from obtained profiles.

Filter design and subsequent signal processing is easy implemented in a computing program Matlab. From the purpose of efficient filter choosing, the analysis of filter type was performed in this program. The usage of highpass filter is obviously because of the information about microtexture related to low wavelengths and thus to high frequencies. Butterworth, Chebyshev, Chebyshev inverse and Eliptic highpass filters were analyzed. The Butterworth filter was chosen because of the frequency response of the Butterworth filter is maximally flat and thus has no ripples in the passband and rolls off towards zero in the stopband. Compared with a both Chebyshev filters and an Elliptic filter, this filter has a slower roll-off and a more linear phase response in the passband than can achieve other investigated filters. Thereafter, the filter transition band was necessary to define by determination of band-boundary frequencies (passband and stopband frequency) (Figure 2).

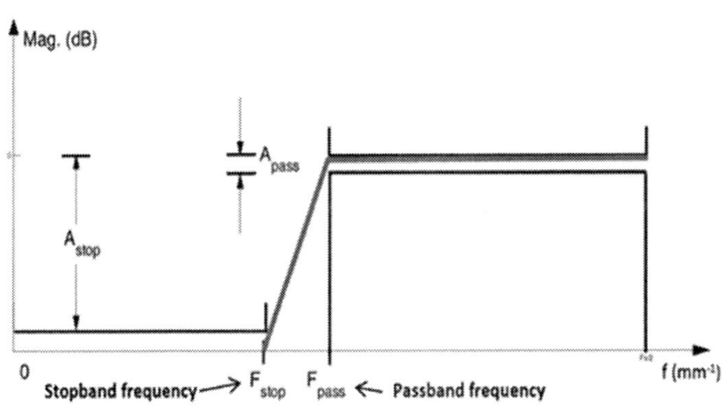

Figure 2. Specification of design. Butterworth highpass filter.

As is shown in the Figure 2, the particular primary profile (Figure 3 black profile *y0*) was filtered by different band-boundary frequencies. The best results were achieved for passband frequency $Fpass$ = 2 mm^{-1} and stopband frequency $Fstop$ = 0.67 mm^{-1} (Figure 3 red profile *y3*). It is equivalent for 0.5 mm and 1.5 mm wavelengths. This filter specification allows removed the wavelengths higher than 0.5 mm from the investigated profile with the most accuracy (e.g. any deformation at the ends of the profile).

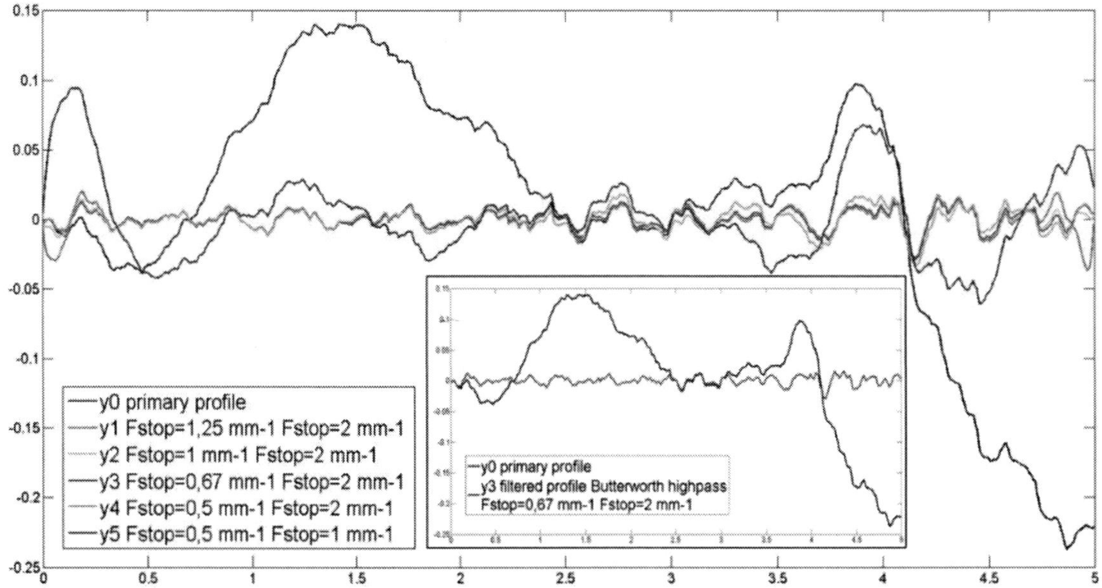

Figure 3. Comparison of resultant profiles filtered with different band-boundary frequencies.

2.2.2. Used statistical texture characteristics. Commonly used statistical surface texture characteristics can be used to evaluate microtexture of aggregate particle from filtered profiles.

These statistical texture characteristics defined in [8,9] presented a group of profile evaluation characteristics related with vertical coordinates distribution of surface profile and reflected basic description of irregularities of investigated profile.

The statistical characteristics mean arithmetic deviation *Ra* (1) and root mean square *Rq* (2) characterize an absolute measure of the surface texture. Characteristics *Ra* gives a very good overall description of height variations but it is not sensitive to small changes in profile whereas characteristics *Rq* is more sensitive to deviations from the main line than *Ra* [10].

$$Ra = \frac{1}{n}\sum_{i=1}^{n}|y_i - \bar{y}| \quad (1)$$

Where: *Ra* = mean arithmetic deviation of a profile (μm), y_i = the individual values of the profile (μm), $\bar{y}$ = average value (μm), *n* = number of points.

$$Rq = \sqrt{\frac{1}{n}\sum_{i=1}^{n}(y_i - \bar{y})^2} \quad (2)$$

Where: *Rq* = root mean square (μm), y_i = the individual values of the profile (μm), $\bar{y}$ = average value (μm), *n* = number of points.

Variance characteristic σ^2 (3) defines a level of high variability.

$$\sigma^2 = \frac{1}{n}\sum_{i=1}^{n}(y_i - \bar{y})^2 \quad (3)$$

Where: σ^2= variance of profile (μm), y_i = the individual values of the profile (μm), $\bar{y}$ = average value (μm), *n* = number of points.

The *Rsk* characteristics (4) describes the symmetry of the heights on the mean line and allows defining so called „positive''surface texture and „negative'' surface texture [11].

$$Rsk = Rq^{-3} . \frac{1}{n} \sum_{i=1}^{n} (y_i - \bar{y})^3 \tag{4}$$

Where: R_{sk} = skewness - asymmetry of the profile (-), Rq = root mean square (μm), y_i = the individual values of the profile (μm), $\bar{y}$ = average value (μm), n = number of points.

The *Rku* characteristics (5) describes the probability density sharpness of the profile. It indicates the presence of disproportionately height peaks or deep valleys and is important for defining a wear conditions.

$$Rku = Rq^{-4} . \frac{1}{n} \sum_{i=1}^{n} (y_i - \bar{y})^4 \tag{5}$$

Where: R_{ku} = kurtosis (-), Rq = root mean square (μm), y_i = the individual values of the profile (μm), $\bar{y}$ = average value (μm), n = number of points.

The MPD characteristics (6) expresses mean depth of profile. The baseline for MPD calculation was chosen 100 μm.

$$MPD = \frac{(M1+M2)}{2} - PP \tag{6}$$

Where: MPD = mean profile depth (μm), M1 = maximum profile value of first half of baseline (μm), M1 = maximum profile value of second half of baseline (μm), PP = average profile value of second half of baseline (μm).

The algorithm for profile filtration process and calculation of above described characteristics was created in the Matlab program. Data processing was performed in two steps. In first step, primary data from the profilometry measurements was filtered to remove the wavelengths that are out of the microtexture range. In second step, the statistical texture characteristics were calculated from the secondary (filtered) data. It was supposed that the filtration process allows more accurately determination of microtexture by this method. Its applicability in the asphalt pavement microtexture investigation was tested on the aggregate grains taken from the pavement surfaces.

2.3. *Data analysis and results*

Each sample of the pavements section consisted of ten aggregate particles. Five aggregate particles represented the sample from the wheel track on the pavement surface and five aggregate particles taken from the core sample after the extraction represented the out of track sample. Six profiles were obtained for particular aggregate particle and statistical characteristics were calculated for the each one. Data in the Table 1 shows the obtained results, where the representative value of investigated microtexture characteristic for each sample was obtained by determination of mean value from thirty values calculated from individual profiles.

The results presented in the Table 1 have shown expressly differences between values determined on the aggregates represent track sample and aggregates represent out of track sample. For better interpretation, the values of *Ra* characteristics are presented in Figure 4. In all samples, the results have shown the degradation of microtexture values due to polishing of aggregate under loading from traffic in relation to the type of used aggregate. It is evident that the considerably lower values of statistical texture parameters were in the road sections, when the limestone was used. And thus it can be argue, that the obtained results confirmed an assumption that the filtration process allows distinguish microtexture levels also in related to type of used aggregate. Obtained results also point to importance of polishing properties of aggregates used in road surfaces.

Table 1. Resultant statistical texture characteristics and *BPN* values.

	Sample	*Ra* (μm)	*σ²* (μm)	*Rq* (μm)	*Rsk* (-)	*Rku* (-)	*MPD* (μm)	*BPN* (-)
Section 1 Andesite	track	7.67	97.34	9.61	0.209	2.87	4.41	55
	out of track	7.84	99.60	9.81	-0.086	2.98	4.69	
Section 2 Andesite	track	8.75	131.66	11.31	0.092	3.53	4.94	64
	out of track	9.23	139.76	11.66	-0.067	3.18	5.22	
Section 3 Limestone	track	1.29	13.00	1.71	-0.193	4.19	1.07	38
	out of track	6.62	73.76	8.47	-0.102	3.26	4.02	
Section 4 Andesite	track	6.79	76.75	4.07	0.057	3.34	4.26	50
	out of track	7.97	118.24	8.61	-0.067	4.10	4.61	
Section 5 Limestone	track	3.23	19.24	4.07	-0.064	2.96	2.07	42
	out of track	5.81	60.10	7.41	-0.084	3.30	3.83	
Section 6 Limestone	track	3.56	21.99	4.64	-0.185	3.91	2.14	37
	out of track	5.29	47.65	6.68	-0.001	3.26	3.39	

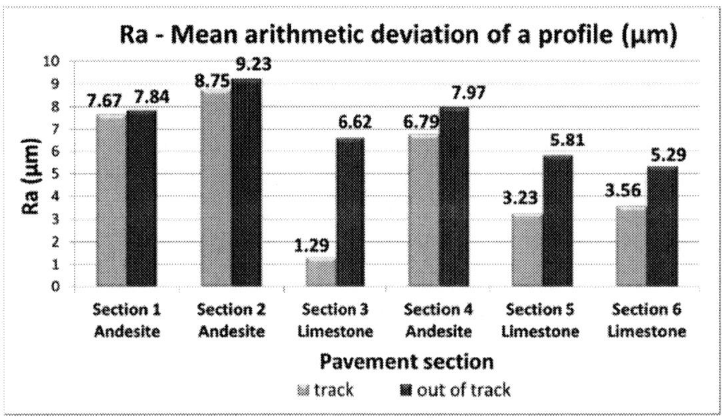

Figure 4. Resultant values of *Ra* characteristics for investigated aggregate samples.

3. Comparison of BPN evaluation with evaluation based on statistical characteristics

The BPN values obtained from measurements on pavements (in the wheel track) were compared with the values of the statistical characteristics obtained from measurements of profiles on the particular aggregate particles taken from the pavement surface (from the sliding path).

The linear regression was chosen, considering of location of values in point charts (Figure 5). The results of comparison are presented in the Table 2, by the coefficient of determination and coefficient of correlation.

Table. 2 Characteristics of regression and correlation analysis – *BPN* and statistical texture characteristics.

	Ra	*Rq*	*σ²*	*MPD*	*Rsk*	*Rku*
Coefficient of correlation r	0.905	0.896	0.917	0.906	0.842	0.348
Coefficient of determination R^2	0.819	0.803	0.841	0.820	0.710	0.121

Data in the Table 2 shows, that some correlation between the *BPN* values and statistical texture characteristics *Ra*, *Rq*, σ^2, *MPD* was found. These results can be useful in order to explain the basic relation between friction and the surface profile proprieties of aggregates. Specifically, higher values of the BPN are related to higher profile heights exclusive of shape properties and vice versa. The coefficient of determination R^2=0.710 between BPN and *Rsk* characteristics implies that *BPN* values is also associated with distribution of surface irregularities within profile because of the *Rsk* characteristics describes the texture by the distribution of peaks and valleys over and under the mean line and allows defining „positive''or „negative'' texture. Thus, from the obtained results is obviously, the polishing process by the tires causes a reduction in the height of the peaks on the aggregate surface with a subsequent reduction of the *BPN*. The lowest correlation between *BPN* and *Rku* characteristics probably related to definition of this characteristics. Whereas statistical characteristics *Ra, Rq*, σ^2, *MPD* and *Rsk* describes irregularities of investigated profile by mean values of profile heights the, *Rku* characteristics describes the texture indirectly by the probability density sharpness of the profile.

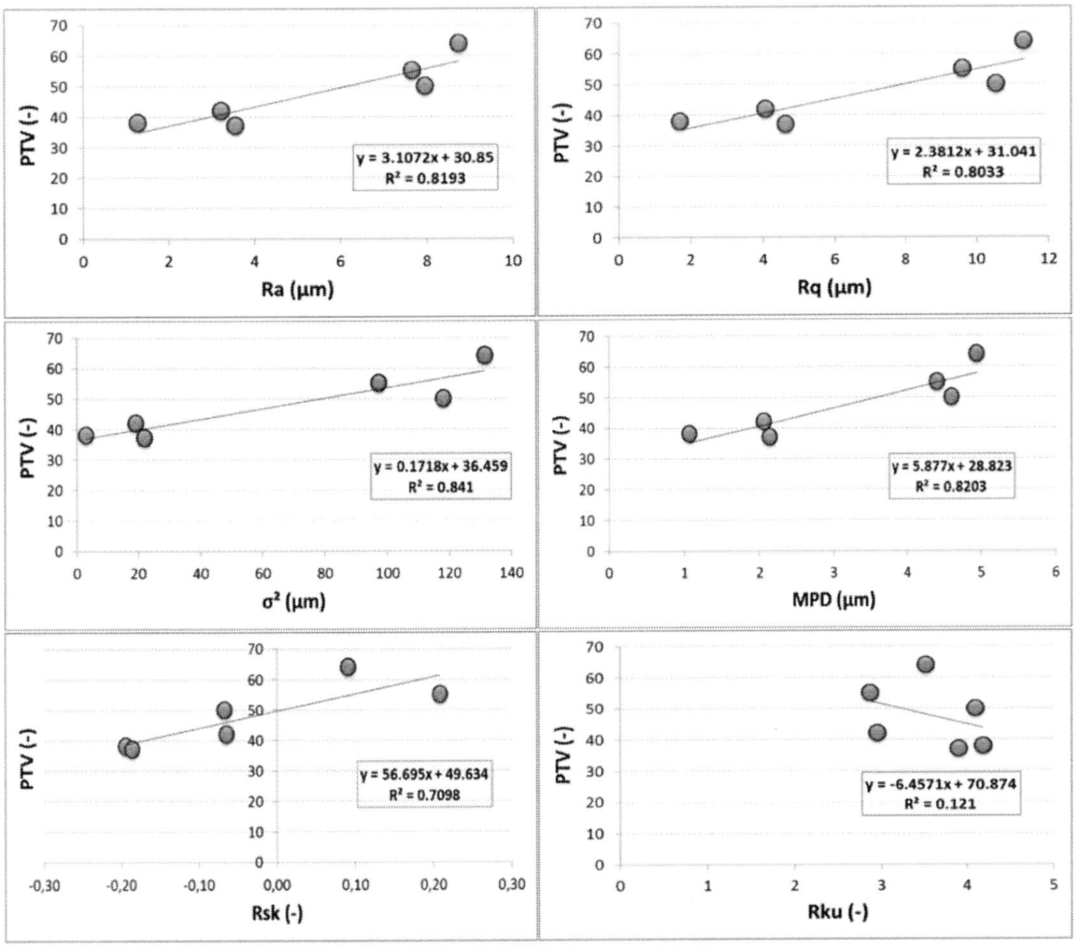

Figure 5. Relation of the *BPN* values with the values of statistical texture characteristics.

4. Conclusions

This paper in short presented results of microtexture measurement. The British pendulum tester measurements have been practiced on selected sections of roads with different asphalt surfaces. Individual grains of aggregates were taken from the surface of each section from the sliding path and also from the core sample after the extraction. The laboratory profilometry measurements have been practiced on these aggregate samples. Subsequently the surface microtexture was investigated based on commonly used texture characteristics and the filtration approach was applied in calculation process.

Based on the results obtained so far it can be argue that the application of filtration approach in the microtexture evaluation based on the profilometry measurements allows more accurately determination of microtexture characteristics and can provide the more representative results in the relation to *BPN* values. From this point of view presented research can provide a methodological recommendation and the implementation of the filtration approach in the further research can be useful for a better understanding of the material proprieties that affect the skid resistant coefficient of the pavement surface.

It is supposed that in the case of implementation of additional measurements even the more relevant correlation between *BPN* parameter and statistical texture characteristics can be expected. In the future research, it is also necessary to select measuring road sections with the available information's about type and composition of asphalt mixtures and the especially about service life and the intensity of traffic.

On the basis of this information's can be possible deal with the degradation of microtexture values in relation to traffic intensity, and mainly to the ability of aggregate be resist to polishing.

Acknowledgement
This research was supported by European project Research Centre of the University of Žilina – Second Phase ITMS: 313011D011

References

[1] Scharnigg K and Schwalbe G 2010 Report on knowledge gaps and proposals for further research concerning optimisation for road surfaces and tyres for skid resistance, rolling resistance and noise emissions *TYROSAFE project* http://tyrosafe.fehrl.org
[2] Hall W et al 2009 *Guide for Pavement Friction* NCHRP Web-Only Document 108 https://www.nap.edu/catalog/23038/guide-for-pavement-friction
[3] European Committee for Standardization (CEN) 2009 Characterization of pavement texture by use of surface profiles - Part 5: Determination of megatexture *(EN ISO 13473-5)*
[4] David K et al 2015 Evaluation of Pavement Safety Performance *Final Report FHWA Publication No. FHWA-HRT-14-065* https://www.fhwa.dot.gov/publications/research/safety/14065/1406 5.pdf
[5] Li S, Noureldin S and Zhu K 2010 *Safety Enhancement of the INDOT Network Pavement Friction Testing Program: Macrotexture and Microtexture Testing Using Laser Sensors* (Purdue University Indiana)
[6] European Committee for Standardization (CEN) 2011 Road and airfield surface characteristics - Test methods - Part 4: Method for measurement of slip/skid resistance of a surface - The pendulum test *(EN 13036-4)*
[7] Lu Q and Steven B 2006 *Friction Testing of Pavement Preservation Treatmens: Literature Review*
[8] European Committee for Standardization (CEN) 1997 Characterization of pavement texture by use of surface profiles - Part 1: Determination of mean profile depth *(ISO 13473-1)*
[9] European Committee for Standardization (CEN) 1997, Geometrical Product Specifications (GPS) - Surface texture: Profile method - Terms, definitions and surface texture parameters *(ISO 4287)*
[10] Sedlaček M et al 2010 Surface Topography Modelling for Reduced Friction *Journal of Mechanical Engineering* **9** pp 674-80
[11] Remišová E, Kováč M and Decký M 2014 The influence of the asphalt mixture composition on the pavement surface texture and noise emissions production *Proc. 14th International Multidisciplinary Scientific GeoConference (Albena)*

The road surface as a source of particulate matter

Daša Fullová[1], Dušan Jandačka[1], Daniela Ďurčanská[1], Adriana Eštoková[2] and Jitka Hegrová[3]

[1]University of Žilina, Faculty of Civil Engineering, Department of Highway Engineering, Univerzitná 8215/1, 010 26 Žilina, Slovakia
[2]Technical University of Kosice, Faculty of Civil Engineering, Institute of Environmental Engineering, Vysokoškolská 4, 042 00 Košice, Slovakia
[3] Transport Research Centre, Líšeňská 33a, 636 00 Brno, Czech Republic

E-mail: dasa.fullova@fstav.uniza.sk

Abstract. Traffic volume is still increasing and has unpleasant impact on the environment and longevity of the pavements. It is also associated with increase in emissions of particulate matter. Slovakia has built up the major part of roads with asphalt surface. Therefore, the research presented in this contribution deals with abrasion of bituminous wearing courses of pavements. The asphalt mixtures of wearing courses are compared in terms of released particulate matter. The samples of asphalt mixtures are abraded in wheel tracking machine DYNA-TRACK. The particulate matter measurements were performed in the laboratory conditions. The experimental laboratory measurements make it possible to sample particulates without contamination from exhaust emissions, abraded particles from vehicles, resuspension of road dust and climate affects. The contribution presents results of measured mass concentrations, chemical analyses of asphalt mixture materials and chemical analyses of particulate matter. The rutted asphalt samples are compared in terms of particulate matter mass concentrations and chemical composition. The contribution presents results of measurements on six trial samples of asphalt mixtures with different composition. The concentrations of metals are subjected to the multivariate statistical analysis (factor analysis) for the identification of sources of particulate matter (bituminous binder, aggregates).

1. Introduction and methodology of experimental measurements

Vehicle motions cause mechanical wearing of the asphalt pavement surface - wearing course by vehicle tires. That leads to gradual fragmentation and abrasion of aggregate. Traffic volume is still increasing and has unpleasant impact on longevity of the pavements and the environment. On the basis of several scientific papers, studies and researches about PM production from road traffic [1], [2], [3], [4] our research is focused on non-combustion emissions - particulate matter from abrasion of the pavement surface. It is very difficult process to identify pavement surface abrasion as a source of particulate matter and separate it from other abrasions and quantify its share of the amount of produced particulate matter in situ. Therefore, PM measurements are performed in laboratory conditions and captured particulates supposed to be without contamination from exhaust emissions, abraded particles from vehicles, resuspension of road dust and climate affects [5].

The aim of the research is to confirm the abrasion of pavement surface as a source of particulate matter and to verify the impact of the composition of asphalt mixture on the PM production. An important phase of the research is to determine chemical elements in the collected particulate matter and to confront it with the representation of these elements in the chemical composition of the asphalt mixture. The research is aimed at finding dependence between the composition of asphalt mixture and collected PM.

The measurements are performed in the laboratory at the Department of Highway Engineering. The asphalt mixture samples of wearing courses are used for laboratory measurements. The used asphalt mixture samples – plates of size 320 x 260 mm and thicknesses of 40 mm are rutted in wheel tracking machine DYNA-TRACK which is used to assess the resistance to rutting of asphalt materials under conditions, which simulate the effect of traffic (Figure 1).

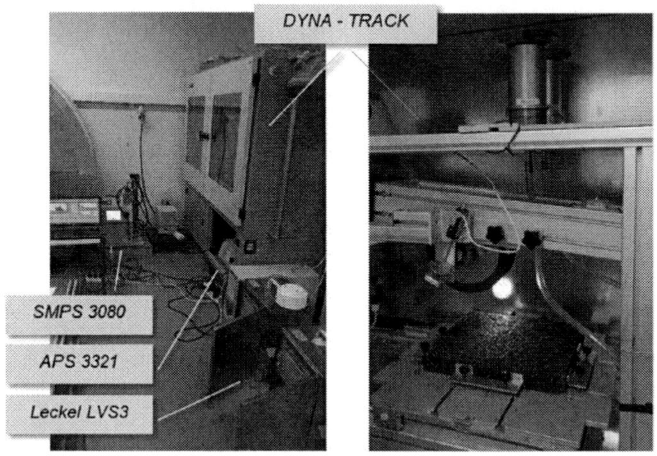

Figure 1. The used machine technology of laboratory tests and rutted asphalt mixture sample.

Air sampling from the inside of wheel tracking machine during the rutting of samples is performed by means of equipment APS 3321 (Aerodynamic Particle Sizer) and SMPS 3080 (Scanning Mobility Particle Sizer) which intercept and distribute particle range from 0.012 to 20 µm and 3 pieces of Leckel LVS3 which capture particulate matter $PM_{2.5}$ and PM_1. Chemical composition tests of materials (aggregates, bituminous binder) of samples were carried out by XRF analyser (X-ray fluorescence spectroscopy) before the rutting. For the experiments the machine SPEKTRO iQ II (AMATEK, Germany) was used. Particulate matter was captured on Nitrate membrane filters for subsequent element determination. The filters were digested in Teflon vessels with HNO_3 at a temperature of 230°C using a high temperature and high pressure Microwave digestion system SW-4 (Berghof, Germany). Samples were transferred and appropriately diluted before element determinations on a triple quadrupole ICPMS (ICPQQQMS, Agilent Technologies, Germany).

2. Results of particulate matter chemical composition

Measurements of asphalt mixture abrasion were carried out on 6 samples. Descriptions of rutted samples and mass concentrations of abraded particulate matter are shown in the following Table 1. The highest average PM mass concentration (13.51 µg/m^3) was measured for the mixture AC 11 with unmodified bitumen CA 50/70 (number 4) and the lowest average PM mass concentration (7.21 µg/m^3) was measured for the mixture AC 11 with unmodified bitumen CA 50/70 (number 1). The mixtures differ in use of aggregate. In the mixture with the highest concentrations the aggregates melaphyre and dolomite were used and in the mixture with the lowest concentrations the siliceous limestone was used.

The sampling of laboratory environment were carried out by means of Leckels before the rutting of samples and after these measurements in order to confirm the difference between the laboratory environment and the environment inside of wheel tracking machine. The difference between PM mass concentrations of laboratory environment and the lowest PM mass concentrations of asphalt mixture is determined at 3.23 $\mu g/m^3$ for particulates of fraction $PM_{2.5}$ and 1.42 $\mu g/m^3$ for particulates of fraction PM_1.

Table 1. Description of rutted asphalt mixtures and average PM mass concentrations.

Number	Asphalt mixture	Bitumen	Aggregate	PM_1 [$\mu g/m^3$]	$PM_{2.5}$ [$\mu g/m^3$]
1.	AC 11 O 50/70, II	CA 50/70	siliceous limestone	4.46	7.21
2.	AC 11 O 50/70, II	CA 50/70	dolomite	7.55	9.04
3.	AC 11 O 50/70, II	CA 50/70	dolomite, andesite	4.22	7.99
4.	AC 11 O 50/70, II	CA 50/70	dolomite, melaphyre	9.23	13.51
5.	AC 11 O PMB 45/80-75, I	PMB 45/80-75	dolomite, melaphyre	8.50	11.83
6.	SMA 11 PMB 45/80-75	PMB 45/80-75	dolomite, melaphyre	4.77	7.11
	Laboratory environment (outside the DYNA-TRACK)			3.04	3.98

For identification of particulate matter originating from the asphalt mixture abrasion the chemical composition of basic materials of mixtures were carried out by XRF analyser. Unmodified bituminous binder CA 50/70 contains approximately 4.8 % of inorganic elements. Sulfur S has the largest representation, it represents 4.19 % (Figure 3). The chemical composition tests were also carried out for all kinds of aggregates used in tested samples. The results have shown that elements silicon Si and calcium Ca have the largest percentage in aggregates. The silicon Si predominates in aggregate melaphyre and andesite and the calcium Ca predominates in aggregate dolomite and siliceous limestone. The elements such as Mg, Al, Fe represent a higher share of identified elements (Figure 2). On the basis of analyses of aggregates and bituminous binder 13 chemical elements (Ca, Si, Mg, Al, Fe, P, S, Cl, K, V, Cr, Mn, and Na) were selected for confirmation of abrasion of asphalt mixture surface.

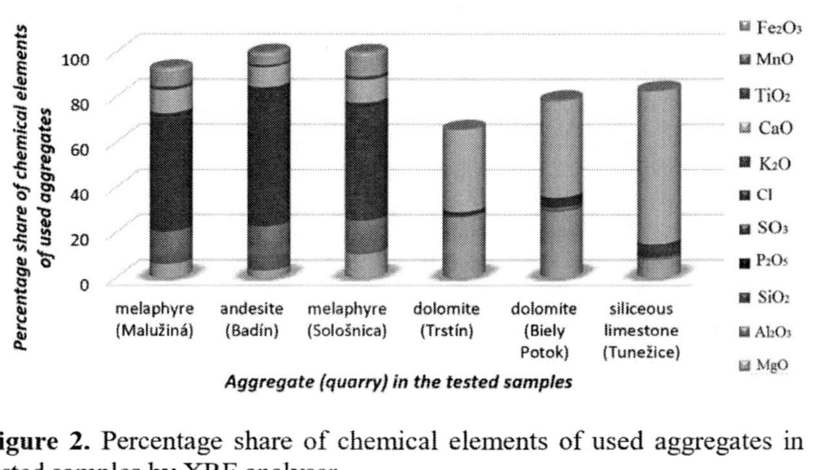

Figure 2. Percentage share of chemical elements of used aggregates in tested samples by XRF analyser.

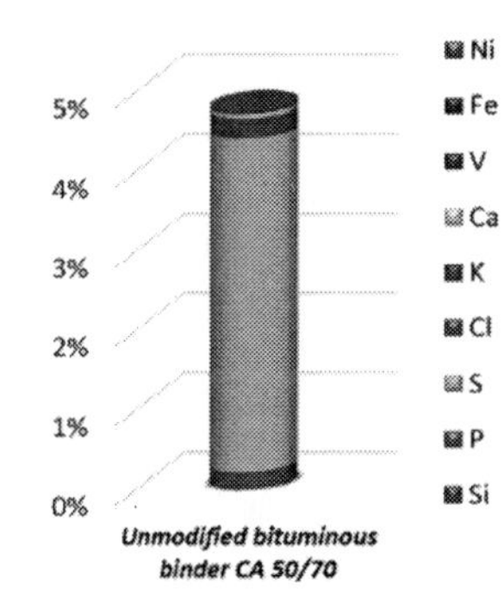

Figure 3. Percentage share of elements of bituminous binder.

Subsequently, exposed filters (51 filters) were analysed for presence of selected elements by means of mass spectrometry with inductively coupled plasma. The Table 2 presents average mass concentrations of selected 13 chemical elements in captured particulate matter $PM_{2.5}$ for each tested asphalt mixture. The elements Ca, S, Cl have the highest representation in the analysed filters. The highest concentrations of selected elements were analysed in the sample AC 11 (number 1), elements represent together 833.34 mg/g of $PM_{2.5}$. Calcium Ca has the highest representation in this sample and its concentration is 546.22 mg/g of $PM_{2.5}$. In this sample of asphalt mixture aggregate siliceous limestone is used. Sulphur with mass concentration 126.15 mg/g of $PM_{2.5}$ has also the highest concentration in this sample. With regard to the chlorine the highest concentration (81.47 mg/g of $PM_{2.5}$) was measured in the sample SMA 11 (number 6).

Table 2. Average mass concentrations of analysed elements in captured particulates $PM_{2.5}$

Number of mixture/ Mass concentrations of analysed elements [mg/g]	**1. AC 11**	**2. AC 11**	**3. AC 11**	**4. AC 11**	**5. AC 11**	**6. SMA 11**
Ca	546.22	117.44	92.53	52.16	68.80	110.01
Si	9.79	6.75	7.69	6.47	14.01	17.47
Mg	7.31	4.01	6.08	3.39	3.43	6.77
Al	7.93	4.51	1.41	0.76	3.39	7.23
Fe	37.63	22.02	12.47	11.04	12.28	26.89
P	2.06	2.16	1.73	0.73	0.46	1.78
S	126.15	60.97	93.60	104.84	57.64	113.50
Cl	62.87	43.94	68.02	34.84	30.86	81.47
K	19.39	10.59	17.75	10.34	12.77	17.65
V	0.04	0.01	0.01	0.03	0.02	0.14
Cr	1.55	0.67	1.22	0.67	0.49	1.70
Mn	1.94	1.76	0.56	0.30	0.71	2.09
Na	10.46	8.35	6.93	3.07	6.91	14.35

On the other hand, the lowest concentrations were measured for the element vanadium V (in range from 10 to 140 μg/g of $PM_{2.5}$) and for elements such as Cr, Mn and P (Table 2).

3. Statistical analysis of measured chemical elements concentrations

Multivariate statistical analyses using Principal component analysis - PCA and Factor analysis – FA were used for statistical assessment [5], [6], [7], [8], [9], [10], [11], [12], [13], [14], [15], [16].

The main objective of using multivariate statistical analysis was to uncover the internal interconnection and the variability of the investigated variables (chemical elements - metals). Identified metals group may characterize some of the partial sources (materials) of asphalt mixtures.

The primary goal of PCA is the transformation of the original characters of x_j, $j=1, \ldots, m$, into a smaller amount of latent variables of y_j. These latent variables possess more appropriate properties: their presence is less significant, they capture and represent almost the entire variability of the original characters, properties and they are mutually not correlated – mutually uncorrelated. Latent variables are known as the main components and they represent linear combinations of former variables: the first main component y_1 describes the greatest part of variability, hence the dissipation, spread of the original data, the second main component y_2 on the other hand the greatest part of dissipation, spread not-contained within y_1 etc..

For the first main component the following relation prevails:

$$y_1 = \sum_{j=1}^{m} v_{1j} x_j \tag{1}$$

where: x_j former character, input variable, j = 1, ..., m,
v_{1j} coefficients of own vectors.

Within the process of FA so called factorial loads are estimated for particular variables (metals) for a generated factor. Factor loads, so to say the expression of correlations between the particular variables and acquired factors. Based on the values of factor loads it is possible to specify a group of variables for each factor, those ones which correlate with it in the closest-possible manner. And vice versa, by means of factor loads, the identified factor is appended with an extent of impact on each of particular variables. The variables with the highest factor loads for a generated factor are to be considered as decisive even when interpreting such a factor. A data matrix serves the purpose of input for calculations, whose lines correlate with particular measurements (objects) and bars of variables, i.e. measured pollutant (character). The variables to be used are those pollutants which are able to specify anticipated sources of pollution.

The fundamental principle of FA lies in the fact that each and every of monitored values Y_j $(j = 1, ..., p)$ may be expressed as a sum of a linear combination of a lesser amount m non-observed (hypothetical) random values $F_1, ..., F_m$ – so called common factors and the further source of variability E_j $(j = 1, ..., p)$ – so called specific (residual) elements. Let us suppose that the following model prevails:

$$Y_j = \sum_{k=1}^{m} \lambda_{jk} F_k + E_j, j = 1, \dots, p, \tag{2}$$

where: λ_{jk} factor load of the k common factor relevant to the j value and of k factor, k = 1,..., m,
F_k the k common factor,
E_j is a random deviation of the exact model, relevant to the one of k value, j=1, ..., p.

Values $Y_1, ..., Y_p$ are standardized, i.e. they have a zero median value and unit variance. In the FA method it is recommended to have at least 5 samples, while the optimum number of samples could reach 20 per each variable [8], [9].

The input data for the purposes of identification of possible PM sources consisted of the concentrations of particular chemical elements, specifically metals (µg/g). The data matrix consisted of 13 variables and 24 objects. Multivariate statistical analyses and source identification were processed for the fraction $PM_{2.5}$.

When it is deemed to be necessary to specify the number of factors before the FA calculation can be run, the PCA was performed beforehand. As a result, we could determine the number of principal components which, to a sufficient degree, account for the variability in the data. As is seen in Figure 4, we are able to fix the number of principal components at 4 (Eigenvalue > 1).

The biplot shown in Figure 5 simplifies our decision-making process over the course of selecting the number of factors for use in FA, where close relationships among several variables can be observed. In this case, there are four collections of variables separated by a significant disparity. That is, there are 4 principal components which may be consecutively used in factor analysis.

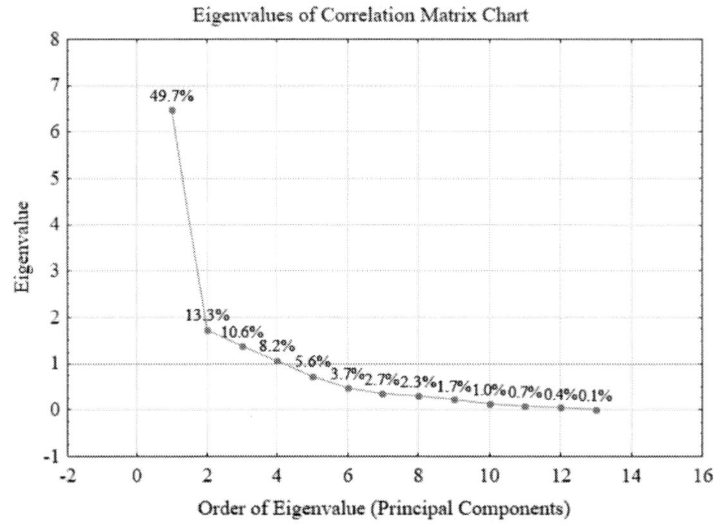

Figure 4. Graph of "foothills" eigenvalues based on PCA.

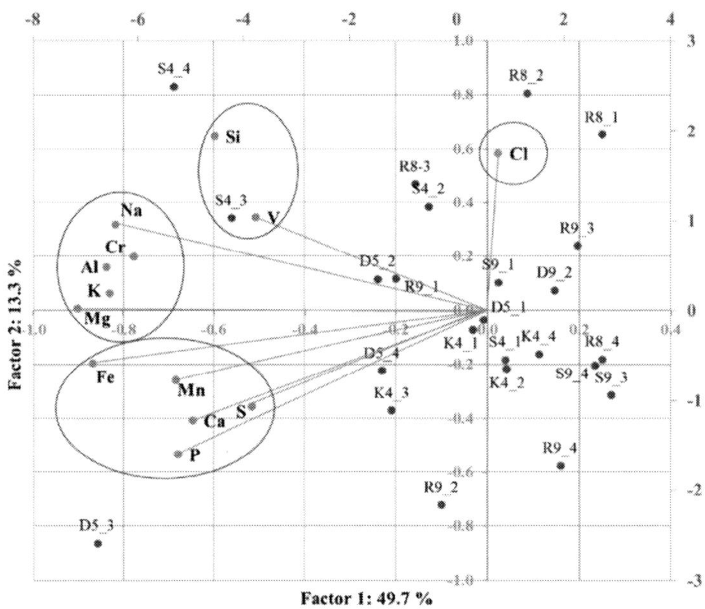

Figure 5. Principal components biplot.

Four factors were selected for factor analysis. These four principal components account for 81.9 % of the total variability in the original data.

The fourth factor was eliminated, which resulted from the consecutive FA using the four factors. The fourth factor was made up primarily of one element, chlorine "Cl", and that is why it represented a specific factor which is at the same time difficult to define and interpret. Three factors account for 76.4 % of the total variability in the original data.

The Varimax model was used for factor rotation. The factor loadings associated with variables may be explained as the correlations between the factors and the variables. They represent the most important unit of information that the interpretation of the factors is based on. Each factor contains information about contributions of selected elements (metals). As the most decisive loading factors, values close to or greater than 0.7 were selected. Based on the representation and presence of elements in particular factors, the following factors may be named (Figures 6, 7, 8). The factors for $PM_{2.5}$ (Figures 6, 7, 8) were identified based on the origins of metals (Figures 2, 3).

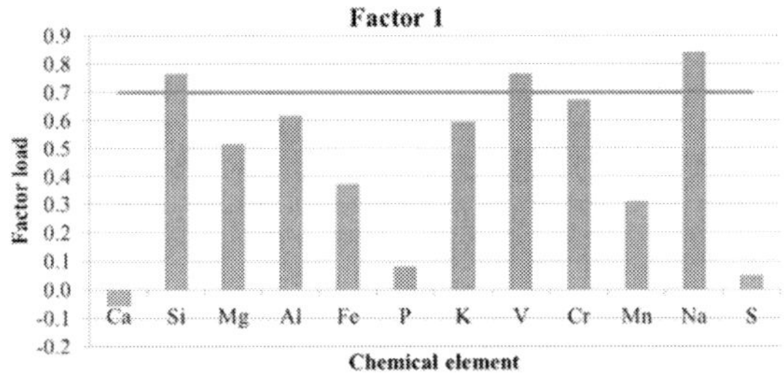

Figure 6. Factor loadings of metals on Factor 1.

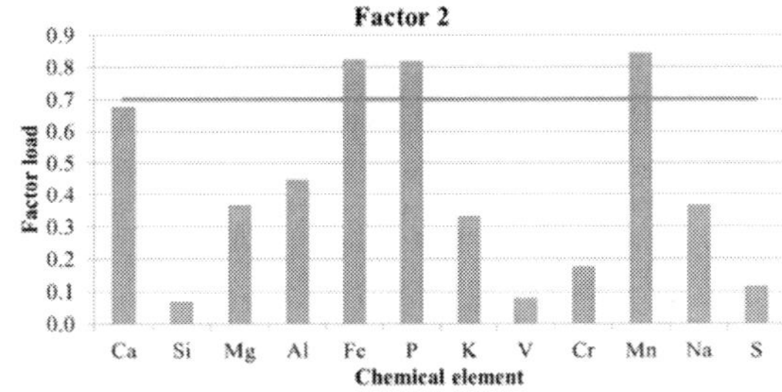

Figure 7. Factor loadings of metals on Factor 2.

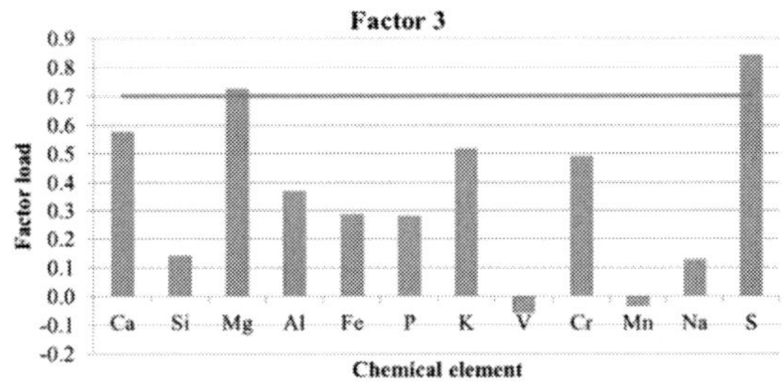

Figure 8. Factor loadings of metals on Factor 3.

Moreover, the variables (elements) with factor loadings of <0.7 do contribute to a given factor by virtue of their weight and may facilitate its naming. Factors obtained via factor analysis of $PM_{2.5}$ were interpreted as Factor 1, representing aggregate (melaphyre, andesite) with marker silicon "Si" (Figure 6), Factor 2, representing aggregate (dolomite and siliceous limestone) with marker calcium "Ca" (Figure 7), and Factor 3, representing bituminous binder with marker sulphur "S" (Figure 8).

4. Conclusion

The highest average mass concentrations of $PM_{2.5}$ were measured for the mixture AC 11 with classic bituminous binder (with mark number 4) and the lowest average mass concentrations were measured for the sample AC 11 with classic bituminous binder (with mark number 1). The mixtures differ in use of aggregate. In the mixture with the highest concentrations the aggregates melaphyre and dolomite were used and in the mixture with the lowest concentrations the siliceous limestone was used.

From chemical analyses of filters the average mass concentrations of selected elements were determined. Calcium Ca has the highest representation in the sample AC 11 with mark number 1 and in this sample the aggregate siliceous limestone is used. The highest concentrations of selected elements were analysed in this sample. Despite of the fact that the sample has the lowest mass concentrations of particulate matter. As resulting, the released particulates from abrasion of mixture surface have similar composition as the basic materials of asphalt mixtures (aggregate, bituminous binder). The first results indicate that it has been found differences in the quality of asphalt mixtures in terms of PM production during the rutting of asphalt pavement surfaces.

Multivariate statistical analysis revealed three meaningful factors (sources of particulate matter) which are characterized by the chemical elements.

The identification of sources based on the presence of metals in the $PM_{2.5}$ fraction, were associated with particular elements: Factor 1, representing aggregate (melaphyre, andesite) (**Si**, V and Na), Factor

2, representing aggregate (dolomite and siliceous limestone) (Ca, Fe, P and Mn) and Factor 3, representing bituminous binder (S and Mg).

Acknowledgements
The paper originated as being supported by means of a grant KEGA 045ŽU-4/2017 Environmental education of engineers for application in public administration

References
[1] Kwak J, Kim H, Lee J and Lee S 2013 Characterization of non-exhaust coarse and fine particles from on-road driving and laboratory measurements *Science of the Total Environment* **458-460** pp 273-82
[2] Pio C, Mirante F, Oliveira C, Matos M, Caseiro A, Oliveira C, Querol X, Alves C, Martins N, Cerqueira M, Camoes F, Silva H and Plana F 2013 Size-segregated chemical composition of aerosol emissions in an urban road tunnel in Portugal *Atmospheric Environment* **71** pp 15-25
[3] Pirjola L, Kupiainen K J, Perhoniemi P, Tervahattu H and Vesala H 2009 Non-exhaust emission measurement system of the mobile laboratory SNIFFER *Atmospheric Environment* **43**(31) pp 4703-13
[4] Ličbinský R, Frýbort A, Huzlík J, Adamec V, Effenberger K, Mikuška P, Vojtěšek M and Křůmal K 2010 Usage of scanning electron microscopy for particulate matter sources identification *Transaction on Transport Sciences* **3** pp 137-44
[5] Haider M, Folkeson L, Conter, Ramsak M, Cholava, Adamec V, Licbinsky R, Durcanska D and Decky M 2009 *Guidelines for the environmental assessment of various pavement types including recommendations to road authorities in NMS, SPENS: Sustainable Pavements for European New Member States* (European commission DG research, Sixth framework programme)
[6] Kachigan S K 1991 *Multivariate Statistical Analysis: A Conceptual Introduction, 2nd Edition*
[7] Manly B F J 2004 *Multivariate Statistical Methods: A Primer, Third Edition* (Boca Raton, USA: CRC press)
[8] Meloun M and Militký J 2006 *Compendium of statistical data processing (in Czech)* (Prague: Academia)
[9] Meloun M, Militký J and Hill M 2012 *Statistical Analysis of Multidimensional Data in the Examples (in Czech)* (Prague: Academia)
[10] Spencer N H 2013 *Essentials of Multivariate Data Analysis* (Boca Raton, USA: CRC press)
[11] Varmuza K and Filzmoser P 2009 *Introduction to Multivariate Statistical Analysis in Chemomentrics* (Boca Raton, USA: CRC press)
[12] Lu X, Wang L, Li L Y, Lei K, Huang L and Kang D 2010 Multivariate statistical analysis of heavy metals in street dust of Baoji, NW China *Journal of Hazardous Materials* **173** pp 744-49
[13] Yang Z, Lu W, Long Y, Bao X and Yang Q 2011 Assessment of heavy metals contamination in urban topsoil from Changchun City, China *Journal of Geochemical Exploration* **108** pp 27-38
[14] Manta D S, Angelone M, Bellanca A, Neri R and Sprovieri M 2002 Heavy metals in urban soils: a case study from the city of Palermo (Sicily), Italy *The Science of Total Environment* **300** pp 229-43
[15] Guo H, Wang T and Louie P K K 2004 Source apportionment of ambient non-methane hydrocarbons in Hong Kong: Application of a principal component analysis/absolute principal component scores (PCA/APCS) receptor model *Environmental Pollution* **129** pp 489-98
[16] Song Y, Xie S, Zhang Y, Zeng L, Salmon L G and Zheng M 2006 Source apportionment of $PM_{2.5}$ in Beijing using principal component analysis/absolute principal component scores and UNMIX *Sciece of the Total Environment* **372** pp 278-86

IOP Conf. Series: Materials Science and Engineering 236 (2017) 012027 doi:10.1088/1757-899X/236/1/012027

Contamination of environment in the road surroudings – impact of road salting on Norway spruce *(Picea abies)* and Scots pine *(Pinus sylvestris)*

Jitka Hegrová[1], Oliver Steiner[2], Walter Goessler[2], Stefan Tanda[2] and Petr Anděl[3]

[1]Transport Research Centre, Líšeňská 33a, 636 00 Brno, Czech Republic
[2]University of Graz, Institute of Chemistry, Analytical Chemistry for Health and Environment, Universitätsplatz 1, 8010 Graz, Austria
[3]Faculty of Environmental Sciences, Czech University of Life, Sciences Prague, Kamýcká 129, Praha 6 – Suchdol, 165 21, Czech Republic

E-mail: jitka.hegrova@cdv.cz

Abstract. A comprehensive overview of the influence of transport on the environment is presented in this study. The complex analysis of soil and needle samples provides an extensive set of data, which presents elemental contamination of the environment near roads. Traffic pollution (including winter road treatment) has a significant negative influence on our environment. Besides sodium and chlorine from winter maintenance many other elements are emitted into the environment. Three possible sources of contamination are assumed for environmental contamination evaluation: car emission, winter maintenance and abrasion from breaks and clutches. The chemical analysis focused on the description of samples from inorganic point of view. The influence of the contamination potential on the sodium and chlorine content in the samples of 1st year-old and 2nd year-old needles of Norway spruce (*Picea abies*) and Scots pine (*Pinus sylvestris*) is discussed. Additional soil samples were taken from each sampling site and analyzed to get insight in the sodium and chlorine distribution. Statistical evaluation was used for interpretation of complex interaction patterns between element concentrations in different aged needles based on localities character including distance from the road and element concentration in soils. This species of needles were chosen because of its heightened sensitivity towards salinization. The study was conducted in different parts of the Czech Republic. The resulting database is a source of valuable information about the influence of transport on the environment.

1. Introduction
Traffic pollution (including winter road treatment) has a significant negative influence on our environment. Besides sodium and chlorine from winter maintenance many other elements are emitted into the environment. In the past lead was known as traffic marker. However, since it has been banned as an anti-knocking additive other elements like platinum group elements (PGEs, from catalysts), antimony (from brakes and clutches), vanadium and nickel (from fuel burning) are typical traffic markers.

Different analytical methods are used for elemental determination of environmental samples. The most common are methods of atomic spectrometry [1]. The used ICPQQQMS instrument has resulted in successful method development to overcome several isobaric, polyatomic and doubly charged interferences in the analysis of environmental samples with complex matrices. The method is suitable for determination of elements at very low concentrations (Platinum group elements – PGE) and elements which are influenced by spectral interferences (PGE, S, P, As, Se).

To investigate a potential influence of traffic pollution (influence of road salting included) on the environment, samples of 1st year-old and 2nd year-old needles of Norway spruce - *Picea abies* and scots pine - *Pinus sylvestris* were collected from two sites differing in traffic loading. Spruce trees were selected because of their abundance and high sensitivity to salinity [2,3]. They are known to be good indicators for urban pollution [3,4]. The 1st and 2nd year old needles were analysed separately. Additionally soil samples were collected at each sampling site. Sodium chloride is the most commonly used de-icing material for road winter treatment. The applied sodium chloride is mobilized by melting snow or ice and rainfall, which leads to its distribution in the soil environment, surface water, geological environment, and groundwater [5].

The aim of this work was the preparation of data for risk assessment of environmental contamination in the vicinity of roads and highways. Soil is the final recipient of a wide range of contaminants and the source for entry into the food chain. The investigation of a potential influence of traffic pollution on the environment attention was focused on all components of the environment.

2. Results and discussion

2.1. Sampling and sample preparation

Sampling was carried out throughout the Czech Republic in the areas with high traffic impact (near the main roads and cities) and the areas with low traffic load (National parks, mountains,...) About one thousand samples of soil, one hundred samples of water and road dust, and three hundred samples of needles were collected during three years.

The study area to investigate the environmental impact of chemical winter road maintenance was chosen to be in the northern part of Czech Republic in the Liberec region. Pine needles were collected in metropolitan area of Prague and surrounding villages. The contaminated spruce needles and soil samples were collected in a distance of 5 m and the uncontaminated samples in a distance of 40 m from the road. The tree health was classified according to the physical appearance (crown density) and color of the needles. Also the contamination potential was estimated. Soil samples were collected over the years 2015 and 2016. Soil samples were collected across the entire country during the years 2013 – 2017.

The different age classes of needles were analyzed separately. The needles were crushed, milled and then extracted by pure water for Cl determination. For the analysis of other elements the samples were digested in Teflon vessels with HNO_3 and H_2O_2 at a temperature of 200°C for 30 min and a pressure of 30 bar using microwave system (SW-4+, Berghof, Germany). The soil samples were sieved, milled and digested in Teflon vessels with Aqua Regia at the same conditions. Samples were appropriately diluted before element analysis With ICPQQQMS. For comparison and Cl determination samples were extracted by pure water 24 hours, then filtered and analyzed. For PGE determination, of HCl was added to stabilize these elements.

2.2. Chemical Analysis

All samples were analyzed with ICPQQQMS (Agilent Technologies, Japan) in different tune modes (collision/reaction, He, H_2, O_2, NH_3). Especially for the detection of PGEs the reaction with NH_3 in MS/MS mode was necessary to achieve low detection limits. As, Se, S and P were analyzed in O_2 MS/MS mode with a mass shift of +16.

The quality of the analytical results was checked via frequent analyses of reference materials (NIST Pine Needles 1575a, QCM – Metranal-33 - Soil, NIST Trace elements in natural water 1640a, ERM – CZ120 – Fine Dust PM_{10} - Like).

2.3. Results

To evaluate the environmental contamination in the vicinity of roads, the methodical scheme was selected:

- Sampling from small amount of selected areas (wider restricted area around the road, with chosen location) with the large number of sampling points (places for sampling in chosen location in different distances from the road).
- Sampling from more areas, with high amount of location
- Sampling high amount of location

Sampling was carried out near the main roads and the roads of 1st and 2nd category.
The results of the elements concentration in the soils were evaluated based on four factors:

- by the distance from the road
  V – road dust from the surface of the road
  K – the nearest sampling point from the road
  E – about five meters far from the road
  B –about twenty meters far from the road
  P – background, about one hundred meters from the road
- by the category of road
  main road
  road of 1st category
  road of 2nd category
- by the level of contamination potential
  1st class – lowest contamination potential, a tree with the lowest possibility of contamination
  2nd class – medium contamination potential, a transitional zone with minor contamination
  3rd class – highest contamination potential, the trees immediately next to the road

The results of the elements concentration in needles were evaluated based on four factors:

- the contamination potential:
  1st class – lowest contamination potential, a tree with the lowest possibility of contamination
  2nd class – medium contamination potential, a transitional zone with minor contamination
  3rd class – highest contamination potential, the trees immediately next to the road
- the health condition:
  1st class – healthy tree
  2nd class – slightly damaged tree
  3rd class – moderately damaged tree
  4th class – damaged tree
  5th class – severely damaged tree
- the distance from the road:
  1st category – samples collected from up to 5 m from the road
  2nd category – samples collected at least 40 m from the road
- the age of needles: first-/second-year

The concentrations for most of the traffic related elements (Na, Ca, S, V, Cr, Co, As, Sb and Pb) increased by 40 % to 60 % from first year needles to the second year needles. Concentration of K and Mg decrease in older needles, because these elements are displaced by Na and Ca from salt from winter maintenance (Figure 1).

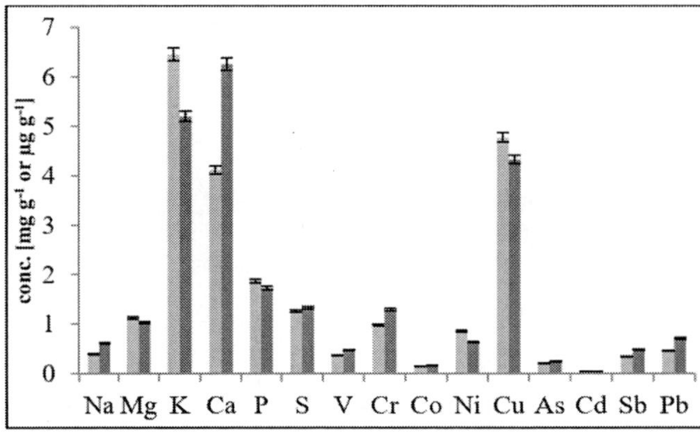

Figure 1. Concentration of traffic related elements in 1st year (light grey) and 2nd year (dark grey) pine needles (Concentration Na, K, Mg, Ca, P, S are in mg g^{-1}; other elements in μg g^{-1}).

Element concentrations found in control samples from "not contaminated area", are typically lower than from all other areas (Figure 2). The antimony concentrations in roadside pine needles show the highest contamination compared to the control group. The behaviour of antimony as typical traffic related element and its interaction with the roadside trees is shown in the Figure 3. The antimony concentrations in roadside pine needles and the vanadium concentrations in urban soils show the highest contamination compared to the control group. The measured element concentrations in needles represent the combination of adsorbed particulate matter and trace elements taken up via the root system of the trees.

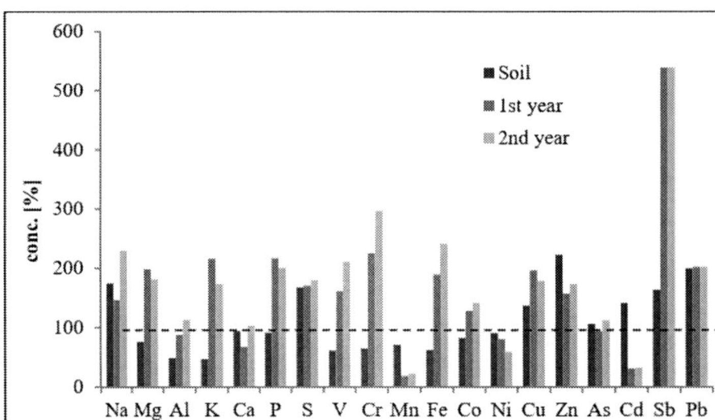

Figure 2. Traffic related element contents from roadside soils and pine needles compared to soils and pine needles from a control area with no direct traffic impact (100%).

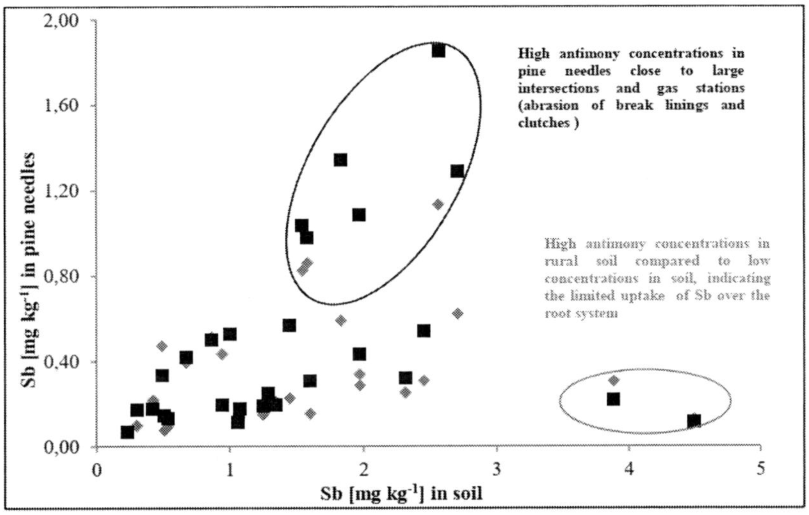

Figure 3. Antimony concentrations in 1st year and 2nd year pine needles and their corresponding soils.

If we compare concentrations of Na in soil samples from different road type side in different distance from the road, we can see a decreasing concentration with an increase of the distance to the road (Figure 4). This confirmed the low level of mobility in the soil and its retention in the vicinity of the road.

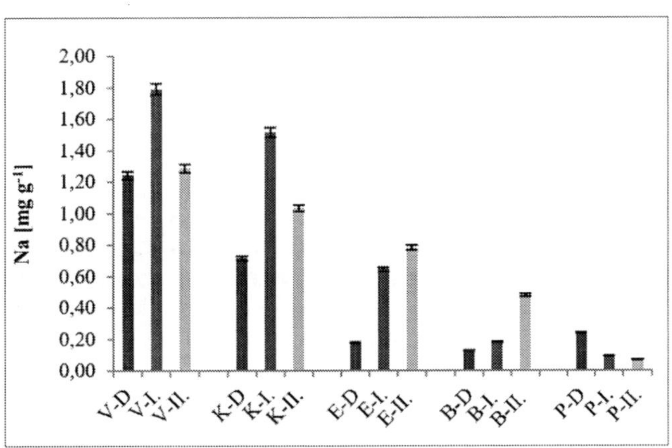

Figure 4. Concentration of sodium in soils near the three type of road in different distance from the roads.

The measured element concentrations in needles represent the combination of adsorbed particulate matter and trace elements taken up via the root system of the trees. The spruce needles and corresponding soil samples show elevated Na concentrations at locations close to the roads where chemical de-icing is performed. Trees from locations with a contamination potential (CP) of 1 show Na concentrations from 0.01 to 0.04 mg g^{-1}, locations with CP 2 from 0.02 to 0.10 mg g^{-1} and CP 3 from 0.12 to 5.0 mg g^{-1}. The Na concentrations in spruce needles directly correlate to the tree health as seen below. In contrast to the control group with comparable CP the Na concentrations in spruce needles close to the road are higher by a factor of 8-10 (Figure 5). In the case of other traffic related elements the trend is opposite. Position of the road in the monitored area has significant influence on amount of elements concentration, in the case of 1st contamination potential, there is the largest accumulation effect (Figure 6).

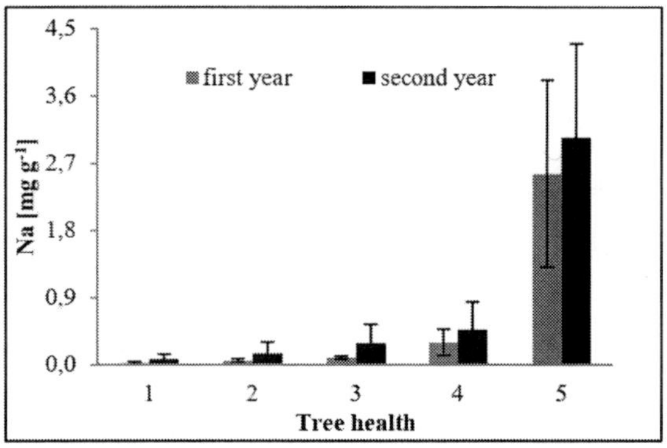

Figure 5. Na concentration in spruce needles from trees with different health stages

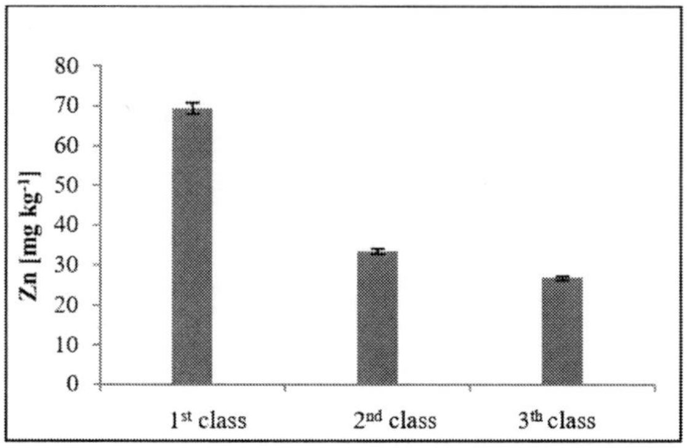

Figure 6. Influence of the contamination potential on total zinc concentration in soil

In contrast to the study of Suoranta, T. et al. [6] the concentrations of Pt, Rh, and Pd were below the detection limits of 1 μg kg^{-1} for urban soil and pine needles.

2.4. Statistical evaluation of the analyses results

Data were evaluated for the two sites differing in traffic loading. Statistical evaluation was used for interpretation of complex interaction patterns between element concentrations in different age needles based on localities character including distance from the road and element concentration in soils. At the beginning, the data was summarized and the basic characteristics for visualization of element content in the needles and soil were calculated.

3. Conclusion

Comprehensive analysis of soil samples collected in the vicinity of roads provides an overview of the effect of transport on soil pollution. The result is an extensive set of data which presents the dependence of the contamination of the environment by selected elements. Results database provides a comprehensive overview of the contents of the traffic related element concentrations in the Czech Republic. Increasing concentrations of selected elements (e.g. Sb, V) in certain locations, where the natural content in soil is low, indicate transport contamination source. Sodium has a higher concentration in non-urban areas (mountains), because of longer winter maintenance. The research into the effects of winter road treatment on selected conifer confirmed the increased sodium and chlorine concentrations in trees growing in the areas with higher contamination potential. A deterioration of health condition of the trees depending on the increasing sodium and chlorine concentration in needles was also observed. The health condition of a tree is an important indicator of contamination. According to the study by Kayama at al. [4] it was also discovered that sodium contents are higher in damaged trees with worse health condition. The distinct sodium concentrations were confirmed in the soil samples. Concentration of PGEs in the monitored areas was below detection limit. The impact of traffic on the concentration of traffic related elements (Sb, V atc.) in soils and the needles is noticeable. The database is a source of valuable information about the impact of transport on the environment.

Acknowledgements

This article was produced with the financial support of the Ministry of Education, Youth and Sports within the National Sustainability Programme I, project of Transport R&D Centre (LO1610), on the research infrastructure acquired from the Operation Programme Research and Development for

Innovations (CZ.1.05/2.1.00/03.0064). The data used for analysis in this paper were obtained in a project funded by Technology Agency of Czech Republic – within the Competence Centre Programme of TA ČR, project n. TE01020168 (CESTI – Centre for Effective and Sustainable Transport Infrastructure).

References

[1] Butler OT et al 2017 Atomic spectrometry update – a review of advances in environmental analysis *Journal of Analytical Atomic Spectrometry* **1** pp 11 - 57

[2] Langen T A and Prutzman H 2006 The effects of chemical road deicers on surrounding boreal forests and lakes *Environmental Impacts of Winter Road Management at the Cascade Lakes and Chapel Pond* (New York State Department of Transportation) pp 15 – 48

[3] Forczek ST et al 1011 Influence of road salting on the adjacent Norway spruce (Picea abies) forest *Plant Soil Environ.* **57**(7) pp 344–50

[4] Kayama M et al 2003 Effects of de-icing salt on the vitality and health of two spruce species, Picea abies Karst., and Picea glehnii Masters planted along roadsides in northern Japan *Environmental Pollut*ion **124**(1) pp 127-37

[5] Howard K W F et al 1993 Hydrogeochemical implications of groundwater contamination by road de-icing chemicals *Journal of Kontaminant Hydrol*ogy **12** (3) pp 245 – 68

[6] Suoranta T et al 2016 Elimination of Interferences in the Determination of Palladium, Platinum and Rhodium Mass Fractions in Moss Samples using ICP-MS/MS *Geostandards and Geoanalytical Research* **40** pp 559–69

Alternative modifications of bituminous binders for mastic asphalt mixtures

Jakub Šedina[1], **Jan Valentin**[1] **and Lucie Benešová**[1]

[1]Faculty of Civil Engineering, Czech Technical University in Prague, Thakurova 7/2077, 166 29 Prague 6 – Dejvice, Czech Republic

E-mail: sedinjak@fsv.cvut.cz

Abstract. This paper focuses on potential benefits of alternative bituminous binders for mastic asphalt mixtures, which were modified by new type of low viscosity additives or activated rubber powder. Paper presents results of laboratory investigation on mixtures with standard bituminous paving grade bitumen 20/30 and mixtures with modified bituminous binders. The reference bitumen 20/30 was modified by *micromilled* activated rubber powder, by a new generation of synthetic waxes (WE-CM20, WE-BM), or by the combination of synthetic wax and *micromilled* rubber powder. Comparison of different mastic asphalt mixtures was based on laboratory testing (indentation test, compressive strength test, bending (tensile) strength test, stiffness modulus test and cyclic compression test). Possible uses of these applications is for example in pavement structures for bridge decks, or for effective sealing of expansion joints on bridges. Mastic asphalt characteristics are compared with selected characteristics of used bituminous binders (complex shear modulus, rotational viscosity, etc.).

1. Introduction

The increasing traffic loading acting on pavement structures not only in the Czech Republic puts generally more and more emphasis on the quality of the asphalt mixture for road construction which should demonstrate improved durability and, primarily, overall asphalt pavement lifetime. At the same time, there are a number of pavement structure components where the above applies even more urgently, for instance also due to difficult installation or the impossibility of compacting a traditional mixture to a sufficient quality level. In many cases, traditional bituminous binders fail to meet more strict performance requirements and their properties have to be modified by appropriate additives or polymers. This has been known in the road construction industry for over 30 years; nevertheless, with the development of a number of new additives or traditional modifier variations, the options have grown wider. Modifications of common paving grade bituminous binders allow influencing asphalt mixture parameters like the processability, load-bearing capacity and durability. Similarly, specific characteristics in need of significant improvement (e.g. resistance to permanent deformation etc.) can be handled as a priority. Bituminous binder modification is common in all types of applications; the general effort strives to improve the parameters of the traditional bituminous binders, thus consequently improving the asphalt mixture quality. The experimental study presented herein focuses on the use of modified binders in mastic asphalt technology, particularly on the modifiers with a potential to influence both the working temperatures and the overall energy requirements of the technology and, on the other hand, on the monitoring of some alternatives to the chemical modification of bituminous binders as such. Due to that, the impact of bituminous binders modified by micromilled activated rubber powder (ARP) has been examined in tests where bituminous binders were partly substituted with secondary material – crumb rubber from processing of old tires in a non-standard format.

Subsequently, bituminous binders are applied to mastic asphalt mixture MA8 where the objective is utilising such mastic asphalt for special applications like narrow pavement lanes close to bridge expansion joints. Although this application is the primary target, the use of such mastic asphalt for large-area structural layers as commonly applied is viable as wellUse.

2. Asphalt Mixture Design

The appropriate combination of crushed aggregate and sand was chosen for mixture MA8 to meet the requirements of technical standards with respect to grading characteristics while being economically viable from the perspective of the quantity of bituminous binder additions; the selected version is marked FSV MA8. The optimal quantity of binder was determined according to CSN 73 6160 [1]. In the case of modification by micromilled active rubber power (ARP), a version with a higher binder content was tested as well since the rubber-modified binders demonstrated poorer workability. The preparation of MA mixtures used aggregate according to the definition stipulated by CSN EN 13043 [2] and CSN EN 13108-6 [3].

Table 1. Design of innovative mastic asphalt mixture „FSV MA8".

Mixture	Bitumen content	Filler	Crushed aggregate 4/8 Zbraslav	Sand 0/8 Uhy	Sand 0/4 Uhy
MA8 FSV	7.40%	22.24%	42.12%	3.70%	24.54%

3. The modifiers or additives used

3.1. Synthetic waxes WE-CM20 and WE-BM

Bituminous binder modification by synthetic waxes is based on the need to modify the current bituminous binders to reduce their working temperatures, or to improve workability of the resulting asphalt mixture (lower temperature, easier application etc.). The mastic asphalt technology per se has high-energy requirements (approx. 2.5 times more energy needed than for compacted asphalt mixtures). The mastic asphalt working temperatures reach up to 250 °C, so the search for a more favourable solution, for instance in the form of wax-modified bituminous binders, is logical – even a slight reduction of working temperatures by 10-15 °C has a rather distinctive economic impact. The use of various wax types for mastic asphalts has been researched since late 1980's, with the primary aim of improving asphalt mixture processability. In the last ten years, it has been applied as a tool to lower working temperatures and reduce emissions, which might be released at higher temperatures.
The synthetic waxes used are innovative products not commonly used in road construction (to improve bituminous binder properties) yet. The product was sourced from a collaborating partner in Germany.

3.2. Micromilled active rubber powder (ARP)

From both the environmental and economical perspective, modification by rubber powder has a considerable potential that has already been demonstrated in the effort to include the material in bituminous binders for standard, compacted asphalt mixtures. With respect to the increasing demand for management of non-renewable sources and with an emphasis on waste recycling, the processing of (not only) used tyres for bituminous binder modification purposes constitutes an interesting option. It must be nevertheless emphasised that from the industrial as well as broader social perspective, the activated material should be tested for reuse in the primary application area – i.e. for rubber products.

The added values of ARP modification of bituminous binders is presently seen in hot mix asphalts (HMA) where such modification positively affects the rheological and mechanical-physical properties of bituminous binders. Asphalt mixtures with modified bituminous binders demonstrate improved resistance to plastic deformation (improved stiffness of the mixture) and increased resistance to frost cracking (behaviour of the binder and, consequently, the mixture in the low temperature range). Some experts also declare a positive effect in the form of a partial improvement of the acoustic parameters, although no major noise reductions can be expected.

The variants depicted below work with a bituminous binder modified by active rubber powder of 0-0.8 mm particle size (adjusted by high-speed milling) with 5 % of a butadiene-based activator. The bituminous binder was modified using the wet process where activated rubber powder is added to the heated bituminous binder (at least 170 °C are required), in our case to the quantity of 15 M% of the bituminous binder. Mixing was carried out in the laboratory with approx. 450 revolutions per minute for at least 30 minutes.

4. Descriptions of the bituminous binders and modifiers applied

The comparison of individual variants involved the standard bituminous binder commonly applied to mastic asphalts – hard paving grade bitumen 20/30 according to CSN EN 12591. The alternative solution and further improvements to common binders consisted of modifying or adjusting the bituminous binders by additives based on synthetic waxes, amide waxes and pulverized (micro-milled) active rubber powder from old tyre processing. The active crumb powder (ARP) used herein showed particle sizes up to 0.8 mm and has been obtained by a specific high-speed milling method as defined by the Czech utility model No. 29199. ARP was supplied by Lavaris s.r.o., which was, in the course of 2016, also involved in the development of a material alternative to the bridge expansion joint damper profile. To improve ARP stability in the bituminous binder, a special butadiene-based activator was added to the entire composition.

Table 2. Characteristics of experimentally assessed bituminous binders.

Bitumen	**Penetration**	**Softening point**	**Penetration index**	**Rotational viskosity @ 6.87 s^{-1}**	
	[0.1 mm]	[°C]	[-]	*135 °C* [Pa.s]	*150 °C* [Pa.s]
20/30	20.5	62.8	-0.4	1.5	0.8
20/30 + 3% WE-CM	17.8	79.9	2.0	1.1	0.6
20/30 + 3% WE-BM	21.9	63.6	-0.1	1.0	0.7
20/30 + 15% CR (APR5AK)	15.0	81.3	1.8	13.8	12.4
20/30 + 15% CR + 2% WE-CM	14.2	87.9	2.5	13.8	11.6

Table 2 indicates the basic characteristics of the modified or doped bituminous binders designed and tested. Besides the conventional tests, the bituminous binders were also described by the penetration index that, aside from auxiliary material characterisation, also provides a simple way to define thermal susceptibility of the bituminous binder. As the workability parameters were important for mixing and further processing of this type of mixture, rotational (dynamic) viscosity was also determined since it could hint at the future behaviour of the mixture with respect to mixing, or any potential reductions in the working temperature and temperature for paving of the mastic asphalt.

What is significant from the point of view of penetration is that the additives applied do not fundamentally modify this characteristic. The synthetic waxes in the test slightly increased the value of penetration while ARP has a stiffening effect, which is completely common for this type of modification. In the case of the softening point, the effect of synthetic wax WE-CM is interesting, raising the value only slightly. The ARP modifier has a more pronounced effect – the softening point is increased considerably. No less important is dynamic viscosity measured by a rotating spindle. For the selected temperature, the effect of the waxes is minimal; it only becomes important from the perspective of mastix in mastic asphalts. On the other hand, the application of micromilled rubber results in an enormous viscosity increase which means deteriorated workability under the temperature concerned, and for the sake of mastic asphalt, it indicates that higher energy might be required under standard working temperatures >200 °C when compared to the reference bituminous binder. The application of wax to an ARP-modified binder obviously reduces dynamic viscosity slightly.

5. Mastic asphalt test specimen preparation

The preparation of MA mixtures governed by CSN EN 12697-35+A1 [7] defines the temperatures of individual components in the mixture as well as the method and time of mixing. In the case presented by this paper, the important issue is manual mixing where we faced a disproportionately longer mixing time along with a rather major problem of keeping the working temperature within the required range. The preparation of the cubic test specimens followed CSN EN 12697-20 [4] in the case of the specimens for indentation test and compressive strength tests. The same method was applied to cylindrical test specimens for the stiffness modulus test according to CSN EN 12697-26 and the cyclic compression strength according to CSN EN 12697-25 to define the deformation behaviour under repetitive stress. The preparation of the beams for the bending tensile strength test utilised re-heated mixture according to ČSN 73 6160 [1] was used. The mixture obtained from the specimens already tested for the indentation number and compressive strength was used to prepare the beams.

6. Workability

The mastic asphalt workability test can be conducted according to ČSN 73 6160 [1], and allows declaring a potential reduction of the working temperatures. Unfortunately, the road construction laboratory of the Civil Engineering Faculty, Czech Technical University, is not sufficiently equipped to perform this test at this moment. Therefore, the individual options are currently only assessed subjectively depending on the demands of manual mixing effort under the selected working temperature, and only serves as an informative indicator. The aggregate and bituminous binder mixture FSV MA8 as designed demonstrated relatively inferior workability where the mixture needed a long time to mix, roughly 10 minutes and more while the standard mixing time should amount to 5 minutes or so. It was also important to pay increase attention to the compaction of the material in the moulds; some specimens in this respect had to be ruled out due to excess air bubbles. From the point of subjective assessment of workability, mixtures with ARP-modified bituminous binders had the poorest behaviour as expected; this was the reason to apply a synthetic wax in combination with ARP to improve workability while mixture properties remain unchanged.

7. Tested MA mixture parameters

The evaluation of suitability of modified binders for MA mixtures depended on the indentation, compressive strength and bending tensile strength tests. The set of basic tests was further accompanied by the findings of the stiffness modulus and cyclic compression tests.

Table 3. Determined characteristics of MA mixtures.

Bitumen		Flexural strength, 100 mm/min	Modulus of elasticity	Compresive strength		Indentation number, 30 min	Indentation number increment, 60 min
		22 °C	*22 °C*	*a=40 mm 22 °C*	*a=70 mm 22 °C*		
		(MPa)	(MPa)	(MPa)	(MPa)	(mm)	(mm)
20/30	7.4%	6.83	245	7.69	6.85	1.33	0.14
20/30+CR	7.4%	6.78	340	9.61	12.27	0.39	0.04
20/30+CR	7.9%	9.35	519	10.01	9.05	0.35	0.25
20/30+CM	7.4%	11.05	485	11.64	8.97	0.95	0.03
20/30+BM	7.4%	10.74	433	11.18	9.26	0.95	0.14
20/30+CR+CM	7.4%	6.33	630	11.86	12.73	0.25	0.02

7.1. Indentation test

The fundamental parameter for mastic asphalt assessment is the indentation test performed on cubes. The process of specimen preparation, storage and testing is described by ČSN EN 12697-20 [4]. For mixtures with aggregate of 0/8 mm grading, the standard stipulates a (recommended) indentation number within the range of 1.0 to 5.0 mm after 30 min (0.6-6.5 mm for verifications); the indentation number increment should not exceed 0.6 mm. The data presented does not comply with one part of the margin conditions. The values measured for mixture FSV MA8 variants with improved or modified bitumens demonstrate lower values after 30 minutes than the minimum indentation number requirement of 1.0 mm for a mixture design according to the ITT (the aforementioned minimum threshold value is required in order to reduce the risk of asphalt mixture brittleness and to prevent cracking). In case the criteria for verification tests are applied, the variants with micromilled ARP modification fail. Despite these results, certain clear dependencies on the choice of bituminous binder and the potential benefits of individual bituminous binder modifications or improvements can be seen. Mixtures modified by micromilled rubber (Figure 4, var. 20/30+CR) demonstrate a distinctive improvement in the indentation number with a reduction of approximately 60 % compared to the reference mixture. Modification by waxes also contributes to lower the indentation number value. Bituminous binder modified by the combination of ARP and synthetic wax showed the best results.

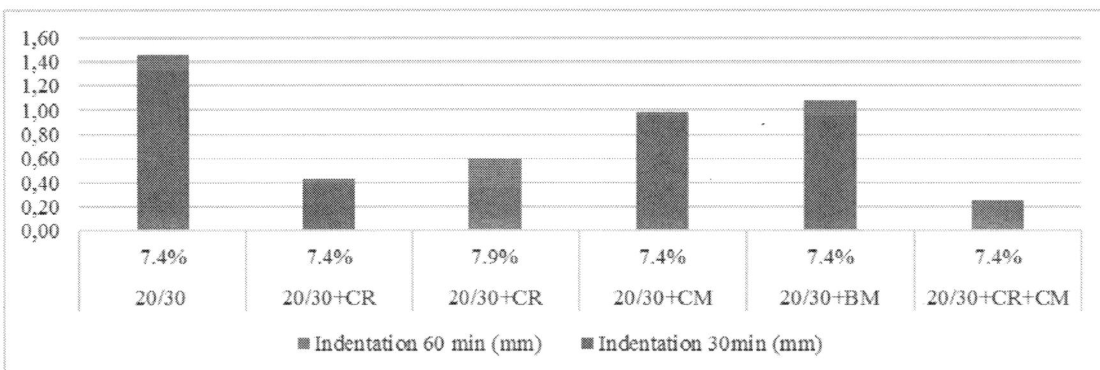

Figure 1. Results of indentation test for mixture variants FSV MA8.

7.2. Compressive strength

Another parameter for the purposes of evaluating suitability of modified binders for mastic asphalt layers is the compressive strength, tested on cubes. The compressive strength was measured both on standard dimension cubes (a=70 mm), and on cubes obtained from the remains of the specimens for the bending beam (flexural) strength test (a=40 mm). The absolute values of the two measurements differed to a point; nonetheless, the trend for comparisons of the individual bituminous binders is similar in both cases. Compared to the reference mixture, the variants with improved or modified binders demonstrate significantly higher compressive strength values. The most preferred variant is, again, the bituminous binder modified by a combination of micromilled rubber and synthetic wax. Even in this case, the modification had a positive effect on the parameters examined. There is a clash with the recommendation given by the existing standards where the resulting compressive strength should fall within the 3.5 – 8.0 MPa range. In this case, too, the threshold value is set to prevent excess stiffness of the material, which is usually linked to increased brittleness, and increased risk of cracking under lower temperatures.

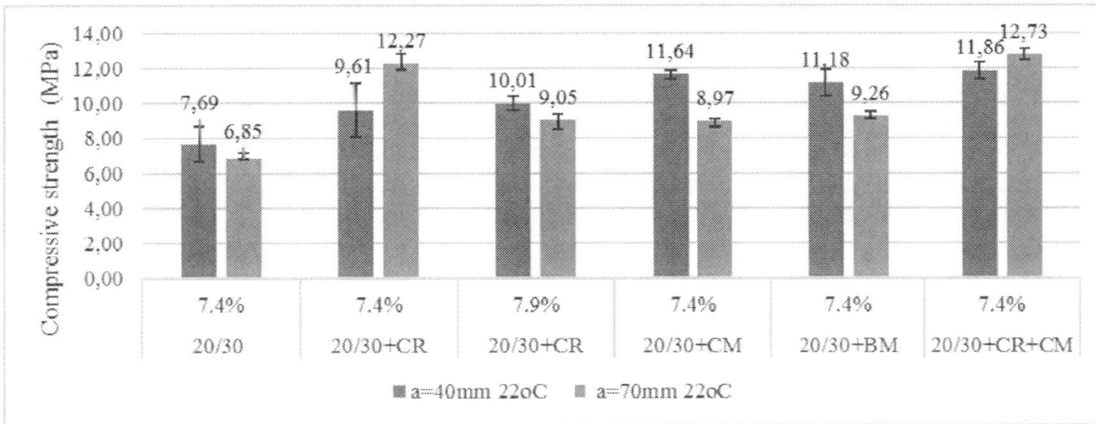

Figure 2. Results of compressive strength test for mixture variants FSV MA8.

7.3. Bending tensile strength

The test of bending tensile strength, or the calculation of the derived modulus of elasticity characteristic, also clashes with the limits set by the applicable national standard where the mixtures with wax-modified bituminous binders exceed the maximum recommended value of 7.0 MPa. If this value is disregarded, bituminous binder modification or additivation achieved an improvement of tensile strength and, when compared to the reference mixture, even the elasticity modulus increased. In this case, mixtures with bituminous binder modified by WE wax achieved the best results. Higher values were also noticed for a mastic asphalt mixture with increased quantities of ARP-modified bitumen. Both mixtures demonstrate an increase in bending tensile strength by approx. 40 %.

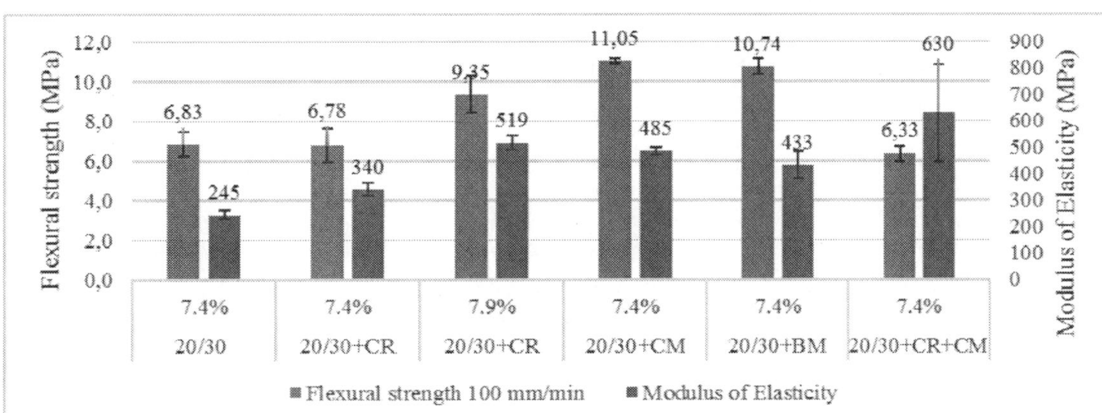

Figure 3. Results of bending beam (flexural) strength test for mixture variants FSV MA8.

7.4. Stiffness modulus

The determination of stiffness is described in ČSN EN 12 697-26 [6]. The test method C was applied, i.e. the non-destructive test of repetitive tensile stress. Generally, the stiffness modulus is not a traditional characteristic to test in mastic asphalt mixtures. In the case of stiffness, we can see the positive impact of ARP-modified bituminous binders where the test specimens demonstrated significantly better results in stiffness testing under higher temperatures – this potentially indicates their suitability for applications where permanent deformation (rutting) should be reduced. A very interesting parameter is thermal susceptibility, which indicates quality of asphalt mixture. Thermal susceptibility is expressed by the ratio of the stiffness modules measured at 0 °C and the stiffness modulus at 27 °C and 40 °C. It is expected, that higher quality of asphalt mixtures is characterized by lower temperature sensitivity. The modified mixtures demonstrated a significant reduction of thermal sensitivity in comparison to the reference mixture. In this case, bituminous binder modification by synthetic waxes had a rather negative effect with respect to the stiffness determined under higher temperatures, and to poorer thermal susceptibility. When the bituminous binder was modified by synthetic wax WE-BM, the higher thermal susceptibility resulted from the high stiffness modulus at 0°C, which acts as the basis for the thermal susceptibility test. From the perspective of stiffness values at higher temperatures, the results were comparable to the reference mixture.

Table 4. Results of stiffness modulus determination for mixture variants of FSV MA8.

Bituminous binder		**Stiffness modulus** 0 °C (MPa)	15 °C (MPa)	27 °C (MPa)	40 °C (MPa)	**Thermal susceptibility** S0/S27	S0/S40
20/30	7.4%	19157	10247	5155	2124	3.7	9.0
20/30+CR	7.4%	19344	13101	8500	5037	2.3	3.8
20/30+CR	7.9%	17597	11738	6612	2998	2.7	5.9
20/30+CM	7.4%	19504	10047	4406	1553	4.4	12.6
20/30+BM	7.4%	20113	10776	5507	2139	3.7	9.4
20/30+CR+CM	7.4%	19504	14857	8363	3826	2.3	5.1

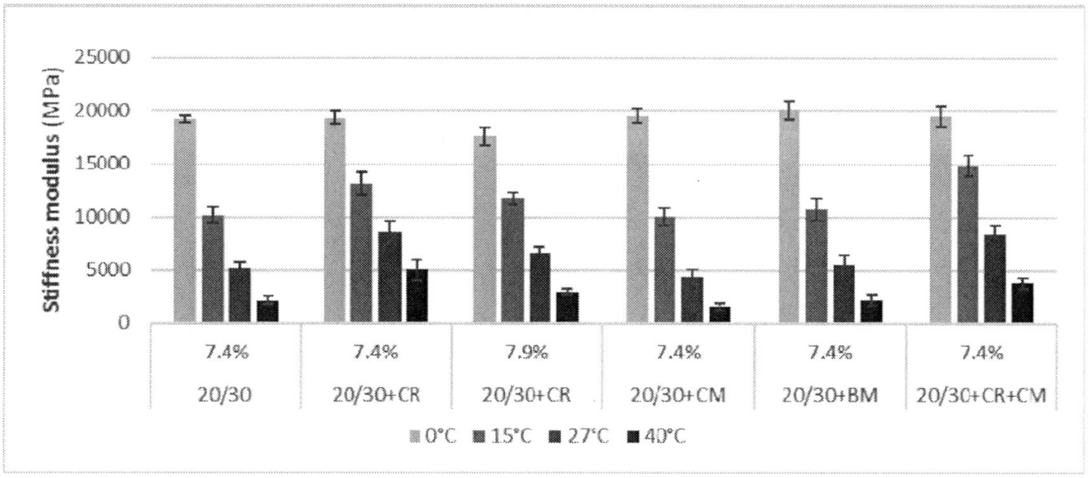

Figure 4. Stiffness modulus determinated for mixture variants of FSV MA8.

7.5. Cyclic compression test

The test method defined in ČSN EN 12697-25 [5] determines mixture resistance to permanent deformation caused by cyclic stress in a cylindrical specimen. In the case of mastic asphalt mixtures, the dynamic indentation number can express sufficiently the permanent deformation according to the aforementioned test. The results obtained from the measurements correspond with the indentation number measurement where the dynamic hardness number is considerably superior for MA mixtures with ARP-modified bituminous binder. This test shows that increased quantities of CRMB binder in mastic asphalt mixtures do not induce any changes in the characteristics tested. Contrastingly, the combination of micromilled ARP and synthetic wax brings on a slight reduction of the characteristic in the case of creep modulus.

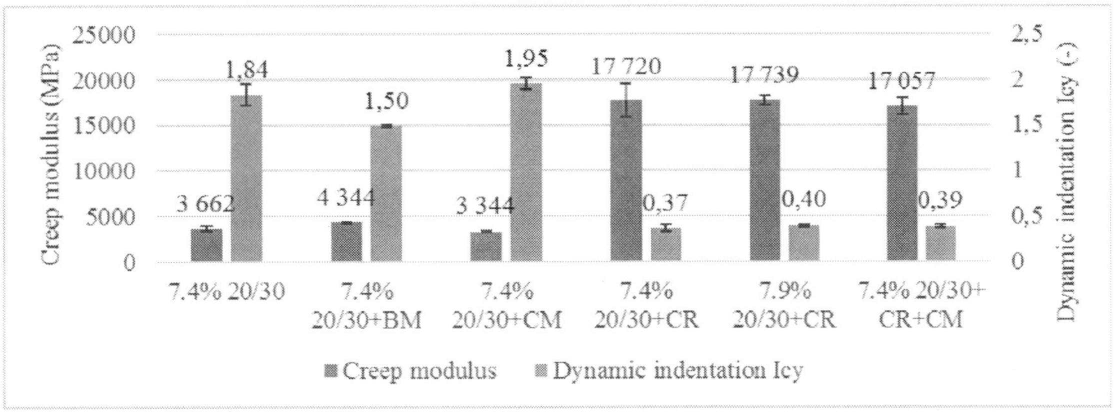

Figure 5. Creep modulus and dynamic indentation number from the cyclic compression test (@ T=60 °C) for mixture variants FSV MA8.

8. Conclusions

The variants of the new MA8 mixture designs with modified or additivised bituminous binders demonstrated an interesting potential where the addition of modifiers positively affects the resulting parameters of assessed mastic asphalt types.

The positive effect of binder modified by micromilled active rubber powder was clearly proven when the mixtures modified thereby demonstrated increased strength characteristics, low thermal susceptibility, high stiffness modulus even under high temperatures, high resistance to cyclic stress and to permanent deformation. Unfortunately, the improved parameters came at the expense of subjectively poorer workability where the mixtures showed higher demands on mixing energy during test specimen preparation.

Bituminous binder modification by a new generation of synthetic waxes, WE-CM and WE-BM, also delivered a benefit in the form of improved strength parameters, particularly bending tensile strength. From the perspective of the stiffness modulus, thermal susceptibility and resistance to permanent deformation, mastic asphalts seem to be more favourably enhanced by WE-BM wax modification where the mixtures with this binder achieved equal or slightly better parameters than the reference mixture. The effect of the waxes on processability, and thus potential working temperature reduction, has not been verified or sufficiently proven, and remains to be examined further.

Acknowledgements
This paper was prepared under the funding scheme of Program Alfa projects by the Technology Agency of the Czech Republic, project no. TA04030714.

References

[1] Český normalizační institut (ČNI) 2008 Zkoušení asfaltových směsí /Testing of bituminous mixtures for roads/ *(ČSN 73 6160)*
[2] Český normalizační institut (ČNI) 2004 Kamenivo pro asfaltové směsi a povrchové vrstvy pozemních komunikací, letištních a jiných dopravních ploch /Aggregates for bituminous mixtures and surface treatments for roads, airfields and other trafficked areas/ *(ČSN EN 13 043)*
[3] Český normalizační institut (ČNI) 2008 Asfaltové směsi – Specifikace pro materiály – Část 6: Litý asfalt /Bituminous mixtures – Material specifications – Part 6: Mastic Asphalt/ (*ČSN EN 13 108-6)*
[4] Úřad pro technickou normalizaci, metrologii a státní zkušebnictví (ÚNMZ) 2012 Asfaltové směsi – Zkušební metody pro asfaltové směsi za horka – Část 20: Stanovení čísla tvrdosti na krychli nebo na válcových zkušebních tělesech /Bituminous mixtures – Test methods for hot

mix asphalt – Part 20: Indentation using cube or cylindrical specimens/ *(ČSN EN 12 697-20)*
[5] Český normalizační institut (ČNI) 2006 Asfaltové směsi – Zkušební metody pro asfaltové směsi za horka – Část 25 Cyklická zkouška v tlaku /Bituminous mixtures – Test methods for hot mix asphalt – Part 25: Cyclic compression test/ *(ČSN EN 12 697-25)*
[6] Úřad pro technickou normalizaci, metrologii a státní zkušebnictví (ÚNMZ) 2012 Asfaltové směsi - Zkušební metody pro asfaltové směsi za horka část 26: Tuhost /Bituminous mixtures – Test methods for hot mix asphalt - Part 26: Stiffness/ *(ČSN EN 12 697-26)*
[7] Český normalizační institut (ČNI) 2007 Asfaltové směsi – Zkušební metody pro asfaltové směsi za horka – Část 35: Laboratorní výroba směsi /Bituminous mixtures - Test methods for hot mix asphalt - Part 35: Laboratory mixing/ *(ČSN EN 12 697-35+A1)*
[8] Valentin J, Mondschein P, Šedina J, Benešová L and Landa F 2016 Lité asfalty s netradičními pojivy *XXI. Seminar Ivana Poliačka 2016*
[9] Valentin J, Mondschein P, Kolařík J and Bureš P 2014 Soubor ověřených technických parametrů nízkoteplotních litých asfaltů pro rozšíření technických podmínek TP 238 *Výzkumná zpráva DVZ 1.1.2* (FSv ČVUT v Praze)

Analysis of laboratory compaction methods of roller compacted concrete

Tomáš Trtík[1], Roman Chylík[1], Petr Bílý[1] and Josef Fládr[1]

[1] Faculty of Civil Engineering, Czech Technical University in Prague, Thákurova 7, 166 29 Praha 6, Czech Republic

E-mail: tomas.trtik@fsv.cvut.cz

Abstract. Roller-Compacted Concrete (RCC) is an ordinary concrete poured and compacted with machines typically used for laying of asphalt road layers. One of the problems connected with this technology is preparation of representative samples in the laboratory. The aim of this work was to analyse two methods of preparation of RCC laboratory samples with bulk density as the comparative parameter. The first method used dynamic compaction by pneumatic hammer. The second method of compaction had a static character. The specimens were loaded by precisely defined force in laboratory loading machine to create the same conditions as during static rolling (in the Czech Republic, only static rolling is commonly used). Bulk densities obtained by the two compaction methods were compared with core drills extracted from real RCC structure. The results have shown that the samples produced by pneumatic hammer tend to overestimate the bulk density of the material. For both compaction methods, immediate bearing index test was performed to verify the quality of compaction. A fundamental difference between static and dynamic compaction was identified. In static compaction, initial resistance to penetration of the mandrel was higher, after exceeding certain limit the resistance was constant. This means that the samples were well compacted just on the surface. Specimens made by pneumatic hammer actively resisted throughout the test, the whole volume was uniformly compacted.

1. Introduction

Roller-Compacted Concrete (RCC) is an ordinary concrete poured and compacted with machines typically used for laying of asphalt road layers. RCC is not significantly different from conventional concrete. Aggregates, cement and water are basic components. Concrete should be sufficiently dry to support the weight of the roller and to avoid excessive deformation of the concrete layer, but moist enough to ensure good placement of cement paste.

Roller-compacted concrete is able to resist severe climatic conditions up to certain degree. In 1999, Pigott [1] made a visual study of RCC structures in Canada and the United States, the age of examined surfaces was from 4 to 20 years. In general, he found that the structures lacked the finest particles, surface scaling reached to the depth of 2 mm. However, the structures were compact and no substantial damage was visible. Therefore he came to a conclusion that RCC can be used for the construction of road infrastructure structures such as parking lots, airfields, temporary roads or landings.

The aim of this study was to assess the suitability of production of RCC samples in laboratory by globally recognized compaction procedure using a pneumatic hammer. This dynamic compaction method was compared with the results of samples obtained by static compaction in loading machine that tried to simulate the same conditions as during rolling without vibratory mode (in the Czech Republic, only static rolling is commonly used).

Samples produced by pneumatic hammer were manufactured in accordance with standards (ASTM C1435 [2], ČSN EN 13286-4 [3]). However, the standards only define the process for the production of laboratory samples, they do not take into account the compliance with the manufacturing process at a construction site. Samples then may have significantly different characteristics than the material produced by rolling and as such they cannot be considered to be representative samples.

The results obtained on both types of laboratory samples were compared with core drills from the construction site.

2. Calculation of compaction parameters

Numerical expression of the degree of compaction can be done using the energy expended during compaction. Various methods use different test specimens and various principles of compaction. It was necessary to normalize the overall compaction energy applied to test specimens.

The first option was to relate the energy to the volume of the specimen. The second option was to relate the energy to the surface area of the tamping plate. As a reference test for determining the energy of compaction has been selected Proctor standard test (ČSN EN 13286-2 [4]). Equations (1) and (2) represent the calculation of the amount of compaction energy relative to the volume of the test specimen and the surface of the tamping plate. Weight of the hammer was 2.5 kg (*m*), hammer was dropped from the height of 0.305 m (*h*). Compaction was performed by 56 strikes (*i*) per layer in each of the three layers (*n*) in which the sample was compacted.

Energy related to the volume of the test sample (Proctor standard test):

$$E_p = \frac{m \cdot g \cdot h \cdot n \cdot i}{V} = \frac{2{,}5 \cdot 9{,}81 \cdot 0{,}305 \cdot 3 \cdot 56}{0{,}002121} = 0{,}59\ \frac{MJ}{m^3} \tag{1}$$

Energy related to the surface of the tamping plate (Proctor standard test):

$$E_p = \frac{m \cdot g \cdot h \cdot n \cdot i}{S} = \frac{2{,}5 \cdot 9{,}81 \cdot 0{,}305 \cdot 3 \cdot 56}{0{,}00196} = 0{,}64\ \frac{MJ}{m^2} \tag{2}$$

The objective was to calculate the time of compaction by pneumatic hammer (*t*) so that the energy introduced into the sample (E) was the same as in the Standard Proctor test. Equations (3) and (4) represent the calculation of the amount of compaction energy relative to the volume of the test specimen and relative to the surface of the tamping plate, respectively. Used pneumatic hammer had the impact energy of $w = 5.9$ J and impact frequency of $o = 48$ impacts per second. The sample was compacted in three layers (*n*).

Energy related to the volume of the test sample (pneumatic hammer):

$$E = \frac{w \cdot n \cdot o \cdot t}{V} = \frac{5{,}9 \cdot 3 \cdot 48 \cdot \mathbf{2{,}5}}{0{,}15 \cdot 0{,}15 \cdot 0{,}15} = 0{,}63\ \frac{MJ}{m^3} \tag{3}$$

Energy related to the surface of the tamping plate (pneumatic hammer):

$$E = \frac{w \cdot n \cdot o \cdot t}{S} = \frac{5{,}9 \cdot 3 \cdot 48 \cdot \mathbf{17}}{0{,}15 \cdot 0{,}15} = 0{,}64\ \frac{MJ}{m^2} \tag{4}$$

Table 1 compares the amount of energy stored according to compaction methods. The aim was to get the same amount of energy for both methods.

Table 1. The amount of energy divided according to compaction methods.

	Energy related to the volume of the test specimen (MJ/m^3)	Energy related to the surface of the tamping plate (MJ/m^2)
Standard Proctor test	0.59	0.63
Pneumatic hammer	0.64	0.64

When the compaction energy was related to the volume of the specimen, 2.5 s compaction time was necessary, when energy was related to the surface area of the tamping plate, 17 s compaction time was required. The results were different. Therefore, the compaction time of cubic samples for the selected pneumatic hammer was determined as an average of both times: 10 s.

For the static compaction method, the same approach could not be used as it would lead to unrealistic compaction force (approximately 240 kN – the molds would not withstand such a force). Therefore, the compaction force was determined according to the stress under the roller (F is the force created by one drum of the roller, A is the estimated contact area between the roller and the surface of RCC layer, both parameters were calculated from the data provided by the manufacturer of the roller):

$$\sigma = \frac{F}{A} = \frac{36{,}05}{0{,}045} = 802\ \text{kPa} \tag{5}$$

Equation (6) calculates the required force for the static compaction method:

$$F = \sigma \cdot A = 802 \cdot 0{,}0225 = 18{,}045\ \text{kN} \rightarrow 19\ \text{kN} \tag{6}$$

3. Production of the samples

The composition of RCC mixture used for the production of samples is given in Table 2.

Table 2. RCC mixture composition.

Compound	kg/m^3
Cement - CEM I 42.5 R	325
Aggregates 0 – 4	1019
Aggregates 4 – 8	631
Aggregates 8 – 16	386
Water	105
Air entraining admixture	0.2
Water retarder	0.35

3.1. Static loading machine

Cubic samples were produced in plastic forms with dimensions of 150x150x150 mm. Total of 21 test samples were produced. Each sample was compacted in three 6 cm thick layers. In the end, final layer was created to receive smooth surface of the sample. The loss of thickness in each layer during compaction was around 1 cm.

For the compaction of test samples in the static loading machine, special compression tool was created (Figure 1) to simulate the same conditions as during rolling without vibratory mode in the laboratory. Basic acting force was F = 19 kN, which was derived in equation (6).

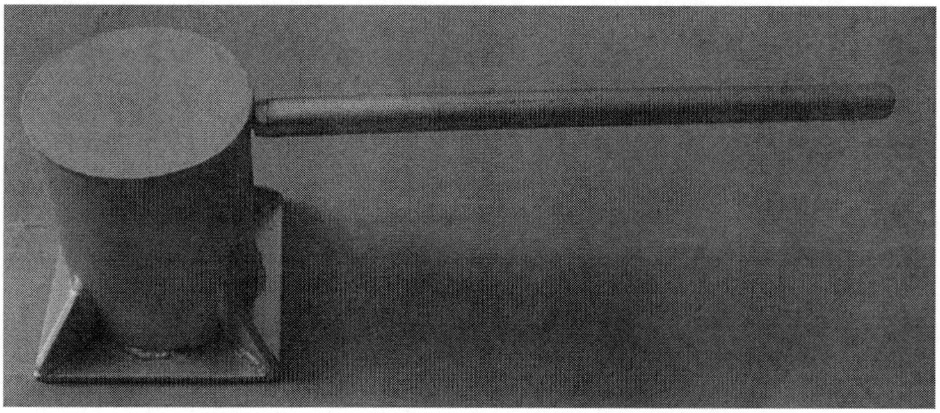

Figure 1. Compression tool with an edge of 145 mm.

3.2. Pneumatic hammer

Production of test samples using a pneumatic hammer is addressed in detail in ASTM C1435 [2]. The standard specifies the requirements for the form and shape of the compaction tamping plate including tolerances. Standard specifies the requirements for the type of the hammer. Minimum power inputs should be 900 W, frequency at least 2000 impacts/min. These important parameters are not listed in the Czech standard ČSN EN 13286-4 [4]. Pneumatic hammer (Figure 2) was chosen to meet the parameters of ASTM C1435 [2].

Cubic specimens were produced in plastic forms in three layers, with the creation of the final layer by tamping plate (Figure 3) Time of compaction for one layer was determined as 10 s, see the explanation in section 2. A total of 21 samples were produced.

To compare the effect of the degree of compaction, another set of samples was produced. In this set, the samples were compacted to the maximum possible degree of compaction; these samples were compacted until no further compaction was possible.

Figure 2. Pneumatic hammer with clamped tamping plate.

Figure 3. Tamping plate with an edge of 145 mm.

3.3. Core drills

Core drills from the parking lot and driveway to the department store were extracted to compare the results of the laboratory samples with the material from the real structure. Weight of dual-drum rollers used during the construction was 7.5 tons, the rollers passed the concrete layer 4 times. The structure was compacted in one layer 20 cm thick. Roller worked only in static mode. The RCC mix composition was the same as in case of the laboratory samples. A total of 14 core drills were extracted.

4. Test results and discussion of results

4.1. Bulk density and compressive strength

For comparison of the various methods of compaction, bulk density was the main evaluation parameter, as an additional aspect the compressive strength was measured. The measured values of bulk density and compressive strength for the different methods are given in Table 3.

Table 3. Bulk density and compressive strength for particular methods of compaction.

	Standard Proctor test	Pneumatic hammer t = 10s	Pneumatic hammer-max	Static loading machine F=19 kN	Core drills
Bulk density (kg/m^3)	2208	2212	2400	2042	2181
Compressive strength (MPa)	(-)	40.2	61.9	14.4	32.3

Core drills had lower compressive strength and lower bulk density compared to the samples compacted by pneumatic hammer. Better result would be probably achieved with the use of dynamic compaction mode or in case that the structure was compacted in more layers [5], but only static rolling in one layer was used by the construction company due to economic reasons.

Static loading machine had the worst results of all the methods of compaction. Basic applied force was low. This confirms the assumption that to achieve comparable results, the acting force would have to be larger, which excludes this method from practical use.

Pneumatic hammer (time of compaction 10 s): The results show almost identical results of bulk density as the standard Proctor test and core drills, which was chosen as the target for verification.

Pneumatic hammer (maximum compaction): The measured results show that it is necessary to determine the parameters of the pneumatic hammer and subsequently to modify the compaction time accordingly. Samples had different characteristics than the material produced on site and samples cannot be considered as representative, which should be their purpose.

4.2. Immediate bearing index (IBI) test

Immediate bearing index expresses the ratio of the force required to push the mandrel by prescribed constant rate to a specified depth of the test sample to the force required to push the mandrel into a reference material [6]. It is therefore a comparative method. The test is used to measure the resistance of the base layers in road construction.

IBI was measured shortly after compaction. The diameter of the penetration mandrel was 50 ± 0.5 mm. The mandrel was pushed with the constant speed of 1.27 ± 0.2 mm / min into the test sample. As reference material was selected crushed limestone located in California (CBR = 100 %).

The test was carried out on three specimens of fresh concrete (immediately after compaction) compacted by different methods. The first specimen was compressed by the force of F = 19 kN in static loading machine. The second specimen was compressed by the force of F = 80 kN in static loading machine. Force was increased four times, because the basic applied force (F = 19 kN) was insufficient. Force of F = 80 kN was chosen due to the stability of forms. The last specimen was compacted by the pneumatic hammer with 10 s compaction time.

Figure 4. Start of the IBI test.

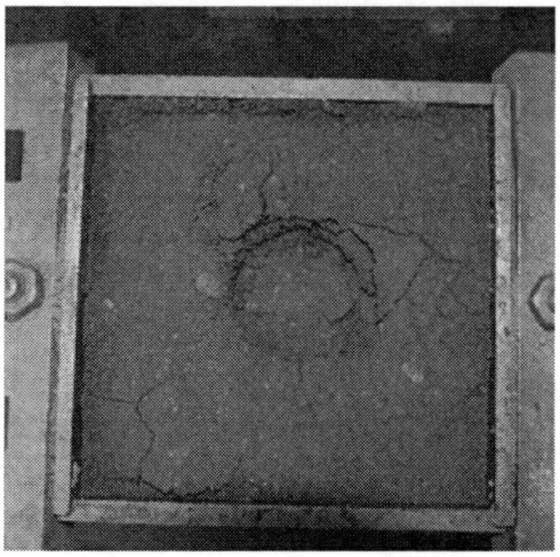

Figure 5. Creation of shear surfaces.

The main purpose of the test was to determine the quality of compaction through the sample height, which the bulk density cannot detect.

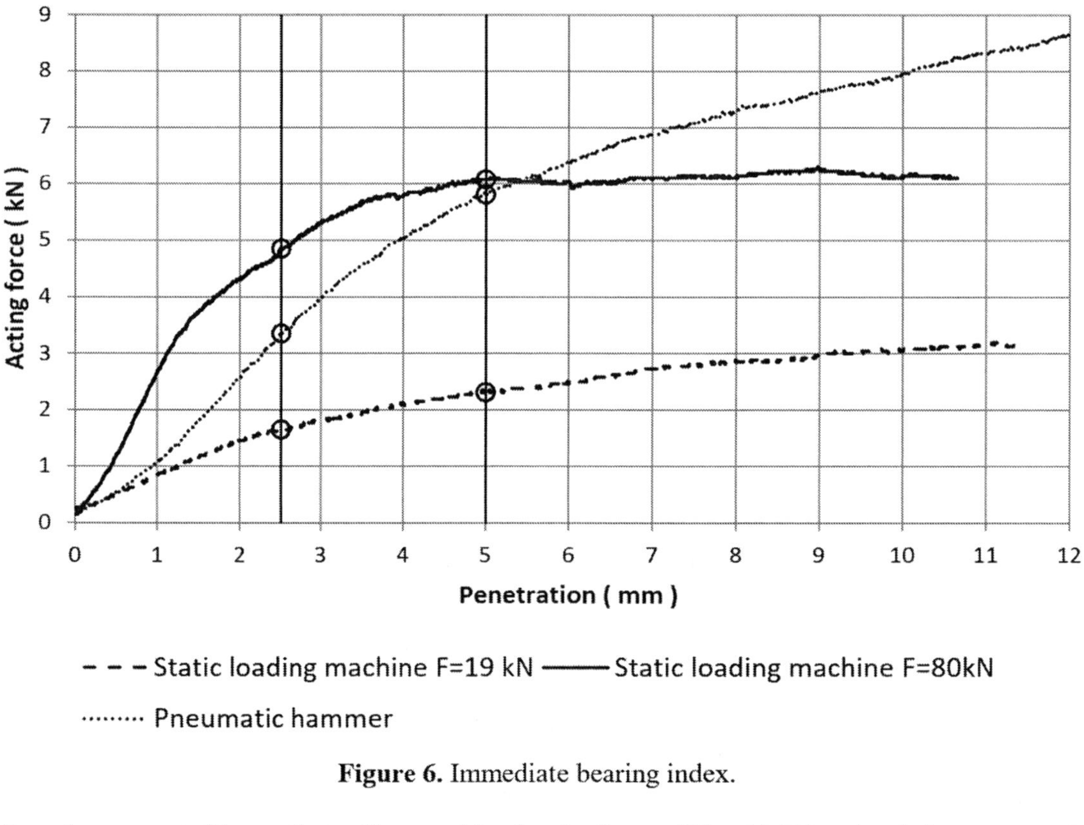

Figure 6. Immediate bearing index.

Sample compacted by static loading machine by the force of F = 19 kN resisted the worst against the penetration of the mandrel (Figure 6).

Sample compacted by static loading machine by the force of F = 80 kN led to a surprising result. After crossing the second point (5 mm penetration), the sample had a constant resistance to the penetration of the mandrel. This behaviour of the specimen can be explained by large compaction just below the surface of the compression tool, where the strongly compacted layer is located. After breaking the surface of the sample and the formation of shear surfaces (Figure 5), the sample had constant resistance to the penetration of the mandrel.

The sample created by pneumatic hammer didn't have the best result of IBI in compared points. However, sample had the best and longest resistance to penetration of the mandrel. This means that the sample was compact throughout the whole depth. Top of the curve was not reached even after 12 mm penetration.

5. Conclusions

The results confirmed the fact that the exact conditions of compaction by pneumatic hammer have crucial influence on the final properties of the sample. Parameters of pneumatic hammer must be taken into account, the time of compaction strongly influences the results. The time of compaction has to be exactly defined in order to obtain representative samples. If the compaction energy introduced into the sample is not regulated, the samples do not correspond to the material produced on the construction site.

The comparison of the results of core drills and samples produced by pneumatic hammer with exactly defined compaction time (10 s) showed that the parameters that influence the compaction were correctly selected. Core drills had similar results of bulk density and compressive strength as samples produced by pneumatic hammer, which confirms the correctness of the calculation.

Static loading machine is unsuitable for practical use. However, the results of samples produced by static loading machine clearly showed the difference between the static and dynamic compaction of RCC. Therefore it can be stated that the production of RCC structures by mere static compaction is unsuitable.

Acknowledgements
This paper was prepared thanks to the support of the Technology Agency of the Czech Republic (TAČR), project "Centre for Effective and Sustainable Transport Infrastructure (CESTI)" (no. TE01020168) and the Science Foundation of the Czech Republic (GAČR), project "Analysis of the relations between the microstructure and macroscopic properties of ultra-high performance concretes" (no. 17-19463S).

References

[1] Piggott R W 1999 *Roller Compacted Concrete Pavements - A Study of Long Term Performance* (Portland Cement Association RP366) pp 17-18
[2] American society for testing and materials 2011 Standard Practice for Molding Roller-Compacted Concrete in Cylinder Molds Using a Vibrating Hammer *(ASTM C1435)*
[3] Úřad pro normalizaci, metrologii a státní zkušebnictví - ÚNMZ 2004 Nestmelené směsi a směsi stmelené hydraulickými pojivy - Část 4: Zkušební metody pro stanovení laboratorní srovnávací objemové hmotnosti a vlhkosti - Vibrační pěch /Unbound and hydraulically bound mixtures – Part 4: Test method for laboratory reference density and water content – Vibrating hammer/ *(ČSN EN 13286-4)*
[4] Úřad pro normalizaci, metrologii a státní zkušebnictví - ÚNMZ 2011 Nestmelené směsi a směsi stmelené hydraulickými pojivy - Část 2: Zkušební metody pro stanovení laboratorní srovnávací objemové hmotnosti a vlhkosti - Proctorova zkouška /Unbound and hydraulically bound mixtures - Part 2: Test methods for the determination of the laboratory reference density and water content – Proctor compaction/ *(ČSN EN 13286-2)*
[5] Portland Cement Association 2004 *Guide Specification for Construction of Roller-Compacted Concrete Pavements* pp 5
[6] Úřad pro normalizaci, metrologii a státní zkušebnictví - ÚNMZ 2005 Nestmelené směsi a směsi stmelené hydraulickými pojivy - Část 47: Zkušební metoda pro stanovení kalifornského poměru únosnosti, okamžitého indexu únosnosti a lineárního bobtnání /Unbound and hydraulically bound mixtures - Part 4: Test method for the determination of California Bearing ratio immediate bearing index and linear swelling */ (ČSN EN 13286-47)*

Precast concrete pavement – systems and performance review

Josef Novak[1], Alena Kohoutková[1], Vladimír Křístek[1] and Jan Vodička[1]

[1]CTU in Prague, Thákurova 7, 166 29 Prague 6 - Dejvice, Czech Republic

E-mail: josef.novak.1@fsv.cvut.cz

Abstract. Long-term traffic restrictions belong to the key disadvantages of conventional cast-in-plane concrete pavements which have been used for technical structures such as roads, parking place and airfield pavements. As a consequence, the pressure is put on the development of such systems which have short construction time, low production costs, long--term durability, low maintenance requirements etc.. The paper presents the first step in the development of an entirely new precast concrete pavement (PCP) system applicable to airfield and highway pavements. The main objective of the review of PCP systems is to acquire a better understanding of the current systems and design methods used for transport infrastructure. There is lack of information on using PCP systems for the construction of entirely new pavements. To most extensive experience is dated back to the 20th century when hexagonal slab panels and system PAG were used in the Soviet Union for the military airfields. Since cast-in-situ pavements became more common, the systems based on precast concrete panels have been mainly utilized for the removal of damaged sections of existing structures including roads, highways etc.. Namely, it concerns Fort Miller Super Slab system, Michigan system, Uretek Stitch system and Kwik system. The presented review indicates several issues associated with the listed PCP systems and their applications to the repair and rehabilitation of existing structures. Among others, the type of manufacturing technology, particularly the position of slots for dowel bars, affects the durability and performance of the systems. Gathered information serve for the development of a new system for airfield and highway pavement construction.

1. General instructions

Cast-in-place concrete pavements have been used for more than 20 years and represent a cost-effective solution for the construction of rigid pavements which provide adequate support for loads exerted by vehicles or airplanes. Long-term traffic restrictions because of extensive and extended lane closures, long curing duration required to reach sufficient strength and the inability to place a material in all weather conditions increase the pressure on the use of systems with enhanced manufacturing technology [1]. Such issues might cause delays and consequently increase the costs related to extended construction time. As a consequence, several PCP systems have been developed up to date and used for the construction of technical structures, particularly for the repair and rehabilitation of roads, parking place and airfield pavements. They are defined as the system of precast panels or elements fabricated or assembled off-site, transported to the building site and installed on a prepared foundation [2]. Such systems do not require field curing for the precast concrete panels as well as require only minimal time for system components to achieve strength before opening to traffic. In comparison with conventional cast-in-place concrete pavements, the potential benefits of PCP systems include better concrete curing conditions and quality concrete, minimal weather restrictions on precast element placement, shorter delay before opening to traffic and elimination of construction related early age failures. Moreover, precast panels can be manufactured in advance and stored until needed. On the contrary, major issues include load transfer between either precast panels themselves or between panels and existing pavement, levelling the panels to avoid bumps at panel edges and costs which is estimated 1.6 to 4 times higher than the cast-in-place repair methods [3].The other report states that the costs related to the precast concrete panels use for repairs is even 7 times higher [4]. The major reason of higher costs for repairs using precast concrete panels is the cost of engineering and the fabrication of the precast concrete panels in small quantities. On the contrary, cost savings associated with shorter time required for repairs and traffic restrictions increase significantly when PCP systems used and consequently slightly compensate the high production costs.

2. Precast concrete pavement systems

There are only a few PCP systems developed particularly for the construction of airfields. Particularly, it concerns a hexagonal-shaped panel system and PAG system invented in the Soviet Union in the 20th century. Since cast-in-situ pavements became more common due to the production costs, PCP systems have been mostly used in order to achieve construction time savings in highway applications with high traffic volume where lane closures are a challenge and for rapid repair and rehabilitation applications at airfield pavements. Considering the current pricing scenarios, the use of PCP systems cannot be justified economically without the benefit of time saving. As a consequence, precast concrete panels are used in two ways - as permanent pavement or temporary replacement for removed panels until concrete can be placed during the next scheduled closure [5]. The most utilized systems include Fort Miller Super Slab System, Michigan system, Uretek Stitch system and Kwik system.

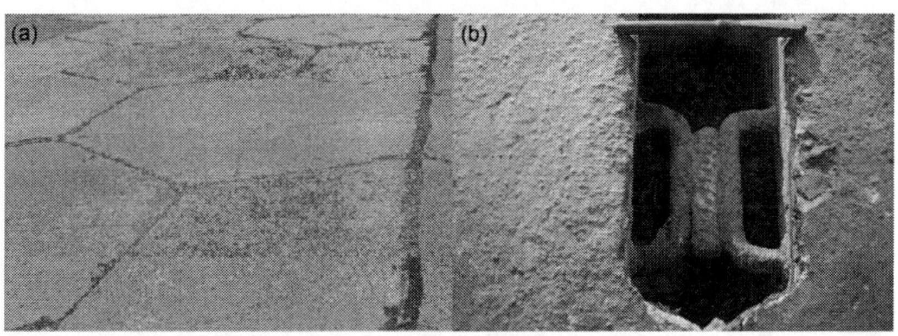

Figure 1. (a) hexagonal-shaped panel system; (b) PAG system - loops of adjacent panels welded together. [6]

2.1. Soviet PCP systems: hexagonal-shaped system and PAG system

In 1931, the first PCP system was developed in the Soviet Union. It was constructed using hexagonal-shaped panels made of plain concrete with no reinforcement (figure 1). The length and thickness of the panels range from 1.2 m to 1.5 m and from 100 mm to 220 mm, respectively, depending on aircraft weight the airfield pavement was designed for [6]. Since no reinforcement was used for the construction of the panels, the system often had problems with spalling. The other issue was associated with the panel rocking caused by the insufficient panel flatness resulted from high temperature gradient, poor production quality or concrete shrinkage.

In 1970, the other system called PAG system was approved as a standard slab system for airfields in the Soviet Union. It consists of longitudinally pre-stressed rectangular panels reinforced also with both a reinforcement mesh and two layers of 12 mm reinforcement bars top and bottom. The size of the panels varies but mostly they were 6 m long, 2 m wide and 140 mm thick. Owing to the panel size and reinforcement used, they exhibited excellent load-carrying capacity. Two lifting loops exposed on each side enabled to manipulate with the panels and install them into a right position. Then, the loops of adjacent panels were welded together (figure 1) and slots filled with mastic. In most cases, the panels were places on sand base since such material provided improved working conditions, low coefficient of friction, uniform support, good drainage and it was also economical and widely available at that time [6].

2.2. Fort Miller Super Slab system

Today, Fort Miller Super Slab System belongs to most utilized PCP systems all over the world. It has been developed for both intermittent and continuous pavement rehabilitation when the replacement of entire mainlines, ramps, intersections and even crosswalks in a series takes only less than 8 hour roadway closures [7]. Precast concrete panels of various sizes are installed directly on a prepared sub-grade with a thin layer of fine bedding material that provides nearly a uniform slab support. The interaction and load transfer between either adjacent precast panels or precast panels and an existing slab is ensured by a unique load transfer system which consists of dowels bars positioned into prepared slots in adjacent panels (existing slab) and grouted (figure 2). In comparison with other systems, the slots are positioned at the bottom of the panels and consequently are not directly exposed to structural and environmental loading. This fact has crucial influence on the structural behavior and performance of the system as pointed out later on. When the panels are placed, the potential air voids underneath the installed panels are filled by bedding grout through bedding grout distribution system bonded to the bottom of the panels with the aim to ensure positive bedding grout distribution [8].

Figure 2. Fort Millar Super Slab system [9]: (a) sub-grade preparation - bedding material; (b) panels installation; (c) interlock systems between adjacent elements; (d) dowel grout pumped into each inverted dovetail slot.

2.3. Michigan system

The Michigan system represents a very similar system which differs only in small but important details. The system, which was refined at the Michigan State University [10], serves for the intermittent pavement rehabilitation. Precast concrete panels are fitted with three or four dowel bars (in each wheelpath) which ensure the load transfer between the adjacent elements [3]. They are placed directly on a prepared sub-grade covered by a flowable fill as a cement-based self-levelling material (figure 3) in such way that the dowel bars fit into slots cut out in an existing slab or adjacent elements [4]. When the precast concrete panels are installed, the slots at the top of the panels and the dowel bars are grouted with a cement-based composite.

Figure 3. Michigan system [3]: (a) sub-grade preparation – flowable fill; (b) panels installation.

2.4. Uretek Stitch system

The Uretek Stitch system has two characteristic features which include a specific lifting (levelling) technique and an interlock system which ensures the load transfer between adjacent elements. When conventional precast concrete panels are placed on a prepared sub-grade, polyurethane foam is injected through portholes in the precast panels in order to provide uniform support and fill air voids as well as to lift the panels to the desire grade [11]. Such a way of lifting represents a unique method also used for levelling sunken or broken portions of earth-supported floors and slabs [12]. After the precast concrete panel placement, a series of thin saw-cut slots (figure 4) is used for fibreglass-reinforced polymeric inserts bonding adjacent elements together. Then, the slots are grouted using a high-density polymer-based composite. However, the use of such bonding system affects the performance of a slab as the movement between joints caused by thermal and moisture changes are not enabled. As a consequence, it is recommended to use this Uretek Stitch system only for small area in order to avoid any damage.

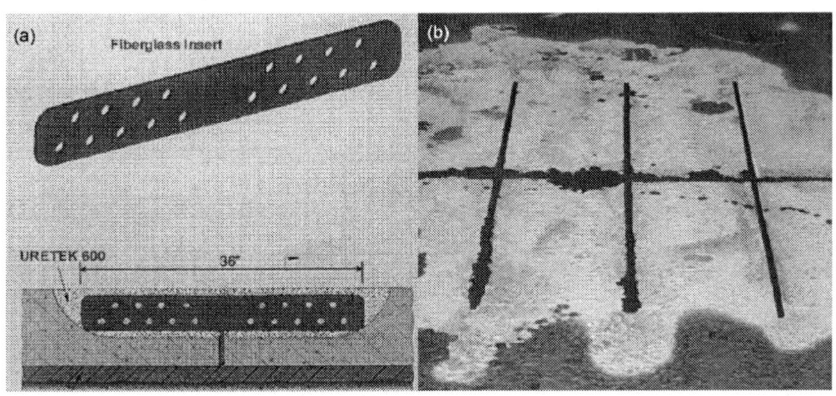

Figure 4. Uretek Stitch system [3]: (a) fibreglass-reinforced polymeric inserts; (b) completed joints.

2.5. Kwik Slab system

The Kwik Slab system represents a precast concrete slab system mainly developed for new construction, extensions or repairs of roadways, bus stops, bridge decks, parking place, airfields, docks, or any project with high demands for the speed of construction and strength of a slab [13]. It is composed of precast concrete panels having a unique interlocking system which allows to forming a slab quickly and easily while maintaining panel reinforcement continuous in two ways with minimum grouting used (figure 5). Each panel has a number of male-type and female-type components in order to easily bond adjacent panels together. The male-type is characterized by a plurality of reinforcing rods extended horizontally through a slab in both ways generally perpendicularly to each other. Whereas the female-type is characterized by steel connector sockets positioned along a side edge which the reinforcing rods are fitted into. As the system is covered by a patent, the detailed description of the Kwik Slab system including Kwik Joint interlocking system is provided in the listed documents [14].

Figure 5. Kwik slab system [13]: (a) female-type and male-type components of interlocking system; (b)steel connector sockets; (c) panel installation; (d) injection of high-strength grout.

2.6. Other systems

The other PCP systems used for either new construction or repairs of existing slabs usually combine various features of the above-listed methods as well as introduce additional features with the aim to enhance manufacturing technology or the performance of a proposed system. For instance, in order to improve levelling technique, several systems use steel bearing plates placed on the sub-grade and screw jack assemblies casted in precast concrete panels at the location of the plates for lifting the panel to the desired elevation (figure 6). Subsequently, air voids underneath are undersealed with a cementitious grout pumped through holes in the panels [4]. There are also systems which consist of precast concrete panels without any dowel bars or dowel slots used for the load transfer between adjacent elements. The panels of such system usually incorporate keyed joints in order to provide some level of load transfer. For instance, in France it has been developed a new PCP system for removable urban pavements. It is composed of hexagonal-shaped precast concrete panels which are easy to place due to their light-weight. As the first application of the system shows, the unique system with low load transfer between adjacent elements seems to be adequate for conventional loading [15].

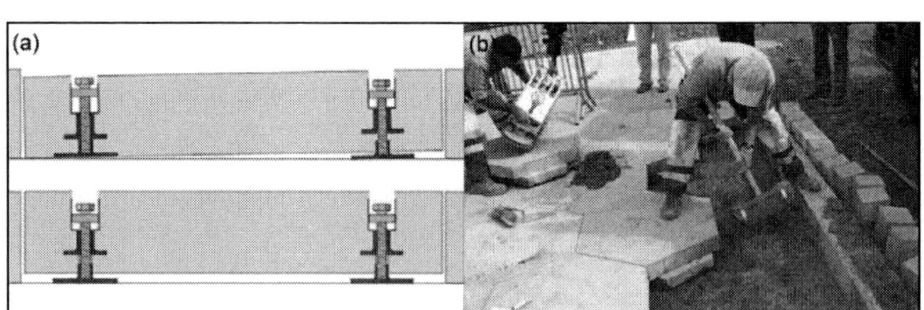

Figure 6. (a) Cast-in screw jack assemblies and (b) hexagonal-shaped precast concrete panels. [15]

3. Performance and experience

More than twenty years experience with PCP systems enables to gather enough information on the performance of the systems when used for either new construction or repairs of existing slabs. Panel cracking, joint spalling, panel surface condition, grout condition, joint elevation difference, joint width measurement and quality of ride belong to main attributes which are monitored during a site inspection. The records on the performance of PAG system are very limited. However, the experimental study conducted by United States Air Force (USAF) shows good performance of the system when used for an airfield subject to C-17 aircraft operations. At the time of site inspection, no sign of cracking or damage to the slabs was observed. However, the C-17 aircraft did break the welded loops. In 2003, the other experimental investigation was undertaken on the system optimized by using enriched limestone materials as a concrete component for the production of panels. The optimized panels were used as reusable temporary access roads for building construction, often in terrible muddy conditions under heavy truck traffic with no complaint of inadequate performance or slab failure. So far, only issues associated with minor concrete cracking and loop separation have come up [6]. Based on the summary of other field testing [16], projects, where precast concrete panels with slots at the bottom (Fort Miller Super Slab system) are used, are usually in very good conditions. In some cases only a few cracks were noted on panels, but it is not considered as concern as steel reinforcement is used in each panel (Tappan Zee Toll Plaza, New York State Jointed Continuous Project; I-66 Ramp, Virginia Jointed Continuous Project). Rarely, dowel slot grout in the joints got separated from the joint face (I-295, New Jersey Jointed Intermittent Repair Project; TH 62, Minnesota Jointed Continuous Project). On the contrary, when precast concrete panels with slots at the top are used the proposed systems exhibit significant damage. For instance, after seven years of using, few panels deteriorated and had to be replaced in the Michigan Project, I-675, where Michigan system was used. The main cause lies in grout spalling at the location of slots (figure 7) which are exposed to structural and environmental loading. The same issue is associated with Uretek Stitch system. Cracking and spalling at the location of stitches belong to the main concerns the system shows [11]. Very likely such failure mode is caused by extreme temperatures during summer as the locked joints inserts enable only a limited in-plane movement of a slab and as a consequence stress concentrated at interlocking system result in crushing of the concrete in the joint area.

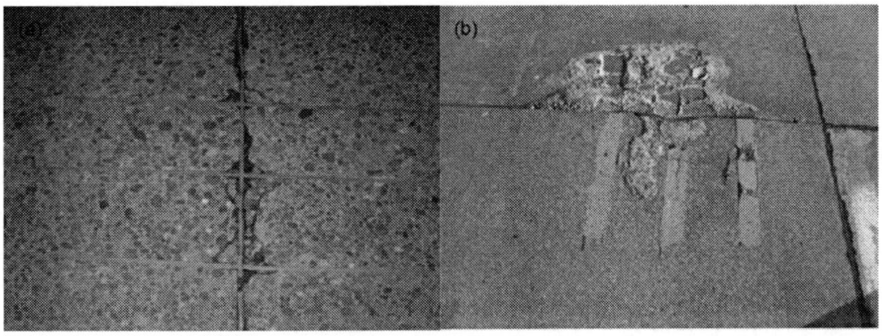

Figure 7. (a) deterioration noted in the Stitch interlocking system and (b) detail of grout spalling at the I-675 project (Michigan system). [16]

4. Conclusion

Better concrete curing conditions and quality concrete, minimal weather restrictions on precast element placement, shorter delay before opening to traffic and elimination of construction related early age failures represent decisive aspects why the PCP systems are used over cast-in-place concrete pavements for the repairs and rehabilitation of existing damaged pavements. Most of the current PCP systems have been developed in USA; namely it concerns Fort Miller Super Slab System, Michigan system and Uretek Stitch system. Their wider utilization is limited by production costs which are estimated 1.6 to 4 times higher than the cast-in-place repair methods. On the contrary, cost savings associated with shorter time required for repairs and traffic restrictions increase significantly when PCP systems used and consequently slightly compensate the high production costs. In terms of the structural behavior and the performance of the listed PCP systems, load transfer between adjacent panels represents a crucial aspect. Concrete spalling and cracking were observed at systems using precast concrete panels with slots at the top surface. Moreover, systems with locked joints inserts enable only a limited in-plane movement of a slab and as a consequence stress concentrated at interlocking system result in crushing of the concrete in the joint area under significant temperature changes. There is also evidence of using PCP systems for entirely new airfields in the Soviet Union during the last century. However, until the latest experimental studies show that the PAG system performs very well. The only issue is associated with the interlocking system when the welded loop got separated and as a consequence the panel joint slightly opened up. The obtained findings play a key role in the ongoing development of a new PCP system for pavement construction which is supposed to be finalized in 2018. Considering the defined issues of the current PCP systems, the main objective of the proposed system is to develop precast panels with keyed joints rather than any dowel bars or dowel slots used for the load transfer between adjacent elements.

Acknowledgements
This work was supported by the grant Development of prefabricated parts intended for the construction of prefabricated airport runways, No. TACR TH02010375, from the Technology Agency of the Czech Republic (TACR), and by the grant Uncertainties on the material level affecting those on the structural level, No. GACR 16-04454S, from the Czech Science Foundation (GACR).

References
[1] Priddy L P, Bly P G and Flintsch G W 2012 Review of Precast Portland Cement Concrete Panel Technologies for Use in Expedient Portland Cement Concrete Airfield Pavement Repairs *TRB Annual meeting* paper No 13-2956
[2] Shiraz T, Ye D and Buch N 2012 Precast Concrete Pavement Technology (Final report: *Modular Pavement Technology – Final Report* (National Academy of Sciences, Michigan, USA)
[3] Reza A S and Haro G 2014 Performance Determination of Precast Concrete Slabs Used for the Repair of Rigid Pavements *Performance Determination of Precast Concrete Slabs Used for the Repair of Rigid Pavements - CAIT-UTC-017 - Final report* (Center for Advanced Infrastructure and Transportation Final Report NJ USA)
[4] Olidis C, Swan D J, Saeed A, Mellerski A C and Hammons M I 2010 *Precast Slab Literature Review Report: Repair of Rigid Airfield Pavements Using Precast Concrete Panels-A State-of-the-Art Review - Final Technical Report* (Air Force Research Laboratory Florida, USA)
[5] Norbert D 2008 *Concrete Pavement Design, Construction and Performance* (New York: Taylor & Francis)
[6] Sapozhnikov N and Rollings R 2007 Soviet Precast Prestressed Construction for Airfields *FAA Worldwide Airport Technology Transfer Conference (Atlantic City, New Jersey)*
[7] The Fort Miller Co, Inc. 2017 *Super Slab* (Brochure, accessed February 28, 2017)
[8] *Super-slab* http://www.super-slab.com/ (accessed February 28, 2017)
[9] The Fortmiller Co., Inc. 2017 *Superslab* http://www.fortmiller.com/pdf/superslab.pdf (accessed February 27)
[10] Tayabji S, Buch N and Ye D 2009 Precast Concrete Pavement for Intermittent Concrete Pavement Repair Applications *National Conference on Preservation, Repair, and Rehabilitation of Concrete Pavements (St.Louis Missouri)* pp 317-34
[11] Anderson K W, Pierce L M and Uhlmeyer J S 2007 *URETEK Stitch-In-Time* (Transport Research Board Washington, USA)

[12] Veikko H 1986 *Method for levelling sunken or broken portions of earth-supported floors and slabs* (USA Patent US4567708 A)
[13] Kwik Slab, LLC *Kwik Slab Brochure* http://aii.transportation.org/Documents/KwikSlabBrochure120406Rev.pdf (accessed February 28, 2017)
[14] Alfred A Y 2006 *Precast concrete slab system and method therefor* (USA Patent US7134805 B2. 14)
[15] Shiraz T and Fugro Consultants 2010 *Precast concrete pavement technology* (Desk scan, American Trade Initiatives)
[16] Shiraz T and Ye D 2010 Performance of Precast Concrete Pavements *7th International DUT-Workshop on Design and Performance of Sustainable and Durable Concrete Pavements (Carmona)*

IOP Conf. Series: Materials Science and Engineering **236** (2017) 012031 doi:10.1088/1757-899X/236/1/012031

Durable high strength cement concrete topping for asphalt roads

Valerii Vyrozhemskyi[1], Kateryna Krayushkina[1] and Nataliia Bidnenko[1]

[1] M.P. Shulgin State Road Research Institute - DerzhdorNDI SE, Peremogy avenue, 57, Kyiv 03113, Ukraine

E-mail: vv@dorndi.org.ua

Abstract. Work on improving riding qualities of pavements by means of placing a thin cement layer with high roughness and strength properties on the existing asphalt pavement were conducted in Ukraine for the first time. Such pavement is called HPCM (High Performance Cementitious Material). This is a high-strength thin cement-layer pavement of 8-9 mm thickness reinforced with metal or polymer fiber of less than 5 mm length. Increased grip properties are caused by placement of stone material of 3-5 mm fraction on the concrete surface. As a result of the research, the preparation and placement technology of high-strength cement thin-layer pavement reinforced with fiber was developed to improve friction properties of existing asphalt pavements which ensures their roughness and durability. It must be emphasized that HPCM is a fundamentally new type of thin-layer pavement in which a rigid layer of 10 mm thickness is placed on a non-rigid base thereby improving riding qualities of asphalt pavement at any season of a year.

1. Introduction

In recent years, road pavement operation conditions have been significantly complicated due to increased traffic intensity and traffic volume. Main roads are exposed to extremely high loads where over 70% of domestic transit is carried out in addition to internal transportation. As a result, traffic intensity on such roads grow at average by 15-20% a year, while on the local roads the growth is within 1-2% [1].

All this leads to deformation and destruction of road pavements.

Pavements which are designed for smaller loads do not withstand current traffic demand that results in deformation and destruction occurrence on the pavement.

It should be noted that the increase in traffic intensity, especially in the number of speed cars, reduces roughness of pavements, i.e. reduces grip coefficient which results in the decrease of traffic safety and leads to emergency situations on the roads.

The experience of leading countries such as USA, Japan, Germany, Sweden, France [2, 3] showed that the use of this type of pavements (international name is HPCM - High Performance Cementitious Materials) will lead to increased roughness, strength and durability of existing asphalt pavements and can be considered as a feasible and economical way of road conditions improvement.

2. Justification of the research related to placement of a thin-layer cement pavement (HPCM)

Thin-layer cement pavement (HPCM) is a strong cement pavement of 8-9 mm thickness reinforced by metal or polymer fiber of less than 5 mm length. Increased grip properties with the car wheel is reached by placing the stone material of 3-5 mm fraction on a concrete surface.

The purpose is to improve roughness and durability of existing asphalt pavement.

Placement of a thin-layer concrete pavement on the existing asphalt pavement is a fundamentally new type of pavement that provides adhesion with existing asphalt pavement and is characterized by increased roughness due to stone material distribution, as well as by waterproofness, crack resistance and durability.

HPCM is a workable type of pavement and it virtually does not involve compaction (roll-on by lightweight rollers). Laying is performed by stackers like placement of guss asphalt mix with simultaneous distribution of stone chippings.

Currently, asphalt pavements are the most widespread out of high performance road pavements [4]. In the USA and CIS countries asphalt pavements make more than 90% of high performance pavements over the total road length carrying over 40% of freights. In Western Europe, asphalt pavement roads make more than 97% of high performance pavements over the total road length.

The researchers [6] believe that roughness is a goodness of fit criterion of asphalt pavement performance properties. It is known that the design speed and traffic safety level are provided at high values of evenness, roughness, wear resistance and grip of the car wheel with the surface of pavement.

One of the up-to-date application directions is a placement of cement pavement over the existing asphalt pavement.

It is known that cement pavements have lower technical characteristics than asphalt pavements in terms of evenness and serviceability but their durability is higher.

Cement pavements have better performance at high loads, especially in summer at high temperatures.

Wearing course of cement mixture ensures better safety and riding comfort for road users and provides sufficient grip values. Experience shows that the combined use of different materials such as cement and asphalt produces a positive effect and provides rigidity and durability of road pavements.

In the project "Long-life surface for busy roads" by the Transport Research Center of the OECD and the International Transport Forum [8] main advantages of HPCM pavement type were determined. Laboratory testing of the HPCM matrix for compressive strength, tensile strength and modulus of elasticity indicated the material can be characterized as High Strength/High Modulus. The results indicate HPCM wearing courses will have good bonding properties as well as durability, confirming these objectives have been achieved.

Testing at medium scale demonstated that a durable bond between asphalt binder course and HPCM can be established, provided the asphalt surface prior to paving of HPCM has been carefully scarified and cleansed. It is also critical for this asphalt to be in the high rage regarding E-modulus and temperature resistance.

The conducted surveys in Ukraine have shown high adhesion of HPCM layer with asphalt layer. Despite complicated traffic conditions, the pavement performs satisfactorily. In this area cracks and other deformations do not occur[9].

Testing has shown that HPCM has great strength and integrity. It is clear that certain requirements need to be met – including a strong and even lower layer and careful embedding of chippings – to ensure maximum performance.

This structure can be used for heavy and intensive traffic on the main roads that connect big cities, industrial centers, airports on the way to cities, etc. Experience on such pavement design application was summarized by Concrete Roads Committee jointly with Nnon-Rigid Pavements Committee established at PIARC (The World Road Association). This pavement design was named as composite (complex) by the Committee; Italian experts called it multifunctional, i.e. each layer in this design performs only its assigned function.

Asphalt pavement is heated because of its dark color. HPCM pavement is lighter by shade, so it isn't heated so much. Thus, asphalt concrete of lower pavement layer doesn't undergo plasticity deformations.

Pavement design including asphalt layer and HPCM wearing course performs as a monolith layer, so the dilatation joints are not arranged.

Considering HPCM's simplicity and workability at placement, such material should be made widespread.

3. Pavement (HPCM) characteristics

Works on the development of a mix design and placement technique for a thin layer cement pavement (HPCM) considering local features and the use of constituent native materials are currently underway in Ukraine.

Constituent materials of cement mix are the following:

- Fibers - fibers of various origins (polymer, metal, glass fiber, basalt fiber);
- Stone materials - granite gravel of 3-5 mm fraction;
- River sand - moisture of 9%;
- Cement - CEM-400, CEM-500 brand with addition of superplasticizers;
- Water for cement preparation.

Due to the fact that works on (HPCM) mix design were performed under an international program but taking into account local features and presence of domestic materials, attention was paid mainly to the choice of a fiber type – a fiber to be used for dispersion reinforcement of cement material.

Types of fibers existing in Ukraine were selected and the studies of their physical and mechanical properties were conducted. They included the following:

- Steel
- Glass fiber (alkali resistant)
- Basalt
- Asbestos (chrysotile)
- Polypropylene
- Polyamide

Basalt fibers have the highest tensile strength and elastic modulus values. Considering that basalt has other positive physical and mechanical properties such as chemical resistance, frost resistance, resistance to aggressive medium (alkaline and acidic) and its fibers placed in concrete articles have ability to limit cracks propagation in concrete, such material was taken for further research.

3.1. Determination of an optimal cement mix design

In the study, the fibers that are chaotically distributed in cement and sand mixture and used as reinforcing additives are considered as the additives regulating physical and mechanical properties of the HPCM layer.

Mix design was determined basing on the results of defining an optimal number of dispersion reinforced additives.

Dispersion reinforced additives were introduced directly into cement and mixed and than sand components of a concrete mix were added to the composite cement.

An optimal concrete mix design per 1 m^3 of concrete is the following:

a) with addition of coarse basalt fibers (CBF)
- cement - 494 kg/m^3
- CBF fiber - 38 kg/m^3

- sand - 1368 kg/m^3
- water - 170 l/m^3

b) with addition of fine basalt fibers (FBF)
- cement - 486.4 kg/m^3
- TBF fiber - 13.6 kg/m^3
- sand - 1400 kg/m^3
- water - 168 l/m^3

3.2. Determination of physical and mechanical properties of the resulting mix design

The research performed at the laboratory of DerzhdorNDI. Comparing the research results with those performed at the laboratory of LCPC (France) showed some identity of mix designs despite the fact that French researchers used metal fibers of d = 0,2 mm, 13 mm length and PVA of d = 0,33 mm and 15 mm length.

It should be noted that basing on DerzhdorNDI's experience of work with dispersion reinforced materials it can be stated that the fibers used at the LCPC laboratory were too big by length that would not provide sufficient workability at the preparation of a cement mixture. In this case uneven distribution of fibers may occur.

Therefore, our studies were conducted with the fibers of not more than 4 mm length.

The optimal concentration of fibers in mixtures was determined experimentally. For coarse basalt fibers it was 2.8% for fine - 4.4% of cement weight. Samples compaction was conducted by vibrocompression. Frost resistant was determined at -5 -20 temperature and thawing in water and in 5% solution of NaCl. Water resistance was determined by the standard method before occurrence of "wet spots" on the samples. Wear was determined in cm after 840 rotations in "LKI-3" circle, impact strength was studied at Page's copra.

Basic physical and mechanical properties of dispersion reinforced cement mixture are shown in Table 1.

Analysis of data presented in Table 1 shows that basalt fibers based additives allow getting cement material of improved strength, density and frost resistance that will positively affect the riding qualities of road pavements.

Table 1. Physical and mechanical properties of materials for HPCM pavement.

Indicators	Measuring unit	Mix types					
		Initial	with addition of coarse basalt fibers CBF	with addition of fine basalt fibers FBF	Steel	Glass fiber (alkali resistant)	Asbestos (chrysotile)
Density	kg/m^2	2200	2400	2600	2700	2600	2500
Porosity	%	2.0	1.0	1.3	1.5	1.1	1.3
Fiber content	%	-	2.8	4.4	3.5	3.0	3.5
Compressive strength	MPa	54.0	68.0	62.0	55.0	61.0	62.0
Tensile strength (axial tensile)	MPa	6.0	14.0	12.0	10.0	11.0	12.0
Water resistance, after 28 days	%	0.8	0.3	0.4	0.5	0.5	0.4
Strength on impact	J/cm^3	2.6	3.2	2.98	2.84	2.75	2.9
Wear, after 840 rotations	cm	0.08	0.04	0.03	0.04	0.04	0.06
Frost resistant coefficient after 200 cycles in water		0.95	1.00	1.0	0.93	0.92	1.0

4. Conclusions

1) Theoretical research, preliminary study of repair work experience in the leading countries of the world and performed laboratory research showed that HPCM pavement is a new promising type of pavement to increase durability and roughness of the existing upper asphalt layer. Such pavements are not widespread in Ukraine yet.
2) The optimal type of fibers to be added to the cement mixture for dispersion reinforcement is basalt (coarse and fine fibers) that are available in sufficient quantities at industrial enterprises of Ukraine.
3) Adding of basalt fibers allows getting a cement mixture with improved physical and mechanical properties and durability; it imparts strength to cement pavement, provides resistance to dynamic load impact, improves tensile resistance, impact strength. Roughness of (HPCM) is ensured by the application of stone chippings.
4) Along with high physical and mechanical properties, dispersion reinforced cement pavements are characterized by rapid growth of strength regardless of temperature and ambient moisture, low water adsorption and high resistance to acids and alkali solutions.

References

[1] Illiopolov S K 2002 Modern issues of improving the reliability of highways *Proc. of scientific and practical workshop New technologies and materials used at road maintenance . The use of cold recycling technology in road reconstruction and repair. (Rostov-on-Don*) (Rostov State Building University) pp 23-31

[2] Verkehrspolitik fuer das Jahr 2000 – Factor fur das Lackiererhandwerk. –Fahrzeug – und Metall *Lackser*. 1989 33, №8. pp 24

[3] Motor Thomas 1989 Regional- und Wirtshaftlichkeitsuntersuchung zur Wiederverwendung von Asphalt *Strasse-+Autobahn* **40**(4б) pp 225-33

[4] Gezentsveya L B ed 1985 *Road asphalt* (Moscow: Transport)

[5] Mozgovoi V V 1998 *Scientific basis of providing temperature crack resistance of asphalt pavements* (Doctoral degree thesis Kyiv) 530 p.

[6] Rudenskyi A V 1992 *Asphalt pavement* (Moscow: Transport)

[7] Marini R 1987 Asfalttipaallysteiden undelleeenkaytto *Tie ja liikaennne* **57**(7-8) pp. 24-6

[8] OECD/ITF 2007 *Long-life surfaces for busy roads* 2007 www.sourceoecd.org

[9] Звіт про виконання науково-дослідної роботи 2011 Розробити рекомендації, технічні умови та технологічні карти на влаштування тонкошарового цементного покриття (HPCM) для підвищення шорсткості і довговічності існуючого асфальтобетонного покриття Київ /Develop recommendations, specification and flow chart on placement of thin-layer cement pavement (HPCM) to increase roughness and durability of existing asphalt pavement/ *Report on R&D work performance* (Kyiv)

IOP Conf. Series: Materials Science and Engineering **236** (2017) 012032 doi:10.1088/1757-899X/236/1/012032

Importance of dowels in transversal joints in concrete pavements

Jiri Grosek[1], Vladimir Chupik[1], Josef Stryk[1] and Ilja Brezina[1]

[1]CDV - Transport Research Centre, Lisenska 33a, 636 00 Brno, Czech Republic

E-mail: jiri.grosek@cdv.cz

Abstract. Concrete pavements are designed for heavy loaded road structures. Their usage brings a number of specific issues. It is necessary to solve them all to ensure that concrete pavements will fulfil their function along the whole design period. One of these issues concerns dowels, which are located in transversal joints. Modelling of load, caused by heavy vehicles, with the use of the finite element method, provides valuable information about the stress condition of concrete pavement. The results of modelling can be verified by measurements or experiments in practice. Dowels and tie bars in jointed unreinforced concrete pavements and the importance of their correct placement, dimensions and material quality on pavement behaviour and lifespan were studied as a part of R&D projects of Technology Agency of the Czech Republic Nos. TA02031195 and TE01020168. The paper presents the experience from the modelling and performed experiments and makes conclusions which are important for the use in practice.

1. Introduction

Dowels are placed in transversal contraction or construction joints of concrete pavements and help to transfer loading between individual slabs. Dowels allow horizontal movement of slabs during contraction in the time period after laying the concrete pavement and at temperature changes. In addition, they prevent different vertical slab movements and occurrence of so-called "steps" during the concrete pavement life span. Transversal joints are critical spots in the structure, when dowels and concrete itself are loaded by the combination of overrunning heavy vehicles and temperature gradient on the slab. Modelling of concrete pavement road structures brings valuable knowledge on structure loading, which is subsequently tested by measurements on testing road segments.

In the Czech Republic, just one steel dowel type with the diameter of 25 mm and length of 500 mm with anti-corrosion coating is used. This dimension of dowels is used for the construction of motorways, where the thickness of the road pavement is 220-300 mm, as well as for the construction of airports, where the thickness of road pavement reaches 400 mm. The existing requirements for dowels in the Czech Republic are specified by standards ČSN EN 13 877-1, ČSN EN 13877-2, ČSN EN 13 877-3 and ČSN 73 6123-1.

National annex ČSN EN 13877-3: 2006 states that dowels used for concrete road pavements CB I must have the minimum diameter of 25 mm and minimum length of 500 mm. The whole length of dowels must be coated with a thin 0.3 mm plastic film, which provides anti-corrosion protection and allow slipping of the dowel in concrete.

IOP Conf. Series: Materials Science and Engineering **236** (2017) 012032 doi:10.1088/1757-899X/236/1/012032

ČSN 73 6123-1 Road building – concrete pavements – Part 1: Construction and conformity assessment states that dowels must be installed in such way that the dowel axis is at the height h/2 minus dowel diameter (where h is concrete pavement thickness), in a single axis, parallel to the surface of the concrete pavement and with the longitudinal axis of the installed traffic lane typically at the distance of 250 mm, or potentially 500 mm. The necessary amount, profile, deployment, and distance of dowels is specified by construction documentation based on the amount of loading of the traffic lane and the merging lane. Unless the construction documentation states otherwise, the distance of dowels on loaded traffic lanes is 250 mm. The distance can be doubled on lightly loaded traffic lanes and hard shoulders.

The following Table 1 contains a list of required dimensions of dowels and horizontal distances for their installation in transversal joints in individual countries where the concrete pavements with cut joints are used.

Table 1. Required dimensions and horizontal distance of dowels.

Country	Dowel diameter (mm)	Dowel length (mm)	Distance (mm)	Pavement thickness (mm)
Czech	25	500	250	-
Germany	25	500	250	-
Austria	25	500	250	-
USA	25	450	300	<200
	32	450	300	200-250
	38	450	300	>250
France	20	400	300	130-150
	25	450	300	160-200
	30	450	300	210-280
	40	500	400	290-400
	45	550	450	410-500
Poland	32	520	300-500	260-300
	40	640	300-500	300-400
	45	800	300-500	>400
Belgium	25	600	300	-
Slovakia	25	500	250-500	≤ 250
	30	500	250-500	> 250

Only few studies focus on examining the contact stress between dowels and the surrounding concrete in jointed concrete pavement. A study performed in West Virginia University confirmed an occurrence of high tensile stress in concrete around dowels and the fact that their size influences the dowel diameter [3]. Another two studies state that serious stress concentration in the surrounding concrete caused by the small diameter and high strength dowel will lead to crush and spall at the edge of the concrete and becomes the main reason for a failure. Based on the multiple diameter analysis, the contact stress at the dowel-concrete interface decreases with the increase of dowel diameter [4, 5]. Therefore, the method of finite elements was selected for the modelling and a testing segment was created, which was to verify the foreign experience.

2. Modelling by finite element method

The calculations were performed at Brno University of Technology, Faculty of Civil Engineering, Institute of Structural Mechanics. The aim of the calculations was to find whether an incorrect position of a dowel in a concrete slab has an effect on tensile stress at the bottom of the concrete slab and on potential occurrence of cracks around the dowel (diameter: 25 mm, length: 500 mm). The calculation took into account 5 variants of "defective" dowel positions (Table 2).

Table 2. Considered variants of defective dowel positions.

Variant	Vertical Translation (mm)	Vertical Tilt (mm)	Description
(V1)	-	-	Basic variant
(V6)	-20	-	Vertical Translation (downwards)
(V7)	-40	-	Vertical Translation (downwards)
(V12)	-	-20	Vertical Tilt
(V13)	-	-40	Vertical Tilt

The calculations were performed with the use of the finite element method in ANSYS software using detailed 3D calculation models. The global calculation model includes the area of road pavement formed by 3x3 slabs (thickness 240 mm). The detailed more accurate model includes the area of transversal joint with five dowels. The global calculation model provides a global field of displacement, field of stress, and boundary conditions for the detailed model. The detailed calculation model provides a local field of displacement and stress around the examined dowel.

Apart from the loading by its own weight and thermal deformation, external loading by a dual-wheel of the total force of 50 kN was inserted into the calculation model. The axis of one wheel crosses the axis of a "defective" dowel. The axis of the other wheel is 344 mm distant in transversal direction. The diameter of contact areas of forces is 240 mm. The wheel contact areas are placed close to the edge of one of the concrete slabs. The axes of forces are 120 mm distant in longitudinal direction from the edge of the contraction joint.

An uneven thermal field was applied to the model of concrete slabs and dowels and tie bars. Linear change in temperature along the slab height was set so that the temperature of the bottom of the concrete slab was 0 °C and the temperature of the top was -7°C. The reference temperature was set to 0 °C. The average axis expansion/shrinking, reference temperature and thermal field correspond with a 3 mm geometric clearance between concrete slabs.

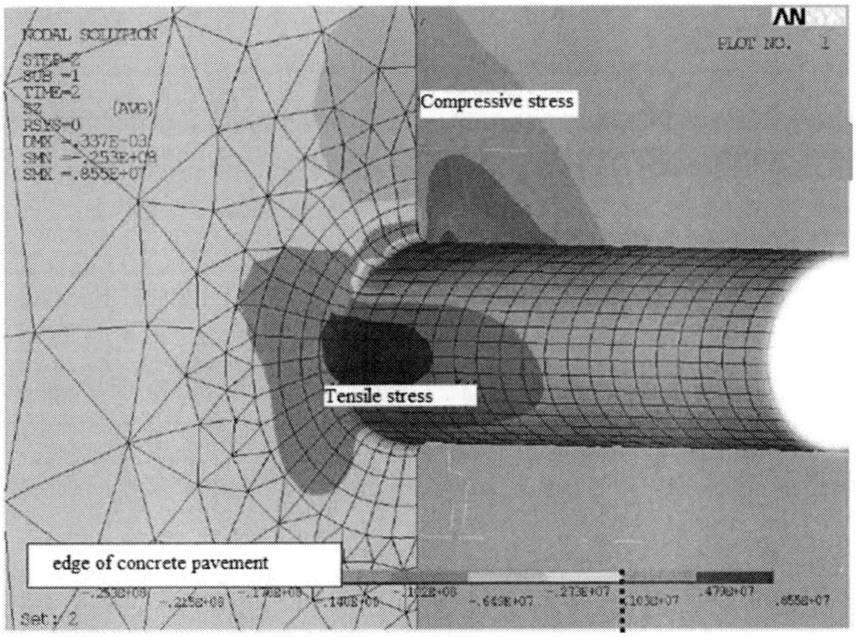

Figure 1. Main tensile stress of concrete around dowel – (V12) variant of defective dowel position.

IOP Conf. Series: Materials Science and Engineering **236** (2017) 012032 doi:10.1088/1757-899X/236/1/012032

The obtained maximum values of tensile stress in concrete slab in close vicinity to defective dowels (σ_3 = 8.0 to 12.0 MPa) make it obvious that the tensile strength of material is significantly exceeded (approximate f_{ct} = 2.3 MPa – derived from the conversion from standard compressive strength) and cracks will occur. Tensile strength was exceeded at all variants of positioning of "defective" dowels. The highest values of tensile stress were found at variants (V12) and (V13), where the dowel is vertical tilted (Figure 1). The results of tensile strength σ_y at the bottom slab edge under defective dowels are influenced by defective dowel positions very little and virtually in all cases the 240 mm slabs range from 1.3 to 1.5 MPa.

3. Measurements in practice on testing field

In order to verify deformations in real concrete slabs, a testing field with concrete pavement was built in Brno. A slab of thickness of 230 mm was laid on a sub-base layer from lean concrete. Dowels on baskets were placed in transversal joints. Their placement on baskets allowed for a setting of different defective position of dowels (vertical and horizontal translation in comparison with the designed position, including vertical tilt). In its premises in Modřice, FIRESTA – Fišer, s.r.o. built a testing site, whose main aim was to construct a single span bridge and an incoming concrete pavement road. The sensors were installed in two transversal joints to the edge of the bottom part of concrete pavement and near dowels with cables running to a green strip of land at the side, where they could be connected to the control measurement unit (Figures 2, 3).

Figure 2. Testing segment Brno-Modřice – installation of strain gauges.

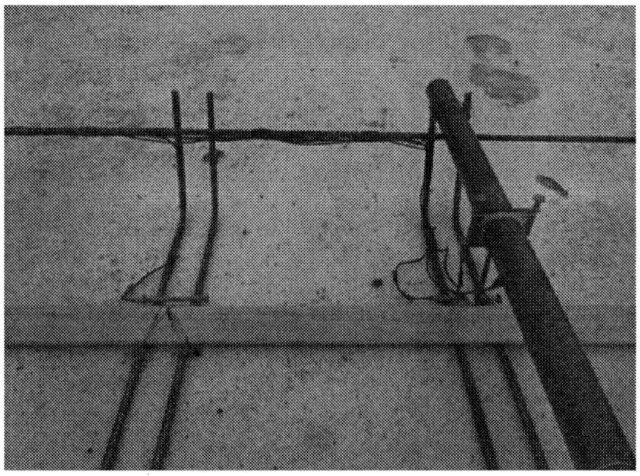

Figure 3. Testing segment Brno-Modřice - installation of strain gauges, detail.

The road construction was designed with a subbase from porous cement concrete, onto which separation woven geotextile was placed. This is to prevent the cement milk from leaking to the porous cement concrete during the concrete pavement installation, which would lead to clogging of this subbase layer with overall drainage effect. The installation of concrete pavement was performed manually, which allowed the installation of baskets with dowels in the area of transversal joints in accurate distance and position. The compacting of concrete pavement was chosen while taking into account the baskets with dowels and sensors as a combination of a hand held concrete immersion vibrator and a vibrating beam on the surface. The concrete on the segment was laid in two longitudinal stripes (Figurer 4, 5).

Figure 4. Testing segment Brno-Modřice – installation of dowels in transversal joints.

Figure 5. Testing segment Brno-Modřice – laying concrete pavement.

Performed tests:

- Measurement of deflection and load transfer efficiency by Falling Weight Deflectometer (FWD) – measurement of deflection under loading in selected points, which simulates a heavy vehicle overrun;
- Measurement of strain in concrete by gauges of PMFL type 50-2LT (with active length of 50 mm) with the aim to study stress around different defectively positioned dowels;
- Measurement of tensile stress at the bottom of slabs – by FWD.

An overview of stress measured in December on 30-day old concrete is shown in Table 4. The average temperature of concrete during the measurements in December was 3.1 °C and negative thermal gradient ranged between -0.01 and -0.29 °C/cm.

Table 3. Measured deformations of concrete loaded by FWD of 115 kN on a 230-mm-thick slab leading to the following stress values (elastic modulus of concrete = 37500 MPa).

Dowel position	Principal tensile stress at the slab bottom (MPa)	Max. tensile stress around dowel (MPa)	Compressive stress around dowel (MPa)
Correct fit	1.23	0.29	-0.77
Vertical Translation (downwards) 20 mm	1.61	0.75	-

The first informative results of the measurement proved suitability of the inbuilt system to monitor stress in the vicinity of dowels. The tests in practice show conformity with theoretical calculations when evaluating tensile stress at the bottom part of the slab. The tests confirm the assumption that stress at the bottom part of the slab under defective dowels is influenced very little by defective position of dowels.

The strain gauges placed around dowels show low values of tensile and compression stress and fail to correspond with theoretical calculations. This discrepancy is caused by the strain gauge length, which exceeds the stress concentration.

4. Recommendations

The development in the field of dowels tends to apply bigger diameters in relation to concrete pavement thickness and their traffic loading. In addition, it is possible to increase horizontal distance of dowels in less busy traffic lanes (fast lane, hard shoulder). Steel dowels must be located in the center of the slab thickness in the same axis, parallel to the road pavement surface and perpendicular to the joint in a tolerance specified by ČSN EN 13877-3.

The recommendations for the use of the numbers of dowels and tie bars of different diameters are mentioned below. The recommendations are based on the experience from abroad, from performed tests, and from modelling. The axial distance of dowels is 250 mm. The distance can be increased, but maximally doubled, (Tables 4 and 5) on less busy traffic lanes (without heavy vehicle traffic) and on hard shoulders. The distance of dowels from slab edges may not be lower than 250 mm. The scheme of this recommendation for the construction of motorway pavements is shown in Figures 6, 7.

Table 4. Recommended dimensions and distance of dowels in transversal joints in concrete pavement.

Dowel diameter (mm)	Dowel length (mm)	Distance (mm)	Pavement thickness (mm)
25	500	250 (500)	≤ 200
25			200-250
30			250-350
35			≥350

Table 5. Recommended dimensions and distance of tie bars in longitudinal joints in concrete pavement.

Tie bar diameter (mm)	Tie bar length (mm)	Distance (mm)	Pavement thickness (mm)
14-16	800	min. 3 pcs (single carriageway)	≤250 mm
16-18		min. 5 pcs (dual carriageway)	>250 mm

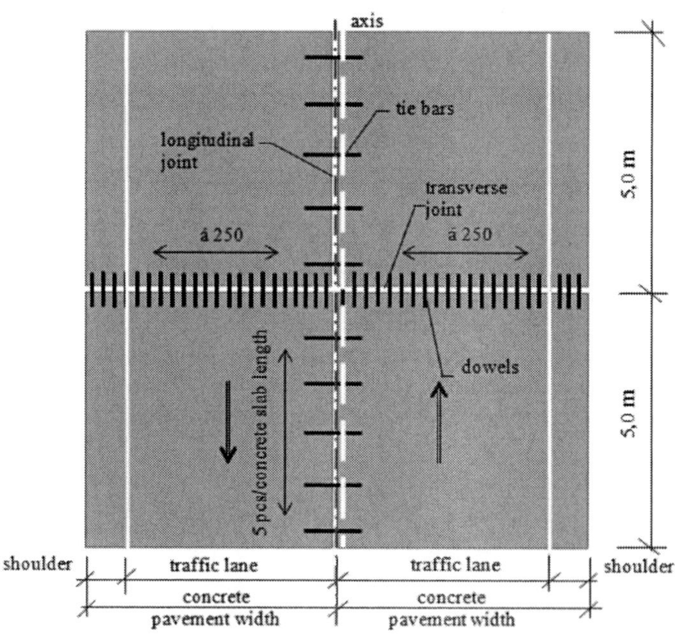

Figure 6. Example of a potential deployment of dowels and tie bars in transversal joints in dual carriageway.

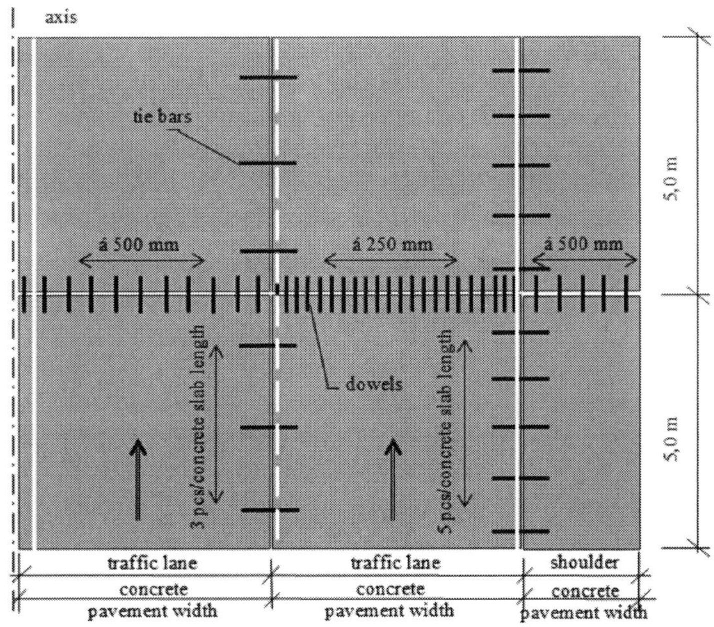

Figure 7. Example of a potential deployment of dowels and tie bars in transversal joints in dual carriageway.

The experience from modelling and testing was applied during the revision of guidelines TP 098 Navrhovanie cementobetónových vozoviek na cestných komunikáciach (Designing Road Concrete Pavements), in which the authors participated.

Acknowledgements
This article was produced with the financial support of the Ministry of Education, Youth and Sports within the National Sustainability Programme I, project of Transport R&D Centre (LO1610), on the research infrastructure acquired from the Operation Programme Research and Development for Innovations (CZ.1.05/2.1.00/03.0064).

The infrastructure used for this work was financed from Technology Agency of the Czech Republic project No. ED2.1.00/03.0064: Transport R&D Centre.

References

[1] Český nomalizační institut 2006 Stavba vozovek – Cementobetonové kryty – Část 1: Provádění a kontrola shody /Construction of roads - Concrete pavements - Part 1: Implementation and Supervision of Conformity *(ČSN 73 6123-1)*

[2] Český normalizační institut 2006 Cementobetonové kryty – Část 3: Specifikace pro kluzné trny /Concrete pavements - Part 3: Specification for dowels to be used in concrete pavements/ *(ČSN EN 13877-3)*

[3] Shoukry S N, William G W and Riad M 2011 Application of LS-DYNA Identifying Critical Stresses Around Dowel Bars *8th Int. LS-DYNA Users Conference*

[4] Li L, Tan Y, Gong X and Li Y 2012 Characterization of Contact Sresses Between Dowels and Surrounding Concrete in Jointed Concrete Pavement *10th Int. Conf. on Concrete Pavements (Québec)*

[5] Nishizawa T, Ozeki T and Kawano H 2012 Surface Stresses of Thick Concrete Pavement Slabs Due To Traffic Loads and Non-Linear Temperature Distributions *10th Int. Conference on Concrete Pavements (Québec)*

[6] Salajka V and Čada Z 2013 *Variant calculation of the influence dowel position on the concrete slabs stress, first series* (Brno, FAST VUT)

[7] Stryk J et al 2013 *Measurement of Deflections and Evaluation of Carrying Capacity of Pavements by Falling Weight Deflectometer - FWD* (Brno, Transport Research Centre).

[8] Zuzulova A 2014 Dowels for Concrete *Research and technical development in road construction XIXth Seminar of Ivan Poliaček with International Participation (Bratislava)* pp 179-84

Assessment of the transport routes of oversized and excessive loads in relation to the passage through roundabout

Jan Petru[1,2], Jiri Dolezel[2,3] and Vladislav Krivda[1]

[1]VSB – Technical University of Ostrava, Faculty of Civil Engineering, Department of Transport Constructions, Ostrava, the Czech Republic
[2]MORAVIA CONSULT Olomouc a.s., Olomouc, the Czech Republic
[3]Brno University of Technology, Faculty of Civil Engineering, Brno, the Czech Republic

E-mail: jan.petru@vsb.cz

Abstract. In the past the excessive and oversized loads were realized on selected routes on roads that were adapted to ensure smooth passage of transport. Over the years, keeping the passages was abandoned and currently there are no earmarked routes which would be adapted for such type of transportation. The routes of excessive and oversized loads are currently planned to ensure passage of the vehicle through the critical points on the roads. Critical points are level and fly-over crossings of roads, bridges, toll gates, traffic signs and electrical and other lines. The article deals with the probability assessment of selected critical points of the route of the excessive load on the roads of 1st class, in relation to ensuring the passage through the roundabout. The bases for assessing the passage of the vehicle with excessive load through a roundabout are long-term results of video analyses of monitoring the movement of transports on similar intersections and determination of the theoretical probability model of vehicle movement at selected junctions. On the basis of a virtual simulation of the vehicle movement at crossroads and using MonteCarlo simulation method vehicles' paths are analysed and the probability of exit of the vehicle outside the crossroad in given junctions is quantified.

1. Introduction

Industry and engineering have a long tradition in the Czech Republic. These sectors produce products with dimensions which can be considered nonstandard. For these products the issue of transport on chosen routes on the roads arises. Routes for excessive transport were previously protected by the Ministry of Transport in terms of dimensions and weight. After year 1992 the routes for oversized loads stopped being protected. Oversized loads consist not only of products, but they also include various construction machinery cranes, etc. that with their dimensions and weight exceed standard allowed maximum limits [1]. Annually the Ministry of Transport of the Czech Republic permits 15 up to 20 thousand of oversized transports, out of which 5,000 transports with respect to the dimensions of the load require extra space to ensure the passage.

More and more resources, both financial and also time are spent in order to ensure the passage of oversized loads. These funds are spent on ensuring technical adjustments to the spatial layout of roads for the passage of individual oversized loads. Such examples include especially building of roundabouts [2,3] with parameters unsuitable for this type of transport, narrowing roads and creating bumps on roads in towns and villages. We can also mention the sc. sewerage transportation in crossroads and junctions and also on fly-over crossings. The problems are also caused by the proximity of lighting, traffic signs and other immobile road elements. Currently, the condition of the road network for usage by oversized load in the Czech Republic is insufficient. Although there exists European directive for transporting excessive loads on the road issued on May 17th, 2006 [4], which recommends to Member States to build a pan-European network of corridors for transporting excessive loads.

2. Critical Points on Roads

Since there are no fixed main routes for transporting of oversized loads in the Czech Republic, they are chosen by the transporter or by the escort. Route selection is based on previous experience and on the assessment of the road condition. This choice depends on the type of cargo, its dimensions and weight. Final approval of the proposed route is granted by the competent authorities [5]. The choice of route itself, due to the passage on roads and bridges, might take several months. Thus, the critical points during the route are level and fly-over crossings of roads, bridges, toll gates, traffic signs, electrical and other lines, etc.

2.1. Traffic Signs

Traffic signs are the most common collision points on the route of oversized load [6, 7]. The signs are located close to the roads, which makes the passage of loads impossible. These signs are very often not possible to be dismantled and it makes the manipulation for ensuring the passage of the transport impossible. If it is possible to dismantle the signs, the problems with anchor bolts of these signs occur. Anchor bolts protrude into the road and it is necessary to cover these places. However, it is not always possible to use the mentioned solution. The problem is caused not only by vertical traffic signs but also by signs which are mounted on the booms over the road. These are mostly newly built or reconstructed pedestrian crossings. [7]

2.2. Immobile Elements on Roads

These elements are primarily lighting, handrails, immobile elements, decorative elements and vegetation. Lighting on roads causes similar problems as traffic signs. Removing of the lighting is very complicated and in most cases it is impossible to pass. Non-removable railings that prevent the passage of cargo and other immobile elements on the route also complicate transport. Not only because of time during the passage but also while selecting the route. In such cases the given part of the road or the intersection becomes impassable for the vehicle. [6, 7]

2.3. Bridge objects

Bridge objects belong among the most critical points on the route of oversized loads. It is because of their limited load capacity but also because of low height clearance. Bridge objects are designed to standardized models of heavy traffic load which do not cover the whole variability of oversized vehicles. For each passage of oversized vehicle on a bridge structure it is necessary to make an individual assessment of static load capacity of the bridge. The height clearance is also connected to the bridges. It is very often inadequate on frequent routes. Since there are not known major routes for excessive loads, it would be useful to ask transporters for any possible requirements which could be used during the designing of new bridges and reconstructing the current bridges. [6, 7]

2.4. Tollgates, power lines and other lines

Height clearance of these objects often limits the passage of loads on individual roads and the assistance of other subjects is often required. The wires of power lines, overhead lines or cables make the transport difficult. In these cases it is necessary to ensure the assistance of experts. The cables are lifted during transport. Sometimes it is necessary to switch off the electricity. [6, 7]

2.5. Intersections

At crossroads and junctions on level crossings the passage of excessive load is very difficult due to often unsatisfactory corner radius of intersections. The excessive transport is usually not taken into consideration during designing or reconstructing roundabouts [8]. Problematic parameters are mainly radius on entrance / exit, geometrical arrangement, the size of the roundabout, or complication with a raised central island and dividing islands [9].

3. Statistical Data for Determination of a Vehicle for Oversized Load and Verification of Passage Through Critical Points

For the verification of the critical points along the route of the oversized load first it is necessary to determine what types of cargos are located on the route. Further, it is also necessary to find out the parameters of trailers which carry the cargo. For this purpose the long-term author's database was used, from it the main route (road of 1st class) was selected. This database contains processed statistics of transports based on the transport time, type of cargo, weight, length, width, height of all set, the beginning and end of the route, a detailed schedule of the route and the number of days of transport.

For the verification of critical points on the chosen route there were used four cargo categories that use this route. A more detailed division into categories is shown in author's doctoral thesis [6]. For the purposes of the article the following categories were chosen:

- Long and heavy transports (generators, transformers, ...).
- Structures (frame structures, bridge structures, steel, wood, or concrete structures, ...).
- Machinery (construction equipment, cranes, ...).
- Mixed category (silos, technological equipment, boats, pools ...).

Individual transports were analysed and divided into a specified range of widths. Figure 1 shows observed values in the specified range according to individual types of transport.

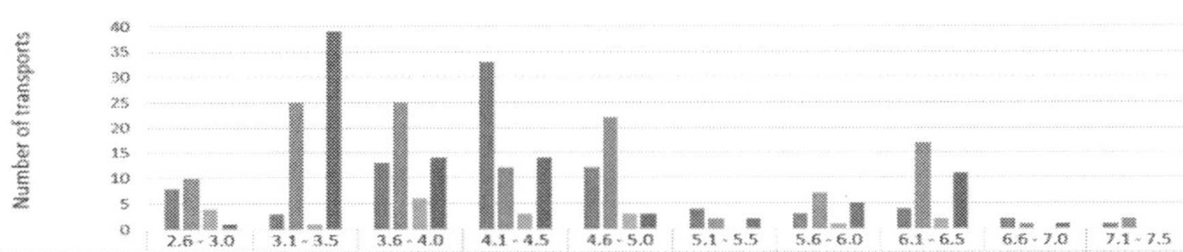

	2.6 - 3.0	3.1 - 3.5	3.6 - 4.0	4.1 - 4.5	4.6 - 5.0	5.1 - 5.5	5.6 - 6.0	6.1 - 6.5	6.6 - 7.0	7.1 - 7.5
Long and heavy transport	8	3	13	33	12	4	3	4	2	1
Structure	10	25	25	12	22	2	7	17	1	2
Machinery	4	1	6	3	3	0	1	2	0	0
Mixed category	1	39	14	14	3	2	5	11	1	0

Width [m]

Figure 1. Number of transports in a range of widths on observed route.

Another observed parameter for verifying the passage through critical points on the route was transport height. Figure 2 shows the resulting observed values within the specified range according to individual types of transport. Transports which were represented the most have between 4.1 to 4.5 meters.

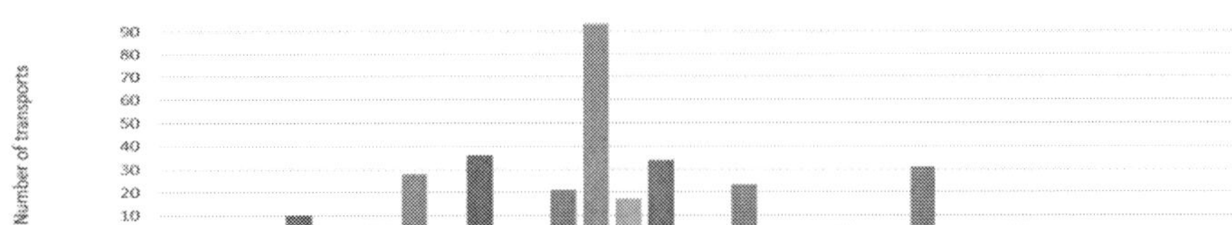

	3.1 - 3.5	3.6 - 4.0	4.1 - 4.5	4.6 - 5.0	5.1 - 5.5	5.6 - 6.0
Long and heavy transport	0	5	21	23	31	3
Structure	0	28	93	2	0	0
Machinery	0	3	17	0	0	0
Mixed category	10	36	34	4	4	0

Width [m]

Figure 2. Number of transports in a range of heights on observed route.

A diagram of clearance gauges was suggested on the basis of the processed statistical data. For the calculated clearance gauge similar requirements are valid as for the clearance gauge in chapter 6.1.2. of CSN 73 6201 Design of Bridges [10].

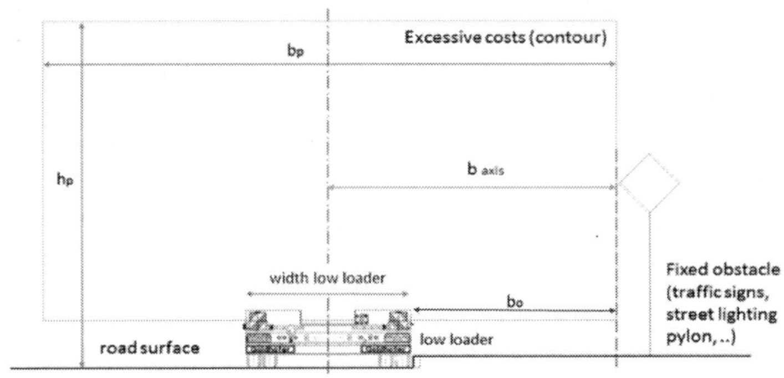

Figure 3. Diagram of clearance gauge including distance from solid obstacles [6,7].

The width of clearance gauge b_p for the passage of oversized loads is shown in Table 1. For individual types of cargo recommended distances are determined. The b_o was calculated as:

$$b_o = b_{asix} - \left(\frac{x}{2}\right) \qquad (1)$$

Table 1 shows a type of transport and for each type the width of used trailer is determined. The width of the trailers was determined for each type of cargo on the basis of types of trailer used in the Czech Republic. The recommended values of b_o were determined on the basis of the calculation. The recommended height h_p is determined for each type of transport. The height h_p of clearance gauge is basis to a free height of the underpass which is listed in section 6.3.2. of CSN 73 6201 Design of Bridges [10].

Table 1. The distance from solid objects and the height area.

Type	The minimum width of used trailers [m]	Recommended b_o [m]	Recommended height h_p [m]
Long and heavy transport	3.00	2.30	6.00
Structures	2.55	2.50	5.00
Machinery	2.75	1,90	4.50
Mixed category	2.55	2.25	5.50
Resulting value	3.00	2.30	6.00

4. Modelling of the Vehicle and its GPS Verification

Based on the analysis and information about parameters, types of vehicles and cargo, it is possible to create the design of the vehicle for verifying trajectories of passages of oversized transports with usage of software AutoTURN. It was necessary to verify such a created model with the help of GPS technology which was installed on the set transporting excessive load (tractor, trailer).

4.1. Modelling of the Vehicle

Two vehicles were created for the verification of the critical points of the route. The first vehicle was created as a test model. On this model the procedure of modelling the car in program AutoTURN was tested. Subsequently vehicle verified with the real vehicle was created.

Figure 4 shows windows of the AutoTURN program that are used for modelling a new vehicle type (set that was tested by GPS technology). Technical parameters were inserted into the program from the manufacturer's data sheets. Other parameters for entering, which were not included in manufacturer's data sheets, were measured additionally. For testing the behaviour of created models of vehicles in AutoTURN program video analysis and data mapping from the transports on roads were used [6, 7].

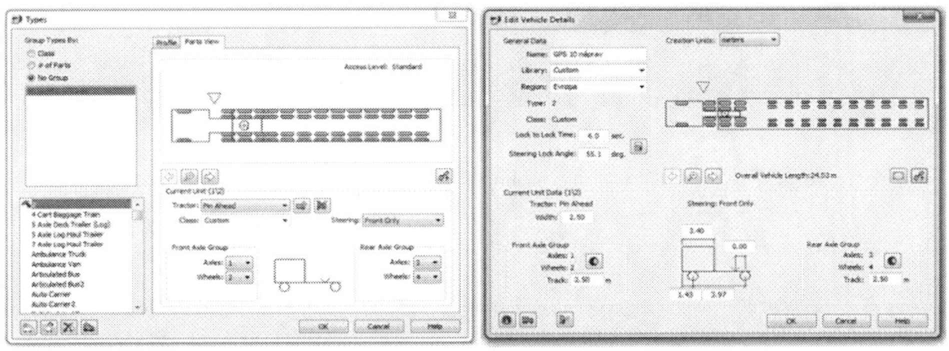

Figure 4. The windows for modelling a new vehicle type in AutoTURN program.

4.2. GPS Measurements and Model Verification

The verification of the created model was carried out on the vehicle with THP axes. GPS technology was installed on this vehicle. For the realization two sets of instruments Topcon HiperGD (GPS observation) and one set of Topcon HiperGGD (observation of GPS / GLONASS) were chosen. Each set consisted of GNSS devices, data recorder for controlling and saving measured data, and a mobile phone. The mobile phone was used for data connection to the source of corrections. TopSURV application was used for recording the measurements, it is a standard application for operation of Topcon. Due to the requirements for measurement accuracy the method of phase measurements with correction in in Real-Time Kinematic (RTK) was chosen.

The fixing the GNSS apparatus was done once to the cabin and twice to the trailer. Fixation and installation of GNSS apparatus is shown in Figures 5 and 6.

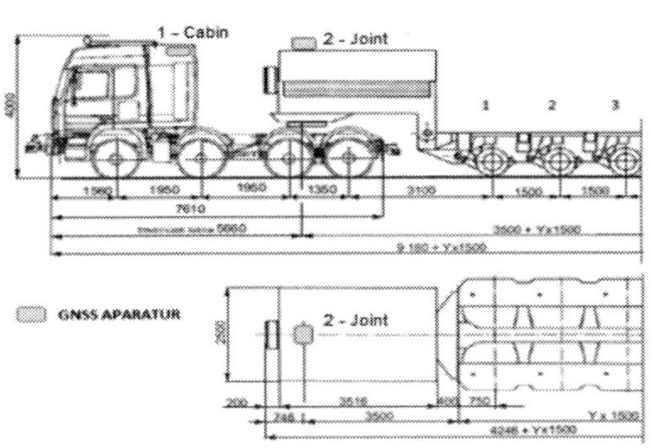

Figure 5. Part of the scheme of GNSS apparatus location on a vehicle transporting oversized loads.

Figure 6. Location of GNSS (2) equipment to trailer with tractor.

Data were processed in ArcGIS Desktop. On the basis of the technical documentation outlines of tractors were calculated from the measured data. Calculations were done by using a combination of formulas of analytical mathematics and analytical tools of ArcGIS Desktop (see Figure 7). For the calculation of the outlines the script for automation of evaluation of measured data was prepared. The result of the processing is – a class of 3D polygon objects (representing a ground plan of the tractor and the trailer) and a class of point objects (representing the outlines of significant edges in the ground plan of the tractor and the trailer).

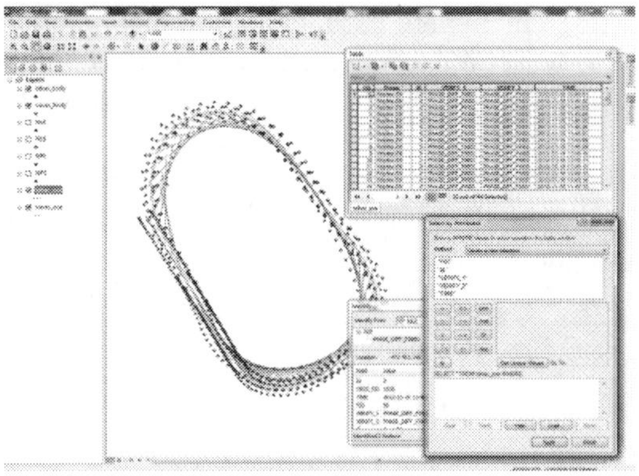

Figure 7. A preview of processing the measurement of GNSS apparatus in ArcGIS program.

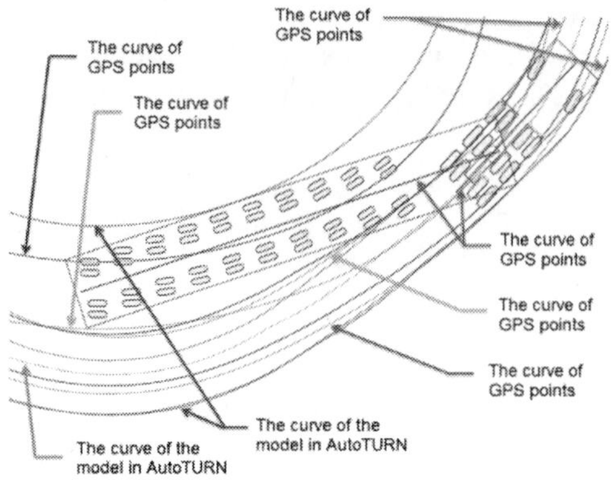

Figure 8. A comparison of the created model in AutoTURN program and GPS measurement.

The newly created vehicle in AutoTURN program was subsequently verified according to the processed and evaluated measurement with the usage of GPS technology. The curve created from the GPS points and the outline curve of the edges of the trailer of newly modelled vehicle in software AutoTURN were compared on the basis of the measurement. Based on the obtained data the conclusion is that the vehicle created in AutoTURN program corresponds to the "real" behavior of vehicle carrying excessive and oversized loads, which was used for the measurement (see Figure 8).

Based on this verified procedure a vehicle for verifying intersections on the route of oversized load was created. The processed statistics from chapter 3 were used for modelling of the vehicle in AutoTURN program.

5. Modelling of Passage of Vehicle and Analysis by the MonteCarlo Method

Based on the modelled vehicle in AutoTURN program a selected intersection on the route of excessive cargo was verified. This intersection is in city Litomyšl. The intersection is located on frequent routes of such cargos. This critical point of the route was deliberately chosen based on many measurements of the author, which were realized at this intersection [4, 7]. Nowadays the intersection is adapted for passage of excessive loads.

Figure 9 shows trailing curves from simulations of passages of the vehicle. For the analysis of the passage 100 realizations at the intersection were done. Figure shows 4 envelope curves of all simulated passages. These curves contain all simulations which were done.

- blue hatch – passage without the need of dismantling the traffic signs, the dividing islands were used for the passage
- grey hatch – passage with the need of crossing the road kerbs and use of unpaved road shoulders and greenery
- red hatch – passage which cannot be realized because of driving on the central island, road shoulder and dismantling the lighting
- purple hatch – passage which was recorded during monitoring of the transport – i.e. transport which was realized through this intersection – during the passage the traffic signs were dismantled, the passage was done across the road kerbs and unpaved road shoulders. Excessive load did not cross the central island of the intersection.

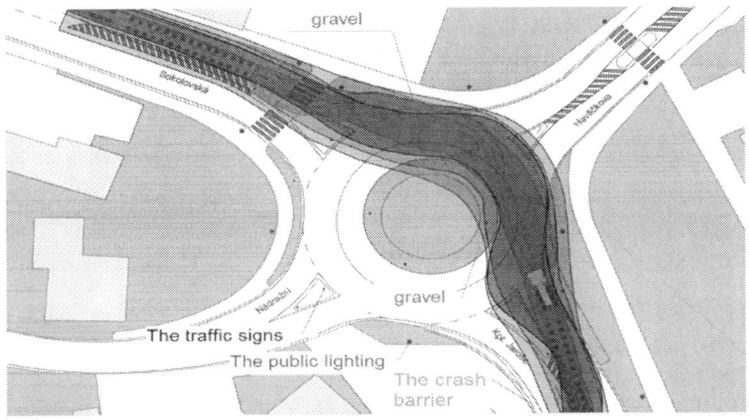

Figure 9. The resulting envelope curves from the simulation of passage through the roundabout in AutoTURN program.

Based on the simulation of movement of oversized vehicle through the intersection the probability of passage was analyzed. For these virtual simulations of the movement of the vehicle through the intersection the MonteCarlo simulation method was used. The trailing curves were analyzed and the probability of exit of the vehicle outside the outline of the intersection during individual passages was quantified. The critical points, such as the distance of the traffic signs, lighting, road kerbs, etc., at this intersection were taken into consideration in this analysis.

Figure 10 shows histogram of evaluation of passage of excessive load through the roundabout. The horizontal axis indicates total width of circulatory roadway which can be used for the passage of the vehicle. Value 0 indicates a half of this width. The vertical axis represents the amount of passages. The blue dashed line indicates a boundary when it is necessary to dismantle the traffic signs and it is necessary to prepare for the passage through this intersection. The black dashed line indicates the necessity of crossing the road kerb (the necessity to underlay the kerbs). The red line indicates the boundary when the passage cannot be realized (there is no possibility to underlay the kerbs, the assistance of many subjects, dismantling the lighting, dismantling the traffic signs is needed).

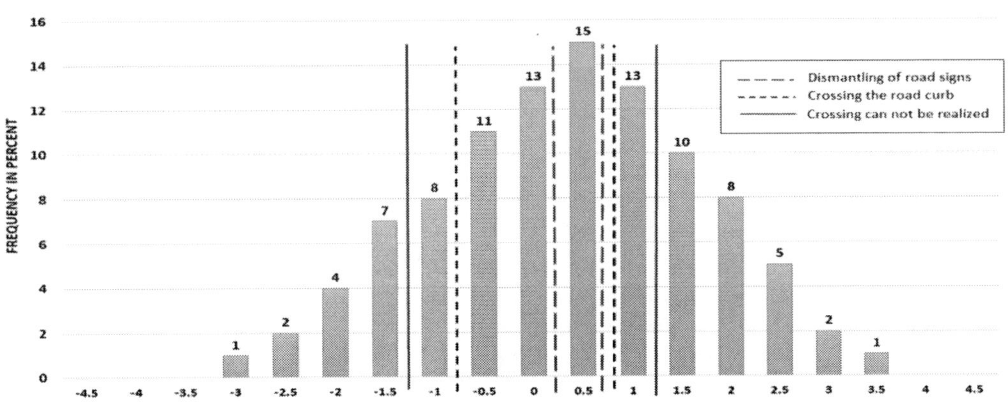

Figure 10. Evaluation of the probability of the passage of excessive load through the roundabout.

6. Conclusion

The article tries to draw attention to the issue of passage of excessive and oversized loads through critical points on the roads. For the purposes of the article one route of excessive cargo was chosen an example. The most common types of transport were evaluated with the help of created database.

Based on the measurements and processed video analyses a passage of vehicle through the selected roundabout was simulated by the usage of a newly created vehicle in AutoTURN program. On the basis of the simulation of vehicle movement through the roundabout the probable passage was analyzed.

From the evaluation of 100 virtual simulations of passage of oversized load through the roundabout arises that only 15 passages can be realized without any adjustments. Further 45 passages can be realized with some adjustments of the route during the passage. And 40 passages cannot be realized.

Based on long-term research and analysis of other critical sites and the results described in this article, indicate that road network in the Czech Republic is not designed for the passage of excessive and oversized load [7,11]. There are many critical points on the main roads of such transports [11, 12]. It causes not only long delays but also financial losses (construction adjustments, rent of other entities for realization of the transport). For these reasons the price of transport (product) increases and this causes reduction of competitiveness of Czech producers. That is why there should be developed main routes for transport of these excessive and oversized loads and there also should be some regulation which would determine suggested parameters of these routes. [6, 7]

Acknowledgements

This paper was supported by MORAVIA CONSULT Olomouc a.s.

The work was supported from sources for conceptual development of research, development and innovations for 2016 at the VSB-Technical University of Ostrava which were granted by the Ministry of Education, Youths and Sports of the Czech Republic.

References

[1] Ministry of Transport of the Czech Republic 2002 *Vyhláška 341/2002 o schvalování technické způsobilosti a o technických podmínkách provozu vozidel na pozemních komunikacích /Notice 341/2002 The approval of technical competence and technical conditions for operating vehicles on roads/*

[2] Mahdalova I and Krivda V 2015 Traffic Safety at Roundabouts in the Czech Republic *Road and Traffic Safety: Practices, Role of Human Behavior and Effective Programs (New York: Nova Science Publisher)* pp 23-48

[3] Brilon W 2005 Roundabouts: A State of the Art in Germany *National Roundabout Conference (Vail, Colorado)*

[4] European Commission - Directorate General for Energy and Transport 2006 *European Best Practice Guidelines for Abnormal Road Transports*

[5] Ministry of Transport Czech Republic 1997 *Zákon 13/1997 o pozemních komunikacích /Act 13/1997 About roads communications/*

[6] Petru J 2014 Nadměrné a nadrozměrné náklady a jejich přepravní trasy /Passage of excessive transportation in the area intersections/ (in Czech) *Sborník vědeckých prací VŠB-TUO* **13**(1) pp 101-8

[7] Krivda V, Petru J, Mahdalova I and Zitnikova K 2016 *Hodnocení stavebních prvků křižovatek s využitím videoanalýzy /Evaluation Intersection Building Elements Using Video Analysis/* (VSB - Technical University of Ostrava) p 184

[8] Giuffre O, Grana A, Marino S and Galatioto F 2016 Passenger Car Equivalent for Heavy Vehicles Crossing Turbo-roundabouts *Transportation Research Procedia* **14** pp 4190-99

[9] Petru J and Zeman K 2013 Analysis of Roundabout Intersections on Routes of Abnormal Loads *Recent Advances in Urban Planning & Construction - Proc. 4th International Conference on Urban Sustainability, Cultural Sustainability, Green Development, Green Structures and Clean Cars - USCUDAR '13 (Budapest)* ed Kung Chang Lee (WSEAS Press) pp 47-53

[10] Úřad pro technickou normalizaci, metrologii a státní zkušebnictví (ÚNMZ) 2012 Projektování mostních objektů /Design of Bridges/ *(ČSN 73 6201)* (in Czech)

[11] Petru J and Mahdalova I 2013 Nadměrná přeprava na pozemních komunikacích /Oversize transport on roads/ (in Czech) *Civil and Enviromental Engineering* **9**(1) pp 63-71

IOP Conf. Series: Materials Science and Engineering **236** (2017) 012034 doi:10.1088/1757-899X/236/1/012034

Capacity analysis of a bypass of roundabouts

Ivan Sedlačik[1] **and Petr Slabý**[1]

[1] Department of road structures, Faculty of Civil Engineering, CTU in Prague; Thákurova 7, 166 29 Praha 6, Czech Republic

ivan.sedlacik@fsv.cvut.cz

Abstract: The capacity of the roads network mainly depends on the capacity of its nodal points - intersections. A connecting branch or a bypass is a lane or lanes inserted between two adjacent branches of a roundabout, providing redirection of vehicles, that would otherwise burden a circular lane. A bypass effect to the capacity of roundabouts, but also other types of level intersections, is undeniable. A connecting branch increases the total capacity of an intersection that takes a part of vehicles performing a manoeuver of the first right turn completely out of an intersection area. Redirecting vehicles reduces delay times at intersections and reduces queues at the entrance to an intersection. Bypasses improve the quality of transport. Limiting for the capacity of bypasses is the point of disconnection from the entrance into the roundabout and the connection point into the exit from the roundabout. Central parts of the bypasses have minimal effects on the capacity. The length of a bypass has to match with the maximum length of a queue of waiting vehicles at a given intensity level. The article deals with analysis of the bypass capacity at the roundabouts.

1. Introduction

The bypass (SV) increases the capacity of a roundabout by adding separate lane (one way road) for turning the vehicles going to the right see [6, 8]. The bigger the demand for the first right turn is, the bigger influence of increased capacity we can see. The situation at the entrance to a roundabout has to enable the early elimination of the vehicles to the bypass. At the same time the smoothly flowing connection from the exit of the bypass to the superior lane must be secured. The capacity of the bypass depends on the capacity of the point of connection and disconnection of the bypass, the central part of the bypass influences the capacity only marginally. The length of the connecting branch should correspond to the maximum length of a queue of vehicles, that is formed in the bypass at a given intensity level. If the length of the connecting branch is short, the queue can reach before the point of the disconnection of vehicles at the entrance, which results in increasing of delay time at the entrance and in total decreasing of quality of transport (UKD) at the intersection. The identical situation can occur at the entrance when it is clogged by vehicles and the queue reaches before the point of disconnection of the bypass.

Empirical evaluation and observing of bypasses was aimed at the simplest but the most used type of the roundabout (roundabout with 1 lane at the circle/1 lane at the exit) with corresponding disposition of bypasses "wedge - wedge" (Figure 1). The given type was also selected because of clear determination of critical time gaps (tg), subsequent time gaps (tf) and minimum time gaps (t0).

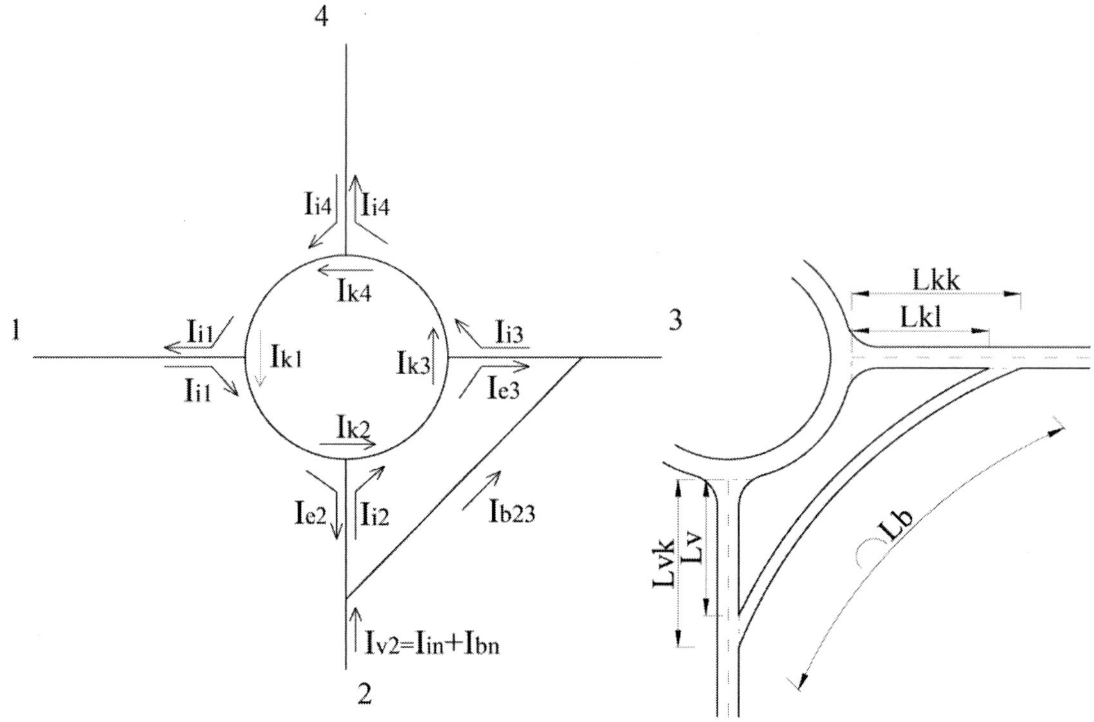

Figure 1. The scheme of intensities, scheme of geometric layout of the bypass - type "wedge -wedge".

2. Capacity of the point of disconnection from the entrance into the roundabout

The capacity of the point of disconnection of the bypass was determined according to the interactive models in the programme PTV Vissim; the tg, tf, t0 values were not detected.

The capacity of the beginning of the bypass mainly depends on the amount of demand for right turn, i.e. intensities percent at the common lane that enter the roundabout (Iej) and intensities of turning to SV (Ibj,j+1). The sum of these intensities makes so called total intensity of an entrance (Ivj). Another important aspect is the position of the start of SV before the border of a roundabout, i.e. Lvk (Figure 1).

From the Figure 2 we can read:

- Whether the connecting branch is necessary, i.e. if the point with coordinates of superior values of intensity on the circle Ikj and entrance Iij is situated above the limit of capacity of an entrance for UKD of level D; in order to make comparison, there is also level B in the figure.
- For the given ratio of intensities Iij a Ibj,j+1 we can read the increase of total capacity Ivj when using the bypass. For the ratio of intensities from 80:20% to 50:50% this increase is approximately 550 - 150 nveh/h, set for UKD of level D and the range of values of intensity at the roundabout 200 – 1200 nveh/h.

This figure can be used when deciding whether the bypass should be built.

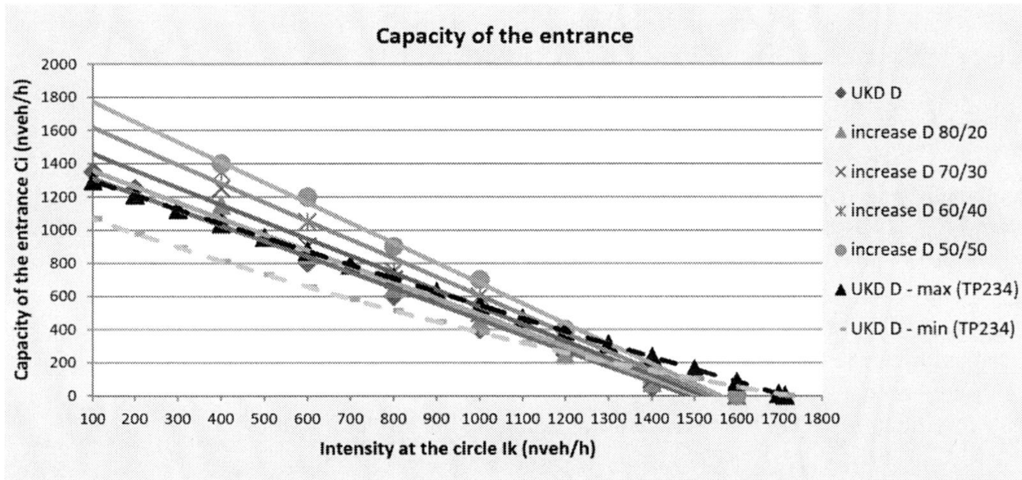

Figure 2. Capacity of the entrance to the roundabout.

The crucial aspect for the capacity of turning is the length of the queue at the entrance to the roundabout. From the Figure 3 we can roughly read the average length of the queue 10m, 20m, 30m, 40m, 100m at the entrance depending on the intensity of vehicles behind the point of disconnection of the bypass and intensity of vehicles at the circle.

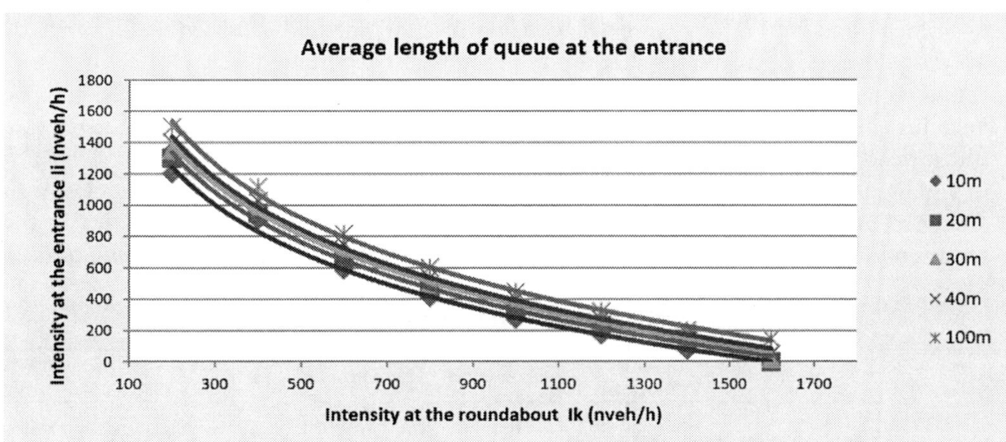

Figure 3. Average length of the queue at the entrance to the intersection.

The length of the queue at the entrance has to be smaller than the distance of the circle and the limit of the point Lv. At the same time the length of the queue that is formed at bypass has to be smaller than the length of bypass, i.e. Lb.

3. Capacity of the connection point into the exit from the roundabout

To be able to determine the conclusion of the capacity assessment of the roundabout it is necessary to verify whether for the intensity of traffic at the entrance the limit value of the medium time of delay (tw) is not exceeded, based on the conditions for the particular levels of UKD according to [1,2,7]:

$$t_w^n \leq t_{w,lim} \tag{1}$$

t_w^n – medium time of delay of a vehicle in the stream of traffic (s)
$t_{w,lim}$ – the longest permissible medium time of delay according to the required level of UKD (s)

For calculation of the capacity of the entrance to the roundabout and the point of bypass connection we use the same formula as for the basic capacity according to [1, 7]:

$$C_i = 3600 * (1 - \frac{\Delta * I_k}{n_k * 3600})^{n_k} * \frac{n_{i,koef}}{t_f} * e^{-\frac{I_k}{3600}*(tg - \frac{tf}{2} - \Delta)} \quad (2)$$

C_i – capacity of the entrance, points of bypass connection (nveh/h)
I_k – intensity of traffic at the roundabout, exit (nveh/h)
n_k – number of lanes at the roundabout, exit
$n_{i,koef}$ - coefficient taking into account the number of lanes at the entrance, bypass
$n_{i,koef} = 1{,}00$ – for one-lane entrances
$n_{i,koef} = 1{,}50$ – for two-lane entrances
t_g - critical time gap (s)
t_f - subsequent time gap (s)
Δ (t0)- minimum time gap between vehicles going on the circle or exit (s)

It is necessary to emphasize that in case of calculating the capacity of the entrance to the roundabout the more precise entrance values according [1] are used. To determine the capacity of the point of bypass connection it is necessary to determine new values.

3.1. Entrance values of the capacity assessment of the connection point
Non - reduced entrance values were set by empiric evaluation of 25 videorecordings of total length of 21 hours (table 1).

- tg (critical time gap) – they were set by using the Raff method a subsequently verified by the Modified method of maximum credibility (Troutbeck) according to [3]
- tf (subsequent time gap) – they were set using two ways: as the average of time gaps between vehicles of side traffic lane and from the departure functions of vehicles at the side stream of traffic according to [4]
- t0 (minimum time gap) – they were set using the departure functions of vehicles at the side stream of traffic according to [4]; it is the biggest time gap between vehicles in the main stream of traffic, that is not yet accepted by the vehicle waiting in inferior stream of traffic.

Table 1. Empirically determined non-reduced entrance values.

	Raff			Departure function			Troutbeck	D	Lkk	Lkl	Lb	Slow vehicles (%)		Intensities (nveh/h)	
	tg (s)	tf (s)	t0 (s)	tg (s)	tf (s)	t0 (s)	tg (s)	(m)	(m)	(m)	(m)	bypass	exit	bypass	exit
Litoměřice 2012	4.3	2.9	2.9	4.5	3.2	2.9	4.2	35	18	11	66	12.38	5.00	366	320
Mělník 2007	4.2	2.7	2.8	4.2	2.8	2.8	4.2	37	25	15	31	15.70	8.02	420	533
Mělník 2014	3.8	2.7	2.4	4.1	2.9	2.7	-	37	25	15	31	14.24	5.77	349	528
Říčany	3.8	2.6	2.5	4.2	3.3	2.5	4.2	38	26	14	29	2.72	3.56	220	366
Šeberov	3.8	2.6	2.4	4.1	3.4	2.4	3.9	33	44	20	71	1.34	3.35	149	537
Svitavy	-	2.9	-	4.3	3.3	2.6	-	45	50	15	59	0.00	0.25	90	396
Jaroměř	4.0	2.6	2.7	3.9	2.7	2.5	-	32	26	14	76	1.48	2.04	337	393
Teplice	4.7	3.0	3.2	4.9	3.9	3.0	-	34	71	23	105	5.23	6.07	153	595

3.2. Capacity of connection point of the connecting branch

Capacity of the bypass connection was determined according to the interactive models in the programme PTV Vissim.

Capacity of the connection point directly depends on the distance of the end of wedge Lkk. The longer the distance Lkk is, the smaller is the critical time gap tg. In practice it means that the driver has better view from the vehicle and therefore affords to accept shorter gap between vehicles in superior stream of traffic. With the help of correlation and under condition that the difference in the capacity of connection point is not bigger than 50 nveh/h, this assumption was found as crucial according to [5].

In practice the connection is usually made using the connecting wedge.

Based on different geometrical dispositions of connection points of observed bypasses, the reductions of entrance values of tg, tf and t0 were set, depending on the distance of the end of connecting wedge Lkk in the range of 10 - 40m (Figure 1). For capacity calculation is not possible to use the values written in [1], as this is another connecting maneuver.

$$t_g = 5 - \frac{L_{kk}}{30} \tag{3}$$

$$t_f = 2{,}7s \tag{4}$$

$$(t0)\Delta = 3{,}2 - \frac{0{,}7}{30} * L_{kk} \tag{5}$$

With increasing distance of the end of wedge Lkk the capacity of connection point of bypass goes up by 200 nveh/h, see Figure n. 4. The figure hereafter also compares the capacity set by simulation programme for Lkk=10m and determined arithmetically according to the formula (2). It is obvious that in the areas of low and high intensities Ie (0 - 400 nveh/h and 1000 - 1300 nveh/h) the simulation deviates more from the calculated values. This can be due to the lack of empirically measured data at low or high intensities.

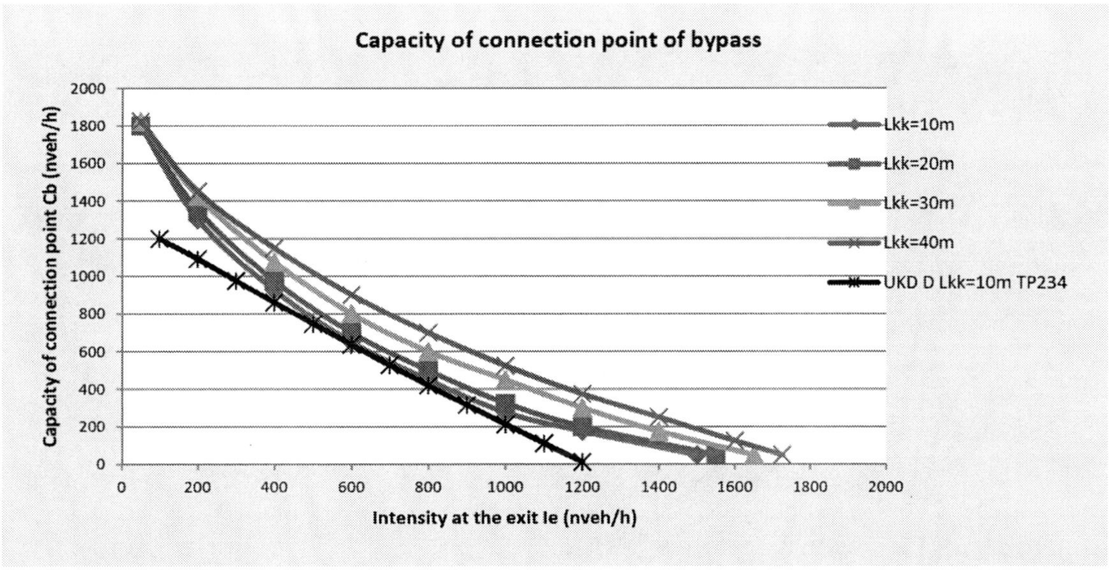

Figure 4. Capacity of connection point of bypass depending on Lkk in the range of 10-40m.

It is always necessary to assess the capacity in relation to the required quality level of traffic UKD. From Figures 5 – 8 we can read the UKD connection points depending on the intensity at the exit Ie, intensity in the bypass Ib and Lkk in the range of 10 - 40m.

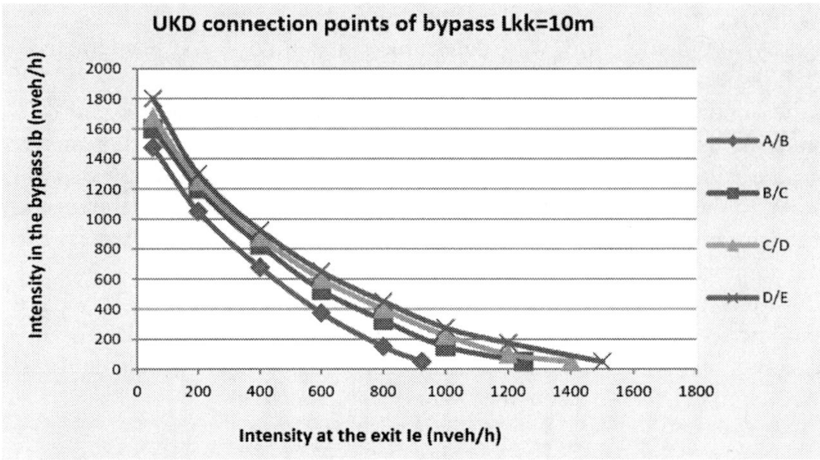

Figure 5. UKD connection points of the bypass Lkk=10m.

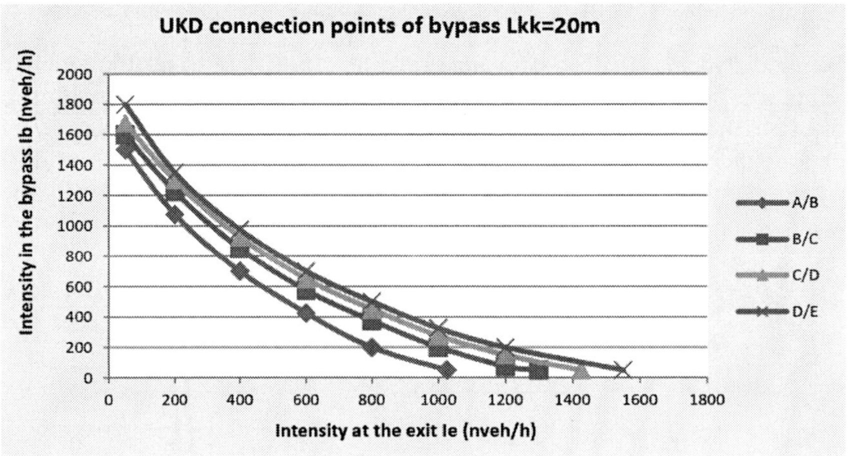

Figure 6. UKD connection points of the bypass Lkk=20m.

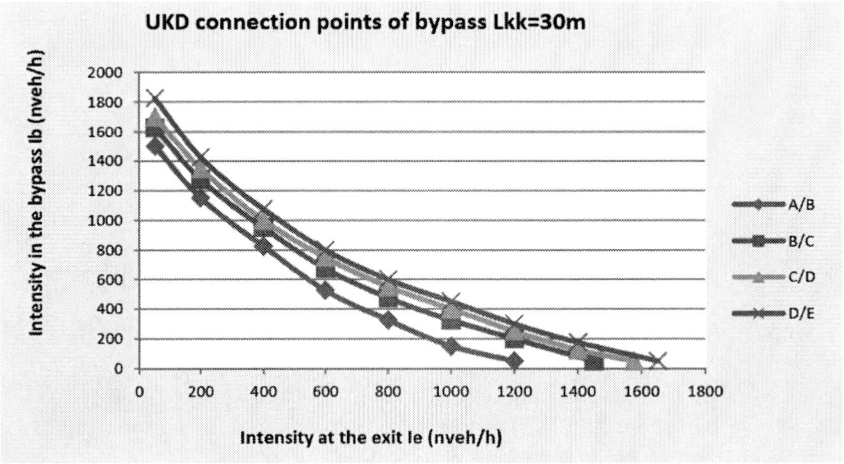

Figure 7. UKD connection points of the bypass Lkk=30m.

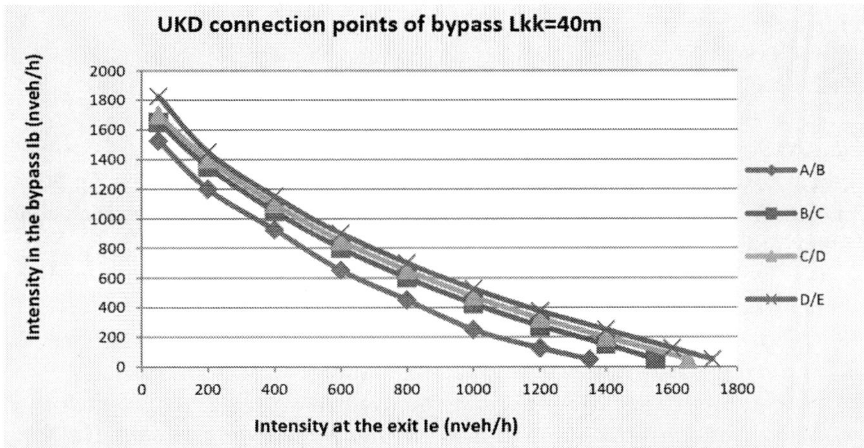

Figure 8. UKD connection points of bypass Lkk=40m.

The important element of capacity assessment of the connection point is the length of queue that is formed in the bypass; the maximum length of queue 10 - 100m depending on the intensities of vehicles and Lkk can be read from the Figure 9.

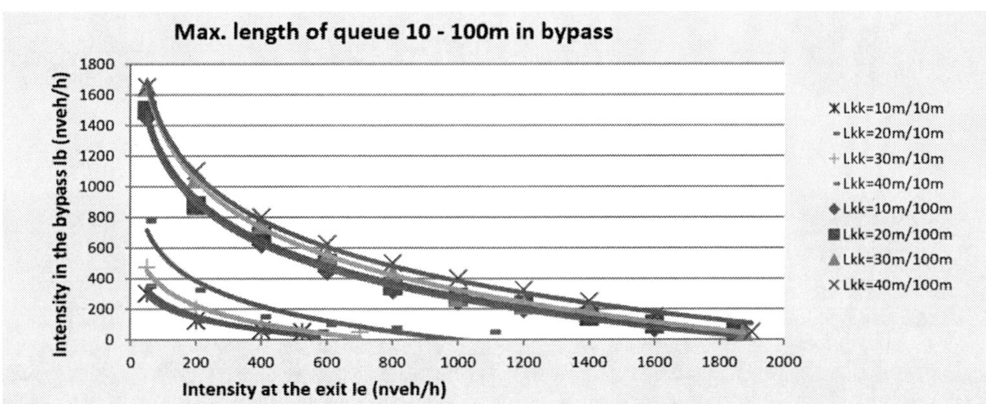

Figure 9. Max. length of queue 10-100m in the bypass depending on Lkk in the range of 0-40m.

4. Conclusion

For effective use of a bypass it is important to design the connection to the exit branch correctly.

For the point of disconnection of the bypass the length of queue of vehicles at the entrance to a roundabout is limiting. If the queue at the entrance reaches behind the point of disconnection of connecting branch, it is necessary to design turning lane or separate lane for entering the bypass; we can read roughly the length of queue from Figure n. 3.

The capacity of connection point is determined apart from intensities of vehicles also by the distance of end of wedge Lkk, that defines how big critical time gap tg a driver is willing to accept. If the bypass is full of waiting vehicles, it is necessary to design the connecting lane or separate lane at the connection point of connecting branch.

The research is aimed at suggesting the complex method of determining the capacity of the bypass as a whole and adding this topic to the rules related to the capacity of roundabouts.

Acknowledgements
This contribution was worked out within the grant Students's grant competition CUT n. SGS14/043/OHK/1T/11.

References

[1] Bartoš L et al 2011 *Posuzování kapacity okružních křižovatek* (EDIP s.r.o.) p 56
[2] Bartoš L et al 2007 *Posuzování kapacity neřízených úrovňových křižovatek, Technické podmínky* (EDIP s.r.o.) p 64
[3] Hála M 2006 Odhadování kritických odstupů na neřízených křižovatkách metodou maximální věrohodnosti *Udržitelná výstavba 2* (ČVUT v Praze) pp 187-96
[4] Lohse D and Schnabel W 2011 *Grundlagen der Straßenverkehrstechnik und der Verkehrsplanung* (transpress VEB Verlag für Verkehrswessen, Berlin)
[5] Slabý P and Ivan Sedlačik 2013 Kapacita spojovacích větví okružních křižovatek *Mariánské Lázně 2013 - Navrhování a posuzování křižovatek na pozemních komunikacích* pp 26-9
[6] Český normalizační institut (ČNI) 2007 Projektování křižovatek na pozemních komunikacích + Změna Z1 *(ČSN 73 6102)*
[7] Forschungsgesellschaft für Straßen- und Verkehrswesen e. V. (FSVG) 2009 *Handbuch für die Bemessung von Straßenverkehrsanlagen (HBS)* p 375
[8] Ministerstvo dopravy - OPK 2005 Technické podmínky: Projektování okružních křižovatek na silnicích a místních komunikacích *(TP 135)*

Influence of roundabout capacity enhancement on emission production

Andrea Kocianova[1]**, Marek Drliciak**[1] **and Eva Pitlova**[1]

[1]Department of Highway Engineering, Faculty of Civil Engineering, University of Zilina, Univerzitna 8215/1, Zilina, 010 26, Slovakia

E-mail: eva.pitlova@fstav.uniza.sk

Abstract. Secondary effects of intersections insufficient capacity in urban areas are negative impacts on environment out of acceleration and deceleration of vehicles moving in long queues. The positive influence of increased intersection performance to reduce delays and queues, as well as negative impacts on the atmosphere is presented in this paper. The case study includes two single-lane roundabouts located close to each other in Žilina. Both roundabouts do not comply with the current traffic loads. This results in long queues and delays lasting not just during the peak hours. The solution to this problem is a new type of roundabout – turbo-roundabout. Capacity characteristics of both the current and new state are determined by microsimulation using PTV Vissim software. Obtained main characteristics of traffic flows are used as inputs to establish emission productions of NOx, CO and HC at the roundabout entries. The paper shows that proposed basic turbo-roundabout provides significant higher capacity performance compared with current state. Waiting times and queue lengths decrease about ten times. Due to this reduction, emission productions decrease about 50-60%.

1. Introduction

Lack of intersection capacity performance causes the overflow waiting times and long queues. This traffic state results in increase of emission production. One of the fundamental solutions is new design of intersections with highest capacity performance. The increase of speed and decrease of density at intersection entries reduced the acceleration and deceleration effects which cause production of emission.

The microsimulation allows evaluate the traffic flow characteristics second by second on base of new design elements. Software PTV Vissim could log several indicators (e.g. speed, density and volumes) from lane segments. Mentioned traffic flow characteristics and known emission factors are used in emission calculation procedure. The data reflect the changes of traffic flow characteristics depending on new design. The comparison of output data shows the positive or negative impacts to emission productions.

The presented case study compares the changes resulting from new layout design of roundabout in Žilina town. The positive influence of increased intersection performance to reduce delays and queues, as well as negative impacts on the atmosphere is presented in this paper.

2. The current and new state

The case study includes two single-lane roundabouts RA 1 and RA 2 located in mutual distance about 160 m (Fig. 1). Traffic problems with queues, long travel times and low speeds daily occur due to the high traffic volumes and un-sufficient capacity of these roundabouts. Additional problem is with two lanes of four-lane road which are merging into the single-lane entry of roundabout RA 1 (entry A). It results in the long delays and queues with length 600 m lasting not just during rush hour (Figure 1).

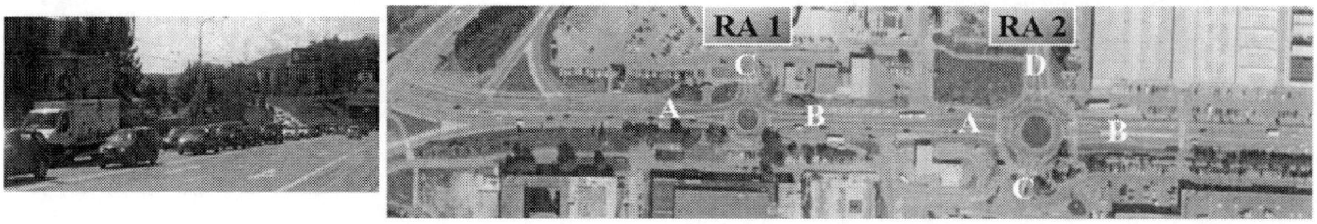

Figure 1. Existing roundabouts RA 1 and RA 2 and queues during rush hour at the entry A of RA 1.

Actual directional distribution of traffic flows at all roundabout entries were ascertained by traffic survey during common working day and it is shown in Figure 2a for the peak hour.

There are plans to cancel the first roundabout RA 1 and then rebuild the existing road to four-lane road along the whole length. There will be designed an uncontrolled intersection instead of RA 1 with main four-lane road, where all left turns will be canceled. The left turn traffic flow from the main entry A will be carried by U-turn around the second roundabout RA 2. The left turn traffic flow from the minor entry C will be carried out of this intersection and solved area. All these factors and also an increase of traffic flow due to unmet transport need at the entry A of RA 1 were considered to set a new traffic flow distribution – see Figure 2b.

There were investigated three scenarios of the case study (see Figure 2):

- Scenario 1 – current state of two existing single-lane roundabouts,
- Scenario 2 – cancelation of the first roundabout and new traffic flow redistribution at second roundabout while the geometry layout of RA 2 maintain,
- Scenario 3 - cancelation of the first roundabout and new traffic flow redistribution at second roundabout that is reconstructed into the basic turbo-roundabout TRA 2.

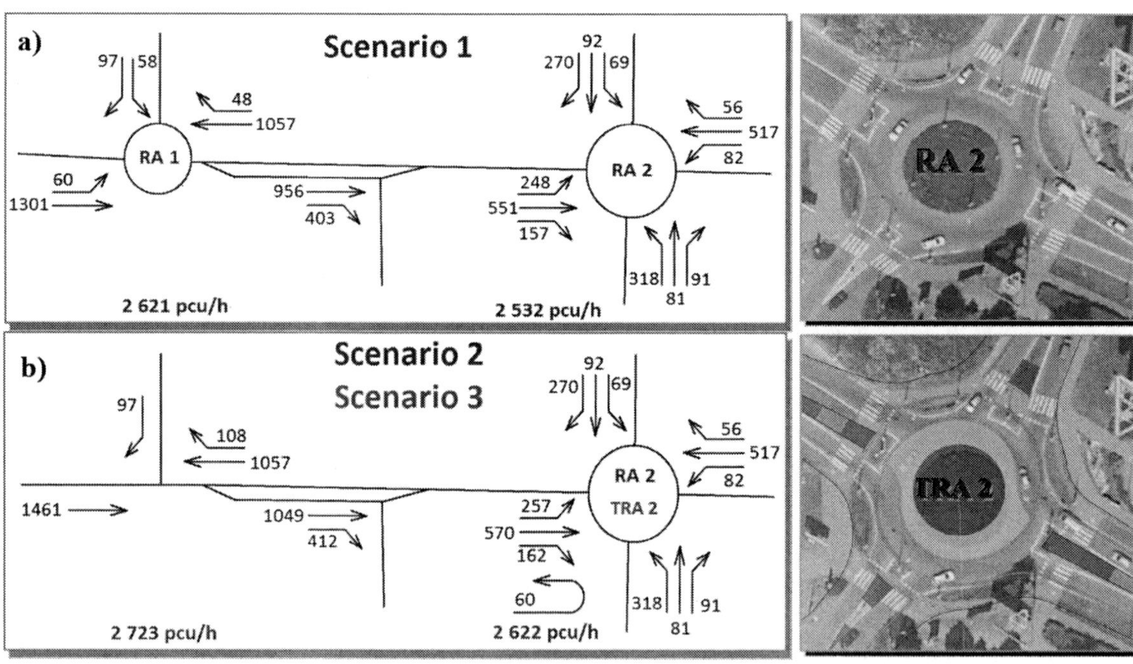

Figure 2. Three investigated scenarios and traffic flow redistributions during the peak hour.

3. Capacity characteristics of investigated scenarios

There were modeled all three scenarios of the case study mentioned above using PTV Vissim software. To set and calibrate simulation of current state Scenario 1 to correspond to the real condition of roundabout, complex field data from survey such as speeds, entry capacities, delays and queue lengths were used.

Values of waiting times (w) and queue lengths (N95) calculated according to the Slovak technical regulation TP16/2015 [2] and obtained from microsimulation are listed for Scenario1 in Table 1. The calibrated current state settings of Scenario 1 were applied at Scenario 2 and Scenario 3 that differ with traffic flow distribution and geometry layout. Capacity characteristic results of Scenario 2 and Scenario 3 calculated according to the Slovak technical regulations TP16/2015 [2] and TP14/2015 [1] and also determined by microsimulations are presented in Tab. 2. Microsimulation time period for evaluation of three scenarios was 3600 seconds; however simulation period was set 4200 seconds with warm up period 600 seconds. The values of microsimulation results are average values of 100 simulation runs with Random seed increment 1 [3].

Table 1. Capacity characteristics of existing roundabouts – Scenario 1.

	RA 1					Scenario 1 - RA 2				
	TP 16/2015			Vissim		TP 16/2015			Vissim	
Entry	v/c (-)	w (s)	N95 (m)	w (s)	N95 (m)	v/c (-)	w (s)	N95 (m)	w (s)	N95 (m)
A	1.15	286.5	633.6	273.6	468.7	0.94	43.9	153.4	44.6	166.2
B	0.90	27.5	125.6	13.0	59.7	1.03	134.0	214.7	138.3	186.8
C	0.39	14.6	11.2	8.3	21.4	0.93	66.0	114.3	81.5	126.2
D	-	-	-	-	-	0.85	42.7	76.4	54.7	83.7

Table 2. Capacity characteristics of existing roundabout RA 2 (Scenario 2) and designed TRA 2 (Scenario 3) with new traffic flow distributions.

	Scenario 2 - RA 2					Scenario 3 - TRA 2					
	TP 16/2015			Vissim			TP 14/2015			Vissim	
Entry	v/c (-)	w (s)	N95 (m)	w (s)	N95 (m)	Lane	v/c (-)	w (s)	N95 (m)	w (s)	N95 (m)
A	1.03	114.2	285.1	133.7	303.9	Left	0.47	6.1	16	5.3	26.2
						Right	0.49	6.6	17	4.9	26.8
B	1.12	267.4	319.2	273.4	253.4	Left	0.45	9.1	15	24.6	25.0
						Right	0.48	10.2	17	3.5	27.2
C	1.04	169.4	193.8	151.5	208.0	Left	0.65	16.5	32	10.1	29.8
						Right	0.10	4.4	2	0.3	0.6
D	0.92	71.1	107.8	62.0	92.2	Left	0.36	9.5	10	4.2	7.4
						Right	0.30	7.0	7	1.6	7.0

Comparison of the roundabout RA 2 performance in three scenarios is presented due to its entries degree of saturation (v/c) in Table 1 and Table 2. Scenario 1 entries degree of saturation approach or even rich value 1.0 which determine Level of Service (LOS) evaluated as F – un-sufficient entry capacity to corresponding volume. Waiting times and queue lengths as capacity characteristics are related to the capacity performance of roundabout and the comparison of their values obtained from microsimulation for three scenarios is shown in Figure 3. It is obvious that new traffic flow redistribution at RA 2 of Scenario 2 affects degree of saturation and therefore both capacity characteristics. The degree of saturation at entries A, B and C overlap value 1.0 resulting in LOS F. Due to that the waiting times arise two even three times and the queue lengths arise about 100 m at all entries except of the entry D. Solution of Scenario 2 capacity problem is Scenario 3 with a new geometry layout of basic turbo-roundabout that provide significant higher capacity performance. In this scenario degree of saturation is lower than 1 with high capacity reserve and waiting times and queue lengths decrease about ten times at all entries.

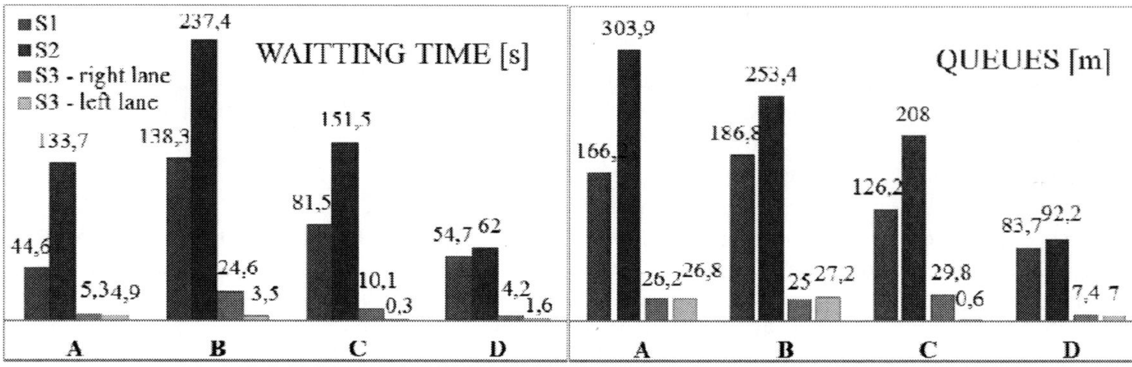

Figure 3. Microsimulation results of three scenarios capacity characteristics.

Since this paper is focus on environmental impact of un-sufficient roundabout capacity, microsimulation models of three scenarios were used to determine not only capacity characteristics but also traffic volumes, speeds and densities, necessary to calculate emission productions.

Density is defined as the number of vehicles per unit length of the roadway and it is related to speed and flow. It reflects traffic conditions on a road network. Occurred queues and long delays cause reduction of speed and increase of density at roundabout entries. Differences between each scenario densities are shown in Figure 4.

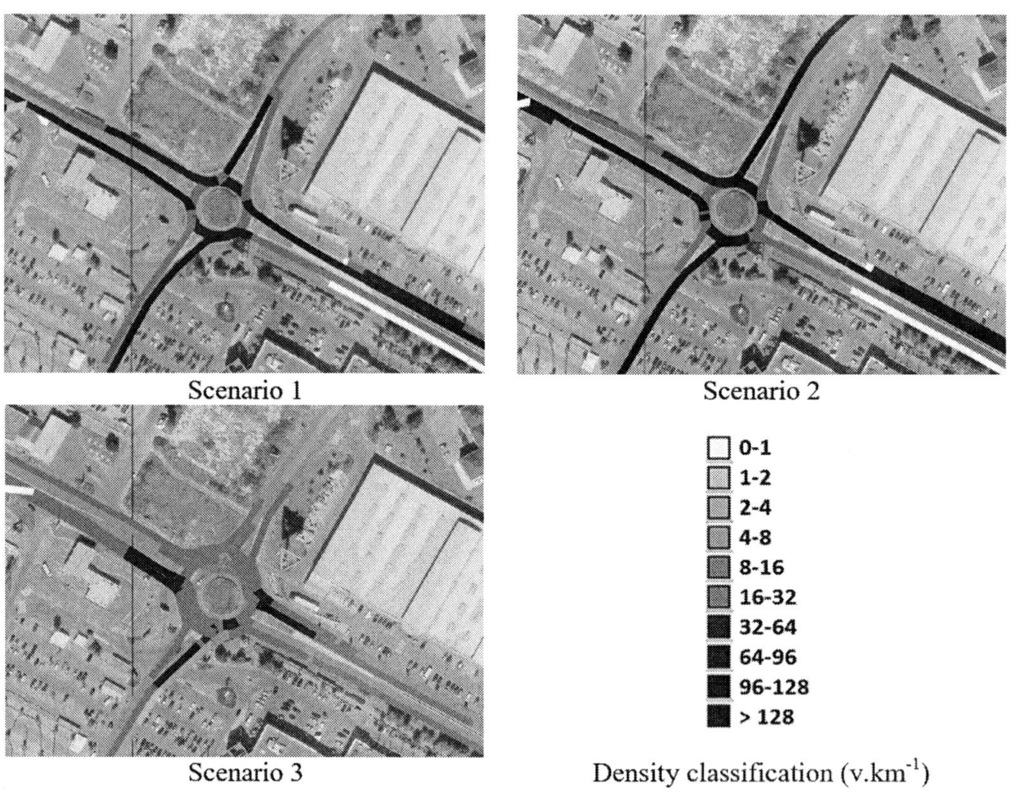

Figure 4. Results of three scenarios densities.

4. Emissions

While capacity and congestion have historically been the major motivating factors of transportation management, new performance metrics have recently garnered attention. In addition to standard traffic metrics, now there is a strong interest in traffic-related emissions in terms of pollutants (e.g. carbon monoxide (CO), hydrocarbons (HC), nitrides of oxygen (NOx), and particulate matter) [4].

It is generally difficult to measure emission from specific mobile sources under real-world conditions due to the mixing and dispersion of emissions as well as contributions from additional sources such as other vehicles, nearby factories and secondary pollutants. These issues are further influenced by environmental factors such as meteorological conditions and topographic features. Some methods of emission measurements, such as tunnel studies, remote sensing, and portable emission monitoring systems, address these problems to some extent. However, these methods, either do not isolate emissions from specific vehicles, or are very limited in testing locations and cannot be practically applied to large sets of vehicles.

One of the ways to quantify traffic-related emissions is the use of emission models. These emissions models are implemented in several softwares which are intended to model in different area scale [5]:

- Macro – e.g. VISUM, CUBE (base on average speed)
- Macro – mesco – e.g. SATURN (base on running speed – three-mode elemental)
- Micro – mesco – e.g. SIDRA INTERSECTION (base on four-mode elemental)
- Micro – e.g. VISSIM, AIMSUN, SIDRA TRIP (base on second-by-second characteristics)

The general formula for emissions calculating is the following:

$$Emission = Traffic\ Activity\ (vkm)\ x\ Emission\ Factor\ (g/vkm) \quad (1)$$

Emission factors are commonly associated with average speed, and researchers often use average speed as a traffic performance measure. It is possible to create these speed-based emission factors (in terms of grams per kilometer) by simply taking the accumulated vehicle activity database and running each individual trip through the emission models. What results is a single emissions value associated with the average speed of the trip.

The used traffic activity characteristics in the case study were evaluated from microscopic model in the PTV Vissim. We can split the trip into smaller segments representing the time spent on freeways, major surface streets, and residential streets. We can then associate an emissions value with the average speed of that particular lane segments on entry lanes of roundabout [3]. Example of speed curve at the entry A is shown in Fig. 5. Low speed in Scenario 2 is due to a long queue at the entry A and a sharp decline in the speed is caused by merging of two lanes into one entry lane 80 m before the single-lane roundabout. A relatively high vehicle speed in Scenario 3 starts to decrease about 130 m before the turbo-roundabout.

In investigated scenarios was used Polutation-emission procedure of PTV Vision software which calculates pollutants CO, HC and NOx, without particulate matter (PM10, PM2.5). This procedure is based on emission factors (*EF*) issued by the Swiss Federal Office for the Environment. For each pollutant, a regression curve with polynomes to the 5th degree is used of the form [3]:

$$EF = a + b.x + c.x^2 + d.x^3 + d.x^4 + f.x^5 \quad (2)$$

Where *a*, *b*, *c*, *d*, *e* and *f* are parameters of the polynome determined separately for different pollutant and *x* is speed in km/h.

The basic speed-emission factors used in the case study are included in emission module of PTV Vision and listed in Table 3. Example of emission factor of CO related to the speed is shown in Fig. 6.

The results of HC, NOx and CO emissions calculated for all scenarios by emission factor and traffic activity are shown in Fig. 7. The mutual comparison is in Table 4.

Table 3. Basic speed-emission factor coefficients of PKW for HC, NOx and CO emissions.

PKW	a	b	c	d	e	f
HC	0.4361	-1.34E-02	1.86E-04	-1.22E-06	3.97E-09	-5.66E-12
NOx	0.1897	5.58E-03	-1.93E-04	2.57E-06	-6.56E-09	-2.07E-11
CO	3.2587	-8.05E-02	6.73E-04	-1.21E-06	-6.39E-09	3.77E-11

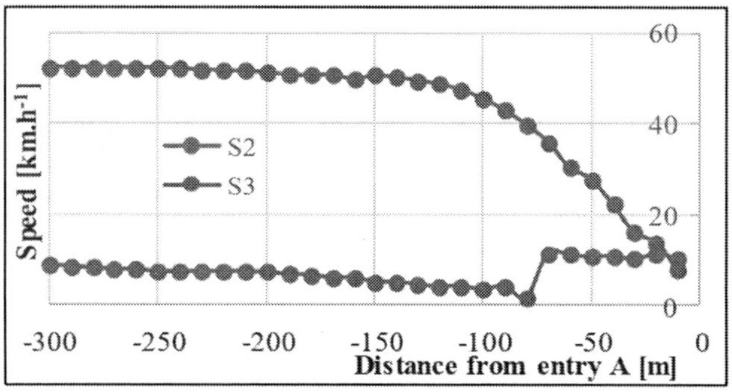

Figure 5. Speed curve of scenario 2 and 3 at entry A.

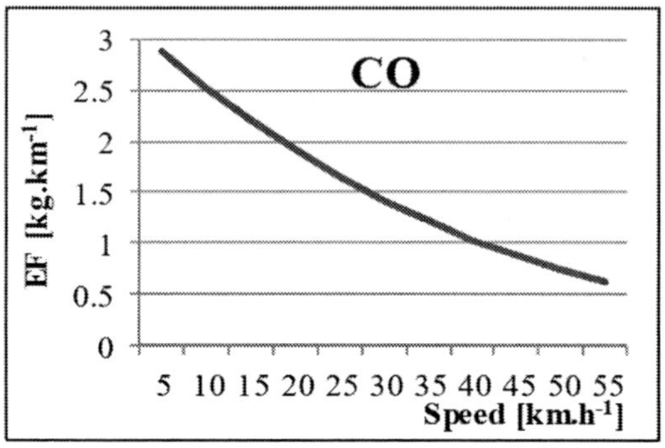

Figure 6. Dependence of emission factor of CO on speed.

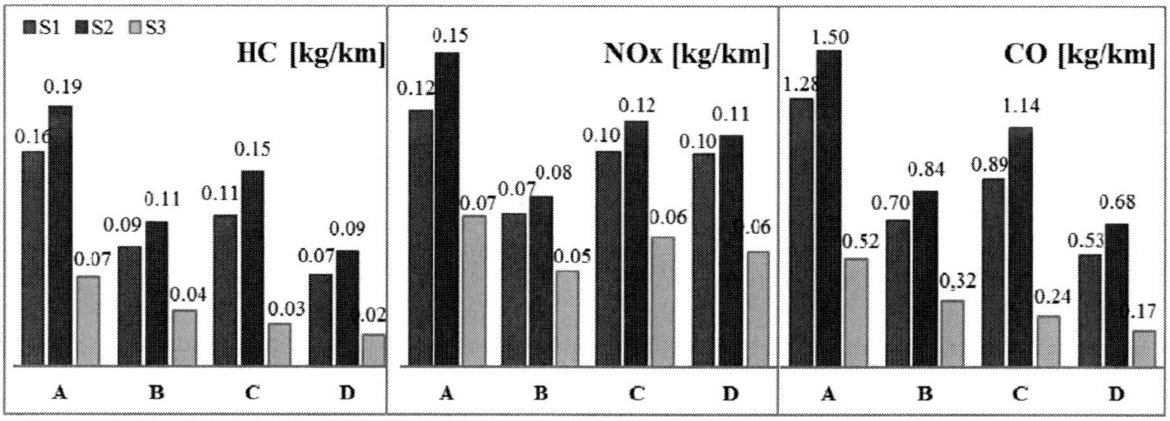

Figure 7. The HC, NOx and CO calculated emissions of three scenarios.

Table 4. Variation of emissions by investigated scenario.

Entry	Scenario comparison	HC	NOx	CO
	S1 (kg.km^{-1})	0.16	0.12	1.28
A	S2 – S1 (%)	+ 17.8	+ 18.4	+ 15.0
	S3 – S2 (%)	- 65.5	- 51.9	- 65.7
	S1 (kg.km^{-1})	0.09	0.07	0.70
B	S2 – S1 (%)	+ 17.5	+ 10.1	+ 16.6
	S3 – S2 (%)	- 61.3	- 44.1	- 62.1
	S1 (kg.km^{-1})	0.11	0.10	0.89
C	S2 – S1 (%)	+ 23.2	+ 12.2	+ 21.8
	S3 – S2 (%)	- 78.5	- 47.0	- 79.0
	S1 (kg.km^{-1})	0.07	0.10	0.53
D	**S2 – S1** (%)	+ 20.3	+ 7.7	+ 21.7
	S3 – S2 (%)	- 73.1	- 50.0	- 74.6

The single-lane roundabout with the new traffic flow redistribution (Scenario 2) causes about 20% increase of emissions compared to Scenario 1. The highest difference was monitored at the entry C with 23.2% increase of HC emissions; contrariwise the smallest difference was monitored at the entry D with 7.7% increase of NOx emissions.

The comparison of Scenario 2 and Scenario3 determine the 60% - 70% decrease of HC and CO emissions. The NOx emissions decrease with 40% -50% at roundabout entries. The highest difference was monitored at the entry C with - 78.5% decrease of HC emissions; contrariwise the smallest difference was monitored at the entry D with - 44.1% decreases of NOx emissions.

5. Conclusions

Estimation of emissions for evaluating traffic conditions is useful for sustainable environmental assessment in traffic design, operations and planning. The modern trends of road design content as capacity issues as well as environmental impacts.

Turbo-roundabouts improve the road network performance. The reductions of waiting times and queues at entries cause the increase of speed and decrease of density. Mentioned positive impact reflects the decrease of emissions.

The presented case study point out the new design of turbo-roundabouts providing higher capacity performance that results in reduction of emissions. The current single-lane roundabout considering the new traffic flow redistribution that is planned in future has the degree of saturation that overlap value 1 resulting in LOS F at entries A, B and C. Due to that waiting times and queue lengths increase at all entries. The change of layout geometry causes the decrease of queues and waiting times about ten times with the same traffic conditions. Also the single-lane roundabout causes the 20% increase of emissions in comparison with the turbo-roundabout with the same traffic flow redistribution that reduces HC and CO by 60% - 70% and NOx values are reduce by 40-50%.

While capacity and congestion are still strong motivating factors of transportation management, emissions have recently garnered attention because they are related to capacity performance of road network as it is shown in this case study. In future Department of Highway Engineering will monitor not only gas particles but also the solid particles [6]. It is necessary to take in consideration also this factor in planning and operating changes in road network.

Acknowledgements

This work was supported by the project "Centre of excellence for systems and services of intelligent transport II" ITMS 26220120050 co-funded from European Regional Development Fund.

References

[1] Ministry of Transport, Construction and Regional Development of the Slovac Republic 2015 *Technical regulations TP 14/2015. Design of Turbo-roundabouts (in Slovak)* http://www.ssc.sk/files/documents/technicke-predpisy/tp2015/tp-14_2015.pdf

[2] Ministry of Transport, Construction and Regional Development of the Slovac Republic 2015 *Technical regulations TP 16/2015. Roads Capacity Calculation (in Slovak).* http://www.ssc.sk/files/documents/technicke-predpisy/tp2015/tp-16_2015.pdf

[3] Planung Transport Verkehr AG - PTV *Visum Installation Manual*

[4] Morris B, Tran C, Scora G and Barth M 2012 Real-Time Video-Based Traffic Measurement and Visualization System for Energy/Emissions *IEEE Transactions On Intelligent Transportation Systems* **13**(4) pp 1667-78

[5] Akçelik R, Smit R and Besley M 2012 Calibrating fuel consumption and emission models for modern vehicles *IPENZ Transportation Group Conference (Rotorua) p 13*

[6] Jandacka D, Durcanska D and Bujdos M 2017 The contribution of road traffic to particulate mater and metals in air pollution in the vinicity of an urban road *Transportation Research. Part D: Transport and Environment* **50**(01) pp 397-408

Road structural elements temperature trends diagnostics using sensory system of own design

Juraj Dudak[1], Gabriel Gaspar[2], Stefan Sedivy[1], Lubomir Pepucha[1] and Zuzana Florkova[1]

[1]Research Center, University of Zilina, Univerzitna 8215/1, 010 26 Zilina, Slovakia
[2]TNtech s.r.o., Rybarska 20, 911 01 Trencin, Slovakia

E-mail: juraj.dudak@rc.uniza.sk

Abstract. A considerable funds is spent for the roads maintenance in large areas during the winter. The road maintenance is significantly affected by the temperature change of the road structure. In remote locations may occur a situation, when it is not clear whether the sanding is actually needed because the lack of information on road conditions. In these cases, the actual road conditions are investigated by a personal inspection or by sending out a gritting vehicle. Here, however, is a risk of unnecessary trip the sanding vehicle. This situation is economically and environmentally unfavorable. The proposed system solves the problem of measuring the temperature profile of the road and the utilization of the predictive model to determine the future development trend of temperature. The system was technically designed as a set of sensors to monitor environmental values such as the temperature of the road, ambient temperature, relative air humidity, solar radiation and atmospheric pressure at the measuring point. An important part of the proposal is prediction model which based on the inputs from sensors and historical measurements can, with some probability, predict temperature trends at the measuring point. The proposed system addresses the economic and environmental aspects of winter road maintenance.

1. Introduction

During winter, many roads located in cold or temperate regions are influenced by severe climatic conditions associated with snow and ice. These conditions have serious consequences on driving conditions, reducing the traffic flow dramatically, because the failure in maintaining roads in winter often leads to road closures. Ice and snow also increase the risk of accidents [1]. In this context, forecasting of road surface and traffic conditions is an important aspect of traffic safety and winter road maintenance. The weather conditions can change quickly, for example, with the onset of snowfall or during rapid temperature variations. A prior knowledge of road weather is important from a public road safety standpoint. Proper consideration of upcoming weather events also helps the road maintenance authorities to attend the roads in an effective and economical manner [2].

Several strategies can be used for winter road maintenance. Among these, the most common are de-icing, in which chemicals are used to melt ice and snow, and anti-icing, a preventive measure that reduces ice by hindering bonds between ice crystals and road pavement or spreading sand to help provide traction [3,4]. The most commonly used de-icing chemical is salt (sodium chloride, NaCl), which usually comes from mined rock salt. It is one of the most cost effective, readily available and efficient chemicals for the prevention and removal of ice and snow and is supplied in a variety of granule size and is sometimes applied in solution. Traditionally this has been spread in dry form; known as dry salt. Another two variants are used; with the addition of some moisture either at manufacture at point of spread resulting in the pre-wetting technique, and the preparation of a brine (salt solution) that is spread as a liquid. The impact of using chemicals for winter road maintenance is a major environmental concern [4]. Studies [5,6] show that soils, vegetation, water, highway structures and vehicles are all affected, and so it is crucial that the de-icing chemicals are used wisely. The growing awareness of the environmental impacts of salt has led efforts to find ways to reduce the amounts entering the environment. There are really only two ways to achieve this. Using something other than salt or optimizing salt usage by applying it strategically [4]. Hence, there is a continual need for improvement in winter road maintenance strategy. One key component of optimized road maintenance decisions is to obtain and use accurate weather information. Weather information can be divided into two temporal categories: observations, which reflect current conditions; and forecasts, which predict future conditions [7]. The largest potential savings to be made in winter maintenance focus upon the prediction of ice formation and snow, and countries prone to icy roads can make significant annual savings [8] by optimizing weather forecasts with salt usage and other operational costs [4].

Thus, it stands to reason, it is necessary to have relevant data about the weather and road surface conditions. They can be obtained from various sources. The most important source of real time meteorological and pavement data are road weather stations (RWS) ensuring automatic data collection and transmission. Currently a wide range technologies of road weather stations are commercially available (e.g. Viasala, Lufft, CrossMet etc.) They are an essential tool for remote monitoring of weather conditions on the roads. The RWS are equipped with variable set of sensors for measuring of meteorological date. They are also equipped with road sensors connected to the station for measuring of road conditions [9]. Road surface conditions are influenced by numerous meteorological, geographical and road parameters, which can produce vast temperature variations across the road section. The most important factors are: air temperature, radiation fluxes, humidity, precipitation amount, wind, topography, properties of the road materials and traffic [10,11]. Their influence on the road is presented in detail in literature [4,12,13].

These RWS are a part of advanced systems known as the road weather information systems (RWIS) and maintenance decision support systems (MDSS). RWIS is a software tool which collects all relevant available data which can be used by dispatchers to support decision making in winter road maintenance management. The RWIS integrates current data from road weather stations, meteorological radars and satellites, cameras, mobile road weather stations, the National Weather Service (NWS) and others. The RWS ensure recording of current conditions in selected locations, but the RWIS can include forecasting module, which will allow for each RWS to generate a point forecast of important parameters in response to the measured data and general weather forecast. General weather forecasts are generally supplied by the NWS, alternatively by a commercial provider. In addition, there are specialized computational models for specialized forecast for road surface. The specialized road weather forecast model is a part of dispatcher maintenance decision support system (MDSS). MDSS as a part of RWIS is focused on special linear weather and road surface status forecast and recommendations on winter maintenance [9]. More information about RWIS and MDSS are presented in [4,9,15]. Various models for predicting meteorological conditions on the road have been developed (e.g METRo [14], RoadSurf [2], SSW model [15] etc.). Based on forecast road condition information obtained be these models, the highway engineer can determine the window of application for anti- and de-icing chemicals to be applied and for snow removal operations to commence. By strategically timing operations the safety of highways can be maintained whilst going some way to optimising the quantity of material applied to the road surface [4].

In this paper is presented the design and implementation of an experimental system in which the desired value of the environment such as temperature of the road are obtained by the direct method, i.e. sensors are placed directly into the road surface. In view of this data acquisition method, the model is suitable for use in a local areas. The data are more accurate than measuring the air temperature using the weather station located near the road.

2. Experimental installation

For the verification of model described in the next section experimental installation of temperature sensors buried in the road surface was created. The sensors are divided into two groups: horizontal layer located in the road surface, and the vertical layer, where sensors are arranged in the vertical direction with a pitch of 20 cm, to the depth of 1 meter. Figure 1 illustrates the location of temperature sensors in the experimental installation. Temperature sensors were made in a special enclosure which protects them against mechanical damage during installation and subsequent use.

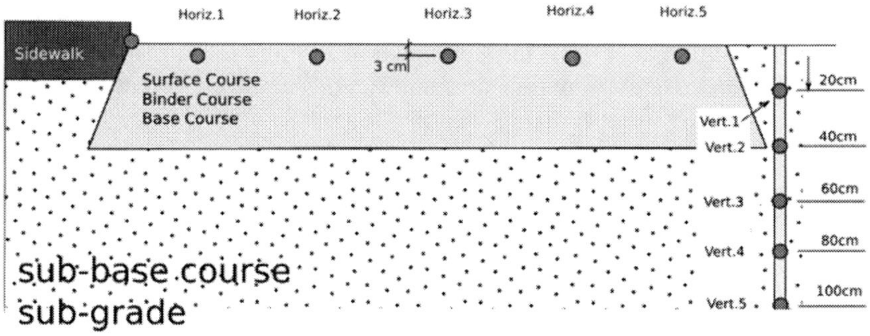

Figure 1. Design of sensors placement.

2.1. *Technical realisation*

For experimental installation following restrictions of the measurement system were defined:

- Robust mechanical enclosures of the temperature sensors to withstand the high temperature of asphalt installation and mechanical stress during the operation of the road, in which the sensors are located.
- Minimizing cabling. Since the system contains more than 10 temperature sensors, it is required to minimize the cabling.
- The temperature measuring range from -50°C to 120°C.
- Simple interface through which it is possible to view and evaluate the measurement results.

The requirement of robust mechanical design was met in the sensor enclosure (Figure 2) made of Ertalon PA66, having the following essential characteristics: melting point 255°C, thermal conductivity 0.28 W/(K.m), the short term operating temperature 200°C and long term operating temperature 80°C.

Requirement of minimizing the cabling was met using the 1-wire bus system, which needs only 2 wires to provide both power and communication with sensors. The selection of the bus is also related to the selection of sensors that are compatible with the selected 1-wire standard. As a suitable candidate which meets the requirement for the range of measured temperatures is the digital thermometer DS18B20.

For the needs of autonomous data collection has been designed and implemented system for autonomous data acquisition. The system consists of server and application parts. The server part includes an operating system service providing repeated measurements at defined time intervals, a relational database for storing measured data and interface for communication with higher-level applications. The hardware requirements for the implementation of the server are set to ensure the functionality even when using minimum hardware configuration.

Figure 2. Installation of temperature sensors in asphalt roads.

3. Evaluation of the soil vertical temperature profile measurements

The object of this section is a description of the road sub-base vertical temperature profile measurement and evaluation of the measured data. The described methods can serve for the creation of predictive models to determine the surface temperature, based on the prior development of temperature and for determination of thermal parameters of the soil. The experiment was designed to measure the horizontal and vertical temperature profiles of an asphalt road. Horizontal thermometers were placed in a line milled into the road surface and fixed using low temperature asphalt mix, vertical thermometer with five sensors was placed on the side of the road.

Data acquisition was controlled by a CNT controller connected to the Internet network and data with a measurement frequency of 1 per minute were saved in the database. Data access and visualization of measurement data continuously is possible through the web interface. Advanced version of the controller allows the connection of local weather station for monitoring air temperature using radiation thermometers at selected heights, solar radiation and wind speed and direction.

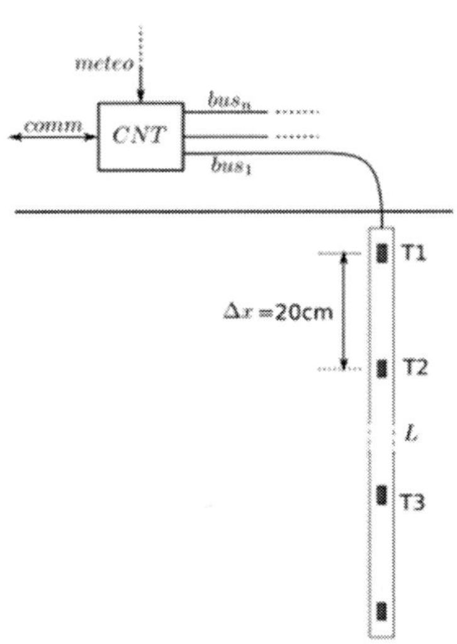

Figure 3. Measurement layout.

3.1. Data processing and results

Data processing design comes from the typical derivation of Fourier equation of heat conduction.

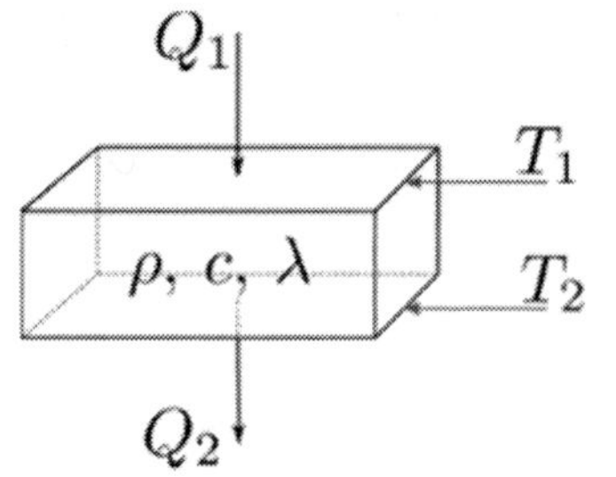

Figure 4. Heat transfer in the segment of environment.

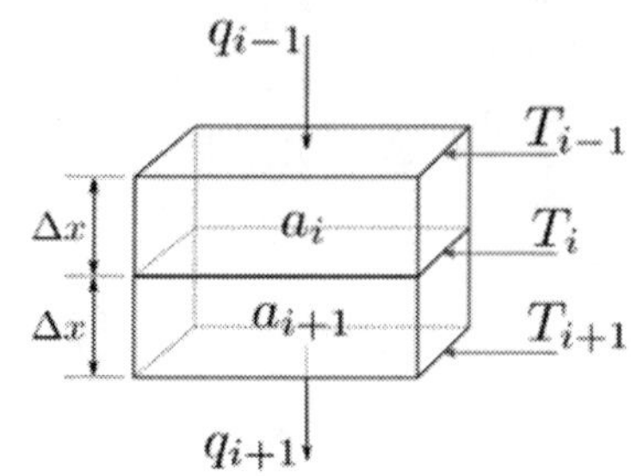

Figure 5. Heat transfer in the segment of non-linear environment.

For heat transfer are valid general equations:

$$(Q_1 - Q_2) = (q_1 - q_2)S\Delta t = -\Delta q S \Delta t \tag{1}$$
$$(Q_1 - Q_2) = mc(T_1 - T_2) = \rho S \Delta x c \Delta T$$

by comparison we get

$$-\frac{1}{\rho c}\frac{\Delta q}{\Delta x} = \frac{\Delta T}{\Delta t} \tag{2}$$

by limiting transition for $\Delta x \to 0$ and $\Delta t \to 0$ we get:

$$-\frac{1}{\rho c}\frac{dq}{dx} = \frac{dT}{dt} \tag{3}$$

Using Fourier law for heat conduction can be after its derivation, excluded immeasurable quantity q from the equation:

$$q = -\lambda\frac{dT}{dx} \quad \to \quad \frac{dq}{dx} = -\lambda\frac{d^2T}{dx^2} \tag{4}$$

by fitting we get:

$$\frac{dT}{dx} - \frac{\lambda}{\rho c}\frac{d^2T}{dx^2} = 0 \tag{5}$$

Value $a = \lambda / \rho c$ is the coefficient of thermal conductivity and expresses how "easy" are temperature differences compensated in given substance. Measurement of the vertical soil temperature profile can be considered as one-dimensional case of heat conduction. In this case, it is possible under constant environmental parameters, obtaining a solution of the equation

$$\frac{1}{a}\frac{\partial T}{\partial t} = \frac{d^2T}{dx^2} \tag{6}$$

In analytic form:

$$T(x, y) = T * er\, f\left[\frac{x}{\sqrt{4at}}\right] \tag{7}$$

The unknown quantity is the thermal diffusion coefficient a, that characterizes the environment in which is the thermal energy spread. From its definition is clear that it is depending on the material parameters of the environment which may not be constant over time.

To experimentally determine the thermal diffusion coefficient, can be used data from the vertical thermometer. For the evaluation we will develop a formula (6) using *central difference formula* into differential form, the indication referred by Figure 5 for linear environment $a_i = a_{k+1} = a$, index k is a time period:

$$T_i^{k+1} = T_i^k + (T_{i-1}^k - 2T_i^k + T_{i+1}^k)\frac{a\Delta t}{\Delta x^2} \quad (8)$$

by fitting we get:

$$T_i^{k+1} = rT_i^k + \left(1 - 2r\right)T_i^k + rT_{i+1}^k\right) \qquad r = \frac{a\Delta t}{\Delta x^2} \quad (9)$$

To verify the functionality of the measurement system, the validity of data and the experimental determination of the coefficient a, we will use a block of the measurement data of Figure 6. T1 correspond to placement Vert.1. in Figure 1 and Figure 3, T2 correspond to Vert.2 etc.

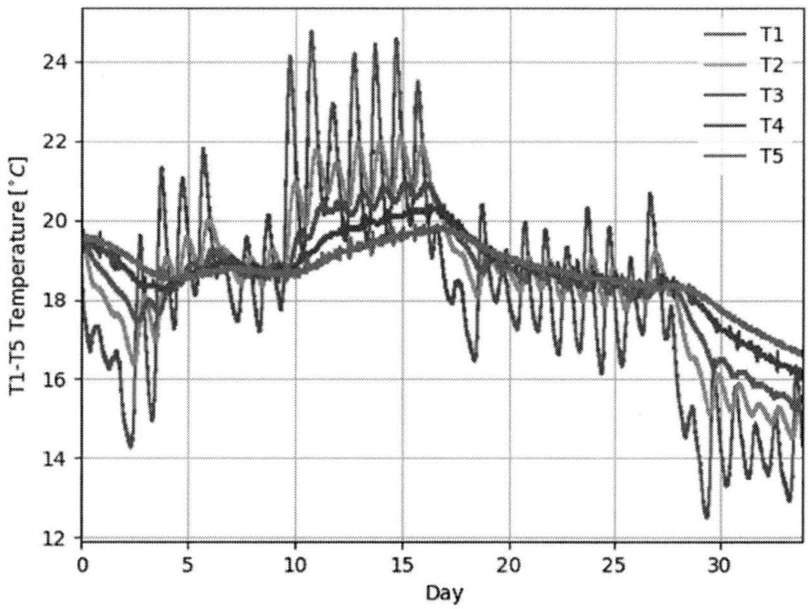

Figure 6. Temperatures chart for autumn 26.8.-29.9.2016.

From the data block we use thermometers T1, T2 and T3. By recursive calculation of the equation in Figure 5 with parameter a we calculate the value T2* and by using least square method we are looking for such a parameter value at which the deviation of calculated from the measured temperature values is minimal. Constant parameter value during the entire monitored period was assumed in the calculation. The result is shown in Figure 7, the detail in Figure 8. For Δx = 20 cm was determined as the optimal value of the thermal diffusion coefficient

$$a = 2.1 * 10^{-7}[m^2s^{-1}], \qquad \Delta x = 20cm.$$

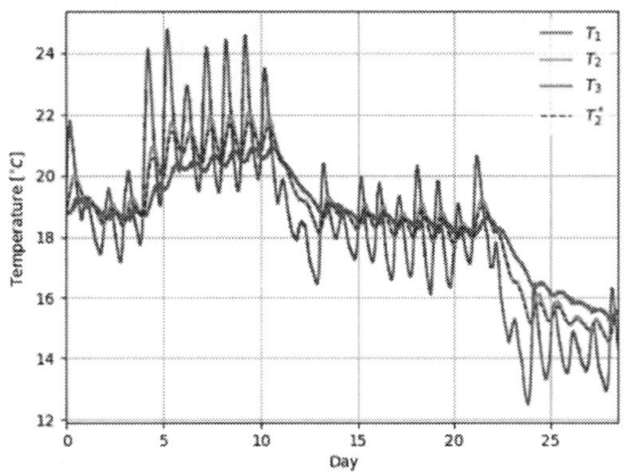

Figure 7. Coefficient a calculation and T2* value according formula (9).

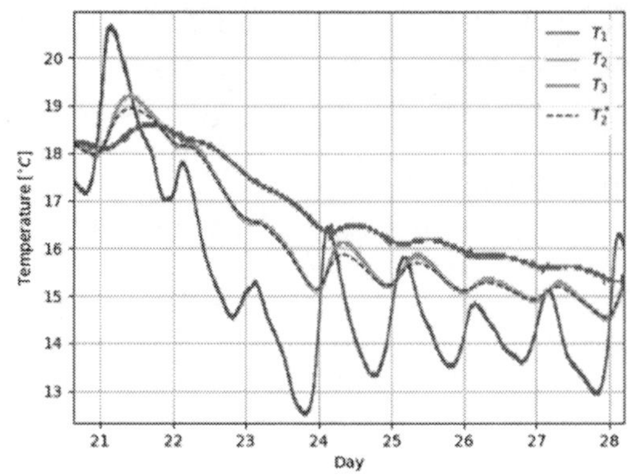

Figure 8. Coefficient a calculation in detail – day 21 - 28 from Figure 7.

Similarly was calculated thermal diffusion coefficient for winter season with temperatures course in Figure 9, the coefficient value was

$$a = 3.1 * 10^{-7}[m^2 s^{-1}], \qquad \Delta x = 20cm.$$

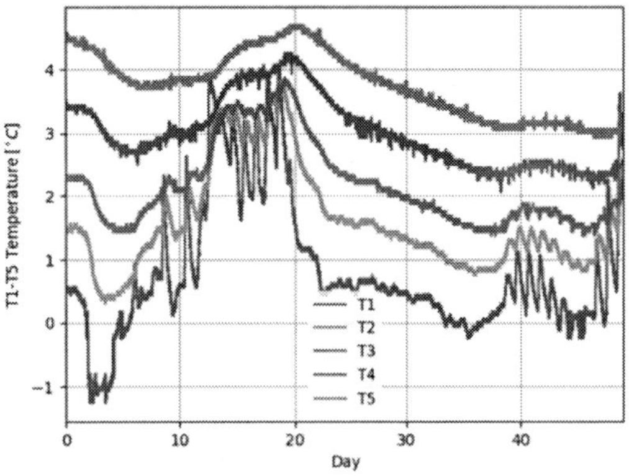

Figure 9. Measured temperatures chart for winter 5.1.-23.2.2016.

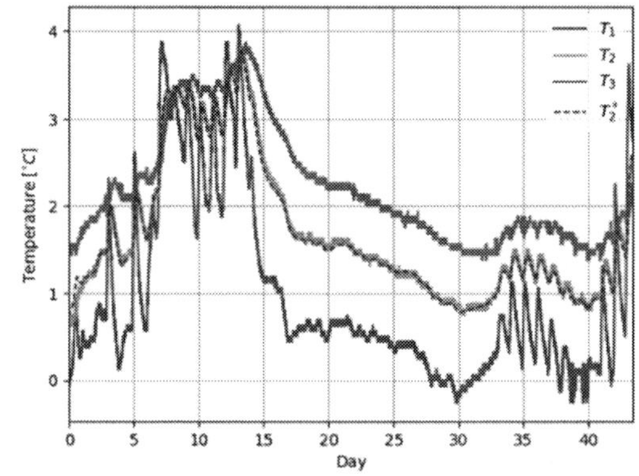

Figure 10. Coefficient a calculation and T2* value according formula (9).

In [16] are given the values of thermal diffusion coefficients for various materials. Water $1.43*10^{-7}$, sand $11.2*10^{-7}$, brick $2.7 - 5.2*10^{-7}$.

Conclusions

It is clear from the above that from measurements of the vertical temperature profile it is possible to determine the thermal diffusion coefficient, by extension of the described method to further depths. From the detailed Figure 8 is obvious that the assumption of constant coefficient is probably not correct, because after a significant change in temperature, likely accompanied by the rainfall, the coefficient changed (day 24-26). Based on fixed marginal conditions (T1, T2), calculated temperature T2* was equal to measured temperature T2 with accuracy better than 0.5°C.

A key factor which influences the thermal diffusion coefficient is the water content in the environment. Clearly, heat transfer will be described as non-stationary and non-linear problem. Knowledge of the current coefficient value is fundamental to determine the road surface temperature, since it determines the speed at which the heat from the road surface transfers to the sub-base or vice versa. Following the above, it is likewise possible that, at a suitable calibration of diffusion coefficient parameters and knowledge of the sub-base (i.e., the structure and composition) could be obtained information by measuring the dynamics of temperature changes on the amount of water and its height profile in the sub-base. Verification of this assumption needs to implement further relevant experiments and verification in real practice.

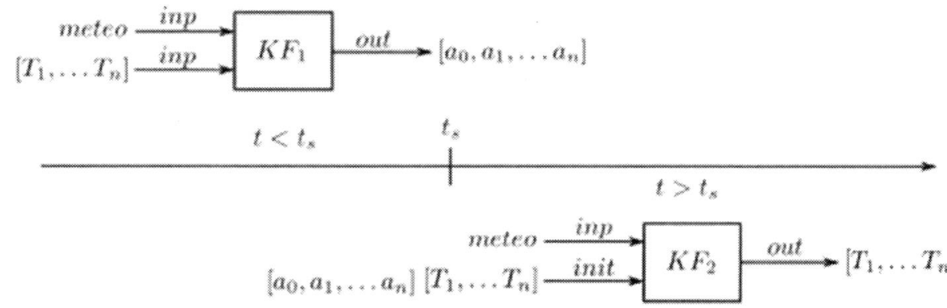

Figure 11. The structure of the predictive model of temperature changes.

Knowledge of the thermal diffusion coefficient and its dependence on the depth provides the basis for creation of the predictive model of temperature changes, the structure of the model is shown in Figure 11. The model consists of two Kalman filters. Filter KF1 calculates into current time on the basis of data flow from thermometers and weather stations (solar radiation, air temperature and wind direction and speed) vector of diffusion values. The actual value of the diffusion coefficient and temperature are used at feedforward simulation in the second filter KF2 as initial values and filter based on knowledge of the course of solar radiation and meteorological forecast estimates a vector of temperatures.

The proposed model, due to its potential can be implemented in MDSS systems and as its part can thus leads to improvment of economic and environmental aspects of road winter maintenance.

Acknowledgements
This research was supported by European project Research Centre of the University of Žilina – Second Phase: ITMS 26220220183

References

[1] Bližňák V et al 2012 The application of METRO model to the Czech road data – preliminary results *Proc. of the 16th SIRWEC conf. (Helsinki)* ID 0077
[2] Kangas M, Heikinheimo M and Hippi M 2015 RoadSurf: a modelling system for predicting road weather and road surface conditions *Meteorological applications* **22** pp 544–53
[3] Berrocal V J et al 2010 Probabilistic Weather Forecasting for Winter Road Maintenance *Journal of the American Statistical Association* **105** pp 522 – 36
[4] White S P, Thornes J E and Chapman L A 2006 *Guide to Road Weather Information Systems SIRWEC* http://www.sirwec.org/documents/rwis_web_guide.pdf
[5] Ramakrishna D M and Viraraghavan T 2005 Environmental Impact of Chemical Deicers - A Review *Water, Air, and Soil Pollution* **166** pp 49–63
[6] Dai H L, Zhang K L, Xu X L and Yu H Y 2012 Evaluation on the Effects of Deicing Chemicals on Soil and Water Environment *Procedia Environmental Sciences* **13** pp 2122-30
[7] Ye Z, Strong C, Fay L, Shi X 2009 *Cost Benefits of Weather Information for Winter Road Maintenance* (Final report Iowa Department of Transportation)
[8] INCA-CE ROAD SAFETY TEAM 2012 *Transregional strategy for the use of nowcasting information in road safety*
[9] Kociánová A 2015 The intelligent winter road maintenance management in Slovak conditions *Proc. XXIV R-S-P seminar: Theoretical foundation of civil engineering* (Procedia

Engineering) **111** pp 410-19

[10] Kršmanc R, Šajn Slak A and Demšar J 2013 Statistical approach for forecasting road surface temperature *Meteorological applications* **20** pp 439–446

[11] Chapman L and Thornes J E 2008 Small-scale road surface temperature and condition variations across a road profile *Proc. 14th Int. Road Weather Conference SIRWEC (Prague)* ID 01

[12] Perry A H and Symons L J 1991 *Highway Meteorology* (London:E & FN Spon)

[13] Kawashima S, Ishida T, Minomura M and Miwa T 2000 Relations between surface temperature and air temperature on a local scale during winter nights *Journal of Applied Meteorology* **39** pp 1570–79

[14] Linden S K and Drobot S D 2010 The Evolution of METRo in a Roadway DSS *Proc. 15th Internat. Road Weather Conference SIRWEC (Quebec)*

[15] Konečný D 2014 Advances in winter maintenance decision support in the Czech Republic *Proc. 17th International Road Weather Conference SIRWEC (Andorra)* ID 040

[16] Blumm J and Lindemann A 2003 Characterization of the thermophysical properties of molten polymers and liquids using the flash technique *High Temperatures-High Pressures* **35/36** pp 627-32

[17] Ďuďák J, Sládek I, Pepucha L and Šedivý Š 2015 The monitoring system of environmental parameters on the road network *Proc. 2th Int. Conf. on Advanced Technology & Sciences (Antalya)* pp 312-14

[18] Fabo P, Skovajsa M and Pepucha L 2016 PSE simulation environment for the real-time simulation and control of mechatronic systems *Proc. 17th Int. Conf. on Mechatronics - Mechatronika* (*Prague)* pp 1-5

Swelling soils in the road structures

Jan Pruška[1] and Miroslav Šedivý[2]

[1]Czech Technical University in Prague, Thákurova 7, 166 29 Prague, Czech Republic
[2]GeoTec-GS, a.s., Chmelová 6, 106 00 Prague, Czech Republic

E-mail: pruska@fsv.cvut.cz

Abstract. There are frequent problems with the soil swelling in the road construction in the past time. This phenomenon is known for decades. This situation is notably given by insufficient knowledge of this problem and difficulties with input parameters describing the swelling process. The paper in the first part proposed regression relations to predict swelling pressure, time of swelling and swelling strain for different initial water contents for soils and improvement soils. The relations were developed by using artificial neural network and QCExpert Professional software (on the data from site investigations by GeoTec-GS, a.s. and experimental data from CTU in Prague). The advantage of the relations is based on using the results of the basic soil tests (plasticity index, consistency index and colloidal activity) as input parameters. The authors inform the technical public with their current knowledge of the problems with the soil swelling on the motorway in the second part of the paper.

1. Introduction

The swelling is a complicated phenomenon affected by a many parameters from mineralogical composition and physically-chemical properties over basic physical property to the influence of the existing surcharge pressure and the type of used material (lime or cement etc.) in the case of soil stabilization.

The conditions needed for a swelling-susceptible soil to swell are the following:

- Stress relief in the soil body (this is always fulfilled for cut slopes) and/or
- water availability in the environment. This may or may not be fulfilled depending on circumstances.

The swelling is tested under laboratory conditions on undisturbed soil samples, usually only linear swelling is tested. Alternative procedure is to measure deformation in time in the oedometer while also measuring the swelling pressure. However, testing in stand with dynamometers where both swelling pressure and deformation are measured in time seems to be the best way to test swelling. The problem is that this test gives the values of the swelling pressure for the given initial state of the soil. But we also have to be interested in the swelling pressure for lower initial water content, because we cannot guarantee that such conditions will not occur in reality. When we focus on the swelling significance in transport engineering and its complexness (influence cut slope stability, destructive effects on pavements, the correct way of soil improvement etc.) the problems with test are clear. The duration of the test is between 160 and 210 hours. Therefore, the laboratory has to be sufficiently equipped for this type of time demanding testing and the cost of tests is also significant. If there are any criteria for swelling in technical standards, it is only for linear swelling. But these values do not give any information about the swelling pressure (deformation is only an accompanying effect). It is shown that the swelling pressure varies with different initial conditions while the deformation may still be the same 3%.

2. Predicting the swelling properties

The swelling is a complicated phenomenon and prediction of the swelling parameters cannot take the account of all effects (especially mineralogical). The direct methods provide physical measurements of swelling pressure and are not suitable for quick identification of expansive soils [1]. For the example testing in stand with dynamometers where both swelling pressure and deformation are measured in time. Because of simplicity usually only linear swelling is tested and so the relative deformation is obtained only. The problem is that this test gives the values of the swelling pressure for the given initial state of the soil (i.e. initial water content). The variations of the shear parameters of soils at swelling with different initial moisture are evident in the Table 1 - there are larger differences in the effective shear parameters for lower initial (immediate) water content (wL = 55%, wn = 18%, Ip = 37%, Ia = 1.61). In the Table 2 it is shown that the swelling pressure varies with different initial conditions (the initial moisture is significant) while the deformation may still be the same 3%.

The indirect methods are based on experience and basic soil parameters and produce quick and reliable identification of expansive soil. There are some empirical equations reported by various author in the literature - summary of some criteria for identifying swell potential is e.g. in [2]. One of the oldest is that of [3]: $S = 3.6x10^5A^{2.44}C^{3.44}$ where S is the swell percent value under 1 psi (6.89 kPa), C is the percent of clay fraction and $A = I_P/C$ where I_P is plasticity index. The other widely used equations were developed e.g. by [4] or [5].

Table 1. Selected geotechnical parameters for the different moisture.

	Immediate moisture w_n (%)		
	12.0	23.0	30.0
Consistency index I_C	1.16	0.86	0.68
Swelling pressure σb	480	285	185
Relative deformation εb (%)	5.50	2.48	1.75
Peak effective friction angle $\phi_{ef,p}$ (°)	21.8	20.1	18.8
Peak effective friction angle at swelling $\phi_{ef,p,s}$ (°)	17.5	18.9	18.0
Difference at friction angle (%)	20.0	6.0	4.2
peak cohesion c_{peak} (kPa)	29.6	23.8	19.9
peak cohesion at swelling $c_{p,l}$ (kPa)	16.3	20.2	17.6
Difference at cohesion (%)	45.0	15.0	11.5

To investigate the relationship between the soil index properties (plasticity index I_P, consistency index I_C, colloidal activity index I_a) and the swelling pressure, time of swelling, and swelling strain it was performed analysis using regression on the large data sets from geological investigation of the many constructions in the Czech and Slovak Republic (tuffaceous clays, mudstones, claystones, chalk marlites and neogene clays). The best equations for swelling pressure, time of swelling and swelling strain are the follows:

- Predicted swelling pressure of the soil:

$$\sigma_{swel} = 0.0552 \cdot I_P^{2.385} \cdot I_C^{1.757} \cdot I_a^{0.397} \quad \text{[kPa]} \tag{1}$$

where: I_P - plasticity index [%], I_C - consistency index [-] and I_a -. colloidal activity index [-].

Note: the higher the water content of the soil (higher the consistency index for given plasticity index and colloidal activity index), the higher the swelling pressure.

- Predicted swelling strain:

$$\varepsilon_{swel} = 0.3 \cdot I_P^{0.2} \cdot I_C^{0.3} \cdot \sigma_{swel}^{0.3} \ [\%] \quad (2)$$

where: I_P - plasticity index [%], I_C - consistency index [-], I_a -. colloidal activity index [-] and σ_{bob} - predicted swelling pressure [kPa].

Note: Considering $\varepsilon_{swel} = \Delta_h/H_v$; if the surface deformation in centimeters is known and the swelling is completed, we may determine (or rather estimate) the thickness of the swelling layer $H = \Delta h/(0{,}01 \cdot \varepsilon_{swel})$.

- Predicted time of swelling:

$$t_{swel} = 0.0013 \cdot H_V \cdot I_P \cdot 78.817 \cdot e^{\left(-4.493/I_C^{1.3}\right)} \ [\text{months}] \quad (3)$$

where: I_P - plasticity index [%], I_C - consistency index [-] and H_v - thickness of the swelling layer [cm].

Note: the swelling in this case is considered with water uptake from both sides. In reality, the water may be available only from one side and the swelling time may therefore be longer.

Table 2. Swelling pressures at relative deformation 3% for different soils.

Natural water content w_n (%)	Water content at liquid limit w_L (%)	Water content at plastic limit w_P (%)	Soil particles < $d_{0.002}$ mm (%)	Plasticity index I_P (%)	Consistency index Ic (-)	Colloidal activity index Ia (-)	Swelling pressure σ_b (kPa)	Relative deformation due to swelling pressure ε_b (%)	Difference between max. swelling pressure and actual value of swelling pressure (-)	Percentual difference between max. swelling pressure and actual value of swelling pressure (%)
32	58.7	18	15	40.7	0.66	2.71	270	3	0	0
23	51.8	16.4	20	35.4	0.81	1.77	238	3	32	11.7
22	52.4	23.8	17	28.6	1.06	1.68	225	3	45	16.7
38	71.6	33.4	59	38.2	0.88	0.6	213	3	57	20.9
20	53.6	24.9	36	28.7	1.17	0.8	200	3	70	26

3. Soil improvements and swelling

Soil modification by lime is very common. Lime significantly reduces swelling but for high values of plasticity index and colloidal activity index it is not sufficient for the reduction of the swelling pressure to negligible values. Because it was not possible to determine the initial water content of the marlstone directly, core samples were extracted from behind the cut slope from depth corresponding to the layers underneath the highway. The samples were tested for swelling after lime addition up to 3%. The results were used in the same manner as for soil without binder (as mentioned in previous sections) to develop equations describing the swelling after lime addition.

- Predicted value of the swelling pressure is given by the equation:

$$\sigma_{swel,ca} = \left(V_{ca} \cdot I_P \cdot I_C \cdot I_a^{0.2}\right) \cdot \left(2.38 \cdot \ln\left(\left(1/V_{ca}^{0.8}\right) \cdot I_P \cdot I_C^{1.2} \cdot I_a^{0.2}\right) - 5.59\right) \ [\text{kPa}] \quad (4)$$

where: I_P - plasticity index [%], I_C - consistency index [-], I_a - colloidal activity index [-] and V_{ca} - volume of lime added [%].

Note: the equation is valid for added lime volume $V_{ca} \geq 0.5$ %.

- Predicted time of swelling for soil with added lime

$$T = 0.00134 \cdot \left[V_{CB}^{5} \cdot H_{v} \cdot I_{P}^{3} \cdot 1.79 \cdot e^{\left(-8.12 \cdot V_{ca}^{0.5} / I_{C}^{0.3} \right)} \right] \text{ [months]} \quad (5)$$

where: I_P - plasticity index [%], I_C - consistency index [-], I_a - colloidal activity index [-] and V_{ca} - volume of lime added [%].

The graph in Figure 1 shows the development of swelling pressure after lime addition in a soil of given parameters. Where the value is negative, the soil is no longer swelling. For initial consistency index $I_c = 1.0$ and swelling layer with lime thickness $H_v = 50$ cm, the swelling time is apparent in the graph in Figure 2.

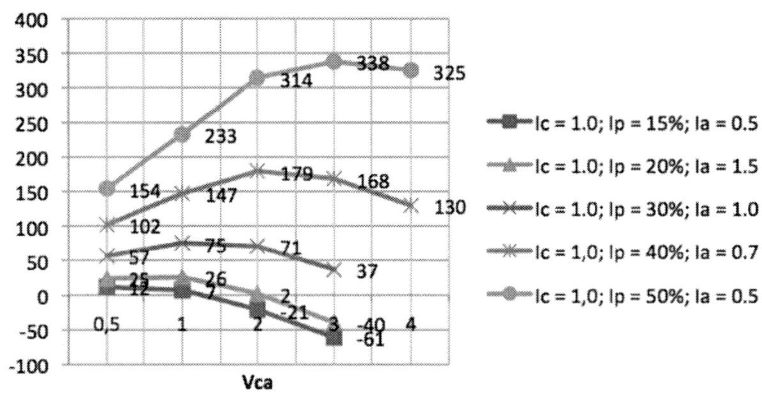

Figure 1. Swelling pressure σ_b development for different volume of lime added V_{ca}.

It is clear that the time of swelling of the soil with lime is proportional to the swelling pressure, as shown in Figure 1. If the soil has high plasticity index, then even after lime addition the swelling pressure remains very significant even though the reduction from the original soil is distinctive.

In order to prevent swelling of the modified material puzzolanic reaction must take place. Its initiation has one condition in the case of lime, the LMO (lime modification optimum). It is a state when the pH reaches 12.45 at a temperature around 25°C. In case this condition is not reached, exothermic reaction (drying of the soil) and flocculation (clustering of clay and silt particles into silt and sand particles) takes place. This happens quite quickly (hours to days) but gives only better physical and mechanical properties of the soil and better workability. At this state the porosity and permeability increases, but it has virtually no influence on the swelling, on the contrary, if future water donation should take place, the swelling conditions would be even better resulting from initial water content. It is known that losely related to the mineralogy of the soil, the LMO is reached at 3% of added lime, for more active minerals of the Ca group (particularly Na montmorillonite) even higher. We know that usual volume of added lime is lower than 3% which is enough to improve physical and mechanical properties of the soil and its workability but not enough to start the puzzolanic reaction leading to soil strengthening and higher resistance to swelling. In the case sufficient volume of the lime was added and the LMO was reached, the water environment is needed. This is the first complication – if we add too much lime, the material is disrupted and the water environment ceases to exist. If all conditions are favourable, exothermic reaction and flocculation follow, then the lime

(CaO) is turned into calcium hydroxide $Ca(OH)_2$. The high concentration of OH ions in alkaline environment starts at low temperature to transform Ca, Na and K ions into solution (more similar to gel) and starts to break the SiO_2 bonds in silicates and aluminates. New needle-shaped minerals grow through the soil structure from the gel, thus strengthening the soil. This is very slow process, according to literature the first signs can be observed after 28 days and full development can be expected after 90 or more days. The process causes decrease in porosity and permeability, the soil strengthens and therefore increases its resistance to swelling. Attention has to be paid to the formation of ettringit which is swelling itself! It was described that for the lime the process is quite complicated and time-demanding, therefore it is suitable to use mixed binders with the ratio 1:1 or with major portion of cement.

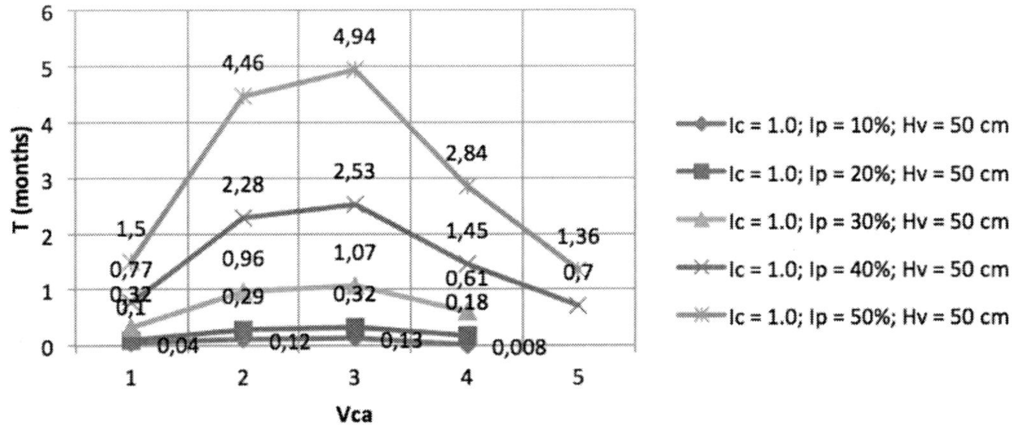

Figure 2. Swelling time T in months for different volume added lime V_{ca}.

4. Impact of swelling

Swelling can be divided on the free and pressure (or resistance). The free swelling is typical for cut slopes where the volume change can be without influence of the structures. In the case of pressure swelling the upper structure (e.g. concrete pavement) acts against the swelling pressure on the soil till reach the state of equilibrium. Generally, the swelling will be stopped in the case of the balanced moisture. The deformation in the free swelling can be four time higher than in the pressure swelling. The value of the swelling pressure can reach 3600 kPa at clays with plasticity index above 80% and low initial moisture. So high pressures can have the significant destruction effects for upper structures or tunnel linings. The swelling has also impact on the soil parameters.

4.1. Damage of the highway

This subsection shows the impact of swelling on the rigid slab of D11 highway in the Czech Republic. Section of the D11 motorway in km 76,000 was opened into use in December 2006. There were discovered instability of slopes on both sides of the highway and cracks in the concrete surface. Figure 3 shows the damage of the rigid highway pavement made of Portland cement concrete and Figure 4. is view on the circa 20 mm vertical shift of the slabs (the slabs have been grinded off). The kinetics of swelling was seen relatively quickly [6].

Figure 3. Disruption of the concrete road structure due to soil swelling, the direction of Prague.

Figure 4. Detail of cracks in the left lane at the 76,000 km in summer 2010. grinded off.

The concrete slabs had to be repeatedly grind due to fault of the surface was several centimeters. The shift of two slabs reached several centimeters and the surface had to be grinded off several times. The slab thickness after construction was 24 cm. After the grinding the shift reached up to 8 cm and the recommended speed was of 80 km/h, instead of the projected speed of 130 km/h. The numerical modelling of the water flow in the subsoil was carried out at CTU in Prague, FCE (Department of geotechnics) using Plaxis software. The FEM model showed that the deep drains at the bottom of the undercut slopes massively saturated road subsoil, even from the middle lane divider (Figure 5). Underlying marls, although very little permeable, has very high capillarity. Well, it results in saturation of the subsoil and intense swelling of subsoil under the road with a totally destructive impacts.

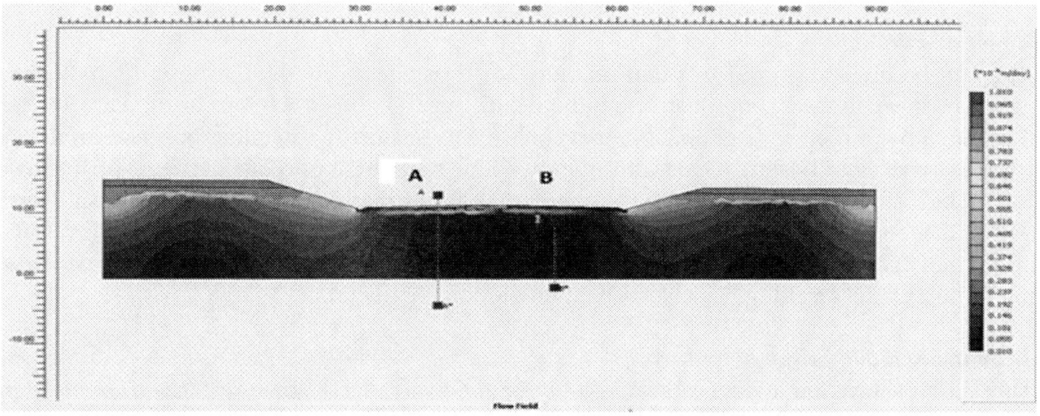

Figure 5. Numerical model of the water flow in the subsoil.

The standards or criteria often requirement only the values of the linear swelling but they tell nothing about the swelling pressures. And the swelling pressures have destructive effects. Deformations are only an accompanying phenomenon. Table 2 shows swelling pressures at relative deformation 3% for different soils. As it is obvious for the various initial parameters of different swelling pressure the deformation is still 3%. Based on this fact, the authors developed classification of cracks development on the road surface (indicative only). The classification is shown in Table 3.

Table 3. Classification of cracks development on the road surface.

Swelling pressure	Evaluation	Asphalt surface	Concrete surface
< 30 kPa	Negligible	De facto no	Virtually impossible
30 – 60 kPa	Increased	little noticeable surface deformation -	They may see the first signs of cracks
60 – 100 kPa	High	Slight undulations	Clearly visible cracks on surface
> 100 kPa	Very high	Significant undulations, depending on swelling pressure	Significant cracks and slip of the broken slabs

As seen from Table 3 the deformation on the surface are visible in the range of 30 to 60 kPa of swelling pressure in the subsoil. Contact stress below the structural layer of road with thickness 65-67 cm (on the subgrade) is about 14 to 16 kPa. If we consider the unit weight of the structural layers and the concrete surface 23 kN/m^3 * 0.7 m, we have about 16 kPa. Not every sublayer has unit weight 23 kN/m^3, some materials have a lower unit weight. If the swelling pressure is 200 kPa or more, then it is obvious the final result. Swelling pressures were greater than 300 kPa on the described part of motorway D11.

5. Conclusion

In this paper, the dependence of the swelling pressure of expansive soils in the Czech and Slovak Republic on the soil properties was examined. Based on the multiple regression analyses made on the extensive experimental activities the reliable correlations have been established for swelling pressure, swelling strain and swelling time with initial water content, initial dry density, plasticity index, consistency index and collodial activity index. Compare of the measured and predicted swelling pressure values indicates a high prediction performance (the correlations coefficient R^2 = 0,88). For the soil modification by lime it was not possible to determine the initial water content of the marlstone directly, so the developed equations describing the swelling after lime addition were presented. The authors believe the presented relations are a suitable alternative for the classical calculations and give quick prediction of swell pressures from easily determined soil properties. There are described the problem with soil swelling on the highway D11 in the second part. The authors also show classification of cracks development on the road surface based on their current knowledge of the problems.

Acknowledgements
The article was processed under financial support by Competence Centres program of Technology Agency of the Czech Republic (TA CR), project no. TE01020168 Centre for Effective and Sustainable Transport Infrastructure (CESTI).

References

[1] Chen F H 2016 *Foundations on Expansive Soils* (Elsevier, Amsterdam)
[2] Yilmaz I 2006 Indirect stimation of the swelling percent and a new classification of soils depending on liquid limit and cation exchange capacity *Engineering Geology* **85** pp 295-301
[3] Seed H B, Woodward R J and Lundgren R 1962 Prediction of Swelling Potential for Compacted Clays *Journal of Soil Mechanics and Foundation Division* **93** pp 325-53
[4] Holtz R D and Kovacs W D 1981 *An Introduction to Geotechnical Engineering* (Prentice - Hall, Inc., Englewood Cliffs, NJ)
[5] McKeen R G 1992 A Model for Predicting Expansive Soil Behaviour *Proc. 7th International Conference on Expansive Soils Vol. 1 (Dallas)* pp 1-6
[6] Šedivý M, Veselý J and Pruška J 2017 Soil Swelling Problems in the highway construction *15th International Multidisciplinary Scientific GeoConference SGEM 2015* pp 681-6

IOP Conf. Series: Materials Science and Engineering **236** (2017) 012038 doi:10.1088/1757-899X/236/1/012038

Tendencies in the development of operational quality of ballasted and ballastless track superstructure and transition areas

Libor Ižvolt[1], Janka Šestáková[1] and Michal Šmalo[1]

[1]University of Žilina, Faculty of Civil Engineering, Department of Railway Engineering and Track Management, Univerzitná 8215/1, Žilina, Slovak Republic

E-mail: michal.smalo@fstav.uniza.sk

Abstract. Department of Railway Engineering and Track Management is cooperating with Slovak Railways (Železnice Slovenskej republiky – ŽSR) on assessment of quality of railway tracks, structures of railway tracks and parts of these structures. One of diagnostics methods is monitoring of track geometry focused on durability of ballastless track quality and its transition area to ballasted track. The diagnostics is carried out by continuous method and results are used for prediction of future degradation of construction. The first part of the paper deals with a brief recapitulation of information about the experimental sections and the methods of diagnostics of track geometry. The second part of the paper is carrying out the analysis and prognosis of development of the quality of construction of the experimental sections, which is a tool for the track superstructure maintenance planning.

1. Introduction

The ballastless track construction is on Slovak Railways network mainly used as railway superstructure construction on bridges, in tunnels and adjacent sections. Diagnostics of ballastless track sections focused on transition areas is used for monitoring of track geometry quality progress presented through quality marks of selected parameters. The quality marks and subsequently the quality numbers are for infrastructure manager useful in process of scheduling track maintenance. The early detection of track quality degradation and subsequent planning of repair interventions in the initial stage of errors in track geometry is essential precondition for achieving an extended lifetime of the structure with the required diagnostics and repair costs and also for safe operation of the railway line.

The experimental section is located near the portals of the newly built tunnel Turecký vrch. This track is a part of V. multimodal transport corridor TNT-T Venezia – Terst/Koper – Ljubljana – Budapest – Užhorod – Ľvov with the branch Va Bratislava – Žilina – Užhorod. The ballastless structure (system Rheda 2000®) was built within modernisation of the railway track Nové Mesto nad Váhom – Púchov, object Nové Mesto nad Váhom – Trenčianske Bohuslavice [1].

The ballastless structure passes though different types of subgrade. It starts on earthwork at the south portal and leads through the tunnel. At the north portal, the ballastless structure continues on two bridges and earthwork. The total length of the ballastless track is 2 280.145 m. A new type of transition area, which using standard superstructure components, carries out between the end of the ballastless structure and the ballasted structure. The structure consists of a longitudinal concrete bed,

20 m long. Each track has its own concrete bed. The concrete beds are dilated longitudinally. The railbed thickness under sleepers is decreasing in the direction to the ballastless track, which causes the increase of the subgrade stiffness. The bed bottom and walls are lined with elastic anti-vibration rubber mats that are supposed to simulate the earthwork soil deformation properties.

Results of the diagnostic measurements are realised until now, enable to create the global forecasting of the quality development of the monitored ballastless track – of the experimental section. The detailed information about the parameters, measuring instruments, diagnostic methods, ways of its evaluation and its partial results were published in [1, 2].

2. Methods of diagnostics

The track geometry diagnostics is carried out by the measuring trolley KRAB^{TM} - *Light*. The measuring system measures the track parameters in an unloaded state. The measured parameters are [3]:

- gauge tolerance *RK* (after calculating the change of gauge *ZR* is also recorded),
- alignment of right rail *SP* (after calculating the alignment of left rail *SL* is also recorded),
- rail top level of right rail *VP* (after calculating the rail top level of left rail *VL* is also recorded),
- superelevation *PK*,
- quasi-twist on a short base (calculated to a quasi-twist on a base of 1.8 m long – $ZK_{1.8}$, 6.0 m long – $ZK_{6.0}$ and 12.0 m long – $ZK_{12.0}$).

From the point of view of the diagnostics evaluation, the monitored section is divided into two subsections [4]:

- section 1.1 (line No.1, the southern portal of the Tunnel Turecký vrch) and 2.1 (line No. 2, the southern portal; both sections are 175 m long; 102.360 000 – km 102.535 000):
- section 1.2 (line No.1, northern portal of the Tunnel Turecký vrch) and 2.2 (line No.2, the northern portal); both sections are 640 m long; km 104.200 000 – km 104.840 000):

The monitoring of the experimental section is executed from July 2012 in a half-year period. By now, 9 measurements have been carried out: one before commissioning (marked as MSO) and 8 operational measurements (marked from PO1 to PO8). Within the initial measurement (MSO) were recorded 40 local faults, 30 of which in the sections with the ballasted track, 9 in the section with the ballastless track and 1 local fault was diagnosed in the transition area of the section 2.1. The contractor carried out after this measurement a repair of the track geometry and a microgeometry repair of the rail heads by grinding [5].

In the second operational measurement (PO2) were for the first time diagnosed the local faults of the vertical profile of the right (*VP*) and the left (*VL*) rail in the transition area of the section 1.1. In the following measurements – PO3 and PO4 – was diagnosed another decrease in quality of the section 1.1, which proves the increased number of the local faults, as well as the decreasing value (increasing) of the general quality number. In November 2014 the railway infrastructure manager executed the maintenance in the transition areas of the sections 1.1 and 2.1. The diagnosed local faults were removed by means of the maintenance. The complex continual diagnostics executed by the measuring trolley KRAB^{TM} - *Light* and the monitoring executed by the railway infrastructure manager by means of the levelling method from December 2014 to August 2015 in the sections 1.1 and 2.1 prove other settlements of the track level. Their values do not occur as the local faults so far because the operational tolerances were not overrun. However, there is expected the formation and occurrence of the local faults within a short time period because of the proved quality decrease tendency of the particular parameters of the mentioned sections. The sections 1.2 and 2.2 in the transition areas do not prove any faults of the values. The quality development tendency of the measurements recorded so far does not imply any probability of their occurrence either in the following measurements [3].

3. The prognosis of the quality development tendency of the construction

The prognosis of the quality development tendency of the particular track geometry parameters is concentrated on the problematical section from the point of view of the track quality. As introduced in the parts 1 and 2, it concerns the transition areas from the ballastless track to the ballasted track in the subsections 1.1 and 2.1. From the data acquired by the measuring trolley were processed the graphs of the real deviations of the track axis alignment (*SK*), gauge (*RK*), gauge change (*ZR*), superelevation (*PK*), the projected value of which is in the mentioned section 0 mm, the twist fault at three measuring bases ($ZK_{1.8}$, $ZK_{6.0}$, $ZK_{12.0}$) and the rail top level of the left and the right line (*VL*, *VP*) in the section from km 102.454 000 to km 102.487 000. A part of this section are the transition areas in both lines from km 102.460 500 to km 102.480 500. In the paper are presented the graphs of real deviations of a selected parameter – the rail top level of the left and the right rail and the development tendencies of this parameter because it is in regard to the maintenance planning of the transition areas the most critical parameter in the long term [5].

In the transition area of the section 1.1 (Figure 1 and Figure 3) detect the tops of the right (after calculation also the left) rail a fault of this parameter on the level of the operational limit overrun defined by the values ±6 mm in PO4 for the right rail , respectively in PO3 and PO4 (the upper limit ±6 mm is overrun) and in PO2, PO3 and PO4 (the lower limit -6 mm) for the left rail . The repair of the track geometry executed in the transition area in November 2014 reduced the deviation values. However, the places with the limiting values of the deviations were remained (the contact of the transition area with the ballastless track and the ballastless track beginning).

In the transition area of the section 2.1 (Figure 5 and Figure 7) was not recorded any overrun of the operational limits of the parameters *VL*, *VP* within the PO1 to PO4. However, the railway infrastructure manager carried out the maintenance in the same time at both tracks. The maintenance executed in the transition area of the track No. 2 did not improve the deviation values of the parameter. It came to the translocation of irregularities and places with the limiting values and there was recorded an increase of the deviation *VL* from -1.5 to -3.5 mm and *VP* from -1.0 to -3.0 mm on the contact of the conventional permanent way with the construction of the transition area.

The prognosis of the further quality development of the diagnosed parameters records various tendencies. The planning of the maintenance and repair works depends on the quality of the parameters which values are fastest impairing. In the case of the experimental sections in the area of portals of the Tunnel Turecký vrch, it is the vertical profile of the left and the right rail in the transition areas of the subsections 1.1 and 2.1. From the tendency lines created by means of the values *VL* and *VP* measured within PO5, PO6, PO7 and PO8 (Figures 2, 4, 6 and 8) is possible to predict that in the section 1.1 will be reached the operational limit (±6 mm) of the parameter *VL* probably in September 2017 and the parameter *VP* in May 2018. In the section 2.1 will the *VL* and *VP* reach the value of the operational limit probably in September 2017, respectively in June 2018.The overrun levels of the operational limits should be detected within the second measurement campaign of the measuring trolley of the railroads in Slovakia (in summer 2017). According to the measurement results is the railway infrastructure manager going to plan the execution of the track geometry repair in the mentioned sections.

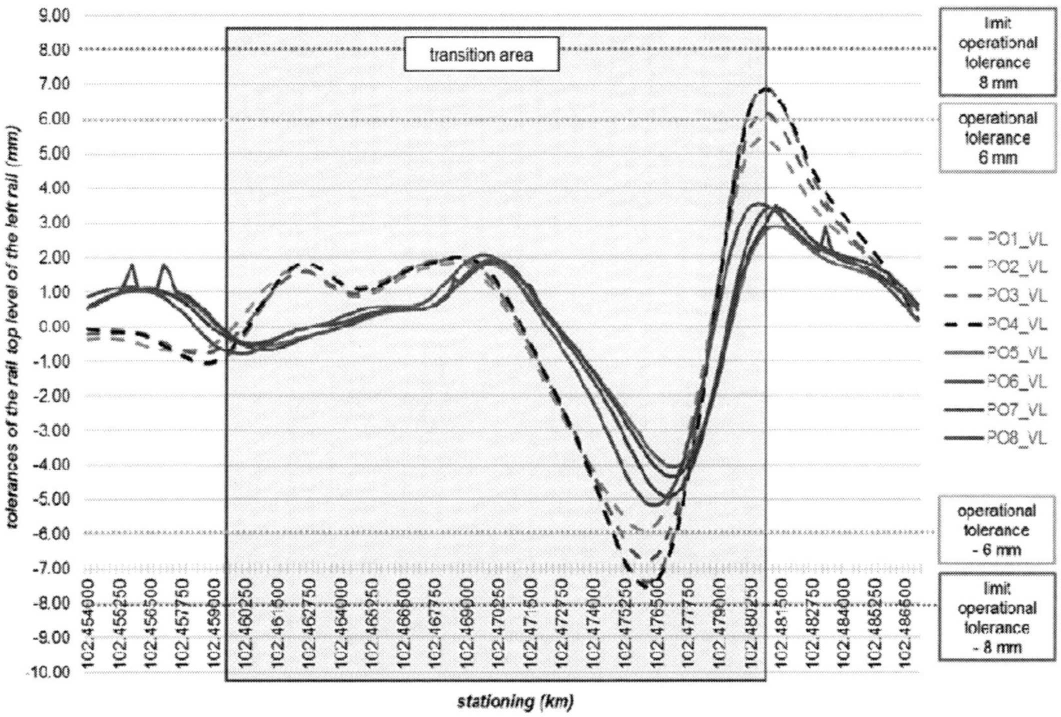

Figure 1. The record of the deviations of the rail top level of left rail (*VL*) in the transition area of the section 1.1.

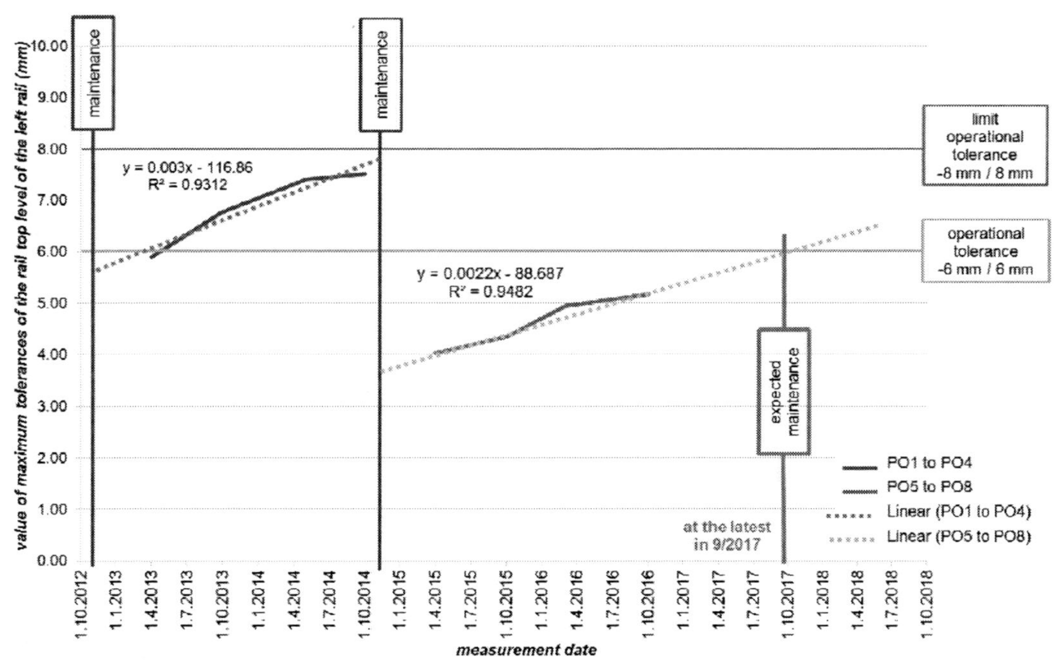

Figure 2. The development tendency of the maximum deviations of the rail top level of left rail (*VL*) in the transition area of the section 1.1.

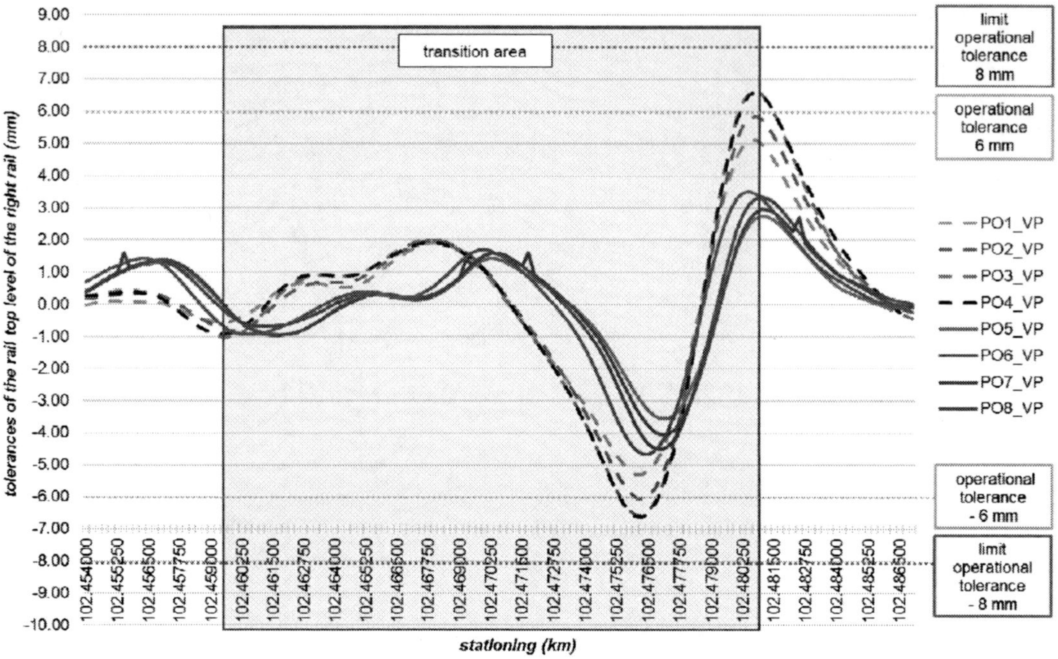

Figure 3. The record of the deviations of the rail top level of right rail (*VP*) in the transition area of the section 1.1.

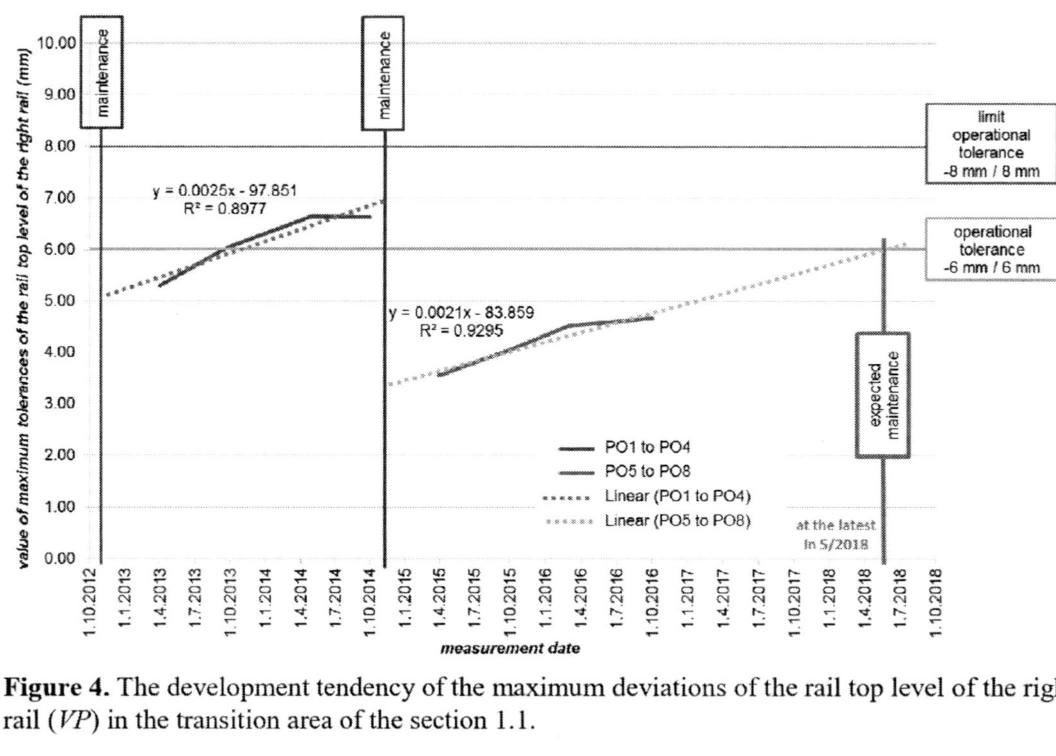

Figure 4. The development tendency of the maximum deviations of the rail top level of the right rail (*VP*) in the transition area of the section 1.1.

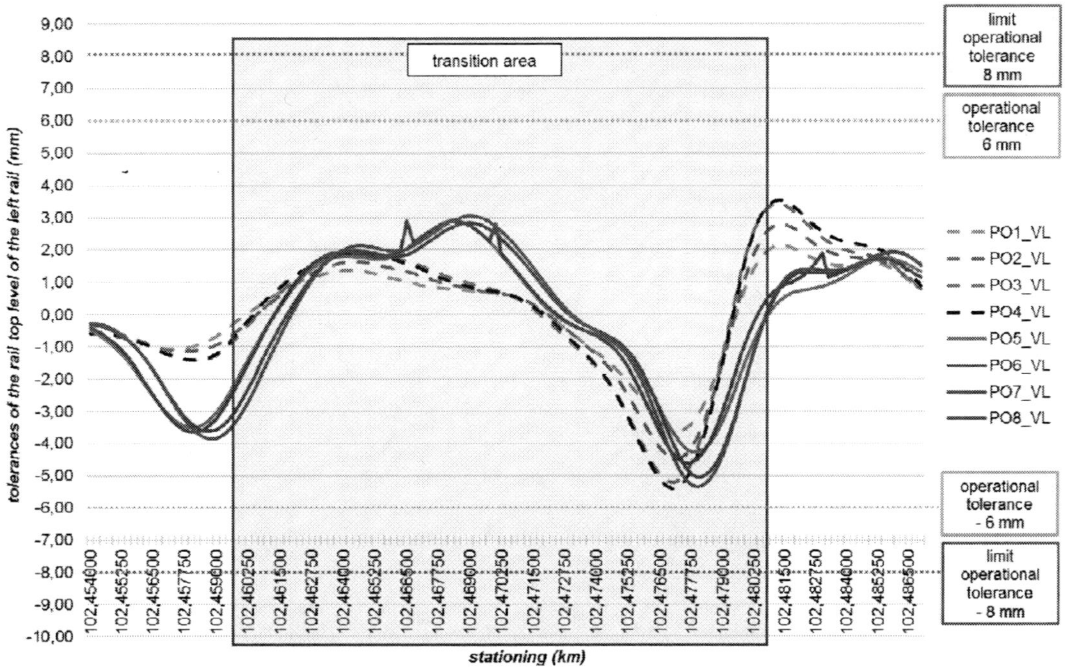

Figure 5. The record of the deviations of the rail top level of left rail (*VL*) in the transition area of the section 2.1.

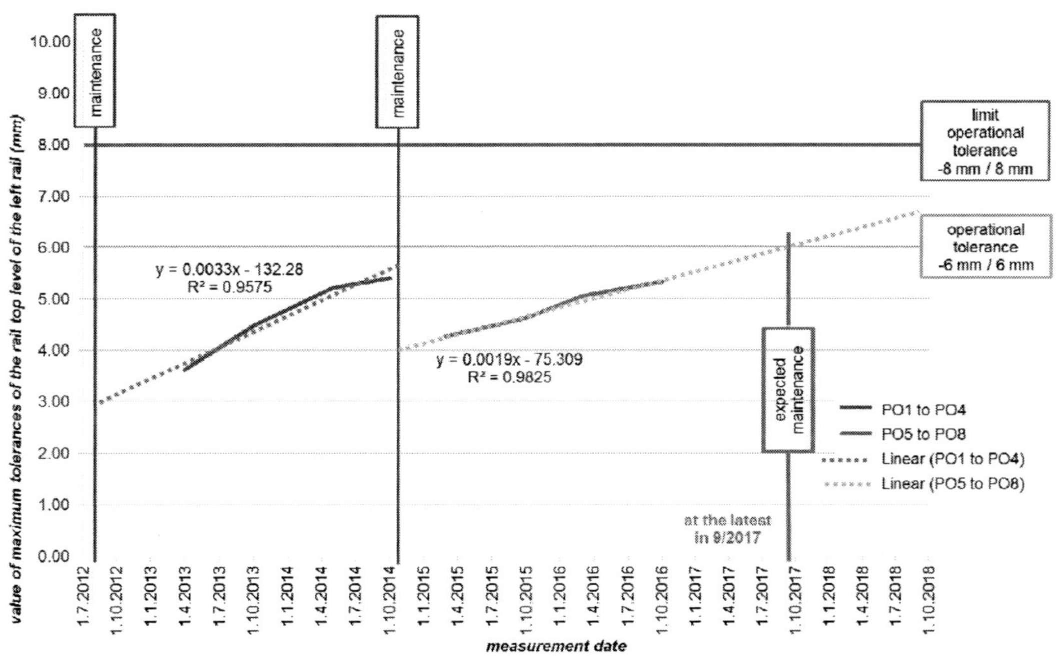

Figure 6. The development tendency of the maximum deviations of the rail top level of the left rail (*VL*) in the transition area of the section 2.1.

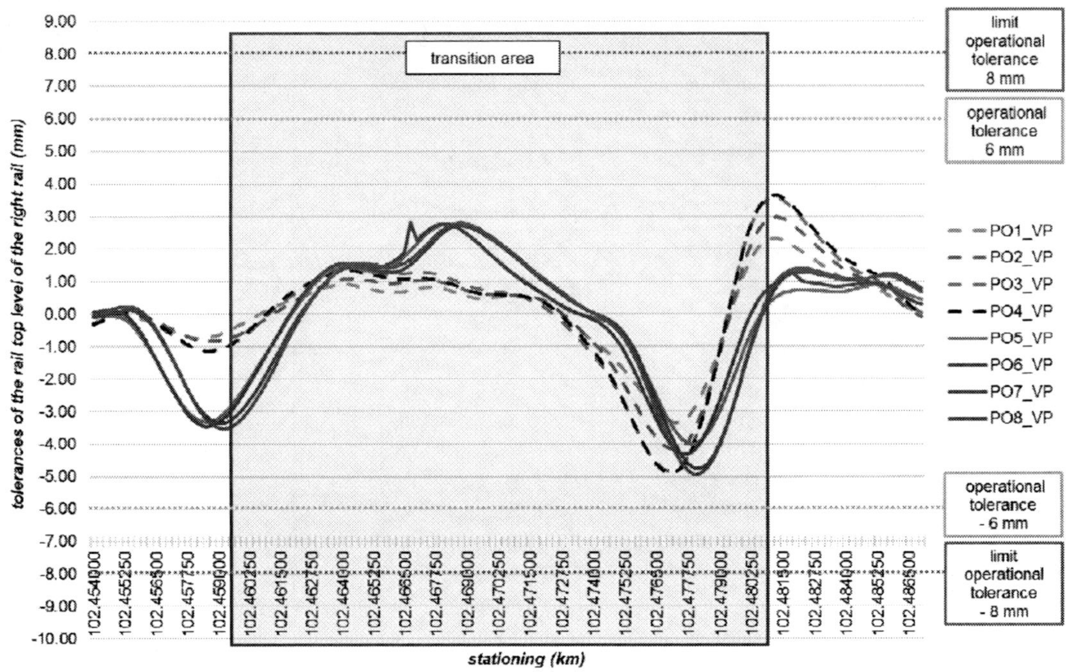

Figure 7. The record of the deviations of the rail top level of right rail (*VP*) in the transition area of the section 2.1.

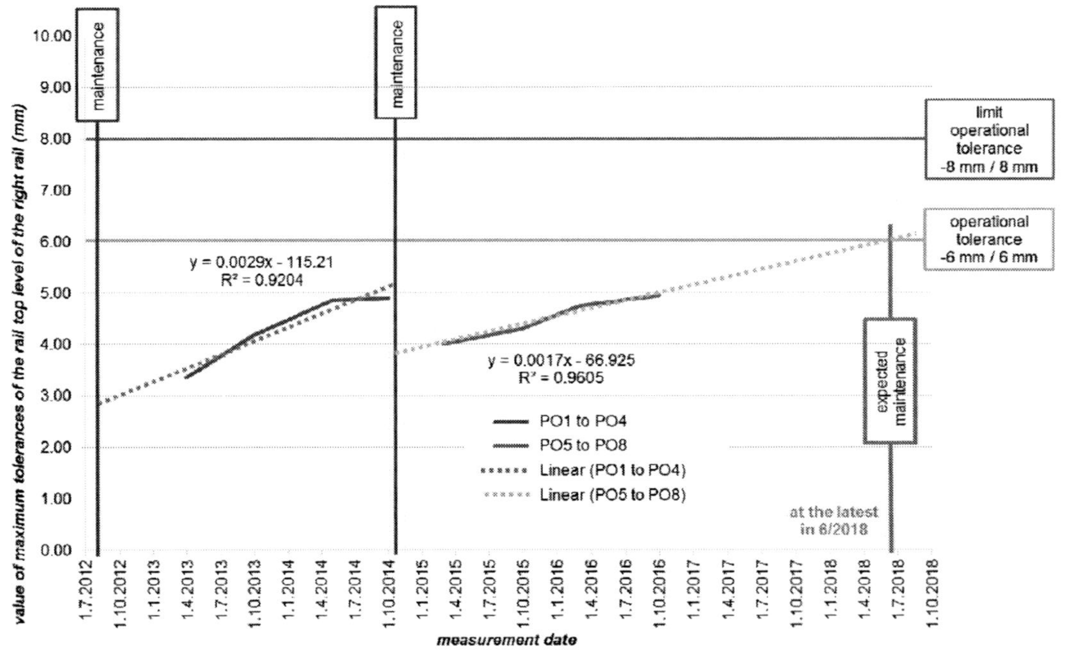

Figure 8. The development tendency of the maximum deviations of the rail top level of the right rail (*VP*) in the transition area of the section 2.1.

4. Conclusion

The monitoring of the ballasted and ballastless track construction with the centre of interest in the transition area between them has been executed for five years in the surrounding of portals of the Tunnel Turecký vrch. Within the continual diagnostics of the track geometry parameters was found out the quality decrease (represented by the increase of the value quality number – *QN*) in the sections 1.1 and 2.1. The quality impairment cause in the largest extent the parameters *VL* and *VP* by the overrun of the operational limits within PO2, PO3 and PO4. The repair after the measurements PO4 caused the improvement, however, it is again recorded the impairment of the measured deviations *VL* and *VP* and consequently *QN*. With regard to the quality development of the sections 1.2 and 2.2 in the elapsed period of the monitoring it is predicted the lasting partial decrease of the construction quality responding to the railway operation without any emergency operation events.

Based on the relative and absolute construction track layout and the achieved results can be stated that the monitored sections with the ballastless track are settled and they do not show any relevant changes of the single parameters among the individual measurements. The sections with the ballasted track show changes, those ones, however, result from the character of this construction. The ballasted track sections require in certain intervals (depending on the operation loading) reparation of the track geometry: tamping and filling-in of the ballast bed and further repairs aiming at the preservation of the quality parameters during the whole lifetime of the structure. However, the measurements of the relative construction track layout have pointed at the irregularities in the transition areas between the ballasted track and the ballastless track especially in the area of the southern portal of the Tunnel Turecký vrch, in the tracks No. 1 and No. 2. These results were subsequently proved by the geodetic survey and also by the outputs of the measuring trolley of the ŽSR. The decreases recorded in these transition areas from the beginning of the measurements resulted into the limit overruns of the quality marks of the rail top level of the rails QM_{VK}, as well as into the appearance of the local errors what led to the realization of the corrective action. This corrective action slightly slowed the error development tendencies, however, the quality of the track geometry of both lines continues in the degradation and the next limit overruns are to be expected already during the second half of the year 2017, out of what will result the necessity to realize another corrective action.

Acknowledgements

The paper contains results of the grant VEGA 1/0275/17 “Application of numerical methods to define the changes of geometrical track position”.

References

[1] Ižvolt L, Šestáková J, Šmalo M and Gocálová Z 2014 Monitoring of the Track Geometry Quality around the Portals of new Tunnel Construction Turecky vrch – Preliminary Results *Communications: Scientific Letters of the University of Zilina* **16**(4) pp 9-20
[2] Ižvolt L, Šestáková J and Šmalo M 2016 Monitoring of parameters of track layout and track geometry of track sections of ballastless track and its transition areas around portals of tunnel Turecký vrch *New Railway Technique* **24**(2) pp 12-16
[3] Šestáková J 2015 Quality of Slab Track Construction - Track Alignment Design and Track Geometry *Civil and Environmental Engineering* **11**(1) pp 2-9
[4] Ižvoltová J, Pisca P, Villim A and Mancovič M 2014 Precision Analysis of Height Measurements Realized on Ballastless Track *Communications: Scientific Letters of the University of Zilina* **16**(4) pp 63-67
[5] Ižvolt L and Šmalo M 2016 Verification and optimization of transition areas of ballastless track in the tunnel Turecký vrch *Proc. CETRA 2016 – 4th International Conference on Road and Rail Transport (Šibenik)* ed S Lakušić et al (Zagreb: University of Zagreb) pp 723-32

Application of geocomposite placed beneath ballast bed to improve ballast quality and track stability

Leoš Horníček[1], Petr Břešťovský[1] and Petr Jasanský[2]

[1] Czech Technical University in Prague, Faculty of Civil Engineering, Thakurova 7, Prague, 166 29, Czech Republic

[2] The Railway Infrastructure Administration, state organization, Dlazdena 7, Prague, 110 00, Czech Republic

E-mail: leos.hornicek@fsv.cvut.cz

Abstract. The article deals with the application of a stabilization hexagonal geocomposite for the improvement of poor stability of railway tracks caused by undesirable migration of fine soil particles from the subgrade into the ballast bed. The establishment of a test railway section on a single-line track situated near Domazlice and its long-term monitoring programme are described. Evaluation is aimed especially at track geometry parameters, the load-bearing capacity of the ballast bed, elastic rail deflection during train passages and the durability of geocomposite's physical properties. The data taken from the test section during five measurement campaigns are compared with both adjacent sections. In one of them, only the ballast bed renovation was carried out, whereas in the second one no intervention was performed at all. The usage of a pioneering geosynthetic product in combination with new trends in ballast bed restoration seems to be an innovative as well as effective solution to analogous problematic spots on railway tracks in the Czech Republic.

1. Introduction

Due to complex geological conditions and gradual track degradation, railway tracks in the Czech Republic with high traffic volumes suffer from the migration of fine-grained soil particles from subgrade layers into the ballast bed and from the appearance of muddy spots. The accompanying phenomenon of the above process is local reduction of the vertical track stiffness, malfunctioning ballast drainage as well as total deterioration of track geometry parameters, and often also reduced load-bearing capacity of the track formation. This situation successively requires either frequent operative repairs (ballast cleaning and tamping) or a single, costly and extensive intervention into the whole roadway. The standard procedure consisting in repairing mud spots by cleaning the ballast bed, adding missing aggregates into the ballast bed and rail tamping, however, usually does not lead to the solution of this problem, but, on the contrary, to its periodic repetition.

An alternative, longer-term solution to the above issue without the necessity of an extensive and costly intervention into the track bed is the application of suitable geogrids or geocomposites placed underneath the ballast bed or in its bottom part, based on the knowledge of the interaction of geosynthetics with granular material [1]. Efficient interlocking of the geocomposite with ballast aggregates leads to the limitation of the horizontal movement of aggregate grains around the geogrid and reduction of permanent vertical deformations of the ballast, which is favourably manifested by increased stability of track geometry parameters, less frequent maintenance cycles and extended service life of aggregates, or reduced degradation of aggregate grains due to frequent tamping [2]. Separation or filtration geotextiles, which are part of the geocomposite, prevent the migration of fine-grained soil particles into the ballast allowing, simultaneously, water passage in both directions to enable functioning ballast drainage in the long-term. In the case of a properly taken measure, economies in material, maintenance costs and costs of traffic limitations can be reached [3].

An efficient interconnection of the geogrid and granular aggregate material via interlocking leads to mechanical stabilization of aggregates within the reinforced zone, which results in the reduction of lateral movements of aggregates and increased stiffness of the layer. In practical terms, it is necessary to place the geosynthetic in a proper depth to secure it from damage during machinery-based maintenance, even though its placement closer to the sleeper would be more efficient [4]. The effective combination of the geogrid aperture size and the aggregate grading is also very important [5].

2. Operational verification in the Czech Republic

The ballast bed stabilization with geogrids and geocomposites in the Czech Republic has been verified since 2008 [6]. Based on a stimulus from the Centre for Effective and Sustainable Transport Infrastructure (CESTI), the Czech Railway Infrastructure Administration, state enterprise, approved the extension of the operational verification, including a combination of a new type of geosynthetic and transversally inclined track formation for efficient drainage of the ballast bed.

The verified technology applies a progressive type of geosynthetic product, which is inserted between the ballast and subballast layer, or directly on the subgrade, e.g. as part of a planned ballast replacement or repair. The inserted geosynthetic – in this case a geocomposite composed of a monolithic hexagonal geogrid and a separation geotextile – should fulfil two objectives at once. One of them is to stabilize the ballast by interlocking aggregate grains with the geogrid, which limits the pushing out of ballast bed grains in the transverse direction, reduces the vertical deformation of the ballast and extends the intervals between track geometrical position adjustments. The second objective is to prevent the penetration of fine-grained soil particles from the subgrade into the ballast bed, which protects the ballast bed from undesirable contamination reducing its ability of draining water from the ballast as well as friction between individual aggregate grains.

3. Description and establishment of a test section

3.1. Test section description

The position of the test section was selected on a non-electrified, single-line track national railway line Plzeň-Jižní předměstí – Česká Kubice in the section of Domažlice-město – Havlovice. Mud spots had been found within the selected section on a long-term basis, which, however, was visibly manifested in 2014 (Figure 1).

Figure 1. Muddy ballast bed.

At the same time, the quality of key track parameters, i.e. track alignment and longitudinal rail height, had significantly deteriorated between 2012 and 2015. An important precondition for choosing this section was the fact that ballast cleaning and renovation ca 30 years after its placement had been planned in this section.

The test section is situated in a straight line, with a gradient of ca 9 ‰, passing over an at-grade crossing. The line speed limit in this section is 90 km/h, a standard gauge of 1435 mm and the line is serviced by mixed traffic. The substructure does not contain any structural layers, therefore, underneath the ballast bed there is just subgrade, consisting of frost-susceptible impermeable soil.

The monitored track section is subdivided into three immediately adjoining subsections. Stabilization geocomposite was placed in a subsection 95 m in length (km 169.850-169.945). This subsection is adjoined on one side by subsection No. 1, 100 m in length (km 169.750-169.850), in which a new ballast bed was installed and tamped. On the other side, it is adjoined by subsection No. 3, 100 m in length (km 169.945-170.045), where no intervention was made, and which, therefore, serves as a reference section.

3.2. Test section establishment

The geocomposite was placed in the test section in June 2015 during ballast replacement using the conventional method of track renewal. In the test section, the track panels were first lifted by means of a truck crane, the remaining ballast bed was extracted down to the track formation level (forming, at the same time, the subgrade) and this subgrade was treated to produce a one-sided gradient of min. 3 % towards the longitudinal ditch.

Two rolls of stabilization geocomposite, 3.8 m wide, with a mutual longitudinal overlap of 50 cm were unrolled onto non-compacted subgrade. The TENSAR TriaX TX190L-GN geocomposite, which complies with the ETA 12/0531 certificate, was chosen. It is composed of a monolithic hexagonal geogrid and non-woven geotextile, both components being mutually interconnected. The geocomposite serves both for the ballast stabilization due to the interlocking of aggregate grains with geogrid apertures (geogrid aperture size corresponds to the ballast aggregate grading of 31.5/63) and for avoiding the migration of fine-grained soil particles from the subgrade into the ballast (low geotextile effective aperture size corresponds to this). The geogrid of polypropylene has a hexagon pitch of 120 mm, an isotropic stiffness ratio of 0.75 and radial stiffness at 0.5% elongation of 540 $kN.m^{-1}$. The effective aperture size of the non-woven polypropylene geotextile is 95 μm.

The geocomposite was placed centrically to the rail axis. A ballast layer of new coarse crushed aggregates with a grading of 31.5/63 in a min. thickness of 35 cm underneath the rail seat was successively laid (Figure 2).

Figure 2. Ballast bed placement on stabilization geocomposite.

The aggregates were brought by lorry and spread by wheel loaders using a so-called cascade method when the aggregates must first be spread in front of the loader in a thickness of min. 15 cm and only then can a machine move over them. The track panels were successively put back, welded and the track position was adjusted into the desired condition by a tamping machine.

4. Test section monitoring

First, initiation measurement was performed on the original state of the construction in June 2015, and, successively, after the geocomposite placement underneath the ballast bed, the 2.5-year monitoring programme was launched, including:

- periodic checking of track geometry parameters by a track geometry car,
- measurement of the load-bearing capacity of the ballast bed and the track bed by means of static and impact plate load tests,
- checks of the degradation rate of geocomposite's utility properties via laboratory testing of geocomposite specimens sampled from trial holes,
- visual checks of the environment at the placed geocomposite level,
- measurement of rail deflection during the passage of trains,
- levelling of the rail string height.

4.1. Measurements performed in 2015 and 2016

Three measurement campaigns were carried out in the test section and both adjoining sections in 2015. The first of them (06/2015) took place on the original state of the roadway immediately before the start of repairs. This campaign covered the measurement of rail deflection during the passage of trains in 4 cross sections, 2 of which were situated in the test section and the remaining 2 in both adjoining sections. Besides, the running plane was levelled and soil specimens were sampled from trial holes made between sleepers. The second measurement campaign (06/2015) was a component part of the test section establishment. Static plate load tests in profiles showing the greatest problems with the track geometry parameters were performed on the modified track formation in the rail axis. Furthermore, impact plate load tests were performed in the rail axis with a spacing of 10 m and at a distance of 0.80 m in both directions. Soil specimens were sampled from the track formation to carry out index laboratory tests. The third measurement campaign (09/2015) took place immediately after the last necessary rail machine tamping. This campaign, as well as all successive measurement campaigns, included rail deflection measurement during the passage of trains in 4 cross sections identical to those used in the first campaign, running plane levelling in 5 cross sections 50 m apart and in the middle of the railway crossing, and impact plate load tests at the rail seat level near the rail string in 15 profiles 10 m apart from each other.

In 2016, two measurement campaigns followed, performed in spring and autumn (04/2016, 11/2016), i.e. at the time when the load-bearing capacity of the subgrade is usually lower. Both campaigns included 4 static plate load tests at the rail seat level in the test section as well as both adjoining sections. As they were performed in the rail axis, railway traffic on the respective rail had to be interrupted for the necessary time. In the fifth measurement campaign, the first sampling of two geocomposite specimens from the space between sleepers in the rail axis took place (Figure 3). The samples were successively sent to the geocomposite manufacturer's laboratory to detect changes in its mechanical properties from the time of placement.

5. Evaluation of results

The test section, as well as both adjoining sections, are regularly monitored in terms of the development of track geometry parameters. In view of the type of track, this happens twice a year, as part of the spring and autumn measurement campaign carried out by the track geometry car's travels. The results achieved in the form of charts (continuous time patterns of measured parameters) and section evaluations (numerical evaluation of track sections using so-called quality parameters –

mathematical recalculation of standard deviations of monitored parameters) represent an important source of data. The results of monitored geometric parameters (particularly alignment and height deviations) and their development over time allow assessing the stability and long-term behaviour of the track bed construction.

Figure 3. Excavated geocomposite specimen.

An important aspect for a serious evaluation of the development of track geometry parameters in the monitored sections was also the assessment of the development of these parameters in the period preceding the stabilization geosynthetic's installation. Figure 4 displays the comparison of charts from the track geometry car immediately before the test section's establishment and ca one year after its establishment. Individual monitored subsections are marked in the figure (section 1 – section with a new ballast bed; section 2 – section with a geocomposite; section 3 – section without any intervention).

The time pattern of parameters from measurement 1/2015 (June) clearly shows that it was a track section complying with operating conditions in terms of observing the track geometry parameter limits. There are no evident spots where the maximum allowable limits were exceeded, but, in a detailed analysis of mainly vertical type parameters, it cannot be labelled as optimum and because of the development of quality parameters, the section required maintenance interventions.

The lower part of Figure 4 shows the course of measurement 1/2016 (May) performed nearly one year later after the geocomposite placement. There is an evident improvement in the course of individual parameters. In the longitudinal height time pattern (labelled VL and VP in the right part of the charts), the occurrence of more significant deviations at interfaces between the monitored sections can be observed. This phenomenon can be attributed to differences in the building interventions and the composition of the track bed construction. The subject of further monitoring, including the overall assessment of the development of track geometry parameters in relation to the other monitored variables in individual sections, is whether these deviations will gradually consolidate or will further develop under operating conditions.

The results of static plate load tests (Table 1) imply that the average load-bearing capacity identified in the rail axis at the rail seat in the test section fitted with a geocomposite was 55.1 MPa in spring and 76.9 MPa in autumn. In adjoining sections, the average values identified were 31.8 MPa, or 34.5 MPa respectively. Thus, a significantly higher load-bearing capacity was measured in the test section, which, moreover, further grew by ca 40 % between both measurement campaigns. A more detailed analysis of the results from the adjoining sections reveals that the load-bearing capacity value in subsection 3 (km 169.960) has remained practically unchanged, while in subsection 1 (km 169.830) the load-bearing capacity has increased by 19.1 %. This difference is caused by the on-going consolidation of cleaned ballast in subsection 1.

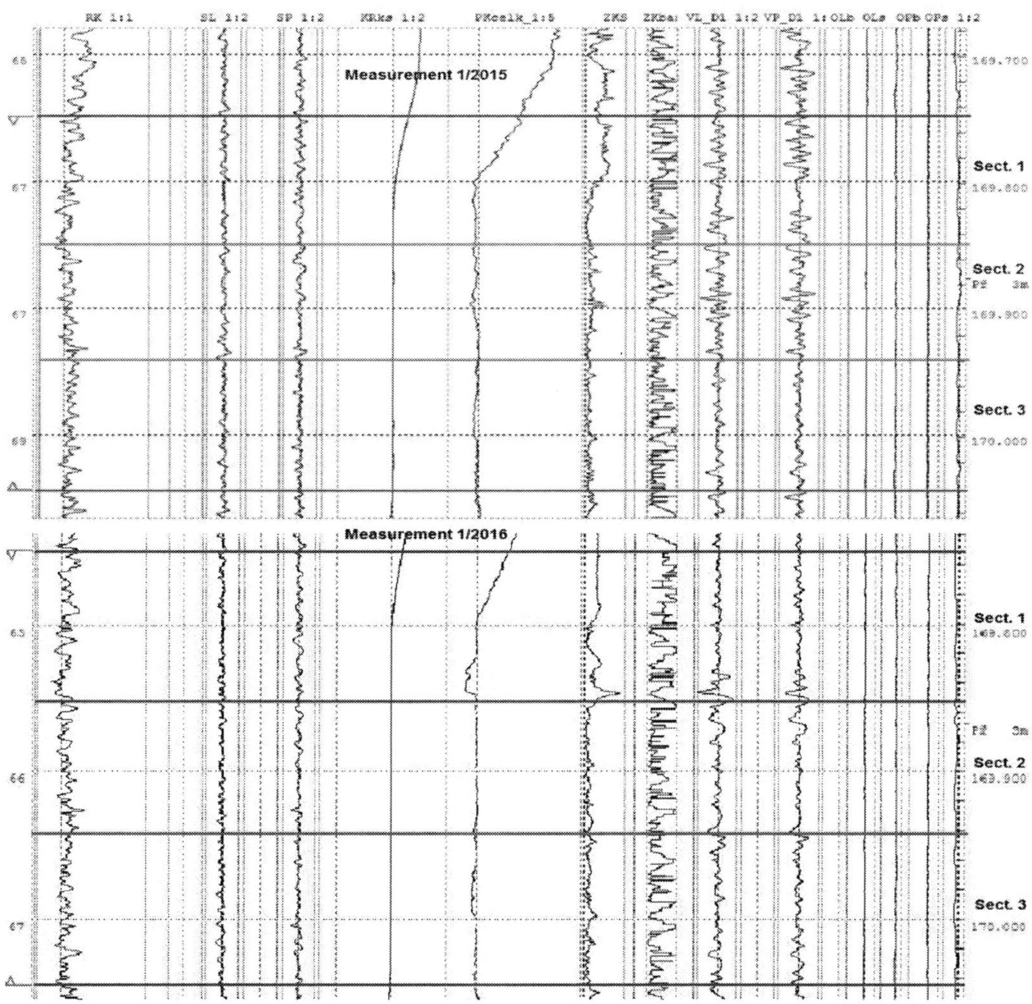

Figure 4. Comparison of track geometry measurements before and after test section's establishment.

Table 1. Results of static plate load tests.

Stationing [km]	E_2 [MPa] 04/2016	E_2 [MPa] 11/2016	Change [MPa]	Change [%]
169.830	25.0	29.7	4.8	19.1
169.860	51.9	76.1	24.1	46.5
169.910	58.3	77.8	19.5	33.4
169.960	38.7	39.2	0.6	1.4

Figure 5 displays the development of the impact modulus of deformation detected during three successive measurement campaigns at the rail seat level near the left rail string. The results imply that the average load-bearing capacity in the section fitted with a geocomposite (41.3 MPa to 44.0 MPa) was higher in all campaigns than in the adjoining sections without a geocomposite (25.8 MPa to 27.2 MPa). The chart also clearly shows a significant change in the load-bearing capacity at the interface of the test section and subsection 1 (km 169.840 vs. km 169.850).

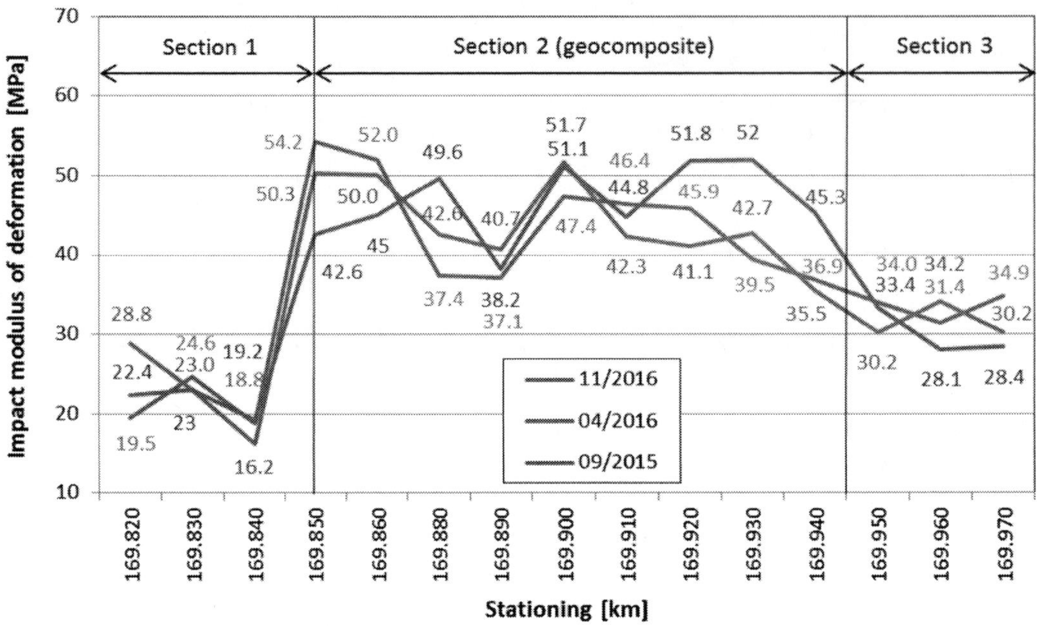

Figure 5. Results of impact plate load tests.

Table 2 presents the average values of the rail string deflection during the passage of trains, identified in 4 characteristic profiles. The measurement took place in spaces between sleepers, on the right rail string for the reason of better access and higher safety during the measurement. The highest values detected after the passage of all trains within an integral time interval lasting always no less than 4 hours in both directions were included in the evaluation.

The results imply that after the geocomposite placement, or after ballast cleaning and successive tamping, a significant reduction in the average deflection in the respective profiles occurred. Since then, the deflection values in profile 1, which falls in subsection 1, have shown a rising trend, while the values in both profiles of the test section have remained at roughly the same level. A significantly lower average speed of trains in the second line of the table is related to the reduction of the line speed limit in the monitored section during the time of ballast consolidation, i.e. until the time shortly after the last tamping.

Table 2. Average values of rail deflection during train passages.

Date of measurement	Average speed [km/h]	Rail deflection [mm]			
		km 169.846	km 169.889	km 169.916	km 169.955
06/2015	59.0	4.65	11.85	3.07	1.52
09/2015	46.3	1.65	1.51	1.43	1.80
04/2016	66.7	3.37	1.16	1.38	1.78
11/2016	63.8	5.92	1.29	1.36	1.25

The continuous measurement of the rail string height by means of levelling allows stating that the ballast bed consolidation process is gradually coming to an end. Whereas the average rail deflection in the test section with the geocomposite measured between the measurement campaigns of September 2015 and April 2016 was 7.2 mm, the difference between April 2016 and November 2016 accounted for only 1.6 mm. Trial holes allowed visual checks of the state of geocomposite placement. It was found in undamaged condition and no penetration of fine-grained soil from the subgrade into the ballast had been detected.

6. Conclusion

Based on the measurements performed in the monitored section to-date, the findings obtained are as follows:

- The geocomposite was successfully fitted under the ballast, or on the transversally inclined track formation. This allows efficient ballast drainage.
- The geocomposite prevented undesirable migration of fine-grained soil particles from the subgrade into the ballast.
- A significant improvement of key track geometry parameters was detected in the section with the geocomposite, deviations are evident at interfaces of sections with a different composition of the construction.
- The load-bearing capacity at the rail seat level identified in both the static and impact plate load test was significantly higher in the test section than in both adjoining subsections.
- A drop and homogenization of rail deflection values during train passages was identified in the section fitted with the geocomposite, which is accompanied by enhanced riding comfort and lower wear of the component parts of the permanent way and substructure.

The monitoring campaign of the section will continue at least by the end of 2017, and a complex evaluation of the efficiency and contributions of the used technology, including its effects on track geometry parameters, will be subsequently made.

Acknowledgements
Research reported in this paper was supported by Competence Centre's programme of Technology Agency of the Czech Republic (TA CR), project Centre for Effective and Sustainable Transport Infrastructure (no. TE01020168).

References
[1] Brown S F, Kwan J and Thom N H 2007 Identifying the key parameters that influence geogrid reinforcement of railway ballast *Geotextiles and Geomembranes* **25** pp. 326-35
[2] Tutumluer E, Huang H and Bian X 2009 Research on the Behaviour of Geogrids in Stabilisation applications *Proc. of Jubilee Symposium on Polymer Geogrid Reinforcement (London)*
[3] Cook J, Dobie M, Buckley J and Bhavsar R 2015 The Use of Multiaxial Geogrids in Rail Trackbed Stabilisation *Proc. Int. Heavy Haul Association Conf. IHHA2015 (Perth)*
[4] Hall C D and Sharpe P 2007 Review of geogrid stabilisation of railway ballast with reference to performance and durability *Proc. Int. Conf. on Railway Engineering (London)* ed M C Forde (Edinburgh) pp 201-5
[5] Indraratna B, Hussaini S K K and Vinod J S 2013 The lateral displacement response of geogrid-reinforced ballast under cyclic loading *Geotextiles and Geomembranes* **39** pp. 20-29
[6] Horníček L, Jasanský P and Nachtneblová K 2014 Long-term monitoring of railway test sections with geocomposites positioned under the ballast bed *Proc. Inter. Conf. on Geosynthetics (Berlín)* ed M Ziegler, G Bräu et al (Essen: Deutsche Gesellschaft für Geotechnik e.V.)

IOP Conf. Series: Materials Science and Engineering **236** (2017) 012040 doi:10.1088/1757-899X/236/1/012040

Determination of the elastic modulus of fly ash-based stabilizer applied in the trackbed

Vít Lojda[1], Martin Lidmila[1] and Marek Pýcha[2]

[1]Czech Technical University in Prague, Faculty of Civil Engineering, Thákurova 7, Prague 160 00, Czech Republic

[2]University of Birmingham, School of Engineering, Birmingham Centre for Railway Research and Education, Edgbaston, Birmingham, B15 2TT, United Kingdom

E-mail: vit.lojda@fsv.cvut.cz

Abstract. This paper describes a unique application of a fly ash-based stabilizer in the trackbed of a railway main line. The key goals of the stabilizer application are to protect the subgrade against the ingress of rain water, to increase the frost resistance and to remediate the natural ground constituted of weathered rock. The stabilizer was designed as a mixture of fly ash, generated as a waste material from coal plants, gypsum, calcium oxide and water. The mixture recipe was developed in a laboratory over several years. In 2005, a trial section of a railway line with subgrade consisting of clay limestone (weathered marlite) was built in the municipality of Smiřice. Since then, periodical measurements including collection of samples for laboratory evaluation of the fly ash-based stabilizer have taken place. Over the time span of the measurements, changes in mineral composition and development of fly ash transforming structures leading to the formation of C-A-S-H gel were detected. This paper describes the experimental laboratory investigation of the influence of dynamic loading on the elastic modulus of fly ash stabilizer samples and the development of permanent deformation of the samples with increasing number of loading cycles.

1. Introduction

Trackbed is an integral part of a railway track structure, designed to and maintained in such a state that it can preserve the line and the level of a railway track. It provides support for sleepers of the track and prevents the failure of the structure due to excessive settlement or shear failure on the formation or subgrade level. In a conventional track structure, trackbed consists of a ballast layer and additional optional trackbed layers. The additional layers are introduced into the trackbed system for two main reasons. First reason is mechanical and is applied when the bearing capacity of subgrade soils is not high enough to support the ballast layer. Various materials can be utilized for the layers with blanketing sand or crushed aggregate/engineering filling being the most common [e.g., 1,2]. Second reason relates to the influence of climate on subgrade soils. At some instances, the mechanical strength of subgrade soils is sufficient to provide the required support but the soils are susceptible to either the ingress of the surface water or the effect of frost cycles or both. In such cases, the trackbed layers might be introduced to insulate the subgrade soils against the water ingress and against the low temperatures. It can be seen that 'traditional' trackbed layer materials as sand or aggregates are usually permeable, i.e. cannot be used as an insulation against the water ingress. Also, the thermal conductivity of sands/aggregates is quite high, which may result in a high required thickness of the trackbed layers if these materials are to be used as thermal insulators [e.g., 1, 3, 4]. To overcome the above stated disadvantages of sands/aggregates, some standards allow for the design of a protective layer made of concrete or asphalt or the usage of geosynthetics [e.g., 1]. While such solutions clearly provide impermeability against the water ingress and usually fulfil the thermal insulation requirements as well, their disadvantage is high cost. Hence, there has been an effort to investigate the possibility of introducing a new construction material for trackbed layers which could provide sufficient water and thermal insulation while reducing the costs required to build the layer.

Part of the research into the new trackbed material has been conducted at the Czech Technical University in Prague. As soon as 2005, Lidmila [4] investigates the possibility of introducing a trackbed layer made of fly ash (pulverised fuel ash) to protect subgrade consisting of clay limestone (weathered marlite) during the reconstruction of a main line track in Smiřice station, Czech Republic. The proposed solution was approved and a unique application of a trackbed stabilizer made of fly ash was built in April 2005. Since 2005 to 2011 and then again from 2014, periodic biannual investigations are conducted on the site. The investigations consist of in-situ Plate Load Test measurements and sampling for further laboratory investigations, e.g., compressive strength, reference density. Details and a summary of the first ten years of the research can be found in [3, 5, 6].

Since the beginning, the research has focused on time-related changes in the material properties of the fly ash layer, e.g. changes in compressive strength, changes in reference density, water regime of the layer in reference to weather seasons and microanalysis of the material. However, recently the research has been expanded to include the behaviour of the material under dynamic loading. The research extension seeks to accomplish two main aims. Because no methodology for dynamic investigation of trackbed materials is currently adopted in the standard system of the railways in the Czech Republic, the first aim has been to develop a suitable methodology for the investigation. The second aim has been to investigate the changes in elastic modulus and the evolution of permanent deformation of the fly ash material under cyclic loading.

This paper presents the methodology adopted for the research along with some preliminary results. The adopted cyclic loading sequence is discussed and results for a representative sample of the fly ash material under the loading are shown. The changes in the elastic modulus and the evolution of permanent deformation with the increasing number of loading cycles are discussed.

2. Methodology

The investigation of the behaviour of the fly ash material was based on laboratory experiments performed on samples extracted from the fly ash trackbed layer. The samples were obtained by core drilling from the layer during the biannual visits in Smiřice. Details of the sample collection can be found in Lidmila & Lojda [5]. After performing the cyclic loading experiment described here, the samples were subjected to a destructive compression strength test. For this reason, sample collection and preparation followed ČSN EN 13286-41 [7] and ČSN EN 13286-43 [8] as closely as practicable.

As mentioned above, because no agreed-on methodology currently exists in the railway standards system in the Czech Republic, the methodology developed for the presented research combines cyclic loading sequences adopted from resilient modulus testing with the laboratory instrumentation adopted from the testing of the elastic modulus. Both parts are described in detail below.

2.1. Loading function

For cyclic loading function, it was intended to subject the material to a loading as close to real traffic loading as reasonably practicable. Details about this problem and review of its solutions can be found e.g. in Brown [9]. Nowadays, suggestions for a cyclic loading function sufficiently close to the real traffic loading are included for instance in BS EN 13286-7 [10] or AASHTO T307 [11]. In terms of frequency and loading shape, AASHTO T307 suggests a loading sequence consisting of a half-sine impulse with the duration of 0.1 s followed by a 0.9 s long period of zero loading. Such loading sequence corresponds well with the real traffic loading under intermediate speeds (70-100 km/h) [e.g., 9]. However, due to the limited sampling rate of the data logger used (see below) the loading sequence adopted for the presented research had to be changed into the form as presented on Figure 1. Also, the loading frame used (see below) requires minimum 'seating force' of 0.2 kN to be applied all the time.

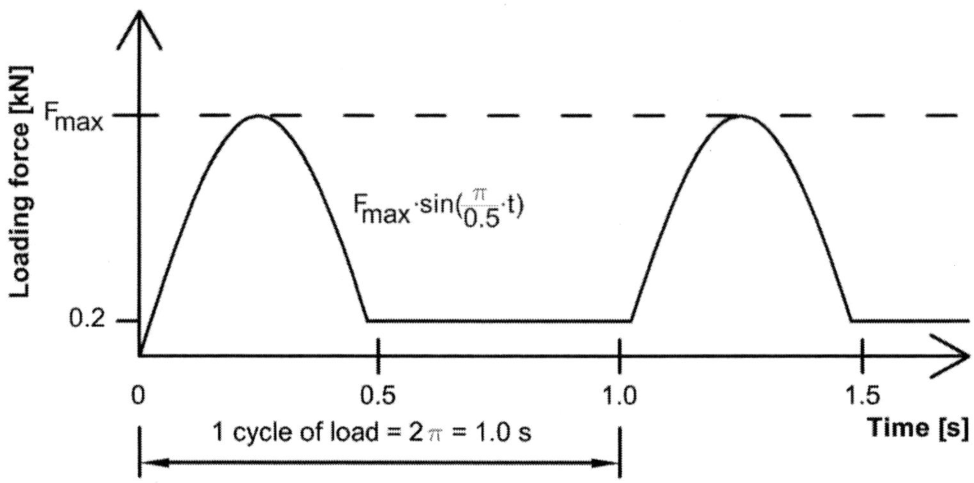

Figure 1. Adopted cyclic loading sequence for the laboratory experiment.

In terms of the maximum load per cycle and the number of loading cycles, BS EN 13286-7 suggests to apply load in five equal steps and to perform 100 cycles in each loading step. However, the primarily focus of the BS standard is on testing of soils in a triaxial apparatus. Hence, the suggested maximum load values were too small to be applicable for the presented research. Instead, based on the previous investigation of the compressive strength of the fly ash trackbed material [3], the maximum load of 15 kN (2 MPa) was adopted for this research. The loading was applied in five steps with the maximum load of 3, 6, 9, 12 and 15 kN respectively and 100 cycles of loading were performed in each step. The applied loading force measured during the experiment for a representative sample is shown on Figure 2.

2.2. *Instrumentation of the laboratory experiment*

The testing of the samples was performed with a loading frame Inova Praha ZUZ 200 1350 with the maximal dynamic load of 160 kN. The contraction of the samples in the vertical direction was measured along the samples by means of three absolute LVDT sensors of displacement Ahlborn FWA025T [12]. In accordance with ČSN EN 13286-43 [8], the displacement sensors were installed in between two metal rings fixed to the top and the bottom of the sample. The applied force as well as the vertical displacement were stored in Ahlborn Almemo 2690-8A data logger [13]. The sampling rate of 20 samples per second was chosen for the data storage, which is the maximum available rate the data logger can provide. The experiment set up is show on Figure 3.

3. Results and discussion

In this paper, only preliminary results of the research are presented. Hence, only results for a representative sample are shown.

Figure 4 shows the measured vertical displacement for all three LVDT sensors. It can be seen that with increasing force the displacement increases into negative values, i.e. the sample contracts. Sensors 1 and 2 seem to be in agreement but Sensor 3 seems to not follow the anticipated trend. The discrepancy may be explained by the method used to fix the steel rings to the samples during the experiment. The rings were fixed to the samples by rectifying screws. Even though the screws had pointed tip, as the cyclic loading progressed, it was observed that the screws were becoming loose and the rings deviated from its original positions slightly. Despite the absolute value measured by Sensor 3 does not have a significant impact on the mean value used for further evaluation, the inconsistency needs to be investigated further in order to refine the applied methodology. Figure 5 shows the mean value of displacement calculated from all three sensors, which is used for further investigation.

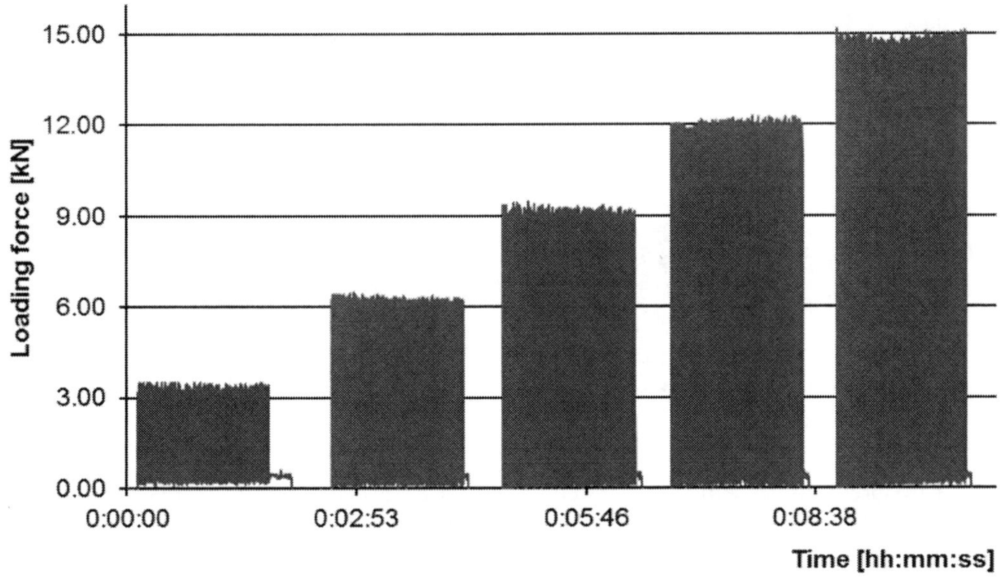

Figure 2. Loading force for the representative sample.

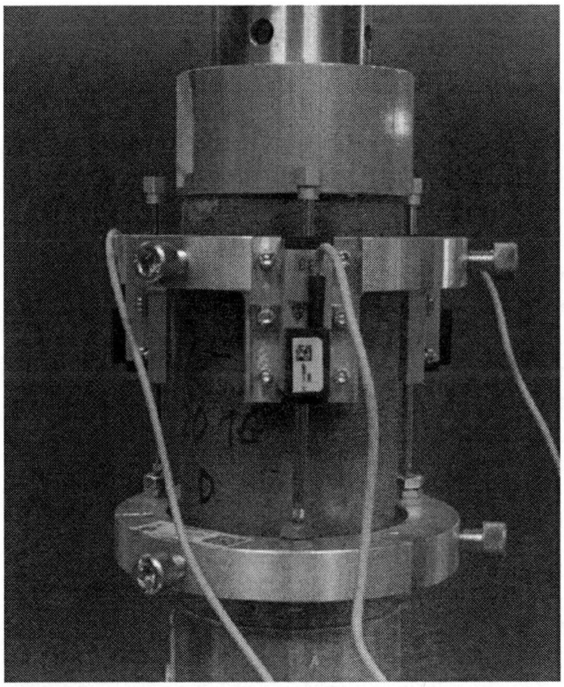

Figure 3. Instrumentation used during the experiment.

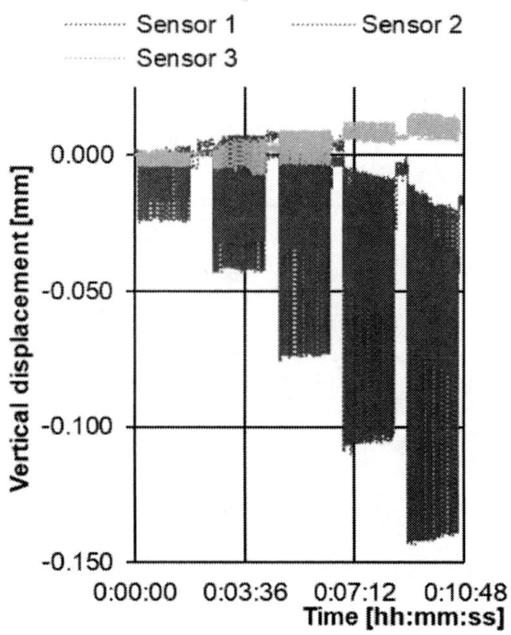

Figure 4. Measured displacement for the representative sample.

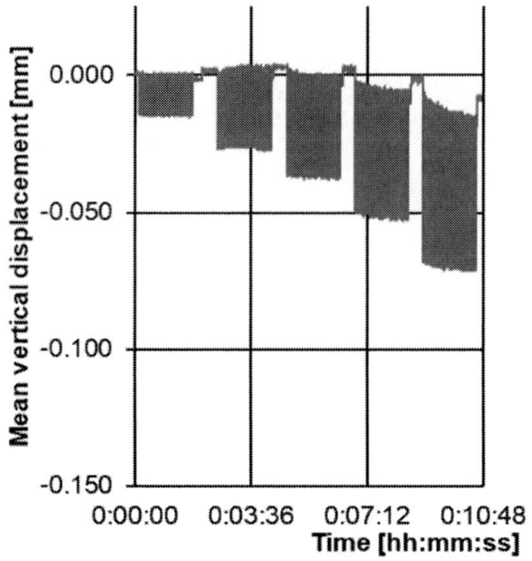

Figure 5. Mean value of displacement for the representative sample.

The applied uniaxial stress in the sample was calculated using the force from Figure 2 and the geometry of the sample. Similarly, the applied strain was calculated using the mean value of displacement from Figure 5. The resulting stress and stain were combined to plot hysteresis loops of the loading cycles on Figure 6.

It can be seen on Figure 6 that hysteresis loops for Step 1 and Step 2 appear quite stable, neither changing their shape nor move in the stress-strain space. This suggests close to pure elastic behaviour of the sample in Steps 1 and 2 of the loading For Steps 3, 4 and 5 it can be seen that the shape of the hysteresis loops is becoming more convex, which is a characteristic change with increasing non-linearity [14]. Also, there is clearly visible ratcheting for Steps 3-5 and the rate of ratcheting seems to increase with the increasing load, which again suggests increasing non-linearity. From Figure 6 it can be derived that the elastic limit for the representative sample is approximately 1000 kPa.

The elastic modulus E was calculated for every loading cycle as a secant modulus with the strain at the end of the loading cycle being adjusted for the influence of the permanent strain. The change of the elastic modulus with the increasing number of cycles of loading is presented on Figure 7.

It can be seen from Figure 7 that for the first two steps of loading the value of the modulus lies between 6000-6500 MPa, a slightly lower value than the value for the remaining steps, which approaches 7000 MPa. The ratcheting is also visible, especially for the last step of loading.

The last investigated characteristic was the evolution of permanent deflection, presented on Figure 8. Due to the anticipated movement of the steel ring (above) it was only possible to calculate the value of permanent strain which is added to the total permanent strain in each loading cycle. Hence, it should be noted that the figure does not show total permanent strain but only its rate.

It can be seen from Figure 8 that the absolute measurement of the permanent strain approaches the resolution limits of the Ahlborn FWA025T sensors. However, when linear trend is calculated from the measured data, the evolution of the permanent strain is practically zero for the first three steps. A non-zero evolution of permanent strain can be observed for Step 4 and 5. Also, the trend for the last two steps decreases, which corresponds well with the ratcheting behaviour observed from the hysteresis loops on Figure 6. The combination of results on Figures 6 and 8 suggests that the last three steps lie in the non-elastic area but under the shakedown limit. Hence, the shakedown limit of the representative sample is higher than 2000 kPa for the representative sample.

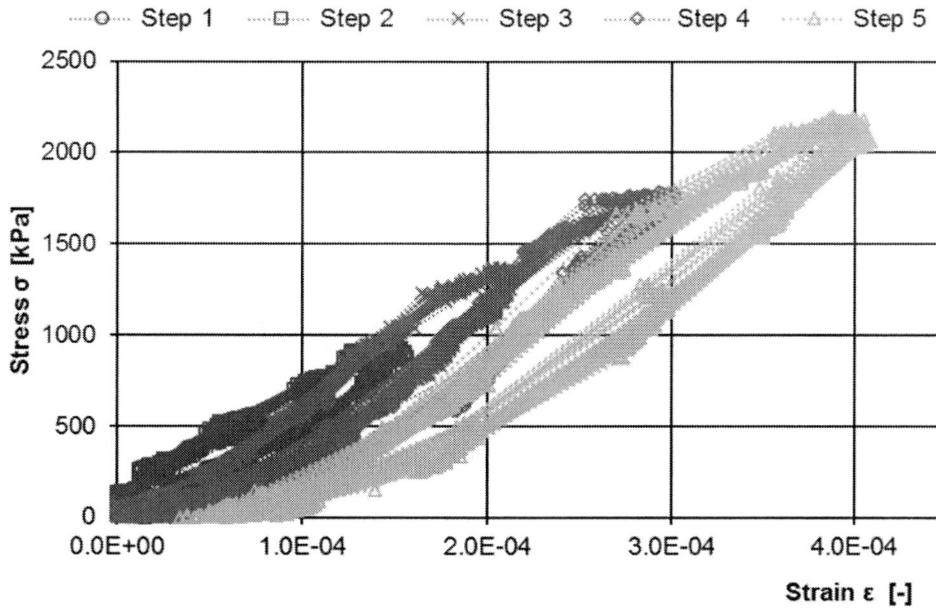

Figure 6. Hysteresis loops for cyclic loading of the representative sample.

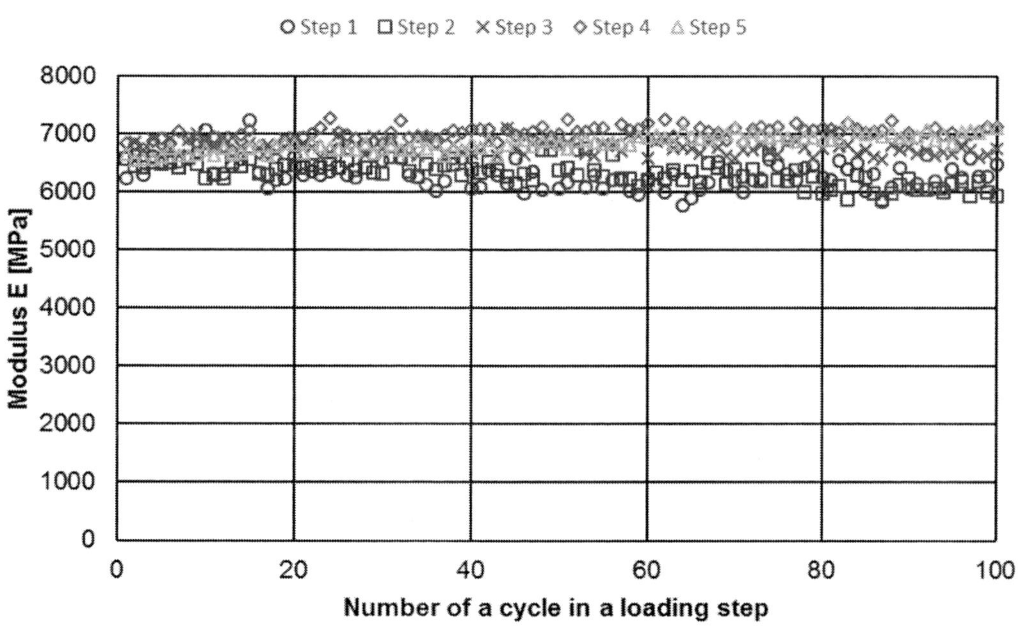

Figure 7. Evolution of the elastic modulus with cycling for the representative sample.

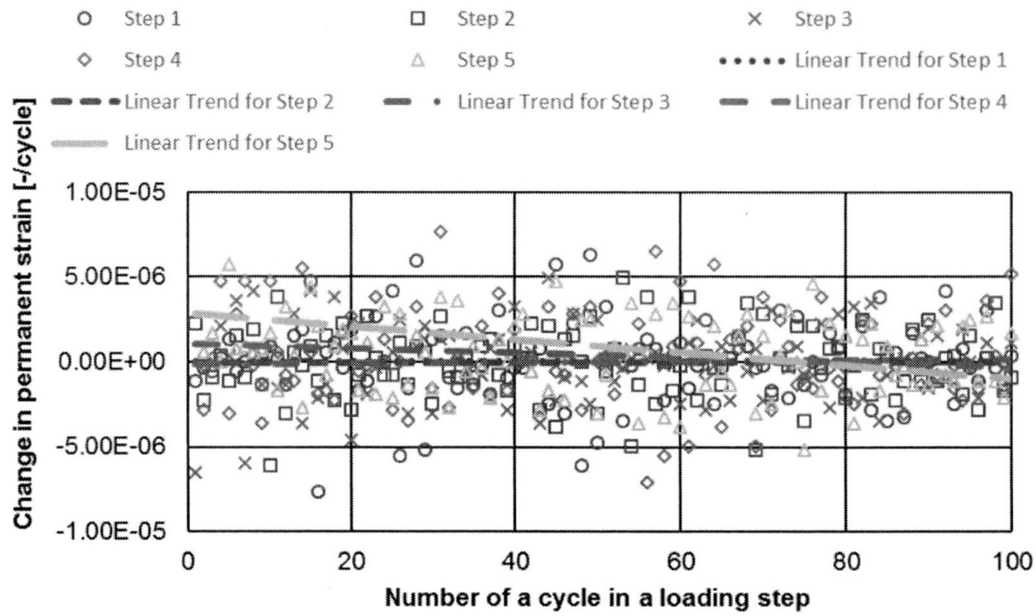

Figure 8. Evolution of the rate in permanent strain for the representative sample.

4. Discussion and Conclusions

In this paper, recent results from the investigation of the behaviour of fly ash trackbed material subjected to cyclic dynamic loading were presented. The methodology adopted for the investigation was discussed and the preliminary results of the investigation for the representative sample of the material were shown. From the presented results, the following conclusive remarks can be made:

For the methodology:

- The fixing of the steel rings to the sample via the rectifying screws shall be further refined to minimize the movement of the rings as the experiment progresses.
- The adopted maximum load of the last cycle seems to lie above the elastic limit, hence it allows to investigate some non-linear properties of the material. However, the load seems to lie below the shakedown limit, which might be important for limit analysis [9]. For this reason, it is recommended to apply additional loading steps until the failure of a sample is reached.
- For similar reasons as above, it is recommended to further investigate the failure point of the sample by continuing the cyclic loading in the last step until the failure of the sample
- The adopted instrumentation does not seem to be sufficient for the measurement of the absolute value of permanent contraction/strain.
- The maximum sampling rate of the data logger used requires significant changes to be made to the loading sequence. In order to use the loading frequency as suggested in e.g. AASHTO T307 a data logger with minimum sampling rate of 100 Hz should be used.

For the behaviour of the fly ash trackbed material:

- The material behaves non-linearly when dynamic loading exceeds the elastic limit.
- The material exhibits ratcheting in the non-linear area. The rate of ratcheting increases with increasing loading.
- The elastic limit for the representative sample seems to be approximately 1000 kPa.

- The elastic modulus of the representative sample lies between 6000-6500 MPa for loading in the elastic area and approaches 7000 MPa for loading in the non-linear area.

For future research, it is recommended to further refine the methodology of the experiment, as suggested above, so that the methodology can be applied on statistically representative number of samples.

Acknowledgements
The research presented in this paper was supported from the Competence Centers programme of Technology Agency of the Czech Republic (TAČR) within the Centre for Effective and Sustainable Transport Infrastructure (CESTI), project number TE01020168. This support is hereby gratefully acknowledged.

References

[1] SŽDC 2008 *S4: Železniční spodek (účinnost od 1. 10. 2008)* (Praha: SŽDC s. o.)
[2] Network Rail 2015 Track Bed Investigation, Design & Renewal *(NR/L2/TRK/4239)*
[3] Lidmila M, Lojda V, Kopecký L and Prošek Z 2015 Fly ash-based stabilizer for railway trackbed *International Journal of Advanced Research in Science and Engineering* **04**(02) pp 91-95
[4] Lidmila M 2005 *Ekologické využití vedlejších energetických produktů v konstrukci pražcového podloží* (CTU in Prague)
[5] Lidmila M and Lojda V 2015 Fly Ash-Based Stabilizer Sampling in Railway Track Bed and Determination of Young's Modulus *J. Polish Miner. Eng. Soc.* **16**(2) pp. 11–16
[6] Lojda V, Zikmund T, Sojka K, Kaiser J, Prošek Z and Lidmila M 2017 Microstructural Analysis of Fly Ash-Based Stabilizer for Track Bed *Key Eng. Mater.* **731** pp 66–73
[7] Czech Office for Standards, Methodology and Testing 2014 Unbound and hydraulically bound mixtures - Part 41: Test method for determination of the compressive strength of hydraulically bound mixtures *(ČSN EN 13286-41)*
[8] Czech Office for Standards, Methodology and Testing 2004 Unbound and hydraulically bound mixtures - Part 43: Test method for determination of the modulus of elasticity of hydraulically bound mixtures *(ČSN EN 13286-43)*
[9] Brown S F 1996 Soil Mechanics for Pavement Engineers *Géotechnique* **46**(3) pp 383–425
[10] British Standards Institution 2004 Unbound and hydraulically bound mixtures - Part 7: Cyclic load triaxial test for unbound mixtures *(BS EN 13286-7)*
[11] AASHTO 1999 Standard Method of Test for Determining the Resilient Modulus of Soils and Aggregate Materials *(AASHTO T307)*
[12] Ahlborn 2017 *FWA xxx T Displacement Sensor Product Sheet* (Holzkirchen, Germany)
[13] Ahlborn 2015 *Almemo 2690-8A Operating Instructions 4.4* (Holzkirchen, Germany)
[14] Pýcha M 2016 *Dynamic Elasto-Visco-Plastic Material Model for Modelling of Light Weight Deflectometer Test - a Uniaxial Preliminary Study* (University of Birmingham)

Impact capacity reduction in railway prestressed concrete sleepers with vertical holes

Chayut Ngamkhanong[1], Dan Li[1] and Sakdirat Kaewunruen[1,2]

[1]Department of Civil Engineering, The University of Birmingham, UK
[2]Birmingham Centre for Railway Research and Education, The University of Birmingham, UK

E-mail: s.kaewunruen@bham.ac.uk

Abstract. Railway prestressed concrete sleepers (or railroad ties) are principally designed in order to carry wheel loads from the rails to the ground as well as to secure rail gauge for dynamic safe movements of trains. In spite of the most common use of the prestressed concrete sleepers in railway tracks, the concrete sleepers are often modified on construction sites to fit in other systems such as cables, signalling gears, drainage pipes, etc. This is because those signalling, fibre optic, equipment cables are often damaged either by ballast corners or by tamping machine. It is thus necessary to modify concrete sleepers to cater cables internally so that the cables or drainage pipes would not experience detrimental or harsh environments. Accordingly, this study will extend from the previous study into the design criteria of holes and web openings. This paper will highlight structural capacity of concrete sleepers under dynamic transient loading. The modified compression field theory for ultimate strength design of concrete sleepers will be highlighted in this study. The outcome of this study will improve the understanding into dynamic behavior of prestressed concrete sleepers with vertical holes. The insight will enable predictive track maintenance regime in railway industry.

1. Introduction

Railway is one of the most efficient and effective modes of transportation, conveying cargo, passengers, minerals, grains, and so forth. Notably, railway prestressed concrete sleepers have been used in railway industry for over 50 years [1-5]. In general, railway sleepers (or called 'railroad ties' in North America) are a main part of railway track structures. The sleepers can be made of timber, concrete, steel or other engineered materials [6]. They have been systemically introduced to railway industry for many decades ago and currently are introduced in almost every railway network in the world. The main duties of sleepers are: (a) to transfer and distribute loads from the rail foot to the underlying ballast bed, (b) to hold the rails at the proper gauge through the rail fastening system, (c) to maintain rail inclination, and (d) to restrain longitudinal, lateral, and vertical movements of the rails and the rail gauge for safe passages of rolling stocks. It is important to note that railway sleepers are a structural and safety-critical component in railway track systems [7-15]. Therefore, the sleeper design and analysis must be ascertained to assure public safety and operational reliability. Critical performance criteria such as static capacity, dynamic strength and ultimate impact loading capacity should be evaluated as each property are mutually important and interconnected. Although the quasi-static capacity is often adopted in design process, the failure modes of the structural component must be known [16-17].

Railway track structures often experience impact loading conditions due to wheel/rail interactions associated with abnormalities in either a wheel or a rail [18]. All static, quasi-static, and impact loads are very important in design and analysis of railway track and its components. Generally, dynamic shock loading corresponds to the frequency range from 0 to 2000 Hz due to modern track vehicles. The shape of impact loading varies depending on various possible sources of such loading, e.g. wheel flats, out-of-round wheels, wheel corrugation, short and long wavelength rail corrugation, dipped welds and joints, pitting, and shelling. Wheel/rail irregularities induce high dynamic impact forces along the rails that may greatly exceed the static wheel load. In all cases, the impact forces are significantly dependent on the train speed. These impulses would occur repetitively during the roll. Loss of contact between wheel/rail, so-called "wheel fly", will occur if the irregularity is large enough, or the speed is fast enough. However, the impact force could be simplified as a shock pulse acting after the static wheel load is removed. The typical magnitude of impact loads from the reviewed cases [4] varies roughly between 200 kN and up to 750 kN, depending on the causes and the traveling speed of train. The durations of such loads are quite similar, varying between 1 and 10 msec. However, the representative values of the first peak (P1) of the forces caused by dipped joints should be about 400 kN magnitude with 1 to 5 msec time duration. For the second peak (P2), the average values are about 80 kN magnitude and 5 to 12 msec time duration. Therefore, it should be taken into account that the typical duration of impact wheel forces varies widely between 1 and 12 msec [5-7]. Note that these actual loading conditions are different to the loading conditions specified in type testing methods of sleeper standards [1, 10]. This is because the type testing methods are commonly used for benchmarking purpose.

Currently, railway track components are often being modified at construction sites to fit with signaling gears, cables, and additional train derailment protections, such as guard rails, check rails, Earthquake protection rails, etc. [19-21]. The practical guideline for crosstie retrofit has not been well established and many attempts were carried out based on trials and errors. Despite a common task in construction site, the behavior of holes and web openings on concrete crossties has not been well documented in open literature. In this manner, it is important to ensure that concrete crossties can be retrofitted and modified for add-on fixture in practice. The emphasis of this study has been placed on the impact capacity of the crossties with vertical holes (e.g. for bolting and fixture of automatic train control system). The insight into these behaviors will not only improve safety and reliability of railway infrastructure, but will enhance the structural safety of other concrete structures.

2. Prediction for ultimate moment capacity

2.1 Modified compression field theory

In this study, the moment curvature has been used to represent the capacity of prestressed concrete sleepers. The moment capacity are predicted by the modified compression field theory using Response-2000 [22]. This theory is capable of predicting the behaviour of reinforced concrete subjected to in-plane shear and normal stresses. The concrete stresses in principal directions along with prestressing steel are considered in only axial direction and uncracked portion will carry on to sustain a load in the analysis [23].

2.2 Effect of strain and loading rates

The strain rate plays an important role in material strengths. Based on the assumption of perfect bond between prestressing wires and concrete, the dynamic material properties of concrete and prestressing wires can be determined as follows [24].

Concrete:

$$\frac{f'_{c,dyn}}{f'_{c,st}} = 1.49 + 0.268\log_{10}\dot{\varepsilon} + 0.035[\log_{10}\dot{\varepsilon}]^2 \quad (1)$$

Prestressing wires:

$$\frac{f_{y,dyn}}{f_{y,st}} = 10^{0.38 \log_{10} \dot{\varepsilon}^{-0.258}} + 0.993 \qquad (2)$$

Where $f_{y,dyn}$ is the dynamic upper yield point stress, $f'_{c,st}$ is the static upper yield point stress of prestressing wires (about 0.84 times proof stress), and $\dot{\varepsilon}$ is the strain rate in tendon.

3. Material properties

In this study, there are 1 prestressed concrete sleeper with full cross section and 4 sleepers with a vertical hole, as shown in Figure 1. The position of the vertical hole is at the middle between the sleeper end and the edge of rail seat. It is interesting to note that the size of the vertical hole are varied from 20 mm to 50 mm.

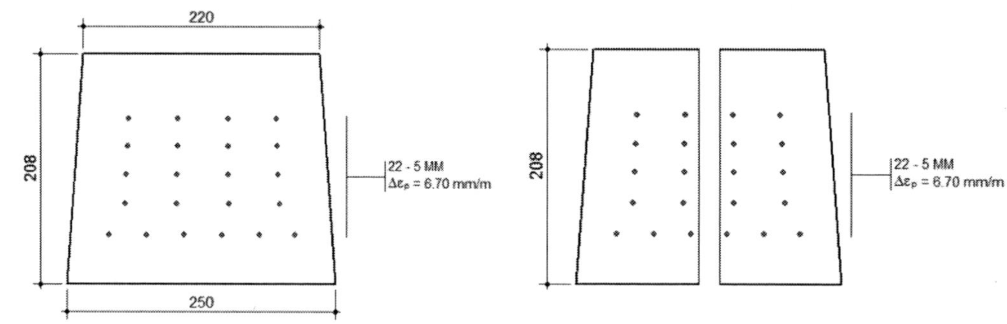

(a) (b)

Figure 1. Prestressed concrete sleepers a) Full cross section b) with vertical hole.

3.1 Static

The dimension and shape of prestressed concrete sleepers are shown in Figure 1.The high strength concrete was used with the design cylinder compressive strength of 55 MPa. The stress-strain curve of concrete derived by Vechio and Collin [23] was used in this study, as shown in Figure 2. The 22-prestressing steels used were the high ultimate strength with rupture ultimate strength of 1860 MPa, as shown in Figure . The initial elastic modulus of prestressing steel was 20000 MPa.

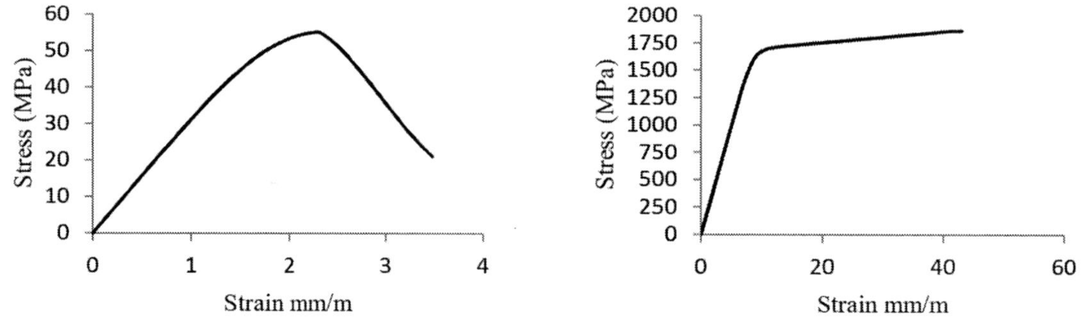

(a) (b)

Figure 2. Stress-strain curve of a) concrete b) steel.

3.2 Dynamic

The prediction of moment capacity has been carried out using the data obtained from the previous experiments [25-26]. It should be noted that the average total duration of impact forces is about 4 ms.

In this study, the strain rate of concrete is varying from 2 ms to 8 ms. It is well known that the dynamic ultimate strain of prestressing steel is about 20x10^3, and the total duration of impact force influencing the steel fibre is roughly from 6 ms to 12 ms. This is because the impact stress wave delays during the stress propagation and will be impeded through concrete [24]. Using Equation. (1) and (2), the dynamic strength of materials can be obtained as the input for the sectional analysis. The 4 pairs of strain rates variations are used in this study, as shown in Table 1.

Table 1. Strain rate variations under impact loading.

Material	A	B	C	D
Concrete	2	4	6	8
Prestressing wires	6	8	10	12

4. Results and discussions

4.1 Static analysis

In general, the first stage of the behavior of material is elastic range when there is no damage in the material and the applied external force is less than the proportional limit. Then, when the moment reaches the proportional yield point, the nonlinear behaviour takes place till the member reaches the ultimate capacity. After that, the curve drops rapidly due to the crushing and spalling of concrete.

The effect of vertical hole in the prestressed concrete sleepers is evaluated by the moment capacity. The obtained results indicate that the vertical hole in prestressed concrete sleepers can affect the moment capacity due to the reduction in cross sectional area. Figure 3 illustrates the moment curvature of prestressed concrete sleepers. It can be clearly seen that the ultimate moment capacity of prestressed concrete sleepers without a vertical hole is about 60 kNm which is more than the moment capacity of all sleepers with a vertical hole.

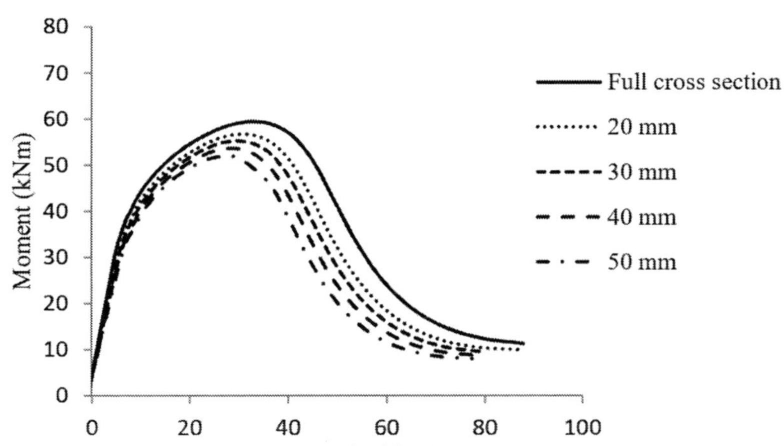

Figure 3. Moment-Curvature comparison between 5 prestressed concrete sleepers.

Table 2 displays the comparison of recorded moment capacity between the prestressed concrete sleepers with and without vertical hole. There are 2 conditions, crack and fail, can be determined from the modified compression field theory. It can be observed that the first crack of the sleepers is between 25-32 kNm and the fail condition is about 51-60 kNm. It can be clearly seen that when the vertical hole increase in diameter, reducing in width and cross sectional area, the moment capacity decrease.

Table 2. Moment capacities of prestressed concrete sleepers under static loading.

Section	Target conditions	Recorded moment capacity (kNm)
Full section	Crack	31.86
	Fail	59.34
20 mm hole	Crack	29.56
	Fail	56.67
30 mm hole	Crack	28.33
	Fail	55.14
40 mm hole	Crack	27.01
	Fail	53.57
50 mm hole	Crack	25.64
	Fail	51.90

4.2 Dynamic analysis

Figure 4 and Table 3 illustrate the moment curvature of the prestressed concrete sleepers under impact loading. It can be clearly seen that the trend of moment-curvature relationship under impact loading are similar to static loading but it is interesting that the ultimate moment capacity is more higher. Apart from that, it is observed that strain rate play a significant role in impact loading and moment capacity. It can be seen that when the strain rate increase, the moment capacity also increase.

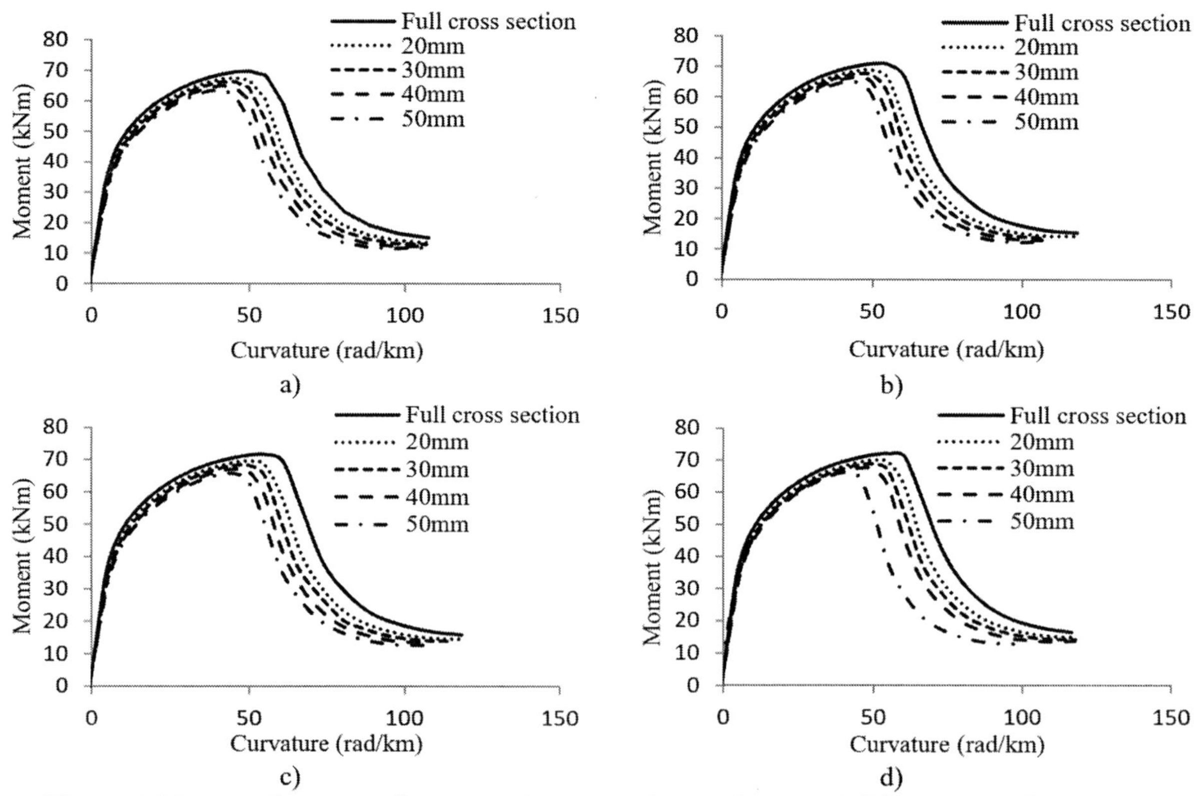

Figure 4. Moment-Curvature of prestressed concrete sleepers between 4 differences strain rate a) Case A b) Case B c) Case C d) Case D.

Table 3. Moment capacities of prestressed concrete sleepers under impact loading.

Section	Target conditions	Recorded moment capacity (kNm)			
		A	B	C	D
Full section	Crack	31.98	32.572	32.92	33.17
	Fail	69.79	70.89	71.50	72.02
20 mm hole	Crack	29.64	30.26	30.59	30.84
	Fail	67.63	68.69	69.49	69.96
30 mm hole	Crack	28.36	28.97	29.33	29.58
	Fail	66.32	67.59	68.17	68.78
40 mm hole	Crack	26.70	27.61	27.97	28.24
	Fail	65.03	66.30	67.04	67.48
50 mm hole	Crack	25.63	26.21	26.56	26.81
	Fail	63.61	64.83	65.61	66.20

Figure 5 illustrates the relationship between percentage of ultimate moment capacities reduction of prestressed concrete sleepers and percentage of area reduction. Ultimate capacities reduction can be calculated from Eq.(3).

$$Ultimate\ moment\ reduction\ (\%) = (M_{full\ cross\ section} - M_{with\ vertical\ hole})x100/M_{full\ cross\ section} \quad (3)$$

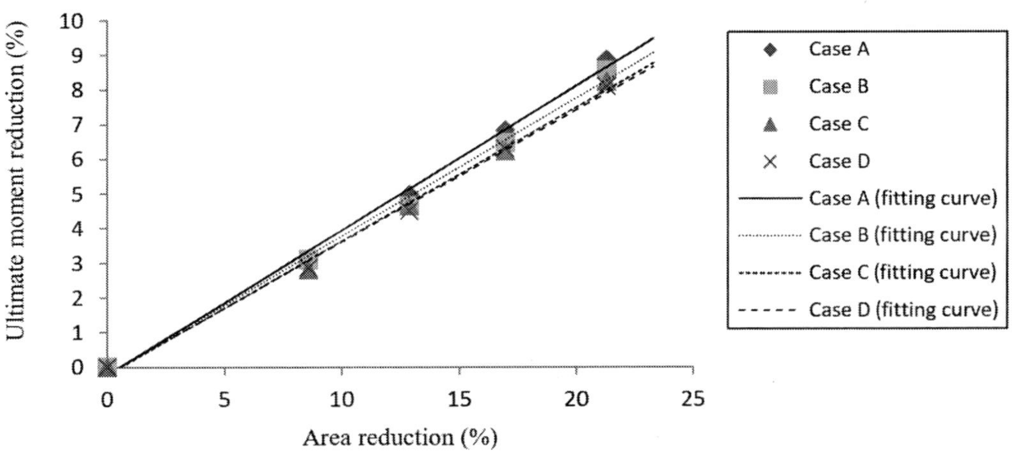

Figure 5. The ultimate moment reduction-area reduction relationship.

The moment capacities reduction show a linear upward trend when the cross sectional area are reduced. This indicates about nearly 10% reduction of moment capacities by about 20% reduction in cross sectional area. Moreover, it is interesting to note that, at the same percentage of area reduction, the moment capacity of sleepers with the lower strain rate is greater than that with higher strain rate.

5. Conclusions

This study is the world first to investigate the effect of vertical holes on the impact capacity of railway prestressed concrete sleepers. It exhibits that the vertical holes undermine strength and impact capacity of railway concrete sleepers. Therefore, it is essentially important for track and rail engineers to assure that the modification or retrofitting of concrete sleepers at construction sites is carried out in a proper manner. By the results obtained from these unprecedented studies, the vertical hole can reduce almost 10 % of dynamic load bearing capacity of the sleepers and such the hole also results in significant reduction of moment-curvature area, resulting in impaired structural ductility. The insight into the impact behavior of the concrete sleepers with vertical holes will enable safer built environments in railway corridor, especially for concrete sleepers whose structural inspection is very difficult in practice.

Acknowledgements
The authors would also like to thank British Department of Transport (DfT) for Transport - Technology Research Innovations Grant Scheme, Project No. RCS15/0233; and the BRIDGE Grant (provided by University of Birmingham and the University of Illinois at Urbana Champaign). The first author would like to express his gratitude for the PhD scholarship provided from the Birmingham Centre for Railway Research and Education and the School of Engineering at the University of Birmingham. The last author gratefully acknowledges the Japan Society for the Promotion of Science (JSPS) for his JSPS Invitation Research Fellowship (Long-term), Grant No L15701, at Track Dynamics Laboratory, Railway Technical Research Institute and at Concrete Laboratory, the University of Tokyo, Tokyo, Japan. The authors are sincerely grateful to European Commission for the financial sponsorship of the H2020-RISE Project No. 691135 "RISEN: Rail Infrastructure Systems Engineering Network," which enables a global research network that tackles the grand challenge in railway infrastructure resilience and advanced sensing in extreme environments (www.risen2rail.eu). We would also like to acknowledge the support from European Cooperation in Science and Technology (EU-COST) Action: TU1404 Towards the next generation of standards for service life of cement-based materials and structures.

References
[1] Standards Australia 2003 Railway track material - Part 14: Prestressed concrete sleepers *(AS1085.14-2003)*
[2] Kaewunruen S and Remennikov A M 2006 Sensitivity analysis of free vibration characteristics of an in-situ railway concrete sleeper to variations of rail pad parameters *Journal of Sound and Vibration* **298**(1) pp 453-61
[3] Kaewunruen S and Remennikov A M 2007 Investigation of free vibrations of voided concrete sleepers in railway track system *Proceedings of the Institution of Mechanical Engineers Part F - Journal of Rail and Rapid Transit* **221**(4) pp 495-507
[4] Kaewunruen S, Minoura S, Watanabe T and Remennikov A M 2016 Remaining service life of railway prestressed concrete sleepers *Proceedings of International RILEM Conference on Materials, Systems and Structures in Civil Engineering (Lyngby)*
[5] Kaewunruen S and Chamniprasart K 2016 Damage analysis of spot replacement sleepers interspersed in ballasted railway tracks *Proceedings of the 29th Nordic Seminar on Computational Mechanics (Chalmers University of Technology, Gotenburg, Sweden)*
[6] Esveld C 2001 *Modern Railway Track* (Delft University of Technology)
[7] Kaewunruen S and Remennikov A M 2013 On the residual energy toughness of prestressed concrete sleepers in railway track structures subjected to repeated impact loads *Electronic Journal of Structural Engineering* **13**(1) pp 41-61
[8] Remennikov A M, Murray M H and Kaewunruen S 2007 Conversion of AS1085.14 for railway prestressed concrete sleeper to limit states design format *Proc. CD-Rom Conf. AusRAIL Plus (Sydney)*
[9] Remennikov A M, Murray M H and Kaewunruen S 2012 Reliability based conversion of a structural design code for prestressed concrete sleepers *Proceedings of the Institution of Mechanical Engineers: Part F Journal of Rail and Rapid Transit* **226**(2) pp 155-73
[10] British Standards Institution 2009 Railway applications. Track. Concrete sleepers and bearers. Prestressed monoblock sleepers. *(BS EN 13230-2:2009)*
[11] Fryba L 1996 *Dynamics of railway bridges* (Thomas Telford Ltd.)
[12] Gamage E, Kaewunruen S and Remennikov A M 2015 Design of holes and web openings in railway prestressed concrete sleepers *Railway Engineering Conference (Edinburgh UK)*

[13] Remennikov A M and Kaewunruen S 2009 Reliability based design of railway prestressed concrete sleepers *Reliability engineering advances* ed Hayworth G I (New York: Nova Science Publishers, Hauppauge)
[14] Kaewunruen S and Remennikov A M 2009 Structural safety of railway prestressed concrete sleepers *Aust. J. Struct. Eng* **9**(2) pp 129–140
[15] Kaewunruen S and Remennikov A M and Murray M H 2012 Limit states design of railway concrete sleepers *Proc. of ICE Transport Journal 2012* **164**(TR1)
[16] Standards Australia 2009 Design of concrete structures *(AS3600)*
[17] Standards Australia 2005 Reliability-based design of structures *(AS5104)*
[18] Remennikov A M and Kaewunruen S 2008 A review on loading conditions for railway track structures due to wheel and rail vertical interactions *Structural Control and Health Monitoring* **15**(2) pp 207-34
[19] Kaewunruen S, Gamage E K and Remennikov A M 2016 Structural behaviours of railway prestressed concrete sleepers (crossties) with hole and web openings *Procedia Engineering* **161** pp 1247-53
[20] Kaewunruen S, Gamage E K and Remennikov A M 2016 Modelling railway prestressed concrete sleepers (crossties) with holes and web openings *Procedia Engineering* **161** pp 1240-46
[21] Gamage E K, Kaewunruen S, Remennikov A M and Ishida T 2017 Toughness of Railroad Concrete Crossties with Holes and Web Openings *Infrastructures* **2**(3)
[22] Bentz EC 2000 *Sectional analysis of reinforced concrete members* (University of Toronto, Department of Civil Engineering)
[23] Vecchio F J and Collins M P 1986 The modified compression field theory for reinforced concrete elements subjected to shear *J Am Concr Inst* **83**(2) pp 219–31
[24] Wakui H and Okuda H 1999 A study on limit-state design for prestressed concrete sleepers *Concr. Lib. Int. JSCE* **33** pp 1–25
[25] Kaewunruen S and Remennikov A 2009 Impact capacity of railway prestressed concrete sleepers *Engineering Failure Analysis* **16**(5) pp 1520-32
[26] Li D, Ngamkhanong C and Kaewunruen S 2017 Influence of vertical holes on creep and shrinkage of railway prestressed concrete sleepers *Int. Conf. on Building up Efficient and Sustainable Transport Infrastructure* (*Prague*)

Influence of vertical holes on creep and shrinkage of railway prestressed concrete sleepers

Dan Li[1], Chayut Ngamkhanong[1] and Sakdirat Kaewunruen[1,2]

[1]Department of Civil Engineering, The University of Birmingham, UK
[2]Birmingham Centre for Railway Research and Education, The University of Birmingham, UK

E-mail: s.kaewunruen@bham.ac.uk

Abstract. Railway prestressed concrete sleepers (or railroad ties) must successfully perform two critical duties: first, to carry wheel loads from the rails to the ground; and second, to secure rail gauge for dynamic safe movements of trains. The second duty is often fouled by inappropriate design of the time-dependent behaviors due to their creep, shrinkage and elastic shortening responses of the materials. In addition, the concrete sleepers are often modified on construction sites to fit in other systems such as cables, signalling gears, drainage pipes, etc. Accordingly, this study is the world first to investigate creep and shrinkage effects on the railway prestressed concrete sleepers with vertical holes. This paper will highlight constitutive models of concrete materials within the railway sleepers under different environmental conditions over time. It will present a comparative investigation using a variety of methods to evaluate shortening effects in railway prestressed concrete sleepers. The outcome of this study will improve material design, which is very critical to the durability of railway track components.

1. Introduction

The railway sleepers (or called 'railroad ties') are a main structural and safety-critical component of railway track structures. The sleepers can be manufactured using timber, concrete, steel or other engineered materials [1]. Concrete sleepers were initially installed on track for many decades ago and at present are being used in almost everywhere in the world. Their major role is to redistribute wheel loads from the rail feet to the supporting ballast bed. Railway track structures are often subjected to impact loading conditions owing to wheel/rail interactions with common defects in either a wheel or a rail [2-5]. Durability and long term behaviours of railway prestressed concrete sleepers depend largely on their creep and shrinkage response. Many investigators have proposed various material models to predict creep and shrinkage but those were mostly based on general reinforced concrete concept. The popular uses of prestressed concrete in long span bridges, stadiums, silos and confined nuclear power plants have led to a concern of practitioners whether those predictive models could be realistically applied to prestressed concrete. Due to high initial elastic shortening in prestressed concrete, the creep and shrinkage effects should be critically re-evaluated in flexural members. This study will investigate methods to evaluate creep and shrinkage effects in railway prestressed concrete sleepers. Comparison between design codes of EUROCODE2 and AS2009-3600 will provide the insight into durability of

IOP Conf. Series: Materials Science and Engineering **236** (2017) 012042 doi:10.1088/1757-899X/236/1/012042

concrete sleepers. The outcome of the project will help rail track engineers to better design and maintain railway infrastructure, improving asset management efficacy [6].

This research aims at providing a principle understanding of creep and shrinkage effects in railway prestressed concrete sleepers. It will review the design criteria of prestressed concrete in Eurocode 2 and Australian Standard 2009-3600 and evaluate creep and shrinkage effects in railway prestressed concrete [7]. Furthermore, the effects among design method above should be compared in order to develop existing codes.

2. Prediction for creep and shrinkage

2.1. Methodology

There are three approaches in order to meet the objectives of the research. The initial step is to review the codes to understand the principles of prestressed concrete design, calculation processes of creep, shrinkage strain, loss of prestress and deflection. Before the calculation, the difference between codes should be known. Then all the calculation will be carried out analytically.

Second step is to analyse obtained data. To investigate the creep and shrinkage effects, sensitivity analysis has been carried out. Comparative study among different codes of practice will be performed.

The final step is to establish the loss of prestress and to calculate shortening of prestressed concrete sleeper.

2.2. Creep Prediction

Neville, A. M (1981) stated the concrete under load that strain increases with time is due to creep. Therefore, creep can be defined as the increase in strain under the sustained stress and it can be several times as large as the initial strain [8]. Creep is a considerable factor in concrete structure. According to Martin, L , creep occurs very slowly after immediate elastic strain has happened[9]. The factors influence creep coefficient following:

(1) Original water content
(2) Effective age at transfer
(3) Effective section thickness
(4) Ambient relative humidity
(5) Ambient temperature

If the load is removed, the strain decreases immediately due to elastic recovery and a gradual incomplete recovery due to creep. This behavior is shown in Figure 1.

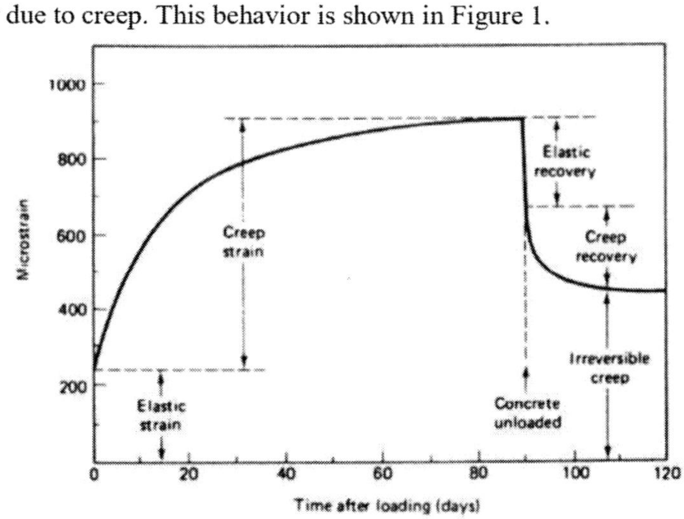

Figure 1 Time dependent creep.

When creep is taken into account, its design effects are always evaluated under quasi-permanent combination of actions irrespective of the design situation considered, i.e. persistent, transient or accidental.

2.2.1. Eurocode 2

The total creep strain ε_{cc} (∞, t_0) of concrete due to the constant compressive stress of σ_c applied at the concrete age of t_0 is given by :

$$\varepsilon_{cc}(\infty, t_0) = \varphi(\infty, t_0) \times \frac{\sigma_c}{E_c} \quad (1)$$

Where (∞,$_0$) is the final creep coefficient, which the value of σ_c does not exceed $0.45f_{ck}$ (t_0). E_c is the tangent modulus.

$$\varphi(\infty, t_0) = \varphi_{RH} \times \frac{16.8}{\sqrt{f_{cm}}} \times \frac{1}{(0.1+t_0^{0.20})} \quad (2)$$

$$\varphi_{RH} = 1 + \frac{1-0.01\times RH}{0.1+h_0^{0.333}}, \quad f_{cm} \leq 35MPa \quad (3)$$

$$\varphi_{RH} = (1 + \frac{1-0.01\times RH}{0.1+h_0^{0.333}}\alpha_1)\alpha_2, \quad f_{cm} > 35MPa \quad (4)$$

$$\alpha_1 = (\frac{35}{f_{cm}})^{0.7}, \quad \alpha_1 = (\frac{35}{f_{cm}})^{0.2} \quad f_{cm} = f_{ck} + 8MPa$$

$$t_0 = t_{0,T}(\frac{9}{2+t_{0,T}^{1.2}} + 1)^{\alpha} \geq 0.5, \quad \alpha = \{-1(S), 0(N), 1(R)\} \quad (5)$$

Where: RH = relative humidity in %, h_0 = 2Ac/u mm, Ac = cross sectional area, u = perimeter of the member in contact with the atmosphere, S, R and N refer to different classes of cement.

The final creep will be larger and final creep coefficient (∞,$_0$) is multiplied by a factor k_σ if the compressive stress applied at the age of t_0 exceeds $0.45f_{ck}(t_0)$ as can happen during prestress transfer process. The Table 1 shows the value of k_σ. The factor k_σ is given by:

$$k_\sigma = \exp[1.5 \times \left(\frac{\sigma_c}{f_{ck}(t_0)} - 0.45\right)] \quad (6)$$

Table 1. Value of $\boldsymbol{k_\sigma}$ in terms of f_{ck}.

$\frac{\sigma_c}{f_{ck}(t_0)}$	k_σ
0.5	1.078
0.6	1.252
0.7	1.455
0.8	1.691
0.9	1.964
1.0	2.282

The creep coefficient at any age *t* can be given by empirical solutions [6, 10-12].

2.2.2. Australian Standard 3600-2009

The creep coefficient at any time φ_{cc} can be determined by:

$$\varphi_{cc} = k_2 k_3 k_4 k_5 \varphi_{cc.b} \tag{7}$$

Where k_2 is the development of creep with time; k_3 is the factor which depends on the age at first loading τ (in days); k_4 is the factor which accounts for the environment; and k_5 is the factor which accounts for the reduced influence of both relative and humidity and specimen size.

For the development of creep with time k_2 can be calculated by:

$$k_2 = \frac{\alpha_2 (t-\tau)^{0.8}}{(t-\tau)^{0.8} + 0.15 t_h} \tag{8}$$

$$\alpha_2 = 1.0 + 1.12 e^{-0.008 t_h}\ ,\ t_h = 2A_g/u_e$$

Where t is any time in days; t_h is the hypothetical thickness; A_g is the cross-sectional area of the member; u_e is the portion of the section perimeter exposed to the atmosphere plus half the total perimeter of any voids contained within the section .

For factor k_3 which depends on the age at first loading τ can be shown as:

$$k_3 = \frac{2.7}{1+\log(\tau)}\ (\text{for } \tau > 1 \text{ day} \tag{9}$$

For the factor k_4 which accounts for the environment:

k_4 = 0.7 for an arid environment
= 0.65 for an interior environment
= 0.60 for a temperate environment
= 0.5 for a tropical or near-coastal environment

For the factor k_5 is given by:

$k_5 = 1.0\ when\ fc' \leq 50MPa$
$k_5 = (2.0-\alpha 3) - 0.02(1.0-\alpha 3) f'c\ when\ 50MPa \leq f'c \leq 100MPa$

Where α_3=0.7/($k_4\alpha_2$). The basic creep coefficient $\varphi_{cc.b}$ is shown Table 2 below:

Table 2. basic creep coefficient.

f_c' (MPa)	20	25	32	40	50	65	80	100
$\varphi_{cc.b}$	5.2	4.2	3.4	2.8	2.4	2.0	1.7	1.5

2.3. Shrinkage Prediction

Bhatt, P (2011) stated that both of creep and shrinkage are influenced by the same parameters. Shrinkage is not an entirely reversible process like creep and it can be also influenced by relative humidity, surface exposed to atmosphere, compressive strength of concrete and types of cement. Shrinkage can be divided by two parts [11]:

(1) Plastic shrinkage: it happens in few hours after concrete placed.
(2) Dry shrinkage: evaporation leads to loss of water.

IOP Conf. Series: Materials Science and Engineering **236** (2017) 012042 doi:10.1088/1757-899X/236/1/012042

2.3.1. Eurocode 2

The total shrinkage strain ε_{cs} can be given by:

$$\varepsilon_{cs}=\varepsilon_{ds}+\varepsilon_{as} \tag{10}$$

Where ε_{ds} is drying shrinkage strain; and ε_{as} is autogenous shrinkage strain.

2.3.2. Australian Standard 3600-2009

The total shrinkage strain ε_{cs} is shown below:

$$\varepsilon_{cs} = \varepsilon_{cse} + \varepsilon_{csd} \tag{11}$$

Where ε_{cse} is autogenous shrinkage strain; ε_{csd} is drying shrinkage strain.

The autogenous shrinkage εcse is given by:

$$\varepsilon_{cse} = \varepsilon'_{cse}(1.0 - \exp\{-0.1t\} \tag{12}$$

$$\varepsilon'_{cse} = (0.6f'_c - 1.0) \times 50 \times 10^{-6}\ (f'_c\ in\ MPa) \tag{13}$$

$$\varepsilon_{csd.b} = (1.0 - 0.008f'_c) \times \varepsilon'_{csd.b} \tag{14}$$

Where $\varepsilon'_{csd.b}$ depends on the quality of the local aggregates and may be taken as 800×10^{-6} for concrete supplied in Sydney and Brisbane, 900×10^{-6} in Melbourne and 1000×10^{-6} in elsewhere.

The drying shrinkage strain ε_{csd} after the beginning of drying $(t-\tau_d)$ can be estimated:

$$\varepsilon_{csd} = k_1 k_4 \varepsilon_{csd.b} \tag{15}$$

Where k_1 is the factor which describes the development of drying shrinkage with time; and k_4 is the factor which accounts for the environment [6, 10-12].

3. Material Properties

The effects of various proposed holes used for estimating creep, shrinkage strain will be evaluated. The fundamental engineering properties of prestressed concrete sleeper used for calculation are based on previous research by Remennikov et al. [2]. The parametric results are generated for comparisons between Eurocode 2 (EC2) and Australian standard 3600-2009 (AS). Figure 2 shows the cross section at rail seat of the prestressed concrete sleepers.

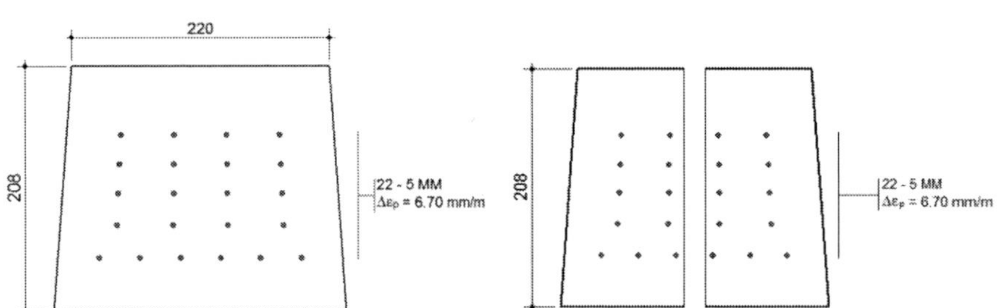

(a) full cross-section (b) with vertical hole

Figure 2. Cross section of railway sleepers.

4. Creep and Shrinkage Evaluations

4.1. Creep evaluations

To investigate relationship between hole size and creep, the 5 cases have been analysed using different characteristic strength (no hole, 20mm dia., 30mm dia., 40mm dia. and 50mm dia.), which are plotted in Figure 3. The data of creep coefficient are calculated by EC2 and AS codes respectively. All the cases are estimated from 1 day up to 36500 days (100 years) in the same conditions (uniform dimension of sleepers, 70% relative humidity, curing for 7 days and first loading at 14 days etc.).

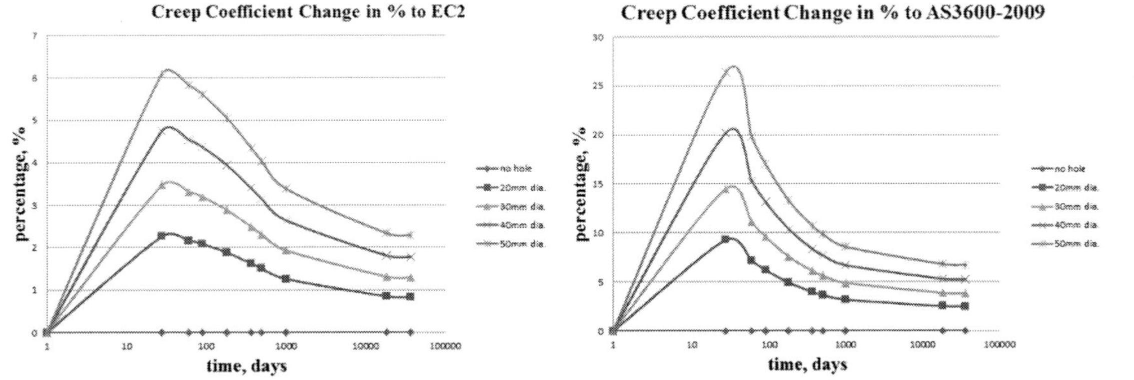

Figure 3. Hole-creep effect.

4.2. Shrinkage evaluations

Figure 4 shows five cases of different size of holes on the shrinkage effect. The data of shrinkage stains are calculated by EC2 and AS3600-2009 codes respectively.

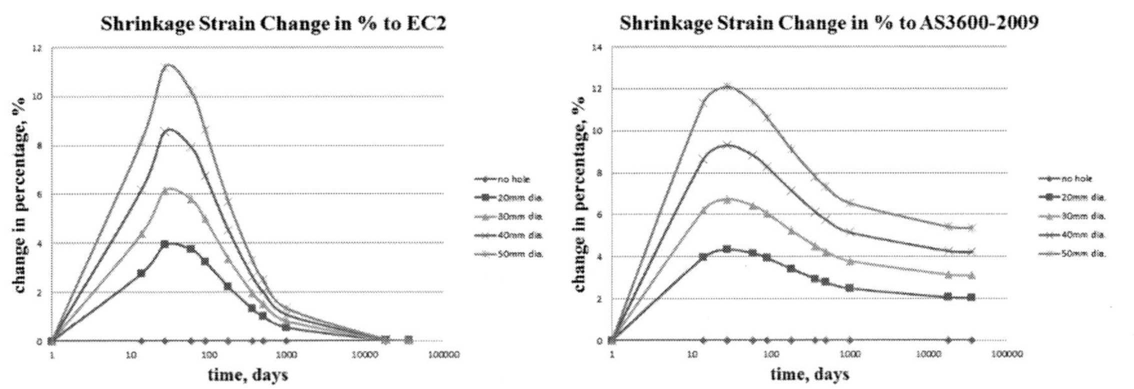

Figure 4. Hole-shrinkage effect.

Based on the sensitive analysis, we found that long-term performance in prestressed concrete sleeper largely depends on geometry. Both of creep and shrinkage are proportional to hole sizes. Furthermore, larger strains result in more loss of prestress in sleeper.

5. Conclusions

There are two main duties for railway prestressed concrete sleepers (or railroad ties) that must successfully perform: first, to carry wheel loads from the rails to the ground; and second, to secure rail gauge for dynamic safe movements of trains. In many cases, inappropriate design of the time-dependent behaviours of railway concrete sleepers due to their creep, shrinkage and elastic shortening responses of the materials affect significantly the rail gauge control. In addition, the concrete sleepers are often modified on construction sites to fit in other systems such as cables, signalling gears, drainage pipes, etc. Accordingly, this study is the world first to investigate creep and shrinkage effects on the railway prestressed concrete sleepers with vertical holes. This paper highlights constitutive models of concrete materials within the railway sleepers under different environmental conditions over time. Comparison has been carried out among a variety of reputable methods to evaluate shortening effects in railway prestressed concrete sleepers. Based on the sensitivity analyses, we found that creep and shrinkage in railway sleeper depends largely on the cross-section. According to the obtained data, the both of creep and shrinkage stains depend on holes' size, which means large diameter of hole could result in more stain in prestressed concrete sleeper. This insight will improve material design and structural restraints, which are very critical to the durability of railway track components.

Acknowledgements
The authors would also like to thank British Department of Transport (DfT) for Transport - Technology Research Innovations Grant Scheme, Project No. RCS15/0233; and the BRIDGE Grant (provided by University of Birmingham and the University of Illinois at Urbana Champaign). The last author is gratefully acknowledge the Japan Society for the Promotion of Science (JSPS) for his JSPS Invitation Research Fellowship (Long-term), Grant No L15701, at Track Dynamics Laboratory, Railway Technical Research Institute and at Concrete Laboratory, the University of Tokyo, Tokyo, Japan. The authors are sincerely grateful to European Commission for the financial sponsorship of the H2020-RISE Project No. 691135 "RISEN: Rail Infrastructure Systems Engineering Network," which enables a global research network that tackles the grand challenge in railway infrastructure resilience and advanced sensing in extreme environments (www.risen2rail.eu). We would also like to acknowledge the support from European Cooperation in Science and Technology (EU-COST) Action: TU1404 Towards the next generation of standards for service life of cement-based materials and structures.

References
[1] Esveld C 2001 *Modern Railway Track* (Delft University of Technology)
[2] Remennikov A M, Murray M H and Kaewunruen S 2012 Reliability based conversion of a structural design code for prestressed concrete sleepers *Proceedings of the Institution of Mechanical Engineers: Part F Journal of Rail and Rapid Transit* **226**(2) pp 155-73
[3] Remennikov A M and Kaewunruen S 2008 A review on loading conditions for railway track structures due to wheel and rail vertical interactions *Structural Control and Health Monitoring* **15**(2) pp 207-34
[4] Remennikov A M and Kaewunruen S 2014 Experimental load rating of aged railway concrete sleepers *Engineering Structures* **76**(10) pp 147-62
[5] Remennikov A M and Kaewunruen S 2015 Determination of prestressing force in railway concrete sleepers using dynamic relaxation technique *ASCE Journal of Performance of Constructed Facilities* **29**(5)
[6] Li D, Kaewunruen S and Robery P 2016 *Effects of creep and shrinkage in railway prestressed concrete sleepers* (University of Birmingham, UK)
[7] Kaewunruen S. and Remennikov A M 2013 On the residual energy toughness of prestressed concrete sleepers in railway track structures subjected to repeated impact loads *Electronic Journal of Structural Engineering* **13**(1) pp 41-61
[8] Neville A 1979 *Properties of concrete 3th edn* (United Kingdom: Longman ELBS)
[9] Martin L and Purkiss J 2006 *Concrete Design to EN 1992 2nd edn* (London: Butterworth-Heinemann)
[10] British Standards Institution 1992 Eurocode 2 - Design of concrete structures: Part 1-1: general rules and rules for buildings *(EN 1992-1-1)*

[11] Bhatt P 2011 *Prestressed concrete design to eurocodes 1st edn* (Abington: Spon Press)
[12] Gilbert R et al 2016 *Design of prestressed concrete to AS3600-2009 2nd edn* (Boca Raton: CRC Press)
[13] Gamage E 2015 *Structural Design Of holes and Web Openings In Pre-stressed Concrete Sleepers* (UK: University of Birmingham)
[14] European Committee for Standardization (CEN) 2003 Actions on Structures; Traffic loads on bridges *Eurocode 1, Part 2 (EN 1991-2)*
[15] Ngamkhanong C, Li D and Kaewunruen S 2017 Impact capacity reduction in railway prestressed concrete sleepers with vertical holes *Int. Conf. on Building up Efficient and Sustainable Transport Infrastructure (Prague)*

IOP Conf. Series: Materials Science and Engineering **236** (2017) 012043 doi:10.1088/1757-899X/236/1/012043

Discrete simulation of railway ballast shear test: spherical and polyhedral grain shapes

Radek Dubina[1] and Jan Eliáš[1]

[1]Institute of Structural Mechanics, Faculty of Civil Engineering, University of Technology in Brno, Veveří 331/95, Brno 602 00, Czech Republic

E-mail: dubina.r@fce.vutbr.cz

Abstract. Dynamical behaviour of railway ballast is important in design and maintenance of railroads. Various modelling approaches are available. The least phenomenological but also the most computational demanding approach is direct representation of every ballast grain and its interaction with other grains in the model. The computational complexity is partially reduced by assuming non-deformable bodies with simplified shapes. Two different shapes of the bodies are considered: spherical and polyhedral. Spherical shapes are advantageous because of their great computational simplicity; however, missing shape information must be compensated by adding phenomenological rolling resistance. These two model variants are used to simulate different shear test. Results of the models are compared to each other and also to the experimental data from the literature.

1. Introduction

An assumption of material homogeneity is generally used to investigate many complex problems in the railway engineering, for example, railway sleepers [1], wheel/rail sliding contact [2] or fatigue of railway axles [3]. Homogeneous continuous formulation in connection with complex phenomenological constitutive relations can be suitable tool for mathematical modelling of the observed phenomena at the macroscopic scale. But with such an approach, it is difficult to take into account phenomena such as dilatancy, pressure sensitivity, compaction, crushing etc. The discontinuous heterogeneous nature plays an important role in the behaviour of granular materials.

The Discrete Element Method (DEM) is an approach that takes an advantage of the granular nature of the material. Nowadays, DEM is one of the most useful and efficient tools to describe the behaviour of discontinuous materials. In the Discrete Element Method, originally proposed by Cundall and Strack [4], every grain is represented by an ideally rigid and independent body which can interact with the other bodies through forces and moments at their contacts. The rigidity of the bodies is motivated by assumption that macroscopic material behaviour is mostly dictated by rearranging of the grains, not by the actual grain deformations. The DEM typically uses explicit integration of the motion equations. The time is discretized time into small time steps Δt of size limited by stiffness and size of the grains. In every time step, it is necessary to detect all contacts between bodies and evaluate contact forces. The computational cost of this operation significantly limits the applicability of the DEM to some practical, large scale problems, where millions of particles are typically involved. The difficulty of contact detection and force evaluation depends on shape of the bodies that represent the real grains of the railway ballast. We consider two common possibilities of shape simplification. The elaborative one is to use convex polyhedral shape which closely corresponds to real shape of ballast grains [5, 6, 7]. The simple one is to use spherical bodies [8, 9]. Other options, not considered here, involve e.g. clumps of spheres [10], ellipsoids [11] or superquadrics [12]. The simple spherical shape requires enhancement or constitutive law to provide sufficient mutual locking of bodies. The enhancement consists in additional phenomenological rolling resistance. The rolling resistance mimics resistance of two grains of complex shape in contact rolling relatively to each other [13 14, 15].

Motivated by difficulties in adjusting the rolling model parameters [16, 17], we are presenting this study. It aims at comparison of performance of polyhedral and spherical shapes in simulation of shear test. The shearing is performed after compaction that is realized via application of large pre-stressing.

2. Derivation of equivalent constitutive relation for spheres and polyhedras

The polyhedral and spherical shape use completely different formulation of constitutive relations. For comparison purposes, material behavior for both models must be approximately equal. Simple relation between material parameters can be established under assumption of ideal polyhedral particle of spherical shape. In this section, we analyze in detail contact of spherical particles using two formulations of the constitutive law, the one for spheres and the one used by polyhedrons. Relations between material parameters for polyhedras and spheres are then established by comparison of these two cases. Superscripts *(s)* and *(p)* are used to distinguish between the constitutive law formulations.

Two spherical particles in contact can mutually interact in different modes by forces and moments that are sketched in Figure 1. The constitutive formulation for spheres is based on Hertz-Mindlin [18, 19] with additional rolling resistance. The polyhedral formulation uses intersecting volume to determine repulsive force [20]. The twist mode is neglected in both cases. All the presented equations are taken from open source platform YADE [21].

For the Hertz-Mindlin contact type, one can express the normal force as

$$\boldsymbol{F}_{\mathrm{N}}^{(s)} = \frac{4}{3} E_{\mathrm{eff}}^{(s)} \sqrt{R_{\mathrm{eq}}}\, \delta^{3/2} \boldsymbol{n} \tag{1}$$

Where $E_{\mathrm{eff}}^{(s)}$ is effective Young's modulus

$$E_{\mathrm{eff}}^{(s)} = \frac{E^{(s)}}{2\left(1-\left(\nu^{(s)}\right)^2\right)} \tag{2}$$

R_{eq} is equivalent radius of spheres

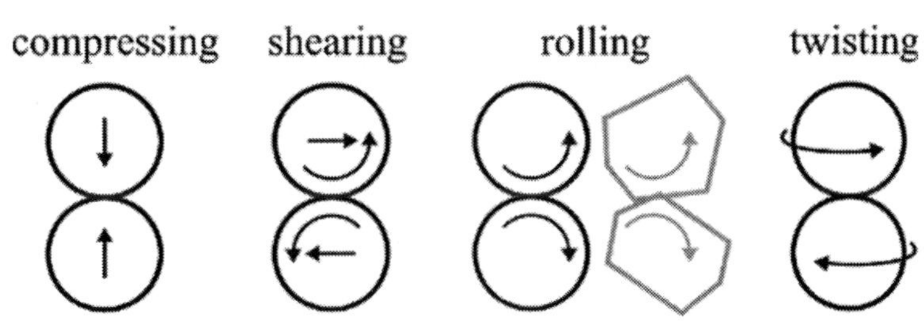

Figure 1. Schematic plot of relative movements of two particles in contact.

$$R_{\mathrm{eq}} = \frac{R_1 R_2}{R_1 + R_2} = \frac{R}{2} \tag{3}$$

IOP Conf. Series: Materials Science and Engineering **236** (2017) 012043 doi:10.1088/1757-899X/236/1/012043

$E^{(s)}$ is Young's modulus, $\nu^{(s)}$ is ration between normal and shear stiffness, $R=R_1+R_2$ is radius of both spheres, δ is mutual overlapping of particles and $\boldsymbol{n}$ is a normal in direction determined by particle centroids. The equations (1)-(3) combined together gives

$$\boldsymbol{F}_{\mathrm{N}}^{(s)}=\frac{4}{3}E_{\mathrm{eff}}^{(s)}\sqrt{R_{\mathrm{eq}}}\,\delta^{3/2}\boldsymbol{n}=\frac{4}{3}\frac{E^{(s)}}{2\left(1-\left(\nu^{(s)}\right)^{2}\right)}\sqrt{\frac{R}{2}}\,\delta^{3/2}\boldsymbol{n}=\frac{E^{(s)}\sqrt{2R}}{3\left(1-\left(\nu^{(s)}\right)^{2}\right)}\delta^{3/2}\boldsymbol{n} \tag{4}$$

For polyhedral shapes, we can express the normal force as

$$\boldsymbol{F}_{\mathrm{N}}^{(p)}=E_{\mathrm{eff}}^{(p)}V_{\mathrm{I}}\boldsymbol{n} \tag{5}$$

with effective Young's modulus for polyhedrons

$$E_{\mathrm{eff}}^{(p)}=\frac{E^{(\mathrm{p})}}{2} \tag{6}$$

The cap of sphere with depth $\delta/2$ has volume

$$V_{\mathrm{cap}}=\frac{\pi\delta^{2}}{24}(6R-\delta) \tag{7}$$

and the volume of the intersection of spheres is then

$$V_{\mathrm{I}}=2V_{\mathrm{cap}}=\frac{\pi\delta^{2}}{12}(6R-\delta) \tag{8}$$

Then the equations (5), (6) and (8) provides

$$\boldsymbol{F}_{\mathrm{N}}^{(p)}=\frac{E^{(p)}}{2}\frac{\pi\delta^{2}}{12}(6R-\delta)\boldsymbol{n} \tag{9}$$

Requiring the same force produced by both constitutive formulation gives us

$$\begin{gathered}\boldsymbol{F}_{\mathrm{N}}^{(s)}=\boldsymbol{F}_{\mathrm{N}}^{(p)}\\ \frac{E^{(s)}\sqrt{2R}}{3\left(1-\left(\nu^{(s)}\right)^{2}\right)}\delta^{3/2}\boldsymbol{n}=\frac{E^{(p)}}{2}\frac{\pi\delta^{2}}{12}(6R-\delta)\boldsymbol{n}\end{gathered} \tag{10}$$

After the modification of equation (10), we can compute required ratio of stiffness parameters, k

$$k=\frac{E^{(p)}}{E^{(s)}}=\frac{8\sqrt{2R}}{\pi\left(1-\left(\nu^{(s)}\right)^{2}\right)(6R-\delta)\sqrt{\delta}} \tag{11}$$

The same approach can be repeated for tangential direction. The expressions for shear directions [21] shall produce the same force.

$$\begin{gathered}\boldsymbol{F}_{\mathrm{T}}^{(s)}=\boldsymbol{F}_{\mathrm{T}}^{(p)}\\ k_{\mathrm{T}}^{(s)}\Delta\boldsymbol{u}_{\mathrm{T}}=k_{\mathrm{T}}^{(p)}\Delta\boldsymbol{u}_{\mathrm{T}}\\ \frac{E^{(s)}\sqrt{2R\delta}}{(1+\nu^{(s)})(2-\nu^{(s)})}=\frac{E^{(p)}\nu^{(p)}}{2}\end{gathered} \tag{12}$$

The shear coefficient $\nu^{(p)}$ can be expressed as

$$\nu^{(p)} = \frac{2\sqrt{2R\delta}}{k\left(1+\nu^{(s)}\right)\left(2-\nu^{(s)}\right)} = \frac{\pi\left(1-\nu^{(s)}\right)(6\mathrm{R}-\delta)\delta}{4\left(2-\nu^{(s)}\right)} \tag{13}$$

Both for spheres and polyhedras, the maximum shear force is limited by Coulomb friction model [21] with additional parameter, φ, called angle of internal friction same for both constitutive formulations.

$$\left\|\boldsymbol{F}_{\mathrm{T}}\right\| \le \left\|\boldsymbol{F}_{\mathrm{T,MAX}}\right\| = \tan\varphi\left\|\boldsymbol{F}_{\mathrm{N}}\right\| \tag{14}$$

The both derived relations are dependent on the radius R and mutual overlapping δ. The dependence on overlapping δ is disadvantageous because it varies during the simulations and has to be approximately chosen. Moreover, the polyhedral particle shapes are far from the spherical shapes. Thus, the derived relations cannot ensure equal results of the simulation. However, they still provide reasonable guess of the values for both models.

The spherical constitutive formulation involves also the rolling resistance. The rolling resistance [13, 14, 15] is an artificial moment arising on the contact of two discrete elements that mimics resistance of two grains of complex shape in contact rolling relatively to each other. It is necessary to use rolling resistance when spherical elements are employed to reproduce behaviour of non-spherical particles such as grains in railway ballast. The model of rolling resistance assumes linear dependence of the rolling moment increment, $\Delta\boldsymbol{M}_{\mathrm{R}}$, on the rolling angle increment, $\Delta\boldsymbol{\omega}_{\mathrm{R}}$, via the stiffness in rolling, k_{R} [Nm] [21].

$$\Delta\boldsymbol{M}_{\mathrm{R}} = k_{\mathrm{R}}\Delta\boldsymbol{\omega}_{\mathrm{R}} \tag{15}$$

The stiffness in rolling is calculated based on parameter α_{R} [-]

$$k_{\mathrm{R}} = \alpha_{\mathrm{R}} E_{\mathrm{eff}}^{(s)} R^3 \nu^{(s)} \tag{16}$$

Similarly to the Coulomb shear, the upper limit is imposed on the rolling moment magnitude.

$$\left\|\boldsymbol{M}_{\mathrm{R}}\right\| \le \left\|\boldsymbol{M}_{\mathrm{R,MAX}}\right\| = \mu_{\mathrm{R}}\left\|\boldsymbol{F}_{\mathrm{N}}\right\| R \tag{17}$$

Parameter μ_{R}[-] controls the maximum rolling moment similarly to the tangent of internal friction angle in shear.

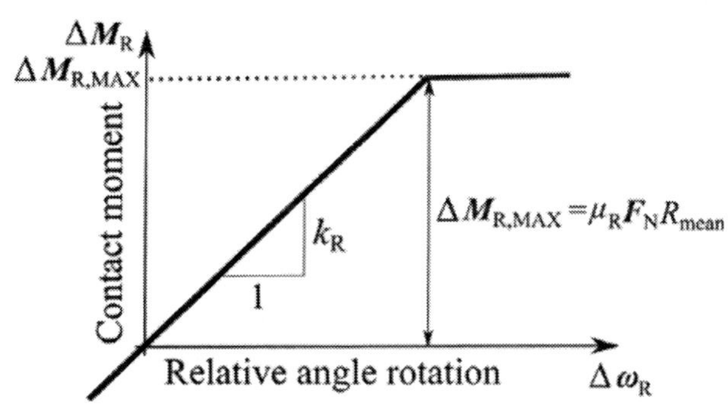

Figure 2. Rolling resistance on the contact between discrete particles.

3. Time step

The estimation of time step is based on the contact stiffness, the material density and particle size. For spherical particles the critical time step is typically considered as

$$\Delta t_{\mathrm{cr}} \leq \frac{R_{\mathrm{eq}}}{\sqrt{E_{\mathrm{eff}}^{(s)} / \rho}} \tag{18}$$

where R_{eq} is the equivalent radius of sphere, $E_{\mathrm{eff}}^{(s)}$ is effective Young's modulus and ρ is the density. For polyhedrons the equivalent volume radius R^{V} is set and the critical time step is consequently estimated.

$$R^{\mathrm{V}} = \sqrt[3]{\frac{V_{\mathrm{I}}}{4/3\,\pi}} \tag{19}$$

$$\Delta t_{\mathrm{cr}} \leq \frac{R^{\mathrm{V}}}{\sqrt{E_{\mathrm{eff}}^{(p)} R^{\mathrm{V}} / \rho}} \tag{20}$$

In case of spheres, the equation (18) is rigorously derived. It ignores the shear and rolling stiffness, that are typically much lower compared to the normal stiffness. The estimation of critical time step for polyhedrons is rather empirical and needs further improvements.

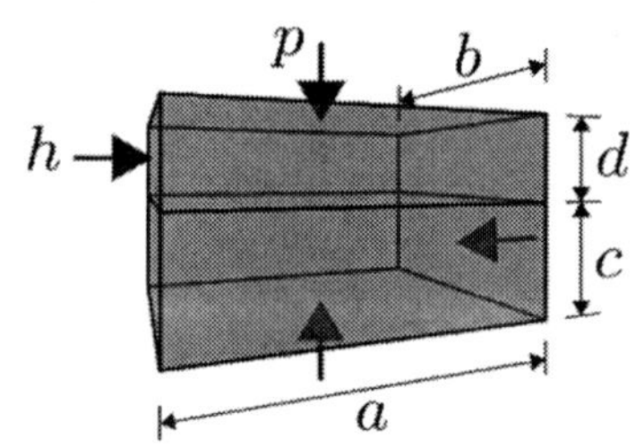

Figure 3. Scheme of the shear test arrangement; The loading is realized by displacement of the lower part of box in horizontal direction, *h*, and surcharge at the top, P.

4. Influence of particle shape in the shear test

The comparison between spherical and polyhedral discrete elements is shown on simulation of the shear test. The experiments published in [22] are used for comparison. Two separate steel boxes are filled by a granular material (the spherical/ polyhedral discrete elements in simulations). A steel loading plate that provides vertical surcharge during shearing is placed on the top. The bottom part of the box is horizontally displaced up to 37 mm, while the upper part remains at the original position. The dimensions of a large scale shear box (Figure 3) were a = 300 mm, b = 300 mm, and c = d = 100 mm. The vertical pressure p, that is kept constant during the loading, is 15 kPa. Before the shear loading starts, the sample is compacted by an incresed vertical pressure. The influence of this compacting pressure is studied, the results are compared with the experiments [22]. Three variants differing in shape of the grains are numerically simulated: (i) spherical shapes of constant radius R, (ii) convex polyhedrons of random geometry of size comparable to spheres with radius R (iii) spheres with radii randomly sampled from a uniform distribution over interval from $0.5R$ mm to $1.5R$ mm. The radius R corresponds to the mean radius R_{50} from the experiment [22], but the experimental sieve curve was not represented. The maximal radius of laboratory tested ballast was 53 mm.

The material a geometrical properties are shown in Table 1. The elastic parameters for the polyhedrons are estimated based the equations (11) and (13) assuming δ=1.42×10^{-4} m. The remaining properties of both particles were identified in previous research or assumed from the literature.

Table 1. The material and geometrical properties for discrete elements used in the shear test. Note the different units for normal stiffness and shear/normal stiffness ratio for spheres and polyhedrons.

	spheres with identical R	polyhedrons	spheres with random R
Normal stiffness E [Pa]/[N/m^3]	70×10^9	$500 \times 70 \times 10^9$	70×10^9
shear/normal stiff. r. ν [-]/[m^2]	0.3	4.0×10^{-6}	0.3
Friction angle φ [rad]	0.6	0.3	0.6
Density ρ [kg/m^3]	2600	2600	2600
Radius R [mm]	17.5	17.5	8.75-26.25
Rolling parameter α_R	0.1	0.0	0.1
Rolling parameter μ_R	0.1	0.0	0.1

Three samples of different initial configuration of bodies are simulated for every level of compaction and for every model variant. Three vertical compacting pressures are considered: without compaction (15 kPa), compaction 200 kPa and compaction 2000 kPa. The obtained results are shown in Figure 4; all the curves are averages over three samples computed. The first row of graphs shows dilatancy - evolution of vertical displacement of loading plate v on horizontal displacement of the bottom box h. Both spherical model variants show similar results; there is an initial sharp peak dependent on compaction level followed by monotonic slow increase of volume. Initially, the sample has to rearrange the grains and the loading plate goes up. The later stable growth is not sufficiently explained yet, but it does correspond to the experimental measurement. The polyhedrons are much more sensitive to compaction – omitting compaction leads to large contraction. For compacted samples, one can see the same compaction dependent peak as in the case of spheres, but followed by almost constant level of vertical shift v. Theoretically, after jumping behind the first peak, the compaction effect should diminish. This can be seen in graphs in the third row, where the curves of the bulk density (total mass of the sample divided by total volumes including pores) are getting close to each other.

The results for shear stress show disorder and large variability among the various samples. This is attributed to relatively low number of particles in the boxes. In case of polyhedrons, the measurement of shear stress exhibits lot of jumps and sharp peaks. The reason is probably inadequate length of the time step Δt. Two polyhedral bodies may come into contact by protruding parallel facets, giving rise to sudden increase in intersecting volume and large repulsive force. The smaller time step may reduce the sudden increase of volume and reduce the jumps in the shear stress. The authors intent to improve the estimation of time step necessary to obtain stable simulation in their future work.

5. Conclusion

The normal and shear stiffness of polyhedral grains were estimated based the comparison of their volumetric constitutive law and Herz-Mindlin constitutive laws for spheres. The calculation of corresponding parameters depends on mutual overlapping of particles δ, which has to be estimated.

Simulations of shear test using polyhedral and spherical particles were performed. The shearing was realized after application of various vertical compaction pressures: no compaction (15 kPa), compaction 200 kPa and compaction 2000 kPa. The polyhedral model yields the highest sensitivity to compaction, however also spherical grains were significantly influenced - polydisperse assemblies more than monodisperse. The simulations did not consider the breakage of ballast grains. Including this phenomena may lead to improvement of the results.

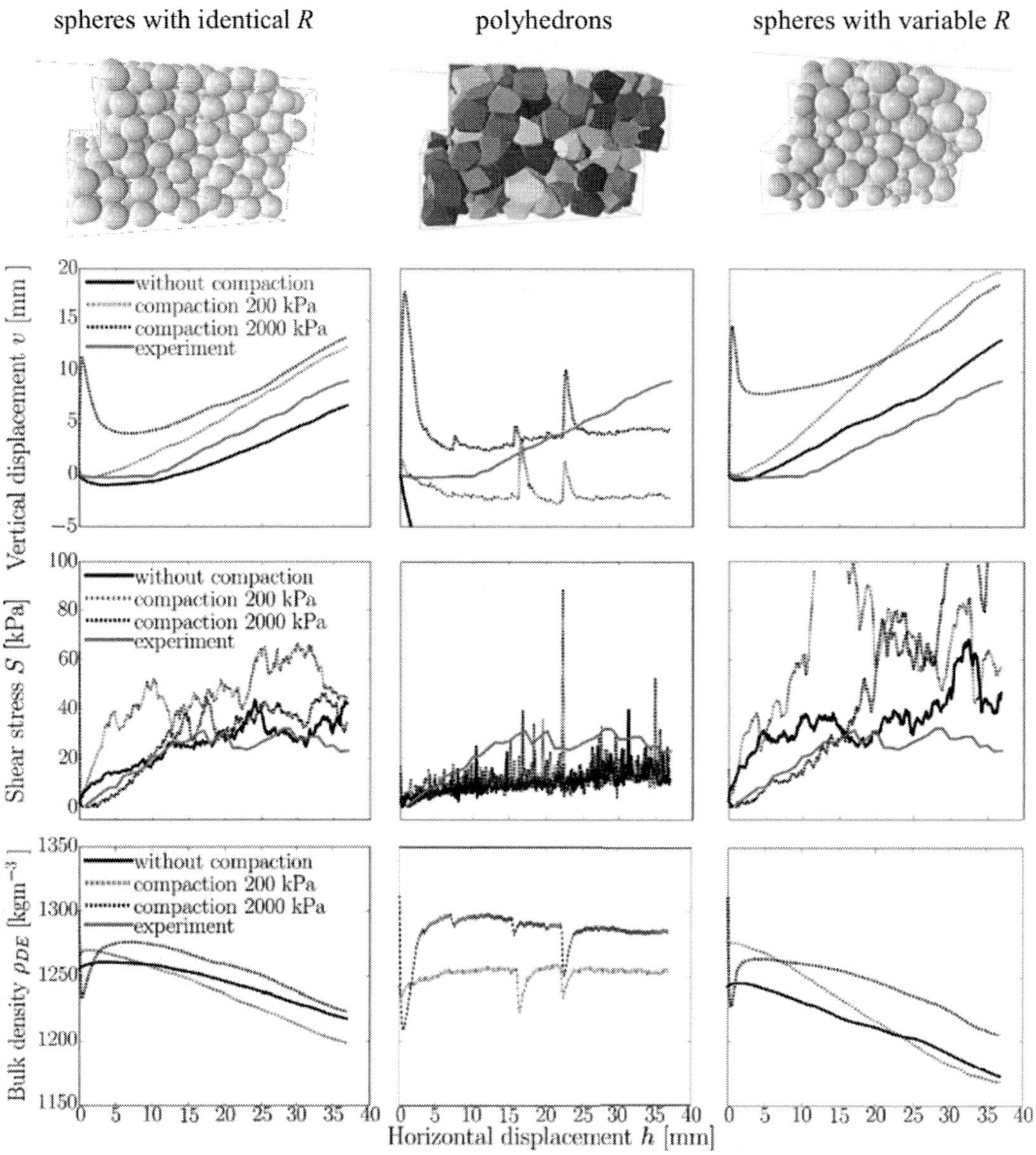

Figure 4. Dependency of vertical displacement v, shear stress S and bulk density ρ_{DE} on the shearing displacement h for different model variants and compaction levels.

Acknowledgements
Research reported in this paper was supported by Competence Centres program of Technology Agency of the Czech Republic (TA CR), project Centre for Effective and Sustainable Transport Infrastructure (no. TE01020168).

References

[1] Bezgin N O 2017 High performance concrete requirements for prefabricted high speed railway sleepers *Construction and Building Materials* **138** pp 340-51
[2] Wu Y, Wei Y, Liu Y, Duan Z and Wang L 2017 3-D analysis of thermal-mechanical behaviour of wheel/rail sliding contact considering temperature characteristics of materials *Applied Thermal Engineering* **115** pp 455-62

[3] Náhlík L, Pokorný P, Ševčík M, Fajkoš R, Matušek P and Hutař P 2017 Fatigue lifetime estimation of railway axles *Engineering Failure Analysis* **73** pp 139-57
[4] Cundall P A and Strack O D L 1979 A discrete numerical model for granular assemblies *Geotechnique* **1** pp 47-65
[5] Hoang T M P and Alart P 2011 A domain decomposition method for granular dynamics using discrete elements and application to railway ballast *Ann. Solid Struct. Mech.* **2** pp 87-98
[6] Ouhbi N, Voivret Ch, Perrin G and Roux G P 2016 Railway Ballast: Grain Shape Characterization to Study its Influence on the Mechanical Behaviour *Advances in Transportation Geotechnics 3rd Int. Conf. on Transportation Geotechnics (Guimaraes)* ed A G Correia et al (Guimaraes: Elsevier) pp 1120 - 27
[7] Lee S J, Hashash M A and Nezami E G 2012 Simulation of triaxial compression tests with polyhedral discrete elements *Computers and Geotechnics* **43** pp 92-100
[8] Bourrier F, Kneib F, Chareyre B and Fourcaud T 2013 Discrete modeling of granular soils reinforcement by plant roots *Ecological Engineering* **61** pp 646-57
[9] Le B D, Dau F, Charles J L and Iordanoff I 2016 Modeling damages and cracks growth in composite with a 3D discrete element method *Composites Part B* **91** pp 615-30
[10] Engenzinger Ch, Seifried R and Eberhard P 2012 A discrete element model predictiong the strength of ballast stones *Computers & Structures* **108-109** pp 3-13
[11] Cleary P W 2009 DEM prediction of industrial and geophysical particle flows *Particuology* **8** pp 10-118
[12] Podlozhnyuk A, Stefan P and Kloss C 2017 Efficient implementation of superquadric particles in Discrete Element Method within an open-source framework *Computational Particle Mechanics* **4** pp 101-18
[13] Ai J, Chen J-F, Rotter J and Ooi J Y 2011 Assesment of rolling resistance in discrete element simulations *Powder Technology* **206** pp 269-82
[14] Belheine N, Plassiard J P, Donzé F V, Darve F and Seridi A 2009 Numerical simulation of drained triaxial test using 3D discrete element modeling *Computers & Geotechnics* **36** pp 320-31
[15] Irazábal J, Salazar F and Oñate E 2017 Numerical modelling of granular materials with spherical discrete particles and bounded rolling friction model. Application to railway ballast. *Computers & Geotechnics* **85** pp 220-29
[16] Dubina R and Eliáš J 2016 Comparison of spherical and polyhedral discrete modelling of railway ballast *2016 Engineering Mechanics Institute Int. Conf. (Metz)* pp 134
[17] Dubina R and Eliáš J 2016 Effect of rolling resistance in DEM models with spherical bodies *Transactions of VŠB – TU Ostrava, Civil Engineering* **16** pp 11 - 18
[18] Hertz H 1882 Ueber die berührung fester elasticher körper (on the contact of rigid elastic solids) *J. reine und angewandte Mathematik* **92** pp 156-171 (translated and reprinted in English in *Hertz Miscellaneous Papers* (Macmillan & Co., London 1896)
[19] Mindlin R D 1949 Compliance of elastic bodies in contact *ASME J Appl Mech* **16** pp 259-68
[20] Eliáš J 2014 Simulation of railway ballast using crushable polyhedral particles *Powder Technology* **264** pp 458-65
[21] Šmilauer V et al 2015 *Yade Documentation 2nd ed* (The Yade Project) DOI http://yade-dem.org/doc/
[22] Indaratna B, Ngo N and Rujikiatkamjorn C 2011 Behavior of geogrid-reinforced ballast under various levels of fouling *Geotextiles and Geomembranes* **29** pp 313-22

Design of high-speed turnouts and crossings

Lukáš Raif[1,2], Bohuslav Puda[1], Jiří Havlík[1] and Marek Smolka[1]

[1]DT – Výhybkárna a strojírna, a.s., Dolní 100, Prostějov 796 01, Czech Republic
[2]Brno University of Technology, Faculty of Civil Engineering, Veveří 331/96, Brno 69 003, Czech Republic

E-mail: raif@dtvm.cz

Abstract. Recently, the new ways to improve the railway switches and crossings have been sought, as the railway transport increases its operating speed. The expectation of these adjustments is to decrease the dynamic load, which usually increases together with velocity, and this influences the comfort of the vehicle passage, the wear of the structural parts and the cost of maintenance. These adjustments are primarily the turnout elements such as the optimized geometry of the turnout branch line by means of transition curves application, which minimizes the lateral acceleration during the vehicle passage through the track curve. The rail inclination is solved either by means of inclination in fastening system, or by machining of the rail head shape, because this ways of adjustment retain the wheel-rail interaction characteristics along the whole length of the turnout. Secondly, it is the crossing with movable part, which excludes the interruption of the running surface and optimization of the railway stiffness throughout the whole turnout length as well. We can see that the different stiffness along the turnout influences the dynamic load and it is necessary to optimize the discontinuities in the stiffness along the whole length of the turnout. For this purpose, the numeric modeling is carried out to seek the areas with the highest stiffness and subsequently, the system of stiffness optimization will be designed.

1. Introduction

Recently, there is an important increase of the operating speed on railways, which leads not only to higher requirements towards the railway track design, but also to higher demands towards the switches and crossings as well. The increase in operating speed brings increased dynamic load and the existence of even a small discontinuity may cause a big force effect. It may affect the railway track quality, i.e. the passengers' comfort during the vehicle passage and the maintenance cost for the infrastructure owner as well. That is why it is necessary to deal with a new design of turnout parts, which can reduce the extent of the dynamic load and this way the passage quality may be increased and the maintenance cost demands decreased at the same time. The high-speed turnouts are equipped with several specific elements. [6] These are the crossings with movable parts or the special turnout geometry used in the branch line using the transition curve. Moreover, the application of the up-to-date setting system is necessary to secure the quick throw-over of the switch rails and the points of the crossings with movable parts into required position. The setting system is equipped with positioning sensors which help to inform the operator quickly, that the operation is properly carried out and the switch rails in the switch and the point in the crossing with movable part are now switched into required position. To allow the easier movement of the switch rails and of the point of the movable part of the crossing, the installation of the roller slide chairs is advisable, to decrease the throw-over resistance as much as possible. The application of the electric heating into the set of switches and into crossing with movable parts under our territorial climatic conditions is essential. It enables the point setting even under the most unfavourable weather conditions without any problem. The frequent application of materials with higher resistance is often carried out in the high-speed turnouts manufacture to decrease the requirements towards the wear of the critically loaded turnout parts and this way of course diminish the claims towards the volume of the maintenance activity as well. [4,5]

2. High-speed turnouts construction elements

Some of the mentioned special elements used in the high-speed turnouts were applied so far and verified under operation when installed into commonly used geometry of the turnout. For the infrastructure manufacturer and for the operator it is useful when the new elements are properly tested and verified within the standard geometry turnout and only then installed into brand-new high-speed turnout. The developed turnouts are in conformance with all Czech and European technical standards and with related specifications, such as TSI (Technical Specifications for Interoperability Relating to the Infrastructure Subsystem). [1]

2.1. Branch line geometry

The geometry parameters of rails situated within the turnout have to follow the requirements of the standard for the proposed velocity in the branch line of the turnout. In particular, it is the standard EN 13 803-2 and the Czech national standard CSN 73 6360-1. [2,3] The turnouts designed for high speed are proposed usually with respect to the following parameters: superelevation deficiency I [mm], abrupt change of superelevation deficiency ΔI [mm] and the rate of change of superelevation deficiency at an abrupt change in curvature as function of time dI/dt. Principally under higher required speed in the turnout branch line it is very useful to design the curved track as a transition curve by clothoid. Thanks to this solution it is possible to design the optimal geometry of the turnout under observing of all parameters required by the given standard. The application of track transition curve by clothoid also diminishes the side impacts from stress values higher then 1,0 $m \cdot s^{-3}$ at values up to 0,8 $m \cdot s^{-3}$ and the dynamic stress within the turnout passage as well, which means lower wear and lower cost of maintenance. The very first turnout in the Czech republic where the clothoidal curved transition branch line will be used is the turnout type J60-1:33,5-8000/4000/14000-PHS-E2.

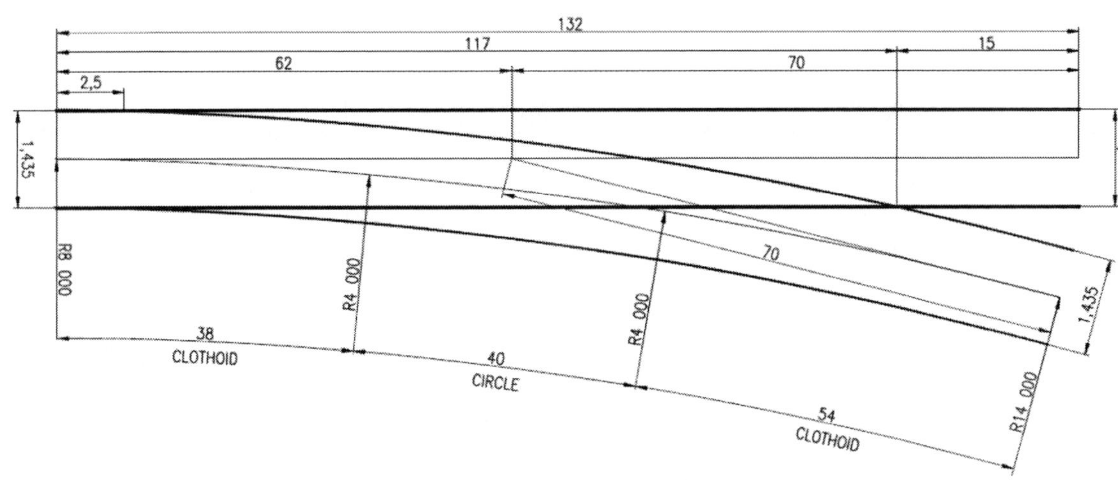

Figure 1. The very first turnout developed in the Czech Republic with unstable curvature of the branch line (type J60-1:33,5-8000/4000/14000-PHS).

IOP Conf. Series: Materials Science and Engineering **236** (2017) 012044 doi:10.1088/1757-899X/236/1/012044

In 2010 began the work on development of the turnout, which would be able to allow vehicle passage at speed 160 km.h^{-1} in the branch line, which is currently the highest possible speed within the Czech and Slovakian Railways as well. Within the proposal of the turnout geometry, the transition curves were used for the first time, which are commonly applied for the high-speed turnouts in the world. Thanks to the branch line special geometry, three different applied radii are introduced directly in the turnout name (see the Figure 1). The first number (8000) indicates the radius of the clothoid osculating circle at the beginning of the turnout. Then follows the clothoid with gradually elevating curvature (or diminishing radius as well) up to radius 4000 meters. Another element is continually joined to the transition curve, which gradually diminishes its curvature towards the osculating radius of the circle 14000 meters at the end of the turnout. If we let the transition curve of the same clothoid parameter A continue until radius ∞ (or the curvature 0 m^{-1}), this point will be exactly the point of inflection in the single crossover assembled from these turnouts with axis spacing 4,75 m. If the axial distance of 5 meters is required, it is possible to follow the turnout by the modified transition curve to reach the inflection motive again.

2.2. Rail inclination

Within the common track the rails laid under inclination 1:20 or 1:40 according to the applied section of rail and the fastening system. On the other hand, the rails in common turnouts are usually fastened vertically (the inclination 1:∞). This change of the rail inclination influences the characteristics of the contact geometry, which may inconveniently influence the running vehicle kinematics and the increased stressing of the turnout structure may occur, mainly at high speeds. That is why in TSI (Technical Specification for Interoperability) this issue is specified and it says that for the main lines with speed higher than 250 km.h^{-1}, the rails must have the inclination. This inclination can be carried out either just by the laying on the inclined base plate, or by the shape of the active part of the rail head profile (see Figure 2). The company DT designed a special shape of the rail head marked as K (1:40), which helps to reach the properties of the contact geometry of rail profile 60 E2 laid in the inclination 1:40 for rail attached in the vertical position. DT also developed fastening system that enables placement of the rail directly in the inclination. This type of turnout has not been applied in the Czech railways network yet. [1,6]

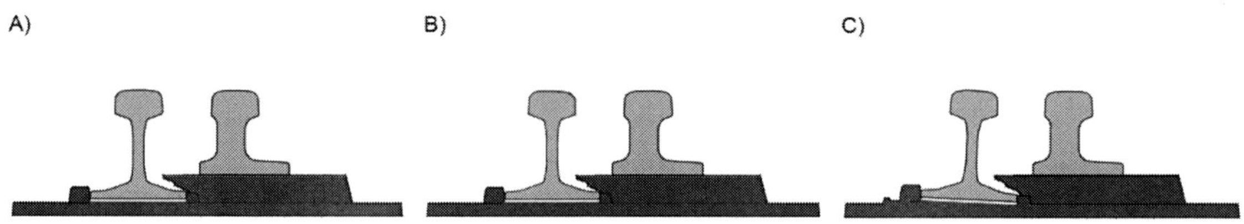

Figure 2. The possibilities how to solve the rail inclination in the slide chair. The rail can be laid either vertically (A), or inclined. The inclination can be carried out by the shape of the active part of the rail head profile (B), or by the inclined fastening system (C).

2.3. Crossing with movable parts

The first important constructional part, which is necessary for high-speed turnouts is the crossing with movable part (Czech abbreviation "PHS"). According to the TSI (mentioned above), the fixed crossing can be applied just up to the velocity 250 km.h^{-1}. Since the high-speed turnouts are designed for operation under higher speeds within the main lines, application of the crossing with movable parts seems to be a necessity. At high speeds, the dynamic stresses coming from vehicles towards the track become more prominent. That is why the elimination of any irregularities on the rail running surface is necessary. Such irregularity is the fixed crossing itself, where the wheel guiding is interrupted and is substituted by the check rail situated into opposite rail. The railway vehicle running on the straight line rail performs the natural sinusoidal movement, coming from the wheel profile shape and from the rail head profile, i.e. from equivalent conicity. At the crossing with movable parts the wheel set is not forced to interrupt this sinusoidal movement, which has advantageous influence towards the passage dynamics and towards the comfort ride for passengers as well.

Two basic types of the common crossings with movable parts are employed. They differ with parts, which are moving. The former, widely used in Europe, it is the movable point rail situated into frame between the wing rails. The latter, not so frequently used, are the movable wing rails and the point is tightly attached. The DT selected the first one from mentioned options, when the crossings with movable parts are applied. The very first single turnout equipped with crossing with movable parts was the turnout type J60-1:12-500-PHS installed in 2003 in Vranovice railway station to be verified under operation. Another turnout with crossing with movable parts was manufactured in 2007. It was the type J60-1:26,5-2600-PHS installed in Poricany railway station. Using of crossings with movable parts was gradually extended on the whole production of the turnouts designated by tangens of turnout angle and designed for application of the rail profile 60 E2. The original construction of the frame for crossing with movable parts was made of flat bottom rails (see Figure 3 B). The production was subsequently extended to the design of welded and even cast frame with optimized shape, which is successfully used for lowest-angle turnouts. The successful development can be found in the applied material as well, because our effort is to deliver materials with increased wear resistance. As to brand-new manufactured frames for crossings with movable parts, we are prepared to apply the bainitic steel, which meets all requirements.

A)

B)

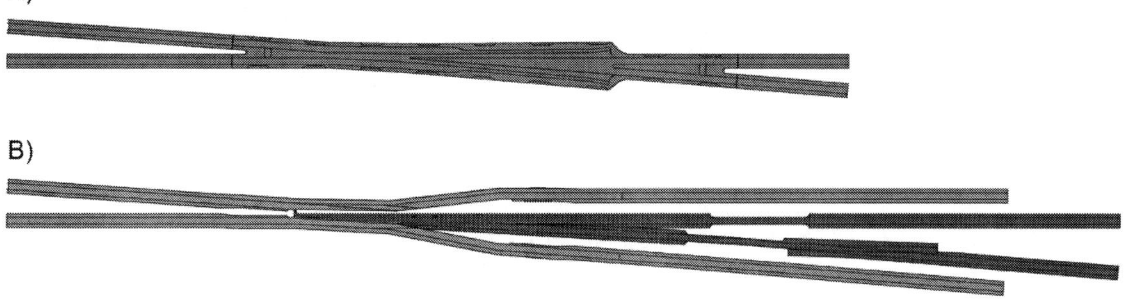

Figure 3. Comparison of the fixed crossing with interrupted running edge (A) and the crossing with movable parts (B).

2.4. Rail track stiffness

The feature, which is given by its construction, is higher stiffness of the turnout, than that of the plain track. This is given by the fact that the turnout structure consists of much more stiff elements fundamentally influencing the stiffness along the turnout. The elements getting the turnout stiff, are usually for instance the switch rails laid onto tough slide chairs (without rail pad), the crossing, check rail, and also the concrete sleepers, because the sleepers installed within the turnout are longer and more robust than those installed within the plain track. The same facts apply for the high-speed turnout J60 1:33,5-8000/4000/14000-PHS and this is our effort to optimize the railway track stiffness across the turnout length by means of the fastening system stiffness adjustment. There is an assumption, that the certain parts of this turnout shall have much higher stiffness than the standard turnout. The basic element principally influencing the high-speed turnout stiffness is the crossing cast frame for movable point rail, which has the robust steel structure. [7]

The stiffness variations across the turnout length fundamentally influence the railway vehicle dynamic behaviour when running across the turnout. The influence of the dynamic effects increases together with higher speed, which means that the way of the high-speed turnouts designing, as long as the running vehicles are supposed to reach speed exceeding 300 $km.h^{-1}$, the turnout variable stiffness across the whole turnout length has to be taken into consideration. At the speeds mentioned above, even lesser variations of the stiffness can cause much bigger loading of some parts, resulting in the quicker degradation of the some parts, consequently sooner intervention of the maintenance staff, including possible worn part exchange. Moreover the occurrence of the quicker destruction of the rail geometry parameters within the turnout can be found. All this can bring the necessity to carry out the adjustment of the rail position within the turnout area more often. The stiffness optimizing, pertinently the surfacing of declined base along the turnout length should bring, besides the other, the diminished cost of maintenance, which is one of bigger requirements in connection with the high-speed turnouts. The knowledge acquired within the dynamic behaviour research of the turnout J60 1:33,5-8000/4000/14000-PHS, which is designed for speed 160 $km.h^{-1}$, will for sure be used within the future development of the further low-angle turnout for high-speed tracks, e.g. for turnout allowing the passage through the branch line with speed 230 $km.h^{-1}$. [6]

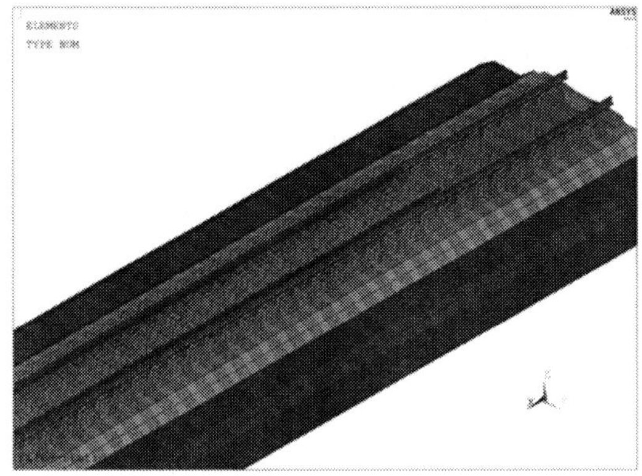

Figure 4 and 5. The high-speed turnout model processed by ANSYS software. The left picture shows the area of crossing, the right picture shows the set of swithes with switch rails.

The first step was to establish the 3D geometric model, which takes account of all important elements of the turnout. All substantial parts were designed under considerations that they can influence the stiffness across the turnout length and then the dynamic behaviour of the whole turnout structure as well. It applies for example to the rails, stock rails, switch rails, frame of crossing including the movable part of the crossing, fastening systems and the sleepers. In some cases the certain simplifications had been adopted, based on the idea to exclude some designing details, which should uselessly increase the needs to disgrace the issue with final elements and could prolong the computing period as well. The further fundamental step was the 3D geometric model import into FEM software ANSYS, in which all important computations were carried out (Figures 4 and 5). After a model file import into ANSYS program, the adding of material characteristics was carried out and the model was equipped with another data, representing the weight features of used parts, which influence the dynamic characterization of the system. Consecutively, the discretionary action towards the final elements was performed and this way the turnout computing model was prepared for the individual static and dynamic computation. The first output is the development of the stiffness expressed by the rail top deflection along the turnout length (Figure 6). From this scene we can assess that the highly stiff part is predominantly the crossing part of the turnout, where after the wheel transition from the curved stock rail onto straight switch rail the prominent increase of stiffness is manifested, thanks to the stiff and flexible pad-free slide chairs. Moreover, we can see the apparent influence of sleepers with changing increased length, which influences the stiffness of the turnout internal rail. According to our assumption, the crossing area also shows high stiffness.

In the further step, the calculation of the dynamic example is carried out by means of application of the wheel forces magnified by the dynamic effect coming from the railway track inequalities and from non-homogeneity in stiffness calculated by the static way. The result shall be the proposal to make the fastening system more flexible to remove the biggest discontinuities.

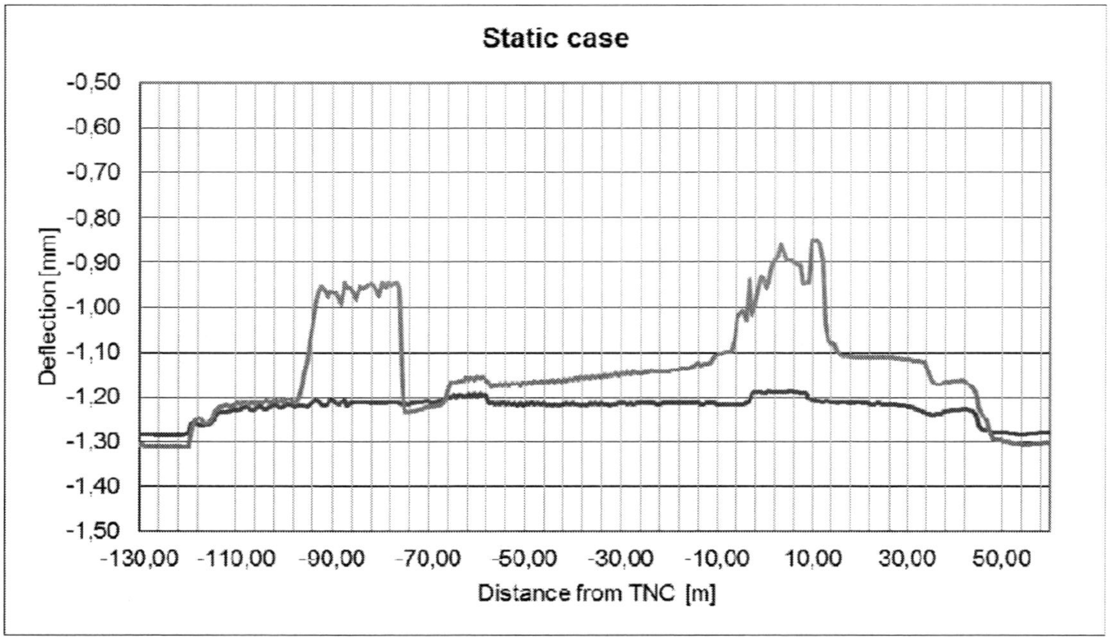

Figure 6. Stiffness diagram shows the vertical deflection within the high-speed turnout. The rail coming across the crossing is of blue colour and the opposite rail is of violet colour. The horizontal axis displays the distance from the theoretical nose of crossing.

3. Conclusion

The development of the high-speed turnouts is a continuous activity which have been carried out for many years. During this activity the innovative elements are developed and usually verified at first within the framework of the standard geometry turnouts. In the process of verification many parameters are usually checked, measured and consecutively assessed. After acquired experience the new elements may be still improved. A new constructional solution can be consecutively enforced into high-speed turnouts. Considering the demanding character of the manufacture and the constructional solution of the high-speed turnout, a functioning sample of the turnout type J60-1:33,5-8000/4000/14000-PHS was manufactured in DT in 2012, at which a number of structural elements, setting mechanisms and the retention systems testing is carried out.

Shortly, the first high-speed turnouts type J60-1:33,5-8000/4000/14000-PHS-E2 shall be installed into Czech railways network in Prosenice railway station. This event will become another milestone within the Czech Republic railway infrastructure development, which leads to increasing the running speed across the turnouts, possibly within the whole adjacent segments of the track. The turnout installations open the door for testing turnouts in the real operation, which will become the source of important knowledge with respect to the correctness of the proposed solution and to its functionality, which should consecutively become an impulsion for further improvement of the design with the aim to install fully verified turnouts into Czech first high speed rails. The high-speed turnout development is naturally still going on and our next goal is to extend our portfolio with turnout which will allow the passage in the branch line by speed 230 $km.h^{-1}$.

Acknowledgements
Research reported in this paper was supported by Competence Centres program of Technology Agency of the Czech Republic (TA CR), project Centre for Effective and Sustainable Transport Infrastructure, no. TE01020168.

References
[1] Český normalizační institut (ČNI) 2008 Konstrukční a geometrické uspořádání koleje železničních drah a její prostorová poloha - Část 1: Projektování *(ČSN 73 6360-1)*
[2] Commission Regulation (EU) 2014 Technical specifications for interoperability relating to the 'infrastructure' subsystem of the rail system in the European Union *(Regulation No 1299/2014)*
[3] European Committee for Standardization (CEN) 2006 Railway applications – Track – Track alignment design parameters. Track gauges 1435 mm and wider. Switches and crossings and comparable alignment design situations with abrupt changes of curvature *(EN 13803-2:2006+A1:2009)*
[4] Puda B 2006 Výhybky pro vysokorychlostní tratě *Seminář Železniční dopravní cesta 2006* pp 81-88
[5] Puda B and Vévoda A 2015 Vysokorychlostní výhybka 1:33,5 pro ŽST Prosenice *Seminář Stroje pro práce na železničním spodku a svršku* pp 59-64
[6] Raif L, Puda B and Žák M 2016 Výhybky pro vysoké rychlosti *Nová železniční technika* **2** pp 70-72
[7] Raif L, Puda B, Havlík J and Smolka M 2015 Výhybky se zpružněnými uzly upevnění v ŽST Ústí nad Orlicí *Silnice železnice* **5** pp 70-72

IOP Conf. Series: Materials Science and Engineering **236** (2017) 012045 doi:10.1088/1757-899X/236/1/012045

Under sleeper pads in switches & crossings

Otto Plasek[1] and Miroslava Hruzikova[1]

[1]Brno University of Technology, Faculty of Civil Engineering, Veveri 331/95, 602 00 Brno, Czech Republic

E-mail: plasek.o@fce.vutbr.cz, hruzikova.m@fce.vutbr.cz

Abstract. Under sleeper pads (USPs) have been installed in the Czech railway lines since 2007 into two trial sections in switches and crossings and in a trial section with the aim to reduce rail pitch corrugation. The basic motivation for the construction of the trial track sections in turnouts was an evaluation of USPs' influence on improvement of track quality as a consequence of ballast bed protection against extreme stress. The paper comprises experience of USPs installation in the Czech Republic in the two trial sections with switches and crossings built in 2007 and 2014. The paper particularly provides an information about USPs applications focusing on the track settlement received by an evaluation of data from the geodetic surveying by precise levelling. Some positive influence of under sleeper pads on the track quality is observed.

1. Introduction

The increase of dynamic effects in railway track structures is connected with the train service speed and axle load increase. An increase of ballast stress is caused not only by the higher dynamic effects but also by the use of concrete sleepers and bearers which are characterized by higher bending stiffness and a relatively small contact area with ballast bed [1]. The dynamic effects occur due to imperfections of a railway track and vertical stiffness variation. The dynamic effects cause changes in sleeper or bearer support in ballast bed that unfavorably influence the quality of track geometry parameters.

A progressive deterioration of the track quality characterized by irregular supports of concrete sleepers or even by voiding sleepers which can vibrate in the ballast bed (so called dancing sleepers) as described, e.g., in [2] and [3], is a consequence of the evidence that track imperfections and higher dynamic effects are closely connected together, [4] and [5].

Under sleeper pads (USPs) which are fixed on the underneath surface of a sleeper decrease the railway track stiffness. Simultaneously the contact area between sleepers or bearers and ballast is significantly higher. Static and dynamic loads on sleepers or bearers are decreased and vehicle–track dynamic system properties are modified. Similarly, furthermore, the transfer of vibrations to a ballast bed is interrupted and damping of vibrations is moved to upper parts of a track, [6] and [7].

The first two trial sections for concrete sleepers or bearers with USPs were built in the Czech Republic in 2007 [8]. The basic motivation for the construction of the trial track section in Plana nad Luznici railway station was an evaluation of the influence of USPs on improvement of track quality as a consequence of the ballast protection against extreme stress. The second trial section was constructed in the Havlickuv Brod – Okrouhlice railway line section with the objective to reduce the rail corrugation development in the curve of small radius. This second trial section is outside of the paper scope. The third trial section with USPs was constructed in Usti nad Orlici railway station in 2014. The line tonnage is approximately 5.6 MGT per year in Plana nad Luznici, while 27,8 MGT in track No.1 and 31,8 MGT in track No.2 in Usti nad Orlici.

Usti nad Orlici station was chosen for testing of more types of the track stiffness modification in switches and crossings (S&C). S&C with special type of rail fastening as well as the USPs are investigated in the station. The variation of different properties of components should homogenize the vertical track stiffness in S&C along the track to reduce dynamic effects caused by abrupt changes of track stiffness. A special type of rail elastic fastening was installed in two selected turnouts and USPs into other two turnouts. The combined solution with the special elastic fastening together with USPs was not used since the dynamic response of such design has not been investigated yet.

Figure 1. Trial sections with under sleeper pads in the Czech Republic.

The paper comprises experience with applications of USPs in the Czech Republic from point of view of track settlement. Here published results are complementary to the evaluation of track settlement and track quality in the sections with USPs, which was published by authors in [9]. An influence of USPs on track quality and influence on sleepers or bearers deflections and vibrations has been monitored for more than nine years.

1.1. Plana nad Luznici trial section

The first trial track section was built in Plana nad Luznici railway station (the 4th Czech railway corridor: Prague – Ceske Budejovice – Linz) in the Czech Republic in 2007. The trial section comprises of a turnout J60-1:12-500-I (the rail 60 E1) with USPs and an adjacent track with and without USPs. The line tonnage ranges from 20 000 to 40 000 gross tons per day, the class of the line according to the maximum admissible axle load and load per unit length is D4 (axle load 22.5 t or load per unit length 8 $t.m^{-1}$).

The assembly of the USPs for the turnout was designed in the finite element model [10] with the aim to reduce abrupt changes of track vertical stiffness. The basic bedding modulus of installed USPs is 0.25 $N.mm^{-3}$ – Getzner SLB 2210. Softer USPs – Getzner SLS 1010 and SLS 1707 – were used in the crossing panel just behind a frog nose and in the area of long bearers behind the crossing. The softer USPs are only in the middle part of the bearers. Transition zones that allow smooth transition of vertical track stiffness between the track with USPs and the track without USPs were designed. The bedding modulus of USPs in the transition zones is 0.30 $N.mm^{-3}$ – Getzner SLB 3007. The length of the transition zone in the adjacent plain track on both ends is 32.4 m, i.e., 54 sleepers. The total length of the track with USPs is 205 m. The pads were installed on the underneath surface of sleepers and bearers by gluing. The neighbouring turnout and the adjacent track were chosen as a comparative conventional section. Continuous welded rail was constructed in the whole track section including both turnouts. The review of the whole trial section is in Figure 2.

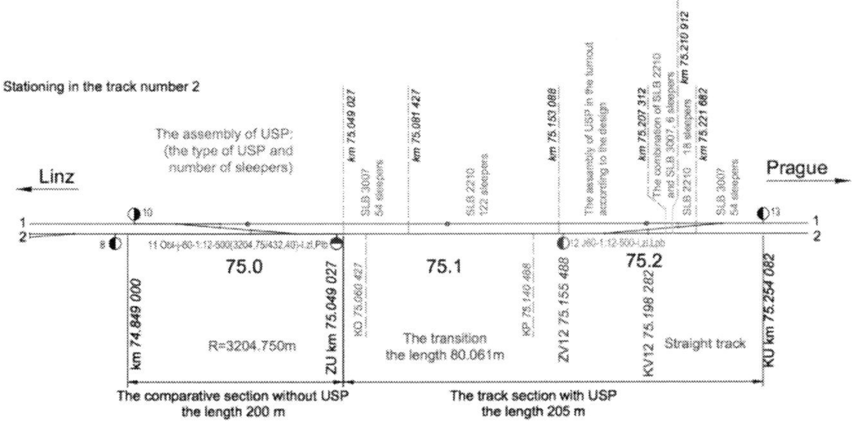

Figure 2. The scheme of the trial track section for USPs in the turnout in Plana nad Luznici railway station.

1.2. Usti nad Orlici trial section

The second trail section with S&C with USPs is located in Usti nad Orlici railway station on the main line Prague – Brno / Ostrava. The traffic load of this track ranges from 80 000 to 130 000 gross tons per day, the class of the line according to the maximum admissible axle load and load per unit length is D4. USPs are installed into two turnouts of different types – J60-1:14-760 and J60-1:12-500-I (both the rail 60 E2). Both turnouts are located on the eastern part of the railway station in the main tracks with V_{max} = 120 km.h^{-1}, for tilting trains (Pendolino CD class 680) V_{max} = 130 km.h^{-1}. The investigated turnouts are: No. 7 J60-1:14-760 in track No.1 and No. 8 J60-1:12-500-I in track No. 2, see Figure 3.

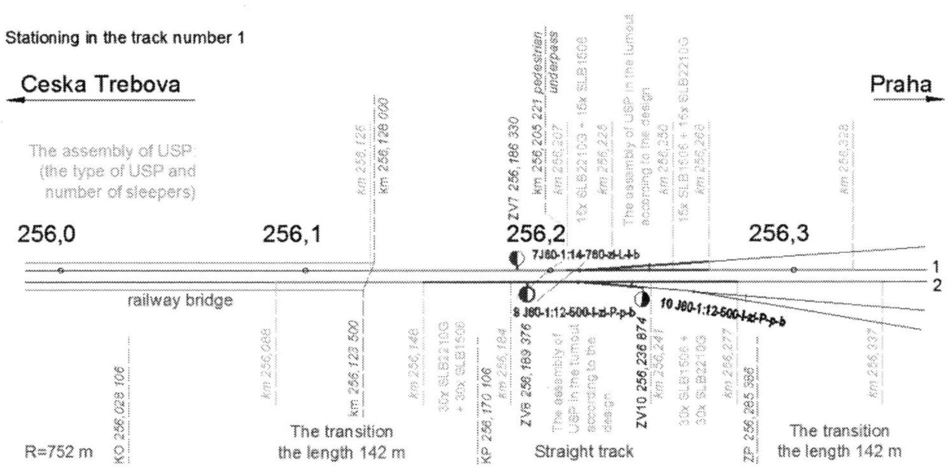

Figure 3. The scheme of the trial track section for USPs in the turnouts in Usti nad Orlici railway station.

The USPs assemblies and stiffness of USPs on particular sleepers or bearers were designed based on the numerical modelling of structures and their static and dynamic analyses. The USPs arrangement was determined regarding the vertical track stiffness homogenization but simultaneously to get the assembly design as simple as possible because of construction work procedures. USPs of 4 different types (bedding moduli) were used in the turnouts, adjacent track including transition zones. The basic bedding modulus of installed USPs is a softer than in Plana nad Luznici, i.e. 0.13 $N.mm^{-3}$ – Getzner SLS 1308G. The softest USPs were installed in the crossings just under the frog nose, the stiffest one in the transition zones between the track with and without USPs. The USPs are installed in the whole length of turnout No. 8, transition zones are located in front of and behind the turnout. The economical design of the USPs assembly was used in the turnout No. 7 where USPs were installed only in the crossing while one transition zone is located between switch and crossing, the second one in the adjacent track behind the crossing. USPs of two bedding moduli were installed into the transitions because of a gradual change of the vertical track stiffness, details of the USPs assemblies were published in [9].

The locality in which the turnouts were installed is rather complicated because of two bridges in the sections vicinity – a pedestrian tunnel and a viaduct over the rivers confluence. Both turnouts are influenced by the transitions of the track from the embankment to the bridge structures. The switch of the turnout No.7 even partially lays on the deck of the pedestrian tunnel structure.

2. Monitoring of USPs trial sections

2.1. Monitoring principles

The trial sections were built with the aim to protect permanent way elements against extremely high dynamic loading. That is why all parameters that could be influenced by the application of USPs are assessed. The following parameters are being observed in all trial sections:

- quality of track geometry parameters;
- track settlements;
- vertical deflections under a running axle;
- vibrations of railway superstructure elements;
- transfer of vibrations to a track vicinity;
- noise propagation to a track vicinity.

This article deals only with the track settlement monitoring assessed by precise levelling. The aim of this assessment is to verify the stability of the support of sleepers or the long bearers with USPs in ballast bed.

2.2. Track level and settlement

The level of running surface of both rails was monitored through geodetic surveying by precise levelling in all trial sections. Rail levels, bracket-type datum mark and other check points were monitored. The distance between the monitored cross sections in the plain track was 6.0 m; the distance between the cross sections in the turnouts and in the immediate adjacent track was 3.0 m. A total of 93 cross sections are monitored in Plana nad Luznici railway station, a total 48 cross sections in the track No.1 and 65 cross sections in the track No.2 in Usti nad Orlici railway station have been investigated.

The track level is evaluated in a relative altitude system. Relative deviations from the optimized track position, which was found out by regression in the initial observation, were calculated. Designed parameters of graduated transition curves were taken into account.

The evaluation of relative geometry parameters was chosen for two reasons. The first reason is the fact that a printing of absolute track level in a chart does not provide an evidence of the track level deviation which is relatively small to the absolute track level. An expression of the deviation relative to the designed track level is influenced by overall changes of the track level in the whole section - settlement or tamping of the track – which is the second reason.

The longitudinal level of the track in both trial track sections was measured twice or occasionally three times every year after the trial sections had been put into operation. If a track geometry correction by tamping procedure had been planned, the levelling was carried out of the track a few days before and after the maintenance work.

Due to a weak subsoil the track sections in the plain track behind the turnouts in Usti nad Orlici had to be tamped in March 2015 in both tracks and in November 2015 in track No.2. In the period between both tamping non-specific maintenance was carried out in track No.1 so this period was removed from the numerical evaluation. Next tamping of both tracks and turnouts is planned for March of 2017. Such maintenance makes the evaluation difficult since only measurements between two consecutive tamping can be taken into account.

The charts in Figure 4 and Figure 5 provide an impression of qualitative assessment of the track settlement in the particular sections. The small and homogeneous settlement is evident in the turnout regions in both tracks in comparison with the adjacent track sections.

The level deterioration and variation is evident in both track behind the turnouts, which is caused by the weak subsoil in this area. The information about the weak subsoil was provided by railway infrastructure manager based on data from the construction work acceptance process (these data are confidential and not available for the research team). Unfortunately, this fact also influences an evaluation of behavior of the track with USPs since the track level in the transition zones follows the level of the adjacent track on the weak subsoil.

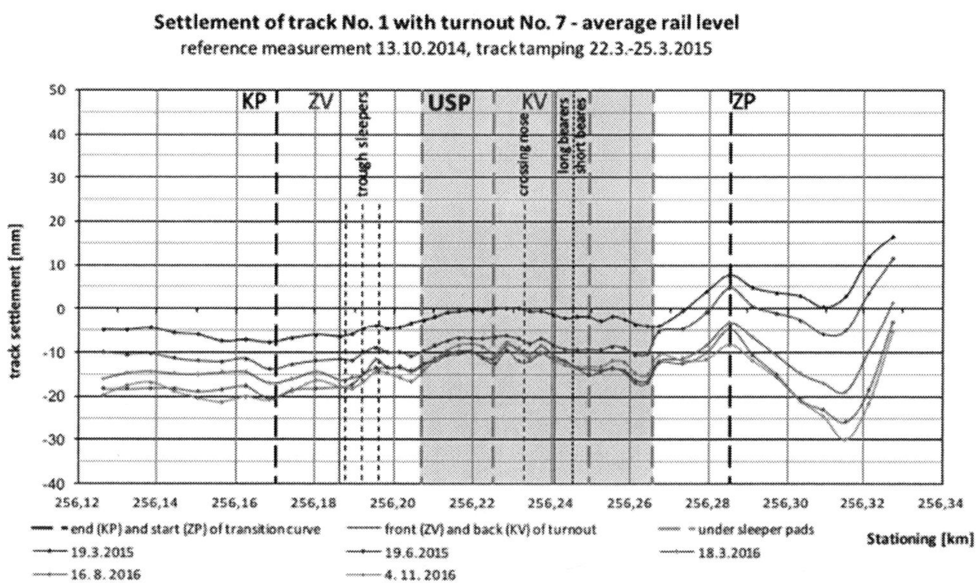

Figure 4. The track settlement in track No.1 Usti nad Orlici after track tamping (March 2015).

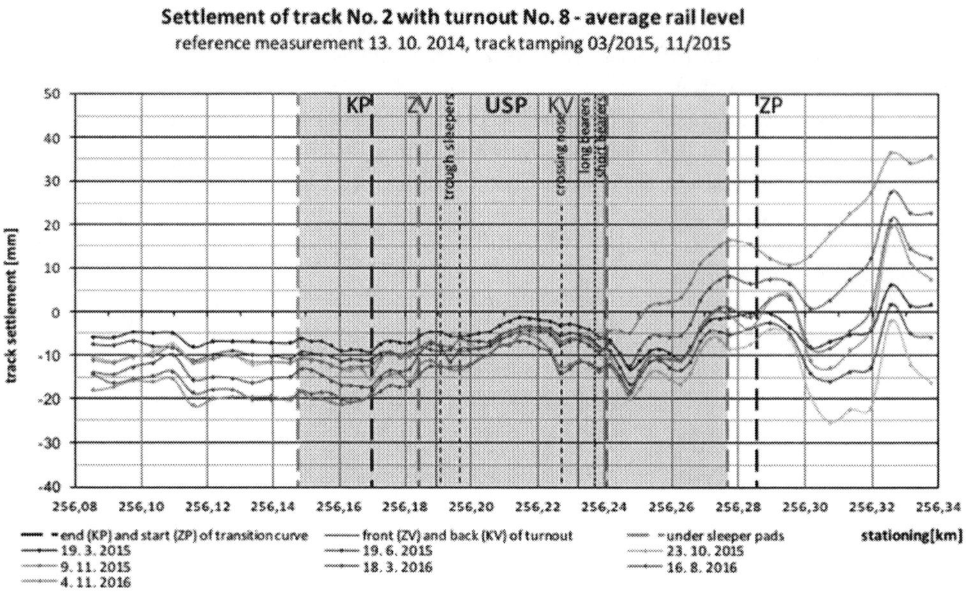

Figure 5. The track settlement in track No.2 in Usti nad Orlici after track tamping (22–25th March 2015).

2.3. Numerical evaluation of track settlement

The settlement progress analysis against line tonnage in both sections for plain track with and without USPs was carried out. The exponential regression of settlement curves was calculated by least square method following the formula:

$$S(l) = L \cdot \left(1 - e^{-\frac{k}{L} l}\right) \quad (1)$$

in which symbols mean:

S(l) ... function of settlement [mm];
l ... line tonnage [MGT ... million gross tons];
L ... limit of settlement [mm];
k ... slope of tangent at $l = 0$ [mm.MGT^{-1}].

The regression parameters used in the formula (1) were calculated by the Newton iterative method using the least square method. The detailed description of a numerical calculation of parameters is provided in [11].

The track settlement in Usti nad Orlici was evaluated in both tracks since November 2015 after the last tamping in track No.2 because no maintenance has not been carried out in both tracks. The evaluation of the track settlement including nonlinear regression is presented on charts in Figure 6 and Figure 7.

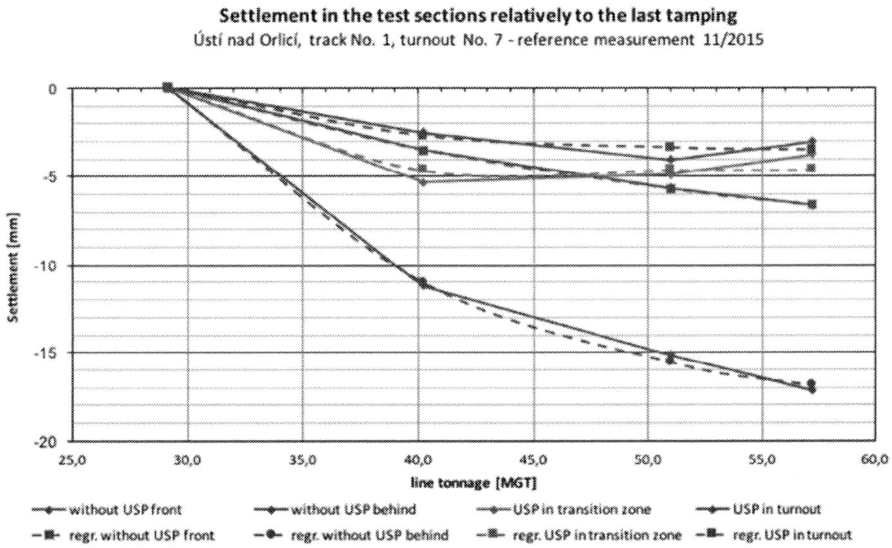

Figure 6. The track settlement progress in Usti nad Orlici after the track tamping 11/2015.

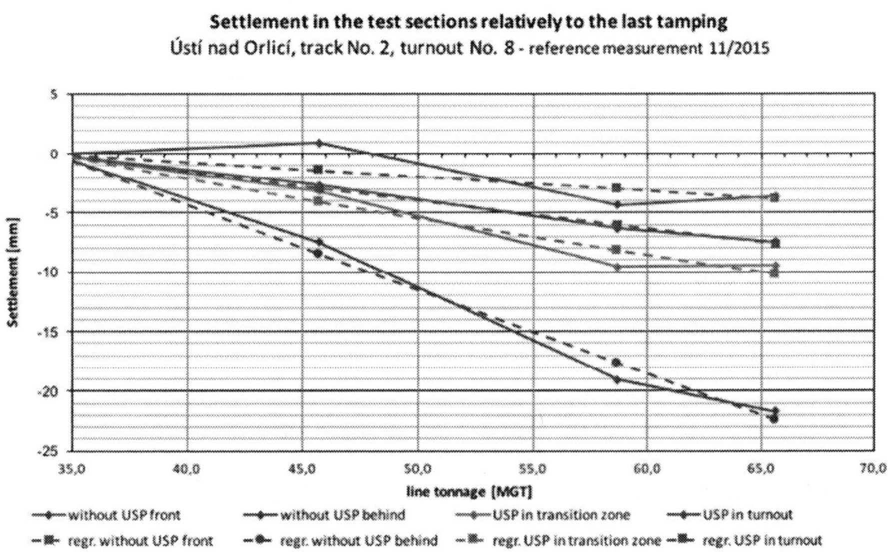

Figure 7. The track settlement in Usti nad Orlici after track tamping (22–25th March 2015).

Table 1. Results of numerical evaluation of the track settlement.

	Usti nad Orlici track No.1, turnout No.7		Usti nad Orlici track No.2, turnout No.8		Plana nad Luznici [11]	
	L [mm]	k [mm.MGT^{-1}]	L [mm]	k [mm.MGT^{-1}]	L [mm]	k [mm.MGT^{-1}]
without USPs front	9.9	0.4	63.3	0.3	10.62	0.02
without USPs behind	19.0	1.5	260	0.7		
USPs in transition zone	4.6	7.7	43.5	0.4	-	-
USPs in turnout	3.6	0.5	95 700	0.1	10.47	0.02

3. Conclusion

Based on the numerical evaluation (Table 1, Figure 6, Figure 7) following facts can be stated:

- the biggest track settlement caused by weak subsoil occurs behind both turnouts in Usti nad Orlici (red lines in Figure 6 and Figure 7);
- the dependence of track settlement on line tonnage is untypically almost linear in track No.2 in Usti nad Orlici so a relatively fast settlement can be expected in this track;
- the track settlement in the track section with USPs in Plana nad Luznici is approximately twice bigger than in the track section with USPs in Usti nad Orlici.

A negative relative settlement (uplift) has been identified in the track sections with USPs, although it is very small, up to 1 mm. This effect has not been completely explained yet. The reason can be the limited precision of measurement and evaluation methods. Because this effect has not been identified for the first time, authors consider also other factors, which could cause this effect, e.g. climate influence on ballast bed and substructure layers. Generally, the conclusion is that USPs in the turnouts in Usti nad Orlici positively influence the track settlement, consequently the track quality, i.e. the main objective for USPs installation is satisfied.

Acknowledgements

This outcome has been achieved with the financial support of Technology Agency of the Czech Republic (TA CR), project TE01020168 Centre for Effective and Sustainable Transport Infrastructure.

References

[1] Plasek O and Hruzikova M 2007 The Use of Under Sleeper Pads as a Protection of Ballast in Plain Track and Turnouts *Akustika* (Studio Akustika: Ceske Budejovice) pp 2-14
[2] Schneider P, Bolmsvik R and Nielsen J C O 2011 In Situ Performance of a Ballasted Railway Track with under Sleeper Pads *Proc. of the Institution of Mechanical Engineers, Part F: J. of Rail and Rapid Transit* **225**(3) pp 299-309
[3] International Union of Railways (UIC) 2009 USP - Under Sleeper Pads - Semelles sous traverses - Schwellenbesohlungen *Summarising Report of UIC Project No. I/05/U/440*
[4] Izvolt L and Kardos J 2011 Influence of Parameters of Railway Track Construction on Vertical Dynamic Interaction Vehicle/Track *Communications - Scientific Letters of the University of Zilina* **3** pp 63-70
[5] Bencat J and Stypula K 2013 Buildings Structure Response Due to Railway Traffic *Communication - Scientific Letters of the University of Zilina* **2** pp 41-48
[6] Leykauf G and Stahl W 2004 Untersuchungen und Erfahrungen mit besohlten Schwellen *EI – Eisenbahningenieur* **55**(6) pp 8-16
[7] Potvin R 2008 Schieneneinsenkung und Schwellenbesohlungssteifigkeit *EI – Eisenbahningenieur* **6** pp 39-44
[8] Hruzikova M, Plasek O, Smutny J, Svoboda R and Salajka V 2009 Trial Track Section with Under Sleeper Pads in the Czech Republic *Proc. of Railway Engineering-2009* (Engineering Technics Press : Edinburgh) pp 58-68
[9] Hruzikova M, Plášek O and Bílek J 2016 Hodnocení zkušebních úseků s podpražcovými podložkami *Věda a výzkum pro práce na železniční dopravní cestě (Děčín)* pp 156-168
[10] Plasek O and Hruzikova M 2007 Design of Under Sleeper Pads for Turnout *Archiwum Instytutu Inzynierii Ladowej* (Poznan: Archives of Institute of Civil Engineering) pp 241-249
[11] Plášek O, HrUzíková M, Svoboda R and Bílek J 2014 Under Sleeper Pads in Railway Track *Komunikácie* **4** pp 27-34

Vibro-acoustic performance of newly designed tram track structures

Ivo Haladin[1], Stjepan Lakušić[1] and Maja Ahac[1]

[1]University of Zagreb Faculty of Civil Engineering, Zagreb, Croatia

E-mail: ihaladin@grad.hr

Abstract. Rail vehicles in interaction with a railway structure induce vibrations that are propagating to surrounding structures and cause noise disturbance in the surrounding areas. Since tram tracks in urban areas often share the running surface with road vehicles one of top priorities is to achieve low maintenance and long lasting structure. Research conducted in scope of this paper gives an overview of newly designed tram track structures designated for use on Zagreb tram network and their performance in terms of noise and vibration mitigation. Research has been conducted on a 150 m long test section consisted of three tram track types: standard tram track structure commonly used on tram lines in Zagreb, optimized tram structure for better noise and vibration mitigation and a slab track with double sleepers embedded in a concrete slab, which presents an entirely new approach of tram track construction in Zagreb. Track has been instrumented with acceleration sensors, strain gauges and revision shafts for inspection. Relative deformations give an insight into track structure dynamic load distribution through the exploitation period. Further the paper describes vibro-acoustic measurements conducted at the test site. To evaluate the track performance from the vibro-acoustical standpoint, detailed analysis of track decay rate has been analysed. Opposed to measurement technique using impact hammer for track decay rate measurements, newly developed measuring technique using vehicle pass by vibrations as a source of excitation has been proposed and analysed. Paper gives overview of the method, it's benefits compared to standard method of track decay rate measurements and method evaluation based on noise measurements of the vehicle pass by.

1. Introduction

Urban railway design is often influenced by the vibro-acoustic performance parameters. This is due to the fact that the railway track resides very close to surrounding structures (residential or business buildings, infrastructure etc.) where people live and work in. In Europe such areas are regulated by standards controlling noise and vibration levels [1], [2] that people are exposed to in their living or working environment. For that reason, light rail, metro and tram tracks have to incorporate certain level of vibration and noise damping components in order to achieve safe, comfortable, steady and quiet operation of railway vehicles.

It is very well known that the dominant source of noise emitted by rail vehicles at speeds greater than 40 km/h and lower than 250 km/h is in fact induced by interaction of wheel and rail at their interface due to unevenness or roughness of both surface of wheel and rail [3]. On the track side, the dominant radiator of noise is the rail that vibrates at high frequencies. It is possible to deal with wheel-rail interaction vibrations by controlling the smoothness of running surface of wheels and rails. Rail grinding can be a challenging and demanding task in collision with retaining level of service on busy urban rail systems running 24/7. Track can however by applying certain damping properties in design phase, absorb a portion of vibrations that would otherwise generate noise in the environment. Such damping properties can reduce noise at its source and reduce or eliminate the need of installing additional noise barriers or other means of noise mitigation measures.

Zagreb Municipality Transit System (ZET) acquired 140 new trams in recent years, thus changing the conditions on the tram network in terms of power demand, loads transferred to the track, wheel-rail interaction mechanism etc. Shortly after the introduction of new low-floor trams, it was observed that this fleet modernization has increased exploitation demands on other components of tram infrastructure [4]. In such conditions the deterioration of tram infrastructure

2. Design of new fastening systems for Zagreb Municipality Transit System (ZET)

The primary role of the rail fastening system is positioning and fixing of the rails and transferring the vehicle load from the rails to the track substructure. Generally, the type and characteristics of fastening system is chosen depending on the required elasticity of the track, planned load and type of rail. On major part of Zagreb tram tracks, grooved rails are discreetly laid on the levelling layer, made of micro synthetic concrete, built on reinforced concrete slab.

Standard fastening system used on a large part of ZET tram network is indirect elastic fastening system with decreased stiffness PPE with 50 MN/m stiffness, developed during the 1990s which do not meet today's highly complex exploitation requirements on Zagreb's tram tracks such as high wheel loads on low-floor trams (more than 3.5 tons per wheel) and stricter environmental noise regulations (adoption of EU regulations [1]). Based these requirements, during the last few years the Faculty of Civil Engineering University of Zagreb has developed five different concepts for new fastening systems. Two of them, with working titles 21-CTT (based on Classic Tram Track) and 21-STT (Slab Tram Track), were chosen for further development. Their characteristics, production requirements and installation methods best match previous experiences in the construction and maintenance of tram tracks in Zagreb.

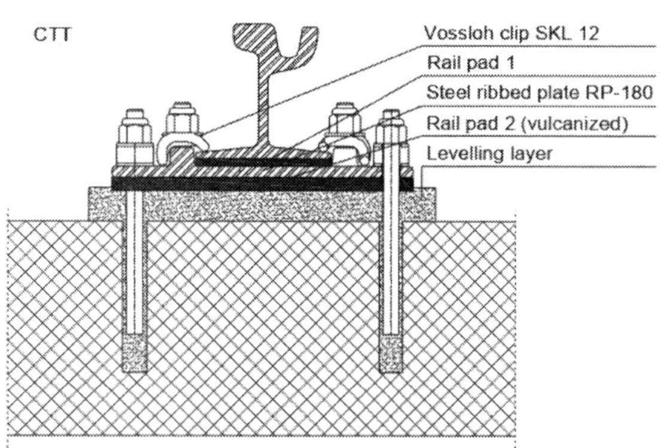

Figure 1. Cross section of 21-CTT fastening system.

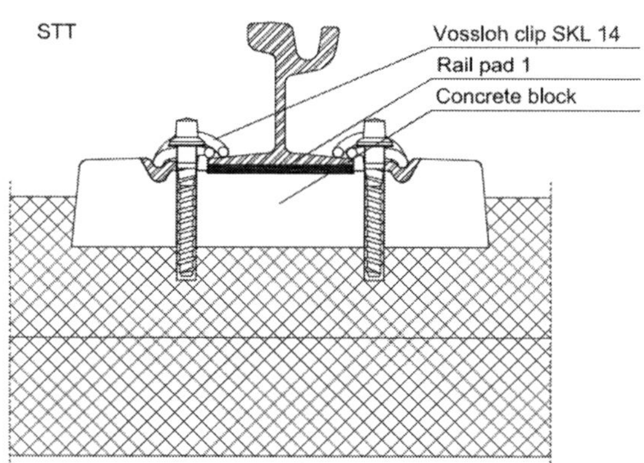

Figure 2. Cross section of 21-STT fastening system .

21-CTT system (Figure 1) places the rails discreetly on the levelling layer, made of micro synthetic concrete, built on reinforced concrete slab. The distance between supports is one meter in the tram rolling direction. The rail foot is supported by neoprene pads placed on ribbed steel plate, laid on the levelling layer. To ensure the durability of individual components and ease the installation during construction and dismantling during the track reconstruction, the underside of the ribbed steel plate is fitted, by vulcanization process, with elastic pad. Vulcanization process enabled production of a compact element electrically isolated from other components (anchors), but also provided for additional elasticity of the whole fastening system.

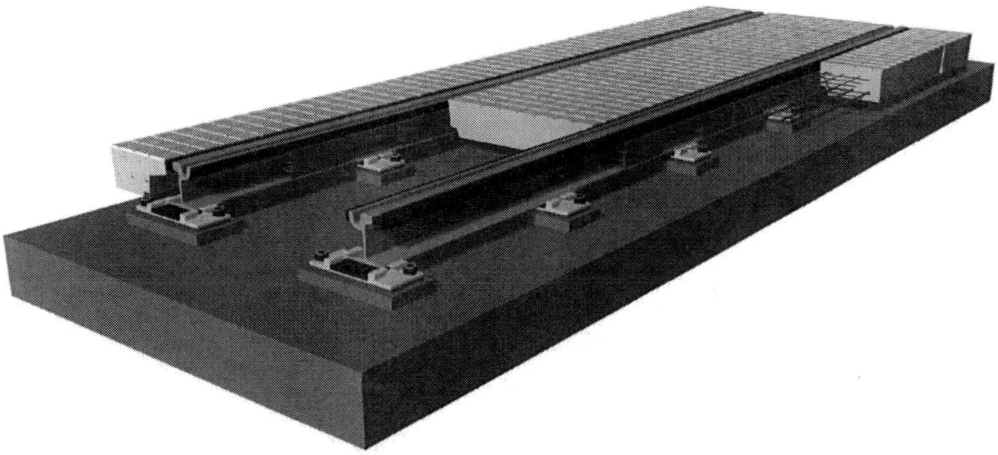

Figure 3. Design model of tram track structure with 21-CTT fastening system.

21-STT system (Figure 2) represents a completely new solution for tram tracks superstructure construction in Croatia. Its main advantage over the previous systems is that the levelling layer (which is quite difficult to construct) are replaced by two block precast concrete sleepers laid on reinforced concrete base slab, one meter apart (Figure 4). Upper reinforced concrete slab is constructed after track horizontal and vertical alignment adjustment. This is a direct elastic fastening system, with one elastic pad between the rail foot and block sleeper.

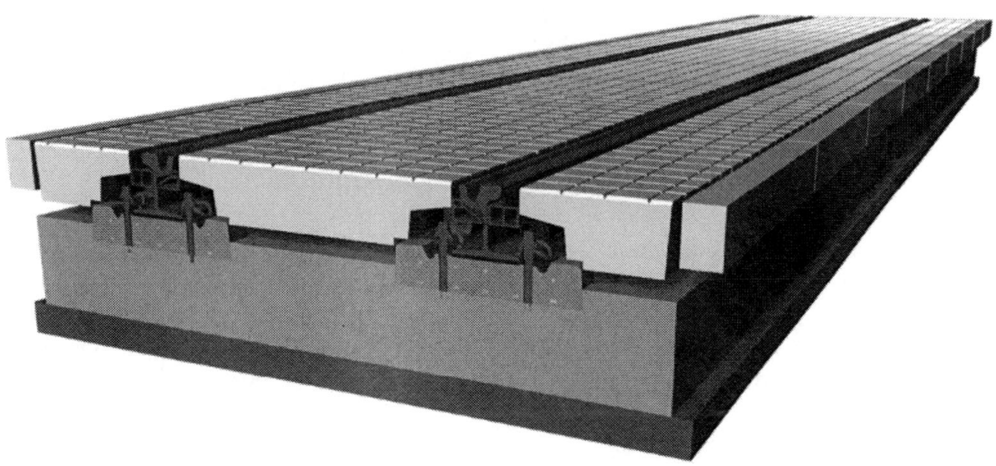

Figure 4. Design model of tram track structure with 21-STT fastening system.

3. Data acquisition at test site

To evaluate the newly designed track structures, test track has been installed in Savska street in Zagreb, providing the opportunity to observe and measure vibro-acoustic behaviour of 21-CTT, 21-STT track systems compared to PPE, a standard track fastening system commonly used on ZET network. Sub-sections of 21-CTT, 21-STT and PPE (each 50 m long) have been installed on 150 m long test track.

The test track incorporates different features for track monitoring:

- strain gauges in bottom zone of RC slab for strain and deformation monitoring,
- auxiliary shafts for easy access to equipment wiring and PC connections,
- small shafts next to the rails for access to the rail web and foot for accelerometer mounting,
- larger revision shafts for visual inspection of fastening systems along the test section.

Each test sub-section has been fitted with 4 strain gauges (two in horizontal, two in lateral direction in respect to a plain through rail running surface, Figure5), 4 accelerometers attached to rail foot and rail web for vibration measurements, Figure 6, and microphones along the track, 2.5 m away from the track axis.

Figure 5. Position of strain gauges prior to concrete casting on 21-STT subsection.

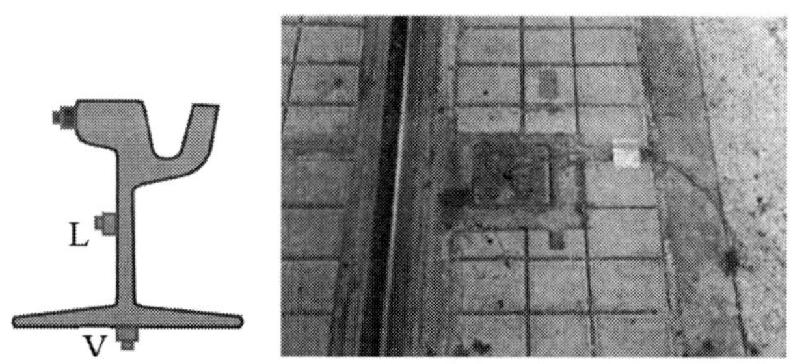

Figure 6. Position of accelerometers on a rail foot (vertical direction) and rail web (horizontal direction), inside a small shaft.

4. Measuring procedure and data analysis

At each sub-section, measurements of strain, noise and vibrations have been conducted using tram TMK 2266 as a source of excitation. The tram has been set to run at a constant speed of 15 km/h and 30 km/h along the three subsections. By using the same tram, under the same running conditions, differences in recorded levels of strain, vibrations and noise should are mostly influenced by track type the tram is running on.

4.1. Strain measurements

Strain reading on the bottom of concrete slab is a good indicator of load distribution through the track structure, influence of dynamic load on the lifespan of structure and possible cracks in the track structure. Data has been collected with sampling frequency of 50 Hz, giving a strain reading every 0.02 sec. Measurement results are shown in Table 1, where the overview of maximum strain readings of track structures is presented (for PPE, 21-CTT and 21-STT) for tram speed of 30 km/h.

By comparing the continuous time record of strain at all subsections (Figure 7) it is evident that 21-STT structure has very good load distribution to lower structure layers. Strain in this structure is around 3 times smaller than the other two observed subsections. It is also visible that structures 21-

CTT and PPE have similar strain measurements, as it was to be expected since similar fastening elements have been used in terms of load distribution.

Table 1. Values of strain reading at the bottom of RC slab.

Maximum strain reading [MPa]						
Section	PPE		21-CTT		21-STT	
Direction	LONG	LAT	LONG	LAT	LONG	LAT
24.11.2014.	**2.38**	**1.87**	**2.89**	**3.33***	**0.56**	**0.73**
04.03.2015.	**3.33**	**2.25**	**2.26**	**1.14**	**0.86**	**1.01**

*sensor is located near a rail weld of poor geometry so additional strain is to be expected

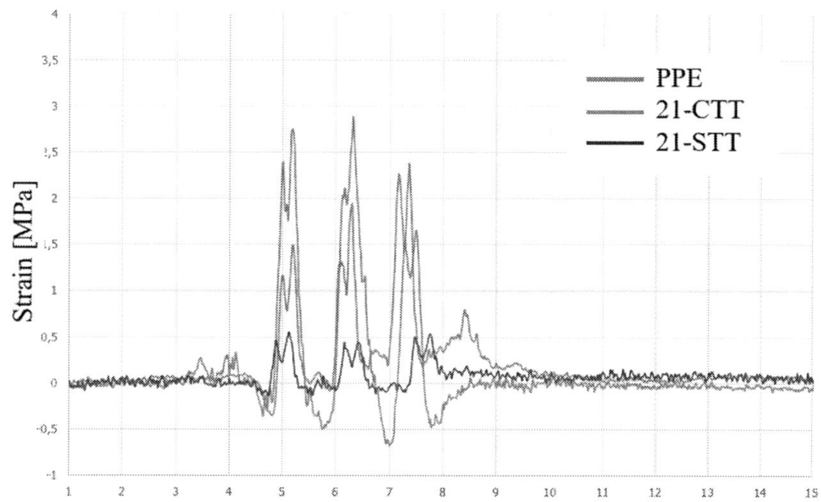

Figure 7. Time recording of strain measurements on three subsections.

4.2. *Vibrations and track decay rate*

Time recording of accelerations of rail in horizontal and vertical direction, under vehicle pass by has been recorded and used as a base for further analysis. Time recording of horizontal and vertical rail vibrations under TMK 2266 tram pass by have been analyzed. To express vibro-acoustic behaviour of a railway track, a parameter of track decay rate has been used. The contribution of noise emitted by rail vibration is a significant component of rolling noise up to 5 kHz [5]. One major factor affecting track noise performance has been identified to be the attenuation rate of vibration on the rail as a function of the distance from the excitation point [6], referred to as track decay rate measured in dB/m. Namely, the longer the section of the rail that vibrates, more noise is emitted to the environment. If the rail vibrations attenuate rapidly along the rail, and high decay rate is achieved, less noise is emitted.

Track decay rate is standardly measured by hammer impact method described in EN 15461:2011 [7] and very effective measurement technique on standard ballasted railway tracks [8]. On tramway tracks with higher decay rate values and different exploitation conditions (running surface shared with road vehicles, less time for conducting measurements), a method of measuring track decay rate from vehicle pass by time recording of vibrations is proposed [9]. Such a method has been adopted, modified and evaluated on the test section in Savska street.

Signal is first transferred to frequency domain and analyzed in 1/3 octave bands with center frequencies from 100 Hz to 5000 Hz. Track decay rate is based on the ratio of recorded rail vibrations around the wheels of the pass-by vehicle (T_{L1}) and vibrations of whole vehicle pass by (T_{L2}), Figure 8.

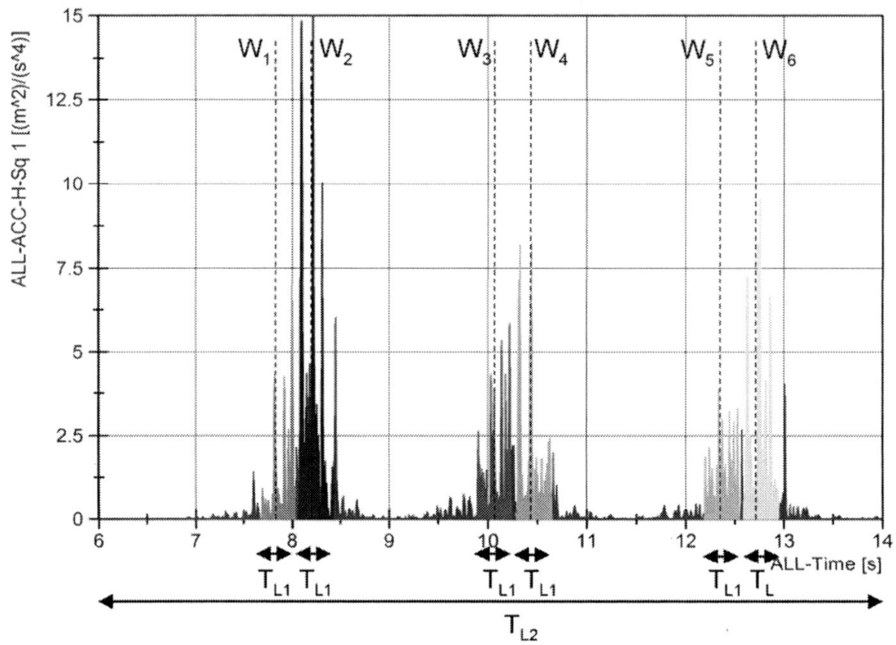

Figure 8. Time recording of vibrations with integration time T_{L1} around each wheel and T_{L2} for the whole tram pass by.

Calculated ratio, $R(f_c)$, of amplitudes around each wheel (A^2_{L1}) over amplitudes for whole tram pass by (A^2_{L2}) described in (1) is used to express track decay rate, TDR (f_c) for certain frequency band (2). Calculated track decay rate is further analyzed through energy iteration method to eliminate the influence of neighboring wheels on calculated decay rates of the observed wheel.

$$R(f_c) = \frac{A^2_{\sum L_1}(fc)}{A^2_{\sum L_2}(fc)} \approx 1 - e^{-\beta L_1} \qquad (1)$$

$$\mathrm{TDR}(f_c) = -\frac{8.686}{L_1}\ln(1 - R(f_c)) \qquad (2)$$

Detailed procedure of measurements is described in [10] while further relation to measured noise levels is analyzed through this paper. Using this method, decay rates on all three test sub-sections have been determined in vertical and lateral direction. Displayed results are averaged values of track decay rate (TDR) of 3 tram pass-bys.

It can be observed from Figure 9 that the vertical track decay rates are very similar on all three test sections in lower frequency range. In frequency range from 1000 Hz to 5000 Hz better performance of vibration attenuation is achieved by 21 CTT fastening system. 21 STT fastening system has the lowest TDR values i.e. less vibration attenuation due to its construction with only one elastic under-rail pad and overall stiffer slab track structure.

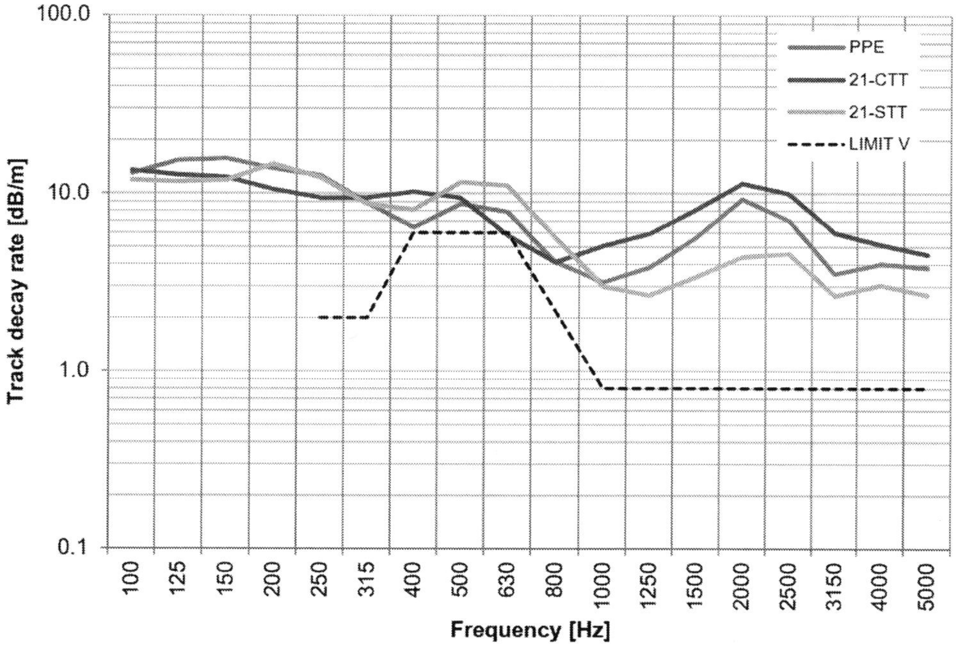

Figure 9. Track decay rate for all sub-sections in vertical direction.

4.3. Noise measurements

As proposed by the European standard [11], time signals of sound pressure recorded during tram TMK2266 pass-by have been expressed as sound pressure level in dB with the reference value $p_0 = 20$ μPa. Equivalent sound pressure level has been calculated for whole vehicle pass by to conduct comparison of different test sections. Equivalent sound pressure level is expressed as

$$L_{pAeq,Tp} = 10\log\left(\frac{1}{T_p}\int_0^{T_p}\frac{p_A^2(t)}{p_0^2}\right)\ [dB] \quad (3)$$

where $L_{pAeq,Tp}$ is the A-weighted equivalent continuous sound pressure level in dB; T_p is the measurement time interval in s; $p_A(t)$ is the A-weighted instantaneous sound pressure at running time t in Pa; and p_0 is the reference sound pressure ($p_0 = 20$ μPa).

Recorded sound pressure levels of 2 pass-by at each test sub-section has been recorded for tram TMK2266 pass-by at 30 km/h and the average value expressed as final equivalent noise level, Table 2.

Table 2. Equivalent sound pressure levels on test section.

Test section	$L_{pAeq,Tp}$ [dB]	SD
PPE	84.5	0.3
21-CTT	80.9	1.4
21-STT	82.9	0.8

It is obvious that under the same conditions (constant speed, low ambient noise, same tram vehicle), newly designed tram tracks perform better from the vibro-acoustic point of view. Noise reduction of 1.6 dB is recorded on Zagreb 21-STT section and 3.6 dB on Zagreb 21-CTT section.

5. Conclusion

Due to high traffic loads, tram tracks in the City of Zagreb are exposed to high stresses, they rapidly degrade and deteriorate. To answer the harsh tram traffic operation conditions, and to optimize maintenance procedures, two new solutions for rail fastening systems were developed, named 21-CTT 21-STT systems. The main objective in developing the new systems was to develop a tram track structure which would be quick and simple to construct, have longer exploitation life, be easy to maintain, and have good exploitation characteristics.

Measurements conducted on test section in Savska street give an opportunity to evaluate the newly designed tracks. The new tracks, especially 21-STT shows improved performance in load distribution. Combined with fast construction (by using pre-fabricated and pre-assembled components) and low construction cost it is a great alternative when considering construction of new tram tracks. For reconstruction of current tram tracks 21-CTT system that is based on standard fastening system typically used in Zagreb, brings an improvement in component availability, stray current insulation as well great noise and vibration attenuation.

Track decay rate determination using vehicle pass-by method has been proven as a very reliable and practical method to measure vibro-acoustic performance of tramway tracks.

Two procedures, track decay rate measurements and noise measurements using vehicle pass-by have confirmed track contribution to overall noise levels is highest on referent track PPE, lower on stiffer 21-STT with direct elastic fastening and significantly lower on 21-CTT.

References

[1] Directive 2002/49/EC of the European Parliament and of the Council relating to the assessment and management of environmental noise, *Off. J. Eur. Communities 2002* Available at http://eur-lex.europa.eu/LexUriServ/LexUriServ.do?uri=OJ:L:2002:189:0012:0025:EN:PDF

[2] European Committee for Standardization (CEN) 2003 Mechanical vibration and shock - evaluation of human exposure to whole-body vibration - Part 2: Vibration in buildings (1-80Hz) *(EN ISO 2631-2:2003)*

[3] Hemsworth B 2013 *Strategies and Tools to Assess and Implement Noise Reducing Measures for Railway Systems (STAIRRS) - Final technical report*

[4] Ahac M 2013 *Mechanical-empirical model for tram track gauge deterioration during exploitation - Doctoral thesis* (University of Zagreb)

[5] Thompson D J and Jones C J C 2000 A Review of the Modelling of Wheel/Rail Noise Generation *Journal of Sound and Vibration* **231**(3) pp 519–36

[6] Jones C J C, Thompson D J and Diehl R J 2006 The use of decay rates to analyse the performance of railway track in rolling noise generation *Journal of Sound and Vibration* **293**(3–5) pp 485–95

[7] European Committee for Standardization (CEN) 2008+2010 Railway applications - Noise emission - Characterisation of the dynamic properties of track sections for pass by noise measurements *(EN 15461:2008+A1:2010)*

[8] Haladin I, Lakušić S and Košćak J 2016 Measuring vibration damping level on conventional rail track structures *Gradjevinar* **68**(6)

[9] European Committee for Standardization (CEN) 2016 Railway applications - Acoustics - Measurement method for combined roughness, track decay rates and transfer functions *(CEN/TR 16891)*

[10] Lakušić S, Haladin I and Ahac M 2016 The Effect of Rail Fastening System Modifications on Tram Traffic Noise and Vibration *Shock and Vibration* **4** pp 1–15

[11] International Organisation for Standardization (ISO) 2013 Acoustics – Railway applications – Measurement of noise emitted by railbound vehicles *(ISO 3095)*

Proposal of an intelligent wayside monitoring system for detection of critical ice accumulations on railway vehicles

Frank Michelberger[1], Adrian Wagner[1], Michael Ostermann[2] and Thomas Maly[2]

[1]Carl Ritter von Ghega Institute for integrated Mobility Research - University of Applied Sciences St. Poelten, Matthias Corvinus-Straße 15, 3100 St. Poelten, Austria
[2]Vienna University of Technology, Institute of Transportation, Research Center of Railway Engineering, Traffic Economics and Ropeways, Karlsplatz 13/230-2, 1040 Vienna, Austria

E-mail: frank.michelberger@fhstp.ac.at

Abstract. At railway lines with ballasted tracks, under unfavourable conditions, the so-called flying ballast can occur predominantly for trains driving at high speeds. Especially in wintertime, it is highly likely that the causes are adhered snow or ice deposits, which are falling off the vehicle. Due to the high kinetic energy, the impact can lead to the removal of ballast stones from the structure of the ballasted track. If the stones reach the height of vehicles underside, they may be accelerated significantly due to the collision with the vehicle or may detach further ice blocks. In the worst case, a reinforcing effect occurs, which can lead to considerable damage to railway vehicles (under-floor-area, vehicle exteriors, etc.) and infrastructure (signal masts, noise barriers, etc.). Additionally the flying gravel is a significant danger to people in the nearby area of the tracks. With this feasibility study the applicability and meaningfulness of an intelligent monitoring system for identification of the critical ice accumulation to prevent the ballast fly induced by ice dropping was examined. The key findings of the research are that the detection of ice on railway vehicles and the development of an intelligent monitoring seem to be possible with existing technologies, but a proof of concept in terms of field tests is necessary.

1. Introduction

Unfavorable conditions and predominantly high speeds of trains on ballasted railway tracks can cause the so called ballast flying. Especially in winter, it is considered to be highly probable that such phenomena are caused by the falling of snow or ice accumulation adhering on the bottom surface of vehicles. As a result of their high kinetic energy, the impact on the superstructure can strike stones out of their cohesion structure. When these disengaged stones gain enough height, collisions with the vehicles underside are possible. On the one hand, such stones may be significantly accelerated and are able to pull out further stones from the ballasted bed. On the other hand, the impact on the vehicle can lead to further detachments of snow or ice accumulations. Due to these mechanisms an avalanche effect can appear, which intensify damages of vehicles (under-floor-area, vehicle exteriors, etc.) and on the infrastructure (signal masts, noise barriers, etc.). Furthermore ballast flying is a serious threat to people which are nearby the railway track (railway staff, pedestrians at platforms or on streets or paths beside the track). Relevant research activities on this topic can be found for example within a project done by Deutsche Bahn AG Systemtechnik, and Bombardier Transportation GmbH in 2007 [1] or within extensive research work in Korea for example by Kwon, H. B. [2]. Although this phenomenon of ballast flying is known for quite some time, due to the seldom occurrence in real life and the concomitant destruction of the initial situation there is no evidence about the concrete circumstances, which induce the avalanche effect. But some factors like the amount of adhesive snow or ice, the speed of the train and the weather conditions of the former trains run can be assumed, which may influence the probability of occurrence. Thus, this paper deals with the new approach of wayside monitoring of trains in operation to identify critical snow or ice formations before they falling off the vehicles.

2. Development

In order to find a solution for this problem, the University of Applied Sciences St. Pölten (Carl Ritter von Ghega Institute for Integrated Mobility), the Vienna University of Technology (Institute of Transportation and the Institute of Building Construction and Technology), the Johannes Kepler University in Linz (Institute of Technical Mechanics) and the Rail Tec Arsenal sat up the project EISMON. With this study, the applicability and meaningfulness of mentioned monitoring approach for identification of the critical ice accumulation was examined.

2.1. Environmental analysis

In a first step, a survey on international documented ballast flying incidents, including the boundary conditions prevailing in each case such as travel speed, weather conditions or vehicle types as well as research of the common practice and development of counter measures were done. These investigations showed that detailed scientific studies on ballast flying induced by ice droppings haven't been made until now in a sufficient number. Nevertheless, some examples of strategies from railway operators were found, how to avoid these incidents. For instance, Deutsche Bahn (DB) uses for their intercity express trains (ICE), which are used in cross-border long-distance service, a wayside de-icing system. Thereby the train drives very slowly over the system and 42 nozzles spray a layer of glycol on their trains, so the ice attachment is minimized up to 80% [3]. In contrast Austrian Federal Railways (ÖBB) count on visual inspections if the weather conditions encourage ice adhesions. When this inspection reveals ice and snow on the train, the speed is limited to a maximum of 160 km/h [4].

In order to achieve a high degree of track and/or vehicle utilization railway operators have to assess the risks as accurately as possible. For this purpose a monitoring system might be beneficial, if it provides detailed and reliable information about the snow or ice attachments and their location on the vehicle underside. To develop such a system, it is necessary beforehand to gain experience of typical characteristics of snow and ice adhesions (e.g. areas which favor adhesions, shapes, degree of impurities, ratio between ice and snow). For this reason investigations on railway vehicles were made during snowy periods in winter. These showed that ice chunks can have considerable dimensions and masses (for instance Figure 1 shows an ice block with dimensions 40 x 20 x 10 cm and mass about 4 kg). Such attachments were predominantly found in the bogie areas and the areas between the vehicles, where many corners and juts can be found. The remaining surfaces of vehicles downsides have a smooth, flat surface. These areas were almost free of snow or ice or, at least, covered with considerably fewer adherences.

In the bogie and inter-vehicle areas the adhesions usually consist of a snow layer in the contact region to the vehicle surface and an iced layer on the top of the snow layer. This structure may be formed due to changing temperatures around the freezing point, where exposed surfaces of snow adherences are subject of an alternating thawing and freezing process (e.g. tunnel passages in winter). In corners and in sheltered areas were mostly found sole snow attachments.

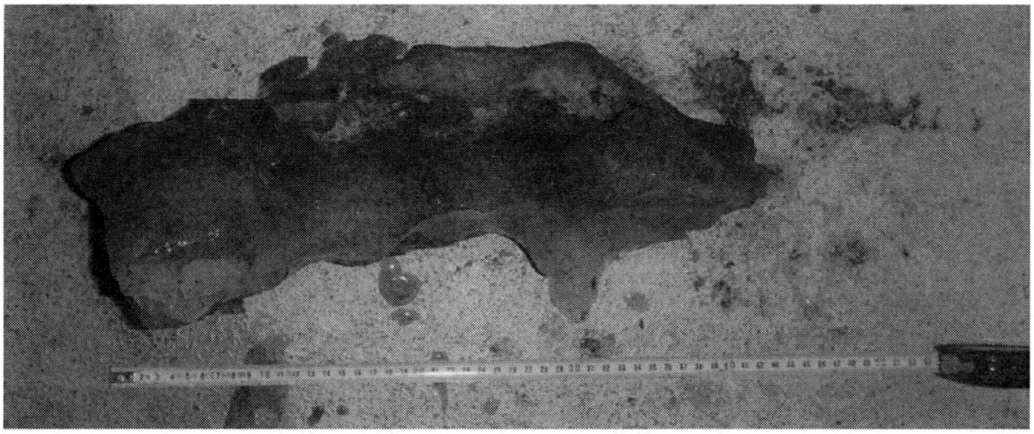

Figure 1. Typical ice block with a mass of 4 kg.[5]

2.2. Mechanical modelling

In a next step to get a deeper understanding of the phenomenon, the basics for a mechanical modelling were executed, mainly through calculation of the flight path from the ice block and the released gravel. For this purpose, input parameters such as the vehicle speed, the diameter of an idealized spherical-shaped, dropped off attachment, the density, the fall height and some more values are solved by setting up a differential equation. It is an interesting note, that in the case of mass equality between the ice dropping and the gravel, it's possible that the gravel hits the vehicle.

2.3. Measurement methods

The future measurement system should be able to detect the volume of the ice or snow accumulation on the train by comparing the measured surface profile with a reference measurement without ice or snow adhesions. If dangerous ice or snow attachments are detected the measurement system has to send a warning to the next operation control centre. Because of the fact that mass equality of snow or ice droppings and gravel could be dangerous (Chapter 2.2.), some requirements in regarding accuracy and spatial resolution were specified. In addition to avoid blocking the line-of-sight between the sensor and the train surface due to snow resuspension the crossing speed is limited to up to 80 km/h. This assumption has to be evaluated in the further development. But because of operational issues the track speed must not be limited by the measurement system. Therefore potential measurement sites are primarily in station areas or other areas where the track speed is limited for example due to narrow curves anyway.

Due to the fact that there are no standard solutions for an automated identification of ice or snow accumulations on trains in operation, a general literature and internet research of applicable contact-free distance measurement methods was made. Through discussions with sensor manufacturers the feasibility of every method was investigated. A comparison of the considered measurement methods was shown through a rating matrix, where every measurement method was evaluated relating to the determined requirements of accuracy and spatial resolution. Based on this matrix the methods laser scanning and light-section appear to be best fitting for this application.

Cameras can additionally be used for verification especially for field tests of a monitoring system development. Thus area scan cameras and line scan cameras were compared separately to the distance measurement methods. Because of the considerably simpler uniform illumination of the recording area, which is necessary for high shutter speeds and in further consequence for sharp images, line scan cameras are preferable.

Up to now these distance measurement methods have never been used with icy or snowy surfaces on fast moved objects like a running train. This means that there is no experience for this measurement task, so the proof of concept for the measurement of ice and snow adhesions must still be provided. To implement a measuring system under real conditions, a test phase with a prototype is therefore necessary. In this phase, the installation of an additional line scan camera to record the train surface is recommended to evaluate the results of the measurement system.

2.4. *Intelligent wayside monitoring system*

The basic idea of the proposed intelligent wayside monitoring system is the combination of different information. The more information can be used, the higher the quality of the result may be. Besides the output of above-described future measurement system, information already available for the infrastructure operator, like train data, infrastructure data or weather data seems to be applicable. Within these information categories, different levels of detail exist. For example train data can also include historic information about the train e.g. how long and where a train stopped before in a depot the night before, on which line the train ran the night or day before and so on. Also useful data of further measurement systems (e.g. temperature or air velocity in tunnels) may be included into the monitoring system. The basic architecture that was developed within the project is demonstrated in Figure 2.

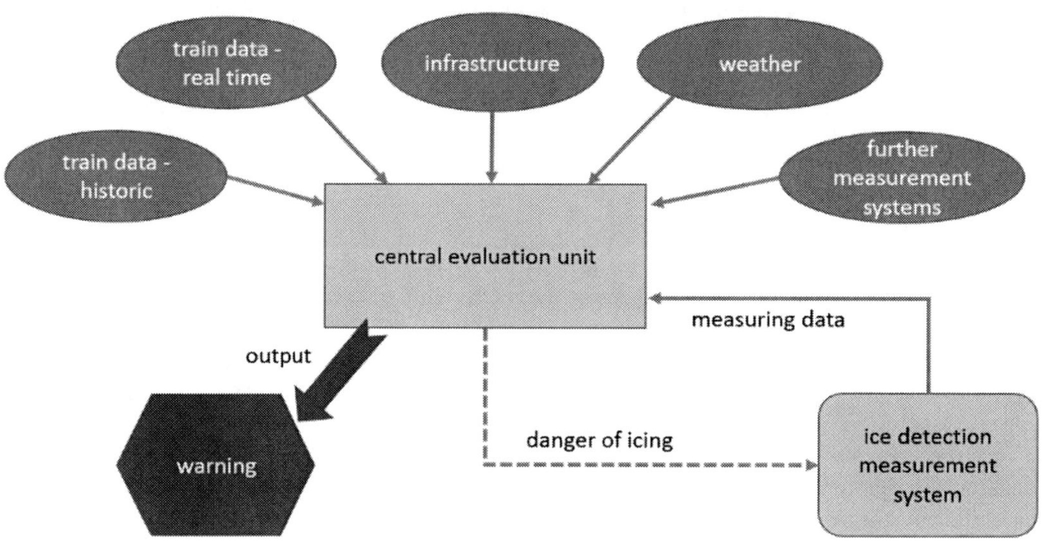

Figure 2. Architecture of intelligent monitoring system. [5]

Within the central evaluation unit the data of the different sources are combined. Mainly for a reduction of maintenance cost due to shorter operating times, the measurement system for ice detection may be deactivated in periods without any danger of icing. For example if there are high temperatures along the complete course of a train in summer, there seems to be less need to measure ice adherences.

The output of the central evaluation are warnings about the detection of suspicious attachments. Based on the strategy of the infrastructure manager these warnings could be an input for the dispatching system where the dispatcher will set suitable measures. But the warning can also be a direct input to a train control system to set further actions automatically (e.g. define a speed limit or stop the train).

3. Problem Analysis

To evaluate the potential and applicability a SWOT analysis was carried out. The main results are illustrated in Figure 3.

Strengths	Weaknesses
• Detection of underfloor is possible • Assessment of ice volume is possible	• Sufficient network of measuring points required • Interfaces for data exchange are necessary • Interpretation of measuring data necessary
Opportunities	**Threats**
• Reduction of train delays • After prototype tests sufficient functionality is realistic	• No detection of ice possible • Algorithm not reliable

Figure 3. SWOT-analysis. [5]

4. Conclusions

As a result of the project it can be concluded that the three-dimensional distance measurement methods laser scanning or light-section seem to be applicable for the detection of snow or ice adhesions on railway vehicles. These data are amongst other necessary as input for an intelligent wayside monitoring system that has the ability to give out a warning or information to the infrastructure manager in order to prevent ballast flying and therefore to avoid accidents or to avoid delays in operation. Measurements of ice or snow on the surface of trains have not been performed under real conditions up to now. As a next step, the development of a prototype measurement system and detailed field test as proof of concept for the measurement as well as for the intelligent monitoring system itself is vital.

Acknowledgements

This study is a result of the research project EISMON, which was funded by the Austrian Ministry for Transport, Innovation and Technology within the program "Mobility of the Future".

References

[1] Kaltenbach H A 2008 *Schlussbericht zum Verbundvorhaben DEUFRAKO-AOA-Aerodynamics in Open Air: Untersuchungen der Schienenfahrzeugaerodynamik zur besseren Beherrschung von Seitenwind und Schotterflug im Hochgeschwindigkeitsverkehr*

[2] Kwon H B and Park C S 2006 An Experimental Study on the Relationship between Ballast-flying Phenomenon and Strong Wind under High-speed Train *The 7th world congress on railway research (Montreal)*

[3] DVV Media Group GmbH (Eurailpress) 2013 *Deutsche Bahn: Neue Glykol-Sprühanlage zur ICE-Enteisung*

[4] Haas R 2016 *Präventionsmaßnahmen gegen Eisbildung – Schotterflug* (University of Applied Sciences St. Poelten, Carl Ritter von Ghega Institute for Integrated Mobility)

[5] Michelberger F et al 2017 *EISMON Projec t- Final report (internal document)*

Assessment of rail long-pitch corrugation

Jan Valehrach[1], Petr Guziur[1], Tomas Riha[1] and Otto Plasek[1]

[1]Brno University of Technology, Faculty of Civil Engineering, Institute of Railway Structures and Constructions, Veveri 331/95, 602 00 Brno, Czech Republic

E-mail: valehrach.j@fce.vutbr.cz

Abstract. The paper focuses on defects of the running surface of the rail, namely the rail corrugation defect and specifically long-pitch corrugation in curves of small radii. These defects cause a shorter life of the rails, greater maintenance costs and increase the noise and vibration pollution. Therefore, it is very important to understand the formation and development of the imperfection of the rails. In the paper, various sections of railway tracks in the Czech Republic are listed, each of them completed with comparison of defect development, the particular track superstructure, rolling stock, axle load, traffic load etc. Based on performed measurements, defect development has been proved as different on sections with similar (or even same) parameters. The paper assumes that a train velocity is the significant circumstance for defect development rates. Assessment of track section with under sleeper pads, which are expected to be the one of the possible ways to suppress the corrugation defect development, is included in evaluation.

1. Introduction

The interaction between wheelset and track in railway applications is a complex issue. Due to the operation on railway infrastructure rail, defects occur. One of those defects is the defect of the running surface of the rail – such as corrugation. All those defects shorten the life length of the rails and increase the maintenance costs and also vibration and noise pollution.

This paper focuses on formation theory of long-pitch corrugation, performed measurements and its evaluation. Long-pitch corrugation occurs mostly in curves of small radii; therefore, such track sections were selected and monitored for over a year in order to be able to understand more thoroughly the long-pitch corrugation development. Then, the aim is to determine the crucial factor, which contributes to the corrugation development.

2. Long-pitch corrugation

The determination of the rail corrugation types is rather complicated. Nevertheless, for the purposes of the research is essential to classify this defect, since there are some similar factors affecting the formation and development of the corrugation. Unfortunately, there is no full agreement in this matter; the view of some infrastructure managers is different from others. As an example, the view of Infrastructure manager of the Czech Republic (SZDC) and International Union of Railways were picked. SZDC issued the standard "SZDC S 67 Vady a lomy kolejnic" (Rail Defects and Rail Breaks), in which the long-pitch corrugation is marked with the code 2201 D [8], whereas the International Union of Railways marks the long-pitch corrugation with code 2202 [6]. The reason for inconsistency is the differing opinion on the relations and connections between the formation and occurrence of the defects. For the purposes of this text, the view of SZDC is adopted, since the author's belief complies more with this approach.

The long-pitch corrugation, sometimes called also as "short waves", is sinuous craggedness of the running surface of the rail. Peak distance of the waves is usually between 8 and 30 cm and the depth of the waves is between 0.1 and 1.2 mm. Typical location, where the long-pitch corrugation could be found, is the lower rail in curves with radii less than 600, sometimes even 700 m [4,5].

Wheel, passing along the rail with corrugations (short-pitch corrugation) or with short wavelets (long-pitch corrugation) gains certain amount of the vertical acceleration. Induced accelerations cause, in contribution with the unsprung mass of the wheelset, additional dynamic forces, which cause vibrations of the track and the vehicle in dependence on periodical changes in acceleration [5].

Negative externalities [5] can be divided into three main categories:

- noise pollution,
- vibration dispersion,
- negative effect on infrastructure.

The influence on the track infrastructure is characterized by the defects such as damaged rail pads, broken concrete sleepers, coach screw snap off, abraded aggregate behind the heads of the sleepers and others.

Figure 1. Long-pitch corrugation defect.

2.1. Corrugation formation theory

The basic principle of movement of wheelset along the curve is shown in Figure 2. In curve with radius r_B, each of the wheels (with wheel radii r_0) travels different length because of the same angular velocity of both wheels.

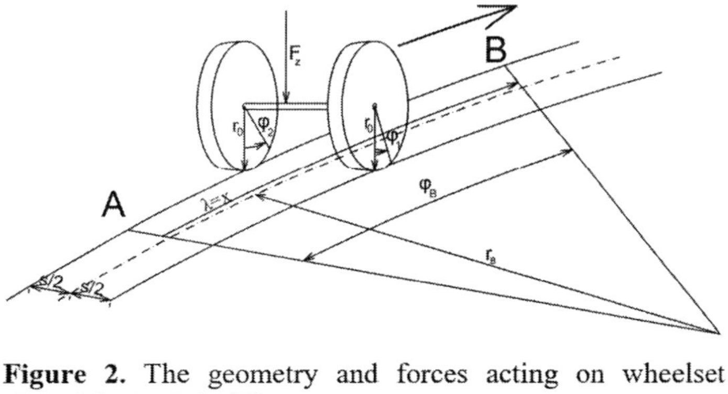

Figure 2. The geometry and forces acting on wheelset altered the basis in [1].

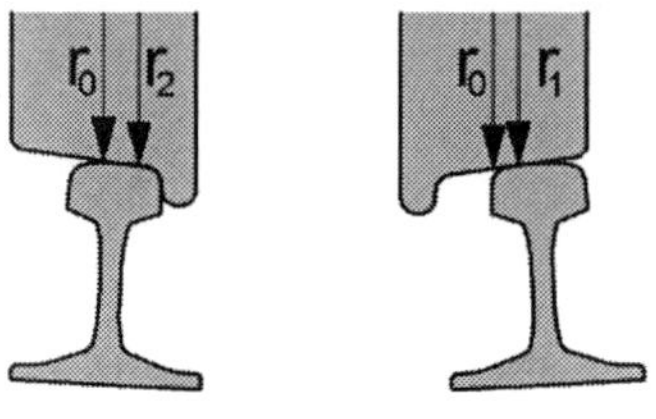

Figure 3. Wheelset position in right-hand curves.

This difference in distances $\Delta\varphi$ corresponds with rotation angles φ_1 and φ_2. Distances reached by each wheel (x_1 for inner wheel and x_2 for outer wheel) can be obtained from arc angle φ_B, radius of the curve r_B and s is distance of rolling circles 1500 mm.

$$x_1 = \left(r_B - \frac{s}{2}\right)\varphi_B \qquad x_2 = \left(r_B + \frac{s}{2}\right)\varphi_B \tag{1) (2}$$

$$x = r_B \varphi_B = \lambda \tag{3}$$

$$\varphi_B = \frac{r_0}{s} \Delta\varphi \tag{4}$$

As both wheels are connected with rigid axle, not allowing one to rotate against the other, the whole wheelset rotates with just one angular velocity. In small radii curves, this forces the inner wheel to slip, since the inner wheel does not have the possibility to dissociate the movement momentum into the actual rotation. The extent of the slip is most likely derived from variables such as rail/wheel friction force depending on relative wheel slip and wheelset attack angle, axel stiffness and others. The length of resulting long-pitch corrugation wave, with vehicle parameters taken into account, can be determined as follows [1]

$$\lambda = \left(\frac{r_0^2 \mu F_Z}{G I_p}\right) r_B \tag{5}$$

where μ is friction coefficient, F_Z is nominal wheel load, G is shear modulus and I_p is moment of inertia.

The vehicle influence describes the part of the formula in parentheses, on the other hand the infrastructure influence is represented by curve radius r_B. Therefore, using this simplified view, from the perspective of infrastructure we could develop a basic idea that a length of the wave λ depends on curve radius.

$$\lambda = f(r_B) \tag{6}$$

The simplified view does not reflect the position of wheelset in curve. To add more complexity to the problem, we need to consider it. Figure 2 explains the implementation of the wheelset lateral position.

The larger the difference between the sizes of the contact circles (r_1 and r_2), the smaller the radius, which the wheelset can pass without the relative in-between spin motion and consequent longitudinal slip of the inner wheel [9].

3. Monitored locations

In cooperation with SZDC (rail infrastructure manager in the Czech Republic), locations suitable for monitoring of the long-pitch corrugation were chosen, preferably with significant occurrence of this defect. In some of the locations, various precautions with various results were applied before in order to slow down the development of the corrugation. The most accentuated precaution was the asymmetric rail grinding [14], the others were the usage of rail fastening with reduced vertical stiffness and with clamps with enhanced fatigue limit, installation of under sleeper pads or the usage of rail lubricators.

Our chosen locations are in most cases double track and thus allow us to compare both tracks in-between each other. The comparison is not always possible, even in two beside laid tracks some parameters differ. The most often limiting parameter is different composition of trains, and the influence of the slope gradient (for one direction uphill, for other direction downhill; both resulting in the different rate of acceleration or deceleration). Furthermore, it would be optimal to monitor the curves as a whole, yet unfortunately, the curves still will not be the same.

For above-mentioned reasons, just a short track sections not exceeding 50 metres are monitored in every location and with only few exceptions more than one measuring places are observed. This approach allows us to compare also the measuring sections within one track and so the heterogeneity of the train composition is ruled out. Particular measuring sections are chosen based on assumption of influence on long-pitch corrugation formation and development. Together with the track alignment, also substructure and superstructure composition is taken into account. The substructure influences track stiffness (with changes along bridges or bedrock), and so does the superstructure (with changes originating in usage of rail fastenings with high elasticity, special elastic rail fastening, clamps with enhanced fatigue limit or under sleeper pads [11].

Based on the criteria mentioned in previous text, the searches in the infrastructure manager databases were conducted. The research then follows two main aims:

- monitoring of the development rate of long-pitch corrugation,
- definition of the factors influencing the development of long-pitch corrugation.

3.1. *Methods of monitoring and evaluating*

Corrugation is measured using device "Salamander" (allowing continuous laser scanning of rail running surface). The device evaluates the measured data according to the standard EN 13231-1 [3,9] in these wavelengths ranges:

- 10 ÷ 30 mm: surface defects of rail head, rail joints,
- 30 ÷ 100 mm: surface defects of rail head, rail joints, rail wear caused by braking,
- 100 ÷ 300 mm: long-pitch corrugation (inner rail in curves),
- 300 ÷ 1000 mm: rail defects persisting from the production of rails.

The evaluation is performed using RMS value (RMS stands for Root Mean Square) [9], which is the basic microgeometry quantity reflecting the long-pitch corrugation. It could be described as moving average of effective values. When assessing, n equidistant samples y_i of calculated sector with defined length L for the point on rail with position x could be obtained from formula

$$RMS(x,L) = \sqrt{\sum_{i=1}^{n} \frac{y_i^2}{n-1}} \qquad (7)$$

The set of RMS values for four wavelength ranges, or wavebands, is computed using this formula. The first and last wavebands are for the purposes of this research expendable, since these wavebands do not reflect the long-pitch corrugation, but other imperfections of the rail. For this reason and on the basis of the chapter 2.1., the only monitored wavebands are 30 ÷ 100 mm and 100 ÷ 300 mm [5,11,13]. The percentage of the samples, which exceed below mentioned limits, can be considered as one of the rail quality indicators.

Table 1. RMS limits.

Waveband [mm]	10 ÷ 30	30 ÷ 100	100 ÷ 300	300 ÷1000
Limit [μm]	4	4	12	40

Together with microgeometry of the rails, other supportive quantities are measured, such as track geometry parameters and velocity of the passing vehicles. Track geometry parameters are measured using the device "Krab", which is able to measure a number of track parameters. For the purposes of the research, just cant and track gauge are considered as relevant in this stage.

The velocity of the passing vehicles is measured with the speed radar, to determine the type of rolling stock and its exact speed on measured section, optical recording device is used. The velocity of the vehicles influences the position of wheelset in curve and it is also strongly believed to be the contributing factor in the corrugation development. In order to describe the expected positions of the wheelsets in curves, the cant excess and the cant deficiency are computed based on the measured speeds, since these quantities allow us to compare the train passages between each other [1]. For this purpose, a weighted average of all speed measurements in a set section was calculated, where the mass of each train was taken into consideration [13].

4. Development rate of long-pitch corrugation

4.1. Initial phase of development

In order to monitor the development rate (formation and development speed) of long-pitch corrugation, it is necessary to start the observation optimally immediately after the renewal of the rails and preventive grinding, or in case of reparative grinding immediately after such grinding. Then it is possible to record the progression from initial condition (e.g. the "flawless" surface of the rails or within maintenance acceptance criteria).

Such measurements were conducted in location Hady, located on the line Brno - Ceska Trebova, between km 161.685 and km 164.485. The first measurement was done before the rails renewal; the second one was performed immediately after the renewal (in order to record the "flawless" state of the rails. Consequently, further measurements were performed with a few months' headways.

The measuring location consists of two sections containing three measuring places, first measuring place is located in front of the tunnel 1 (curve A), the second one is located in front of the tunnel B (curve B) and the third one is located in tunnel B. For the whole measuring location, only the track No. 1 is monitored. Considering the common structure and geometrical parameters of the track (curve radii 283 m and 261 m, cant 123 mm in both curves, fastening W14, concrete sleepers B 91S etc.) and the railway superstructure, all three measuring places do not exhibit significant differences.

Corrugation development in all wavebands is shown in the Figure 4, Figure 5 and Figure 6.

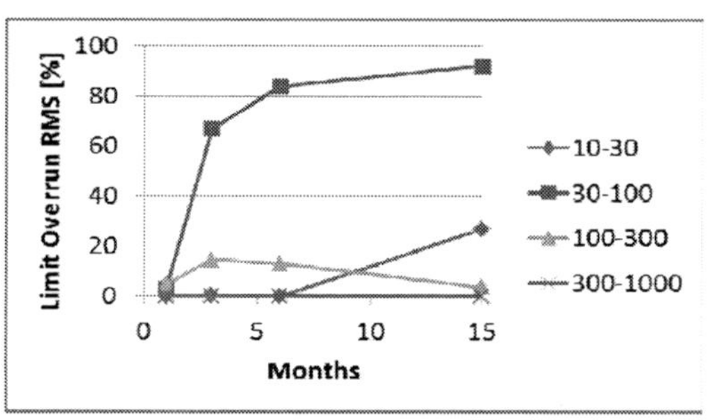

Figure 4. RMS values for measuring point in the tunnel 2.

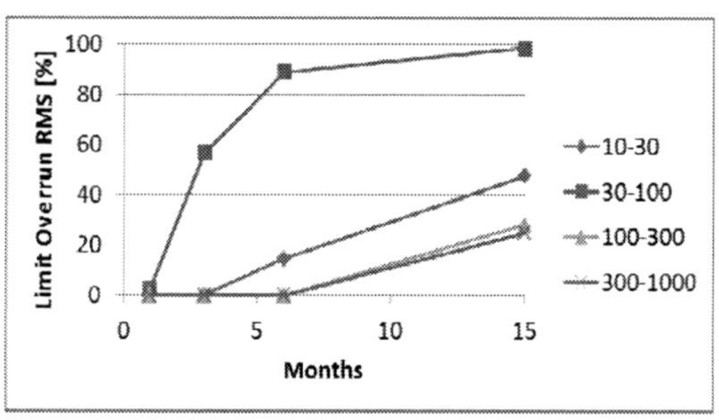

Figure 5. RMS values for measuring point in front of the tunnel 2.

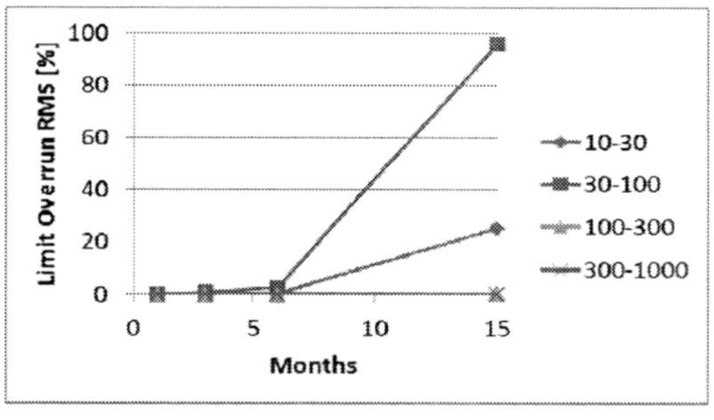

Figure 6. RMS values for measuring point in front of the tunnel 1.

As it could be seen in all the Figures, after 15 months the RMS values in waveband 30 ÷ 100 mm are reaching 100%. In case of the curve B (Figures 4 and 5), the progression in this waveband is for both measuring places alike. On the other hand, there is significant rise in RMS values between sixth and fifteenth month after the rail renewal in curve A. This sudden change in development of the defect is not yet satisfactory explained, in the present time we believe the change of previous sleepers B 91S to B 91T could be the factor. Or, to be more precise – not the change of the sleepers type, but the changing process itself (the interference into the superstructure).

When evaluating only the first six months, the curve B (in front of the tunnel 2) embodies the weighted cant deficiency I = 87 mm. The curve A (in front of the tunnel 1) shows the weighted cant deficiency I = 57 mm.

4.2. Developed defect phase

For development rate monitoring in areas, where more developed defect takes place, the location Havlickuv Brod was chosen. Four curves (two in each of the tracks) are monitored in the line No. 324 Havlickuv Brod – Kutna Hora between the km 224.394 and km 225.150 in an approach to the station Havlickuv Brod from the direction Kutna Hora. The location consists of a double track (numbers 1 and 2) with consecutive reverse compound curves (marked A and B). Since all of the curves are in some regards different, it is possible to compare them between each other. For instance, in curve A the cant differs in tracks No. 1 a No. 2, in curve B one of the tracks is equipped with under sleeper pads and the other one is not, also in some curves the trains are passing with the constant speed, and in others the trains are decelerating.

The railway superstructure in the location Havlickuv Brod consists of the rails 49 E1, fastenings E14 with clamps Skl40 and W14 with clamps Skl14, the sleepers B 91S/1 are on both tracks. The difference between used fastening systems could not be considered as a cause of the change in the defects, since the replacement of the fastening system was the result of the development of the defects and should have been the resolution of the problem [13].

Based on conducted measurements, the RMS values ordinarily exceed 100%. For this reason, it is not possible to monitor the development of the defect in time under the terms of above mentioned RMS value calculation. Thus, in following figures the vertical axis does not stand for RMS values in percent, it is transferred to the scale from zero to one in order to be able to compare values magnitudes. Slight alteration was performed to reach this possibility – the limits from table 1 had to be changed, which will allow us to monitor the development furthermore. The downside of this modification is the need for careful examination of the values, otherwise the results could be misrepresented (some of the RMS values could be neglected unintentionally). In addition, the impossibility to compare the results from this location to others could be considered as a disadvantage.

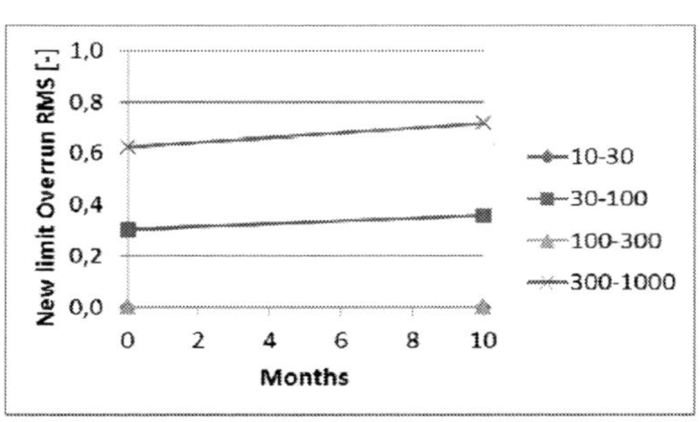

Figure 7. RMS values for measuring point in track No.1 in curve A (E = 2 mm).

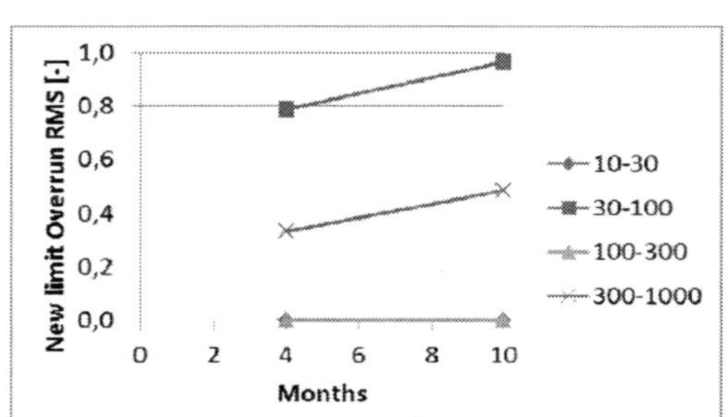

Figure 8. RMS values for measuring point in track No.1 in curve B (E = 72 mm).

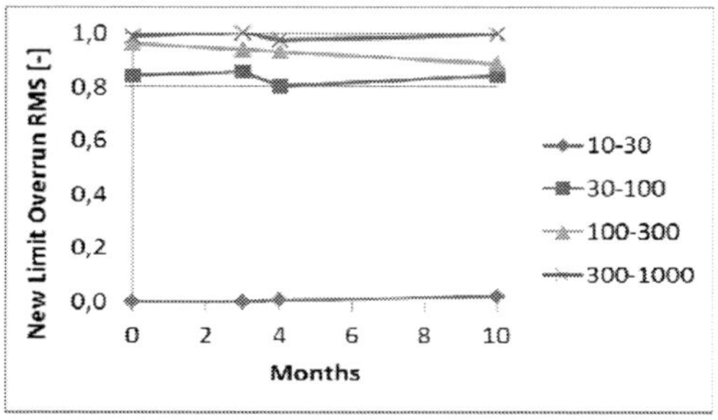

Figure 9. RMS values for measuring point in track No.2 in curve A ($E = 27$ mm).

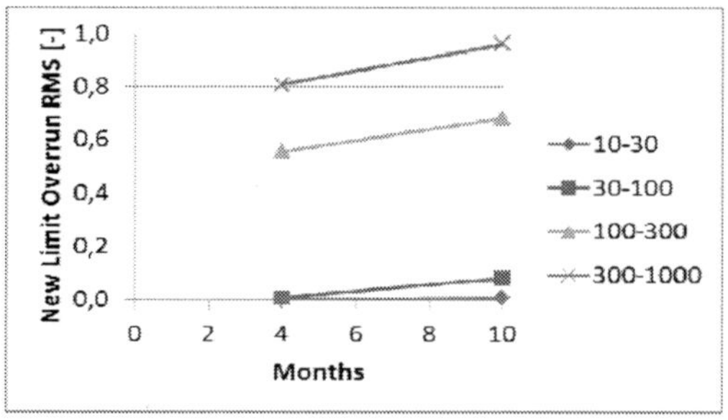

Figure 10. RMS values for measuring point in track No.2 in curve B ($E = 57$ mm).

When evaluating only the wavebands representing the long-pitch corrugation on Figures 7 and 8, then both curves in track No. 1 show significantly lower RMS values than in track No. 2. Higher values in waveband 30 ÷ 100 mm in curve B of a track No. 1 could be the result of the braking effect, which is usually present in this waveband [5]. In comparison of both wavebands in all four curves with cant excess values (specified in figure titles), we could presume that in curve A of the track No. 1 the lower RMS values are the consequence of a cant excess $E = 2$ mm. In case of the curve B of the same track, where the highest cant excess values ($E = 72$ mm) were found, the most probable factor is believed to be the influence of the under sleeper pads. According to [7], under sleeper pads reduce the generation of vibrations and noise emissions in these frequencies.

The figure 9 shows, that after the certain state the decrease in development rate is present for some of the wavebands. The reason of this phenomenon has not been explained yet and it is a subject of further investigation.

4.3. Presumed superposition

In order to understand the development process of long-pitch corrugation as a phenomenon, the authors believe, that it is possible to superpose the initial and developed phases of the defect progress. Such combination of the phases will show the whole life cycle of the defect and what is more important – it is able to mirror *the changes* in the development. The author's idea of presumed theoretical development of long-pitch corrugation is shown in Figure 11. Every curve displays different rate of development. The steeper the curve, the more raging the progress is and the less favourable the state of the track is.

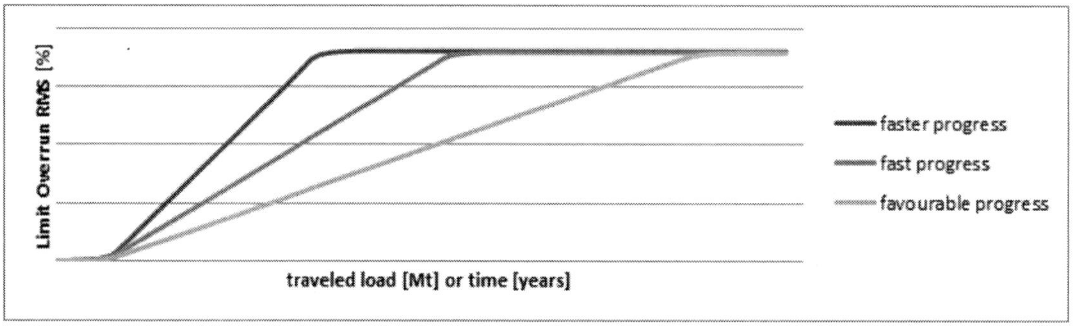

Figure 11. Presumed theoretical development rate of long-pitch corrugation.

5. Conclusion

Longitudinal and lateral motion of the wheelset in curve is a complex issue. Many variables have to be taken into account, and in some cases, negative effects can come up. In curves of small radii, it is especially corrugation defect on inner rail running surface. This defect is easily recognisable due to increased noise and vibration levels.

The paper aims on the assessment of long-pitch corrugation in curves of small radii. Several locations were determined as suitable track sections for the long-pitch corrugation defect measuring.

The measurements results indicate that cant excess and cant deficiency can be one of the most significant factors supporting the development of long-pitch corrugation. In addition, it is showing that the development rate of the corrugation might be, under right circumstances using preventive measures, decelerated and thus it is possible to extend the life cycle of the rails.

The research will be furthermore focused on investigations and should imply more than development progress of the corrugation. A detailed investigation of the factors and their influence should take place. When done, it would be possible to predict future development of the defect, plan the maintenance works and result in improving the quality of the railway transport.

Acknowledgements

Research reported in this paper was supported by Competence Centres program of Technology Agency of the Czech Republic (TA CR), project Centre for Effective and Sustainable Transport Infrastructure (no. TE01020168) and Student's grant tender of the Brno University of Technology.

References

[1] Berthel H and Gerber U 2009 Schienenfehler und Schienenschleifen *Proc. Int. Conf. on Železniční Dopravní Cesta (Děčín)* pp 143-8

[2] Český ormalizační institut (ČNI) 2008 Konstrukční a geometrické uspořádání koleje železničních drah a její prostorová poloha - Část 1: Projektování /Geometrical characteristics of railway tracks – Part 1: Layout/ *(ČSN 73 6360-1)*

[3] Úřad pro normalizaci, metrologii a státní zkušebnictví (ÚNMZ) 2012 Železniční aplikace - Kolej - Přejímka prací - Část 3: Přejímka reprofilace kolejnic v koleji /Railway applications – Track – Acceptance of works – Part 3: Acceptance of reprofilling rails in track/ *(ČSN EN 13231-3)*

[4] Esveld C 2001 *Modern railway track* (Zaltbommel: MRT-Productions)

[5] Funke H 1986 *Rail Grinding* (Berlin: Transpress)

[6] International Union of Railways (UIC) 2002 Rail defects *(UIC Code 712 R)*

[7] Plasek O and Hruzikova M 2006 The Contribution of Under Sleeper Pads to Improving of Track Geometry Parameters *Proc. Int. Conf. Pozemne komunikácie a dráhy, Environmentálna kapacita pozemných komunikacií a dráh (Herlany)* (Kosice: Technical University of Kosice)

[8] Správa železniční dopravní cesty s.o. (SŽDC) 2017 Vady a lomy kolejnic (in Czech) /Rail defects and breaks/ *(Regulation SZDC S67)*

[9] Scholle S 2015 The Possibilities of Recording Cars TUDC for Rail Defects Detection *Vyhybky a vyhybkove konstrukce. Kolejnicove vady. (Brno)*

[10] Valehrach J 2015 Long-pitch Corrugation *Proc. Int. Conf. Young Scientist 2015 (Jasna)* (Kosice: Technical University of Kosice)

[11] Valehrach J 2015 Long-pitch Corrugation – Theory and Praxis *Proc. Int. Conf. Juniorstav 2015 (Brno)* (Brno: Brno University of Technology)

[12] Valehrach J 2016 Long-pitch Corrugation and the Corrugation Measurement *Proc. Int. Conf. Juniorstav 2016 (Brno)* (Brno: Brno University of Technology)

[13] Valehrach J and Smid J 2016 Monitoring of Track Sections with Long-pitch Corrugation *Proc. Int. Conf. Interoperability of Railway Transport - IRICoN 2016 (Praha)* (Praha: Czech Technical University in Prague)

[14] Zelenka J and Kohout M 2010 Posibilities to Influence Running and Guiding Behaviour of Rolling Stock by Railway Track Parameters *Proc. Int. Conf. Zeleznicni dopravni cesta 2010 (Pardubice)* (Praha: Sprava zeleznicni dopravni cesty, statni organizace) pp 60-67

IOP Conf. Series: Materials Science and Engineering **236** (2017) 012049 doi:10.1088/1757-899X/236/1/012049

Seismic impact of the railway on the geotechnical constructions

Martin Stolárik[1] and Miroslav Pinka[1]

[1]VŠB – Technical University of Ostrava, 17. listopadu 15/2172, Ostrava-Poruba 708 33, Czech Republic

E-mail: martin.stolarik@vsb.cz

Abstract. Nowadays, the focus on more ecological means of material and persons transport is still higher. Big loads can be transported on railways more effectively and with lower environment impact than on roads. The geotechnical structures are inherent parts of railway infrastructure, such as embankments, sides of notches and, of course, tunnels, foundation constructions of buildings or pillars of bridges and the others geotechnical constructions (e.g. retaining walls, culverts, transition area of bridges). By train pass, vibrations are caused and these vibrations are relayed to the soil. These vibrations can make adverse impact to surrounding objects and to technologies placed in. This so far uncared-for influence gets into the foreground by current trend of everyday life technical equipment increasing. The article introduces different kinds of geotechnical structures and the influence of by-passing railway transport on their constructions and surroundings. The data are evaluated in the amplitude and frequency domain.

1. Introduction

Nowadays there is emphasis on more environmentally-friendly ways to transport material and personnel than road transport. Rail can move large loads more effectively and with less environmental impact in particular. In the Czech and Slovak Republics there is a very dense railway network, which was started to be built in the mid-19th century and is still being modernized. An inseparable part of the transport routes also include geotechnical structures such as embankments, cuts, the transition zones of bridges, but of course also tunnels and foundation structures of the surrounding buildings.

Passing trains produce vibrations that are transmitted to the ground [1, 2, 3]. The vibrations propagate through the rock environment and decrease with increasing distance [4]. In the near-surface sediments the vibrations may even be amplified; resonant vibrations are induced here and the vibration period is extended as a result of free vibration of the near-surface sediments. The resulting vibrations, transmitted through natural ground or artificial soil body, may have an adverse effect on both the surrounding objects, on people living in these buildings, as well as on the technology located in them which can be very sensitive to vibration. With the current trend of the technical equipment of everyday life on the increase, this so far neglected effect comes to the fore. It is therefore important to realize the experimental measurements to obtain values representing these vibrations. The main task of these measurements is to determine the maximum velocity values (or acceleration), the parameters of waveforms and frequency spectra of the measured signals. The resulting data is then very important for civil engineers who, based on the measured values, can specify the effects on the existing technical condition and possible susceptibility to damage [5]. The impact of such vibrations on buildings or on the population is then assessed by the relevant technical standards and regulations [6]. The effect of vibrations in the vicinity of the railway line can be also determined using numerical modelling of the actual geological structure, vibration and eventually buildings. The current models of the effects of vibrations on buildings are based on numerical methods such as FEM - "Finite Element Method", or BEM - "Boundary Element Method", while the building of relevant models is very often based on the results of experimental seismic measurements [7, 8].

This article presents the results of two experimental seismic measurements in the context of theoretical research on the dynamic impact of rail transport on various types of geotechnical structures. Specifically, it is the dynamic response of the secondary tunnel lining to the train's passing through the neighbouring tunnel tube and the response of the reinforced concrete walls. Subsequently a concrete example of a practical solution to the problem dealing with the same issues will be presented.

2. Instrumentation

All measurements were taken using Gaia2 and Gaia2T seismographs with ViGeo2 sensors (both of them are manufactured by Vistec Prague). The Gaia2 or Gaia2T seismograph is a three-channel seismic station with a dynamic range of 138dBp-ps and with the triggered and continuous digital data recording option. Time synchronization is provided by a GPS module. ViGeo2 is a compact, active, short-period, three-component and velocity seismometer for field and station applications. The seismometer includes three mechanical vibrating systems (sensors) with a natural frequency of 2 Hz and the frequency range is 2 Hz to 200 Hz. SWIP (Waves Seismic Interpretation Program), supplied as standard with Gaia seismometers by Vistec Prague, was used to process seismic data. This program enables to process a seismic signal both in the amplitude region and in the frequency domain. In the amplitude region, the processing software does not allow for the recalculation of vibration amplitude values to physical units [$mm.s^{-1}$], so the vertical axes are in quantizing levels [cnt] in all the wave pattern figures. Conversion formula: 1 cnt = $2.975.10^{-6}$ $mm.s^{-1}$. Figures of all wave patterns show top down the vertical component, marked SHZ, the horizontal radial component SHN (directed in the direction in which the tunnel runs) and the transverse SHE; the horizontal axis is time in seconds (the axes are with the same amplitude and time scales).

3. Measurement 1 - seismic response of the secondary tunnel lining

During the reconstruction of the Jablunkov Tunnel II the operation of trains was diverted to tunnel I, which was temporarily operated. After an incident in 2009 that led to the collapse of the entire tunnel profile from the 70th to 160th tunnel meter (TM), a series of measures was taken [9]. In the blocked part mining was renewed, in the northern, unburdened part of the tunnel the secondary lining has already been installed. On this lining, seismic response measurement of the response of the invert of the secondary lining to a passing train in tunnel I was realized. Four velocity seismometers were set on the invert in a profile parallel to the longitudinal direction of the tunnel, and the radial horizontal axis (SHN) was oriented in the direction of the structure and the horizontal transverse axis (SHE) perpendicular to the axis of the temporarily operated tunnel I.

The largest response was observed from a passing freight train of thirty-four full four-axle wagons pulled by two locomotives. In Figure 1 the wave pattern of recording of the whole train can be seen, and in Figures 2 and 3 there is a detail of wave recording of the captured maximum amplitude on sensor 1 together with the corresponding frequency spectrum. Maximum velocity values captured on sensors 1 to 4 are shown in Table 1.

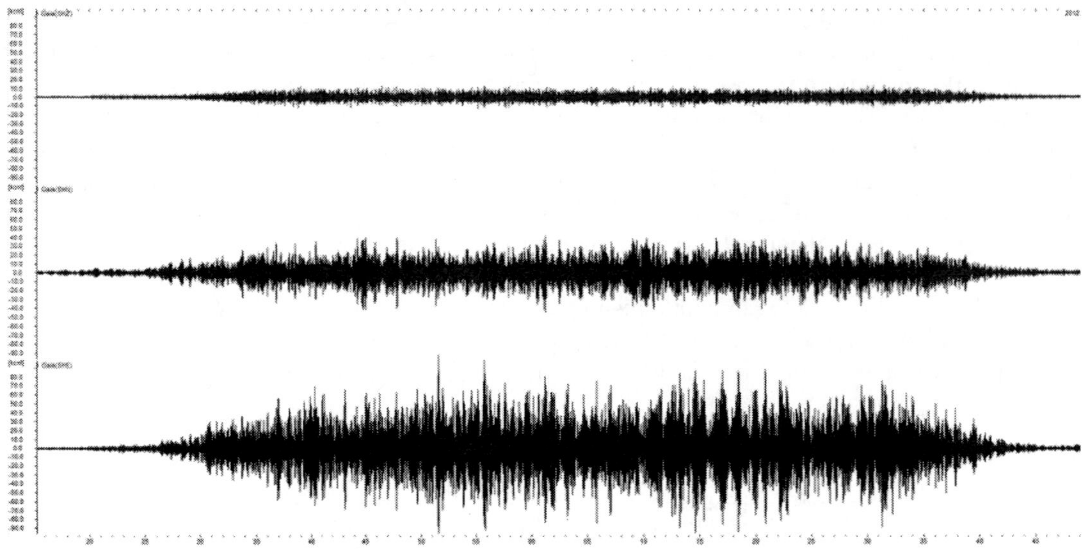

Figure 1. Wave pattern of recording of the whole train.

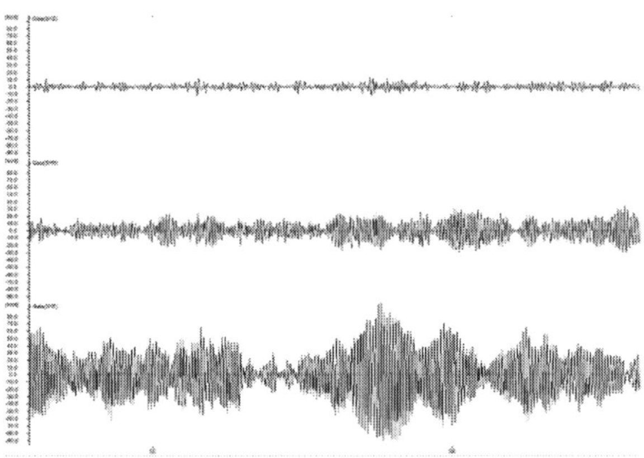

Figure 2. Detail of wave pattern of recording.

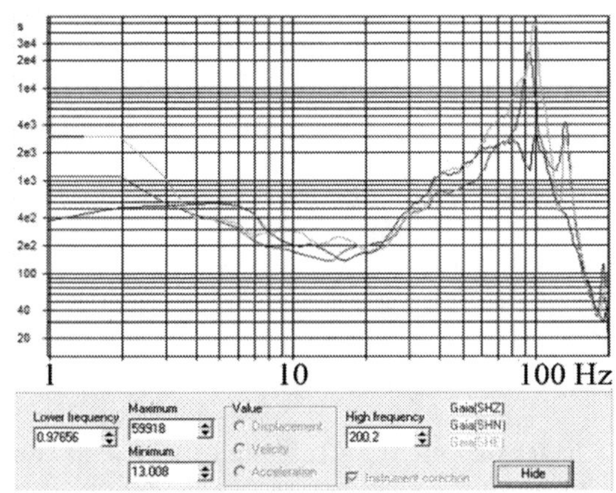

Figure 3. The corresponding frequency spectrum.

Table 1. Maximum amplitudes of velocity – measurement 1.

Sensor	Maximum velocity amplitude of oscillation [mm.s^{-1}]		
	SHZ	SHN	SHE
1	0.037	0.135	0.312
2	0.025	0.112	0.174
3	0.031	0.146	0.157
4	0.054	0.224	0.250

4. Measurement 2 - seismic response of a reinforced concrete angled wall

The second presented measurement is the response of a reinforced concrete angled wall in Ostrava - Svinov, Prodloužená Bílovecká. At a distance of fifteen metres from the nearest operated rail, measurements were realized both in the foot of the wall (sensor 1) and in its crown (sensor 2) and another station was also located on the far side of the road (sensor 3, distance from the nearest operated rail around thirty-four metres), which is built on an artificial ground body stabilized by the angled walls. All velocity seismometers are oriented with their radial horizontal axis (SHN) perpendicular to the track axis and horizontal transverse axis (SHE) horizontally with the axis of the track. From the recorded train sets (closest rail - CityElefant, electric unit series 460, and rail at a distance of twenty-two or twenty-six metres - RegioJet, SC Pendolino, express train, an empty freight train) the largest response observed against all expectations, was from the CityElefant. Figures 4-6 present wave patterns or records of CityElefant and frequency spectra obtained from individual stations. The maximum amplitudes of velocity are shown in Table 2.

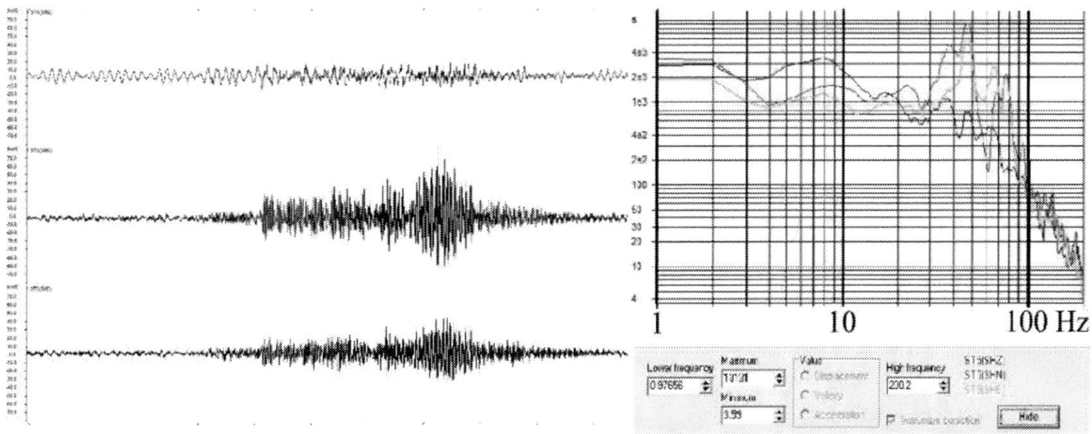

Figure 4. Wave pattern and frequency spectrum of recording at the foot of the angled wall (sensor 1).

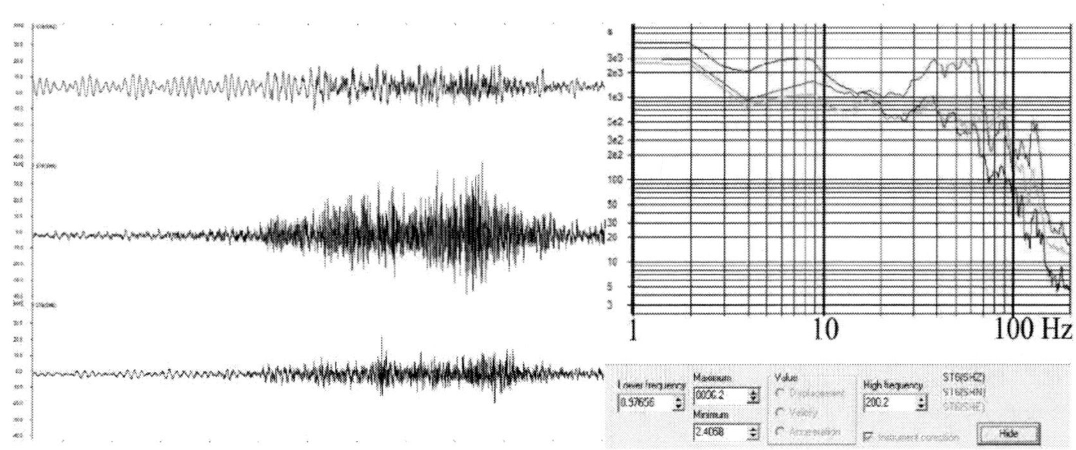

Figure 5. Wave pattern and frequency spectrum of recording at the crown of the angled wall (sensor 2).

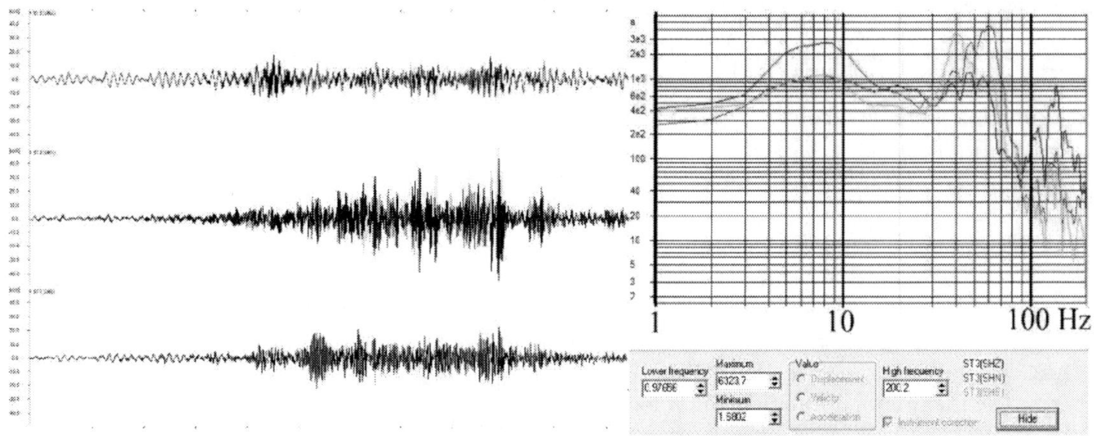

Figure 6. Wave pattern and frequency spectrum of recording at the remote side of the communication (sensor 3).

Table 2. Maximum amplitudes of velocity – measurement 2.

Sensor	Maximum velocity amplitude of oscillation [mm.s^{-1}]		
	SHZ	SHN	SHE
1	0.060	0.243	0.131
2	0.052	0.128	0.062
3	0.056	0.146	0.073

5. Measurement 3 - seismic response of a reinforced concrete floor

In the following section, seismic measurements are presented which dealt with the specific practical task when a technology very sensitive to vibrations should be located in a production hall, and in close proximity outside the hall works a railway was operated at a distance of twenty metres. The aim of the measurements was to assess the maximum possible seismic load and frequency of journeys per twenty-four hours. A velocity seismometer was placed on the concrete floor of the empty hall at the location where the equipment should be installed in the future, with the radial horizontal axis (SHN) being orientated perpendicular to the axis of the railway. Figure 7 is a wave pattern of recording at the time from 9 am to 11 pm with distinct peaks representing various passages of trains. In the background among others, the noise caused by the technology in the next production hall can be traced. The following figure shows the wave pattern of the phenomenon with the maximum captured velocity amplitude and the corresponding frequency spectrum, where a significant peak in the 50-55 Hz region represents the just mentioned technological background process.

From the wave pattern of the entire twenty-four hour recording the maximum velocity amplitude of oscillation was deducted for the six largest phenomena (Table 3). Note: the time indicated in the tables is CET. The end result was a chart showing the time distribution of the seismic load for a given location (Fig. 9).

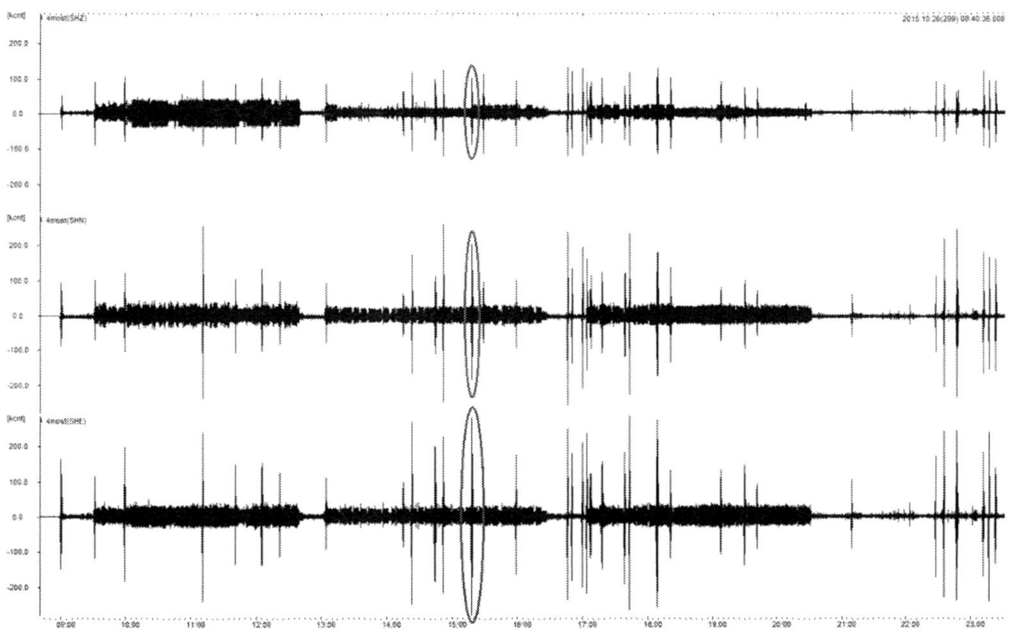

Figure 7. Wave pattern of recording at the time from 9 am to 11 pm with recorded maximum amplitude – red color.

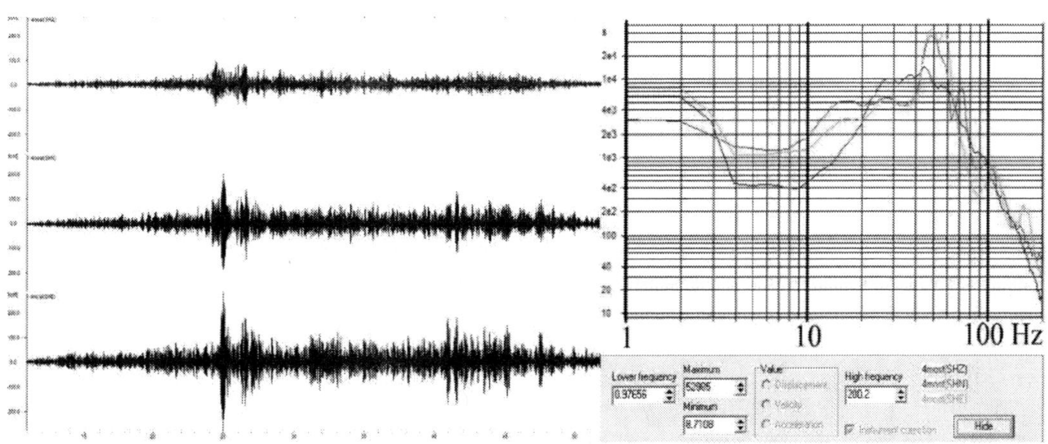

Figure 8. Wave pattern and frequency spectrum of recording of the phenomena 3.

Table 3. Maximum amplitudes of velocity – measurement 3.

Evaluation phenomena number	Time captured of the phenomena	Maximum velocity amplitude of oscillation [mm.s^{-1}]		
		SHZ	SHN	SHE
1	11:11 am	0.279	0.756	0.714
2	2:51 pm	0.366	0.768	0.672
3	3:18 pm	0.339	0.601	0.848
4	4:47 pm	0.387	0.762	0.741
5	5:42 pm	0.342	0.693	0.827
6	7:27 am	0.202	0.726	0.720

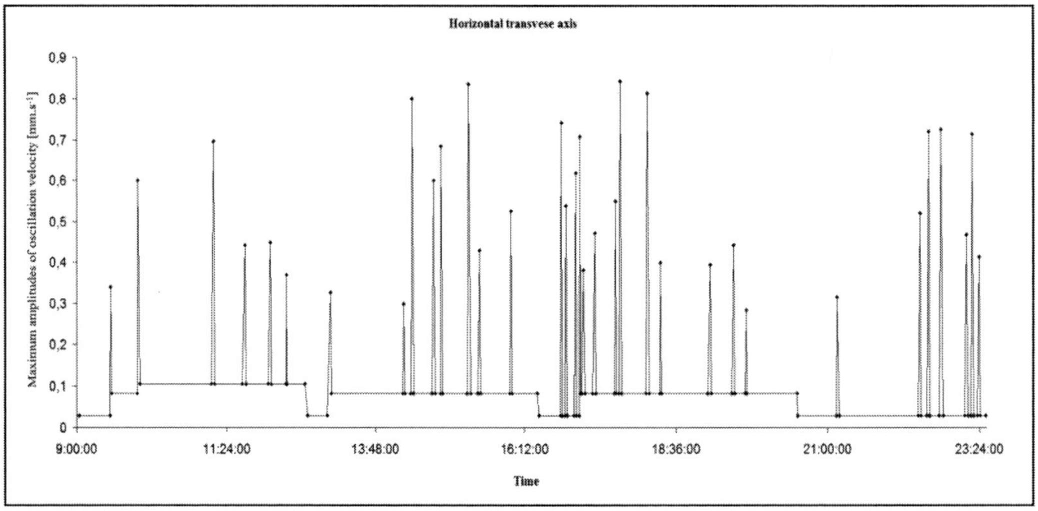

Figure 9. Distribution of the seismic load for a given location in the time period from 9 am to 11 pm – horizontal transverse axis (SHE).

6. Conclusion

In this paper we presented the results of two experimental seismic measurements carried out in the framework of theoretical research on the dynamic influence of rail transport (and transport in general) on various types of geotechnical structures. Specifically, the dynamic response of the secondary lining of the train's passing through a side tunnel tube and the response of the reinforced concrete angled wall loaded with dynamic effects of rail traffic were mentioned. Following this a specific example of the problem in practice of dealing with seismic loads was presented, which could jeopardize the smooth operation of industrial technology very sensitive to vibration. The results of this measurement ultimately proved the unsuitability of the location selected for the location of the equipment.

The research interest is obviously much broader. Within this extensive study dealing in general with the dynamic effects of traffic, some results both in terms of seismic experimental measurements [10] and in terms of mathematical modelling of seismic effects [11] were already presented within the workplace. The issue is solved comprehensively and as shown by the results of measurements three (response of reinforced concrete floors in the production hall) in the article, it is necessary to monitor and analyse not only vibrations with stronger intensity which can lead to damage to buildings, but also vibrations with an intensity that is significantly lower.

Acknowledgements
The paper was prepared with the support of the Competence Centers of the Technology Agency of the Czech Republic (TAČR) within the project Centre for Effective and Sustainable Transport Infrastructure (CESTI), project number TE01020168.

References
[1] Kaláb Z 2015 Vibrations by moving trains: Case study *15th International Multidisciplinary Scientific Geoconference SGEM 2015* ed J P Burg, D Y Pushcharovsky et al (Albena: STEF92) pp 855-62
[2] Kouroussis G, Connolly D P, Alexandrou G and Vogiatzis K 2015 The effect of railway local irregularities on ground vibration *Transportation Research Part D: Transport and Environment* **39** pp 17-30
[3] Bian X, Jiang H, Chang Ch, Hu J and Chen Y 2015 Track and ground vibrations generated by high-speed train running on ballastless railway with excitation of vertical track irregularities *Soil Dynamics and Earthquake Engineering* **79** pp 29-43
[4] Connolly D P, Kouroussis G, Woodward P K, Alves Costa P, Verlinden O and Forde M C 2014 Field testing and analysis of high speed rail vibrations *Soil Dynamics and Earthquake Engineering* **67** pp 102-118
[5] Sanayei M, Maurya P and Moore J A 2013 Measurement of building foundation and ground-borne vibrations due to surface trains and subways *Engineering Structures* **53** pp 102-111
[6] Kouroussis G, Florentin J, Conti C, Verlinden O, Connolly D P 2014 Building vibrations induced by railways: An analysis of commonly used evaluation standards *21st International Congress on Sound and Vibration* ed Crocker M J (Beijing: IIAV) pp 375-382
[7] Nejati H R, Ahmadi M and Hashemolhosseini H 2012 Numerical analysis of ground surface vibration induced by underground train movement *Tunnelling and Underground Space Technology* **29** pp 1-9
[8] Hall L 2003 Simulations and analyses of train-induced ground vibrations in finite element models *Soil Dynamics and Earthquake Engineering* **23** pp 249-330
[9] Aldorf J, Ďuriš L, Hrubešová E and Vojtasík K 2011 Jablunkov tunnel collapse *WTC 2011,* ed Särkkä P, Tolppanen et al (Helsinki: ITA-AITES) pp 1443-51
[10] Kaláb Z and Hrubešová E 2015 Evaluation of seismic effect of traffic-induced vibrations Acta *Montanistica Slovana* **20** pp 33-7
[11] Hrubešová E, Stolárik M, Lunáčková B, Pinka M and Petřík T 2015 Model analysis of the dynamic effects of railway traffic, depending on the selected determining factors *15th International Multidisciplinary Scientific Geoconference SGEM 2015*, ed J P Burg, D Y Pushcharovsky *et al* (Albena: STEF92) pp 465-70

IOP Conf. Series: Materials Science and Engineering **236** (2017) 012050 doi:10.1088/1757-899X/236/1/012050

Acoustic energy propagation around railways

Petra Cizkova[1]

[1]CTU in Prague, Faculty of Civil Engineering, Thakurova 7, Praha 6, 166 29, Czech Republic

E-mail: petra.cizkova@fsv.cvut.cz

Abstract. The article deals with the issues of acoustic energy propagation around railways. The research subject was noise emission spreading into the surroundings during the passage of trains over a directly travelled steel bridge construction. Noise emissions were measured using direct measurements in the field. The measurements were performed in two measurement profiles. The noise exposures A *LAE* measured near the steel bridge construction were compared against the noise exposures A *LAE* captured on an open track. From the difference of these data, the noise level of the steel bridge structure was determined. Part of the research was to evaluate the effect of the reconstruction of the railway track superstructure on the acoustic situation in the given section of the railway track. The article describes the methodology of measurements, including the processing and evaluation of measured data. The article points out the noise levels of the steel bridge construction and assesses changes in the acoustic situation after the reconstruction.

1. Introduction

This article addresses the propagation of acoustic energy in the surroundings of an open railway track compared to an area around a steel bridge construction. It is generally assumed that noise emissions arising on a trafficked bridge construction will reach higher values than on a classic track section. This, in particular, refers to steel railway bridges, which represent a relatively frequently designed structural type.

Foreign literature sources state that the increment in the sound pressure level A on steel bridge constructions with directly trafficked bridge decks with bridge sleepers ranges from +3 to +16 dB [1]. Thus, in acoustic terms, bridge constructions represent a major local source of noise on the railway track. There are two main causes of the growth in noise emissions. The first cause are vibrations of the bridge construction, which successively emits noise, caused by the passage of a train unit. The second cause is the noise emitted by the rail itself. Its intensity relies, among others, on the rail fastening method used on the bridge. [2, 3, 4] In this study, the investigated subjects are a railway bridge and its adjoining straight track section. The whole studied section is 468 m in length. The difference in elevation is 1 m (a gradient of 2 ‰).

2. Solved territory

The solved territory is situated near the settlement of Sány in the Nymburk District. The territory is crossed by a single-rail railway track No. 505A Choceň – Týniště n. O. – Velký Osek. The track incorporates a steel railway bridge structure across the Cidlina River. The load-bearing part of the bridge is a welded steel truss construction with a trough bridge deck. The trough bridge deck consists of cross girders, longitudinal girders and a steel plate. The cross girders and longitudinal girders are welded plate girders. Rigid rocker bearings are mounted on concrete support O01 in the direction of Velký Osek, while expansion roller bearings are used on support O02 in the direction of Chlumec nad Cidlinou (Figure 1). The project included the measurement of sound pressure levels before and after the railway track reconstruction. Figure 2 shows the passage of a freight train unit across the steel bridge construction. [5]

Figure 1. Application of bearings on the bridge.

Figure 2. Track section.

3. State before and after reconstruction

In spring 2015, the reconstruction of the railway track superstructure was carried out. Within the reconstruction, the ballast graded 32/63 was completely cleaned, the R 65 jointed flat-bottom rail with visible corrugation (Figure 3) was replaced with a new jointless flat-bottom rail 60E2 (Figure 4) along the whole length of the solved territory.

Figure 3. Straight track section before reconstruction.

Figure 4. Straight track section after reconstruction.

SB6 concrete sleepers with ribbed baseplates and ŽS4 rigid clips were replaced with B91S concrete sleepers with W14 elastic baseplateless fastening on the straight track section. Before the reconstruction, the rail was jointed, and after the reconstruction is was fitted as jointless.

A transition section, 17.5 m in length, is inserted between the adjoining track section and the closing wall of the bridge construction. The track skeleton construction before the reconstruction consisted of two wooden sleepers with the K type rail fastening and the R 65 flat-bottom rail (Figure 5). During the reconstruction, the rail was replaced and the Skl 24 elastic clip on a ribbed baseplate was newly used (Figure 6).

Figure 5. Transition track section before reconstruction.

Figure 6. Transition track section after reconstruction.

The load-bearing part of the railway bridge, 48 m in length, is composed of a welded steel truss construction with a trough bridge deck. During the reconstruction, the R 65 jointed rail was replaced with the 60E2 jointless flat-bottom rail on a ribbed baseplate using the Skl 24 elastic clips instead of the original Skl 12. The flooring of channelled plate was completely replaced (Figures. 7 and 9). Part of the superstructure are L-shaped bridge guards (Figure 8).

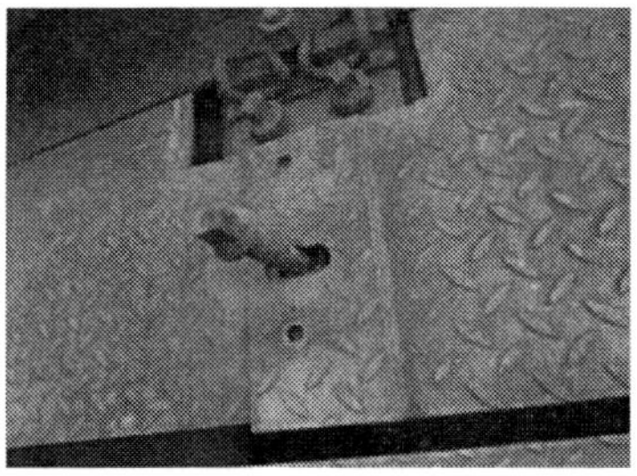

Figure 7. Railway bridge before reconstruction.

Figure 8. Railway bridge after reconstruction, L-shaped bridge guards.

Figure 9. Railway bridge after reconstruction.

4. Measurement methodology

After the reconstruction, the railway track No. 505A was put back into service on 1.6.2015. Technical measurement pursuant to ČSN EN ISO 3095 Railway Applications – Acoustics – Measurement of Noise Emitted by Wheeled Vehicles [6] was carried out on the railway track before and after its reconstruction. The measurement was aimed at finding out the effect of the reconstruction on the sound situation in the surroundings of the bridge construction and, at the same time, identifying the value of the noise emission increase due to the steel bridge construction. The measurement was performed simultaneously at two measurement points. The position of the points before the reconstruction corresponded by their location to the position after the reconstruction (Figure 10). The measurement point No. 1 (MP1) was situated in the straight track section on the embankment (stationing of km 3.984). The measuring microphone of the first noise meter was placed 7.5 m from the rail axis, at a height of 1.2 m above the running plane (Figure 11). The measurement point No. 2 (MP2) was situated in the stationing direction, at a level of ¾ of the bridge construction length (km 4.452). The measuring microphone of the second noise meter was placed outside the bridge construction, at a distance of 7.5 m from the rail axis, at a height of 1.2 m above the running plane, i.e. 3.55 m above the terrain level (Figure 12). The measured variables were time-related developments of the equivalent continuous sound pressure level A $L_{Aeq(t)}$ with reading steps of 1 s.

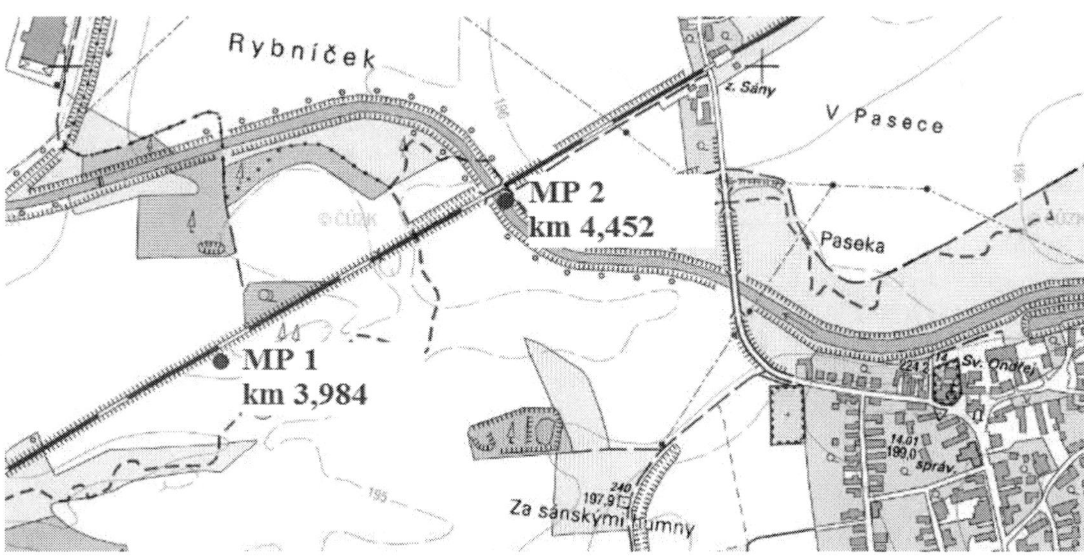

Figure 10. Position of measurement points.

5. Direct field measurement before reconstruction

The measurement before the reconstruction was performed simultaneously at both measurement points on 11. 6. 2014. A total of 20 passages of train units in both traffic directions were measured. The speed of train passage during the crossing was measured. The climatic conditions at the time of measurement complied with the requirements of the ČSN EN ISO 3095 standard [6]. The expanded combined measurement uncertainty was identified according to the procedure described in the document METHODOLOGICAL INSTRUCTION for the measurement and evaluation of noise in non-working environments [7]. For this type of measurement, the expanded combined measurement uncertainty according to the above document reaches a value of ± 2dB [7].

6. Direct field measurement after reconstruction

On 23.10.2015, a total of 15 passages of train units in both traffic directions was measured. The speed of train passage during the crossing was measured. The climatic conditions at the time of measurement complied with the requirements of the ČSN EN ISO 3095 standard. The expanded combined measurement uncertainty was identified according to the procedure described in the document METHODOLOGICAL INSTRUCTION for the measurement and evaluation of noise in non-working environments [7]. For this type of measurement, the expanded combined measurement uncertainty according to the above document reaches a value of ± 2dB [7].

Figure 11. Position of sound level meter, MP1.

Figure 12. Position of sound level meter, MP2.

7. Processing of measured data

The measured data were processed using the B&K Type 7820 Evaluator programme. Individual passages of train units were evaluated by selecting the time development sections in which the equivalent continuous sound pressure level of the noise (emitted by the monitored train unit) had exceeded the value $L_{Aeq(1s)} \geq 60$ dB. This criterion was chosen based on a sufficient distance from the residual sound recorded during the measurement. To compare the noise emissions arising by the passage of a train unit along the straight track section on the embankment (MP1) and across the bridge construction (MP2) noise exposures A L_{AE} were generated in the Evaluator programme.

8. Evaluation of results

To gain a more accurate informative value the noise exposures A L_{AE} of individual passages were standardized according to velocity and the number of rail cars. The reference values for velocities and the number of rail cars were identified separately for each type of train. The units were divided into fast trains, freight trains and expresses.

Table 1 below displays the increments in the sound pressure levels on the bridge construction against the adjoining section for the situation before and after the reconstruction.

Table 1. Comparison of $L_{AE,\ norm.}$ at measurement points MP1 and MP2 before and after reconstruction.

	$L_{AE,\ norm.}$ [dB] before reconstruction			$L_{AE,\ norm.}$ [dB] after reconstruction		
	MP1	MP2	MP2 - MP1	MP1	MP2	MP2 - MP1
All trains	109.9	116.9	7.0	98.0	112.1	14.1
Fast trains	110.5	116.5	6.0	96.4	110.6	14.2
Freight trains	113.7	121.8	8.1	103.7	117.8	14.1
Expresses	101.2	108.3	7.1	93.0	105.3	12.3

The growth in the level LAE, norm in the bridge construction area is in a range of 6 - 8 dB before the reconstruction and 12 – 14 dB after the reconstruction. At measurement point MP1 – on an open track, the reconstruction resulted in a significantly higher sound pressure level attenuation than at measurement point MP2 – around the bridge construction (Table 2). The higher sound pressure level increment on the bridge construction after the reconstruction could have been due to e.g. the replacement of a jointed rail with a jointless one. The bridge construction was fitted with a jointless rail before the reconstruction as well.

Table 2. Reduced sound pressure levels at measurement points MB1 and MB2 due to reconstruction.

	$L_{AE,\ norm.}$ [dB]	
	MP1	MP2
All trains	11.9	4.8
Fast trains	14.1	5.9
Freight trains	10.0	4.0
Expresses	8.2	3.0

9. Conclusion

Foreign literature states that the increment in the sound pressure level A on steel bridge constructions with directly trafficked bridge decks with bridge sleepers ranges from +3 to +16 dB [1]. The carried out sound measurement revealed a growth in noise emissions on the bridge construction against a track section ranging between 6 - 8 dB before the reconstruction and 12 – 14 dB after the reconstruction. The above reconstruction reduced the sound pressure levels at MP 1 – on a straight track section by 8 – 14 dB against MP 2 – around the bridge construction where the attenuation only accounted for 3 to 6 dB. The completed reconstruction had a positive effect particularly on the straight track section where a significant sound pressure level attenuation was recorded. The attenuation at MP1 after the reconstruction could be caused by the replacement of a corrugated flat-bottom rail with a new rail. A steel bridge construction is an element on the railway track very noisy on its own, for this reason, the rail replacement was not manifested by such a significant reduction in the sound pressure levels as on a straight track. The conclusion drawn from the obtained data is the need to focus on potential applications of noise attenuating elements directly in the steel bridge construction.

Acknowledgements
Research reported in this paper was supported by Competence Centres program of Technology Agency of the Czech Republic (TA CR), project Centre for Effective and Sustainable Transport Infrastructure (no. TE01020168).

References

[1] Thompson D 2009 *Railway Noise and Vibration* (Oxford: Elsevier)
[2] Poisson F and Margiocchi F 2006 The use of dynamic dampers on the rail to reduce the noise of steel railway bridges *Journal of Sound and Vibration* **293** pp 944-52
[3] Alten K and Flesch R 2012 Finite element simulation prior to reconstruction of a steel railway bridge to reduce structure-borne noise *Engineering Structures* **35** pp 83-88
[4] Bonnett C F 2005 *Practical railway engineering* (London: Imperial College Press)
[5] Pilíková E 2014 *Measurement of sound pressure levels in the surroundings of a bridge structure* (Czech Technical University in Prague, Faculty of Civil Engineering)
[6] Úřad pro normalizaci, metrologii a státní zkušebnictví 2014 Akustika - Železniční aplikace - Měření hluku vyzařovaného kolejovými vozidly / Acoustics - Railway applications - Measurement of noise emitted by railbound vehicles/ *(ČSN EN ISO 3095)*
[7] Ministry of Health and Main Health Officer of The Czech Republic 2005 Metodický návod pro měření a hodnocení hluku v mimopracovním prostředí /Methodological instruction for the Measurement and Evaluation of Noise in Non-Working Environments/ *(HEM-300-11.12.01-34065)*

Standardization of quality control plans for highway bridges in Europe: COST Action TU 1406

Joan R Casas[1] and Jose Campos e Matos[2]

[1]UPC-BarcelonaTech, Jordi Girona 1-3, Barcelona 08034, Spain
[2]UMinho, Campus Azurem, Guimaraes, Portugal

E-mail: joan.ramon.casas@upc.edu

Abstract. In Europe, as all over the world, the need to manage roadway bridges in an efficient way led to the development of different management systems. Hence, nowadays, many European countries have their own system. Although they present a similar architectural framework, several differences can be appointed. These differences constitute a divergent mechanism that may conduct to different decisions on maintenance actions. Within the roadway bridge management process, the identification of maintenance needs is more effective when developed in a uniform and repeatable manner. This process can be accomplished by the identification of performance indicators and definition of performance goals and key performance indicators (KPI), improving the planning of maintenance strategies. Therefore, a discussion at a European level, seeking to achieve a standardized approach in this subject, will bring significant benefits. Accordingly, a COST Action is under way in Europe with the aim of standardizing the establishment of quality control plans for roadway bridges.

1. Introduction

An efficient transportation network is essential for the modern society from the economic, societal and environmental point of view. Today, it is a challenge for operators to manage road infrastructures under their responsibility in an efficient way, meeting the present and future needs of the community they serve. Therefore, quality control plans have to be implemented by the operators in order to serve the increasing public demands with the always limited budget available for maintenance.

For this purpose, the authorities need to produce an asset management plan, which should not only define the goals to be achieved by exploiting the roadway bridge network, but that should also identify the investment needs and priorities based on life-cycle cost criteria. As a result, many European countries have their own system. Although they present a similar architectural framework, several differences can be appointed, for example, with regard to the condition assessment procedure and the use of life-cycle cost schemes. These differences constitute a divergent mechanism that may conduct to different decisions on maintenance actions, although affecting to the same transnational traffic.

Within the roadway bridge management process, the identification of maintenance needs is more effective when developed in a uniform and repeatable manner. This process can be accomplished by the evaluation of performance indicators, improving the planning of maintenance strategies. Therefore, a discussion at a European networking level, seeking to achieve a standardized approach in this subject, will bring significant benefits. Accordingly, COST Action TU-1406 aims to standardizing the establishment of quality control plans for roadway bridges. In this context, a first step would be the establishment of specific recommendations for the assessment of roadway bridges, namely, used methods for the quantification of performance indicators. A set of reference time periods for these assessment actions should be also presented. A second step would be the definition of standardized performance goals. Finally, a guideline for the establishment of quality control plans in roadway bridges would be developed. Moreover, the concept of sustainable roadway bridge management, involving the evaluation of environmental, economic and social performance indicators during the whole life cycle, is also highlighted.

In the following, these aspects are fully developed.

2. Objectives and main beneficiaries

The main ambition of the Action is to develop a guideline for the establishment of QC plans in roadway bridges, by integrating the most recent knowledge on performance assessment procedures with the adoption of specific goals [1,2]. This guideline will focus on bridge maintenance and life-cycle performance at two levels: (i) performance indicators, (ii) performance goals. The possibility to incorporate new indicators related to sustainable performance will also be considered. By developing new approaches to quantify and assess bridge performance, as well as quality specifications to assure expected performance levels, bridge management strategies will be significantly improved, enhancing asset management of ageing structures in Europe.

In order to reach this main general aim, the following more specific objectives/deliverables have been considered [1,2]: (i) to systematize knowledge on QC plans for bridges, which will help to achieve a state-of-art report that includes performance indicators and respective goals; (ii) to collect and contribute to up-to-date knowledge on performance indicators, including not only technical indicators but also environmental, economic and social ones; (iii) to establish a wide set of quality specifications through the definition of performance goals, aiming to assure an expected performance level; (iv) to develop detailed examples for practicing engineers on the assessment of performance indicators as well as in the establishment of performance goals, to be integrated in the developed guideline; (v) to create a data basis from COST countries with performance indicator values and respective goals, that can be useful for future purposes; (vi) to support the development of technical/scientific committees.

The target groups and end users who will exploit the outcome of this Action are: (i) public/private owners, as their assets will be maintained in an upscale level; (ii) operators, as standardized procedures for reducing maintenance costs, guaranteeing the same quality-level, will be introduced; (iii) design and consultant engineers, as the assessment of roadway bridges performance will be established in a uniform way, according to the developed guideline; (iv) equipment and software companies, as a new perspective will be given, regarding the most suitable equipment and software for the assessment of roadway bridges; (iv) academics and research engineers, as they will take an advantage of their involvement in the guideline preparation; (v) students, as they will benefit from COST tools (e.g. training schools) and from the contact with different stakeholders involved in this Action; (vi) relevant European, international and national associations, with which the main outcomes of this Action will be shared; (vii) standardization bodies and code writers, which will benefit from the developed guideline.

3. Performance indicators database

It is known that management systems are supported in QC plans which, in turn, are supported by performance indicators. Therefore, it is highly important to analyse such indicators in terms of used assessment frameworks (e.g. what kind of equipment and software is being used), and in terms of the quantification procedure itself. These are the objectives of working group 1 of the Action, with the aim to define:

(a) Technical indicators: the goal is to explore bridge structures performance indicators, in the course of international research cooperation, which captures the mechanical and technical properties and its degradation behaviour. Moreover, environmental condition, natural aging, and quality of material regarding to determined indicators will be investigated and evaluated in their meaningfulness. These considerations, however, also include service life design methods, aimed at estimating the period of time during which a structure or any component is able to achieve the performance requirements defined at the design stage with an adequate degree of reliability. Based on the input information quality (mainly concerning the available degradation models), it is possible to distinguish among deterministic methods, usually based on building science principles, expert judgment and past experience, which provide simple estimations of service life, and probabilistic methods;

(b) Sustainable indicators: in addition to technical performance indicators, which characterize the ultimate capacity as well as serviceability conditions, environmental based sustainability indicators will also be formulated. These variables characterize the environmental impact of a structure in the course of its total life cycle, expressed in terms of total energy consumption, carbon footprint (CO2 emissions), raw materials balance, etc. These indicators can be separated into direct and indirect, where the former are related to the construction/maintenance itself and the latter are caused e.g. as consequence of limited functionality;

(c) Other indicators: other sustainable indicators, economic and social based, may be used to evaluate bridge performance. These indicators, based on the technical performance of a structure, capture additional aspects that may influence the decision process and typically represent the discounted (accumulated) direct or indirect costs associated with construction and maintenance. Summed up over the full life-time, they represent part of or the full life-cycle costs. They can, in the context of multi-objective optimization, be understood as a weighting scheme to arrive to a single objective function to be minimized.

The determination of performance indicators for bridge structures from European countries and its harmonization on a European level is complex, extensive, and time consuming. These facts were confirmed in processing WG 1 "Performance indicators" of the COST TU1406. There are the following findings and important aspects associated with the Performance Indicator Survey Processes that was undertaken at the beginning of the project, as presented in Figure 1 [3,4]:

- A complete translation of codes or guidelines as used by owners and operators from the national language to international European format has been considered as unnecessary, since only some pages are devoted to the subject of interest (performance indicator, performance goal,...).
- The nomination of a responsible to collect the relevant parts of existing guidelines and translate them to English turned out to be much more effective. The responsible person must have good knowledge and expertise on inspection/assessment of existing bridges in order to identify the relevant parts
- A request for replying the questions in the questionnaire, and for up-loading the relevant parts of the document, both the original and the translated versions was regarded as very significant. It supports to objectify the language translations, since (a) it was revealed that many times the same operation or concept has different English translations or wording, and (b) to avoid subjectivity in some way.
- Because of the objective to propose enhancements to the existing practice of performance assessment by the different owners and showing recent advances and new performance indicators two types of documents are asked for: operator documents (actually in use by the different Agencies in the form of guidelines or recommendations) and research documents.
- Due to the different languages used across Europe and the different formats of both type of documents (guideline or research oriented) it was decided to nominate in each country several persons with the main tasks of contacting owners and operators of highway bridges asking for available documents in practice (operational performance indicators), for the preparation of tutorials for the screening of documents, processing screened documents, fill-in the data base and finally analyze the data base and to obtain the main results and conclusions. Those persons were also responsible to identify the research groups in each country and ask them to

provide information about new proposal still in the researching phase for performance indicators (research performance indicators).

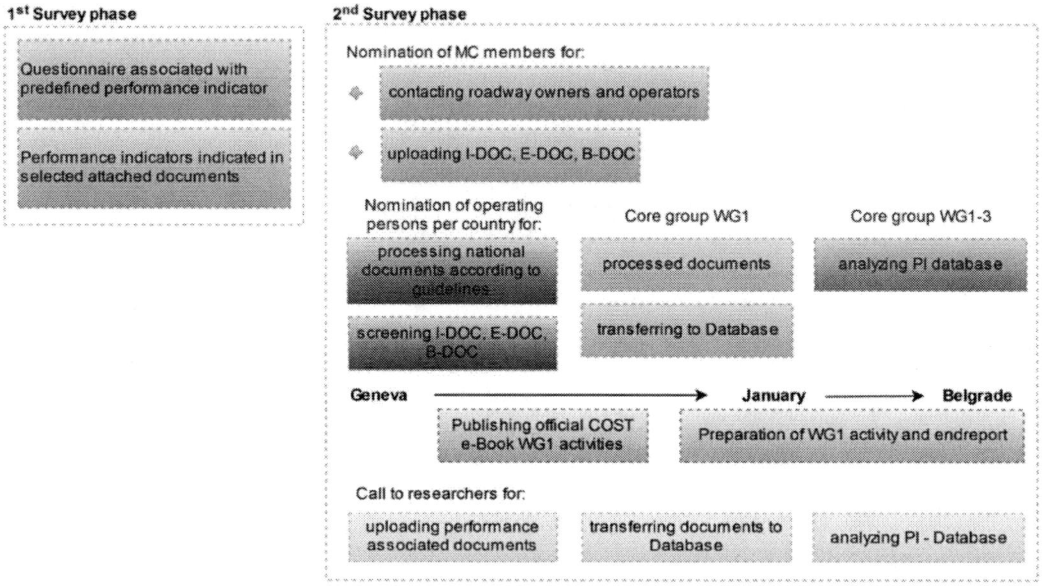

Figure 1. Technical survey on performance indicators (I-DOC inspection document, E-DOC evaluation document, B-DOC background document).

The final result of WG1 has been the publication of a report on these performance indicators and how they have been categorized and homogenized [5]. The core of the survey process for the key performance indicators (KPI's) and performance indicators (PI's) is given in Figure 2. The COST countries must choose beforehand the relevant documents (e.g. inspection, evaluation, research etc.) from which the PI's and KPI's and related information are going to be extracted. To support this process, a user interface is necessary. Here, it must be acknowledged that the amount and level of information varies between documents, even in those of the same type. Thus, one of the main requirements in the survey is to allow an unrestricted data input.

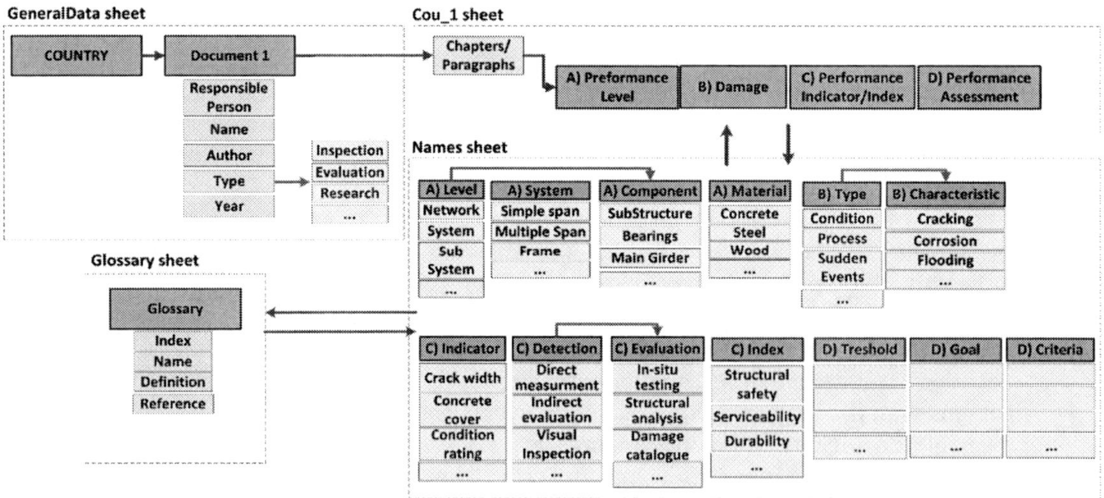

Figure 2. Database - core of the survey process.

After collecting the input from different countries, based on surveying of inspection and evaluation documents related to bridge maintenance, assessment and management, it was concluded that results are partly heterogeneous with a number of overlaps. Therefore, a critical overview of contributions from different countries, with respect to the content and definitions, was necessary in order to homogenise the terms to be used within the PI database. Figure 3 shows the main steps of this procedure.

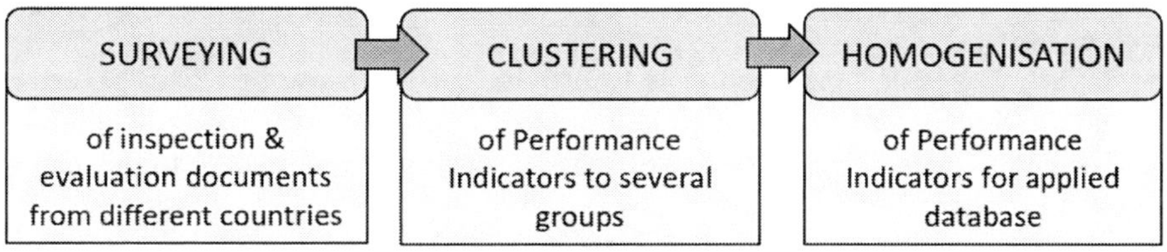

Figure 3. Procedure for the homogenization of PIs of applied database.

Collecting and surveying of research-based performance indicators, in order to reveal those that are already applicable in practice as well as those in whose development is worth investing, is envisaged to improve existing performance assessment methods within bridge maintenance systems and consequently the management of roadway bridges at the European level. As mentioned previously, indicators related to scientific achievements in, for example, testing and monitoring, dynamic behaviour and reliability of bridge structures should be included and continuously developed.

Collecting of research-based indicators is still an ongoing process through which several important questions need to be answered in order to extend the operators database:

- What is the type of indicator?
- Is there any related mathematical formulation?
- What are the intentions of this indicator, where is it to be applied?
- What is the threshold related to performance goal?
- What is the level of its maturity within the research?
- Through which type of case study is it verified?

Table 1. Definitions of parameter readiness level.

Ranking	(PRL) Parameter Readiness Level	Definition
1	basic principles observed	The **principles underlying** the parameter are known
2	**parameter** concept formulated	The parameter is applied in **analytical studies**
3	experimental proof of concept	**Analytical and experimental studies (indoor)** performed on a **laboratory scale on a component level** to validate analytical predictions
4	**parameter** validated in laboratory	**Experimental studies are** performed in laboratory on **a reduced scale model** of the structure/asset to produce a database for which estimate the parameter
5	**parameter** validated in laboratory in simulated environment	**Experimental studies** performed in controlled laboratory (or outdoor) on **a large model** of the structure/asset **reproducing real environmental conditions** to produce a database for which estimate the parameter
6	**parameter** demonstrated in relevant environment	Experimental studies performed **on a real structure/asset**
7	**parameter** demonstrated in operational environment	**Performance goals** are defined
8	system complete and qualified	**Testing protocols** are defined
9	Actual system proven in operational environment	**Decisions** on possible interventions in a bridge (repair, maintenance,...) are made

To this end, Table 1 is prepared to check for the maturity/readiness of the proposed indicator. Level 9 means maximum maturity: the performance indicator is ready for practical use.

Namely, answering this type of questions will help not only to identify research indicators, but also to decide which operational indicators are the most important and significant within the collected database from the actual state of practice in different countries.

4. Performance goals

The objective of WG2 (performance goals) is to provide an overview of existing performance goals for the indicators previously identified in WG1 and to develop technical recommendations which will specify the performance goals. These goals will vary according to technical, environmental, economic and social factors.

Performance goals are usually defined at different levels, from high-level strategic decisions to low-level, object-specific objectives. Further information on this WG can be found in [6]. This paper presents structure and basic ideas for the development of a guideline document linking different aspects of performance goals and bridge performance. It is explained there how bridge performance goals should be set as a multi-objective system, taking into account different aspects of bridge and network performance. The basis of the approach is the work developed so far in the Netherlands. Netherlands' Rijkswaterstaat (RWS) has developed RAMS SHEEP concept, which additionally to RAMS (Reliability, Availability, Maintainability and Safety) defines the following aspects: Security, Health, Environment, Economics and Politics [7], where each criterion is defined as follows:

- Security: related to the safety of a system regarding to vandalism and unreasonable human behavior.
- Health: being related to physically, mentally and socially defined aspects.
- Environment: concerns the physical environment requirements.
- Economics: regarding the relationship between cost and value.
- Politics: concerning political-administrative and social requirements.

However, practical indicators for quantification of these aspects within the risk assessment procedure are still lacking.

In Australian Roads documents [8,9], the following objectives are identified: Functionality (minimization of traffic delays), Safety (safe for intended use), Aesthetics (maintenance of an acceptable appearance), Sustainability (no backlog of repairs, and the workload remains at a manageable level), and Economic (maintenance is based upon lifecycle cost analysis).

Multi-criteria decision-making (MCDM) provides a systematic approach to combine these inputs with benefit/ cost information and decision-maker or stakeholder views to rank the alternatives. MCDM is used to identify and quantify decision-maker and stakeholder considerations about various (mostly) non-monetary factors in order to compare alternative courses of action [10]. Hierarchy structure for linking multi-objective bridge performance goals, covering most of the previously mentioned aspects with performance indicators is required. Possible result of multi-criteria assessment of different bridge maintenance alternatives is shown in Figure 4, which can be finally used for a decision making about the optimal maintenance or design alternative.

Alternatively, the multiple performance criteria can be combined into a so-called utility function, in which all the criteria are brought into a single scale. In order to transform the various out into a single (mostly monetary) scale it is necessary to establish weight factor for the individual types of criteria. Some of the weight factors are available in some countries (for example weight factor for traffic delays, noise, injuries etc.), depending on the selection of criteria, some weight factor may still need to be developed. In the development of the weight factors the starting point can be taken in the qualitative approach mentioned above, from which the apparent relative weight can be deducted. Once the possible outcomes have been brought to a single scale, the best decision can be found as a formal optimized decision process, in which option with the maximum "utility" shall be selected as the recommended decision.

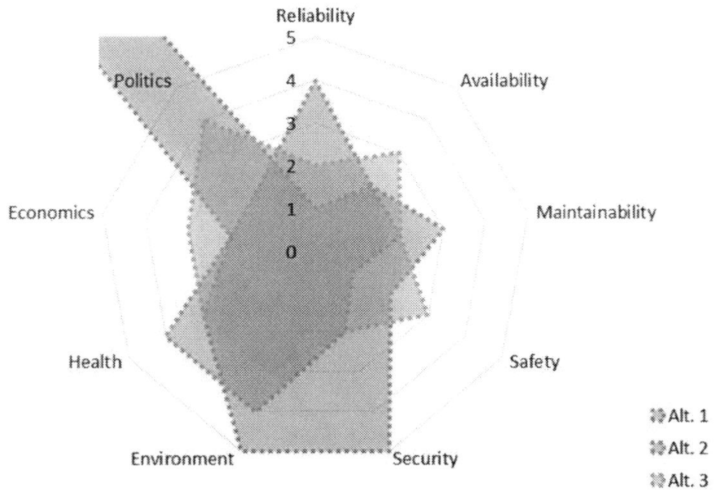

Figure 4. Spider plot as a possible result of multi-objective assessment of different maintenance alternatives against different performance aspects.

One of the main challenges in future research is how to quantify performance goals other than technical, and how to link strategic level to the performance requirements on the project level. Network or even societal goals tend to be rather broad in their definition. Furthermore, there is often no exclusive relationship between performance indicators set at a lower level and goals at a higher level. An important notion is that in many countries, the main focus of bridge management is still the condition assessment of the particular objects or elements thereof.

5. Quality Control Plans

Based on the results of WG 1 and WG 2 as well as on survey of existing approaches in practice, the objective of WG3 (Quality Control Plans) is to provide a methodology with detailed step-by-step explanations for establishment of QC plans for different types of bridges. The QC plan has to relate performance goals, which are user / society related, e.g.: Traveling time, Traffic allowance, Safety level, Comfort / Serviceability. In terms of bridges the quality can be defined as degree to which the bridge performance fulfils the performance goals. Given that the road bridges serve primarily their users by providing them a fast and safe crossing, one should try to derive performance goals from their desires and expectations. The users desire:

- maximum (unrestricted) weight allowance
- maximum (unrestricted) clearance
- minimum (zero) fatalities and/or injuries due to bridge collapse
- maximum (24/7) availability

The general public demand

- minimum (no) noise
- minimum (no) pollution and sometimes
- good (spotless) visual appearance i.e. no signs of deterioration

The owner and/or operator, however, strive to spend as little money as possible for the maintenance and operation. These conflicting desires and demands need to be resolved in a political process and thus implicitly define the performance goals. The resolution is carried out by the following stakeholders:

- Users, which are represented by government bodies, interest groups, transport corporations, car industry, etc.
- General public, which is represented by neighborhood associations, environmental groups, professional associations, etc.
- Owner, which is a government body itself, an agency controlled by the government, a concessionaire, private corporation, etc.

The main challenges in the development of a quality control plan for highway bridges are the following:

- Quality assurance and control is important in all industrial fields and construction industry is no exception. Bridge management is a part of construction industry activities that ensures that the bridge attains the desired service life and quality level. Important parts of any BMS are quality assurance and control plans that must provide the basis for quality data acquisition and processing. Reliable data related to the condition of the structures is essential for all further processes, however ensuring continuously the quality of this data is a demanding task, as the data acquisition is mainly based on periodic visual inspections.
- Data processing and decision models, on the other hand, are dependent on preselected criteria and decision methods. A well-designed decision model will yield reliable result through the lifetime of its use.
- The third component of quality assurance and control in bridge management is maintenance planning. Bridges differ in size, condition, structural design, materials used etc., therefore these specific features need to be acknowledged in the maintenance plans. Designing unique plan for every bridge is irrational and designing a general plan for all bridges impossible. Additionally, optimizing maintenance for individual bridge may worsen optimization on the network level and viceversa. A compromise has to be made in order to implement a quality plan successfully into practice.

The general framework to evolve from performance indicators to key performance indicators and to quality control plans is shown in Figure 5. Further information on this WG can be found in [11].

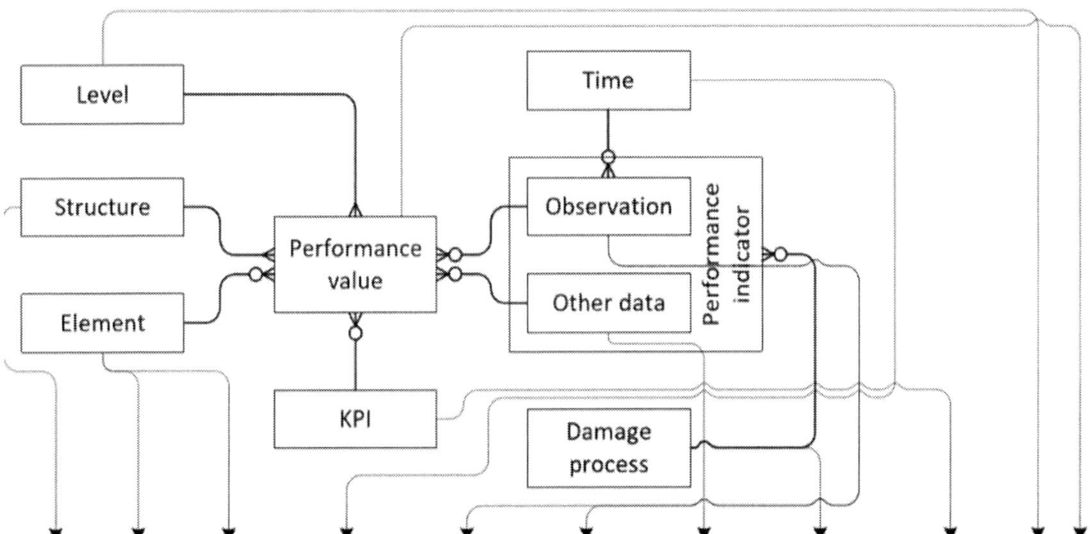

Single Girder Bridge	Beam	Reinforced concrete	*Time of the latest inspection*	Crack	0.4 mm	Construction year	Repeated overloading	Safety, Reliability, Security	C	2
Single Girder Bridge	Beam	Reinforced concrete	*Time of the latest inspection*	Spalling	$1.0\ m^2$	Construction year	Corrosion of Reinforcement	Safety, Reliability, Security	C	3
Single Girder Bridge	Beam	Reinforced concrete	*Forecast in 10 years*	Spalling	2.0	Construction year	**Corrosion of Reinforcement**	Safety, Reliability, Security	C	4

Figure 5. Road-map in use in WG3.

Acknowledgements
This article is based upon the work from COST Action TU-1406, Quality specifications for roadway bridges, standardization at a European level (BridgeSpec), supported by COST (European Cooperation in Science and Technology).

References

[1] Matos J C, Casas J R and Fernandes S 2016 COST Action TU 1406 Quality Specifications for Roadway Bridges - BridgeSpec *Proc. 8th Int. Conf. on Bridge Maintenance, Safety and Management (Foz do Iguaçu)* (London: Taylor and Francis)

[2] Matos J C 2016 An overview of COST Action TU1406, Quality Specifications for Roadway Bridges (BridgeSpec) *Proc. 5th Int. Symp. on Life-Cycle Civil Engineering (Delft)* (London: Taylor and Francis)

[3] Casas J R 2016 European Standardization of Quality Speci-fications for Roadway Bridges: an Overview *Proc. 8th Int. Conf. on Bridge Maintenance, Safety and Management (Foz do Iguaçu)* (London: Taylor and Francis)

[4] Strauss A, Vidovic A, Zambon I, Dengg F, Tanasic N and Campos J C 2016 Performance indicators for roadway bridges *Proc. 8th Int. Conf. on Bridge Maintenance, Safety and Management (Foz do Iguaçu)* (London: Taylor and Francis)

[5] Straus A, Mandic A, Campos J and Casas J R 2016 *WG1 Technical Report: Performance indicators for roadway bridges of Cost Action TU1406* (COST Brussels)

[6] Stipanovic I, Hoj N and Klanker G 2016 Multi-objective bridge performance goals *Proc. 5th Int. Symp. on Life-Cycle Civil Engineering (Delft)* (London: Taylor and Francis)

[7] Ministry of Infrastructure and the Environment (RWS) 2012 *Framework for performance analysis civil structures* (in Dutch, Rijkswatertsaat, Analysekader vaste kunstwerken)

[8] Austroads 2009 *Guide to Asset Management, Part 6: Bridge Performance* (Sydney, Australia)
[9] Austroads 2013 *Bridge Management Using Performance Models* (Sydney, Austroroads) p 175
[10] Kabir G, Sadiq R and Tesfamariam S 2014 A review of multicriteria decision-making methods for infrastructure management *Structure and Infrastructure Engineering* **10** pp 1176–1210
[11] Hajdin R 2016 Quality control plans on roadway bridges *Proc. 8th Int. Consf. on Bridge Maintenance, Safety and Management (Foz do Iguaçu)* (London: Taylor and Francis)

Thermo-mechanical simulations of early-age concrete cracking with durability predictions

Petr Havlásek[1], Vít Šmilauer[1], Karolina Hájková[1] and Luis Baquerizo[2]

[1] Czech Technical University in Prague, Faculty of Civil Engineering, Department of Mechanics, Thákurova 7, 166 29 Praha 6, Czech Republic
[2] LafargeHolcim R&D, 95, Rue du Montmurier, 38291 St Quentin Fallavier, France

E-mail: petr.havlasek@fsv.cvut.cz

Abstract. Concrete performance is strongly affected by mix design, thermal boundary conditions, its evolving mechanical properties, and internal/external restraints with consequences to possible cracking with impaired durability. Thermo-mechanical simulations are able to capture those relevant phenomena and boundary conditions for predicting temperature, strains, stresses or cracking in reinforced concrete structures. In this paper, we propose a weakly coupled thermo-mechanical model for early age concrete with an affinity-based hydration model for thermal part, taking into account concrete mix design, cement type and thermal boundary conditions. The mechanical part uses B3/B4 model for concrete creep and shrinkage with isotropic damage model for cracking, able to predict a crack width. All models have been implemented in an open-source OOFEM software package. Validations of thermo-mechanical simulations will be presented on several massive concrete structures, showing excellent temperature predictions. Likewise, strain validation demonstrates good predictions on a restrained reinforced concrete wall and concrete beam. Durability predictions stem from induction time of reinforcement corrosion, caused by carbonation and/or chloride ingress influenced by crack width. Reinforcement corrosion in concrete struts of a bridge will serve for validation.

1. Introduction

The term *mass concrete* describes elements in which high thermal gradients may lead to cracking. Binder content and initial concrete temperature belong to the most critical factors; even concrete members that are only 0.5 m thick may be susceptible to thermal cracking [1]. Significant tensile stresses may develop, leading to early-age thermal cracking, reducing concrete durability [2].

Models for stand-alone temperature prediction in hardening concrete belong among the most traditional concrete models. They consider either adiabatic temperature rise or general thermal boundary conditions [3]-[5]. Humidity transport plays no major role in massive concrete elements; hence only temperature field needs to be calculated. Thermal models can be widened for multiscale formulation, recognizing the role of cement paste for hydration [3]. Precise results can be obtained by first fitting hydration models to semi-adiabatic measurements on small samples and then by upscaling them to a structural scale [6],[7].

Several complex thermo-chemo-mechanical models were formulated during the last decades. They include prediction of stress, often considering concrete creep, autogenous shrinkage, drying shrinkage, and they may predict early-age cracking and crack width [8],[9]. The number of input parameters presents the bottleneck of simulations, varying broadly over cements, cement types, local practices and boundary conditions. Nevertheless, growing computer power and calibrated material laws allow simulating various concrete structures. The simulations of the massive concrete structures are in the focus of this paper. These simulations use a thermo-(chemo-)mechanical model implemented in a open-source finite element package OOFEM [10] .

2. Thermal model

Temperature field on a block is predicted from heat balance equation

$$-\nabla^T q(\mathbf{x}) + \overline{Q}(\mathbf{x},t) = \rho(\mathbf{x}) c_V(\mathbf{x}) \frac{\partial T(\mathbf{x},t)}{\partial t} \tag{1}$$

using further Fourier's law on an isotropic material

$$q(\mathbf{x}) = -\lambda(\mathbf{x}) \nabla \mathrm{T}(\mathbf{x}) \tag{2}$$

where $T(x,t)$ stands for temperature, $Q(x,t)$ for heat power from hydrating cement (W/m^3), $\rho(x)$ for concrete density (kg/m^3), $c_V(x)$ for concrete thermal capacity (J/kg/K), and λ for heat conduction (W/m/K). Equation (1) can be further complemented with initial conditions and several other boundary conditions including convection. Here, thermal radiation is neglected.

The heat power $Q(x,t)$ originates from hydrating cement paste according to Figure 1. The figure defines material scale with an affinity hydration model that captures reaction kinetics. Since the affinity model depends strongly on temperature, coupling with structural scale needs to be carried out. The model uses a strong coupling with a predictor-corrector scheme to achieve balance on both scales.

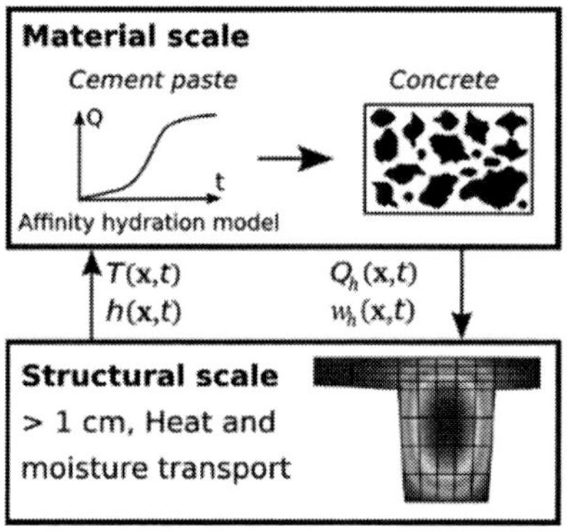

Figure 1. Multiscale hydration model with material and structural scales.

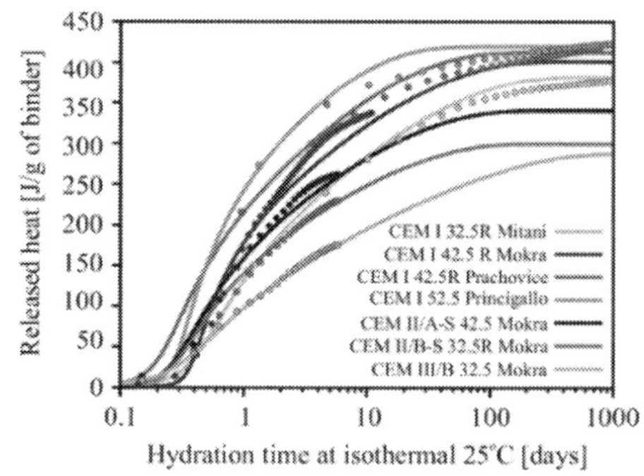

Figure 2. Performance of affinity hydration model on selected cements.

Affinity hydration model is the heart of the multiscale formulation. The evolution of hydration is approximated with a four-parametric function

$$\tilde{A}_{25} = B_1 \left(\frac{B_2}{\alpha_\infty} + \alpha \right) (\alpha_\infty - \alpha) \exp\left(-\overline{\eta} \frac{\alpha}{\alpha_\infty} \right) \tag{3}$$

where $\tilde{A}_{25}$ is the affinity at isothermal temperature 25°C, and B_1, B_2, α_∞, α are the four parameters for calibration. Isothermal calorimetry is a perfect method how to obtain these parameters. Figure 2 shows released heat of common cements [6]. As described in Figure 1, temperature influences hydration rate and vice versa. Temperature-dependent kinetics is approximated with Arrhenius equation

$$\frac{1}{Q_{h,pot}}\frac{\partial Q_h}{\partial t}=\frac{\partial \alpha}{\partial t}=\tilde{A}_{25}\exp\left[\frac{E_a}{R}\left(\frac{1}{T_{25}}-\frac{1}{T}\right)\right] \tag{4}$$

where $Q_{h,pot}$ is the potential hydration heat (J/g), E_a the activation energy and R the universal gas constant. This multiscale heat transfer model has been used several times for validation with excellent results on structures with volume from 0.01 m^3 up to 1050 m^3 [6].

3. Mechanical model

The mechanical part combines extended B3 creep model [11] with an isotropic damage model. The incremental constitutive equation reads

$$\Delta\sigma_{eff}=\bar{E}\mathbf{D}_V(\Delta\varepsilon-\Delta\varepsilon''-\Delta\varepsilon_{sh,aut}-\Delta\varepsilon_T) \tag{5}$$

where $\Delta\sigma_{eff}$ is an increment of effective stress, $\bar{E}$ incremental elastic modulus, $\Delta\varepsilon$ strain increment, $\Delta\varepsilon''$ strain increment due to creep, $\Delta\varepsilon_{sh,aut}$ an increment of autogenous shrinkage, $\Delta\varepsilon_T$ thermal strain increment [11],[16]. $\mathbf{D}_V$ stands for a unit elastic stiffness matrix. Using the concept of damage mechanics, the nominal stress is computed as

$$\sigma=(1-\omega)\sigma_{eff} \tag{6}$$

where ω is a scalar damage. Cracking is initiated when the effective principal tensile stress exceeds the tensile strength of concrete. Exponential softening is assumed with a crack-band approach to maintain objectivity with respect to the finite element size [10]. Evolution of compressive strength, tensile strength and fracture energy is adopted from ModelCode 2010 [15] and the user can directly enter concrete strength at 28 days. Autogenous shrinkage is controlled by maximum asymptotic value and the evolution can be specified according to ModelCode 2010 [15] or B4 model [16]. The crack width w is evaluated as

$$w=L_{ch}\omega\varepsilon_{I,red} \tag{7}$$

where L_{ch} stands for element length projected in direction of the first principal stress and $\varepsilon_{I,red}$ is the first principal strain decreased by thermal and autogenous strain.

A staggered approach is adopted; the thermal part of the analysis passes temperatures to the mechanical part. A user can define the end of setting time (4 hours by default) and the beginning of mechanical analysis (8 hours by default).

4. Thermo-mechanical validations

4.1. Temperatures on Sri Lanka's foundation caps

A new commercial/residential building Access Tower II with 28 floors has been constructed during 2014-2016. There have been several pile cap foundations, the biggest one with dimensions of 16.0 × 8.0 × 2.5m. Technical standards specified maximum concrete temperature of 70°C with compressive strength over 45 MPa at 28 days. After concrete mix optimization, 395 kg/m^3 of fly ash cement was used. A pile cap with dimensions of 7.15 × 7.15 × 2.5 m was validated via gage T3 which is located 125 mm from the concrete surface as indicated in Figure 3. The maximum measured temperature in T1 was of 70.1°C while the model predicted 72.7°C. Thermal gages served also in other two pile caps with slightly different dimensions, yielding maximum temperature differences 2.2 and 1.6 °C.

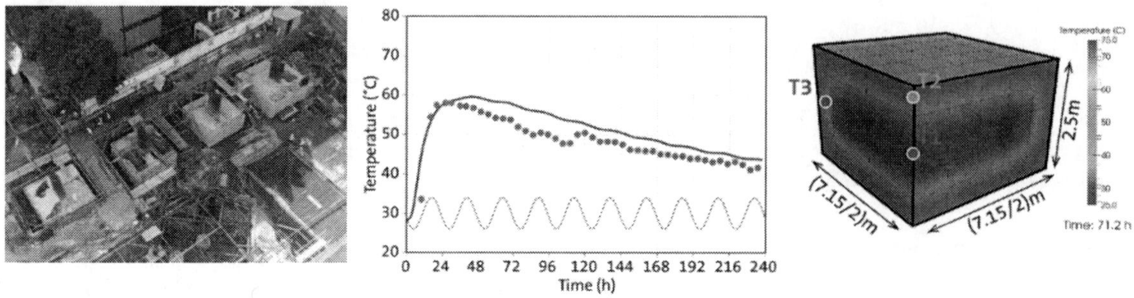

Figure 3. Top view of foundation caps, temperature validation in gage T3 and temperature field in a characteristic cap.

4.2. Simulation of Oslo's restrained wall

The Norwegian Public Roads Administration developed a "low-heat" concrete with a minimal risk of early-age cracking with a forerunner experiment for Bjørvika submerged tunnel project running between 2005 and 2012. A reference concrete SV 40 with CEM I 52.5 404 kg/m^3 and silica fume 20 kg/m^3 was tested from a series of small-scale lab tests to a restrained wall [17]. The first test was a small semi-adiabatic calorimeter using a concrete cube with 247 mm edge wrapped in EPS foam.

The cement used in this test had slightly higher initial kinetics than that of similar cement in the database [6]; hence, the kinetic was slightly accelerated to match core temperature in the adiabatic calorimeter, see Figure 4. A restrained wall with the same concrete composition was built in Oslo, see Figure 5. In particular, the right wall utilized low-heat concrete with a minimal risk of early-age cracking and high water tightness, whereas the left wall utilized SV40 concrete for comparison. Both walls were cast on a hardened raft concrete slab [17] and the left wall is validated.

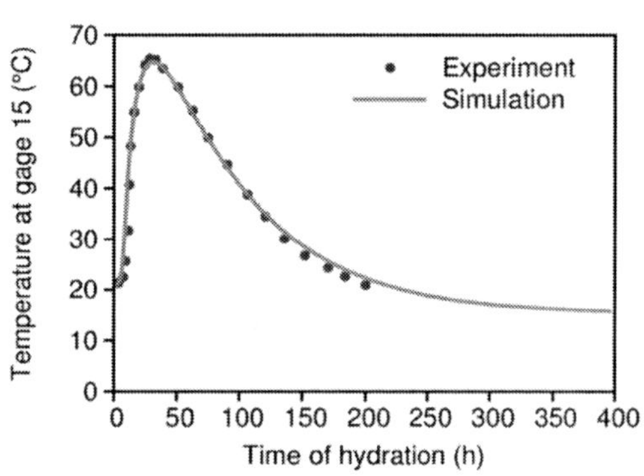

Figure 4. Core concrete temperature is a semi-adiabatic calorimeter.

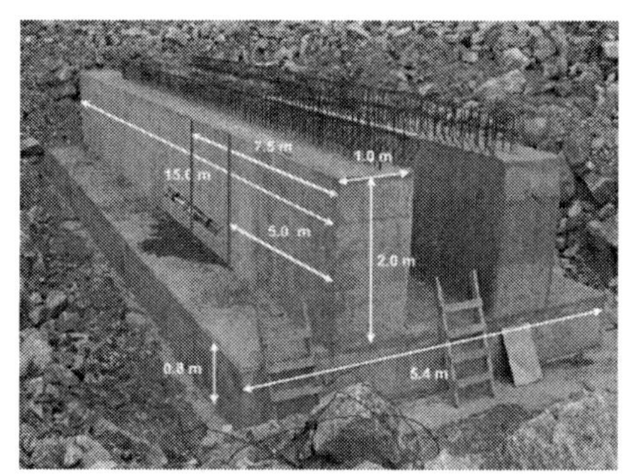

Figure 5. A restrained wall in Oslo [17].

Thermal simulation is carried out on a wall with a cross section of 1.0 × 2.0 m. The length was shortened to 6.5 m to decrease computing time. The next reduction came from symmetry so the modeled block size shrank to 0.5 × 2.0 × 3.25 m.

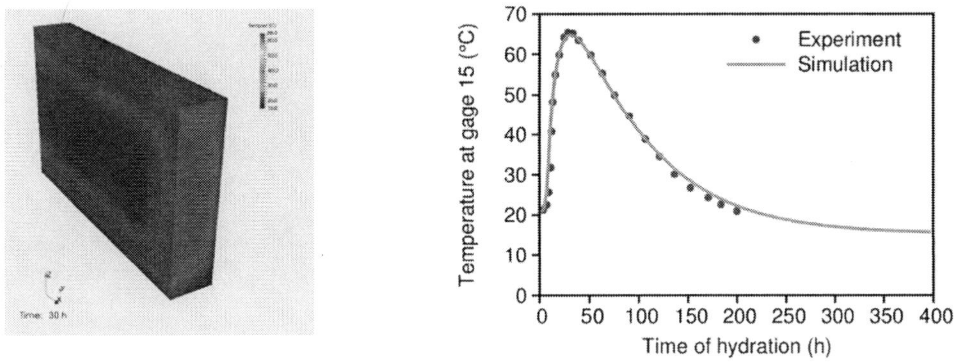

Figure 6. Temperature field during maximum temperature at 30 hours (left) and validation (right).

A fixed temperature of 15°C was assigned to the bottom surface and fluctuating ambient temperature ranging 12-18°C was prescribed to the remaining surfaces. Heat transfer coefficients set at 4.50 W/m^2/K acted on vertical surfaces covered with a formwork while the top was covered with a foil yielding a heat transfer coefficient of 13.50 W/m^2/K. Figure 6 shows temperature validation in gage 15, located in the middle of the wall and 1.2 m from the bottom. Formwork was removed after 216 hours from mixing time and the formwork stripping down was neglected in the simulation.

The mechanical part used B3 creep model for concrete with Microprestress-Solidification extension [11]. A mean compressive concrete strength was set at 77 MPa, ultimate autogenous shrinkage -180 με, and the coefficient of thermal expansion taken $10 \cdot 10^{-6}$ K^{-1} as the average [17]. Reinforcement ratio was estimated as 0.002 which had a small effect on mechanical behavior but improved convergence. It was assumed that the hardened concrete was fully restrained at the bottom of the simulated wall.

The simulation shows that first cracks on the surface appeared already 12 hours after casting but they closed at 60 hours due to surface cooling. The majority of the cracks appeared after 70 hours on the interface with the hardened concrete base where a high shear stress leads to tension on inclined planes. Figure 7 shows cracks and principal stress at t=300 h, where the widest cracks attain width of 0.58 mm. The validation of horizontal strain measured by vibrating wire strain gage is in agreement with experimental data, see Figure 7.

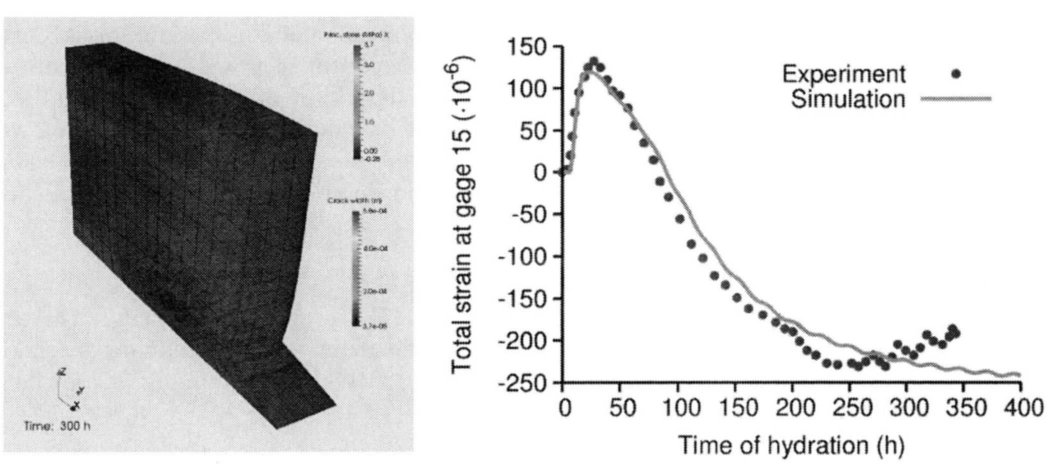

Figure 7. First principal stress and cracks at 300 hours. Validation of horizontal strain.

5. Extension for durability

Presence of cracks accelerates both carbonation and chloride ingress. Carbonation depth can be computed from Papadakis and Tsimas's model [18], which reads

$$x_c = \sqrt{\frac{2D_{e,CO2}CO_2}{0.218(C+kP)}}\sqrt{t} = A_1\sqrt{t} \tag{8}$$

and the acceleration due to crack width can be computed with Kwon and Na's model [19]

$$x_c(t) = (2.816\sqrt{w}+1)A_1\sqrt{t} \tag{9}$$

where w is the crack width (mm). Note that the presence of a 0.3 mm wide crack increases carbonation depth by a factor of 2.54. This also means that the induction time is 6.46 times shorter compared to a sound concrete.

The model for chloride ingress is based on Kwon et al. [19] as 1-D transient problem with initially chloride-free content

$$C(x,t) = C_S\left[1 - erf\left(\frac{x}{2\sqrt{D_m(t)f(w)t}}\right)\right] \tag{10}$$

where C_S is the chloride content at surface (kg·m^{-3}), D_m is the averaged diffusion coefficient at time t (mm^{2}s^{-1}), x is the position from the surface in mm and $f(w)$ gives the acceleration by cracking (equals to one for a crack-free concrete). Based on recent results, the following scaling function is proposed (Kwon, Na et al. 2009)

$$f(w) = 31.61w^2 + 4.73w + 1 \tag{11}$$

where w stands for crack width (mm). The presence of a 0.3 mm crack increases the mean diffusion coefficient by a factor of 5.26. In reality, crack width evolves and incremental solution needs to be formulated. Several examples for carbonation and chloride ingress were already presented in [20].

6. Conclusions

Performance of thermo-(chemo)-mechanical model was demonstrated on several mass concrete structures. Generally speaking, temperature predictions show excellent accuracy, with the error falling usually below 3°C. Mechanical part needs more accurate definition of boundary conditions and contains several parameters in the constitutive laws. Current simulations disregard drying which would become more important in thinner concrete elements. We demonstrated reasonable predictions on a restrained wall with crack localization and extension to durability considering carbonation or chloride ingress.

Acknowledgements

We acknowledge collaboration with LafargeHolcim R&D, providing valuable temperature measurements and technical details for validations. The financial support from the Technology Agency of the Czech Republic under the project CESTI TE01020168 is also gratefully acknowledged.

References

[1] Neville A M 1997 *Properties of Concrete* (London: Wiley)
[2] Weiss W, Yang W and Shah S 1999 Factors influencing durability and early-age cracking in high strength concrete structures *Proc High Performance Concrete: Research to Practice (Farmington Hills, MI)* pp 387–409

[3] Bentz D P, Waller V, and de Larrard F 1998 Prediction of Adiabatic Temperature Rise in Conventional and High-Performance Concretes Using a 3-D Microstructural Model *Cement and Concrete Research* **28** pp 285–297

[4] Estrada C F, Godoy L A, and Prato T 2006 Thermo-mechanical behavior of a thin concrete shell during its early age *Thin-Walled Struct* **44** pp 483–495

[5] Park K-B, Jee N-Y, Yoon I-S and Lee H-S 2008 Prediction of Temperature Distribution in High-Strength Concrete Using Hydration Model *ACI Mater. J.* **105** pp 180–6

[6] Leal Da Silva W, Šmilauer V and Štemberk P 2015 Upscaling semi-adiabatic measurements for simulating temperature evolution of mass concrete structures *Materials and Structures* **48** pp 1031-41

[7] Briffaut M, Benboudjema F, Torrenti J-M and Nahas G 2012 Analysis of semi-adiabiatic tests for the prediction of early-age behaviour of massive concrete structures *Cement and Concrete Composites* **34** pp 634–641

[8] Gawin D, Pesavento F and Schrefler BA 2006 Hygro-thermo-chemo-mechanical modelling of concrete at early ages and beyond Part I: hydration and hygro-thermal phenomena *International Journal for Numerical Methods in Engineering* **67** pp 299–331

[9] Gawin D, Pesavento F and Schrefler BA 2006 Hygro-thermo-chemo-mechanical modelling of concrete at early ages and beyond Part II: shrinkage and creep of concrete *International Journal for Numerical Methods in Engineering* **67** pp 332–363

[10] Patzák B 2012 OOFEM - an object-oriented simulation tool for advanced modeling of materials and structures *Acta Polytechnica* **52** pp 59–66

[11] Jirásek M and Havlásek P 2014 Microprestress-solidification theory of concrete creep: Reformulation and improvement *Cement and Concrete Research* **60** pp 51–62

[12] Hellmich C, Mang HA and Ulm F-J 2001 Hybrid method for quantification of stress states in shotcrete tunnel shells: combination of 3D in situ displacement measurements and thermochemoplastic material law *Computers & Structures* **79** pp 2103–15

[13] Jendele L, Šmilauer V and Červenka J 2014 Multiscale hydro-thermo-mechanical model for early-age and mature concrete structures *Advances in Engineering Software* **72** pp 134–146

[14] Ulm F-J and Coussy O 1998 Couplings in early-age concrete: From material modeling to structural design *International Journal of Solids and Structures* **35** 4295–4311

[15] Fédération Internationale du Béton 2010 Model Code 2010 *FIB Bulletin International Federation for Structural Concrete* **55**

[16] RILEM Technical Committee TC-242-MD 2015 Model B4 for creep, drying shrinkage and autogenous shrinkage of normal and high-strength concretes with multi-decade applicability *Materials and Structures* **48** 753–770

[17] Ji G 2008 *Cracking risk of concrete structures in the hardening phase: Experiments, material modeling and finite element analysis* (NTNU Trondheim)

[18] Papadakis V G and Tsimas S 2002 Supplementary Cementing Materials in Concrete Part I: Efficiency and Design *Cement and Concrete Research* **32** 1525–1532

[19] Kwon S J et al 2009 Service Life Prediction of Concrete Wharves with Early-aged Crack: Probabilistic Approach for Chloride Diffusion *Structural Safety* **31** 75-83

[20] Šmilauer V, Jendele L and Červenka J 2013 Prediction of Carbonation and Chloride Ingress in Cracked Concrete Structures *Proceedings of the Fourteenth International Conference on Civil, Structural and Environmental Engineering Computing Stirling* (Civil-Comp Press Ltd) pp 1-12

Improving resistance of high strength concrete (HSC) bridge beams to frost and defrosting salt attack by application of hydrophobic agent

Jiri Kolisko[1], Lukáš Balík[1], Michaela Kostelecka[1] and Petr Pokorný[1]
[1]Klokner institute, CTU in Prague, Solinova 7, Prague 6, 16608, Czech Republic

E-mail: jiri.kolisko@cvut.cz

Abstract. HSC (High Strength Concrete) is increasingly used for bearing bridge structures nowadays. Bridge structures in the Czech Republic are exposed to severe conditions in winter time and durability of the concrete is therefore a crucial requirement. The high strength and low water absorption of HSC suggests that the material will have high durability. However, the situation may not be so straightforward. We carried out a study of the very poor durability of HSC concrete C70/85 used to produce prestresed beams 37.1 m in length to build a 6-span highway bridge. After the beams were cast, a production control test indicated some problems with the durability of the concrete. There was a danger that 42 of the beams would not be suitable for use. All participants in the bridge project finally decided, after extensive discussions, to attempt to improve the durability of the concrete by applying a hydrophobic agent. Paper will present the results of comparative tests of four hydrophobic agents in order to choose one for real application and describes this application on construction site.

1. Introduction

The problem of concrete durability is very important for transport structures in the Czech Republic. Our experience has shown that the application of HSC concrete leads to new problems in this area. The high strength and low water absorption of HSC suggests that the material will have high durability. However, the situation may not be so straightforward. In 2010-2011, we carried out a study of the very poor durability of HSC concrete C70/85 used to produce prestresed beams 37.1 m in length to build a 6-span highway bridge (see Figure 4). There was a danger that 42 of the beams would not be suitable for use. After the beams were cast, a control test indicated some problems with the durability of the concrete. This problem is not easy to solve. All participants in the bridge project finally decided, after extensive discussions, to attempt to improve the durability of the concrete by applying a hydrophobic agent.

The action of hydrophobic agents is a too broad issue to be discussed in detail in a short paper. More about this topic can be found in [1], [2], [3], [8]. The principle of hydrophobic treatment involves increasing the capillary tension of the materials and increasing the wetting angle of the water by forming a thin layer (film) of a hydrophobic agent on the inner surface of the pores of the construction materials. This layer must be extremely thin, invisible to the eye, in order to avoid reducing the diameter of the pores, which would clog them and block the vapour permeability. Surfaces treated with a hydrophobic agent do not allow water to create a continuous surface and moisten the surface. The water runs down in a bullet-like shape.

2. Experimental verification of the four hydrophobic agents

The functional behaviour of hydrophobic agents applied to concrete is specified in EN 1504-2 [4]. This standard is followed by testing standards defining a series of tests. The most important tests are:

- Test of drying rate EN 13579 [5]
- Test of water absorption through an impregnated surface EN 13580 [6]
- Test of the resistance of an impregnated surface to alkali effects EN 13580
- Determination of the depth of penetration of hydrophobic agent EN 1504-2 [4]

Concrete cubes (100 mm edge) prepared from concrete with water cement ratio w/c = 0.45 were produced for a drying test, a test of water absorption, and a test of resistance to alkalis. Cubes for depth of penetration test were prepared from concrete with water cement ratio w/c = 0.7. The test specimens were impregnated according to the instructions of the producer of the hydrophobic treatment agent. The depth of the dry (light) area (see Figure 1) was considered as the effective depth of the hydrophobic agent. Results of individual tests and normative standards are stated in table 1.

Table 1. Results of executed tests

Parameter		Test label of hydrophobic treatment agent			
		A	B	C	D
	Agent basis	oligomeric siloxanes	silanes	oligomeric siloxanes	silanes
Consumption of agent for test samples [g/m^2]		450	420	400	320
Coeff. of drying speed [%]	Measured	48.3 %	56.2 %	25.0 %	57.5 %
	Standard [4] req.	class I: >30 %; class II: >10 %			
Absorption relation [%]	Measured	5.9 %	5.0 %	18.6 %	4.0 %
	Standard [4] req.	< 7,5 %			
Absorption relation after influence of alkali [%]	Measured	12.4 %	9.1 %	36.3 %	7.9 %
	Standard [4] req.	< 10 %			
Penetration depth [mm]	Measured	1-1.5 mm	4-6.5 mm	0 mm	8 - 9 mm
	Standard [4] req.	class I: < 10 mm; class II: ≥ 10 mm			

Figure 1. Broken impregnated test specimens. Penetration depth of four various hydrophobic treatment agents. From the left A, B, C, D.

Only agents B and D pass all the normative criteria, see table 1. In addition, the depth of penetration is higher when using these agents (4-6.5 mm for agent B and 8-9 mm for the agent D). Agent A did not pass the test of long-term durability after the test specimens were exposed to alkali effects, most probably due to the small depth of penetration, which was only 1-1.5 mm. Agent C did not pass any of the normative regulations, and no depth of penetration was measured there. When this agent was applied, it formed only a film on the specimen surface.

3. Laboratory test of the effect of agent on the resistance of concrete to frost, water and defrosting salt

Within the wider experimental program, the effect of hydrophobic agent D was tested on the change in the resistance of the concrete against the attack of frost and defrosting salt solution. The result of the test with agent D is presented in the following text. The test was carried out in accordance with Czech standard ČSN 731326 [7] – method C similar to CDF method according specification CEN/TS 12390-9. Test procedure contains 5 mm layer of 3% salt solution on the tested surface. Than the air temperature varies from +5°C (3 hours include temperature change) to -18°C (3 hours include temperature change). The result of the test is the amount of spalled material from the tested surface (see Figure 2). The investor's requirement for XF4 environment grade is 1000 g/m^2 spalling of the surface material after 75 cycles of test method C. The surface of the tested concrete samples (cylinders) prepared during the standard production of non-air entraining high strength concrete (HSC) C 70/85 XF4 was used for the comparative test. This concrete was used to produce prestressed bridge beams for a new highway bridge. Oscillation of the compressive strength during production was 90-100 MPa. The results of production tests for this concrete indicated significantly impaired resistance of the concrete to frost attack and defrosting salt attack.

Figure 2a. Start of the durability test on cores drilled from beams.

Figure 2b. Massive surface deterioration (more than 3 kg/m^2 of spalled material) after 50 cycles of test C according ČSN 731326 [7].

The degradation found with the high-strength concrete was rather unusual, as there was none of the gradual spalling of individual flakes that is usually found due to the effect of frost and defrosting agents, but series of cracks were found in the structure (see Figure 3).

In order to avoid any impact of surface processing during the production of cylinders (150/300 mm), the resistance test was carried out on the cut surface of the samples. Total 6 disks about 50 mm in height (3 treated + 3 reference) were prepared from three cylinders. Using a brush, agent A was applied to the tested surface of three cylinders. Three layers of agent was applied according to producer recommendation. The total average consumption of the agent on all cylinders was 390 g/m^2. Three cylinders were used as a reference sample. The test started very soon = 5 days after the agent application. Not as much time was left as is usually required by the suppliers to allow the agent to penetrate into the surface of the concrete.

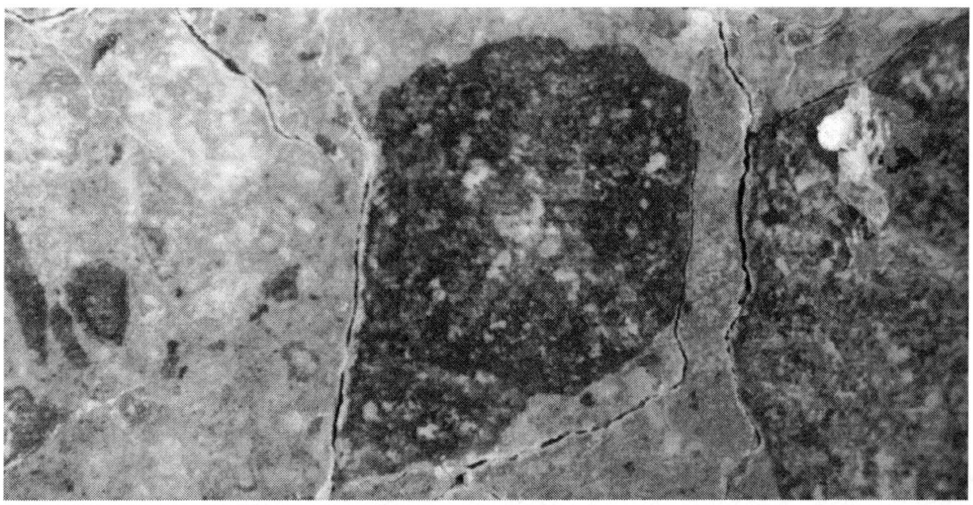

Figure 3. Characteristic pattern of cracks on the HSC concrete surface as a result of frost atact. Magnification 15 x.

Table 2. Result of the frost and defrosting salt attack test - test method C according to ČSN 731326.

Sample No.	Bulk density [kg/m^3]	Surface water absorption 15 min. [g/m^2]	Spalling material after freeze thaw cycles [g/m^2] 25	50	75	100	125
1 – ref	2490	164	300	874	1594	2069	2266
2 - ref	2450	98	344	1207	1763	2282	2588
3 - ref	2470	82	0	0	795	2007	2347
Average	2470	115	215	693	1384	2120	2400
1 - A	2480	55	0	0	0	16	33
2 - A	2540	28	0	0	0	39	189
3 - A	2480	28	0	0	0	33	156
Average	2500	37	0	0	0	30	126

The results of the durability test were satisfactory and very promising, and the investor decided to use this technology to improve the durability of the beam concrete.

4. Practical application of the hydrophobic agent on bridge beams

Agent type A achieved the best results in the laboratory tests. This agent was chosen for field application to improve the protection of the concrete beams of a six-span beam bridge (see Figure 4). Forty-two precast beams 37,1 m in length were produced from HSC concrete C 70/85 that was not air entrained and was suspected of poor frost resistance.

The field application comprises several steps. First of all, the technology for applying the agent was tested for applying the required amount of the agent (min. 400 g/m^2 of wet hydrophobic agent). A small field test was arranged in 2010 (May - September), almost 1 year before final application.
During the field pre-test, we determined:

- The number of layers needed to apply 400 g/m^2 of the agent (Figure 5).
- The reference surface water absorption, in order to provide reference information for the control program during final application to the beams (Figure 7).

A reference measurement of surface conductivity, in order to evaluate the quality of the hydrophobic layer (Figure 6).

Figure 4a. Six-span beam bridge = one side of two separate highway bridges. Span 38.33 m.

Figure 4b. Right - cross section of a pre-stressed bridge beam. One span consists of 7 beams.

Figure 5. Pre-test application test to get the required amount of agent on the surface.

Figure 6. Surface conductivity test.

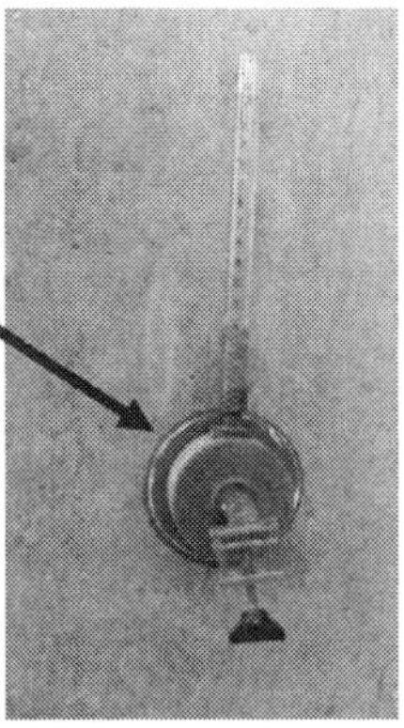

Figure 7. Surface absorption test – Karsten flask.

Afterwards (September 2011 – one year later), the agent was applied on 42 beams (see Figure 8) and control testing (amount, conductivity, surface absorption tests) was also carried out.

Figure 8. Application of the hydrophobic agent on the bridge beams.

The amount of the agent that was applied was controlled by comparing the weight of the applied material with the treated area. Comprehensive results of a surface conductivity test after the agent had been applied are presented in Figure 9.

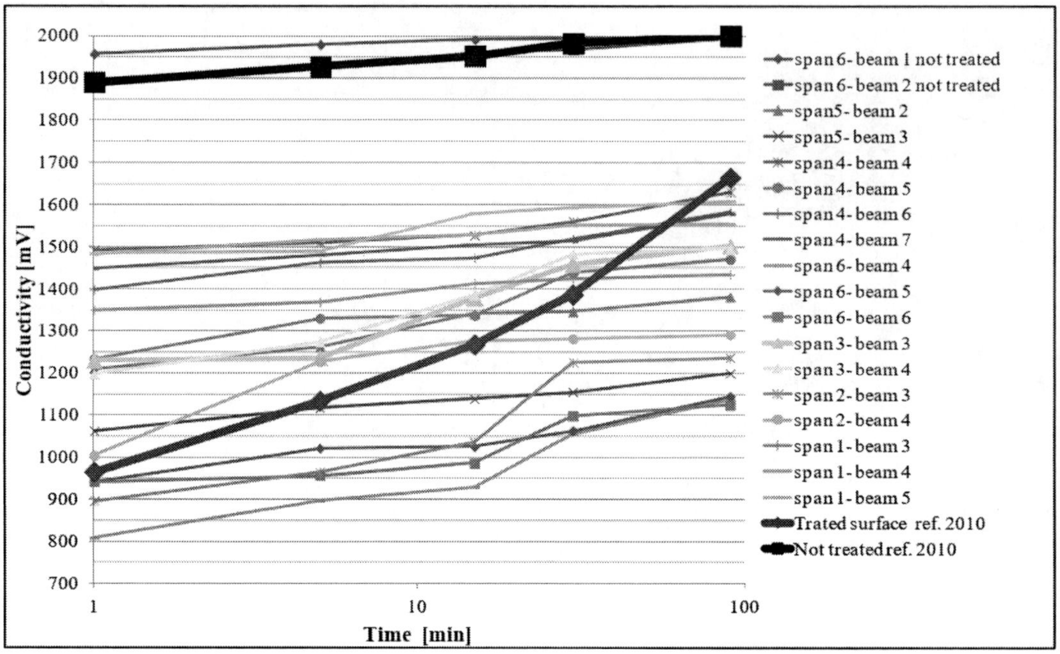

Figure 9. Comprehensive results of the control conductivity test on the bridge beams. Results on non- treated beams is mentioned as non treated.

The results of the surface water absorption test (Karsten flask 120 min./200 mm water column) are presented in Table 3.

Table 3. Result of water surface absorption - Karsten flask 120 min./200 mm water column.

Measured span	Number of measurements	Mean of surface absorption g/m^2 versus time [min.]		
		0 min	60 min	120 min
Span 1 – treated	3	0	0	0
Span 2 – treated	2	0	0	89
Span 3- treated	2	0	125	246
Span 4- treated	4	0	0	122
Span 5- treated	4	0	47	176
Span 6 – treated	3	0	50	226
Reference untreated surface	4	0	354	972

During application of the hydrophobic agent, the control tests showed that:

- The amount of agent that was applied was controlled by comparing the weight of the applied material with the treated area. In this way, it was proved that required amount of 400 g/m^2 of the hydrophobic agent was applied.
- Comprehensive results of the surface conductivity test after the agent had been applied are presented in Figure 9. Less conductivity with a shallower slope of the graph indicates better results i.e. better agent. It is seen that the application meets the reference values acquired during the field pre-test in 2010, and this was counted as a satisfactory result.
- The reference value for the water surface absorption of the treated surfaces was determined during the pre-test as <350 g/m^2. Table 3 shows that the control results meet the reference values obtained during the field pre-test in 2010, and this was counted as a satisfactory result.

5. Conclusion

There has been relatively little practical field experience in the Czech Republic with the effect of protecting concrete on the long-term behaviour of hydrophobic impregnations. Our comparative test of the resistance of concrete surfaces has proved that hydrophobic agent has a remarkable positive impact (even unsatisfactory parameters according to EN 1504-2 [4]) on the resistance of the concrete surface against the impact of frost, water, and defrosting salt attack. However, long-term durability and effectiveness is very important when applying hydrophobic agent. Resistance against the impact of weather conditions is mainly related to the ability to penetrate as deep as possible into the structure of the concrete, in order to slow down any degradation of layers caused by weather conditions as much as possible. At the same time, there is a very important effect of an alkaline environment on the resistance of hydrophobic agent against degradation.

The practical application of a hydrophobic agent on the surface of concrete C 70/85, XF4 will provide relevant data on the efficiency of surface treatment with a hydrophobic agent. Treated concrete is five years old after agent application and the bridge structure is now under long-term supervision. Visual inspection of bridge does not show any sign of surface deterioration and concrete defects until now. We are discussing with bridge administrator possibility of surface testing to check reliability and durability of hydrophobic treatment. If the results are satisfactory, we can expect an increased number of applications of surface protection with hydrophobic agents in the Czech Republic.

Acknowledgements

This paper was prepared in the frame of project GACR No. 15-10591S Impact of long-term environmental conditions on the hydrophobic properties of the surface layers. Project is financially supported by Grant Agency of Czech Republic GACR.

References

[1] De Vries J and Polder RB 1997 Hydrophobic Treatment of Concrete *Construction and Building Materials* **11**(4) pp 259-65

[2] Dobiáš D and Kolísko J 2009 Possibilities of hydrophobic treatment of silicate materials *Conference - Redevelopment and Reconstruction of Buildings (Brno)* pp 37-41

[3] Kolísko J and Hromádko J 2010 Durability of concrete of bridges and tunnels *Textbook Highway D8 – concrete constructions*

[4] Úřad pro technickou normalizaci, metrologii a státní zkušebnictví (ÚNMZ) 2006 Products and systems for protection and repairs of concrete structures – Definitions, requirements, quality control and conformity assessment – Part 2: Systems of concrete surface protection (*ČSN EN 1504-2)*

[5] European Committee for Standardization (CEN) 2003 Products and systems for the protection and repair of concrete structures - Test methods - Drying test for hydrophobic agent (*EN 13579)*

[6] European Committee for Standardization (CEN) 2003 Products and systems for the protection and repair of concrete structures - Test methods - Water absorption and resistance to alkali for hydrophobic agents (*EN 13580)*

[7] Úřad pro technickou normalizaci, metrologii a státní zkušebnictví (ÚNMZ) 1985 Stanovení odolnosti povrchu cementového betonu proti působení vody a chemických rozmrazovacích látek (Test of concrete against attack of water, frost and defrosting salt) (*ČSN 731326)*

[8] Pernicova R and Ticha P 2016 Experimental method of measuring the efficiency of hydrophobic surface layer of concrete *EAN 2016 - 54th Int. Conf. on Experimental Stress Analysis (Copenhagen)*

IOP Conf. Series: Materials Science and Engineering **236** (2017) 012054 doi:10.1088/1757-899X/236/1/012054

Experiments on fibre orientation in UHPC

Jan L. Vítek[1], David Čítek[2] and Robert Coufal[3]

[1]Metrostav a.s., and CTU in Prague, Koželužská 2450/4, 180 00 Prague 8, Czech Republic
[2]CTU in Prague, KI, Šolínova 7, 166 08 Prague 6, Czech Republic
[3]TBG Metrostav, s.r.o., Koželužská 2246/5, 180 00 Prague 8, Czech Republic

E-mail: vitek@metrostav.cz

Abstract. Distribution and orientation of fibres in fibre reinforced composites are essential for the structural performance of elements and structures. The paper deals with experimental research where this phenomenon was investigated. During construction of the first large footbridge, many experiments were made, which should guarantee the quality of the UHPC segments. Later the problem was investigated more in detail, when a technology of hollow core slabs was developed. Three basic experiments were carried out where the direction of pouring of concrete and the structural response were compared. Standard beams which were cast in horizontal and vertical positions, small beams were cut from the hollow core slab in different directions and finally thin slabs were cast in horizontal and vertical directions and their load carrying capacity was tested. Reduction of flexural strength in different directions may vary significantly.

1. Introduction

Ultra high performance concrete is a high strength composite which tends to be widely used in concrete structures. Since the practical application requires the reinforcement of high strength cementitious matrix by fibres (steel or non ferrous) some authors use the abbreviation UHPFRC. The fibres should guarantee that the brittleness of the composite will be limited, and that the material will be able to carry a tensile force also after cracking. It is generally expected that the fibres are distributed uniformly in the material, which is understood that they are oriented in space uniformly and also that the number of fibres is similar in individual sections of the structural element. However, the experience from the production showed that the fibres are not distributed uniformly; their orientation is dependent on the way of production. If a self-compacting UHPC is used, which is quite often, the fibres tend to find their location in dependence on the flow of concrete into the formwork. The paper deals with some experiments which show the differences in the behaviour of structural elements in dependence on their production.

2. Footbridge in Čelákovice

The footbridge over the Labe River in Čelákovice was the first large structure, where the load carrying structure was made from UHPC in the Czech Republic [1]. The UHPC was produced from local materials. Its compressive strength was about 150 MPa. The cable stayed bridge still has the largest span of the cable stayed structure in the Czech Republic (156 m) and it is shown in Figure 1. The bridge deck (3 m wide) is made of segments which have two longitudinal beams which are connected by a thin slab (only 60 mm deep) reinforced by ribs in the distance of about 1 m. The slab has not any reinforcement with exception of fibres. At the time of production of segments (2013), there were executed many tests [2] in order to develop a reliable way of casting of individual segments so that the fibres would be distributed uniformly. The distribution of fibres was checked on segments which were cut and the fibres were observed on the cut surfaces. The distribution appeared to be rather uniform (Figure 2). The structural performance was verified by mechanical tests. The load carrying capacity of the slab was tested by a point load acting on the produced segment. The results were rather surprising; the load carrying capacity was significantly higher than that expected. The load carrying capacity in punching of the slab 60 mm thick was 350 kN (average from 4 tests). The distribution of fibres was observed on the slab after the punching test, and the fibres can be seen in all the area subjected to fracture (Figure 3).

Figure 1. Completed footbridge over the Labe River in Čelákovice.

Figure 2. Distribution of fibres in the top part of the longitudinal beam.

Figure 3. Opening in the slab after the punching test.

Tests executed prior the construction and during the construction of the footbridge brought valuable information and also proved that the applied UHPC and the technology of casting were suitable and reliable. It was also shown that it would be difficult to guarantee the quality of the footbridge without the executed tests. Some tests showed that there are some reserves, and some dimensions could be even smaller. However, the completely new material was applied and a conservative approach to the design was desirable. The experience after almost 3 years of the footbridge operation exhibits that the accepted decisions were correct.

3. Hollow core slabs

Hollow core slabs were designed as a convenient part of different structures. The slabs which were tested were 300 mm thick and had triangular openings. The segments are assumed to be 2 m long, and casting in vertical position was proposed, in order to get all surfaces of a very high quality. In order to verify the structural performance of the slabs and also a production technology, models of the hollow core slab were produced. The dimensions of the models were 1.5 x 2.0 x 0.3 m, i.e. the thickness and the length of the model were identical to the dimensions of the segment, the width of the model was smaller (Figure 4). During the production, it was verified that the segments can be successfully produced and the formwork of the openings can be reliably removed after several hours after casting. This time is dependent on the curing of concrete. If UHPC is cured at normal conditions, the formwork of the internal openings may be pulled out after 6 hours, if heating of the segment is used, then the time can be reduced to 4 hours. At any case it is possible to produce one segment per day in one mould.

Figure 4. Model of a hollow core slab.

If a thin slab is cast in horizontal position, most of the fibres are usually oriented in horizontal position. Casting of the segment in vertical position indicates a question what is the orientation of fibres in top and bottom slab and partitions between the openings in the model of a hollow core slab, which are rather thin – about 30 to 45 mm.

The models were tested in bending in longitudinal and in transversal directions. Since no reinforcement or prestressing was applied the load carrying capacity of the models was determined by the tensile properties of UHPC. The UHPC was reinforced with high strength steel fibres. The amount of fibres was 2%. The testing is rather expensive; therefore the models were cut into smaller specimens which were subjected to loading tests. The compressive and tensile strengths were determined from laboratory tests executed on cubes, cylinders and beams. Also the tests in direct tension were executed on small cylindrical cores (∅60 mm and length 120 mm). The flexural strength was measured on two kinds of beams. The beams (150 x 150 x 700 mm) were cast 1. in the horizontal position and 2. in the vertical position. The loading scheme was the same, 4 point bending test (span 600 mm) with two point loads located in the distance of 200 mm. The beams cast in horizontal position exhibited larger flexural strength. Basic material parameters are summarized in the Table 1.

The fibre orientation in the flanges of hollow core slabs was investigated. Small beams of dimensions 40 x 40 x 160 mm were cut from the model of the hollow core slab. The beams were cut in the horizontal and vertical directions and also in the inclined direction of 45° with respect to the direction of casting.

Table 1. Material parameters of the applied UHPC.

	Mean value	Coeff. of variation [%]
Compressive strength – cylinders ∅150 mm x 300 mm [MPa]	140	3,9
Flexural strength – beams cast in horizontal position [MPa]	15.8	4.8
Flexural strength – beams cast in vertical position [MPa]	13.8	4.7
Direct tensile strength – cores ∅60 mm x 120 mm [MPa]	4.8	12.4
Modulus of elasticity [GPa]	46.3	0.9

The beams were subjected to 3 point bending test (Figure 5 – span 100 mm) and the results were compared. The measured flexural strengths of the beams cut from the experimental model are described in the Table 2. The results of 6 beams cut in each direction and tested in 3 point bending test, clearly identify that the direction of casting is rather important. The bending strength of the small beams cut in the direction parallel with the openings (i.e. vertical direction when compared to the direction of casting), is rather small (13.3 MPa) and the scatter is large. The beams cut in horizontal direction (with respect to the direction of casting) exhibited higher flexural strength (19.9 MPa). It was interesting, that the beams cut in the direction 45° showed the highest flexural strength (22.7 MPa) as well as the smallest scatter.

Figure 5. Small beams cut from the hollow core slab – 3 point bending test.

Table 2. Flexural strength of the small beams.

Spec. No.	Flexural strength [MPa]		
	Parallel	Perpendicular	Inclined 45°
1	18.6	22.6	20.9
2	9.4	17.9	24.8
3	16.2	15.8	24.5
4	18.0	25.9	21.6
5	8.6	19.6	26.5
6	8.8	17.6	17.8
Mean value	**13.28**	**19.90**	**22.68**
Relative mean value [%]	**66**	**100**	**114**
Coeff. of variation [%]	**36**	**19**	**14**

4. Tests on thin slabs

Small slabs only 30 mm thick were cast in different positions, in a horizontal position and in a vertical position. The slabs had dimensions 800 x 800 mm. Some slabs were cut into strips 800 x 200 mm and tested in 4 point bending. Other slabs were simply supported along the perimeter and loaded by a point load in the middle of the slab.

4.1. Strips from slabs

At the horizontally cast slabs, the direction of sawing was not specified. The slabs which were cast in vertical position were cut into strips either vertically or horizontally. The strips were subjected to 4 point bending test. The span of the strips was 730 mm (Figure 6).

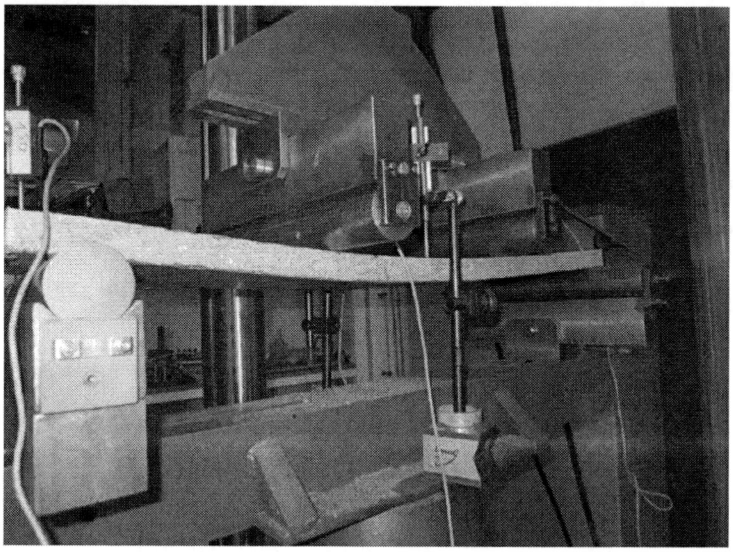

Figure 6. 4-point bending test of the slab strip.

Table 3. Load carrying capacity of the strips [kN].

Strip no.	Horizontally cast slab	Vertically cast slab	
		Horizontal strips	Vertical strips
1	2	3	4
1	3.73	3.57	2.75
2	4.06	4.05	2.36
3	4.01	4.99	2.56
4	3.75	4.62	2.43
5	4.95	---	---
6	5.11	---	---
Mean value	**4.27**	**4.31**	**2.53**
Relative mean value [%]	**100**	**101**	**59**
Coeff. of variation [%]	**14.2**	**14.6**	**6.8**

The tests were controlled by the deflection at the midspan. The ultimate load carrying capacity of the individual strips is summarized in the Table 3. The results completely proved that the direction of casting is important. The fibres tend to get into the horizontal position in all arrangements. The strips from horizontally cast slabs and horizontal strips from vertically cast slabs exhibited very similar results. The fibres seem to be in the direction of the span and they are able to carry the horizontal forces and thus contribute to the resistance of the sections subjected to bending. On the other hand, at the vertically cast slabs only smaller amount of fibres remained in vertical position, which was in the direction of the span of vertical strips. It may be a reason why the load carrying capacity of vertical strips (column 4 in the Table 3) is smaller than that of other strips. However, it is rather strange, that the coefficient of variation is significantly smaller than that at horizontally cast slabs.

5. Conclusions

The casting technology has an effect on the fibre distribution in structural elements. The fibres tend to occupy rather the horizontal position than the vertical position related to the casting direction. This phenomenon was observed on different elements. The experiments showed that in vertical direction the flexural strength can drop to 60% of that in horizontal direction (with respect to the pouring of concrete). In case of the hollow core slab the highest strength was achieved in the direction of 45° (with respect to the pouring of concrete). If the effect of orientation was investigated on standard beams, the effect of fibre distribution and orientation was significantly smaller. The fibres are oriented specifically in dependence on the pouring of concrete and on the shape of the structural element. It can be recommended to verify the fibre distribution and orientation on actual structural elements, since the laboratory specimens or similar elements need not provide a realistic figure on the fibre efficiency.

Acknowledgements

The supports of the Technological Agency of the Czech Republic (project No. TE 01020168), of the Grant Agency of the Czech Republic (project No. 16-04454S) and of the Ministry of Industry (project no. FR TI3/531) are gratefully acknowledged.

References

[1] Coufal R, Vítek J L and Kalný M 2016 The first large application of UHPC in the Czech Republic *First International Interactive Symposium on UHPC -2016 (Des Moines)*

[2] Vítek J L, Coufal R and Čítek D 2013 UHPC - Development and Testing on Structural Elements *Concrete and Concrete Structures 2013 (Žilina)* (University of Žilina) pp. 218-23

IOP Conf. Series: Materials Science and Engineering **236** (2017) 012055 doi:10.1088/1757-899X/236/1/012055

Lightweight HPC beam OMEGA

Michal Sýkora[1], Petr Jedlinský[1] and Jan Komanec[2]

[1] EUROVIA Services, s.r.o., Národní 138/10, 110 00 Praha 1, Czech Republic
[2] Pontex spol. s r.o., Bezová 1658, 147 14 Praha 4, Czech Republic

E-mail: petr.jedlinsky@eurovia.cz

Abstract. In the design and construction of precast bridge structures, a general goal is to achieve the maximum possible span length. Often, the weight of individual beams makes them difficult to handle, which may be a limiting factor in achieving the desired span. The design of the OMEGA beam aims to solve a part of these problems. It is a thin-walled shell made of prestressed high-performance concrete (HPC) in the shape of inverted Ω character. The concrete shell with prestressed strands is fitted with a non-stressed tendon already in the casting yard and is more easily transported and installed on the site. The shells are subsequently completed with mild steel reinforcement and cores are cast in situ together with the deck. The OMEGA beams can also be used as an alternative to steel - concrete composite bridges. Due to the higher production complexity, OMEGA beam can hardly substitute conventional prestressed beams like T or PETRA completely, but it can be a useful alternative for specific construction needs.

1. Context – What was the problem to be solved?

Making ever longer spans of bridge superstructures from precast beams is a common practice. One major problem is the growing weight of the beams which complicates both transport and assembly on site. This problem is even bigger in case of difficult terrain on site and/or when assembly is done from greater distances, e.g. with the crane placed behind the abutment.

The proposed solution is based on an extensive use of high performance concrete (HPC), decreasing the beam weight very significantly. It not only helps to solve the usual problem of excessive weight of beams but has also a potential for improvements in overall economics of the structure, both during construction and long term maintenance costs. For example, transport can be arranged for pairs of beams on one truck, reducing transport costs and CO_2 emissions; service life can be extended thanks to increased durability of the material [1] etc.

2. What solution did we find?

The proposed "OMEGA" beam solves the problem of beam weight quite effectively. The cross section is a thin walled shell trough made of pre-tensioned pre-stressed high performance concrete in the shape of inverted letter Ω (Figure 1). This concrete shell has also all ducts and tendons for post-tensioning installed in the precast yard. After the transport and assembly of beams on site, the mild steel reinforcement is placed and the core of the beams as well as the composite deck with optional end crossbeams is cast in situ.

The "OMEGA" beam can be effectively used also as an alternative to the steel beams used in standard steel-concrete composite bridges.

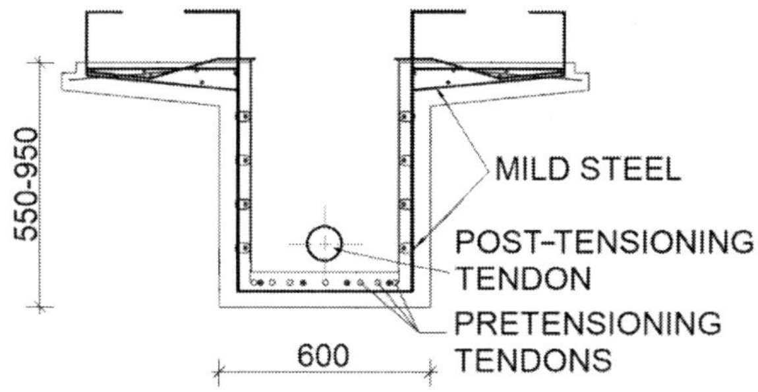

Figure 1. "OMEGA" beam – typical cross section.

3. Production of the beams

At first, the excessive weight of currently used precast beams was defined as the main problem to be solved. The target parameters were defined along with corresponding main dimensions of the new beam type, taking into account the practical technological possibilities in contemporary precast yards.
The next step was the static analysis of proposed beams in all stages of the construction and service life according to the applicable codes. Also, a design of formwork for the first testing beams was prepared by formwork contractor Doka (Figure 2).

Figure 2. Formwork of the beam.

Figure 3. Test of the transport equipment cast into the surface of the beams.

Figure 4. Detail of the beam end region.

Figure 5. Omega beam – general view.

The first beam cast served to verify the technological methods used, including the transport equipment (Figure 3).

In the next phase, a group of four beams was cast (Figures 4, 5) and a set of loading tests was conducted on a selected beam.

4. Loading tests

The results of the static load-bearing test of a selected beam showed very good correlation with the calculated values (Figures 6, 7 and Table 1).

The basic requirement, i.e. no cracks while casting the composite 'in-situ' part of cross section, was also perfectly met, no temporary supports are needed on site. The total deflection during the static load test was measured together with other parameters.

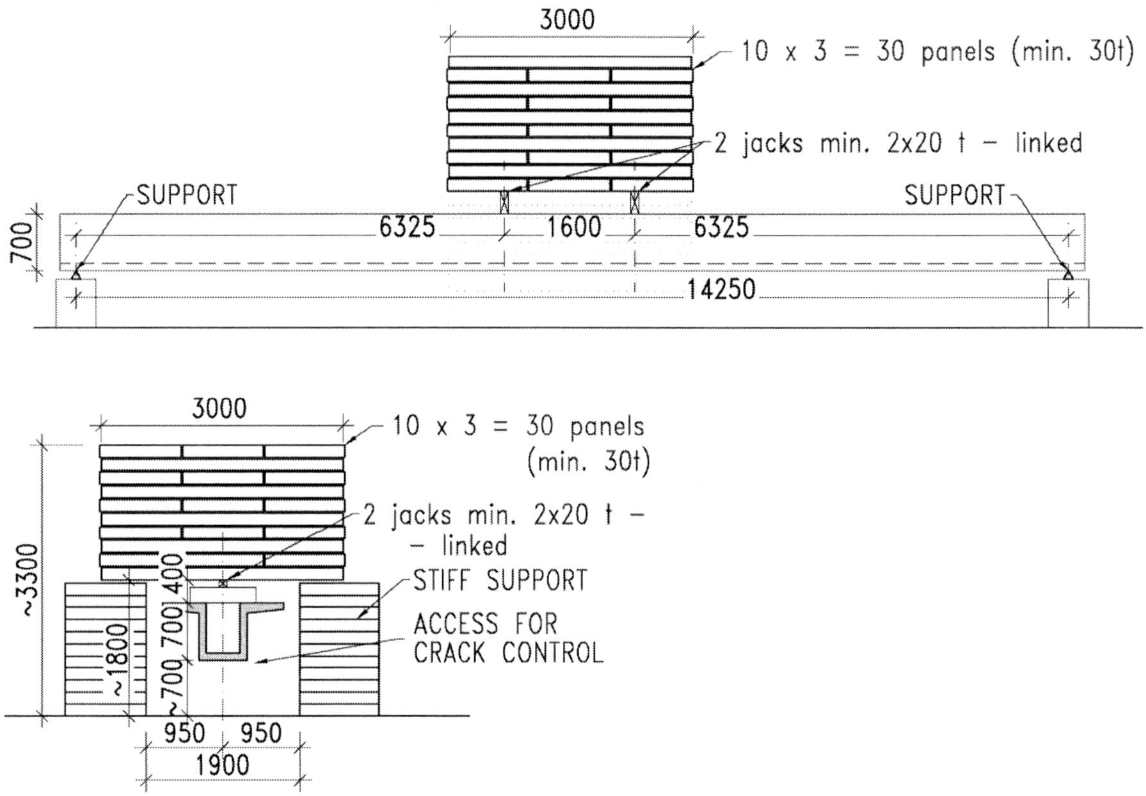

Figure 6. Loading test, general arrangement.

Table 1. Estimated deflections during the experiment.

Load step	Load [%]	Force [kN]	Deflection [mm]
Theoretical ULS	121	2x 172	-
Test ended	100	2x 142	~ 40
Concrete tensile strength	78	2x 111	24
Decompression	63	2x 89	19
Core and deck cast	60	2x 85	16

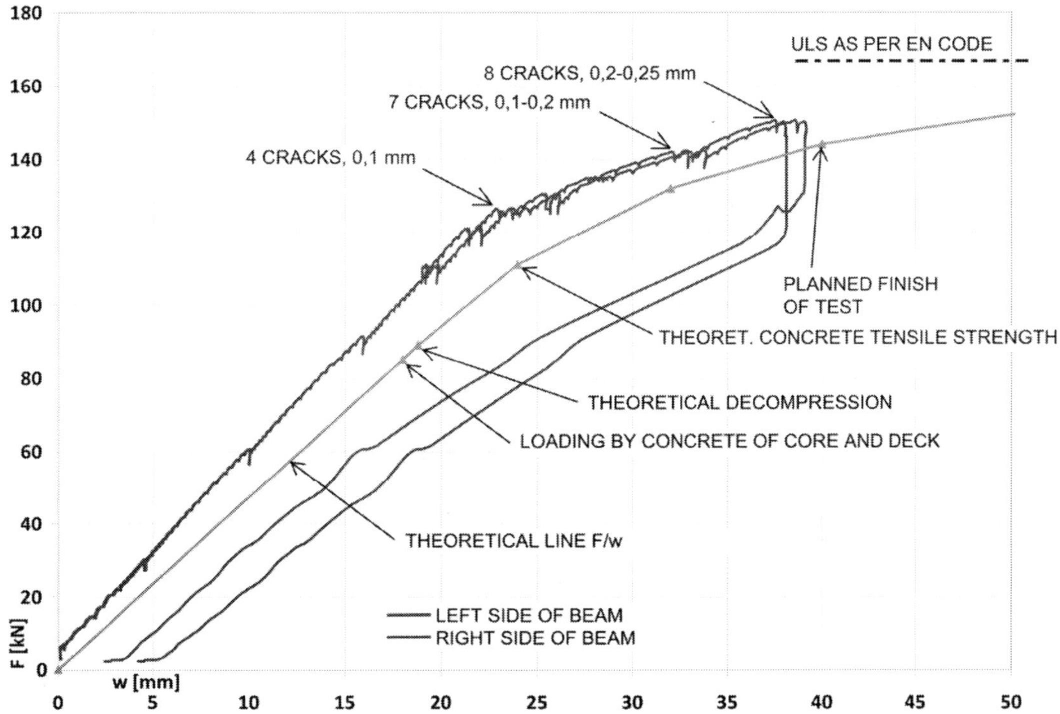

Figure 7. Loading test, theoretical and measured deflections during test.

A test of the bond between the precast part of the cross section and the subsequently cast in situ core of the beam was carried out (Figures 8, 9, 10). The core on the left side (the side of active anchor) was separated from the composite deck to eliminate the effect of the deck. The prestressing tendon consisted of 19 strands Y1860-S7-15.7.

In the gap between both cores a set of displacement sensors was installed for measuring a possible movement of the core relative to the shell. The force in the tendon was gradually increased in steps of 400 kN with a time step of 1 minute. The test was finished after the maximum force 4.1 MN was reached. The results of the test showed that only elastic deformations occurred (Figures 11, 12) and sufficient bond was achieved.

Figure 8. Test of bond.

Figure 9. Test of bond, displacement sensors (LVDT displacement sensors).

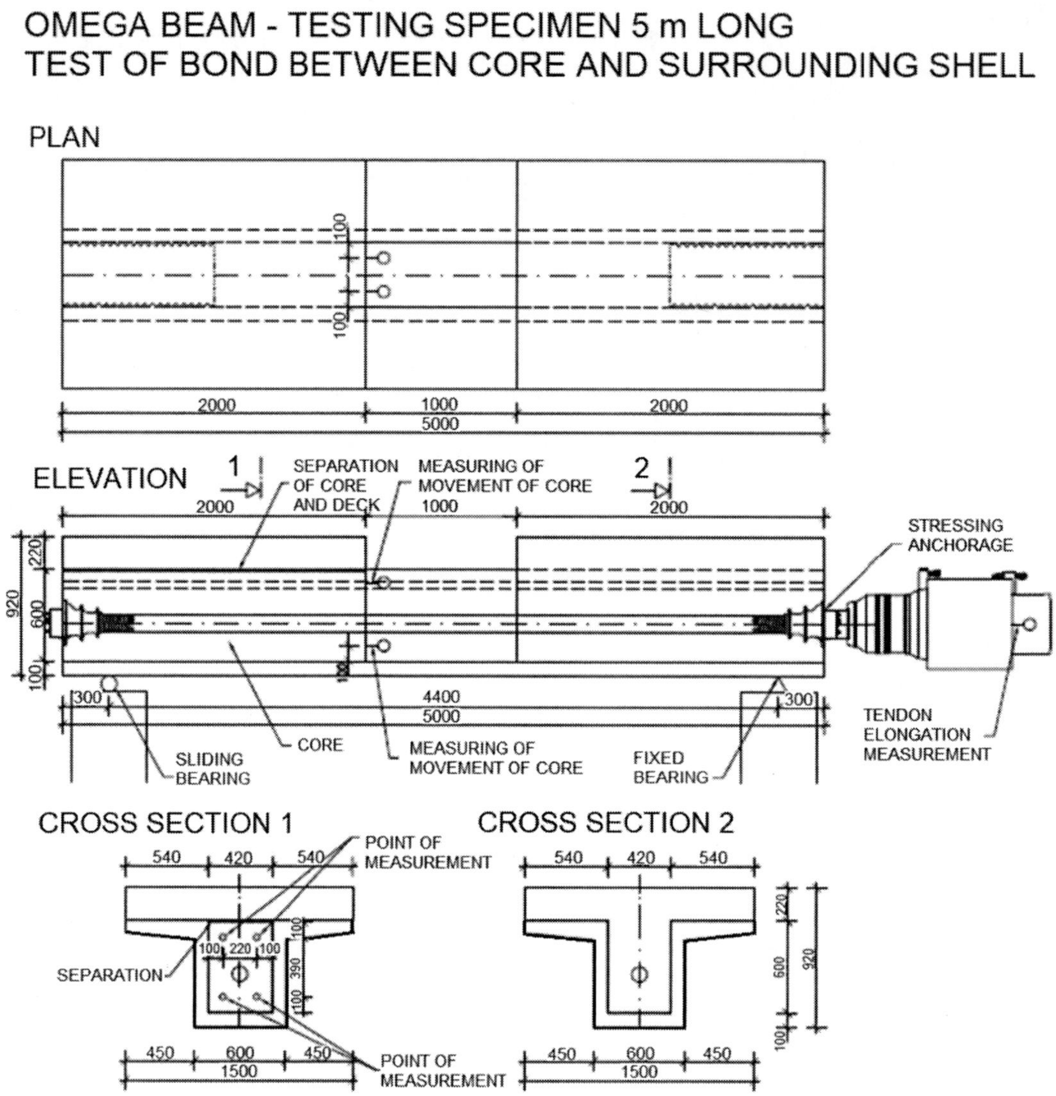

Figure 10. Test of bond between core and shell – general arrangement.

The resulting product meets all requirements defined at the first stage. Therefore, the development of the new type of beam is considered finished. The idea of a ligtweight precast concrete beam was fulfilled and the design is currently protected as a registered design no. 29825 at the Czech Office of Industrial Intellectual Property. A registration of invention was also filed under no. PV 2016-410.

The article of Petr Bílý from ČVUT (Czech Technical University, Prague) called "FEM simulation of static loading test of the Omega beam" [2] presents the numerical simulation of static loading test of the beam.

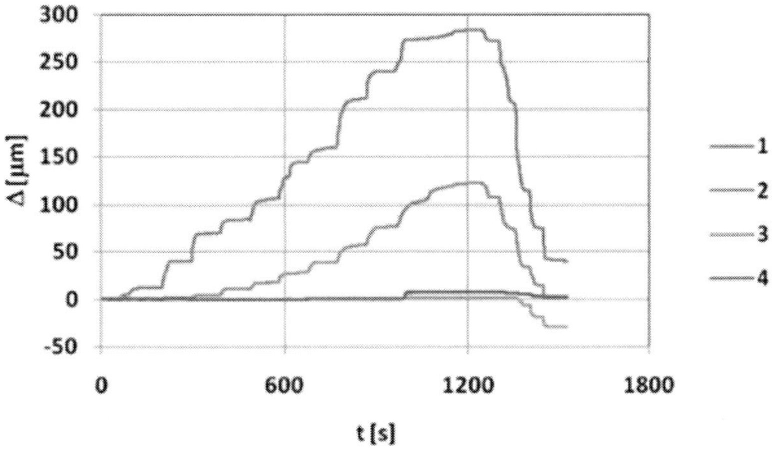

Figure 11. Test of bond – displacement measured on particular sensors. 1: Bottom, right; 2 – bottom, left; 3 – top, right; 4 – top, left.

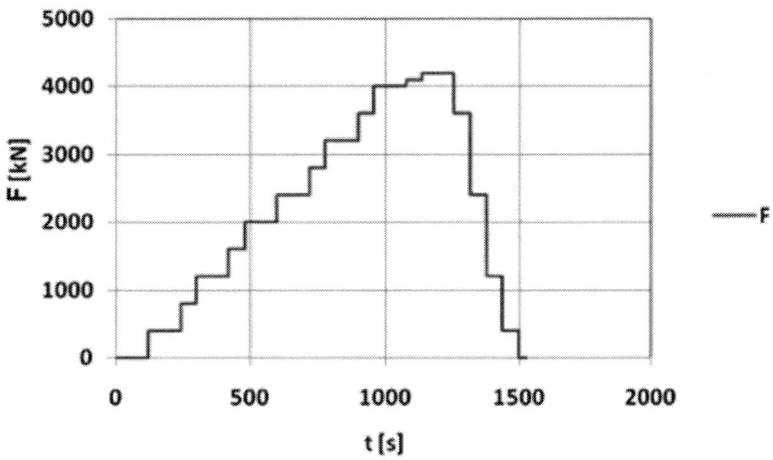

Figure 12. Test of bond – prestressing force applied.

5. Wider adoption of the innovation

The described OMEGA beams will be seen first in the situ use on the project of a utility bridge in a quarry of the Eurovia kamenolomy company located at the village Třebnůška in the Czech Republic. Two of the beams will be equipped with strain gauges which will allow further investigation and measurements to be done both during the construction period and in service.

Assembly of 4 OMEGA beams was carried out on above mentioned bridge (Figures 13, 14) on 2.3.2017. Typical cost comparison is shown in Tables 2 and 3.

Table 2. Comparison of weights for various dimensions of beams.

Beam depth	[mm]	550	650	750	850
Span	[m]	12	15	18	21
T – beam	[t]	14.7	20.7	27.6	35.5
Omega beam	[t]	6.4	8.6	11.0	13.7

Table 3. Comparison of cranes required for various beams.

Single span	Ω15	T15	Ω21	T21	Ω24	T24
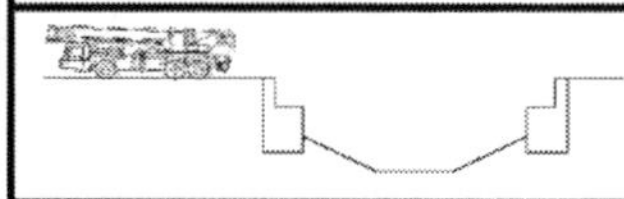	LTM 1055	LTM 1200	LTM 1100	LTM 1200	LTM 1200	LTM 1300
	810 EUR	3890 EUR	1670 EUR	3890 EUR	3890 EUR	5470 EUR
2 span	**2 x Ω21**	**2 x T21**	**2 x Ω27**	**2 x T27**		
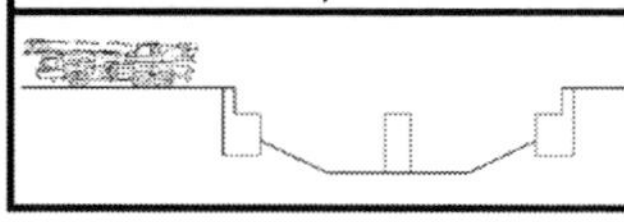	LTM 1200	LTM 1500	LTM 1500	LTM 1500*	* Including 2 assembly positions	
	5520 EUR	7470 EUR	9260 EUR	11260 EUR		

The cost comparison is done on a fictitious site with travel distance 50 km. For 2xT27 T beams it is necessary to move the crane behind the other abutment due to beam weight, thus further increasing the costs. The cost estimate for crane transfer behind the other abutment is EUR 3.000. For conversion, the ratio 27 CZK/EUR is used. Further savings could be achieved by transport of 2 OMEGA beams on one truck.

Figure 13. OMEGA beams used on site.

Figure 14. Finished bridge in Třebnůška Quarry.

6. Conclusions

The newly proposed shell beam, able to withstand all loads during construction, offers some financial savings in transport and construction costs. Feasibility of the design was verified on a real bridge in Třebnůška, Czechia, finished in May 2017 (Figures 13 and 14).

Acknowledgements

This paper was prepared thanks to the support of the Technology Agency of the Czech Republic (TAČR), project "Centre for Effective and Sustainable Transport Infrastructure (CESTI)" (no. TE01020168).

References

[1] Fládr J, Bílý P and Vodička J 2016 Experimental Testing of Resistance of Ultra-high Performance Concrete to Environmental Loads *Proceedia Engineering* **151** pp 170-176

[2] Bílý P, Kohoutková A and Jedlinský P 2017 FEM simulation of static loading test of the Omega beam *IOP Conference Series: Materials Science and Engineering (MSE) – Proceedings of BESTInfra 2017 conference* (Prague, Czech Republic) to be published

Tresfjord Bridge – a human friendly and traffic efficient structure

Kristian B Dahl[1], **Aja Anta Magerøy Tønnessen**[1] **and Lars I Toverud**[1]
[1]Multiconsult ASA, Nedre Skøyen vei 2, N-0213 Oslo, Norway

E-mail: lars.toverud@multiconsult.no

Abstract. The E136 Tresfjord Bridge opened in October 2015, and crosses the Tresfjorden on the west coast of Norway. It is a concrete bridge with a total length of 1290 m, consisting of 19 viaduct spans, 60 m each, and a FCM (free cantilever method) main span of 160 m. The E136 is one of the most important transportation routes in the county of Møre and Romsdal and starts in Ålesund, and passes along Tresfjorden to Åndalsnes. The existing road is very narrow with speed limit of 60 km/h and characterizes by many accidents involving cars and people. The traffic flow is approximately ca 2500 vehicles a day, of this is 25% heavy vehicles. Those transport fresh salmon from the breeders in the fjords along the coast. To try to decrease the transportation time is very important for the fresh salmon. The bridge reduces the distance between Ålesund and Åndalsnes by 13 km. The speed limit is now 80 km/h, and with much less risk for accidents since there are separate lanes for cars and pedestrians over the whole bridge. This means that the bridge represents a human friendly and traffic efficient structure to the benefit for the people and the region.

1. Introduction

The Tresfjord Bridge which was opened in October 2015 and crossing Tresfjorden on the north-west coast of Norway, is a concrete bridge with a total length of 1290 m, consisting of 19 spans, 60 m each, and a FCM (free cantilever method) main span of 160 m, see Figure 1. The bridge has a box cross section with a maximum height of 9.2 m and a total width of 13.5 m, with 2 traffic lanes plus 1 pedestrian lane. The navigational channel is 60 m wide and 32 m high.

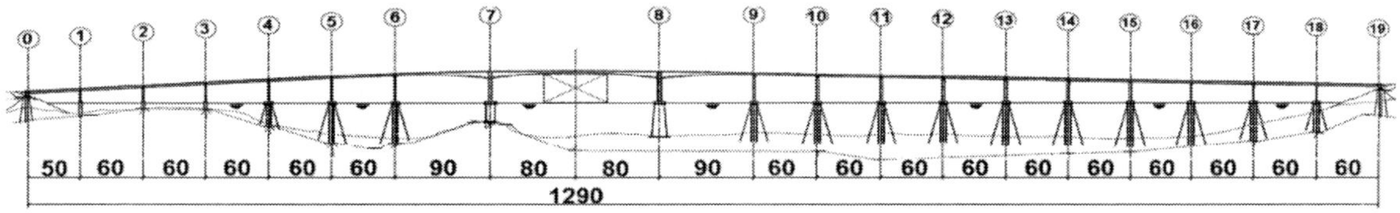

Figure 1. Tresfjord Bridge with total length 1290 m, built 2012-2015 at the west coast of Norway.

IOP Conf. Series: Materials Science and Engineering **236** (2017) 012056 doi:10.1088/1757-899X/236/1/012056

2. Foundations

2.1. Pile foundations

The foundations for the approach spans, a total of 13, are founded on pile groups of steel pipe piles ø1220, reinforced and concreted. Each group consists of 12-18 piles, typical 16 piles, see Figure 2. The maximum depth in the fjord is approximately 40 m. The piles are driven 5-6 m down in the very stiff moraine. The pile driving equipment was fixed to a steel jacket, supported by 4 steel legs standing on the seabed, thus making a very stable working platform for the pile driving. The piles were driven at an elevation 5:1. When the driving were completed the piles were fixed to a steel frame and cut at elevation +0.50. A bottom formwork was installed and a 400 mm thick concrete slab was concreted. This slab was the support for a precast concrete box which acted as a formwork for the pile cap. When installed the piles were reinforced with prefabricated reinforcement cages and concreted. There after the pile caps were reinforced and concreted.

Figure 2. Steel piles Ø1200 (left), and precast walls for pilecap installed by floating crane.

2.2. Caissons

The main span is founded on two caissons, one on bedrock, Axis 7, and one on soil seabed, consisting of dense, ground moraine, Axis 8.

2.2.1 Caissons in axis 7. The Caisson in Axis 7 is placed directly on rock at elevation -21. After underwater blasting the rock surface has been leveled by using AUV-concrete. Upon the concrete slab 3 circular foundations ø=2000 are established. The caisson is shaped like a cylinder with diameter 9.0 m and a wall thickness 0.45 m. The fender slab in top of the caisson has an outside diameter of 12 m and a height of 3.5 m. The total height is 21.5 m, see Figure 3.The caisson is constructed on a barge anchored at a quay not far from the site. After reaching a certain height, the barge was lowered until the caisson was floating. The caisson was completed up to the top when submerged in the water. The finished structure was then towed to the final position and lowered on the pre-prepared foundations by ballasting with water. The opening between the bottom of the caisson and the concrete slab was filled with concrete through pre-installed pipes. The caisson was then filled up with water saturated gravel before casting of the fender slab. 8 vertical steel pipes ø200 was embedded in the wall. Through these pipes holes were drilled 14 m down in rock in order to establish anchor for the rock anchors, 12ø0.6.

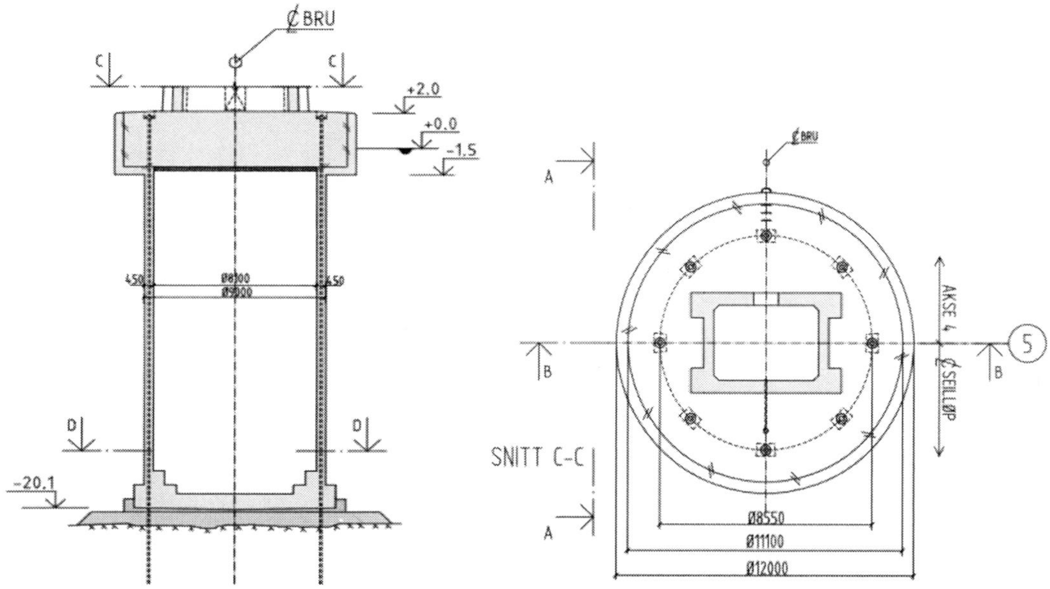

Figure 3. Caisson axis 7.

2.2.2 Caissons in axis 8. The deepest foundation, see Figure 4, in the FCM (free cantilever method) main span has a water depth of approximately 38 m. The thickness of the soil is approx. 20 m down to rock. Since the soil is very dense, it is impossible to excavate down to rock. Thus, the foundation is placed on the seabed after excavated approx. 2 m. In Norway, this is the first large caisson founded directly on soil seabed. The caisson resembles a GBS (gravity-based structure) similar to offshore structures. This, in conjunction with its size and general complexity, required a comprehensive design procedure.

The bottom slab consists of posttensioned lightweight aggregate concrete and has a diameter of approximately 24 m, whereas the shaft diameter ranges from 9 -15 m. The fender slab has a diameter of 12 m and a thickness of 3 m, constituting a monolithic base for the bridge pier. The caisson is filled with water saturated gravel. The weight of the structure is 3900 tonnes. It was designed with the in-house code MultiCon which is a postprocessor specifically developed for large concrete offshore structures. The construction took place on a barge at a quay a short distance from the bridge site, see Figure 5. When a specified part of the caisson was finished, the barge were lowered until the caisson was floating, and then the rest of the caisson was constructed. The caisson was 40 m from top to bottom. During the construction period it was stabilized with water, then afterwards it was towed to the site and placed into position with only 65 cm deviation from the correct position. On beforehand the seabed had been excavated in order to remove loose material and to be sure that the foundation surface was stiff. After excavating the top layer, a steel frame with 3 supports was installed and levelled, one of the supports were entrusted with guide in order to facilitate the positioning of caisson. The work on the bottom had to be performed by divers, the allowable working time at that depth was 30 min, and then it was back in the compression chamber for 12 h. Hence, it was a very time-consuming work. When in position on site the fender slab was reinforced and casted.

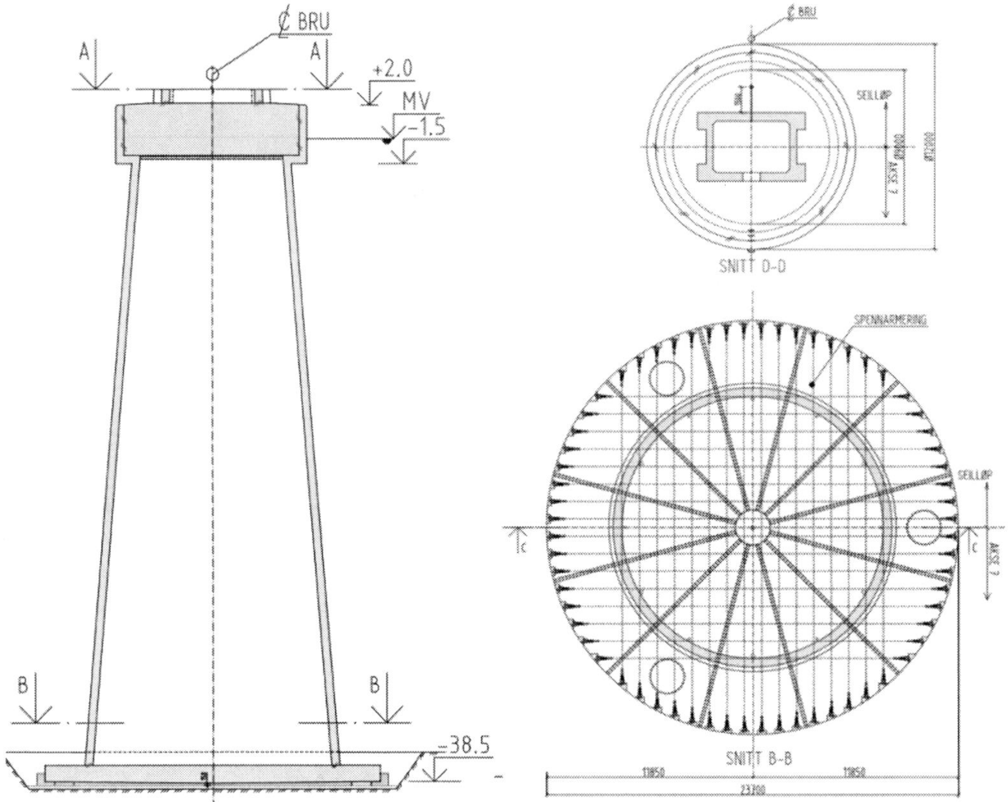

Figure 4. Caisson axis 8.

Figure 5. Caissons constructed at barge by quay.

3. Approach spans

The superstructure is a single box section with total width 13.5 m, web spacing 5.4 m, girder depths constant 3.0 m, span lengths 60 m. It was a span by span construction with a underslung gantry which used approx. 3 weeks on each span, see Figure 6. In each web it is 6 tendons 18ø0.62 and additional 6 tendons 31ø0.62 at each support. The superstructure is supported by rectangular shaped columns 1.6x5.2 m and 2.4x5.2 m respectively and height from 11 to 29 m.

Figure 6. Approach spans with MSS.

4. Main Bridge

The main bridge was constructed by the balanced cantilever method. The depths of section is varying from 3.0 m to 9.2 m for the main bridge. The thickness of the bottom slab varies from 0.35 m to 0.80 m, thickness of the webs from 0.35 m to 0.40 m. Form travelers were used to construct the free cantilevers, and typical section has a length of 5 m, produced on a weekly sequence, see Figure 7. In the deck slab it is maximum 52 tendons 19ø0.62, and in the bottom slab in center span it is 30 tendons 18ø0.62. Piers with box section 4.0 x 5.4 m and wall thickness 0.5 m support the superstructure. The piers are constructed with a climbing scaffolding system.

Figure 7. Balanced cantilever method.

5. Environmental effects

Route E136 is one of the most important transportation routes in the county of Møre and Romsdal. It starts in Aalesund, passes by Vestnes, go into the bottom of Tresfjorden, then out to Vikebukt and further on to Åndalsnes, through the valley of Romsdal before heading on to Oslo and Europe. The existing road along the fjord is characterized by a lot of spread housing alongside the road, many accidents involving cars and people, speed limit of 60 and 70 km/h, and is a very narrow road with a width in some places of 5.8 m. There are also several places along the existing road along the fjord where there are avalanches, with downfall of stones. The traffic flow is approx. ca 2500 ÅDT, of this is 25% heavy vehicles. The heavy vehicles transport mostly furniture from the factories around Aalesund and fresh salmon from the breeders in the fjords along the coast. These items are transported directly to the markets in Europe. It is very important to try to shorten the transportation in time, especially for the fresh salmon. The bridge shortens the distance between Aalesund and Åndalsnes by 13 km, while the distance between Vikebukt and Vestnes will be reduced by 22 km. The speed limit will be 80 km/h, instead of 60 km/h -/70 km/h, and with much less risk for accidents since there are separate lanes for cars and pedestrians over the whole bridge. All this means that the bridge represents a human friendly and traffic efficient structure to the benefit for the people and the region.

Figure 8: Completion of the key element in midspan.

6. Cost benefits

Total cost for the bridge inclusive roads is 800 mill NOK (100 mill EURO (2010)). Consequences (for a period of 25 years):

- Reduced travel time 22 min
- Net value for society +490 mill NOK
- Reduced transportation cost for private sector 1230 mill NOK
- Reduced transportation cost for business sector 640 mill NOK
- Change in CO2-utslipp -4000 ton pr year
- Reduced accidents cost 80 mill NOK each year
- Reduced dead rate from car accidents 0.12 pr year

7. Conclusion

The completed Tresfjord Bridge, shown in Figure 9, reduces the distance between the two local towns Ålesund and Åndalsnes by 13 km, and offers low risk for accidents since there are separate lanes for cars and pedestrians. Moreover, the speed limit is increased. All in all, the bridge represents a human friendly and traffic efficient structure to the benefit for the people and the region.

Figure 9. The completed Tresfjord Bridge.

References

[1] Dahl K B, Toverud L I and Brekke D E 2015 Tresfjord Bridge – foundation of main span on 40m caisson on soil seabed *Multi Span Large Bridges* **2015** (CRC Press, Taylor & Francis Group) pp 283-84

Strengthening of bridges by post-tensioning using monostrands in substituted cable ducts

Ladislav Klusáček[1] and Adam Svoboda[1]

[1]Institute of Concrete and Masonry Structures, Faculty of Civil Engineering, Brno University of Technology, Veveří 331/95, Brno 602 00, Czech Republic

E-mail: klusacek.l@fce.vutbr.cz

Abstract. Post-tensioning is suitable, reliable and durable method of strengthening existing engineering structures, especially bridges. The high efficiency of post-tensioning can be seen in many applications throughout the world. In this paper the method is extended by a structural system of substituted cable ducts, which allows for significantly widening application of prestressing so it's convenient mostly for application on beam bridges or slab bridges (built in years 1920 – 1960). The method of substituted cable ducts is based on theoretical knowledge and technical procedures, which were made possible through the development in prestressing systems, particularly the development of prestressing tendons (monostrands) and encased anchorages, as well as progress in drilling technology. This technique is highly recommended due to minimization of interventions into the constructions, unseen method of cable arrangement and hence the absence of impact on appearance, which is appreciated not only in case of valuable historical structures but also in general. It is possible to summarise that post-tensioning by monostrands in substituted cable ducts is a highly effective method of strengthening existing bridges in order to increase their load capacities in terms of current traffic load and to extend their service life.

1. Introduction

When strengthening by post-tensioning, the prestressing tendons must be placed inside a cross section of a bridge structure. It is possible to use the free space between beams in case of monolithic reinforced concrete beam bridges built between 1920 and 1960. U-shaped beam (parapet) bridges do not provide such an option and there is not any free space in case of slab bridges as well. However, there is still a large amount of these structures in the Czech Republic, around 6000 according to databank data.

The possibility of placing prestressing tendons to substituted cable ducts is a construction method based on additional drilling of cable ducts into the concrete of existing bridges. It is then possible to place the prestressing system tendons both into the existing bridge slabs and the existing parapet beam bridges. This construction method is very suitable for beam systems of bridges wherein part of the post-tensioning is placed into concrete beams and part into the free space between them using the substituted cable ducts.

The following article text describes this method in detail and includes examples of successful application.

2. Substituted cable ducts, saddles and anchorage areas

2.1. Implementation of substituted cable ducts

The substituted cable ducts can be applied by construction drilling. It is possible to create very long cavities of relatively small diameters (40–52 mm) and use them for the insertion of prestressing tendons and strands.

When creating the substituted cable ducts in the concrete of existing bridge, it is a drilling of long drilling holes of appropriate diameters suitable for passing the prestressing tendons assembled from either individual strands (separate monostrands) or tendons made of 3, 4 or 6 monostrands. In terms of the existing technology, diamond and hammer drilling using carbide drill bits can be used for such drilling.

Diamond drilling technologies are based on grinding the material by appropriately shaped grinding alloy cutting edges. Industrial diamonds in the form of particles of a few tenths of a millimeter are dispersed in light metal alloy. The drilling segments of a diamond drill are shown in Figure 1. Drills for this technology are made in the shape of hollow metal cylinders; they are fitted with diamond segments on the side intended for drilling in the structure (see Figure 1b). When drilling, a hollow drill is pushed towards a structure with a relatively large force exerted by an auxiliary stand, the segments graze the structure material and the exposed diamond particles grind the drilled material. Drive units are electric or hydraulic. The power consumption of electric drive units are between 1.8 kW up to 4 kW, i.e. small input powers common for small construction site machinery. The drills are supplied in lengths from 300 to 500 mm; the drills of up to 1000 mm can be produced as special order.

Figure 1a. Drilling segments of a diamond drill bit.

Figure 1b. Diamond drill bit.

Hammer drilling technologies with solid carbide drill bits are modifications of classical drilling technology with helix drill bits. Augers are equipped with carbide cutting edges – appropriately shaped and grinded plates from sintered carbides. Figure 2a) shows the drill bit fitted with four cutting edges. Figure 2b) shows a typical carbide spiral drill bit. Small-mechanization machines are designed to drill holes for installations, openings, wall plugs, etc. They are intended for manual handling with no exceptions, which means that workers hold a machine tool – drill in their hands and drill holes without any special demands on the accuracy of drilling trajectory. Hammer drilling technology drive units are mainly electric, exceptionally hydraulic. The power consumption of electric drive units lays between 0.8 kW up to 2.0 kW, i.e. small input powers again, common for small construction site machinery. In terms of power, they are classified by the energy of one impact which is measured in Joules. Impact energy ranges from 4 to 20 J. The augers are supplied in lengths from 150 to 800 mm, in diameters from 8 to 40 mm; it is possible to have drill extension shafts without a spiral of up to 1500 mm produced as a special order.

Figure 2a. Carbide drill bit with four cutting edges.

Figure 2b. Carbide drill bit.

The application of substitute cable ducts requires drilling holes into the concrete of bridge construction in significantly longer lengths than those made by serially supplied equipment and tools. Special positioning drilling equipment – drill support [1] is used for this purpose. Its main parts are rails and drill carriage; see Figure 3. The carriage allows defined fixing of drilling machinery, power transmission for exerting the necessary thrust while drilling. The guide rods along with the carriage allow the extension of drilling shafts. The purpose of the equipment is a guidance of drilling machine, a reduction in drilling machinery operator's fatigue to an acceptable level and a maximum achievable increase in the accuracy of the trajectory of drilled cable ducts with the possibility of setting any trajectory in a space by individually adjustable vertical angle and horizontal angle (usually an angle to the longitudinal axis of supporting structure). The drill support can be coaxially fitted by a diamond and hammer drilling technology which may even be interchanged while applying one cable duct. It can be mounted advantageously to the rehabilitated structure itself or to the auxiliary steel structure. Figure 4 shows deployment of the drill support with diamond drilling technology; Figure 5 shows the location of individual monostrands in the made duct. The cable ducts of 52 mm in diameter and at a length of 6 m, exceptionally even 10 m, can be normally drilled in this way.

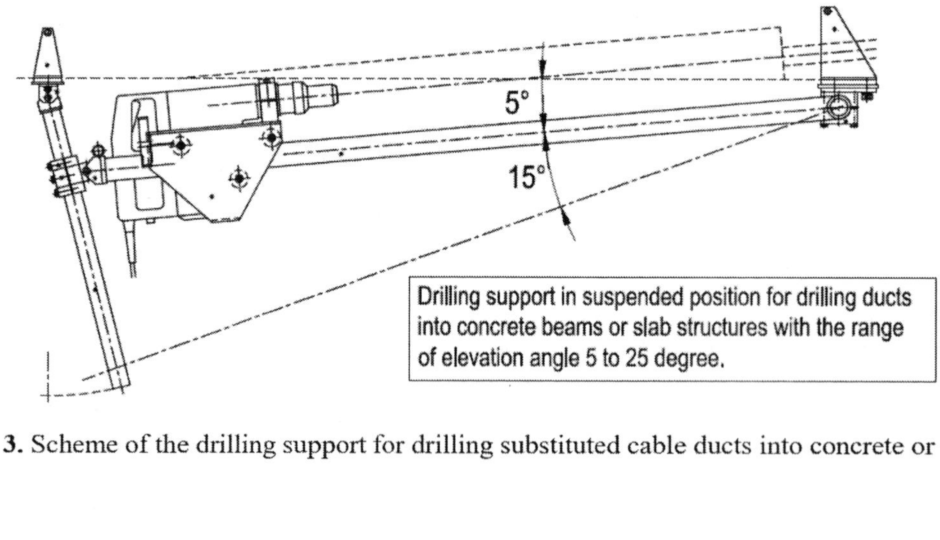

Figure 3. Scheme of the drilling support for drilling substituted cable ducts into concrete or masonry.

When using carbide impact drilling technology for drilling cable ducts, the extension System 6 by Joran is successfully used. HILTI or MILWAUKEE currently uses more precise extension systems. Extension rods have a length of 500 mm or 1000 mm, couplers with patented lock which transfers the impact energy almost without losses from the drill to the drill bit are used for connection. It does not matter how many couplers are used, the system operates with acceptable productivity up to the lengths of about 4 m; the optimal use is for ducts of up to 2.5 m in length. Unit price per 1 m' of drilled duct using the carbide hammer drilling technology is significantly lower than using the diamond technology; it indicatively ranges from 30 to 50% of the diamond technology price. It is due to less wear and lower entry price of impact tools and their much wider application in other construction sectors.

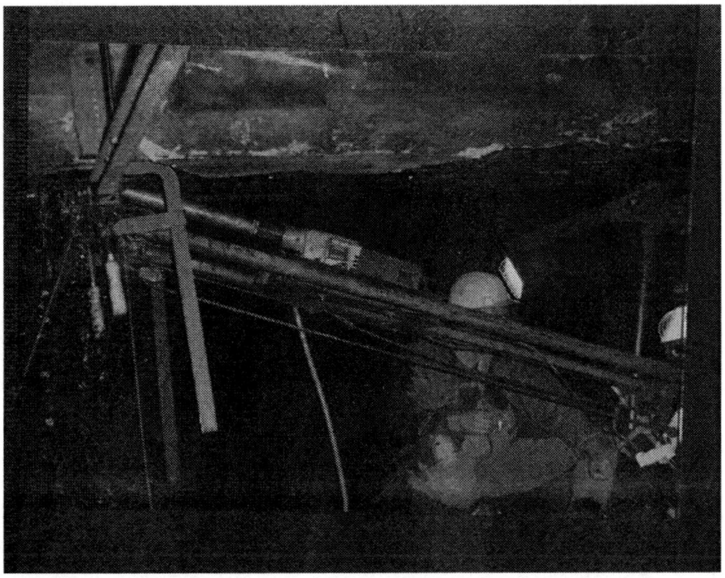

Figure 4. Drilling of substituted cable duct in parapet bridge beam using drilling support.

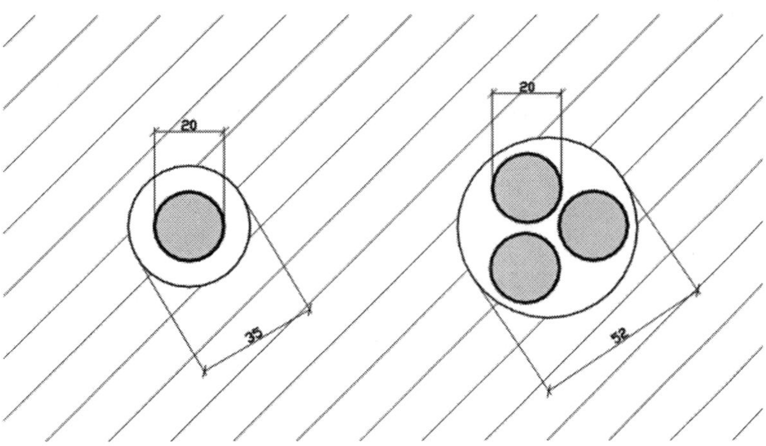

Figure 5. Geometry of substituted cable ducts for single monostrand and tendon with 3 monostrands.

The drill support can be fitted in any position for drilling ducts. The article authors elaborated and verified the preferred method of fitting from the bottom to a structure when drilling cable ducts in reinforced bridge structures – beams and slabs. It is generally possible to perform oblique cable ducts top-down or bottom-up in these structures. The gaps between the existing concrete reinforcement are used to guide the monostrands (Figure 6). During the top-down process, it would be necessary to lead the drilled duct accurately to 'fit' into the gap between the reinforcement at the bottom structure surface; particularly in case of the rehabilitated reinforced concrete beams, it is regarded almost an impossible task demanding on the precise geodetic survey of the drilling hole axis due to local irregularities in the structure. In contrast, with the bottom-up process, it is possible to find the suitable gaps in advance, to prepare a drilling start as advantageous as possible, to correct plan projection of the cable axis according to local conditions. Concrete reinforcement is usually placed relatively sparsely near the upper surface (in bridge deck) and the problem of 'fitting' into the gap is almost eliminated. A significant advantage associated with the drilling from the bottom approaches this: gravity facilitates ejection of drilled material from the hole which remains permanently clean while drilling. Similarly, flushing water flows away by gravity. However, the decisive advantage of this process is its independence from the traffic restriction on the rehabilitated bridge structure. If the drilling of cable ducts is done from the bottom, most preparatory works for the rehabilitation by tensioning may be carried out with ongoing traffic, thus significantly shortening the necessary time of lockout the bridge. From experience, investors strongly prefer this variant wherever they deal with the strengthening of a bridge with ongoing traffic.

Figure 6. Substituted cable ducts in the gaps between existing reinforcement.
Left – possible variants of the directions for single monostrand.
Middle – cable ducts ϕ 52 mm in suitable gap between reinforcement.
Right – prestressing tendons with 3 monostrands in cable ducts.

2.2. Saddles

The saddles (deviators) are auxiliary parts of the structure designed to alter cable trajectory. Their purpose is to enable the change of cable trajectory and transfer the resulting radial forces into the structure while maintaining the standard conditions prescribed for prestressing reinforcement, its cover and concrete in contact with the reinforcement. Cross-section of the monostrand must not lose its circular shape and a primary protection cannot be broken. One of the sources giving instructions in the Czech Republic was the standard [2] requiring the following, inter alia, in Annexe E:

- Transverse stress of monostrand in the saddle should not exceed 600 kN/m' of the strand (due to bending and lateral compression of the prestressing reinforcement to prevent loss of strand profile circularity, thereby increasing the risk of fatigue);
- Lateral stress of monostrand on protective plastic cover should not exceed 100 kN/m' of the strand (due to denting the cover and losing passivation lubricant);

- Monostrand in the saddle should pass through a bent steel tube which should ensure the position of the strand in the saddle. Its radius should be at least 3 m (requirement ensuring the circularity of strand profile).

These requirements are established primarily to ensure the durability of the prestressing reinforcement at the increased stress during the passage through deviator where the effects of local bending, partial stretching and the tension arising therefrom are added to the overall tension in the reinforcement by prestressing forces. However, these requirements (especially the passage through the tube) are formulated rather for the deviators of new structures with free external prestressing tendons, their application for strengthening existing bridges leads to structural complications.

Saddles made directly in concrete or masonry of the reinforced structure are part of the substituted cable duct method. This modified process of making deviators was verified by numerous applications. It is based on the above requirements, although the saddle construction is more adapted to the nature of work while strengthening. The basic idea is to create a deviator directly in the structure concrete to eliminate complicated transformation of radial forces by auxiliary weldments, for example. The saddle is created using the technology of diamond cutting and partial cutting. When creating it, it is always assumed that the corresponding cable duct has already been made. The saddle is done in the following steps:

1. Lining cuts at the sides of saddle. They are carried out manually in planes parallel with the plane defined by a prestressing tendon polygon. The width between the cuts is equal to the width of cable duct, e.g. 40 mm for monostrands, 60–80 mm for 3strand (4strand) tendons composed of monostrands. This deviator shape ensures that the prestressing tendon (cable) cannot slide to another position. Therefore, it fulfils the function of the tube according to [2] Annexe E.
2. Supporting saddle surface. It is done manually by undemanding cutting; material loosened by lining cuts is collected. It is prepared according to a metal template designed by a designer which defines the saddle shape (the shape of a part of a circular arc). Inequalities up to ± 5 mm can be achieved in the deviator shaped in concrete.
3. Alignment of bearing surface. The bearing surface is aligned – smoothed by high-strength thixotropic anchor mortar. Saddle reinforcement is used to check the shape.

Figure 7. Saddle in the concrete of bridge beam with saddle reinforcement.

4. Fitting saddle reinforcement. The aligned bearing surface of the deviator is reinforced by load distribution reinforcement. It is generally sufficient to use structural steel (e. g. S 235). The strap of steel (see Figure 7) has the same width in the concrete as the diameter of cable duct,

the thickness of 5-8 mm. It also replaces the function of the tube in the deviator according to [2] Annexe E particularly by ensuring the smooth shape of the radial bearing surface and distributing theoretically line load of structure concrete by radial forces to the surface load. This solves the contact problem. Smooth bearing surface without any roughness is necessary for long-term protection of the prestressing tendon plastic cover (to avoid cutting through the plastic cover or tube).

5. Reprofiling the saddle area. After prestressing the structure, the areas of saddle are smoothed out (reprofiled). For this purpose, additionally established concrete covering layer for the cables on the bottom face of the bridge beams is used.

The saddles for monostrand and 3strand tendons were developed and experimentally tested in this manner. Designed and implemented radii r_s = 1200–1500 mm met the requirements, buckling of tendons was avoided and plastic monostrand HDPE covers were not damaged. The radii are therefore feasible smaller than the ones for new buildings with free cables required by the standards [2, 4]. Design of the saddle is done by establishing its radius and testing if the allowable radial stress on strand protecting cover is not exceeded. The overall static solution of the strengthened bridge structure is decisive, which determines the angle of cable trajectory behind the saddle relation to its previous direction.

2.3. Anchorage areas

The latest views on the protection of prestressing tendon passing through anchor require hermetically coated anchors, including bearing plates with leak proof connection of the cover of monostrand and anchor. After prestressing, such an anchor is covered with a plastic sheath and the entire space is injected with the same passivation material as was used in the actual tendon.

Monostrand anchor system by DYWIDAG has a planar bearing surface and is suitable for subsequent installation (Figure 8). It consists of a compact anchor in plastic cover, a sleeve which seals the transition from the monostrand to the anchor, and a cover sheath which ensures closure of the self-locking anchoring three-jaw system after anchoring and cutting the single strand off. It can be underlaid with ordinary locksmith-made bearing plates or weldments according to the particular shape and needs. Figure 8 shows examples of use of this system for anchoring tendons made of several monostrands.

Figure 8. Examples of anchoring tendons using encapsulated system DYWIDAG.
Left – Anchoring of 3strand tendon with tetrahedron load distribution weldment.
Right – Anchoring two 3strand tendons with pentahedron load distribution weldment.

3. Arrangement of prestressing tendons with examples

The substituted cable duct method is especially suitable for slab bridge structures. It can be suitably applied when strengthening in longitudinal direction but also for strengthening in transverse direction. It is very suitable for simple and continuous reinforced concrete slab structures. It can also be adapted to slab structures assembled from lengthwise precast prestressed segments.

Reinforced concrete slabs have a flat surface and there is no suitable structure gap for additional trajectory of prestressing tendons. Therefore, there is no other possibility of lifting cables in the longitudinal direction than using the substituted cable ducts drilled through the slab. At the same time, the original concrete reinforcement of reinforced concrete slabs (longitudinal) with respect to drilling ducts is placed relatively sparsely, usually every 100 to 160 mm in the transverse direction. Sometimes, the single tendons – individual monostrands are sufficient, multiple strand tendons in ducts of larger diameters are used for medium-span slabs, especially for economic reasons (drilling smaller number of ducts is also faster and reduces the extent of preparatory work). The ducts for each monostrand and cables can be appropriately placed between the original concrete reinforcement, without interrupting it, thus eliminating undesirable weakening. The ducts do not need to be at regular intervals. Of course, designing of reconstruction must be preceded by proper structure diagnostics. The tendons can be lead near the lower face of the rehabilitated slab and the structure provided with additional concrete covering layer, or it is possible to cut grooves into the original cover layer using the diamond technology and place the tendons into them. The grooves are then puttied using high-strength cement repair mortar. Figure 9 shows the arrangement of 3strand tendons suitable for strengthening the continuous slab bridges. This arrangement was used when strengthening a continuous slab structure with a span of L = 2 x 22 m.

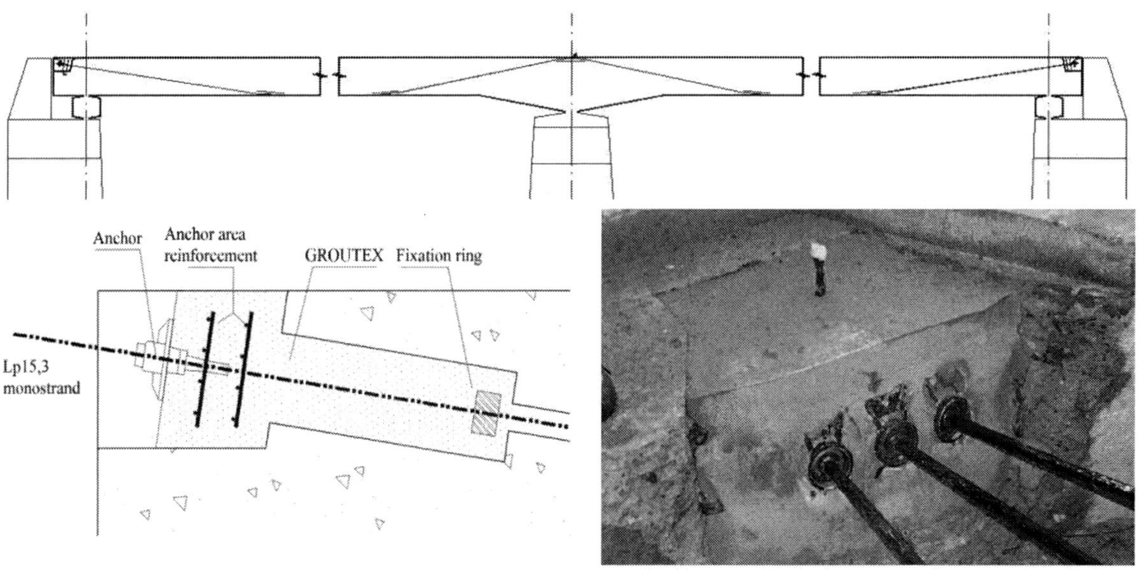

Figure 9. Arrangement of 3strand tendons in continuous slab structure with a span of L = 2 x 22 m.
Top – scheme of the trajectory 3strand tendons in slab structure.
Left bottom – anchoring area in concrete with widening duct outfall.
Right bottom – anchoring area with encapsulated anchors.

When strengthening slab structures, anchorage areas can be done directly into the concrete of the existing slab (this is the most commonly suggested method) or they may also be done (particularly at lower strength of the original slab concrete) in concrete of composite slab, if used. In case of continuous structures, it is necessary to design and implement two-way saddles over internal supports to change direction of additional cables from one (rising) duct to another (decreasing). A necessary condition is the adherence to a prescribed accuracy of drilling ducts (compliance with the deviations of up to ± 30 mm on the duct length is usually prescribed). After making the ducts, the saddles over supports are directionally adjusted to the outfall of ducts. The deviators (over the internal supports) themselves are actually double normal saddles used on the bottom surface of the structure, again with complying radii r_s = 1.2 to 1.5 m.

Parapet beam bridges have two main beams pulled over the road and the bridge deck is supported by cross beams. The cross beams connect the two main beams forming a half-frame in the transverse direction which provides spatial rigidity of the structure. These bridges can be advantageously strengthened by post-tensioning using the substituted cable ducts method in both the longitudinal and transverse direction. The span of these structures is usually between 15 to 25 m.

Figure 10 shows the shape and arrangement of the prestressing system when strengthening the U-shaped beam bridge with variable height of the main beam on a III class road with a span of l = 14.1 m. The main beams were post-tensioned by two tendons with four monostrands, cross beams by one tendon with two monostrands. This created the conditions for carrying up the weight of additionally composite slab. This is also example of anchoring with encapsulated anchoring system DYWIDAG CPS with the primary protection in anchors (already standardly used method of anchoring in bridge structures). Secondary protection was ensured by usual concreting of anchor cellars. Figure 11 shows the details of tendon anchors in the main beam and the cross tendon anchors on the sides of the main beam.

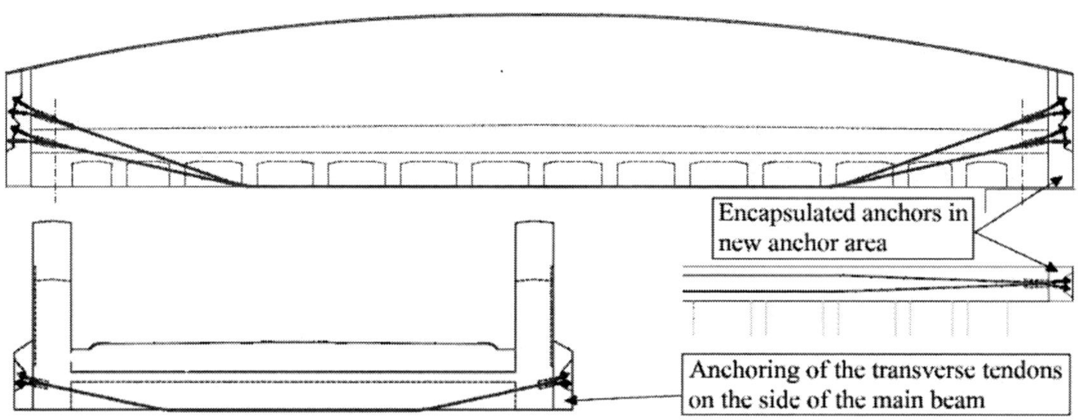

Figure 10. Trajectory and arrangement of prestressing system in parapet beam bridge.
Top – tendons trajectory in the main parapet beams (2 tendons with 4 monostrands).
Left bottom – tendons trajectory in transverse direction (2 tendons with single monostrand).
Right bottom – distribution of monostrands in the main parapet beam (ground plan).

Strengthening the parapet beam bridges by post-tensioning using the substituted cable ducts system generally brings several benefits:

1. Unlike bonded reinforcement, which is activated after being loaded, therefore not participating in the transfer of forces from permanent load, introducing prestressing into a construction balances a significant portion of the internal forces resulting from permanent load; thus, it effectively improves the condition when the structure is not subjected to imposed load and the necessary reserve for the transfer of traffic load effects is created.
2. Increase in load capacity by this method is significant, usually 200 up to 300 %; it is almost several times higher effect than when using bonded reinforcement where, from experience, the improvement by ca. 30 % can be achieved.

3. Cracks caused by static or dynamic loads in strained parts of reinforced concrete beams significantly accelerate the corrosion process of reinforced concrete. Bringing stress forces by prestressing leads to partial or even complete closure of these cracks, thus extending the durability of concrete structures against corrosion.
4. Most of the work associated with this technology can be done without interrupting the traffic on the bridge or with its partial restriction.
5. When strengthening bridges by prestressing, we use the entire width of the prestressing level interval. When strengthening beam bridges, it usually reaches the values of $\lambda = 0.15$ to 0.25.
6. In case of bridges which appear irreparable in terms of structural stability or other methods which cannot achieve the required load capacity and when demolition and construction of a new building occurs, the required parameters can be achieved for a third to half the price of new building using this method of increasing the load capacity.

Figure 11. Anchor areas of parapet beam bridge.
Left – disposition of new anchor area on the face of existing parapet beam bridge.
Middle – actual implementation of cast in situ anchor area block on the face of parapet beam bridge.
Right – implementation of cast in situ anchor areas for transverse tendons on the side of parapet beam.

The design and implementation of reinforcement must emphasize the fact that the structural strengthening is part of the overall reconstruction and rehabilitation of the bridge. It must create conditions for the reliability and durability of the chosen design. Therefore, it is necessary to consider and suggest the following measures:

1. Thorough preparation of degraded concrete of the entire bridge (surface layers, including plaster), first mechanically and then using high-pressure waterjet with rotary nozzle.
2. Careful cleaning of the exposed and corroded steel reinforcement in the entire existing bridge structure, first mechanically and then with high-pressure waterjet with a one-nozzle (a point) tool.
3. Carrying out the protection of steel reinforcement by silicate materials.
4. Implementation of own post-tensioning of the bridge construction by both using prestressing and composite slab.

5. Application of adhesive layer in the entire area of rehabilitated bridge concrete components.
6. Performing coarse and fine reprofiling of load bearing bridge structure.
7. Fitting inner and upper surfaces of U-shaped beams by unifying and protective coating.

Using prestressing tendons in substituted cable ducts can strengthen the simple and continuous reinforced concrete beam structures in other way as it was described in [5, 6, 7, 8]. In the case of simple supported constructions (mostly simply placed grids), the ducts and the prestressing tendons are arranged in the appropriate shape which is similar to shape of bending moment diagram resulting from permanent load; in the case of continuous structure as shown in Figure 12. The continuous structures can be effectively prestressed by continuous lifted cables tensioned from both anchors. They may be complemented by not-continuous cables anchored in the composite slab and anchoring blocks (cast in site) which are placed between the original beams.

Figure 12. Strengthening of continuous beam bridge using substituted cable ducts method.
Top – trajectory of tendons of 1. system (in ducts and on the bottom surface of beams) and trajectory of tendons of 2. system (free unbonded tendons in mid span).
Bottom – examples of outlets of substituted cable ducts ϕ 52 mm in gaps between steel reinforcement of beam.

Friction coefficient when prestressing monostrands which is applied for this arrangement only in the saddles has a value of 0.06 to 0.1. This value was repeatedly verified during the prestressing of thus strengthened structures by comparing the calculated and the achieved sizes of stretching monostrands. Friction is not applied in the straight sections of cable trajectories (the monostrands here are mostly almost in line from saddle to saddle, i.e. in the air and without friction).

The substituted cable ducts can be drilled into the structure very precisely and with minimal damage to the original beam reinforcement. This requires the construction site diagnostics of beam reinforcement, providing data on the most appropriate gap for placing the duct. The designer specifies the position of theoretical points only in the longitudinal direction and accurate outfall of ducts across the width of beams is left for the building process. In many cases, the concrete of reinforced structure in the region of the lower edge of beams is affected by corrosion in the extent that the concrete cover dropped away and the deployment of the reinforcement is obvious. In other cases, it is possible to remove the original cover in the place of future deviators as the cables will be protected by anchored concrete cover after the prestressing and the surfaces of the entire structure are typically rehabilitated by rehabilitation layers. Figure 12 below shows the examples of duct outfalls between the reinforcement of beams of strengthened continuous beam bridge construction. They show that it was always possible to find a suitable gap and to prepare the substituted ducts without or with minimal damage.

4. Summary

The substituted cable ducts method is especially suitable for slab, parapet beam and simple and continuous beam bridge structures. It can be applied when strengthening in longitudinal direction but also for strengthening and structural rehabilitation in transverse direction. It enables efficient life extension of existing bridges which do not meet the current requirements without reconstruction.

Acknowledgements

This paper has been worked with the support of the program Competence Centre of Technology Agency of the Czech Republic (TAČR) within the Centre for Effective and Sustainable Transport Infrastructure (CESTI), project number TE01020168.

This paper has been created during the solution of Specific junior research FAST-J-17-4545 "Analysis of the interaction of unbonded prestressing tendon and protective cover in deviators with smaller radii".

References

[1] Klusáček L 2009 *Strengthening of structures using post-tensioning by cables in substituted cable ducts and slots* (Brno University of Technology, Faculty of Civil Engineering)
[2] Czech Office for Standards 1993 Design of prestressed concrete bridge structure *(ČSN 73 6207)*
[3] European Committee for Standardization (CEN) 2006 Design of concrete structures. General rules and rules for buildings *Eurocode 2, Part 1 (EN 1992-1-1)*
[4] European Committee for Standardization (CEN) 1997 Design of concrete structures. General rules - Structures with unbonded and external prestressing tendons *Eurocode 2, Part 5 (ENV-1992-1-5)*
[5] Petrangeli M, Usai G and Zoppis E 2009 Bridge Repair by External Prestress: The Gibe Crossing in Ethiopia *Sustainable Infrastructure. Environment Friendly, Safe and Resource Efficient - IABSE Symposium (Bangkok)*
[6] Recupero A and Spinella N 2012 Method for Extending Life of Existing Bridge: A Case Study. *8th Int. Conf. on Concrete in the Low Carbon Era (Dundee)*
[7] Recupero A, Spinella N, Colajanni P and Scilipoti C 2014 Increasing the Capacity of Existing Bridges by Using Unbonded Prestressing Technology: A Case Study *Advances in Civil Engineering* **2014**
[8] Bouhaloufa A, Kadri T, Zouaoui S and Belhacene A 2015 Strengthening of Bridges by Additional Prestressing 2015 *17th Int. Conf. on Sustainable Design and Construction Engineering - ICSDCE 2015 (Paris)*

Effective way to reconstruct arch bridges using concrete walls and transverse strands

Ladislav Klusáček[1], Robin Pěkník[1] and Radim Nečas[1]

[1]Brno University of Technology – Faculty of Civil Engineering, Veveří 95, 602 00, Brno, Czech Republic

E-mail: klusacek.l@fce.vutbr.cz

Abstract. There are more than 500 masonry arch bridges in the Czech Road system and about 2500 in the Czech Railway system. Many of them are cracked in the longitudinal (span) direction. The barrel vaults are separated by the cracks into partial masonry arches without load bearing connection in transverse direction. These constructions are about 150 years old and they are also too narrow for the current road system. This paper presents a strengthening method for masonry arch bridges using transverse post-tensioning. This method is very useful not only for strengthening in the transverse direction, but widening of masonry arches can be taken as secondary effect especially in case of road bridges. Several bridges were successfully repaired with the use of this system which seems to be effective and reliable.

1. Failures of the arch bridges in the longitudinal direction

Some masonry arch bridges were built with cylindrical vaults. The type of vault is usually a half of the circle or a part of the circle in the longitudinal direction. For calculation of their load bearing capacity a simple frame model is used (Figure 1). There are only normal forces and bending moments acting on the cross section. There are no forces in the transverse direction in this case. But real vaults are separated by cracks in the longitudinal direction creating more sections (Figure 2). These cracks are typically situated in the central area of the vaults. They are also situated near the left and right edges in a distance about 1.5 m.

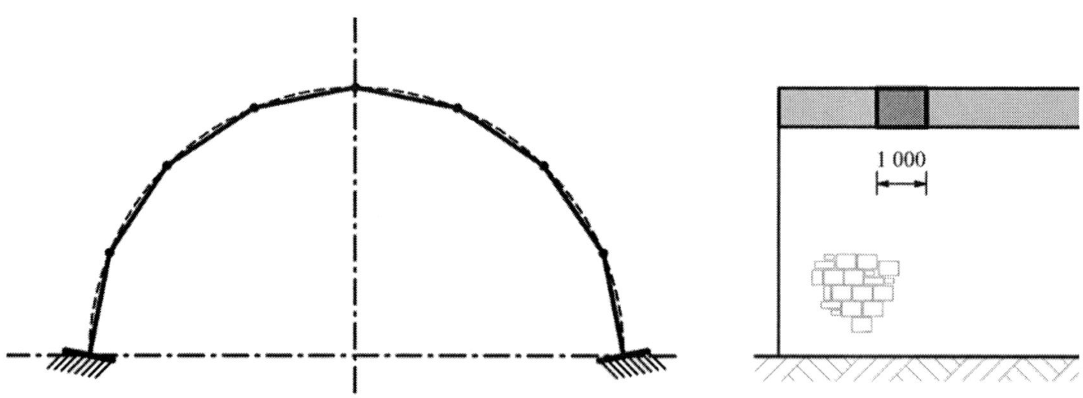

Figure 1. Frame model of a masonry cylindrical vault.

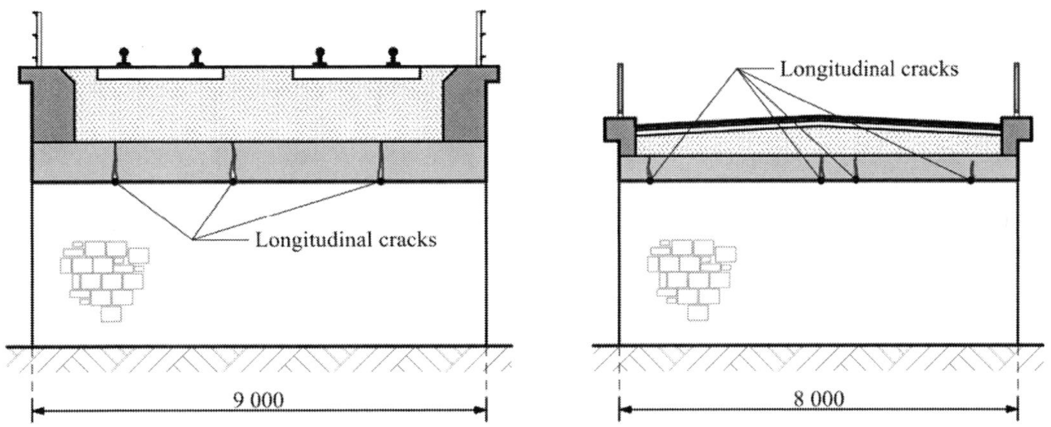

Figure 2. Railway (on left side) and road (on right side) masonry arch bridges; barrel vaults were cracked in the longitudinal direction.

A shell model can be used for explaining the central cracks. Deformation of the vault using the shell model is shown in Figure 3. We obtain internal forces (normal force and bending moment) in both directions: in the longitudinal direction (axis X) and in the transverse direction (axis Y) too. Because of very low tensile strength of masonry in the transverse direction cracks can appear in the central part of vaults. Explanation of the existence of cracks near edges is not so clear. There are also horizontal forces due to traffic load in these areas. These cracks are large, their widths were observed from 10 to 30 mm. Sometimes their size is about 100mm and this part of vault is fully separated from the face (spandrel) wall. In this case the structure is unstable and the bridge should be closed to traffic.

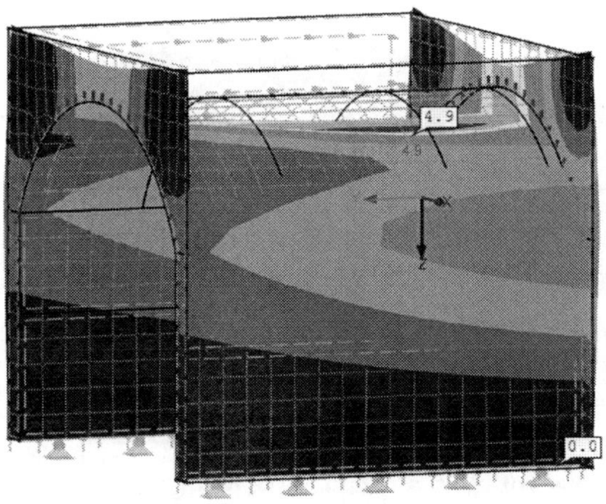

Figure 3. Deformation of the shell model due to live load.

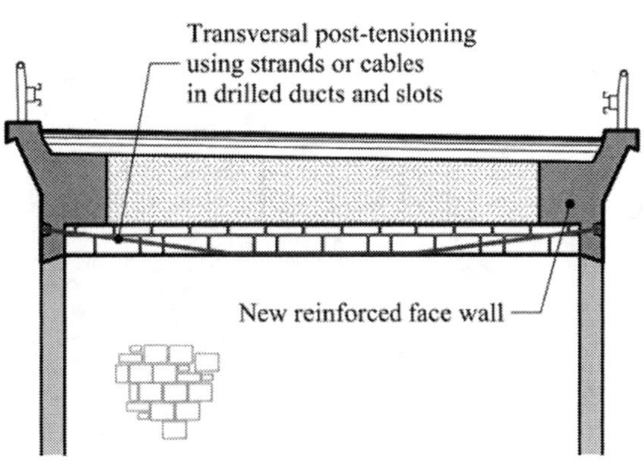

Figure 4. Arrangement of transversal post-tensioning.

2. Design of strengthening using transverse post-tensioning

Structural strengthening is based on an analysis of the vaults using the shell model. Tensile forces in the masonry barrel vault are taken by transverse post-tensioning (Figure 4). Post-tensioning is achieved by transverse cables. Cables usually consist of monostrands, each with two or three pieces. They are pushed into drilled cable ducts and slots prepared in advance in the masonry. Diamond drilling and sawing technology should be used. The special drilling support has to be used for preparing of cable ducts. Anchorages are positioned into new widening reinforced concrete (RC) walls. Prestressing forces are distributed by them to the crown and supporting abutment walls, so no crushing of stones or brick masonry due local pressure takes place. New RC face walls can also be used for widening of the bridge. This is the second benefit of this method and can be used in case of the road bridges. Widening can be large because the span of the cantilever can be designed up to 3m length without difficulties. Stability of the new face walls is blocked by forces in the anchorages. Those prestressing forces should be taken from force equilibrium in horizontal direction. Active forces are the horizontal part of earth pressure due to traffic load and embankment over the vault multiplied by safety coefficient 1.5 or the partial safety coefficient design method can be used. Passive forces are in fact prestressing forces in the anchorages. A check of the stress in masonry vault in horizontal direction is also necessary. According to the experience a value 20% of strength of masonry in perpendicular direction can be taken into account. The example of a real road bridge reconstruction using described method is given in the pictures below (Figures 5, 6 and 7).

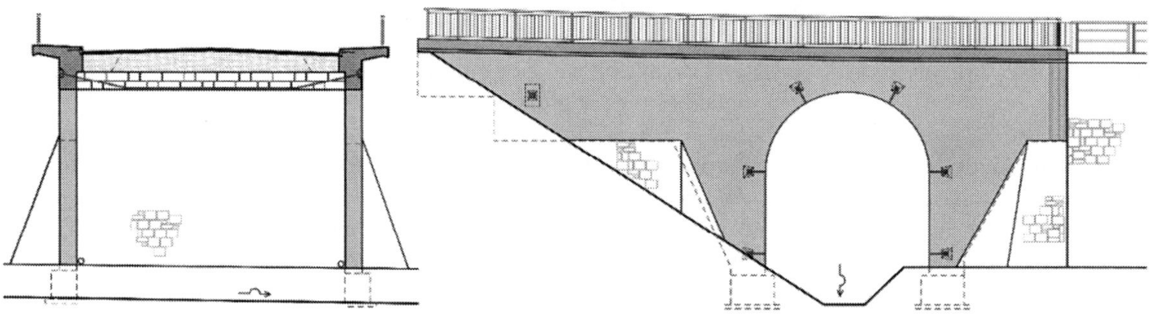

Figure 5. Example of design reconstruction of a road bridge using transversal post-tensioning with RC face walls.

Figure 6. Collapse of the old masonry face wall due to water, frost and ice.

3. Example of the reconstruction of the road stone masonry arch bridge

This road bridge was built in 1880. It is situated on 1st Class road No14. The main structure is a barrel vault with face walls. Span of the vault is 6.6 m (Figure 5). Construction comprises sandstone masonry.

Longitudinal cracks similar to those shown in Fig. 1 and buckling of one face wall were observed in 2002. The buckling became so extensive through the years that finally a part of face wall collapsed (Fig. 6). The traffic over the bridge was stopped. Design for reconstruction using strengthening of this vault by transverse post-tensioning was made in 2003. It is based on transverse cables with anchorage in the new RC face walls. Widening of the bridge was achieved as a secondary effect. The reconstruction was made in 2003 by MADOS MT ltd., the post-tensioning with drilled cable ducts was made by Mitrenga-stavby ltd. Further pictures show the bridge after repair works (Figures 7 and 8).

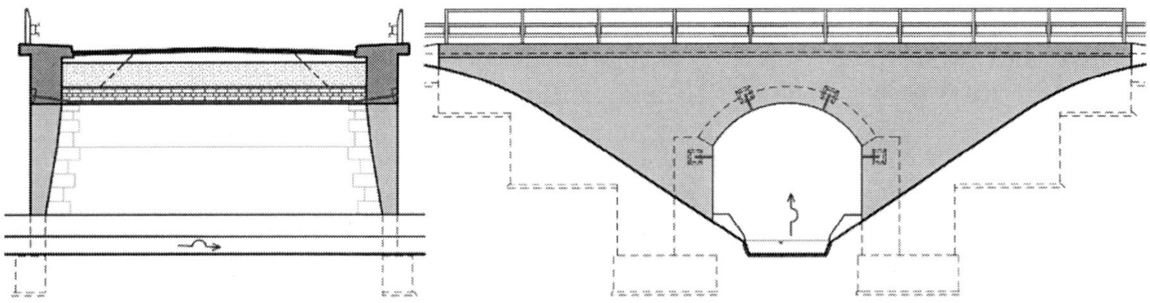

Figure 7. Plan of the structure measured during prestressing; left - cross section of the vault with cables; right - view of new face wall with anchorage ducts for four cables.

4. Process of the post-tensioning

This method of repair of the masonry arch bridge is a typical example of the process of post-tensioning. Cable ducts and slots were prepared before new RC walls as the first stage of reconstruction. Walls were built in the second stage and anchorages areas were prepared in concrete too. In the third stage the cracks were grouted by mortar injections and cavities over the vault were filled through drilled holes too. The post-tensioning was installed after concreting. The cables were tensioned to a magnitude of 180 kN with 5 minutes relaxation using single-cable prestressing jack of a range up to 200 kN. The cables were tensioned from one side and the extension of individual cables was measured during the process. Tensioning was applied carefully in 20 kN steps simultaneously increasing of the force in all cables of one arch. After post-tensioning the cable ducts were injected with common cement mortar and the slots were grouted up.

5. Example of the reconstruction of clay brick masonry arch road bridge

The vault is cylindrical with its center line in the shape of a circular segment from masonry 450mm thick. Lime mortar was used, partly flushed away, of strength 0.2 to 0.4 MPa. The span was 4.5m, width of the vault 10m, length of newly realized face walls 18.4m. The plan of the structure is in Fig. 7.

The main reason for the repair was buckling of both face walls and cornices with subsequent occurrence of longitudinal cracks 20-30mm wide between the vault and walls downstream and about 10mm upstream. In relation to the above mentioned type of defects the strengthening was designed in such a way that new RC walls were concreted against the original face masonry walls. RC walls were stabilized by transverse prestressing. Load bearing capacity was set in the longitudinal direction (suitable for Class A) without decreasing the calculated values caused by defects in the transverse direction. Face walls were stabilized by transverse prestressing with unbonded cables. The cables comprised three strands Ls 15.7 mm NPE (protected prestressing cables against corrosion, manufacturer Austria Draht).

Original stone vault was provided with cable ducts, diameter 52mm for setting of prestressing cables. Cable ducts formed a frame with inclination 10 degrees to the longitudinal radial plane. They were drilled from the vault towards the faces. On the upper surface of the stone vault the cables were set in slots 70 × 50 mm. New RC walls were concreted against the original faces and remaining parts of the face walls. Defects of the original structure are shown in Figure 8.

Measurement of horizontal deformation of the structure was carried out in three reference lines 2.2m long. The reference lines were made of steel anchors fixed in the vault masonry and steel pipes. Deformations were captured by inductive displacement sensor WETA 2mm with a range 0.001mm. Prestressing force was measured by a tester Proceq 200 kN. Deformation was continually monitored by a computer. The temperature during monitoring was practically stable +5°C.

Prestressing forces were induced evenly into the vault so that influence of uneven prestressing on displacement of concreted walls was avoided. First, the cable No. 1 was stressed in all four three-strand cables, followed by the cable No. 2 and then No. 3. Prestressing of the vault was carried out by a prestressing jack in 20 kN steps up to the designed prestressing force. It was possible to apply the full prestressing force in the cables (180 kN).

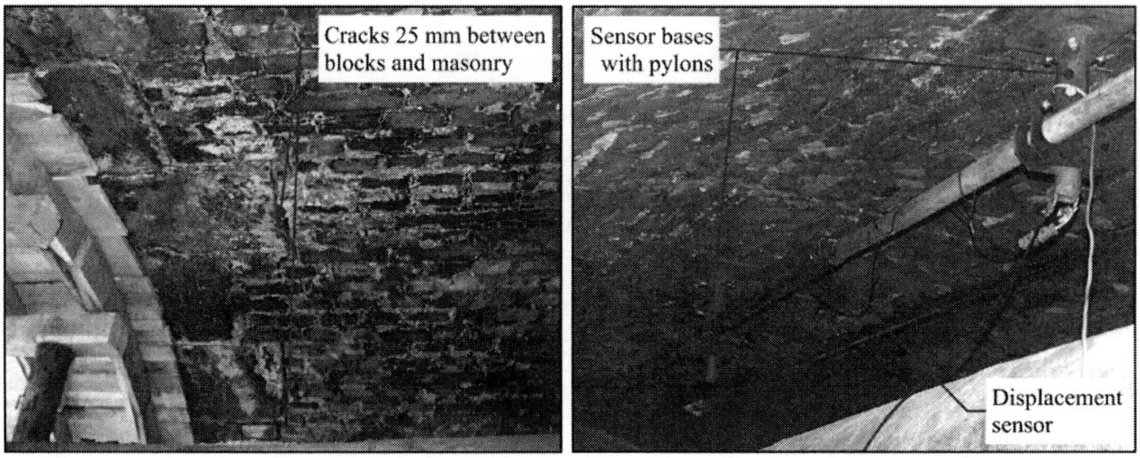

Figure 8. Defects in vault and sensor base; left - cracks between face wall blocks and masonry caused by buckling; right – sensor base in the crown with displacement sensor.

6. Deformation of masonry walls

Compression in the walls increased evenly in all sensor reference lines in time. During prestressing masonry was compressed immediately after prestressing was induced, after anchoring of each cable the increase tended to stop. After prestressing was completed (last cable was prestressed) the increase in compression stopped fully.

Relation between the deformation (compression) of masonry and magnitude of prestressing force is in Figure 9.

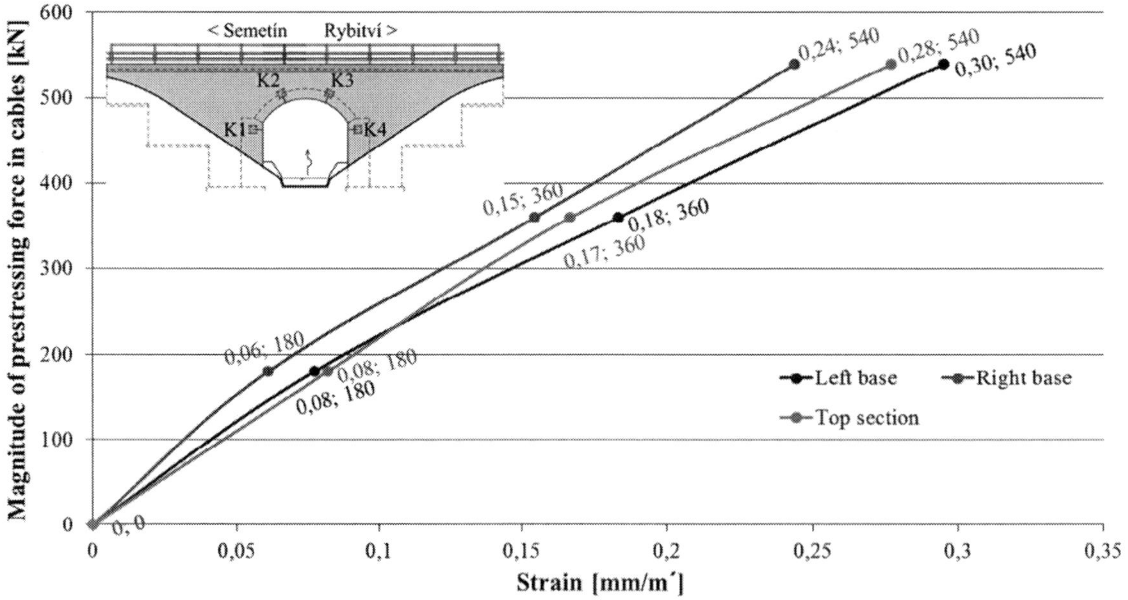

Figure 9. Relative compression of vault masonry during horizontal prestressing.

Following conclusions may be induced from the resulting relation:

1) It was possible to stress the cables up to the predicted and designed force 3 x 180 kN = 540 kN in one cable. Such forces were sufficient in relation to stability of the structure described in Paragraph 2. With such forces no negative impact occurred (cracks, crushing etc.). The concreted structure remained still without movement. The masonry was compressed practically linearly, i. e. prestressing was performed in elastic part of the masonry behaviour. It means that original longitudinal cracks in vault (between brick vault and original stone face walls) were properly filled with grout from the statical point of view, because no displacement occurred when there was no prestressing force. Subsequent grouting can reliably through drilled holes fill cracks and cavities that were created by defects in original vault.
2) Left side of the vault (left abutment area) was slightly softer – and it was compressed more. The difference is 10 % compared with the average value; and it is practically of no significance. Vault masonry com-pressed regardless of cohesion between vault and consolidated infill (reconstruction was carried out during operation without removal of infill) and without visible influence of newly realized walls on original walls and infill. Such influence is thus not big and with relatively small magnitude of compression only stiffness of vault and abutment in longitudinal direction may be important.
3) Average max. compression reached in springings was 0.27 mm/m′; in the crown 0.28 mm/m′. It is almost constant compression in the transverse direction upwards and proves good design of prestressing forces and also good realization of the reconstruction. Deformation of the structure is in compliance with assumptions given in Paragraph 2 and it is possible to consider them as confirmation of this system of reconstruction of bridge vaults (Figure 10).

Figure 10. View into the barrel vault; the picture shows sand stone masonry of the vault and cable slots after completion.

7. Example of the reconstruction of the clay brick masonry arch railway bridge

The bridge carries a double railway line Brno – Česká Třebová. Due to longterm influences of operation the vault was separated by longitudinal cracks into two independently acting parts under each track with third edge vault belt 1.2 m wide which was about to collapse. First repair was carried out in 2014 – new face wall statically secured by transverse prestressing. A set of arm amplifiers located in rail track No. 1 and 2 axes provided technology for deformation measurement.

The superstructure of the railway vault bridge is formed by segment brick vault with a clearance 7.540 m and a rise of 1.970 m. The springing is 2.100 m above the pavement and the crown is 4.220 m above the road. Thickness of the vault is about 0.840 m; this value comprises in lower part about 35 – 60 mm of torcrete. The repairs of both vault faces were formed by new RC face walls (Figure 11) which transfers the induced transverse prestressing into the whole vault section. Stabilization of separated parts of segment vault was achieved by induction of prestressing in the transverse direction by four cables. All masonry in the vault was grouted by cement grout in a matrix 600x600 mm.

Figure 11. Monitored bridge vault, new cast-in-situ walls.

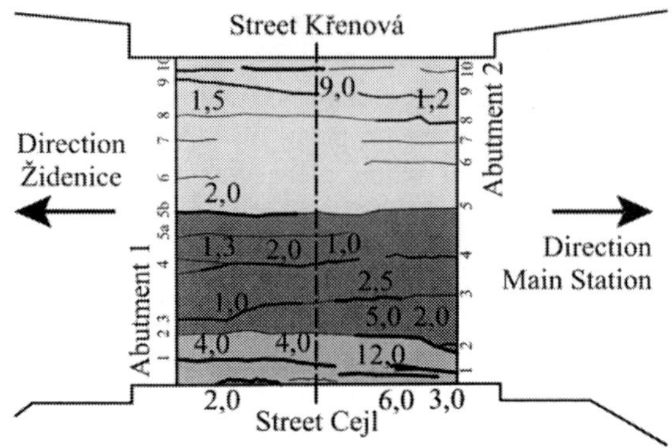

Figure 12. Location of cracks and division of vault into three separate parts.

Larger cracks divided the vault into three independent belts (Figure 12) which of course influenced the whole structure. The edge belt had also visibly dropped when compared with other parts of the vault. Diagnostics by probes confirmed that the cracks extended into masonry and that the masonry was saturated with water. The cracks were found in both abutments in approx. vertical direction. In abutment 1 there were mainly hair cracks, mostly below the springing, whereas in abutment 2 they were up to 2.0 mm wide and extended up to ¾ of the height from the springing. Other defects were corresponding to the age of the bridge. Surface defects were visible on both face walls especially below non-functional drainage.

8. Method of strengthening the structure

The main aim was to stabilize the structure and restore cohesion between separated parts in transverse direction. As a logical step separated parts of segment vault were stabilized by prestressing in the transverse direction. After grouting of cracks the prestressing in the whole structure was induced with the help of four prestressing cables through new face walls. Transverse continuity is supposed to increase stiffness of the whole structure and thus joint influence of the vault in the transverse direction for live load transfer.

9. Joint influence measurement principle

9.1. Measurement technology

Deformation measurement was carried out in three places along the axis of the vault, similarly to measurement before the repair. (Figure 13) [1]. A set of arm amplifiers was placed in the same anchorage places. For the purpose of this paper data from arm amplifiers placed below track axes were also used. (RZ1 and RZ2).

9.2. Data collection for measurement

For deformation measurement i.e. strain under traffic load a train with engine 560 „Pantograf" (Figure 14) with weight of 37t was used. (more precisely – values measured under the first axle of the engine in the direction of traffic). Measured and recorded passages of this train were eight in 2013 and four in 2014. During measurement the direction and tracks were monitored. Sampling frequency was 20 Hz in 2013 and 50 Hz in 2014 (Figure 15). Frequency 50 Hz sufficiently documents all vibrations of the vault during train passages.

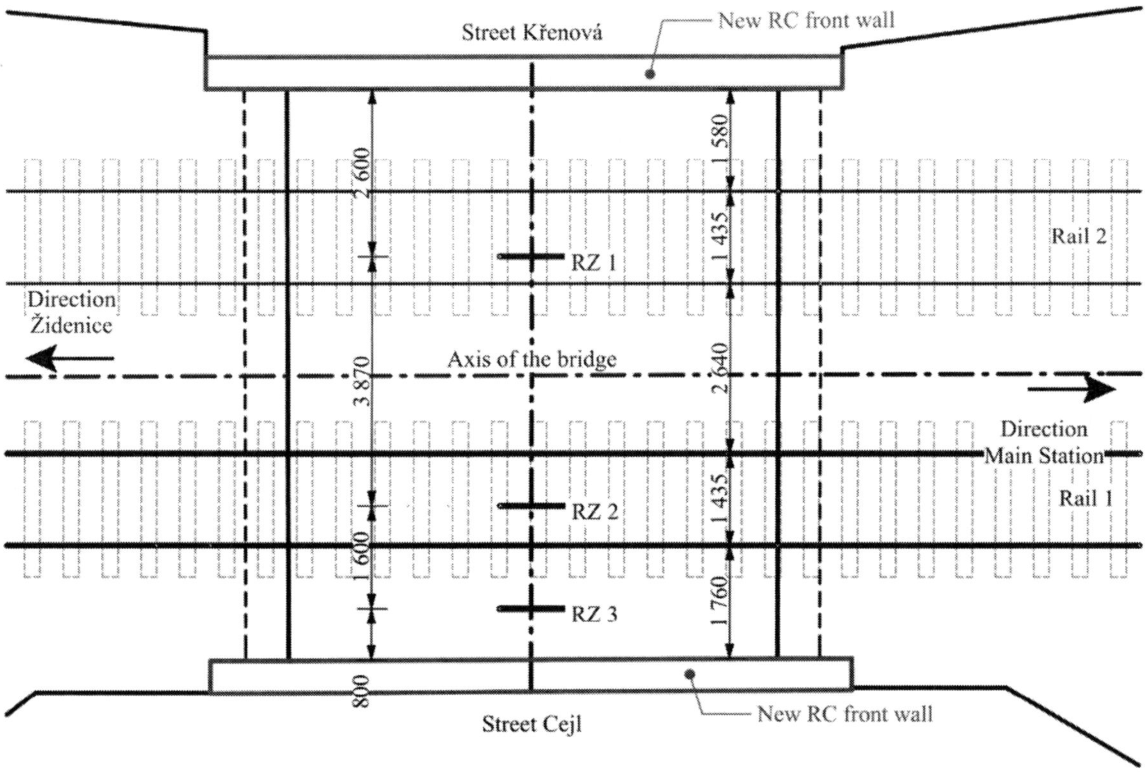

Figure 13. Positions of measuring basis RZ.

Figure 14. Train with engine 560 „Pantograf".

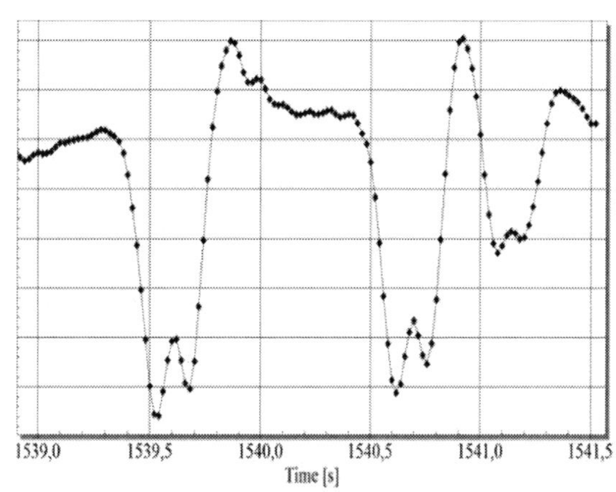

Figure 15. Train passage – sampling frequency 50 Hz.

10. Assessment of the measurement

Passages in the same direction and on the same track were compared; the comparison was based on data measured by arm amplifiers RZ1 a RZ2. Measured data were subsequently converted into strain on the crown both face (RZx-D) and back (RZx-H) sides of the vault. Values of strain are clearly shown in the following table (Tab.1).

IOP Conf. Series: Materials Science and Engineering **236** (2017) 012058 doi:10.1088/1757-899X/236/1/012058

Table 1. Values of strain.

Year	Train	Time	Train on rail nu.	RZ1-H	RZ1-D	RZ2-H	RZ2-D
2013	Pantograf	1435	1	-174.10	-33.48	-81.79	8.26
2013	Pantograf	5850	2	-14.86	10.94	430.57	-48.95
2013	Pantograf	6450	1	-141.52	-30.96	-83.98	8.67
2013	Pantograf	7875	2	-28.59	10.16	428.45	-51.61
2014	Pantograf	1540	1	-63.54	-10.59	-30.19	8.15
2014	Pantograf	3155	2	-12.33	2.59	-79.33	8.28
2014	Pantograf	5100	1	-68.36	-8.98	-27.87	6.69
2014	Pantograf	6780	2	-10.77	3.15	-89.53	10.51

10.1. Passage on track 2 above RZ1 (Figure 16)

From the measured data after passage on Track 2 above the arm amplified RZ1 it is clear that the vault was before and after the repair fully compressed. However, after the repair the strain decreased on the back side from its minimal value -167.67 μm/m to -68.36 μm/m and on the face side from its minimal value -34.68 μm/m to -10.59 μm/m. Proportionally the stiffness increase in the vault in transversal direction can be described for the back face as 2.45:1 and face side as 3.27:1. It can be stated that stiffness of the vault has increased three times in track 2 above the arm amplifier.

Similarly during passage on Track 2 the strain on the face side of the vault almost did not change in the place of Track 1 above the arm amplifier RZ2, on the back side of the vault it decreased from its minimal value of -86.31 μm/m to -30.19 μm/m, i. e. ratio 2.86:1.

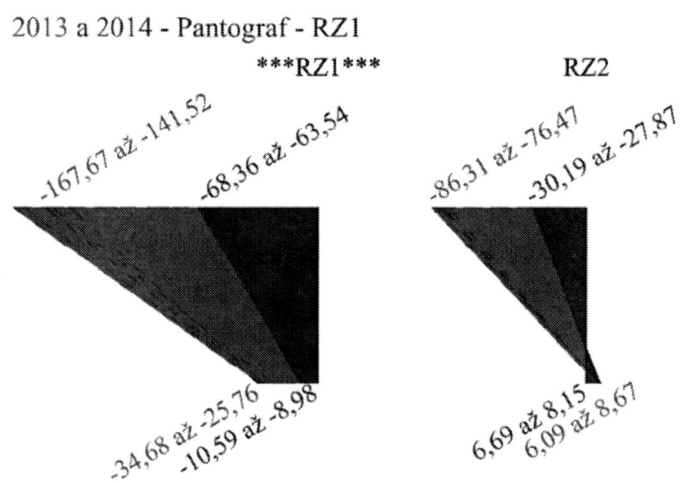

Figure 16. Strain of the vault in the location of RZ1 a RZ2 during passage above RZ1.

10.2. Passage on Track 1 above RZ2 (Figure 17)

From the measured data after passage on Track 1 above the arm amplifier RZ2 it is clear that the vault was before and after the repair on the back side in compression and on the face side in tension. After repair the strain on the back side decreased from measured minimal value -28.59 μm/m to -2.33 μm/m. Strain on the face side decreased from 10.94 μm/m to 3,15 μm/m. Proportionally the stiffness increase in the vault in transversal direction can be described for the back face as 2,32:1 and face side 3.47:1. It can be stated that stiffness of the vault has increased in the place of RZ1 three times.

During the passage on Tack 1 above RZ2 the strain on the face side decreased from measured maximal value of 51.61 μm/m to 10.51 μm/m, i.e. ratio 4.91:1. On the back side of the vault the strain decreased from minimal value of -430.57 μm/m to -89.53 μm/m, i. e. ratio 4.81:1.

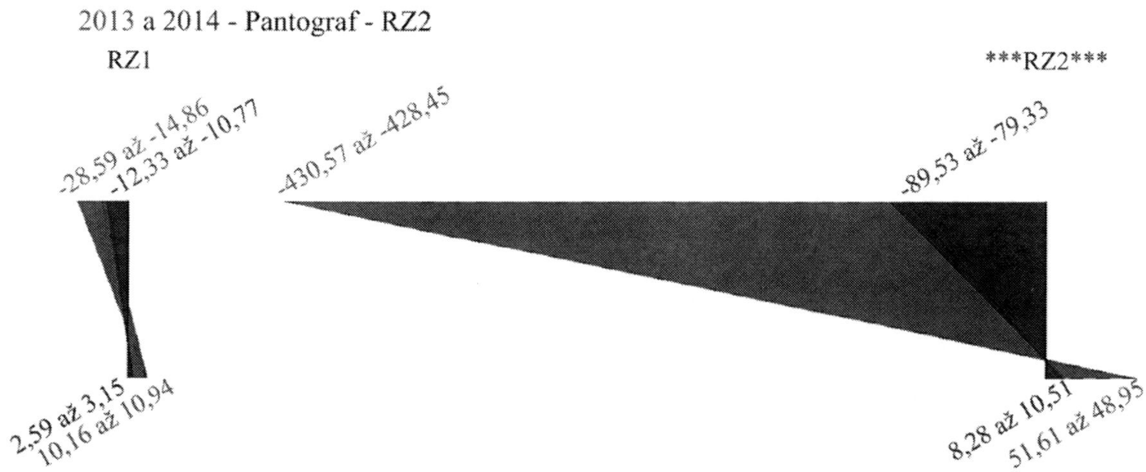

Figure 17. Strain of the vault in the location of RZ1 and RZ2 during passage above RZ2.

11. Conclusion

Described method of strengthening and widening masonry arch bridges with barrel vaults using the transverse post-tensioning with new RC walls has a lot of advantages. It is simple and very effective too. The repair costs are about 40% of the new construction. If the quality of stone is satisfactory, then this method may be recommended.

Acknowledgements
This paper has been prepared under project No. LO1408 "AdMaS UP – Advanced Materials, Structures and Technologies", supported by Ministry of Education, Youth and Sports under the „National Sustainability Programme I".

References

[1] Klusáček L 1993 Podélné trhliny v klenbách z kusových staviv /Longitudinal cracks in masonry vaults/ *Časopis Beton a zdivo /Journal Concrete and Masonry/*
[2] Klusáček L 2005 Zesilování betonových a zděných konstrukcí předpětím /Strengthening of concrete and masonry structures with post-tensioning/ *Sborník příspěvků konference Zesilování betonových a zděných konstrukcí /Proc. Conf. Strengthening of concrete and masonry structures Brno/* (Česká betonářská společnost - ČBS) pp 113-119
[3] Klusáček L and Pěkník R 2016 Reconstruction of Arch Bridges using Concrete Walls and Transversal Strands *2. Brückenkolloquium Beurteilung, Ertüchtigung und Instandsetzung von Brücken - Tagungschandbuch* (*Esslingen, Deutschland*) pp 203-206
[4] Klusáček L and Strnad J 2011 Rekonstrukce klenbových mostů betonovými čely a příčnými předpínacími kabely /Reconstruction and strengthening of masonry arch bridges using concrete face walls and transversal prestressing cabels/ *Časopis Silnice a železnice /Journal Roads & Railways/* **2** pp 23-29
[5] Požár M, Klusáček L and Nečas R 2015 Výsledky stabilizace železniční mostní klenby příčným předpětím /Results of stabilization of the railway vault bridge using transversal post-tensioning/ *Sborník příspěvků 20. Mezinárodní sympozium mosty /Proc. 20th Symposium Bridges 2015 – Brno/* (SEKURON s.r.o.) pp 249-254

Special heavy plates and steel solutions for bridge building

Tobias Lehnert[1]

[1]DILLINGER, Werkstr.1, 66763 Dillingen, Germany

E-mail: tobias.lehnert@dillinger.biz

Abstract. In many European countries infrastructure, -road as well as railway infrastructure-, needs intensive investments to follow the growing demands of mobility and goods traffic. Steel or steel composite bridges offer in this context viable and very sustainable solutions. Due to its unlimited recyclability steel can in general be seen as the ideal material for such sustainable constructions, but especially when designers or fabricators exploit the nowadays available possibilities of steel industry very cost-efficient and remarkable constructions are realizable. This paper will highlight some of these newest developments in heavy plates for bridge building. For example, for small span railway bridges the so-called thick plate trough bridges have proven to be a favourable concept. Very heavy plates with single plate weights up to 42 t allow building these bridges very efficiently out of one or very few single plates. Another interesting development is the so-called longitudinally profiled plates which allow a varying plate thickness along the actual loading profile. As last point the rising entry of higher strength steels in bridge building will be discussed and it will be shown why thermomechanically rolled plates are the ideal solution for these demands.

1. General instructions

In many European countries infrastructure, -road as well as railway infrastructure-, needs intensive investments to follow the growing demands of mobility [1] and goods traffic. Steel or steel composite bridges offer in this context viable and very sustainable solutions [2]. Due to its unlimited recyclability steel can in general be seen as the ideal material for such sustainable constructions. Especially when the designers or fabricators take advantage of what the steel industry offers them nowadays very cost-efficient and remarkable constructions are realizable. This paper will highlight some of these developments in heavy plate production, adjusted to the needs of modern bridge building. For example, the so-called thick plate trough bridges have proven to be a very favorable concept for small span railway bridges [3]. Very heavy plates with single plate weights up to 42 t allow building these bridges very efficiently out of one or very few single steel plates. Another interesting development is the so-called longitudinally profiled plates which allow a varying plate thickness along the actual loading profile. As last point the benefits of high thickness thermomechanically rolled plates (TM-Steel) will be discussed which offer the ideal solution as higher strength steels in bridge building gain more and more ground.

2. Heavy single plate weights for thick plate through bridges

Europe's rail infrastructure is showing its age. Bridge refurbishment projects - and new bridges - are needed urgently at many locations. The continent's politicians have recognized this and have set up programs of investment in many European countries. In Germany, for example, over 800 rail bridges are to be (re)constructed by 2019 [4]. These will, in the main, be short-span bridges, such as rail overbridges, for example. Nowadays not only the "classical" criteria, such as construction costs and project completion time, but also, more and more frequently, other important factors, such as the sustainability and durability of these bridge structures, are taken into account in investment decisions. In most cases, construction must be implemented on the existing system, i.e., "in traffic", for which reason the rapidity with which bridges can be completed is now playing an ever more important role. Trough bridges featuring thick deck plates have proven to be the ideal bridge design for such short-span overbridges (Figure 1).

Figure 1. Trough bridge with thick deck plates ready for installation.

The special benefit of this type of trough bridge is the fact that the deck plate consists of ideally one single thick steel plate (possibly also more plates with weld joints), with the advantage that it is possible to dispense with crossbeams underneath the track plate. Web plates are in many cases welded on to stiffen the trough walls.

The advantages of thick-plate trough bridges:

- High vehicle headrooms and clearances are possible, thanks to the lower structural height compared to conventional trough bridges (with crossbeam thickness 360-500 mm).
- Short traffic-interruption times ("possessions"), easy fabrication and fast installation thanks to high level of prefabrication.
- Reduced noise emissions and thus less noise nuisance, thanks to the heavy foundation plate.
- Good maintenance conditions thanks to the flat and homogeneous bridge undersides.
- Life-cycle assessment benefit, plus sustainability [3].

Grades S275NL/ML or S355NL/ML in accordance with DIN EN 10025, Parts 3 and 4 are generally used for the deck plate (plate thickness ~ 100 mm). Plates of particularly great widths are popular here, in order to avoid joints in the deck plate. Where a deck plate joint is unavoidable, the weld should be positioned in length direction at the one-third point, referred to bridge length, and must be tested for its fatigue strength.

Thanks to the large range of dimensions and formats available, steel heavy plates make it possible to avoid, or at least reduce to minimum, transverse and longitudinal joints in thick-plate trough bridges, also achieving, simultaneously, an improvement in fatigue performance. Grades S275NL/S355NL can, for example, be supplied in maximum plate weights of up to 42 t and in plate widths of up to 4,750 mm for such applications, while Grades S275ML/S355ML are available up to 35 t and 4,600 mm.

The following maximum plate lengths are available for a 100 mm thick plate in Grade S355NL, assuming a maximum plate weight of ~ 42,5 t (Figure 2).

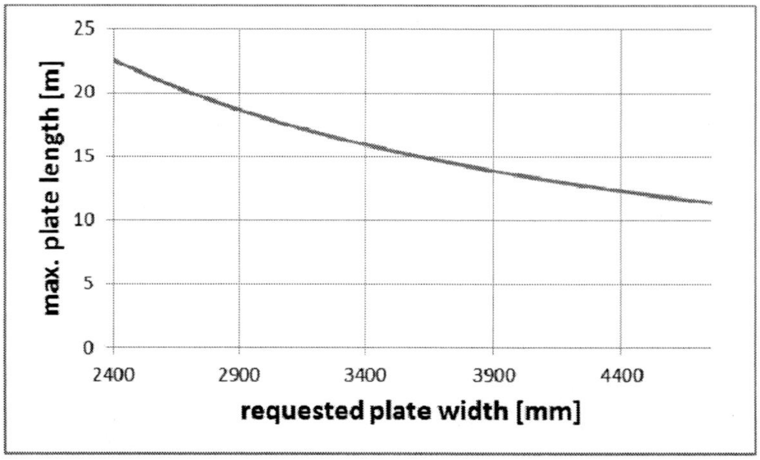

Figure 2. Possible plate length for a 100 mm thick S355NL with different plate widths.

3. Longitudinally profiled plates

Another interesting product of the steel making industry is the longitudinally profiled (LP) plate. By perfect control of the rolling parameters during rolling the thickness over the length of a plate can be varied to a certain extent and a thickness profile can be obtained [5]. Figure 3 gives an overview of the possible profile types. A segment defined as plate part which has either the same thickness or the same slope has to be at minimum 1000 mm long, but a longitudinally profiled plate can generally have up to 5 such segments. Newest developments for the production of such LP plates could increase the possible slope of up to 10 mm/m. These new possibilities offer a greater freedom to the engineering in choosing the plate profile desired and adjusted to the load situation for various applications, e.g. bridge girders.

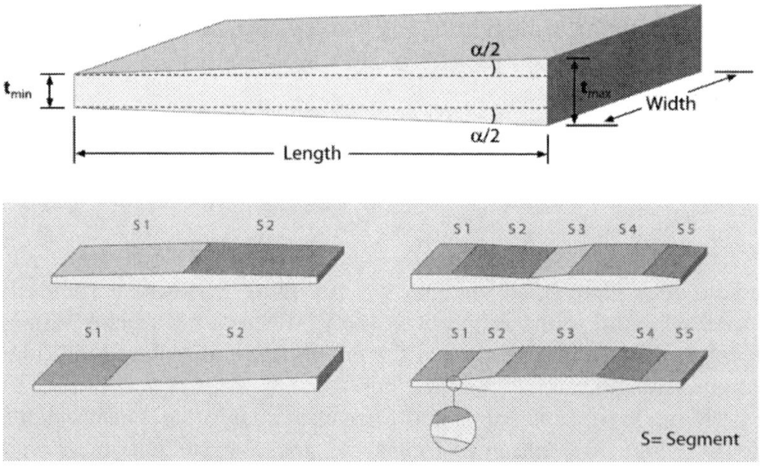

Figure 3. Overview of possible profiles for longitudinally profiled plates (LP plates).

Similar to the usage of high strength steels the economic and ecologic benefit of this very special type of steel plates lies mainly in an achievable material weight reduction. The tunable plate thickness gives the opportunity to save excess material in regions where the thickness is not needed from static calculations. Furthermore the welds in e.g. a bridge girder can be moved to positions where the fatigue is less governing the design (Figure 4). Given that the usage of LP plates can offer a cheaper solution by representing less excess material and less welding costs with possibly improved fatigue properties. This consideration implies only the fabrication cost and is not involving the inherent reduction of transportation cost and energy which is associated with such a lower component weight.

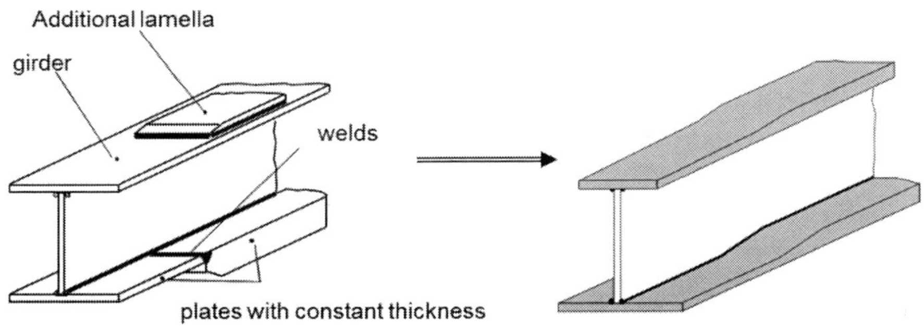

Figure 4. Possible applications of LP plates in a bridge girder.

4. Thick high strength plates, thermomechanically rolled

Higher strength steels, as S460 with a minimum specified yield strength of 460 MPa, are gaining more and more ground in international bridge building due to its great benefits regarding weight savings and achievable cost reductions. Depending on the load situation significant reduction in the cross-section is possible by stepping into a higher strength class. If for example comparing the rated values for S355 and S460 (fy = 315 MPa at 100 mm plate thickness to fy = 430 MPa at 73 mm plate thickness), the change towards a higher strength steel allows a plate thickness reduction of around 30 % (not taking into account fatigue effects which are depending on the exact construction detail) (Figure 5).

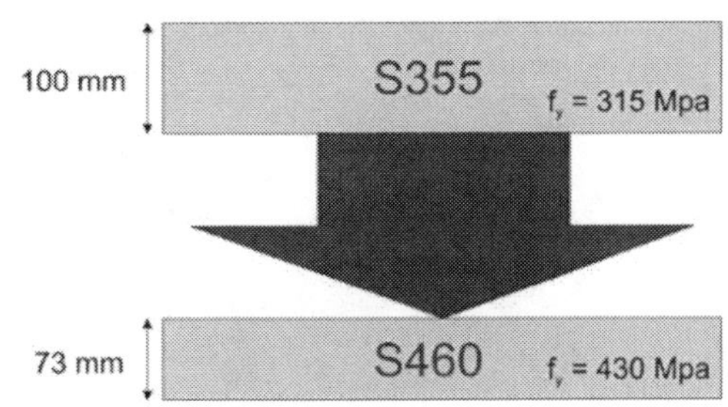

Figure 5. Possible reduction of plate thickness when using higher strength steel leads to savings in material consumption [6].

Beside the reduced material amount these reduced cross sections impact the cost and eco-balance of a construction in multiple ways [7]. First, the lower component weight enables bigger assembly units and thereby faster assembly plus optimized transportation. The hence lowered transport energy expenses (for example less truck transport) add a relevant sustainable aspect to the high economic efficiency of high strength steels in bridge building.

But the biggest impact on cost reduction is coming from the fact that the thinner steel plates lead to a considerable reduction of welding costs. Due to the geometrical situation the weld seam volume is a quadratic function of the plate thickness and therefore the enabled savings in weld material and welding time are disproportionately high with decreasing plate thickness (Figure 6).

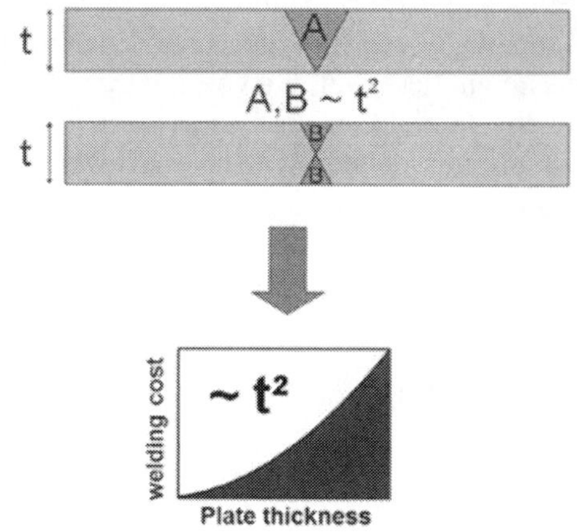

Figure 6. Realizable cost savings in welding dependent on plate thickness.

To achieve such "higher" strength steel grades (with yield strength > 355 MPa) different production routes are possible [8]:

1. Alloying: higher carbon equivalents and therefore decreased weldability
2. Thermomechanical rolling: excellent weldability, mainly up to 500 MPa
3. Quenching and Tempering: in special cases used up to 690 MPa in bridge building

The big advantage of using thermomechanically rolling as a production route for achieving higher strength steel grades is the remaining excellent weldability, while the other processes tend to an increased carbon equivalent and by that to a reduced weldability.

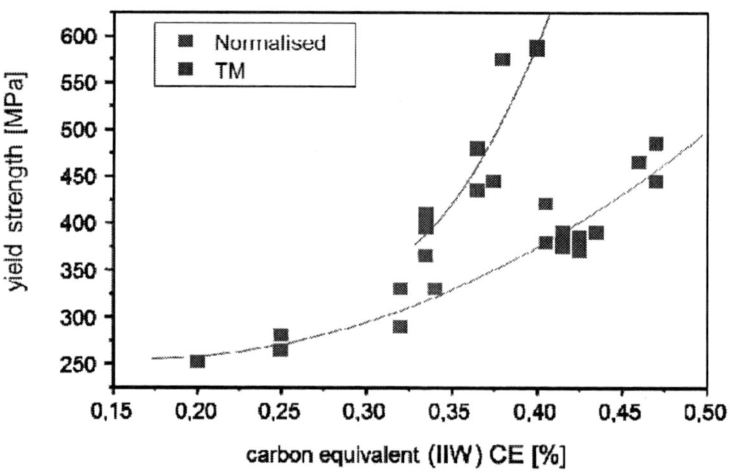

Figure 7. Yield strength as a function of alloying, expressed by the carbon equivalent CE (IIW).

Thermomechanically rolled plates need less carbon and alloying material compared to its normalized equivalent to achieve the same strength (Figure 7). Their low alloying contents are therefore the basis for the major advantage of TM-steels, their excellent weldability. This is even more evident when talking about higher strength steels, as the gap between the Carbon equivalent (CE(IIW) of a normalized steel and a TM-steel increases with yield strength.

Up to recently the big advantages thermomechanically rolling offer especially for higher strength plates, as e.g. S460M/ML, were only available for plate thickness up to 120 mm. Surpassing this limit lead to the situation that the fabricator had to choose between a normalized steel, as S460NL, which has a high CET and needs special care during welding and a quenched and tempered steel which has an indeed acceptable CET value but higher cost due to more energy intense production route.

The highest thickness for thermomechanically rolled plates is mainly limited by the availability of strong rolling forces as well as of semi-finished steel products with appropriate product thickness in order to allow a minimum degree of thickness deformation between the slab and the final heavy plate. This is of crucial importance to achieve the beneficial properties of TM-Steels. Thus EN 10025-4, as standard for TM plates in steel constructions, has defined such steels also only up to a thickness of 120 mm. Nevertheless, newest developments in steel manufacturing, e.g. very thick slabs up to 500 mm possible due to the invest in a new innovative continuous caster, are now able to push this production-driven limit to 150 mm. In order to allow maximum usability in bridge constructions these thick TM-plates fulfill the same mechanical requirements as a 120 mm thick plate according to EN 10025-4.

With this new development now also for plates exceeding 120 mm an excellent weldable and cost-efficient steel grade alternative is available for the fabricator. The significantly improved weldability of TM-steels can be verified when comparing the carbon equivalent CET, as common measure of weldability, for the different alternatives of S460 (Table 1). CET was computed with Eq. 1 [9].

$$CET = C + (Mn + Mo)/10 + (Cr + Cu)/20 + Ni/40 \quad (1)$$

Table 1. Typical CET values for S460 steels in 140 mm.

Steel grade	typical CET /%	typical CEV /%	max. CEV /% acc. to EN 10025
S460NL	0,34	0,50	0,55
S460QL	0,31	0,42	0,50
S460ML	0,26	0,41	-

Figure 8. Higher strength thermomechanically rolled steel (S460ML) in bridge construction, Botlek Bridge in Rotterdam, Netherlands (Copyright SEH Engineering).

5. Conclusions

By developing and refining modern steel concepts, the steel producers can significantly support the efficiency and design of a modern sustainable steel bridge. Mainly the potential savings these modern steel grades enable, are associated with a reduction of fabrication time (less welding, less time for preheating or transportation) or also with reduced energy and material consumption (weight

IOP Conf. Series: Materials Science and Engineering **236** (2017) 012059 doi:10.1088/1757-899X/236/1/012059

reduction). Therefore positive effects on the environment are immanent. This combination of effects can help the steel fabricators reaching the ambitious goals of high profitability, safety and sustainability in modern bridge construction.

References

[1] Zinke T, Ummenhofer T, Pfaffinger M and Mensinger M 2014 The Role of Traffic in Sustainability Assessment of Road Bridges *The 8th Int. Conf. on Bridges in Danube Basin* ed E Petzek, R Băncilă pp 255-263

[2] Kuhlmann U, Maier P, Pfaffinger M et al. 2014 Sustainability Assessment of Bridges – Recent German Research Results *JGBS München* pp 7-8

[3] Breunig S and Kuhlmann U 2015 Improvement of Fatigue Resistance of Transverally Bended Longitudinal Fillet Welds by High Frequency Mechanical Impact Treatment *Steel bridges: Innovation & New Challenges 2015* ed N Yardimci, A B Aydöner, H Y Gür'es pp 615-624

[4] Deutsche Bahn 2016 *Leistungs- und Finanzierungsvereinbarung: Infrastrukturzustands- und-entwicklungsbericht 2015*

[5] Richter K and Schmackpfeffer H 1988 Herstellung von LP-Blechen und deren Verwendung im Brückenbau *Stahlbau* **57**(2) pp 33-38

[6] Willms R 2008 High strength steels in steel construction: Application and processing. *Proc. 5th European Conference on Steel and Composite Structures EUROSTEEL (Graz, Austria)* ed R Ofner, D Beg et al

[7] Stroetmann R 2011 High strength steel for improvement of sustainability *Proc. 6th European Conf. on Steel and Composite Structures EUROSTEEL (Budapest, Hungary)* ed L Dunai, M Ivanyi et al

[8] Samuelsson A and Schröter F 2005 High performance steels in Europe *Use and Application of High-Performance Steels for Steel Structures* (IABSE Structural Engineering Documents) chapter 5 pp 99-146

[9] Uwer D and Höhne H 1991 Charakterisierung des Kaltrißverhaltens von Stählen beim Schweißen *Schweißen und Schneiden* **43**(4) pp 195-199

IOP Conf. Series: Materials Science and Engineering **236** (2017) 012060 doi:10.1088/1757-899X/236/1/012060

Dynamic response of footbridges with tuned mass dampers

Jiří Máca[1]

[1]Czech Technical University in Prague, Faculty of Civil Engineering, Department of Mechanics, Thákurova 7, Praha 6, CZ-166 29, Czech Republic

E-mail: maca@fsv.cvut.cz

Abstract. The increase of vibration problems in modern footbridges shows that footbridges should no longer be designed for static loads only. Not only natural frequencies but also damping properties and pedestrian loading determine the dynamic response of footbridges and design tools should consider all of these factors. Footbridge vibrations don't cause usually structural problems, but if the vibration behaviour does not satisfy the comfort criteria, changes in the design or damping devices could be considered. The most popular external damping devices are viscous dampers and tuned mass dampers (TMD). The paper presents the basic principles of optimal TMD configuration and design procedure. The efficiency of TMD is demonstrated on the example of a footbridge prone to vibrations induced by pedestrians. It is shown that if the TMD is tuned quite precisely the reduction of accelerations can be very significant.

1. Introduction

Modern footbridges are very often lightweight and flexible structures, where the first natural frequencies of vibration may fall close to dominant frequencies of the dynamic excitation due to walking or running. Such bridges are susceptible to vertical as well as to horizontal vibrations leading to a resonant response characterized by high levels of vibration and a dynamic design is necessary.

The increase of vibration problems in modern footbridges shows that footbridges should no longer be designed for static loads only. But fulfilling the natural frequency requirements (given in many codes) restricts footbridge design: very flexible, lightweight structures, such as stress ribbon bridges, cable-stayed or suspension bridges may not satisfy these requirements. Moreover not only natural frequencies but also damping properties and pedestrian loading altogether determine the dynamic response. Design tools should consider all of these factors.

Although footbridge vibrations don't cause usually structural problems, they can induce some uncomfortable sensation, and so many codes establish maximum acceptable values of acceleration. Provided that the vibration behaviour due to expected pedestrian traffic is checked with dynamic calculations and satisfies the required comfort, any type of footbridge can be designed and constructed. If the vibration behaviour does not satisfy some comfort criteria, changes in the design or damping devices could be considered.

2. Proposed design procedure

The design procedure should satisfy two slightly contradictory requirements: to be simple and accurate. The proposed design procedure should contain the following steps (cf. [5]):

- Determination of fundamental frequency limit values in horizontal and vertical directions ($f_{v,limit}$, $f_{h,limit}$) above which the vibration limit state is satisfied. The proposed limit value for vertical fundamental frequency is $f_{v,limit}$ = 5 Hz, the limit value for horizontal fundamental frequency $f_{h,limit}$ = 2.5 Hz.
- Evaluation of footbridge natural frequencies in vertical and horizontal direction (f_v, f_h).
- Comparison of footbridge natural frequencies to the limit values.

If the frequencies are higher than the limit values the serviceability limit state in relation to vibration is satisfied, otherwise the verification procedure will be continued according to the following steps:

- Determination of the acceleration limit values ($a_{v,limit}$, $a_{h,limit}$) to ensure pedestrian comfort. The proposed limit values for acceleration in vertical and horizontal directions are $a_{v,limit}$ = 0.7 m/s^2 and $a_{h,limit}$ = 0.2 m/s^2, respectively. These limit values are suggested in most of the codes.
- Determination of pedestrian loads (determination of load models). Three different types of human motion are commonly considered to model the dynamic loads applied by pedestrians, namely walking, running and rhythmic jumping. The pacing frequency f_p and the pedestrian forward speed v_p are two parameters that play a fundamental role in terms of the characterisation of the excitation.
- Evaluation of maximum accelerations (a_v, a_h) of footbridge for each load model.
- Comparison of footbridge maximum accelerations and limit accelerations.

If the accelerations are higher than the limit values, changes the vibration characteristics of the footbridge (natural frequencies) or damping devices will be considered and the design procedure is repeated.

3. Tuned mass damping

To avoid undesirable vibrations of the structure it is a good idea to install tuned mass dampers (TMDs) on the footbridge to dissipate the energy from one or more modes. A TMD is often a much more lucrative solution when compared to changing the natural frequencies of the structure.

The theory of how a TMD works, and how to determine the optimal characteristics are summarized in this chapter. The idealized system of a TMD is shown in Figure 1. This configuration is characterized by a mass m_d attached to the main structure, having the modal mass m_0 by a spring having the stiffness k_d and a viscous damper with the parameter c_d. The frequency of the structure is characterised via the modal stiffness k_0 and the internal damping of the structure is c_0. When the structure vibrates, a relative motion of the attached mass is generated. The velocity of this relative motion enables the damper to dissipate energy from the system and thereby comprehend large response while the structure is exposed to a harmonic force close to its natural frequency.

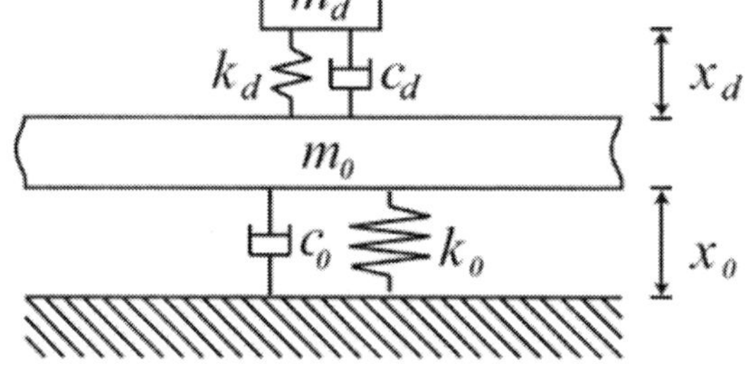

Figure 1. Idealized system of a TMD.

3.1. *Optimal damping configuration*

When installing a TMD the following steps should be considered to obtain the optimal damper properties:

- The position of the damper must be selected, so it can perform optimally.
- The frequency of the damper must be tuned.
- The optimal damping level must be determined

The TMD must be designed to prevent undesirable response due to specific mode. The energy dissipated by the TMD is proportional to the relative velocity of the deck. As the relative velocity is related to the maximum displacement, the TMD should be placed to the point, where the vibration mode has the largest modal displacement.

TMDs are normally tuned so that the two peaks of the damped system frequency response curve have the same dynamic amplification, when expressed in terms of displacements.

If the damping is too low, undesirable dynamic amplification will occur near the two undamped resonance frequencies because the dampers dissipate too little energy. On the contrary, if the damping to high, the relative motion of the damper will be to low resulting in a configuration where damper and structure acts as one unit.

3.2. *Design procedure*

The design procedure may be as follows (cf. [1]):

1. Choice of TMD mass m_d, based on the ratio μ to the structural modal mass m_s ($\mu = m_d / m_s$). Typical values of the mass ratio can range from 0.01 to 0.05.
2. Calculation of optimum TMD frequency ratio, expressed by the ratio δ between the TMD's frequency f_d, and the system's frequency f_s ($\delta = f_d / f_s$)

$$\delta_{opt} = \frac{1}{1+\mu} . \quad (1)$$

3. Calculation of optimum TMD damping ratio ξ_{opt}

$$\xi_{opt} = \sqrt{\frac{3\mu}{8(1+\mu)^3}} . \quad (2)$$

4. Calculation of the TMD constants:

Natural frequency $$f_d = \frac{f_s}{1+\mu} ; \quad (3)$$

Stiffness $$k_d = (2\pi f_d)^2 m_d ; \quad (4)$$

Damping $$c_d = 2\sqrt{k_d m_d} \xi_{opt} . \quad (5)$$

Important remarks on TMD design:

- The frequency of the TMD shall be tuned quite precisely.
- The compliance with the optimum damping is less important.
- TMDs are most effective when the damping of the structure is low.
- It is not worth increasing the mass ratio too much, for large mass ratios, the amplitude of the TMD oscillations is reduced.

- The exact tuning of the TMD occurs experimentally; therefore great care should be paid to construction details.
- The performance of a TMD is extremely sensitive to frequency de-tuning, which can occur as consequence of slight frequency changes associated with pedestrian loads or with modifications within the structure during its lifetime; therefore it is of interest to evaluate the TMD efficiency for an estimated range of frequencies.

4. Example - Footbridge with TMD

The footbridge is a simply supported beam made of lightweight concrete. The dimensions of the beam were chosen to make it prone to vibrations induced by pedestrians. A TMD is mounted at the midspan.

Parameters of the footbridge:

Mass	$m_s = 5300$ kg
Stiffness	$k_s = 870$ kN/m
Damping ratio	$\xi_s = 0.015$
Natural frequency	$f_s = 2.04$ Hz

Parameters of the TMD:

Mass	$m_d = 250$ kg
Mass ratio	$\mu = m_d / m_s = 0.047$
Natural frequency (cf. Eq.(3))	$f_d = f_s /(1+\mu) = 1.95$ Hz
Stiffness (cf. Eq.(4))	$k_d = (2\pi f_d)^2 m_d = 37.5$ kN/m
Damping ratio (cf. Eq.(2))	$\xi_{opt} = \sqrt{\dfrac{3\mu}{8(1+\mu)^3}} = 0.124$
Damping (cf. Eq.(5))	$c_d = 2\sqrt{k_d m_d}\,\xi_{opt} = 759$ kg/s

One 95 kg-heavy person crosses the footbridge with a frequency of 2 Hz, which is approximately equal to the first natural frequency of the bridge. The length of the step is 0.70 m. The excitation force is given in Figure 2. Other types of analytical force models can be found in the literature: time-domain models (deterministic and probabilistic force models) and frequency-domain models – for a detailed review cf. [2] and e.g. [4]. The suitable model of mutual interaction between human gait and elastic bridge has been developed in [3].

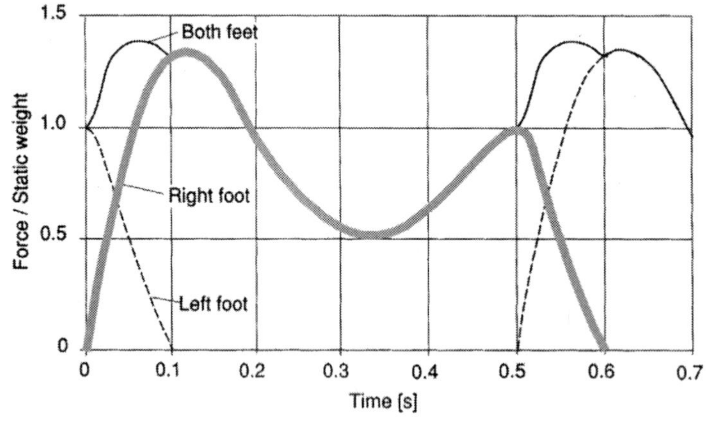

Figure 2. Excitation force.

Figure 3 shows the time history of the displacement at midspan with locked TMD, acceleration is shown in Figure 4. The results with free TMD are given in Figure 5 and Figure 6.

The effect of the TMD can be easily seen in Fig. 3 and Fig. 5 for displacement time histories and in Figures 4 and 6 for acceleration time history. The maximum acceleration at midspan reduced from 1.6 m/s^2 (locked TMD) to 0.35 m/s^2 (free TMD), which corresponds to a permissible value 0.7 m/s^2.

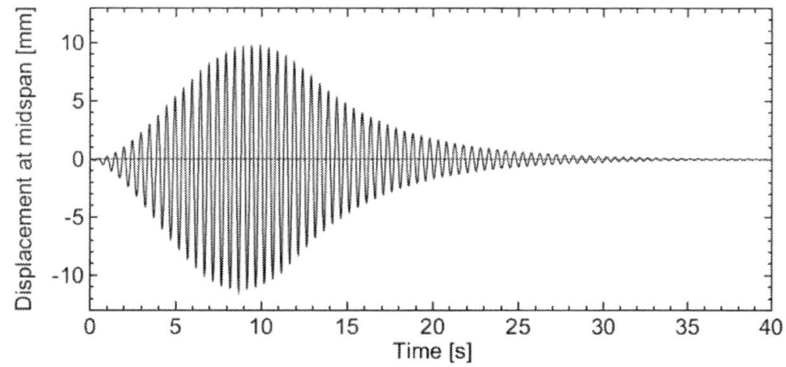

Figure 3. Displacement at midspan with locked TMD.

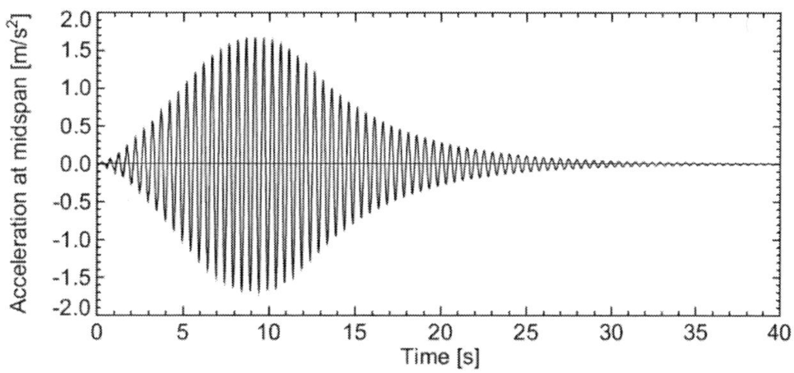

Figure 4. Acceleration at midspan with locked TMD.

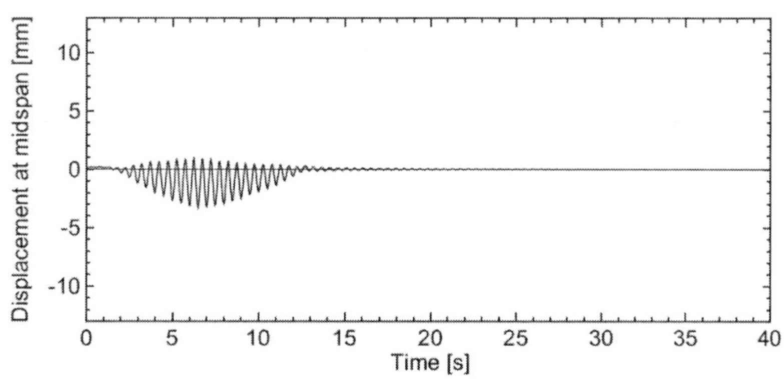

Figure 5. Displacement at midspan with free TMD.

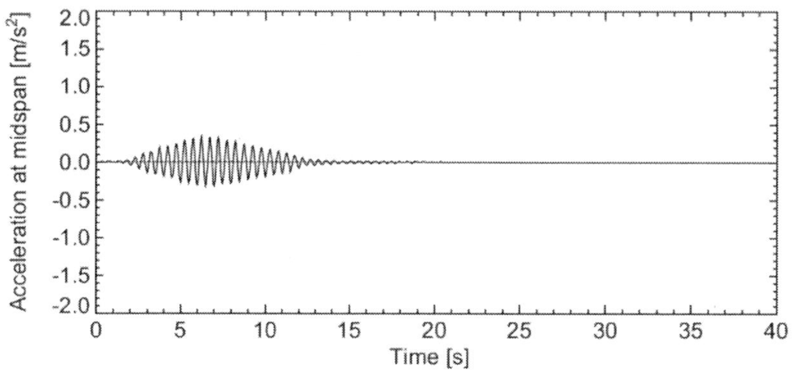

Figure 6. Acceleration at midspan with free TMD.

5. Conclusions

Not only natural frequencies but also damping properties and pedestrian loading determine the dynamic response of footbridges and design tools should consider all of these factors. Footbridge vibrations don't cause usually structural problems, but if the vibration behaviour does not satisfy the comfort criteria, changes in the design or damping devices could be considered.

The control of the vibration response in a footbridge implies the introduction of modifications, which can include variation of the mass, frequency or structural damping. For an already constructed structure, the simplest approach is based on the increase of the structural damping, which can be usually achieved either by implementation of external damping devices. The most popular of these are viscous dampers and TMDs.

The paper presents the basic principles of optimal TMD configuration and design procedure. The efficiency of TMD is demonstrated on the example of a footbridge prone to vibrations induced by pedestrians. Is has been shown that if the TMD is tuned quite precisely (especially its frequency) the reduction of accelerations can be very significant.

Acknowledgement

The author gratefully acknowledges support from Technology Agency of the Czech Republic (TAČR), project TE01020168 *Centre for Effective and Sustainable Transport Infrastructure* (CESTI) and CTU SGS project *Development and application of advanced algorithms for numerical analysis and modeling of structures and materials.*

References

[1] Research Fund for Coal and Steel 2007 Human induced Vibrations of Steel Structures *Design of Footbridges*

[2] Živanovič S, Pavic A and Reynolds P 2001 Vibration serviceability of footbridges under human-induced excitation: a literature review *Journal of Sound and Vibration* **279** pp 1-74

[3] Máca J, Valášek M, Machač J and Vampola T 2010 Dynamic Interaction between Human Gait and Footbridges *Proc The Tenth Int. Conf. on Engineering Computational Technology (Valencia)* ed B H V Topping, J M Adam *et al* (Stirlingshire, Scotland: Civil-Comp Press) CDROM pp 1-13

[4] Bachman H and Ammann W 1987 *Vibrations in Structures Induced by Man and Machines* (IABSE Structural Engineering Document 3e)

[5] Grandić I Š 2015 Serviceability verification of pedestrian bridges under pedestrian loading *Tehnički vjesnik* **22** pp 527-537

Reserves in load capacity assessment of existing bridges

Jan Žitný[1] and Pavel Ryjáček[1]

[1]Czech Technical University in Prague, Thákurova 7, 166 29 Praha 6, Czech Republic

E-mail: jan.zitny@fsv.cvut.cz

Abstract. High percentage of all railway bridges in the Czech Republic is made of structural steel. Majority of these bridges is designed according to historical codes and according to the deterioration, they have to be assessed if they satisfy the needs of modern railway traffic. The load capacity assessment of existing bridges according to Eurocodes is however often too conservative and especially, braking and acceleration forces cause huge problems to structural elements of the bridge superstructure. The aim of this paper is to review the different approaches for the determination of braking and acceleration forces. Both, current and historical theoretical models and in-situ measurements are considered. The research of several local European state norms superior to Eurocode for assessment of existing railway bridges shows the big diversity of used local approaches and the conservativeness of Eurocode. This paper should also work as an overview for designers dealing with load capacity assessment, revealing the reserves for existing bridges. Based on these different approaches, theoretical models and data obtained from the measurements, the method for determination of braking and acceleration forces on the basis of real traffic data should be proposed.

1. Introduction

The Czech Republic has one of the most dense and oldest railway networks in the world. Its history reaches back to the beginning of the 19th century. Due to the several historical and political reasons its development stagnated in the second half of the 20th century. Despite the tradition of railway transportation and the size of the railway network, it quickly became outdated in comparison to western neighboring countries. During the early stages of the railway construction, the steel bridges were widely used to span long and medium distances. Apart from their practical purpose, these bridges have often great architectural and historical value. Nowadays, the needs of transportation rapidly grow and the modern codes reflect these needs by introducing new traffic models and approaches for calculation of load capacity of existing bridges. In the Czech Republic, the Eurocodes are used for the design and assessment of bridge structures. Eurocodes are often too conservative for the historical steel bridges, which are in many cases part of the local railway tracks with lower transportation importance.

One of the load actions, which often limit old bridges designed according to historical norms, are braking and acceleration forces. Although these forces do not usually have impact on the overall load capacity of the main load-bearing superstructure, the cross-beams and horizontal bracing members do not usually satisfy the structural checks.

The aim of this paper is to review the methods for braking and acceleration forces determination. Even though majority of European countries use Eurocodes for design and assessment of new bridges, many national authorities for railway transportation have their own approaches for load capacity calculation of existing bridges. These methods are either based on national historical norms or at least adjust the Eurocode models.

2. The development of braking and acceleration forces

In the following paragraphs is the summary of historical norms used in the Czechoslovakia and the Czech Republic. This chapter outlines the models used for calculation of acceleration and braking forces.

Table 1. The list of historical norms used in the Czechoslovakia and the Czech Republic until Eurocode introduction. One example of load model is shown for all codes.

Designation	Original name	English translation	Period	Load Model
-	Nový mostní řád	New Bridge Regulations	1904-1923	2x119t+3.7t/m
-	Československý mostní řád	Czechoslovak Bridge Regulations	1923-1937	2x150t+8.0t/m
ČSN 1230	Jednotný mostní řád	Unified Bridge Regulations	1937-1941	2x150t+8.0t/m
-	Výnos ministerstva dopravy	Ministry of Transport Proceeds	1941-1950	2x175t+8.0t/m
-	Smernice pro navrhování mostů	Bridge Design Guidelines	1950-1953	2x175t+8.0t/m
ČSN 73 6202	Zatížení a statický výpočet mostů	Loads and Structural Analysis of Bridges	1953-1968	2x192t+7.6t/m
ČSN 73 6203	Zatížení mostů	Load Actions on Bridges	1968-1987	2x192t+7.6t/m
ČSN 73 6203	Zatížení mostů	Load Actions on Bridges	1987-2004	UIC 71

Until the ČSN 73 6203 – Zatížení mostů (Load actions on bridges) which was used between year 1987 and 2004, all previous norms had assumed the braking forces as 10% of the overall weight of the whole train. This was limited to bridges in open track and stations with maximum horizontal allignment of 10‰. The dynamic coefficient was introduced and used since the Jednotný mostní řád (Unified Bridge Regulations) 1937-1941.

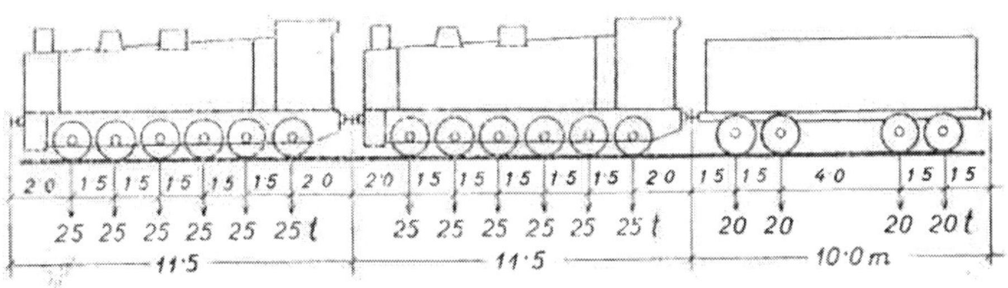

Figure 1. The load model from the Czechoslovak Bridge Regulations.

In the ČSN 73 6203 – Zatížení mostů (Load actions on bridges) 1987-2004 new concept of braking and acceleration forces calculation was implemented. Acceleration forces were introduced acting against the train direction of travel, where braking forces act in the direction of travel. The friction coefficient is used for calculation of both, braking and acceleration forces. The coefficient depends on the type of acting force, the used connections of the rail and the presence of rail expansion joints, stiffness of bearings and the loaded length of the bridge. The friction coefficient is based on the theoretical model ČSD – Frýba theory [1] and the results of vast measurements made under the UIC ORE D 101 research in 1967 focused on the braking and acceleration forces.

3. The current standards and national codes

3.1. EN 1991-2: Eurocode 1: Actions on structures – Part 2: Traffic loads on bridges

This norm should be used in all countries in European Union for the determination of traffic load actions on bridges. Acceleration and braking forces in this code match those in *UIC 774-3 R – Track/bridge interaction, recommendations for calculations*.

The acceleration and braking forces are calculated from the loaded length of the bridge for load models LM71 and heavy train load models SW according to following equations.

$$Q_{lak} = 33[kN/m]L_{a,b}[m] \leq 1000[kN] \quad (1)$$

$$Q_{lbk} = 20\ [kN/m]L_{a,b}[m] \leq 6000[kN] \quad (2)$$

The characteristic values are not multiplied by dynamic coefficient or reduction factor. The values are multiplied by load classification factor. In the case of specified traffic on the track, the acceleration and braking forces can be taken as 25% of the acting axel load. For bridges, shorter than 40 m, the forces in bearings may be reduced depending on the type of rail connections and expansion joints.

3.2. RIL 804 – Railway bridges and other engineering structures – construction and maintenance

The value of acceleration and braking forces in the German code is estimated according to the EN 1991-2. However, the reduction factor for of the bearing load depending on the rail connections and expansion joints is defined up to 300 m. The reduction factor varies between 0.5 and 0.9.

The *RIL 805 – Safety of existing railway bridges* defines the lower upper limit for acceleration forces on the non-electrified sections of the track.

3.3. SIA 269 – Maintenance of structures – load actions

This swiss norm adjusts the *SIA 261 – Load actions on structures* for existing bridges. The value of acceleration forces is not modified. However, the braking forces and their upper limits differ for various categories of tracks based on the traffic density, and they are up to 10% percent lower, than Eurocode.

3.4. TDOK 2013:0267 – Calculation of load capacity of existing bridges

The acceleration and braking forces are calculated according to this Swedish norm as 1/7 of the vertical load. The forces are limited to 4000 kN for the load models TLM 1 and TLM 2 (axle load of 300 kN) and to 6000 kN for the load model TLM 3 (axle load of 350 kN).

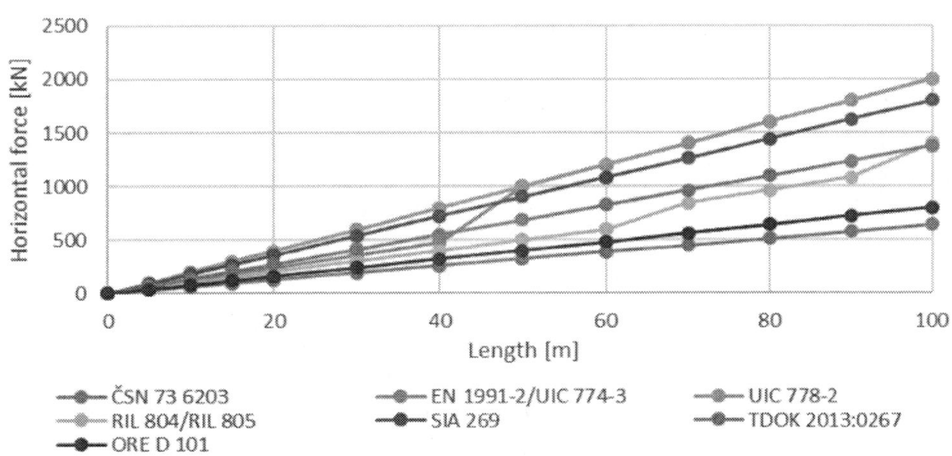

Figure 2. Braking forces according to different current and historical codes.

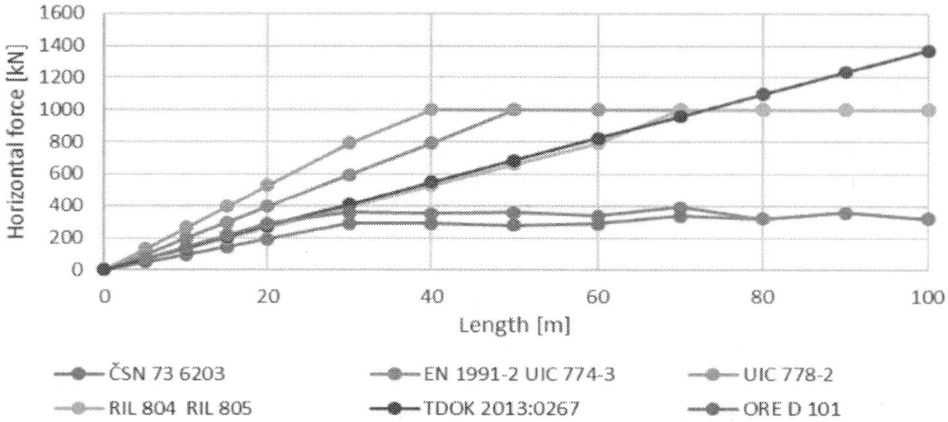

Figure 3. Acceleration forces according to different current and historical codes.

3.5. Action comparison
If we compare the acceleration and braking forces, we can clearly see their value and compare in the Figure 2 and 3. We can see very high variability of the actions.

4. Braking tests results

4.1. ORE D 101 Braking and acceleration forces
The team of specialists consisting of representatives of different European national railway authorities was established in the 1967. The main goal was to solve the issue of estimation of braking and acceleration forces and their distribution over the superstructure, substructure bearings and rails. The assumed variable parameters were: a) structural type of the bridge, b) material of the superstructure, c) bridge span, d) type of bearings, e) railway superstructure and e) rail connections and expansion joints.

One of the outputs is the theoretical model ČSD – Frýba [1]. In the model, the superstructure is defined as a beam element supported at its ends. The rail is replaced by another beam element with free ends continuously connected to the bearing structure and surrounding area by springs. The stiffness of the springs consists of stiffness of connecting elements and stiffness of the rail superstructure. The horizontal load is calculated from the vertical load and friction coefficient. The resulting quantities are forces in rails, bearing structure and bearings.

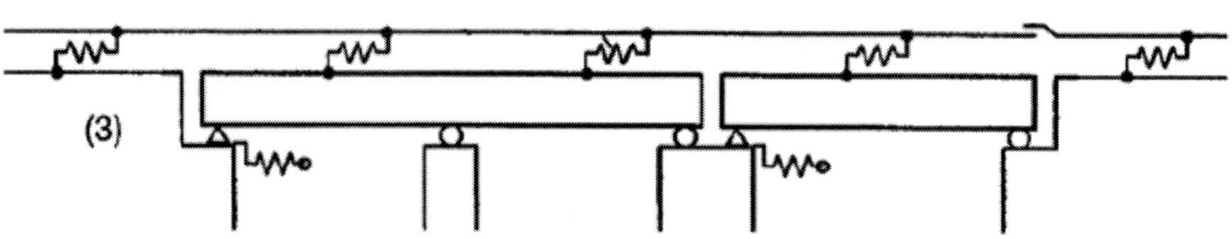

Figure 4. The chart of the quasi-static ČSD – Frýba model.

The most important findings of ORE D 101 research are:

- The maximum braking force occurs shortly before the stopping of the train. The acceleration force grows together with the train thrust.

- The horizontal forces are due to the friction transferred to the rail, bridge, bearings, substructure and surrounding earth. The distribution of forces to different structural members depends on the type of rail connections and expansion joints, length of bridge and type of bearings.
- The influence of structural type, material and rail superstructure has almost no impact on the value of braking and acceleration forces.
- The greatest influence on the value of braking and acceleration forces has the type of the train and adhesion between the braking or accelerating vehicle and the rail.

The value of the vertical force coefficient differs under the given parameters from 0.04 to 0.25. For the most bridges under 50 m the acceleration is the dominant effect and for the bridges over 50 m it is braking.

4.2. The braking test on Losí Bridge
The Losí Bridge is a four-span continuous girder bridge with orthotropic deck and rail bed, placed in the Czech Republic. The bridge transfers a double track. The bridge is supported by pot bearings. The braking test was performed in April 2016. Four electric locomotives Class 230 Laminátka were used for the braking test, each of weight 85 tons. The configuration on the bridge was divided into three cases, single locomotive, four on one track and two on each track. The maximum measured braking force was equal to 17.5 of the acting vertical force [5].

4.3. The braking test on Chotoviny Bridge
The Chotoviny Bridge is a Langer beam bridge transferring double track with welded rails and supported by pot bearings, placed in the Czech Republic. The Langer beam is 100.5 m long. The braking test was performed in 2015, using one locomotive HV740 with twelve Faccpp 9-429.0 wagons and assembly of two HV749 and two HV771 locomotives.

Table 2. The results of braking and acceleration test on Chotoviny bridge [3].

Assembly	Weight [t]	Braking force [kN]	Acceleration force [kN]	Braking force - ratio	Acceleration force - ratio
HV740+Faccpp	1104	525	205	5%	2%
HV771+749	381	920	305	24%	8%

4.4. The braking test on Oskar Bridge
The Oskar Bridge is an arch bridge with box girders, orthotropic deck and rail bed, placed in the Czech Republic close to Austria. There are two same bridges. The arch span is 97.5 m long and carries one track. The bridge us supported by spherical bearings. The bridge is a subject of long term monitoring which started in 2015 and is planned to continue until 2018. The braking test was performed by assembly of two Siemens Taurus ES64U2 locomotives and ten Falls wagons. The maximum measured braking force was equal to 12% of the total weight of the assembly [7].

5. Case study – assessment of Červená Bridge
The Červená Bridge is the perfect example of historical steel railway bridge struggling with calculation of load capacity according to Eurocodes. The bridge was built in 1889 and is a part of the railway track connecting two cities in the southern Czech Republic, Tábro and Písek. The bridge is 254.2 m long with three main spans of 84.4 m, the bearing structure is formed by two truss girders with overall height of 9.9 m and system of cross and longitudinal beams supporting intermediate bridge deck. The rail expansion device is installed on both ends.

Figure 5. The Červená Bridge.

Figure 6. The model of Červená Bridge.

The load capacity calculation was made by SUDOP PRAHA a.s. [4] and the load capacity estimation of this bridge was also the topic of thesis "Static and dynamic behavior of the Červená railway bridge" [6]. In order to investigate the utilization of all structural elements, accurate model was created in both cases. First, the capacity was calculated using the LM71 load model according to Eurocode. The design capacity of the cross beam at the abutment, due to braking forces, was exceeded more than eight times. Different approach was chosen to prove the bridge can at least carry the current traffic. The bridge is located on a local track and therefore, the real traffic model was used. The load train consists of five vehicles each with four 180 kN axels and the braking force was assumed of 20% of the vertical load. Even though, the utilization level of the abutment cross beam was still around 2.1, there was a significant improvement.

The results from the load capacity assessment were taken and used for a simple exhibition of other approaches of braking force calculation. The bridge was assumed to be loaded by LM71 load model from the Eurocode. The safety and combination factors were preserved according to the EN 1990. Only the approach of the braking load model was changed. The results are shown in Table 3.

IOP Conf. Series: Materials Science and Engineering **236** (2017) 012061 doi:10.1088/1757-899X/236/1/012061

Table 3. Červená Bridge cross beam utilization level.

Norm	ČSN 73 6203	EN 1991-2 UIC 774-3	UIC 778-2	RIL 804 RIL 805	SIA 269	TDOK 2013:0267	ORE D 101
Braking Force [kN]	2847	5084	5084	4576	4576	3486	2644
Stress – traffic [MPa]	1002.3	1647.6	1647.6	1500.9	1500.9	1186.7	943.7
Stress – overall [MPa]	1176.5	1821.8	1821.8	1675.1	1675.1	1360.9	1117.9
Utilization [-]	5.45	8.44	8.44	7.76	7.76	6.30	5.18

Even though the cross beam does not still satisfy the design checks, the table shows very significant impact of the brake/acceleration model on the overall load capacity of the bridge. Also, the conservativeness of the Eurocode is evident.

6. Real traffic model proposal

The real traffic data on the whole railway network of the Czech Republic were provided by the Czech railway maintenance authorities SŽDC. The data contain of traffic during June and November on all track sections owned by SŽDC. These months were chosen due to lower cargo traffic during summer months and higher cargo traffic at the end of every year. Therefore, combined data from these two months should be sufficient to affect the whole year and to form the accurate real traffic model.

The data provide information about the type of locomotive, the total number of axels, weight and length of the whole train assembly. The evaluation of these data can be used to determine the year frequencies of train assemblies of certain type on each track per year. Combined with the data about locomotive braking and acceleration forces, the normative theoretical models and real in-situ measurements, real traffic model can be created for the whole Czech railway network. Evaluation of the data for every single track of the railway network would be too complex. Therefore, the track classes can be used for such a purpose. The Czech railway network is divided into seven classes by SŽDC based on the amount of gross transported tons per year. Such dividing criterion can accurately secure separation of local tracks with low traffic importance used only for public transportation purposes and heavily used tracks for cargo transport.

7. Conclusion

The review of historical and current used national codes has provided useful information about the determination of braking and acceleration forces. Together with the data obtained from several in-situ measurements of these actions it creates a clear idea how Eurocodes can be conservative especially for load capacity calculation of existing bridges. Moreover, the Eurocodes consider the design service life for bridges of 100 years, where some of the old bridges are assumed to be in service for maximum ten to twenty years.

In order to be able to preserve these bridges and not to lower their allowable capacity, the braking and acceleration forces effects should be in some cases reduced. The combination of real traffic loading and reduced braking forces can lead to significant improvements in load capacity assessment and historical bridges such as Červená Bridge can be used for another several years before the necessary funding is secured for either the bridge reconstruction or the structure replacement.

For such purpose the real traffic model should be created, providing the useful information about the trains and frequencies on each of the seven railway track classes in the Czech Republic.

Acknowledgements
Research reported in this paper was supported by Competence Centers program of Technology Agency of the Czech Republic (TA CR), project Centre for Effective and Sustainable Transport Infrastructure (no. TE01020168).

IOP Conf. Series: Materials Science and Engineering **236** (2017) 012061 doi:10.1088/1757-899X/236/1/012061

References

[1] Frýba L 1978 *Brzdné a rozjezdové síly na železničních mostech* (Výzkumný ústav železniční v Praze)

[2] Ryjáček P, Macho M, Stančík V and Polák M 2016 Deterioration and assessment of steel bridges *8th International Conference on Bridge Maintenance, Safety and Management IABMAS 2016* pp 1188-95

[3] Ryjáček P, Brůna M and Karmazín K 2016 The Behaviour of the Bridges with the MW Steering Bar *Procedia Engineering* **156** pp 395-402

[4] SUDOP PRAHA a.s. 2015 *Statický přepočet mostu km 41,791 trati Tábor - Písek včetně návrhu řešení opravy*

[5] Krúpa J 2017 *Chování Losího mostu při účinku brzdných sil* (CTU in Prague)

[6] Vachutka T 2016 *Statické a dynamické chování mostu Červená* (CTU in Prague)

[7] Werunský M 2017 *Chování šikmého mostu s řídící tyčí MW při brzdných silách* (CTU in Prague)

[8] Žemličková L 2004 *Ekvivalentní rozkmit napětí železničních mostů* (CTU in Prague)

IOP Conf. Series: Materials Science and Engineering **236** (2017) 012062 doi:10.1088/1757-899X/236/1/012062

Instrumentation and monitoring of segmental post-tensioned girders

Roberto Gomez[1], J Alberto Escobar[1] and Hector Guerrero[1]

[1] Structural Engineering Department, Institute of Engineering, UNAM, 04510 Mexico City, Mexico

E-mail: rgom@pumas.ii.unam.mx

Abstract. As a result of the failure of prefabricated segmental girders during their construction, and in order to evaluate the structural safety of the superstructures of three bridges, different activities of instrumentation, monitoring and mathematical modelling were performed. This paper presents the first stage of a study describing the tasks of instrumentation and monitoring of stresses in an isolated precast post-tensioned beam during the process of application of the tension loading, and before being placed in its final position. The main goal of this task was to confirm the transmission and distribution of stresses at segmental joints. Also, ambient vibration tests were carried out in different phases of tensioning, as well as in the final position of the girder. Recorded stories of accelerations are presented and fundamental periods of vibrating of the isolated girder are derived. The results of the measurements will be used for the calibration of mathematical models and simulation of different scenarios of loads and thus assess the structural safety of the different sections of the superstructures of the three bridges in question.

1. Introduction

During the manufacture of post-tensioned prefabricated segmental concrete girders, 53m long, three of them failed in their shear key joints during the application of the post-tension load, while the rest do not show currently any kind of incipient form of instability. Most of the prefabricated beams have been already placed in the different spans of three bridges located along a new beltway in Mexico.

A technical opinion about the failure of the girders was requested. For this purpose a study was proposed to assess the behaviour of new beams and superstructures, to be built with the same type of construction system, under different types and levels of static and/or dynamic loading.

With this in mind, it was necessary to define an instrumentation scheme of the isolated girders during construction and of some spans of the superstructures. This paper only presents the instrumentation activities of some beams, representative of the superstructures, with the objective of measuring deformations, displacement and camber during their tensioning; parameters that could provide valuable information and help to determine probable causes of failure, as observed. Also, frequencies and periods of vibrating of some isolated girders were derived based on ambient vibration tests. These parameters were measured only during the tensioning.

Additionally, mathematical models were elaborated in order to reproduce the experimental data and collect reliable analytical information to predict the behaviour of the bridge under certain events of loading. A design review of the superstructure is not the aim of this paper.

2. The superstructures

Each span of the superstructures of the bridges studied is comprised of 5 posttensioned concrete beams of 2.4 m height, with 4 transversely symmetrical cables of pre-stress. Figures 1 and 2 show a typical cross section of the superstructure and the position of cables in a girder, respectively. Each girder is composed of five segments which are concreted in a special yard near the construction site; shear keys are provided for each segment which are assumed to be cast in the same order as they will be placed in their final position.

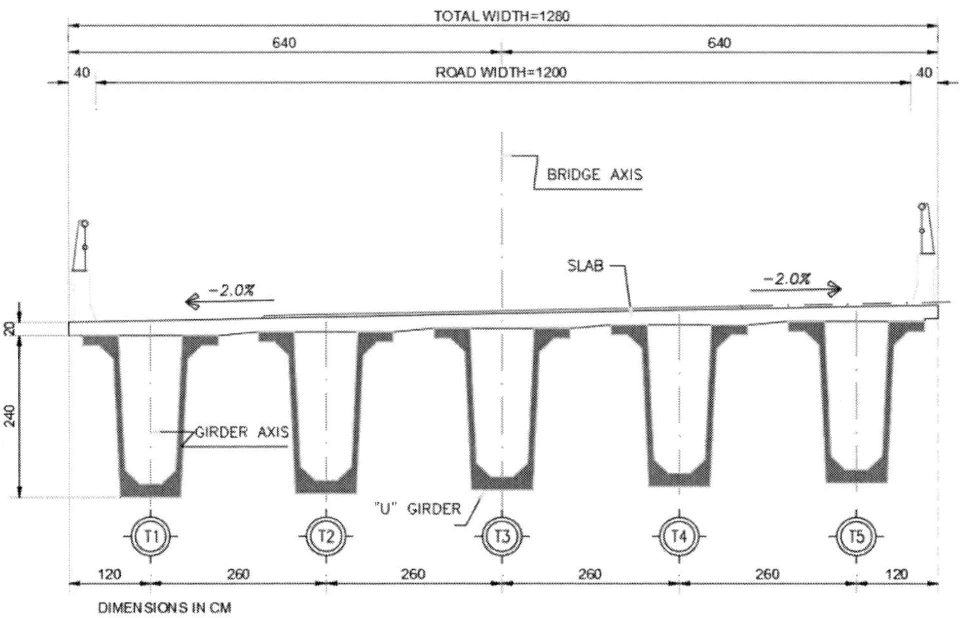

Figure 1. Typical cross section of the studied superstructures.

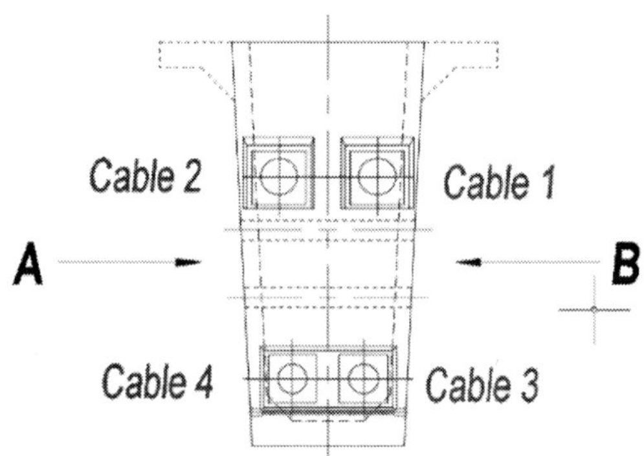

Figure 2. Position of cables/tendons for post-tensioning.

The sequence of tensioning was applied as follows: first, cable 1 is tensioned up to 50% of its design load; immediately after, cable 2 is tensioned up to 100% of its design load; cable 1 is then completed to its 100% design load. Then, tendon 3 is tensioned up to 50% of its load followed by the application of 100% of presstress to cable 4. The tensioning devices are returned to cable 3 in order to complete 100% of its design tension. The post-tension loads were: 384.7 t, 384.7 t, 305.8 t, and 308.5 t in cables 1 to 4, respectively.

Once the five segments are finished, they are transported separately and placed on appropriate temporary supports in a special area just by the approach of the bridge in construction. In this area, the post-tension load is applied. Afterwards, girders are launched to their final position in each span of the bridge.

3. Instrumentation and monitoring

An instrumentation scheme was designed for monitoring strains and stresses during posttensioning of the girders. The scheme comprised electrical strain gauges, signal conditioning cards, communication devices and wiring as well as an industrial type automatic recording software [2].

Based on previous analyses and the identification of stresses in the beams, a total of 12 sensors were placed in each beam studied. Only the joints with the maximum compression stresses were instrumented placing a sensor (strain gauge) on each side at the bottom and top of the same joint, on both sides of the girder. In the vertical direction, sensors were placed at 0.30m and 2.0m from the bottom surface of the girder; the distance between sensors on both sides of a joint was 0.50m.The aim was to monitor the transfer and distribution of stresses between segments. Figure 3 shows the location and nomenclature of each of these sensors for a specific beam.

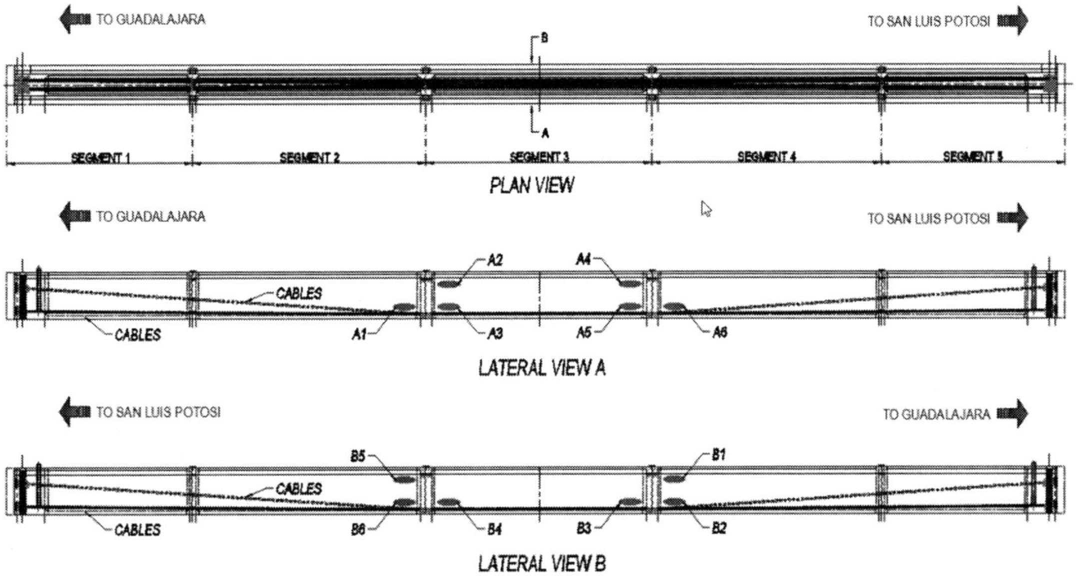

Figure 3. Instrumentation scheme for segments of beam TR7, Don Viejito bridge.

4. Results

4.1 Distribution of stresses

Below are shown some results associated to one of the instrumented girders (labelled as TR7), as a product of the tensioning of the cables. Table 1 shows a summary of the maximum values recorded. According to design calculations, no tension was to be expected during the application of the post-tension load. Also, it was expected an even distribution of stresses at each side of the matching joints.

As shown in the table, no tensions were registered in any sensor. Also, observing the magnitudes for the maximum values for pairs of sensors located on opposite sides of the joints: A1 - A3, A5 - A6, B2 -B3 and B4 -B6 it is shown that they are comparable. For example -18.43MPa and -20.1MPa for sensors B4 and B6, respectively. The difference may be attributed to the position of the sensors in the longitudinal direction. The same applies for the rest of pairs of sensors. Additionally, when comparing values of stresses at sensors on different joints but in the same longitudinal symmetric position, like joints B1-B5, B4-B3, A2-A4 and B2-B6, the results are better comparable confirming the assumption of an even distribution of stresses.

Table 1. Maximum strains and stresses recorded during tensioning of beam TR7.

Sensor	Initial condition		Final condition	
	μm/m	σ (MPa)	μm/m	σ (MPa)
A5	0	0	-592	-19.02
A6	0	0	-501	-16.1865
B4	0	0	-573	-18.4428
B6	0	0	-624	-20.1105
A1	0	0	-487	-15.696
A3	0	0	-458	-14.715
B2	0	0	-621	-20.0124
B3	0	0	-562	-18.1485
A2	0	0	-109	-3.5316
B1	0	0	-114	-3.7278
A4	0	0	-96	-3.1392
B5	0	0	-125	-4.0221

Figure 4 shows the stories of strains and stresses recorded during the tensioning of the 4 tendons in the sequence described above. Each graph shows, on the horizontal axis, the elapsed time during tensioning, and, on the vertical axis the recorded strain (left graph) in sensors, or equivalent stresses (right graph). Regarding these plots, the first two rows are associated to sensors at the bottom of the U girder, and the last row to sensors at the top of the girder. It is shown that shapes and magnitudes of stresses are similar; and all values increase as long as the load is increased.

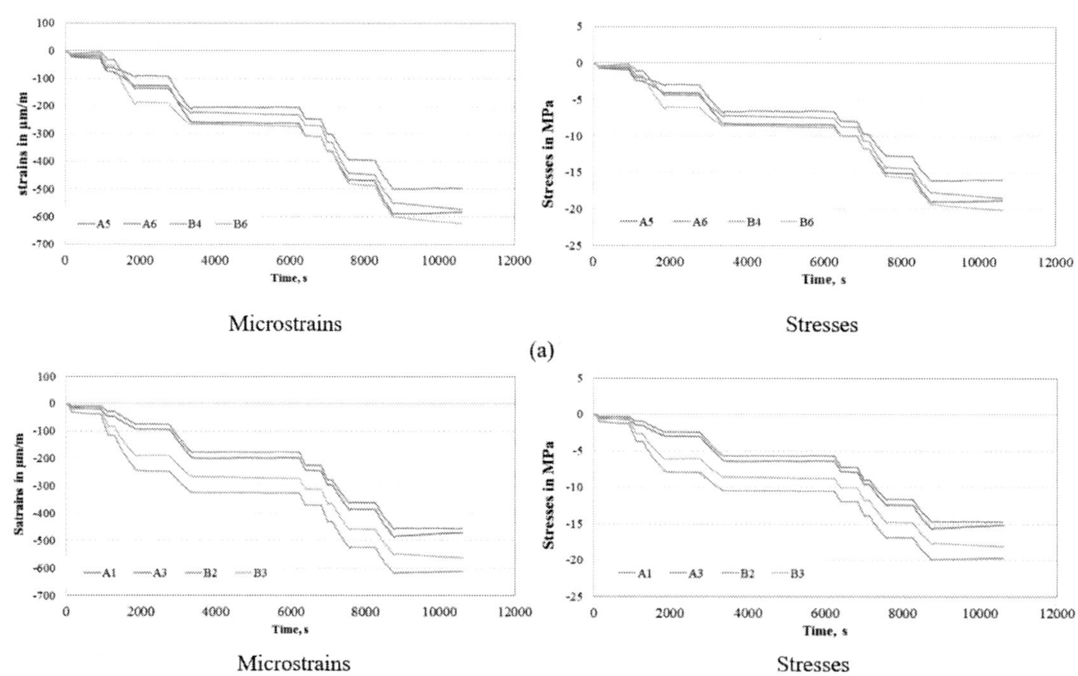

Figure 4. Time histories of microstrains and stresses, girder TR7.

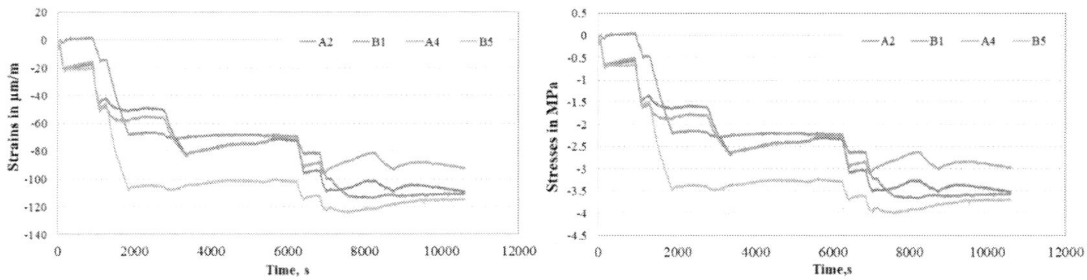

Microstrains Stresses

(c)

Figure 4. (Cont.) Time histories of microstrains and stresses, girder TR7.

Also, it is worth to note that the sensors at the top of the joints (A2-A4 and B1-B5) present the lowest values of the compression stresses, as expected.

4.2 Vibrational frequencies

In order to experimentally determine the dynamic properties of the post-tensioned girders before being placed in the superstructure of the bridge, during the six stages of tension of the four cables, ambient vibration tests were carried out with the purpose of obtaining frequency and natural vibration periods, as well as modal shapes. In particular, the results shown in this paper are focused on the measurements obtained from the fourth stage of tension of the cables up to the final tension of these elements (sixth stage). The ambient vibration tests comprised the recording of accelerations in three orthogonal directions: longitudinal (L), transverse (T) and vertical (V), at three different locations on the girders.

Figure 5 shows schematically the locations where accelerometers were placed during the tensioning of the cables of girder TR7. In order to clarify this matter, a photograph of an accelerometer already placed (position 2) on the top of the web of the "U" girder is included in this same figure. In this photograph the tendons at the bottom of the girder may be observed, as well.

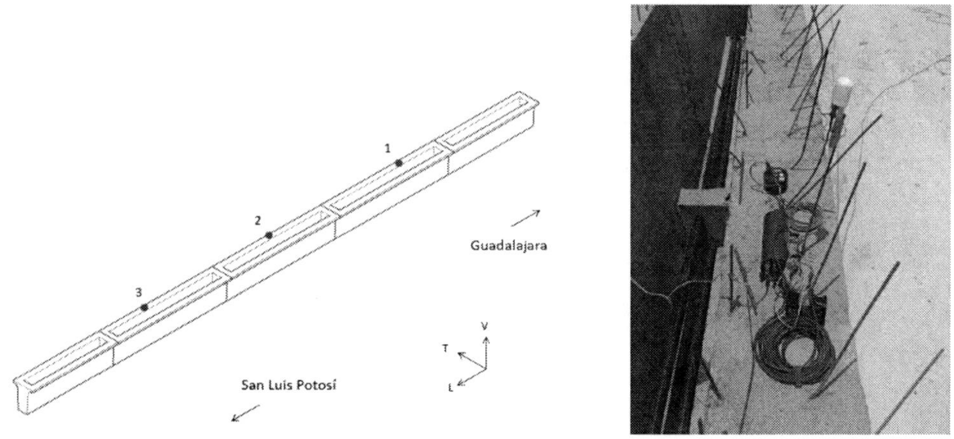

Figure 5. Schematic representation of location of sensors for ambient vibration of "U"girder TR7.

From the vibration program proposed for each of the post-tensioned beams, selected acceleration records were processed by means of a spectral analysis based on pairs of signals [1], from which the spectral density, coherence, phase angle and transfer functions were calculated.

Table 2 shows the identified periods of vibration for the different stages of post-tensioning mentioned. These values were derived from sets of functions like those shown in Figure 6, in accordance with a procedure based on the identification of frequencies associated to peaks on the spectral density function and correlation of values in the coherence function. Another check with the same frequency values is performed for the phase angle function, which quantifies the synchrony of both signals. In this way, the frequencies/periods associated to fundamental and higher modes of vibration of a structure may be identified.

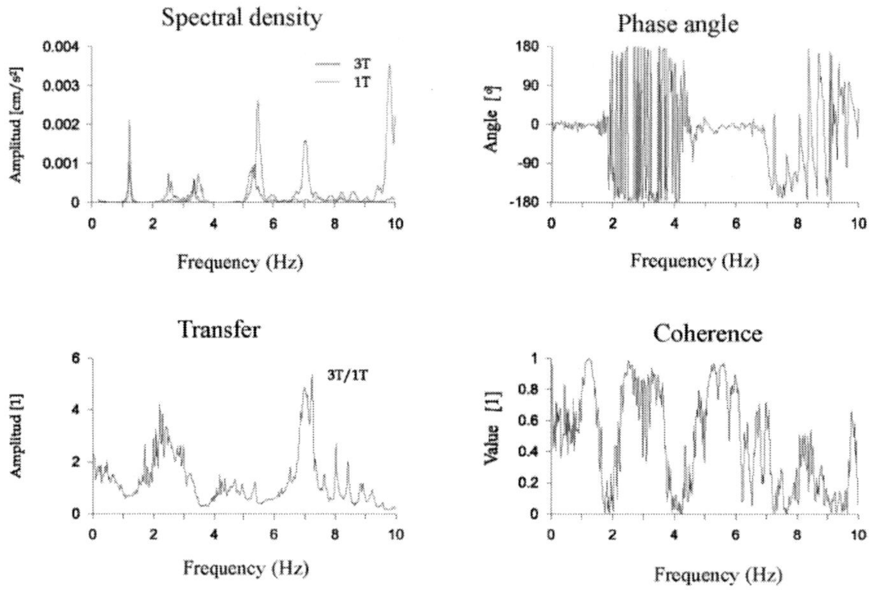

Figure 6. Spectral density, phase angle, transfer and coherence functions at locations 1 and 3; T component, 4th stage of tensioning of cables, girder TR7.

Table 2. Natural frequencies of vibration of girder TR7, after the 4th, 5th and 6th stages of posttensioning.

Mode	Component	Freq. [Hz] Stage 4	Freq. [Hz] Stage 5	Freq. [Hz] Stage 6
1	T	1.221	0.831	0.830
2	L-V	3.345	1.855	1.855
3	T	3.491	2.173	2.148
4	V	8.228	6.714	6.738

The identified frequencies presented in Table 2 show that the first modes of vibration, vertical and longitudinal, are coupled, a circumstance which can be attributed to the conditions of support at the ends of the girder, and to the sequence of loading. Another issue that is observed in the table is the effect of the post-tensioned load: up to certain magnitude of the applied load, the girder is supported on several points (joints) changing the supported length and in consequence the frequency. During the final stages of load, the girder is supported only at the ends. For the purposes of calibration of a mathematical model of the girder, attention should be focused in reproducing the first vertical mode and the first and second lateral modes of vibration.

5. Mathematical modelling

As mentioned, three-dimensional solid finite element models of several isolated girders were developed in order to be calibrated with experimental results. The calibration was performed using SAP2000 program [3] and the final post tension forces (stage 6). All models used the same mechanical nominal characteristics of materials and considered the same effect of pre-stress cables and ducts.

Figure 7 shows a 3D view of the mathematical model of a segment used to conform a typical girder.

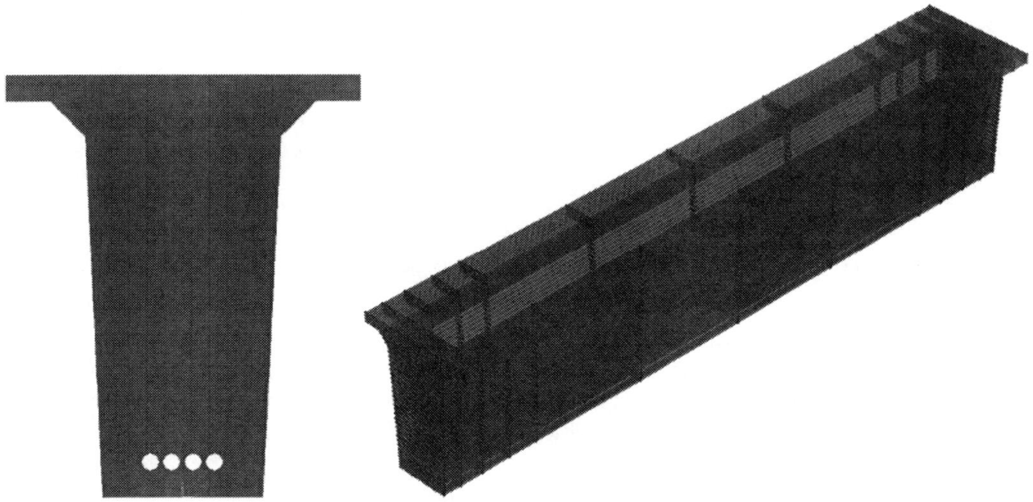

Figure 7. Mathematical model of a TR7 girder segment.

In the following table a comparison between the experimental and theoretical results are presented.

Table 3. Comparison of experimental and numerical vibrational periods for a type TR7 girder.

Direction	Transversal (s)		Vertical (s)	
Mode	1	2	1	2
Experimental	1.205	0.466	0.539	0.148
Mathematical model	1.133	0.522	0.568	0.147
Relative error	-6.0%	12.0%	5.4%	-1.0%

On the other hand, finite element results of axial stresses for the last post tension stages (3-6), showed only compression stresses in the whole cross section of the girders. Additionally, when comparing the maximum numerical stress at point B6, its magnitude was -22.57MPa and the one obtained experimentally was -20.1MPa, which is a fair approximation considering the difficulties involved in developing the model of the girder. The same situation was observed when comparing the maximum numerical and experimental camber: 78.3 vs 80mm, respectively.

6. Final comments

The results of the first stage of a study developed to instrument and monitor different joints of segmental girders for three bridges were presented. The objective was to assess the transfer and distribution of stresses at different construction joints of the girders. The results of this stage included the recording of strains during the application of the post-tensioning load on the girders before being placed in their final position, as well as ambient vibration tests.

Regarding the results obtained with the strain gauges, a good behaviour was observed in spite of some minor misalignments and geometry flaws that were observed prior to the monitoring of the strains. A good transference and distribution of stresses was recorded. The results do not reveal any possible deficiency in the structural stability of the girders studied.

Although some coupled vibration modes were perceived, results of the ambient vibration study showed that these segmental girders have a global behaviour that does not differ significantly from the one assumed for a regular girder cast in a single element. It is expected that these behaviour will not have any effect on the global behaviour of the complete deck: girders and slab. Some changes in the frequency values were observed between different stages of tension.

Acknowledgements

Thanks to O. Rosales, M. A. Mendoza, and L.M. Arenas who were in charge of the instrumentation and monitoring of the girders. L.M. Arenas also helped during the preparation of the final version of this paper.

References

[1] Bendat J S and Piersol A G 1989 *Random data: analysis and measurements procedures* (New York: Wiley Interscience)

[2] Gómez-Rosas E, Gomez R, Arenas-Garcia L M, Mendoza-Garcia M A, Rosales-Gonzalez O N and Escobar-Sanchez J A 2012 *Instrumentation and monitoring of six road bridges by means of a measurement system developed at the Institute of Engineering of UNAM (in Spanish)* (VII Mexican Society of Instrumentation Congress, Culiacan, Sinaloa, Mexico)

[3] CSI 2009 *SAP2000 Version 18.0.1, Integrated Finite Element Analysis and Design of Structures* (Computers and Structures Inc. Berkeley: California)

Steel bridge in interaction with modern slab track fastening systems under various vertical load levels

Vojtěch Stančík[1], Pavel Ryjáček[1] and Miroslav Vokáč[2]

[1]Czech Technical University in Prague, Faculty of Civil Engineering, Department of Steel and Timber Structures, Thakurova 7, Prague 6 16629, Czech Republic
[2]Czech Technical University in Prague, Klokner Institute, Department of Experimental Methods, Solinova 7, Prague 6 16608, Czech Republic

E-mail: vojtech.stancik@fsv.cvut.cz

Abstract. In modern slab tracks the continuously welded rail (CWR) is coupled through the fastening system with the substructure. The resulting restriction of expansion movement causes significant rail stress increments, which in the case of extreme loading may cause rail failures. These interaction phenomenon effects are naturally higher on a bridge due to different deformation capabilities of the bridge and the CWR. The presented contribution aims at investigating the state of the art European direct fastening system that is suitable for application on steel bridges. Analysis involves experimental determination of its nonlinear longitudinal interaction parameters under various vertical loads and numerical validation. During experimental procedures a two and a half meter long laboratory sample equipped with four nodes of the Vossloh DFF 300 was tested. There have been checked both DFF 300 modifications using the skl 15 tension clamps and the low resistance skl B15 tension clamps. The effects of clamping force lowering on the interaction parameters have also been investigated. Results are discussed in the paper.

1. Slab track in interaction with bridge

Ballastless track technologies can be conveniently used for constructing shallow steel bridge decks with reduced self-weight. On a steel bridge the ballastless track is formed by connecting the continuously welded rail (CWR) to the orthotropic steel deck directly through the special direct fastening system. Naturally in the CWR the longitudinal forces are rising because of the restriction of its expansion movement. On a bridge there is even more significant stress increment caused by different deflections of the bridge compared to CWR (see Figure 1).

In the case of ballastless tracks the crucial component, describing the transmission of longitudinal forces, is the fastening system. For performing a sufficiently reliable bridge/track interaction analysis it is necessary to investigate the longitudinal nonlinear stiffness of the fastening system. However values of this input parameter are quite uncertain. In the scientific literature and national codes various values of the ballastless track longitudinal resistance r_x can be found. Fundamentals in this field of study were laid by prof. Frýba and are summarized in [1]. Frýba conducted several in-situ experiments on different bridge deck types, involving the ballastless tracks. In order to obtain the values of r_0, prof. Frýba introduced analytical solution. By solving systems of differential equations considering various values of r_0 he was able to find the most fitting value to the experimental results. However in this case only the parameters for unloaded track were investigated. When assessing the longitudinal bridge/track interaction bridge designer must prevent reaching rail stress additional increment limits of 72 MPa in compression and 92 MPa in tension caused by live loads. These limits have been calculated for 60 E2 rail and for other rail types need to be recalculated. More recently Geissler and Freystien released a study [2] where they proposed higher stress increment limits when the bridge is arranged appropriately. Amongst other they demonstrated that ballastless tracks may resist higher stress increments in compression thanks to their rigid transversal resistance.

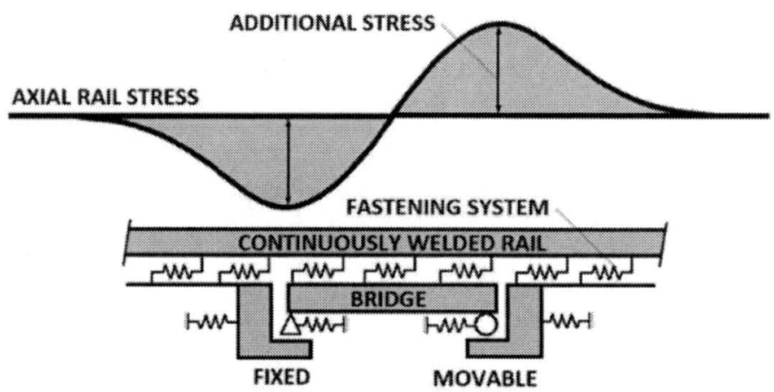

Figure 1. Bridge rail interaction phenomenon.

Regarding the modelling of interaction functions, general recommendations are given by UIC-774-3R [3]. Some important parameters may be found amongst other in German national codes [4] or Dutch national codes [5], while the current version of European code [6] doesn't involve recommendations for ballastless tracks. For the purposes of proper interaction analysis each node of fastening may be modelled using the nonlinear spring (Figure 3.), where the stiffness K_x refers to the recalculated value of bilinear longitudinal resistance (Figure.2) and K_z refers to vertical stiffness of elastic pads and tension clamps. Variables marked with capitals refer to one node only, while the small letters denote values recalculated for one meter of railway track length. In the before mentioned codes, values of plastic longitudinal resistance r_0 varies between 30 and 40 kN/m for unloaded track load cases, respectively 50 and 60 kN/m for loaded track load cases. According to codes the plastic resistance is reached when the relative bridge/track displacement meets the limit u_0, which may be considered with the values in range between 0.5 and 1 mm. Thus the longitudinal stiffness k_x is significantly higher compared to standard ballasted tracks. In his publication [7] Esveld suggested slightly higher value of r_0 = 47 kN/m for unloaded slab tracks. Other experiments brought different results. For example evaluation of in-situ monitoring on a newly built bridge equipped with DFF 300 and the low resistance tension clamp skl B15 [8] indicated rather lower values of slab track longitudinal resistance r_0 = 25 kN/m. Besides that, research study carried out at the CTU Prague [9] also proved that interaction functions are temperature dependent.

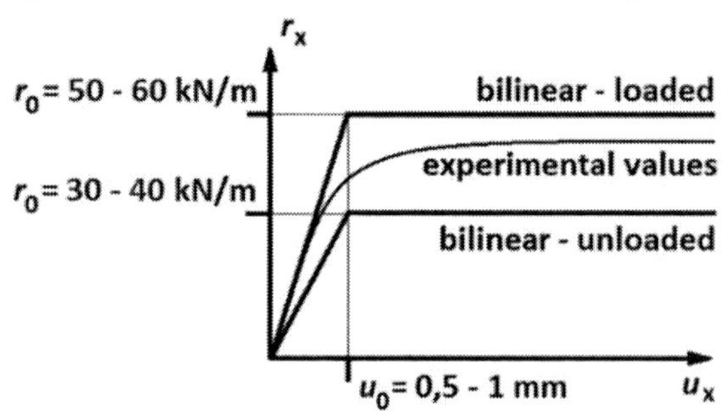

Figure 2. Variations of slab tracks nonlinear longitudinal coupling functions according to [3,4,5].

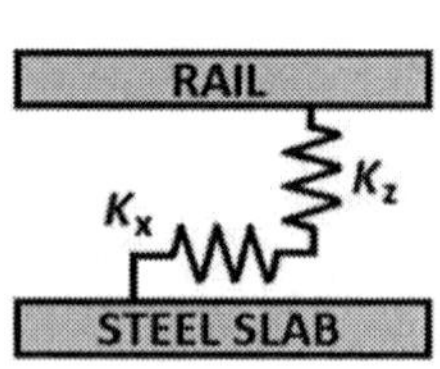

Figure 3. Standard model of the bridge/track coupling interface.

2. Experiment

The aim of experimental analysis was to determine longitudinal interaction parameters of DFF 300, the state of the art direct fastening system, under various vertical loads. To meet the goals, an experiment was conducted in the laboratory of CTU Klokner Institute. Tested sample was assembled from an asymmetric 2.65 metres long steel beam with wider upper flange and four nodes of DFF 300 fastening system. In vertical direction the sample was supported by two sliding supports to achieve the sample planned spanning of 2.5 metres. Horizontal deflection was restricted by anchoring the sample with the M30 steel bars. Each DFF 300 fastening node was assembled according to Vossloh requirements using the steel base plate, elastomeric intermediate plates, steel distribution plate and the hard plastic pad that was placed under the rail toe. Testing involved both the Skl 15 tension clamp and the low resistance Skl B15 tension clamp.

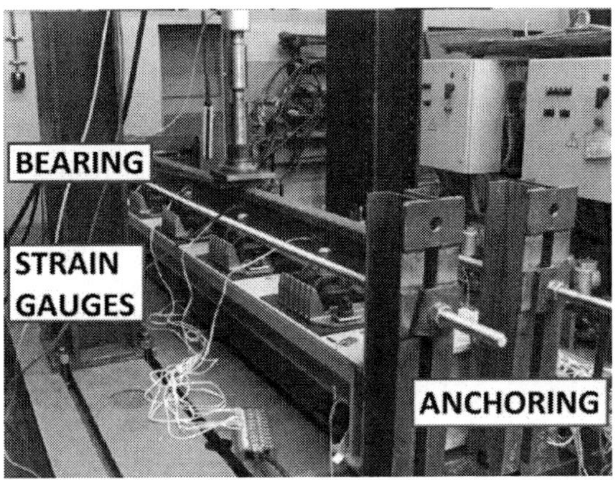

Figure 4. Stress measurement on the small scale sample.

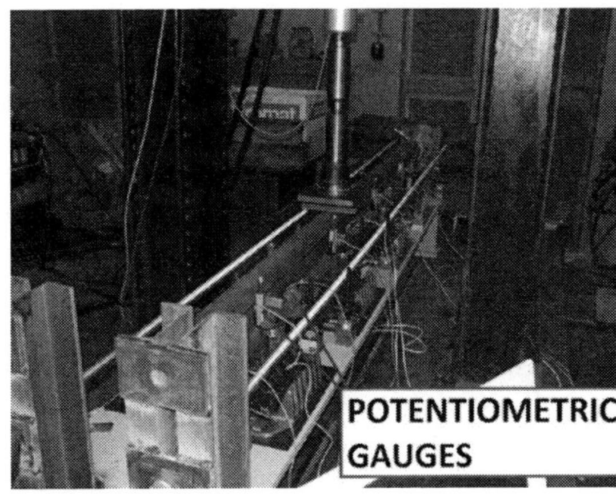

Figure 5. Longitudinal and vertical deflection measurement on the small scale sample.

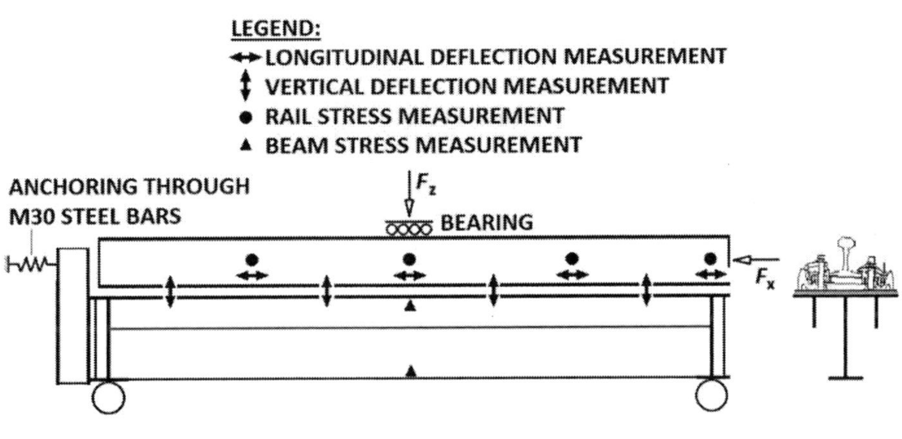

Figure 6. Experiment layout.

During the testing procedure the rail and beam stress, so as the longitudinal and vertical rail deflection were measured. The rail strain gauges were placed in the intermediate sections between the fastening nodes alike the horizontal potentiometric displacement sensors. In vertical direction the deflections were measured directly at the location of each fastening node. Arrangement of gauges is shown in Figure 6. Experimental procedures were based on EN 13481-1 code [10], nevertheless some

IOP Conf. Series: Materials Science and Engineering **236** (2017) 012063 doi:10.1088/1757-899X/236/1/012063

modifications had to be adopted, because the regulation doesn't involve recommendations for testing under vertical loads. Procedure started with vertical loading F_z. After reaching the prescribed vertical load level the rail was loaded horizontally with the force F_x. When the force increment approached the plastic resistance level, the horizontal loading was quickly reduced and plastic straining was observed. After two minutes of displacement development, the sample was horizontally and subsequently vertically unloaded. Four load cases were designed to determine the fastening interaction functions. In the first load case the interaction parameters for unloaded track were observed, while the other three load cases focused on determining interaction parameters under vertical pressure. Thus the sample was gradually tested under 0, 40, 80 and 125 kN vertical loads. Whole procedure was repeated twice. First the fastenings were equipped with Skl 15 and subsequently with Skl B15 tension clamps.

3. Numerical analysis

Generally the brief review of scientific literature showed that the longitudinal resistance of the direct fastening systems is a quite uncertain input parameter that depends on friction between the layers of different materials used within the fastening node as it depends on the elasticity of the tension clamps. Along with the verification of experimental values the numerical analysis aimed at investigating the effects of different tension clamp types and the clamping force lowering on longitudinal resistance. Effects of clamping force lowering are sometimes being discussed as the possible reason for discrepancies between the measured values of r_0. For that case a modified model of the bridge/track coupling interface had to be developed.

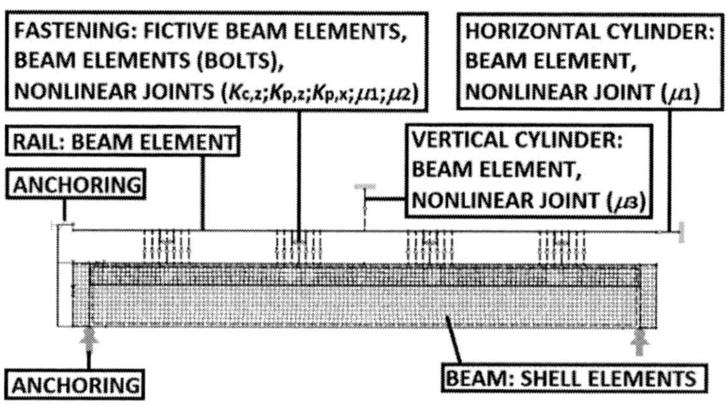

Figure 7. Numerical model description.

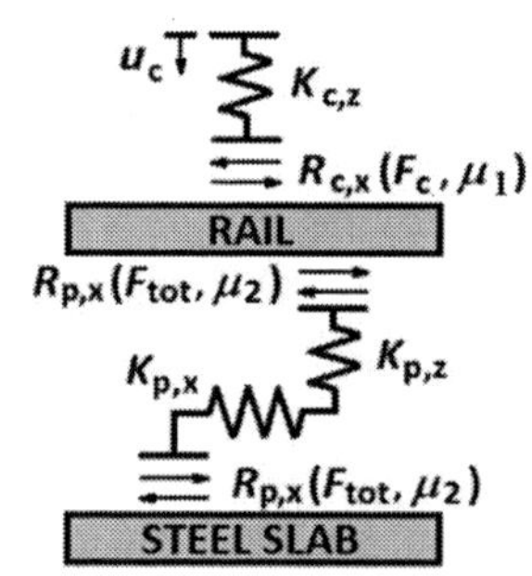

Figure 8. Modified model of bridge/track coupling interface.

The global model for finite element analysis (FEA) was created in Dlubal RFEM software and is described on Figure 7. Steel plates of the asymmetric supporting beam were modelled using shell elements while the rest of the components, thus the rail, hydraulic cylinders and fastening bolts, were modelled using the Euler-Bernoulli beam elements. All the necessary eccentricities were introduced through fictive rigid beams. The most challenging task was to adopt coupling parameters of the fastening system. This was solved using the nonlinear joints at the endpoints of the corresponding beam elements. Main principle of each fastening node model is presented on Figure 8, where u_c is the nominal deflection of tension clamp caused by the fastening bolt tightening, $K_{c,z}$ is the vertical stiffness of the tension clamp, $R_{c,x}$ is the friction force between the clamp and the rail, $R_{p,x}$ is the friction force between the pad and steel parts of the fastening, $K_{p,z}$ is the vertical stiffness of the fastening elastic parts and $K_{p,x}$ is the longitudinal stiffness of the fastening elastic parts. Independent elements for tension clamps and the rest of the fastening components have been used. Therefore $R_{c,x}$ depends on the clamping force F_c and the steel-steel friction coefficient μ_1, while $R_{p,x}$ depends on the total vertical force F_{tot} (1) and the steel-elastomer friction coefficient μ_2.

$$F_{tot} = F_z + F_c \tag{1}$$

Table 1. Input parameters for FEA.

clamp	u_c (mm)	F_c (kN)	μ_1	μ_2	μ_3	$K_{p,z}$	$K_{c,z}$	$K_{p,x}$
Skl 15	20	22	0.1	0.55	0.05	Figure 9.	Figure 10.	rigid
Skl B15	20	12.6	0.1	0.55	0.05	Figure 9.	Figure 10.	rigid

Models of hydraulic cylinders were involved in the FEA so that the friction forces between the cylinders and the rail could be adopted. Mainly the coefficient of rolling friction μ_3 modelling the cylinder resistance against the rail longitudinal displacement was important during the FEA process to improve its accuracy. Also by using the cylinder models the loading could be controlled by horizontal displacement increments while the output was the total resistance force. Otherwise the plastic resistance values could be hardly determined by FEA. An input parameters setting is shown in Table 1. Nonlinear stiffness of the fastening components were taken from [9] and from the Vossloh DFF 300 guide [11]. It has to be mentioned that some assumptions had to be made when modelling vertical stiffness of the clamp skl B15, since there isn't presented a full description of the deflection – force behaviour in the guide.

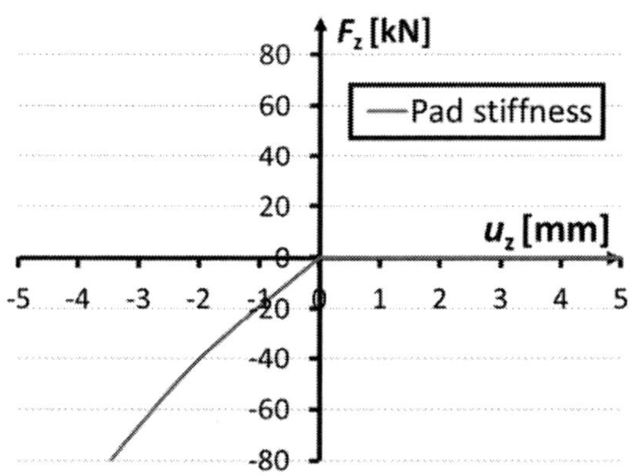

Figure 9. Vertical stiffness of elastic pad [9].

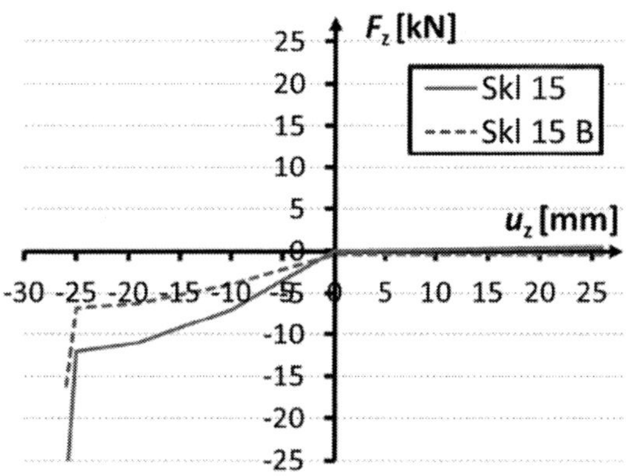

Figure 10. Vertical stiffness of clamps [11].

4. Results

As mentioned before the experiment was inspired by EN 13481-1 [10]. However this regulation describes testing of one fastening node only, therefore the evaluation procedure was slightly modified. During the evaluation process some difficulties had to be dealt with. Obtained functions of longitudinal resistance had slightly disjointed character because of sudden irregular slipping in particular fastening nodes. Because of that a polynomic regression function was used to substitute the elastic part of the graph and a mean value was used to evaluate the plastic resistance r_0 (2). Moreover maintaining simultaneous constant vertical pressure and very small horizontal force increments during plastic displacement development was quite challenging task. For similar purposes controlling the loading procedure by displacement should be recommended. Also it may be stated that longer sample is more appropriate for experimental analysis of longitudinal interaction behavior. That is because the longer rail doesn't rotate in the fastening nodes that much and some effects of eccentricities and second order are minimized compared to the only one fastening node testing procedures. After all no latter mentioned effects may occur in the real track.

$$r_0 = \frac{1}{n}\sum_{i=1}^{n} r_{0,i} \quad \text{where} \quad r_{0,i} > r_{x,yield} \tag{2}$$

$$r_x = 2 \cdot R_x/(4 \cdot s) \tag{3}$$

$$k_x = r_x(u_{x,max} - u_{x,res})/(u_{x,max} - u_{x,res}) \tag{4}$$

$$k_x = r_x(u_{x,yield})/u_{x,yield} \tag{5}$$

Longitudinal resistance r_x of the track was determined using the formula (3) where R_x stands for total cumulated resistance of four fastening nodes and s stands for axial spacing between the nodes. Using this formula a mean value of longitudinal resistance is obtained and recalculated for one meter length of the slab track. Values of unloaded track longitudinal stiffness k_x were evaluated using formula (4), where $u_{x,max}$ is the maximum measured displacement, $u_{x,res}$ is the residual displacement after unloading and $r_x(u_{x,max} - u_{x,res})$ is the resistance at the fictive yield point. The yield point term isn't recognized by the standards, it refers to the point where the longitudinal rail deflection increment starts to grow significantly higher due to rail slipping. Formula (4) was adopted from [10] and is appropriate to use in the case of indistinct yield point, while for the loaded track load cases, where the yield point is quite obvious, formula (5) had been used. Here the variable $u_{x,yield}$ indicates the displacement at the yield point and $r_x(u_{x,yield})$ indicates the resistance at the yield point.

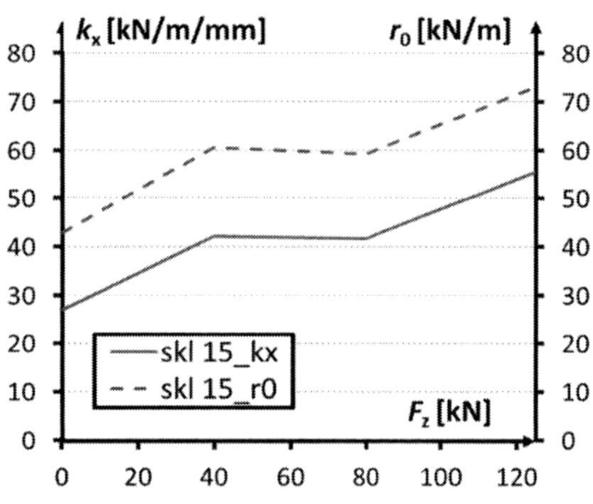

Figure 11. Experimental longitudinal stiffness and plastic resistance dependence on vertical load level for DFF 300 (Skl 15).

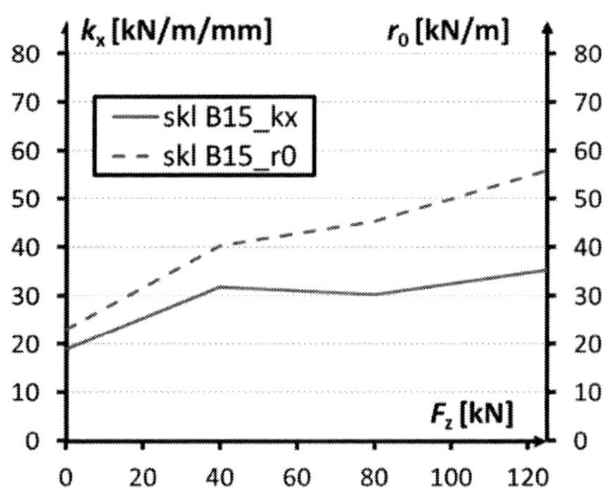

Figure 12. Experimental longitudinal stiffness and plastic resistance dependence on vertical load level for DFF 300 (Skl B15).

Figures 11 and 12 introduce the dependence of longitudinal stiffness and resistance on vertical loading. Table 2. introduces comprehensive comparison between experimental interaction parameters r_0, k_x and corresponding numerically simulated parameters $r_{0,F}$, $k_{x,F}$. Besides some exceptions the numerical results showed quite good accuracy, meaning the modified coupling interface model is suitable for this type of analysis. Major part of discrepancies is caused due to uncertainties in the input material parameters for FEA presented in table 1. Results demonstrate that the tension clamp is an important component, which significantly affects interaction parameters. By observing the longitudinal stiffness results it is quite interesting that the determined values are significantly lower than recommended in the codes [3,4]. Limit values of elastic deformation u_0 were found to lie in interval between 1.2 to 1.6 mm which rather matches in situ measured values proposed in [8]. By observing the plastic resistance results it may be concluded that the unloaded track values for Skl 15 clamp correspond to the standard recommended values quite well. The same remark may be stated for the loaded track load cases, where the vertical load level is below 80 kN, referring to common rail vehicles. In the case of heavily loaded track (F_z = 125 kN) the track resistance value 72.5 kN/m is considerably higher than usually recommended value 60 kN/m. Latter mentioned heavily loaded track value refers to the recalculated impact of UIC 71 load model. When exploring the results of the track equipped with Skl B15 tension clamp it may be stated that the longitudinal resistance values have diminished significantly. This applies to the unloaded and quite surprisingly also to the loaded track load cases.

Table 2. Experimentally and numerically determined interaction parameters.

F_z (kN)	clamp	r_O (kN/m)	k_x (kN/m/mm)	$r_{O,F}$ (kN/m)	$k_{x,E}$ (kN/m/mm)	$r_O/\ r_{O,F}$	$k_x/\ k_{x,F}$
0	Skl 15	42.9	27.1	41.4	30.3	1.04	0.89
40	Skl 15	60.6	42.2	56.5	37.6	1.07	1.12
80	Skl 15	59.3	41.8	61.5	40.8	0.96	1.02
125	Skl 15	73.0	55.5	72.5	43.83	1.01	1.27
0	Skl B15	23.0	18.9	25.4	19.1	0.91	0.99
40	Skl B15	40.3	31.9	45.2	29.7	0.89	1.07
80	Skl B15	45.4	30.3	48.9	31.0	0.93	0.98
125	Skl B15	55.9	35.2	61.4	32.5	0.91	1.08

Effects of clamping force lowering are demonstrated in Table 3, where u_c' is the final clamp deflection after the fastening bolt loosening, F_c' is the final clamping force after the fastening bolt loosening and similarly $r_{O,F}$' is the numerically obtained plastic longitudinal resistance after the fastening bolt loosening. To quantify these effects the validated modified coupling interface model (Figure 8.) was applied. Generally speaking it has been proven that the clamping force lowering may cause track resistance reduction. However results show that the influence isn't of big significance, at least when considering the highly elastic Skl 15 and Skl B15 tension clamps. Due to their good elastic capacity large clamp deflection change is needed for considerable clamping force lowering and subsequently for considerable longitudinal resistance reduction. Important question is how the deflection in fastening screws really reduces and what deflection change is realistic to achieve. On the other hand if we consider more rigid older types of tension clamps the effects of clamping force lowering will be considerably higher.

Table 3. Effects of clamping force lowering.

F_z (kN)	clamp	u_c' (mm)	u_p'/ u_p	F_c' (kN)	F_c' / F_c	$r_{O,F}$' (kN/m)	$r_{O,F}$'/ $r_{O,F}$
0	Skl 15	20	1.00	11.0	1.00	41.4	1.00
0	Skl 15	19	0.95	10.5	0.95	41.1	0.99
0	Skl 15	18	0.90	10.0	0.91	39.5	0.95
0	Skl 15	17	0.85	9.6	0.87	37.7	0.91
0	Skl 15	16	0.80	9.1	0.83	36.1	0.87
0	Skl 15	15	0.75	8.7	0.79	34.4	0.83
0	Skl B15	20	1.00	6.3	1.00	25.4	1.00
0	Skl B15	19	0.95	6.0	0.95	24.9	0.98
0	Skl B15	18	0.90	5.8	0.92	24.3	0.95
0	Skl B15	17	0.85	5.5	0.87	23.3	0.92
0	Skl B15	16	0.80	5.3	0.84	22.5	0.89
0	Skl B15	15	0.75	5.0	0.79	21.5	0.85

5. Conclusions

Performed study provided with useful information for practical bridge design. Detailed overview of the analysis remarks is given below:

- Modified numerical model of the coupling interface demonstrated quite good accuracy for simulating longitudinal interaction functions for ballastless tracks on a steel bridge. Using this model it is possible to predict longitudinal resistance of direct fastening, however precise knowledge of the friction coefficients and material input parameters is needed for reliable analysis.
- Dependence of longitudinal interaction parameters on vertical load for DFF 300 fastening system equipped with both the Skl 15 and the Skl B15 tension clamps is proposed in the paper. It may be stated that the tension clamp type significantly affects longitudinal interaction. Low resistance clamp Skl B15 is very convenient not only for lowering of track resistance during unloaded track load cases, but even in loaded track load cases.
- Longitudinal stiffness of the evaluated fastening system was determined to be rather lower than recommended in valid codes [3,4] while the values of plastic resistance quite matched the recommended ones.
- Effects of the clamping force lowering don't seem to be very significant. For greater influence the clamp deflection lowering would have to reach values about five millimetres. Reason for this good behaviour is the high elasticity of both tension clamps Skl 15 and Skl B15. On the other hand if the older types of tension clamps with lower elastic capacity were applied, the influence of clamping force lowering would be much higher.

Acknowledgements

Research reported in this paper was supported by Competence Centre's Program of Technology Agency of the Czech Republic (TA CR), project Centre for Effective and Sustainable Transport Infrastructure (no. TE01020168).

References

[1] Frýba L 1996 *Dynamics of railway bridges 2nd ed.* (London: ICE Publishing)
[2] Freystein H and Geissler K 2013 Rail/Bridge interaction on Steel Bridges with Examples *Steel Structures* **82**(2) pp 78-86 [in German]
[3] Union Internationale des Chemins de fer (UIC) 2001 Track/bridge interaction, Recommendations for calculations (*UIC 774-3R)*
[4] Freystein H 2010 Rail/Bridge interaction – State of the Art and Examples *Steel Structures* **79**(3) pp 220-31 (in German)
[5] Regulations for Structures 2003 Appendix 6 Part 1: Loads (in Dutch)
[6] European Committee for Standardization (CEN) 2003 Actions on Structures; Traffic loads on bridges *Eurocode 1, Part 2 (EN 1991-2)*
[7] Esveld C 2001 *Modern Railway Track 2nd ed.* (Zaltbommel: MRT-Productions)
[8] Ryjáček P and Vokáč M 2014 Long-term monitoring of steel railway bridge interaction with continuous welded rail *Journal of Constructional Steel Research* **99** pp 176-86
[9] Stančík V and Ryjáček P 2017 Effects of biaxial and thermal loads on rail-slab track interaction behavior *Advances and Trends in Engineering Sciences and Technologies II* ed M Al Ali and P Platko (Boca Raton: Taylor & Francis Group) pp 835-40
[10] European Committee for Standardization (CEN) 2012 Railway applications; Track; Test Methods For Fastening Systems; Part 1: Determination Of Longitudinal Rail Restraint *(EN 13146-1)*
[11] Vossloh Fastening Systems GmbH 2014 Highly elastic rail fastening for high speed and conventional rail; The flexible single support point for slab track *System DFF 300*

IOP Conf. Series: Materials Science and Engineering **236** (2017) 012064 doi:10.1088/1757-899X/236/1/012064

SHS and RHS stainless steel slender members loaded by compression and bending interaction

Břetislav Židlický[1] and Michal Jandera[1]

[1]Department of Steel and Timber structures, Czech Technical University in Prague, Thákurova 7, Prague, 166 29, Czech Republic

E-mail: bretislav.zidlicky@fsv.cvut.cz

Abstract. Behaviour of stainless steel slender members loaded by interaction of axial compressive force and bending moment is investigated in this research. Square hollow sections (SHS) made of austenitic stainless steel grade are considered. An initial numerical parametric study in FE software Abaqus is given and its results are compared to the existing design procedures and design standard rules. The investigated parameters are mainly the column slenderness, section slenderness, ratio between the applied bending moment and axial compressive force and the moment distribution along the member. The necessity of having additional design rules for stainless steels is firstly demonstrated on the values of interaction factors k_y which are significantly higher for stainless steel members due to the material non-linearity with decreased stiffness even at lower stress levels. As an alternative, the General Method is used for comparison to the Abaqus GMNIA model results. The limitation of the method when used for members of non-linear material behaviour is shown and a safe modification of design procedure is suggested.

1. Introduction

Stainless steel alloys utilization as a structural elements increased, mostly due to their corrosion resistance, mechanical properties, aesthetic appearance and easy maintenance, in the last few decades. Nevertheless, it is still restrained by the high initial investment which is, partially, caused by the insufficient background of stainless steel structures behaviour. Design procedure of stainless steel members loaded by the interaction of compressive force and bending moment provided by stainless steel code EN 1993-1-4 [1] is mostly based on the carbon steel code 1993-1-1 [2]. That leads to some inaccuracies. Many researches tried to develop this procedure to be more accurate and efficient in the last years. 1993-1-1 [2] providing another design approach called General method applicable for beam-columns. The procedure is not commented in [1] and therefore there is no limitation regarding to material behaviour. This paper is primarily focused on the evaluation of the General Method for stainless steel square hollow section (SHS) members and some improvement is suggested. Further, new design approach developed by Ou Zhao [3] and current design approach provided by stainless steel [1] are evaluated too.

2. Investigated members

As was mentioned before, only SHS member research is presented in this paper. Cross-sections 80x3 and 80x5 mm were considered. Member lengths are various between 1000 mm and 7000 mm. The same cross-section members also being tested and will be presented on the conference as well. The austenitic stainless steel grade 1.4404 was selected. The material properties were obtained through the tensile tests recently. Due to high strains occurred in the cross-section corner parts during cold-working, tensile tests were performed for corner and flat coupons separately. Location of the test coupons providing Figure 1.

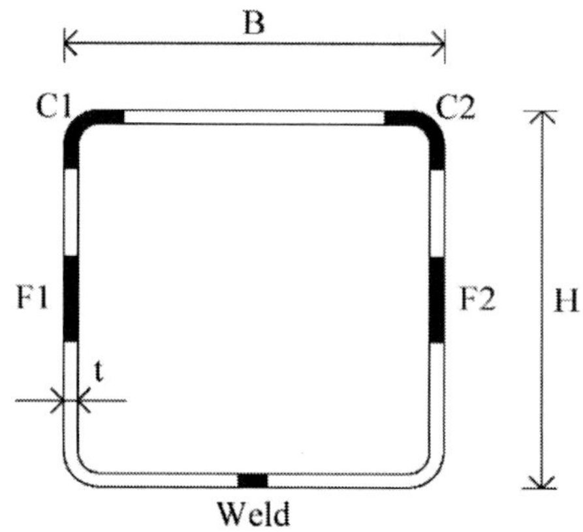

Figure 1. Location of flat and corner coupons.

MTS Qtest 100 kN electromechanical testing machine with all data recorded at 5 HZ using the SPIDER data acquisition system with CATMAN32 data acquisition software was used to all coupon tensile tests. The tests were controlled by the strain. To obtain smooth stress-strain diagram curve, the accepted strain rate for testing up to 1.5 % strain was 0.007 % per second, followed by 0.2 % per second until fracture. In the research, flat coupon behaviour was considered for the whole cross-section. Effect of the corner region different properties was shown as negligible on the interaction in [4]. Member lengths and slenderness information are summarised in Table 1. Table 2 providing material properties information. Where E_0 is the initial young modulus of elasticity, $\sigma_{0.2}$ is the 0.2% proof stress, $\sigma_{1.0}$ is the 1.0% proof stress and σ_u is the ultimate stress.

Table 1. Member lengths and slenderness.

Section	**L [mm]**	1000	1500	2000	2300	2500	3000	3500	4000	4500	5000	6000	7000
80x3	$\bar{\lambda}$	0.64	0.76	1.01	1.16	1.26	1.51	1.77	2.02	2.27	2.52	3.02	3.53
80x5	$\bar{\lambda}$	0.58	0.87	1.15	1.32	1.44	1.73	2.01	2.30	2.58	2.87	3.44	4.02

Table 2. Material properties.

Section	E_0 [MPa]	$\sigma_{0.2}$ [MPa]	$\sigma_{1.0}$ [MPa]	σ_u [MPa]
80x3	183200	397.1	465.3	627.6
80x5	190500	448.1	516.2	627.4

3. Numerical modelling

In order to obtain the real member behaviour, 3D model using GMNIA, geometrically and materially non-linear analysis with imperfections, was created in software Abaqus. The model will be validated soon on the planned tests. The same model was successfully validated in [4].

3.1. Considered loading

The research is focused on the compressive and bending interaction. Only uniform moment distribution along the member length was considered. The loading values are based on the axial compressive force N_{Ed} and member column resistance $N_{b,Rd}$ ratio labelled as n_b. Values of n_b are shown in Table 3.

Table 3. Considered n_b values.

n_b	0	0.3	0.5	0.7	0.8	1.0

For the initial prediction of the bending moment, the interaction formula provided in [3] was used. The bending and column resistances were obtained from Abaqus. The interaction formula is given by equation (1) and the interaction factor formula according to [3] by equation (2).

$$\frac{N_{\mathrm{Ed}}}{N_{\mathrm{b,Rd}}} + k_{csm}\frac{M_{\mathrm{Ed}}}{M_{\mathrm{b,Rd}}} = 1 \tag{1}$$

$$k_{csm} = 1 + D_1(\bar{\lambda} - D_2)n_{\mathrm{b}} \leq 1 + D_1(D_3 - D_2)n_{\mathrm{b}} \tag{2}$$

where N_{Ed} is the design compressive force,
M_{Ed} is the design bending moment,
$N_{\mathrm{b,Rd}}$ is the buckling resistance,
$M_{\mathrm{b,Rd}}$ is the bending resistance,
n_{b} is the N_{Ed} to $N_{\mathrm{b,Rd}}$ ratio,
k_{csm} is the interaction factor according to [3],
$\bar{\lambda}$ is the non-dimensional slenderness,
D_{i} are the constants according to stainless steel grade and they are given by Table 4.

Table 4. Parameters of the interaction factor, D_{i}.

Grade	D_1	D_2	D_3
Austenitic	2.0	0.30	1.3
Duplex	1.5	0.40	1.4
Ferritic	1.3	0.45	1.6

3.2. GMNIA model

The numerical model was created in software Abaqus. Four-noded shell element with reduced integration S4R was used. Two Reference Points situated on each member end in the centroid, rigidly coupled to the shell edged, ensured pin-ended boundary conditions.

To obtain the critical load with the corresponding eigenmodes the linear buckling analysis was made firstly. The first buckling mode was a flexural buckling mode as the sections are quite stocky. Local buckling mode was always some higher mode. Both local and global modes were used for the imperfection shape definition in the following calculation. The global imperfection amplitude was considered as L/1000 whereas the local imperfection amplitude was calculated according to Dawson and Walker [4] formula given by equation (3).

$$\omega_0 = \gamma\left(\frac{\sigma_{0.2}}{\sigma_{cr}}\right)t \tag{3}$$

where γ is the material constant, for stainless steel is equal to 0.023,
$\sigma_{0.2}$ is the 0.2% proof stress,
σ_{cr} is the cross-section critical stress,
t is the thickness of the structural element.

Separate model was used to section compression resistance N_{Rd} calculation. The same model was equipped by the additional supports in the middle of flat parts of the cross-section along the whole member. That prevent the local buckling phenomena. Only local imperfection was considered.

3.3. GNIA model
The General Method is based on the geometrically non-linear analysis of the structure with member imperfection (GNIA). Therefore, simplified 2D model, using B21 beam element, was created in software Abaqus. SHS cross-sections were simplified into the box cross-sections with modified wall thickness in order to keeping the same moment of inertia. The model was considered elastic using the elasticity modulus given in Table 2. The section resistance according to General Method is given by equation (4).

$$\frac{N_{Ed}}{N_{Rd}} + \frac{M_{Ed}^{II}}{M_{Rd}} \leq 1 \tag{4}$$

where N_{Rd} is the section compressive resistance,
M_{Rd}^{II} is the maximum of the bending moment along the beam calculated by GNIA.

For the section resistance in compression and bending, values from the 3D GMNIA model were used. The member imperfection differs from those in chapter 3.2 as it should cover also the effect of material non-linearity. The initial geometric imperfection e_0 was calculated by equation (5).

$$e_0 = \alpha(\bar{\lambda} - \bar{\lambda}_0)\frac{M_{Rd}}{N_{Rd}} \tag{5}$$

where $\bar{\lambda}$ is the non-dimensional slenderness,
$\bar{\lambda}_0$ is the plateau length,
A is the characteristic of the buckling curve.

4. Evaluation

4.1. Ou Zhao
Evaluation of the Ou Zhao method [3] is presented in this chapter. Assessment is focused on the interaction factor *k* calculated by [3] and obtained from the numerical GMNIA model. The comparison is shown in Figure 2 for SHS 80x3 mm and in Figure 3 for SHS 80x5 mm. Curves represent interaction factor according to [3] and dots are values obtained from numerical study.

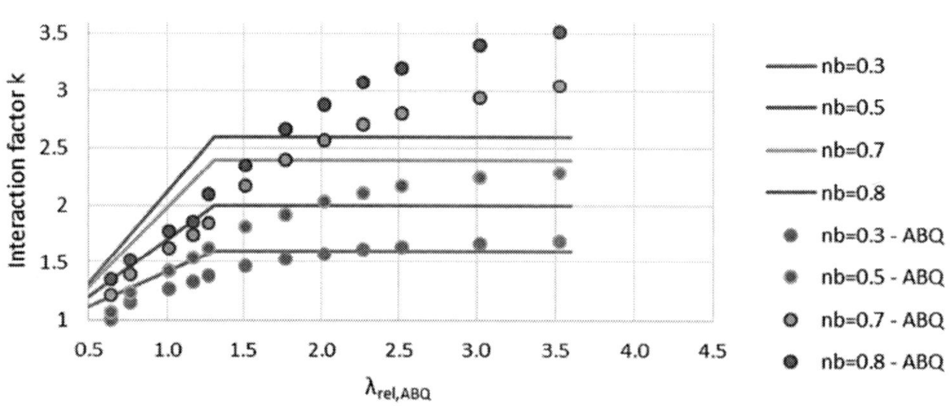

Figure 2. Results for SHS 80x3 according to Zhao and numerical model.

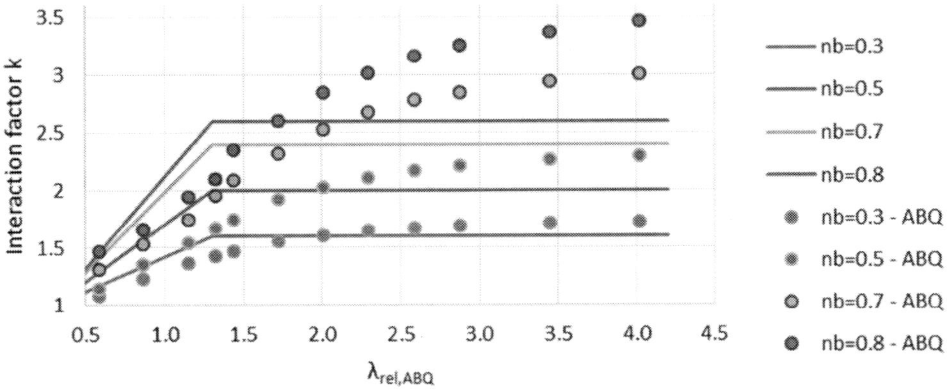

Figure 3. Results for SHS 80x5 according to Zhao and numerical model.

In the comparison to the numerical model, the initial part of the formula seems to be conservative for $\bar{\lambda} \leq 1.7$ but the upper bound of the factor is unsafe, especially in the cases of dominant compressive force. That could be caused by a different cross-section used (in [3] RHS members were used) or more probably, by different material behaviour of the considered austenitic stainless steel.

4.2. EN 1993-1-4

Design procedure provided by Eurocode [1] is evaluated in this section. The interaction factor is calculated by equation (6). The comparison is shown in Figure 4 for SHS 80x3 mm and in Figure 5 for SHS 80x5 mm. Curves represent again the formula of the interaction factor k [1] and dots are values obtained from the numerical study.

$$1{,}2 \leq \quad k_y = 1 + 2(\bar{\lambda} - 0{,}5)\frac{N_{\mathrm{Ed}}}{N_{\mathrm{b,Rd}}} \quad \leq 1{,}2 + 2\frac{N_{\mathrm{Ed}}}{N_{\mathrm{b,Rd}}} \tag{6}$$

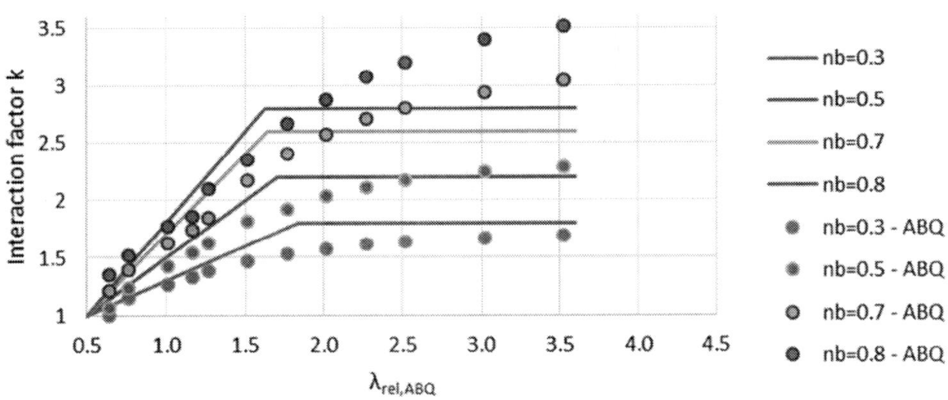

Figure 4. Results for SHS 80x3 according to EN 1993-1-4 and numerical model.

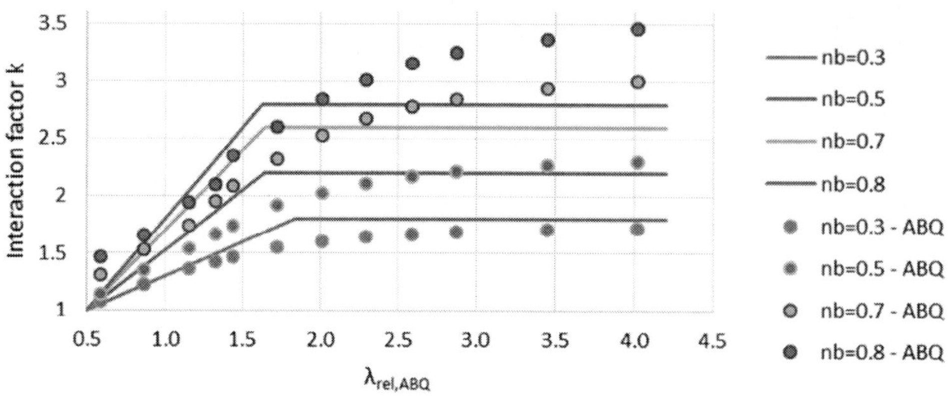

Figure 5. Results for SHS 80x5 according to EN 1993-1-4 and numerical model.

This approach is mostly safe for $\bar{\lambda} \leq 2{,}0$. However, the method has quite good results in the case of uniform bending moment along the member, which is this case. For other moment distribution, it becomes too conservative as there is no factor considering the moment distribution along the member.

4.3. General Method

Evaluation of the General method is conducted through the ratio of compressive force $N_{Ed,ABQ}$ obtained from GMNIA shell model and $N_{Ed,GNIA}$ obtained from equation (4) based on the GNIA beam model. Because both SHS 80x3 mm and SHS 80x5 mm indicates similar results, the assessment is shown together in Figure 6. Values greater than one indicate safe results.

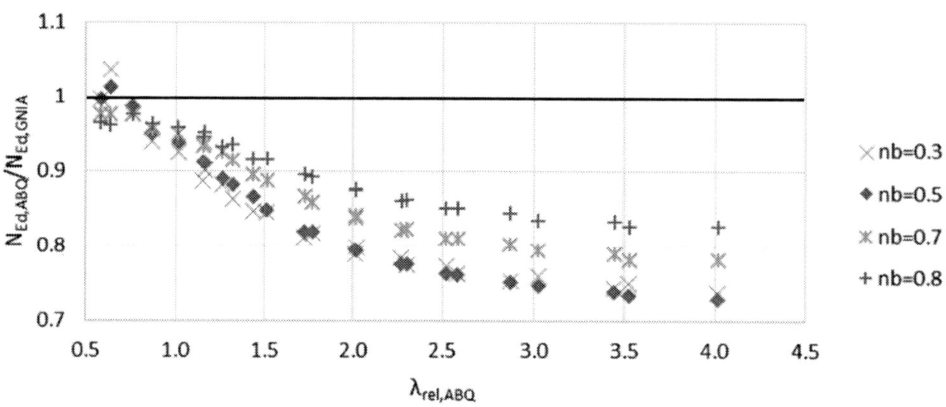

Figure 6. Comparison of the General Method.

Almost all results are on the unsafe side, except members with very low slenderness. The unsafe results are probably caused by the simplification of the material behavior, which is considered in the General Method. That could leads to the neglecting of second order effect resulting from the material non-linearity, especially for very slender members and significant contribution of bending moment loading to high stress level.

Therefore, a modification was suggested. Initial young modulus of elasticity E_0 has been replaced by the secant modulus of elasticity for stress level equal to the yield strength $E_{s,0.2}$. Equation (7) was used to calculation of $E_{s,0.2}$. Results of this modification are shown in Figure 7.

$$E_{s,0.2} = \frac{\sigma_{0.2}}{\frac{\sigma_{0.2}}{E_0} + 0.002} \tag{7}$$

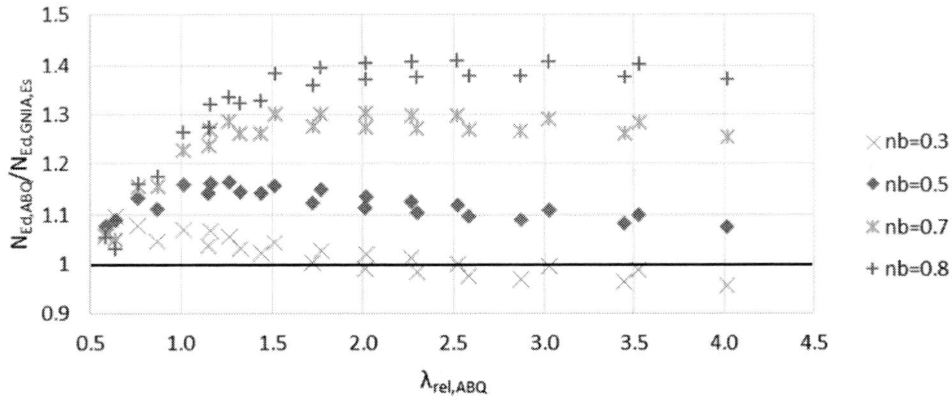

Figure 7. Comparison of the General Method $E_{s,0.2}$ modification.

For each n_b line of results, SHS 80x3 mm is represented by the upper row of points and SHS 80x5 mm by the lower row of points. The results are mostly on safe side, except very slender members loaded mostly by bending moment. However, in the case of dominant compressive force the results are very conservative.

In order to eliminate over-conservative results, another modification was suggested. The secant modulus $E_{s,M}$ was calculated based on the stress caused by the bending moment. The evaluation is presented only for SHS 80x3 mm through the Figure 8.

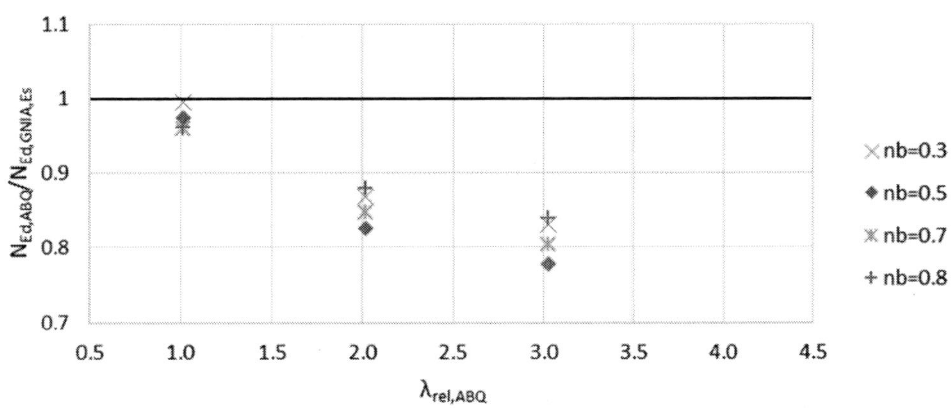

Figure 8. Comparison of the General Method $E_{s,M}$ modification.

As could be seen, all results are on the unsafe side again. So, the calculation with the secant modulus equal to the yield stress level is the most suitable modification, currently.

5. Conclusions

The main objective of this research is to study behaviour of stainless steel slender members loaded by the compressive and bending interaction. The General Method provided by EN 1993-1-1 [2] evaluation is presented in this paper, mainly. Otherwise, design approach provided by stainless steel code EN 1993-1-4 [1] and approach developed by Ou Zhao [3] are compared too.

3D shell element numerical model using GMNIA was created in software Abaqus to obtain realistic members behaviour. Then, simplified 2D beam model using GNIA was made in order to produce the General Method results.

The design procedure developed by Ou Zhao [3] provide safe results for members with low non-dimensional slenderness. The method become very unsafe with increasing non-dimensional slenderness, especially in the cases of dominant compressive force. That could be caused by different cross-section type considered in [3] but most likely by different stainless steel material. EN 1993-1-4 [1] results are very similar with unsafe results as well.

The General Method evaluation shown very unsafe results, therefore it is not suitable for materials with non-linear behavior in the current form. Two modifications of this method were demonstrated. Replacement of the initial modulus of elasticity E_0 in the global analysis by the secant modulus of elasticity equal to the yield strength $E_{s,0.2}$. That leads to safe, usually very conservative results, except for very slender members loaded mostly by bending moment. In order to eliminate over-conservative results, second modification considering secant modulus of elasticity calculated through the stress caused by bending moment $E_{s,M}$ was presented. The results are very unsafe again, so this modification is useless. Further investigation of this method is under way to obtain more accurate results.

Acknowledgements
The support of the Czech Science Foundation grant 17-247695 "Nonlinear stability and strength of slender structures with nonlinear material properties" is gratefully acknowledged.

References
[1] European Committee for Standardization (CEN) 2006 General rules – Supplementary rules for stainless steel *Eurocode 3, Part 1-4 (EN 1993-1-4)*
[2] European Committee for Standardization (CEN) 2005 General rules and rules for buildings *Eurocode 3, Part 1-1 (EN 1993-1-1)*
[3] Zhao O 2015 *Structural Behaviour of Stainless Steel Elements Subjected to Combined Loading* (University of London, Faculty of Civil Engineering)
[4] Ares M R 2016 *Structural behaviour of stainless steel beam-columnsubjected to combined loading* (University of Barcelona/Prague, Faculty of Civil Engineering)
[5] Dawson R G and Walker A C 1972 Post-buckling of geometrically imperfect plates *Journal of the Structural Division ASCE* **98** pp 75-94

Behaviour of several fatigue prone bridge details

Petr Kubiš[1] and Pavel Ryjáček[1]

[1]Czech Technical University in Prague Faculty of Civil Engineering, Thákurova 7, Prague 6,166 29, Czech Republic

E-mail: kubispe1@fsv.cvut.cz

Abstract. Three fatigue welded bridge joints analysed in this work are the alternative details of the bottom flange connection. This construction detail is mainly used for the erection connection for steel and composite bridges. If applied in the place, where live load is significant, the fatigue becomes the main design criterion. The detail category is thus very important factor. The aim of this paper is to analyse the possibilities of the improving the behaviour of this detail, by various methods. First solution is to modify the shape of the cope hole to the elliptic shape. Second option is to use the "Olemutz" fully welded detail. This detail is often used in bridge designing despite there is no exact information about the fatigue category, and doubts of the performance exists. "Olemutz" is a long web plate slit that is filled by the double bevel weld after the execution of the bottom flange weld. The last detail is the elliptic cope hole filled by the plate-cap welded into an empty hole. The geometry is the same, as in the first case. The conclusion of the numerical analysis and the pilot fatigue experiments is discussed with several practical recommendations for designing.

1. Introduction

Fatigue design takes place in every bridge assessment and the fatigue issue could affect the whole dimension of the main bridge structure. There are many available fatigue details which are verified and recommended in codes, however not all details are covered.

The main purpose of this paper is to analyze the possibilities of the improving the behavior of three specific fatigue details - "Olemutz", the elliptic cope hole and the elliptic cope hole filled by the welded plate-cap.

The first part describes the parametric study, which was performed in order to optimize the shape of the cope hole fatigue detail. The second part contains experimental pilot tests of the fatigue life of all studied fatigue details.

Bridge designers widely use EN 1993-1-9 [1] for fatigue assessment, where the methodology is based generally on nominal stresses approach. The fatigue strength is then divided into the categories according to the stress range for 2.10^6 load cycles, that the detail can withstand. The same fatigue details are also described in IIW recommendations [2]. Besides, there is a definition how to determine Hot spot stresses or Effective notch stresses. For the Hot spot method, the mesh performance, mesh density, element usage, and hot spots (points for integration stresses) location are given as well. Numerical Hot spot method is also appropriate in a combination with the experimental hot spot modification. Highly advanced modeling containing notch effect of weld is summarized in the chapter 2.2.4. Effective Notch Stress. A recommendation from this publication was used for numerical and experimental study of the elliptic cope hole detail.

A valuable source of knowledge provides master thesis written by Mohsen Heshmati [3]. There is a very similar cope hole detail study that provide series of new shapes that improve a standard using circular R=50mm hole. A portion of effort is put into observing a combination of shared and normal stresses.

The post welded treatment is also an alternative way how to increase a fatigue life of the cope hole. There is an experiment that follows this topic in the part [3]. Illustrative Figure 1 shows how burr grid of weld toe looks on the specimen. According to [3], the test results show an increase of 190% in terms of fatigue strength for cope-holes treated with Burr Grinding. The same conclusions can be found in Sung-Min Choi et al. [4].

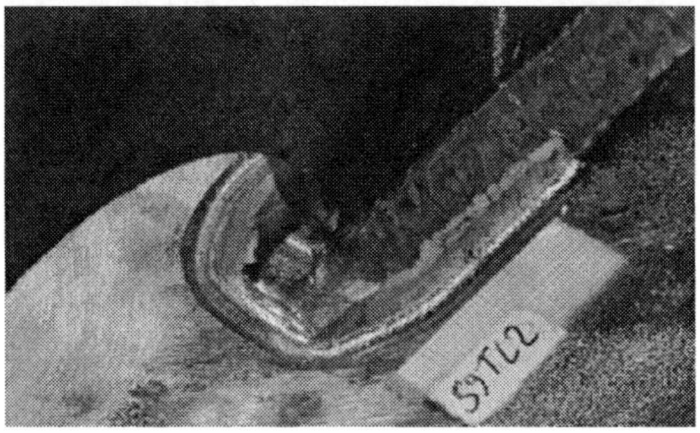

Figure 1. The cope hole weld toe treated with Burr Grinding [3].

2. Studied details

The common solution for railway bridges is to divide them in the middle, allowing two erection parts to be easily transported to the site and welded only with one weld on the site, as shown in the Figure 2. However, the connection of the tensile lower flange is a subject of mainly live load stresses and thus the fatigue loads. To perform a transversal butt weld well even in a location of the web, the cope hole is usually placed, as shown in the Figure 2. The highlighted part shows the area that is exposed to the dynamic load and sensitive to the fatigue. Also, fatigue can limit the whole design of this structural element.

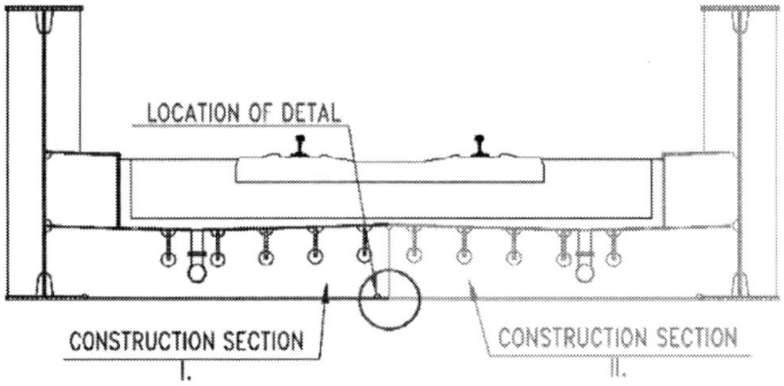

Figure 2. An example of the application of studied details. Cross section of the railway bridge with the lower deck.

Generally, three solutions were studied here.

- The standard fatigue detail with the cope hole R=50mm of category 71. Here, the detail was improved in order to increase the fatigue detail with usage of modern FEM tools.
- An alternative solution, used by designers, is called "Olemutz", see the Figure 3. However, standard codes or literature give no information about this fatigue detail. Designers apply this detail to aim higher fatigue strength but without any exact knowledge. A significant

longitudinal weld of a narrow slit could seriously affect the zone around weld where the fatigue failure could begin. The slit allows the transverse butt weld of the flange to be welded.

- The last detail represents an "improvement" of the previous detail. After welding of the flange weld, the cope hole is filled and welded by the shaped plate. The cope hole exists no more here, and detail seems to increase its category. However, combination and amount of welds, and related internal stresses creates doubts of this assumption, although designers use this procedure sometimes.

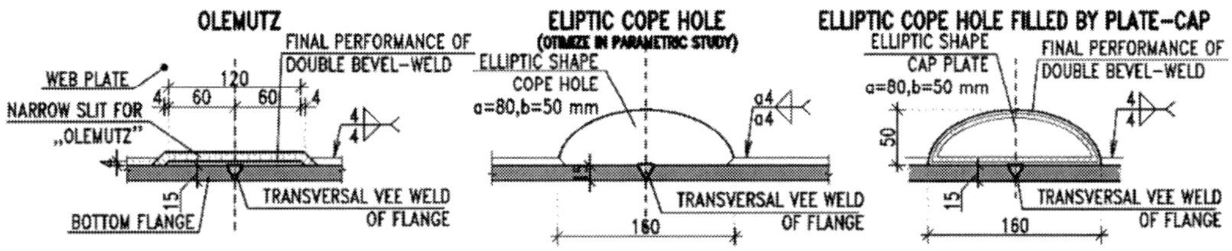

a) b) c)

Figure 3. Investigated fatigue details a) Elliptic cope hole, b) "Olemutz", c) Elliptic cope hole filled by the welded plate.

The following chapters describe the parametric analysis of the cope hole and the preliminary experimental work focused on these three details.

3. Study of cope hole shape

3.1. Parametric analysis

The first stage of the study is focused on a parametric study of the optimization of the cope hole shape. The typical solution a standard cope hole with a radius R=50mm. According to IIW and [1], this fatigue detail has a prone spot in a toe of weld. The aim of this study is to try to use different shapes of the cope hole to redistribute local stresses due to geometry of detail. The studied shapes are an oval, a parabola and an ellipse. The parametric study follows these main parameters: width (W) and height (H). Two general cases are established. For the situation Analysis I – the parameter width (W) is changeable and height (H) is fixed on the value 50mm. The Analysis II follows the second parameter - height (H) - and there are used changeable shapes of ellipse and circle with vertical moving center, see Figure 7.

Hot spot method is an appropriate life fatigue assessment method for this type of problem. This method could take into the account the local geometry of the detail in the sufficient accuracy. Despite that it excludes a notch effect of weld. Hot spot stress σ_{hs} is extrapolated from the spots that are in a specific position.

It is necessary to avoid the stress peaks near the flange and the web connection because of the unsuitable meshing. Only if the prescribed, verified procedure of modeling and generating of mash near welded detail is used, the exact hot spot stresses could be reached.

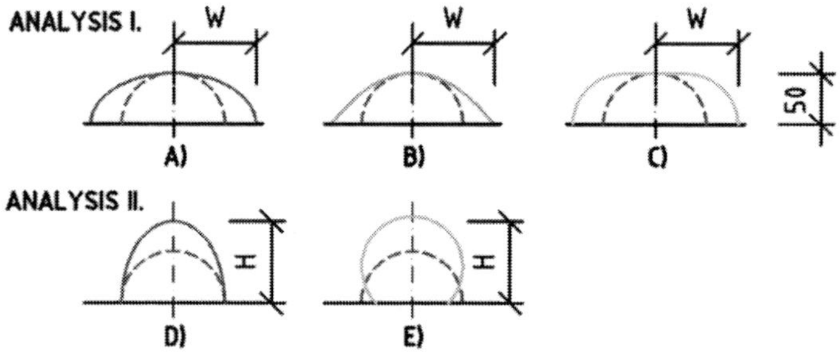

Figure 4. The selected shapes and parameters for parametric study, A) ellipse B) parabola 2° C) oval D) ellipse E) Circle with a moving centre; Red line define a reference circle R=50mm.

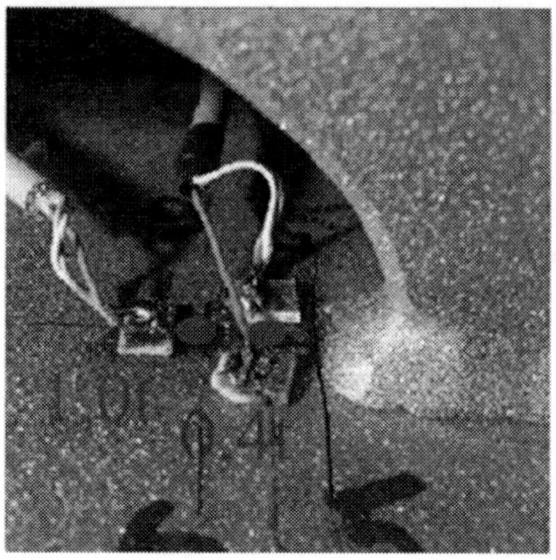

Figure 5. The decomposition of hot spots near the weld. The figure describes hot spot application on the sample.

The Ansys FEM software and linear elastic analysis was used here. The spot stresses were taken from nodes in distances 0.4 t and 1.0 t from weld toe. Below is an equation (1) for the hot spot stress σ_{hs} extrapolation.

$$\sigma_{hs} = 1.67.\sigma_{0.4t} - 0.67.\sigma_{1.0t} \tag{1}$$

An important condition was an appropriate generation of the mesh because the highlighted nodes in Figure 5 should the keep the same distance from the weld toe in all details. Size of element in numerical model is 3.2 mm.

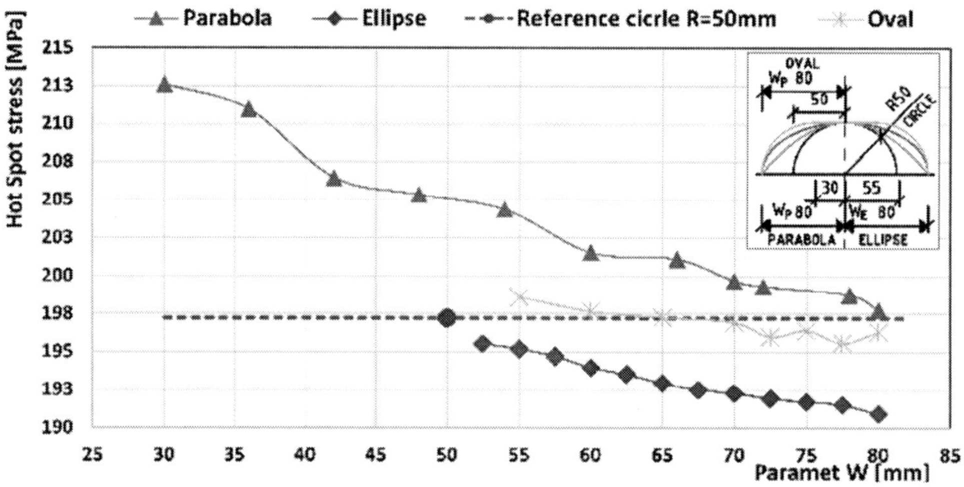

Figure 6. Results of Analysis I.

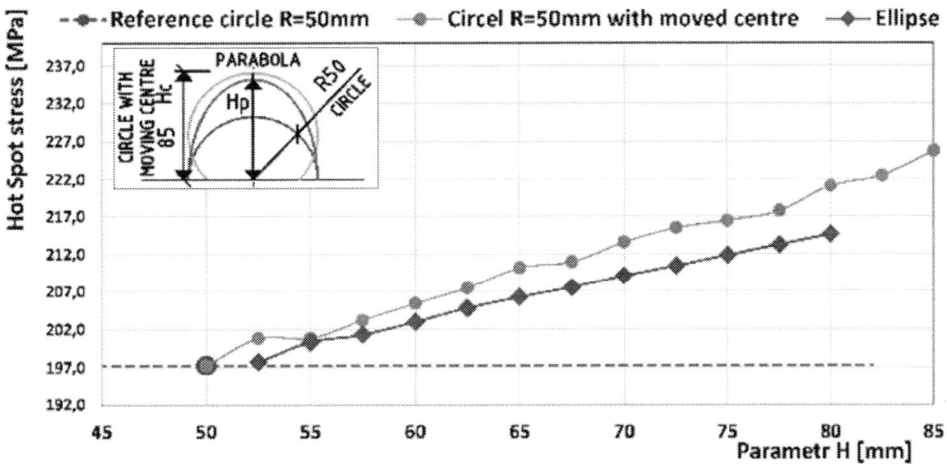

Figure 7. Results of Analysis II.

Table 1. Results of the parametric study.

Shape of cope hole	*Nominal stress N [MPa]*	*Hot spot stress H [MPa]*	*Factor N/H*	*Predicted fatigue life [cycle]*	*Fat. life comparison [%]*
reference circle	117.3	195.82	1.670	266 358	-
parabola 2°	117.3	197.79	1.686	258 478	-2.96
ellipse	117.3	190.92	1.628	287 406	7.90
oval	117.3	195.69	1.668	266 897	0.2

3.2. Experimental hot spot analysis

Hot spot method is also suitable for an experimental measurement and stress-strain evaluation. The equations (2) and (3) are used to evaluate the hot spot stress σ_{hs}. The location of hot spots set in an experimental and numerical application is shown in the Figure 5. The aim of the experiment is to verify FEM model used in the chapter 3.2. and also to get the detail category of the modified cope hole. Totally six weld toes were measured by strain gauges in twelve spots.

Figure 8. Experimental assembly.

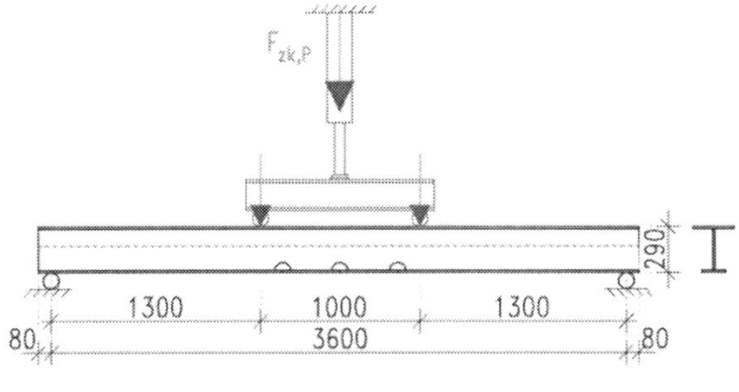

Figure 9. Scheme of the tested beam.

$$\varepsilon_{hs} = 1.67.\varepsilon_{0.4t} - 0.67.\varepsilon_{1.0t} \quad (2)$$

$$\sigma_{hs} = E.\varepsilon_{hs} \quad (3)$$

The fatigue load specification is given here:

- nominal stress range $\Delta\sigma_n$=117.3MPa in tension,
- no shear stress applied,
- load frequency 4 Hz.

3.3. Conclusion of the cope hole study

The summary of obtained numerical and experimental results is in the Table 2. The comparison of the numerical and experimental hot spot stresses provides a good accuracy together and it is indicating that the numerical model used for the parametric analysis was created well and reflects real values. The fatigue life was evaluated according to the mean of the SN curve FAT 137. This curve is recalculated from characteristic SN curve FAT 100, that is recommended in IIW [2]. Also, real fatigue life that was measured for our specimen is shown here.

Table 2. Determination of fatigue life.

Approach of fatigue life determination	***Numerical model***	***Experimental strain gauge***	***Experiment results***
Method based on	SN-curve FAT137	SN-curve FAT137	Cycle counting
Hot spot stress [MPa]	187.8	180.6	-
Fatigue Life [cycles]	778 367	875 165	1 077 335

The parametric analysis provides these conclusions:

- From the range of the tested shapes, an ellipse with a=80 and b=50mm was chosen as an efficient alternative of the cope hole. However, the improvement is quite small, 7.9% improvement in comparison with standard cope hole R=50mm.
- The wider narrow shapes seem to behave better than the tall narrow ones.

4. Experimental tests of the fatigue details

In the last parts of the paper, a pilot full scale fatigue experiment will be presented. The size of the specimen is taken to be comparable with the standard beams used for bridges. The tested details are introduced in the chapter 2. For each type of the detail one specimen was fabricated. The results should check and compare behaviour of these chosen fatigue details, as a pilot experiment, that cal lead to the further wider study. An ultrasonic test and a dye penetrant inspection were used to check the weld defects.

4.1. The experiment specification

The layout of the test is shown in the Figure 8 and Figure 9. A simple supported beam is loaded by a hydraulic device by four-point bending. This layout forms a constant internal moment in the central part of the beam, where three same fatigue details are located without any shear stresses.

The specimens were tested in separate time period and because of the time limitation, the stress ranges were not the same. Despite this, constant stress range was ensued in each single test. Experiments were finished when fatigue crack reached 70 mm.

IOP Conf. Series: Materials Science and Engineering **236** (2017) 012065 doi:10.1088/1757-899X/236/1/012065

4.2. Discussion of the test results

Main results of the experiment are summarized in the Table 3, where the approximate fatigue categories are calculated. It is obvious, that "Olemutz" reaches the best fatigue behavior. If we accept assumption that this experimentally determined fatigue category is the mean fatigue category, characteristic fatigue category would be FAT 93. It shows that whole detail is limited by transverse butt weld (of category FAT90 according [2]). This assumption is supported by fact that fatigue crack began and progress exactly in the butt weld, as shown in the Figure 10 and 11.

Table 3. Results of the experiment.

Fatigue detail		*Stress range* $\Delta\sigma_n$ *Load [MPa]*	*Experiment Number of cycles*	*Experimentally determined Corresponding fatigue category*
"Olemutz"	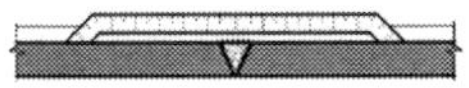	225	360 134	127
Elliptic cope hole	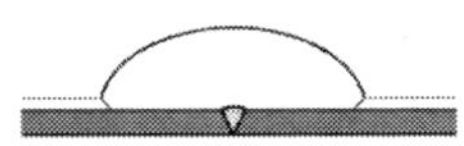	117.3	1 077 335	95
Elliptic hole filled by plate cap	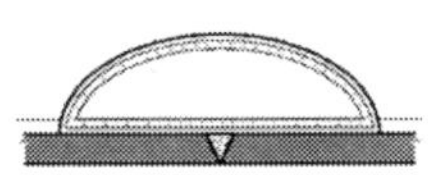	201	215 288	96

The results also show that filling of the cope holes by welding of additional plates does not improve the fatigue behavior, compared to the cope hole itself.

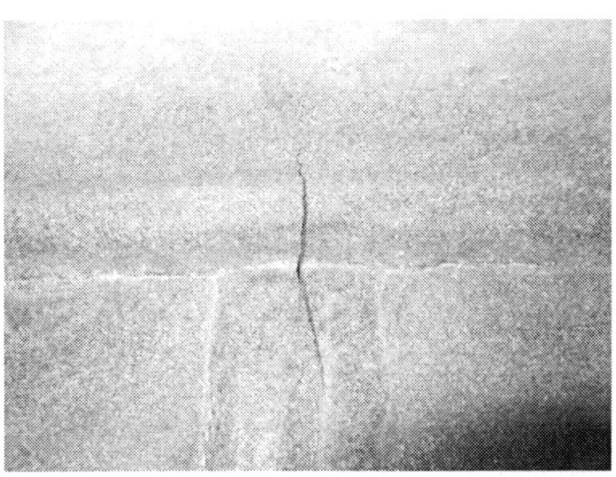

Figure 10. Fatigue crack through butt weld of "Olemutz" detail.

Figure 11. View on the lower flange with crack of "Olemutz" detail.

Figure 12. Cross section of the beam with "Olemutz". Cut in the section of crack.

5. Conclusion

The results of the parametric study shows, that the possibilities of the improvement of the cope holes by the geometry modification exist, however are quite small. The elliptic shape with the dimension of main axis a=80mm and b=50mm provides 7.9% improvement of the fatigue life.

The second part of the paper shows the experimental tests of two fatigue details in the full-scale test. It is obvious, that "Olemutz" reached the best fatigue behavior. The results also shows, that filling of the cope holes by welding of additional plates does not improve the fatigue behavior, compared to the cope hole itself.

All of those results can be directly applied in the bridge engineering for the design of the fatigue sensitive details.

Acknowledgements

Research reported in this paper was supported by Competence Centers program of Technology Agency of the Czech Republic (TA CR), project Centre for Effective and Sustainable Transport Infrastructure (no. TE01020168). The authors thanks for the fabrication of the samples to Bilfinger MCE Slaný s.r.o.

References

[1] European Committee For Standardization 2006 Design of steel structures: Fatigue (*EN 1993-1-9)*

[2] Hobbacher A 2008 *IIW document: Recommendations for Fatigue Design of Welded Joints and Components* Doc. XIII-2151r4-07/XV-1254r4-07 (Paris: International Institute of Welding)

[3] Mohsen H 2012 *Fatigue Life Assessment of Bridge Details Using Finite Element Method: Master's Thesis* (Gothenburg: Chalmers University Of Technology)

[4] Choi S-M, Tateishi K and Hanji T 2013 Fatigue strength improvement of weld joints with cope hole *International Journal of Steel Structures* **13**(4) pp 683–90

FEM simulation of static loading test of the Omega beam

Petr Bílý[1], Alena Kohoutková[1] and Petr Jedlinský[2]

[1]Faculty of Civil Engineering, Czech Technical University in Prague, Thákurova 7, 166 29 Praha 6, Czech Republic
[2]EUROVIA Services, s.r.o, Národní 138/10, 110 00 Praha 1, Czech Republic

E-mail: petr.bily@fsv.cvut.cz

Abstract. The paper deals with a FEM simulation of static loading test of the Omega beam. Omega beam is a precast prestressed high-performance concrete element with the shape of Greek letter omega. Omega beam was designed as a self-supporting permanent formwork member for construction of girder bridges. FEM program ATENA Science was exploited for simulation of load-bearing test of the beam. The numerical model was calibrated using the data from both static loading test and tests of material properties. Comparison of load-displacement diagrams obtained from the experiment and the model was conducted. Development of cracks and crack patterns were compared. Very good agreement of experimental data and the FEM model was reached. The calibrated model can be used for design of optimized Omega beams in the future without the need of expensive loading tests. The calibrated material model can be also exploited in other types of FEM analyses of bridges constructed with the use of Omega beams, such as limit state analysis, optimization of shear connectors, prediction of long-term deflections or prediction of crack development.

1. Introduction

When an unusual new type of a load-bearing element for bridges is designed with the use of a non-traditional material, it is sometimes the case that a static loading test of 1:1 scale model is performed before practical application to verify the expected behavior of the element. Such tests are usually quite expensive and therefore it is suitable to use their results for calibration of numerical models that will allow avoiding the large-scale tests in the future. A properly calibrated numerical model can fully replace the static loading test.

In this paper, the process of numerical modeling of the static loading test of so called Omega beam (for more information, see paper "Lightweight HPC beam OMEGA" by Sýkora et al. published in the same proceedings as this paper [7]) is described and the results of the model and the test are compared. Omega beam is a precast prestressed high-performance concrete (HPC) element with the shape of Greek letter omega. Omega beam was designed as a self-supporting permanent formwork member for construction of girder bridges. It exploits high durability of HPC [6]. FEM software ATENA Science version 5.3.4 [1] was used for numerical modeling.

2. Static loading test

A prototype of Omega beam (see Figures 1 and 2) was produced and subjected to static loading test in the age of 17 days. The test took place in the precast concrete production plant. The load was applied to the beam through the steel crossbars loaded by hydraulic presses propped against a group of concrete panels (see Figure 3). The total weight of the panels was 30 tons, the panels rested on concrete supports.

Figure 1. View of the end part of the Omega beam.

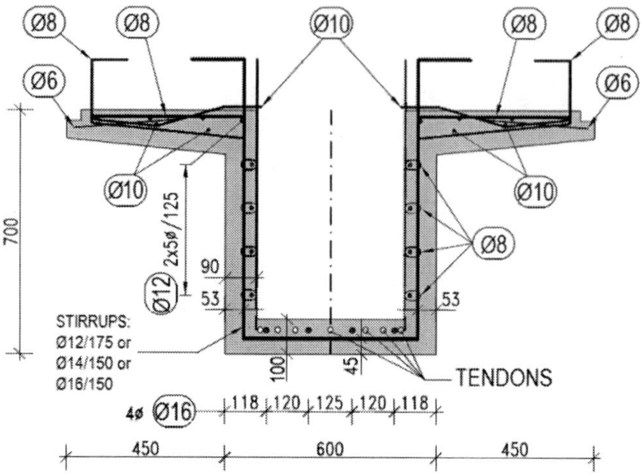

Figure 2. Section of the Omega beam with description of the reinforcement.

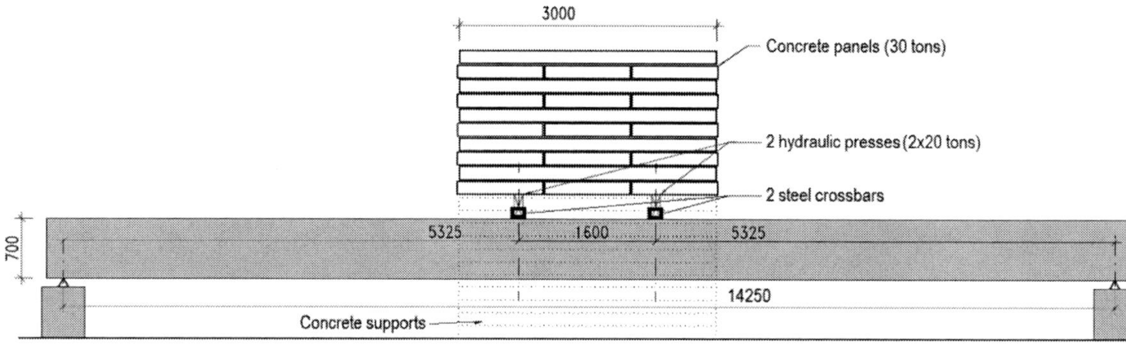

Figure 3. Scheme of the static loading test of the Omega beam.

The schedule of the static loading test is given in Table 1. The first load step simulated the loads from in-situ concreting of the monolithic part of the bridge structure. The second one represented the decompression of the lower part of the cross-section. The third one corresponded to the anticipated moment of reaching the tensile strength of concrete, i.e. the moment when the first cracks occur and the non-linear behavior of the structure begins. It was planned to interrupt the test before reaching the theoretical ultimate load-bearing capacity as the prototype was meant for exploitation in a real structure. The loads in particular steps were calculated by hand calculations performed according to ČSN EN 1992-1-1 [2].

Table 1. Load steps of the static loading test.

Load step no.	Load step	Load [kN]	Load [%]
1	In-situ concreting	2 × 85	60
2	Decompression	2 × 89	63
3	Tensile strength	2 × 111	78
4	End of the test	2 × 142	100
-	Theoretical capacity	2 × 172	121

3. FEM model

The simulation of the loading test was performed in ATENA Science version 5.3.4 [1]. Creep module was used to include the creep and shrinkage of concrete between the production of the prototype and the static loading test. Newton-Raphson calculation method was applied. The model was prepared in GiD 11 preprocessing tool.

3.1. Geometry

To optimize the computation process, symmetrical quarter of the beam was modelled (Figure 4). The model was supported by a hinge in the middle of the supporting plate (orange colour in Figure 4), horizontal constraints were applied in the direction perpendicular to the section planes (green and yellow colour in figure 4) to simulate the influence of cut parts of the structure. The load was introduced through a part of the crossbeam (red arrow in Figure 4). Regular hexahedral mesh with five elements in thickness of the Omega beam was used, the length of the elements was 50 mm in the central part of the beam and 115 mm in the rest. Each bar of the longitudinal reinforcement was divided in 100 linear finite elements, each stirrup was divided in 10 linear finite elements.

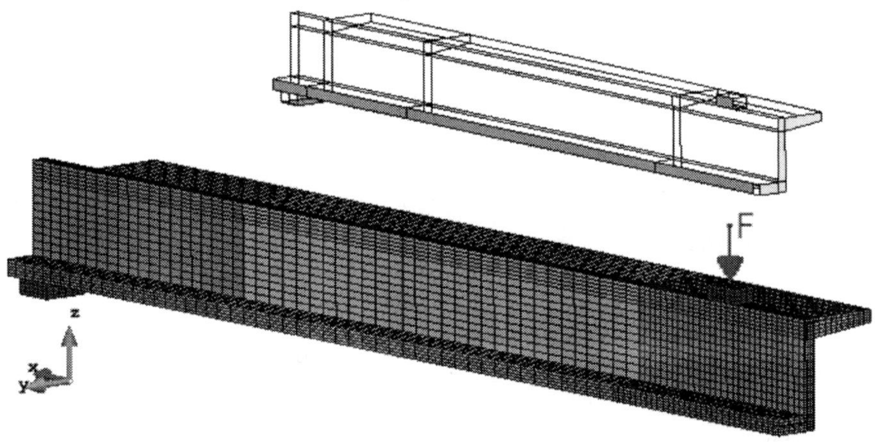

Figure 4. Model of the Omega beam. Lower part: FEM mesh and the load. Upper part: boundary conditions (explanation in the paragraph above).

3.2. Materials

Mean cylindrical compressive strength of concrete f_{cm} = 68.2 MPa was obtained from the results of laboratory tests. Other properties were calculated according to the formulae given in the *fib* Model Code 2010 [3], see Table 2. Concrete class C60/75 XF2, XD1 was used.

Material model CC3DNonLinCementitious2 was utilized for the basic deformation behavior of the material. The model combines constitutive models for tensile (fracturing) and compressive (plastic) behavior. The fracture model is based on the classical orthotropic smeared crack formulation and crack band model. It employs Rankine failure criterion. The hardening/softening plasticity model is based on Menétrey-Willam failure surface [4].

To model the time-dependent behavior of concrete, material model CCmodelB3Improved based on B3 model created by Bažant and Baweja [5] was exploited. The parameters used in the model are summarized in Table 3. This model considers autogenous and drying shrinkage, basic and drying creep, temperature effect on creep and coupling between creep and fracture. B3 model was also used to calculate the time development of material properties of concrete from the properties in the age of 28 days.

The structure was reinforced by B500 B bars and St 15.7-1660/1860 tendons with very low relaxation. Mean material parameters were used in the model (see Tables 4 and 5). CCReinforcement material model with bilinear stress-strain diagram was exploited for embedded reinforcement bar elements.

Table 2. Mechanical properties of concrete

Property	Label	Value	Unit	Formula [3]
Mean cylindrical compressive strength	f_{cm}	68.2	MPa	-
Mean elastic modulus	E_{cm}	40.76	GPa	$21.5(f_{cm}/10)^{1/3}$
Mean axial tensile strength	f_{ctm}	4.4	MPa	$2.12\ln(1+f_{cm}/10)$
Relative fracture energy	G_F	156	Nm^{-1}	$73f_{cm}^{0.18}$
Tension stiffening factor	F_t	0.4	-	-
Bulk density	ρ_c	2285	$kg.m^{-3}$	-

Table 3. Material parameters for B3 model.

Property	Value	Unit
Cement type	CEM I 52.5 R	-
Cement content	457.6	$kg.m^{-3}$
Water to cement ratio	0.41	-
Aggregate to cement ratio	3.70	-
Effective thickness of the beam	0.094	m
Relative ambient humidity	70	%
Cross-section shape factor	1.0	-
Curing conditions	air	-
Curing time	2	days

Table 4. Mechanical properties of reinforcement bars.

Property	Label	Value	Unit
Mean yield limit	f_{ym}	550	MPa
Elastic modulus	E_s	200	GPa
Ultimate strength	f_{um}	594	MPa
Ultimate elongation	ε_{um}	5	%
Bulk density	ρ_s	7850	$kg.m^{-3}$
Bar diameters	Ø	6 – 16	mm

Table 5. Mechanical properties of tendons.

Property	Label	Value	Unit
Mean yield limit	f_{pym}	1826	MPa
Elastic modulus	E_p	193	GPa
Ultimate strength	f_{pum}	1972	MPa
Ultimate elongation	ε_{pum}	3.5	%
Bulk density	ρ_s	7850	$kg.m^{-3}$
Cross-sectional area of 1 tendon	A_{p1}	150.5	mm^2

3.3. Loads

The model was loaded in several steps according to Table 6. The forces in steps 3 – 6 represented the static loading test described in section 2. As one quarter of the beam was modelled, the applied loads are equal to one quarter of the total load in the test. Some values of the forces are slightly different from those in Table 1 as the real process of the test did not agree exactly with the plan (the cracks occurred later). The forces in steps 3 – 8 were applied in several increments in order to reach the convergence of the solution process. The model was loaded up to the ultimate limit state (ULS).

The self-weight was applied to the model by means of the defined bulk densities of the materials (concrete, reinforcement).

The prestress was introduced as the initial strain of tendons $\varepsilon_{p,0}$. The value was calculated using the Hooke´s law and considering the final stress of tendons after straining $\sigma_{p,0} = 1462$ MPa, anchorage slip $p = 2.4$ mm and total length of the tendons 37 557 mm:

$$\varepsilon_{p,0} = -\frac{\sigma_{p,0} - \frac{p}{L} E_p}{E_p} = -\frac{1462 - \frac{2,4}{37557} \cdot 193000}{193000} = -7.511 \cdot 10^{-3} \tag{1}$$

The loss due to elastic shortening of concrete after application of the prestress was modeled by the software. The same applies to the progress of creep and shrinkage of the structure, including the respective losses in prestress. The loss due to relaxation of tendons was calculated by hand according to [2] and entered to the model as a reduction of the initial strain of tendons $\Delta\varepsilon_{p,r} = 1.38 \times 10^{-4}$.

Table 6. Load steps

Step no.	Load step	Time [days]	New applied loads	Number of increments [-]	Total force in the end of the step [kN]
0	Production, curing	0.0 – 2.1	-	-	-
1	Prestressing	2.1 – 2.2	Self-weight, prestress	5	-
2	Aging, losses in prestress	2.2 – 17.0	Creep, shrinkage, losses in ps.	5	-
3	In-situ concreting	17.0 – 17.01	0.5 × 85 kN	5	0.5 × 85
4	Decompression	17.01 – 17.02	0.5 × 4 kN	5	0.5 × 89
5	Tensile strength	17.02 – 17.03	0.5 × 36 kN	10	0.5 × 125
6	End of the test	17.03 – 17.04	0.5 × 24 kN	5	0.5 × 149
7	Theoretical capacity	17.04 – 17.05	0.5 × 23 kN	5	0.5 × 172
8	ULS	17.05 – 17.065	0.5 × 75 kN	15	0.5 × 247

4. Results and discussion

The load-displacement curves obtained from the static loading test and the numerical model are presented in Figure 5 and Table 7. Very good overall agreement between the test and the basic material model with the parameters defined before (denoted as ATENA) was reached.

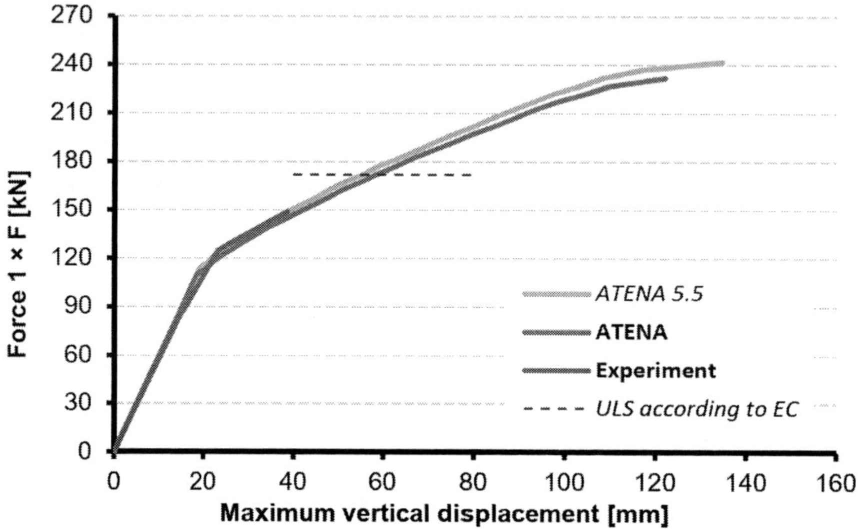

Figure 5. Comparison of load-displacement curves.

Table 7. Comparison of displacements from the experiment and the models.

Step no.	Load step	Total force in the end of the step[kN] 0.5 × F	Maximum vertical displacement [mm]		
			Experiment	ATENA	ATENA 5.5
3	In-situ concreting	0.5 × 85	14.6	14.2	14.2
4	Decompression	0.5 × 89	15.5	14.8	14.8
5	Tensile strength	0.5 × 125	23.4	26.1	23.7
6	End of the test	0.5 × 149	38.6	41.2	38.6
7	Theoretical capacity	0.5 × 172	-	58.4	55.4
8	ULS	0.5 × 232	-	121.9	108.3
9	ULS ver. 5.5	0.5 × 242	-	-	134.6

Certain disagreement is evident at the beginning of the non-linear behavior of the model. The first cracks occurred in the model as early as at the load of F = 111 kN, while no cracks were observed during the test until F = 125 kN. The width of the first cracks in the model was less than 0.02 mm. Cracks that thin would have no practical meaning in the real structure, but they influenced the stiffness of the very sensitive numerical model.

An attempt was made to further improve the affinity of the curves in the non-linear area by varying material parameters (tensile strength, fracture energy, tensile stiffening factor). The best result was reached when the tensile strength of concrete was increased to 5.5 MPa (the curve denoted as ATENA 5.5 in Figure 5). The tensile strength of the material was not measured directly by tests. This means that the real tensile strength could have been higher than the theoretical one considered in the basic model. The value of 5.5 MPa is less than 95 % quantile of tensile strength of C60/75 concrete ($f_{ctk,0.95}$ = 5.7 MPa according to [2]), therefore it is realistic. However, there is no experimental evidence to justify such a high tensile strength and the value is not very likely according to the experience of the authors. Therefore, the ATENA 5.5 model is not recommended for further simulations.

The development of cracks is documented in Figures 6 – 11.

- The first cracks were observed during the test at the load of F = 125 kN when four cracks of approximately 0.1 mm appeared in the central area of the beam (Figure 6). At the same load, six cracks of 0.06 – 0.08 mm occurred in the corresponding area of the model (Figure 7).

- At the load of F = 141 kN, seven cracks of the width between 0.1 and 0.2 mm were registered during the test. No new cracks exceeding the width of 0.1 mm appeared in the model, the width of the six cracks in Figure 8 was between 0.1 and 0.12 mm.
- At the end of the test, the beam was disturbed by eight cracks 0.2 – 0.25 mm wide, while there were eight cracks approximately 0.1 – 0.15 mm wide in the model (Figure 9).
- Crack pattern at the theoretical ULS is in the Figure 10, the widest cracks had 0.2 mm.
- The extensive disruption of the beam at the ULS of the model is represented by Figure 11, with the width of the cracks reaching 0.5 mm.

As a result, it can be said that the crack pattern of the model corresponds to the reality, but the model tends to underestimate the width of the cracks and overestimate the number of the cracks at the same time. The calculated crack widths are dependent on the finite element mesh size. Theoretically, better agreement between the cracks in the model and the test could be reached by increasing the size of the elements in the *x* direction, but this could lead to problems with convergence of the solution process. Therefore the amendment was not applied.

The ultimate limit state of the ATENA model was reached at the load of F = 232 kN. At this moment, the highest compressive stress in concrete in the middle of the beam was σ_{cc} = 45 MPa, the tendons were slightly behind the yield limit (σ_p = 1850 MPa) and the reinforcement bars were exactly on the yield limit (σ_s = 552 MPa).

Further small increase of the applied load accompanied by plastic deformation of the reinforcement and large vertical displacements of the model would be possible, up to the moment of reaching the ultimate strength of one of the materials. The Newton-Raphson method used for the calculation did not allow simulating this situation; the Arc-Length method is not available in the Creep module of ATENA Science software. In any case, finding the load and displacement at the final ULS would have little practical meaning.

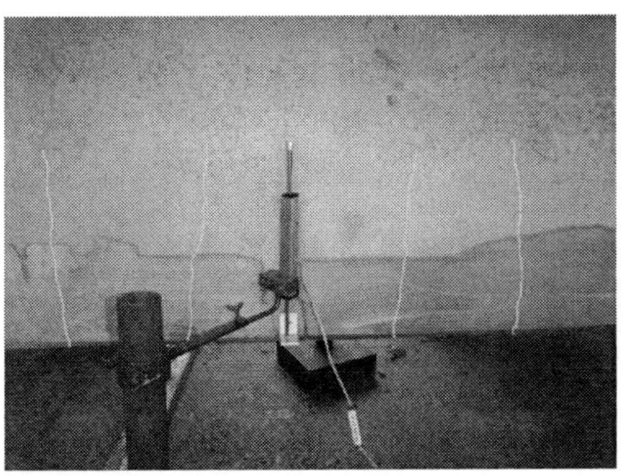

Figure 6. First cracks in the test (4 × 0.1 mm, F = 125 kN).

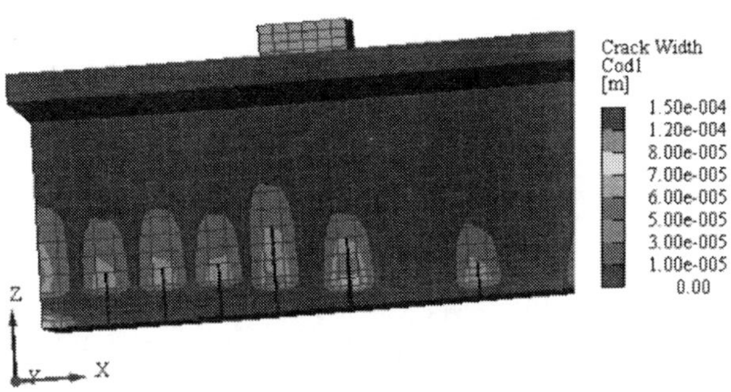

Figure 7. Cracks in the model at F = 125 kN. Width is between 0.06 – 0.08 mm.

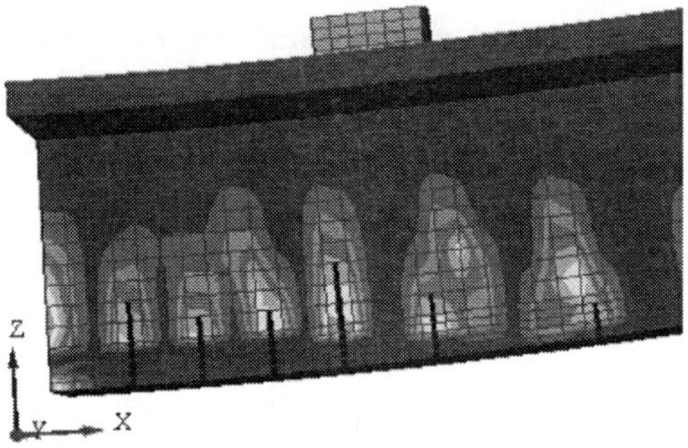

Figure 8. Cracks in the model at F = 141 kN. Width is between 0.1 – 0.12 mm. The same legend as in Figure 7 applies.

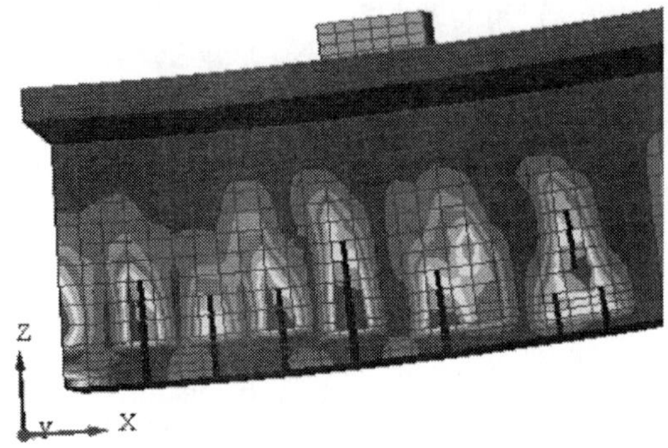

Figure 9. Cracks in the model at F = 149 kN. Width is between 0.1 – 0.15 mm. The same legend as in Figure 7 applies.

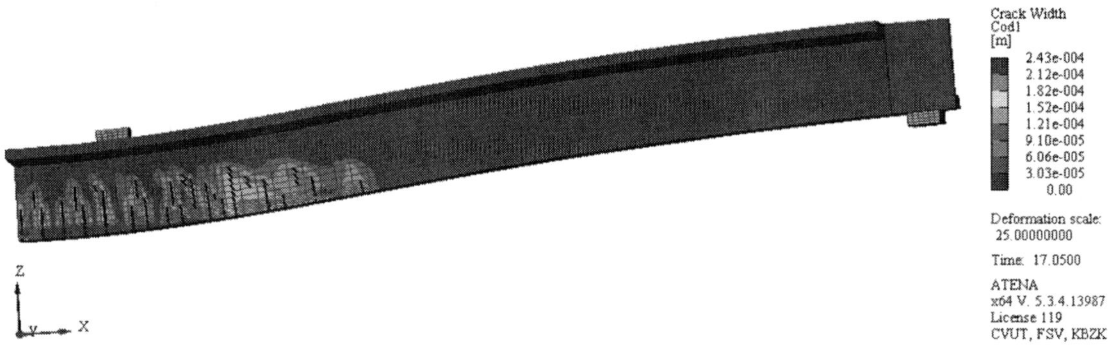

Figure 10. Deflection and cracks in the model at F = 172 kN (theoretical ULS). Width of the cracks is between 0.1 – 0.2 mm.

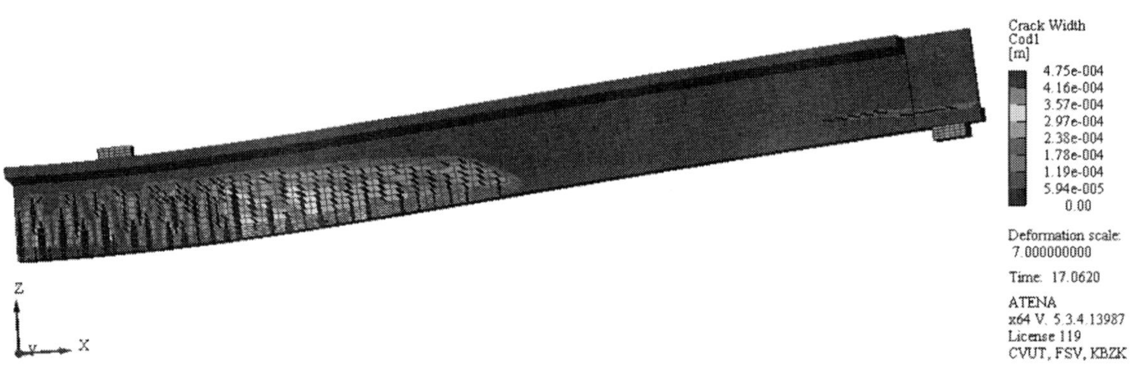

Figure 11. Deflection and cracks in the model at F = 232 kN (ULS of the model). Width of the cracks is between 0.1 – 0.5 mm.

5. Conclusion

Very good overall agreement of load-displacement curves from the experiment and the FEM model was reached. Further improvement of the model would be probably possible if the real tensile strength of the material was known. The crack patterns obtained from the model corresponded to the reality, but the model tended to underestimate the width of the cracks and overestimate the number of the cracks at the same time.

The calibrated model can be used for design of optimized Omega beams in the future without the need of expensive loading tests. The calibrated material model can be also exploited in other types of FEM analyses of bridges constructed with the use of Omega beams, such as limit state analysis, optimization of shear connectors, prediction of long-term deflections or prediction of crack development.

Acknowledgements
This paper was prepared thanks to the support of the Technology Agency of the Czech Republic (TAČR), project "Centre for Effective and Sustainable Transport Infrastructure (CESTI)" (no. TE01020168), and the Science Foundation of the Czech Republic (GAČR), project "Analysis of the relations between the microstructure and macroscopic properties of ultra-high performance concretes" (no. 17-19463S).

References
[1] Červenka Consulting 2016 ATENA x64 Version 5.3.4. Available online at <http://www.cervenka.cz/products/atena>
[2] Český normalizační institut 2006 Design of concrete structures - Part 1-1: General rules and rules for buildings *Eurocode 2, Part 1-1 (ČSN EN 1992-1-1)*
[3] Walraaven J et al. 2013 *fib Model Code for Concrete Structures 2010* (Berlin: Ernst&Sohn)
[4] Pisano A A 2012 An Algorithm Approach for Peak Load Evaluation of Structural Elements Obeying a Menetrey-William Type Yield Criterion *Electronic Journal of Differential Equations* **167** pp 1-9
[5] Bažant Z P and Baweja S 2000 Creep and Shrinkage Prediction Model for Analysis and Design of Concrete Structures: Model B3 *Adam Neville Symposium: Creep and Shrinkage – Structural Design Effects* ed A Al-Manaseer (Farmington Hills, Michigan, USA: ACI)
[6] Fládr J, Bílý P and Vodička J 2016 Experimental Testing of Resistance of Ultra-high Performance Concrete to Environmental Loads *Proceedia Engineering* **151** pp 170-176
[7] Sýkora M, Jedlinsky P and Komanec J 2017 Lightweight HPC beam OMEGA *IOP Conference Series: Materials Science and Engineering (MSE) – Proceedings of BESTInfra 2017 conference* (Prague, Czech Republic) to be published

IOP Conf. Series: Materials Science and Engineering **236** (2017) 012067 doi:10.1088/1757-899X/236/1/012067

Analysis of the connection of the timber-fiber concrete composite structure

Milan Holý[1], Lukáš Vráblík[1] and Vojtěch Petřík[2]

[1]CTU Prague, Faculty of Civil Engineering, Department of Concrete and Masonry Structures, Thákurova 7, 166 29 Praha 6, Czech Republic
[2]Obermeyer Helika, a. s., Beranových 65, 199 21 Praha 9, Czech Republic

E-mail: milan.holy@fsv.cvut.cz

Abstract. This paper deals with an implementation of the material parameters of the connection to complex models for analysis of the timber-fiber concrete composite structures. The aim of this article is to present a possible way of idealization of the continuous contact model that approximates the actual behavior of timber-fiber reinforced concrete structures. The presented model of the connection was derived from push-out shear tests. It was approved by use of the nonlinear numerical analysis, that it can be achieved a very good compliance between results of numerical simulations and results of the experiments by a suitable choice of the material parameters of the continuous contact. Finally, an application for an analytical calculation of timber-fiber concrete composite structures is developed for the practical use in engineering praxis. The input material parameters for the analytical model was received using data from experiments.

1. Introduction

Slender, aesthetic and durable constructions can be achieved by combining timber and fiber reinforced concrete (FRC). Especially an application of the ultra-high performance fiber reinforced concrete (UHPC) enables to design very slender decks, an architectonic surface of the concrete without a need of any special treatment. The structural behavior of timber-(fiber) concrete composite members is governed by the shear connection between timber and concrete in general.

2. Experiment basis

The push-out shear tests were carried out in order to determine characteristics of the connection between timber and fiber concrete members. Specimens (see Figure 1) were formed as a solid timber beam of strength class C24 coupled in five cases with a FRC slab of thickness 60 mm, and in four other cases with a UHPC slab of thickness 35 mm. The connection between the timber and fiber-concrete parts was carried out by special coupling screws TCC diameter 7.3 mm and length 150 mm embedded at an angle of 45 degrees to the timber beam. The comparison of relevant material properties of the used fiber-concretes see in Table 1.

A record of the push-out shear test represents a relationship between the connection slip and the transferred force on the contact area. A start-up inclination of the record diagram caused by pushing of supports was eliminated for purposes of the numerical simulation. This was achieved by moving asymptotes that pass through the origin of the coordinate system. Such modified test records transferred to the stress-slip diagram are shown in Figure 2.

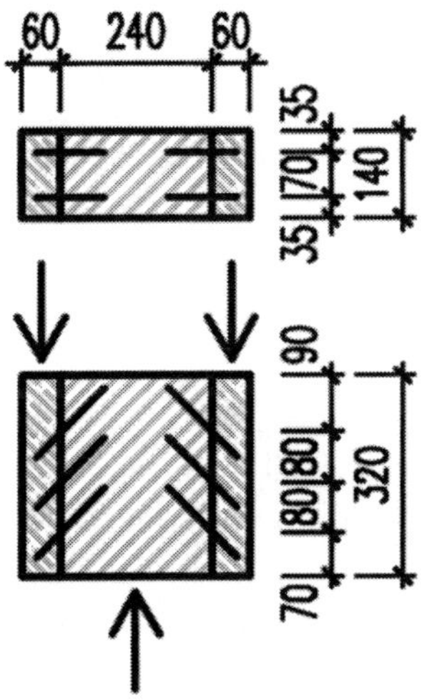

Figure 1. Dimension of specimen and direction of loading.

Table 1. Material properties of the used fiber-concretes.

Material properties	FRC	UHPC
Compressive strength (MPa)	45	150
Flexural strength (MPa)	4,4	14
Equivalent tensile strength (MPa)	1,6	5,4
Modulus of elasticity (MPa)	29016	50600

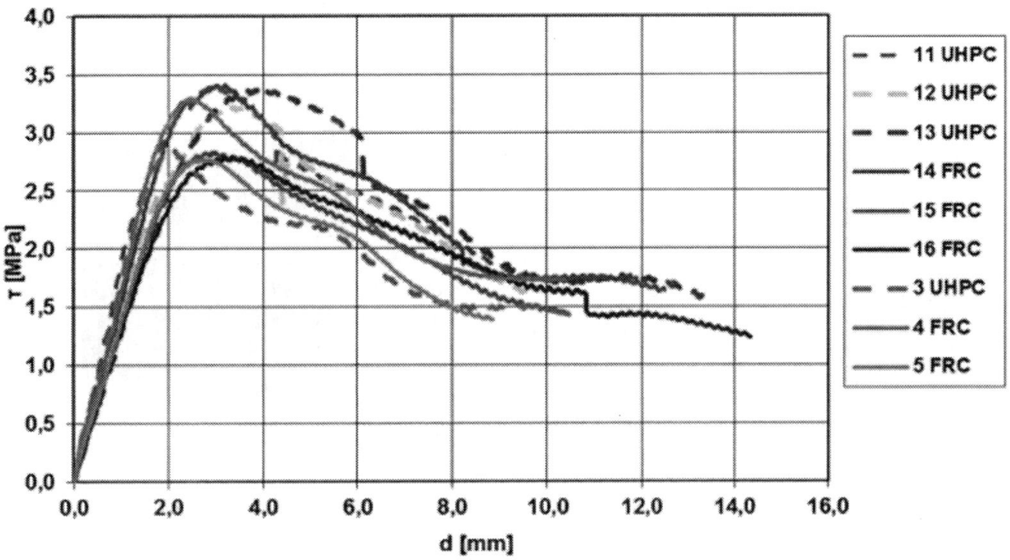

Figure 2. Stress-slip diagram for modified records of the push-out shear tests.

Results of the push-out shear tests show that a different thickness and different material properties of fiber-concrete components FRC x UHPC does not affect the structural behavior of the connection (specifically regarding a tangential stiffness, a shear strength and a strain softening), because the failure of the connection generally occurs in the area of the timber beam, the concrete remains undamaged.

3. Simulation of experiment

An implementation of correct material models of the connection is essential in terms of relevancy of the analysis of the timber-concrete structures. If the connection strain remains in the linear elastic range until the timber or concrete members fail, a linear-elastic behavior of the connection may be assumed. However, if the connectors reach their load carrying-capacity, a nonlinear behavior of the composite structure should be considered.

3.1. Material model

The push-out shear tests were simulated with a nonlinear numerical analysis in program ATENA (Červenka Consulting, s.r.o.). A continuous contact spread over area of the contact between both composite members was applied, because the continuous contact is more suitable by use in the complex models than a model of the connection based on discrete coupling elements, which leads to a high number of finite elements and means a high computational difficulty.

The material model of the contact in program ATENA is based on the Mohr-Coulomb theory. The model of the contact is defined by parameters representing real physical properties of the contact on the one hand. That constitute a shear strength (cohesion) c, a tensile strength f_t and a coefficient of friction φ. On the other hand, additional parameters, which are basic and minimum values of the axial and tangential stiffness K_{nn}, $K_{nn,min}$ and K_{tt}, $K_{tt,min}$, are defined for numerical purposes.

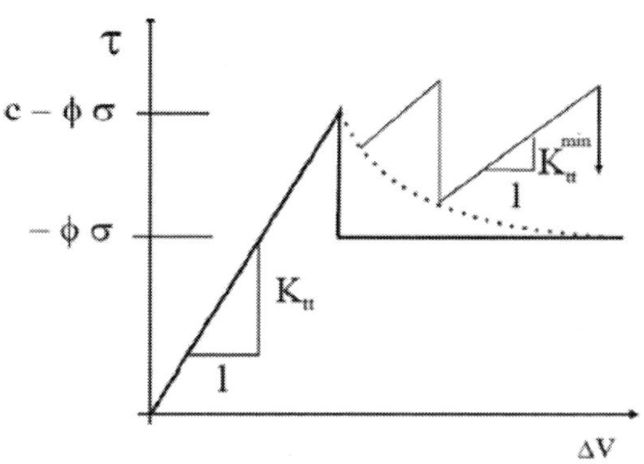

Figure 3. Schema of material model used for continuous contact spread over area in program ATENA.

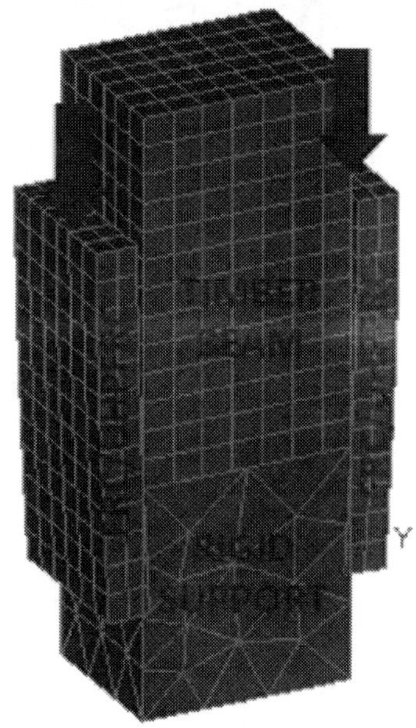

Figure 4. Screenshot of simulation of the push-out shear test in program ATENA – visible deformation of the connection caused by eccentricity between load and support.

The contact surface is primarily subjected to shear in the push-out shear test, therefore the shear material parameters have a major influence in this case. A basic value of the shear stiffness determines the start inclination of asymptote of the slip-stress diagram. A value of the shear strength determines the apex of the slip-stress diagram. The shear softening can be approximated by a function defining a decrease of the cohesion depending on increasing displacement. The coefficient of friction between timber and concrete has an insignificant effect in the case of push-out shear test, because the contact surface is not pressed by external load. Although, the coupling elements are installed with an inclination of 45 degrees. Some axial stresses in the contact surface occur due to an eccentricity between the support and the introduced force. The upper part of the connection is pulled and the bottom part is pushed (see Figure 4). The coefficient of friction was considered with a conservative value 0.2.

It would be necessary to perform a tensile test of the connection (an additional experiment) to determine the axial stiffness and the tensile strength of the contact relatively exactly. However, the tensile strength value is usually about half of the shear strength value. Therefore, the tensile strength was derived from the shear strength. The decrease of the tensile strength (a tensile softening) should approximately correspond to the shear softening. It is recommended that the axial stiffness should be the same as the shear stiffness according to the experience of the authors of the ATENA software.

3.2. Adjustment of parameters

Basically, it can be used more ways how to achieve a good compliance of the simulation with the experiment. One approach is based on the assumption that the concrete and timber do not deform, they behave as rigid bodies. Any deformation depending on the introduced force is simulated exclusively by the material model of the contact in this case. Although, this model approximates the push-out shear test very well, it is not suitable for further application in complex models, because of its tendentious interpretation of the determining material parameters (modulus of elasticity E and Poisson's ratio ν of the timber and concrete members must be adjusted).

It is more apt to implement the material characteristics E and ν of the composite members unchanged and vary only the parameters of the connection. However, the calculation converges more difficult and do not reach such accordance with the experiment - see "corrugations" in stress-slip diagram (see Figure 5).

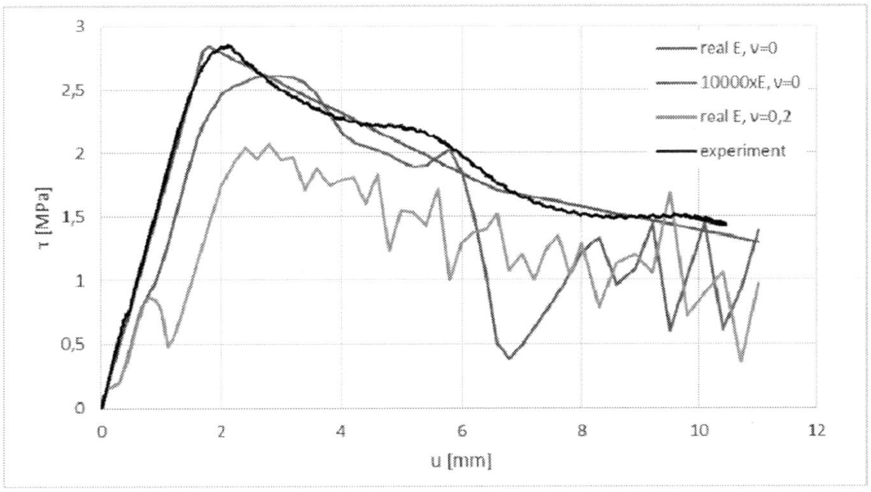

Figure 5. Effect of an adjustments of modulus of elasticity and Poisson's ratio of coupled materials on the simulation of the push-out shear test.

It was found that it is possible to eliminate the "corrugations" in stress-slip diagram also by reducing of the axial stiffness of the contact (see Figure 6). The smooth curve in stress-slip diagram is achieved by reducing of the axial stiffness with respect to the tangential stiffness and the model approximates the tests very well. Similarly to the experiments, a certain "jump" can be observed at the beginning of loading. The start-up inclination of the experiment record diagram caused by pushing of supports was eliminated for the purposes of numerical simulation. It is apparently, that the part of the gradual start-up inclination is caused due to a deformation of the material of composite members.

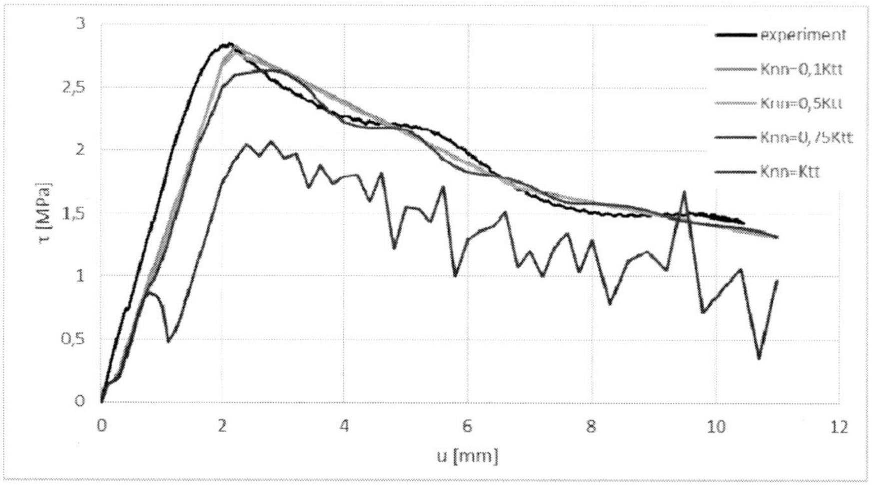

Figure 6. Effect of an adjustment of the normal stiffness on the simulation of the push-out shear test.

4. Complex models

4.1. Linear elastic analysis

The ultimate strength of the timber and the concrete members is mostly reached before it is achieved a nonlinear part of the stress-slip diagram. The connection can be advantageously considered as linear elastic in this case.

A simplified design method for mechanically jointed beams according to Annex B of Eurocode 5 is usually used for a design purposes. This simplified design method called "γ-method" is based on the differential equation for the partial composite action with continuous constant slip stiffness. It was developed for simply supported beams subjected to loads giving a bending moment varying sinusoidal, because for such case the differential equation of the partial composite action has a simple analytical solution. However, the γ-method is sufficiently accurate also for loads giving a bending moment varying parabolic and for a slip stiffness varying according to the shear force.

A program resulting from a need of the design profession for easy and quick design and optimizing of the dimensions of the composite timber–fiber concrete structures was created in MS Excel. The program works on the basis of the γ-method. The input material characteristics of the connection received directly from the experiments are included. It is planned to create a database of the material characteristics of the connection for different arrangements of coupling elements.

4.2. Verification of nonlinear material model

A verification of the material model of the connection was made on the simply supported beam with a T-shaped cross-section loaded with a uniformly distributed load. Adjustment of the material model of the connection was made primarily in program ATENA 2D. The results were compared to 3D model in ATENA 3D and to linear calculation on the basis of the γ-method.

The T-shaped cross-section (see Figure 7) was composed of timber beam of structural timber class C24 with width 14 cm and height 24 cm coupled with fiber concrete slab of a compressive strength of 45 MPa and modulus of elasticity of 29 GPa of width 94 cm and height 6 cm. A linear elastic material model was used for timber and FRC composite members. Connection between timber and fiber-concrete members was carried out by special coupling screws TCC diameter 7.3 mm and length 150 mm embedded at an angle of 45 degrees to timber beam. Span of simply supported beam was 4 m.

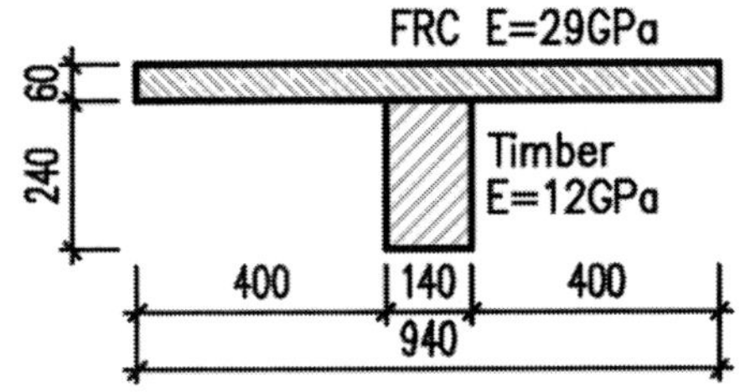

Figure 7. Dimension of T-shaped cross-section.

A change of the axial stiffness and the tensile strength of the connection has not a decisive influence on the structural behavior of the bent structure (it is logical, because a vertical load presses the joint). The curves for variants $K_{nn} = K_{tt}$, $K_{nn} = 0{,}5K_{tt}$, $K_{nn} = 0{,}1K_{tt}$, $0{,}5f_{t0}$, $2f_{t0}$ overlap each other in the load – deflection diagram and differ from each other in the order of hundredths (see Figure 8).

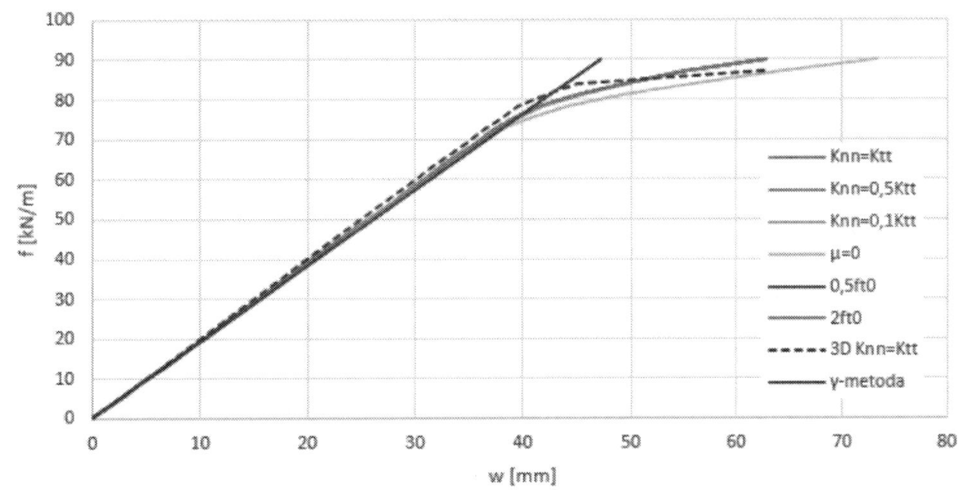

Figure 8. Load-deflection diagram of simply supported beam with uniformly distributed load for cross-section in the middle of span. Comparison of calculation with several material parameters of connection in ATENA 2D with ATENA 3D and γ-method.

In the opposite to the simulation of the push-out shear tests, the coefficient of friction μ has a considerable influence on the structural behavior of the bent beam (see Figure 8, default value $\mu = 0.2$, yellow variant $\mu = 0$). Friction takes an effect after reaching of the shear strength c.

The ultimate strength of the timber beam and the UHPC/FRC deck has been reached before it was achieved a non-linear part of stress-slip diagram.

5. Conclusion

An implementation of correct material models of the connection is essential in terms of relevancy of the analysis of the timber-concrete structures. If the connectors reach their load carrying-capacity, a nonlinear behavior of the composite structure should be considered. It was approved by use of the nonlinear numerical analysis in program ATENA (Červenka Consulting, s.r.o), that it can be achieved a very good compliance between results of numerical simulations and results of the experiments by a suitable choice of the material parameters of the continuous contact. It was verified that the material parameters of the connection model derived from the experiment could be effectively applied in a complex model for elements subjected to bending.

If the connection strain remains in the linear elastic range until the timber or concrete members fail, a linear-elastic behavior of the connection may be assumed. A program based on the γ-method for easy and quick design and optimizing of the dimensions of the composite timber–fiber concrete structures was created in MS Excel. The input material characteristics of the connection received directly from the experiments are included. It is planned to create a database of the material characteristics of the connection for different arrangements of coupling elements.

Acknowledgements
Special thanks for expert advice and consultation belongs to Ing. Radomír Pukl, CSc. and Ing. Jan Červenka Ph.D. (both Červenka Consulting, s.r.o). This paper was written in the context of TAČR number TH02020730.

References
[1] Caldová E, Blesák L, Wald F, Kloiber M, Urushadze S and Vymlátil P 2014 Behaviour of timber and steel fibre reinforced concrete composite constructions with screwed connections *Wood Research* **59**(4) pp 639-60
[2] European Committee for Standardization (CEN) 2006 Eurocode 5: Design of timber structures – Part 1-1: General - Common rules and rules for buildings *(EN 1995-1-1)*
[3] Glaser R 2005 *Zum Kurz- und Langzeitverhalten von Holz-Beton-Verbundkonstruktionen* (TU Cottbus)
[4] Holý M, Petřík V and Pukl R 2016 Idealizace modelu spřažení dřevo-vláknobetonových kompozitních konstrukcí /The idealization of the connection model for timber-fiber concrete composite structures/ *Beton - Technologie, Konstrukce, Sanace* **16**(2) pp 18-21
[5] Holý M 2016 Idealizace modelu kontaktu kompozitních dřevo-vláknobetonových konstrukcí a jeho použití v komplexních modelech /The idealization of the connection model for timber-fiber concrete composite structures and its use in complex models/ *PhD Workshop 2016* (CTU Prague)
[6] Petřík V and Philipp N 2013 Dřevo-vláknobetonové kompozitní konstrukce - experimentální program /Timber-fibre concrete composite structures – experimental program/ *Beton - Technologie, Konstrukce, Sanace* **13**(1) pp 70-73
[7] Petřík V, Philipp N, Křístek V and Půlpán M 2011 Full-scale testy kompozitních konstrukcí FRC-dřevo, UHPFRC-dřevo *18. Konf. Betonářské dny (Praha)* pp 337-42
[8] Moshiri F, Gerber C, Valipour H R, Shrestha R and Crews K I 2013 The predictive model for strength of inclined screws as shear connection in timber-concrete composite floor *Materials to Structures: Advancement Through Innovation - Proceedings of the 22nd Australasian Conf. on the Mechanics of Structures and Materials* pp 1059-64
[9] Nežerka V 2010 *Timber-Concrete Composite Structures* (CTU Prague)
[10] Symon D, Persaud R and Stanislaus H 2010 Strength of inclined screw shear connections for timber and concrete composite construction *The Structural Engineer* **88**(1) pp 25-32
[11] Šlapka P 2013 Vyhodnocení protlačovacích zkoušek a stanovení materiálových parametrů modelu rozetřeného spřažení a následná numerická analýza chování spřahovacího trnu *Phd Workshop 2013* (CTU Prague)

Fracture properties of concrete specimens made from alkali activated binders

Hana Šimonová[1], Barbara Kucharczyková[1], Libor Topolář[1], Vlastimil Bílek, Jr.[2] and Zbyněk Keršner[1]

[1]Brno University of Technology, Faculty of Civil Engineering, Veveří 331/95, 602 00 Brno, Czech Republic
[2]Brno University of Technology, Faculty of Chemistry, Purkyňova 464/118, 602 00 Brno, Czech Republic

E-mail: kersner.z@fce.vutbr.cz

Abstract. The aim of this paper is to quantify crack initiation and other fracture properties – effective fracture toughness and specific fracture energy – of two types of concrete with an alkali activated binder. The beam specimens with a stress concentrator were tested in a three-point bending test after 28, 90, and 365 days of maturing. Records of fracture tests in the form of load versus deflection (*P–d*) diagrams were evaluated using effective crack model and work-of-fracture method and load versus mouth crack opening displacement (*P–CMOD*) diagrams were evaluated using the Double-*K* fracture model. The initiation of cracks during the fracture tests for all ages was also monitored by the acoustic emission method. The higher value of monitored mechanical fracture parameters of concrete with alkali activated blast furnace slag were achieved with substitution blast furnace slag by low calcium fly ash in comparison with substitution by cement kiln dust.

1. Introduction

Currently, the binders based on ordinary Portland cement are the most commonly used in manufacturing of concrete. At the same time, the immense emphasis is put on environment protection through reductions of CO_2 of which large quantities are produced by cement industries. Therefore, the effort to reduce CO_2 emissions is leading to develop some alternative binders among them belongs also alkali activated binders during the recent decades. These binders are formed by mixing of some aluminosilicate based material, such as blast furnace slag (BFS) or fly ash (FA), with an alkaline activator and water [1, 2]. The specific disadvantages of BFS and FA with regard to alkaline activation may be partially overcome by their mixing [3–5]. Therefore, this paper focuses on the determination of fracture parameters of concrete based on alkali activated binder which blends BFS and FA. Also a concrete based on alkali activated BFS blended with by-pass cement kiln dust (CKD) was tested.

This paper is focused on the determination of fracture parameters of above mentioned types of alkali activated binders based concrete and on their evolution in time. The records of three-point bending fracture tests of specimen with initial stress concentrator were evaluated using effective crack model, work-of-fracture method and the Double-K fracture model. The initiation of cracks during the fracture tests was also monitored by the acoustic emission method.

2. Experimental details

Two different concrete mixtures with alkali activated binder were prepared. The both of them contained BFS, which was substituted by low calcium FA (50 %) in the first case and by CKD (25 %) in the second case. Details, regarding to material composition, specimens preparation and their curing can be found, for example, in [6].

The six specimens of the nominal dimension 75 × 75 × 295 mm were prepared from each mixture and tested in three-point bending at the age of 28 days. The three slab specimens (75 × 295 × 600 mm) were also prepared from each mixture and tested by vacuum test method. Two specimens of the nominal dimensions of 75 × 75 × 295 mm were cut from each slab (from undamaged end part of slab after vacuum tests were performed). One half of these specimens was tested in three-point bending at the age of 90 days and the second half at the age of 1 year.

The fracture experiments were conducted on beams specimens with initial stress concentrator made by a diamond blade saw before testing. The depth of initial edge notch on the bottom side of specimen was approximately 1/3 of specimen depth. The span length was 245 mm. If displacement increment loading of specimen is performed, it is possible to record the load versus displacement (*F–d*) and the load versus crack mouth opening displacement (*F–CMOD*) diagram during testing. For illustration, the *F–d* diagrams for both types of concrete made from alkali activated binder at the age of 90 days, with well recorded descending part, are shown in Figure 1.

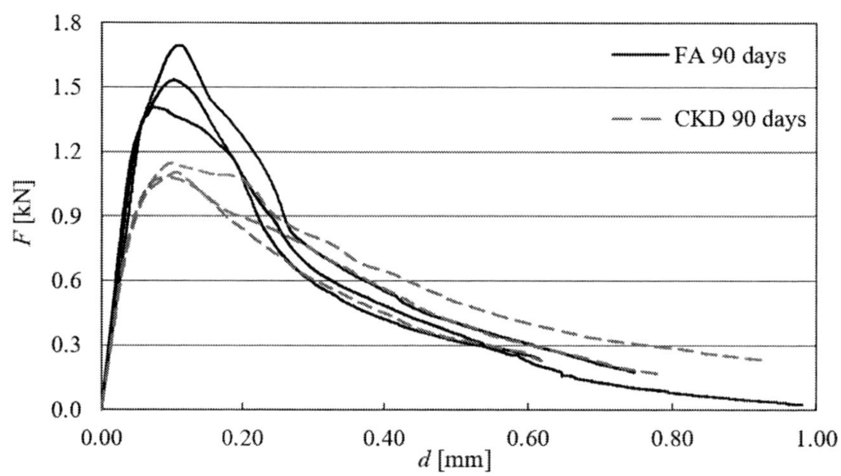

Figure 1. *F–d* diagrams of specimens tested at three-point bending at the age of 90 days.

The initiation of cracks during the fracture tests for all ages was also monitored by the acoustic emission (AE) method. The AE activity was recorded during the fracture tests. The four AE sensors were attached to the surface by beeswax. The AE signals were recorded by measuring equipment DAKEL XEDO with four acoustic emission sensors IDK-09 with 35 dB preamplifier.

The compressive strength informative values were also determined on the fragments remaining after the fracture experiments had been performed.

3. Methods

3.1. Effective crack model and work-of-fracture method

At first, almost linear part of *F–d* diagram is used to estimate modulus of elasticity value. Afterwards, the set of parameters is determined using effective crack model: critical effective crack length, effective fracture toughness and (using the known modulus of elasticity value) effective toughness [7].

The specific fracture energy value is determined from whole *F–d* diagram using the work of fracture value [8].

3.2. Double-*K* fracture model

The Double-*K* model [9] contains two material parameters; the first is called the initiation stress intensity factor K_{Ic}^{ini}, which defines the onset of cracking, and the second the unstable fracture toughness K_{Ic}^{un}, which defines the onset of unstable cracking or failure. These parameters characterize the tested materials in more detail and proceed from the combination of linear fracture mechanics principles and the concept of cohesive forces acting on the faces of the fictitious (effective) crack increment. In addition to the specimen dimensions and notch depth, other input data for the Double-*K* fracture model was taken from the above mentioned *F–CMOD*, namely maximum load F_{max} and its corresponding critical crack mouth opening displacement $CMOD_c$, and load F_i read from the linear part of diagram and its corresponding crack mouth opening displacement $CMOD_i$. The *F–CMOD* diagrams (part until peak load) of both tested materials at the age of 1 year are shown in Figure 2.

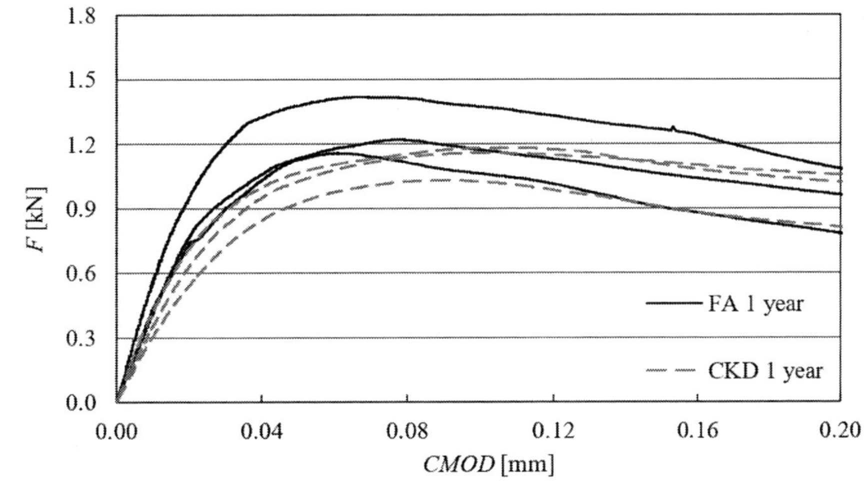

Figure 2. Part until peak load of *F–CMOD* diagrams of specimens tested at three-point bending at the age of 1 year.

In this case, the unstable fracture toughness K_{Ic}^{un} was numerically determined first, followed by the cohesive fracture toughness K_{Ic}^{c}. When both of these values were known, the following formula was used to calculate the initiation fracture toughness K_{Ic}^{ini}:

$$K_{Ic}^{ini} = K_{Ic}^{un} - K_{Ic}^{c}. \quad (1)$$

Details regarding the calculation of both the unstable and the cohesive fracture toughness values can be found in, e.g. [9, 10]. The bilinear variant [11] of the softening function was used in calculations where the tensile strength value was estimated based on the measured compressive strength value suggested by the relevant documentation [12] and the critical crack opening displacement was determined based on the fracture energy value obtained from above mentioned *F–d* diagrams.

3.3. Acoustic emission method

To describe acoustic emission signals formed during the three-point bending test, the attention is focused on the number, duration, amplitude and energy of AE signals [13]. The number of AE events corresponds to the material ability to resist fracture. The occurrence of a large number of cracks generates a rather large number of events prior to fracture. The duration of an AE signal is the time difference between the crossing of the first and last threshold. The amplitude is the highest measured voltage in a waveform. This is an important parameter in AE inspection because it determines the detectability of the signal. Signals with amplitudes below the operator-defined limit are not recorded. Energy of AE signals is defined as an area under the envelope of the rectified linear voltage time signal from the transducer [14].

4. Results and discussion

The arithmetic mean and coefficient of variation of resulting mechanical fracture and acoustic emission parameters of both sets of concrete specimens at the age of 28, 90 days and 1 year are summarized in Table 1: bulk density γ, informative value of compressive strength f_c, modulus of elasticity E, effective crack elongation a_e–a_0, effective fracture toughness K_{Ice}, specific fracture energy G_F, unstable fracture toughness K_{Ic}^{un}, toughness ratio K_{Ic}^{ini}/K_{Ic}^{un}, i.e. the ratio of initiation fracture toughness to unstable fracture toughness, load ratio F_{ini}/F_{max}, i.e. the ratio between the force at the beginning of stable crack propagation from an initial stress concentrator and the maximal force corresponding to peak of *F–CMOD* diagram, and maximum crack tip opening displacement w_c. Mechanical fracture parameters are supplemented with parameters of AE signals, namely number of AE events N_{AE}, duration of AE signals D_{AE}, amplitude of AE signals A_{AE} and energy of AE signals E_{AE}.

Table 1. Mean values of selected parameters (coefficients of variation in %).

	FA			CKD		
	28 days	90 days	1 year	28 days	90 days	1 year
γ [kg/m^3]	2010 (0.9)	2010 (0.2)	2000 (0.4)	2230 (0.6)	2220 (0.3)	2200 (0.1)
f_c [MPa]	22.4 (2.6)	25.4 (9.8)	24.7 (5.1)	18.3 (6.3)	23.4 (9.4)	23.6 (8.8)
E [GPa]	9.6 (8.7)	7.6 (9.9)	9.4 (10.9)	5.3 (10.4)	6.0 (5.4)	6.6 (6.1)
a_e–a_0 [mm]	12.0 (22.5)	12.1 (12.0)	14.1 (8.7)	14.4 (23.7)	14.8 (11.0)	16.7 (6.0)
K_{Ice} [MPa·m$^{1/2}$]	0.579 (13.6)	0.610 (11.3)	0.553 (12.5)	0.404 (18.2)	0.497 (8.0)	0.558 (7.5)
G_F [J/m^2]	118.2 (22.0)	134.4 (18.2)	103.1 (22.8)	90.6 (19.4)	127.3 (16.9)	107.0 (25.4)
K_{Ic}^{un} [MPa·m$^{1/2}$]	0.529 (13.1)	0.680 (15.4)	0.531 (14.8)	0.382 (14.2)	0.521 (11.6)	0.537 (17.1)
K_{Ic}^{ini}/K_{Ic}^{un} [–]	0.411 (26.8)	0.311 (30.0)	0.201 (11.6)	0.139 (29.5)	0.097 (27.2)	0.129 (41.0)
F_{ini}/F_{max} [–]	0.582 (21.3)	0.547 (34.0)	0.330 (16.2)	0.224 (25.1)	0.184 (35.9)	0.235 (31.8)
w_c [mm]	0.223 (21.5)	0.233 (17.8)	0.183 (22.0)	0.177 (17.8)	0.233 (16.5)	0.194 (21.9)
N_{AE} [–]	91 (1.7)	170 (1.2)	188 (1.3)	30 (3.1)	87 (2.0)	218 (0.9)
D_{AE} [µs]	1365 (0.1)	1434 (0.2)	1310 (0.1)	1376 (0.4)	1386 (0.2)	1500 (0.2)
A_{AE} [mV]	2717 (0.3)	2900 (0.2)	2677 (0.3)	2466 (0.8)	2645 (0.3)	2571 (0.2)
E_{AE} [10^{-3}V·s]	62.9 (0.5)	124.8 (0.5)	132.0 (0.3)	29.2 (2.7)	69.8 (0.9)	180.0 (0.6)

As can be seen from Table 1, higher value of monitored mechanical fracture parameters of concrete with alkali activated BFS were achieved with substitution BFS by low calcium FA in comparison with substitution by CKD. The modulus of elasticity value is about 30–45 % lower in dependence on specimens age for CKD concrete in comparison with FA concrete. The lowest value was obtained for CKD concrete at the age of 28 days. Similar results were obtained for fracture toughness value determined by two different methods: this value is almost the same at age of 1 year for both type of concrete with alkali activated binder. In case of the ratio expressing the resistance to stable crack propagation the value is lower even about 70 % in case of CKD concrete and the value is also decreasing with age of specimens, it is almost 50 % decrease in case of FA concrete. The specific fracture energy value is also lower for CKD concrete, the trend is similar as in case of fracture toughness.

In case of AE parameters higher values were also achieved with substitution BFS by low calcium FA in comparison with substitution by CKD. In case of duration and amplitude of AE signals differences are until 10 %, whereas number of AE events and energy of AE signals shows significant differences. These two parameters were increased two times with age of specimen for FA concrete, in case of CKD concrete the increase is more than six times. It can be presumed that higher value of N_{AE} and E_{AE} indicates better mechanical properties of concrete thanks to a better bond of the matrix. The cracks of larger size generate a larger value of AE amplitude, from this point of view arises that the largest size of cracks were for specimens at the age of 90 days for both types of concrete.

5. Conclusions

The aim of this study was to quantify the influence of fly ash (FA) and by-pass cement kiln dust (CKD) on mechanical fracture parameters of alkali activated BFS concrete and their comparison for different age of specimens.

The mechanical fracture parameters of alkali activated binder based concrete were higher in the case of substitution BFS by FA compared to CKD. With exception of resistance to stable crack propagation, other monitored parameters were increased with age of specimens. The slight decrease of some parameters of FA concrete was monitored at the age of 1 year.

The AE technique was used to monitor the specimens failure during the experiments. It is evident, that different types of cracks generate different AE signals. In summary, the results of the experiments conducted as part of this project can be used to infer properties of materials as well as the specific characteristics of micro-cracks. The properties of the micro-cracks can ultimately be linked to the overall behavior of the building materials under stress.

Acknowledgements

This outcome has been achieved with the financial support of the Competence Centre programme of the Technology Agency of the Czech Republic (TACR) within the project Centre for Effective and Sustainable Transport Infrastructure (CESTI), project no. TE01020168, and with financial support from the Czech science foundation, project no. GA17-03670S Development of shrinkage reducing agents designed for alkali activated systems.

References

[1] Provis J L and van Deventer J S (eds) 2014 *Alkali activated materials: state-of-the-art report, RILEM TC 224-AAM* (Dordrecht: Springer)

[2] Shi C, Krivenko P V and Roy D 2006 *Alkali-activated cements and concretes* (London: Taylor & Francis)

[3] Marjanović N, Komljenović M, Baščarević Z, Nikolić V and Petrović R 2015 Physical–mechanical and microstructural properties of alkali-activated fly ash–blast furnace slag blends *Ceramics International* **41** pp 1421–35

[4] Chi M and Huang R 2013 Binding mechanism and properties of alkali-activated fly ash/slag mortars *Construction and Building Materials* **40** pp 291–98

[5] Aydin S 2013 A ternary optimisation of mineral additives of alkali activated cement *Construction and Building Materials* **43** pp 131–38

[6] Bílek V Jr, Šimonová H, Havlíková I, Topolář l, Kucharczyková B and Keršner Z 2017 Alkali Activated Binders Based Concrete Specimens: Length Change and Fracture Tests *Solid State Phenomena: Materials Structure & Micromechanics of Fracture VIII (Brno)* ed P Šandera (Switzerland: Trans Tech Publications) **258** pp 623–6

[7] Karihaloo B L 1995 *Fracture Mechanics and Structural Concrete* (New York: Longman Scientific & Technical)

[8] RILEM TC-50 FMC Recommendation 1985 Determination of the fracture energy of mortar and concrete by means of three-point bend test on notched beams *Materials & Structures* **18** pp 285–90

[9] Kumar S and Barai S V 2011 *Concrete Fracture Models and Applications* (Berlin: Springer)
[10] Reinhardt H W, Cornelissen H A W and Hordijk D A 1986 Tensile tests and fracture analysis of concrete *Journal of Structural Engineering* **112** pp 2462–77
[11] Petersson P E 1981 *Crack growth and development of fracture zone in plain concrete and similar materials Report No. TVBM-1006* (Lund Institute of Technology)
[12] Červenka V, Jendele L and Červenka J 2014 *ATENA Program documentation – Part 1: Theory* (Prague: Cervenka Consulting)
[13] Grosse C U and Ohtsu M 2008 *Acoustic Emission Testing: Basics for Research – Applications in Civil Engineering; with contributions by numerous experts* (Heidelberg: Springer)
[14] Sagar R V 2009 An experimental Study on Acoustic Emission Energy and Fracture Energy of Concrete *Proc. of the National Seminar & Exhibition on Non-Destructive Evaluation (Tiruchirappalli, India)* pp 225–8

IOP Conf. Series: Materials Science and Engineering **236** (2017) 012069 doi:10.1088/1757-899X/236/1/012069

Push-out tests and evaluation of FRP perfobond rib shear connectors performance

Ludvik Kolpasky[1] **and Pavel Ryjacek**[1]

[1] CTU in Prague, Thakurova 7, Prague 6 166 29, Czech Republic

E-mail: ludvik.kolpasky@fsv.cvut.cz

Abstract. The behavioural characteristics of FRP (fibre-reinforced polymer) perfobond rib shear connector was examined through push-out tests in order to verify the applicability for pedestrian bridge structure. The aim of this study is to determine interaction between high performance concrete slab and handmade FRP plate which represent web of the composite beam. Combination of these modern materials leads to structural system with both great load bearing capacity and also sufficient flexural stiffness of the composite element. Openings cut into the GFRP plate at a variable spacing allow GFRP reinforcement bars to be inserted to act as shear studs. Hand lay-up process can increase suitable properties of FRP for connection by perfobond rib shear connectors. In this study, three push-out tests on fiber-reinforced polymer were performed to investigate their shear behaviour. The results of the push-out tests on FRP perfobond rib shear connector indicates great promise for application in full scale structures.

1. Introduction

Bridges play a central role as an essential component of the infrastructures network as they do not only cover a large investment volume but also guarantee the smooth transport of goods and passengers. Compared to building, bridges are especially long living structures with scheduled service life of at least 100 years [1]. Most bridges suffer from inadequate maintenance and therefore lot of them must be repaired more times during their life cycle than it is necessary. Some of them end up being unrepairable and must be removed before their planned service life. Clients start to demand sustainable structures which can survive the whole life cycle without any crucial defects and related high maintenance costs.

Although the traditional material for this purpose is steel, the maintenance expenses are forcing the bridge operators to look for more durable and maintenance free solutions. The main goal is to replace steel elements that require service and coating, by more durable FRP systems. The advantages are especially high durability and no additional expenses.

1.1. Performed research on composite connection

One of the first research of composite FRP-concrete beam was carried out by Descovic [2]. Beam specimens composed of a rectangular FRP tube in the tensile region and plain concrete in the compression zone. In their work shear connectors between upper flange and concrete slab was provided by applying epoxy resin before pouring the concrete. This solution was not sufficient, because the beam failed suddenly by debonding of concrete slab. As a conclusion of this experimental study there was a demand for more robust shear connectors.

Next study was focused on feasibility of composite connection by steel bolts [3]. This connection system showed some level of ductility, but the initial stiffness was very low in comparison with the epoxy adhesive joint.

Cho performed tests on pultrude square hollow sections FRP tube connected by concrete wedge and coarse sand coating [4]. Connection by coarse sand coating resisted higher maximum load, but the failure was brittle. The advantage of the connection by concrete wedge was some level of ductility. In the fatigue flexural tests two variants of connections were tested. Tests of connection by coarse sand coating and mixed shear connection system (coarse sand coating together with concrete wedge) were performed. Coarse sand coating can resist fatigue load only at low level loading. The mixed shear system performed the remarkable fatigue resistance for identical fatigue loading.

To obtain the influence of wedge diameter and spacing they performed additional pull-out tests of the FRP pultrude plates embedded in concrete block [5]. Pultrude profiles were lubricated to exclude the bonding effect. From the extensive research the discrete spring model was derived and proposed for the concrete dowel action including post-failure frictional effects, considering the diameter of the hole and spacing between them as parameters. The resistance induced by concrete dowel action increase practically linearly to the area of the rib hole. The number of rib holes did not increase the resistance linearly due to imperfect reallocation of load after occurrence of failure caused by low stiffness of FRP.

Pultrude profiles are reinforced by fibers mostly in the longitudinal direction which leads into very low transverse strength. In FRP made by hand lay-up process the amount of fibers and their direction can be adjusted in order to increase suitable properties. This paper focused on the development of composite FRP-HPC beam, which consists of pultrude lower flange, hand lay-up FRP webs and high performance concrete slab. The modulus of elasticity is low, for common profiles between 17 – 30 GPa which implies that the serviceability limit state is much more important here, than in other usual materials, especially due to deflection requirements and local failures in connections. The disadvantage of low modulus of elasticity is compensated by the use of concrete part in the composite cross-section.

2. Cross-section of proposed beam

For this study we propose following composite cross-section (see Figure 1 below). Composite box girder using trapezoidal FRP box made by hand lay-up process. FRP box consist of glass fibers in the whole section and additional carbon fibers in the bottom flange. Webs are embedded to the HPC slab and form the composite action.

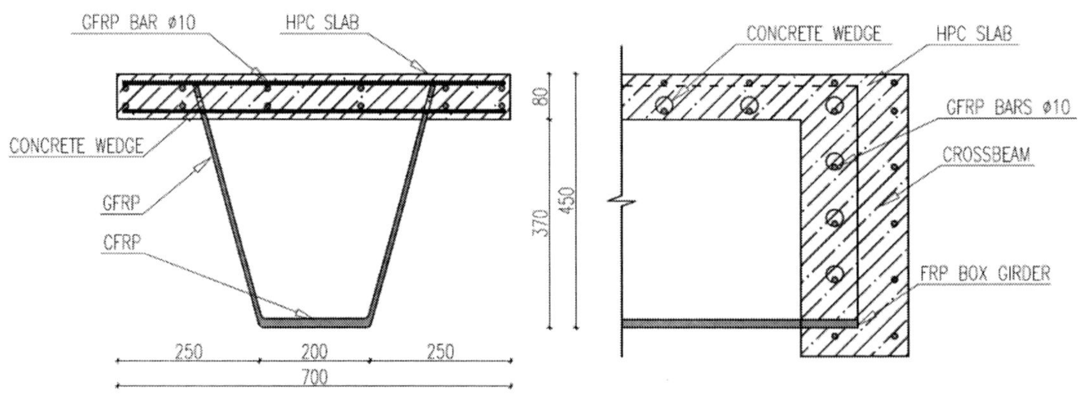

Figure 1. Proposed composite FRP-HPC cross-section.

Concrete is used as the top layer due to its high compressive strength. The use of HPC (high performance concrete) should limit the initiation and development of cracks. HPC has significant role in aggressive environment to ensure high durability. Hand lay-up FRP plates are used as webs due to their good properties in transverse direction. The second reason for this choice is that the high pultrude I profiles are still not available in supply of manufacturers.

3. Setup of the push out tests

Three push-out tests were performed in order to study the behavior of composite system and to compare the influence of diameter of holes and their spacing. The shear performance of rib shear connectors was examined using the 40 a 50 mm diameter with spacing of 250 and 300 mm as it is shown in Table 1. The specimen is illustrated in Figure 2. The load is transmitted by steel fixture to the FRP plates.

FRP plates with dimensions 580 x 540 mm were connected to steel fixture by 16 bolts of diameter 30 mm. In order to minimize the effect of slip between steel fixture and FRP plates the holes were drilled with the same diameters as bolts. Holes with 50 mm or 40 mm diameter were drilled into each side of the plate. After connection of the FRP plate to the fixture, GFRP bars were installed and fixed in the position by steel wire. On the bottom side of the FRP plates foam blocks with thickness of 30 mm were inserted to eliminate the effect of supporting the frontal side of the FRP plate.

Specimen was placed into the formwork and the concrete was poured to the form and vibrated on the shaking table. The concrete was cured indoors for 28 days.

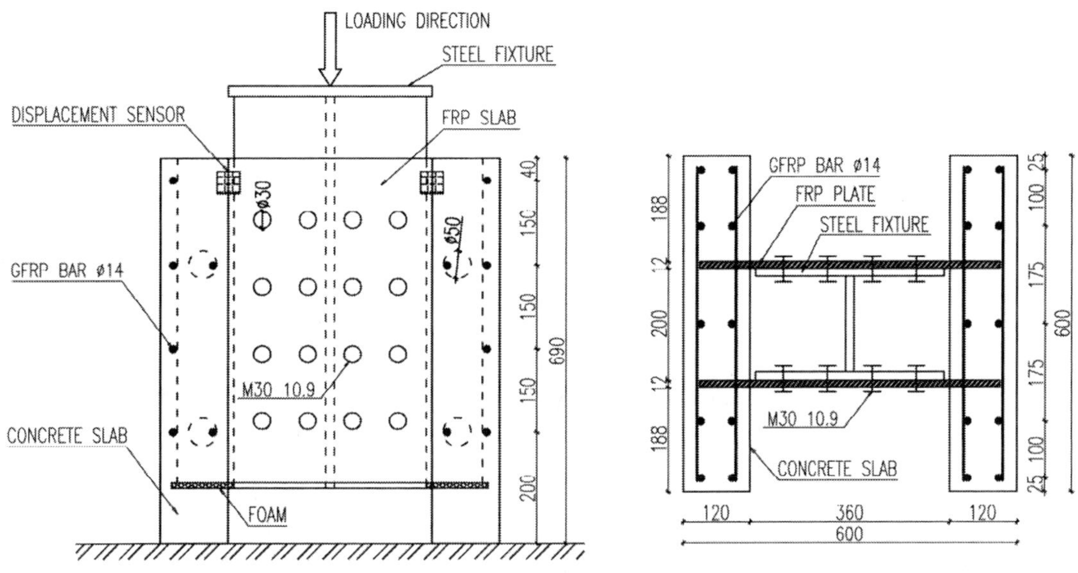

Figure 2. Shape of push out test specimen S1.

Table 1. Parameters of the push out test specimens.

	Diameter of rib hole (*mm*)	Number of rib holes (*mm*)	Spacing of rib holes (*mm*)	Thickness of the slab (*mm*)
S1	50	2	300	120
S2	50	2	250	120
S3	40	2	250	110

4. Material specifications

4.1. Concrete
In order to obtain concrete strength class C70/85 for the test specimen, the concrete was mixed and casted in a laboratory. The properties of concrete were determined by six cube tests and three tensile bending strength tests at the time of push out testing. The average compressive strength and the tensile strength were 91.7 MPa and 6.8 MPa, respectively.

4.2. GFRP plates
The handmade 12 mm thick FRP plates were supplied by Tatragrate, Slovak republic. Tensile characterisation of the material was carried out in accordance with ASTM D 3039. The strip specimens were cut out from the FRP plate used in push-out specimen. From the strip specimens the maximum failure stress and stiffness of the material were determined. The material properties are the same in longitudinal and transverse direction due to used textile reinforcement. An average tensile strength is 120 MPa and tensile modulus 8.5 GPa. To obtain the tensile modulus extensometer Epsilon 3542 was used.

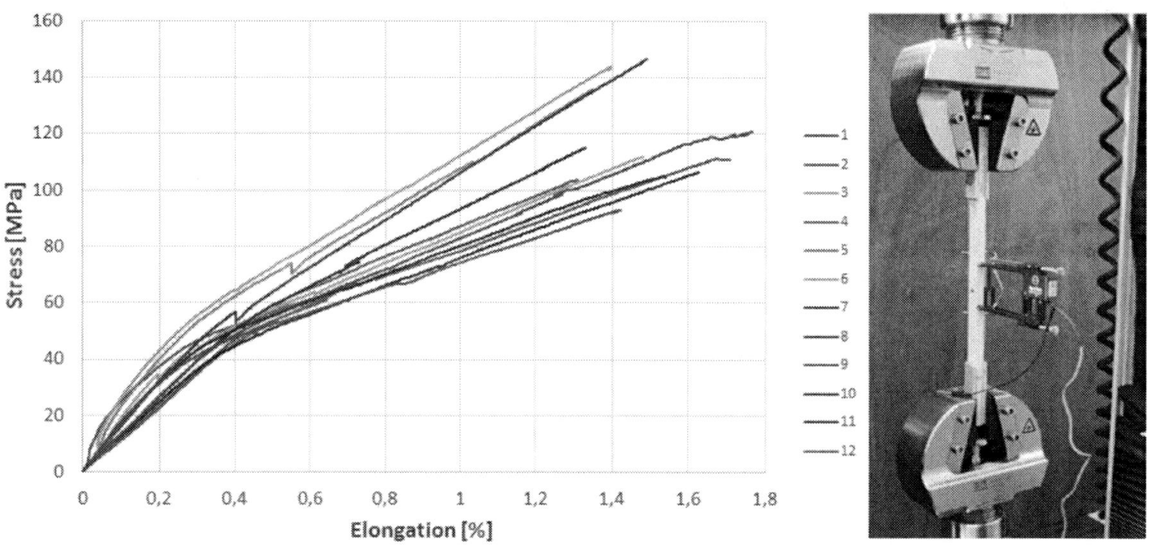

Figure 3. Stress-elongation curve of tensile test specimens and view of the test.

4.3. GFRP bars
GFRP pultruded bars of 14 mm diameter, consist of E-glass fibers with carbon fibers in the middle of the cross section, were supplied by Prefa kompozity, Czech Republic, and were used as shear connectors in this study. These bars with sand coating are commercially available as reinforcement for concrete structures. Tensile tests were provided by the producer according to ACI 440.3R-04. Bars had average tensile strength, modulus and ultimate strain of 1050 MPa, 67 GPa and 2.14 %, respectively.

5. Experimental results
Tests were performed using a hydraulic jack and loading was applied through displacement control at speed of 0.2 mm/min. Four displacement sensors were installed at each composition of FRP plate and concrete slab to measure the relative slip. According to EN 1994-1-1 the specimens were loaded 25x till 40% of the estimated maximum force and then loaded to the failure.

The maximum force 1000 kN of the jack was reached without failure. The curve S1_1 shows that loading. The following specimens were tested on hydraulic jack with maximum loading force of 2000 kN. Figure 4 shows the comparison of load-relative slip curves of three specimens. In the second loading case of first specimen the stiffness was lower due to debonding between FRP plate and concrete slab during the first loading.

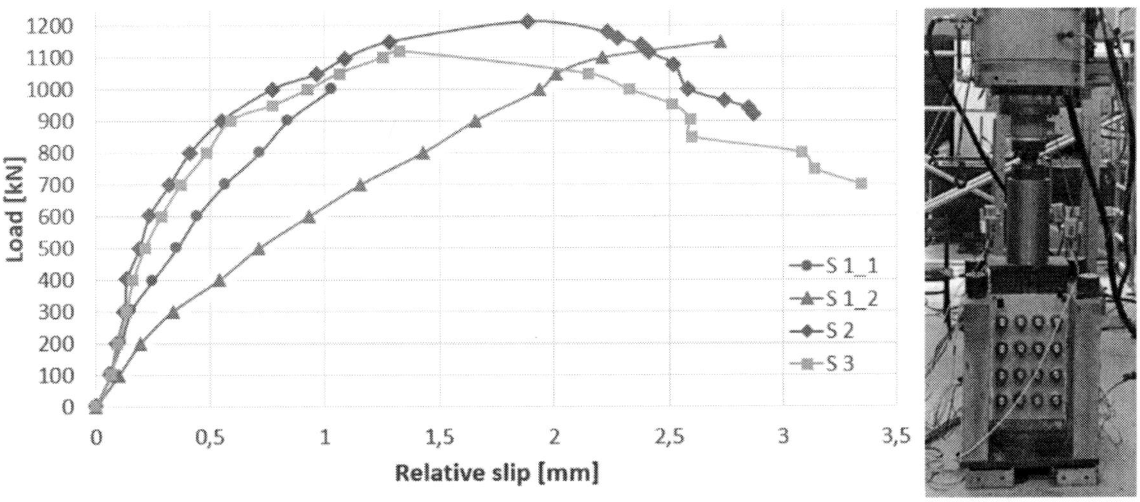

Figure 4. Load-relative slip curve of push out test specimens and view of the test.

Table 2. Summary of load and displacement of push out test.

	Diameter of rib hole (*mm*)	Number of rib holes (*mm*)	Spacing of rib holes (*mm*)	Thickness of the slab (*mm*)	Ultimate load (*kN*)	Displacement at ultimate load (*mm*)
S1	50	2	300	120	1150	2.37
S2	50	2	250	120	1214	2.78
S3	40	2	250	110	1119	3.34

The experiment demonstrated that there were no significant differences between the three specimens. As can be seen in Table 2 the ultimate force was between 1119 – 1214 kN with a maximum difference of a 8.5 %. Figure 4 shows that all three specimens had a high initial stiffness of 2.43 x10^{-6} kN/m. Near the load 1000 kN the loud cracking of glass fibers was heard. When the shear resistance was reached, the shear crack developed from the upper rib hole in slope of 45 degrees. When the load gradually dropped by 20%, next shear crack appeared.

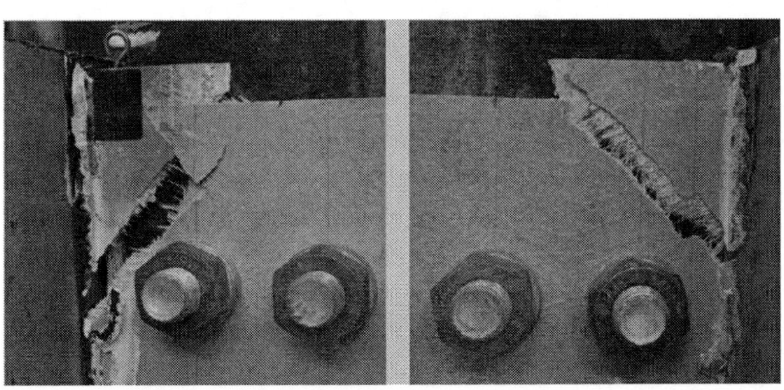

Figure 5. Shear cracks of the push out test specimen.

To explore the failure of the composition, specimens were cut in longitudinal direction of the web. Figure 6 shows the failure modes of the specimen S3. The upper rib hole failed by shear crack combined with tensile crack. The bottom rib hole failed by crushed concrete wedge and delamination of the FRP web.

Figure 6. View of the failure of the composite connectors.

Before the composite concrete slab was casted, on the FRP webs were installed strain gauges, 12 pieces on each web. The distribution of strain gauges on the web is apparent on the Fig. 7.In the vicinity of each corner wedge were used 3 strain gauges: 1 in horizontal direction, one in vertical direction and 1 in 45 degrees inclination along expected shear crack. Strain gauges HBM 1-LD20-6/120 with high strain were used on the specimens.

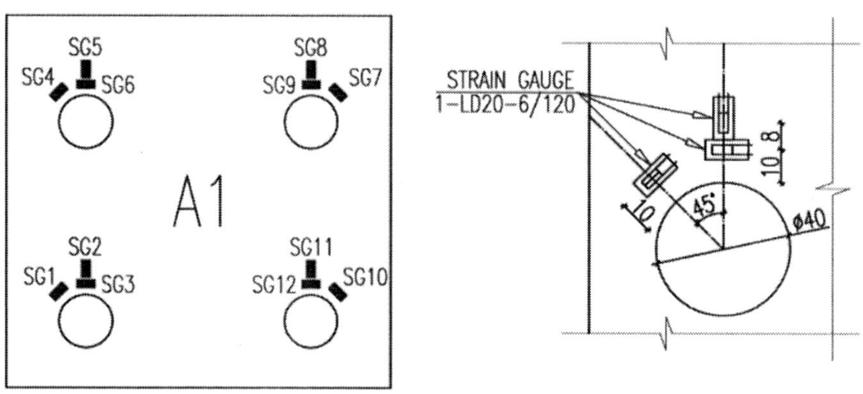

Figure 7. Location of the strain gauges on the FRP plate A1.

The strain-stress diagrams for all gauges of the specimen 1, FRP plate 1A are displayed on Figure 8.

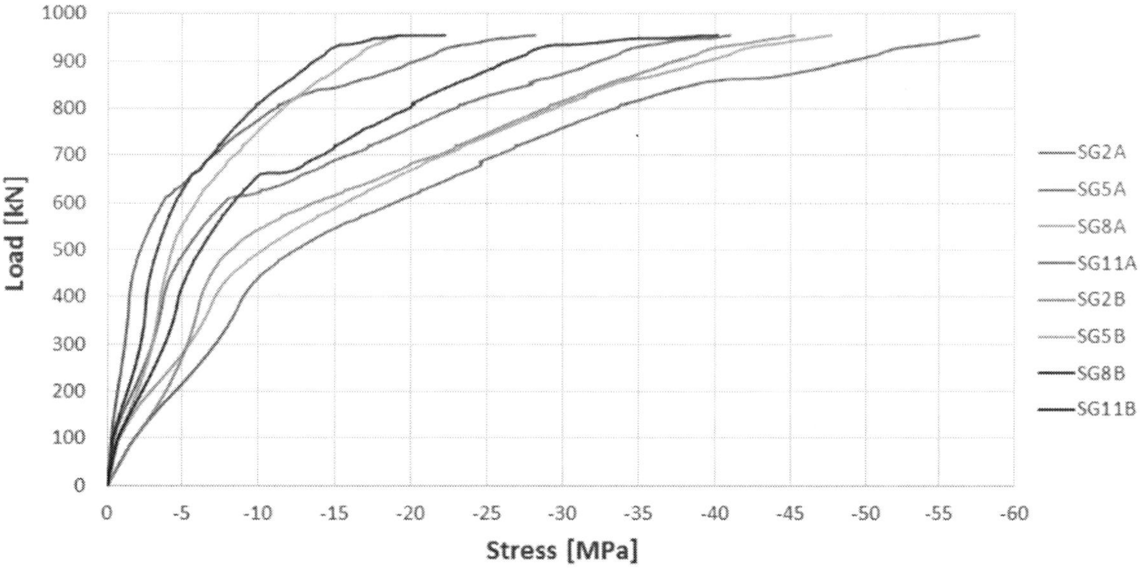

Figure 8. The strain-stress diagrams for vertical gauges of the specimen 3.

From the stress distribution in all wedges it is apparent, that the specimens were not loaded symmetrically as one slab is loaded more than the other one. Figure 8 shows the stresses from the strain gauge positioned vertically above the opening for concrete dowel. The graph shows the uneven distribution of load force for each concrete dowel. From the linear part of the working diagram was calculated load distribution between the individual dowels. To determine the elastic resistance of one dowel, the magnitude of stress was determined when nonlinear behaviour first appeared. From this value was calculated the magnitude of effective acting force on each dowel. Values are shown in Table 3.

Table 3. Summary of distribution, stress and effective acting force.

	Specimen 1			Specimen 2			Specimen 3		
	Distribution (%)	Stress (MPa)	Force (kN)	Distribution (%)	Stress (MPa)	Force (kN)	Distribution (%)	Stress (MPa)	Force (kN)
SG2A	16.3	-12.23	116.4	13.4	-7.87	75.6	9.7	-10.46	82.1
SG5A	7.33	- 8.26	78.6	11.1	-8.34	80.0	4.3	- 4.96	38.7
SG8A	4.0	-6.03	57.4	8.9	-7.76	74.5	15.4	-8.49	66.6
SG11A	7.2	-7.71	73.4	9.0	-5.54	53.2	20.1	-11.54	90.5
SG2B	24.4	-11.38	108,3	11.7	-7.88	75.6	18.6	-10.9	85.5
SG5B	18.0	-6.94	66.0	9.9	-8.62	82.7	10.8	-8.27	64.9
SG8B	13.0	-6.99	66.5	15.5	-10.38	99.6	6.4	-6.31	49.5
SG11B	9.7	-10.29	98.0	20.5	-9.27	88.9	12.2	-6.31	49.5

For the first specimen is the mean value of the elastic performance 83.1 kN, standard deviation 21.75 and 5% quantile 45.1 kN. For the second specimen is the mean value of the performance 78.8 kN, standard deviation 13.35 and 5% quantile 55.5 kN. For the third specimen is the mean value of the performance 69.3 kN, standard deviation 15.31 and 5% quantile 42.5 kN.

6. Conclusions

A reliable connection between FRP and concrete slab is important to prevent sudden and brittle failure. The use of perfobond rib shear connectors can reach this robustness. This study demonstrated the feasibility of composite connection by perfobond rib shear connectors between concrete slab and hand lay-up FRP plates. Concrete wedge together with GFRP bar achieved great properties in performed push out tests.

The effect of diameter of the holes was not observed throughout the tests as all specimens failed by shear crack or debonding. The hand lay-up FRP plates may be further enhanced by adding suitable reinforcement. Additionally, a design with more holes with smaller spacing should lead to more even stress distribution and increase the overall performance of the tested specimens.

Due to small modulus of FRP, composite beam may have problems with serviceability limit states and especially in deflection. Therefore the behavior of the connection should be rigid. Solution of connection with rib shear connectors reached ductile behavior but the debonding between surface of FRP and concrete slab occurred before the maximum load was reached. In order to increase the initial stiffness of the connection, coarse sand coating of the FRP surface can be applied.

Due to uneven distribution of acting force it is apparent that the strength of composite connection between FRP and concrete has even greater load bearing capacity than was determined by performed test with 2 FRP web specimens. The composite connection is therefore evaluated as strong link in the structure and therefore future research will focus on flexural strength of the composite girder.

Acknowledgements

Research reported in this paper was supported by Competence Centres program of Technology Agency of the Czech Republic (TACR), project Centre for Effective and Sustainable Transport Infrastructure (no. TE01020168).

References

[1] Mensinger M, Pfaffinger M and Schnell W 2012 Sustainability Assessment of Road Bridges *IABSE Congress Report* **18**(13) pp 1173-80
[2] Deskovic N, Meier U and Triantafillou T 1995 Innovative Design of FRP Combined with Concrete: Long-Term Behavior *Journal of Structural Engineering* **121**(7) pp 1079-89
[3] Correia J, Branco F and Ferreira J 2009 Flexural behaviour of multi-span GFRP-concrete hybrid beams *Engineering Structures* **31**(7) pp 1369-81
[4] Cho K, Park S Y, Kim S T, Cho J-R and Kim B-S 2010 Shear connection system and performance evaluation of FRP concrete composite deck *KSCE Journal of Civil Engineering* **14**(6) pp 855–65
[5] Cho J, Park S, Cho K, Kim S and Kim B 2012 Pull-out test and discrete spring model of fibre-reinforced polymer perfobond rib shear connector *Canadian Journal of Civil Engineering* **39**(12) pp 1311-20

Assessment of the transportation route of oversize and excessive loads in relation to the load-bearing capacity of existing bridges

Jiří Doležel[1,3], Drahomír Novák[1] and Jan Petrů[2,3]

[1]Faculty of Civil Engineering, Brno University of Technology, Brno, Czech Republic
[2]MORAVIA CONSULT Olomouc a.s., Olomouc, Czech Republic
[3]Faculty of Civil Engineering, VSB – Technical University of Ostrava, Ostrava, Czech Republic

E-mail: dolezel.j@fce.vutbr.cz

Abstract. Transportation routes of oversize and excessive loads are currently planned in relation to ensure the transit of a vehicle through critical points on the road. Critical points are level-intersection of roads, bridges etc. This article presents a comprehensive procedure to determine a reliability and a load-bearing capacity level of the existing bridges on highways and roads using the advanced methods of reliability analysis based on simulation techniques of Monte Carlo type in combination with nonlinear finite element method analysis. The safety index is considered as a main criterion of the reliability level of the existing construction structures and the index is described in current structural design standards, e.g. ISO and Eurocode. An example of a single-span slab bridge made of precast prestressed concrete girders of the 60 year current time and its load bearing capacity is set for the ultimate limit state and serviceability limit state. The structure's design load capacity was estimated by the full probability nonlinear MKP analysis using a simulation technique Latin Hypercube Sampling (LHS). Load-bearing capacity values based on a fully probabilistic analysis are compared with the load-bearing capacity levels which were estimated by deterministic methods of a critical section of the most loaded girders.

1. General instructions

Transportation routes of oversize and excessive loads are currently planned in relation to ensure the transit of a vehicle through critical points on the road [1]. Critical points are level-intersection of roads, bridges etc. Bridge construction has to be designed in relation to the assumed reference period and the reliability level to resist all possible load combinations that might appear during its usage time.

In the Czech Republic load bearing capacity of new and existing read bridges is assessed according to the Czech technical standard ČSN 73 6222 [2]. Before the determination of load bearing capacity, a bridge inspection must be carried out and consequently, a real condition of the structure must be taken into account. If the load bearing capacity of concrete bridges is assessed with respect to the ultimate as well as the serviceability limit states, than the limit states of decompression (for prestressed concrete bridges) and the crack width (for prestressed and reinforced concrete bridges) should be verified.

There are three basic types of load bearing capacity which are defined by that norm. Normal load bearing capacity (V_n) is defined by the load model LM1 according to the norm EN 1991-2 [3] and is represented by a continuous load and a two-axle vehicle. Reserved load bearing capacity (V_r) is represented by a six-axle vehicle and exceptional load bearing capacity (V_e) is defined by a nine-axle vehicle. Transportation of oversize and excessive loads is assessed individually for each vehicle and its corresponding load. Minimal recommended values of the load-bearing capacity of existing bridges after their reconstruction are listed in Table 1.

Table 1. Minimal recommended values of the load-bearing capacity of existing bridges after their reconstruction.

Group of roads according to ČSN EN 1991-2	Load bearing capacity type		
	Normal (V_n)	Reserved (V_r)	**Exceptional (V_e)**
1	32 t	80 t	**180 t**
2	22 t	40 t	-

Assessment of reliability of existing bridge structures can be performed by deterministic approach, which requires a design action effect E_d to be smaller than a design resistance R_d of the construction, or by stochastic approach, when calculated failure probability P_f is lower than the required failure probability $P_{f,t}$ for the given reference period.

An equivalent term to the failure probability is the reliability index β that is a commonly used measure of reliability of existing bridge structures. The reliability index β is formally defined in terms of the probability of failure P_f as:

$$\beta = -\Phi^{-1}(P_f) \tag{1}$$

where $-\Phi^{-1}(.)$ is the inverse function of the standardized normal probability distribution.

Values of target reliability index β_t and corresponding probability of failure $p_{f,t}$, that are for selected limit states presented in Table 2, can be specified in more details depending on the estimated residual lifetime of a bridge, on the consequences of damage, or considering the economic, social and ecological consequences, e.g. Czech technical standard ČSN ISO 13822 [4] and ČSN EN 1990 [5].

For common bridges on highways and roads class I. and II., the value of the reliability index of ultimate limit states (ULS) is $\beta_t = 3.8$ and of the serviceability limit states (SLS) is $\beta_t = 0.0$ (limit state of decompression and limit state of cracking).

Table 2. Target values of reliability index.

Limit state type	β_t	$p_{f,t}$
Serviceability limit states:		
- **reversible**	**0.0**	-
- irreversible - small consequence of damage	1.3	9.6×10^{-2}
- irreversible - medium consequence of damage	1.5	6.7×10^{-2}
- irreversible - high consequence of damage	2.3	1.1×10^{-2}
Ultimate limit states:		
- very small consequence of damage	2.3	1.1×10^{-2}
- small consequence of damage	3.1	9.7×10^{-4}
- **medium consequence of damage**	**3.8**	$\mathbf{7.2\times.10^{-5}}$
- high consequence of damage	4.3	8.5×10^{-6}

2. Load bearing capacity assessment of single-span post-tensioned composite bridge

A bridge with an ID 55-046 built in 1950 bridges a road class I over the railway. The bridge is considered as a single-slab span object with a prefabricated supporting structure and a monolithic substructure. Superstructure of the bridge is a simple beam consisting of 12 prefabricated and precast prestressed concrete girders MPD4 (10 intermediate) a MPD3 (2 outlying). Length of the superstructure is 20.50 m, length of the bridging is 17.50 m and width of the supporting structure is 11,0 m. Detailed diagnostic survey of the superstructure didn't show any damages of girders MPD4 a MPD3 as a result of exceeding neither the ultimate limit state nor the serviceability limit state. Corrosion of the reinforcement and prestressed tendons wasn't recorded either.

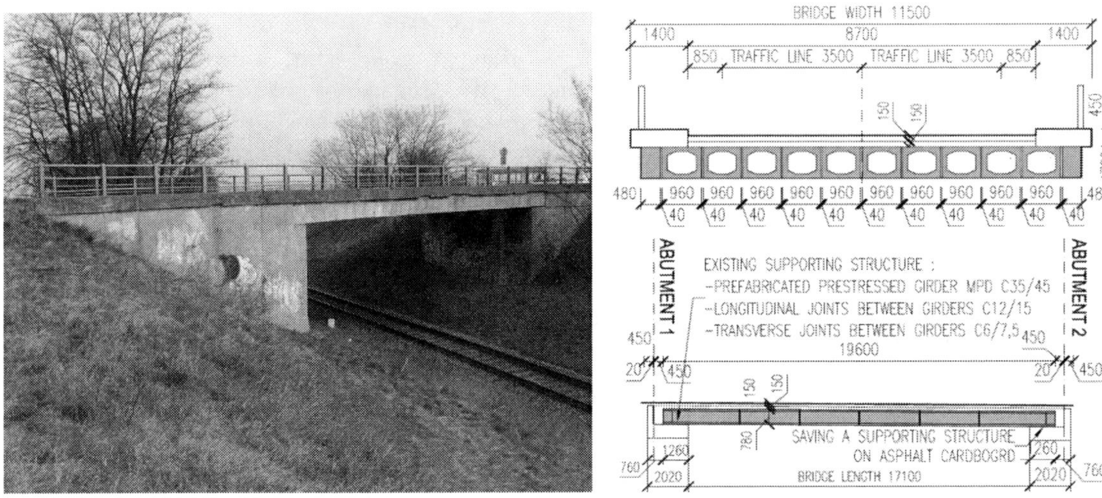

Figure 1. Side view of analyzed bridge, transversal section and longitudinal section of the bridge.

Bridge girders MPD3 and MPD4 were designed for continuous load of pavements of intensity 0.6 t/m^2 in combination with a crawler vehicle having the weight of 60 tons according to the provisional directive for the design of bridges (year 1945). Load bearing capacity of superstructure was determined for the bridge before its reconstruction. Oversize vehicle was represented as twelve-axle THP type having the width of 3.0 m (see Figure 2).

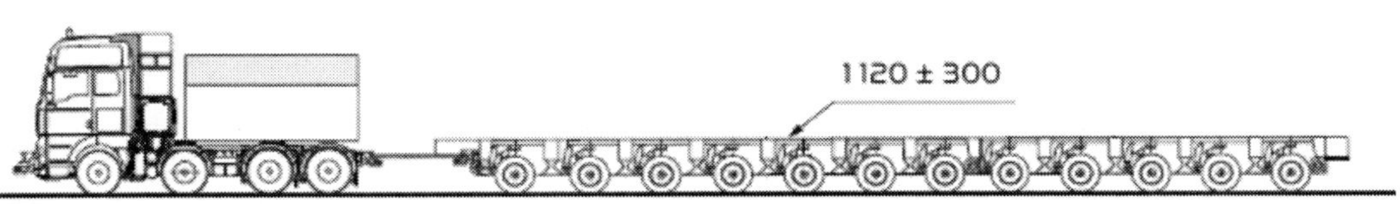

Figure 2. Twelve-axle THP type vehicle.

2.1. Full-probabilistic analysis

Probabilistic analysis of resistance and action can be performed by numerical method of Monte Carlo-type of sampling, such as LHS sampling method. Results of this analysis provide random parameters of resistance and actions, such as mean, standard deviation, etc. and the type of distribution function for resistance. Load-bearing capacity function S is estimated according to the aggregate statistics of random simulations of structure response with a vector of random variables

$$S = s(f_c, f_{ct}, E_c, f_{s,y}, f_{s,p}, \theta_R \ldots, g_0, g_1, V_i, \theta_E) \tag{2}$$

Material parameters, permanent loads g_0, g_1 and model uncertainties of resistance θ_R and uncertainties of model load effects θ_E are considered as random variables. Random traffic load V_i is assumed to be deterministic. Design value of the load-bearing capacity $V_{i,d}$ is estimated as

$$P(S \leq V_{i,d}) = \Phi(-\beta_t) \tag{3}$$

2.1.1. Numerical model. By assumption of forming ideally rigid slab from individual girders MPD3 a MPD4 thanks to their transverse prestressing, plain FEM model was created using the computer program ATENA 2D [6]. Concrete parts of each girder were modelled as a material model 3D Non Linear Cementitious 2, prestressed reinforcement was modelled as a discrete member and soft reinforcement of the girder individual segments was modelled as smeared. In both cases the reinforcement material was modelled by bi-linear working diagram with hardening. Simple beam placing on abutments was assumed. Structure model was loaded by its dead load, transverse prestressing effects and other permanent loads. Loading scheme for assessment of normal load-bearing capacity was represented by a continuous load and a two-axle vehicle. Loading schema for assessment of reserved load bearing capacity was represented by six-axle vehicle and in case of the exceptional load bearing capacity, the model of twelve-axle THP type vehicle was assumed.

Loading was simulated by increasing of deformation or by loading force until a limit state was observed which corresponds to the loading level at limit state of decompression (D), the limit state of crack formation (T) and the limit load-bearing capacity (U), see Fig. 2.

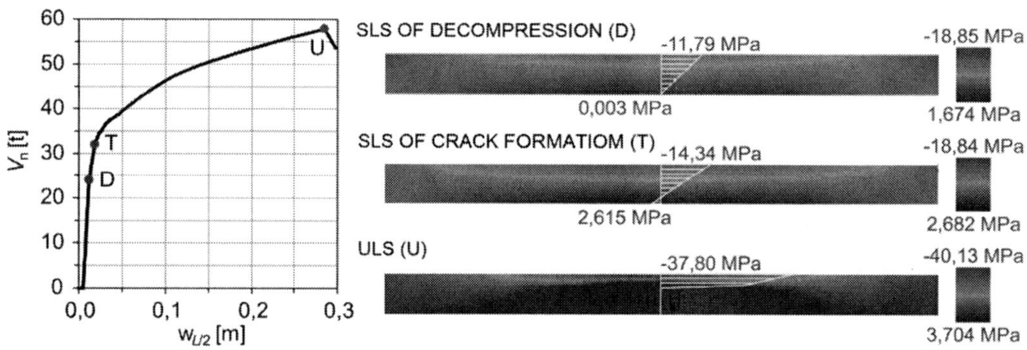

Figure 3: Diagram of normal load-bearing capacity – deflection; direct stress at the supporting structure.

2.1.2. Stochastic modelling of basic variables. Stochastic parameters of basic variables were defined using software FReET [7] according to Joint Committee on Structural Safety [8], including model uncertainties [9], and these were updated based on the material properties of concrete and reinforcement obtained from diagnostic survey.

32 random simulations were generated using stratified Latin Hypercube Sampling (LHS) method, which is capable to cover space of random variables in terms of relatively small number of sample [10]. As random variables material parameters of concrete as well as parameters of reinforcement, prestressed tendons and value of secondary dead load were chosen. The self-weight of the structure a prestressed force was also randomized using concrete mass density. Definitions of random input variables by their probability density functions (PDF), mean values and coefficients of variation are summarized in Table 3. The statistical correlation between material parameters of concrete and reinforcement was also considered and imposed with respect to formerly performed tests and JCSS recommendations using simulated annealing approach [11]. Finally, traffic load for determination of load bearing capacity was defined using deterministic value of load according to valid loading schemes introduced in current Standards.

Table 3. Definition of input parameters for assessment of load bearing capacity.

Value		Unit	PDF	Mean	V_X
Parameters of concrete girder MPD					
Young's modulus	E_c	[GPa]	Lognormal 2 par.	37.2	0.10
Tensile strength	f_{ct}	[MPa]	Lognormal 2 par.	3.30	0.15
Compressive strength	f_c	[MPa]	Lognormal 2 par.	43.4	0.08
Fracture energy	G_f	[N/m]	Lognormal 2 par.	8.25×10^{-5}	0.15
Mass density	γ_c	[kN/m3]	Normal	23.80	0.03
Parameters of concrete transverse joints between girders					
Young's modulus	E_c	[GPa]	Lognormal 2 par.	34.0	0.15
Tensile strength	f_{ct}	[MPa]	Lognormal 2 par.	2.81	0.35
Compressive strength	f_c	[MPa]	Triang.	19.1	<29.8;8.5>
Fracture energy	G_f	[N/m]	Lognormal 2 par.	4.78×10^{-5}	0.25
Mass density	γ_c	[kN/m3]	Normal	23.80	0.08
Parameters of prestressing tendons					
Young's modulus	E_p	[GPa]	Normal	190.0	0.03
Ultimate strain	$\varepsilon_{p,lim}$	[-]	Normal	0.05	0.07
Yield strength	$f_{p,y}$	[MPa]	Normal	1248.0	0.03
Ultimate strength	$f_{p,u}$	[MPa]	Normal	1716.0	0.03
Parameters of reinforcement					
Young's modulus	E_s	[GPa]	Normal	200.0	0.07
Ultimate strain	$\varepsilon_{s,lim}$	[-]	Normal	0.05	0.07
Yield strength	$f_{s,y}$	[MPa]	Lognormal 2 par.	462.1	0.07
Ultimate strength	$f_{s,u}$	[MPa]	Lognormal 2 par.	581.4	0.07
Load and prestressing force					
Secon. dead load	g_1	[kN/m2]	Normal	6.600	0.05
Load model	V_e	[tuny]	Det.	180	-
Force in time t = 60 y.	P_t	[MN]	Normal	$P_{m,t}$	0.09
Model uncertainties					
resistance R	$\theta_{R,M}$	[-]	Lognormal 2 par.	1.00	0.10
load E	$\theta_{E,M}$	[-]	Lognormal 2 par.	1.00	0.10

Table 4. Final values of normal, reserved and exceptional load bearing capacity.

Limit state	Load bearing capacity by full-probabilistic method			Load bearing capacity by deterministic method			Load bearing capacity from bridge documentation (2012)		
	V_n [tons]	V_r [tons]	V_e [tons]	V_n [tons]	V_r [tons]	V_e [tons]	V_n [tons]	V_r [tons]	V_e [tons]
decompression	25	80	172	21	71	163	25	48	-
crack formation	32	103	220	23	79	181			
ultimate	38	129	298	25	85	197			

2.1.3. Probability analysis of load bearing capacity of superstructure. Load-bearing capacity of the superstructure was estimated considering the reliability index β_t accoring to the Table 2. Final values of normal, reserved and exceptional load bearing capacity determined by the full-probabilistic analysis are summarized in the Table 4 and they are compared with values of load bearing capacity based on deterministic assessment of the critical section of the most stressed girder MPD3 and MPD4, as well as with load bearing capacity values listed in the bridge documentation from the year 2012.

Recommended values of maximum axle load for passage of the twelve-axle THP type vehicle was specified to 14 tons for the limit state of decompression and for the ultimate limit state, the pressure on a single axle was set to 25 tons.

3. Conclusions

In conclusion, it has been proved that probabilistic methods in combination with nonlinear FEM analysis represent an effective and practical tool in cases of evaluation of load bearing capacity and reliability of existing structures. It has been demonstrated that probabilistic approach is less conservative and loads to slightly higher capacities than the deterministic one, which is mostly applied using current Standards. However, for more detailed assessment of structures, more information is necessary to reduce model uncertainties and reach the most exact results therefore information from detailed diagnostic surveys should be used to improve the computational models.

Acknowledgements

This paper has been worked out under the project No. TA04030713 provided by the Technology Agency of the Czech Republic (TAČR).

References

[1] Petrů J, Kludka M, Křivda V, Mahdalová I and Zeman K 2015 Verification of census devices in trasprtation research *Acta Polytechnica* **55**(6) pp 415-421

[2] Úřad pro normalizaci, metrologii a státní zkušebnictví - ÚNMZ 2013 Zatížitelnost mostů pozemních komunikací /Load bearing capacity od road bridges/ *(ČSN 73 6222)*

[3] Úřad pro normalizaci, metrologii a státní zkušebnictví - ÚNMZ 2005 Eurokód 1: Zatížení konstrukcí – Část 2: Zatížení mostů dopravou /Eurocode 1: Actions on structures - Part 2: Traffic loads on bridges/ *(ČSN EN 1991-2)*

[4] Úřad pro normalizaci, metrologii a státní zkušebnictví - ÚNMZ 2014 Zásady navrhování konstrukcí – Hodnocení existujících konstrukcí Bases for design of structures - Assessment of existing structures *(ČSN ISO 13 822)*

[5] Úřad pro normalizaci, metrologii a státní zkušebnictví - ÚNMZ 2004 Eurokód: Zásady navrhování konstrukcí /Eurocode: Basis of structural design *(EN 1990)*

[6] Červenka V, Jendele L and Červenka J 2012 *ATENA Program Documentation – Part 1: Theory* (Červenka Consulting, Praha, Czech Republic)

[7] Novák D, Vořechovský M and Rusina R 2012 FReET – Program Documentation: User's and Theory Guides, Version 1.5. (Brno/Červenka Consulting, Czech Republic) http://www.freet.cz

[8] Joint Committee on Structural Safety (JCSS) 2016 *Probabilistic Model Code* (Zurich) http://www.jcss.byg.dtu.dk/

[9] Schlune H, Plos M and Gylltoft K 2011 Safety formats for nonlinear analysis tested on concrete beams subjected to shear forces and bending moments *Engineering Structures* **33**(8) pp 2350-56

[10] McKay M D, Beckman R J and Conover W J 1979 Comparison of three Methods for Selecting Values of Input Variables in the Ananlysis of Output from a Computer Code *Technometrics* **21**(2) pp 239–45

[11] Vořechovský M and Novák D 2009 Correlation control in small-sample Monte Carlo type simulations I: A simulated annealing approach *Probabilistic Engineering Mechanics* **24** pp 452-62

Measures for the reduction of sinter formations in tunnels

Gerhard Harer[1]

[1]GEOCONSULT ZT GmbH, Hölzlstraße 5, A-5071 Wals, Austria

E-mail: gerhard.harer@geoconsult.eu

Abstract. A considerable part of the maintenance costs of tunnel structures is related to the inspection, maintenance and repair of the drainage system. The drainage system of tunnels is frequently clogged with Calcium precipitates. Cleaning and water conditioning are cost-intensive for operating companies. Apart from the direct costs associated with inspection, maintenance and repair works of the drainage system indirect costs are generated, such as by the blocking of the tunnel while inspection, maintenance or repair or by the reduction of the permitted operation speed. Sintering and clogging of the drainage systems is mainly caused by dissolution of cement minerals in concrete and mortar and/or by inadequate design and construction of the drainage system and/or grubby workmanship. With long-term studies and in-situ experiments in Austria traffic tunnels the specific input factors for sinter mechanism have been identified and appropriate counter measures could be defined. In particular modified mix designs for shotcretes and mortars have proven to bring a significant beneficial effect. By means of constructional measures and by the application of hardness stabilizers a further reduction of hard deposits inside the drainage system is achievable. The paper will deal with the specific aspects and will propose adequate counter measures.

1. Introduction

In the recent past in Austria a large number of traffic tunnels have been constructed or are still under construction. In particular long railway tunnel projects, such as the Koralm tunnel (l = 33 km) or the Semmering Base tunnel (l = 27 km), but also long road tunnels, such as the Gleinalm tunnel (l = 8 km) required new considerations in regard to the simplification and reduction of maintenance.

The drainage system of tunnels is frequently clogged with Calcium precipitates. Assessments on railway tunnels have shown that around 15 – 20 % of the entire inspection and maintenance works are in relation to the drainage system. For road tunnels similar dimensions are well known, leading to high direct costs for inspection and maintenance of the drainage system, e.g. for motorway tunnels between 5,000 and 10,000 EUR per kilometer tube and year. Apart from the direct costs associated with inspection, maintenance and repair works of the drainage system indirect costs are generated, such as by interfering the traffic or blocking of the tunnel while inspection, maintenance or repair. These works have high negative impact on the operation, the operational costs as well as on the capacity of the tunnels. Thus, from the early stages of the design of a tunnel project special effort has also to be made to find solutions to minimize these costs and negative interferences.

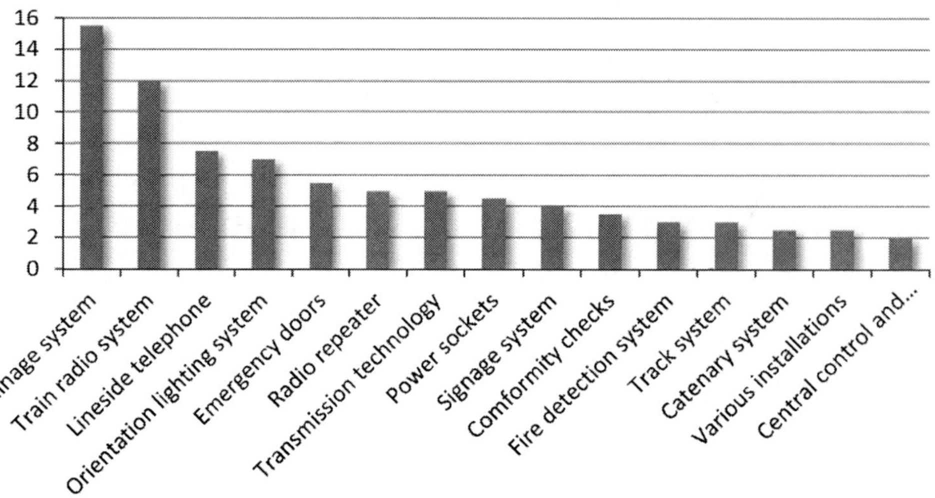

Figure 1. Average percentage of inspection, cleaning & maintenance works for various installations in Austrian railway tunnels (after [1]).

2. Sinter mechanism in tunnels

The phenomenon of iron precipitations will not be discussed in this article. In case of carbonate dissolution in tunnel drainage systems mostly Calcium carbonates precipitate from aqueous solutions, which are generated from local groundwater and the reaction with the shotcrete of the tunnel construction [2], [3]. Accordingly, the chemical composition of the drainage solutions depends on the evolution of the local groundwater as well as on the chemical composition and reactivity of the materials used for construction.

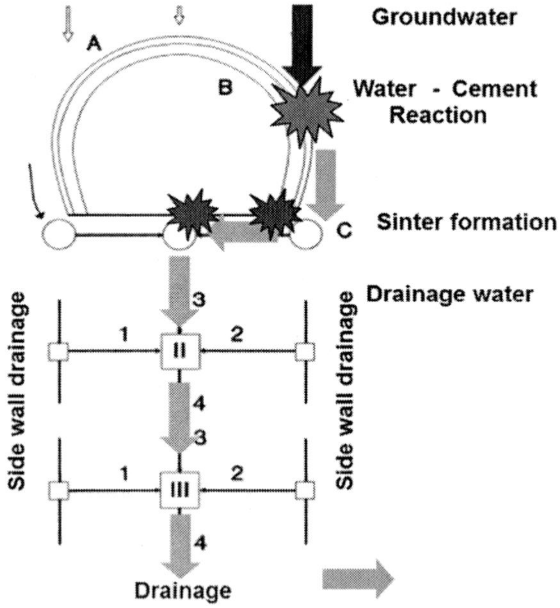

Figure 2. Simplified tunnel drainage concept and groundwater – construction interaction (after [3]).

Frequently the precipitates consist of the minerals Calcite ($CaCO_3$), Aragonite ($CaCO_3$) and also Brucite ($Mg(OH)_2$). The general mechanisms of interaction are described by [3] as follows.

2.1. Dissolution of limestone

The drainage solutions are gained from groundwater, which dissolves carbonate minerals from the natural rocks. The dissolution of carbonates is promoted by the uptake of carbon dioxide from the Earth´s atmosphere. Calcium carbonates dissolve according to the overall reaction

$$CO_{2(gas)} + H_2O + CaCO_3 = Ca^{2+} + 2HCO_3^- \quad (1)$$

The dissolution usually results in slightly alkaline Calcium bicarbonate solutions. The concentration of total dissolved carbonate strongly depends on the primary content of carbonic acid and the occurrence of solid carbonates.

2.2. Reactions at the shotcrete

As the groundwater gets into contact with the shotcrete of the tunnel, complex mineral water reactions appear. Simplified, the Portlandite of the shotcrete dissolves which results in an increase of the Calcium concentration and pH. Drainage solutions can reach pH values up to 13 and more. On the other hand, Calcium ions can be fixed by the precipitation of Calcium carbonate and sulphate minerals directly at the shotcrete. Newly formed sulphate minerals at the shotcrete may comprise Gypsum, Ettringite, Monosulfate and Thaumasite, depending on the availability of aqueous sulphate ions. In highly alkaline solutions Brucite can be precipitated.

2.3. Mechanisms of carbonate sinter formation

Calcite precipitation from the drainage solutions can be induced by the dissolution of Portlandite in the primary groundwater at the shotcrete. The dissolved Calcium and carbonate ion can be precipitated according the equation

$$Ca(OH)_2 + 2HCO_3^- + Ca^{2+} = 2CaCO_3 + 2H_2O \quad (2)$$

as supersaturation with respect to Calcite is reached. However, once the solutions come into contact with the Earth's atmosphere, the internal partial pressures of CO_2 in the solutions control the CO_2 exchange with the atmosphere and ongoing carbonate precipitation.

Figure 3. Typical sinter formations (in whitish colour) on the surface of a shotcrete lining.

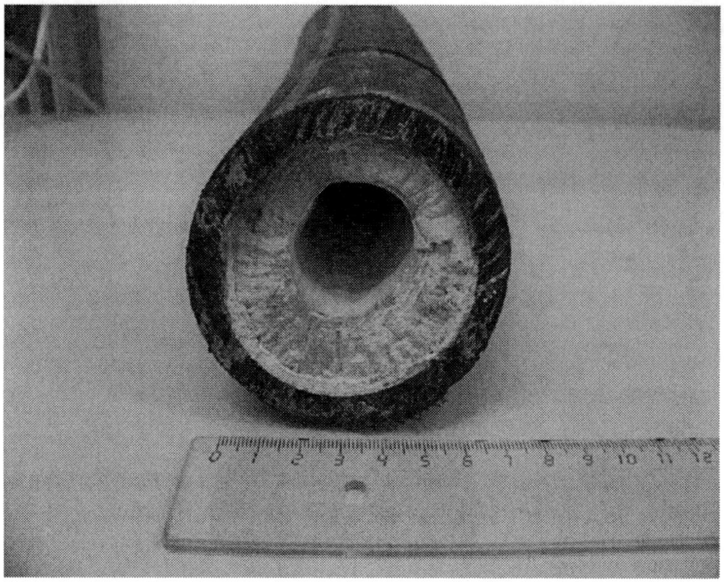

Figure 4. Sinter crusts inside a drainpipe (after [3]).

Thus, carbonate precipitation in drainage systems may cause serious problems due to the reduction of the cross section of drainage tubes (Figure 4) and the pollution of receiving streams by suspended carbonates and ongoing sinter formation.

3. Impact factors and counter measures

Consequently, the main impact factors on sintering are:

- the composition of ground water,
- the materials of tunnel structure,
- as well as constructional details.

To reduce the negative effects of sintering right from the early stages of a project special attention shall be paid towards these impact factors.

3.1. Composition of ground water

As stated above quantity, the chemical composition of ground water as well as the saturation respectively under-saturation of ground water has high influence to the dissolving behavior. Therefore it is recommended to analyze the prevailing groundwater conditions also regarding to their sinter-relevant aspects. These aspects shall be considered as early as of the first ground investigation campaigns and can be continued over all periods of design and construction.

On the example of the Koralmtunnel it has been shown that by means of specific hydrogeological monitoring, carried out also subsequently to the construction, zones with high and zones with low sinter potential could be identified [3] and specific counter measures could be defined on the basis of this knowledge.

The ground water can be treated by means of inhibitors, e.g. Polyaspartic acid or Polysuccinimid with the effect that the precipitation mainly occurs in a colloidal matter not forming hard crusts in the drainage system. On the other hand several examples have shown that the overdose of inhibitors or the application of inappropriate products may cause negative effects. This might lead to an overfertilization of waters with the result of a significant growth of aqueous slime and felt.

The application of supply stones had been tested in several projects. Also their functionality depend on specific project respectively groundwater conditions. Extensive tests in this regard have been carried out for the Gotthard Base tunnel [4] leading to the development of a tailor made hardness stabilizer system.

On the example of the Austrian Lainzer railway tunnel the application of supply stone showed no convincing results. After intensive investigations it was described [5] that under the specific conditions of this project the most effective solution was to treat the groundwater by means of the liquid dosing of diluted hydrochloric acid instead of the abovementioned organic acids. The reaction of the hydrochloric acid with partly formed sintering crusts leads to the dissolution of the calcite.

The best knowledge of the sinter relevant input parameter of the groundwater conditions, therefore, is mandatory. Specific hydrogeological monitoring programs shall therefore be mandatory parts of the project. System, application and dosing of inhibitors strongly depend on the project specific groundwater conditions and shall be provided as a tailor made solution for each project.

3.2. Custom-made concretes and mortars

Carbonate sinter formation in tunnels are, as stated above, closely related to the groundwater-concrete interaction. Sinter formation can significantly be reduced by an adequate construction and by using tailor-made construction materials. Lab experiments show that shotcrete with low contents of Portlandite and sulphate active minerals as well as with low permeabilities should be used for a sustainable reduction of carbonate sinter formation. In case of shotcretes these modifications have already been incorporated in the national guidelines [1].

Based on the study results for the Koralm tunnel appropriate recipes for shotcrete under construction site conditions are developed using low hydraulic cement content to reduce Portlandite formation during the hydration, alkali-free setting accelerators, and proper conditions to avoid fissure formation.

One suitable composition was obtained by using around 280 kg/m^3 of CEM I 52.2R, 140 kg/m^3 of AHWZ (pre-processed latent hydraulic acting material such as fly ash), and about 7 wt.% of an Aluminium sulphate containing setting accelerator. Results from water-shotcrete interaction experiments confirmed that such tailor made shotcrete can significantly lower the liberation of Ca^{2+} and OH^- ions into the solution. The Ca^{2+} liberation can be reduced by about 65% (!) by the application of the above recipe in comparison to common shotcrete compositions. Advanced shotcretes with respect to improved workability or specific fields of application are meanwhile applied (e.g. 320 kg/m^3 of CEM I 52.5R, 110 kg/m^3 of AHWZ, 40 kg/m^3 of steel fibers or 320 kg/m3 of CEM I 52.5R, 102 kg/m3 of AHWZ), showing also excellent early strength behaviors.

Already the optical impressions of the shotcrete lining constructed with such modified shotcretes shows a significantly improved visual appearance.

Figure 5. Visual appearance of shotcrete lining without sinter efflorescence inside the Koralm tunnel lot KAT1.

IOP Conf. Series: Materials Science and Engineering **236** (2017) 012071 doi:10.1088/1757-899X/236/1/012071

Segments applied in shield-TBM tunneling can be considered to be uncritical in respect to dissolution due to their dense concrete structure and their long curing time. However, also in mechanical excavations considerable amounts of backfill grout mortars are in use. These mortars (in case of a drained tunnel system), as well as mortars used for grouting and rock bolting still provide a significant sinter potential. Verbal reports from the Swiss Lötschberg tunnel or from HSR Frankfurt – Köln report of at least area by area heavy sinter deposits due to heavy grouting measures.

Measures for further reduction of the sinter potential of mortars analogous to the above mentioned shotcrete modifications have to be consequently further developed. Manufactured products already available on the market, such as TIWOFILL mortar, show satisfying developments, however, further investigations, such currently running in accordance with the Semmering Base tunnel construction indicate that there is still a remaining space for further improvements.

3.3. Constructional details

Calcite precipitation from the drainage solutions can be induced by the dissolution of Portlandite in the groundwater at the shotcrete. If a supersaturated solution comes into contact with the Earth's atmosphere, the internal partial pressures of CO_2 in the solution control the CO_2 exchange with the atmosphere and ongoing carbonate precipitation.

The mechanisms of Calcite precipitation in the drainage solutions show that the drainage system should be designed to minimize the CO_2 exchange with the atmosphere. Wide and long-lasting contact with air as well as turbulent flow of solutions has to be avoided. In addition an easy accessibility for inspection and cleaning of the drainage system shall be part of a proper design. In case of the Koralm tunnel the main drainage pipe is located directly under the centerline of the track, allowing easy rail bound inspection and cleaning [1], [3].

By using filter gravel with a reduced addition of cement (less than 200 kg/m^3) additionally a reduction of nutrient supply can easily achieved.

These measures have to be accompanied by proper workmen ship, avoiding steps, kinks or blockages of the flow path. By such an approach the precipitation of Calcium carbonate can be significantly reduced.

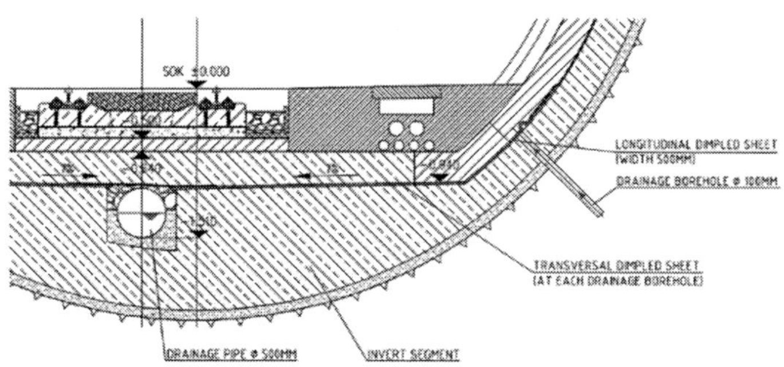

Figure 6. Invert detail of the Koralm tunnel with laminar flow path and only one central main drainage (after [3]).

In case of remaining precipitations these might be treated with scale inhibitors, e.g. hardness stabilization, considering the sensitivity of such counter measures. Once carbonate sinter is formed in the drainage system, a proper drainage design for the discharge of drainage solution and for simplified cleaning actions is essential for cost-efficient maintenance of the tunnel system. The alignment of pipelines shall be smooth, with short shaft distances, sufficiently large drainage pipes with sufficient space for sinter formation, as well as smooth-walled pipe surfaces, which do not show any damage.

4. Conclusion

The drainage systems of tunnels are frequently clogged with calcite precipitates. Cleaning and water conditioning are cost-intensive for the operating companies. Multiannual investigations and in-situ experiments in the recent past in Austria and other European countries have shown that the dissolved Calcium in the drainage solutions derives from the groundwater and especially from the dissolution of Portlandite at the shotcretes and mortars. By using custom-made shotcretes sinter formation can be significantly reduced. Matched to the specific groundwater conditions these custom-made shotcretes are meanwhile successfully applied. Analogous developments have also to be made with further construction materials, e.g. grouting mortars.

With a design of flow paths avoiding turbulent flows and less contact to Earth's atmosphere the precipitation of Calcium carbonate can be significantly reduced, too.

The consideration of these measures provide an important contribution for the significant reduction of maintenance and repair works of tunnel drainage systems, yielding to less interference in operation and a significant reduction of operational costs.

Acknowledgements

The author would like to thank ÖBB Infrastruktur AG and ASFINAG for their studies and the provision of data. I would also like to offer my special thanks to Martin Dietzel from Technical University of Graz and his team for their contribution and their valuable technical support.

References

[1] Harer G 2008 Planerische Vorkehrungen zur Erzielung eines erhaltungsarmen Entwässerungssystems beim Koralmtunnel *Proc. Symposium Darinagesysteme im Tunnelbau; Design, Versinterung und Instandhaltung (Graz)* pp 1-14

[2] Gamisch T and Girmscheid G 2007 *Versinterungsprobleme in Bauwerksentwässerungen* (Bauwerk Verlag)

[3] Dietzel M, Rinder T, Leis A, Reichl P, Sellner P, Draschitz Ch, Plank G, Klammer D and Schöfer H 2008 Koralm tunnel as a case study for sinter formation in drainage systems - precipitation mechanisms and retaliatory action *Geomechanics and Tunnelling* **4** pp 271-78

[4] Fabbri D, Ferrari A and Keiper K 2008 Härtestabilisation - Erfahrungen aus dem Baulos Bodio des Gotthard Basistunnels *Proc. Symposium Drainagesysteme im Tunnelbau (Graz)* pp 143-61

[5] Stur M, Ottner F, Schachinger T and Wriessnig K 2015 Lokale Beseitigung von Versinterungen in Bergwasserdrainagen *http://www.tunnel-online.info*

[6] Austrian Society for Concrete and Construction 2009 *Guideline Sprayed Concrete*

IOP Conf. Series: Materials Science and Engineering **236** (2017) 012072 doi:10.1088/1757-899X/236/1/012072

Geotechnical risk and earth structures

Ivan Vaníček[1], Daniel Jirásko[1] and Martin Vaníček[2]

[1]Czech Technical University in Prague, Faculty of Civil Engineering, Geotechnical Dept. Thákurova 7, 166 29 Prague 6, Czech Republic

[2]SGeosyntetika Ltd, Nikoly Tesly 3, 160 00 Prague 6, Czech Republic

E-mail: daniel.jirasko@fsv.cvut.cz

Abstract. There is a general acceptance that the complexity of each geotechnical design should correspond to the expected risk. The attention of the paper is therefore focused on the closer specification of the geotechnical risk – risk with which the design and realization of geotechnical structure is connected. The special attention with respect to this geotechnical risk is devoted to the earth structures.

1. Introduction

There is a general feeling that the geotechnical structures design is connected with higher risk than other civil engineering structures, especially when the concrete or steel are the main construction materials. Steel and concrete are typical man-made construction materials and the designer of concrete or steel structures has a great chance that the defined properties in the phase of the design will be very close to the properties in the realized structure. On the other hand the geotechnical structures are connected with natural material, soil, rock, properties of which are strongly variable and our opportunity to check their properties are limited on extremely small volume of the real volume which is part of the realized structure. From the three main types of geotechnical structures – foundation structures, underground structures and earth structures, the last one are more sensitive to the discussed problem as natural environment is not only in the interaction with the realized structure but this material is also the main construction material. There is a general acceptance that the complexity of each geotechnical design should correspond with expected risk. This fact is explicitly expressed also in Eurocode 7 Geotechnical design, however the closer specification is indirect via classification of the geotechnical structures into three Geotechnical Categories. Therefore the paper is focused on the closer specification of the geotechnical risk – the risk with which the design and realization of geotechnical structure is connected. Special attention is devoted to the geotechnical risk for earth structures, as the problem is there even more problematic.

2. Risk generally

All human activities are facing many different factors and influences, both internal and external that make it uncertain as to whether, when and the extent to which they will achieve their objectives. The effect of this uncertainty is called risk. To make it more likely that their objectives are achieved, the detection and understanding to this risk is needed and, if it seems prudent, necessary or beneficial, to modify those risks. With respect to the organization these principles are defined by ISO/CD 31000

"Risk management – Principles and guidelines" where are also specified such terms as "risk management process", "risk assessment", "risk identification" or "risk analysis".

In geotechnical engineering, for cases connected with natural hazards, closer specification in this direction was proposed e.g. by [1], [2] or [3]. They are working with specifications as probability of natural hazards occurrence, with expected cost of the total loss of the element, respectively with vulnerability of the element in risk, as under some circumstance not total loss of the element at risk can occurred.

It is obvious that most problematic in this expression is specification of H – probability that a particular danger (threat) occurs within a given period of time. For natural hazards, as floods, slope stability of natural slopes, there are many historical, statistical evaluations, so that the estimation can be roughly specified. However the similar approach for the geotechnical structures design cannot be used as the estimation of all negative factors including their frequency is practically impossible. In this case the geotechnical risk evaluation can be done first of all in more general level, however with the utilization of terms (processes) specified in ISO/CD 31000.

3. Risk according to EC 7

Problems with EC 7 design in general are described e.g. by [4] or [5]. EC 7 in section 2 Basis of geotechnical design recommends as the first step to distinguish between light and simple structures and other geotechnical structures. For these light and simple structures it is possible to ensure that the minimum requirements will be satisfied by experience and qualitative geotechnical investigations, with negligible risk.

This statement is very important as allows the design on the base of up to date experiences – know how – without necessity to calculate limit states.

On the other side EC 7 distinguishes very large or unusual structures, the design of which should follows the basic principles of EC 7 however often needs alternative provisions and rules to these in EC 7.

EC 7 is using terms Geotechnical categories and simple structures are falling into 1st GC, very large or unusual structures into 3rd GC. However not only character of structure but also character of ground conditions (geological and geotechnical conditions) have to be evaluated, when terms as "straightforward" for 1st GC or "unusual or exceptionally difficult" for 3rd GC are used.

Therefore the EC 7 concentrates on conventional structures. However for such conventional structures the risk can be in very wide range and so closer specification is needed, which are presented further. If to the above mentioned character of structures and character of geological and geotechnical conditions the impact of failure of the proposed structure on human lives and environment generally is included, we are obtaining three characterizations with three different degrees:

- Demandingness of the structure – simple, middle, very demanding
- Complicacy of ground conditions – simple, modest, very complicated
- Impact of failure of the proposed structure on human lives and environment generally - practically negligible, moderate (conventional), very high.

With respect to the above mentioned segmentation, into 1st GC are falling structures for which all characterizations are simple and into 3rd GC structures for which at least one characterization is very complicated, e.g. [5]. After that we recommend to divide the 2nd GC into three subgroups, whether one, two or all three characterizations are falling into middle range – 2nd GC with low risk, with middle risk and with high risk.

This closer characterization is very important from the view of selection of characteristic values of geotechnical parameters, values, which are used during the design with the help of calculation model. EC 7 states that either standard tables of characteristic values, related to soil investigation parameters, or characteristic values evaluated from the measured results via statistical interpretation can be used, always as very cautions estimate. These two possibilities strongly affect the demands on geotechnical investigation as for the first case the index properties are sufficient for the classification, on which standard tables are based, [6]. Utilization of standard tables we recommend for 2nd GC with low risk, respectively the statistical evaluation for 2nd GC with high risk (and of course for 3rd GC as well). In ordinary praxis very often the number of measured results is deficient for good statistical evaluation, and therefore the final cautious estimate is based on both, measured and table values - this case can be recommended for 2nd GC with middle risk.

4. Risk based on the uncertainties of the main steps of the geotechnical structure design and performance

Another risk evaluation can be based on the uncertainties associated with four main phases of the design and performance, as is briefly shown in Figure 1 [7]. Here the "geological model" represents a geometrical model of the geological environment; "geotechnical model" specifies for geological model geotechnical data obtained during field and lab tests and subsequently, after cautious estimation, the characteristic values used for the structure design; "calculation model" specifies the basic approach to the geotechnical structure design and "execution" (construction) of the geotechnical structure – specifies the construction technology.

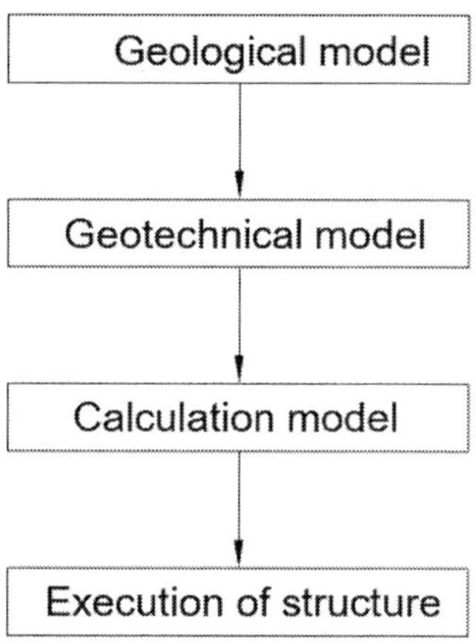

Figure 1. Main phases of the design and performance of the geotechnical structure.

Any uncertainties are connected with each of this phase and risk assessment can be based on credibility of each ones. For example – the credibility of the geological model depends on:

- Seriousness of the geological environment; its anisotropy, non-homogeneity, irregularity of discontinuities; generally speaking, the more problematic this geological environment, the greater the risk connected with the design and performance is;
- Actual state of exploration of this geological environment; e.g. during earlier steps of site investigation and construction implementation;
- Extend of the ground investigation and its quality;
- Skill of the persons responsible for the site investigation interpretation.

5. Specificity of earth structures

The designer of earth structures does not need to know only the properties of natural ground, on which the fill is situated, but also properties of fill alone. As far as the specification of ground properties is more complicated process than for man-made construction materials as steel, concrete, the specification of fill properties is even more complicated. Therefore we can speak about specific risk connected with earth structures. In this case, the designer has specified in advance the properties of fill material, which are used during the design, [8], [9]. This specification is based on the laboratory compaction tests, for fine grained material mostly on Proctor test, for coarser material with index of density. For samples compacted in laboratory, for the consequently defined range, the tests of strength, deformation and permeability are performed and after that from this received set of results the characteristic values of the geotechnical parameters for fill are selected as cautious estimate. The specification of demanded properties for the contractor is in most cases indirect, not with the help of mechanical-physical properties, but via dry density and moisture content, which are subsequently controlled on the site. Therefore, only in limited cases the designer is receiving a feedback that the properties are as demanded, e.g. permeability for sealing layer of dams or sanitary landfills.

Logical scheme of the Geotechnical Design Report (GDR), as specified in EC 7, is for earth structures shown on Figure 2. Geological model have to be presented not only for the ground but also for borrow pit. GIR – is Ground (Geotechnical) Investigation Report. Geotechnical model is divided into two parts:

- Geotechnical site (ground) model - representing really obtained results of the individual tests (measured or derived values) and the executor of GIR is responsible for it;
- Geotechnical design model – representing characteristic values used during the design and designer is responsible for it.

The comparison of these two "Geotechnical models" will show very clearly what values were selected for the calculation from values obtained during the phase of investigation.

Another specificity of earth structures compared with most of other structures is so called 2D effect. For earth structures the 2D effect prevails, either for earth structures of transport engineering - motorways, railways so for earth structures of water engineering – channels, small dams, dykes along rivers etc. Two basic problems are playing more important role here:

- Distance between individual borings (investigation points), namely with respect of interpretation of the geological model between these two investigation points. It is obvious that closer spacing between investigation points will help to decrease the potential risk that some untypical changes will not be disclosed in advance.
- The potential failure in relatively small part of the whole structure will cause total collapse of the functionality of such structure, as is typical case for dykes along river. Gash of width about 10 m can cause a complete loss of functionality e.g., 10 km long dyke.
- Therefore some conservatism shall be accepted in such cases, even we will know, that about 99.9 % of such structure is safe, somewhere nearly extremely. The discussion between investor and designer is needed, as relatively small amount of money added to the geotechnical investigation can decrease the probability of missing weakest section and in such way to reduce conservative approach to the selection of characteristic geotechnical parameters. The combination of the weakest part in subsoil combined with weakest part for the fill cannot be neglected.

Last note to earth structures is connected with sustainability approach, as there are a great opportunities, [10]. Smart earth structures can safe agricultural land, natural aggregates or energy but in comparison with classical solutions they are connected with higher risk.

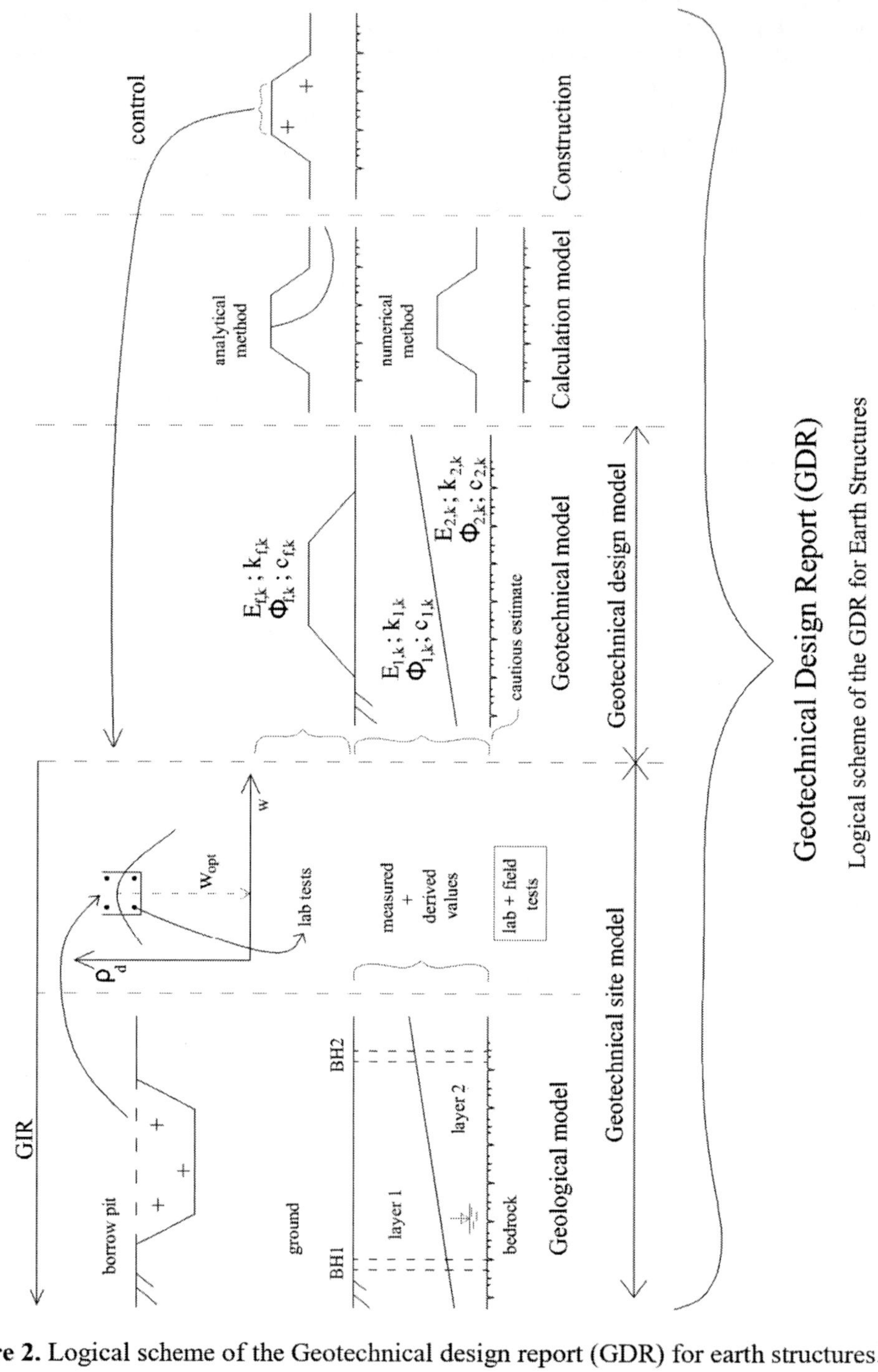

Figure 2. Logical scheme of the Geotechnical design report (GDR) for earth structures.

6. Conclusion

Word meaning "risk" is very wide and a certain guidance can be found in ISO/CD 31000 "Risk management – Principles and guidelines". Risk evaluation for structures affected by natural hazards, as e.g. landslides, floods, can be evaluated when probability of occurrence is known, however for general case it is very vague, as the estimation of all negative factors including their frequency is practically impossible. Nevertheless the closer specification of risk is needed as the complexity of each design should be strongly connected with this risk. In this case the geotechnical risk evaluation can be done first of all in more general level, with risk estimation, risk assessment, taking into account basic steps during which the geotechnical design has to pass. Two possibilities are therefore mentioned by the authors, which are however very close to each other. The first is utilizing demandingness of the structure, complicacy of ground (geological and geotechnical) conditions and the impact of failure of the proposed structure on human lives and environment generally. The second possibility is based on the evaluation of uncertainties or credibility of individual phases of the geotechnical design and performance as geological, geotechnical and calculation models plus execution of structure. Specificity of earth structures are expressed and recommended logical scheme of the Geotechnical design report presented.

Acknowledgements

The paper was prepared with financial support of the research project TE01020168 of the Technology Agency of the Czech Republic CESTI – Centre for effective and sustainable transport infrastructure.

References

[1] Lee E M and Jones D K C 2004 *Landslide risk assessment* (Thomas Telford London)
[2] Kalsnes B, Nadim F and Lacasse S 2010 Managing Geological Risk *Proc. 11th IAEG Congress (Auckland, New Zealand)* pp 111-126
[3] Vaníček I 2011 *Sustainable Construction* (CTU Press Prague) 163 p
[4] Bond A and Harris A 2008 *Decoding Eurocode 7* (London : Taylor and Francis) 598 p
[5] Frank R, Bauduin C, Driscoll R, Kavvadas M, Krebs Ovesen N, Orr T and Schuppener 2011 *Designers guide to Eurocode 7 Geotechnical Design* (Thomas Telford London) 216 p
[6] Vaníček I 2015 Earth Structures – sensitivity of the design to the characteristic geotechnical parameter values determination. Chapter in: Jubilee Volume – A. Anagnostopoulos *50 years of service at the National Technical University of Athens* ed M Kavvadas pp 543-52
[7] Vaníček I 2011 Risk in geotechnical engineering and profession prestige *Proc. 3rd Int. Symposium on Geotechnical Safety and Risk (Munich)* ed N Vogt, B Schuppener, D Straub, G Brau (BAW Bundesanstalt für Wasserbau Karlsruhe) pp 3-9
[8] Vaníček I and Vaníček M 2008 *Earth Structures in Transport, Water and Environmental Engineering* (Springer) 637 p
[9] Vaníček I and Vaníček M 2013 Modern Earth Structures of Transport Engineering *Procedia Engineering* **57** pp 77-82
[10] Vaníček I, Jirásko D and Vaníček M 2013 Geotechnical engineering and protection of environment and sustainable development *Proc. of the 18th IC SMGE Challenges and Innovations in Geotechnics (Paris)* (Presses de l'école nationale des ponts et chaussées) pp 3259-62

IOP Conf. Series: Materials Science and Engineering **236** (2017) 012073 doi:10.1088/1757-899X/236/1/012073

Thermal load of the final lining of traffic tunnels

Lukas Duris[1], Josef Aldorf[1] and Jiri Geryk[2]

[1]Technical University of Ostrava, Ludvika Podeste 1875/17, Ostrava 70833, Czech Republic
[2]Inset Ltd., Rudna 21,Ostrava 70030, Czech Republic

E-mail: lukas.duris@vsb.cz

Abstract. The final lining of mined tunnels using NATM is connected with the need to ensure long-term durability of the construction. The interaction between the secondary lining and surrounding rock mass is ensured by contact with the primary lining. Static load depends already on the interaction between tunnel lining and rock mass. Loading is given by geotechnical conditions, the size of the excavation, construction progress and very important is weather conditions. In this paper presents the results of long-term temperature measurements of the final lining of traffic tunnel. Measurement at selected tunnel pass over ten years. The monitoring results should be available for future design of the final lining of tunnel construction.

1. Introduction

The design of the lining of traffic tunnels has a number of requirements that the lining must fulfill over the lifetime of the tunnel. Any interruption of the traffic always has far-reaching socio-economic consequences. Many of these requirements, such as the completion of the load development and stabilization of the rock mass with the tunnel lining are fulfilled after successful completion and commissioning of the tunnel. There are some factors that in the short term are of little weight, but as their effect gradually accumulates, they can significantly affect the lifespan of the structure in the long term. Representative of these factors are periodic stress changes in tunnel linings due to temperature variations in tunnels. In countries such as the Czech Republic, with four seasons characterized by different temperatures and also by temperature fluctuations during the day, this phenomenon can not be ignored and must be included in the design when assessing the life of the tunnel lining.

Geological arrangement, direction, height and layout of the tunnel route, but also the range of projects and machinery potential of contracting companies in the Czech Republic are the reasons why in recent times the absolutely decisive majority of our mined tunnels are designed on the principles of conventional tunnelling using the "New Austrian Tunnelling Method" (NATM). The NATM system uses two types of lining to ensure stability. The primary lining is usually a layer of shotcrete with welded mesh, supplemented by bolts. This lining is temporary and secures the mined space during the construction. At this stage, the deformations usually stabilize and the stress in the massif regroups. The secondary lining is a continuous supporting structure carrying the entire applied load throughout the life of the structure. The final lining is generally composed of monolithic concrete or reinforced concrete structures, consisting in a cross section of the upper vault, the side (base) blocks and possibly bottom vault. Unfortunately, our modern tunnels so far only rarely use final lining just of plain concrete. Abroad, the use of plain concrete is commonplace.

Pouring of concrete of the individual structural elements of the lining usually uses the continuous method in sections; in the bottom vault into the formwork system, the top vault into hydraulically controlled mobile steel form. The length of working widths vary with the size of the transverse profile, the horizontal alignment and length of the form used, from about 5 m to about 15 m. Each ring of the final lining acts statically on its own.

The final lining is nowadays already designed using modern computer tools. Internal forces and deformations of the lining are calculated with numerical models using the finite element method, where the whole geological environment is simulated including the progress of the excavation and insertion stages of both the primary and the secondary lining. Static calculations are carried out mainly in areas with the most adversely loaded profiles, taking into account the number of load cases and combinations thereof, including:

- Own weight
- Geostatic load of the rock environment
- Hydrostatic groundwater pressure (in the case of closed insulation)
- Creep and shrinkage of the final lining
- The effect of temperature (cooling/warming)
- Technological load, etc.

Based on the above, it was decided to determine how in reality the existing tunnels are stressed or loaded and whether their actual behaviour is consistent with the assumptions in the static calculations. Therefore, in the context of a research project within the geotechnical monitoring, selected devices were placed in the final or temporary tunnel lining to monitor relative stress and temperature; strain vibrating string gauges. The strain gauges are today used to monitor, for example, these tunnels in the Czech Republic:

- Valík (motorway D5)
- Panenská (motorway D8)
- Libouchec (motorway D8)
- Klimkovice (motorway D47)
- Mrázovka (City Ring Road in Prague)
- Dobrovského Tunnels in Brno
- Tunnel complex Blanka (City Ring Road in Prague)
- Vítkovské (railway tunnels of the New Connection in Prague)
- Cable tunnel Vltava (Prague)
- Jablunkov (railway tunnel)

The paper presents the results of measurements on the tunnel lining in Klimkovice. The automatic measuring system was installed as the first in the Czech Republic in the Klimkovice tunnel, built in 2006, during the installation of the secondary lining. Systems for measuring temperature and stress in the final lining then also became to be more used for newly-built tunnels to gain more accurate knowledge of the behaviour of the final lining during operation of the structure. The measuring system has been in operation for over nine and a half years and still keeps collecting the necessary data from two measuring profiles. [1]

Secondary lining measurements are common. We can see examples of tunnel measurements that were presented in [5, 6].

2. Measurement system on the Klimkovice Tunnel

In 2008 the next section of the motorway D47 - Lipník nad Bečvou - Bohumín was put into operation, designated as construction 4707. On this section, less than 10 km long, is also the Klimkovice Tunnel -

one of the major civil engineering structures of the motorway. The tunnel itself is 1,080 m long and is located near the municipality of Klimkovice at a distance of about 1 km from the building of Klimkovice spa. It passes here underneath the ridge between the villages of Klimkovice and Hýlov. If the tunnel had not been built, the area would be inconsiderately and quite drastically divided by the busy motorway and its traffic. The underground route of the motorway will allow development of this territory in the future.

The tunnel is routed in two separate tunnel tubes A and B; both tunnel tubes have two lanes. Both tunnels have the same spatial parameters of the one-way two-lane tunnel T9.5 category, and same layout. The lane in the tunnel has a unilateral cross slope and width of 9.50 metres between curbs. The clearance above the ground is 4.80 m. Two-sided walkways have a width of at least 1.10 m. The tunnel has a clear width of 12.204 m. The gross cross area of the tunnel, including the technologically required overcut, is 120.17 m^2. Because of the routing of the motorway in the terrain configuration and geological conditions, the tunnels were mined in longer stretches of rock massif; the shorter sections next to the portals were excavated in open trenches with subsequent backfilling. The excavated sections of the single-tunnel construction have reverse umbrella waterproofing and the vault strips sit on longitudinal footings. The mined sections are double-walled construction tunnels with closed intermediate waterproofing.

The tunnel was mined using the new Austrian tunnelling method with the application of shotcrete, honoring the NATM principles, and with a horizontal articulation. In the first step, the full width of the top heading was mined with the excavation area of about 65 m^2 and a minimum height of 6.10 m of the vault crown. The lower profile opening was further spatially divided into three parts - the lower bench in two steps (right, left) and excavation of the invert. The primary lining supporting the excavation is shotcrete with reinforcing steel mesh, lattice girders and bolts. Construction of the lining distinguishes 5 types of protection depending on the nature of the geological environment.

The final lining of the Klimkovice tunnel is designed as reinforced concrete C30/37. The vault is reinforced with armature. The basic length of the individual section (tunnel block) is 12 m. The lining of the Klimkovice tunnel has already claimed one first - to increase the fire resistance of concrete vaults concrete filled with polypropylene fibers was used. This solution was selected because the coarser fractions of aggregate in the concrete were greywacke. Fire resistance tests confirmed the correctness of the selection of concrete type – the reinforced concrete of the secondary lining passed the tests [2].

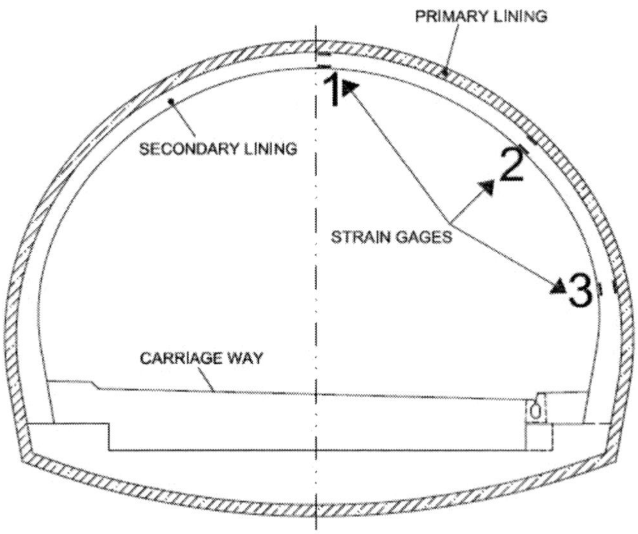

Figure 1. Tunnel cross-section with monitoring equipment.

During the construction of the secondary lining, a controversy occurred regarding the thermal load of the tunnel lining. For this reason, temperature sensors were installed at different distances from the portal in two cross sections of the tunnel B for measuring temperature and lining strain. Furthermore, temperature sensors were installed to measure air temperature in the tunnel and in the rock 1.0 m behind the lining.

In each profile there are 12 such sensors. The sensors were placed in three positions (see Figure 1). At each position they were fitted to the upper and lower edge of the reinforcement in the radial and tangential direction. In each measurement profile a sensor for measuring air temperature is also mounted. The measurement system is automatic. The reading and collection of data is automatic. The intervals of reading are every four hours.

The first measured section was in the block labelled B90, approximately 55 m from the eastern portal. The record of temperature measurements is shown in Figure 1. Block B90 is the first strip from the portal on the transition between excavated and mined part of the tunnel. The second profile was located 190 metres from the entrance portal in the tunnel strip labelled B78.

3. The temperature measurement on the Klimkovice Tunnel

As mentioned, the design of the tunnel lining is always complex due to the effect of a large number of load cases and combinations thereof. Among these influences are the temperature changes, which reflect the climatic conditions outside the tunnel. This paper will evaluate a single profile that is close to the portal. This profile is influenced more by external factors, as indicated by the current measurement and the difference of readings between the two profiles is up to 8 °C (air temperature). This paper will assess the measurements over the full nine years. Overall temperature during this period is recorded in Figure 2. The recorded temperatures are those on the three levels of the tunnel vault, in the outer and inner face of the lining, and the air temperature and the temperature in the rock. The air temperature is measured at the crown where the thermometer is just below the top of the crown. Rock temperature is also monitored at two levels, both in the top of the crown and in the side of the vault. The respective sensor is always placed in a one-metre-deep borehole. The course of the air temperature is the most interesting, as it obviously varies according to the time of day and the season. Its extremes fall to both times of the year (summer and winter). This temperature is faithfully copied by the temperature in the lining in all three positions. The seasonal trends act of course on the temperatures in the rock. Its maximum and minimal manifestations are less than the air temperature in the tunnel, but again respond to external influences. Of course the temperature in the rock shows some lag compared to the air temperatures. Maximum, minimum and average values are shown in Table 1 below. Over twenty-nine thousand entries were evaluated. The highest air temperature was measured in August 2013 (34.15 °C). Usual summer extremes amounted to thirty degrees Celsius. The lowest temperatures were achieved in February 2012, when the temperature in the tunnel fell to -15.3 °C. This was due to long-term low temperatures, which lasted for over twenty days. In winter, temperatures routinely fall below freezing. This usually happens in January and February. The average air temperature in the tunnel is about 13°C. The temperature in the rock reacts to external weather influences and the temperature varies from 17 C to 4°C throughout the year. In summer, extreme temperatures in concrete are about 20% lower than the air temperature inside. During winter extremes the difference is greater. The measurement shows that the highest or the lowest air temperatures are not identical with the temperatures in the lining at the same time.

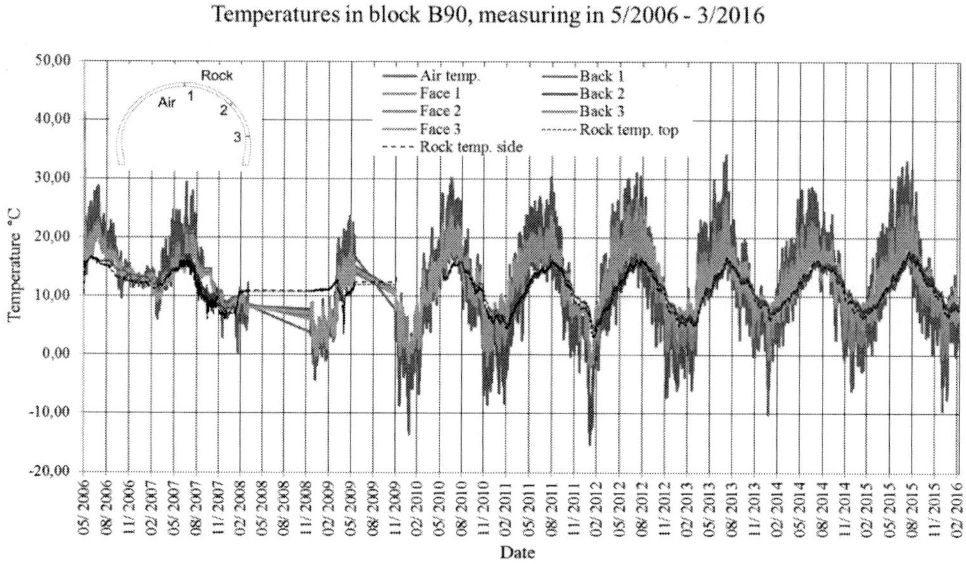

Figure 2. Course of temperatures in the concrete lining.

Table 1. Measured temperature.

	Air	Back 1	Face1	Diff.	Back 2	Face 2	Diff.	Back 3	Face 3	Diff.	Rock
Maximum	34.1	27.4	24.9	4.6	27.5	24.9	6.0	27.0	24.3	5.7	17.6
Minimum	-15.3	-1.4	-2.0	0.0	-1.1	-1.9	0.0	-2.0	-2.0	0.0	4.1
Average	13.2	13.1	13.2	0.5	13.1	13.1	0.7	12.9	12.9	0.9	12.0

Table 1 also shows the temperature differences. This means the difference between the face and the reverse side (back) of the structure. Strain gauges are placed on the reinforcement on both sides of the cross section, and therefore we are able to determine a simple linear temperature gradient across the cross section. This is very important in terms of the structural design and stress of construction due to temperature influences. The load is not determined by the highest or lowest temperature attained, but rather by the temperature gradient between the reverse and face of the structure. The maximum and minimum value is always set for the difference at the time of reading, not from the maximum and minimum values. This temperature gradient is not uniform for all three sensors positions in the vault. At the crown, the differences are minimal. 70% of the measured values are within 0.5 °C. The concrete thickness is 0.35 m here. In the heel the vault is extended to 0.7 m. At the position of sensors two and three, only 65% and 50% of the measured values respectively fall within 0.5 C. The temperature difference from 0.5° to 1.0°C is roughly around 17%. The highest value of the difference that has been reached is 6 °C, and only rarely. This maximum was recorded for sensor position 2 (see Table 1). Minimum values are limited to zero, because the difference was determined from the absolute value of the difference between the face and the reverse. A zero value means that at that time both temperatures were the same.

The significance of the temperature gradient has already been mentioned. Another reason for monitoring the temperatures at both surfaces of the lining is the German railway directive DS 853 [4]. This directive prescribes what temperature differences must be considered in the design of the final lining of tunnels. The directive takes only the tunnel length (the distance from the portal) and the season (summer, winter) into account. It has been found that the temperature gradient is also influenced by the thickness of the lining and the distance from the portal. The prescribed temperature differences are absurdly large, and completely diverge in comparison with the measurement results.

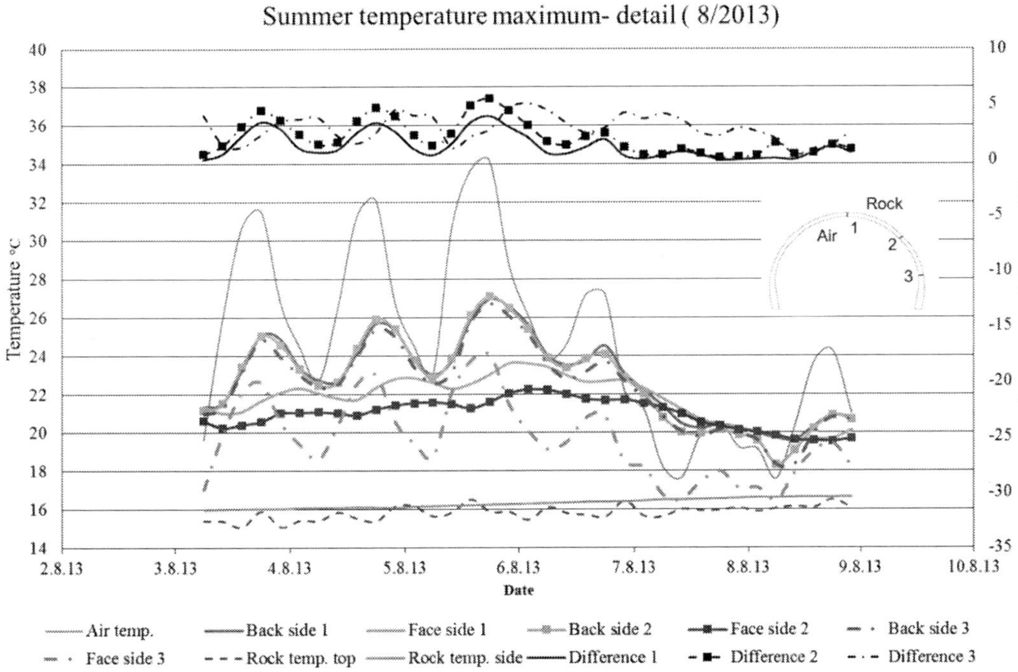

Figure 3. Detail of temperatures during extreme summer.

To illustrate the daily temperature course, two detailed graphs for a short period (5 days) were compiled. These graphs show the extreme temperatures of the air in the tunnel in summer and winter. All three sensor positions in the vault, the air temperature and the temperature in the rock were always evaluated. Furthermore, the differences in temperature between the face and the reverse of the structure were determined.

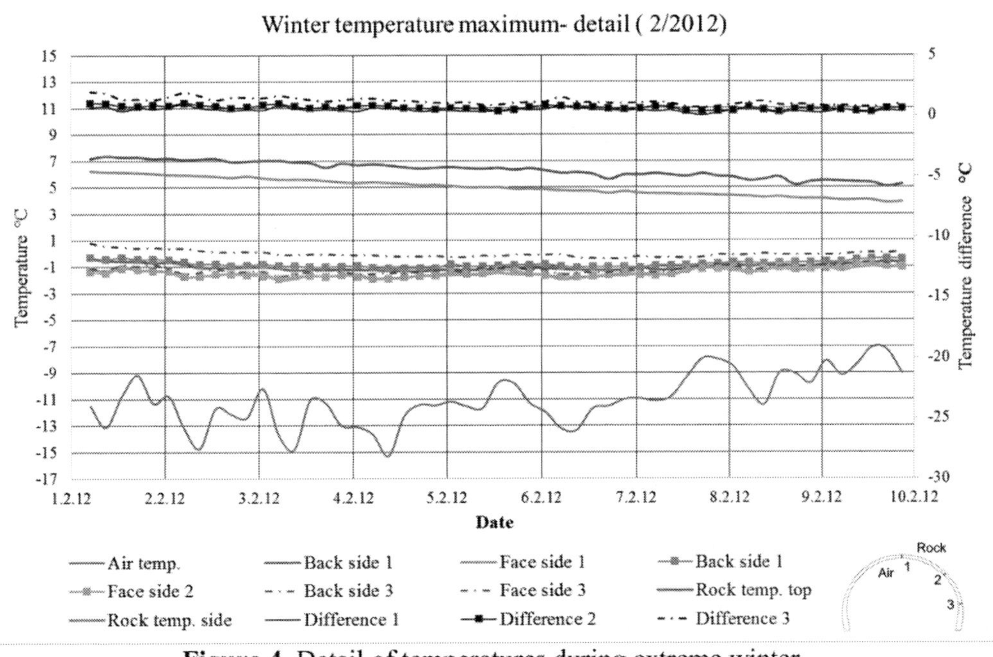

Figure 4. Detail of temperatures during extreme winter.

IOP Conf. Series: Materials Science and Engineering **236** (2017) 012073 doi:10.1088/1757-899X/236/1/012073

Figure 3 shows the recorded detail of summer. The air temperature climbed briefly to a maximum of 34 C, but the other days were cool, which is also reflected in the course of temperatures. The difference between the air temperature and the highest temperature in the structure is 4 °C. The courses of temperature at the outer edge are almost identical; the differences are greater on the reverse of the lining. The courses no longer overlap and it is possible to see the differences by the sensor position or the thickness of the lining. The temperature in the rock is constant (16°C) and the difference between the outer surface of the lining and the rock is up to 6°C. When the temperature of the air drops, the temperature in the lining reacts and the temperatures of the reverse and the face equalize. This effect is particularly evident from the difference in temperatures at the top of the tunnel vault. These results show that the structure is not heated to the air temperature in the tunnel and this temperature acts only briefly and is not able to affect the lining too much. Extreme weather will have only a minimal effect and for a very short time. The temperature gradient in the structure is affected mainly by the thickness and position of the cross section of the vault (the upper part is heated more). The temperature in the structure does not respond immediately to changes in outside temperature, and may even have a higher temperature than the ambient air. This is caused by accumulation of heat and partial heating of the back part of the rock, which has no possibility to quickly dissipate the heat from the structure, i.e. the structure is cooled from one side only.

Figure 4 is a detailed curve for winter. The winter temperatures demonstrate a significant difference from the rail directive. For extreme recorded values the air temperature was -15 °C, and also for long periods (3 days). Despite this long-term cooling, no significant drop in temperatures in the tunnel lining occurs. The lowest temperatures on the face are around -2 °C, which is a 13 °C difference from the air temperature. The temperature in the rock gradually decreases due to the long lasting frosts. Its temperature does not drop below +4 °C, which causes heating of the underside of the secondary lining. The reverse side is cooled to 0 °C. Temperature differences in the wall (temperature gradient) in these winter temperatures are far lower than anticipated by the rail directive; within two degrees Celsius. This finding is of great significance in terms of stress of the structure due to climatic conditions. The combination of loads with winter temperatures in the static calculation is much more unfavourable than at summer temperatures. It is very different to consider a temperature gradient of 10 °C or 2 °C when calculating the load of the structure (see [3]).

The second measurement profile is located 190 m from the portal. The course of temperature practically corresponds to temperatures in the first measurement profile. The difference of the average air temperature in the tunnel between the profiles is 0.3 °C. Virtually the same maximum values during the same period were achieved. The difference is in winter when the temperature difference is up to 3 °C. Winter temperatures are higher deeper in the tunnel than closer to the portal. The temperatures are virtually identical in both profiles, and then the findings for block B90 apply also for the second measurement profile.

4. Conclusion

Summary of findings from measurement:

- Maximum, minimum and average values obtained in each measurement nodes, the air temperature at the ceiling and surrounding rock were determined.
- Temperature differences between the face and the reverse of the lining and the maximum, minimum and average values were determined. Based on evaluation of the frequency of temperature differences it has been found that the majority of differences is within 0.5 °C with the frequency of over 70%.
- For the assessment of extremes the effects of hydration of concrete were eliminated.
- The temperature in the rock does not fall below freezing in the winter; the annual temperature difference in the rock is 13 °C.

In terms of measuring temperatures in the tunnel the extreme values of these temperatures were obtained, which brought a lot of new knowledge. In particular, the lining responds to the course of the ambient temperatures, but also does not drop to the temperature of ambient air at very low temperatures. Temperature differences between the reverse and the face are less than projected by the rail directive. The temperature differences are a function of the thickness of the reinforcement, position in the vault and distance from the portal. The temperature is also affected by the temperature of the rocks.

The measurement of temperature and relative deformations in the lining of the Klimkovice Tunnel clearly demonstrated their importance and necessity. Monitoring allows tracking the effects of climatic conditions in a tunnel lining. The acquired knowledge significantly contributes to a more realistic look at these temperature effects. It even brings verification of the assumed static solutions in terms of the loading process of the lining.

Acknowledgements
The paper was prepared with the support of the Competence Center of the Technology Agency of the Czech Republic (TAČR) within the project Center for Effective and Sustainable Transport Infrastructure (CESTI), project number TE01020168.

References
[1] Šourek P, Vítek J, Aldorf J and Ďuriš L 2008 Měření deformací a teplot na definitivních tunelových ostěních *Tunel* **17**(4) pp 70-76
[2] Pechman J et al 2008 Klimkovický tunel uveden do provozu *Tunel* **3** pp 18 - 25
[3] Ďuriš L and Aldorf J 2011 Vliv ochlazení na tunelové ostění *Geotechnika* **14**(3) pp 22-25
[4] DB Netz – Deutsche Bahn Gruppe 2002 *Richtlinie 853, Eisenbahntunnel planen, bauen und instand halten*
[5] Zhao X and Qiu H 2007 Application of fiber Bragg grating sensing technology to tunnel monitoring Yanshilixue Yu Gongcheng Xuebao/Chinese *Journal of Rock Mechanics and Engineering* **26**(3) pp. 587-93
[6] Iida H, Iura T, Konishi S, Ono T, Koyama Y and Mizuhara K 2006 Internal strain measurement method of the extruded concrete lining *Tunnelling and Underground Space Technology* **21**(3-4)

Assessment of tunnel´s face support pressure

Eva Hrubesova[1] and Lukas Duris[1]

[1]VSB-Technical University of Ostrava, Ludvika Podeste 1875/17, Ostrava 70833, Czech Republic

E-mail: eva.hrubesova@vsb.cz

Abstract. The evaluation of support pressure ensuring tunnel´s face stability during excavation using tunnel boring machines is one of the fundamental problems in tunnelling. This paper presents various approaches to this problem. The analytical models, based either on the principle of limit equilibrium or the limiting analysis (low boundary or upper boundary limiting methods), requires considerable simplification (circular tunnel, homogeneous rock mass etc.). But, on the other hand they are less time-consuming in comparison with the numerical models based on the finite element method. This paper describes the comparison of results of minimum face pressure obtained on the basis of different calculation methods with respect to the specific geometrical and material characteristics of the problem. Geotechnical software Midas GTS NX (based on finite element method) was used for presented numerical analysis. The paper also provides a comparison of modelling results with the results of the face pressure corresponding to the real excavation of the tunnel.

1. Introduction

One of the fundamental tasks in a tunnel excavation using tunnel boring machines is determination of the optimum face support pressure, which ensures the safety of the excavation itself, helps to minimize a surface settlement and ultimately prevents the collapse of the entire tunnel. The value of the face support pressure generally depends on many factors, primarily on the size of the tunnel, the depth below the surface, cohesion and internal friction angle of the rock environment, unit weight of the soil environment, permeability of the rock environment, machine type, excavation speed, depth of groundwater level, etc.

Research concerning the determination of the optimum tunnel face pressure is globally conducted by experimental methods, including centrifugal tests (e.g Kamata [9], Kimura, Mair [10]), as well as by various mathematical (analytical and numerical) methods [3].

2. Basic outline and principles of calculation methods to determination of the face pressure

The methods to calculate pressure on the tunnel face differ in their mathematical approach, calculation assumptions, the degree of conservatism of results, required computing time and calculation efficiency, which is a very important factor, particularly when practical problems are solved.

Analytical calculation methods to determine the minimum tunnel face pressure are either based on the limit equilibrium methods (Broms and Bennmark [4], Krause [11], Jancsecz and Steiner [8], Anagnostou and Kovári [1], Broere [3], Carranza-Torres [5]), or the limit analysis method (Atkinson and Potts [2], Davis, et all. [6], Leca and Dormieux [12]). Although derived analytical methods to determine the minimum tunnel face pressure are usually derived under quite restrictive conditions, mainly relating to the homogeneity of the rock environment, circular cross-section of the tunnel, hydrostatic state of primary stress, disregarding the nature of the lining behind the face in the excavated area, etc., they are less time consuming in comparison with numerical methods.

Analytical calculation methods to determine the minimum face support pressure based on limit analysis can be divided into two basic groups. The first group of methods is based on the so-called "lower bound theory". These methods allow for the determination of a statically acceptable solution for minimum face pressure, that satisfies both balance and rock failure conditions. The basic underlying assumptions for these types of methods imply, that they allow obtaining conservative value of the minimum tunnel face support pressure.

The second group of computational methods is based on the so-called "upper bound theory", which assumes kinematically acceptable solution subject to a certain accepted rock failure mechanism. Methods of this type assume that, given a certain rock failure mechanism, the compatibility of displacements between individual blocks of material in a rock failure process is not breached. Unlike solutions based on lower bound theory, these computational approaches to obtain a kinematically acceptable solution allow obtaining a non-conservative value of face pressure.

Computational schemes, which foresee the emergence of a particular rock failure mechanism before the face, are based on the Horn model [7], which replaces the circular shape of the tunnel face with a square shape, and approximates the corresponding rock failure area before the tunnel face using rigid blocks – wedge and the follow-up prism (chimney).

3. Characteristics of the numerical model for calculation of the face support pressure

For the comparative analysis of the various approaches to determining the required minimum face pressure, the authors of this paper used the numerical model of an excavated tunnel of a circular cross-section. Both software Midas GTS NX and Plaxis 3D were used for numerical analysis. These codes were developed for the stress-strain and stability analysis of geotechnical problems and are based on a numerical finite element method. Mentioned software allow the simulation of linear and nonlinear behavior of soil, provide to model both hydrostatic and hydrodynamic effects of water in soils and the interaction between structures (reinforcement, anchors and wall) and the soil.

The basic model dimensions 70x50x75 m were considered (see Figure 2). The basic tunnel diameter $D = 5$ m was considered at different depths $h = 10$, 20 or 30 meters below the surface. The ratio between the diameter of the tunnel and the depth (h/D) achieves value 6, 4 and 2 respectively. In the next part of research next variants of diameters $D = 8$, resp. 10m were considered. The length of the excavated tunnel was considered 25 meters. Due to a more objective comparison with analytical methods, the calculations assume the hydrostatic state of primary stress, i.e. the coefficient of lateral pressure is equal to one. Influence of ground water was not considered in the calculation. The input material properties of the rock environment and concrete structure are given in Table 1. Angle of internal friction was considered from 25° to 35° with a step of 5°. The model used the final tetrahedron-shaped elements. Elastic- perfectly plastic Mohr-Coulomb constitutive model of soil was applied in the model. The concrete lining of thickness of $d = 35$ cm was characterized by elastic constitutive model. This lining will be installed immediately in one step after excavation (no gap between the excavation and lining installation was assumed). The standard boundary deformational conditions (rigid box) were assumed.

For simulation of stabilizing pressure at the face the pressure on full face was considered. The values of pressure were changed from 10 to 800 kPa. The first calculation stage involved deactivation of the volume elements inside the tunnel and activation of the shell elements corresponding to the tunnel lining. In the first stage the full pressure load on the face was activated. Each loading phase was solved separately.

Table 1. Properties of the rock environment and concrete structure.

	Model	E (MPa)	ν	γ (kN/m^3)	c (kPa)	φ (°)
Soil	M/C	45	0.35	20	2	from 25 to 35
Concrete	Elastic	20 000	0.2	24	-	-

4. Assessment of minimum tunnel face support pressure obtained by analytical and numerical methods

To the assessment of minimal tunnel face pressure both analytical and numerical methods were applied. Figure 1 shows the comparison of the results of the selected analytical approaches to determining the minimum stabilization face pressure. Equations were published earlier [11, 14, 15]. Due to the practically non-cohesive nature of the investigated rock environment (cohesion is equal to 2 kPa), methods working with undrained shear strength c_u are not considered in the presented study. Analyzed methods don´t reflect the impact of tunnel depth.

Based on the results of the calculations, it is possible to formulate the following conclusions:

- The range of the minimum face support pressure for all the compared analytical methods is 20 to 80 kPa for internal friction angle φ = 20°, values of 15 to 50 kPa for φ = 30° and 10 to 38 kPa for φ = 40°.
- The most conservative values of the minimum face support pressure result from the Krause method, considering the quadrant rock failure mechanism.
- In terms of the face support pressure, the least conservative analytical method is the Krause method considering hemispherical rock failure mechanism.
- The model indicates more significant dependence of the face support pressure on the internal friction angle in the case of lower values of the internal friction angle. [15]

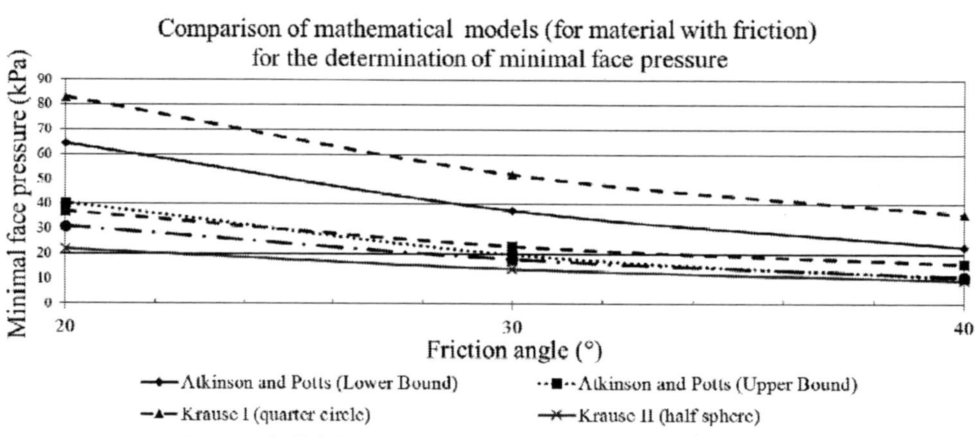

Figure 1. Comparison of the values of minimum face support pressures determined by analytical computational methods.

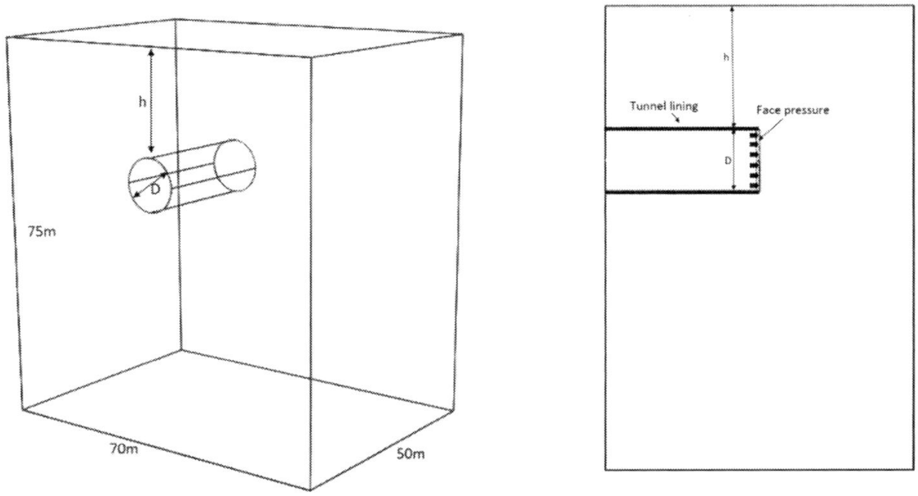

Figure 2. Conceptual model.

Three different tunnel diameters D were assumed (D = 5, 8 or 10 m respectively) in the performed numerical analysis. Moreover, the variant friction angle between 25 and 35° was assumed. Depth h of the tunnel below the surface was considered 20 m.

Based on the numerical model, all alternative depths, internal friction angles and face loads were assessed for horizontal face displacements corresponding to face extrusion into the space behind the face.

Figure 3 illustrates the dependence of the maximum horizontal face displacements depending on the face pressure for three different values of excavation diameter D at the same tunnel depth of 20 m. Friction angle of 30° was considered. For the diameter D = 10 m under the same face pressure of 50 kPa the face displacements are increased 3.5 times in comparison with the diameter D = 5 m. For the higher value of face pressure the corresponding face displacements are practically identical independently on the diameter. The face pressure, which causes such a qualitative change in the developments of horizontal displacements and more significant face extrusion, is considered the required minimum face pressure. Based on this approach, the minimum face support pressure determined by the model for D = 10, 8 and 5 m reaches 150 kPa, 100 kPa and 75 kPa respectively. Face pressure of approximately 600 kPa leads to virtually no signs of face extrusion, with horizontal displacements being zero. At pressures exceeding 600 kPa, the calculation model shows developments of horizontal face displacements in the opposite direction (the face is pushed back by the pressure).

Figure 4 illustrates the dependence of the maximum horizontal face displacements depending on the face pressure for three different values of internal friction angle and identical tunnel depth below the surface of 30m. This dependence allows us to conclude that, depending on the internal friction angle, up to a certain face pressure, the horizontal face displacements are not very significant and there is no significant extrusion into the free space behind the face. The minimum face support pressure reaches 60 kPa for the highest considered internal friction angle of 35°, approximately 75 kPa for φ= 30° and 100 kPa for φ = 25°.

Figure 5 documents the range of the deviatoric strain for safety calculation related to the face support of 10 kPa, the tunnel diameter of 5 m, the tunnel depth of 20 m and friction angle of 30°. Range of maximum deviatoric strain zone in front of the face is about 3.8 m and the shape of this zone corresponds to the Horn´s scheme (including chimney) [7]. Angle of the wedge from the vertical position is about 37.5°, range of this zone above the roof (in the overburden of the tunnel) is 6 m (approximately the same as the tunnel diameter).

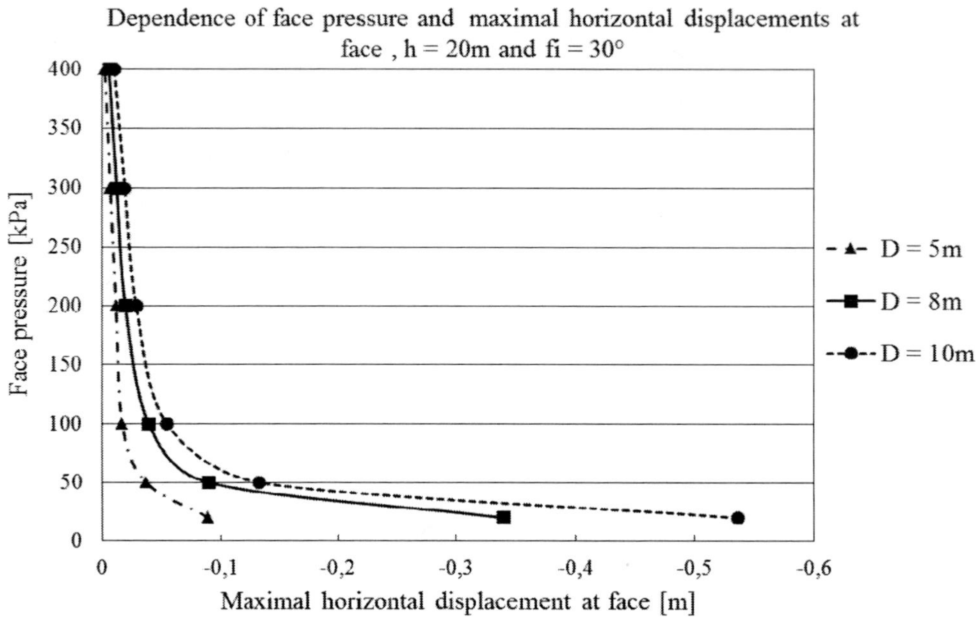

Figure 3. Dependence of face pressure and maximal horizontal displacements at face for various tunnel diameter.

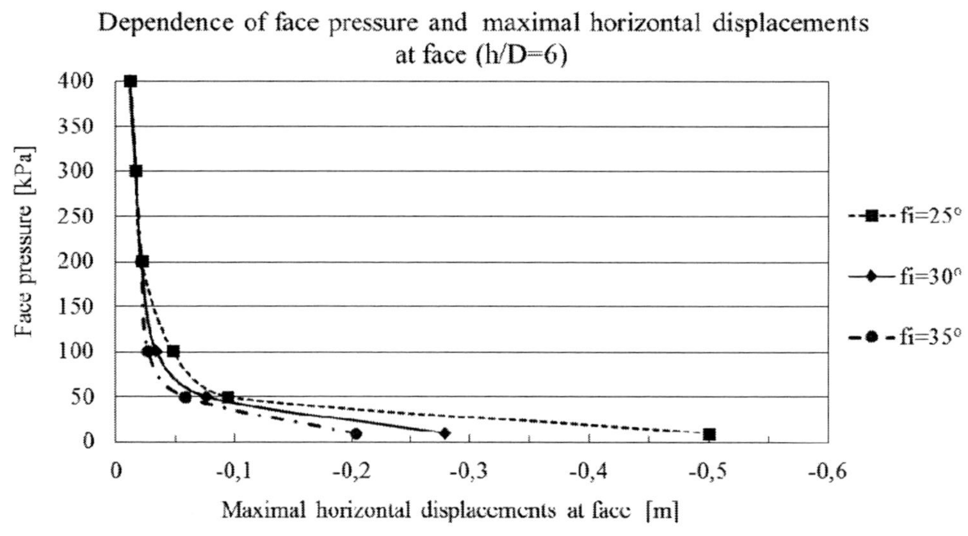

Figure 4. Dependence of face pressure and maximal horizontal displacements at face for various friction angle.

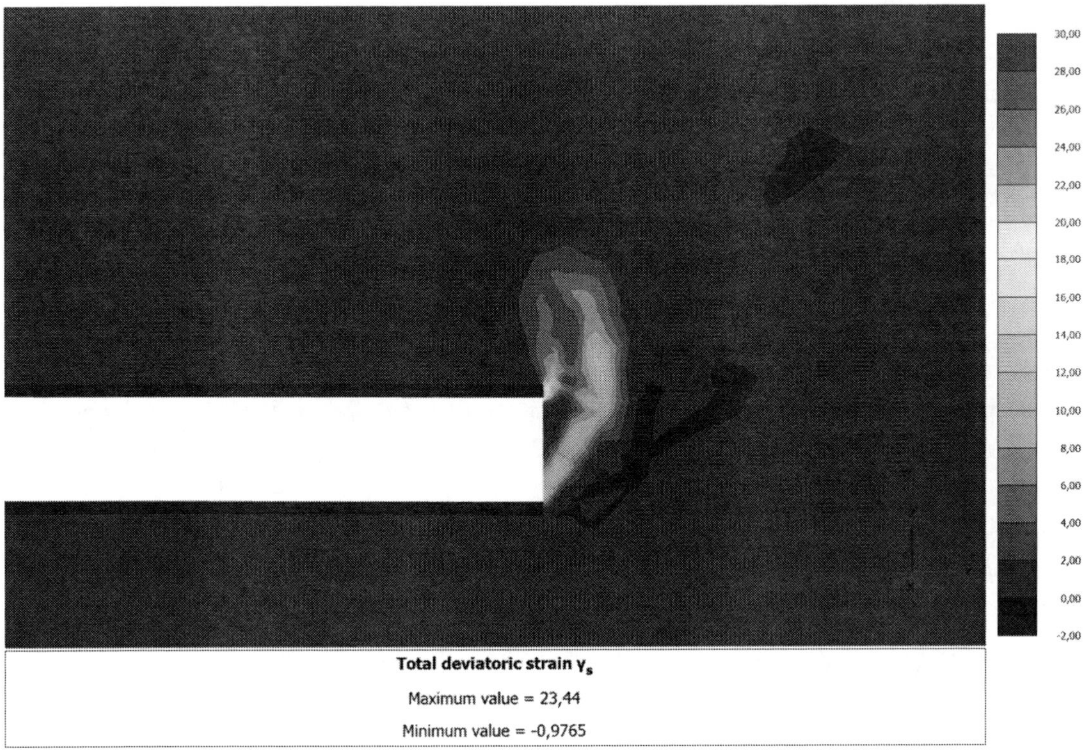

Figure 5. Pattern of total deviatoric strain in front of the face (PLAXIS).

5. Conclusion

The results of analytical computational models and numerical 3D model created in MIDAS GTS (finite element method) indicate minimum face pressure values during the excavation of the specified tunnel with circular cross-section excavated by full tunnel face. Using the finite element method for a depth of 30 m, the given model task results in 20% higher minimum face support pressure compared with the most conservative analytical Krause method (which, however, disregards tunnel depth). For the tunnel depth of 20 m, the values of the maximum face support pressures are comparable for the most conservative analytical method as well as the finite element method. For the tunnel depth of 10m, the value of the minimum face support pressure is smaller in comparison with the most conservative analytical method, but the value is greater than the value determined by the least conservative analytical method. The zone of maximum deviatoric strain, resulted from the 3D PLAXIS calculation, confirms the shape of Horn´s failure area in front of the face.

When formulating these conclusions, account must also be taken of the fact that the finite element method is a deformation method, and calculation thus depends not only on strength parameters of the soil environment, but also on deformation parameters (Young's modulus and Poisson's ratio), which are not even included in the analytical methods to determine the minimum support face pressure. However, although the quantitative values of face deformations (equivalent to horizontal displacements on the face) can be expected to naturally be different for different values of the Young's modulus of the soil environment, they would not be very different qualitatively. To verify these particular conclusions, it is necessary to conduct further, more extensive numerical analysis and comparison with a real face pressure during an actual tunnel excavation.

Acknowledgements
The paper was prepared with the support of the Competence Centers of the Technology Agency of the Czech Republic (TAČR) within the project Center for Effective and sustainable transport infrastructure (CESTI), project number TE01020168.

References
[1] Anagnostou G and Kovari K 1996 The face stability of slurry- shield driven tunnels *Tunneling and Underground Space Technology* **9**(2) pp 165-74
[2] Atkinson J H and Potts D M 1977 Stability of a shallow circular tunnel in cohesionless soil *Geotechnique* **27**(2) pp 203–15
[3] Broere W 2001 *Tunnel Face Stability & New CPT Applications* (TU Delft)
[4] Broms B B and Bennermark H 1967 Stability of clay at vertical openings *ASCE Journal of Soil Mechanics and Foundation Engineering Division* **93** pp 71-94
[5] Carranza-Torres C 2004 *Computation of factor of safety for shallow tunnels using Caquot's lower bound solution - Technical Report to Geodata S.p.A* (Itasca Consulting Group, Minneapolis)
[6] Davis E H 1980 The stability of shallow tunnels and underground openings in cohesive material *Geotechnique* **30**(4) pp 397-416
[7] Horn M 1961 Horizontaler Erddruck auf senkrechte Abschlussflächen von Tunneln /Horizontal earth pressure on vertical tunnel fronts/ *Landeskonferenz der ungarischen Tiefbauindustrie (National Conference of Hungarian Civil Engineering Industry*) (Translation into German by STUVA, Düsseldorf)
[8] Jancsecz S and Steiner W 1994 Face support for large mixshield in heterogeneous ground conditions *Proc. Tunnelling '94* (London: Chapman & Hall) pp 531-50
[9] Kamata H and Mashimo H 2003 Centrifuge model test of tunnel reinforcement by bolting *Tunnelling and Underground Space Technology* **18** pp 205-12
[10] Kimura T and Mair R J 1981 Centrifugal testing of model tunnles in soft clay *Proceedings of the 10th ICSMFE Vol 1* pp 319-22
[11] Krause T 1987 Schildvortrieb mit flüssigkeits- und erdgestützter Ortsbrust *Mitteilung des Instituts für Grundbau und Bodenmechanik* (Technische Universität Braunschweig)
[12] Leca E and Dormieux L 1990 Upper and lower bound solutions for the face stability of shallow circular tunnels in frictional material *Geotechnique* **40**(4) pp 581-606
[13] Mollon G, Dias D and Soubra A 2009 Two new limit analysis mechanisms for the computation of the collapse pressures of circular tunnels driven by a pressurized shield. *Proc. 2nd International Conference on Computational Methods in Tunnelling* (Ruhr University Bochum)
[14] Vermeer P A, Ruse N and Marcher T 2002 Tunnel heading stability in drained ground *Felsbau* **20**(6) pp 8-18
[15] Hrubešová E, Duriš L, Melicha A and Jaafar M 2017 Determination of the tunnel face stabilizing pressure using numerical and analytical methods *Advances and trends in Engineering Sciences and Technologies II* pp 767-72 (Taylor & Francis Group)

Ovčiarsko tunnel – interpretation of the results of the geotechnical monitoring

Jakub Ondráček[1]

[1]Jakub Ondráček, Mlýnská 425/70, Brno 602 00, Czech Republic

E-mail: jakub.ondracek@geotechnika.cz

Abstract. The Ovčiarsko twin-tube tunnel is part of the development of the Hričovské Podhradie – Lietavská Lúčka section of the D1 motorway, running southwest of the Žilina town. The D1 motorway overcomes the detached mass of an overthrust sheet and the western adge of the Paleogene fill of the Žilina Basin through the Ovčiarsko tunnel. The 2T – 8 category is designated for the Ovčiarsko tunnel with the total length of the northern tunnel tube 2360 m and southern tunnel tube 2367 m. the tunnel orientation to the cardinals is on the west-east axis, the marking of tunnel portals and tunnels tubes is distinguished according to it. The design speed is 100 km/h, respectively 90 km/h under worsened weather conditions. Parts of the tunnel are three cross passages passable for vehicles and five cross passages passable for pedestrians, serving as protected escape routes. This report describes geotechnical monitoring, construction technology and procedures with respect to complicated geological conditions. The geological conditions require the operative approach of the project owner, contractor, designer and geologist to the optimum solution for the running of the tunnel construction.

1. Introduction

Slovakia's location is important for all countries in Europe, especially as a transit country through which the lead one of the major pan-European corridors, Corridor V (Venice - Trieste / Koper - Ljubljana - Budapest - Uzhgorod - Lviv). The construction corridor in tunnels Ovčiarsko near the town of Žilina, on the D1 motorway in the section Hričovské Podhradie - Lietavská Lúčka. D1 is the most significant and longest Slovakian motorway, upon completion of which connects Bratislava with Záhor border crossing on the border with Ukraine.

The tunnel orientation to the cardinals is on the west-east axis, the marking of tunnel portals and tunnel tubes is distinguished according to it. The total length of the southern tunnel tube (JTR) on the axis is 2367 m, the mined part is 2320 m long (Figure 1). The route is on two reverse curves with the minimum radius of 739 m. The alignment rises from the western portal on a 1,70 % gradient. The total length of the northern tunnel tube (STR) on the axis is 2360 m, the mined part is 2300 m long. The alignment is again on two reverse curves, with the minimum radius of 777,5 m and rises on a 1,68 % gradient. The tunnel being proposed is of the 2T – 8.0 category with unidirectional traffic. The design speed is 100 km/h, respectively 90 km/h under worsened weather conditions. Parts of the tunnel are three cross passages passable for vehicles with the 3,5x3,6 m cross-sections and five cross passages passable for pedestrians with the 2,0x2,4 m cross-sections, serving as protected escape routes. Construction commencement is 1/2014 and completion 1/2018. Geological documentation and geotechnical monitoring of tunnels excavation provides SG Geotechnika, Inc. and GEOFOS, Ltd.

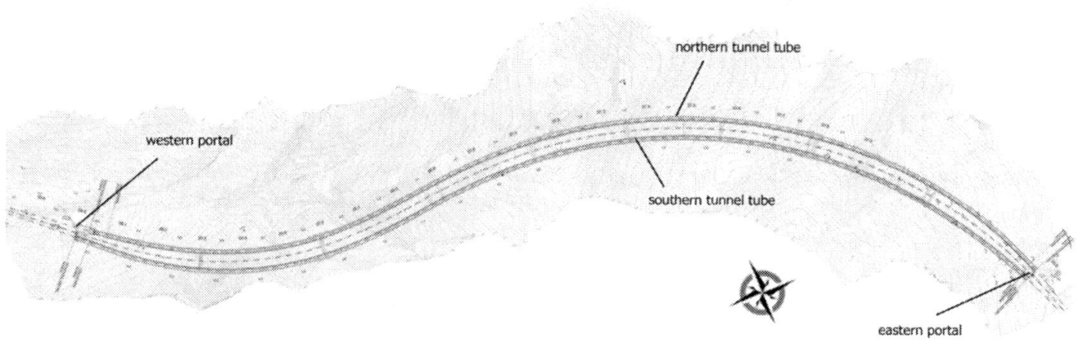

Figure 1. Alignment of the tunnel.

2. Engineering geological conditions

2.1. The geological conditions

The rock mass of the Ovčiarsko tunnel is formed and represented by two basic fault units (Figure 2):

The Manín unit – represented by Mesozoic flysch series of layers formed by series of layers of calcareous clay, marl slate, calcareous claystone, sandstone, and sandy limestone. The rock mass is very tectonically faulted and represents Outer Western Carpathians.

The Inner Carpathian Paleogene, which is represented by flysch series of layers of the Paleogene complex (Inner Western Carpathian). The Paleogene complex is represented by Hričov-Žilina Paleogene (Paleocene), which is represents blocks of limestone and formation of conglomerates, sandstone, and claystone. The younger Paleogene complex (Eocene) is forms by formation of claystone, sandstone, and conglomerate (Súlov and Domanín formation). The youngest complex (young Eocene) is represented by Žilina formation with typical claystone-sandstone rock mass with "colorful claystone" [1]

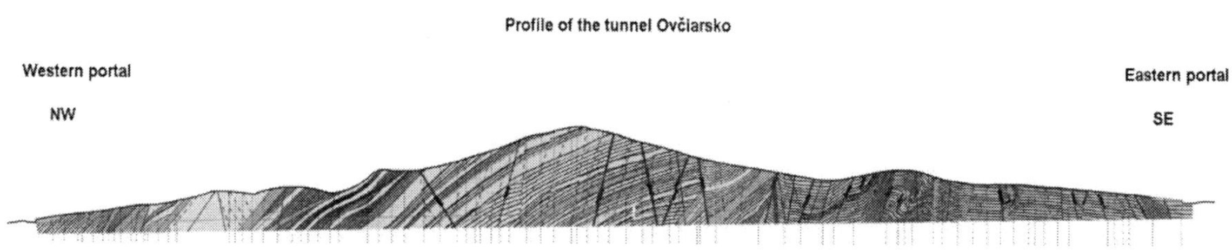

Legend: Quartery: deluvial caly; Mesozoic: marl and marlite, limestone and sandy limestone; Paleogene: sandy and colorful claystone, formation of conglomerate, sandstone and claystone, Súlov conglomerate, red claystone, sandy claystone; Tectonic: regional faults, sublocal faults, alpine tectonic faults

Figure 2. Geology profile of the Ovčiarsko tunnel [2].

2.2. Hydrogeological conditions

Hydrogeological conditions are subject of the climatic conditions together with geomorphological, drainage conditions, geological and tectonic structure, and other factors (reforestation, etc.). of these factors, it depends on the proportion of infiltrated rainfall. And also, its verified by exploration gallery, which is situated in the location of the northern tunnel tube, thus fulfilling the function of drainage and evacuating water in a gravity way toward the western portal, where it is continually pumped out. In the remaining section of the tunnel hydrogeological conditions were verified by drilling work, which determined groundwater levels. During the excavation from the western portal were larger inflows of groundwater linked to tectonic interfaces and transitions lithological units (usually claystone and conglomerates), but there are no significant groundwater flows into the tunnel driven from the western portal. However, there are significant groundwater flows into the tunnel driven from the eastern portal. The Paleogene complex (Žilina formation) make hydrogeological massif even more significant with the continuous hydrogeological collector layer type with circulation of groundwater mainly in zones of fracture and faulting. The existing drainage system was supplemented by additional horizontal drainage holes bored to the massif south of the southern tunnel tube in the water-bearing zones.

3. Tunnel design

Excavation of tunnel takes place on the bases of the principles of the New Austrian Tunnel Method (NATM), this is a cyclical way of excavation. The so-called "horizontal" excavation sequence design consists of the top heading, bench, and invert. The tunnel is driven from both portals; uphill from the western portal and downhill from the eastern portal. The excavation support consisting of two lining layers – the primary lining from shotcrete and the secondary lining from reinforced concrete. The intermediate waterproofing layer between the two liners is formed by geotextile and a plastic waterproofing membrane. The excavation operations are organised depending on the type of the rock mass, either by mechanical excavation (tunnel excavator) or using the drill and blast.

3.1. Primary tunnel lining

There are 16 excavation support classes which proposed for the excavation of the road tunnel. In comparison to the tender documents, excavation support classes VT4/2 was added for the excavation in the best geotechnical conditions (conglomerates and sandstone, figures 3). Excavation support classes (table 1) VT 7MP, VT 7/1, VT8/1, and VT8/1MP were proposed for the portal area of the portal and VT 6/5, VT 6/4, VT 6/2, VT 6/1, and VT 7/2 for the bad geotechnical conditions and for the fault zones, and VT 5/2 NZ and VT 6/2 NZ for the emergency bay. The following elements are used for restoring the ground mass stability disturbed by the excavation [3]:

- shotcrete – C 25/30 grade, with the strength gain on the J2 curve
- lattice girder is designed for the top heading and bench primary lining
- steel mesh with 6x6 /150x150, 8x8/150x150, 8x8/100x100 mm and steel grade B 500 B
- hydraulically expanded rockbolts 3 m and 4 m long – 150 kN
- grouted rockbolts anchors – Ø28 mm, the length of 3, 4 and 6 m – 250 kN, B500B
- self-drilling rockbolts R32; 4,6 and 8 m long– 280 kN
- self-drilling spiles 4 m long – R32, 9 m long – R51 and 4 m and 9 m long
- canopy tube pre-support from grouted steel tubes Ø 114/6.3, 18 m long
- cement grout – strength of 25 MPa, cement R32.5 or AM25

Figure 3. Excavation of the southern tunnel tube with excavation support class VT4/2 - massive conglomerates without tectonic deformation (Súlov formation).

Table 1. Geological conditions with excavation support class.

Excavation support class	Value of RMR/quality of the rock mass	Geological conditions
8MP	0-20/very poor	Formation of the heavy weathering claystone (eastern portal)
7MP	0-20/very poor	Formation of the heavy weathering siltstone, claystone, and sandstone (western portal)
7/2	0-20/very poor	Weakly to moderate weathering siltstone and claystone
6/5	24-30/poor	Calcareous claystone with sandy limestone, moderate to heavy weathering
6/4	28-35/poor	Sandy limestone, claystone, and siltstone, weakly to moderate weathering
6/2	33-40/poor	Formation of claystone, siltstone, and sandstone (calcareous sandstone – sandy limestone), weakly to moderate weathering
5/2	Min. 45/fair	Conglomerates, sandstones, siltstones, claystone (flysch series), healthy to weakly weathering
4/2	≥50/fair	Massive conglomerates with sandstone and claystone, healthy to weakly weathering

3.2. Secondary tunnel lining

The secondary tunnel lining is designed as a 0.3 m thick reinforced concrete structure from C30/37-XF4, XC3 concrete (up to the distance of 150 m from portals) and C25/30-XF2, XC3 concrete (in the internal tunnel section). Concrete is to be cast behind a movable formwork. The length of the expansion blocks is 12.5 m. The lining is reinforced with steel mesh, bar reinforcement and auxiliary lattice girders used for the realisation of the reinforcement of individual blocks of the secondary lining. B500B steel is used for all reinforcement of the secondary lining. The installation of the secondary lining is possible only after the primary lining deformations stabilise.

3.3. Tunnel waterproofing

The tunnel tube lining system is designed as a sandwich structure consisting of a primary lining and secondary lining. A drainage and waterproofing layer consisting of min. 900 g/m^2 geotextile and min.

2 mm thick waterproofing membrane is inserted between the two liners. The umbrella-type waterproofing system is designed for the tunnel. The invert is not provided with waterproofing. The drainage layer of the waterproofing evacuates groundwater to the DN 200 mm side drainage system. Drainage water is directed from cleaning niches to a collector sewer located under the left-hand traffic lane.

4. Geotechnical monitoring

Principles of the NATM consist in controlling of the deformation process surrounding the excavation, which include geotechnical documentation of the rock mass, lining deformations and extensometer measurement, measurement of stress in primary lining and contact primary lining-ground, measurement of water outflow from the exploratory gallery, physical-chemical analyses of water, supervision of warning states (value "A") and hydromonitoring (measurement of groundwater level). And, there is monitoring in the area of the portals. The following measurement activities are performed in the area. 3D measurement of deformations, measurement of stress in anchors, automatic surveying system on the eastern portal face, inclinometer measurements and inclinometer deformation measurements.

There are 20 (STR) and 14 (JTR) geotechnical block that are characteristic with the following principles. On the bases of the RMR classification [4], excavation support class and geological with tectonic situation of the rock mass and geotechnical situation.

We can divide the geotechnical blocks to the three elementary parts on the bases of the rock mass age and tectonic deformation (section with significant and local deformation and section without deformation). The first section is a mass of Mesozoic rock (claystone, limestone, and sandstone) with very tectonic deformation and steep layers. General course of the layers is NE-SW (orientation of plunge is to NW with 45-60 °) and tectonic is NE-SW and SE-NW (with orientation to NW and NE with 60-80 °). The next section is Paleogene complex of claystone, sandstone, and conglomerates. There is section with or without tectonic deformation and the best of geotechnical block (excavation support class 4/2) for the conglomerates (Súlov rock formation) with the length of the block 147.3 m. The general course of the layers and faults is almost same as previous section. Formation of folding claystone and sandstone (there are a minimum of sandstone, figures 4) represents Paleogene complex of Žilina formation. The orientation of the layers is NE-SW (with orientation to NW and 10-30 °) and faults with NW-SE trend (SW and NE plunge with 60-80 °).

Figure 4. Paleogene complex (Žilina formation) of subhorizontal claystone, siltstone, and sandstone layers.

The value of the supervision of warning states are high level of safety, allowable state changes and critical condition (Critical state was achieved exceeding 125 % of the limit value "A".)

Extensometer measurements are used to analyse the behaviour of the rock mass around the excavation interact with the primary lining. Extensometer is measured relative deformation between the anchor point and head extensometer. Extensometer are designed with a rod of glass fibber guided in a protective tube, especially for reading mechanical and hydraulic anchor point. In each main measurement profile of the excavation is build threesome radially arranged three-stage extensometers with Tread depth 3.6 and 9 m.

Measurement of stress on the primary lining-ground contact using pressure measurement boxes. The purpose is to determine the load bearing capacity of the primary lining deforming rock mass. In each main profile measurement needs to be installed five pressure measurement boxes distributed around the perimeter of the excavation. The principle of measure of stress in the primary lining using the same method as previously measurement.

Tension in the secondary lining is measured indirectly by detecting the vibration of strain using tensiometer mounted on the rod of concrete reinforcement at the outer and inner surfaces. Vibration tensiometer consist of two parts a steel wire stretched between two retaining blocks and measuring the drive coil. Deformation of the structure will produce mutual shift the retained strain gauge blocks, changing the length respectively. tension steel wire, which is measured as the natural frequency of oscillation of the excitation. Part of each tensiometer is a thermometer. Vibration tensiometer is installed in the main measurement profile (10 respectively 12 vibration Sensors).

Lining deformation measurement is carried out in three dimensions by routine geodetic surveying using total stations. Optical reflector targets are installed at sections along the tunnel (every 10 to 20 meters). There are typically 5 reflectors, which are installed in each section: upper vault (crown), mid vault, and at the wall sides.

Table 2. Limits of the lining deformation measurement.

Excavation support class	Top heading deformation (mm)
8MP	130
7MP	130
7/2	120
6/5	80
6/4	60
6/2	35
6/1	30
5/2	20
4/2	20

5. Conclusion and discussion

The tunnel excavation was completed in August 2016 and now work is continuing the secondary lining. Geotechnical monitoring of Ovčiarsko tunnel was continuously carried out under the approved project geotechnical monitoring. The basic objective of the monitoring was to compare the identified engineering geological and geotechnical conditions with the assumptions of standard project.

Based on continuous geological documentation of the individual advance, the engineering geological and geotechnical characteristics of the massif were evaluated for the determination of the excavation support class. The most important criterion for determining the equipment class was the point assessment of the rock mass according to the classification system RMR and QTS, which are based on parameters such as strength of intact rock material, drill core quality (RQD index), spacing and condition of discontinuities, ground water, etc.

The observed geotechnical conditions in the tunnel were almost the same as the predicted geological conditions in the presupposition. The main differences were in percentage of the support classes, percentage of the clay, sandstones and conglomerates and groundwater flows into the tunnel. During the excavation of the tunnel, generally better excavation support class were defined than assumed. The excavation was mostly done in the support class 4/2 (almost 45 %) with massive conglomerates and sandstones.

The most unfavourable excavation conditions were in portal areas (highly weathered and decomposed rock mass) and chainage 432.83 to 527.38 tunnel meter of the northern tunnel tube (significant representation of tectonically damaged rocks and very favourable discontinuity orientation). The tectonic zones were mostly bound to the intersection of the sandstone (conglomerates) and claystones with tributaries of groundwater into the tunnel tube (water inflows to 0.1 l/s).

The overall hydrogeological assessment of portals and tunnel routes was favourable. There was no correlation or influence of the underground water of the basal conglomerates on a spatially extensive hydrogeological structure with a great water-management significance between near villages. The inflows of groundwater complicated the excavation through a change in the geological characteristics (tectonic zones and lithostratigraphic layer). The most unfavourable hydrogeological conditions were in the eastern portal. In the sedimentary sediments, accumulated water, which negatively influences the geotechnical properties of the rocks. Artificial drainage improved the situation.

The maximum of top heading deformation in the western portal were 4.1-35.9 mm (STR, VT 4/2-5/2), 6.0-29.7 mm (JTR, VT 4/2-5/2) and 9.6-94.7 mm (STR VT 6/2), 17.1-104.5 mm (JTR, VT 7MP-8/1) in the eastern portal. And maximum of lining deformation was 5.1-46.8 mm (STR), 3.8-26.0 mm (JTR) in the wester portal and 2.9-32.4 mm (STR), 10.8-87.4 mm (JTR) in the eastern area. Now all deformations are in the state of permissible changes.

As the results of the geotechnical monitoring show, the excavation in the southern tunnel tube from the eastern portal was very complicated (weather and tectonic condition, very unfavourable of discontinuity orientation and presence of groundwater). Whole area of the eastern area represented very unfavourable geotechnical conditions.

Laboratory testing of the properties of the rock material has shown its extreme heterogeneity and the resulting complications in its construction use. The highest quality material (conglomerates, which consist predominantly of dolomite with a small proportion of other minerals). Use them for more demanding construction purposes (concrete or underlying road surface) is limited by the low frost resistance of this material. Flysch rocks are hampered by the frequent and irregular layers of conglomerates, sandstones and claystones. It is possible to characterize the materials as suitable for use in the lower and central portions of the backfill, respectively as a lesser quality part of the sandwich type.

Despite the bad geotechnical conditions during the excavation, the tunnel will be opened in the assumed term in the first quarter of 2018, the Ovčiarsko tunnel will become part of the traffic junction on the D1 motorway and will significantly contribute to the improvement of the traffic situation in Žilina region.

References

[1] Mahel' M 1980 Pribradlové pásmo, charakteristika a význam *Mineralia* **12** pp 193-207
[2] Matějček A, Mitter P and Panek M 2014 Inženierskogeologický profil 1-1´masívom prieskumnej štôlne (Žilina) *Dokumentácia pre stavebné povolenie*
[3] Zwilling R et al 2014 Dokumentácia pre realizáciu stavby: Tunel Ovčiarsko. Část' primárne ostenie (Bratislava-Basler & Hofmann Slovakia s. r. o.)
[4] Bieniawski Z T 1989 *Engineering rock mass classifications: a complete manual for engineers and geologists in mining, civil and petroleum engineering* (New York: Wiley-Interscience)

IOP Conf. Series: Materials Science and Engineering **236** (2017) 012076 doi:10.1088/1757-899X/236/1/012076

Hierarchical stochastic model of terrain subsidence during tunnel excavation

Tomáš Janda[1], Jiří Šejnoha[1] and Michal Šejnoha[1]

[1]Czech Technical University in Prague, Faculty of Civil Engineering, Department of Mechanics, Thákurova 7/2077, 166 29 Praha 6, Czech Republic

E-mail: tomas.janda@fsv.cvut.cz

Abstract. In this contribution the Bayesian statistical method is applied to assess the expected probability distribution of the terrain subsidence in the course of tunnel excavation. The approach utilizes a number of simplifying assumptions regarding the system kinematics to arrive at a very simple model with just a few degrees of freedom. This deterministic model together with the intrinsic uncertainties of its parameters and measurement inaccuracies are used to formulate the stochastic model which defines a distribution of the predicted values of terrain subsidence. Assuming the measured data to be fixed, the stochastic model thus defines the likelihood function of the model parameters which is directly used for updating their prior distribution. This way the model parameters can be incrementally updated with each excavation step and the prediction of the model refined.

1. Introduction

Bayesian statistics and especially Bayesian parameter inference is a powerful tool for assessing uncertainty of complex systems. To understand its concepts we need to change our perception of probability. While in case of the observable quantity the probability density expresses how often the given value occurs (frequentist interpretation of probability), in case of unobserved quantity the probability density expresses how much we believe in that particular value (Bayesian interpretation of probability). Although the difference may seem subtle, it allows us to attribute the probability not only to measured data but also to model parameters.

Then, if there is a stochastic relationship between the unobserved quantities and observed data we can reappraise our prior knowledge about the unobserved quantities, i.e. their prior probability densities, and obtain a rationally updated knowledge – their posterior probability density.

The main advantage of this approach is its mathematical soundness: if our model (stochastic relations) is correct then the posterior belief is the most rational one given the prior belief and the observed data. On the other hand, one must be prepared that the integral statistics of posterior distribution such as the mean value, standard deviation or quantiles can be computed analytically only for a limited category of models. Posterior distribution obtained with models of an arbitrary structure has to be analyzed numerically via Markov chain Monte Carlo methods such as the Metropolis–Hasting algorithm, Gibbs sampling or the Hamiltonian Monte Carlo algorithm. Since these algorithms are implemented in several statistical programs we can concentrate only on the definition of the stochastic model and the prior distribution of its parameters.

This contribution presents a simple stochastic model including a simplified deterministic description of the evolution of settlement during tunnel construction as well as various input and measurement uncertainties. The parameters are updated based on data from monitoring in one tunnel cross-section. Then the settlement in the other cross-section is predicted both with and without taking the previously observed data into account.

2. Deterministic model of sequential excavation of driven tunnel

In a typical engineering application the stochastic approach usually involves a deterministic model of the idealized reality. In our case the model predicts the evolution of displacements in the soil body induced by successive excavation of a driven tunnel. The model is founded on the assumption that the displacement field and the subsidence trough can be approximated with rather simple shape functions controlled by a small number of parameters. The complete formulation of the model goes beyond the scope of this paper so that only the basic model principles are provided. Further details can be found in [1].

The model predicts the increment of vertical displacements following a single excavation stroke of length b as shown in Figure 1. Further we assume that the increment of displacements in segment II will vanish with the distance, i.e. the shape function $f(z)$ decreases in segments I and III, see Figure 2. The distribution of the displacement increment in the $x \times y$ plane is controlled by similar shape functions displaced in Figures 3, 4 and 5.

As has been shown in [1] the spacial distribution of the vertical displacement $v(x, y, z)$ is completely defined by five model degrees of freedom (equivalent to nodal displacements in standard FEM formulation, see a considerably more complex approach taking also into account the interaction of upper and underground structure [2]) V_1, V_2, V_3, ΔV_1 and ΔV_2, see Figure 5, derived as a sum of an infinite series of their incremental values. The proposed kinematic relations then enable formulating the overall stiffness matrix and the load vector through the standard application of the principal of virtual work. The load vector due to the excavation forces is computed from the known geostatic stress.

As displayed in Figure 3, the model includes three layers of homogeneous material. If the real geological profile consists of more layers, the Young modulus in the particular model layer is computed as an average of the values in the profile sub-layers weighted by their thicknesses.

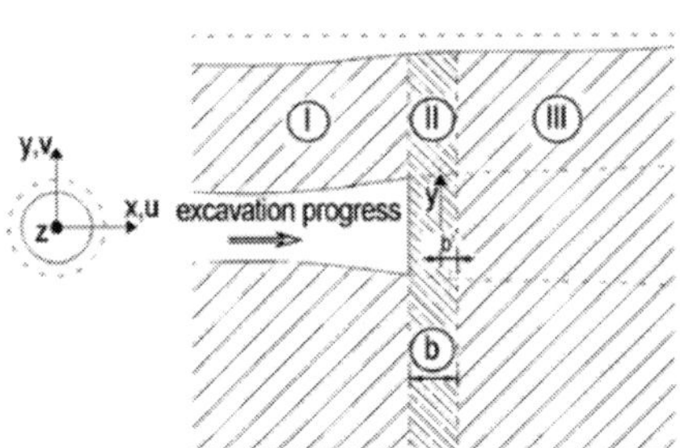

Figure 1. The model is divided to three segments in the longitudinal direction. The increment of the vertical displacement induced by the excavation of the segment II is computed at any point.

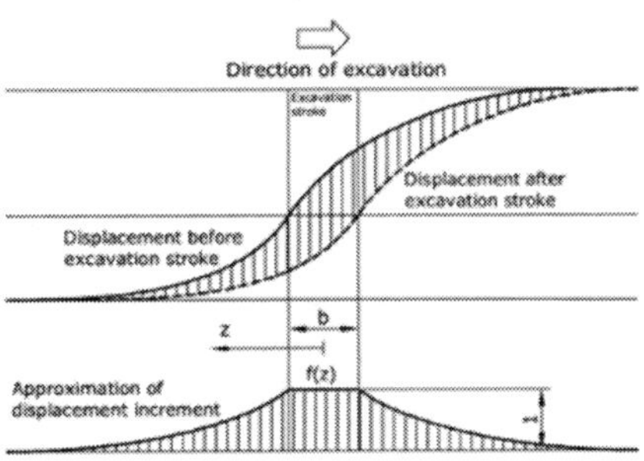

Figure 2. Shape function $f(z)$ defines how the displacement increment evolves in the longitudinal direction. The rate of decrease of the function $f(z)$ is controlled by parameter α_z.

From the perspective of stochastic modeling, the deterministic model described above is abstracted as a black-box function taking the material parameters (Young modulus, Poisson ratio and density in each model layer), geometry (thicknesses of the model layers, the tunnel radius and the length of excavation step) and shape function parameters as an input and projecting them to the displacement

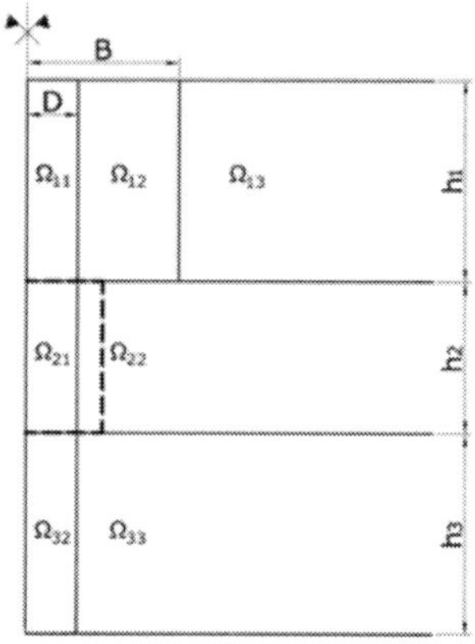

Figure 3. In $x \times y$ plane the model is divided into several domains each governed by convenient shape function.

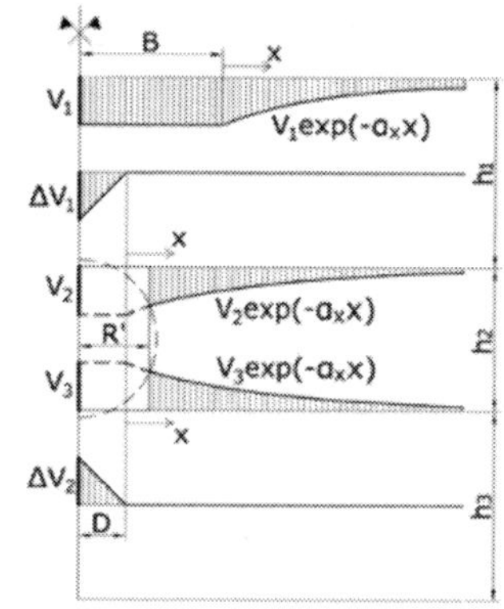

Figure 4. Shape functions in the x direction. The parameter α_x defines the rate of decrease of the displacements.

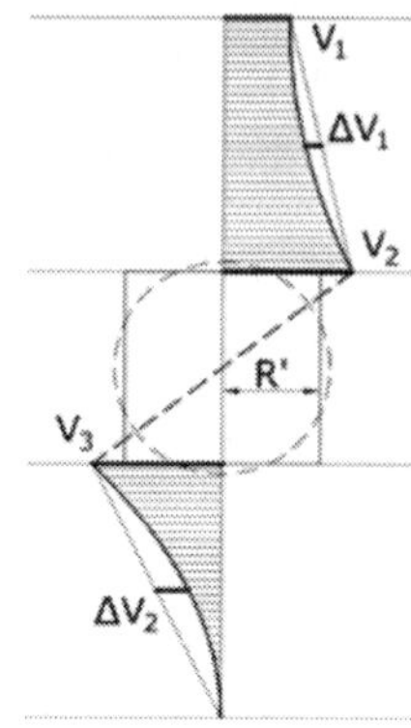

Figure 5. The approximation in the vertical direction is controlled by auxiliary quantities ΔV_1 and ΔV_2.

field $v(\boldsymbol{x}) = v(x, y, z)$, where $\boldsymbol{x}$ is measured relatively to the position of the tunnel face, see Fig 1. For our particular case, this can be schematically written as

$$\left.\begin{array}{ll} \text{Soil profile:} & \langle E_i, \nu_i, \rho_i, t_i \rangle, \quad i = 1, 2, 3 \\ \text{Geometry:} & h_1, h_2, h_3, R', b \\ \text{Shape parameters:} & \alpha_x, \alpha_z, D, B \end{array}\right\} \xrightarrow{\text{Model}} v(\boldsymbol{x}). \tag{1}$$

For a specific tunnel with a given geometry and a given soil profile some of the input parameters are known while the other might remain unknown or at least uncertain. The choice which parameters are known (constant) and which uncertain (assigned suitable prior distribution) is solely a matter of engineering judgment. In our practical example we limit our attention to five parameters only collected in vector $\boldsymbol{b} = \{E_1, E_2, E_3, \alpha_x, \alpha_z\}^{\mathsf{T}}$. The remaining parameters are assumed to be known. The deterministic mapping function that projects the the vector of unknown variables on the vertical displacement then simplifies as above schema. For a

$$\boldsymbol{b} \xrightarrow{g_{\text{model}}} v(\boldsymbol{x}). \tag{2}$$

In such a case uncertainty of the displacement field is attributed to uncertainty of Young's moduli and the shape function parameters α_x and α_z. The details on how the model is used are given in Section 4.

3. Stochastic model of measured data

The above deterministic function is part of the stochastic model which takes uncertainty of the model parameters and the measurement error into account. According to the Bayes theorem the posterior joint probability distribution of the model parameters is proportional to the product of likelihood function and the prior joint probability distribution

$$p(\boldsymbol{\theta}|\boldsymbol{y}) \propto p(\boldsymbol{y}|\boldsymbol{\theta})p(\boldsymbol{\theta}) \tag{3}$$

where vector $\boldsymbol{\theta}$ denotes parameters of the stochastic model and $\boldsymbol{y}$ represents the vector of observed data. Note, that the parameters $\boldsymbol{\theta}$ are not directly observable, whereas the data $\boldsymbol{y}$ can be measured. In fact, in the above relation the data $\boldsymbol{y}$ are assumed to be already known and therefore fixed leaving both sides of Eq. (3) independent of $\boldsymbol{y}$.

Let's suppose that the vertical displacement in the vicinity of the tunnel tube will be measured at n points with known coordinates (x_j, y_j, z_j) where $j = 1 \dots n$. We denote the measured values as v_j and express our belief about probability distributions by the following set of relations

$$v_j \sim \mathcal{N}(\overline{v}_j, \sigma_v), \tag{4}$$

$$\overline{v}_j = g_{\text{model}}(\boldsymbol{b}(\boldsymbol{x}_j)) \tag{5}$$

$$E_i \sim \mathcal{U}(E_{i,min}, E_{i,max}), \quad i = 1, 2, 3 \tag{6}$$

$$\alpha_x \sim \mathcal{U}(\alpha_{x,min}, \alpha_{x,max}), \tag{7}$$

$$\alpha_z \sim \mathcal{U}(\alpha_{z,min}, \alpha_{z,max}), \tag{8}$$

$$\sigma_v \sim \mathcal{U}(\sigma_{v,min}, \sigma_{v,max}). \tag{9}$$

Relation (4) implies that the measured displacements v_j are normally distributed around the theoretical mean value $\overline{v}_j$ with standard deviation σ_v. This way the measurement error and the modeling uncertainty are jointly covered by a simple stochastic relation. The theoretical mean values $\overline{v}_j$ are a direct outcome of the deterministic model g_{model} as prescribed by (5). These two relations form the model likelihood function. Finally, relations (6)–(9) define the prior probability distributions of the top-level model parameters. In particular, the parameters $E_1, E_2, E_3, \alpha_x, \alpha_z$ and σ_v are described by uniform prior distributions with specified limits.

4. Example and results

Two cross-sections of the road tunnel Blanka of average radius $R = 5$ m constructed in Prague are chosen to provide real-world data for the stochastic model. Before the construction of the tunnel a borehole was drilled in each monitored section. The soil profile has been documented and the borehole was equipped with extensometers to allow for monitoring of the vertical displacement developed during the tunnel construction. A simplified stratigraphy of the two monitored sections denoted as J22 and J24, 101 m apart, is provided in Tables 1 and 2. The ranges of the Young modulus were adopted from the ČSN 73 1001 standard for each of the soil class. The stroke size b was set equal to 2 m. The remaining model parameters equal to $D = 4$ m and $B = 10$ m were derived from an identification procedure outlined in [1].

The displacement evolution monitored only at extensometer J22 on terrain and just above the tunnel crown, i.e. at a depth of 15 m, were used in the inference procedure. In particular, the measurements of vertical displacements from seven different positions of the tunnel face relative to the monitored section were adopted. As shown in Figure 6 these were taken from the interval -82 m where the tunnel face has not yet reached the monitored profile to 85 m where the tunnel face has already passed the extensometer. The total number of measurements in Eq. (5) thus amounted to $n = 14$.

The ranges of the uniform prior distribution for the Young moduli in the soil layers were taken from Table 1 and the parameters α_x and α_z were assumed to be uniformly distributed between 0.01 and 1. The prior distribution for the standard deviation of the measurement error σ_v was also uniform with ranges from 0.1 to 5 mm.

Table 1. Soil profile in borehole (J22).

i	Depth [m]	Soil type	E_i [MPa]	ν_i [–]
1	0–4.6	Clay (F3)	3–15	0.35
2	4.6–7	Gravel (G3)	80–100	0.25
3	7–20	Slate (R3)	60–300	0.2

Table 2. Soil profile in the borehole (J24).

i	Depth [m]	Soil type	E_i [MPa]	ν_i [–]
1	0–5.5	Clay (F3)	3–15	0.35
2	5.5–8.8	Gravel (G3)	80–100	0.25
3	8.8–18.5	Slate (R4)	40–120	0.25

The hierarchical model given by equations (4)–(9) was specified in a dialect of the BUGS language [3] and run in JAGS program [4]. Due to the limited range of operators and functions available in the JAGS

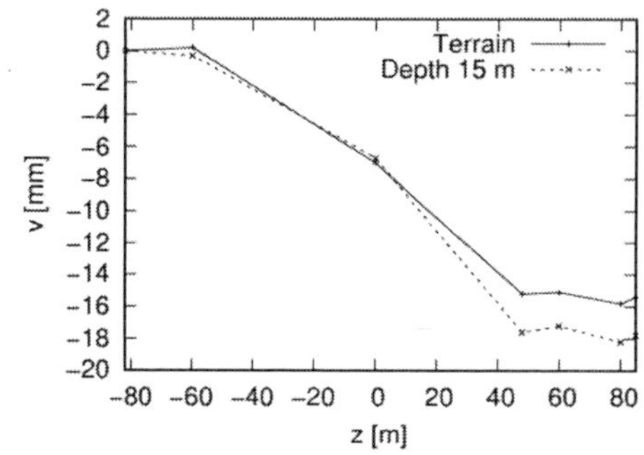

Figure 6. Development of the measured vertical displacement v on terrain and at the depth of 15 m with the position of the tunnel face z.

program the formulation of the deterministic function g_{model} needs to be approximated. The Taylor series in the form

$$\begin{aligned}\overline{v}_j &= g_j(\boldsymbol{b}(\boldsymbol{x_j})) \approx g_j(\overline{\boldsymbol{b}}) + \sum_{r=1}^{5} \frac{\partial g_j}{\partial b_r}|_{\boldsymbol{b}=\overline{\boldsymbol{b}}}(b_r - \overline{b}_r) \\ &+ \frac{1}{2}\sum_{r=1}^{5}\sum_{s=1}^{n} \frac{\partial^2 g_j}{\partial b_r \partial b_s}|_{\boldsymbol{b}=\overline{\boldsymbol{b}}}(b_r - \overline{b}_r)(b_s - \overline{b}_s) + \ldots \qquad (10)\end{aligned}$$

is a suitable tool for such approach. In what follows, the model is further reduced to a constant and the linear term approximating $g_{\text{model}}(E_1, E_2, E_3, \alpha_x, \alpha_z)$ at $\overline{E}_1$ = 8 MPa, $\overline{E}_2$ = 90 MPa, $\overline{E}_3$ = 120 MPa, $\overline{\alpha}_x$ = 0.1 and $\overline{\alpha}_z$ = 0.1.

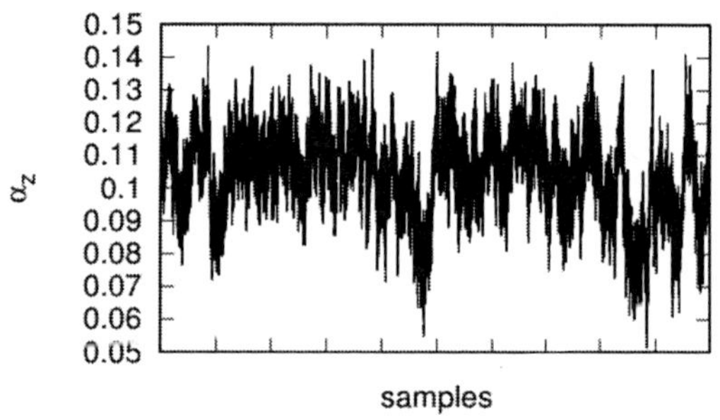

Figure 7. Markov chain of the generated samples of the parameter α_z.

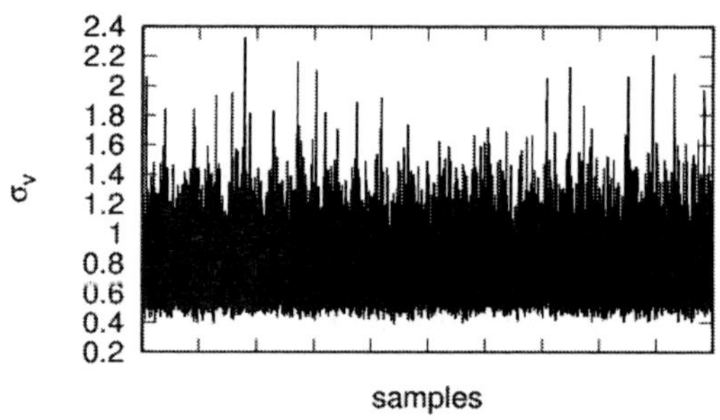

Figure 8. Markov chain of the generated samples of the parameter σ_v.

The program uses Gibbs sampling to generate the samples of the model parameters according to the joint posterior distribution. The burn-in period of 1000 samples, which were discarded, was followed by generating two chains of 50000 samples. For illustration, the Markov chains for parameters α_z and σ_v are displayed in Figures 7 and 8. Both chains show sufficient mixing of the generated samples indicating proper settings of the Gibbs sampling algorithm. Therefore, the generated histograms should converge to the theoretical continuous distributions.

The histograms computed from the generated Markov chains of the top-level model parameters are displayed in Figures 9–14 along with their prior uniform distributions. The differences of the prior (uniform) distributions and the posterior distributions (visualized in form of histograms) tell us how the

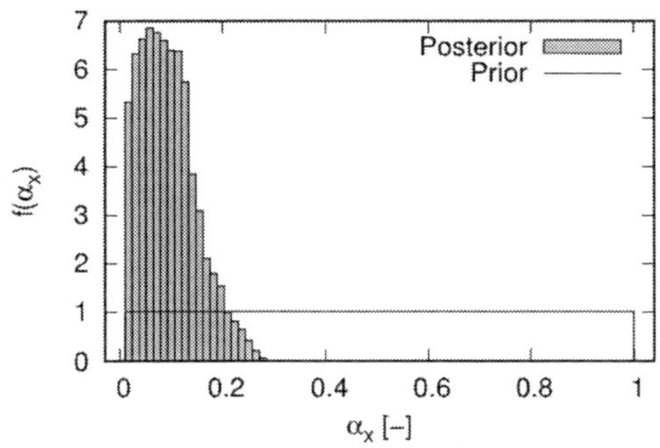

Figure 9. Prior and posterior distribution of parameter α_x.

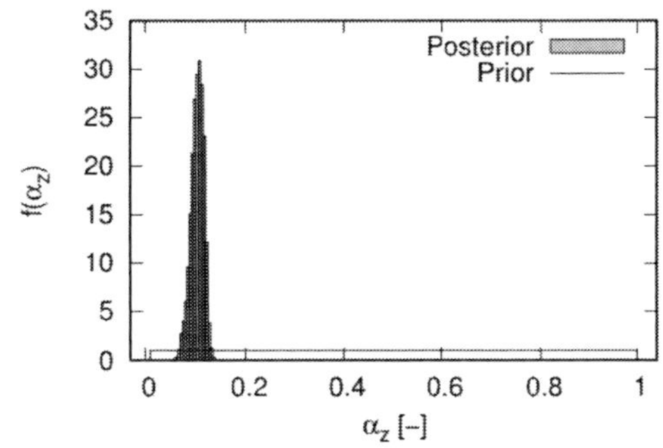

Figure 10. Prior and posterior distribution of parameter α_z.

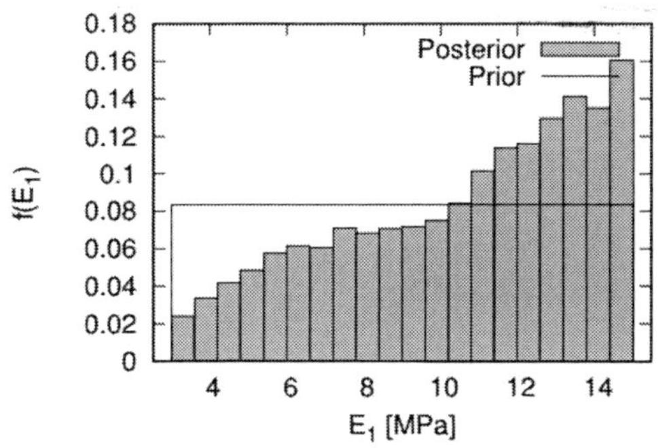

Figure 11. Prior and posterior distribution of the Young modulus of clay layer E_1.

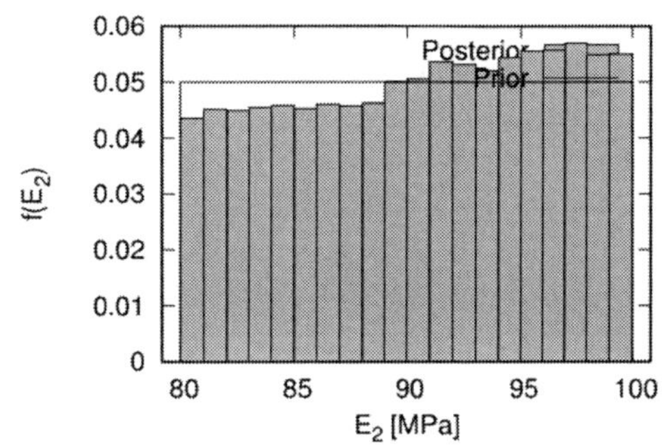

Figure 12. Prior and posterior distribution of the Young modulus of gravel layer E_2.

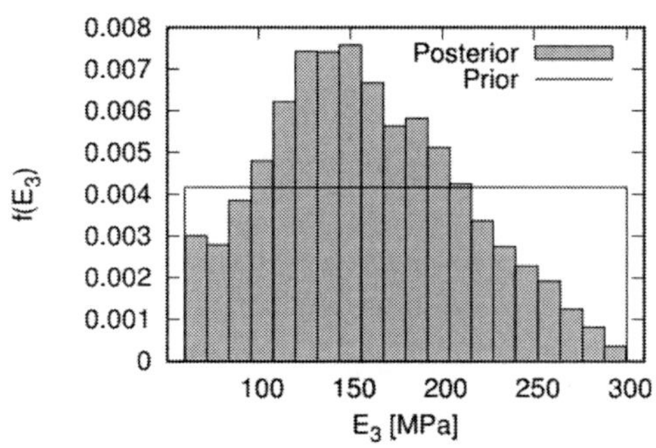

Figure 13. Prior and posterior distribution of the Young modulus of rock layer E_3.

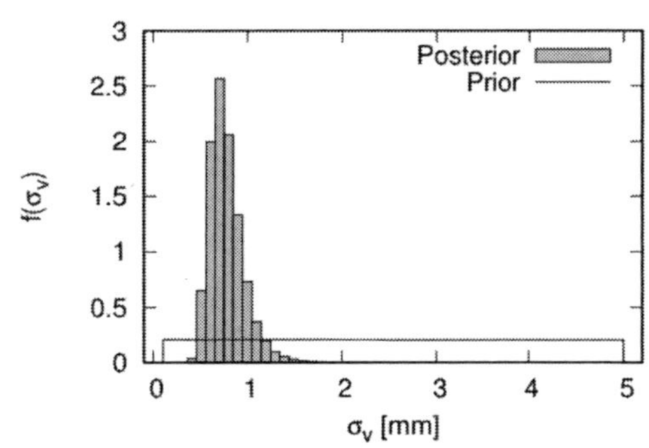

Figure 14. Prior and posterior distribution of the standard deviation of measurement error of vertical settlement σ_v.

prior knowledge about the value of the parameter changed after the data were observed. These graphs suggest some partial conclusions. Clearly, the algorithm managed to identify the shape parameters α_x, α_z, Figures 9 and 10, together with the measurement error in Figure 14 relatively well. It also appears, perhaps even intuitively expected, that the direct impact of the gradually excavated mid section on the settlement measured at the selected points is not significant. Additionally, the results plotted in Figure 14 may draw one to the conclusion that the original soil classification might be erroneous.

This also fully uncovers the advantage of the Bayesian inference when adopted full scale analysis incorporating in-situ measurements rather then just measurements from small scale experiments.

The principal objective of the application of the Bayesian inference is to improve original predictions given new (updated) information about the model parameters. In this paper, this is demonstrated by providing predictions of the settlements measured at a different section than was the one exploited in the updating process. The predicted settlements were found assuming:

(i) A vague knowledge of model parameters specified by their prior distribution with no knowledge of their potential statistical dependence.

(ii) Improved (updated) knowledge of model parameters represented by their joint probability distribution function available from the already generated Markov chain.

In the fist case variability of the settlement was predicted using simple Monte Carlo method with individual outcomes of vector $\boldsymbol{b}$ sampled from previously adopted statistically independent uniform prior distributions. In the second case, these outcomes were adopted directly from the Markov chains of parameters obtained within the inference procedure. More specifically, only the updated parameters E_1, E_2, α_x, α_z and σ_v were used in the prediction. The updated distribution of the parameter E_3 was not used here as it represented the Young modulus of the rock layer of class R3 in section J22 while the rock layer in J24 section is classified as R4, i.e. more fractured. Due to this subtlety, a separate prior uniform distribution with the range 40–120 MPa was considered for the Young modulus of the rock layer in section J24. The range was again taken from the ČSN 73 1001 standard. Note that the measured settlements from this cross-section were not used within the updating procedure.

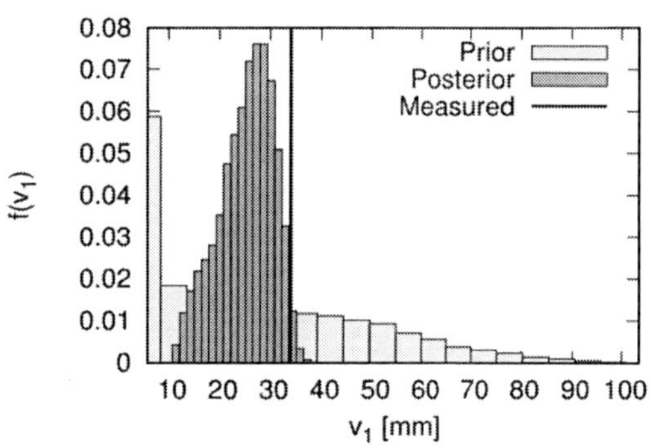

Figure 15. Prior predictive and posterior predictive distribution of the final vertical displacement at extensometer J24 at the terrain.

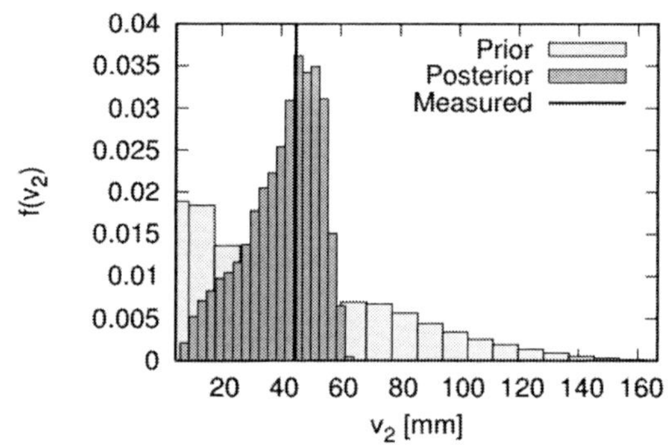

Figure 16. Prior predictive and posterior predictive distribution of the final vertical displacement at extensometer J24 at a depth of 15 m.

The settlements derived from prior distributions (Prior) and the posterior joint probability distribution (Posterior) in section J24 are shown in Figures 15 and 16. They correspond to the tunnel face being at a distance of 74 m passed section J24. Apart form the distributions the figures show also the values of the actual measured vertical displacements. The 98% confidence interval computed from the posterior predictive distribution (approach ii) of the total settlement on the terrain is 12.6 mm – 34.6 mm, and 10.3 mm – 60.1 mm for the displacement at a depth of 15 m, respectively. This is a more useful piece of information when compared to the significantly wider confidence interval for prior predictive distributions (approach i) that are 4.4 mm – 85.1 mm for terrain, and 1.9 mm – 134.4 mm for the displacement at 15 m depth, respectively. Both measured values, i.e. 34 mm at the terrain, and 45 mm at the depth of 15 m, respectively, fit into the prior as well as the posterior predicted confidence intervals.

5. Conclusions

The presented analysis demonstrated that the originally quite wide range of the admissible values of the model parameters is significantly narrowed when the observed data are taken into account. Moreover, the updating procedure reflects the structure of the stochastic model. Due to the model structure some of the parameters are refined substantially while the distribution of the other parameters may remain practically unchanged. This is natural because some of the parameters do not influence the type of the observed data and thus the knowledge of the data does not refine their prior distribution. The updated set of parameters was used in the stochastic model for the prediction of the yet unobserved data, i.e. the settlements evolution in a particular tunnel cross-section. Recall that the purpose of the stochastic model is not to predict the settlements with a maximum accuracy but to predict its probability distribution. We are generally interested in the distribution of the predicted data, i.e. their confidence interval and not the single most probable value.

Nevertheless, the predictions obtained by the stochastic model outlined in this paper can be further improved in several ways. First and foremost, the prediction are improved with additional observed data. This does not only mean to monitor more sections but also to collect more types of the data such as the displacement sideways to the tunnel tube.

Additionally, the deterministic model used within the stochastic description can itself be improved. As for the reduced model for successive excavation, the introduction of the nonlinear material would make the analysis more realistic and allowing us to take the soil strength into account. Finally, the approximation of the model by means of the Taylor series (10) could be expanded by adding the higher order terms.

For reference, a similar formulation based on the finite element method and a nonlinear material is formulated in [5]. When comparing these two approaches, the following peculiarities deserve mentioning:

(i) The effectiveness of the newly proposed approach exploiting the simplified deterministic model for the prediction of the surface settlement.

(ii) The robustness of the Gibbs sampling algorithm allowing us to reliably capture the statistical dependence of the input parameters (components of vector $\boldsymbol{b}$).

Acknowledgments

The financial support of the project no. TE01020168 provided by TA ČR is gratefully acknowledged.

References

[1] Janda T, Šejnoha J, Srnec V and Šejnoha M 2014 Modeling of soil structure interaction during tunnel excavation considering uncertainties *Proceedings of the Ninth International Conference on Engineering Computational Technology* (Civil-Comp, Ltd.)

[2] Šejnoha M, Janda T and Šejnoha J 2016 Tunnel 2D3D: FEM program for modeling tunnel-structure interaction *13th International conference on undeground construction Prague 2016* ed Hilar M, Butovič A, Škopková I and Zlámal J (JMP Praha)

[3] Lunn D, Spiegelhalter D, Thomas A and Best N 2009 The BUGS project: Evolution, critique and future directions *Statistics in Medicine* **28** 3049–3067

[4] Plummer M 2003 JAGS: A program for analysis of bayesian graphical models using gibbs sampling https://www.r-project.org/conferences/DSC-2003/Proceedings/Plummer.pdf accessed: 2017-01-30

[5] Špačková O, Janda T, Novotná E, Šejnoha M, Jarušková D and Šejnoha J 2011 *Computational Models for Tunnel Risk Assessment* (Czech Technical University in Prague)

Supervision of tunnelling constructions and software used for their evaluation

Aristotelis Caravanas[1] and Matous Hilar[1,2]

[1]Czech Technical University in Prague, Thakurova 7, Prague 166 29, Czech Republic
[2]3G Consulting Engineers s.r.o., Zeleny pruh 95/97, Prague 140 00, Czech Republic

E-mail: ari.caravanas@tunnelsupervision.com

Abstract. Supervision is a common instrument for controlling constructions of tunnels. In order to suit relevant project's purposes a supervision procedure is modified by local conditions, habits, codes and ways of allocating of a particular tunnelling project. The duties of tunnel supervision are specified in an agreement with the client and they can include a wide range of activities. On large scale tunnelling projects the supervision tasks are performed by a high number of people of different professions. Teamwork, smooth communication and coordination are required in order to successfully fulfil supervision tasks. The efficiency and quality of tunnel supervision work are enhanced when specialized software applications are used. Such applications should allow on-line data management and the prompt evaluation, reporting and sharing of relevant construction information and other aspects. The client is provided with an as-built database that contains all the relevant information related to a construction process, which is a valuable tool for the claim management as well as for the evaluation of structure defects that can occur in the future. As a result, the level of risks related to tunnel constructions is decreased.

1. Introduction

Tunnels are being constructed in an environment whose parameters are explored in details only locally priory to the commencement of construction. Therefore, changes to the expected geotechnical conditions are not rare when constructing a tunnel. Contractors adopt an excavation method and rock support as per given conditions in order to construct a tunnel in time manner and quality as per the contract's conditions. The final product, a tunnel, must serve without defects for its entire service life.

There is a large amount of data related to tunnel excavation, geotechnical conditions and rock support obtained a during tunnel construction period. These data are essential not only for the prompt evaluation of rock mass conditions, the determination of excavation and rock support measures and payments, but lately they form a base for the claim management as well as the analysis of possible defects of the structure.

The defects of a tunnel structure or the necessity of its improvement may appear during the operation of a tunnel. The knowledge of an excavation process, geotechnical conditions, used rock support and any complications that happened in the course of tunnelling are vital for choosing a correct remedial action. The later analyses of the conditions are more difficult because tunnel lining covers the surrounding rock mass. The experienced clients incorporated an obligation to provide a database of as-built information into their contract with the supervision and/or the contractor.

IOP Conf. Series: Materials Science and Engineering **236** (2017) 012077 doi:10.1088/1757-899X/236/1/012077

2. Supervision scope of work

Supervision work in tunnelling projects, based on the contract between clients and supervisions, can include, but is not limited to, the below listed tasks:

- Check of quality and quantity of contractor's tunnelling work including excavation process and installed rock support
- Regular evaluation of tunnel excavations and contractor's performance
- Geological mapping
- Determination of excavation classes and/or instructions of rock support
- Check of health and safety and compliance with environmental regulations
- Quality tests of construction material
- Design check, proposal of design improvements according to actual conditions; exceptionally designing
- Check of the contractor's technical documentation including method statements, as built drawings etc. and proposing improvements
- Financial matters – regular check of payment applications, claim management and quasi-arbitrator
- Client's representation
- Regular and ad hoc reports and analyses, Final Report
- Holding meetings

Some of the above listed tunnel supervision activities are described in more details in the following chapters. Tools used by a supervision to perform the tasks are presented, including the example of an on-line software system specialized in tunnel supervision purposes.

3. Excavation classes

According to contract conditions the supervision role can include the mapping of geological conditions, the determination of excavation classes and the check/control of rock support to be installed. It is of extreme importance to carry out this work properly because of health and safety reasons, technological reasons and not least economic reasons.

Clients pay contractors for the actual provided work as per contract conditions. In case of tunnel excavations the payments are often made as a cost per linear meter of a tunnel for a particular excavation class. The excavation classes are specified by their excavation methods and used rock support according to geotechnical conditions. There are two kinds of classes – expected (prior to construction) and actual (as built). The construction schedule is based on the expected excavation classes. The difference between expected excavation classes and actual excavation classes has significant influence on both project cost and schedule. Refer to Table 1 for an example which presents the difference between actual and expected excavation classes of a 4km long railway tunnel in the Czech Republic and its influence on project cost.

When the actual geotechnical conditions lie within the limits specified in the contract, it is usually relatively easy to agree on a difference/change between expected excavation classes and actual excavation classes. The impact on construction schedule and cost is then automatically evaluated according to predefined conditions. However, when geotechnical conditions are outside the limits of the contract, then the agreement on additional work and its payment is sometimes difficult to reach, taking into account the fact of not predefined influence on construction schedule and cost.

In both cases – the actual geotechnical conditions are a) within or b) outside the expected limits – proper unambiguous specification of the excavation classes needs to be clearly stated in the contract. There are usually more aspects with the impact on determination of excavation classes. The determination of excavation classes must be carried out by experienced personnel in tunnelling.

Unfortunately, sometimes the definition of excavation classes is vague, ambiguous and unthorough. Furthermore, sometimes the specification of expected geotechnical conditions and their limits is incomplete and vague and/or with mistakes. It is wise to define expected geotechnical conditions and their limits in a special document. Such a document is often called Geotechnical Baseline Report (GBR) [2]. However, it is often a case that GBR or a similar document is not provided, especially when projects are located outside Western Europe and North America.

Table 1. Example: The difference between actual and expected excavation classes in a 4km long railway tunnel and the approximate impact on project cost.

Excavation class	Length [m]	Approx. cost for excavation class per linear meter [EUR]	Approx. cost per excavation class [EUR]	Approx. construction time based on bid advance speed [days]	Daywork - additional approx. cost for extended time [EUR]
1 - expected	650	2,413	1,568,450	33	---
1 - actual	*950*		*2,292,350*	*48*	---
2 - expected	1,641	2,849	4,675,209	99	---
2 - actual	*658*		*1,874,642*	*40*	---
3 - expected	868	2,855	2,478,140	60	---
3 - actual	*200*		*571,000*	*14*	---
4 - expected	169	4,466	754,754	12	---
4 - actual	*316*		*1,411,256*	*22*	---
5 - expected	784	4,870	3,818,080	63	---
5 - actual	*1,988*		*9,681,560*	*159*	---
Sum - expected	4.1 km	---	13,294,633	266	585,674
Sum - actual			*15,830,808*	*282*	
Add. cost in total			**estimate: + 3,121,849 EUR (+ 23.5%)**		

4. Construction schedule

The supervision shall monitor and evaluate the actual progress of tunnel excavations and should continuously compare it with the scheduled progress. It is an advantage when the supervision is capable of predicting of the future progress of excavations based on statistical methods. Such methods use data from the already excavated sections of tunnels and take into account the effects of actual encountered geological conditions, delays, maintenance periods, learning curve etc. It is a demanding task which needs a special software but when available, then the supervision is able to keep the construction schedule under tight, continuous and realistic control.

An example of construction schedule of a 4km long railway tunnel is shown in Figure 1. The excavation was commenced on February 2015. For the first three months the actual excavation progress was very slow. The status of excavation is related to the 30th of September 2015, when approximately half of the tunnel was completed. The chart presents the comparison of the actual tunnelling progress (2114m excavated) with the originally planned progress where planned excavation classes were used. The expected progress (2887m) – was updated for the actually encountered geological conditions, i.e. excavation classes. The predicted future progress is shown as an area of probabilistic distribution with the breakthrough date shifted from November 2015 to March 2016.

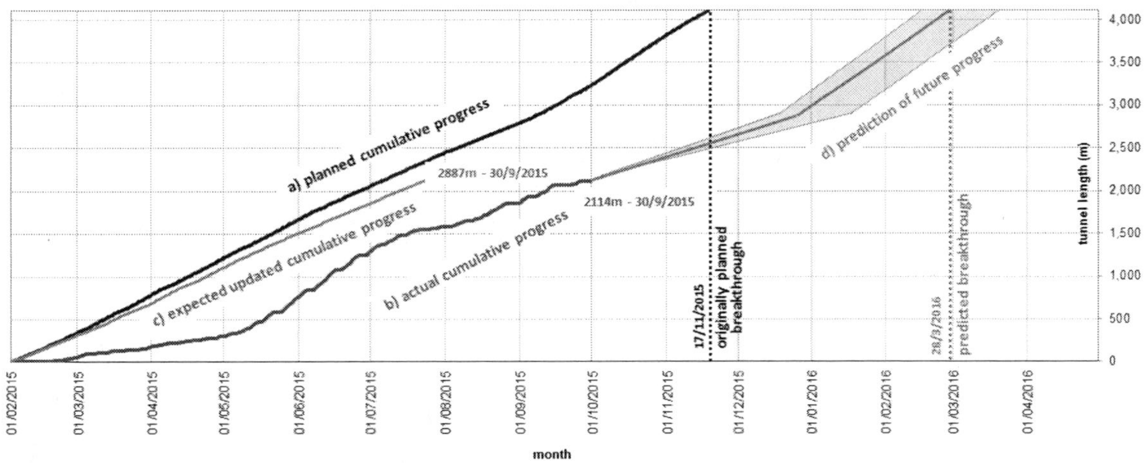

Figure 1. Construction schedule: a) black – originally planned excavation progress based on expected excavation classes; b) red - actual excavation progress shown till the end of September 2015; c) green – expected updated excavation progress based on actual excavation classes and advance rate as per bid; d) blue – probabilistic prediction of future excavation progress based on previous excavation data.

Figure 1 shows that that the actual excavation progress b) was delayed in comparison with the originally planned progress a) by 86 days (horizontal offset), which represented 1080m (vertical offset). However, due to aggravated geotechnical conditions, represented in higher/heavier excavation classes with slow advance rate, the progress was delayed by 20 days, which is a horizontal difference between the curves a) and c). The rest of the delay, i.e. 66 days, is on the account of unforeseen geotechnical conditions, unexpected events, learning curve, contractor's delays including breakdowns and extensive maintenance etc.

When these delays are claimed by a contractor, it is up to a supervision to confirm or determine their owner. Having known the high cost of daywork – 66 days were equal to approx. 2.5 million EUR in the above mentioned project – it is of the critical importance to follow and record the activities and delays in details during the tunnelling.

Due to the amount of data it is necessary to use a special software application. One of such applications is mentioned in Chapter 7. The application analyzes events, i.e. activities and delays, for any time in the past. The Figure 2 presents a simple chart showing partition of main activities and delays during construction of a 14km long headrace tunnel in Iceland [1]. The tunnel was excavated by a 7.6m diameter hardrock open type TBM, see Figure 3. A noticeable amount of time was spent on the activity No. 7) Utilities, which included repairs of a conveyor belt. As the tunnel conveyor was getting longer, breakdowns in the conveyor belt happened more often and thus the excavation progress became considerably slower. The breakage of the conveyor belt was caused by an insufficient number of installed boosters. Having identified and quantified the true cause of the deceleration of the excavation progress, the issue could be further addressed.

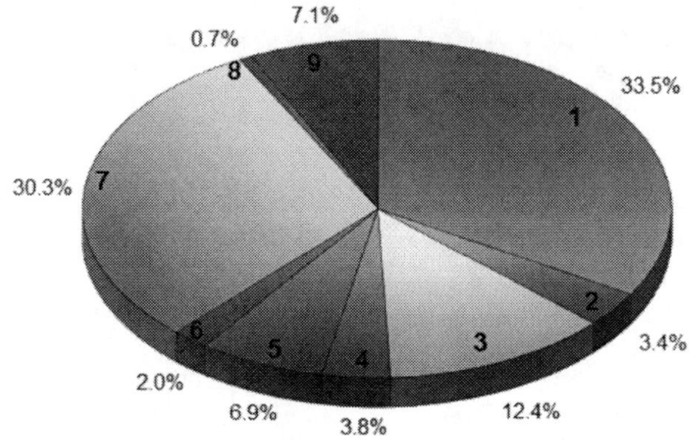

Figure 2. Actual TBM utilization after boring 14km long headrace tunnel.
1 – TBM boring, i.e. utilization – 33.5%
2 – TBM re-gripping
3 – installation of rock support
4 – cutter change and check
5 – TBM maintenance and breakdown
6 – back-up
7 – utilities including installing and repairing of tunnel conveyor – 30.3%
8 – probing
9 – others (downtime, holiday, grouting, etc.)

Figure 3. A 7.6m diameter hardrock TBM prior to the excavation of a 14km long headrace tunnel in Iceland.

5. Quality of tunnel structure

A tunnel structure should be constructed as specified in the design in order to fulfil its operational function during its entire service life. The final structure sometimes more or less varies in comparison with the design. There can be more reasons for the variation, such as for example the change in supplied materials, the different procedure of excavation, etc. It is necessary to have all the changes properly documented including the reasons for changes with all relevant records.

In cases when the variation in construction is not agreed, it is required to issue a special form called in many countries "Non-conformance Report" (NCR). The tunnelling industry regularly uses the NCR as a way to keep the track of deviation and work that fails to meet specifications and standards. A NCR should minimally include the information on what exactly went wrong and in which place, what the main reasons were, why the work did not meet specifications and what remedial action should be taken. A NCR is open until the time when the remedial works are undertaken and agreed, after which the NCR is closed. All NCRs shall be closed until a tunnel is handed over for operation to a client.

The remedial actions that are needed for closing NCRs are often time consuming and expensive. Refer to Table 2 and Figure 4 that present an example of time lost and money wastage due to the collapse of rock support and surrounded rock mass during the excavation of a 28km long headrace tunnel in Pakistan. The event happened because the contractor did not comply with rock support specifications. Having all relevant information including NCRs properly recorded the supervision was able to defend the client against the later false contractor's claim for "suddenly worsen geotechnical conditions". It is interesting to note that the direct cost of remedial measures is often much less expensive when compared with the estimated income from a project operation.

Table 2. Collaps consequences.

Item No.	Description	Additional time [day]	Daywork – approx. cost for additional time [EUR]	Approx. direct cost for work [EUR]
1	Stoppage of excavation progress because of collapse and remedial works. Tunnel on critical project path. Two tunnel faces stopped and one partially affected.	22	520,000	---
2	Additional installed rock support – shotcrete, rockbolts, wire mesh, grouting etc.	---	---	63,000
Direct cost in total [EUR]		---	**583,000**	
For information purposes: Commencement of hydropower project operation delayed. Estimated price for generated electricity during 22days of hydropower project operation based on 90 EUR/MWh.		---	27.9 million	

Figure 4. An excavator buried by collapsed rock support and surrounded rock mass. The photo taken during time when rock was still falling and shotcrete lining was cracking. An excavator operator was pulled out alive.

Depending on the reasons remedial actions could be paid solely by a contractor, a client or expenses can be shared between them. It is a supervision who has to guide this process. Contractors usually try to include the provided remedial works into claims regardless true reasons. Therefore, it is of a paramount importance to keep all the necessary relevant information related to NCRs. Furthermore, contractors are often directly penalized for any issued NCR according to contract conditions.

There can be plenty of information and documents rělated to any single NCR. Speaking of tunnels that are often several kilometres long and constructed for a period of couple of years, the amount of gathered data to be stored is vast. This imposes a necessity to involve software database solutions so that we could analyse the data efficiently even many years after the construction of a tunnel is finished.

6. Claim management

According to Reference [4] a construction claim can be defined as a request by either party of the contract, usually the contractor, for compensation for the damages caused by the failure of the other party to fulfil his part of obligations or for the change in expected conditions as specified in the contract. The compensation is usually in the form of an additional payment or an extension of time.

Construction claims are one of the most worrying and unpleasant events of a project. Hard competition forced contractors to submit their bids with minimum profits or worse underbid. In addition, projects are becoming more complex and risky. This put additional burden on contractors to construct increasingly sophisticated and risky projects with less resources and profits. As a consequence, the number of claims within the construction industry increases.

A large majority of claims involve delays in construction schedule. As could be seen in the examples above, refer to Tables 1 and 2, the cost for one daywork is significant. Therefore, it is of an extreme importance to understand what was happening in the tunnel all the time, as well as to know the true reasons for it. The causes of delays must be well described and the "owner" of a delay identified. When it comes to assessing claims, information and knowledge are vital in order to make the correct decision.

Special software applications, such as for example [3], are able to analyse activities and delays for any period of time and help to manage claims.

7. Software for tunnel supervision purposes

There exist only few on-line software applications or databases specialized in the tasks of tunnel supervision. The Tunnel Supervision [3] is one of such on-line applications which provides project participants with essential data from tunnelling. The Tunnel Supervision application manages and analyses data related to for example:

- Excavation classes
- Geological conditions and underground water
- Rock support and tunnel lining details
- Construction schedule
- Activities and delays
- Site diaries
- Site instructions
- Non-conformance reports
- Payments and claims
- Tests

The data of various fields of expertise are integrated and accessible all the time through one uniform environment. Routine reports are generated automatically, special reports are provided and statistics calculated as required. Refer to Figures 1 and 2 for examples where the application provided important information.

The application has been developed to provide support for:

- Certifying payments, defending or supporting claims
- Answering specific questions on excavation progress and construction schedule
- Passing data and information, i.e. enhancing collaboration in the team

- Reporting and graphic presentations
- Producing statistics as required
- Refining data for a wide spectrum of tasks that underground work generates

8. Conclusion

The construction schedules and costs of tunnel projects are often exceeded. Aggravated geological conditions are regularly proclaimed by contractors as the reasons for extra time and cost, although the true reasons often consist of combinations of factors, such as delays due to contractor's faults, construction mistakes and errors in design, poor contract conditions, poor geotechnical survey and worsen actual geotechnical conditions. The supervision should expose the true factors including mistakes, try to find and suggest solutions and do as much as possible to avoid mistakes. The supervision plays a very important role in tunnel construction industry, contributes to finish the project in required quality with time schedule and cost matching contract conditions.

During tunnelling construction a database of gathered data shall be established and managed in order to provide information and analysis. The input data include information such as for example geotechnical conditions, actual and expected excavation classes, installed rock support quantity and quality, other payment items, activities and delays during construction, deformation monitoring, photo documentation, etc. at any given place. Due to the amount of data and number of users with different rights to access them, it is wise to use a special on-line software application which provides ad hoc and precise analysis on the fly. A software database should be handed over to the client when the construction of a tunnel is finished. The client can use cross connected data and analysis any time e.g. for later maintenance or in case of some later structure defects.

Acknowledgements

This paper was prepared thanks to the support of the Technology Agency of the Czech Republic (TAČR), project "Centre for Effective and Sustainable Transport Infrastructure (CESTI)" (no. TE01020168).

References

[1] Caravanas A 2009 Tunnels for Kárahnjúkar Hydropower Project *Tunel* **1** pp 22-31
[2] Klee L and Hruška D 2013 How to efficiently allocate the risk of unforeseeable physical conditions in underground construction? *Tunel* **3** pp 51-62
[3] www.tunnelsupervision.com
[4] Hadikusumo B and Tobgay S 2015 Construction Claim Types and Causes for a Large-Scale Hydropower Project in Bhutan *Journal of Construction in Developing Countries* **15** pp 49-63

Calibration of an advanced material model for a shotcrete lining

Juraj Chalmovský[1], Martin Závacký[1] and Lumír Miča[1]

[1]Department of Geotechnics, Faculty of Civil Engineering, Brno University of Technology, Veveri 95, 602 00 Brno, Czech Republic

E-mail: chalmovsky.j@fce.vutbr.cz

Abstract. Proper choice of a constitutive model is an essential part of any successful application of numerical methods in geotechnical engineering. In most cases, attention is paid to the soil constitutive model. For structural elements, such as tunnel linings, retaining structures etc. elastic constitutive models are often used. These material models however do not involve many aspects of a real structural behavior such as limited tensile and compressive strength, strain softening and time dependent behavior during service life of a construction. In the proposed paper, an application of the novel constitutive model for shotcrete (Schädlich, Schweiger, 2014) is presented. The paper is focused at the process of determination of input parameters values of this model based on performed laboratory test. Section of the primary collector network in Brno was chosen for the purpose of obtaining shotcrete lining samples.

1. Tested material and laboratory tests

For described calibration of the material model sprayed concrete from lining of collector network in Brno was selected. The particular section of primary collector is situated under Leitnerova street (near the shaft Š 13 A) and which was constructed in 1980's. Sensors of geotechnical monitoring has been installed there so the sampling was easier and desirable right there. A drill core with diameter of 23 cm through whole thickness of the lining was obtained at first. Smaller samples for laboratory testing were made afterwards from the first core.

Amount of the material for testing was limited. Also, a lot of steel bar reinforcement was involved in the big drilled core. Thus, smaller diameter for core drilling of laboratory test samples was selected. Three types of test procedures were done:

- Splitting tensile strength
- Uniaxial compressive strength
- Triaxial compressive strength

1.1. Splitting tensile strength

Parameter of tensile strength is required in the shotcrete material model. Tensile strength was determined by three point bending test in other calibration of the material model [1]. Lack of the testing material lead us to decision to obtain this parameter by splitting tensile test. Cylindrical samples with diameter 55 mm were prepared. Requirement of the standard [4] to height-to-diameter ratio at least 1.0 was satisfied with 55 mm height of the sample. The test was driven by rate of loading at 190 N/s. A sample after test is shown in Figure 3.

As the result of splitting test was obtained 3.4 MPa in tensile strength for the sprayed concrete. This value is an average of three individual tests with standard deviation 0.12 MPa, hence the result can be considered as reliable.

Figure 1. Sample after failure in uniaxial compressive test.

Figure 2. Hoek cell during triaxial test.

1.2. Uniaxial compressive strength

Due to considerable amount of reinforcement bars in the concrete, as it was mentioned before, was selected diameter of 38 mm for core drilling of laboratory samples. The diameter of the aggregate used in the concrete was estimated to fraction less than 8 mm. Ratio of aggregate diameter to cored diameter was less than maximally allowed 1:3 according standard [5] so the selection of the drill core diameter was appropriate. To satisfy criterion of height-to-diameter ratio 2.0 was given requirement on samples height equal to approximately 76 mm.

Figure 3. Sample after splitting tensile test.

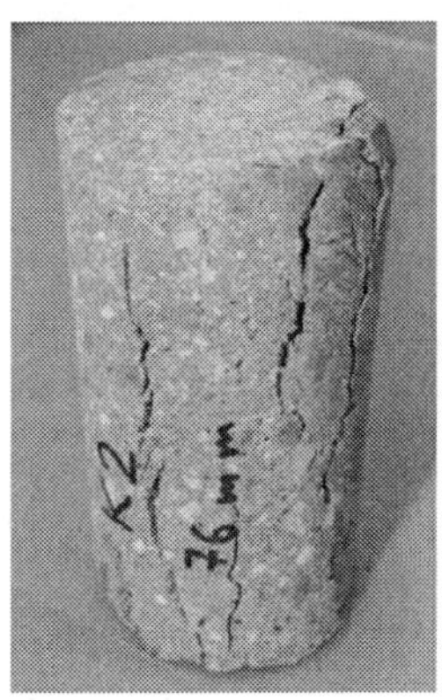

Figure 4. Sample after uniaxial compression test.

Figure 5. Sample after triaxial test at σ_1 = - 8 MPa with diagonal shear crack.

The compression test was governed by deformation with rate 0.5 μm/s. For illustration see Figure 1. Higher deformation rates were also tested, but behavior of the concrete was more brittle in other cases. Smaller increments of deformation caused more ductile response of samples. Thus, it was considered as suitable for obtaining calibration parameters of the shotcrete material model.

Results from one uniaxial compressive test were selected as representative with further relations to triaxial tests. Peak stress level was reached -44.1 MPa as peak strength.

1.3. Triaxial compressive strength

The same dimensions and axial loading conditions were used for triaxial compressive tests as for the uniaxial testing. Hoek cell was used for applying lateral (radial) pressure (see Figure 2) and LVDT sensors were used for deformation control. Convention of "-" for compression and "+" for tension stresses is used in this paper. Thus, σ_1 represents the highest stress level and σ_3 represents the lowest stress level with taking into account the signs.

Three different lateral pressures were applied gradually. Reached peak strength values are shown in Table 1 below. Post-peak softening of the material was observed in all three cases. The sample 3 after test is shown in Figure 5.

Table 1. Peak axial stress σ_3 reached in triaxial tests at corresponding lateral pressure σ_1.

Sample	σ_1 (MPa)	σ_3 peak (MPa)
1	- 2.0	- 56.0
2	- 4.0	- 63.5
3	- 8.0	- 84.2

2. Model calibration

Process of determination of input parameters values for the Shotcrete model (SC) developed by [1,2] is presented in the following section.

2.1. Brief description of the Shotcrete model

The SC model belongs to a group of elasto-plastic constitutive models involving strain hardening and softening. The following main features are involved in the SC model:

- Mohr-Coulomb yield surface for a deviatoric loading and Rankine yield surface for a tensile loading regime
- Strain hardening and softening in tension and compression
- Regularization during the softening regimes in order to avoid mesh dependency
- Increase of the shotcrete stiffness and strength with time
- Decrease of the shotcrete ductility with time
- Creep and shrinkage strains

Determination of input parameters related to the hardening during loading is presented in the paper. The stress – strain curves for loading in compression and tension are shown in Figures 4 and 5.

$H_c = \varepsilon_3^P/\varepsilon_{cp}^P$ and $H_t = \varepsilon_1^P/\varepsilon_{tu}^P$ are normalized hardening – softening parameters in compression and tension, ε_3^P is a minor plastic strain, ε_{cp}^P is a plastic peak strain in uniaxial compression, ε_1^P is a major principal plastic strain and ε_{tu}^P is a plastic ultimate strain in uniaxial tension. Further details about the material model can be found in [1,2]. The list of relevant input parameters is stated in the Table 2.

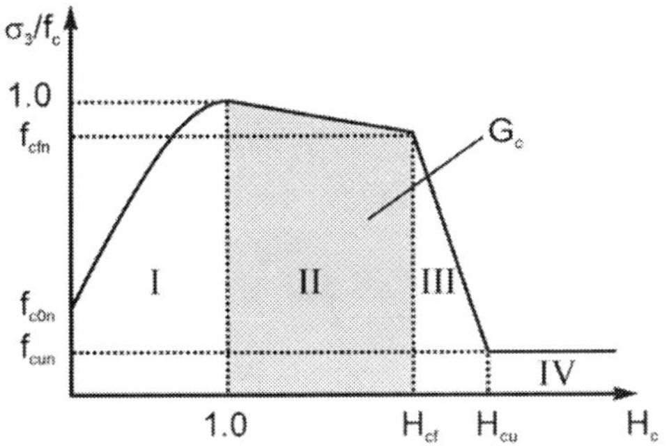

Figure 4. Normalized stress strain curve in compression (taken from [2]).

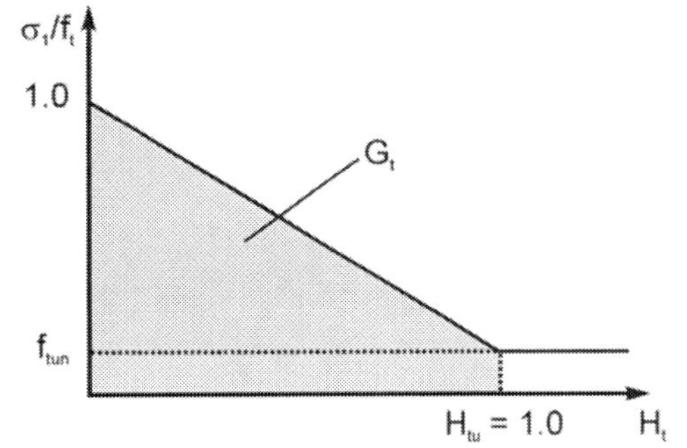

Figure 5. Normalized stress strain curve in tension (taken from [2]).

Table 2. Input parameters related governing strain hardening

Parameter	Unit	Description
E_{28}	MPa	Young´s modulus of cured shotcrete
ν	-	Poisson´s ratio
$f_{c,28}$	MPa	Uniaxial compressive strength of cured shotcrete
$f_{t,28}$	MPa	Uniaxial tensile strength of cured shotcrete
f_{c0n}	MPa	Normalized initially mobilized strength
$\varepsilon_{cp}{}^{p}$	-	Uniaxial plastic failure strain
a	-	Parameter governing increase of ε_{cp} with p´
φ_{max}	°	Maximum friction angle

2.2. Model calibration

The uniaxial tensile strength $f_{t,28}$ = 3.4 MPa was derived from the performed splitting tests. The parameter *a* governing the shotcrete ductility and the maximum friction angle φ_{max} were derived from series of three triaxial tests described in the previous section. For determination of the remaining parameters it is possible to use the uniaxial test. Measured and calibrated stress strain curves are shown in Figure 6. Calibration was performed on a stress point level using the Plaxis 2D 2016 software package. Care must be taken when comparing post-peak behavior with experimental data due to inhomogeneity such as shear bands and transverse cracks occurring in the sample during post-peak strain softening. This requires a complete 3D FE model of a laboratory test. The final model parameters are listed in Table 3.

Table 3. Final set of input parameters

E_{28} [MPa]	ν [-]	$f_{c,28}$ [MPa]	$f_{t,28}$ [MPa]	f_{c0n} [MPa]	$\varepsilon_{cp}{}^{p}$ [-]	a [-]	φ_{max} [°]
13500	0.2	45	3.4	0	0.0015	18	40

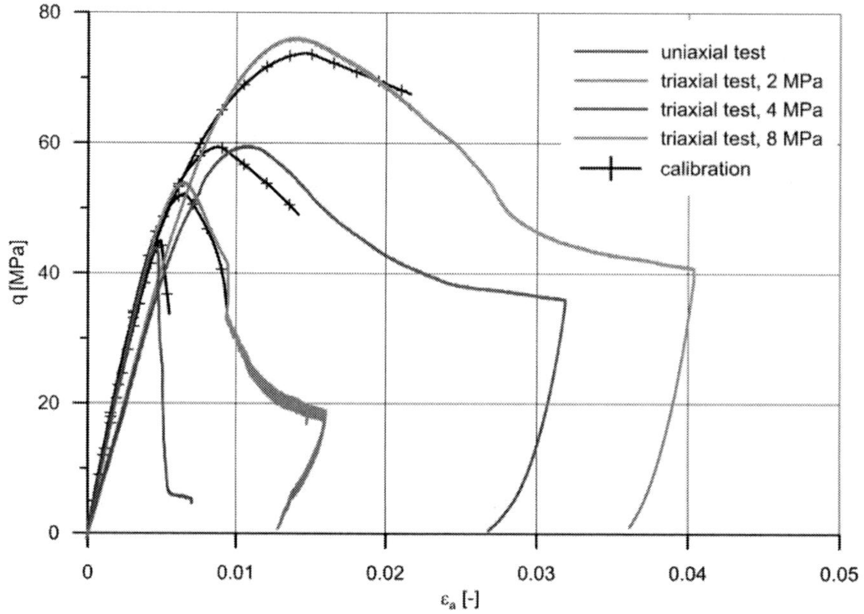

Figure 6. Stress – strain curves: measurements and calibration uniaxial test, triaxial tests with confining pressures -2, -4 and -8 MPa.

3. Results evaluation

3.1. Performed laboratory tests

The compressive strength and shotcrete ductility is increasing and the rate of strain softening is decreasing for higher confining pressures. Differences between the peak and residual stresses are lower for higher confining pressures. All these aspects are in accordance with other experimental results (f. e. [3]). The initial stiffness for the triaxial tests with confining pressures -4 and -8 MPa is slightly lower in comparison with the remaining test which is in contrast with theoretical assumptions. This discrepancy might be caused by an initial sample disturbance or a slightly different original position of the sample in the lining.

3.2. Calibration of the SC model

SH model correctly predicts the following features of shotcrete behaviour:

- Nonlinear stiffness decrease during strain hardening
- The compressive strength and ductility increase with increasing confining pressure
- The decrease of the rate of strain softening with increasing confining pressure. This aspect is however possible to analyse only from a qualitative point of view. A quantitative prediction would require a complete 3D model.

4. Conclusions

The computed stress – strain curves are in reasonable match with the laboratory tests. The discrepancy described in the previous section is not taken into account in the calibration. Slightly lower compressive strength was reached for the test simulation with the confining pressure -8 MPa. A linear Mohr-Coulomb yield surface is implemented in the model. The failure surface for rocks and concrete is however not linear which might cause differences especially for higher ranges of confining pressures ranges. From an engineering point of view the level of compliance is however satisfactory.

Acknowledgements
This research was financially supported by the research programme "Centra kompetence", Technological agency Czech Republic, No. TE01020168 and research grant No. FAST-S-15-2743.

References
[1] Schädlich B, Marcher T, Schweiger H F and Saurer E 2014 Application of a novel constitutive shotcrete model to tunnelling *ISRM Regional Symposium - EUROCK (Vigo)* ed L R Alejano, A Perucho, C Olalla and R Jimenez (London: Taylor & Francis group) pp 799-804
[2] Schädlich B, Schweiger H F 2014 A new constitutive model for shotcrete *Numerical Methods in Geotechnical Engineering* **1** pp 103-108
[3] Von R A 1992 *Softening of concrete in compression* (University of Technology Eindhoven)
[4] European Committee for Standardization (CEN) 2009 *Testing hardened concrete - Part 6: Tensile splitting strength of test secimens (EN 12390-6)*
[5] European Committee for Standardization (CEN) 2009 *Testing concrete in structures – Part 1: Cored specimens – Taking, examining and testing in compression (EN 12504-1)*

IOP Conf. Series: Materials Science and Engineering **236** (2017) 012079 doi:10.1088/1757-899X/236/1/012079

Probabilistic approach to damage of tunnel lining due to fire

Jiří Šejnoha[1], Jan Sýkora[1], Eva Novotná[1] and Michal Šejnoha[1]
[1]CTU in Prague, Thákurova 7, Prague 160 00, Czech Republic

E-mail: novotna.eva@fsv.cvut.cz

Abstract. In this paper, risk is perceived as the probable damage caused by a fire in the tunnel lining. In its first part the traffic flow is described as a Markov chain of joint states consisting of a combination of trucks/buses (TB) and personal cars (PC) from adjoining lanes. The heat release rate is then taken for a measure of the fire power. The intensity λ_f reflecting the frequency of fires was assessed based on extensive studies carried out in Austria [1] and Italy [2, 3]. The traffic density AADT, the length of the tunnel L, the percentage of TBs, and the number of lanes are the remaining parameters characterizing the traffic flow. In the second part, a special combination of models originally proposed by Bažant and Thonguthai [4], and Künzel & Kiessl [5] for the description of transport processes in concrete at very high temperatures creates a basis for the prediction of the thickness of the spalling zone and the volume of concrete degraded by temperatures that exceed a certain temperature level. The model was validated against a macroscopic test on concrete samples placed into the furnace.

1. Introduction

High temperatures produced by a fire may cause damage to structures, particularly to the tunnel lining. This sub-structure then would be deteriorated not only by concrete spalling but also by a drop of strength in the volume stricken by the temperatures substantially acceding a hundred degrees Centigrade.

The large-scale fires (Mont Blanc tunnel, Tauern tunnel, Saint Gotthard road tunnel) are not typical and cannot be incorporated in the scope of standard statistics-based approaches, and as such, must be analyzed ad hoc. On the other hand, there exists a considerable volume of data on standard–size incidents which may be exploited in a risk analysis of fire in road tunnels and will be briefly discussed in the sequel. The risk analysis monitors both probabilistic and economic viewpoints. The probability of a fire is closely related to the probability of a traffic incident, which could be not only the consequence of a collision of two or more vehicles but also any event causing the vehicles' inflammation (e.g. overheating of the engine, brakes, leakage of fuel, etc.) Fire incidents in road tunnels are prevailingly caused by electrical and/or mechanical defects of vehicles.

In order to address the fire risk assessment as a whole, this paper is composed of two parts. In the first part, the probabilistic concept of the fire appearance in a road tunnel will be outlined. To this end, the findings of two extensive studies carried out in Austria [1] and Italy [2, 3] will be utilized in this study.

The second factor affecting the extent of damage due to a fire in a tunnel and corresponding financial costs is, barring the probability of fire occurrence, the evolution of the fire characterized by the following attributes: (i) the evolution rate (slow after smoke vs. explosive), (ii) the peak of the heat release rate (HRR).

As predicted in [1], the spontaneous ignition of TBs was followed by a slow evolution (80 [%]) and the explosion was triggered out in 20 [%] of cases. In the case of TBs, the fire of the cab must be distinguished from the fully developed fire of the whole vehicle. Due to insufficient data, HRR and the frequency of the fire occurrence have been predicted in an expert way by means of event tree analysis [1]. More details about the traffic flow and the fire itself are summarized in Section 2. Based on the data about the traffic flow and fire characteristics, a simple probabilistic model is outlined in Section 3.

In the second part of this paper, the data on fires will be used to analyze the degree of degradation affecting the tunnel lining exposed to high temperatures. In quest for a cogent yet effective model, a special combination of approaches originally proposed in [4, 5] is presented in Section 4. The report [6] on the results of the Runehamar tunnel fire tests sets forth simple yet reliable formulas for the description of the phenomena accompanying a fire. It becomes a basis for the hygro-thermo-mechanical model for the prediction of damage due to a fire. Next, a case study under simplified assumptions demonstrates a prediction of the risk of a fire in a road tunnel, see Section 5.
Conclusions and possibilities of further work on the fire risk analysis are outlined in Section 6.

2. Available statistical data as a basis of a probabilistic model

The results of studies carried out in Austria and Italy serve not only as representatives of fire occurrences in Europe but also as a guideline for their prediction that provides the data allowing us to assess the rate of fires as the number of fires per 10^9 [veh.km] travelled in tunnels, see Tables 1 and 2.

Table 1. Frequency of fire occurrence (the number of fires per 10^9 [veh.km] in Austria, source [1]).

Type of vehicle	The number of fires per 10^9 [veh.km]	Self-ignition	[%]	Subsequent	[%]
All	6.5	6.0	92	0.5	8
PC	4.2	3.6	86	0.6	14
TB	25.0	24.3	97	0.7	3

Table 2. Frequency of the fire occurrence λ_f [fires (veh.km)$^{-1}$] (the number of fires per 10^9 [veh.km] in Italy, source [2]).

Type of vehicle	Type of tunnel	Frequency of fire occurrence
All	2 lanes	33.3
	3 lanes	40.8

The results in Table 2 were calculated from the following formula:

$$\lambda_f [\text{fires (veh. km)}^{-1}] = \frac{N_f}{365\{\text{AADT}\}L_{\text{ref}}} \tag{1}$$

where N_f [fires year^{-1}] is the average annual number of fires monitored in tunnels with their total length L_{ref} [km] and the average annual daily traffic AADT [veh.day^{-1}] (see Tab 3).

Table 3. Summary statistics of the characteristics of tunnels with recorded fire occurrence, source [2]

Type of tunnel	N_f	$L_{ref} = \Sigma L_i$ [km]	AADT [veh.day^{-1}]
2 lanes	9.5	46.2	16931
3 lanes	5.0	8.3	24247

The second characteristic of a fire is the maximum value of the heat release rate Q_{max} [MW], which is a measure of the fire power. The HRR per square meter of exposed fuel is generally found for most vehicles to be in a very narrow range (0.27 [MWm^{-2}] - 0.4[MWm^{-2}]), see [7].The resulting fire power depends on the critical scenarios that give the most significant contribution to the overall risk. In this paper, we used as an illustrative example the Austrian model which considers three representative scenarios of fires (5[MW], 30[MW], 100[MW]). The distribution of Q_{max}[MW] was obtained in [1] for trucks and buses (TB) as a combination of databases and expert assessment exploiting an Event Tree Analysis (ETA).

The results of ETA are a suitable point of departure when estimating the probability mass function of Q_{max} [MW] conditional on the fire developed on one truck/bus, p_{Q1} (q_1). This function is displayed in Fig. 1(a). If two trucks/buses are involved in the incident giving rise to a fire, the resulting HRR is a composition of both HRRs, i.e. $Q = Q_1 + Q_2$, see Figure 1b.

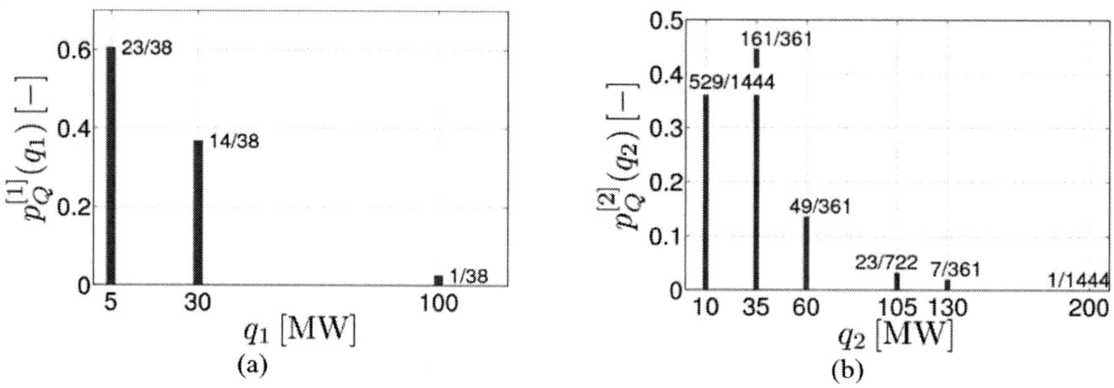

Figure 1: (a) Probability mass function of Q_{max} [MW] (one truck/bus) (b) probability mass function of Q_{max} [MW] (two trucks/buses).

3. Probabilistic model for the prediction of fire risk in road tunnels

In this chapter a pragmatic probability-based model will be proposed. In doing this, the following factors will play a decisive role in the model: (i) the probability of a certain configuration of vehicles involved in the fire accident, (ii) the probability that the subsequent fire will ensue a collision (see Tab. 1), (iii) the distribution of heat developed in a given configuration of vehicles engulfed by the fire (Figs. 1(a) and (b)), and (iv) the damage of the tunnel lining as a result of the fire with a known HRR.

The probable configurations of vehicles potentially involved in a fire incident in the traffic flow with two lanes are sketched in Fig. 2, where V stands for the empty space (Void).

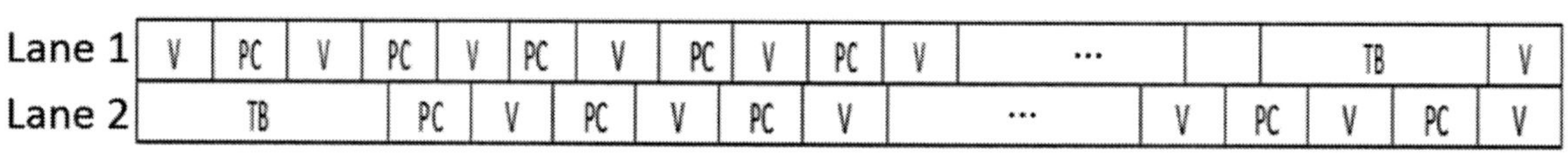

Figure 2. Two-lane traffic flow.

For two stationary (homogeneous) independent parallel traffic flows, the probabilities of joint states, e.g. (V,V), (PC,V) etc., can be simply calculated as the product of the probabilities of individual states in adjoining lanes.

Hence

$$P_l = P_i^1 P_i^2, \quad i,j = \mathrm{V} \equiv 0, \mathrm{PC} \equiv 1, \mathrm{TB} \equiv 2, \qquad l = 1, \ldots, 9. \tag{2}$$

The aforementioned simplistic approach will now be enhanced based on the Markov chain model. This model describes the evolution of the probabilities of joint states at a given position x_0 and at discrete times. Recall that a random sequel $\{ X_n : n = 0, 1, 2, \ldots \}$ is the Markov chain if

$$P(X_{n+j} = j | X_n = i, X_{n-1} = i_{n-1}, \ldots, X_0 = i_0) = \\ = P(X_{n+j} = j | X_n = i). \tag{3}$$

If the transition probabilities (3) do not depend on n, the Markov chain is called homogeneous. The probabilities expressing the distribution of the Markov chain at a time step n

$$p_i(n) = P(X_n = i), \quad i = 0, 1, 2, \ldots, n = 0, 1, .. \tag{4}$$

In case of a non-homogeneous chain it holds

$$p_i(n+1) = \sum_{(j)} p_j(n)\, p_{ji}(n), \qquad i, j = 0, 1, 2, .., n = 0, 1, \ldots \tag{5}$$

The individual states, $0 \equiv$ V, $1 \equiv$ PC, $2 \equiv$ TB, can be combined to obtain $3^2 = 9$ initial joint states. Admitting the possibility of switching lanes increases the number of joint states to a total of fifteen states. The three of the six supplementary states are displayed in Figure 3. They demonstrate the newly emerged joint states, the process of a TB changing lanes, $1 \rightarrow 2$, has been phased in.

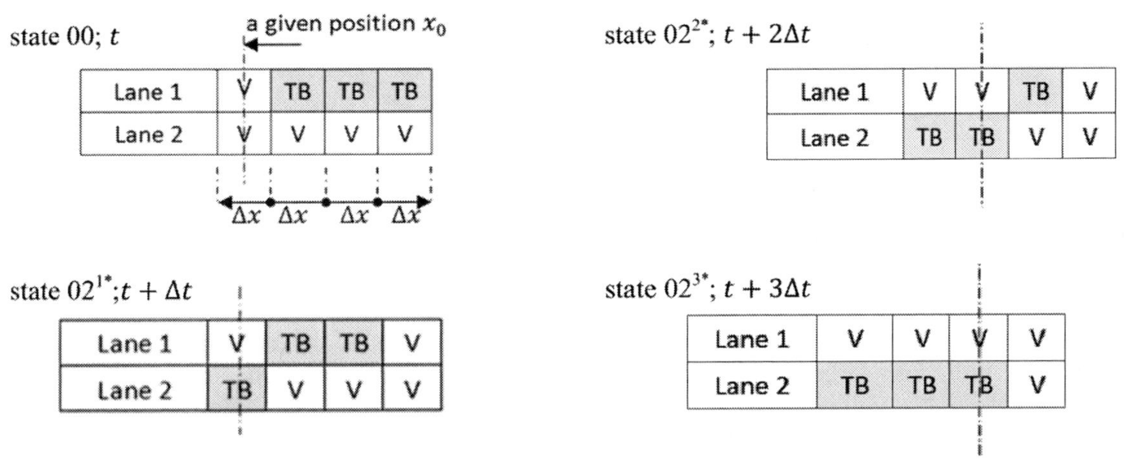

Figure 3. The process at a time instant $t = x/c$ in which a TB pulls from lane 1 to lane 2; c is the velocity of traffic flow.

In what follows, individual states n will be identified with time t and Eq. (5) will be formally rewritten into this matrix form:

$$\boldsymbol{p}(t + \Delta t) = \boldsymbol{P}^{\mathrm{T}}(t) \,.\, \boldsymbol{p}(t) \tag{6}$$

The current entries of the transition probability matrix $\boldsymbol{P}(t) = [p_{ij}(t)]$ need to be up-dated at the end of the previous step.

The entries of the column matrix $\boldsymbol{p}(t)$ correspond to nine initial joint states 00, 01, 02, …. , 22 supplemented with six additional states 02^{1*}, 02^{2*}, 02^{3*}, 20^{1*}, 20^{2*}, 20^{3*} reflecting the switching of lanes.

As the illustrative example, let us consider a two-lane traffic flow with a configuration depicted in Fig. 2. To be specific, let in the initial state there be 8 PCs and 1 TB detected in lane 1, and 8 PCs and 2 TBs in lane 2. The best likelihood estimates now can be used to obtain the transition probabilities. The whole transition probability matrix cannot be presented here because of the limited scope of the paper.

The two models of switching lanes are assumed in this illustrative example: (i) any PC ≡ 1 switches both lanes with the same probability $p_1^{12}= p_1^{21}= 0.5$ and, analogously, as to a TB $p_2^{12}= p_2^{21}= 0.05$, (ii) the probabilities of switching the lanes are different, namely $p_1^{12} = 0.5$, $p_1^{21} = 0.8, p_2^{12}= 0.05$, $p_2^{12}= 0.1$.

The non-stationary solution was carried out based on randomly chosen four non-zero initial probabilities: $p_{11}(0) = 0.64, p_{12}(0) = p_{21}(0) = 0.16, p_{22}(0) = 0.04$.

The probabilities of joint states in the initial phase strongly depend on the initial conditions and may swiftly vary and differ from each other. As an example, the evolution of the probabilities of joint states 02 and 12, respectively, are displayed in Fig. 4a, b considering distinct assumptions about transition probabilities with which the vehicles switch lanes 1 and 2.

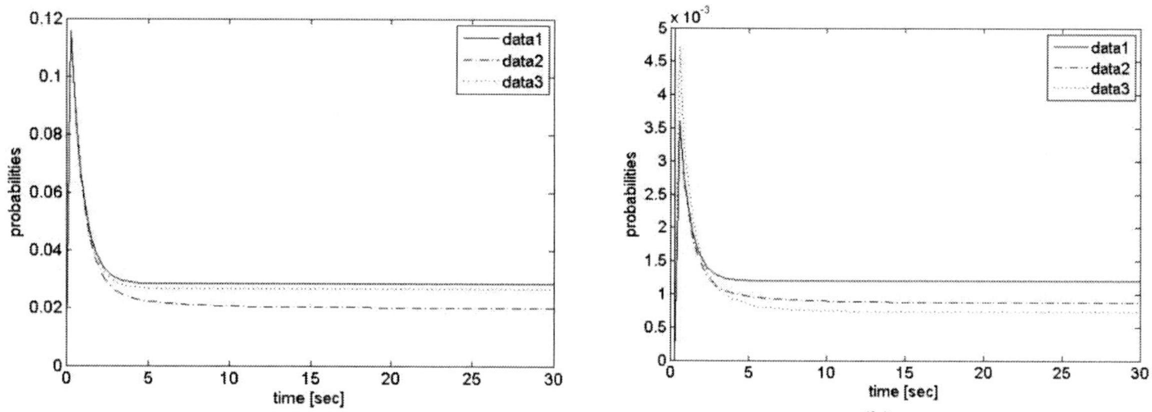

(a) (b)

Figure 4. The evolution of the probabilities of joint states (a) 02 and (b) 12 (data1- no switching lanes; data2 - switching lanes with the same probabilities; data3 – switching lanes with divers probabilities)

In what follows, the fire incidents are regarded as components of the Poisson process with the intensity λ_f. As by far a largest number of fires are attributed to self-ignition, the rate of fires per 10^9 [veh.km], see Eq. (1), will be taken for the intensity of the process. For a more detailed risk analysis, the total counts should be split at least into two sub-categories involving one TB and/or two TBs. In more detailed modeling, series configurations may be allowed for as well.

Denote the probabilities of states conditional on the fire accident as $P_f^{[i]}$, $[i] = 1, 2$, and express the respective fire intensities as

$$\lambda_f^{[i]} = \kappa^{[i]}\lambda_f \,, \qquad \sum_{i=1}^{2} \kappa^{[i]} = 1. \tag{7}$$

The probability mass function of heat, Q, developed during a fire, $p_Q(q)$, has already been discussed in Section 2. In the optimal case, functions $p_Q^{[i]}(q)$ would be desirable for all possible configurations of vehicles. However, the reduction of these configurations into two sets of the utmost importance seems to be acceptable in this pragmatic model, so only the two functions displayed in Figs. 1 (a) and (b) will be adequate.

On the other hand, we should not disregard the fact that, in general, the parameters $P_f^{[i]}$, $\kappa_{[i]}$, λ_f may vary along the length of the tunnel.

Considering that risk is the probable damage and with regard to the chain rule of conditional probabilities, a general formula allowing us to predict the average risk of a fire in road tunnels takes this lucid form

$$RISK = T\{\mathrm{AADT}\} \sum_{[i]} \sum_{(q)} p_Q^{[i]}(q) \int_L C^{[i]}(q,x) P_f^{[i]}(x)\, \kappa[i](x) \lambda_f(x) \mathrm{d}x, \tag{8}$$

where T [day] is the period of time for which the risk is calculated, L [km] is the length of the tunnel, $C^{[i]}$ is thhe cost that must be expended to eliminate the damage caused by a fire. It should be pointed out that $C^{[i]}$is a non-linear function of q and, therefore, any simplifications of Eq. (8), no matter how tempting, cannot be recommended. This variable is also a function of x due to the varying effect of a fire along the tunnel. Putting the accent on the technological aspects of damage, the cost $C^{[i]}$ in Eq. (8) may be alternatively substituted by the volume of the damaged concrete lining to be restored. We should further remember that AADT is the annual average daily traffic per direction [veh.day^{-1}] and as such it is held constant in Eq. (8).

4. Hydro-thermo mechanical model for the prediction of fire risk in road tunnels

The mathematical formulation of the material model consists of two governing equations representing the conservation laws of heat and mass. Their detailed description along with the corresponding boundary and initial conditions as well as numerical solution can be found in [8]. In the same source the model parameters have been identified based on a macroscopic test carried out on large scale panels made from fly ash concrete reinforced with FORTA-FERRO fibers and loaded in furnace (RWS fire curve). The results of numerical simulations are compared with those obtained from measurements in Figure 5.

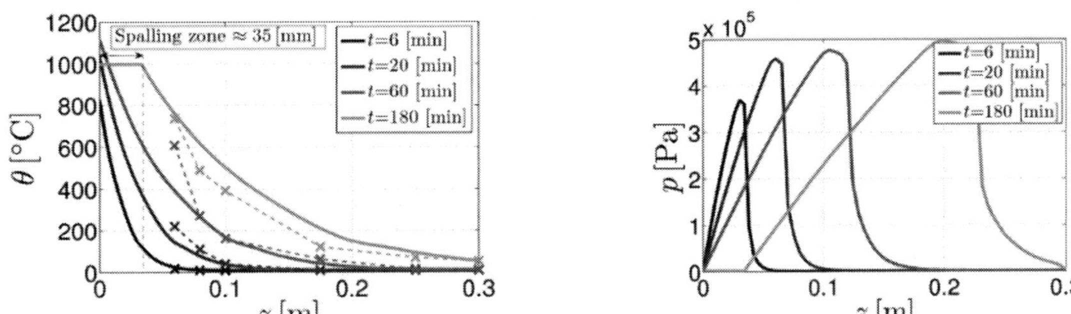

Figure 5. Numerical result for an optimized set of material parameters in different time steps: (a) Evolution of temperature, (b) evolution of pore pressure.

5. Risk assessment of fire in a road tunnel, case study

In this section we adopt formulas presented in [6] and references therein. The maximum ceiling temperature $\theta_{\mathrm{ceil,max}}$ [°C] in a specific region is expressed as

$$\theta_{\mathrm{ceil,max}} = \begin{cases} \theta_{\mathrm{reg,1}} & \theta_{\mathrm{reg,1}} < 1350[°\mathrm{C}], \\ 1350 & \theta_{\mathrm{reg,1}} \geq 1350[°\mathrm{C}], \end{cases} \theta_{\mathrm{reg,1}} = 17.5 Q^{\frac{2}{3}} H_{\mathrm{eff}}^{-\frac{5}{3}}, \tag{9}$$

where H_{eff} [m] is the effective height of the tunnel and Q [MW] is the heat release rate. This term is then used in the definition of the ceiling temperature $\theta_{\mathrm{ceil}}(t,x)$ [°C], which is a function of location and time as

$$\theta_{\mathrm{ceil}}(t,x,z) = c_{\mathrm{time}}(t) c_{\mathrm{loc}}(x)\, c_{\mathrm{loc}}(z) \theta_{\mathrm{ceil,max}} \,. \tag{10}$$

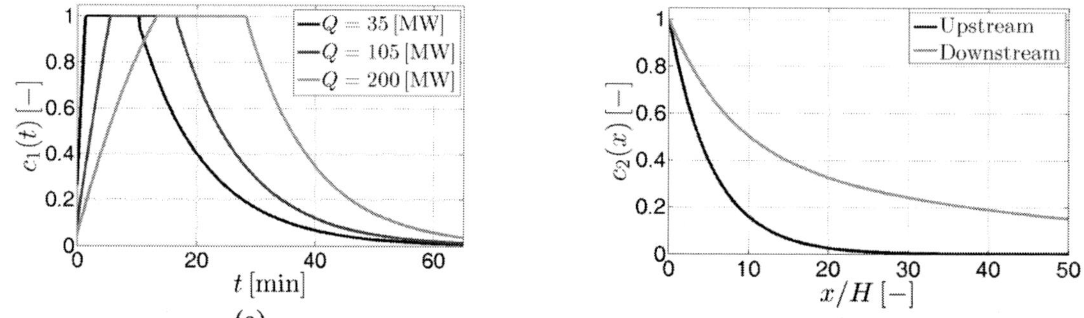

Figure 6. Evolution of dimensionless coefficients: (a) c_{time} [−], (b) c_{loc} [−] .

The evolutions of c_{time} [−] for typical values of Q [MW] and c_{loc} [−] used as example in this study are shown in Figure 6. The 3D nature of the problem is captured by using a parabolic function of vertical coordinate z to describe the distribution of the gas temperature along the tunnel lining.

The typical values of Q (5, 10, 30, 35, 60, 100, 105, 130, 200[MW]) representing the power of fire due the ignition of one and/or two vehicles, respectively, were considered to take into account the selected scenarios in two-lane road tunnel. The damaged volum [m^3] of the tunnel lining at the end of simulation (after 3 hours) is displayed in Figure 7 (a). In this study the volume V_{dam} [m^3] is limited to the material volume loss when the temperature exceeded a limit value $\bar{\theta}$ given the specified value of Q. It consists of the volume V_{spal} [m^3] destroyed by spalling, and the volume V_{deg} [m^3] degraded through the passage of high temperature. Both volumes constitute a material basis of risk analysis.

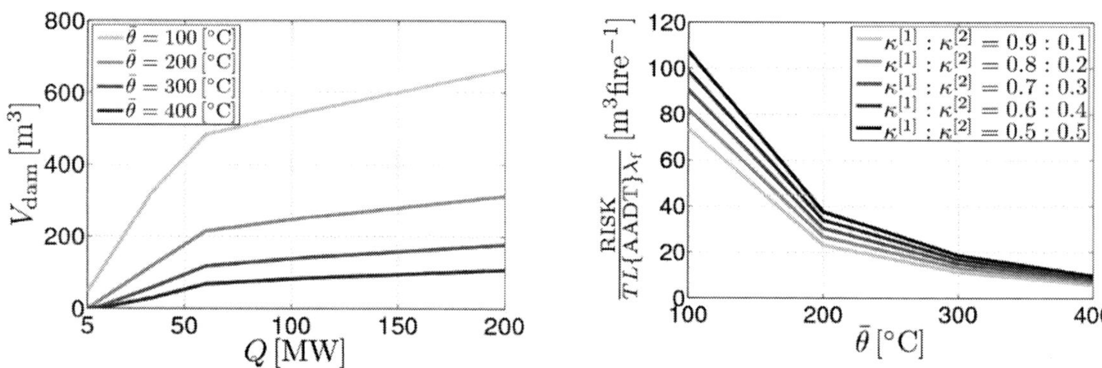

Figure 7. (a) Volumes of the tunnel lining degraded by spalling and high temperatures exceeding a given limiting value $\bar{\theta}$ [°C], (b) random damage relative to the number of fire incidents.

The right hand side of Eq. (8) is graphically represented in Figure 7(b) and can be regarded as the damage randomly caused by a single fire accident. To be specific let us consider a two-lane road tunnel, 2 [km] long, with AADT= 17000 [veh.day^{-1}], see Table 3, and T = 1 [year] = 365 [days].

Finally, make a tentative guess of the fire intensity $\lambda_f = 30 \cdot 10^{-9}$ [fires (veh.km)$^{-1}$], see Table 2 and permit $\bar{\theta}$ = 150 [°C] as the maximum temperature the tunnel lining is able to support without an essential repair.

It follows from Eq. (8) and Fig. 7(b) that *RISK* ranges from 13 to 20 [m^{3}year^{-1}]. Considering that a total volume of the lining struck by a fire along a 520 [m] long stretch is V = 3463 [m^3], we arrive at the value of risk relative to this volume $RISK_{rel}$ = (0.38 ÷ 0.58) [%] of damaged concrete per year.

6. Conclusions

In this paper, we proposed a pragmatic model for the risk analysis of road tunnels engulfed by a fire. In the probabilistic description of the traffic flow the Markov chain model, non-homogeneous in general, appeared to be a suitable tool to cover this problem. Admitting the possibility of switching lanes increases the number of joint states, for which the Markov chain must be developed. Apparently, the probability of changing lanes must be known or reasonably estimated. Our numerical experiments with varying initial conditions (see Section 3) imply that the traffic flow becomes homogeneous (stationary) within 300 – 400 m from the initial position (semaphore). This fact considerably simplifies the assessment of the probable damage due to a fire.

The probabilistic modelling of the heat release rate is another factor worthy of improvement. The probability mass functions of the respective HRRs strongly depend on the scenarios that must be taken into account when estimating the consequences of a fire.

Last but not least, the fire intensity, λ_f, should not be omitted in this discussion. Insufficient data on the type of vehicles involved in the fire incident do not allow for the fire intensity of joint states to be grasped sufficiently well. Moreover, we have inadequate knowledge of the variability of λ_f along the tunnel, but taking this quantity for a constant seems to be a reasonable compromise.

All things considered, the proposed model is applicable for risk analyses in sufficiently long tunnels and in all cases where the traffic flow is already homogeneous. It will be further augmented while estimating damage caused by fire incidents at the proximity of portals. Except for exploiting the methods based on CFD, the expert approaches to the prediction of the upstream/downstream gas flows are close at hand.

Acknowledgements

The financial support of the project no. TE01020168 (Center for Effective and Sustainable Transport Infrastructure – CESTI) provided by TA ČR is gratefully acknowledged.

References

[1] ASFiNAG Wienn 2013 *Auswertung der ASFINAG-Tunnelbrandstatistik 2006-2012*
[2] Caliendo C and Guglielmo M L D 2012 Evaluation of traffic and fire accidents in road tunnels, and a cost-benefit analysis *International Journal of Civil Engineering Research* **3**(3) pp 201–222
[3] Caliendo C, Guglielmo M L D and Guida M 2013 A crash-prediction model for road tunnels *Accident Analysis & Prevention* **55** pp 107–115
[4] Bažan Z P and Thonguthai W 1997 Pore pressure in heated concrete walls: theoretical prediction *Magazine of Concrete Research* **31**(107) pp 67–76
[5] Künzel H M and Kiessl K 1996 Calculation of heat and moisture transfer in exposed building components *International Journal of Heat and Mass Transfer* **40**(1) pp 159–167
[6] Ingason H, Lönnermark A and Li Y Z 2011 *Runehamar tunnel fire tests* (SP Technical Research Institute of Sweden)
[7] Hurley M J 2015 (editor in chief) 2015 *SPFE Handbook of fire protection engineering, 5th edition* (Springer)
[8] Sýkora J, Jarušková D, Šejnoha M and Šejnoha J Fire risk analysis focused on damage of the tunnel lining, submitted to *Fire Safety Journal* (Elsevier)

IOP Conf. Series: Materials Science and Engineering **236** (2017) 012080 doi:10.1088/1757-899X/236/1/012080

Application of fire and evacuation models in evaluation of fire safety in railway tunnels

Kamila Cábová[1], Tomáš Apeltauer[2], Petra Okřinová[2] and František Wald[1]

[1]Department of Steel and Timber Structures, Faculty of Civil Engineering, Czech Technical University in Prague,Thakurova 7, Prague 16629, Czech Republic
[2]Brno University of Technology, Faculty of Civil Engineering, Veveri 331/95, 602 00 Brno, Czech Republic

E-mail: kamila.cabova@fsv.cvut.cz

Abstract. The paper describes an application of numerical simulation of fire dynamics and evacuation of people in a tunnel. The software tool Fire Dynamics Simulator is used to simulate temperature resolution and development of smoke in a railway tunnel. Comparing to temperature curves which are usually used in the design stage results of the model show that the numerical model gives lower temperature of hot smoke layer. Outputs of the numerical simulation of fire also enable to improve models of evacuation of people during fires in tunnels. In the presented study the calculated high of smoke layer in the tunnel is in 10 min after the fire ignition lower than the level of 2.2 m which is considered as the maximal limit for safe evacuation. Simulation of the evacuation process in bigger scale together with fire dynamics can provide very valuable information about important security conditions like Available Safe Evacuation Time (ASET) vs Required Safe Evacuation Time (RSET). On given example in software EXODUS the paper summarizes selected results of evacuation model which should be in mind of a designer when preparing an evacuation plan.

1. Introduction

Historically, fire safety has been restricted by prescriptive standards. These prescriptive codes including many requirements are based on traditional fire testing techniques. Using of these codes may lead to more expensive and inefficient solutions. In many countries the performance based design is widely accepted as a natural and inseparable evolution of fire engineering. Fire engineers or structural designers may use knowledge of latest research to optimize their design, not only the cost and life safety, but also environmental impact, in practice. In fire safety engineering, natural evolution is associated with an increase of exact knowledge about the fire dynamics. Development of computational technology, numerical analysis and mathematical modelling has also contributed to change the approach. Using all benefits of the knowledge and computer development, fire safety in railway tunnels is investigated by application of fire and evacuation models.

During a fire in a tunnel there are three-dimensional flows which are influenced by buoyancy effects. In addition, the flow velocity and length scales are sufficiently large that the flow is generally turbulent. At the fire, a combustion process takes place, and so there are chemical reactions going on, producing soot particles and combustion products at high temperatures [1]. Part of the heat is transferred by radiation and part by convection to the tunnel walls. Generally, the problem of fire safety in tunnels increases with their length. In case of railway tunnel fire dynamics is influence by high longitudinal flow due to natural ventilation and also supported by moving trains (known as "piston effect"). The problematic of fluid dynamics in case of fire in railway tunnels is therefore very complicated. However it can be solved numerically, when correct method is applied. The degree to which qualitative agreement is achieved varies with the modeller and the situation.

2. Design fire models for tunnels

In the design phase fire safety is evaluated from the length of a tunnel of 350 m, which is considered as the length with higher risk for human safety. Temperature resolution and development of toxic gases in a tunnel are together with visibility considered as the most important parameters affecting the safety of people involved in tunnel accidents. Gas temperature may be evaluated based on the temperature curves and simple empirical equations. However, the aim is to determine the exact behavior of the fire and the spread of toxic gases in the tunnel, which corresponds to reality, and to find the balance between the tunnel safety equipment components and their contribution to improvement of traffic safety at the site. In this case it is advantageous to use a numerical simulation. Findings from the numerical simulation of fire and smoke development enable to improve models of human evacuation during fires in tunnels. Together with the optimization of the elements of safety equipment of tunnels and evacuation scenarios the use of numerical solutions may help to improve emergency procedures of IRS in the case of tunnel fires or to assess the behavior of materials of tunnel lining under fire even with sudden cooling caused by the intervention of firefighters.

3. Numerical modelling of fire

Computational Fluid Dynamics (CFD) modelling is a numerical approach to representing fluids that divides a fluid domain into a number of smaller sub-domains, resulting in the generation of a mesh of cells (control volumes). Three-dimensional, time-dependant partial differential equations of conservation of mass, momentum and energy transfer and conservation of species are written for each control volume based on fundamental equations of fluid dynamics, thermodynamics, chemical reactions and mechanics. Appropriate initial boundary conditions are then applied to find numerical solutions to these equations. Currently available CFD models include sub-models for solution of burning and heat transfer, so they provide a framework for including all phenomena which are present during fire in a tunnel into a calculation.

One of the freely available numerical code, Fire Dynamics Simulator (FDS), developed by National Institute of Standards and Technology (NIST), is frequently used for simulation of fires. It is a numerical solver to simulate a flow and movement of fluids, also caused by burning. FDS solves numerically a form of the Navier-Stokes equation appropriate for low-speed, thermally-driven flow with an emphasis of the transport of heat and smoke from fires [2]. The code consists of several independent models, for example model of burning. To interpret data obtained from FDS postprocessor Smokeview is used.

Despite the huge advantage of numerical simulations, a level of reliability and accuracy of numerical predictions must be always evaluated. Therefore, when using numerical simulation verification and validation is always recommended. Verification and validation examples of prediction of gas temperature resolution and smoke distribution during the fire in a railway tunnel are presented in [3].

Considering tunnel fires, the thermo-fluid-dynamics behaviour is affected by several internal and external factors, such as the barometric pressure at the portals, the tunnel slope, the set-point of the ventilation system, and others [4]. Reliability of computer predictions of such complex system has been evaluated in many studies. Ventilation velocity and backlayering distance was validated in [5] and [6]. In [5] and [7] local field data as velocity and gas temperature in the vicinity of the fire source were predicted and validated to experimental results. Based on results of these and other studies Colella stated [8] that CFD method can be used to produce sufficiently accurate results of flow pattern and temperature gradient located in close areas of the fire. However, as it was mentioned above reliability of numerical solution and accurateness of results have to be checked.

4. Numerical modelling of passengers evacuation

Applied simulation tool in this case is EXODUS, developed by a team Fire Safety Engineering Group at the University of Greenwich. The model geometry is filled with network nodes that are normally spaced at a constant interval of 0.5 m. Each node is connected with neighbours and represents elementary space that can be occupied by one person only (see Figure 1). Any change in cell size has noticeable effect on the model behaviour and can be carried out only in well-justified cases.

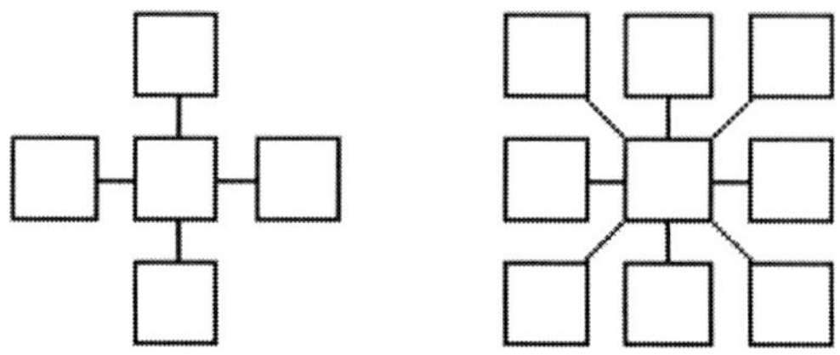

Figure 1. Elementary space represented by nodes with spacing 0.5 m.

Model also takes into account the particular passenger's parameters describing the age profile, reaction time and movement speed [9]. If it is necessary to apply Czech national standards, it is possible to use the inputs according to CSN 73 0802 [10].

Time before the movement is equally or more important than the movement time and includes activities difficult to calculate. *Time to recognize the danger* is the period from the start of the alarm the moment when people begin to respond to an alarm triggered. During this time, people continue with an activity before the alarm appeared. Variability of this time interval is very high, generally ranges from several seconds to many minutes and is different also among individuals in the same space. The *time of response to the danger* lasts from the moment a person realizes formation of critical situations, until the moment when they decide for a particular method of evacuation strategy. Among the typical activities in this period include luggage collecting, searching of other family members, informing others or assessing the appropriate escape routes. For the purpose of the model was therefore assigned persons reaction time in the interval 0-30 s with normal distribution.

5. Application to a railway tunnel

5.1. Fire model

A 1747 m long tunnel with an arc shape cross-section of 6.97 m width and 5.33 m high is undertaken to numerical study in FDS. Computational domain of dimensions 1747 m x 6.97 m x 5.33 m consists of five meshes. Size of mesh is determined according to rules given in [2]. Cell size in axis (x, y, z) equals to (0.345 m, 0.348 m and 0.355 m). In the region of fire source the grid is refined to half size, in axis (x, y, z) cell size equals to (0.182 m, 0.174 m, 0.178 m). Detail of meshing in the location of fire source is shown in Figure 2. The total number of cells is 1 530 600.

The tunnel lining is formed by 0.4 m thick layer of concrete with density of 2200 kg/m^3, conductivity of 1.3 W/mK and specific heat of 1.02 kJ/kgK. Tunnel portals are opened throughout the whole cross-sectional area (in FDS code plane surface type VENT is OPEN). In the tunnel there is negligible natural gas flow as the high difference of both portals is very small. Initial gas flow velocity is therefore set to 0 m/s. Gas temperature before fire ignition is considered as 10°C.

Fire scenario of a 20 MW fire in a passenger train wagon is simulated by time dependant heat release rate corresponding results of fire test of passenger train wagon described in [11]. The heat flux is released through four window openings, each of area of 1.44 m^2, situated 2 m above the ground level. The fire (areas releasing the heat flux) is located in the third wagon from the total of five wagons. The train stopped after 935 m after entering its portal (stands between 810 m and 935 m of the tunnel length). In FDS burning is simulated by mixed-fraction ratio of polyurethane (defined by fractions of carbon, hydrogen, oxygen, nitrogen and soot yield particles - 'C_6.3 H_7.1 O_2.1 N_1.0 SOOT_YIELD_0.10'). Development of rate of heat release is based on t-quadratic curve for ultra-fast fire.

In the model there are sensors to measure high of smoke layer ('SMOKE LAYER HEIGHT'), temperature of hot gas layer and temperature of gas layer below the smoke level ('UPPER TEMPERATURE' a 'LOWER TEMPERATURE'). These sensors are situated in 50 m span sections in the axis of the tunnel tube and in the position of 0.6 m from the side lining, which should control the area of safety evacuation path.

In FDS turbulence model is applied by Smagorinski formulas of large eddy simulation (LES) with coefficient Cs equals to 0.2. Heat transfer by radiation is applied by 100 discrete angles.

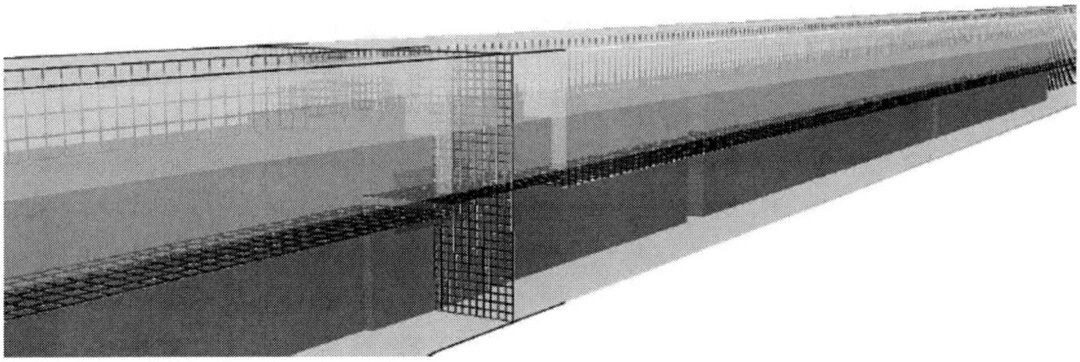

Figure 2. Meshing of the model.

5.2. Discussion on selected results of fire model

As it was expected, after fire ignition layer of hot smoke gases accumulated below the tube arch. The tunnel linings cooled hot gases down and smoke sank to lower level of the tunnel. Visualisation of smoke originated from the fire of passenger train wagon in 2 min and 5 min is given in Figures. 3 - 4. In 10 min from the ignition the entire tunnel cross section is full of smoke. The speed of smoke spread is slowing down closer to the portals. In 20 min smoke reaches a distance of 340 m from ignition sources.

Temperature of hot smoke layer calculated in the axis of the fire source reaches the peak value of 800°C. Figure 5 illustrates the comparison of calculated hot smoke layer temperature in the numerical model and temperature curves which are usually used in the stage of design. From the figure it is obvious that numerical model gives lower temperatures. Hydrocarbon temperature curve as well as RABT ZTV temperature curve for trains gives too conservative estimation.

Figure 6 shows decrease of smoke layer high along the tunnel length in 10 min. In the diagram three curves are introduced – curve of smoke layer high calculated by numerical model in the axis of evacuation path (0.6 m from the tunnel linings), curve of smoke layer high calculated in the axis of the tunnel cross-section and a curve indicating the level of maximal smoke layer decrease according to ČSN 73 7508. In the figure proves that the calculated high of smoke layer in the axis of evacuation path and also in the axis of the tunnel is lower than the level of 2.2 m. This level is considered as the maximal limit for safe evacuation. From the view of safe evacuation temperature of 80°C should be also controlled in the high of 2 m above the evacuation path.

IOP Conf. Series: Materials Science and Engineering **236** (2017) 012080 doi:10.1088/1757-899X/236/1/012080

Figure 3. Visualization of smoke development during passenger train fire in 2 min.

Figure 4. Visualization of smoke development during passenger train fire in 5 min.

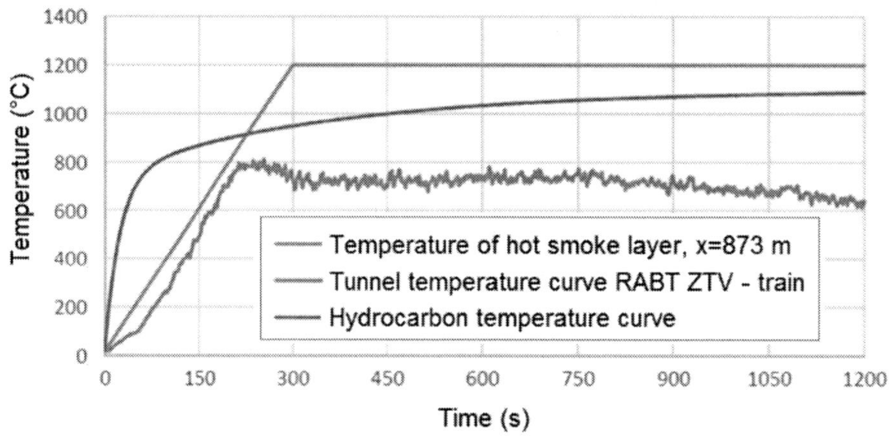

Figure 5. Comparison of temperature of hot smoke layer from FDS with simplified temperature curves.

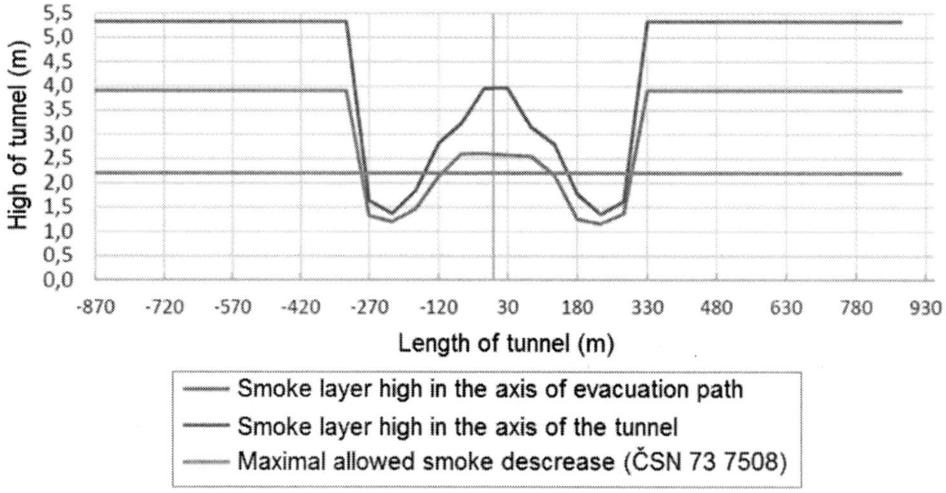

Figure 6. Smoke layer high above the tunnel floor during passenger train fire in 10 min.

5.3. Selected results of passenger's evacuation model

The key advantage of using simulation tools in such situations is the ability of statistical processing of the results of several tens or hundreds of the same simulation with random sweep key parameters and the initial position of persons, while evacuation experiment can be performed at the best for only a small number of repetitions. The sensitivity analysis of the results is also quite easy. Comprehensive evaluation of the model behavior is therefore beyond the scope of this text, so the results with more general impact on safety were selected.

The simulation results show, moreover, that the course of evacuation significantly affects the order of the vehicle carriage. People escaping from the first and last carriages are very significantly restricted in movement by persons that are no longer found on the sidewalk, and have traveled in wagon closer to the set. On the Figure 7 there is a comparison during the evacuation of people from the first and the third carriage, wherein it is clearly evident slowing the process of evacuation of the first wagon in the second half of the time interval when the sidewalk in front of the carriage is already full of persons mainly from other carriages.

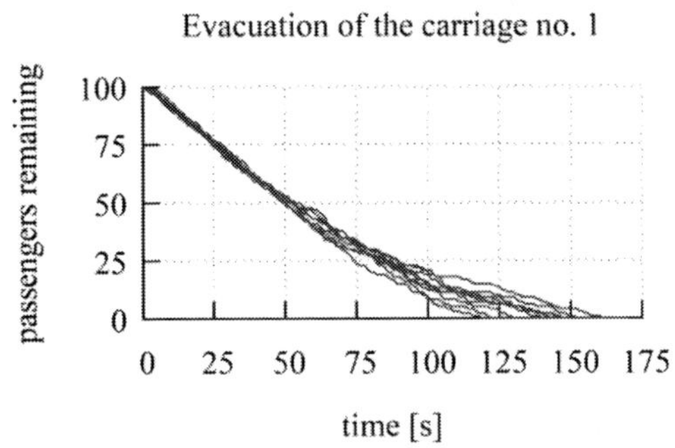

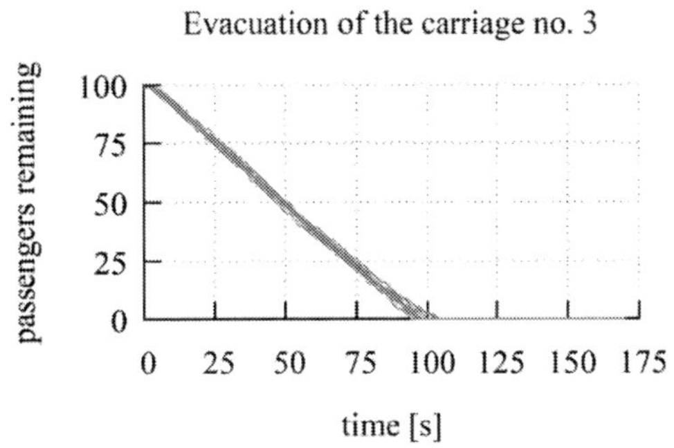

Figure 7. Influence of the order of the vehicle carriage on evacuation of wagon no. 1 and no. 3. Each variant shows results of 10 separate simulations of the model.

From an overall comparison of Figure 8 it is then clear that the time differences of evacuation can vary between different carriages significantly. Rather than a specific time of the evacuation of the wagon it is precisely this fact that generally applicable and it is important to have it in case of evacuation plan in mind.

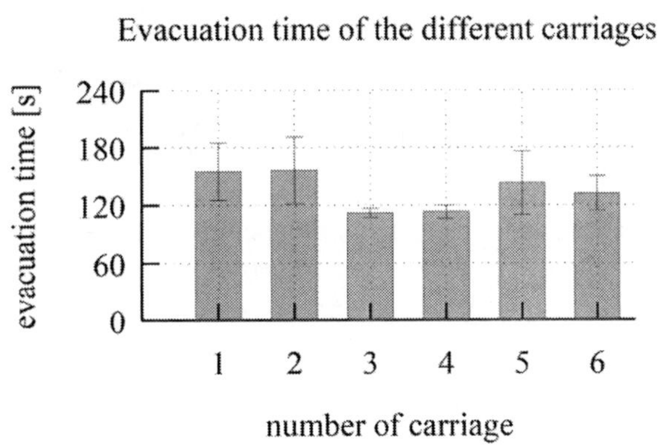

Figure 8. Evacuation time of each carriage. For each wagon the mean value of the time required for evacuation and the standard deviation of this time as the result of 10 simulations is shown. It is obvious that the evacuation time is significantly different and depends on the carriage position.

High density of people, which can lead to serious injury or death even without the presence of high temperatures and toxic fumes is also an important security risk during the evacuation process. Density of people usually shown through so-called Level of Service and using the six-scale A-F, where A is the ideal stage, grade D or higher is already regarded as an area of increased risk of injury. On the Table 1 the particular pedestrian densities for different environments and levels of services are shown.

Table 1. Level of service according to Fruin [12].

Sidewalks		Stairs		Waiting areas		LoS
Min	Max	Min	Max	Min	Max	
0	0.308	0	0.538	0	0.828	A
0.308	0.431	0.538	0.718	0.828	1.076	B
0.431	0.718	0.718	1.076	1.076	1.538	C
0.718	1.076	1.076	1.538	1.538	3.588	D
1.076	2.153	1.538	2.691	3.588	5.382	E
2.153	-	2.691	-	5.382	-	F

The situation is modeled in Figure 9, which captures the area around the middle carriage no. 4. On the picture the pedestrian density is described with equidistant time step 30 s. The color scale corresponds to levels of quality pedestrian traffic, while the red shows the worst situation of people with a density of over 2,153 persons.m^{-2}. The results can be identified as being critical points around the evacuated wagons on the sidewalk and in niches.

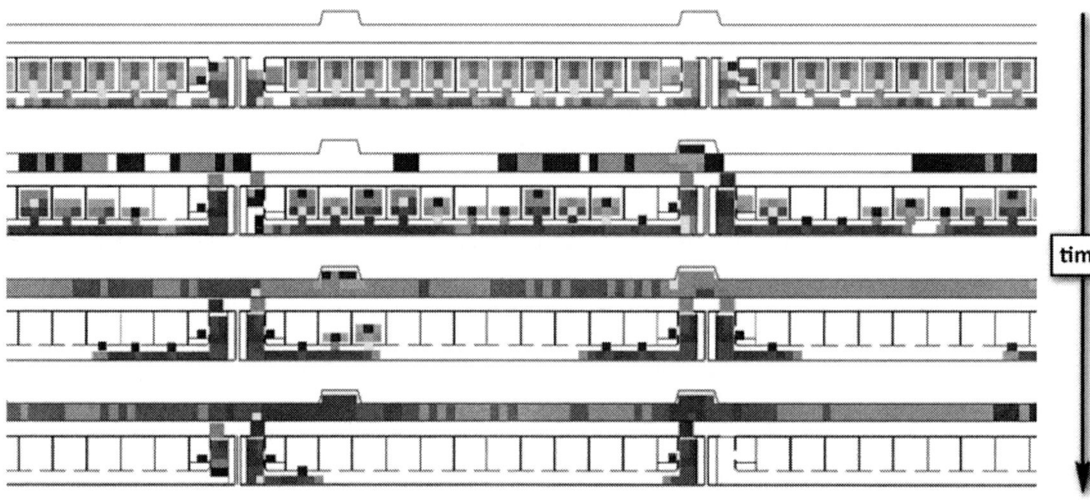

Figure 9. Displaying the pedestrian density around the carriage no. 4. Especially in the later stages evacuation is critical due to increase of the density of people evacuated around cars and thus the risk of injury is increased.

6. Conclusion

Commission Regulation (EU) no. 1303/2014 from November 2014 concerning the technical specification for interoperability relating to safety in railway tunnels rail system (TSI) has brought many new facts. These facts also include the possibility to develop an evacuation plan with respect to the above described safety risks. The models of evacuation and fire are the optimal tools.

Using these models, it is possible at present to analyse the evacuation process and its results are substantially wider than it is using classical engineering calculations based on prescriptive standards.

In the same way as evacuation models, correctly applied CFD models may reproduce the qualitative behavior of fire in each tunnel. The degree, to which qualitative agreement is achieved, is necessary to find out by the aid of verification and validation process. Accurate prediction of fire and smoke spread in tunnels on the bases of physical properties is a promising approach how to stress the human safety and fire design economy.

Acknowledgements
Work on the research presented in this paper is supported by project no. TE01020168 – Centre for Effective and Sustainable Transport Infrastructure, in the frame of Competence Centres program of Technology Agency of the Czech Republic (TA CR).

References

[1] Rhodes N 2005 CFD modelling of tunnel fires *Handbook of Tunnel Fire Safety* (London: ICE Publishing) p 329-46
[2] McGrattan K et al 2016 Fire Dynamics Simulator (Version 6.4) *Technical Reference Guide* (NIST Special Publication 1019-6 US Government Printing Office) p 86
[3] Cábová K, Apeltauer J and Wald F 2016 Verification and validation of a numerical model of fire and smoke development in a railway tunnel *Acta Polytechnica* **56**(6) pp 432-9
[4] Beard A and Carvel R (ed) 2005 *Handbook of Tunnel Fire Safety* (London: ICE Publishing)
[5] Vauquelin O and Wu Y 2006 Influence of tunnel width on longitudinal smoke control *Fire Safety Journal* **6** vol 42 pp 420-6
[6] Van Maele K and Merci B 2008 Application of RANS and LES field simulations to predict the critical ventilation velocity in longitudinally ventilated horizontal tunnels *Fire Safety Journal* **8**(43) pp 598-609
[7] Vega M G, Arguelles Diaz K M, Oro J M F, Ballesteros T R and Morros C S 2008 Numerical 3D simulation of a longitudinal ventilation system: Memorial Tunnel case *Tunelling and Underground Space Technology* **5**(23) pp 539-51
[8] Collela F, Verda V, Borchiellini R and Rein G 2005 One-dimensional and multi-scale modelling of tunnel ventilation and fires *Handbook of Tunnel Fire Safety Chapter 17* (London: ICE) pp 365-90
[9] RIMEA 2009 Richtlinie für Mikroskopische Entfchtungs-Analysen
[10] Úřad pro technickou normalizaci, metrologii a státní zkušebnictví (ÚNMZ) 2009 Požární bezpečnost staveb - Nevýrobní objekty *(ČSN 73 0802)*
[11] White N 2010 *Fire development in passenger trains* (Victoria University, Melbourne)
[12] Fruin J. J. 1987 *Pedestrian Planning and Design* (Elevator World, Inc.) p 206

Site investigation as tool for elimination of natural hazard impact on construction project

Pavel Pospíšil[1] and Alexandr Rozsypal[2]

[1]VSB – Technical University of Ostrava, Faculty of Civil Engineering, 17. listopadu 15, Ostrava-Poruba, 708 33, Czech Republic
[2]Brno University of Technology, Faculty of Civil Engineering, Veveří 95, Brno, 602 00, Czech Republic

E-mail: pavel.pospisil@vsb.cz

Abstract. Natural hazards and their impact on traffic structures represent the most risk factor during the construction phase and maintenance of traffic structure. The article deals with the process of site investigation focused on elimination of natural hazards impact on the structure. It is described the importance of gradual assessment especially of rock environment conditions in relation to specific traffic structure. Significant role plays cooperation between foundation designers and investigation workers especially for right classification of geotechnical category and final determination of characteristic values for documented rock formations. Positive and negative role of standardised approaches included in various standards are also.

1. Introduction

High-quality construction process especially in transportation engineering guaranties quality final structure as road/highway, railroad, pipeline etc. Transportation structures has been both very demanding from the point of view of design and construction and vulnerable by natural hazards, accidental and deliberate hazards in all phases of construction and utilization [1].

It is essential to understand basic concepts of the risk analysis and risk management theory including terminology.

A hazard (threat) is a potential threat of an undesirable event occurring in the future. It is a state of endangering by the undesirable event occurrence and resulting in a consequential damage origin. The hazard is a dimensionless parameter. The hazard has not been established as a legal concept.

A risk scenario is a manner in which the undesirable event may occur, circumstances, and conditions, under which the undesirable event occurs, and its development in time.

A hazard source is a factor, which is the hazard originator, which may be an insufficiently investigated rock massive, human factor, untested construction technology, exceedingly daring design of the structure, rapidly changing outer environment at the area of interest, external influences, such as poor weather conditions, or even danger of natural disaster, and the like.

An undesirable event (occurrence) is an aggregate process, an event, the occurrence of which will cause a material or immaterial damage or any other financially expressible detriment to the risk bearer.

A damage (harm) is a legal concept. The damage is a particular material/immaterial detriment caused to the injured party (the risk bearer). They are unforeseen costs related to the surmounting of the undesirable events or removing consequences of their occurrence, lost property benefit, and the like. The final consequence is mostly a financial loss, which has already arisen, or which is most likely to arise after the undesirable event occurred.

Natural hazards threaten all engineering projects. It is necessary to identify all potential natural hazards in the first phase of the ground investigation.
Natural hazards including: Earthquakes (including tsunamis), volcanoes, floods and droughts, windstorms, including hurricanes and tornados, erosion, ground displacements, including landslides, frost heave, and settlement, avalanches, icing, chemical threats (degradation of concrete structures), wildfires, geomagnetism.

Other important concepts of a risk assessment is vulnerability of the particular structure and the system performance. Vulnerability includes the potential for physical damage and loss of life with respect to: Physical facilities, functional systems, environment, administrative/financial activities, human safety.

System Performance includes the consequences resulting from system damage or disruption as measured by: Capital and revenue losses, service disruption and downtime, casualties, hazardous materials release and environmental damage.

In addition to scientific works dealing with natural hazards and their impacts on the natural and urbanized environment [2], [3], [4], [5] and project titled "The Spatial Effects and Management of Natural and Technological Hazards in Europe - ESPON 1.3.1" [6] has been solved as a part of the European Union scientific programs ESPON, which dealt with a territorial assessment of specific regions of all European Union countries in aspect of a possible negative impact caused by the natural and technological hazards.

From the point of view of the natural hazards, especially southern and south-eastern EU countries may be potentially affected by endogenous hazards (earthquake and volcanism).

The main natural hazards in the central European regions affecting structures during their construction and lifetime are certain properties and behaviour of the rock environment and its changes in the course of time (ground displacement). Moreover, also exogenous influences, such as extreme precipitation causing the rock environment erosion, or possibly floods. Recently, drought has also started becoming a problem, affecting the quality of certain types of the rock environment (contraction of some of the rock masses). Storms, avalanches, icing, chemical hazards are local, not regional influencing factors. Furthermore, these events are very random ones. It is geomagnetism that plays a very specific role. The Earth´s geomagnetic field has been continually changing and may affect electronic instruments, and thus have a negate impact especially on the transportation and traffic management in the future. A general classification of the natural hazards for the specific EU regions is shown in Figure 1.

2. Elimination of natural hazards

The natural hazards elimination is a complex process, which must constitute an integral part of every construction project being solved, especially in transportation engineering. This can be achieved through a consistent application of the risk analysis and risk management in the process of the engineering project designing and execution.

2.1. Risk assessment

The risk analysis is a sequence of risk identification and quantification procedures. The risk analysis does not include a decision-making process focused on the risk reduction. It includes a parametric analysis of various possibilities, which can be taken into consideration to reduce the particular risks. (Which tools are taken into account to reduce the probability of the undesired event occurrence? Which actions and under what conditions can be applied to reduce the consequences of the undesired events occurrence, i. e. damage and loss? – and the like). The risk analysis also deals with the analysis of circumstances, under which potential hazards may arise, which result the undesired events, and with their development. It is focused on the assessment of outer as well as random factors, which may influence the course of the undesired events.

IOP Conf. Series: Materials Science and Engineering **236** (2017) 012081 doi:10.1088/1757-899X/236/1/012081

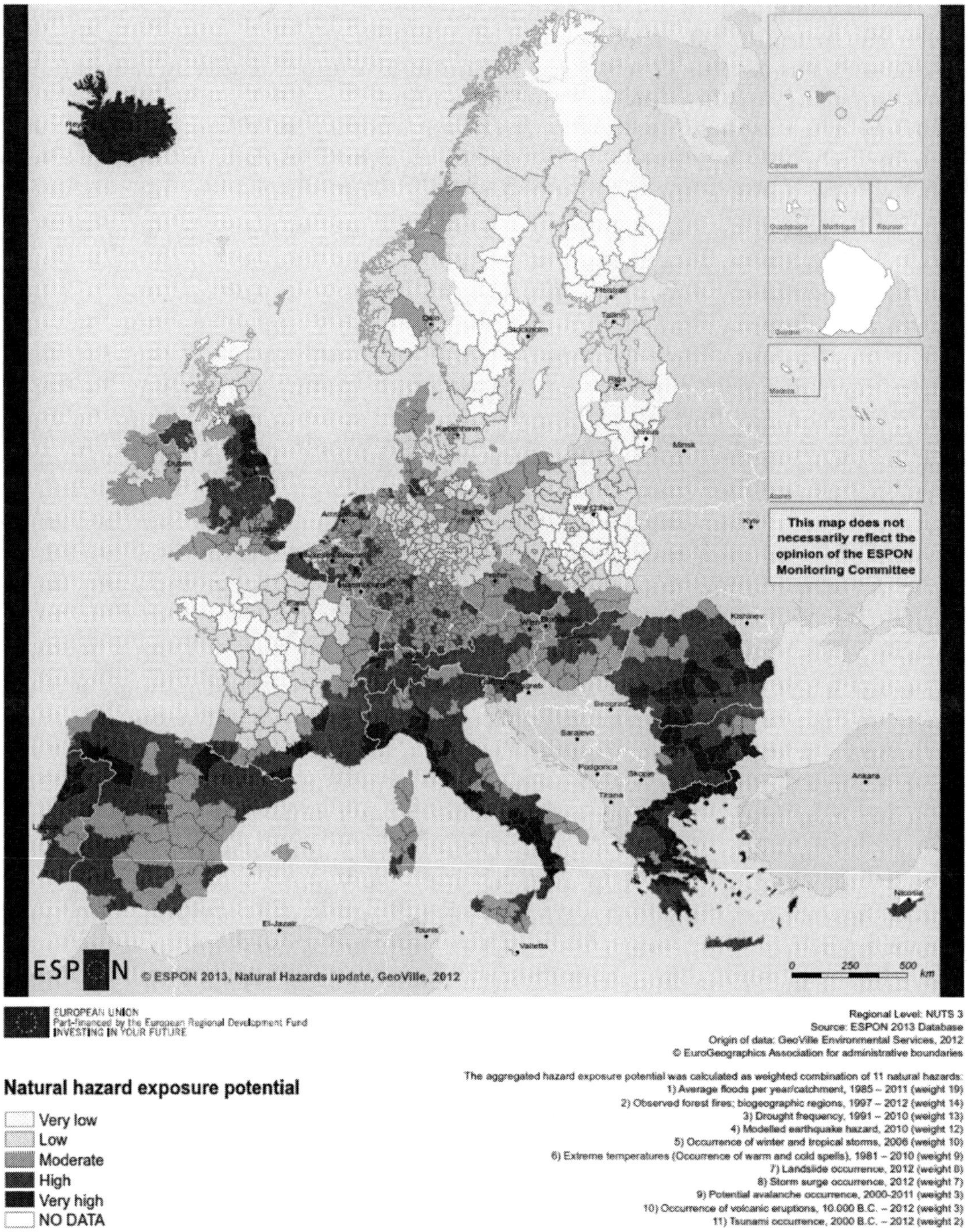

Figure 1. Natural hazard exposure potential. (Source. EU - ESPON 2006 Programme 1.3.1 [6])

The evaluated results of the risk analysis allow for the application of the risk management procedures at the project preparation and above all at the construction. The risk management is the base of optimal variants selection for the design solution and technological procedures in further construction process.

2.1.1. Risk identification. The risk sources identification is a procedure resulting in a list of all possible undesirable events, which may occur, and a description of circumstances, under which they may happen. Basically, it is the list of all hazards, which may be taken into account, and a description of scenarios. The risk identification concept is sometimes also indicated as a risk specification.

2.1.2. Risk quantification. The risk quantification is understood to be a procedure, which results in the risk expression in financial or other physical units. The risk quantification is based on determination or estimation of probability of occurrence its impact (Figure 2). The risk quantification is done either directly by calculation, or by expert estimations.

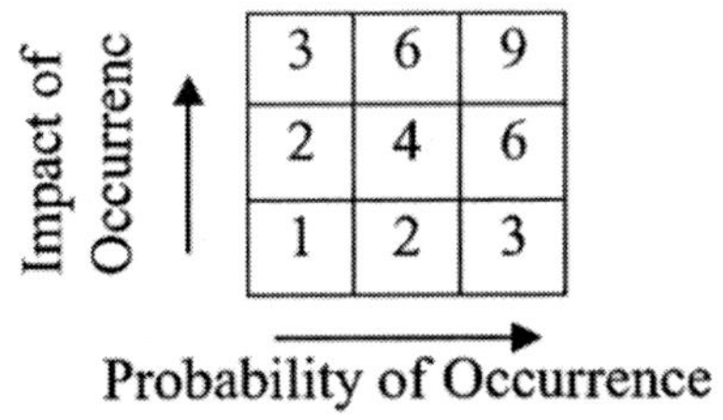

Figure 2. Risk analysis matrix [7].

The risk analysis finally results in assessment of system performance. The approach is shown in Figure 3.

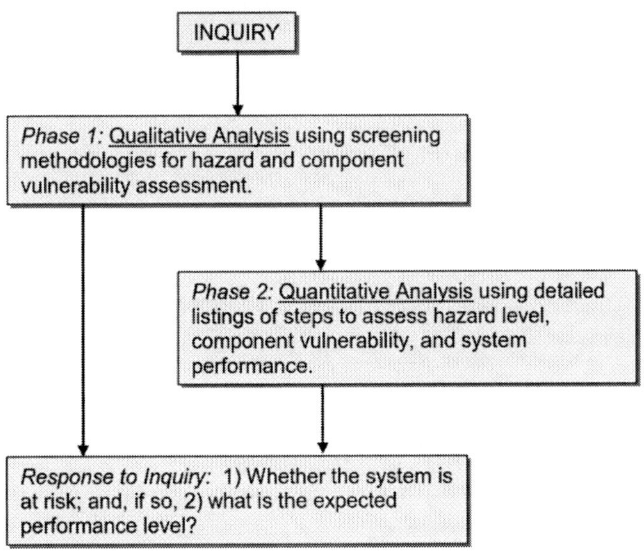

Figure 3. Two-phase approach to performance assessment [8].

2.2. Risk management (risk control)

The risk management is the risk analysis completed with a decision process aiming at the risk reduction to an acceptable level. It is a decision-making process based on the risk analysis result. At the beginning, the decision-making is to define the risk management goals and the risk management strategy. The latter determines the decision-making principles, goals, and criteria. The basic goals are as follows: Protect public safety, maintain system reliability, prevent monetary loss, prevent environmental damage.

A part of the risk management strategy is the acceptable risk level definition for specific risk bearers. The goal of the risk management is to reduce the risk amount under the acceptable level. The risk management is typically applied to decision-making processes comprising an uncertainty. (Under the conditions when all circumstances influencing the result are not sufficiently known, or they cannot be unambiguously quantified).

2.3. Specific features of decision-making process at risk management in construction engineering

The risk management at the preparation and construction of structures is typical of decision-making under the conditions of uncertainty, when several, or all variables influencing the decision are only partially or insufficiently known. The decision-making is therefore executed at a higher or lower uncertainty rate.

The use of deterministic methods to solve the problems related to the risks is therefore limited and requires a special carefulness.

That is why a probability approach is needed for most of the problems solved in relation to the risk management of construction of structures of various types.

The risk analysis and risk management are integral parts of the preparation and construction of every structure. They condition the structure's safe and, at the same time, economical construction. They allow for the construction using elements of the observation method when the construction progress is adjusted to the real behaviour of the system "rock environment (foundation ground) versus structure".

The risk management minimizes the situations, which may result in origination of emergency events, minimizing also the possible consequences of such events. Thus, the occupational safety is enhanced, and extra works, extra costs, and construction term extension due to the emergency events arisen are reduced to minimum. The risk management requires the application of the observation method and the geotechnical monitoring.

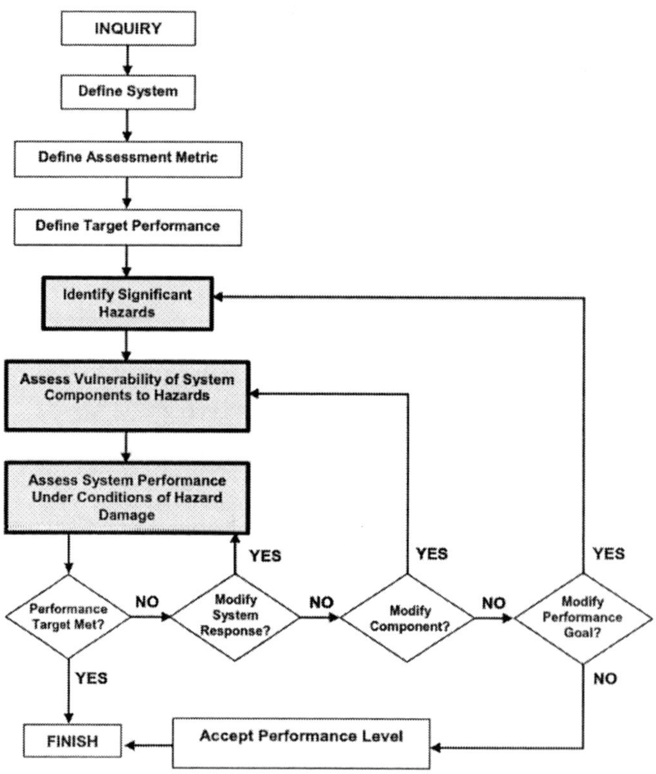

Figure 4. Decision-making process for assuring system performance goals [8].

3. Investigation

Understanding the site and subsurface characteristics of the project site are another key aspect of geo-engineering risk identification. Typically, there are unique risks associated with particular geologic formations or structures such as, presence of sinkholes in karst terrain, and rock gouge and slickensides along fault zones. Regional and local ground water conditions are another key risk factor that must also be considered. Valuable knowledge of the geologic and ground water conditions for a particular site can generally be attained through literature research which usually includes previous geotechnical and ground water studies, geologic and agricultural maps, journals, and research studies. Mining maps, historical data, topographic surveys and aerial photos are other sources used to aid in the identification of potential geo-engineering risks especially those associated with previous man-made features e.g. resulted from mining activities [9], [10].

It is widely recognized that risk identification is best achieved through knowledge and experience. In assessing geo-engineering risks, experience can extend beyond that of the agency's geotechnical engineers or engineering geologists. Local experience of individual contractors, field inspectors, geotechnical design consultants and even adjacent landowners have the potential for providing valuable insight in identifying geo-engineering risks associated with a particular project site [7].

A complex and correct execution of the ground investigation at the area of interest is extremely demanding. It is only the correctly executed ground investigation that is able to eliminate the natural hazards, which may in the future negatively affect the engineering project being designed or under construction. The basis is the creation and gradual improvement of the rock environment 3D model – engineering-geological model.

3.1. Gradual assessment

The effectively executed ground investigation should be divided into several interconnected stages. The terrain works should be preceded by a desk study of archive materials available, then an orientation and preliminary investigation should follow. It should be also followed, for complex structures or structures designed in a complex rock environment, by a detail ground investigation or also by an additional ground investigation)

3.2. Role of cooperation between designer and investigator

Very important is to provide for the information transfer between the ground investigator and the designer.

3.2.1. Determination of geotechnical category. It is the geotechnical category determination at the investigation beginning which should result from the cooperation of the investigator of the area of interest and the engineering project designer. The geotechnical category should be refined as needed in the same way as the selection of the ground investigation methods and refinement of the engineering and geological model.

3.2.2. Determination of characteristic values of the rock environment. In the course of the ground environment the basic engineering and geological types of rock formations and their detail characteristics should be defined on the 3D engineering and geological model. At the end of the ground investigation, derived values and designed characteristic values should be determined of relevant parameters in the meaning of the ground investigation goals and purpose in possible cooperation with the engineering project designer.

4. Positive and negative role of standardised approaches (Standards)

The rock environment ground investigation can be standardized for many engineering tasks. In EU the Eurocode 7-1 a 7-2 have been approved and published, which methodically describe the procedures of the ground investigation. With regard to partial differences both in legislation and especially in composition, properties, and behaviour of the rock environment it is nevertheless necessary to respect national annexes of the Eurocodes and national standards. In case of the ground investigations at the territory of the Czech Republic it is also necessary for the traffic structures to respect, in addition to the Eurocode, the ČSN 736133 and the newly approved ČSN P 731005. The standardized procedures, however, cannot describe all situations at the specific area of interest. This can be evidenced by the ground investigations and technological procedures executed at the construction of highways in the Czech Republic, such as the rolling D1 highway near Ostrava or the north Bohemian D8 highway buried in a landslide.

5. Conclusion

The ground investigation results should be handed over to the designer in an "understandable" form where an instructive verbal description of all significant parameters and the rock environment behaviour will be supplemented with clearly arranged tabular analyses and measurements results together with a presentation of the 3D model of the rock environment at the area of interest in the form of a digital model (e. g. AutoCad solid) or in the form of an appropriately chosen 2D sections; preferably in the form of such files that could be further used for creation of the geotechnical model simulating the rock masses behaviour under certain marginal conditions.

Problems still remain in certain cases in the ground investigation works organized by investors; their insisting on a possibly cheapest and fastest investigation prevents the observation of fundamental procedures of the quality and quantity risk analysis of the specific rock environment for the particular engineering project. The experience from abroad suggests that more advantageous is a design-build construction method employing an experienced contractor capable of better organization of activities and allocation of finance.

The use of the design-build (DB) procurement process has become increasingly popular in recent years especially in the transportation industry. The allocation of responsibility to a single source (design-builder) gives an agency the ability to define total project costs and to decrease the time for project procurement and delivery. It also reduces the potential for disputes and claims that are commonly associated with the traditional design-bid-build (DBB) projects. The DB contractor has more flexibility in managing costs and schedule during both the design and construction phases [7].

Acknowledgements

The work was realized under the support of project No. TE01020168 by Competence Centers program of Technological Agency of the Czech Republic entitled Centre for Effective and Sustainable Transport Infrastructure (CESTI).

References

[1] ASCE Council on Disaster Risk Management 2006 *Infrastructure risk management processes natural, accidental, and deliberate hazards* ed Taylor C and VanMarcke E p 296
[2] Hyndman DW and Hyndman DW 2011 *Natural hazards and disasters* (Brooks/Cole, Cengage Learning)
[3] Münchener Rückversicherungs-Gesellschaft 2011 *NATHAN World Map of Natural Hazards*
[4] ASCE Council on Disaster Risk Management 2011 *Quantitative risk assessment (QRA) for natural hazards* ed Uddin N, Ang H S A p 87
[5] Vinciguerra S and Bernabâe Y eds 2009 *Rock physics and natural hazards* (Birkhauser Basel) p 427
[6] ESPON Project 1.3.1 2006 *The Spatial Effects and Management of Natural and Technological Hazards in Europe - Executive Summary (on line)* ed Schmidt-Thomé P p 309
[7] Juang C H, Phoon K K, Puppala A J, Green R A and Fenton G A 2011 Geotechnical Risk Assessment and Management *Proceedings of Georisk 2011* (ASCE Press)
[8] American Lifelines Alliance 2005 *Guideline for Assessing the Performance of Electric Power Systems in Natural Hazard and Human Threat Events* p 56

[9] Harat A, Rapantova N, Grmela A and Adamczyk Z 2015 Impact of mining activities in the upper silesian coal basin on surface water and possibilities of its reduction *Journal of Ecological Engineering* **16**(3) pp 61-9

[10] Malucha P and Rapantova N 2013 Impact of underground coal mining on quaternary hydrogeology in the czech part of the upper silesian coal basin *13th SGEM Geoconference on Science and Technologies in Geology, Exploration and Mining* **1** pp 507-14.

IOP Conf. Series: Materials Science and Engineering **236** (2017) 012082 doi:10.1088/1757-899X/236/1/012082

Practical example of the infrastructure protection against rock fall

Daniel Jirásko[1] and Ivan Vaníček[1]

[1] Czech Technical University in Prague, Faculty of Civil Engineering, Geotechnical Dept. Thákurova 7, 166 29 Prague 6, Czech Republic

E-mail:daniel.jirasko@fsv.cvut.cz

Abstract. The protection of transport infrastructures against rock falls represents for the Czech Republic one of the sensitive questions. Rock falls, similarly as other typical geo-hazards for the Czech Republic, as landslides and floods, can have negative impact on safety and security of these infrastructures. One practical example how to reduce risk of rock fall is described in the paper. Great care is devoted to the visual inspection enabling to indicate places with high potential to failure. With the help of numerical modelling the range of rock fall negative impact is estimated. Protection measures are dealing with two basic ways. The first one utilize the results of numerical modelling for the optimal design of protection measures and the second one is focused on the monitoring of the rock blocks with high potential of instability together with wire-less transfer of measured results. After quick evaluation, e.g. comparison with warning values, some protection measures, mostly connected with closure of the potential sector, can be recommended.

1. Introduction

In the field of Transport Infrastructure some priorities during last years were defined, e.g.

- FEHRL Vision 2025 for Road Transport in Europe [1];
- ECTP reFINE (2012) [2].

Basically all are emphasizing three aspects: sustainability, availability and affordability.

Availability is putting an accent on increase of infrastructure capacity not only for current but also for expected changes in future, e.g. from the weather change point of view. Therefore the interaction of transport infrastructure with natural hazards as e.g. landslides, rock falls, floods are studied very intensively – e.g. European project INTACT, as well by authors: [3].

Therefore the paper is focused on one such example, describing the infrastructure protection against rock fall.

The slopes above the road II/102 in section Strnady - Štěchovice represent a substantial risk for the traffic on the road from its construction in 1st mid-20th century [4]. An increasing number of cases of falling stones and blocks on the road have been observed in recent period. Fortunately it was without any casualties. Road management and maintenance office in Kladno decided to solve this alarming situation. They contacted the Department of Geotechnics of the Faculty of Civil Engineering, CTU in Prague with a request for assessment of the risk and recommendations for its elimination.

The main aim of the assessment was determination of the risk areas which was based on our own site survey. The second aim was to propose risk reduction measures, together with the proposed long-term monitoring of the selected rock objects (the places with the highest potential risk), using wireless data transmission system with connection to the warning system.

2. Description of the site

Assessed section is 12.45 km long, and there were four defined surveyed subsections with the high risk, Figure 1. These are the sections: Strnady, Vrané nad Vltavou, Štěchovice 1 (Žižkův vrch) and Štěchovice 2 (Šlemín - Hradišťko).

Rock slopes above the dam Vrané nad Vltavou was evaluated as a riskiest part - Figures 2 and 3. Vltava river valley runs parallel to the direction of layers having tendency ca 40-50° to the SE. This tendency is identical with the inclination of most studied slope and forms the majority of its surface.

A typical characteristic of this section are four clearly visible smoothed rock surfaces - "plates" of shale. Inclination of the layers is oriented down toward the valley. A huge rock falls of volcanic tuffs occurred along these shale plates in the past as a result of excavation of the foot of the slope for road construction in the twenties of the last century. Principle and chronology of the landslide is shown in Figure 4 [4].

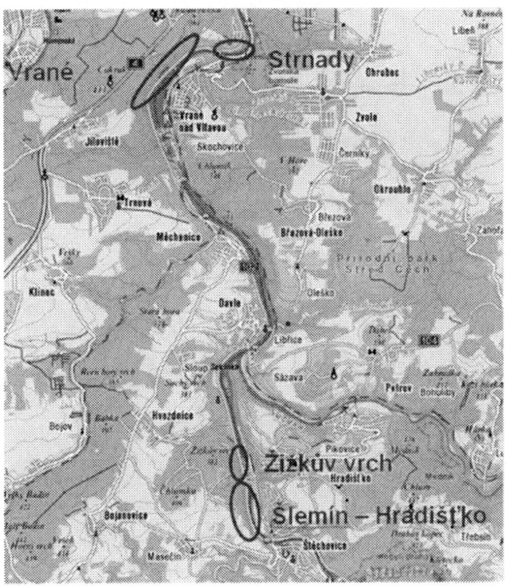

Figure 1. Assessed sections.

Figure 2. Vrané nad Vltavou – section with highest risk above the dam.

In addition to this typical feature there are large rock blocks of various sizes with various stage of weathering situated on the main scarp of the landslide above a rock plates – Figure 5. These blocks represent a significant risk of fall on the road. Progressive weathering occurs along parallel discontinuities which can result in possibility of their loosening and subsequent movement or even fall down into the valley.

Figure 3. View on the section with highest risk - Vrané nad Vltavou section above the dam reservoir.

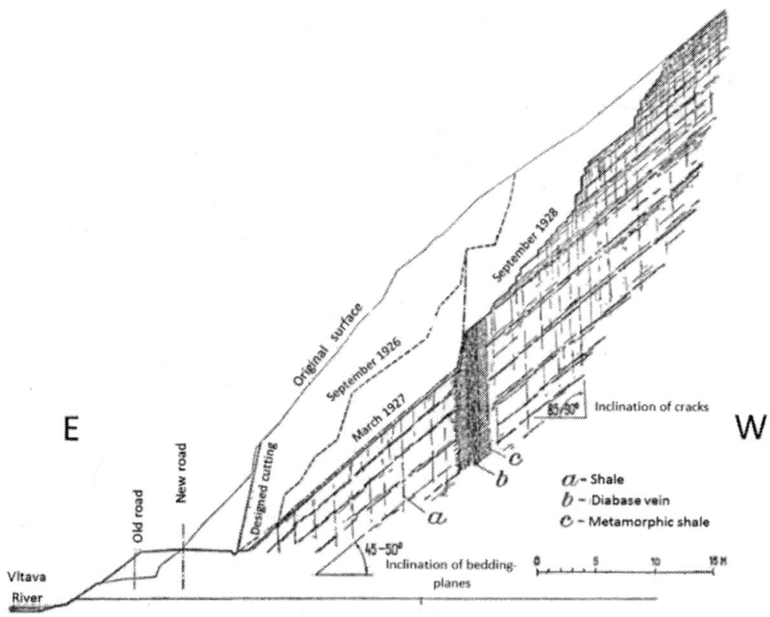

Figure 4. Principle of landslide occurred in Vrané nad Vltavou [4].

Figure 5. View on main scarp above rock plates.

3. Risk reduction

Evaluated section of most risky rock slopes along the road II/102 Strnady - Štěchovice is in indifferent equilibrium. It is almost impossible to guarantee required slope stability (usually expressed as a factor of safety F = 1.3 to 1.5) by cutting and reduction of steepness of the slopes, because it would mean a huge excavation of rock mass, which would also affect land in the upper edge of the slope.

The solution to this potential risk can have several approaches, depending on the acceptable risk. This risk is given by to direct threat of falling released rocks from small fragments to large blocks, which directly threaten the traffic on a road at the foot slope or nearly situated buildings and water structures. Recent cases where the fall of loose blocks and their collision with vehicles occurred mean that a danger is very high.

One extreme solution is an approach whose main premise is to implement such remedial measures, which would effectively not allow direct impact of falling blocks with the road. It is obvious that this solution can reduce the risk to ensure a high level of safety for several decades on one hand, but also implies the extremely high costs of these remedial measures. But 100% safety can be never reached.

The second extreme solution is based on reduction of remedial measures to the minimum with maximum attention devoted to the limiting the direct collisions of persons and vehicles with falling rocks. This solution has its limits in terms of monitored blocks, where the drivers would be informed in advance about released block by some warning system (traffic lights, gate).

The approach chosen by authors is based on the finding of some optimum between these two extreme solutions. For sections where the potential danger is less, (for example with respect to the small size of possible loosed fragments, or in case where toe of the slope has some retention area) was chosen first passive approach. This approach involves both limiting the movement of loose particles down the slope, and its capture just before the impact on the road. For large blocks (of the order of 10 m^3 and more) with the highest potential risk in the event of its release, it is recommended the second approach, i.e. monitoring of behavior of these blocks (especially inclination and displacement in time). For cases with higher potential of release and with relatively demanding measures to dissipate its energy the disintegration or anchoring is recommended (usually for size of blocks 5-10 m^3).

4. Proposed measures for risk reduction

Because of variable character of the observed area five types of remedial and protection measures were proposed: retaining reinforced walls, covering of weathered rock surface by mesh and anchoring, dynamic barriers, removal of unstable blocks and monitoring.

Effectiveness of designed protective measures should be evaluated by simulation in rock fall program. The program allows simulations the trajectories of the falling blocks. Designed block with a certain weight, shape and modulus of elasticity is released from a certain part of the slope. Its kinetic energy has to be safely captured by proposed barriers while falling blocks must not jump the barrier. Example of rock fall simulation in GeoRock 2D program for selected sections of Vrane nad Vltavou site is shown in Figure 6.

4.1. Retaining reinforced walls

Retaining reinforced earth walls are recommended in places where there is an enough space between the toe of the slope and the road. Larger stones may be a partially disintegrated during falling and the smaller particles can be captured by retaining walls. We recommended reinforced soil with using old tires, especially on the side toward the slope, which should reduce the impact of fallen rock blocks. Schematic cross section is shown in Figure 7. It seems to be a very effective and low cost measure with great potential for energy absorbing. Therefore, this measure should be implemented wherever the morphological conditions allow it. Moreover, it is possible to increase total protected height by vertical steel fence with fixed wire mesh. This fence is especially very useful for capture of smaller rebounding rock fragments and stones.

IOP Conf. Series: Materials Science and Engineering **236** (2017) 012082 doi:10.1088/1757-899X/236/1/012082

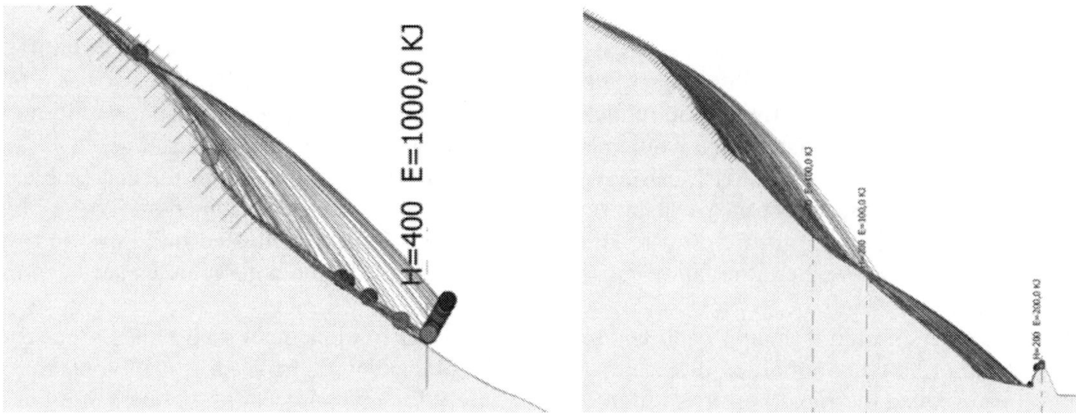

Figure 6. Example of the effectiveness of the proposed measuresof by rock fall simulation.

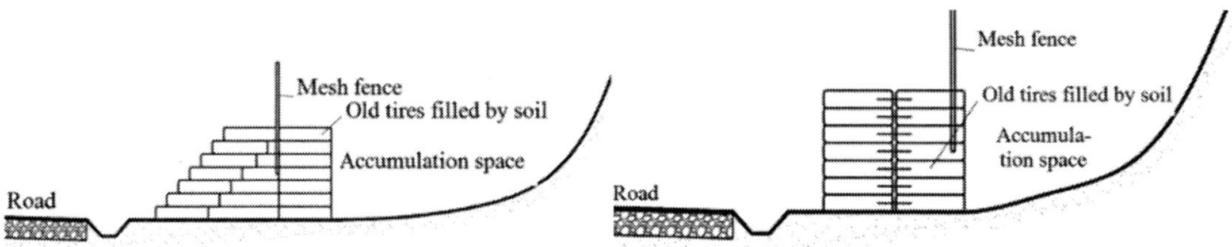

Figure 7. Schematic cross sections of retaining reinforced earth wall with old tires facing.

4.2. Covering by mesh and anchoring
In principle selected parts of the rock slope is covered by steel wire mesh or high strength geogrid anchored by bolts or short anchors into the rock. The main purpose of this measure is reducing the movement of weathered fragments and even larger boulders of the order of up to 1 m^3 to hold them in place during the first stage of the weathering and fix their position until further weathering into smaller pieces. After that weathered particles are targeted removed. This measure is easily realizable on a flat surface slope of the slab character. The measure requires regular inspections and releasing the captured material.

Negatively inclined rock structures that are bounded by discontinuities can be stabilized by anchors or rock bolts. Drilling of boreholes for bolts or anchors will be done by using climbing techniques. All steel elements have to be provided with suitable anticorrosive coating.

4.3. Dynamic Barriers
Dynamic barriers represent special approach. They are most suitable on the places where spontaneously falling boulders and large debris may fall down. In principle dynamic barrier consists of high strength steel mesh which is connected to the steel vertical anchored beams. The connection is made by using brake elements that capture the deformations, absorb energy and stop the block. These barriers represent an effective but demanding system which requires gradual release of captured blocks in order to restore full functionality of the barrier. Therefore, it is recommended only for places where there is a real danger of rock fall, which can impact the road. In our case we assumed that the proposed dynamic barriers have been able to keep the size of boulders up to 1-2 m^3.

4.4. Disintegration of rock (mechanical loosening)

Disintegration is very suitable for very danger rock blocks which are unstable and their fall could be expected soon. Blast is usually controlled with micro-second delay and with steel mesh protection, preventing wide spread of loose parts. This measure solves the most problematic cases, and thus significantly reduces the potential risk of falling blocks to the road. The procedure requires safety measures; work on the slope is not easy with respect of difficult access. Hydraulic wedges and lifting bags are most commonly used for mechanical disintegration.

4.5. Monitoring – warning system

The purpose of monitoring which is associated with the warning system can be divided into several follow-up phases. In principle, monitoring is observation of deformations. Deformation can be understood as a result of cyclic weathering processes - such as the influence of water freezing in rock cracks, crevices, the influence of temperature changes which cause increase or reduction of rock volume, as well as the process of reduction of shear strength on potential slip surfaces. Shear strength is a function of shear displacement. In the first phase shear strength increase with the displacement, but after crossing the so-called peak strength, shear strength can be reduced on shear surface. This second phase may occur as a gradual acceleration of deformation in time, with resulting shear strength excess and followed with sudden shear drop off. The basic character of the expected deformation is illustrated in Figures 8 and 9, when the first case is classical shear and second case is overturning.

Figure 8. Block with expected translation movement.

Figure 9. Block with expected rotation movement.

5. Phases of the monitoring

The following phases of the monitoring are proposed:

- Regular visual inspections of the documentation points or other points that were photographically documented and briefly described. This stage is a rough assessment of the visually observable changes;
- Regular visual inspection of remediation measures - assessment of the status of remedial measures, the volume of captured loose debris, boulders and recommendations for their release, to restore the accumulation space;
- Geodetic observations in the recommended time intervals to monitor relatively large areas on a slope for selected observation points. The aims are to identify places with excessive movements, to assess their development in time, to recommend the installation of additional documentation points or implementation of other remedial measures;

- Instrumentation and monitoring of selected blocks with deformation measuring (displacements of particular blocks, widening of cracks) with using crackmeters and inclinometers with high resolution. Time intervals can be selected depending on the potential danger. It is possible to define critical conditions for increasing deformation rate and provide wireless transmission of measured data to a desk of responsible person.

5.1. Description of continuous monitoring methods using MEMS sensors with wireless data transmission

For monitoring of rock slopes where there is very limited access and where there is a necessity to perform long term monitoring, is worthwhile to use wireless data transfer. This approach allows connecting various types of sensors used for monitoring via the wireless network. Once the data are recorded they are sent to the central point of the wireless sensor network and afterwards all measured data from a short period of time are together transferred again wirelessly via internet on the server for further processing.

The authors of this paper have a good experience with similar type of monitoring which is designed, installed and operated in the Prague metro [5]. Processing and analysis of data can be operated automatically and measured data can be displayed on a web interface.

At the same time it is possible to create a warning alarm system, which defines the critical value of measured deformation. In the event that the measured value of deformation is higher than critical value, system will automatically send a warning e-mail and SMS messages to the addresses of all persons responsible for safety management of the road. This system can be extended to include an automatic traffic lights and gates on the road. Scheme of the continuous measurement system using MEMS sensors with wireless data transmission is shown in Figure 10.

Wireless technology works in non-licensed band of 2.4 GHz on the ZigBee platform that operates on Intel chips. The whole solution of wireless data collection from individual measuring points that can connect both analogue and digital sensors is supplied by Crossbow Company. Everything is just a question of interconnection of relevant chips (interfaces) for sensing (data collection) with the chip for wireless communication. The heart of the whole system is an embedded computer running the Linux operating system which communicates with the gateway point of wireless ZigBee sensor network via Ethernet or USB interface. This embedded PC acts as well as temporary storage of monitored data before they are sent to the server in the office. Each network point is necessary to program before its installation for specific task, position and connected sensor in the wireless sensor network.

The results can be continuously evaluated and it is possible to define the limit values of deformation, respectively time development of deformation. In case of exceeding of limit deformations, warning system will automatically signal on the road (speed reduction signal, an increased risk of falling rocks, locking the gate etc.), while a group of experts will visit the section on which signal was registered. To check the automatic sensors classic strain gauges with a lower degree of accuracy will be installed as a comparative measurement. It is a classical form of dual control.

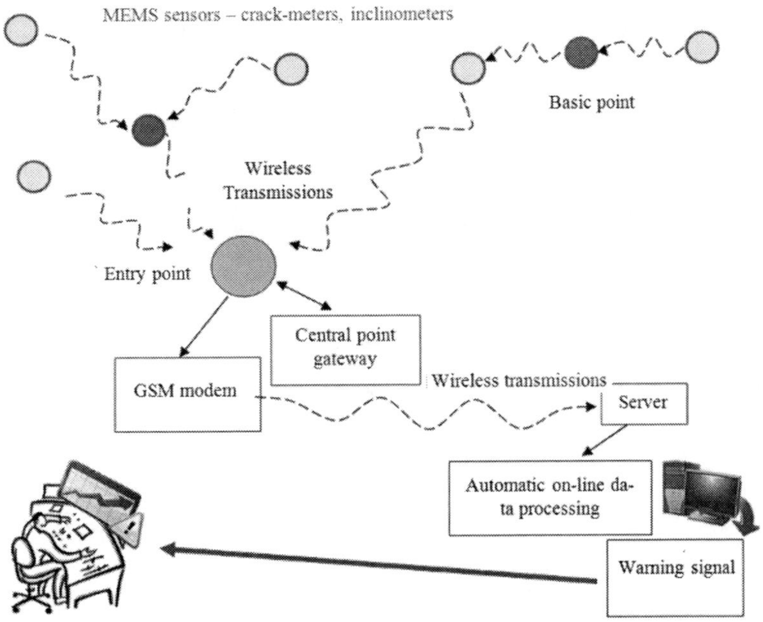

Figure 10. Scheme of monitoring system using MEMS sensors and wireless network.

6. Conclusion

The risk associated with a rock fall on the road is currently very topical and much discussed not only in the Czech Republic but also worldwide. Defining and quantifying the risks is very difficult, since the risk depends on many factors and specific condition of particular site.

Paper is therefore focused on rock slope above the road from Prague to Štěchovice. Most dangerous places were described on the base of personal investigation. After that different countermeasures were discussed. Final recommendation for risk reduction is based also on the numerical modeling of blocks up to the size of 2 m^3. With the help of numerical modeling we specified protection barrier at the slope toe utilizing waste tires.

One of the most important measures for reducing the risk of the site are regular visual inspections by experienced geotechnical engineer and using of modern methods of monitoring system connected to the warning signal.

Acknowledgements

The paper was prepared with financial support of the research project TE01020168 of the Technology Agency of the Czech Republic CESTI – Centre for effective and sustainable transport infrastructure.

References

[1] http://www.fehrl.org/
[2] ECTP reFINE 2012 Research for Infrastructure Network in Europe initiative – Building up Infrastructure Network of a Sustainable Europe *Strategic targets and expected impacts* www.ectp.org
[3] Vaníček I 2011 *Sustainable Construction* (CTU Press, Prague) 163 p.
[4] Záruba Q and Mencl V 1957 *Engineering Geology* (ČAV Prague) p 425
[5] Soga K, Vaníček I and Gens A 2011 *Micro-Measurement and Monitoring System for Ageing Underground Infrastructures* (CTU Press Prague)

Static design of steel-concrete lining for traffic tunnels

Karel Vojtasik[1], Marek Mohyla[1] and Eva Hrubesova[1]

[1]VŠB – Technical University of Ostrava, 17. listopadu 15/2172, Ostrava-Poruba 708 33, Czech Republic

E-mail: karel.vojtasik@vsb.cz

Abstract. Article summarizes the results of research focused on the structural design of traffic tunnel linings that have been achieved in the framework of a research project TE01020168 that supports The Technology Agency of Czech Republic. This research aim is to find and develop a process for design structure parameters of tunnel linings. These are now mostly build up by a shotcrete technology. The shotcrete is commonly endorsed either with steel girders or steel fibres. Since the installation a lining structure is loaded while strength and deformational parameters of shotcrete start to rise till the setting time elapses. That's reason why conventional approaches of reinforced concrete are not suitable. As well as there are other circumstances to step in shown in this article. Problem is solved by 3D analysis using numerical model that takes into account all the significant features of a tunnel lining construction process inclusive the interaction between lining structure with rock massive. Analysis output is a view into development of stress-strain state in respective construction parts of tunnel lining the whole structure around, including impact on stability of rock massive. The proposed method comprises all features involved in tunnel fabrication including geotechnics and construction technologies.

1. Introduction

Contemporary tunnel lining of subterranean works build by New Austrian Tunneling Method (NATM) is double wall system. It consists of two particular structures an outer primary lining and the inner secondary lining. The both are separated by a thin insulating layer of a plastic membrane which prevents groundwater to permeate into the inner lining.

The both lining are essential in stabilizing rock ground. The first one starts stabilizing rock ground around excavation and provides safe space for execution all of tunnel compounds, inclusive of the secondary lining installation. The secondary lining steps in operation later after the primary lining loses bearing capacity due not being protected against aggressive groundwater.

Static response of the primary lining built from shotcrete and steel frame structure is marked by variable working diagram consequently in accord with setting time of shotcrete. Initially it is in operation the steel frame structure only and instant stiffness of the lining is not high. Then lining stiffness grows and eventually settles as shotcrete becomes solid with time. This quality is welcomed because it allows rock ground relaxation that is indispensable to achieve the lowest rock ground load on a lining structure. Installation of the insulating layer and a followed construction of the secondary lining structure built from reinforced concrete require more time. The both can be performed only for steady rock ground excavation which is achieved by assistance of a primary lining.

Interaction of primary and secondary lining is only a temporary act limited to an initial period of lining operation after it's completion. The primary lining is not protected against the impact of the aggressive groundwater which limits the life time of a primary lining. The necessary response to stabilize the rock ground is gradually transposed from primary to secondary lining. So the bearing capacity of a secondary lining must be sufficient to avoid a the next rock ground deformation when the transfer is accomplished and the secondary lining must hold the excavation space steady forever. Construction period of the primary lining is a decisive event for stabilizing of the rock ground. The rigid working diagram of a primary lining implies a higher value of stabilization response. A design intention of the primary lining is to achieve a minimum response.

2. Design approaches

Design procedures for lining of subterranean works are not explicitly included in the current legislation of the Czech Republic. Eurocode EC7 [1] that focuses the design of geotechnical structures does not include construction of tunnel lining. Eurocode EC2 [2] which addresses the design of concrete building structures, can be used for the design of the secondary lining structure on condition that this is made from concrete or reinforced concrete.

The EC2 takes not notice of a primary lining structure built from shotcrete and steel frame arch inclusive of rock ground interaction. Static activity the primary lining structure, however, is crucial for the steadiness of the rock ground and the amount of the permanent load put on the secondary lining.

Design and evaluation of the lining of subterranean works made from concrete and steel elements are analyzed in the manual by Bartak, Horejsi and Zapletal [3]. In the manual proposed method uses partial coefficients modifying the input calculation parameters, especially those ones that implement phenomena inherent in rock ground. An other contribution to the solution of lining presents a work by Zapletal [4].This author assumes a design of the lining as a composite steel and concrete structures. The evaluation of a variable elastic modulus of steel shotcrete cross-section of lining investigates work by Rott [5].

Most contemporary structural designs of subterranean work linings is performed on available commercially distributed computer programs. These programs are based on theories of mechanics of continuous materials (Midas NX, Plaxis) or discontinuous materials (UDEC). These programs provide respective elements to stand for rock ground as well as lining structure. There are several options and elements to represent a lining structure.

Authors of this contribution are developing design method to set up lining parameters systematically for more than twenty years. They designed the original method of homogenizing heterogeneous cross-section of the lining structures assembled from steel girder combined with layers of shotcrete. Lining solution is based on the theory of cooperating rings and homogenizing a heterogeneous steel shotcrete cross-section of the lining [6]. Unlike the evaluation of a variable elastic modulus of steel shotcrete cross-section at work [5] this time the homogenization algorithm is based on an analytical model for calculation of stress strain state in a multi-layer circular ring, which was formulated by Bulytschev [7]. This analytic model uses the theory of analytic functions of complex variables based on Kolosov's functions for complex potentials. Next development in solution was implementing the analytical model of lining cross-section with a numerical model standing for rock ground [8]. This option extends the static analysis of a lining structure for possibility to take better off an account on geomechanical circumstances and to carry out calculations for alternative excavation scheme, construction technology and mechanical parameters of lining. It depicts an interaction of rock ground and lining on an interface plane that they met on. It is apt to relate dependency of rock ground stress-strain behavior on lining parameters and vice versa. The final stress-strain state on the interface plane makes the load secondary lining.

Last proposal of design of subterranean work lining parameter animates 3D numerical model. This enables to integrate rock ground constitutive structures, primary stress-strain in rock ground, mechanical properties of the rocks, lining construction phases, i.e. geometric conditions and material properties that describe lining construction .

3. Numerical 3D design

The following example documents a solution by 3D numerical model of a simple primary lining structure of oval shape that supports a rock ground excavation. This computes the internal forces in the primary lining constituents concrete shell and steel girders. The computation sets the necessary primary lining response to maintain the rock ground around it in steady conditions not to collapse. The necessary lining response provides initially only the primary composite lining. Due not being protected against ground water this lining dilapidates and the response is transposed to the secondary lining made from reinforced concrete lining. Design and assessment of reinforced concrete lining structure is carried out separately from the 3D model according to structural mechanics for known discreet load specified with 3D model. The design conditions also allowable deformation of secondary lining that has not to go beyond regulated value. Calculation of the primary tunnel lining were made by Midas MX.

Figures 1 and 2 show a 3D computational model of rock ground and primary lining and a close up view of the steel shotcrete cross-section modelling of the primary lining.

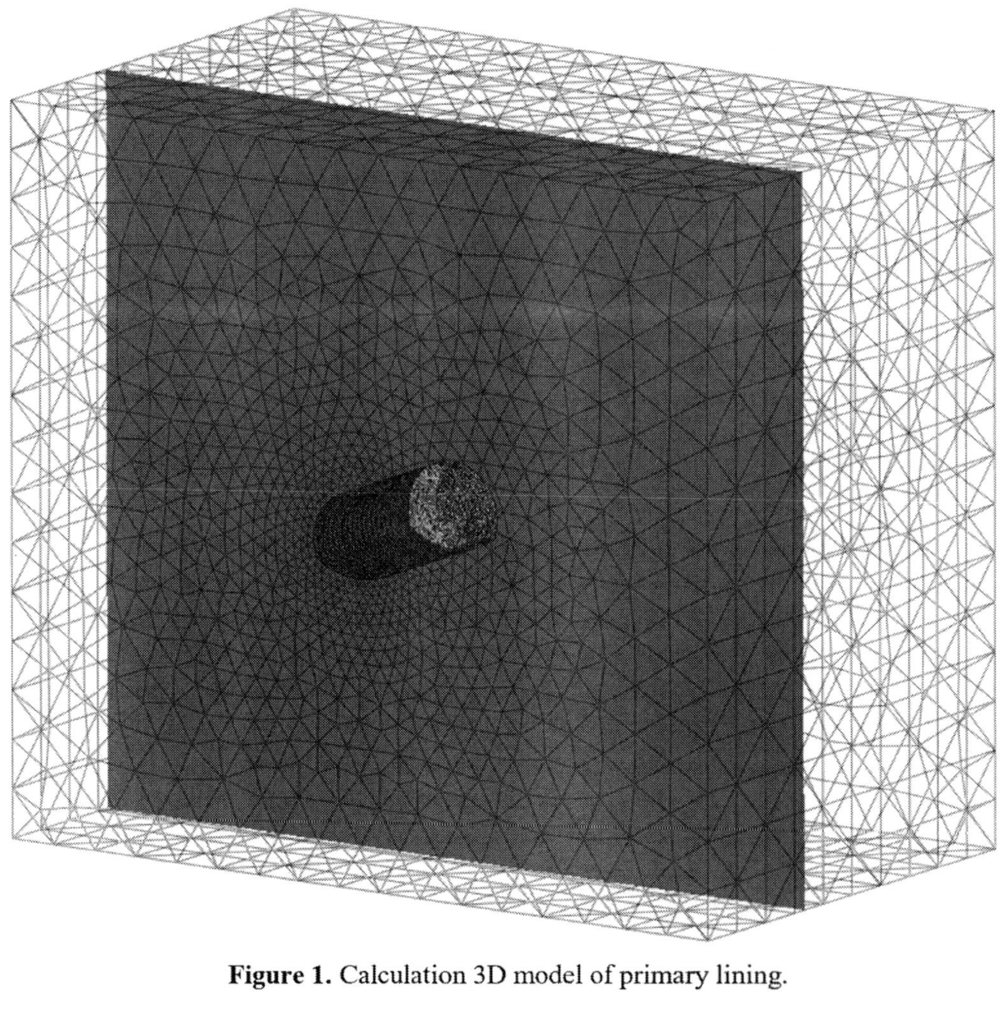

Figure 1. Calculation 3D model of primary lining.

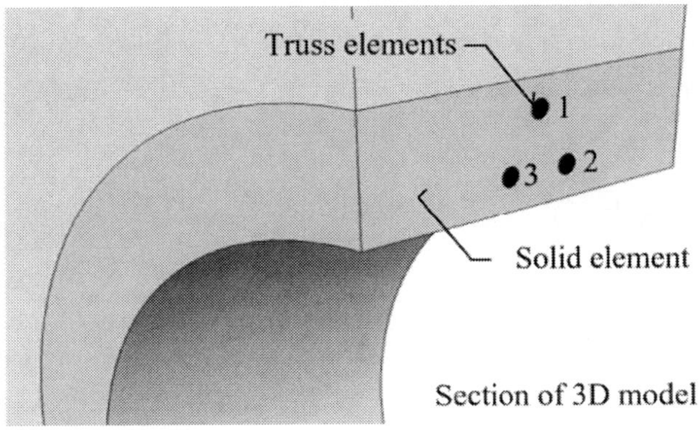

Figure 2. The modelling of a steel shotcrete cross-section.

Figure 3 shows calculated values of stresses along the primary lining perimeter in the steel elements.

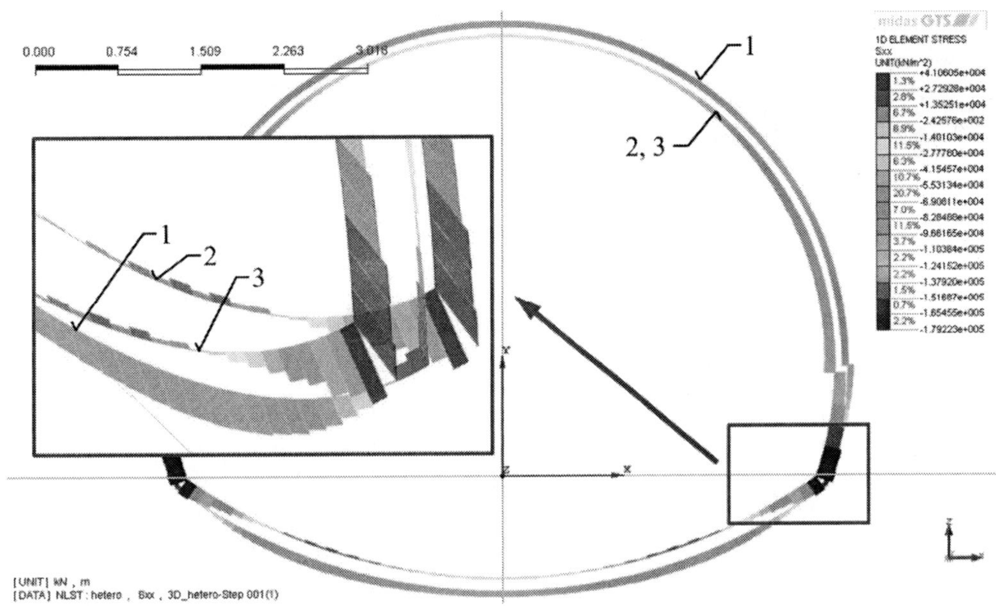

Figure 3. Depiction of stresses along a perimeter the primary lining in steel elements.

Figure 4 shows mean stress in the rock ground around the excavation and a curve of the mean stress development along perimeter of a lining on the interface plane with rock ground.

4. Conclusions

Structural analysis of tunnel lining by mathematical 3D model allows comprehensive inclusion of all relevant circumstances come in on tunnel construction. These include rock ground constitutive structures, primary stress strain in rock ground, mechanical properties of the rocks, lining construction phases, mechanical parameters and working diagram.

Significant feature of 3D numerical model is that it deals the interaction of both rock ground and lining structure through constitutive tress-strain relations that other contemporary approaches miss.

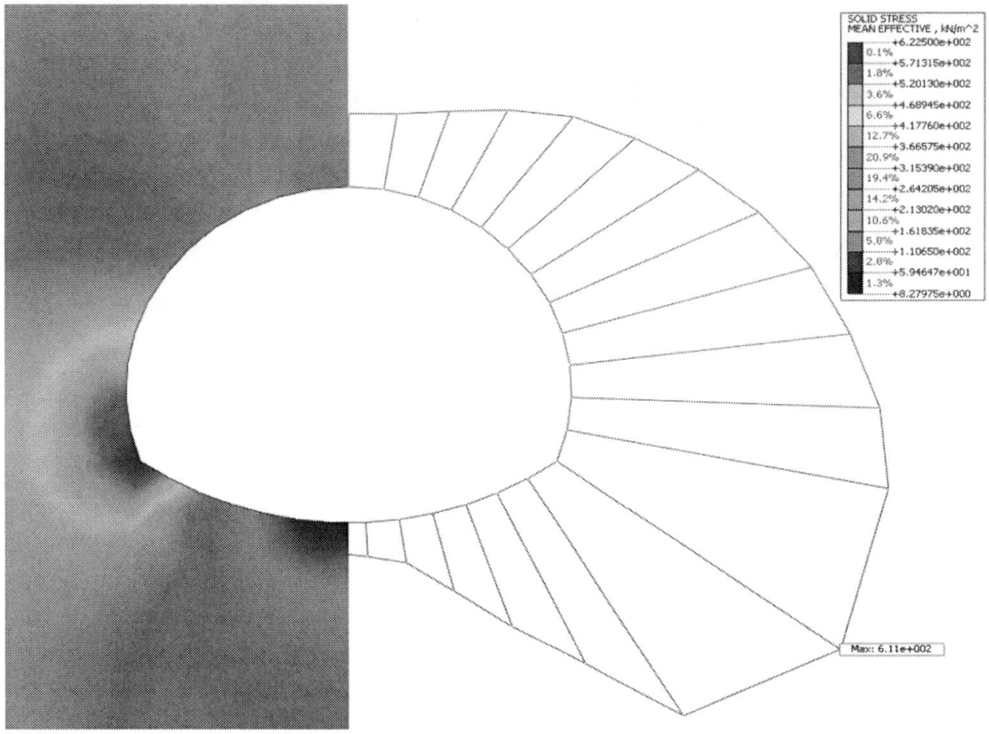

Figure 4. The curve of the mean stress along perimeter of a lining on the interface plane with rock ground.

Acknowledgements

The paper was prepared with the support of the Competence Centers of the Technology Agency of the Czech Republic (TAČR) within the project Centre for Effective and Sustainable Transport Infrastructure (CESTI), project number TE01020168.

References

[1] European Committee for Standardization (CEN) 2004 Geotechnical design -Part 1 *Eurocode 7, Part 1 (EN 1997-1)*

[2] European Committee for Standardization (CEN) 2004 Design of concrete structures *Eurocode 2, Part 1-1 (EN 1992-1-1)*

[3] Bartak J, Horejsi J and Zapletal A 2010 *Recommendations for the processing of static calculations of tunnels according to EC* (Czech Tunnelling Association ITA-AITES)

[4] Zapletal A 2007 Structural model of a steel-concrete composite lining *Tunel* **4** pp 68-71

[5] Rott J 2014 Homogenisation and modification of composite steel-concrete lining, with the modulus of elasticity of sprayed concrete growing with time *Tunel* **3** pp 53-60

[6] Vojtasik K, Hrubesova E, Mohyla M and Duris L 2013 An analytical model of a system of the concentric rings to state the stress in the material heterogeneous cross section of a steel shotcrete tunnel lining *Proc. Int. Conf. on EURO:TUN 2013 Computational Mehtods in Tunneling and Subsurface Engineering* ed G Meschke, J Eberhardsteiner et a*l* (Aedificatio Publishers) pp 205-211

[7] Bulytschev N C 1982 *Mechanika podzemnych sooruženij* (Moskva NEDRA)

[8] Mohyla M, Vojtasik K, Hrubesova E and Zdvorak J 3D numerical analysis of primary tunnel lining *Proc. Int. Conf. on 15th International Multidisciplinary Geoconference and EXPO, SGEM 2015* ed STEF92 Technology Ltd. pp 3-8

Design, construction and conditions of the application of unreinforced concrete final lining in conventionally driven tunnels

Jan Faltýnek[1], Jiří Hořejší[1], Libor Mařík[1] and Pavel Růžička[1]
[1]HOCHTIEF CZ a.s., Plzeňská 16, 150 00 Prague 5, Czech Republic

E-mail: jan.faltynek@hcohtief.cz

Abstract. The way to an economic design in the final lining in conventionally driven tunnels lies in structural analysing based on the actually encountered geotechnical conditions. Regarding reinforced concrete structures, many standards and regulations applicable to designing and building structures and taking them over by the client before their commissioning and before the end of the warranty period respectively exist in the Czech Republic. If the local conditions allow it, it is possible to design the final lining as an unreinforced concrete structure. In such a case it is necessary to take the differences into consideration in the structural design and in the possibilities of the lining behaviour and to set criteria for taking over the lining allowing for its use. Setting too stringent criteria for cracking can lead to an increase in the contract price, either because of the necessity for reinforcing the lining or because of the fact that the contractor reduces the risk by incorporating the assumed cost of repairs into the total cost. The paper describes basic differences in the approach to reinforced concrete and unreinforced concrete linings, the possibilities of limiting formation of cracks by means of the concrete mix design, by selection of the technological procedure of the work and the method of curing after stripping. The text contains a comparison of criteria for assessing the surface of an unreinforced concrete lining with criteria in foreign regulations.

1. Introduction

The use of unreinforced concrete final lining is currently usually associated with conventionally driven tunnels, where the primary lining is formed by shotcrete and the final lining is cast-in-place. The NATM prevails in the Czech Republic in the area of conventionally excavated tunnels. All tunnels in the modern tunnelling history, with the exception of the Březno tunnel driven using the Mechanical Pre-cutting Method, were driven using the NATM principles. The NATM application was made possible only after the revolution in 1989, which opened borders and gave the green light to technologies which had not been used until then. Designers gradually gained access to high performance computer technology and program equipment, construction companies access to mechanical equipment and building materials and both groups jointly gained the possibility to draw on foreign experience.

In the context of the NATM it is often mentioned that it is a tunnelling method allowing operative responding to the actually encountered geotechnical conditions and expending only such financial means that are necessary for adhering to the quality requirements and requirements for construction work safety. To make it possible, it is necessary to create space not only for the above-mentioned modern technological equipment and for obtaining the knowledge and experience necessary, but also to create the technical and legal framework which would allow both parties to the contract to carry out changes during the construction legally and operatively. The excavation of a tunnel and the method of stabilising the excavation is what we mostly understand under the term "optimisation of a technical solution with respect to geotechnical conditions". Not less important role in terms of laboriousness, minimisation of risks during works execution and the final amount of investment costs is played by the technical solution of the final lining. Even in this case is it possible to speak about the NATM as an observational method. It is possible to create a mathematical model approximating realistic conditions on the basis of the results of geotechnical measurements conducted during the underground excavation, condition surveys of the faces of individual excavation advance rounds and responsible determination of geotechnical parameters of the ground mass, thus obtaining the inner forces required for the design of the amount of reinforcement or for the application of an unreinforced concrete lining.

2. Differences in designing reinforced concrete and unreinforced concrete linings

2.1. Two approaches to the function of primary and final linings

No unambiguous opinion exists among the technical public on the load-carrying function of the primary and final (secondary) linings. Supporters of the basic load-carrying function of the primary lining stand on one side. They assign it, in cooperation with the ground ring stabilised with anchors, the capability of transmitting ground pressure throughout the tunnel design life. In such a case, the final lining fulfils only the facing and aesthetic role because it is loaded only by its own weight and effects of volumetric changes. Only in the cases of pressure insulated tunnels they have to resist hydrostatic pressure as well. The Pisárky tunnel in the Czech Republic was constructed under these assumptions. High-durability fibreglass anchors were used for reinforcing the ground ring. Supporters of the theory of total degradation of the primary lining stand on the other side. In their opinion, the primary lining loses its load-carrying function earlier than the assumed life period ends and all loads have to be carried by the final lining. A fundamental issue of durability of the primary lining is the effect of aggressive environment not only on the individual components of the primary lining (shotcrete, supporting frames and welded mesh), but also on the elements reinforcing the ground ring in the excavation surroundings. In the Czech Republic the latter approach is usually applied to designing the final lining and the final lining fully takes the loading from the primary lining that is considered to be temporary. An answer regarding the degree of degradation of the primary lining after tens of years can be obtained from a research conducted currently in Austria, where the second tubes of tunnels, which were opened to traffic over 30 years ago, are driven in order to improve operation safety. It is possible to collect samples of the primary lining during the building of cross passages or repairs of linings of existing tunnel tubes and analyse them. Recent results have proved that the speed of degradation of the primary lining shotcrete is far from the assumptions and both the concrete and steel keep fulfilling their load-carrying function [10]. An answer regarding the durability of shotcrete linings can, in addition, be found on numerous Scandinavian tunnels, where cast-in-place secondary lining is virtually unused. The required lining quality remains even after many years despite direct exposure to effects of the environment from the side of both the ground mass and the road space. In the text below we will deal with unreinforced concrete final lining as the only load-carrying element providing stability to a tunnel throughout its life (100 years).

2.2. Principle of designing reinforced concrete linings

When a lining is being designed and criteria for the unreinforced concrete lining are being determined, it is necessary to take into account the differences in the static action of an eccentrically compressed reinforced concrete or an unreinforced concrete structure. In case of the reinforced concrete lining, cracks start and develop at adequate exploitation of tensile reinforcement in adjacent concrete because the tensile strength of concrete is exceeded already at a small strain of tensile fibres in the section. For that reason, an assumption is introduced that concrete in tension is excluded and tensile stresses in the section are transmitted solely by reinforcing bars. However, a linear material model is sufficient for the calculation of inner forces for which the lining is being designed; it means that the stiffness values of lining sections are not affected by the formation of cracks and the full (initial, unreduced) values enter the calculation. The resultant position of the normal force can be under these conditions even outside the section boundaries.

2.3. Principle of designing unreinforced concrete lining

A different situation is in the case of an unreinforced concrete lining. If the unreinforced concrete lining is modelled using a linear material model, the unavoidable result of the design is an uneconomically great thickness of the lining. In such a case the insufficiently sophisticated calculation model where the lining is designed according to the tensile strength of concrete is blamed for the unsatisfactory result. Tensile strength is smaller by an order of magnitude than compressive strength. However, the compressive strength is in reality crucial for the bearing capacity. For that reason it is necessary to proceed to calculation with a non-linear model of the lining, where the cracks and potential plastification of the sections are taken into account. The problem can be solved, for example, using the FEM with planar elements or even using a simpler beam-based model of the lining.

In the models where total exclusion of tension is assumed in the area of the crack, the resultant of normal stresses always lies inside the section and the bearing capacity exclusively depends on the compression strength of concrete. It leads to a thinner lining, thus also to a more economic design.

In the method of calculation of unreinforced concrete lining outlined above, the results are arrived at by means of an iteration process. The state of stress and deformations in the section is calculated under predefined assumptions. For example, on a beam-based model of a structure with a non-linear model of the lining, a linear distribution of stress is assumed in the compression zone of the section; during additional loading, the design compression strength of concrete is first reached in the outermost fibres and, subsequently, the phase of gradual plastification of the material starts. It ends by reaching the ultimate limit state of the section. The assumption of planar section deformation is adopted for the development of deformations.

In the text above it is possible to find some differences in the approach to the structural design of a reinforced or unreinforced concrete lining. However, some differences must be applied even in the phase of evaluation of success of the lining the design. The assessment of cracks in the lining is an important issue related to this statement.

2.4. Difference in the approach to the width of cracks in reinforced and unreinforced concrete linings

One of the criteria for assessment of function and building quality of the lining is the width of cracks and their number in a certain area of the tunnel profile (e.g. the top of arch, a tunnel sidewall). Here it is necessary to note that the formation of a crack caused by direct (static) loading does not have to be considered as a defect or failure. Rather the contrary, the origination of cracks is a usual phenomenon arising from the principles and adopted assumptions regarding the static action of a reinforced concrete or only unreinforced concrete structure. Nevertheless, there are some aspects which must be monitored in this respect and they are different for both of the lining types.

In case of reinforced concrete linings, the crack width is limited and the values which are not to be exceeded can be found in respective standards and regulations. The main purpose of limiting the crack width is in this case to protect the steel reinforcement and to prevent its corrosion. Limit values are recommended for various types of the environment in which the lining may be located and for those kinds of substances which can threaten the reinforcement. When these prescribed values are fulfilled and the concrete cover is sufficient, it is possible to expect that the penetration of harmful matters into the area of the reinforcement will be prevented and neither the load-carrying function nor the durability of the structure, which is planned for 100 years in the case of tunnel structures, will be disturbed. Another criterion for limiting the width of cracks is the potential unfavourable influence on the structure appearance.

Another situation occurs in the case of an unreinforced concrete lining. The requirement for the undisturbed load-carrying function of the section remains the same, but this function is ensured by the concrete itself as the only structural element. The transition of forces by means of steel reinforcement ceases to exist in this case and the criteria relating to cracks and determined for reinforced concrete lose their significance. The main requirement for the function of an unreinforced concrete lining is the safe transition of forces in the concrete section, which has to be ensured mainly by the sufficient depth of the compression zone corresponding to particular stress conditions in the section. The resulting crack width therefore can be different and greater under various stress states in comparison with a reinforced concrete section. Just as with reinforced concrete, this width may be limited by requirements for the structure appearance.

In general, the possibility of designing the tunnel lining as an unreinforced structure depends on several factors. Above all, it is a reliable assessment of the geological conditions. The possibility of applying unreinforced concrete linings only in favourable geological and geotechnical conditions is accepted even in foreign regulations. It is assumed that reinforced concrete is used in the cases of geological faults with an impact on the magnitude and different intensity of loads acting around the tunnel circumference. Another factor important for a successful design is the geometry of the tunnel excavation cross-section and the shape of the lining. It is especially in the case of unreinforced concrete that the success of the design is very sensitive to this parameter. But it is obvious that only seldom is it possible to adapt the shape of the lining purely to the needs of statics. The convergence of the iteration process and the final position of the normal force inside the section are achieved in the calculation of a lining which is statically satisfactory. If the equilibrium is not achieved in the calculation (in the case of reinforced concrete it is possible to solve it by additional reinforcement), the only way of remedy available is changing the shape of the lining, designing a greater section thickness or a change from unreinforced to reinforced concrete lining.

It is particularly important in the case of unclosed tunnel profiles to thoroughly determine the boundary conditions to describe the contact of the vault and footings. In such a case, we consider it to be correct if the tunnel profile is analysed comprehensively, i.e. jointly with the lining vault and the footing.

3. Principles of designing concrete mix formula for unreinforced concrete lining

From the aspect of material, the formation of cracks in unreinforced concrete is predestined first of all by the low tensile strength of the material. It is a property which can be changed only with difficulty without the use of steel reinforcement. The means for limiting the formation of cracks caused by volumetric changes during the initial hours after casting of concrete is a correct design of the concrete mix formula with low hydration heat and slow strength gain. The reduction of the hydration heat magnitude can be achieved by adhering to several rules. Reducing the hydration heat is positively affected by the low content of cement ranging from ca 250 to 280 kg/m^3 and by the use of suitable cement with low content of C3A, the use of fly ash as a concrete improving admixture (ca 50–80 kg/m^3) and the design of such a mixture which reaches the maximum temperature difference up to 15K during laboratory tests on insulated cubes. The initial temperature of the mixture should be moderately low (13 – 18 °C) and it is recommended that the fresh concrete temperature should not be higher than 25 °C. It is quite common abroad that a combination of several different cement types is used to achieve optimum values of the mixture being designed for unreinforced concrete linings. In addition, it is necessary to choose low water content within the interval of ca 170 – 190kg/m^3 to increase the proportion of fine particles – cement, cinder, fine aggregate < 0.125 mm to the volume > 370 kg/m^3, to limit the amount of silt particles, to provide 4/8 mm fraction aggregates in the recommended amount of ca 3-5 %, to maintain the w/c ratio at 0.63 as the maximum and to use aggregates with a low coefficient of thermal conductivity as much as possible. The use of aggregate fraction D = 32 mm is recommended in the case of an unreinforced concrete lining.

Even the low concrete strength at formwork stripping is important as far as the formation of cracks is concerned. With respect to the fact that the final lining concrete is usually cast after the stabilisation of the primary lining deformations and the final lining is not loaded by ground pressure, it is not necessary from the static point of view for the structure to exhibit the required terminal strength value already after 28 days. It is possible to expect that the value of loading assumed by the calculation will

IOP Conf. Series: Materials Science and Engineering **236** (2017) 012084 doi:10.1088/1757-899X/236/1/012084

act on the lining later, after the complete degradation of the primary lining, if ever. It is advantageous for the mixture design if it is possible to use a "slow" concrete formula with low hydration heat and the terminal compressive strength is reached after 56 or even 90 days.

4. Assessing the quality of the surface of unreinforced concrete lining and contractual criteria

4.1. Causes of the formation of cracks and their influence on the function of linings

An unreinforced concrete tunnel lining is still rather a rarity than a generally established practice in the Czech Republic. An often discussed issue is the number, orientation, depth and width of cracks in a lining. We can divide the cracks into several types, assign significance to these types and, with a certain degree of simplification, even the cause of their origin. The first cause of their formation lies in volumetric changes of concrete - shrinkage, temperature changes during concrete casting due to hydration heat and, during the tunnel life, due to the temperature changes corresponding to the seasons. Another cause of the formation of cracks lies in the static loads acting on the structure.

Cracks due to volumetric changes of concrete are either vertical cracks usually on the tunnel sides, which can in some cases run across the whole lining vault, or directionless "cobwebby" cracks in the whole area of the structure. An example of a vertical shrinkage crack is presented in Figure 1.

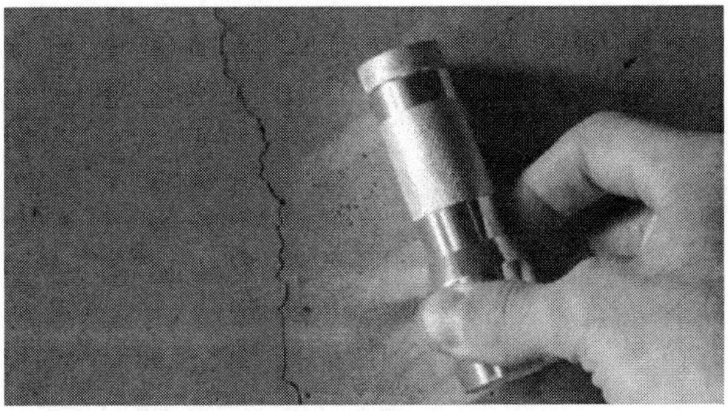

Figure 1. Vertical crack on the side of the lining caused by concrete shrinkage (photo courtesy of Libor Mařík).

Cracks due to static loading are usually oriented in the direction of the tunnel axis (horizontal). They are usually created in the vault crown; they can appear even on the tunnel sides, depending on the loading intensity and proportion between the vertical and lateral pressure. In reality, the two causes of the formation of cracks cannot be strictly distinguished one from the other. The size of a crack due to a static action is affected by the state of stress in the lining due to concrete shrinkage and vice versa.

The approach to assessing cracks from the point of view of the lining functionality is also double. First of all, it is necessary to distinguish the crack direction. Vertical cracks do not fundamentally influence the static action of the lining. It is possible to imagine their function as splitting of the lining into shorter blocks, the static function of which does not change in the transverse direction. The main criterion in this case is aesthetic. It is a matter of an agreement between the contractor and the client about the client's idea of the lining surface appearance. On the contrary, cracks due to static loading may directly affect the static function of the lining. When they start, an imperfect hinge develops in the lining, the statical schema of the structure changes as well as the distribution of inner forces, mainly bending moments. With the normal force almost unchanged, the compression zone of the lining section is reduced. In such a case it is necessary to assess the structure and determine the allowable width and depth of the crack on the basis of a structural analysis. Apart from assessing the compression zone of the section it is necessary to assess also the danger of secondary cracks formation in the lining perpendicularly to the direction of radial cracks, which may cause falling of the parts thus damaged from the lining. It is dangerous first of all with respect to the operation safety. This phenomenon is described in Reference [9].

4.2. Assessing the quality of unreinforced concrete linings in the Czech Republic

Whilst in Austria and Germany the problems of assessing cracks are known and contained in directives and regulations, no regulation defining the relationship between the client and the contractor during the final acceptance at the end of the warranty period existed in the Czech Republic until the end of 2015. In a better case, the allowable parameters of the lining surface were defined in tender documents, whilst, in a worse case, the rules were set later, in the course of the construction. In such a case, the contractor takes the risk of an undefined scope of potential rehabilitation work. In case of a responsible attitude, the contractor evaluates this risk and incorporates it in the bid for execution of the works, thus the investment cost is unnecessarily increased.

In case of railway tunnels this situation still exists, in case of road and motorway tunnels it changed after the issuance of a review of Technical Specifications of the Ministry of Transportation No. 18 – Concrete Structures and Bridges in January 2016. The regulation allows use of unreinforced concrete as a structural material and contains not only criteria for the allowable formation of cracks, but also requirements for the quality of the unreinforced concrete lining surface. Cavities (bubbles) with closed surface and maximum diameter of 25mm or maximum area of 5cm2 and maximum depth of 10mm are permissible on the surface of an unreinforced concrete lining. An example of these cavities is presented in Figure 2.

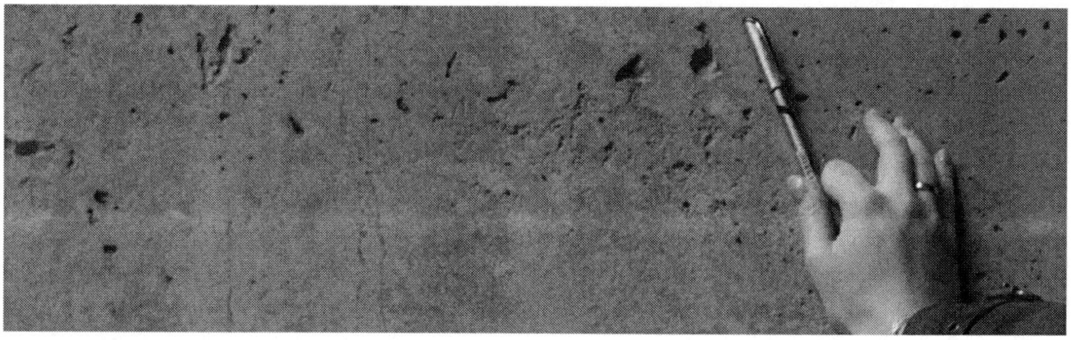

Figure 2. Surface cavities in the lower part of the lining (photo courtesy of Libor Mařík).

Cavities in the lining surface have to be treated if one of their dimensions (diameter, area, depth) exceeds the respective limit. The repair of concrete surface in the repair design, for example by grinding, if possible, is better and more durable than a thin layer of a repair material. Such a layer may cause a far greater damage if it breaks away from the substrate and falls on a passing vehicle than an imperfect surface of the lining. It applies first of all to the area at the tunnel vault crown above the roadway. Breaking of concrete at the edges of construction or expansion joints and/or in the vicinity of cracks is assessed from the point of view of its location in the lining. If a crack of unknown depth is located at a distance of 200mm or smaller from a construction or expansion joint or another closest crack, it is necessary to assume that a part of the lining concrete can break off. If the crack and the construction or expansion joint and/or more cracks form a continuous closed shape delimiting the possibility of breaking away or loosening a part of concrete (see Figure 3), the crack width (including a non-structural crack) will be monitored during regular inspections at least once a year, unless the width exceeds 0.2mm.

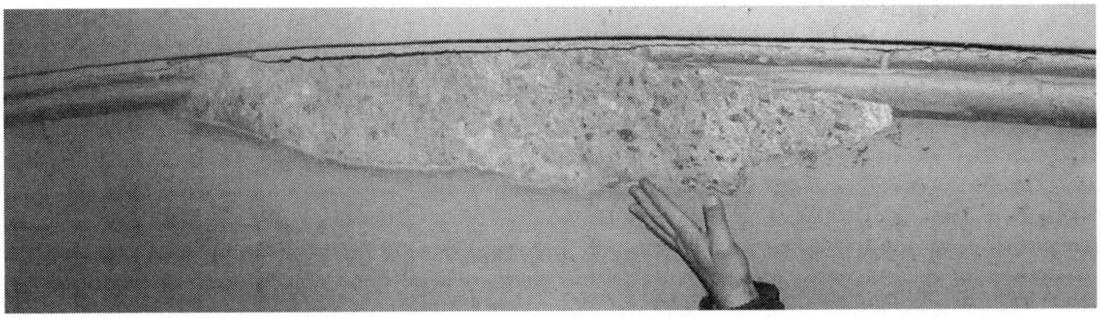

Figure 3. Loosened part of the lining at the joint between concrete casting blocks (photo courtesy of Viktor Petráš).

At the width ranging from 0.2 to 0.8mm the crack will be monitored during extraordinary inspections 1x in 3 months and, at the width over 0.8mm, it is necessary to design and implement measures (for example a repair of the particular section of the lining). Spalls and loosened parts of concrete, first of all those loosened from the edges of construction or expansion joints and/or cracks, are considered to be a defect preventing safe operation and have to be repaired before opening to traffic.

In the case of unreinforced concrete lining the maximum depth of structural cracks is defined by the requirement for the minimum compression zone depth of the concrete section. It has to be determined on the basis of a structural analysis for the particular case. The depth of non-structural cracks in a reinforced concrete lining is not limited and can run throughout the lining thickness. Cracks in linings of all of the above-mentioned types must not pass across anchors to concrete serving, for example, to fix tunnel equipment. In the case of the unreinforced concrete lining it is necessary to fill cracks when their full load-carrying capacity and usability of the structure are not proved and/or when their dimensions exceed the permitted values presented in Table 1.

Table 1. Maximum permitted values for cracks in unreinforced concrete linings [1].

Maximum permitted values of monitored parameters of defects of unreinforced concrete linings		Unit	Criterion at final accep-tance	Criterion at the end of warranty period	Criterion at the end of design life
Structural cracks	Width of cracks (horizontal)	[mm]	0.5	1	1.5
	Depth of cracks	[mm]	To be determined by structural analysis		
	Vertical displacement (perpendicularly to tunnel axis)	[mm]	1	2	3
	Horizontal displacement (perpendicularly to tunnel axis)	[mm]	1	2	3
	Number of cracks (horizontal) in a 12m long section	[pcs]	3	4	5
Non-structural cracks	Width of cracks (horizontal)	[mm]	2	3	3
	Depth of cracks	[mm]	Up to the full lining depth		
	Number of cracks in a 12m long section (vertical, running across the whole vault)	[pcs]	2	2	2

4.3. Examples of criteria from foreign practice

In the Czech Republic, the unreinforced concrete lining is at the moment found in the Pisárky tunnel in Brno, the Libouchec tunnel on the D8 motorway linking Prague and Dresden, both tubes of the Vítkov railway tunnels in Prague and the Olbramovice, Tomice I and Sudoměřice tunnels on the fourth railway corridor leading from Prague to Linz. The formation of cracks is being monitored in these tunnels without significantly negative experience. However, long-term monitoring is not available yet.

The relative lack of experience with the use of unreinforced concrete linings in the Czech Republic in comparison with the Alpine countries, where economically designed tunnels are necessary for securing the transport infrastructure, probably leads to the determination of more detailed criteria for assessing the cracks in linings. Above all, it is defining the number of cracks per concrete casting block, but also detailed defining the time to which the crack width is related. An advantage is a transparent relationship between the client and the contractor in potential claim enforcing procedures. In foreign regulations the criteria are set in a substantially simpler way because of the fact that experience shows that use of an unreinforced concrete lining is not a risk from the point of view of the tunnel operation or even securing its stability.

The Austrian regulation [3] does not consider cracks due to volumetric changes or mechanical effects up to the width of 1mm to be defects. When the width is greater, the cracks have to be assessed in terms of durability of the lining, tightness and static function. First of all, this regulation defines possible defects of reinforced concrete and unreinforced concrete lining including photodocumentation. It, in addition, prescribes the way of their treatment.

According to a German regulation DB853 [7] it is necessary to treat unbranched radial cracks wider than 2mm, longitudinal cracks in the sidewalls or in the vault crown wider than 1.5mm and crescent cracks in the vicinity of construction joints and anchoring elements wider than 0.5mm by injecting grout into them. In comparison with this regulation the criterion for the permissible width of crack of 0.5mm contained in the regulation [1] for horizontal structural cracks at the moment of the final acceptance by the client appears to be very strict. It may be limiting when the decision about use of reinforced or unreinforced concrete is being made, thus negatively affecting the cost of the works. It would be more appropriate to set the same criterion as that for the end of the warranty period or, as in the case of assessing the crack depth, to require a static assessment. The reason is the possibility of formation of secondary cracks, which is described in Reference [8]. Another reason is the fact that when a crack does not start immediately after stripping, it is usually caused by the action of ground pressure on the structure.

5. Measures restricting surface defects and formation of cracks in unreinforced concrete lining

In the case of a reinforced concrete lining it is possible to a certain degree to control the width and depth of a crack by a suitable design of reinforcement. In the case of unreinforced concrete lining it is impossible to prevent from the formation of cracks in economically designed cross-sections. From the aspect of the static function and usability of the structure it is even unnecessary, unless the crack parameters exceed the statically or contractually defined limits. Nevertheless, the formation of cracks can be limited during the concrete casting process by designing a suitable formula, adhering to construction principles, a suitable concrete casting procedure and curing of the concrete lining after stripping.

Measures designed to limit the formation of cracks in an order without considering the weight of importance:

1. Maintaining the geometry of the primary lining surface smooth and unbroken as much as possible (without toothing, edges);
2. Using intermediate waterproofing or a separation membrane between the primary and final linings;
3. Reducing the time lag between the casting of concrete for footings/invert and the lining upper vault to a technologically possible minimum (best between 2 and 3 days, acceptable up to 7 days);
4. Using concrete mixture with low hydration heat;
5. Using concrete mixture with slow gain of strength in the initial period after casting the concrete;

6. Stripping the lining block when the concrete strength is low (1.5 MPa to maximum 3 MPa), with the stripping time of ca 12 hours;
7. Curing of concrete after stripping with focus on restricting drying and moderating the thermal shock (temperature conditioned travelling scaffolding);
8. The use of tunnel form travellers with good thermal conductivity skin (steel);
9. Low initial temperature of concrete mixture ranging from 13 °C to 18 °C, under controlled conditions max 25 °C can be acceptable;
10. Recommended length of a concrete casting block up to 12.5 m.

5.1. Permissible deviations from the theoretical thickness of lining

An unreinforced concrete lining is usually used in tunnel sections located in ground mass allowing excavation with longer advance rounds or the application of a primary lining without support frames. As a result, technologically conditioned overbreaks and larger deviations from the designed excavation geometry occur. The overbreaks within allowable tolerances are backfilled with cast-in-place concrete of the final lining. When unreinforced concrete is used, it is necessary to avoid too rapid changing of the thickness and creating toothing in the primary and final linings, which could introduce undesirable components of the tangential stress into the lining. When intermediate waterproofing is used, the conditions for its installation mostly meet also the requirements for the smoothness of the inner surface of the primary lining. Another limiting criterion securing fluent changing of the final lining thickness is the permissible deviation from the design thickness. It is, for example, according to the Austrian directive on the final lining, 100% of the theoretical thickness. With respect to the fact that the theoretical thickness of an unreinforced concrete lining usually ranges from 200mm to 300mm, the maximum real thickness of the final lining is up to 600mm. All irregularities failing to meet the two criteria are filled with shotcrete prior to the installation of the waterproofing membrane and casting of the final lining concrete. This way it is guaranteed that the real thickness of the final lining will not vary sharply and the variable loading by hydration heat in the concrete casting phase and shortly after stripping will not be a cause of excessive formation of cracks.

5.2. Measures on formwork

From the point of view of operation safety it is important to maintain integrity of the structure in the area of joints between concrete casting blocks. They are carried out either as butt construction joints or as expansion joints containing an elastic strip. An important role is played also in this case by the flatness of the surface of the concrete casting block front end, which should be ideally smooth. In general, all construction joints between the blocks must be perfectly cleaned up. The creation of a crack having the shape of a flat parabola in the vault crown (see Figure 4) is a frequent defect.

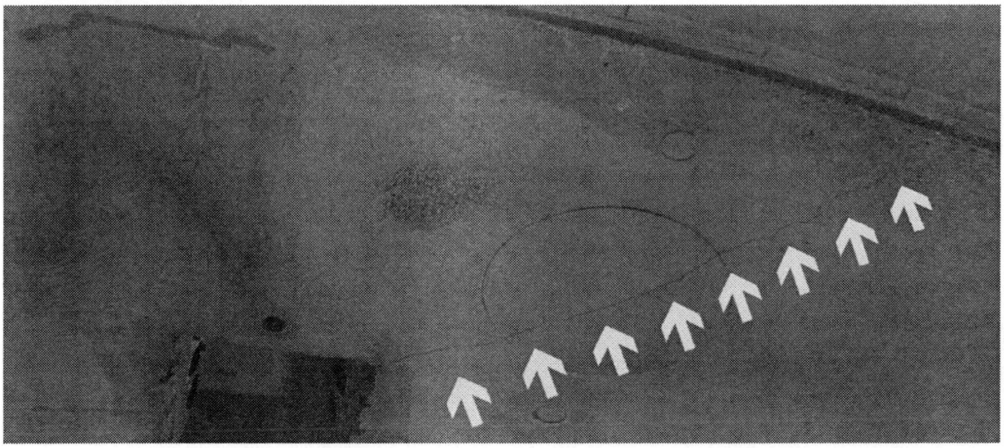

Figure 4. Crescent crack at the vault crown (photo courtesy of Libor Mařík).

It is caused by rough handling of the tunnel form traveller, where a part of the skin leans against the freshly stripped block cast in the previous step. In combination with the unfilled top of the vault, this procedure leads to damaging the top of the vault. A measure for preventing this phenomenon is careful handling of the tunnel form traveller and installation of a soft strip on the contact steel sheet at the end of the form traveller to reduce the local load acting on the traveller.

For achieving the required appearance of the lining surface it is necessary to pay attention to preparation of the formwork surface. The formwork structure must be designed to be sufficiently stiff and watertight. All external vibrators installed on the formwork must be fixed in a way ensuring that the vibration is transmitted first of all to the concrete casting area. It is desirable that the quality of processing all joints in the formwork skin is high. The most suitable surface material for the formwork skin is steel. Prior to the casting of concrete, the formwork must be cleaned up and treated with a suitable separation spray, which ensures that the external surface of concrete does not adhere to the form and smooth stripping is possible. A separation spray must be applied uniformly in thin layers on the whole surface and must be chosen with respect to the necessity of preventing physical or chemical damage to the concrete surface. An important role is played by the compatibility of the separation spray with subsequently applied curing sprays or coloured and protective coats applied to the already finished secondary lining.

5.3. Measures regarding transport of concrete mixture and casting concrete

The overall quality of the lining is influenced by the production, transport and placement of concrete behind formwork. If possible, the temperature of fresh concrete should range from 13 °C to 18°C. Fresh concrete for casting the upper vault with the temperature lower than +10 °C and higher than +25 °C is not considered to be suitable for unreinforced concrete linings. It is therefore necessary to respond flexibly to seasonal fluctuations of temperatures and humidity conditions. Protection against external climatic effects must be ensured during the transport of a fresh concrete mixture. In addition, it is necessary to choose suitable concrete workability time taking into consideration the distance of the batching plant from the construction site and to place concrete on site in time. A rule generally applicable to pouring concrete mixture behind formwork is that the maximum concrete casting rate should be 2 m/h and the horizontal spacing of concrete casting gates should not be greater than 3 m. It is necessary to take care of continual supply of concrete mixture to the construction site so that undesired construction joints are not formed.

5.4. Measures regarding technological procedure and stripping

The formation of both vertical and horizontal cracks can be limited by a suitable technological procedure of casting the concrete. Vertical shrinkage cracks running from the foundation approximately up to the centre of the upper vault can be best limited by a short time lag between casting of the concrete footings and the upper vault. This can be achieved by a suitably selected procedure of the work in the streamlined system of the construction of the final lining. A general recommendation is reducing the time lag between casting the concrete footings and casting the tunnel vault so that it does not exceed 48 hours. This way, an approximately identical shrinkage is achieved in both parts of the structure. The horizontal cracks which mostly appear on a part or the whole length of the concrete block shortly after stripping are not caused by ground pressure. The final lining concrete is attached to the primary lining usually under the condition of stabilisation of deformations and when a predefined very low rate of the deformation increase is achieved. The most important cause of the origination of the horizontal cracks is delayed stripping, when the concrete strength is already too high. According to the Austrian directive [2] it is recommended that unreinforced concrete tunnel linings are stripped when the concrete compression strength reaches 1.5 to 3MPa as a maximum, during ca 12 hours after casting the concrete. Strength values at stripping higher than 3MPa are indicated in the directive as a case with increased risk of the formation of cracks. It is guaranteed by stripping the structure in the moment when the compressive strength is still low that the "young" concrete of the block being stripped will induce sufficiently high compressive stresses in the lining through its own weight, which will contribute to timely closing of originating cracks.

5.5. Measures regarding curing after stripping

It is necessary to duly cure each concrete structure after stripping and secure the required climatic conditions during the concrete casting work with respect to the resultant quality of the surface. The airflow velocity in the tunnel should be reduced to 1m/s as a maximum and the humidity of air should not drop under 90%. The formation of cracks can be significantly limited by eliminating the thermal shock after stripping the formwork.

Temperature conditioned travelling scaffolds are successfully used apart from the standard measures, such as closing the portals, reducing airflow velocity or applying spray against drying. It is another traveller in the overall final lining production flow. It is deployed just behind the travelling form. Its length ranges from two up to three lengths of the concrete casting block, which means that in a model example with casting 5 blocks in 7 days ca 34 hours are allotted to one cycle and each block is protected after stripping for 68 up to 102 hours. During this time a significant part of hydration heat fades away, thus the risk of the cracks formation due to rapid cooling is minimised.

The structure of the temperature conditioned travelling scaffold differs depending on the possibility of curing. The simplest scaffolds consist only of a steel structure covered with a thermally insulating membrane, which maintains natural temperature and humidity in the gap between the just stripped concrete and the scaffold structure.

The concrete curing travelling scaffold standing in front of the Považský Chlmec motorway tunnel on the D3 motorway near Žilina, Slovakia (see Figure 5) is presented as an example. It is a light-weight steel structure covered with a thermally insulating membrane. Inflatable rims serving to seal the space between the inner surface of the lining and the cover of the traveller are presented in Figure 6. The length of the curing scaffold is 25m, which corresponds to the length of two concrete casting blocks.

Figure 5. Steel structure of the temperature conditioned travelling scaffold (photo courtesy of Vladimír Prajzler).

Figure 6. Inflatable rims and thermally insulating cover of the temperature conditioned travelling scaffold (photo courtesy of Vladimír Prajzler).

More complex temperature conditioned travelling scaffolds have water and air brought to the system and create a cold mist by means of spraying jets. In this way sufficient moisture of the concrete surface is secured. The most complex temperature conditioned travelling scaffolds are fully controlled and they allow regulating the humidity and temperature in the gap between the temperature conditioned travelling scaffold and the concrete surface or, as the case may be, to actively cure the concrete structure with steam. The temperature conditioned travelling scaffold structure is in the case of these scaffold types in addition insulated, for example by a polystyrene layer. All parameters of curing can be set and regulated as needed, depending on the changing external conditions.

Nevertheless, in any case the maximum temperature at the surface of the curing concrete structure should not exceed 45°C.

6. Conclusion

Unreinforced concrete has its irreplaceable position as a material for final linings of conventionally driven tunnels. In suitable geotechnical conditions it represents a way to reduce investment costs with the required quality maintained. In addition, minimisation of the risk of damaging the intermediate waterproofing layer, which the placing of reinforcement undoubtedly is, and the complete exclusion of problems with concrete carbonation, can be considered as a bonus. Whilst the formation of cracks and surface irregularities plays no important role in terms of maintenance and the static function, it is necessary to pay special attention to filling of the top of the vault and possibly to creation of cracks in the vicinity of construction or expansion joints between concrete casting blocks, which could pose a threat to the operation safety.

As in the beginnings of using the NATM when it was necessary to start thinking „outside the box“ and to cope with the idea of the self-supporting capacity of rock mass and begin to perceive ground as a building material, it is necessary to respect properties of unreinforced concrete, first of all, the low tensile strength and the formation of cracks associated with it. A different approach in thinking must be applied in all stages, from the lining design through the construction up to the takeover by the client. The designer must take into account the non-linear behaviour of the structure as a whole and assess the existence of cracks in terms of the lining function; the contractor must adapt the concrete mixture design, the technological procedure for building and curing; the client must define realistically possible quality requirements. Only through synergy of all of the above-mentioned components an economic design of the structure can be reached and the NATM principle can be confirmed as an observational method capable of responding to the actually encountered conditions not only in the field of securing the excavation stability, but also in the solution of final lining.

Acknowledgement

This paper was prepared thanks to the support of the Technology Agency of the Czech Republic (TAČR), project "Centre for Effective and Sustainable Transport Infrastructure (CESTI)" (no. TE01020168).

References

[1] Ministerstvo dopravy České Republiky odbor pozemních komunikací 2016 Betonové konstrukce a mosty *Technické kvalitativní podmínky staveb pozemních komunikací*

[2] Österrechische Bautechnik Vereinigung 2012 *Richtlinie Innenschalenbeton*

[3] Österrechische Vereinigung für Beton- und Bautechnik 2009 *Richtlinie Bewertung und Behebung von Fehlstellen bei Tunnelinnenschalen*

[4] Deutscher Ausschuss für unterirdisches Bauen (DAUB) – Arbeitskreis: Unbewehrte Tunnelinnenschalen 2007 *Empfehlungen zu Ausführung und Einsatz unbewehrter Tunnelinnenschalen*

[5] Bundesanstalt für Straßenwesen, Brücken- und Ingenierbau 2013 *Kriterien für die Anwendung von unbewehrten Innenschalen für Straßentunnel - Heft B92*

[6] Bundesanstalt für Bauwesen 2007 *Zusätzliche Technische Vertragsbedingungen und Richtlinien für Ingenieurbauten ZTV-ING, Teil 5 Tunnelbau, Abschnitt 1 Geschlossene Bauweise*

[7] DB Netz AG 2013 *Richtlinie DB853 Eisenbahn Tunnel planen, bauen und instand halten*

[8] Kupfer H B, Kupfer H and Steckbaner A 1984 *Tragfähigkeit von schlanken Druckgliedern aus unbewehrtem Beton* (Springer-Verlag)

[9] Pöttler R 1993 *Die unbewehrte Innenschale im Felstunnelbau – Standsicherheit und Verformung im Rissbereich Beton- und Stahlbetonbau*

[10] Vogel F, Sovják R, Holčapek O, Mařík L and Šach J 2015 Experimental Study of Primary Lining Tunnel Concrete after Thirty Years of Operation *Mechanics and Materials* **732** pp 403–6

IOP Conf. Series: Materials Science and Engineering **236** (2017) 012085 doi:10.1088/1757-899X/236/1/012085

The ultimate limit state of the underground circular tunnel segment lining

Václav Ráček[1] and Jaromír Zlámal[2]

[1]CTU in Prague, Faculty of Civil Engineering, Dept. of Concrete and Masonry Structures, Thákurova 7/2077, 166 29 Prague 6 – Dejvice, Czech Republic
[2]POHL cz, a.s., Nádražní 25, 252 63 Roztoky, Czech Republic

E-mail: vaclav.racek@fsv.cvut.cz

Abstract. Circular tunnel segment lining with staggered joints in track tunnels of Prague Underground was found to be on the ultimate limit state; static analysis (mathematical modeling) is used to determine the causes which led to such situation. This situation is signalized by cracks and related deformations; lining load-limit coefficient can be used to determine the cause. Analysis is performed in the form of parametric study, where the variables are the values of geotechnical figures, the rigidity of the lining with staggered joints and the load of the lining. This paper focuses on analysis of reinforced concrete segmental lining Ø5.3 / 5.8 m (5 + 1 element) and cast iron Ø5.1 / 5.5 m (9 + 1 element). Parametric study using coefficient of loading limit for both of these cases in the Prague Underground leads to fast and relatively easy determination of the cause of the reaching of the ultimate limit state.

1. Introduction

In the track lines of Prague Underground, the circular tunnel lining (concrete or cast iron) with shifted staggered joints reached the emergency conditions of the load bearing capacity several times; it is a task of the static analysis (mathematic modelling) to assess the source of this emergency situation.

When the circular lining reaches the state of load bearing capacity, usually there is not enough time to elaborate complicated safety measures, especially if the reason of such changes is not known. Proposal for safety measures of the tunnel lining is usually based on the mathematical model analysis. The static model is prepared to correspond with real state as best as possible.

This article deals with analysis of selected reinforced concrete segmental lining Ø5.3/5.8 m (6 members) and cast iron lining Ø5.1/5.5 m (10 members). The goal of the parametric study is to assess the load bearing capacity of the lining and its margins. The reinforced concrete segmental lining was indicated with hair-like cracks parallel to the longitudinal track axis; the hypothesis was that the cracks were results of the impact pressure of the end of the shield in curve with small radius of the track. The cast iron lining was diagnosed with significant openings in abutting joints in the place, where lining was mounted on the concrete basement in the starting pit. Cast iron lining was backfilled, but compaction of the dusting was inadequate and after short time of the utilization the deformation arose.

IOP Conf. Series: Materials Science and Engineering **236** (2017) 012085 doi:10.1088/1757-899X/236/1/012085

2. The load bearing capacity analysis method

Designing the structure, we try to satisfy two basic but often contrasting conditions. We want to have the structure economical and safe too. Underground structures are loaded with continuous stress which creates its deformation. This state can be described mathematically in terms of structural mechanics. The durability of the structure is defined by three factors – placing the structure, load applied on the structure and natural environment surrounding the structure.

Static computation reveals the bearing-capacity of the lining considering rock-support interaction of different quality. For exact assessment of the load-bearing capacity, we use static parameters M_i, N_i, T_i and then we can set measures to stop the additional destruction of lining.

Let P_{Umax} denote the limiting bearing-capacity of the cracks, P_{Lmax} be the limiting loading capacity of the tunnel lining (Figure 1), and P_i, P_j be the discreet loadings in points i, j for the calculated static values M_i, N_i, T_i and M_j, N_j, T_j, then

$$P_i(t) = f(M_i . N_i . T_i);\ P_j(t) = f(M_j . N_j . T_j) \tag{1}$$

If he limiting bearing-capacity of the tunnel lining is described by the function $P_{Umax} = f(M_U, N_U, T_U)$, then for time ($t$) we can simply write equation

$$P_{Umax}(t) = f(A_b . A_s . f_c . f_t) = f(M_{Ut} . N_{Ut} . T_{Ut}) \tag{2}$$

where A_b is the sectional area of concrete, A_s is the sectional area of steel reinforcement, f_c is the compressive strength of concrete, f_t is the strength of steel in tension.

We compare values of the discrete loading P_i a P_j and the bearing-capacity of the cracks described by equation (2). If $P_i(t)$ is less than $P_{Umax}(t)$ then we conclude in equation

$$P_i(t) < P_{Umax}(t) \tag{3}$$

We state that tunnel lining suits limiting bearing capacity of cracks. If loading $P_j(t)$ surpasses $P_{Umax}(t)$, then it results into in equation

$$P_j(t) > P_{Umax}(t) \tag{4}$$

We state that the limiting bearing capacity of cracks is not satisfactory and construction is not suitable. The coefficient of the bearing-capacity of the lining s is defined as the ratio of $P_{Umax}(t)$ and $P_j(t)$

$$\frac{P_{U\max}(t)}{P_i(t)} = s_i \tag{5}$$

The bearing-capacity of the tunnel lining is then calculated by the coefficient s_i on the whole perimeter of the tunnel lining and in the places where the value of s_i (5) is smaller than 1, $s_i < 1$, the construction is not satisfactory due to limiting bearing capacity of cracks . If the construction has more places with coefficient $s_i < 1$, then measures for prevention of increased cracks and deformations have to be adopted.

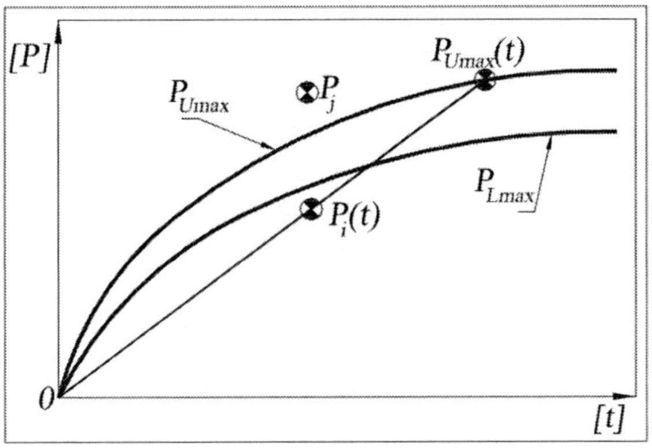

Figure 1. The limiting bearing-capacity and the limiting loading of the tunnel lining.

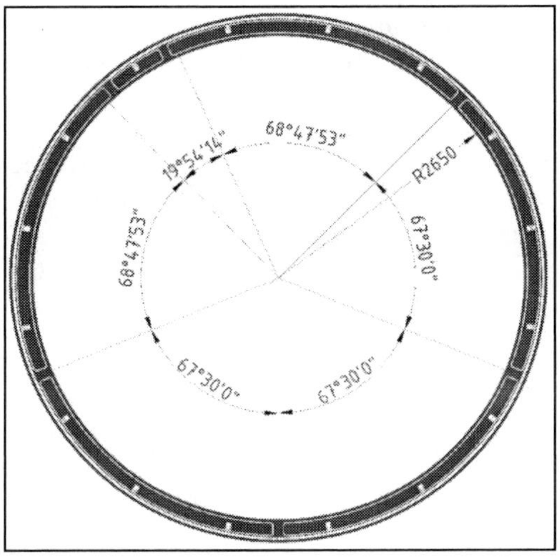

Figure 2. Tunnel lining Ø5.3/5.8 m of Prague underground.

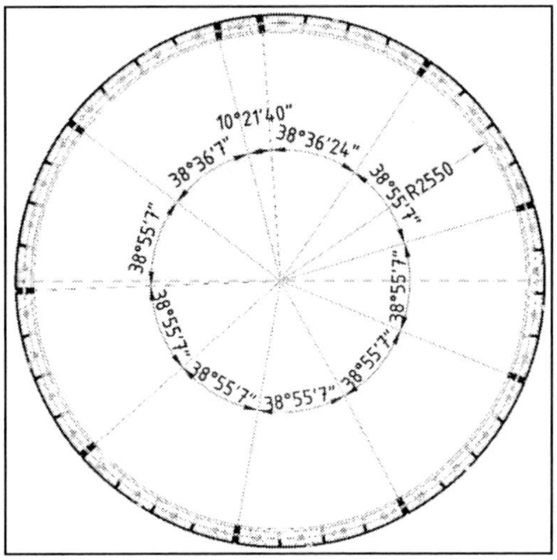

Figure 3. Tunnel lining Ø5.1/5.5 m of Prague underground.

3. Analysed tunnel linings

In the parametric study, the coefficient of bearing capacity of two tunnel linings of Prague Underground, reinforced concrete segmental lining Ø5.3 / 5.8 m (5 + 1 member) and cast iron segmental lining Ø5.1 / 5.5 m (9 + 1 member) is assessed. In both cases shifted abutting joints were used and allowed limited deformations were kept. In the reinforced concrete segmental lining, hairline cracks parallel with longitudinal track axis arises. It has been supposed that the cracks were results of the impact pressure of the end of the shield in curve with small radius of the track.

Cast iron lining was assembled in open pit on the concrete basement of ca 4 m width and the pit backfilled by the soil showed significant openings of the abutting joints after commencing of operation of the underground.

Reinforced concrete segmental lining has outside diameter of 5.8 m and inside diameter of 5.3 m. Thickness of the lining is 0.25 m, width of the ring is 1.5 m, C50/60 XA1 concrete is used. Tubings are reinforced by steel (ØR6 – ØR12) and connected by bolts with dowels in abutting and radial joints. Shape of the ring is in Figure 2.

The cast iron lining Ø5.1/5.5 m (9+1 member) is made from СЧ24-44, width of flange 195 mm and thickness of shell 20 mm. Thickness of the abutting and radial shells are 27 – 36 mm. In abutting shell of tubing there are 4 openings of Ø 30 mm, radial shell has 6 openings of Ø 30 mm placed shifted to the radial joint axes.

Ring is assembled of 7 tubings Н-3-Л, two tubings of С-2-Л and one key tubing of К-2-Л. Weight of the whole ring is 5280 kg. According to the standard СНиП П-В.3-72 - Стальные конструкции-Нормы проектирования (Steel structures – designing code), cast iron СЧ24-44 has flexural compressive strength of 210 MPa, the flexural tensile strength is 80 MPa and shear strength is 60 MPa.

4. Polygonal computation method

Tunnel linings are modelled as polygons with many joints. Computation is made by general deformation method. Load of the rock on the lining is modelled as active loading in polygon vertices and passive counteraction of the rock is modelled by the Winkler's springs. This method is universal and suitable to replace every shape of the construction and different geotechnical parameters. The computation gives us rather good information about inner forces in the lining, but doesn′t provide information about strength in arbitrary place in the rock. We only know reaction and stresses by which rock interacts with the lining. Results of computations depend on the boundary conditions, thus parametrical studies have to be conducted with different geotechnical parameters, values and shape of the loading. Stiffness (E-modulus of the concrete) is altered in the case of the reinforced concrete considering age and quality of the concrete.

The calculation runs in iterative cycles; in the first iterative cycle, all radial rock imitative rods are in function, and the calculations run until all tensile rods are removed and all pressed radial rods are in function only. The springs imitating rock have area and length value of 1. Their tightness relates to the E_o deformation modulus of the rock. The characterization of the elastic rods can be written in the form

$$\overline{K}_i = K_{(\sigma)} \frac{l_i + l_{i+1}}{2} b \tag{6}$$

where l_i is the length of the member of the polygon and b is the width of the ring. The coefficient of elastic resistance of rock $K_{(\sigma)}$ depends not only on geotechnical parameters of the rock, but also on the shape of the construction and is usually determined according to B. G. Galerkin equation.

$$K_{(\sigma)} = \frac{E_o}{R(1+\nu)} \tag{7}$$

Let us assign the deformation and the movement of i joint U_i , V_i , θ_i and internal forces of i joint X_i, Y_i, M_i and outside loading $\underline{X}_i$, $\underline{Y}_i$. Then we can write the columnar matrix $\{P_i\}$ for the internal calculation parameter, and columnar matrix $\{P_i\}$ for outside loading, where

$$\{P_i\} = \begin{Bmatrix} U_i \\ V_i \\ \theta_i \\ X_i \\ Y_i \\ M_i \end{Bmatrix}; \qquad \{P_i\} = \begin{Bmatrix} 0 \\ 0 \\ 0 \\ \underline{X}_i \\ \underline{Y}_i \\ 0 \end{Bmatrix}; \tag{8}$$

We include the reactions of the supports as the internal forces for i joint. Adding the deformation of elastic supports leads to the changes of the stiffness matrix coefficient $[K_i]$ which affects computation parameters. In general, load can be applied to any joint (joints have separate numbering, denoted with j). Matrix notation of the equation for calculating parameters for the joint n, including known boundary conditions, is in the form

$$\{P_n\} = \prod_{i=n+1}^{1}[K_i]\{P_o\} + \sum_{j=1}^{n}\prod_{i=n+1}^{j+1}[K_i]\{\underline{P}_i\} \tag{9}$$

$\{P_o\}$ is column matrix of boundary calculation parameters.

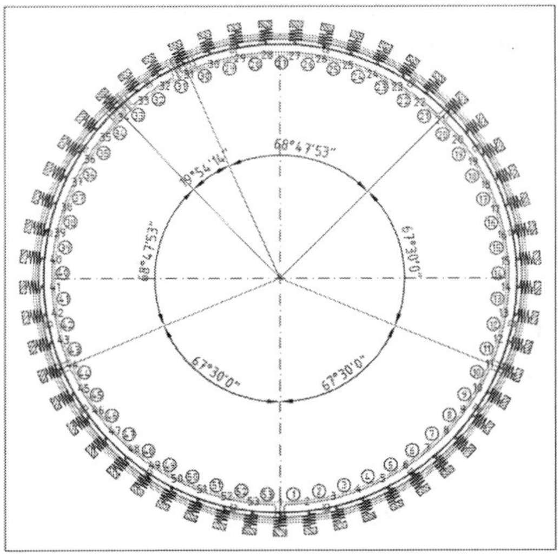

Figure 4. Static scheme of the lining Ø5.3/5.8 m Prague underground.

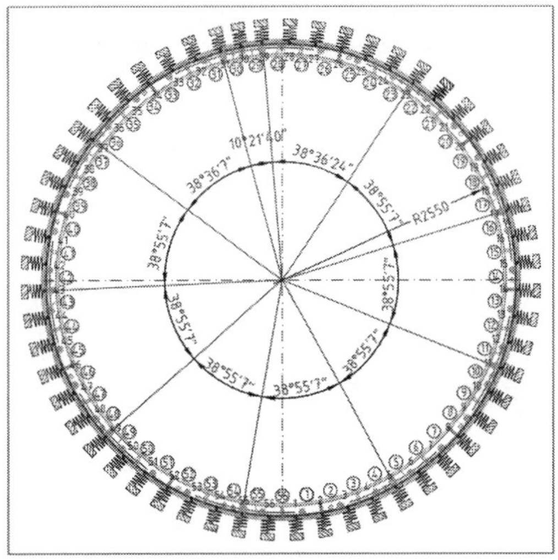

Figure 5. Static scheme of the lining Ø5.1/5.5 m Prague underground.

5. The ultimate load-bearing capacity analysis

5.1. Fenner-Pacher curve for reinforced concrete lining assembled by earth pressure balanced shield and for balanced shield and for cast iron lining assembled by "erector"

Self-support of the rock, which can be considered in technological classification I – III by QTS index (Tesař 1990), is in most cases time-dependent value. After boring, rock is capable to hold its weight itself for a limited time. Thanks to rock deformation, loading on the lining decreases and rock ring is activated – rock arch appears. If the equilibrium is achieved after boring of the face without any lining of the intrados and without breakdowns, rock is fully self-supporting and time dependent development is considered in the rock with rheological property only.

If equilibrium is achieved only by lining assembling after boring of the face, self-supporting of the rock relates to value of time dependent radial deformation of the bored face.

Fenner – Pacher curve illustrates changes of the radial stress σ_r, where partial self-support of the rock is used, resisting the weight of the rock in the head. Weight-loaded rock is deformed in the radial way by the value *u*, which is considered as rheological property of the rock (in many cases rheological property of the rock is short-termed) which is time dependent, i.e. $u = u\ (t)$.

$$u = f\,(\tau,\ \Gamma,\ r,\ R,\ \rho) \tag{10}$$

where τ – shear strength of the rock, γ – weight of the rock, r – radius of the circular boring, R – radius of the bearing rock ring, ρ – rheological property of the rock. Fenner – Pacher curve is given as

$$\sigma_r = f\,(\sigma_{ro}\ .\ \sigma_{rmin}\ .\ u) \tag{11}$$

where σ_{ro} is the highest value directly after assembling tight lining into bored face, σ_{rmin} is the lowest value of the strength σ_r.

One part of the roof $\sigma_o = \gamma h$ is carried by the arch in the roof, second part is acting as loading on the lining, which is consequently transposed as support to deforming bored face (value σ_r). Fenner –

IOP Conf. Series: Materials Science and Engineering **236** (2017) 012085 doi:10.1088/1757-899X/236/1/012085

Pacher curve can only be fully applied on bored face and tensioned lining that are symmetrical along vertical axis. Fenner – Pacher curve in asymmetric profiles is different for every point of lining. Fenner – Pacher curves can be obtained in each discrete point of lining for pre-selected ratio of the vertical and horizontal loading and particular elasticity/deformation modulus of rock by FEM mathematical modelling. Value of the loading with corresponding value of deformation of the bored face in the moment of the lining assembling would be obtained by using such curves. Fenner – Pacher curves for the reinforced concrete lining assembled by earth pressure balanced shield and for cast iron lining assembled by „erector" are shown in Figure 6. Curves differ significantly, earth pressure balanced shield doesn't create big deformations of the rock, while using cast iron lining assembled by „erector" leads to bigger deformations. Smart deformation of the cast iron lining was ceased (see Figure 7) by grouting as depicted in Figure 6.

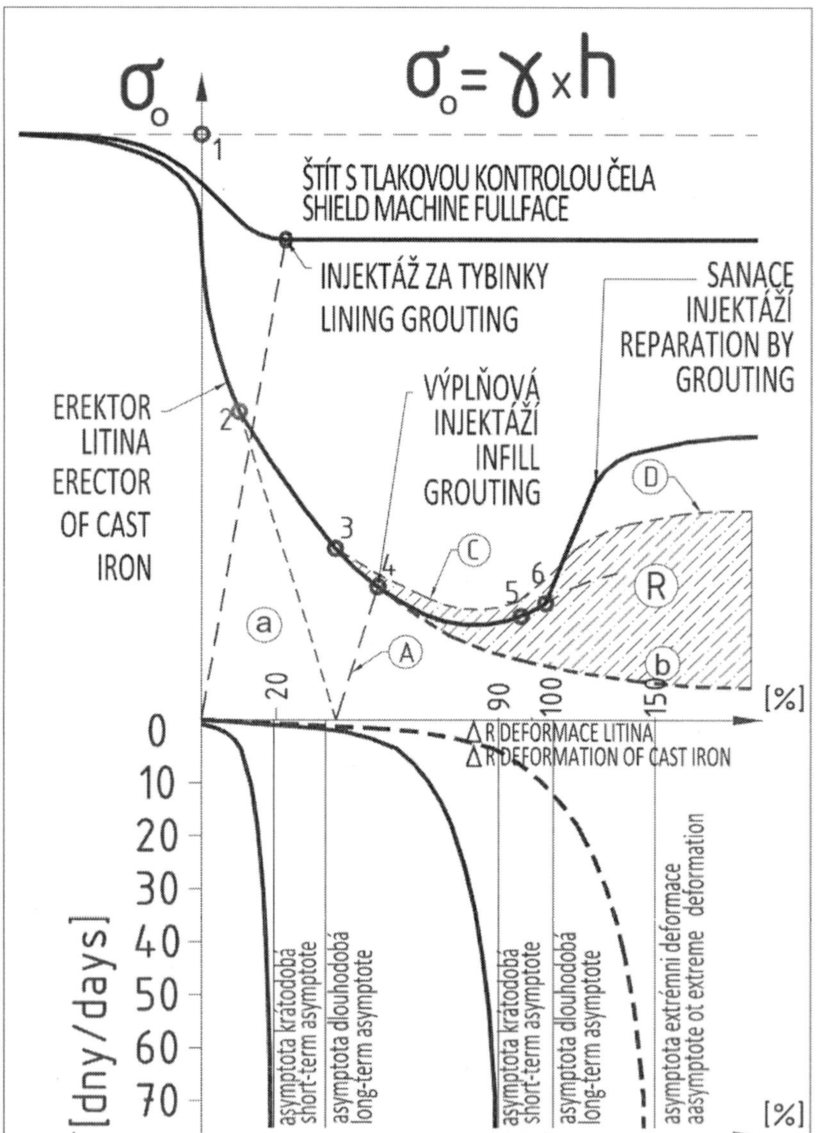

Figure 6. Deformation and stress curve for „erector" and shield. a – elastic rock deformation, b – plastic rock deformation, A – cast iron assembling, C – line of "rock-lining" state, D – state in long term equilibrium, R – area of loosened rock, 1 – value of original tenseness, 2 – beginning of the elastic plastic state of the rock, 3 – assembling of the lining, 4 – beginning of the activated lining after grouting, 5 – achievement short term state of equilibrium, 6 – beginning of the over passing of the limited bearing capacity of the lining, ΔR increment of deformation %.

5.2. Prague underground cast iron lining 5.1/5.5 m in inadequate dusting

Cast iron lining Ø5.1/5.5 m was mounted on the concrete basement in the starting pit 14.0 m deep. Design required compaction of the dusting especially on the side of the lining. Quality of the compaction required by design wasn't delivered and during operation the deformation and opening joints in upper part of lining appeared. Clearance profile for railcars wasn't affected. Tunnel lining state is depicted in Figure 7.

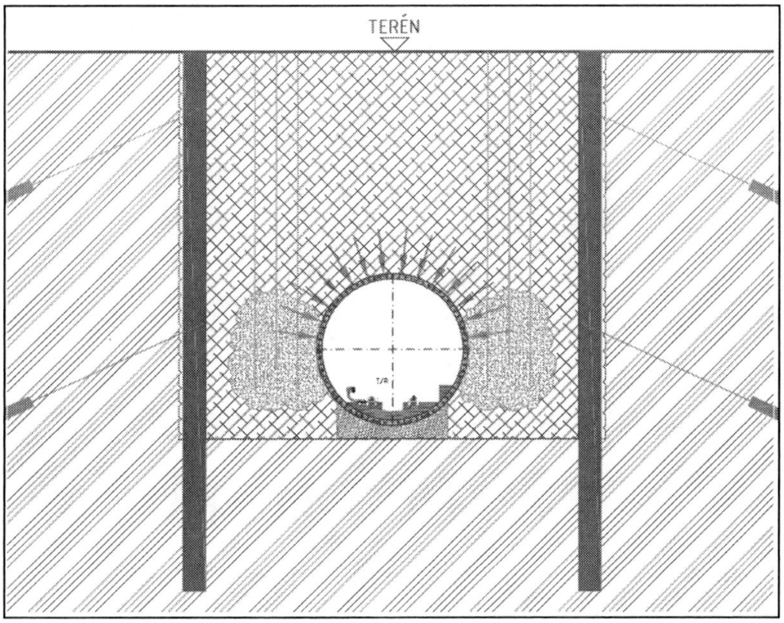

Figure 7. Underground track tunnel's cast iron lining mounted on concrete basement with inadequate dusting on the sides.

After indication of openings in joints of track tunnel, the grouting of the sides of cast iron lining was made (Figure 7). Deformations and opening of the joints immediately stopped. Calculation of the ring composed from tubings connected in abutting joints by numerous screws (radial joints are shifted) can be carried out as construction without joint. Sir Allen Muir Wood [4] uses for such case equation where moment of area is reduced:

$$I = I_s + I_n \cdot \left(\frac{4}{n}\right)^2 \tag{12}$$

where I = reduced area-wise moment, I_s = area-wise moment of the force transmission zone, I_n = area-wise moment of complete section, n = number of segments $n > 4$, (small keysegment not counted), $I_s \ll I$ lining of tubings with shifted radial joints and screws in abutting joints.

Parametric study analysis of the limiting deformation bearing capacity led to calculation of coefficients of loading limiting bearing capacity s_i with deformation modulus of rock E_o = 5, 50 a 100 MPa and with reduced area-wise moment according [4], equation (12). Results are depicted in Figure 8. Value of the limiting bearing capacity $s = 1$, (the bearing capacity of the lining isn't reached) is depicted by grey solid line.

If the coefficient of loading limiting bearing capacity s_i is smaller than 1, openings in joints are registered. It means the bearing capacity of the lining is not sufficient at these spots and it is necessary to take mitigating action to ensure lining stability. Most dangerous is state with E_o = 5 MPa deformation modulus, value s_i reached s_{5MPa} = 0.24 – 0.32; s_{50MPa} = 0.44 – 0.64 and s_{100MPa} = 0.62 –

0.81. Analysis indicated several spots where coefficient of limiting bearing capacity s_i is smaller than 1, or close to this value. Openings in joints facing to the rock wall aren't visible, but must be considered at the time of stabilization of lining. If the coefficient of limiting bearing capacity s_i exceeds 1, the lining has satisfactory load-bearing capacity.

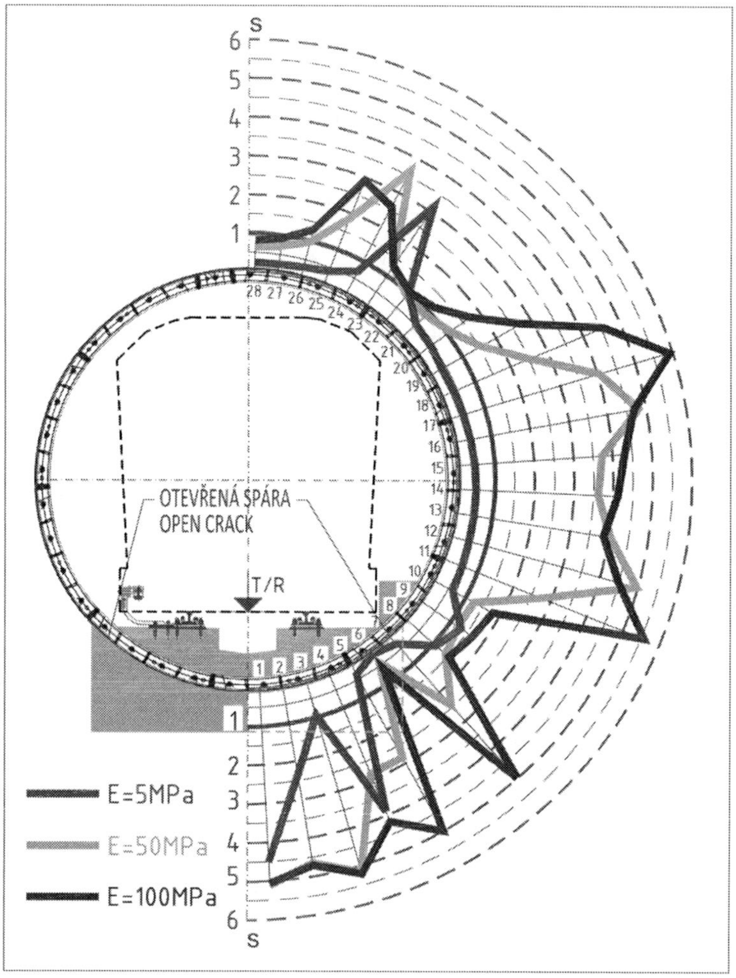

Figure 8. Limiting bearing capacity of cast iron lining assembled in pit on the concrete basement.

5.3. Reinforced concrete segmental lining Ø5.3/5.8 m of Prague underground assembled by earth pressure balanced shield

Hairline cracks parallel to longitudinal track axis were found on the reinforced concrete segmental lining Ø5.3/5.8 m in Prague Underground in the altitude of horizontal axis of the lining. Two hypothesis of the cracks origin were proposed. One theory explained the origin of the cracks as a result of the impact pressure of the end of the shield in curve with small radius of the track. Minimum radius of the track tunnels is R = 650 m and contact of just the end of the shield with reinforced concrete lining is expected in this theory. Second theory explained the origin of the cracks as a result of hydraulic pressure on abutting joints of the lining in the time of shield moving. Such force is often asymmetric and in some cases pressure on tubing can reach extreme values.

5.3.1. Impact pressure of the end of the shield on reinforced concrete segmental lining. Lets assume load of the reinforced concrete segmental lining Ø5.3/5.8 m at the end of the shield according to Figure 9. The contact between the end of the shield and reinforced concrete lining is the most probable case in curve with small radius of V.A metro line which is 650 m. According to the information from the construction site, the direction divergence of the shield face never exceeds 20 mm from the designed track.

Analyzing the contact of the shield end with lining with divergence of the shield face of 20 mm in radius 650 m, we found four theoretical positions of the shield, see Figure 10. It is assumed that contact of the shield end with lining takes place when shield rotation is deviated from designed track by $\alpha = 0°26'25''$.

Shield face is deviated by 20 mm from designed track inside radius and shield end is in contact with reinforced concrete lining outwards of curve – option 1. Force on lining will be calculated as result of this state. For the geometry of the lining loaded by hydraulic pressure see Figure 9. The only force present is the force at contact between reinforced concrete lining and the shield end. Earth pressure force on the shield face is not included. Tubing's ring is loaded with the grouting pressure of 3bars and forces of hydraulic pressure on abutting joints till the tubings are let out of shield.

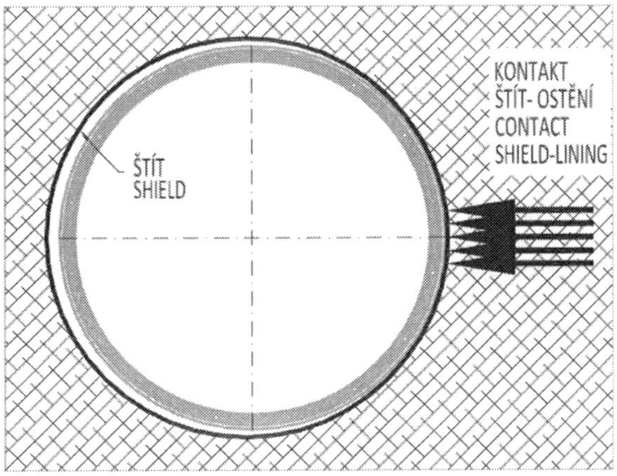

Figure 9. Contact of the shield end with reinforced concrete lining Ø5.3/5.8 m.

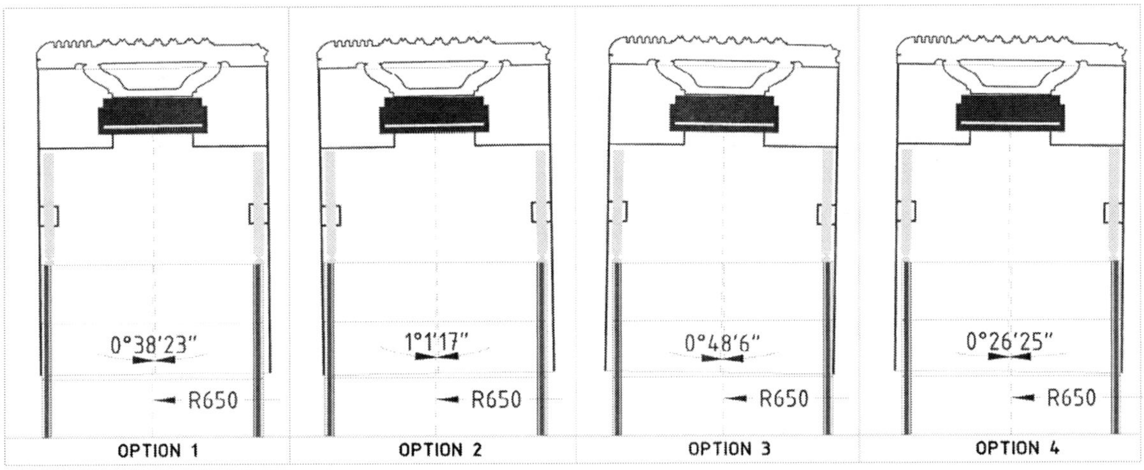

Figure 10. The shield maximum possible rotation in curve R = 650 m, resulting in divergency of the shield face of 20 mm from designed track.

Option 1 – shield face is 20 mm deviated horizontally inwards the radius from designed track. Shield end is in contact with reinforced concrete lining outwards of curve.

Option 2 – shield face is 20 mm deviated horizontally outwards the radius from designed track. Shield end is in contact with reinforced concrete lining outwards of curve.

Option 3 – shield face is 20 mm deviated horizontally inwards the radius from designed track. Shield end is in contact with reinforced concrete lining inside of curve.

Option 4 – shield face is 20 mm deviated horizontally outwards the radius from designed track. Shield end is in contact with reinforced concrete lining inside of curve.

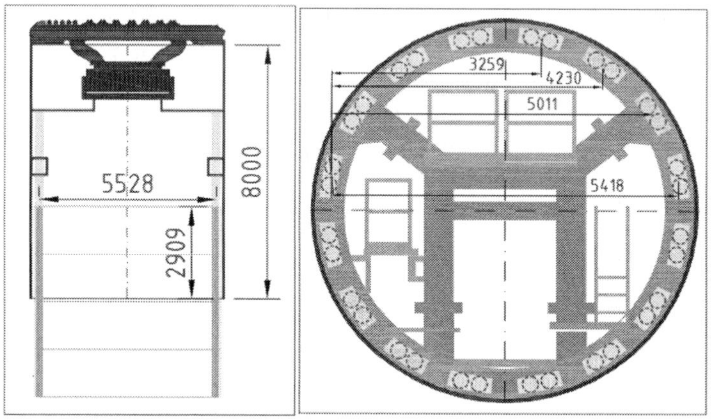

Figure 11. Computation of maximum force which the shield end is acting on lining Ø5.3/5.8 m.

When radius is bored, 8 couples of hydraulic presses push on lining with maximum force $N_8 = 8 \times 2.43 = 19.46$ MN. We can state

$$\Sigma(N_i \times d_i) = P_s \times 2.909 \quad (13)$$

where d_i is the distance of forces. Shield end is acting on the lining by force (simplified approach):

$$P_s = \frac{2.433 \times 2(3.259 + 4.23 + 5.011 + 5.418)}{2.909} = 12.319 MN \quad (14)$$

Head of the track tunnel is 15 – 35 m and if hairline cracks appeared immediately after moving ring from the shield, analysis would not include geostatic pressure. Thus maximum theoretical loading of shield end is $P_{suma} = 12.319$ MN.

Calculation of the cracks in loading lining by shield end according Figure 9 was carried out for three types of loading derived from maximum theoretical loading, which cannot be obviously achieved. Ring loading chosen was where $P_{suma} = 12.319$ MN.

Table 1. Loading by shield end.

$P_{1/20} = P_{suma}/20 = 0.61595$ MN	
$P_{1/30} = P_{suma}/30 = 0.41063$ MN	where $P_{suma} = 12.319$ MN
$P_{1/40} = P_{suma}/40 = 0.307975$ MN	

Table 2. Size of the cracks.

Name of file	N_{Ed} [kN]	N_{Rd} [kN]	M_{Edy} [kNm]	M_{Rdy} [kNm]	Use ID [%]	Calculation ID	Width of cracks w [mm]	Use w_{max} [%]	Review w_{max}
B12M24	-403.75	-13222.57	128.69	142.13	90.6	OK	0.695	> 300	NOK
B12M25	-268.95	-13222.57	86.36	128.79	67.1	OK	0.468	234	NOK
B12M26	-200.30	-13222.57	64.76	121.90	53.2	OK	0.352	176	NOK

Analysis results in conclusion that all three types of loading are satisfactory, but none satisfies limiting bearing capacity of cracks. The maximum allowed size of cracks is w_{max} = 0.2 mm. All analyzed cases breach such limit, thus we can state that origin of cracks as result of shield end contact on lining is possible.

5.3.2. Reinforced concrete lining loading by hydraulic press. Moving shield ahead, hydraulic presses set on last assembled tubings ring. In total shield is pressing on the ring by 16 couples of hydraulic cylinders, with angle 22.5° between each two couples. Each couple is pressing on the lining through steel plate (transferring sheet) with dimensions of 153 x 635mm, with technological maximum load on one steel plate up to force 2.433 MN. This force is not acting in the center of the tubing abutting joint (axis of tubing and axis of hydraulic press cylinder is shifted by 18.5 mm), so loading on tubings is eccentric.

The most loaded tubing is loaded with maximum three couples of hydraulic presses, resulting in force of

$$N_3 = 3 \times 2.432849 = 7.3\ MN \tag{15}$$

All input data described above were applied in calculation of tubings by program ATENA (Červenka Consulting), where crack development was compared for different values of loading (size of cracks can be shown according to chosen preferences).

Task was modelled with nonlinear material model (3d_nonlinear_cementitious), fracture energy was derived from ModelCode 1990.

Boundary conditions of static stage of tubing model were set to follow real situation as close as possible. Loading was applied to the centre of each of the three steel plates.

Complete analysis gives more detailed data of cracks with size > 0.15mm with overall loading ≈ 7.3 MN, which are depicted on Figure 12. Cracks in both cases are similar to cracks observed on lining in V.A track tunnel.

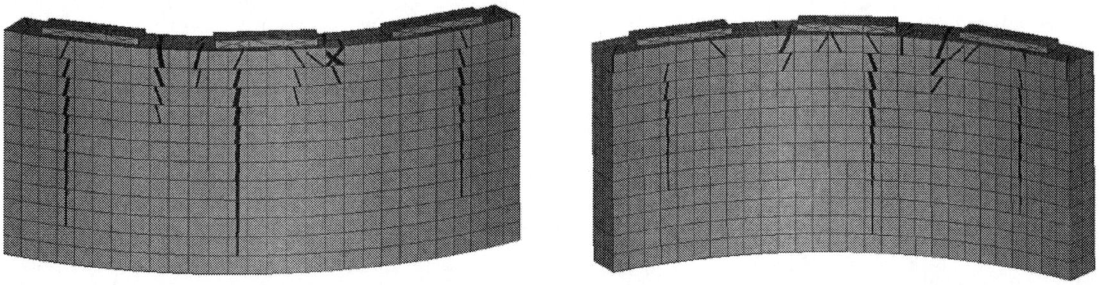

Figure 12. Cracks > 0.15 mm by overall loading ≈ 7.3 MN applied on tubing (2.4 MN/1 couple press).

6. Conclusions

The coefficient of the bearing capacity of the lining *s* describes reserves in the bearing capacity of the lining and also shows the point where the ultimate state will be reached first. Cast iron lining has spots where cracks are faced outwards to rock and this method could find such spots. Analysis made for deformation modulus of rock E_o = 5, 50 a 100 MPa and reduced area-wise moment by Muir Wood A. M. is shown in Figure 8, where one can simply identify values smaller than limiting coefficient of bearing capacity $s = 1$. Computation shows that most dangerous state is with deformation modulus E_o = 5 MPa.

In reinforced concrete lining of tubings Ø5.3/5.8 m in V.A line of Prague Underground, hairline cracks in the altitude of horizontal axis of the lining parallel to longitudinal track axis were identified early after assembling. As contribution to discussion about origin of the cracks, analysis of the hypothetical origin of cracks was made. We compared hypothesis of cracks origin as a result of horizontal force from shield end acting at time of moving shield in direction radius of track R = 650m and hypothesis of crack origin as a result of pressure of shield's hydraulic press cylinders. The result achieved were not decisive: none of both hypotheses is deemed absolutely correct. Cracks could be caused by relatively small force of the shield end and cracks > 0.15 mm could also appear after applying force of 7.3 MN on tubing as depicted in Figure 12. As no decisive result is available, both theories may explain the cause of cracks and it is recommended to study this problem more deeply on both by mathematical and laboratory models.

Acknowledgements

This article was made possible by financial support from „Centrum kompetence Technologické agentury České republiky (TAČR)" program, project number TE01020168 (CESTI).

References

[1] Aldorf J 1992 *Mechanika podzemních konstrukcí* (VŠB – Technická universita Ostrava)
[2] Bulychev N S 1982 *Mechanika podzemnykh konstrukciy* (Moscow: Nedra)
[3] Daushvili A P 1972 *Raschot tonnelnykh obdelok v matrichnoi forme* (Moscow: Transport)
[4] Muir Wood A M 1975 The Circular Tunnel in Elastic Ground *Geotechnique* **25**(1) pp 115-127
[5] Muskhevishvili N I 1966 *Nekotorye osnovnye zadachi matematicheskoj teorii uprugosti* (Moscow: Nauka)
[6] Kovář I et al 1973 *Projekt I: C–1 Provozní úsek, Sokolovská-Kačerov* (DP hl. m. Prahy – Metroprojekt)
[7] Metroprojekt Praha a. s. 2015 *Projekt V: A Dejvická–Motol* (Metroprojekt Praha a. s.)
[8] Széchy K 1966 *The Art of Tunnelling* (Budapest: Akadémiai Kiadó)
[9] Závora K 1979 *Metro – Ostění ražených tunelů – Sborník průřezů a dílců* (DP hl. m. Prahy-Inženýring dopravních staveb a.s.)
[10] Závora K 1974 *Metro – Ostění ražených tunelů – Typizační směrnice* (DP hl. m. Prahy-Inženýring dopravních staveb a.s.)
[11] Comité Euro-International du Béton 1993 *CEB-FIP MODEL CODE 1990: Design Code* (London: T. Telford)

IOP Conf. Series: Materials Science and Engineering **236** (2017) 012086 doi:10.1088/1757-899X/236/1/012086

New drainage tunnel of the tunnel Višňové – design and excavation

Igor Jurík[1], Ladislav Grega[1], Jozef Valko[1] and Peter Janega[2]

[1]URANPRES, spol. s r.o., Čapajevova 29, Prešov 080 01, Slovakia
[2]KAMI PROFIT, s.r.o., Príjazdná 7/A, Bratislava 831 07, Slovakia

E-mail: jurik@uranpres.sk

Abstract. The actual pilot tunnel dated to the period of geological and hydrogeological survey, is designed as a part of the tunnel Višňové, which is located at the section of the D1 motorway Lietavská Lúčka – Višňové – Dubná Skala in Slovakia. Drainage tunnel will be used for the drainage of the main tunnel tubes, where the maximum inflow from the eastern portal is greater than 250 $l.s^{-1}$. Overlapping of the initial pilot tunnel with the profile of the southern tunnel tube led to the demolition of the portal sections of the pilot tunnel during the excavation of main tunnel tubes. These sections were replaced by new drainage tunnels, with the lengths of 288.0 meters from west portal and 538.0 meters from eastern portal, to ensure access from both portals. The new drainage tunnel is excavated under the level of the two main tunnel tubes. Drainage pipes with a diameter of 250 mm will be installed from cleaning niches in the main tunnel tubes to the new drainage tunnel.

1. Introduction to tunnel Višňové

Tunnel with a total length of 7424.9 meters in the northern tunnel tube and 7462.0 meters in the southern tunnel tube is a challenge for contractor from technical, logistical and design point of view. Due to complicated geological and hydrogeological conditions, this tunnel belongs to the most difficult line structures in the history of engineering works in Slovakia.

Višňové tunnel consisted of two separate tunnel tubes, each for one driving direction. According to the width category of the tunnel, the Višňové tunnel belongs to the category of 2T 7,5. Pursuant to Slovak technical standard – STN737507, it is characterized as a long tunnel. The speed limit is 100 km/h.

Portal sections were performed by using new Austrian tunnelling method. For continuity of works, full face profile method ADECO-RS was and is still currently used. The advances of excavation and section types are determined by the designer, based on the actual geological conditions in tunnel [1], [3], [4].

2. The purpose of new drainage tunnel

The initial pilot tunnel (PT), dated to the period of geological and hydrogeological survey of 1998-2002, was redesigned to be used as a part of tunnel construction.

Mentioned PT is designed to perform drainage function for both tunnel tubes. The horizontal alignment of the main tunnel is adjusted, so that the pilot tunnel will be located approximately

IOP Conf. Series: Materials Science and Engineering **236** (2017) 012086 doi:10.1088/1757-899X/236/1/012086

in the middle between tunnel tubes. Vertical alignment is performed with the position of tunnel tubes above of the new drainage tunnel.

Because of mutual interference between PT and southern tunnel tube (STT) in the portal sections (both west and east portal), it was designed to proceed to the partial liquidation of the PT in length of 538.0 meters from the east portal (EP) and 288.0 meters from the west portal (WP).

These sections had to be replaced with new parts of drainage tunnel, which will be connected to the original PT and will provide continuous access from both portals.

The primary lining of new drainage tunnel will be used as a permanent structure, which serves to ensure the stability of the excavation and to limit the deformations of rock mass during excavation works and the use of these objects [1], [3], [4].

3. Geological settings

For excavation of the new drainage tunnel from WP, two basic types of rock mass were identified (Figure 1). First one is complex of Flysh formation (until 90.0 meters from portal) build with Paleogene rocks – claystones, siltstones, sandstones, which was moderately to highly weathered and tectonicaly disturbed. Second rock mass type was made up of a complex of Mesozoic rocks – dolomites, dolomitic limestones, limestones and clayed limestones. The rocks are moderately weathered to unweathered, often faulted.

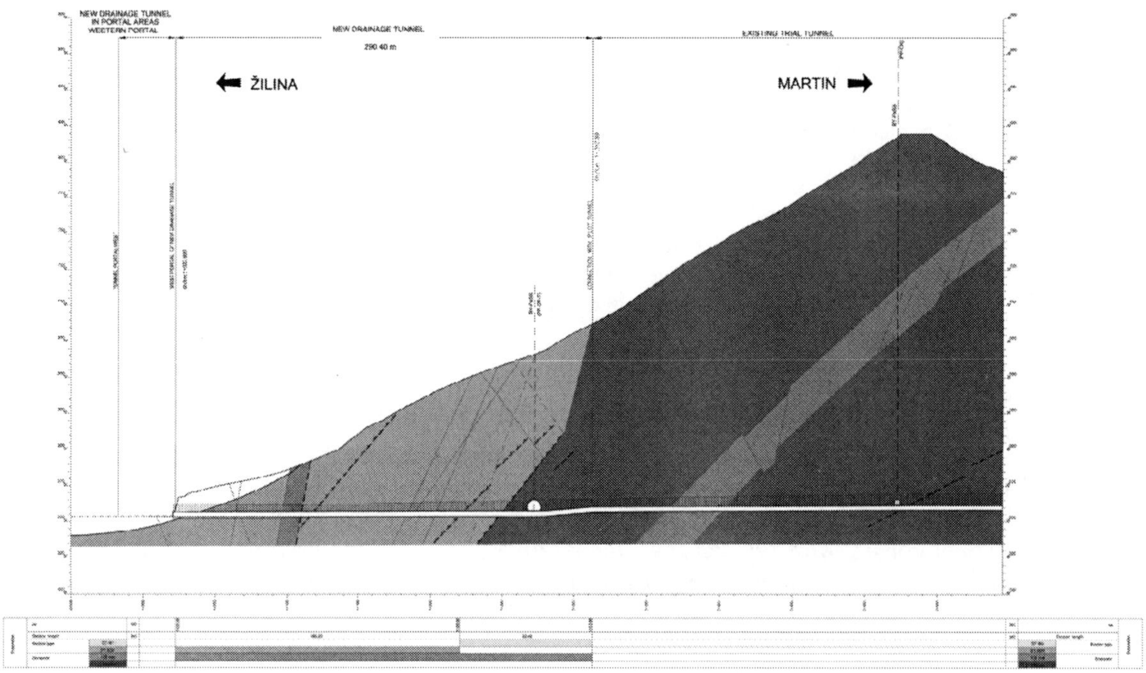

Figure 1. West portal – longitudinal profile and proposed section types [1].

Rock mass from EP (Figure 2) is formed exclusively with basement rocks – granodiorites, pegmatites, tonalites. Rock mass was strongly desintegrated with very small values of RQD, also very often strongly weathered especially in close to portal sections [1], [3], [4].

4. Construction of new drainage tunnel

The excavation of relocation of drainage tunnel at the portal sections was carried out by full face method, known as ADECO-RS from WP and also from EP. From east portal, the relocation of the drainage tunnel was excavated upwards with rising of circa 1.0 %. At the chainage of 8+060.47 km, the rise increases to 9.0 %, until the chainage of 7+933.62 km, where the new drainage tunnel is intended to connect to initial pilot tunnel. From west portal, the new drainage tunnel was excavated with rising of circa 0.2 % until the chainage of 1+279.96km, on following section of 33.0 meters, the rising increases to 9.0 % until the chainage of 1+312.99km, where the new drainage tunnel will be again connected to initial pilot tunnel.

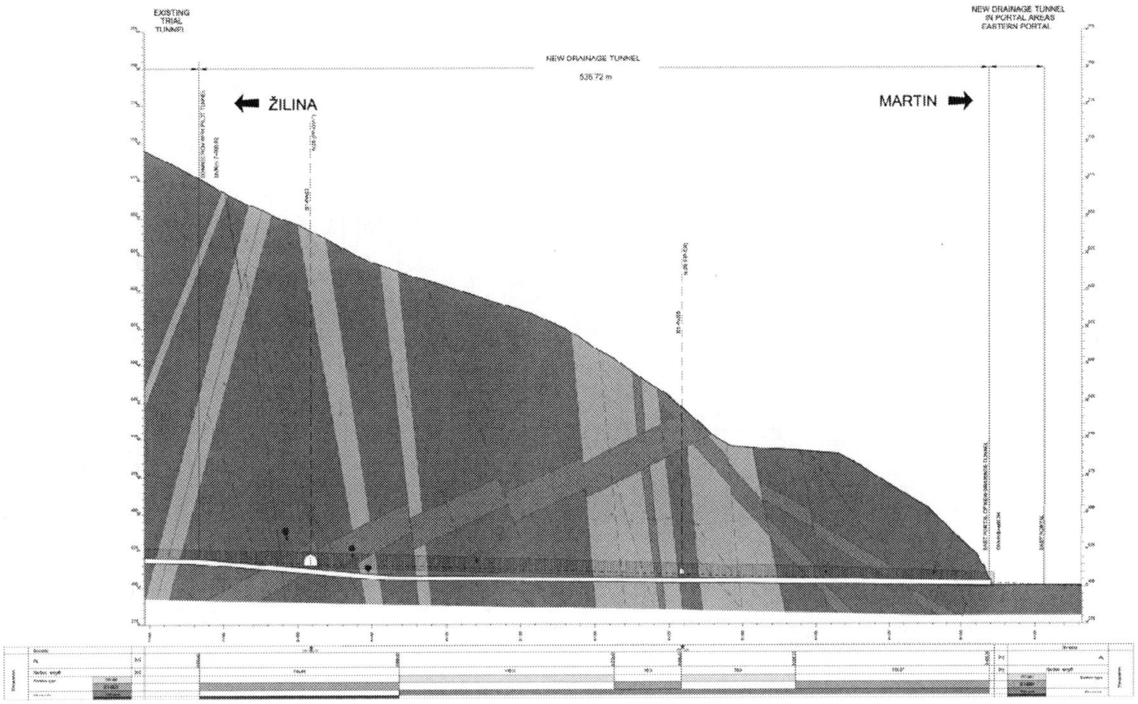

Figure 2. East portal – longitudinal profile and proposed section types [1].

The situation at WP and EP as designed is pictured in Fig.3 and Fig.4. In Fig.5, the typical cross section with secondary lining and the position of new drainage tunnel is pictured.

The excavation is conducted as a full profile, and immediately after the cut, the reinforcement of primary lining is installed according to corresponding section type. For the reinforcement, two section types were designed, DT-B0V a DT-B0. Additionally, for initial portal parts, specific section type which is based on DT-B0V was designed. Simultaneously, the excavation under the cross passage No.1 at WP was conducted according to special design, which is described below.

The mechanical means and drill&blast works were used during the excavation. Blasting works were carried out mainly in rigid rock.

IOP Conf. Series: Materials Science and Engineering **236** (2017) 012086 doi:10.1088/1757-899X/236/1/012086

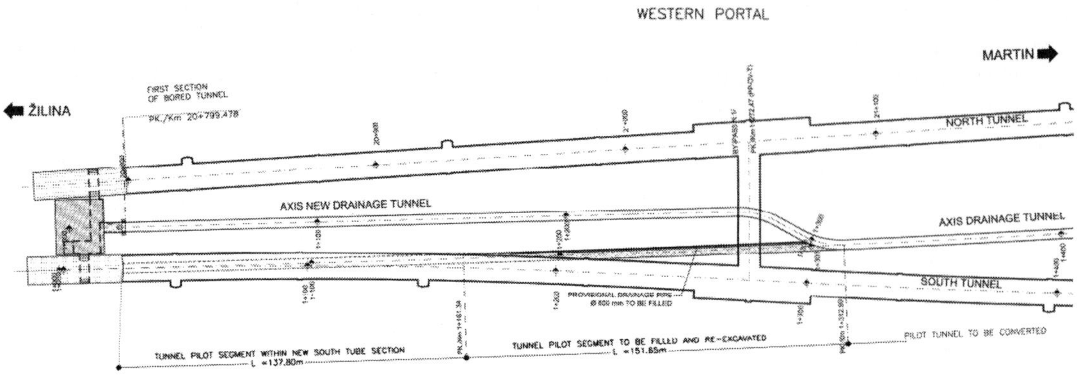

Figure 3. New drainage tunnel and its connection to initial pilot tunnel at west portal [1].

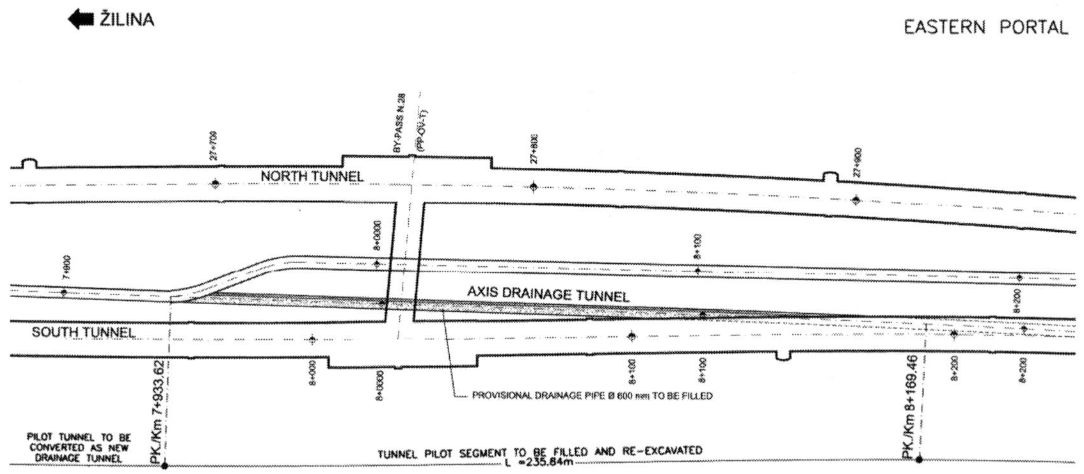

Figure 4. Part of new drainage tunnel and its connection to initial pilot tunnel at east portal [1].

Mechanical means of excavation were intended for parts with appropriate geotechnical conditions, or for parts, where strict measures has to be respected, for instance connection of the tunnels or restrictions regarding the seizmical impact, or under conditions, where blasting could possibly disrupt the rock directly around the face.

The machinery used during the excavation is limited by the dimensions of the tunnel, so basically small machinery has to be used. That was small excavator with pick hammer, robot for shotcreting, mobile mixer for concrete transport into the tunnel, skid-steer loader, special mining loader with bigger bucket, special mining dumper and drill rig with one boom and a platform. The ventilation system was based on suitable fan with flexible air duct of diameter of 600 mm.

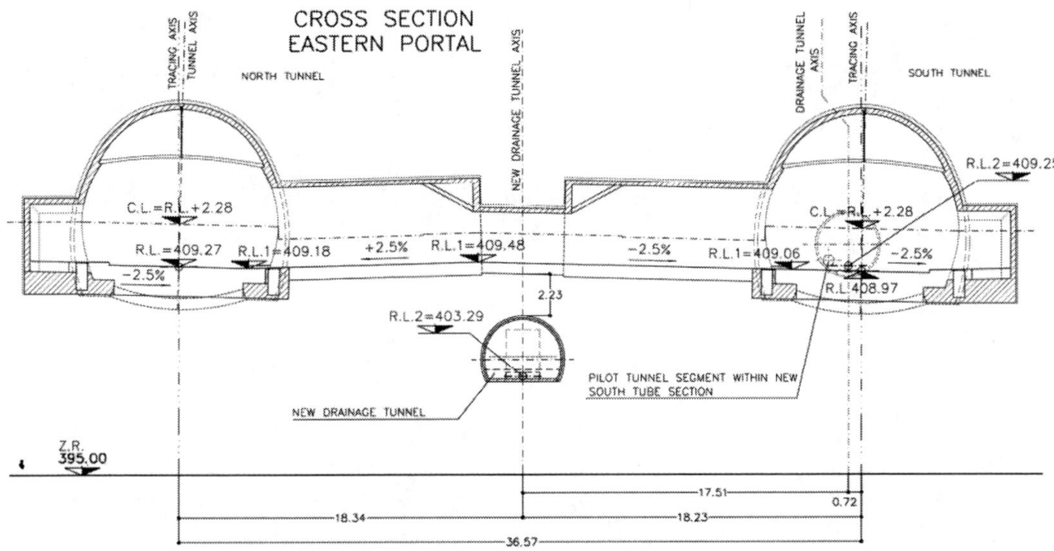

Figure 5. Typical cross section of secondary lining at east portal (view from the tunnel towards portal) with the position of the drainage tunnel [1].

5. Section types, connection to initial pilot tunnel, interference with the cross passage No.1 at west portal and installation of temporary drainage pipes at east portal

For the excavation of new drainage tunnel, two section types as mentioned in previous chapter were designed, DT-B0 and DT-B0V (Figure 6), where mainly at portal areas, the section type with reinforcement made of micropile umbrella DT-B0V was used. The rounds in these sections were defined as maximum of 1.0 meter for section type DT-B0V and maximum of 1.5 meter for section type DT-B0. The reinforcement within both section types includes steel ribs HEA120, wire mesh 150 mm x 150 mm of diameter 6 mm (this is partial modification compared to initial design, where shotcrete with steel fibers was intended to use) and finally, concrete of type C30/37 applied by shotcreting.

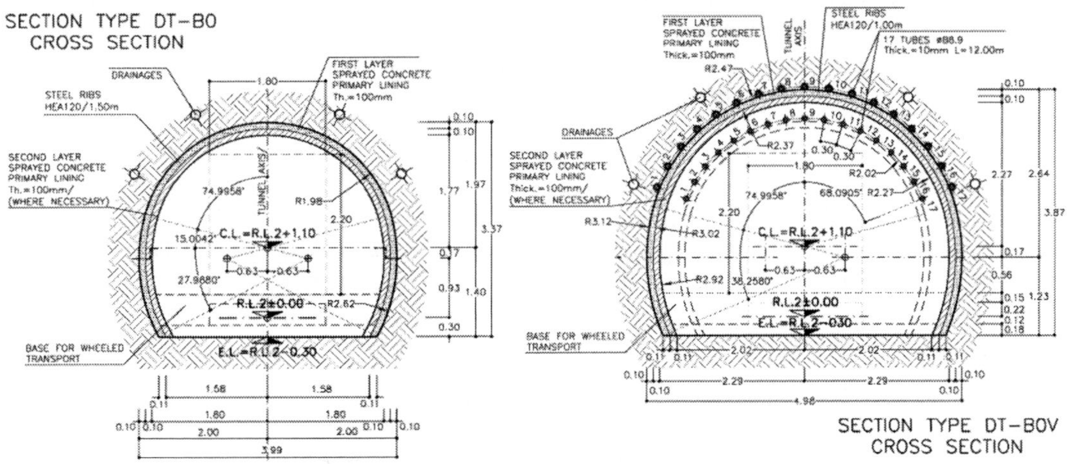

Figure 6. Comparison of cross section for DT-B0 and DT-B0V, indicating difference in reinforcement for both section types of new drainage tunnel [1].

Figure 7. Drilling of micropile umbrella under the section DT-B0V in first meters of excavation at east portal of tunnel Višňové [2].

The works on excavation of new drainage tunnel started in september of 2015 and after the excavation of circa 220.0 meters, the excavation at EP started due to some operational reasons, where until the august of 2016, some 535.0 meters were excavated (Fig.7). After the excavation, the installation of temporary drainage pipe DN500 into the invert of new drainage tunnel was carried out along its whole length (Figure 8 and Figure 9).

In february of 2017, the interrupted works at WP continued by passing the cross passage No.1, in the invert of which the reinforced concrete slab was installed. In the section between the chainage of 1+266.33 km and the chainage of 1+279.96 km, the new drainage tunnel is interfering with cross passage No.1, which obviously connects NTT with STT of main tunnel. As it is stated in the documentation, the distance between the bottom of the cross passage No.1 and the new drainage tunnel is less than 2.0 meters, thereby distinctive technical solution to this issue was designed to cope with mutual interference (Figure 10). The actual connection of both tunnels, even on WP did not avoid the filling of some part of the tunnel with filling concrete in the volume of circa 400 m^3 (Figure 11).

6. Conclusion and current state of new drainage tunnel

Currently, the completion of excavation at west portal in length of circa 15.0 meters is prepared, as the curing of concrete filling in its part is necessary for proper and safe continuing in excavation. Consequently, at the east portal, the connection of drained water from initial pilot tunnel into the temporary drainage pipe which was placed into new drainage tunnel in september and october 2016, will be done. In the context of initial pilot tunnel, the remediation of the primary lining, as well as subsequent temporary ventilation of main tunnel tubes through the pilot tunnel, which would require excavation of vertical shafts is planned in near future.

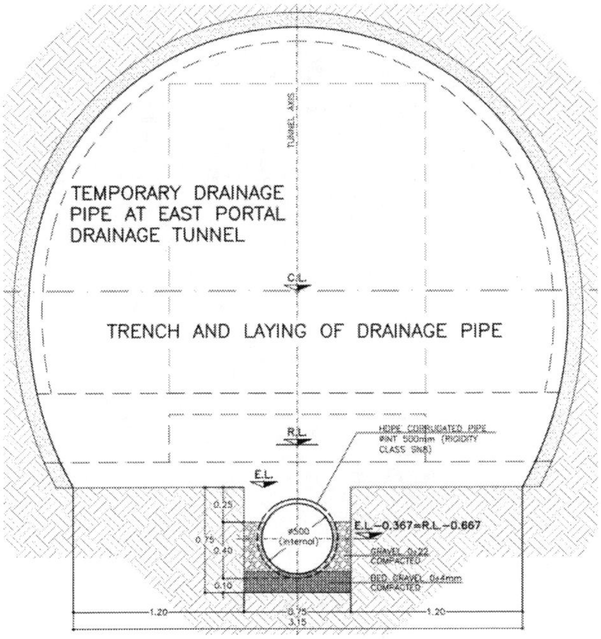

Figure 8. Drawing of crosssection of new drainage tunnel, with detail for technical solution of drainage pipe DN 500 laying into the invert of the new drainage tunnel [1].

Figure 9. Picture of temporary drain pipe DN 500 installation from section, where the excavation was stopped. In the upper part, the initial pilot tunnel is visible, also as the shotcreted pipe in which drained water is collected nowadays [2].

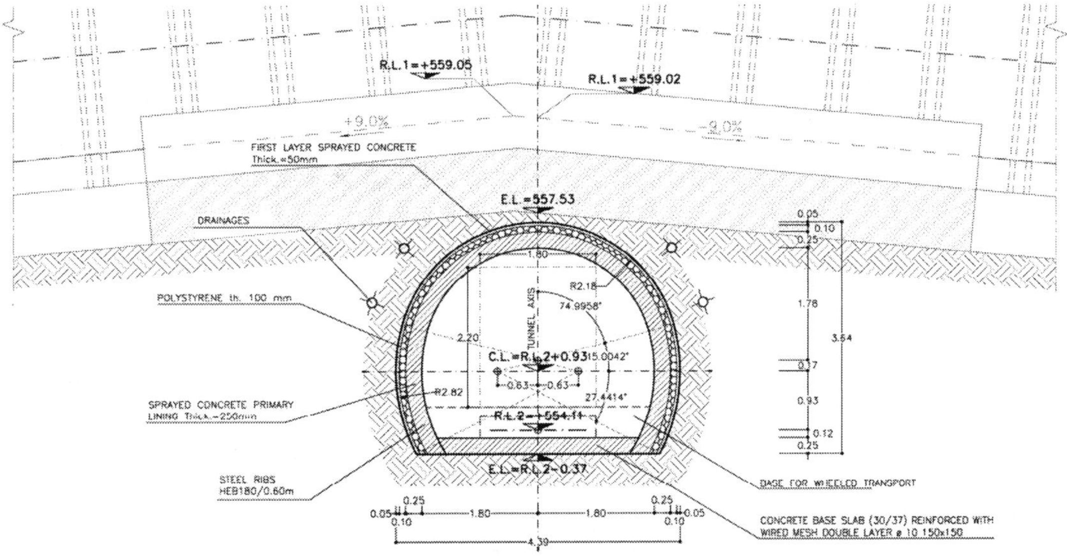

Figure 10. Cross section of interference of new drainage tunnel with cross passing No.1 at west portal of tunnel Višňové [1].

Figure 11. Formwork in pilot tunnel before filling up its short section with concrete [2].

Acknowledgements
Authors wish to acknowledge encouragement and financial support from managing director at URANPRES, spol. s r.o. – Ing. Ján Jaško, and also specially would like to thank to Dr. Alessandro Citterio and to main contractor Salini Impregilo – Dúha Joint Venture.

References

[1] Cassani G 2015 *Drainage tunnel portal sections excavation – DRS* (Rocksoil S.p.A., responsible designer)
[2] Grega L, Jurík I, Udič M, Trúchly S and Vancák I 2015-2017 *Monthly reports* (Višňové-Dubná Skala, archive of Uranpres spol. s r.o., subcontractor for Salini Impregilo-Dúha Joint Venture)
[3] Jurík I, Udič M, Valko J and Grega L 2016 Supplementary engineering-geological survey and following excavation *Underground Construction Prague* www.uconline.com
[4] Valko J, Janega P, Grega L, Jurík I, Udič M and Kaňuščák M 2015 Tunel Višňové *Inžinierske stavby* **5** pp 15-18

Spray-applied waterproofing membranes: effective solution for safe and durable tunnel linings?

Barbora Pisova[1,2] and Matous Hilar[1,3]

[1]Czech Technical University in Prague, Thakurova 7, Prague 166 29, Czech Republic
[2]Duha a. s., Prievozska 4D, Bratislava - mestska cast Ruzinov 821 09, Slovakia
[3]3G Consulting Engineers s.r.o., Zeleny pruh 95/97, Prague 140 00, Czech Republic

E-mail: bara.pisova@seznam.cz

Abstract. What is the perfect tunnel lining? Cost efficient, easy and fast to build with acceptable environmental impact? How to construct a watertight and safe tunnel lining? Would it be possible to apply a waterproofing system directly onto the rock face just after the tunnel face opening? This might be the system of the future enabling all concrete applied to the rock face to remain permanent. For now though, we would like to focus on an optimisation and examination of currently available technologies and materials, such as tunnel linings with the use of spray-applied waterproofing membranes. In this paper, the failure mechanisms of a tunnel lining with a spray-applied waterproofing membrane are described, the behaviour of spray-applied waterproofing membrane under various conditions (dry, moist, wet) is challenged and the possibilities of interface numerical modelling are presented. Tunnel lining design is mainly dependent on the geological and hydrological conditions in the considered area. The application of tunnel linings with spray-applied waterproofing membrane in both hard rock and soft ground tunnelling, are studied.

1. Introduction

To build an efficient and sustainable tunnel structure, tunnel designers must come up with a functional and reliable solution of how to fight water and deliver a safe and durable design to the client. There are various waterproofing measures such as drainage systems, plastic sheet waterproofing membranes, watertight concrete or other methods available to serve this purpose. Until now, plastic sheet waterproofing membranes have represented the most commonly used waterproofing system with all technical standards and design approaches well-known and frequently used. On the other hand, innovative solutions such as spray-applied waterproofing membranes have been introduced and heavily discussed within the profession during the last few years.

Manufacturers of spray-applied waterproofing membranes report that thanks to the use of their material, the tunnel lining can be designed thinner and so the overall cost can be reduced [1]. Material, time and cost savings are major motivations to deeply look into this problem and evaluate whether such proclamations are reasonable or if some limitations exist.

This paper is limited to the use of EVA-based (ethylene vinyl acetate) spray-applied waterproofing membranes.

2. Application of tunnel lining with spray-applied waterproofing membrane in hard rock and soft ground tunnelling

If we look at the largest projects where spray-applied waterproofing membranes have been used we can observe some similarities. No matter if the tunnels were excavated in soft ground or hard rock, the ground had to exhibit, at least from the short-term perspective, low permeability.

Since the spray-applied waterproofing membranes can be manufactured only by applying onto a surface without active water ingress, it is crucial that the primary lining remains without water ingress until the whole sandwich lining (spray-applied waterproofing membrane and secondary lining) is built. However, even if this is achieved, the question of whether the primary lining ever gets saturated or what the humidity conditions acting on the spray-applied waterproofing membrane are remains difficult to answer. A Doctoral Thesis of Karl-Gunnar Holter (Norway) shows an example of water/humidity conditions and presents real in-situ measurements of moisture conditions of the ground surrounding a tunnel lining. KG Holter states in his study [2]: "Recent reported testing of membranes for waterproof SCL (sprayed concrete lining) have not included the moisture condition and moisture properties of the constituent materials in the lining. Testing of moisture properties of membrane and concrete materials has yet to be included in guidance for design and testing of spray-applied membranes."

To get samples from an already existing tunnel lining, test them and carry out a back analysis is possible. However, it would be almost impossible to get such data before a tunnel gets designed, respectively before a tunnel gets excavated. Prediction of real moisture conditions of a future tunnel lining is a challenge and is discussed further in chapter 4.

3. Failure mechanisms of a tunnel lining with a spray-applied waterproofing membrane

The key parameter of a tunnel lining with spray-applied waterproofing membrane is the bond between the membrane and the layers of concrete (primary and secondary lining). The tensile bond strength is not only dependent on the roughness and cleanliness of the substrate but also on the relative humidity. The values reported in various papers and manufacturer's technical sheets vary but for example ITA (International Tunnelling Association) refers to the min. value of 0,5MPa [3].

Bond is closely connected to elongation (crack-bridging capacity) of the membrane. The elongation is highly dependent on the w/c ratio and relative humidity.

Recently, it has been discovered that although contractors are advised to use a regulating layer in order to make the application of the spray-applied waterproofing membrane easier [3], the designer of the particular project should be aware of what kind of material is going to be used. Regulating layers might significantly influence performance of the whole system. If parameters of the regulating layer are not specified by the designer and contractors are allowed to choose according to their own will, the client might end up with a run-out-of-control system. The issue of regulating layer can be assumed as a good example of delicacy of the sandwich tunnel lining.

3.1. De-bonding between primary lining and spray-applied waterproofing membrane

Failure within the primary lining (or even between the primary lining and regulating layer) is not considered.

EVA-based waterproofing membranes are dissolvable in water and it is already well-known that their behaviour very much depends on the w/c ratio of the mix and their water exposure during their lifetime. Even though some researchers claim that direct contact with water is not a realistic case, since there is a high level of uncertainty such cases should be studied. Presence of water at the interface between the membrane and the primary lining might have a detrimental impact on the bond. Once the waterproofing de-bonds from the primary lining, the lining cannot be considered as a composite lining.

3.2. Failure within the spray-applied waterproofing membrane

Two types of failure can potentially occur within the membrane; mechanical or chemical failure. The mechanical failure is represented by loss of the membrane's expected mechanical properties or its rupture. The chemical stability of the membrane should be verified in terms of the chemistry of the ground water in contact with the membrane, or potential influence of other chemicals that might be added to the original mix of a spray-applied waterproofing membrane during construction.

It was reported in Tunnelling Journal this year [5] that "two colours of waterproofing are good". It is explained that it is much easier for the contractors to guarantee the right thickness and covering if two layers of the membrane with two different colours are used.

The issue might be that even if a non-harmful dye is used for colouring the membrane, it should be verified that the colour does not wash off from the membrane and does not influence its microstructure. It is important that the contractors do not make such change, such as adding a vegetable dye to the spray-applied waterproofing mix without consultation with specialists.

3.3. De-bonding between primary lining and spray-applied waterproofing membrane

The bond between the primary lining and the waterproofing membrane can be easily tested during construction by an in-situ pull-off test and the membrane can be easily repaired after by spraying or painting over the tested area.

This is not applicable to the bond between the membrane and the secondary lining though. Once the secondary lining is applied (either sprayed or cast), it is impossible to test the bond without complicated and expensive destructive tests. However, since there is no water expected to act on the interface between the membrane and the secondary lining, the risk of de-bonding of the secondary lining is much less likely than in the case of the interface between the primary lining and the membrane.

4. Behaviour of spray-applied waterproofing membranes under various conditions

As already stated, water changes the properties and behaviour of the spray-applied waterproofing membranes. It is not only necessary to provide the right water content during production of the membrane but also to evaluate and predict the relative humidity/saturation conditions in which the membrane will be placed. These conditions might change during the lifetime of the membrane (tunnel lining). The season and durability of the primary lining is an issue on its own.

It is believed that significant effects of the moisture state of the material on its physical properties are related to the nature of the material EVA polymer colloid containing many long-chain polymers, which on curing lose moisture, allowing the polymers to coagulate into a film, [6]. On exposure to liquid or vapour phase moisture, the spray-applied waterproofing membrane re-absorbs water. This appears to have an effect on physical properties of the membrane, including bond strength and elasticity. On absorption of water, the tensile strength of the membrane reduces significantly, while its elongation at break increases. Only concrete of exceptionally good quality would be expected to resist water penetration for the full tunnel service life. If the tunnel lining includes areas of 'poor' quality concrete, water penetration could occur in these areas within a matter of weeks. In reality, it is likely that the time taken for full penetration of water through the primary SCL lining will vary over a significant range, depending on the relative proportions and distribution of the varying 'quality' of concrete achieved in construction. Should significant cracking and 'honeycombing' of the concrete occur, full penetration is likely to be rapid (achieved in months/years) but, for the time being, remain localised to those areas.

Generally, it is considered likely that full saturation of the primary sprayed concrete lining will not be achieved in the lifetime of the structure. However, a test programme including curing times of each layer of the sandwich structure should be elaborated together with a schedule of design works if the design should be verified with real measured in-situ data.

IOP Conf. Series: Materials Science and Engineering **236** (2017) 012087 doi:10.1088/1757-899X/236/1/012087

5. Interface numerical modelling

All the above mentioned leads to the necessity of careful consideration of what the critical tunnel lining – interface properties – configuration is. It is expected that the shrinkage effect of the secondary lining will be increased, because the secondary lining will be restrained by the bond properties to the spray-applied waterproofing membrane, compared to a system with plastic sheet waterproofing membrane that allows for slipping interface. On Figure 1, a shrinkage crack in the secondary lining can be observed.

Figure 1. Shrinkage crack in secondary lining [2].

As already pointed out by Jiang Su and Michal Uhrin [6] and Pisova [7] the usual design approach taking into account compression-only action between the tunnel lining and the waterproofing membrane might not be the most conservative case and the shear and tension interface properties should be examined.

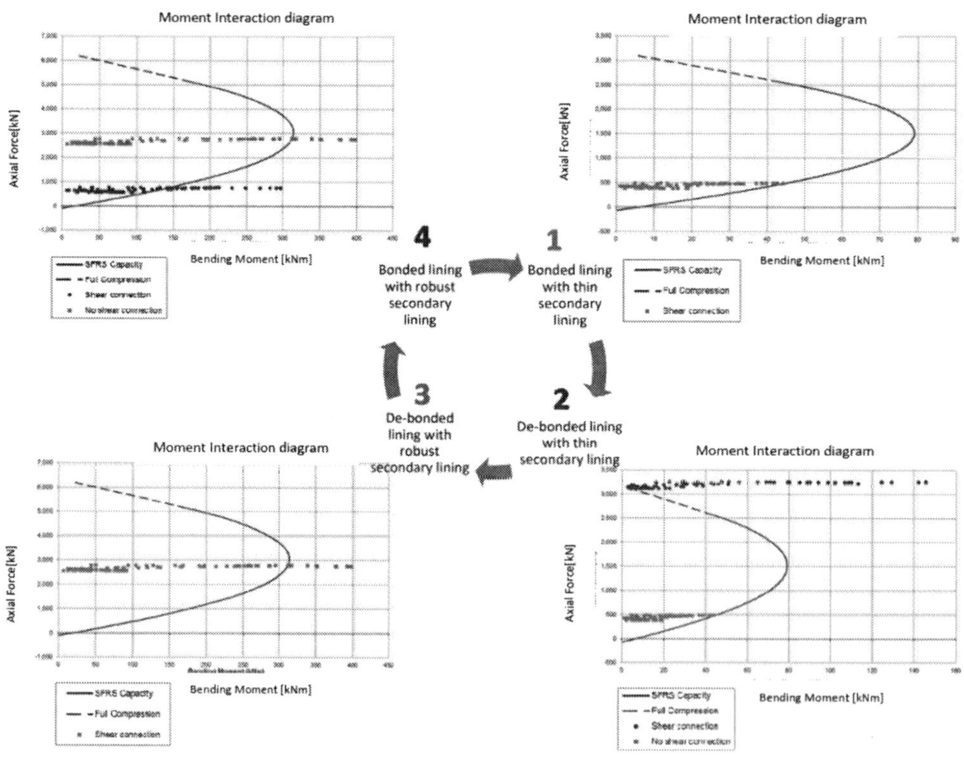

Figure 2. Numerical modelling chart.

On Figure 2, design approach of a sandwich tunnel lining with the spray-applied waterproofing membrane is presented. Analysis of secondary lining with no tensile and shear bond might lead to design of a robust tunnel lining with no savings whatsoever. Also, there is a risk that even though the bond is not considered in the numerical model, it does not necessarily mean that the bond does not exist in the real structure. On the other hand, design of a thin shell secondary lining that is considered to be bonded to the primary lining through the waterproofing membrane is a risk as well, because if the secondary lining de-bonds it might not be able to carry the acting loads on its own.

6. Conclusions

The interface properties have significant impact on the load sharing between the components of the sandwich lining with spray-applied waterproofing membrane. The task for the designers is to find a way of how to define the interface properties and specify the boundary conditions with confidence. The aim is to find universal rules applicable to any ground conditions so that tunnel lining in both soft ground and hard rock can be easily designed and built. In case of the use of spray-applied waterproofing membranes, it has been proven that designers with on-site experience are much more capable to understand the delicacy and complexity of a tunnel lining with spray-applied waterproofing membranes.

Acknowledgements

Financial support from the Technology Agency of the Czech Republic (TAČR) grant TE01020168 (CESTI) is gratefully acknowledged.

References

[1] www.master-builders-solutions.basf.com
[2] Holter KG 2015 *Properties of waterproof sprayed concrete tunnel linings: A study of EVA-based sprayed membranes for waterproofing of rail and road tunnels in hard rock and cold climate* (Norwegian University of Science and Technology, Faculty of Engineering)
[3] ITAtech 2013 ITAtech design guidance for spray applied waterproofing membranes *ITAtech Report 2*
[4] Lemke S 2014 Critical review of spray-applied membrane in the field of tunneling in comparison to sheet membranes: advantages and open questions *Proc. World Tunnel Congress 2014 - Tunnels for Better Life (Iguassu, Brazil)*
[5] Tunnelling Journal 2017 Sprayed membranes: Five things you should know *Tunnelling Journal* **2** pp 28-29
[6] Su J and Uhrin M 2016 Primary-secondary lining interactions for composite sprayed concrete lined tunnels using sprayed waterproofing membrane *Proc. Int. Conf. World Tunnel Congress 2016 (San Francisco, USA)*
[7] Pisova B 2015 Spray applied waterproofing membranes for tunnel linings – fashion or innovation? *Proc. 43rd Conference Foundations Brno 2015 (Brno, Czech republic)*
[8] Šíma J, Krčmář O, Matzner R, Dvořák J and Cháma L 2011 Spray-applied waterproofing membrane Masterseal 345 – successful alternative solution *Tunel* **2** pp 4-11
[9] Holter K G and Zámečník M 2009 Composite sprayed concrete waterproof tunnel linings using spray applied waterproofing membrane *Tunel* **2** 41-47
[10] Holter K G nad Zámečník M 2009 Modern pre-injections in underground construction for water ingress reduction and ground improvement. *Tunel* **2** pp 34-40
[11] Barták J, Hilar M and Pruška J 2013 Development of underground construction engineering in the Czech Republic *Geomechanics and Tunnelling* **2** pp 101-12
[12] Stehlík E 2013 Prague Metro construction – past, present and future *Tunel* **1** pp 69-82
[13] Powell D, Thomas A and Hilar M 2004 The Role of numerical modelling in tunnel design *Tunel* **1** pp 25-8
[14] Kochánek M and Korejčík J 2009 Mined Tunnels on Prague Metro Line VA *Tunel* **3** pp 13-20
[15] Korejčík J and Růžička J 2007 Mined tunnels on the metro line A extension *Tunel* **4** pp 10-14

IOP Conf. Series: Materials Science and Engineering **236** (2017) 012088 doi:10.1088/1757-899X/236/1/012088

Recommendations and strategies for using reclaimed asphalt pavement in the Flemish Region based on a first life cycle assessment research

Wim Van den bergh[1], Patricia Kara[1], Joke Anthonissen[1], Alexandros Margaritis[1], Geert Jacobs[1] and Karolien Couscheir[1]

[1]University of Antwerp, Faculty of Applied Engineering, EMIB research group, Groenenborgerlaan 171, 2020 Antwerp, Belgium

E-mail: wim.vandenbergh@uantwerpen.be

Abstract. In Flanders, using Reclaimed Asphalt Pavement (RAP) is allowed in asphalt mixes for base layers. Primary economic and secondary laboratory-measured mechanical properties are given as justification for higher amounts in specific mixes. However, one should evaluate the performance of these mixes on long-term by environmental impact of the production until end-of-life. In this paper recommendations and strategies for using RA, based on current research, are discussed in a broader perspective such as using a carbon-footprint tool and warm-mix asphalt production in the Flemish Region. The paper aims to a wide discussion by reporting several outcomes of laboratory research, statistics and practical application in order to set a general strategy for the road engineering sector in the Flemish Region.

1. Introduction

Reclaimed Asphalt Pavement (RAP) is a typical by-product in the field of asphalt road rehabilitation these days but recycling of RAP is not a new technique since it was stimulated due to the oil embargo in 1973 and following sharp rise in all construction costs. In Europe, the use of RAP is common practice with some countries having more than 30 years of experience, such in the Netherlands [1], Flanders (Belgium) [2], Germany and Denmark [3]. Nowadays, RAP is considered as the most recyclable material in the construction industry and exceed typical recycling volumes for glass and paper [4]. According to the EU Waste Framework Directive 2008/98/EC, a recycling target of 70% for non-hazardous construction and demolition waste (including asphalt waste) should be achieved by 2020 [5]. Roughly 50 million tons of RAP are produced in Europe each year with over 70% of it being reused for pavement surfaces [6]. However, the asphalt recovery industry in some countries faces the problem of excessive material storage due to the limited quantities of RAP that can be reused in new mixes [7]. Several construction materials are composed by RAP, of which the most used are: asphalt mixes, cement-treated base and loose base materials. Most commonly, RAP is recycled in new asphalt mixes, since RAP is composed by asphalt components, but except it contains aged binder. The recycled hot asphalt mixes are made by several methods such as cold addition, pre-heated addition by a parallel drum and/or using rejuvenators. Reuse of this material is described in the EN 13108-8 standard [8], including material classification and testing.

2. Using RAP in Belgium

In Flanders, the Flemish waste policy is determined in the Flemish Waste Decree and its implementation order, the Order of the Flemish Government for the Establishment of the Flemish Regulations relating to Waste Prevention and Management (VLAREA) [9]. RAP is recognized as a secondary material since late eighties and currently, RA is accepted as a full material within the closed-cycle strategy as part of a circular economy.

In Belgium, exact data of the use of RAP are not available, however, for contractual reasons, the impartial certifying body COPRO monitors the quality of materials and the use of it for the production of asphalt mixes and bound base materials. According to the annual report of the European Association of Asphalt Manufacturers [10], it is estimated that 72% of an 1.5 million ton production of RAP are recycled annually in Belgium in hot mixes. Data from COPRO (year 2016) show that 0.95 million ton is recycled in asphalt, 0.64 million ton is used in foundations (base material loose and cement-treated).

In Belgium different rules for the addition of RAP are defined in the regional regulations (Flanders, Walloon and Brussels-Capital Region) and different works (public, private works or works with a certified mix). Depending on the material, regulations are described, mostly referring to European Standards, in regional technical standards. There is no official database available for RAP characterization in Flanders. Partly information can be obtained from the Flemish Road Agency (FRA) and COPRO. Currently, two new approaches for increasing the use of RAP are being developed: a sustainable selection model for the application of reclaimed asphalt granulate in road design (SSMARAGD) [11] and an IT-architecture for archiving the production and paving process of asphalt in order to provide data for later reclaiming activities [12].

3. Use of RAP in concrete

The use of RAP in concrete mixes for roads can be accepted as a new, even superior application than loose material for base construction. A number of researches have investigated the influence of RAP on the mechanical properties of concrete. Most of these studies have shown that RAP has a negative influence on the properties of the hardened concrete [13-15]. Nevertheless, using RAP in concrete could be useful to appoint non-useful RAP for asphalt, e.g. third generation RAP, lean RAP, etc. in a durable justified situation. Two studies elaborated in EMIB research group at University of Antwerp could be mentioned for that reason:

- In a first study [16] 4 concrete mixes were designed for cycle paths (compressive strength of 42.5 MPa) and tested, containing a percentage of 0; 25; 35 and 50 of RAP 0/20: slump test; modulus of elasticity; compressive strength test; density and air content were performed. The concrete mixes were designed with equal grading, cement and water content. In this phase, no optimization was done. An overall conclusion is that the properties are decreasing when adding RAP into a concrete mix. The compressive strength decreases from 26% up to 42%. It is estimated that 25% of the aggregates could be replaced by RAP to start optimizing concrete mixes.
- In a second study [17] the use of RAP in Portland cement concrete for roads (maximal 4 million ESA) was researched as alternative for virgin aggregates. The influence of RAP 0/20 on the mechanical properties of fresh and hardened concrete was evaluated. In order to determine this influence, an extensive laboratory study was performed on two types of mixes. The first type of mixes contained 25 or 35 percent RAP and the W/C - ratio was decreased to 0.45 and 0.42, while the reference mix had a W/C - ratio of 0.47. The second part of the experimental program focused on the influence of different RAP sizes used: the influence of a *fine and a coarse* RAP was evaluated. In these mixes 25 percent of the virgin aggregates were replaced by RAP. Only the mix proportion of the *fine* and *coarse* RAP varied. The RAP passing the 4 mm sieve was used as a replacement for the fine aggregates. The RAP retained on the 4 mm sieve was used to replace the coarse aggregates. Based on the test results and

discussion, the following conclusions can be drawn as follows: i) when RAP is added to the mixes the workability of the fresh concrete decreases due to the high absorption and internal friction of the RAP; also it is noticed that the *fine* RAP has a greater effect on the workability than the *coarse* RAP; ii) the RAP mixes show a systematic reduction in compressive strength with respect to the control mix; this is probably due to the reduced adhesion between the RAP in the mortar. Reducing the W/ C - ratio is an effective way the increase the compressive strength of the RAP containing mixes. Finally, it is observed that concrete mix holding only *fine* RAP, shows the least reduction in strength, iii) concrete made with RAP displays a lower dynamic modulus of elasticity and freeze - thaw resistance than the control mix, which can be improved by lowering the W/C – ratio, iv) RAP still absorbs water which must be taken into account for the calculation of the W/C ratio.

4. Use of RAP in asphalt mixes

4.1. Directives and technical guidelines

Strict guidelines on the nature of the reclaimed material (size distribution, bitumen content, filler content, bitumen viscosity or hardness, etc.) are enforced in the asphalt industry to guarantee good quality end materials. Homogeneity in terms of size, mineral type, binder type etc. is the most crucial quality criteria of RAP. In the ideal case, RAP is coming from only one source and one layer. Often this is not possible and RAP is mixed together from different pavement and locations. EN13108-8 standard leaves open to the countries to define a maximum particle size for RAP [18].

Issues related to the use of reclaimed binders are: variability of binder in RAP can generally be addressed by homogenizing the RAP stockpiles, hardness and level of recovered binder activation, compensation for the ageing process in the design of the added virgin binder, recyclability of any polymer modifiers, recyclability of asphalt with legacy materials, such as asphalt with coal tar or asbestos [10].

In Flanders, the use of RAP is described in the Flemish Standard version 3.1, TRA64 and TRA13 of COPRO, related to EN13108-8. Base course mix may contain unlimited content of homogenous RAP, when preheated at least at temperature of 110°C, but it is limited to 20% for EME-mixes and for mixes with cold-added RAP. All mixes must fulfil equal performance independent the RAP-content. Currently, mixes for surface layers may not contain RAP. Since it is allowed to use high RAP-content, a qualitative handling of the RAP is set.

4.2 Inventory of RAP and RAP-containing mixes in Belgium

In Belgium a slow increasing trend of using RAP is observed (see Figure 1). This means that the prohibition of using RAP in surface layers does not affect its use. A detailed analysis, based on data from COPRO, illustrates a consistency in grading and penetration (see Table 1), which is likely because the asphalt mix compositions in Flanders have hardly changed during the survey period.

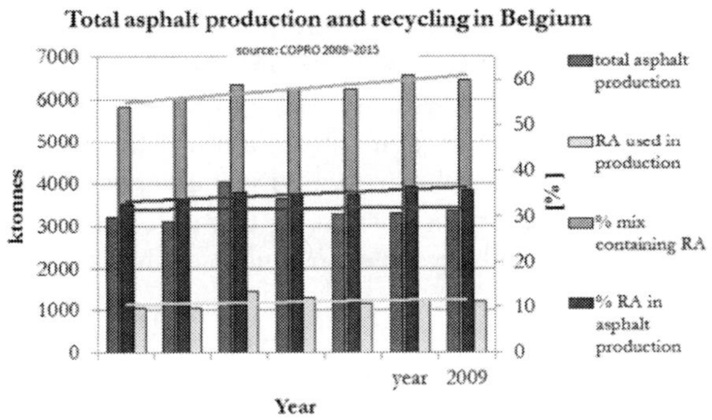

Figure 1. Data for asphalt recycling in Belgium [source: COPRO asbl].

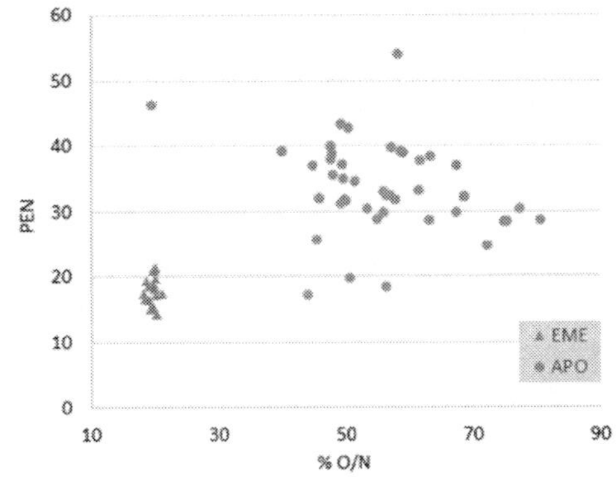

Figure 2. Penetration value of binder to ratio old/new binder in asphalt mixes [Data source: FRA 2015-2016].

Table 1. Binder penetration value of RAP stockpiles (n=number of selected stock piles [dmm]).

Year		n	Minimum	Maximum	Mean	Std. Deviation	Variance
2008	Penetration	95	11	31	20.7	4.5	19.9
2009	Penetration	105	11	38	20.6	4.6	21.2
2010	Penetration	96	13	46	20.6	4.8	23.3
2011	Penetration	104	11	36	19.5	4.6	20.8
2012	Penetration	117	12	37	19.8	4.0	15.9
2013	Penetration	98	14	41	20.0	4.7	21.9
2014	Penetration	111	13	45	20.0	5.7	32.9
2015	Penetration	101	13	30	18.8	2.8	7.7

Recycling techniques, which use old and milled asphalt to produce new asphalt, have been in use for over 30 years [4]. By the year 2010, it has been reported that hot asphalt mixes produced in Belgium, contained between 20% and 50% of RA [19]. Van den Kerkhof stated that at low recycling rates ($\leq$ 20%), there are not much technical problems to use RAP. When higher levels of reuse are to be reached (50% or even more), a good product quality control and management is imperative to guarantee the conformity of production of the new bituminous layer. RAP is currently used in most APO (these are AC-mixtures for base layers) and EME mixes. The RAP/new-binder ratio for EME-mix is limited to 20%; for APO no limit is set. Practical reasons to limit can be found in the technical restrictions on the asphalt plant where a minimal quantity is necessary in the parallel drum to provide overheating. In Figure 2, a summary is given for the binder properties of EME and APO-mixes (registered by Flemish Road Agency 2015-2016). It is clear that APO-mixes contain even more than 50% of RAP-binder.

4.3 Mechanical properties of RAP-containing mixes in Belgium

In [20] and [21] a survey was done on the performances of asphalt mixes containing RAP. In these studies data from 65 technical justification notes (TJN) of asphalt mixes for base layers were analyzed. The RAP-binder content ranged from 0 to 76%. The mechanical properties rutting, fatigue and stiffness were compared. The study concluded that, using Spearman's rho correlation test, a statistically significant, but weak correlation was found for % old/new binder, rutting and for stiffness: the resistance to rutting and the stiffness of the mix increase if % old/new binder was higher.

Considering that all mixes fulfilled the requirements of the European standards, the mix was optimized whether RAP was used or not, such as using a softer binder to fulfill binder classes. When the dichotomous analysis was performed (0% RAP and >%RAP), it was observed that mixes containing RAP showed higher resistance to rutting and fatigue. No strong or robust correlation could be found between these mechanical parameters and %RAP, however the study revealed no adverse influence when using RAP in hot asphalt mixes. Other studies mentioned both positive as negative effects. It is recommended to analyze all these studies into one statistical study in order to find relevant parameters to adjust the mix design, if appropriated.

Since the use of RAP in surface layers is prohibited by SB250 v2.2 and later versions, questions arise whether the use of RAP in only base layers is too restricted; since there is still more RAP on the market. In order to start a research program, two typical mixes for surface layers were tested using RAP [22]. A dense asphalt AC-10, containing 5.6% binder (penetration 53 dmm) is used as reference. Mixes were produced containing 0% and 40% RAP. A softer binder (penetration 76 dmm) was used in order to compensate hard binder of the RAP. For this type of mix only %voids, wheel rutting and resistance to water sensitivity was tested. Compared to the reference mix, using 40% RAP was beneficial to rutting resistance and ITSR. SMA mix was also tested, using 20% and 30% RAP illustrating that RAP has a positive influence on the results for stiffness, rutting, ravelling and water sensitivity. Based on the research results, it is recommended to re-evaluate the current exclusion of the use of RAP in surface layer mixes and to initiate new test methods in order to evaluate mix long term performances, e.g. ravelling tests on aged specimens. Moreover, the prohibition for using RAP in surface mixes was set allowing up to 50% RAP, which is too high for these mixes under high environmental influence.

Some remarks must be pointed when RAP is used in mixes for surface layers. In Europe, the standard EN 13108-8:2005 for reclaimed asphalt establishes that if RAP content is higher than 10% for surface layers and 20% for base layers, a logarithmic blending law for penetration and a linear blending law for softening point should be applied to select the proper virgin binder to use. The Flemish approach with recycling RAP is based on the pen-rule in order to assign the new binder type, because of its simplicity and practicality [1]. A second issue is the use of polymer modified bitumen in mixes. The blending of old and new modified binder is very complex, maybe even case-dependant. There is a lack of guidance in this technology. Developing a proper binder blend's design between RAP binders and virgin materials is the first step for designing mixes with higher content. Currently, different approaches are being followed to carry out this task in different countries. In the U.S., for high RA contents (>20%), NCHRP Report 452 [23] described a particular procedure to obtain blending charts assessing high, intermediate and low critical temperatures of the blend of RAP and virgin binder [24]. Zhou et al. [25] developed a balanced RAP mix design for high RA content mixes for surface layers based on changing the binder content of the mix to optimize the maximum density, with 35% RAP content mixes designed with their methodology in different locations. The overall conclusion from the study was that high RAP mixes can have better or similar performance to virgin mixes, but they must be well designed following appropriate mix design methods. The use of RAP in surface layers might influence the RAP-handling into extra quality checks, e.g. types of aggregates and binder performance at a lower temperature.

The healing mechanism is one of the most important factors with significant impact on the performance of asphalt pavements. Investigators have expressed their belief that the healing mechanism not only can provoke total or partial restoration of the damage but can also extend the pavement service life [2, 26]. Currently there is no established test or procedure in order to study this phenomenon. In order to investigate this mechanism, researchers focused on the factors that mainly effect healing. Chemical and mechanical characteristics could provide an insight on the healing phenomenon and the properties that mostly trigger it [27]. Although, healing is a mechanism that can mainly effect positively the mixes service life, many researchers expressed their doubt about the effect of RAP on this phenomenon. A research conducted by Huurman [28] indicated that the shift factor of 4, established in the Netherlands as healing factor for bituminous mixes, can be reduced to 1.4 when

60% of RAP is initiated. Clearly this addition of RAP can limit the healing mechanism, in an unknown extend until now.

Since healing tests are long-lasting and still under discussion, a new method to evaluate the healing property of an asphalt mix is necessary. A method performed with a Dynamic Shear Rheometer allows to compare mechanical properties for different types of mortar on cylindrical mortar samples, as first described for fatigue tests in [28] and further explored for healing tests by means of a procedure using continuous (10Hz, 15°C) and discontinuous (3s loading, 9s rest; stress-controlled) tests in [2]. The research showed that the healing property of 100% RAP mortar decreased significantly, compared to laboratory-aged mortar and virgin mortar. Since the healing factor is used as a constant for Flemish road design (factor 7.1) for each asphalt mix, an overestimation of the factor can result in faster deterioration of the road construction. It is recommended to quantify the healing factor for asphalt mixes when RAP is used, in order to optimize the mixes is such a way that the fatigue*healing is at least equal than for mixes without RAP.

5. Life cycle assessment of asphalt pavements

In Belgium, some steps were taken into the direction of a carbon-footprint tender including a carbon calculation tool [29], although it was found that there were many significant differences occurred comparing this tool and a detailed LCA study using Simapro software [20] such as different emission factors and the major contributor to the environmental impact is fossil depletion from raw material bitumen and transport. Important conclusions from the recent LCA study [21] are: i) LCA is very case-dependant and holds still a lot of uncertainties, ii) in the specific case, a standard rehabilitation of a dual carriage road by toplayer and baselayer using 0% and 44% of RAP-binder, the two impact categories related to climate change (human health and ecosystems) are together, with 47%, the main contributors to the single score impact. Transport by lorry (47%, including upstream processes) and heat generation (18%) mainly cause the climate change impact, iii) using RAP leads to environmental benefits mainly coming from the avoided amount of crushed gravel, limestone powder and the total amount of bitumen. Impact categories climate change human health (29%), particulate matter formation (19%) and fossil depletion (19%) are the main contributors to this environmental benefit, iv) the benefits only stand when the same service life of the asphalt mixes with and without RAP is obtained.

In general, much more analysis should be done concerning the transport aspect of asphalt mixes, the environmental impact of the use of additives - even rejuvenators - in combination with the service life for asphalt mixes with RA. This service life must be seen as larger than only the fatigue life, e.g. also the resistance to rutting and healing mechanisms. When the service life is theoretically decreased by 20% (e.g. simplified as e6 80 µS instead of 100µS), all environmental benefit is gone when the pavement thickness is increased to compensate this inferior property. This means that using RAP at a high content is only beneficial when also the performance of the mixes containing RA is equal, since transport of material is dominant. The above discussed and accepted means that mixes containing RAP can be designed and produced with equal performance. It is also concluded that the distance between asphalt plant and the work site is an important factor considering the environmental impact of work. Another way to work with reduced environmental impact of asphalt mixes, is the use of reduced production and compaction temperatures. An LCA comparative calculation of a standard mix, a mix produced at decreased temperature and a mix containing 50% RA tend to a conclusion that using RAP is more effective than decreasing the temperature by 30°C. During the project Re-Road [30] the life cycle benefits were investigated from recycling versus using warm mix asphalt (WMA), and demonstrated that even at low recycling rate of 15%, the recycling benefits outweigh those achieved by reducing temperature from 165 to 130 °C. Discussion about the performance and technical production challenges of these WMA mixes in Flanders, leads currently to an on-hold situation.

6. Strategies by conclusions for optimizing the use of RAP in Flanders

In Flanders, high recycling rates of RAP in asphalt mixes for baselayer are used. However, data showed that not all RAP available on the market is used in these mixes, which is currently accepted as one of the best recycling application in Flanders. In order to optimize the use of RAP in a sustainable way, the next steps should be taken into account:

i) The LCA-method is a new tool giving more insight in the environmental effect of production methods, transport and the use of by-products or enhancers (rejuvenators, waxes). Still there is a lack of data and consensus in order to calculate absolute impact values. Comparative studies can be done with this tool and showed that the use of RAP is beneficial to the use of warm asphalt mixes without RAP.

ii) The asphalt mixes currently used, all meet the mechanical requirements, tested in laboratory. In order to evaluate the effect of RAP in situ, the initiation of an inventory of pavements, of which the mix composition and the traffic volume is monitored, is strongly recommended. An important parameter is absent: the healing factor is not well-studied and is related to the use of RAP. It is recommended to test a selection of the mixes also on their healing capacity combined with the fatigue performance: an LCA-study stated that all environmental benefits are counteracted when RAP decreases the service life performance of the mix. Future comparative studies should only take into account optimized mixes, assessing the environmental impact of that optimization.

iii) A statistical analysis taking into account all mechanical performance and mix compositions would provide insight in the effect of RAP on the mechanical durability performance.

iv) The step to use RAP in mixes for surface layer is self-evident for limited recycling rates e.g. 25% taking into account additional quality checks of the RAP used and mix performance.

v) The blending process of mixes using polymer modified binder, rejuvenators, multiple recycled mixes or at higher recycling rates still needs more research. The current pen-rule does not cover the required performance expected: when new binders are used in asphalt mixes, these binders are extensively tested for a wide range of temperatures and ageing conditions; when e.g. 70% of the binder consists of RAP-binder, only the penetration value must meet the binder class. Durability performance on long term or other conditions are neglected.

vi) The use of RAP in concrete for road constructions, is at this moment not beneficial taking into account the mechanical performance. Only low contents can be used. An LCA study can give the environmental information to decide its applicability.

vii) The LCA-study demonstrated that transport of materials is one of the most dominant factors in the asphalt process. An effective RAP management tool for appointing the best recycling process for a geographical area, on the plant or even recycling in situ could be in this case a promising technology. It is recommended for larger projects to incorporate an environmental tool for comparative studies of the available technologies.

Acknowledgements

The authors would like to thank COPRO asbl and the Flemish Road Agency for providing data and test facilities. The authors would also like to acknowledge the support of Belgian Road Research Centre and the Flemish Industry and the financial support of the Flemish Government.

References

[1] Van den bergh W and van de Ven MFC 2009 Hot recycling of asphalt: experiences in Flanders and the Netherlands *Proc. ENVIROAD 2009 (Warsaw)*
[2] Van den bergh W 2011 *The effect of ageing on the fatigue and healing properties of bituminous mortars* (Delft University of Technology)
[3] Nielsen E 2015 Recycling in high trafficked roads in Denmark *NABin seminar (Oslo, Norway)*

[4] European Asphalt Pavement Association – EAPA 2011 *Fit for tomorrow – Asphalt paving: safe, comfortable, sustainable*
[5] Miliutenko S, Bjorklund A and Carlsson A 2012 Opportunities for environmentally improved asphalt recycling: the example of Sweden *Journal of cleaner production* **43** pp 156-65
[6] Falchetto A C, Moon K H and Wistuba M P 2014 Microstructural analysis and rheological modelling of asphalt mixtures containing recycled asphalt materials *Materials* **7** pp 6254-80
[7] Coppola L, Kara P and Lorenzi S 2016 Concrete manufactured with crushed asphalt as partial replacement of natural aggregates *Materiales de Construcción* **6** (324)
[8] European Committee for Standardization (CEN) 2005 Bituminous mixtures – Material specifications – Part 8: Reclaimed asphalt *(EN 13108-8)*
[9] Public Waste Agency of Flanders (OVAM) 2008 *VLAREA – Vlaams Reglement inzake asvalvoorkoming en – beheer* (Brussels)
[10] European Asphalt Pavement Association 2014 *Asphalt in figures*
[11] Van den bergh W 2015 Towards more use of ICT in road engineering *STA Annual conference and awards ceremony (Brussels)*
[12] ROAD IT 2017 *Efficient process management by the implementation of IT in the asphalt road construction* (University of Antwerp, Belgian Road Research Centre, TetraFondsproject IWT150166)
[13] Huang B, Li G and Shu X 2005 Laboratory investigation of portland cement concrete containing recycled asphalt pavements *Cement and Concrete Research* **35**(10) pp 2008-13
[14] Hossiney N, Bergin M and Tia M 2010 Concrete Containing RAP for use in concrete pavement *International Journal of pavement Research and Technology* **3**(5) pp 255-85
[15] Brand A S, Al-Qadi I, Roesler J and Shangguan P 2012 *Fractionated Reclaimed Asphalt Pavement (FRAP) as a Coarse Aggregate Replacement in a Ternary Blended Concrete Pavement* (University of Illinois, USA)
[16] Voorspoels J 2015 *Technical development of concrete pavement containing reclaimed asphalt pavement I* (University of Antwerp, Belgium)
[17] De Dooy B and Huygens D 2015 *Technical development of concrete pavement containing reclaimed asphalt pavement II* (University of Antwerp, Belgium)
[18] Tebaldi G et al 2012 Classification of recycled asphalt (RA) material *Proc. 2nd International Symposium on Asphalt Pavement and Environment (France)* p 14
[19] Van den Kerkhof E 2012 Warm waste asphalt recycling in Belgium – 30 years of experience and full confidence in the future *Proc. 5th Eurasphalt & Eurobitume Congress (Istanbul, Turkey)* pp A5EE-137
[20] Anthonissen J, Braet J and Van den bergh W 2016 Effect of RAP on the mechanical properties of registered asphalt mixtures in Flanders *The 5th Int. Symp. on Life-Cycle Engineering (IALCCE 2016) (Delft)* (DRC Press/Balkema, Leiden) pp 399-405
[21] Anthonissen J 2017 *Bituminous Pavements in Flanders: Quantifying the effect of RAP on the environmental impact* (University of Antwerp, Belgium)
[22] Van den bergh W, Seghers D, Anthonissen J, Vuye C, Benor T and Haulet J-Ph 2016 Recycling reclaimed asphalt in Flanders: state of art 2014 *Proc. 6th Eurasphalt & Eurobitume Congress (Prague, Czech Republic)*
[23] McDaniel R and Anderson R M 2001 Recommended use of reclaimed asphalt pavement in the superpave mix design method: technician's manual *NCHRP Report 452* (TRB Washongton USA) p 58
[24] Presti D L, Del Barco Carrion A J, Airey G and Hajj E 2016 Towards 100% recycling of reclaimed asphalt in road surface courses: binder design methodology and case studies *Journal of cleaner production* **131** pp 43-51
[25] Zhou F, Hu S, Das G and Scullion T 2011 *High RAP mixes design methodology with balanced performance - Report 0-6092-2* (Texas Department of Transportation, USA)
[26] Karki P, Li R and Bhasin A 2015 Quantifying overall damage and healing behaviour of asphalt

materials using continuum damage approach *International Journal of Pavement Engineering* **16**(4) pp 350-62

[27] Ayar P, Moreno-Navarro F and Rubio-Gamez M C 2016 The healing capability of asphalt pavements: a state of the art review *Journal of Cleaner Production* **113** pp 28–40

[28] Huurman M 2007 *Lifetime Optimisation Tool – Main Report 7-07-170-1* (Laboratory of Road and Railway Engineering Delft The Netherlands)

[29] Anthonissen J, Van Troyen D, Braet J and Van den bergh W 2015 Using carbon dioxide emissions as a criterion to award road construction projects: a pilot case in Flanders *Journal of Clean Production* **102** pp 96-102

[30] Wayman M, Andersson-Skold Y, Bergman R, Huang Y, Parry T, Raaberg J and Enell A 2012 Life cycle assesment of reclaimed asphalt *Project Report 7FP Re-Road: End of life strategies of asphalt pavements*

IOP Conf. Series: Materials Science and Engineering **236** (2017) 012089 doi:10.1088/1757-899X/236/1/012089

Polish experience with cold in-place recycling

Bohdan Dołżycki[1]

[1]Gdansk University of Technology, Department of Highway Engineering, Narutowicza str. 11, 80-233 Gdansk, Poland

bohdan.dolzycki@wilis.pg.gda.pl

Abstract. Deep cold in-place recycling using cement and asphalt emulsion has been used for reconstruction of existing roads since the beginning of the 1990s. This paper describes the first Polish requirements for mineral-cement-emulsion mixtures. As requirements stated for the strength of the mineral-cement-emulsion mixtures were quite high, most of the mixtures were designed using high amount of cement and aggregate added for the improvement of gradation. The paper also presents selected results of the visual assessment of existing pavements with bases made of mineral-cement-emulsion mixtures. The mixtures used in bases were designed according to the first Polish requirements. On all of the sections numerous reflective transverse cracks were observed. Most of them resulted from very high requirements stated for the strength. Last part of the paper describes the changes which were introduced to the Polish requirements on the basis of conducted field investigations and other past experiences of using mineral-cement-emulsion mixtures in Polish climatic conditions. Newly developed instruction liberalized the requirements stated for base materials and significantly reduced the amount of cement.

1. Introduction

In the 1990s the road system in Poland was underinvested and nearing the limits of its designed fatigue life. The political changes brought a rapid increase in heavy traffic, which resulted in significant degradation of the basic road system. Soon the roads rutted and showed multiple fatigue cracks. The road administration was in need of a technology that would allow for a quick and relatively cost-effective rehabilitation of the deteriorated roads.

One of the solutions that were put into practice in Poland was to use deep cold in-place recycling to mill the distressed pavement into a base made of mineral-cement-emulsion mixture (for the sake of convenience from now on referred to as an „MCE mixture", which is a very popular term in Poland). In order to facilitate the widespread use of the mineral-cement-emulsion mixtures the first Polish requirements [1] were introduced in 1997. They were soon revised, and it was the 1999 version [2] that was widely used and served as a basis for reconstruction of several thousand kilometers of roads in Poland.

This paper describes the Polish experience with the introduction of deep cold in-place recycling as well as the changes that were introduced after the observations from the first years of use of the MCE mixes in Poland were gathered.

2. The Polish requirements from 1999

The Polish requirements [2] assumed the preparation of MCE mixes using reclaimed pavement materials of the following types: mineral-bitumen, mineral-tar, mineral-cement, or only aggregate from existing pavements. The aggregate mix found in materials used for MCE mixes was required to be continuously graded, reaching 31.5 mm or 63 mm particle size, and lie within the range of good grading defined by the limiting curves. Crushed aggregate was added in order to improve the gradation. Asphalt emulsion and cement were used as binding agents. The mentioned materials were widely used in MCE mixes. The recommended cement content was within the range from 1.5% to 4% if an asphalt reclaimed material was used and from 1.5% to 7% in case of tar-asphalt or tar reclaimed material. The recommended asphalt emulsion content was within the range from 3% to 5,5%.

The primary requirements set in the Polish guidelines [2] were as follows:

- Marshall stability at the temperature of +60°C:
  - for traffic categories KR3 through KR6 (medium and heavy traffic) 8.0 – 20.0 kN
  - for traffic categories KR1 and KR2 (low traffic) 4.0 – 20.0 kN
- Marshall flow at the temperature of +60°C:
  - for all traffic categories: 1.0 – 3.5 mm
- Air voids for all traffic categories:
  - for specimens compacted with a Marshall hammer, 2 x 75 impacts, in a perforated compaction mold, cured for 28 days at room temperature: 9.0 – 16.0%
  - for specimens pressed with a hydraulic press, constant force of 100 kN for 5 minutes, cured for 7 days at room temperature: 5.0 – 12.0%

Those requirements were meant to facilitate the design of MCE mixes without reference to uncommon test methods. The assessment guidelines for MCE mixes were compatible with capabilities of a typical road laboratory at the time.

3. The need to introduce changes to the requirements for MCE mixes

After several years of use of MCE mixes it turned out that pavements with bases of this type were prone to transverse cracking. Assessment of chosen sections of roads reconstructed with the use of MCE bases showed that in most cases the transverse cracks appeared in quantity of several cracks per kilometer. They were not very numerous, but their number was increasing with time. Observations showed as well that the pavements were generally even, without major distress, with sporadic edge cracks. Fatigue cracking was observed after 5 to 10 years. Typical cracks can be seen in Figures 1, while a specimen cored out at the crack is shown in Figure 2.

Assessment of various sections allowed for defining the problem, which was the occurrence of transverse cracks in roads with bases made of mineral-cement-emulsion mixtures. Information about the composition of the built-in MCE mixtures was obtained in order to establish the probable cause of the problem. The information was gathered both from the road administration and from contractors. In total, mix compositions were obtained for 56 road sections built using this technique. The information served as the basis for detailed analysis of the mineral-cement-emulsion mix formulas used. The quantity of every component in the mixtures was analyzed.

It was found out that among the 56 road sections built using cold in-place recycling:

- 20 sections showed cement content in the mix within the range of 2.5 – 3.9%,
- 24 sections showed cement content in the mix equal to 4%, that is the maximum value according to the technical requirements [2] for MCE mixes using asphalt reclaimed material,
- 6 sections showed cement content in the mix within the range of 5.0 – 5.8%,
- 6 sections showed cement content in the mix equal to 7%, that is the maximum value according to [2] for MCE mixes using tar reclaimed material.

The above set of data shows that there is a tendency to use relatively high amounts of cement. Often the cement content is comparable with the one found in cement stabilization, cement bound mixtures or lean concrete. It is the main factor that results in construction of rigid bases that are prone to cracking.

Figure 1. Transverse crack in the national road No. 7.

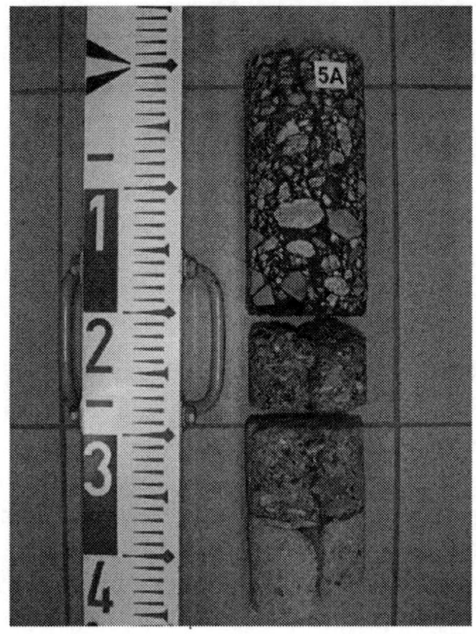

Figure 2. A specimen cored out at the crack in the national road No. 7. The crack propagates through the entire pavement.

Among the 56 road sections built using cold in-place recycling:

- 7 sections had emulsion content in the mix higher than 3%,
- 12 sections had emulsion content in the mix lower than 3%,
- 37 sections had emulsion content in the mix equal to 3%, that is the minimum emulsion content according to the Polish requirements [2].

The above comparison shows that the MCE mix bases constructed in Poland contain too little asphalt emulsion, and the strength requirements are met mainly by means of adding high quantities of cement.

It was found out that among the 56 road sections built using cold in-place recycling:

- 13 sections had new aggregate content of up to 30%,
- 34 sections had new aggregate content between 31% and 50%,
- 7 sections had new aggregate content between 51% and 70%,
- 2 sections had new aggregate content above 71 %.

The above breakdown shows that the Polish MCE mix bases contained relatively high quantities of added aggregate – often high-quality crushed aggregate. In most cases its content was higher than 30% of the MCE mix. In 9 cases new aggregate constituted over 50% of the mixtures. It is even hard to consider those as conventional MCE mixtures, since the reclaimed material – and not added aggregate – should have been their primary ingredient.

The analysis of the obtained data about road sections with MCE bases along with the assessment of the state of the pavement on chosen sections showed that:

- Transverse cracks occurred in the majority of sections with MCE mix bases. While not very high in quantity, they showed that there does exist a problem with cracking in pavements with MCE mix bases.
- It is difficult to define the cause of cracking precisely, but all evidence suggested that the observed cracking should be qualified as reflective cracks. Nevertheless, there is a dose of probability that they were caused by low temperature.
- The excessive content of cement in the MCE mixtures should be noted as the probable cause of the distress. The content of cement used in the MCE mixes was not far from the content associated with typical cement bound materials.
- The content of asphalt emulsion used in the MCE mixtures was very low and often does not exceed 3%.
- The MCE mixes contained high quantities of added aggregate, which in many cases disqualifies them in terms of economic viability.

The practical application of the requirements deviated from the basic idea of the mineral-cement-emulsion mixtures, which were originally meant as a flexible course constructed with the highest possible quantity of the material reclaimed from the existing structure. In search for the possible cause, it was found out that the guidelines [2] set relatively high strength requirements, which in practice translated into the use of high quantities of cement as well as new aggregate with good parameters. Emulsion was often added only in order to enable the product to pass as a mineral-cement-emulsion mixture. Such state of affairs forced the road administration to verify the requirements for MCE mixtures.

4. International experience

The first step in the verification of the requirements [2] was an extensive study on design and placement of MCE mixtures worldwide. Information on various countries was gathered, including Germany, England, America, Norway and Sweden. The literature study [3–14] allowed for formulation of the following conclusions:

- The MCE mix design is performed in various ways. Differences occur on every stage of the design procedure, including selection of materials as well as assessment of the produced mixture.
- The choice of aggregate mix is most frequently based on a range of good gradation defined by limiting curves. The space between them is wide and defined in a manner that ensures proper compaction of the mix.
- Minimalization of cement content in MCE mixtures is advised. Requirements in many countries limit the cement content to 1–2%. Where higher cement content is acceptable, the requirements state that the MCE mix should be designed in such a way that the bituminous binding dominates over the hydraulic binding.
- The basic parameter used for MCE mix assessment is the indirect tensile strength. Sometimes the stiffness of the mix is tested as well.
- Moisture susceptibility is tested very often. This parameter is defined on the basis of decrease in strength of specimens subjected to conditioning cycles in comparison with reference specimens. It is particularly significant in countries with high levels of soil moisture and frost.

The comparison between requirements in Poland [2] and in other countries implied the following fields where change was desirable:

- For the gradation of the mineral-cement mix, the Polish guidelines gave limiting curves that defined a narrower range than in most countries.
- An important element that the Polish guidelines were lacking was the assessment of moisture susceptibility of the MCE mixtures.
- In Poland the assessment of mechanical properties of the mix was based primarily on Marshall stability and flow tested at the temperature of +60°C. Tests performed at this temperature did not reflect the real conditions under which the built-in MCE mix functioned. Possibility of relinquishing the Marshall tests needed be taken into consideration. A different criterion – one that was used worldwide – was to be considered instead, that is the indirect tensile strength.
- The Polish guidelines did not take into account the actual stiffness of the produced mixture, which led to placement of mixtures that were too stiff.

5. New requirements for mineral-asphalt mixtures

The past experience with MCE mixes implied that there was a need to revise the requirements considering their production and placement. The literature study served as a source of information on requirements in other countries. Assessment of the practice in Poland showed the areas that needed change. In 2013 the new instruction for the design and placement of MCE mixes [16] was published, encompassing the in-place mixing technique as well as mixing in a stationary plant. The new requirements took into account:

- experience in terms of design, production and placement from various countries,
- widespread use of the material requirements based on the European standards,
- standardized test methods based on the European standards,
- Polish experience from the past practice with MCE mixtures as well as its implications.

The preface to the new requirements includes, among others, the following statements:

- The nature of an MCE base is dual. Depending on the combination of binding agents and the resulting stiffness, MCE mixtures can be divided into two types:
- mixture with dominating bituminous binding – flexible base,
- mixture with dominating hydraulic binding – rigid base.
- Through selection of type and quantity of the binding agents it is possible to achieve different MCE mixtures, whose performance is similar to hydraulically bound bearing courses or to asphalt bearing courses.
- When designing an MCE mixture, one should aim at achieving a dominating bituminous binding in order to minimalize the risk of reflective cracking.
- The mineral-cement-emulsion mixtures designed in accordance with these rules can be freely used in reconstruction and construction of roads subjected to traffic categories from KR1 through KR4 and are compatible with the rules defined in the Polish catalog of typical pavements as well as in the Polish catalog of pavement reinforcement rehabilitation and repair.

In terms of properties of the materials used for MCE mixtures the following requirements were set:

- The reclaimed material is classified as either mineral-bituminous (i.e. mineral-asphalt, mineral-tar or mixed), mineral-cement or mineral. The reclaimed material cannot contain foreign or organic contaminants. It is necessary to determine the type of binder used in it as well as its ratio of bound vs. unbound material.

- Any new aggregate added to the mineral-cement-emulsion mix should be compliant with the PN-EN 13242:2013 standard.
- Asphalt emulsion of the C60B5R type should meet the requirements of the PN-EN 13808 standard.
- Portland cement of type CEM I or multi-ingredient Portland cement of type CEM II, classes 32.5 and 42.5, should be compliant with the PN-EN 197-1 standard.
- Water should meet the requirements stated in the PN-EN-1008 standard.

Even if the material properties meet the above requirements, it may happen that the produced mix does not achieve the required parameters. The design procedure of the MCE mix should be repeated then with different ingredients or modified ingredient proportions.

The mix design process was divided into several successive steps. They were ordered as follows:

- Design of the mineral mix, which may contain reclaimed material plus new aggregate or only reclaimed material. The gradation of the mix should be continuous. Maximum particle size should not exceed 31.5 mm, but there is a 15% tolerance margin for particles retained on the first sieve.
- Choice of the binding agents. Combination of the binding agents should be determined in accordance with the following general rule:
    - asphalt emulsion: from 2% to 6% (mass content),
    - cement: from 1% do 4% (mass content).

The desired combination of binding agents is the one where the cement content is as low as possible, which minimalizes the risk of reflective cracking.

- Proper design of the MCE mixture should be carried out according to the following procedure:
    - Initial determination of the materials planned for use in the mineral mix.
    - Choice of the binding agents.
    - Determination of the optimum liquid content.
    - Determination of the quantity of water that should be added in order to achieve the optimum liquid content.
    - Forming of mineral-cement-emulsion mix samples for determination of physical and mechanical properties of the mix.
    - Storage of the samples during the setting process.
    - Performance of the tests necessary for determination of physical and mechanical properties.
    - Determination of the MCE mix formula.

Each of the above steps is described precisely in the requirements. The designed MCE mix should meet the requirements given in Table 1.

- The requirements give relatively precise rules for placement and acceptance of the MCE mix courses. An MCE base should be built in such a manner that guarantees its homogeneity and compaction without any visible weak, damaged or non-homogeneous areas. The requirements for the compacted MCE course are given in Table 2.

The above requirements were introduced in practice in 2014. After a transition period, the majority of new projects are currently realized based on the new requirements.

Table 1. Requirements for the designed MCE mixture

Property:	Required values:	
	Traffic KR1 and KR2 (low traffic)	Traffic KR3 and KR4 (medium traffic)
Content of air voids	8% to 18% (vol.) max.14% (vol.)[1)]	8% to 15% (vol.) max.12% (vol.)[1)]
Indirect tensile strength, T = + 5°C, after 7 days, [MPa]	0.40–0.90	0.50–1.00
Indirect tensile strength, T = +5°C, after 28 days, [MPa]	0.60–1.40	0.70–1.60
IT-CY stiffness modulus, T = +5°C, after 28 days, [MPa]	1500–4000	2000–5000
Remaining indirect tensile strength after storage in water, no less than [%]	70	80

[1)]For reclaimed materials that include tar.

Table 2: Requirements for the compacted MCE course

Parameter	requirement	
	Traffic KR1 and KR2 (low traffic)	Traffic KR3 and KR4 (medium traffic)
Relative compaction	≥ 98%	≥ 98%
Air voids	≤ 15% vol.	≤ 12% vol.
Base course bearing capacity:		
• Secondary deformation modulus E_2	$E_2 \geq 140$ MN/m^2	$E_2 \geq 180$ MN/m^2
• Dynamic deformation modulus E_{vd}	$E_{vd} \geq 70$ MN/m^2	$E_{vd} \geq 90$ MN/m^2

6. Summary
The Polish experience showed that improperly defined requirements for MCE mixtures may lead to production of an overly stiff mix, resulting in transverse cracking of the pavement. The change to the requirements was meant to introduce a tendency to design MCE mixtures that are less rigid. The first cases of their usage are promising.

When producing an MCE base it is necessary to make it as flexible as possible in order to minimalize the risk of reflective cracking which may appear in such pavements.

For the last few years interest in deep cold recycling has been increasing in Poland. Apart from the fact that the road administration used the technique more often, especially on local roads, the scientists have been carrying out research on a greater scale [17, 18]. The research is in accord with the international tendency to use and test mixtures that include materials recovered from existing pavements [18, 19, 20].

Acknowledgements
The author would like to thank the Polish General Directorate for National Roads and Motorways for help and support in his research.

References
[1] Zawadzki J and Matras J 1997 Warunki techniczne wykonania warstw podbudowy z mieszanki mineralno-cementowo emulsyjnej metodą recyklingu na miejscu *IBDiM* **53**
[2] Zawadzki J, Matras J, Mechowski T and Sybilski D 1999 Warunków technicznych wykonania warstw podbudowy z mieszanki mineralno-cementowo emulsyjnej (MCE) *IBDiM* **61**
[3] Forschungsgesellschaft für Straßen- und Verkehrswesen Arbeitsgruppe 2005 *Mineralstoffe im Straßenbau Merkblatt für Kaltrecycling in situ im Straßenoberbau*
[4] Yoshida T and Noda E 2001 Technical guidelines for In-situ recycling of base course in Japan *Proc. 1 st International Symposium on Subgrade Stabilization and in situ Pavement*

Recycling Using Cement (Salamanca, Spain)
[5] Theyse H, Long F, Harvey J H and Monismith C L 2004 *Discussion of deep in situ recycling* (Pavement Research Center, University of California Berkeley and University of California Davis)
[6] Marti M and Mielke A 2002 *Synthesis of Asphalt Recycling in Minnesota* (Minnesota Local Road Research Board)
[7] Kandhal P S and Rajib B M 1997 Pavement recycling guidelines for state and local governments *Federal Highway Administration Report FHWA-SA-98-042*
[8] Merrill D, Nunn M and Carswell I 2004 A guide to the use and specification of cold recycled materials for maintenance of road pavements (Report TRL611)
[9] PIARC 2003 *Pavement Recycling. Guidelines for in-place recycling with cement, in-place recycling with emulsion or foamed bitumen, hot mix recycling in-plant*
[10] Wirtgen GmbH 2004 *Wirtgen Cold Recycling Manual*
[11] Jostein M 2000 The use of cold bitumen stabilized base course mixes in Norway *Proc. Seminar Baltic Road Association B2*
[12] Carswell I, Ellis S J and Hewitt A *2008 Design and specification for sustainable maintenance of roads using cold recycling techniques* (PIARC)
[13] Milton L J and Earland M 1999 *Design guide and specification for structural maintenance of highway pavements by cold in-situ recycling* (TRL Crowthorne)
[14] PARAMIX Project 2004 *Road Pavement Rehabilitation Techniques Using Enhanced Asphalt Mixtures*
[15] Dolzycki B 2013 *Instrukcja projektowania i wbudowywania mieszanek mineralno-cementowo-emulsyjnych* (Politechnika Gdańsk, Generalna Dyrekcja Dróg Krajowych i Autostrad)
[16] Dołżycki B, Jaczewski M and Szydłowski C 2017 The Influence of Binding Agents on Stiffness of Mineral-cement-emulsion Mixtures Proc. Int. Conf. on Modern Building Materials, Structures and Techniques (Vilnius) pp 239-46
[17] Iwański M and Chomicz-Kowalska A 2016 Application of the foamed bitumen and bitumen emulsion to the road base mixes in the deep cold recycling technology Baltic Journal of Road And Bridge Engineering 11(4) pp 291-301
[18] Grilli A Bocci E and Graziani A 2014 Influence of reclaimed asphalt content on the mechanical behaviour of cement-treated mixtures Road Materials and Pavement Design 14(3) pp 666-78
[19] Godenzoni C, Graziani A and Perraton D 2017 Complex modulus characterization of cold-recycled mixtures with foamed bitumen and different contents of reclaimed asphalt Road Materials and Pavement Design 18(1) pp 130-50
[20] Valentin J, Čížková Z, Suda J, Batista F, Mollenhauer K and Simnofske D 2016 Stiffness Characterization of Cold Recycled Mixtures Transport Research Procedia 14 pp 758–67
[21] Polski Komitet Normalizacyjny 2013 Kruszywa do niezwiązanych i związanych hydraulicznie materiałów stosowanych w obiektach budowlanych i budownictwie drogowym PN-EN 13242
[22] Polski Komitet Normalizacyjny 2010 Asfalty i lepiszcza asfaltowe. Zasady klasyfikacji kationowych emulsji asfaltowych (PN-EN 13808)
[23] Polski Komitet Normalizacyjny 2012 Cement. Część 1. Skład, wymagania i kryteria zgodności dotyczące cementów powszechnego użytku (PN-EN 197-1)
[24] Polski Komitet Normalizacyjny 2004 Woda zarobowa do betonu. Specyfikacja pobierania próbek, badania i ocena przydatności wody zarobowej do betonu w tym wody odzyskanej z procesów produkcji betonu (PN-EN-1008)

IOP Conf. Series: Materials Science and Engineering **236** (2017) 012090 doi:10.1088/1757-899X/236/1/012090

Utilisation of metallurgical by-products in road construction in the Czech Republic

František Kresta[1]

[1]SG Geotechnika a.s., branch Ostrava, 28.října 150, 702 00 Ostrava, Czech Republic

E-mail: frantisek.kresta@geotechnika.cz

Abstract. Metallurgical by-products, primarily blast furnace slag and steel slag, have ranked among important alternative sources of fill as well as of material for the structural layers in highways. Main hazards of metallurgical by-products are closely connected to their chemical and mineralogical composition and they can be resulted in volume changes. Fears from possible deformations similar to the D47 motorway meant that metallurgical by-products were excluded from several public tenders of road construction. Comparison of blast furnace slag, steel slag and other metallurgical by products parameters allow us to define the most hazardous material as steelworks waste. Linear swelling of steelwork waste achieves more than 40% at 75°C and swelling pressure was higher than 1.5 MPa. Compositional heterogeneity of steelworks waste makes it difficult to establish the long-term behaviour of this material. At the present time we cannot ascertain which maximum values can be reached by deformation and what are the swelling pressures acting on the material while the volume changes are in progress.

1. Introduction

Metallurgical by-products, primarily blast furnace and steel slag, have ranked among important alternative sources of fill as well as of material for the structural layers in highways.

The properties of aggregates based on blast furnace or steel slag are determined to a decisive degree by the process employed by the specific ironmaker or steelmaker in question, or as the case may be, the specific producer of a nonferrous metal.

Such metallurgical by-products tend to be termed 'slags' not just by laymen but even by the professional public. An accurate description and the use of correct terminology in designating the material are indispensable for determining the marginal conditions of its application as well as in retrospect, for analyzing any defects or failures of structures caused by volume changes of the metallurgical by-products.

The terms and definitions applicable to the various metallurgical by-products are in Czech republic listed in Technical Specifications (TP) 138 as well as technical requirements of their utilization in road construction [9].

Air cooled blast furnace slag was historically the first used in earthworks and as aggregate. Utilisation of steel slag was occurred at the end of 20th century. Non-ferrous metal slags were only tested and their utilisation in mass scale is more limited. Finally, so-called steelwork waste was after several trial tests used as fill in embankments of the D47 motorway in Czech republic. The knowledge

IOP Conf. Series: Materials Science and Engineering **236** (2017) 012090 doi:10.1088/1757-899X/236/1/012090

about this material was very limited especially its long term behaviour. However, it was certified as a product and therefore it was used without its detailed study [3].

Main hazards of metallurgical by-products are closely connected to their chemical and mineralogical composition and they can be resulted in volume changes.

Pavement deformation of the D47 motorway does not constitute Czech Republic's first example of damage to an engineering structure due to volume changes of metallurgical by-products, but since this problem became widely known it drew greater attention to the problems of volume stability of the metallurgical by-products.

However, fears from possible deformations similar to the D47 motorway meant that metallurgical by-products were excluded from several public tenders of road construction.

Figure 1. Pit KSJ22 (km 155.876 of the D47 motorway, right lane) with visible deformed layers of the active zone (capping layer) resulting by the volume changes in underlying steelwork waste in embankment.

2. Marginal conditions of metallurgical by-products utilisation

The boundary conditions that have to be met in order to allow for the application of metallurgical by-products within the construction limits of highways are in Czech Republic stated in Technical Specification (TP) 138; they can be defined as follows.

The maximum admissible sulfur content is 1.5% where there is contact with concrete structures. This is a general requirement that follows from standards for concrete structures. In most cases this parameter is met by the metallurgical by-products.

The free lime content as per Technical Specification (TP) 138 must be lower than 4.5%

Volume stability

- Blast furnace slag - this is established by the disintegration test conducted by steaming in an autoclave (as per Technical Specification TP 138). The test is conducted on the 8-16 mm fraction, at the temperature of 105°C and the pressure of 0.20 MPa, for the duration of 2 hours. The disintegration index must be lower than 5%.

- Steel slag - this is established by the expansibility test (as per ČSN EN 1744-1). The fraction subjected to the test is 0-22 mm, and the test is carried out at the steam temperature of 100°C. Test duration is 24 hours for LD slags and 168 hours for BOF and EAF slags. Expansibility must be lower than 5%.

Materials which are defined more unambiguously and materials which are more homogeneous (blast furnace and steel slags) can be applied within the construction limits of highways, provided that the aforementioned prerequisites are met. General recommendation is to use the coarser-grained fractions (above 32 mm), inasmuch as the major portion of the volume changes can be attributed to the fine-grained fractions (usually, 0-4 mm).

Steelworks waste as raw material represents a distinctly regional by-product. Its use was widespread in the Ostrava region only.

The determining properties of steelworks waste include:

- heterogeneity;
- volume instability;
- variable mineralogical and chemical compositions;;
- difficulty of predicting its long-term behavior.

The properties listed above are interrelated and interacting. Volume instability can be attributed to the variable mineralogical and chemical compositions. The components of the material do not lend themselves to macroscopic separation. Particularly the fine-grained fractions that determine volume stability are constituted of mixtures of steel slag, refractories, and other materials. It can be stated that in steelworks waste material, each and every grain is different [5].

The volume changes associated primarily with the hydration and carbonation processes are long-term changes, as has been demonstrated by the results of experimental laboratory tests conducted at higher temperatures on saturated samples.

Also, compositional heterogeneity of steelworks waste makes it difficult to establish the long-term behavior of this material. At the present time we cannot ascertain which maximum values can be reached by deformation and what are the swelling pressures acting on the material while the volume changes are in progress. Further research and laboratory testing will be necessary, and this will be mostly of an experimental character.

Steelworks waste cannot be recommended without reservation for use within the construction limits of highways. In view of the negative experience acquired, it is improbable that this material will ever be used again. Similar conclusions were also arrived at by experts from Germany, France, the U.S. and other countries where the use of secondary by-products of metallurgical, except for blast furnace and steel slags, is prohibited [10], [11].

3. Volume stability of metallurgical by-products

Volume stability of metallurgical by-products has not been systematically studied in Czech republic so far. Tests and criteria used in Germany were adopted for blast furnace slag aggregates. Similar situation was also in case of steel slag. However, a question of long-term development of properties has not been analysed. Also the comparison of short term tests (autoclave tests, steam chamber tests) and long term tests (swelling in the CBR mould) have not been assessed. Last but as not least topic is chemical composition of metallurgical by-products and its influence of their mechanical properties.

Juckes discussed the various types of tests employed to determine the volume stability of slag based aggregates and compared them with the on-site behaviour of steel slag which often was diametrically different from the laboratory results [2]. The expansibility tests (as per EN 1744-1) or the hot water bath tests give results that better approximate the behaviour of the steel slag at actual construction sites. On the other hand, the test which gave the best result in situations where steel slag was used for unbound layers was the long-term CBR swell test.

The measurement of volume changes in laboratory conditions is limited. Short time tests under higher temperature and pressure (e.g. autoclave test) provide results very quickly; however, it is complicated to correlate them with behaviour of material under normal conditions. Methodological limits represent difficulties, too. Autoclave test is carried out on the fraction 8/16 and according to the present specifications it is valid for blast furnace slag aggregate only (see appendix A of TP138) [9]. Steam test according to the EN 1744-1 [7] is used only in case of steel slag aggregate and test is carried out for fraction 0/22.5. General short term test for any material has not been installed, yet.

Long term tests than remains to analyse volume changes. They are applicable for any material (both natural and artificial origin). It is the swell test in the CBR mould according to the EN 13286-47 [8]. A disadvantage of this test is fact that volume changes are slow and results are obtained during several months, even years. Comparison of test results of steelwork waste swelling under standard and non-standard conditions is shown in Figures 8 and 9.

The volume changes of water saturated samples tested at the temperature of 75°C amounted to 27.4% after 122 days for the samples compacted by 100% Proctor Standard energy, and 43.1% after 188 days for the sample compacted by Proctor Modified energy (Figure 2). The volume changes of steelwork waste sample under normal temperature and pressure after 5 years achieve 8.5% and still the trend to stabilization has not been observed (Figure 3).

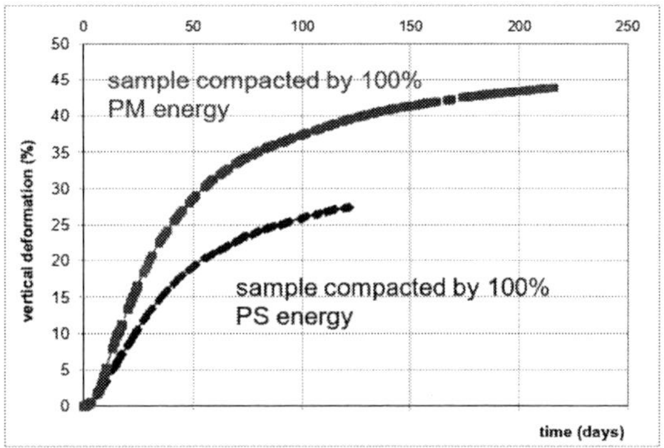

Figure 2. Progression with time of the increment of vertical deformation of steelworks waste samples soaked with water at the temperature of 75°C.

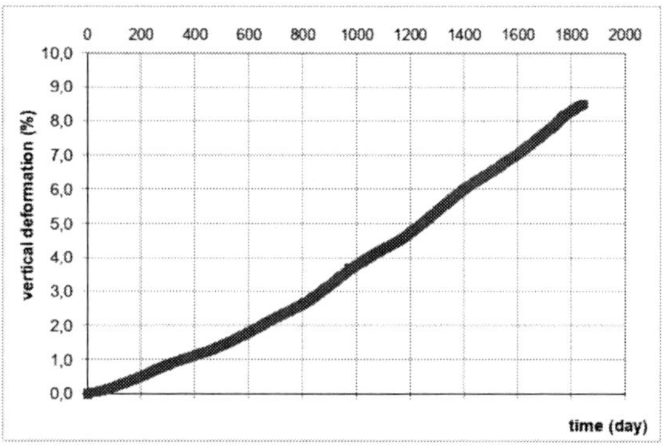

Figure 3. Progression with time of the increment of vertical deformation of steelworks waste samples soaked with water under normal pressure and temperature (period of 30.1.2012-15.2.2017).

Swelling pressure values of steelwork waste were measured in case of experimental samples only. The swelling pressure value of 1.548 MPa obtained for the sample from the D47 motorway after 48 days at 70°C was higher than the values obtained from the autoclave test conducted at the pressure of 357 kPa and the temperature of 137°C for 1 hour (max. 1.28 MPa) presented by Wang [6] for samples of BOF steel slag.

It has to be pointed out, however that we still lack the values of correlation between swelling pressure values obtained in environments that accelerate the changes of volume (at higher temperatures) and those that would be obtained in tests performed at standard temperature and pressure.

Another problem is to estimate in which stage of volume changes a steelwork waste material is. The velocity of mineralogical and phase changes is well described in case of pure minerals. In case of such heterogeneous mixture which is represented by the steelwork waste, it is not possible definitely to set any prognosis.

Mineralogical analyses before and after autoclave tests were carried out in case of the Ikea shopping centre samples to find out changes in mineralogical association. It was observed that after autoclave tests a share of calcite increase comparing with share of portlandite. However, it was not possible to quantify mineralogical changes after autoclave tests to predict behaviour of this material in the future [4].

Figure 4. View of a sample compacted by 100% Standard Proctor energy at the temperature of 75°C after 65 days water at the temperature of 75°C.

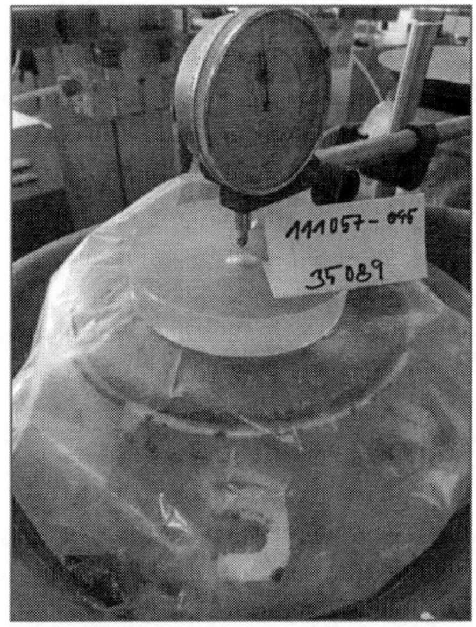

Figure 5. View of sample of steelworks waste samples soaked with water under normal pressure and temperature (period of 30.1.2012-15.2.2017).

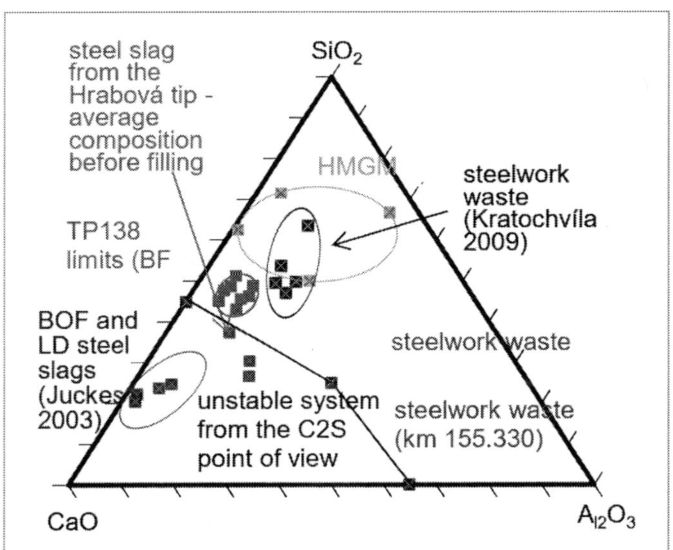

Figure 6. Composition metallurgical by-products, including the steelwork waste, HMGM [1] (*Hüttenmineralstoffgemischen*) in the phase diagram CaO-SiO_2-Al_2O_3. The unstable area from the point of view of C2S disintegration.

Unfortunately, chemical composition of metallurgical by-products is not able to answer if these products are stable or not. Recommended ranges of content of SiO_2, Al_2O_3, CaO and MgO are defined in the TP138 for blast furnace slag, only [9]. However, some experts used these recommended values for all metallurgical by-products and deduced from it that materials outside these boundaries are non-stable. This statement is not right. E.g. steel slag usually contains higher amount of CaO and it is mostly stable. Statements concerning volume stability derived only from chemical composition are very confusing and they may lead to wrong conclusions.

4. Conclusions

From the analyses performed so far it follows that the cardinal error committed in the case of the D47 motorway was the use of a material (steelworks waste) in respect of which no experience was available from other applications within the construction limits of a scope comparable to those of highways.

Results of tests show that although deformations lasted for many years their stabilization has not been definitely proven. Therefore we are not able to predict how many years deformations will be lasted. Values of swelling pressure are so high which show results of tests under non-standard conditions (temperature of 75°C), when there were measured values so far unpublished, but also consequences of volume changes in constructions where steelwork waste was used.

This very expensive experience, because the remedial works of steelwork waste usage will be very expensive, maybe increase the cautiousness in case of utilization of unknown and untested materials. On the other side probably the opposite extreme will not occur when good secondary material (blast furnace or steel slag) will be excluded from constructions only therefore that they have its origin in metallurgical production.

It is necessary to stress that well described and properly tested metallurgical by products are valuable alternative materials used in earthworks.

There are many questions, however, we do not know if they will be anytime answered because research financing to understand properties of this material is not actual in comparison of legal disputes.

References

[1] Jaschke K et al 1998 *Merkblatt über die Verwendung von Hüttenmineralstoffgemischen, sekundärmetallurgischen Schlacken sowie Edelstahlschlacken im Straßenbau, Forschungsgesellschaft für Straßen und Verkehrswessen* (FGSV Verlag GmbH Köln)
[2] Juckes L M 2003 The volume stability of modern steelmaking slags *Mineral processing and Extractive metallurgy* **112**(3) pp. 177-97
[3] Kresta F 2012 *Secondary materials in highway construction* (VŠB TU Ostrava)
[4] Kresta F 2014 Metallurgical by-products in earthworks, hazards of their utilisation *Advanced Materials Research* **1020** pp 98-109
[5] Kresta F 2015 Steelwork waste – non-standard metallurgical by-product *International Journal of Advance Research in Science and Engineering* **14**(02) pp 5-16
[6] Wang G 2010 Determination of the expansion force of coarse steel slag aggregate *Construction and Building Materials* **24**(10) pp 1961-66
[7] European Committee for Standardization (CEN) 2010 Tests for chemical properties of aggregates - Part 1: Chemical analysis (*EN 1744-1*)
[8] European Committee for Standardization (CEN) 2010 Unbound and hydraulically bound mixtures - Part 47: Test method for the determination of California bearing ratio, immediate bearing index and linear swelling (*EN 13286-47*)
[9] Ministry of Transport of Czech Republic 2011 Technické podmínky: Užití struskového kameniva do pozemních komunikací /Technical specification: Utilisation of slag aggregate in road construction (*TP138*) (in Czech)
[10] US Department of Transport Federal Highway Administration 1998 *User Guidelines for Waste and Byproduct Materials in Pavement Construction* (*FHWA RD-97-148*)
[11] FGSV Verlag GmbH Köln 2013 Technische Lieferbedingungen für Böden und Baustoffe im Erdbau des Straßenbaus, Zusätzliche Technische Vertragsbedingungen und Richtlinien für Erdarbeiten im Straßenbau (*TL BuB E-StB 07*)

Potential use of fly ash to soil treatment in the Morava region

Lucia Bulíková[1], František Kresta[1] and Martin Rochovanský[1]

[1] Faculty of Civil Engineering, VŠB – Technical University of Ostrava, Ludvíka Podéště 1875/17 708 33 Ostrava-Poruba, Czech Republic

E-mail: lucia.bulikova@vsb.cz

Abstract. Soil treatment by binders is a standard technology and leads to optimal utilization of excavated soils in road constructions. Soil treatment is controlled in the Czech Republic by EN 14227-15 and Technical Requirement TP 94. Soil treatment using fly ash has not been performed in the Czech Republic, although there is a sufficient normative base. Fly ash produced by burning of hard coal in the Moravian region was tested as a potential binder. Fly ash samples were mixed with loess loams (CI). Tested siliceous fly ash of class F (ASTM C618) did not showed hydraulic properties but it showed positive effect on reducing maximum dry density of mixtures, increasing the IBI value (Immediate bearing index) and decreasing tendency to volume changes when the amount of fly ash was increased. The results of laboratory tests demonstrate the possibility of using fly ashes as a binder for soil treatment.

1. Introduction and background

Treatment and stabilization of soils change the properties of soils to improve their engineering utilisation. Bulk density, moisture content, plasticity, susceptibility to volume changes, shear strength and bearing capacity are the most improving properties of treated soil [1]. Hydraulic road binders, which are used in soil treatment, contain mainly cement and/or lime. Recommended alternative to traditional binders becomes fly ash [2, 3].

Fly ash is an inorganic product of combustion of pulverized coal in power and heating stations. Fly ash consists of small particles of siliceous glass and it is collected by electrostatic precipitators flue gases. Basic mechanical properties of fly ash, that are required for use in engineering, are maximum dry density, granularity, pozzolanicity and absorbability. Mineralogical composition and unburned particles content belong to its important chemical properties [3, 10].

Research on the potential use of fly ash and ash from thermal power stations began in Europe during World War II in the UK. Fly ash was first used as an admixture in concrete to reduce the consumption of Portland cement. Later, the fly ash produced by combustion of coal has been marked as PFA (Pulverized Fly Ash), thus distinguishing it from derivatives produced by combustion of other of types material. PFA has been introduced in the construction industry as a material for construction of embankment of roads [4]. By the end of the 20th century, fly ash was used in road construction only for the construction of embankments and for levelling adjacent surface near its source. Then, the use of fly ash in civil engineering projects increased substantially in the Czech Republic, for example, in 2009, 100 000 tonnes of fly ash were processed per year during the construction of the D11 motorway [2,5].

Fly ash is used in geotechnical applications closely connected with road construction to increase strength properties of soils, to control contraction and swelling of soils, to reduce water content in soils and to modify a particle size distribution [1].

IOP Conf. Series: Materials Science and Engineering **236** (2017) 012091 doi:10.1088/1757-899X/236/1/012091

2. Standards and requirements

Soil treatment by fly ash in the Czech Republic is governed by the European standard EN 14227-15 Hydraulically bound mixtures - Specifications - Part 15: Hydraulically stabilized soils, which was adopted in 08/2016, Technical Requirement TP 94 - Soil treatment, valid since 11/2013 and Technical requirements TP 93 - Design and construction of highway structures using fly ash and ash, valid since 02/2011. Regulations contain technical requirements for materials, material testing, requirements for the construction of road layers, effects of soil treatment and environmental aspects.

3. Materials and methods

Fly ashes from the Morava region were tested as potential binder to soil treatment. First one (FAO), produced by Arcelor Mittal Ostrava power station was collected from the tip near its source. Second one (FAP), was produced in the Heating station of Přerov. This fly ash was deposited on tip in Předmostí until 1996 and then the tip was reclaimed. Fly ash samples were collected from depth of two meters below surface. Both fly ash samples were produced by burning of coal from the Ostrava-Karvina coalfield. Tested soils were sampled from the sites of planned construction of roads in the Morava region, soil SO from the construction of "the Road I/11 Ostrava – the elongation of the Rudná street" and soil SP from the construction of "the D1 Motorway Přerov – Lipník nad Bečvou".

Figure 1. Sampling of materials.
a) FAO Ostrava Bartovice
b) SO Ostrava the I/11 road
c) FAP, SP Předmostí, the D1 motorway

Laboratory tests were carried out both on untreated and treated soils. Soils were mixed with 10% and 15% of fly ash. Several sets of chemical and mineralogical analyses were performed on samples, including leaching of the FAP sample. Leaching of heavy metals is one of the important environmental aspects. The limit values are set in the TP 93. Fluorescence Spectrometry and Scanning Electron Microscopy were carried out in the ICT VŠB TU Ostrava - Institute of Clean Technologies.

The Proctor Standard test sets a maximum dry density and optimal moisture of untreated soils, treated soils and fly ashes. Then values of immediate bearing index (IBI), and California bearing ratio (CBR) after 96 hours of soaking and three days of curing were determined. Minimal value of IBI for utilisation of treated soils as fill is 10% (see TP 94). Volume changes (swelling) represent a crucial parameter of fly ash - they limit its utilisation as fill in road construction. The maximal allowable linear swelling is 3%. This value is measured in CBR mould after 3 days of curing in wet conditions, according to the TP 93. All tests were carried out under laboratory conditions.

4. Results and discussion

Chemical analysis was performed to classify fly ash and determine its reactivity. The analysis showed that SiO_2, Al_2O_3 and Fe_2O_3 represented more than 70% content. The CaO content was 5% maximum (Table 1). Fly ash samples were classified as class F - siliceous (ASTM C618). Siliceous fly ashes are characterized by low volume changes and higher pozzolanic activity.

Table 1. Chemical composition of fly ash samples.

Element	FAO (%)	FAP (%)
SiO_2	42.70	51.28
Al_2O_3	24.53	30.78
Fe_2O_3	4.57	6.29
CaO	4.89	1.61
MgO	0.66	0.80
SO_3	0.95	0.08
K_2O	2.39	2.98

Electron microscope snaps of fly ash FAO from the AMO power station and sample of the SO soil are shown in Figures 2 and 3. Spherical particles of fly ash microspheres and mica particles were observed. Microspheres are formed during combustion in the temperature range of 1200-1500°C and they influence fly ash properties (density, porosity etc.). Soils were represented by loess loams (clays of low to medium plasticity CL-CI).

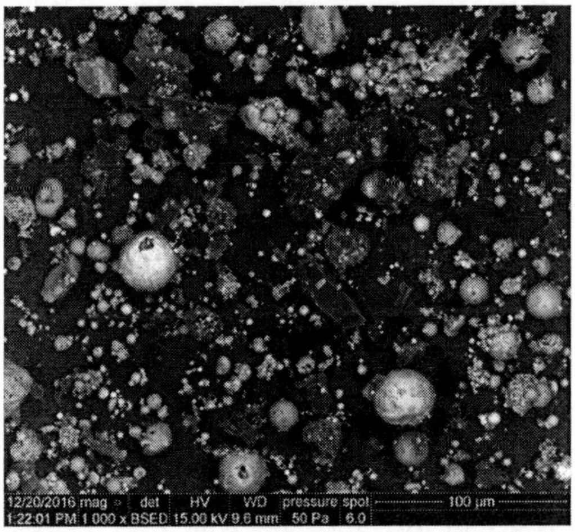

Figure 2. Siliceous fly ash (the FAO samples).

Figure 3. Loess loam (the SO samples).

Results of leachate in case of the FAP sample showed, that this fly ash is suitable in roads construction. Any element does not exceed the limit set by the Czech specification TP 93 and all values are several times lower than the limit ones (Table 2). The highest content of the leachate occurred in case of As, Co and V.

Table 2. Results of leaching test of the FAP sample.

Element	Max (mg/l)	Meas. (mg/l)
Ag	0.1	< 0.005
As	0.1	0.0140
Ba	1	0.0230
Be	0.005	< 0.005
Pb	0.1	< 0.01
Cd	0.005	< 0.003
Cr celk	0.1	0.0070
Co	0.1	< 0.01
Cu	1	< 0.01
Ni	0.1	< 0.006
Hg	0.005	< 0.0003
Se	0.05	< 0.012
V	0.2	0.0170
Zn	3	< 0.01
Sn	1	< 0.02

The Proctor Standard test set an optimal moisture w_{opt} of soils in range of 15 - 17%. Maximum dry density of mixture of soil and FAO was ρ_{dmax}= 1780 kg.m^{-3} (mixture with 10% FAO). In case of the mixture of soil and 10% of FAP a maximum dry density was ρ_{dmax} = 1670 kg.m^{-3}. Measured values of all samples are showed in Table 3. Addition of fly ash caused reduction of the maximum dry density of the mixtures about 100 kg.m^{-3} under the same optimal moisture.

Table 3. Results of the Proctor Standard tests.

Sample	w opt (%)	ρ max (kg/m³)
FAO	36	1080
SO	16	1850
SO+10% FAO	16	1780
SO+15% FAO	15	1740
FAP	45	890
SP	16	1740
SP+10% FAP	17	1670
SP+15% FAP	17	1630

All tested mixtures exceeded minimal value of IBI (min = 10%) for utilisation of treated soils as fill (Figure 4.). The IBI value immediately after compaction was 11% and 16% for untreated soil SP respectively SO. The greatest change was observed in case of both mixtures with 15% of fly ash. The IBI values increased to 15% and 21%. Fly ash alone did not cause the hydraulic behaviour of mixtures. CBR values after soaking of samples with FAP were not considerably different. Measured CBR values were 11% - 15%. CBR values increased by approximately three times, from 4% to 11%, in case of samples with FAO (Figure 5). Hydraulic binder is recommended to add to mixtures to increase the IBI and CBR values. Hydraulic binder will be an activator and subsequently will initiate pozzolanic properties of fly ashes. It will cause the ion exchange reaction between soil particles and fly ash, and it will be resulted in the increase of strength properties of treated soils. All samples fulfil requirements of Czech specification regarding value of volume changes. The curves indicate decreasing tendency to swell when the amount of fly ash is increased (Figure 6.). Linear swelling of mixture with FAO ranged between 0.8 and 1.6%. Critical values occurred in mixtures with FAP, values were 1.8% - 2.8%.

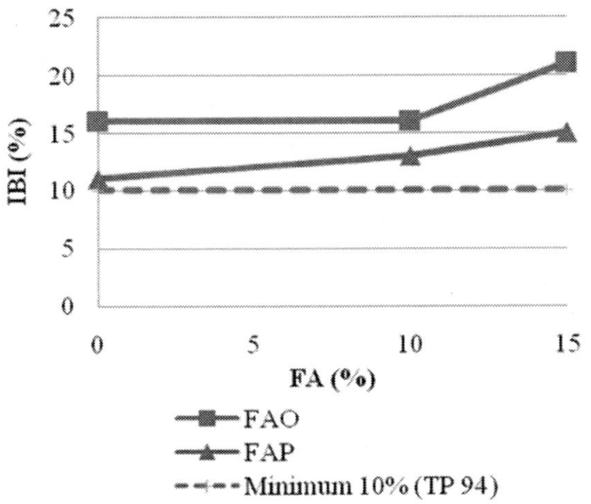

Figure 4. Dependence of IBI values of the mixtures of soil and fly ash.

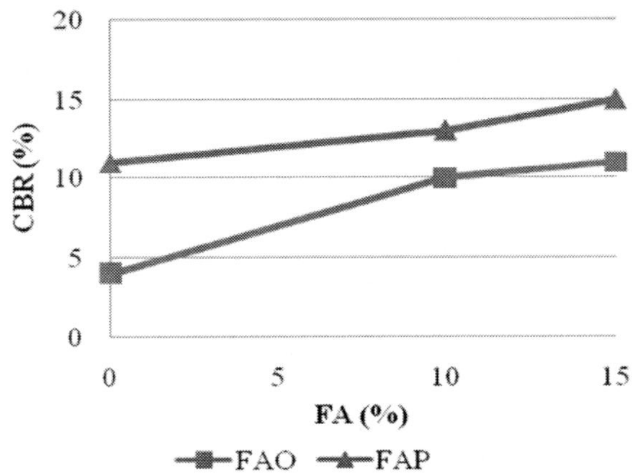

Figure 5. Dependence of CBR values after soaking of the mixtures of soil and fly ash.

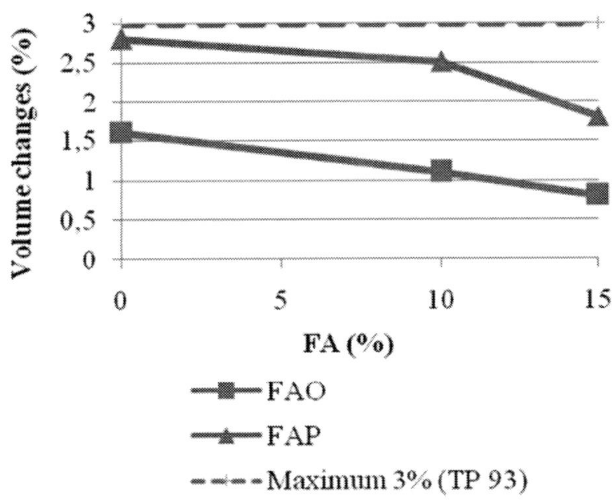

Figure 6. Changes of volume for soil treated with siliceous fly ash.

5. Conclusions and recommendation to future work

Laboratory tests show the potential of utilisation of siliceous fly ash from the Morava region in road construction. Based on the results from executed tests fly ash can be used in mechanical treatment of soils (tests were carried out with loess loams) without admixture of any hydraulic binder. In the further research it is recommended to verify parameters of mixtures with different lime content. The next step would be to check the mixture properties in large-scale tests on construction site using the appropriate machinery. Fly ash from the Předmostí tip will be used as fill in embankments of the D1 motorway. In 2017 is planned starting of earthworks. However, fly ash is designed to be used as fill only. Its utilisation as a binder is not expected.

Acknowledgement
Financial support through the Student Grant Competition VŚB-TUO - Project No. SP2017/155 is gratefully acknowledged.

References
[1] American coal ash association 2003 *Fly ash facts for highway engineers* (Washington DC: Federal Highway Administration) pp 47-52
[2] Kresta F 2013 *Secondary materials in highway construction* (Ostrava: Editorial Centre VŠB-TU Ostrava) pp 54-66
[3] Mezencevová A 2003 Possibilities of utilizing power plant fly ashes *Acta Montanistica Slovaca* **8** pp 146-51
[4] Sear L K A 2001 *Properties and use of coal fly ash* (London: Thomas Telford Ltd)
[5] Kresta F 2013 Fly ash utilisation in road construction in Czech Republic *Proc. 5th symp. on ash, slag and waste landfills in power plants and mines (Brno)* (Subotica) pp 118-25
[6] Kresta F 2015 Fly ash from the Předmostí a Prosenice tips - potential source of fill for the D1 Motorway section Přerov – Lipník nad Bečvou 2015 *Proc. Conf. Fly ash in civil engineering 2015 (Brno)* pp 141-51
[7] European Committee for Standardization (CEN) 2011 Unbound and hydraulically bound mixtures, Part 2 *(EN 13286-2)*
[8] European Committee for Standardization (CEN) 2012 Unbound and hydraulically bound mixtures, Part 47 *(EN 13286-47)*
[9] European Committee for Standardization (CEN) 2016 Hydraulically bound mixtures, Part 15 *(EN 14227-15)*
[10] Ministry of Transportation of the Czech Republic 2011 Technické podmínky - Návrh a provádění staveb pozemních komunikací s využitím popílků a popelů /Design and construction of highway structures using fly and ash/ *(TP 93)*
[11] Ministry of Transportation of the Czech Republic 2013 Technické podmínky - Zlepšení zemin /Soil treatment/ *(TP 93)*

IOP Conf. Series: Materials Science and Engineering **236** (2017) 012092 doi:10.1088/1757-899X/236/1/012092

Utilization of the waste from the marble industry for application in transport infrastructure: mechanical properties of cement pastes

Zdeněk Prošek[1,2], Jan Trejbal[1], Jaroslav Topič[1], Tomáš Plachý[1] and Pavel Tesárek[1]

[1]Czech Technical University in Prague, Faculty of Civil Engineering, Thákurova 7, 166 29 Praha 6, Czech Republic
[2]Czech Technical University in Prague, UCEEB, Trinecka 1024, 273 43 Bustehrad, Czech Republic

E-mail: zdenek.prosek@fsv.cvut.cz

Abstract. This article is focused on the mechanical testing of cement-based samples containing a micronized waste marble powder used as replacement of standard binders. Tested materials consisted of cement CEM I 42.5 R (Radotín, Czech Republic) and three different amounts of the marbles (25, 50 and 70 wt. %). Standard bending and compressive tests of the prismatic samples having dimensions equal to 40 × 40 × 160 mm were done in order to reveal an influence of marble amount on flexural and compressive strength, respectively. Moreover, the dynamic modulus of elasticity and dynamic shear modulus were examined and compared after 7 and 28 days of mixture curing.

1. Introduction

Recycled materials are commonly used as an alternative to non-renewable resources. Unfortunately, these materials have significant problems with variability. As a result of this drawback, the usage of these materials is limited only to the fields of transport infrastructures [1]. The utilization of waste or recycled materials have been spelled out throughout the different researches, as showed e.g. in [2, 3].

Unlike the mentioned research works, we focused directly on the usage of waste marble sludge. The sludge is generated during the block cutting or the finishing process of stone materials. Generally, the production of waste may cause a wide range of environmental problems, for example as follows: air pollution, water pollution, visual impacts and loss of flora and fauna [4]. The amount of this waste was estimated for the area of The Shaq El Thoaban (Egypt) to one million tons per year [5] and for area of Estremoz Anticline (Alentejo, Portugal) to 600 thousand tons per year [6]. Both mentioned areas have approximately the same production of marble stone. On the other hand, the technical, operational and management practices in the industry are less developed in area Shaq El Thoaban. Moreover, they have contributed significantly to greater waste generation. Small amount (from 10 to 20 %) of waste generated by marble industry is used in construction. Remaining part is refused and accumulated in mounds of marble waste close to the extraction quarries [7].

The marble sludge is the most interesting part of the waste because of its almost nil usage. There are several works dealing with the use of marble sludge in cement composites. It was found that adding of small amount of marble sludge into the cement composites may improve their mechanical properties [8]. As pointed out in [9], compressive strength of cement samples (w/c ratio 0.50) containing 5.0, 7.5, 10 and 15 % of marble dust as cement replacement was increased by 7 %, 4 %, 5 % and 14 %, respectively.

Grains of marble sludge clump together into lumps and thus grains of marble sludge are not homogeneously distributed through the mixture. The use of high-speed mill from Lavaris Ltd. May be an option to limit or eliminate this effect. The resulting product is micronized marble powder, which has better mechanical properties than marble sludge [10]. High-speed milling process separates all the grains of marble sludge. The high-speed milling process has been successfully used for micronization of recycled concrete [11].

2. Materials and samples

The tested mixtures consisted of Portland cement CEM I 42.5R (Radotín, Czech Republic) and micronized marble powder with fractions 0 – 40 μm. The micronized marble powder had an average grain and mean grain size of 9.74 and 4.59 microns, respectively. 90 wt. % of grains were smaller than 29.67 microns. The micronized marble powder was obtained from the processing of limestone marble in the Czech Republic as unsorted waste (Beroun, Jež Ltd.). Micronized marble powder was micronized using a high-speed mill from Lavaris Ltd. (Libčice, Czech Republic). The mixtures were designed to maintain the same workability as the fresh reference mixture. The consistency of the mixtures was determined by means of a flow expansion test after 10 (d_{10}) and 20 (d_{20}) impulses (lift up several centimeters and then dropped), (Table 1). The water ratio is defined as water to cement and marble powder weight ratio. Cement was replaced by the waste micronized marble powder by the amount of 25, 50 and 75 wt. percent (Table 1).

Mixtures prepared according to Table 1 were cast in the rectangular molds having inside dimensions equal to 40 × 40 × 160 mm. After casting, these specimens were kept in the molds for 24 h at room temperature of 22 °C. Each set consisted of 6 samples made of the same mixture. After demolding, these specimens were stored in a laboratory environment at 22 ± 1 °C and relative humidity 50 ± 2 % for 28 days.

Table 1. Composition of the mixtures.

Set/ Material	Cement (g)	Marble powder (g)	Water ratio (-)	Flow expansion test d_{10} (mm)	Flow expansion test d_{20} (mm)	Bulk density (kgm^{-3})
C100	3000	0	0.35	160	190	1919 ± 9
C75	2250	750	0.31	157	190	1902 ± 4
C50	1500	1500	0.31	155	190	1760 ± 15
C25	750	2250	0.32	162	195	1668 ± 10

3. Experimental methods

In the initial stage of research, the marble powder was examined from the perspective of material composition by means of microanalysis. The tests were performed using a ZEISS Merlin electron microscope with EDS at the University Centre for Energy Efficient Buildings CTU in Prague. The output of the EDS was spectrum frequency of X-ray signal in the individual energy windows characterized by peaks (Figure 1). Individual peaks corresponded to the individual elements, whose height was proportional to the concentration of the sample element.

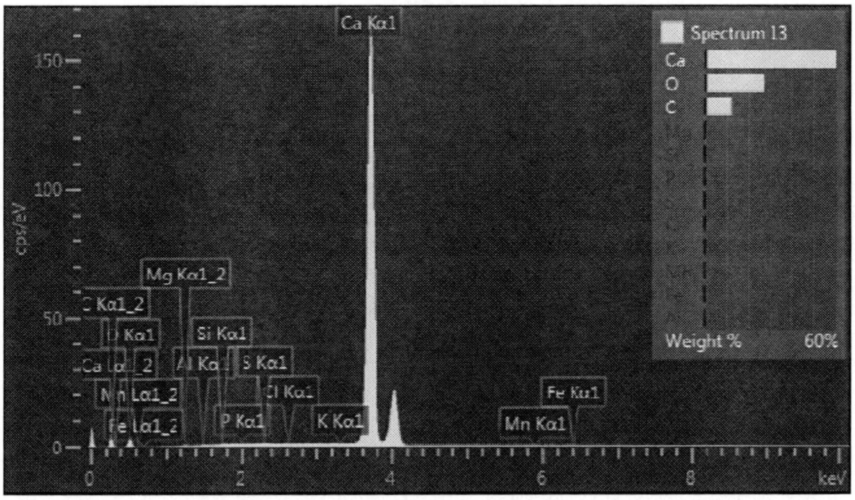

Figure 1. EDS spectrum of the sample.

The values of unit weight, dimension and dynamic modulus of elasticity of the specimens were determined before the destruction test (the compressive and the flexural strength test) at the 7th and the 28th day. The non-destructive resonance method was used for the determination of the dynamic shear modulus and the dynamic modulus of elasticity. Six specimens having dimensions equal to 40 × 40 × 160 mm were tested. Measurements of the dynamic modulus were carried out using the measurement set from Brüel & Kjær. By using measuring assembly and software PULSE LabShop, the basic natural frequencies were evaluated. As the next step, the standard equations for the prismatic sample were used in order to find out the dynamic modulus of elasticity and dynamic shear modulus.

Dynamic modulus of elasticity was determined by measuring the basic longitudinal natural frequency of the samples. The acceleration transducer was located in the center of the vertical side perpendicular to the length of the sample and the impact hammer strikes were carried out in the middle of the opposite side. The sample was supported by a soft and elastic pad at the midpoint of its length, which is the fundamental longitudinal nodal point.

The measurement of the dynamic modulus of elasticity was conducted by measuring of basic flexural natural frequency of the samples. In this case, the acceleration transducer was placed on the edge of the top side of the sample and the impact hammer strikes were carried out on the edge of the same side of the sample. The sample was supported by two soft and elastic pads at the fundamental bending nodal lines, which were located at the distance of 0.224 × length of the sample from each end of the sample.

Dynamic shear modulus was examined by measuring of basic torsional natural frequency of the sample. The acceleration transducer was located in the corner of the vertical side parallel to the length of the sample and the impact hammer strikes were carried out diagonally on the same side. The sample was supported in the same manner as in the case of measuring of longitudinal natural frequency of the samples.

The values of the compressive and the flexural strength of the specimens were determined at the 28th day. Six specimens for each series were used for the determination of the flexural strength on beams of dimensions equal to 40 × 40 × 160 mm and the uniaxial compressive test was performed on the broken halves of the specimens with effective dimensions of 40 × 40 × 40 mm. The load was applied at a constant rate of 0.1 mm/s in the case of the three-point bending test and 0.3 mm/s for the compressive test by using the Heckert loading frame, model FP100. The distance between supports for the three-point bending test was equal to 100 mm.

4. Results and discussion

The microscopic analysis determined the elemental composition of the waste material (micronized marble powder). The results were slightly distorted due to the carbon presence sputtered onto the sample surfaces (necessary technological step needed for the analysis). The results showed that the samples were composed of 98 wt. % $CaCO_3$. Remaining 2 wt. % consisted of minor components, of which the most important was $MgCO_3$. According to the composition, waste micronized marble can be used as an additive to blended cement.

The dimensions of all rectangular molds were measured before casting the mixture. These values were used to calculate approximate shrinkage. Samples C25 and C50 exhibited approximately the same shrinkage, concretely 0.40 ± 0.02 % and 0.37 ± 0.05 %, respectively. On the other hand, C75 and C100 reached shrinkage equal only to 0.21 ± 0.06 % and 0.19 ± 0.06 %, respectively. Increased shrinkage of samples was caused by the addition of fine particles of micronized marble powder characterized by the high specific surface area. This negative effect of powder can be eliminated by using plasticizers.

The results from non-destructive testing (Figure 2 and Figure 3) showed differences between values of modulus at the 7th day and 28th day. Differences between the values were in the size of the standard deviations. It can be stated that between 7th and 28th days, a minimal change of the dynamic modulus of elasticity and dynamic shear modulus occurred. The samples C75 had the largest value of the dynamic shear modulus, namely 7.82 ± 0.13 GPa (Figure 2) and dynamic modulus of elasticity, namely 19.19 ± 0.23 GPa (Figure 3).

The results from the destructive tests are shown on Figure 4 and Figure 5. Samples C50 reached the largest value of flexural strength, namely 3.91 ± 0.41 MPa (Figure 4). The other samples had approximately the same value of flexural strength. Differences between the values did not exceed the size of the standard deviations. Similar trends as the dynamic modulus of elasticity were observed in the compressive strength (Figure 5). The samples C75 had approximately the same value of compressive strength as reference samples (C100), namely 74.52 ± 9.14 MPa.

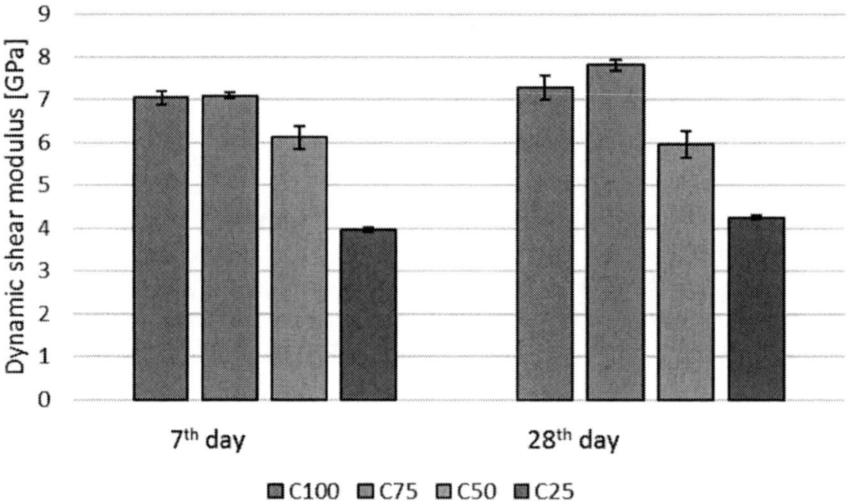

Figure 2. Comparison of the dynamic shear modulus (with standard deviation).

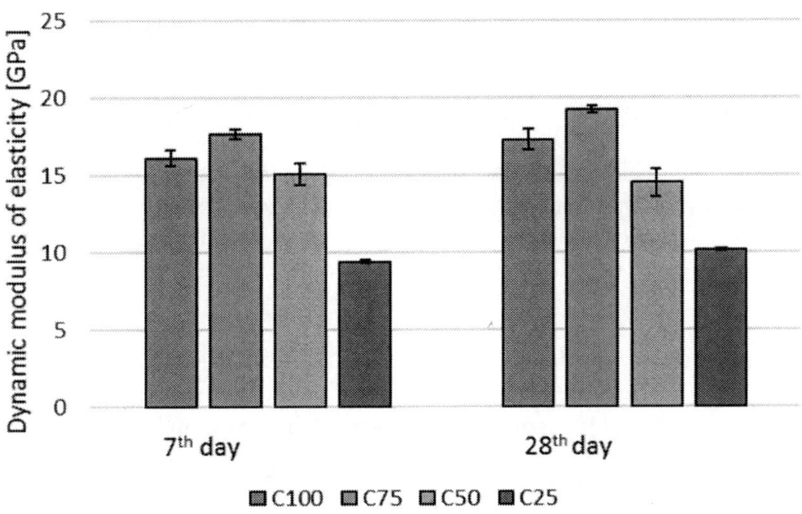

Figure 3. Comparison of dynamic modulus of elasticity (with standard deviation).

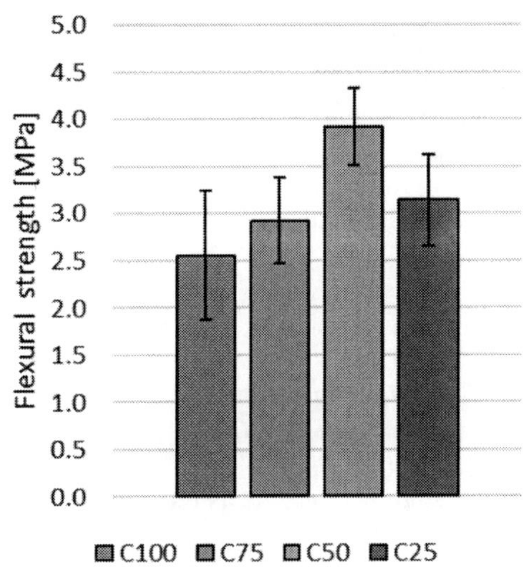

Figure 4. Comparison of flexural strength (with standard deviation).

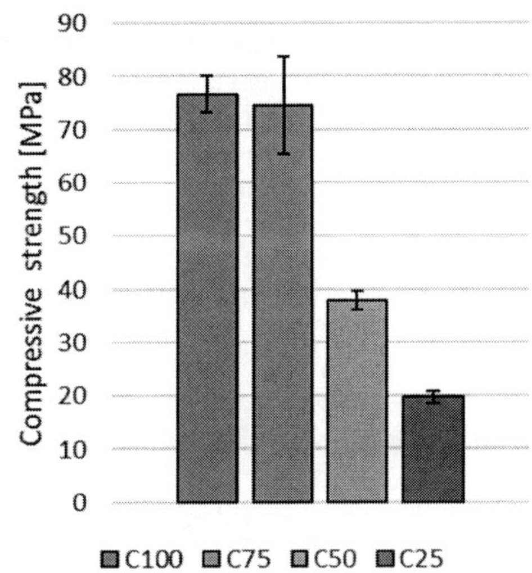

Figure 5. Comparison of compressive strength (with standard deviation).

5. Conclusions

This work was focused on the influence of the micronized waste marble powder on mechanical properties of cement pastes. The researched cement pastes were composed of Portland cement and micronized marble powder in an amount of 0, 25, 50 and 75 weight percent. Based on the results, it can be concluded that:

- The average value of dynamic modulus of elasticity and dynamic shear modulus stayed unchanged between 7 and 28 days.
- For 50 wt. % of the waste micronized marble powder, flexural strength increased by approximately 50 %. For 25 wt. %, compressive strength decreased by approximately 3 %, probably due to its microfiller properties (filling pores by grains of micronized marble

powder). For 75 wt. %, compressive strength decreased by approximately 75 %, probably due to small amount of portland cement.
- The compressive strength of cement pastes made with 25 % of micronized marble powder and cement was found to be either comparable or less than reference samples.

It can be clearly said that the investigated materials have the potential for economic savings because of their almost zero material cost. Therefore the authors consider them to be very promising for the construction of transport infrastructure. In the future, the authors will focus on determining properties such as freeze-thaw resistance and resistance to deicing chemicals.

Acknowledgements

This project was supported by Technology Agency of the Czech Republic (TA ČR) – Competence Centers under No. TE01020168 Centre for Effective and Sustainable Transport Infrastructure (CESTI) and by the Ministry of Education, Youth and Sports within National Sustainability Programme I, project No. LO1605 and by Czech Technical University in Prague under No. SGS project SGS16/201/OHK1/3T/11. The authors also thank the Center for Nanotechnology in Civil Engineering at the Faculty of Civil Engineering, Czech Technical University in Prague.

References

[1] Lidmila M et al 2013 Utilization of Recycled Binder for Improvement of Subsoil under Railway Sleepers *Applied Mechanics and Materials* **486** pp 323-6
[2] Šeps K and Vodička J 2014 Fibre Reinforced Concrete with Recycled Concrete and Insulation Material STERED *Applied Mechanics and Materials* **1054** pp 262-6
[3] Corinaldesi V and Moriconi G 2009 Behvaiour of cementious mortars containing different kinds of recycled aggregate *Construction and Building Materials* **23** pp 289-94
[4] Ibrahim M S, Badr E A, Asker S A and El-galad N M 2011 Health Risk Assessment of Marble Dust at Marble Workshops *Nature and Science* **9** pp. 144-54
[5] Alzboon K K and Mahasneh K N 2009 Effect of Using Stone Cutting Waste on the Compression Strength and Slump Characteristics of Concrete *Proceedings of World Academy of Science: Engineering & Technolog* **51** pp 267-72
[6] Lopes L and Martins R 2012 Marbles from Portugal *Naturstein* online publication p 1-19
[7] Bapna R K 2002 Marble Waste Minimization *Department of Mines & Geology Newsletter* **23** p 4
[8] Binici H et al 2008 Durability of concrete made with granite and marble as recycle aggregates *J. Mater. Process. Technol.* **208** pp 299-308
[9] Aliabdo A A et al 2014 Re-use of waste marble dust in the production of cement and concrete *Constr. Build. Mater.* **50** pp 28-41
[10] Prošek Z, Šeps K, Topič J and Tesárek P 2016 Effect of particle size distribution waste micronized marble powder on the mechanical properties of the cement composite *EAN 2016 - 54th International Conference on Experimental Stress Analysis (Srní)* ed Department of Mechanics, Faculty of Applied Sciences, University of West Bohemia
[11] Prošek Z et al 2016 Influence of recycled concrete composition on its elastic stiffness *Key Engineering Materials* **677** pp 288-91

IOP Conf. Series: Materials Science and Engineering **236** (2017) 012093 doi:10.1088/1757-899X/236/1/012093

Mechanical properties and durability of crumb rubber concrete

Roman Chylík[1], Tomáš Trtík[1], Josef Fládr[1] and Petr Bílý[1]

[1] Faculty of Civil Engineering, Czech Technical University in Prague, Thákurova 7, 166 29 Praha 6, Czech Republic

E-mail: roman.chylik@fsv.cvut.cz

Abstract. This paper is focused on concrete with admixture of rubber powder, generally called crumb rubber concrete (CRC). The inspiration was found in Arizona, where one of the first CRCs has been created. However, Arizona has completely different climates than Central Europe. Could we use the crumb rubber concrete on construction applications in the Central European climate too? The paper evaluates the influence of the rubber powder on material characteristics and durability of CRC. CRCs with various contents of fine and coarse crumb powder were compared. The tested parameters were slump, air content, permeability, resistance of concrete to water with deicing chemicals, compressive and splitting tensile strength. The tests showed that workability, compressive strength and permeability decreased as the amount of rubber increased, but the air content increased as the rubber content increased. Photos of air voids in cement matrix from electron microscope were captured. The results of laboratory tests showed that admixture of rubber powder in concrete could have a positive impact on durability of concrete and concurrently contribute to sustainable development. Considering the lower compressive strength, CRC is recommended for use in applications where the high strength of concrete is not required.

1. Introduction

Increasing amount of waste materials in landfills has become a huge problem in the last decades. Fortunately, considerable amount of waste is recycled in developed countries. This includes for example glass, plastics, paper, electronic devices and worn tires. Constantly growing number of manufactured tires (1 billion in 2016, expected 1.2 billion in 2030) and permanently increasing number of worn tires in landfills created a problem of reuse of the worn tires.

Recycling is carried out in several ways. Worn tires can be incinerated at high temperatures or crushed to rubber powder. The difference between shredding at normal temperature and shredding at temperature below zero (cryogenic shredding) is in quality of rubber powder. Cryogenic shredding leads to higher grinding fineness, but it is more expensive.

The use of shredded tires as an additive to concrete can not only help to solve the problem of further exploitation of worn tires, it can also have positive effect on concrete properties. On this subject, several studies were created. The results of the tests showed that an additive of rubber powder could have a positive influence on deformability [1] and durability [2]. Frankowski [13] studied the influence of crumb rubber on reinforced concrete for structures. The results of his research showed that the addition of rubber powder in reinforced concrete improved the resistance to cracking, shock

IOP Conf. Series: Materials Science and Engineering **236** (2017) 012093 doi:10.1088/1757-899X/236/1/012093

wave absorption and acoustic wave damping and lowered the weight and thermal conductivity of the material.

This paper is focused on workability, air content, compressive strength and splitting tensile strength of concrete with different amounts of added rubber powder and on the resistance of crumb rubber concrete to water with deicing chemicals. The values of concrete with an additive of rubber powder (crumb rubber concrete, abbreviated as CRC) were compared with concrete mixture without an additive (control mix).

2. Experimental program

Concrete mixtures with various amounts and particle sizes of crumb rubber were tested. Rubber powder used in this research was obtained by mechanical shredding at normal temperature. Basic content of rubber powder was 40 kilograms per cubic meter and supplementary content of rubber powder was 80 and 120 kilograms per cubic meter. Concrete mixture contained fine (0-1 mm) and coarse (1-3 mm) rubber particles in several ratios. Rubber powder was not used as a replacement of aggregate, it was added to the basic compounds. Concrete mix compositions are shown in Table 1.

Table 1. Mix compositions.

Compound	Quantity in control mix (kg/m^3)	Quantity in CRC mix (kg/m^3)
Cement CEM I 42,5 R	410.0	410.0
Water	200.0	200.0
Fine aggregate 0-4	840.0	840.0
Coarse aggregate 4-8	340.0	340.0
Coarse aggregate 8-16	620.0	620.0
Stachement 2180	0.82	0.82
Shredded rubber	0.0	40.0 (80.0, 120.0)

All tests were performed with basic content of rubber powder, test of air content, compressive strength and splitting tensile strength were also performed with supplementary content.

Example of designation of concrete mixes: 40 - RUB - 0/100 = 40 kg/m^3 of rubber powder, 0 % of fine rubber / 100 % of coarse rubber.

3. Tests of fresh concrete

3.1. Slump test

The tests were performed according to [10] for content of rubber powder 40 kg/m^3. Ratios of fine and coarse rubber are shown in Table 2.

3.2. Air content

The tests were performed according to [11] for content of rubber powder 40, 80 and 120 kg/m^3. Ratios of fine and coarse rubber are shown in Table 2.

Table 2. Air content and slump test.

Mix	Slump [mm]	Air content [%]
REF	70	2.1
40 – RUB - 0/100	35	2.7
40 – RUB - 20/80	45	3.0
40 – RUB - 40/60	35	3.6
40 – RUB - 60/40	28	4.0
40 – RUB - 80/20	16	5.0
40 – RUB - 100/0	25	7.2
80 – RUB - 0/100	---	5.5
80 – RUB - 100/0	---	8.5
120 – RUB - 0/100	---	6.1
120 – RUB - 100/0	---	7.9

3.3. Results

The tests showed that incorporation of rubber powder in concrete mix caused lower workability and higher air content in fresh concrete. The surface of rubber particles caught molecules of the air which entered the fresh concrete. Fine rubber has higher specific surface, therefore the fine rubber caught more air molecules than the coarse rubber. This is the reason why mix 40-RUB-100/0 had higher air content than other mixes containing 40 kg/m^3 of rubber. The surface of the rubber particles can catch molecules of water too. Higher amount of the fine rubber particles caught more water and therefore reduced the workability.

4. Mechanical properties of hardened concrete

The compressive strength of concrete specimens was determined according to [8] after 7, 28 and 45 days for 40 kg/m^3 (Figure 1) and after 7 days for 80 and 120 kg/m^3 of rubber powder in concrete mixes. Measurements were performed on 150 mm cubes. Measured values are shown in Table 3.

Table 3. Compressive strength.

Mix	7 days [MPa]	28 days [MPa]	45 days [MPa]
REF	38.6	48.6	49.3
40 – RUB - 0/100	30.4	34.2	34.0
40 – RUB - 20/80	30.9	33.8	34.8
40 – RUB - 40/60	27.8	34.8	32.1
40 – RUB - 60/40	28.3	31.7	32.2
40 – RUB - 80/20	26.6	31.4	35.6
40 – RUB - 100/0	25.7	31.8	30.4
80 – RUB - 0/100	23.6	---	---
80 – RUB - 100/0	20.8	---	---
120 – RUB - 0/100	16.5	---	---
120 – RUB - 100/0	20.2	---	---

This research confirmed the results of other works focused on added rubber powder to concrete mixes [3] and [4]. Decrease of mechanical resistance has several reasons. The first one is the lower adherence between the cement matrix and the particles of rubber. This effect could be eliminated by chemical treatment of rubber particles before the mixing, as recommended by [5] and [6]. Chemical pretreatment of rubber improves adherence and mechanical resistance compared with rubber concrete without pretreatment. The next reason is higher content of air voids in cement matrix.

The splitting tensile strength was determined according to [9] for different amounts of rubber powder and only for 0/100 and 100/0 ratios of fine and coarse rubber. Measurements were performed on 150 mm cubes. Measured values are shown in Table 4 and Figure 2.

Table 4. Splitting tensile strength for 40 kg/m^3.

Mix	Strength [MPa]
REF	2.5
40 – RUB - 0/100	1.9
40 – RUB - 100/0	1.8
80 – RUB - 0/100	1.7
80 – RUB - 100/0	1.5
120 – RUB - 0/100	1.3
120 – RUB - 100/0	1.2

The results showed that rubber powder has the same influence on the splitting tensile strength as on the compressive strength. The lower adherence and higher air content caused the decrease of the splitting tensile strength of concrete mixes containing rubber powder. Compressive strength and splitting tensile strength of concrete containing rubber powder has been investigated by several authors [3, 4] with similar results.

5. Resistance of concrete to water with deicing chemicals

Having in mind the influence of rubber powder to content of air voids in fresh concrete mixes, it was expected that the rubber will also influence the resistance to water with deicing chemicals. Previous studies have confirmed positive effect of added rubber powder to freezing and thawing durability [7]. A part of this research was focused on freezing and thawing combined with the effect of deicing chemicals.

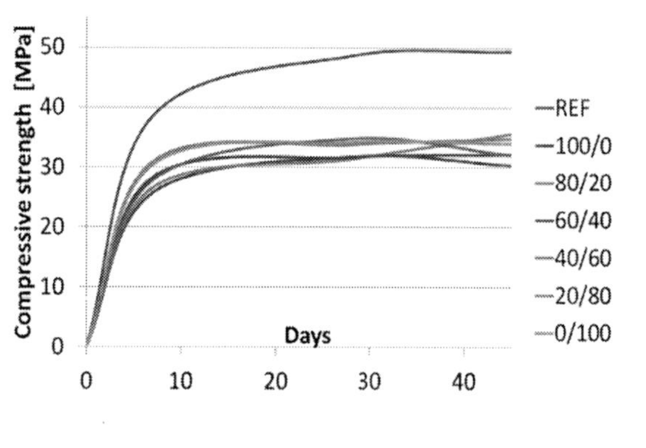

Figure 1. Compressive strength for 40 kg/m^3 of rubber powder.

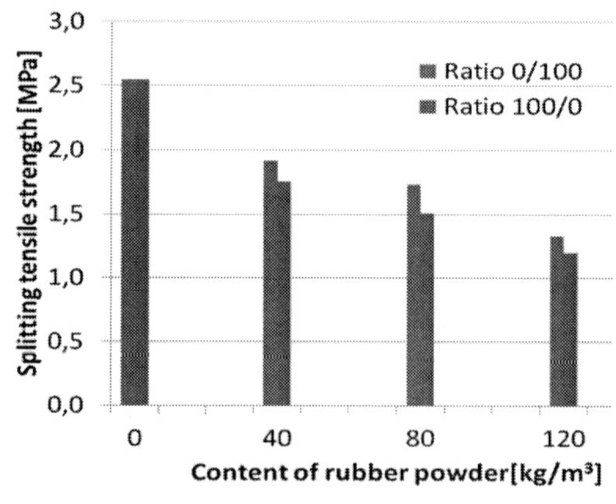

Figure 2. Splitting tensile strength.

The tests were performed for mixes containing 40 kg/m^3 of rubber powder in ratios 20/80 and 100/0. Measurements were conducted according to [12] on 150 mm cubes. Figures 3 and 4 show the mass of scaled material after certain numbers of freezing and thawing cycles. The limits were reached after 32 cycles for mix 40-RUB-20/80 and after 129 cycles for mix 40-RUB-100/0.

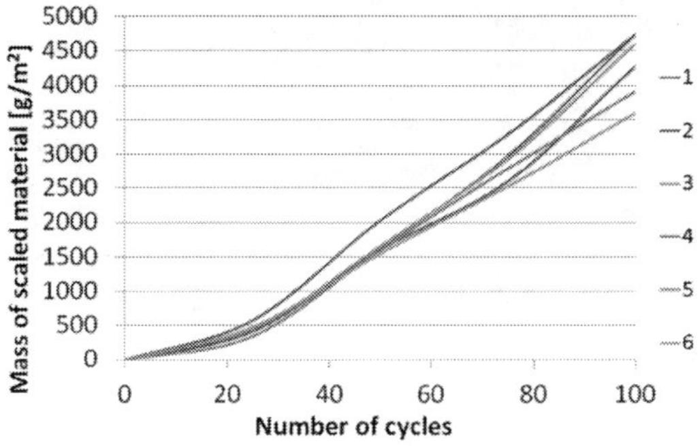

Figure 3. The result for mix 40-RUB-20/80.

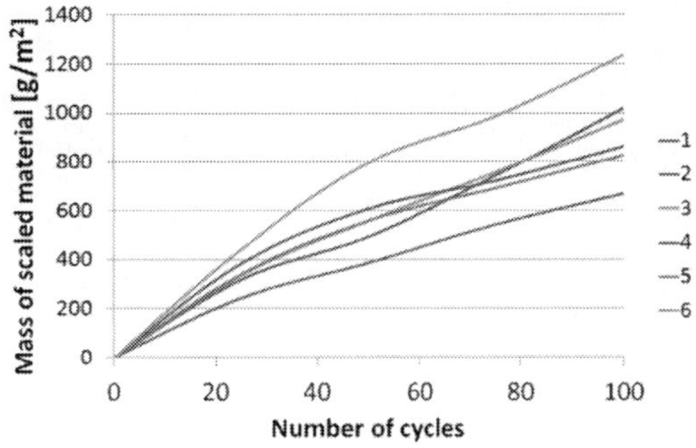

Figure 4. The result for mix 40-RUB-100/0.

The limit value of the mass of scaled material was 1000 g/m^2. This value was determined for exposure class XF4. Both tests were performed for 100 freezing and thawing cycles. For mix 40-RUB-100/0, the limit value was not achieved even after 100 cycles. Therefore the limit value was calculated by linear extrapolation.

The results indicated that incorporation of fine rubber powder to concrete mixes had a highly positive influence on the resistance to water with deicing chemicals. The air voids created additional space for expansion of ice and salt crystals. Optimum size and spacing of voids was confirmed by microscopic analysis.

6. Microscopic analysis

The photos from electron microscope were created for mixes 40-RUB-0/100, 40-RUB-100/0. The images were captured by primary electrons for all mixes, and also by secondary electrons for mixes with additive of rubber powder. The photos proved that the sizes and spacing of pores were better in the mix containing finer rubber particles; this corresponds with the results of the tests of resistance to water with deicing chemicals.

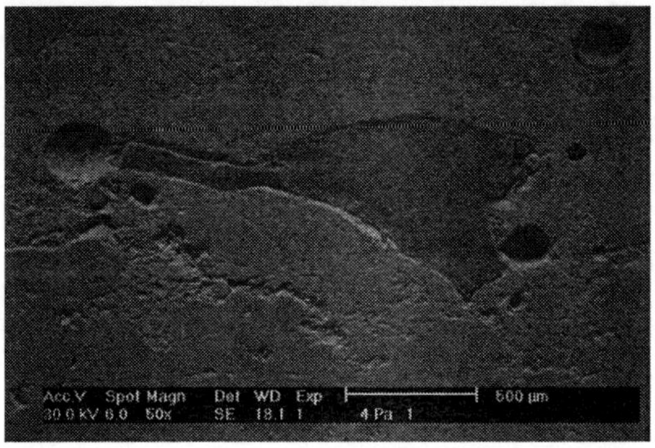

Figure 5. The photo of coarse rubber particle in cement filler by secondary electrons. Zoom - 50x.

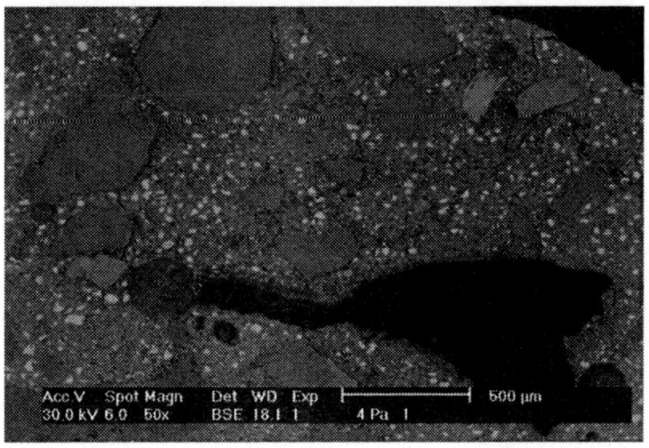

Figure 6. The photo of coarse rubber particle in cement filler by primary electrons. Zoom - 50x.

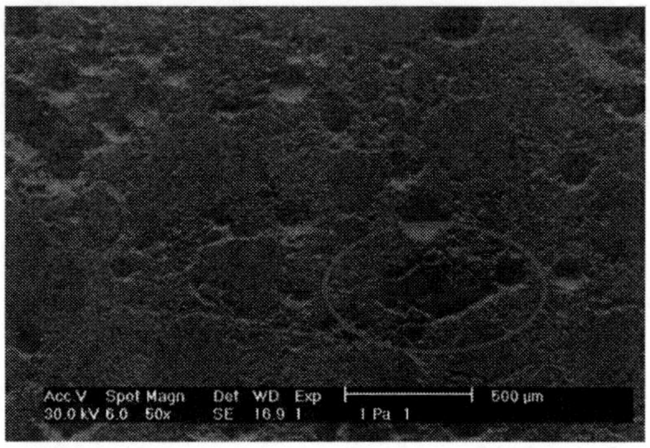

Figure 7. The photo of fine rubber particle in cement filler by secondary electrons. Zoom - 50x.

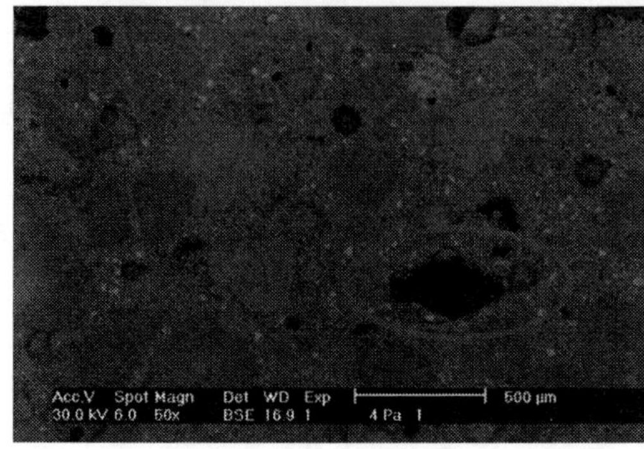

Figure 8. The photo of fine rubber particle in cement filler by primary electrons. Zoom - 50x.

7. Conclusion

Several tests of concrete with admixture of rubber powder were performed. The results indicated that the incorporation of rubber powder to concrete mixes had significant effect on the properties of fresh and hardened concrete. Laboratory tests included slump test, air content test, compressive strength, splitting tensile strength tests and test of resistance to water with deicing chemicals.

The results of macroscopic properties showed the following:

- As the content of rubber powder increased, compressive strength and splitting tensile strength decreased.
- In mixes containing 40 kg/m^3 of rubber powder, the compressive strength generally decreased with increasing ratio of fine rubber powder to coarse rubber powder in earlier stages, After 45 days, the strength of all the mixes was almost the same.
- Fine rubber particles were significantly more efficient in increasing the air content than the coarse ones. The effect of the total amount of added rubber powder was less significant than the effect of fineness.
- Incorporating of rubber powder to concrete mix has also influence on workability. Workability decreased, as the amount of fine rubber increased.

Microscopic analysis performed for two selected mixes confirmed that fine rubber particles contributed more efficiently than coarse rubber particles to creation of optimal size and spacing of air voids in hardened concrete. This is a reason why the mix containing fine rubber particles had better resistance to water with deicing chemicals.

Acknowledgements

This paper was prepared thanks to the support of the Science Foundation of the Czech Republic (GAČR), project "Analysis of the relations between the microstructure and macroscopic properties of ultra-high performance concretes" (no. 17-19463S).

References

[1] Topçu U B 1995 The Properties of Rubberized Concrete *Cement and Concrete Research* 25(2) pp 304-10

[2] Benazzouk A and Queneudec M 2002 Durability of Cement-Rubber Composites under Freeze Thaw Cycles *Proceedings of International congress of Sustainable Concrete Construction (Dundee)* pp 355-62

[3] Bonnet S 2003 Effet de l'Incorporation des Granulats Caout-chouc sur la Résistance à la Fissuration *Rencontres Universitaires de Génie Civil (Des Mortiers)*

[4] Garros M 2007 *Composites Cimentaires Incorporant des Granulats Caoutchouc Issus du Broyage de Pneus Usagés: Optimisation de la Composition et Caractérisation* (Université Paul Sabatier, Toulouse)

[5] Li Z, Li F and Li J S L 1998 Properties of Concrete Incorporating Rubber Tyre Particles *Magazine of Concrete Research* **50**(4) pp 297-304

[6] Abdel-Wahab M and Abdel-Hameed E 2001 Concrete Using Rubber Tyre Particles as Aggregates *Proc. of the Int. Symp. Recycling and Reuse of Used Tyres* pp 251-59

[7] Savas B Z, Ahmad S and Fedroff D 1996 Freeze-thaw durability of concrete with ground waste tire rubber *Transportation Research Record* **1574** pp 80-88

[8] European Committee for Standartization (CEN) 2009 Testing hardened concrete; Compressive strength of test specimens Part 3 (*EN 12390-3)*

[9] European Committee for Standartization (CEN) 2010 Testing hardened concrete – Part 6: Tensile splitting strength of test specimens *(EN 12390-6)*

[10] European Committee for Standartization (CEN) 2009 Testing fresh concrete - Part 2: Slump-test *(EN 12350-2)*

[11] European Committee for Standartization (CEN) 2009 Testing fresh concrete - Part 7: Air content – Pressure methods *(EN 12350-7)*

[12] Český normalizační institut 2003 Stanovení odolnosti povrchu cementového betonu proti působení vody a chemických rozmrazovacích látek *(ČSN 73 1326 Z1)*

[13] Frankowski R 1995 Rubber-Crumb-Reinforced Cement Concrete *US Patent 5391226*

IOP Conf. Series: Materials Science and Engineering **236** (2017) 012094 doi:10.1088/1757-899X/236/1/012094

Influence of increasing amount of recycled concrete powder on mechanical properties of cement paste

Jaroslav Topič[1], Zdeněk Prošek[1] and Tomáš Plachý[1]

[1]Czech Technical University in Prague, Faulty of Civil Engineering, Thákurova 7, 166 29 Praha 6, Czech Republic

E-mail: jaroslav.topic@fsv.cvut.cz

Abstract. This paper deals with using fine recycled concrete powder in cement composites as micro-filler and partial cement replacement. Binder properties of recycled concrete powder are given by exposed non-hydrated cement grains, which can hydrate again and in small amount replace cement or improve some mechanical properties. Concrete powder used in the experiments was obtained from old railway sleepers. Infrastructure offer more sources of old concrete and they can be recycled directly on building site and used again. Experimental part of this paper focuses on influence of increasing amount of concrete powder on mechanical properties of cement paste. Bulk density, shrinkage, dynamic Young's modulus, compression and flexural strength are observed during research. This will help to determine limiting amount of concrete powder when decrease of mechanical properties outweighs the benefits of cement replacement. The shrinkage, dynamic Young's modulus and flexural strength of samples with 20 to 30 wt. % of concrete powder are comparable with reference cement paste or even better. Negative effect of concrete powder mainly influenced the compression strength. Only a 10 % cement replacement reduced compression strength by about 25 % and further decrease was almost linear.

1. Introduction

The need for construction and maintenance of the infrastructure takes a large amount of funds and natural resources and yearly produces a great amount of construction and demolition waste based on concrete and asphalt. Asphalt materials can be directly processed on site with the help of mobile recycling units and incorporated into new structural layers. It is mainly because the material is homogenous and does not need much effort to sort and modify. In railway structures the concrete is mostly contained in sleepers or slabs for tram tracks. Sleepers and boards can be reused but recycling on the site is more complicated because they contain reinforcement [1].

Sorted waste based on concrete and asphalt can be applied to new asphalt mixtures [2,3]. Old concrete is in its simplest form used as an alternative to crushed stone backfilling and soil treatment. It can also be used as sub-base layers of the roads [4,5] or unpaved road surfaces [6,7]. Recycled concrete is also used as fine and coarse aggregate in new concrete [8,9,10]. Problematic part of recycling process is remaining concrete powder < 1 mm, which is mostly composed of old cement paste. One option is to use it as fillers or partial binder replacement in cement composites.

In this paper the high speed mill was used to crush old cement grains and exposed their non-hydrated core. Exposed non-hydrated core can hydrate with water again and therefore can be used as partial cement replacement. Also particles of old cement are smaller after milling and thus can be used as microfiller. This way of using rest of recycled concrete could help to reduce the amount of accumulating construction waste. Based on previous research, the amount of recycled concrete powder (RCP) was limited to 50 wt. % [11]. Replacement of the cement behind this level is inefficient and unfeasible.

2. Materials and samples

The samples were made of the Portland cement CEM I 42.5 produced in Radotín, the Czech Republic. Recycled concrete powder was produced by LAVARIS Ltd. by using a high speed mill. The old concrete was obtained from old railway sleepers of type PB2 and SB8. Six mixtures with 0%, 10%, 20%, 30%, 40% and 50% of RCP, by weight, were prepared (Table 1). W/b ratio was different due to deteriorated workability caused by increasing amount of RCP. Each set contained 6 prismatic samples having dimensions of 40 × 40 × 160 mm. The samples were removed from their casts after 2 days. After that the samples were cured for 28 days in laboratory conditions at the temperature of 21 ± 2 °C and relative humidity of 50 ± 5 %. After 28 days the testing of the samples was performed.

Table 1. Composition of the mixtures.

Mixture	Cement [g] (CEM I 42.5 R)	RCP [g]	Water/mixture ratio	Slump flow test [mm]
CEM (ref)	1500	0	0.35	130 ± 5
RCP 10 (10 wt. %)	1350	150	0.36	130 ± 5
RCP 20 (20 wt. %)	1200	300	0.37	130 ± 5
RCP 30 (30 wt. %)	1050	450	0.38	130 ± 5
RCP 40 (40 wt. %)	900	600	0.39	130 ± 5
RCP 50 (50 wt. %)	750	750	0.40	130 ± 5

3. Experimental methods

The dynamic Young's modulus was determined by using the impact resonance method, based on nondestructive measurements of the basic natural frequency of longitudinal vibration of the sample (Figure 1) and the basic natural frequency of flexural vibration of the sample (Figure 2) of the samples. The dynamic Young's modulus was evaluated by basic longitudinal natural frequency of the samples according to the standard equations for prismatic sample [12]. The basic flexural natural frequency of the samples was used for verification of results. Weight and dimension of each sample was measured before testing. The samples were supported by a soft elastic pad at the nodal points.

Figure 1. Configuration of measuring the basic natural frequency of longitudinal vibration of the sample by impact resonance method.

Figure 2. Configuration of measuring the basic natural frequency of flexural vibration of the sample by impact resonance method.

The measurement of dynamic Young's modulus was carried out by a measuring assembly Brüel & Kjaer containing the measuring station type 3560-B-120, the acceleration transducer type 4519-003, the impact hammer type 8206 and a control notebook. The measuring station recorded excitation signal and response signal. Using the Fast Fourier Transform, the signals were transformed from the time domain to frequency domain and graph of Frequency Response Functions (FRF) as a ratio of a response and an excitation force in the frequency domain was evaluated by Software PULSE LabShop version 14.0.1. From the FRFs the appropriate basic natural frequencies were evaluated (Figure 3).

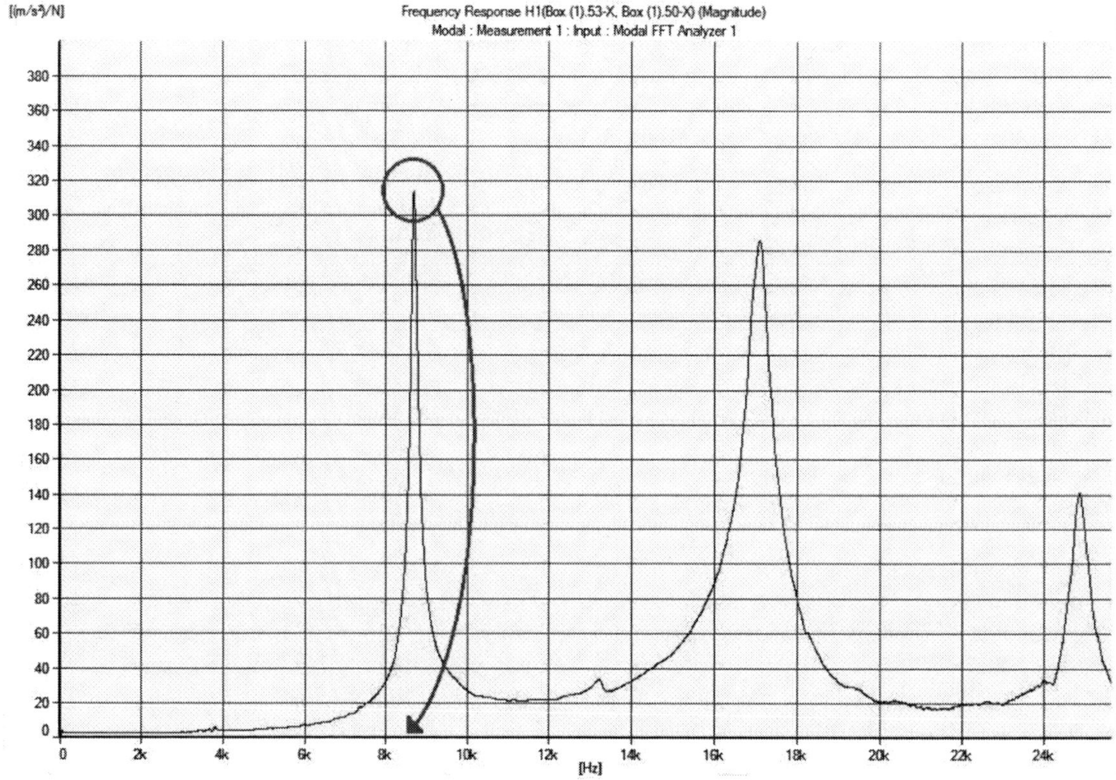

Figure 3. Graph of Frequency Response Functions (FRF) obtained by Software PULSE LabShop version from measurement of the basic natural frequency of longitudinal vibration of the sample.

The flexural and compressive strength were determined on the 28 days old samples using the Heckert device, model FP100. The testing was displacement controlled at a constant rate of 0.1 mm/s in the case of three-point bending and 0.3 mm/s during the compression test. The distance between supports during the three-point bending test was equal to 100 mm. The uniaxial compressive test was performed on the broken halves of the specimens with effective dimensions of 40 × 40 × ~80 mm.

4. Experimental results

Bulk density was evaluated before measurement of dynamic Young's modulus. As can be seen in Table 2, bulk density decrease is directly proportional to the increasing amount of recycled concrete powder. Bulk density of the samples with 50 % of RCP decreased by about 18 %.

Table 2. Bulk density of the samples with different cement replacement.

Samples	Bulk density [kg/m^3]	Samples	Bulk density [kg/m^3]
CEM (ref)	1919 ± 9	**RCP 30** (30 wt. %)	1790 ± 9
RCP 10 (10 wt. %)	1851 ± 1	**RCP 40** (40 wt. %)	1709 ± 4
RCP 20 (20 wt. %)	1839 ± 7	**RCP 50** (50 wt. %)	1566 ± 3

The dynamic Young's modulus of the samples with RCP up to 20 % was within standard deviation comparable with reference sample (Figure 4). Dynamic Young's modulus of the samples with RCP above 20 % decreased with increasing amount of RCP. Modulus of samples with 50 % of RCP was about 38 % lower compared with reference sample.

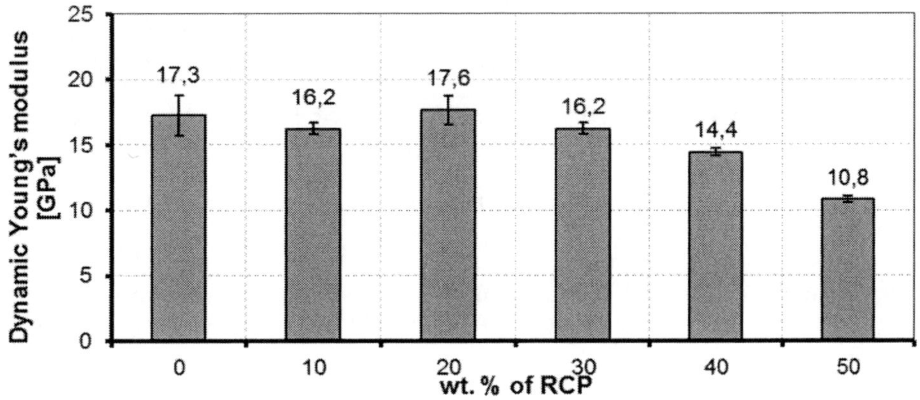

Figure 4. Dynamic Young's modulus of cement paste with increasing amount of RCP.

In Figure 5 the results of compression strength are shown. Samples with 10 % of RCP exhibit lower compressive strength by about 35 % compared to reference samples. But with 20 % replacement, the compressive was lower by only about 25 %. The amount of RCP about 20 % was high enough to reach improvement of compression strength due to microfiller and binder effect and also sufficiently low to limit the decrease of strength caused by lower amount of cement and higher porosity. Compression strength of the samples with RCP above 20 % decreased with increasing amount of RCP.

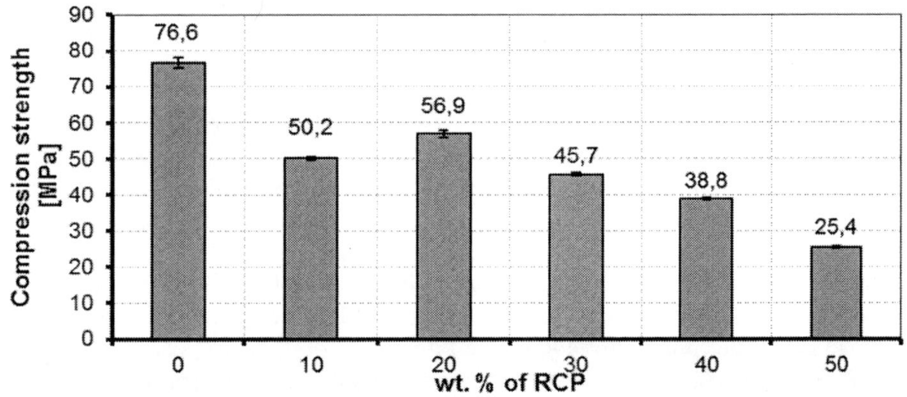

Figure 5. Compression strength of cement paste with increasing amount of RCP.

Results of flexural strength (Figure 6) show that replacement of 10 to 20 % of cement by RCP cause increase of strength by about 25 to 40 %. But high standard deviation needs to be considered. It was likely caused by the same influences as in case of compression strength. All of this contributed to a reduction in cohesiveness of the cement paste and led to increase of flexural strength. Samples with RCP above 20 % were comparable with reference samples.

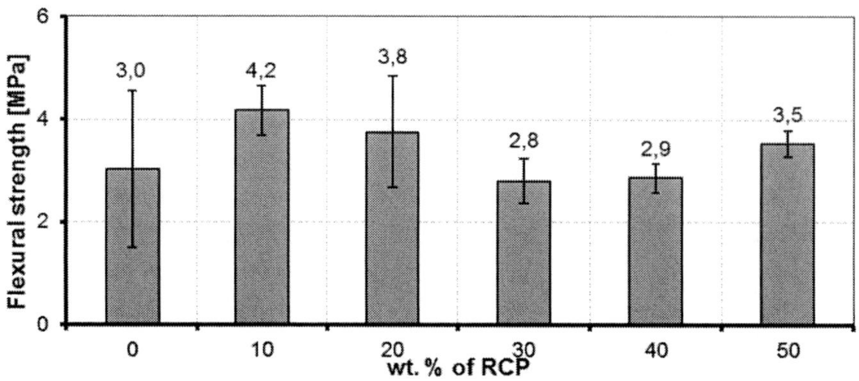

Figure 6. Flexural strength of cement paste with increasing amount of RCP.

5. Discussion

This study evaluated the effects of recycled concrete powder (RCP) on mechanical properties of cement paste. Samples of cement paste contained 0 to 50 wt. % of RCP. From presented results can be summarized those findings.

Bulk density of samples decreased with increasing amount of RCP mostly because of higher water/mixture ratio that caused higher porosity. Partial negative influence can also be attributed to lower density of old cement paste in the RCP. Decrease of dynamic Young's modulus of the samples with > 30 wt. % of RCP was related to lower modulus of elasticity of the old cement paste and higher porosity caused by higher water to binder ratio that cause mentioned higher porosity. But on the other hand, the microfiller and binder effect of RCP retained the dynamic Young's modulus of the samples with RCP up to 20 % on the same or slightly higher level than in case of reference cement paste. Compression strength of samples decreased with increasing amount of RCP, but samples with 20 wt. % of RCP showed better results than samples with 10 wt. % of RCP. Samples with 10 to 20 wt. % of RCP had higher flexural strength than reference cement paste, samples with more than 20 wt. % of RCP had comparable flexural strength as reference samples. According to results of dynamic Young's modulus, compression and flexural strength, the optimal rate of replacement was defined around 20 to 30 wt. % of RCP. Such replacement level allows some improvements due to microfiller and binder effect of RCP and also allows to minimize the negative influence of higher porosity caused by higher water to binder ratio.

The paper focused on using recycled concrete powder and eventually brick powder [13,14] confirmed possibility of using recycled concrete powder as partial cement replacement. Compared with the paper by Šeps and Broukalova [15], the trend of decreasing bulk density according to higher amount of micro-ground recycled concrete was similar. Also the compressive and flexural strength had similar trend. In cited paper, the values of all investigated properties were higher than in this paper. This difference was probably caused by curing in water that provided better conditions during hardening. Also better compaction of samples during molding could have resulted in higher bulk density that primary influenced compressive strength.

Other authors used recycled concrete powder as heat treated binder [16,17] or as replacement of natural ingredients in the mixture for cement production [18]. Those methods require large amounts of energy and produce high CO_2 release although less than in the case of conventional cement clinker.

Only methods using recycled concrete in its raw form can be considered as sustainable and environmentally friendly. Using RCP in geopolymer based binder [19] or as microfiller and partial cement replacement belongs to these sustainable methods.

6. Conclusions

This study evaluated the effects of recycled concrete powder (RCP) on mechanical properties of cement paste. Samples of cement paste contained 0 to 50 wt. % of RCP. From presented results can be summarized those findings:

- Bulk density of samples decreased with increasing amount of RCP mostly because of higher water/mixture ratio that caused higher porosity and partially because of lower density of old cement paste in the RCP.
- Decrease of dynamic Young's modulus of the samples with > 30 wt. % of RCP was related with lower modulus of elasticity of the old cement paste and higher porosity caused by higher water/mixture ratio.
- Compression strength of samples decreased with increasing amount of RCP, but samples with 20 wt. % of RCP showed small improvement.
- Samples with 10 to 20 wt. % of RCP had higher flexural strength than reference cement paste.
- Optimal rate of replacement was defined around 20 to 30 wt. % of RCP. Such replacement allows some improvements due to microfiller and binder effect of RCP and also allows to minimize negative influence of higher porosity caused by higher water to binder ratio.

Acknowledgements

The authors wish to express their gratitude for financial support to Technology Agency of the Czech Republic (TA ČR) – Competence Centers under project No. TE01020168 Centre for Effective and Sustainable Transport Infrastructure (CESTI), Czech science foundation GAČR under project No. 17-06771S and to the Faculty of Civil Engineering, Czech Technical University (CTU) in Prague under SGS project No. SGS16/201/OHK1/3T/11. The authors also thank to the Center for Nanotechnology in Civil Engineering at the Faculty of Civil Engineering, CTU in Prague.

References

[1] Lidmila M et al 2014 Utilization of Recycled Fine-ground Concrete from Railway Sleepers for Production of Cement-Based Binder *Applied Mechanics and Materials* **486** pp 323-326

[2] Lee Ch-H, Du J-Ch and Shen D-H 2012 Evaluation of pre-coated recycled concrete aggregate for hot mix asphalt *Construction and Building Materials* **28** pp 66-71

[3] Pasandín A R and Pérez I 2015 Overview of bituminous mixtures made with recycled concrete aggregates *Construction and Building Materials* **74** pp 151-161

[4] Barbudo A et al 2012 Statistical analysis of recycled aggregates derived from different sources for sub-base applications *Construction and Building Materials* **28** pp 129-138

[5] Engelsen Ch J et al 2012 Field site leaching from recycled concrete aggregates applied as sub-base material in road construction *Science of the Total Environment* **427-428** pp 86-97

[6] Vegas I et al 2011 Pre-normative research on the use of mixed recycled aggregates in unbound road sections *Construction and Building Materials* **25** pp 2674-82

[7] Jiménez R J et al 2012 Utilisation of unbound recycled aggregates from selected CDW in unpaved rural roads *Resources, Conservation and Recycling* **58** pp 88-97

[8] Silva R V, Brito J and Dhir R K 2014 Properties and composition of recycled aggregates from construction and demolition waste suitable for concrete production *Construction and Building Materials* **65** pp 201-217

[9] Agrela F et al 2011 Limiting properties in the characterization of mixed recycled aggregates for use in the manufacture of concrete *Construction and Building Materials* **25** pp 3950-5

[10] Evangelista L and Brito J 2007 Mechanical behavior of concrete made with fine recycled

concrete aggregates *Cement and Concrete Composites* **29** pp 397-401

[11] Topič J et al 2016 Replacement of cement with finely ground recycled concrete: Influence on mechanical properties *Applied Mechanics and Materials* **825** pp 69-72

[12] American Society for Testing and Materials (ASTM) 2006 Standard Test Method for Dynamic Young's Modulus, Shear Modulus, and Poisson's Ratio by Impulse Excitation of Vibration *(E1876-01)*

[13] Lidmila M et al 2014 Utilization of Recycled Fine-ground Concrete from Railway Sleepers for Production of Cement-based Binder *51nd International Conference on Experimental Stress Analysis* pp 323-6

[14] Liu Q et al 2014 Investigation of using hybrid recycled powder from demolished concrete solids and clay bricks as a pozzolanic supplement for cement *Construction and Building Materials* **73** pp 754-763

[15] Šeps K and Broukalová I 2014 Properties of cement based composite with fine ground recycled concrete *Advanced Materials Research* **1000** pp 110-3

[16] Shui Z et al 2008 Rehydration reactivity of recycled mortar from concrete waste experienced to thermal treatment *Construction and Building Materials* **22** pp 1723-29

[17] Florea M V A et al 2014 Activation of liberated concrete fines and their application in mortars *Construction and Building Materials* **50** pp 1-12

[18] Shoon J et al 2015 Fines extracted from recycled concrete as alternative raw material for Portland cement clinker production *Cement & Concrete Composites* **58** pp 70-80

[19] Ahmari S et al 2012 Production of geopolymeric binder from blended waste concrete powder and fly ash *Construction and Building Materials* **35** pp 718-729

IOP Conf. Series: Materials Science and Engineering **236** (2017) 012095 doi:10.1088/1757-899X/236/1/012095

Innovative neuro-fuzzy system of smart transport infrastructure for road traffic safety

Anna Beinarovica[1], Mikhail Gorobetz[1] and Anatoly Levchenkov[1]

[1]Industrial Electronics and Electrical Engineering Institute Riga Technical University, Azenes street 12–1, Riga, LV-1048, Latvia

E-mail: anja19892@inbox.lv

Abstract. The proposed study describes applying of neural network and fuzzy logic in transport control for safety improvement by evaluation of accidents' risk by intelligent infrastructure devices. Risk evaluation is made by following multiple-criteria: danger, changeability and influence of changes for risk increasing. Neuro-fuzzy algorithms are described and proposed for task solution. The novelty of the proposed system is proved by deep analysis of known studies in the field. The structure of neuro-fuzzy system for risk evaluation and mathematical model is described in the paper. The simulation model of the intelligent devices for transport infrastructure is proposed to simulate different situations, assess the risks and propose the possible actions for infrastructure or vehicles to minimize the risk of possible accidents.

1. Introduction

The amount of used transport vehicles is growing very speedily. This causes the bigger amount of road accidents. That is why transportation safety is one of the most studied subjects nowadays. Road accidents can lead to terrible consequences. Safety is one of the first priority tasks in transport domain. A lot of road accidents and crashes caused by driver factor. Driver factor depends on the mood, physical and psychical condition of the driver [1]. A lot of scientists are interested in inventing the intelligent system, which could reduce the human factor by smart analyzing the situation on the road, making the decisions to avoid dangerous situations and transmitting the solution from the infrastructure to the vehicles [10].

Different researches about road traffic safety are made. Neural network is proposed for the traffic sign recognition process [2], for the road surface traffic sign detection [3], risk evaluation [4] etc. Simple fuzzy severity estimation model has a comparable performance to more complicated systems such as the CoTEC (CoOperative Traffic congestion detECtion) [9]. That proves the actuality of the chosen topic.

In the previous works, the system and the algorithm for anti-collision system reducing human factor has been developed and tested [5]. Also intelligent transport safety system was investigated and designed by paper authors [6]. This research is a development of the innovative neuro-fuzzy system of smart transport infrastructure.

The main advantage of the proposed system is that system is using real-time data makes the decision about speed change instantly.

2. Problem formulation

The main purpose of the proposed system is to evaluate the collision possibility and to make a decision about speed change of the vehicles.

The goal of current research is to develop the system for risk evaluation.

The following tasks are defined and solved:

1) to define the structure and functions of the system,
2) to develop the mathematical model and multiple criteria target function for collision possibility minimization,
3) to develop the adaptive algorithm of the system functions,
4) to develop the computer model and simulate the developed algorithm. To compare results before and after neural network training process,
5) to implement the model and the algorithm for microcontroller based embedded devices as a functional prototype of the system.

3. Neuro-fuzzy system model

In this paper, neuro-fuzzy control proposed to be taken as intelligent control method. While fuzzy logic provides an inference mechanism under cognitive uncertainty in reactions to the transport situation danger level, computational neural networks offer exciting advantages, such as learning, adaptation, fault tolerance, parallelism and generalization.[7].

Each neural network input pair is made by two parameters:

- The distance of the vehicle till the trajectories crossing point di;
- Average speed of the vehicle vi.

According to the number of vehicles, neural network has n outputs, and as a result the speed change for the vehicle is output.

Input neurons of output layer has wij, weights and bj shifting.

Each j-th neurons of the ouput layer generates j-th vehicles speed change Δvj.

Values Δv1 ... Δvn are given to the self-training multi-criteria target function F, that evaluate efficiency of the output according the collision possibility P and vehicles common speed change ΣΔvi.

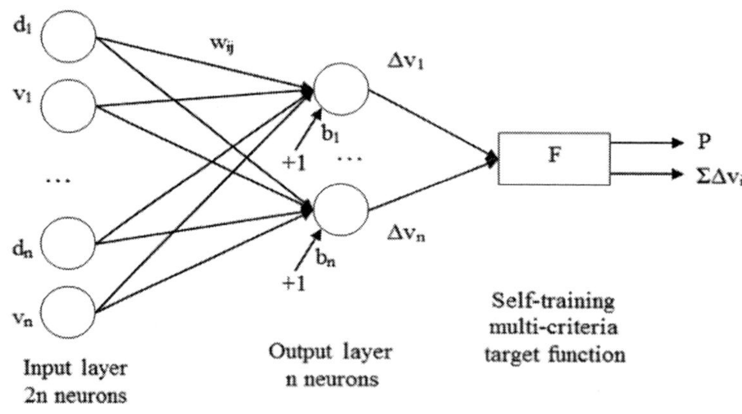

Figure 1. Neural network structure.

General fuzzy logic structure, shown in the Figure 1, consists of input as a risk level of recognized objects, membership functions for risk assessment, rule database for selection of actions and defuzzification functions for the level of the activity. In system identification, rule-based fuzzy models are usually applied. In these models, relations among variables are described by means of if–then rules with fuzzy predicates, such as "if the collision possibility is high, then the car speed must be changed" and "if the car speed was changed, but collision possibility still is high, then train speed must be changed".

Fuzzy sets are defined through their membership functions (denoted by μ) which map the elements of the considered universe to the unit interval [0, 1]. The extreme values 1 and 0 denote complete membership and non-membership, respectively, while a degree between 0 and 1 means partial membership in the fuzzy set. A particular domain element can simultaneously belong to several sets (with different degrees of membership).[8] In Figure 2, for instance, 45% of risk belongs to the set of high risk with membership 0.2 and to the set of medium risk with membership 0.8. We can suppose which action must be chosen for the crash prevention after defining the level of the risk in percentages. This gradual transition from membership to non-membership facilitates a smooth outcome of the reasoning (deduction) with fuzzy if–then rules, in fact a kind of interpolation.[8]

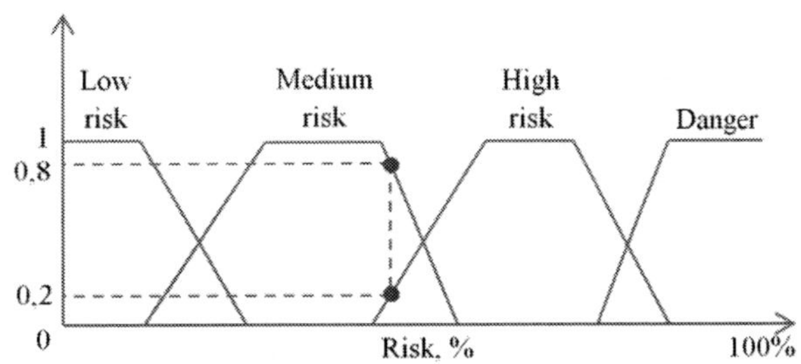

Figure 2. Partitioning of the risk domain into four fuzzy sets.

The neuro-fuzzy system of smart transport infrastructure for road traffic safety system consists of movable and stationary controllable components of the system, i.e. transport units and infrastructure – level crossings (Figure 3).

The system contains centralization software components working in the web environment such as a web-server, database, web interface, dynamic system models and control and optimization modules for long term scheduling and global optimization.

There are communication components ensuring data transmission to or between the components, such as satellite positioning system - GPS, radio frequency modules – RF.

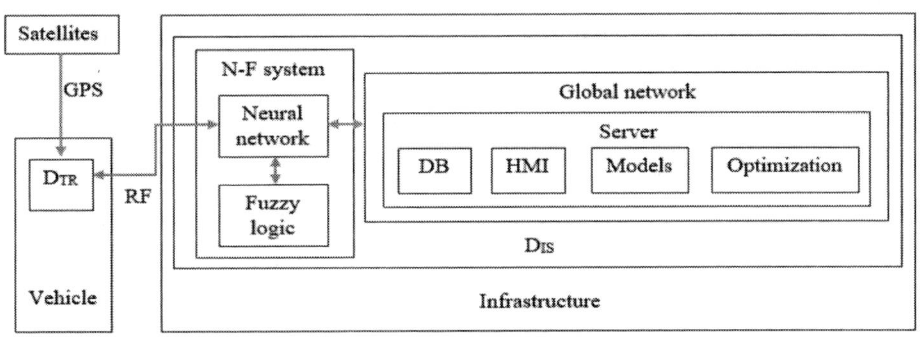

Figure 3. Structure of the system.

Each component has embedded electronic device Di, which performs specific function to ensure prevention of collisions between vehicles.

D_{TR} – the devices for transport units obtain the position, calculate the motion parameters and communicate with other devices. The bidirectional energy flow monitoring and measurement [11] and controlling of the energy flow of the drive [12] of the vehicle can be and additional source of data for the system installed in D_{TR}.

D_{IS} – devices of infrastructure objects, such as stations, level-crossings etc. control the anti-collision function. Consist the block of Global network and developed neuro-fuzzy system, for recognizing the collision possibility and possible speed change for the dangerous situation prevention.

4. Mathematical model and target function

The mathematical model is represented with following sets:

$U \subset (U_1, \ldots, U_n)$ – set of transport units as a subsets of different types, where for different transport safety task it could be:

$U^1 = (U^1_1, \ldots, U^1_{n1})$ – subset of railway transport units

$U^2 = (U^2_1, \ldots, U^2_{n2})$ – subset of road vehicles

$U^3 = (U^3_1, \ldots, U^3_{n3})$ – subset of aerial vehicles etc.

$P = (p_1, p_2, \ldots, p_c)$ – set of infrastructure objects, where the collision of vehicles, e.g. for railway transport it could be level-crossings, switches, etc. For this research, the crossing section is assumed as a short straight segment of the route or trajectory.

The geographical coordinates of all crossing of possible routes or trajectories of transport units are defined by these sets:

$$\chi_b^P=\{\chi_b^{P_1},\chi_b^{P_2},\ldots,\chi_b^{P_c}\},\ \psi_b^P=\{\psi_b^{P_1},\psi_b^{P_2},\ldots,\psi_b^{P_c}\}, \quad (1)$$

$$\chi_e^P=\{\chi_e^{P_1},\chi_e^{P_2},\ldots,\chi_e^{P_c}\},\ \psi_e^P=\{\psi_e^{P_1},\psi_e^{P_2},\ldots,\psi_e^{P_c}\}, \quad (2)$$

Where

$\chi_b^{P_i}$ – latitude of the beginning point of the crossing sector

$\psi_b^{P_i}$ – longitude of the beginning point of the crossing sector (crossing)

$\chi_e^{P_i}$ – latitude of the ending point of crossing sector

$\psi_e^{P_i}$ – longitude of the ending point of the crossing sector

c – number of trajectories crossing point

t^P_{vest} – safe closing time of each trajectories crossing point $p \in P^2$

Neural network can be used as one of the immune system tool for the situation improvement on the crossing point and transportation process optimization.

There is no information either the output value is correct or not, that is why there is no possibility to use error back propagation algorithm.

Random sequential delta law self-training algorithm and target function was developed for the neural network training.

Function of the optimization is defined by two criteria:

- Collision possibility P with the aim of minimizing;
- Changes of the vehicles' speed $\Sigma\Delta vi$ with the aim of minimizing.

First criterion is connected with safety. The situation considered to be dangerous if trajectories of two transport vehicles has a common crossing point, and exists a possibility, that transport vehicles will arrive at the crossing point of their trajectories at the same time. This situation is actual for each transport type.

Second criterion is connected with the transport traffic specifics. For example train traffic precision is very important for railway transportation operations, because train delays causes obstacles for other trains. That is why timetable is very important for such transport vehicles as trains and trams. Also time of departure and arrival is important for the public transport. That is why it is necessary to make minimal speed changes of such type of vehicles.

Based on the individual weighted criteria, the target function was developed:

$$F(\Delta v) = \begin{cases} P = \max(P_{IJ}) \rightarrow \min \\ \sum \Delta v_i \rightarrow \min \end{cases} \tag{3}$$

Where,
Δv – changes of the speed of vehicle, solution of the task;
P – maximal collision possibility [5,6];
P_{IJ} – each i-th vehicle collision possibility with each j-th vehicle;
Δv_i – change of speed of i-th vehicle.

5. Algorithm

Developed algorithm consists of such steps:
Initialization:

- Index of training group element e = 1;
- Chosen for the correction neurons sn = 1;
- Chosen for the correction weight sw = 1;
- Retraining = false.

STEP 1. Take the element $e = \{d_1^e, v_1^e, d_2^e, v_2^e\}$ from the training set.
STEP 2. Read the x_{min} and x_{max} parametres, that limit neuron network output.
STEP 3. Calculate output neurons adder values:

$$\Sigma_j = (\Sigma_{i=1}^{2n} x_i * w_{ij}) + b_j \qquad j = \overline{1..n} \tag{4}$$

STEP 4. Generate output layer neuron output value by positively and negatively saturated linear activation function:

$$\Delta v_j = \begin{cases} x_{min}, & \Sigma_j \leq x_{min} \\ \Sigma_j, & x_{min} < \Sigma_j < x_{max} \\ x_{max}, & \Sigma_j \geq x_{max} \end{cases} \tag{5}$$

STEP 5. Save the previous valuation, if it exists P^{iep}, $\Sigma\Delta v_i^{iep}$.
STEP 6. Evaluate the solution, that was found, using the target function $[P, \Sigma\Delta v_i] = F(\Delta v)$.
STEP 7. If $P > P_{lim}$ or $\Sigma\Delta v_i > \Sigma\Delta v_{i\,lim}$, then go to STEP 8.
STEP 8. If the last element of the training set is not reached $e \neq e_{max}$,
then e=e+1 and go to STEP 1,
else if there is no need to retrain the network then FINISH,
else e = 1 and go to STEP 1.
STEP 9. Weight correction occurs sequentially:
If (sn ≠1 and sw ≠ 1) or ($P^{iep} < P$ and $\Sigma\Delta v_i^{iep} < \Sigma\Delta v_i$), that means if the element isn't first and the result is worse than it was before, then return to the previous weight value $w_{sw,\,sn} = w_{sw,\,sn} - k$ and choose another weight
If sw > 2n, then sn = sn + 1, else sw = sw + 1.
If sn > n, then sn = 1, sw = 1.
STEP 10. The correction value is generated, as random number:

$$k = \text{random}(-1000, 1000)/10000 \tag{6}$$

STEP 11. Weight correction is done:

$$w_{sw,\, sn} = w_{sw,\, sn} - k \tag{7}$$

STEP 12. If weight correction was done, than neural network must be retrained. Retrain = true.

STEP 13. Go to STEP 3.

6. Neural network training experiment

For the experiment target function was used, but neural network is trained to make a decision about the speed change, to prevent the collision between vehicles.

If common amount of vehicles is n, then neural network has paired input number 2n. Neural network input and output neuron amount is dynamic, because the amount of vehicles can be changed.

Neural network structure is given in the Figure 1.

For the experiment such a situation was chosen: one train; one bus; trajectories of the train and bus have crossing point.

In this situation neural network consists of 4 input and 2 output layers.

Parameters taken for the training set are given in the Table 1.

Each element from the set is sent to the neural network input layer during training process. When changes of speed Δv1 and Δv2 for the train and bus are found, these values are evaluated by target function.

Neural network is already trained if:

- Collision possibility is reduced minimum to 0.005;
- Average speed change for the train is not bigger than 3 kmh.

Table 1. Neural network training set.

Number of the training set element	Bus distance till the trajectories crossing point	Train distance till the trajectories crossing point	Bus average speed	Train average speed
1	400	600	40	60
2	390	580	40	60
3	320	550	40	65
4	290	540	40	70
5	260	520	40	75
6	230	500	40	80
7	210	480	40	85
8	200	460	40	85
9	190	440	40	90
10	175	420	40	90

Two solutions x_{min}= -10, x_{max} = 0 un x_{min} = -15, x_{max} = 5 were used for each element of the training sets.

7. Neural network self-training experiment

Another experiment was made using the system based on the Waspmote microcontroller. System can evaluate the situation and give an advice to the train or bus driver to change the speed for the prevention dangerous situation.

For the experiment following situation was defined:

- In the road infrastructure is imbedded developed equipment. When car is entering the radio communication channel earshot, it began to send information about its speed, coordinates and trajectory to the infrastructure embedded device.
- When train is entering the radio communication channel earshot, it began to send information about its speed, coordinates and trajectory to the infrastructure embedded device.
- Infrastructure embedded device, which receives data from vehicles makes up neural network based system, which is trained by the target function, find out how to reduce the speed and minimize collision possibility.

For the experiment a single car and train are approaching the trajectory crossing point. The devices of car and train are transmitting the data to the infrastructure embedded device. Initial car distance is 500 m, and train distance is 1000 m, train speed is 80 kmh, but the car speed is 40 kmh. After evaluation of such situation, the collision possibility is 0.469. After data was received, collision possibility P, Train speed change Δv and target function is calculated.

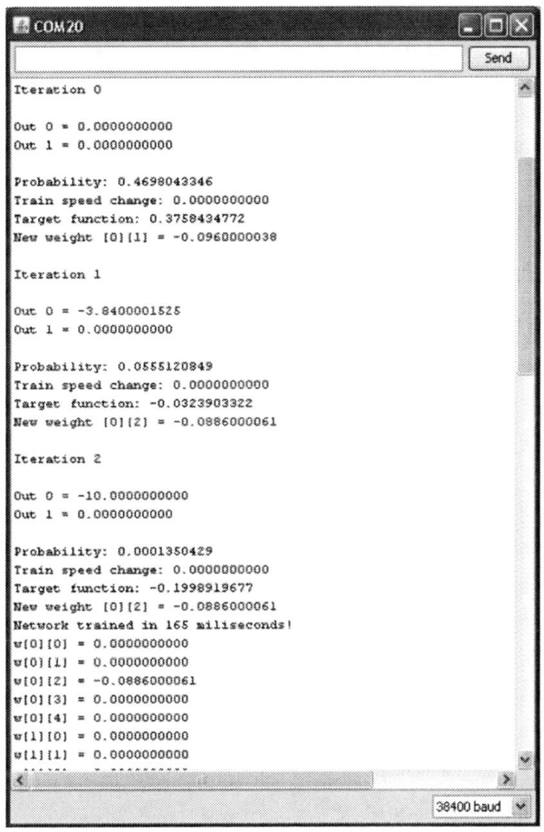

Figure 4. Self-training of the system.

In the Figure 4. self-training system is shown. At the beginning collision possibility P was 0.4698043346, and after neural network training process it was reduced to 0.0001350429. Network was trained in 165 miliseconds.

8. Conclusions

Analysis of the experiments' results allows concluding following. It's necessary to limit the data output, leaving only the most important data. Limited memory and processing power is the reason for controller to realize the algorithm in a simple form.

The results of research shows that neural network and fuzzy logic can be widely used in traffic safety area, because of its possibility of training and learning and reduced amount of input parameters.

Experiment shows that at the beginning, collision possibility P was 0.4698, and after neural network training process and speed change it was reduced to 0.0001. Experiment took 165 miliseconds, system is working fast and it is very important because of the task specific.

Convolutional neural network wasn't proposed in this study, but authors continue making the research about road traffic safety and plan to develop the CNN based system for the object recognition and collision prevention by using the convolutional neural network and making the decision about how to avoid a dangerous situation on the road.

Acknowledgement
The research leading to these results is performed in a frame of ARTEMIS Joint Undertaking project No. 621429.

References

[1] De Nadai S, D'Inca M, Parodi F, Benza M, Trotta A, Zero E, Zero L and Sacile R 2016 *Enhancing safety of transport by road by on-line monitoring of driver emotions* (University of Genoa DIBRIS – Department of Informatics, Bioengineering, Robotics and System Engineering Genoa, Italy)
[2] Abedin Z, Dhar P and Deb K 2016 Traffic Sign Recognition Using SURF: Speeded Up Robust Feature Descriptor and Artificial Neural Network Classifier *Proc. 9th Int. Conf. on Electrical and Computer Engineering (Bangladesh)* pp 198-201
[3] Qian R, Liu Q, Yue Y, Coenen F and Zhang B 2016 Road Surface Traffic Sign Detection with Hybrid Region Proposal and Fast R-CNN *Proc. 12th Int. Conf. on Natural Computation, Fuzzy Systems and Knowledge Discovery* pp 555-9
[4] Yil Z, Wang Y, Tian D, Lu G and Xia H 2010 A Road Safety Evaluation Method Based on Clustering Neural Network *Proc. Int. Conf. on Optoelectronics and Image* pp 108-111
[5] Levchenkov A. and Gorobetz M 2014 An Evolutionary Algorithm for Reducing Railway Accidents Caused by Human Factors *Proc. Second International Conference on Railway Technology: Research, Development and Maintenance (UK)*
[6] Levchenkov A, Gorobetz M and Mor-Yaroslavtsev A 2012 Evolutionary algorithms in embedded intelligent devices using satellite navigation for railway transport *Infrastructure design, signalling and security in railway* ed Xavier Perpinya pp 395-420
[7] Fuller R 2001 *Neuro-Fuzzy Methods for Modeling & Fault Diagnosis* (Lisbon)
[8] Simonyan K and Zisserman A 2014 Two-Stream Convolutional Networks for Action Recognition in Videos *Proc. Int. Conf. on Neural Information Processing Systems*
[9] Nassar L and Karray F 2016 Fuzzy Logic in VANET Context Aware Congested Road and Automatic Crash Notification *Proc. Int. Conf. on Fuzzy Systems – FUZZ IEEE (Vancouver)* pp 1031-7
[10] Nikolajevs A and Mezitis M 2016 Level crossing time prediction *Proc. 57th Int. Scient. Conf. on Power and Electrical Engineering of Riga Technical University - RTUCON (Latvija)* pp 199-202
[11] Apse-Apsitis P, Avotins A and Ribickis L 2015 Bidirectional DC/AC energy flow measurement *Proc. 5th Int. Conf. on Power Engineering, Energy and Electrical Drives - POWERENG* pp 465-68
[12] Stana G, Apse-Apsitis P and Brazis V 2014 Virtual energy simulation of induction traction drive test bench *Proc. 2nd Workshop on Advances in Information, Electronic and Electrical Engineering - AIEEE* pp 1-6

IOP Conf. Series: Materials Science and Engineering **236** (2017) 012096 doi:10.1088/1757-899X/236/1/012096

Joint road safety operations in tunnels and open roads

Adewole Adesiyun[1], Antonio Avenoso[2], Kallistratos Dionelis[3], Liljana Cela[4], Christophe Nicodème[5], Thierry Goger[1] and Carlo Polidori[6]

[1]Forum Des Laboratoires Nationaux Europeens De Recherche Routiere Boulevard de la Woluwe 42, 1200,Brussels, Belgium
[2]European Transport Safety Council, Avenue des Celtes 20, 1040, Brussels, Belgium
[3]European Association with tolled motorways, bridges and tunnels, Rue Guimard, 15 - B-1040, Brussels, Belgium
[4]South East Europe Transport Observatory ,Omladinskih brigada 1, PO Box 14, 11198 Belgrade, Serbia
[5]The European Union Road Federation, 6 Place Stéphanie, 1050,Brussels, Belgium
[6]Italian Association of Road Safety Professionals, Via Bergamo 3 00198 Roma, Italy

E-mail: c.polidori@libero.it

Abstract. The objective of the ECOROADS project is to overcome the barrier established by the formal interpretation of the two Directives 2008/96/EC and 2004/54/EC, which in practice do not allow the same Road Safety Audits/Inspections to be performed inside tunnels. The projects aims at the establishment of a common enhanced approach to road infrastructure and tunnel safety management by using the concepts and criteria of the Directive 2008/96/CE on road infrastructure safety management and the results of related European Commission (EC) funded projects. ECOROADS has already implemented an analysis of national practices regarding Road Safety Inspections (RSI), two Workshops with the stakeholders, and an exchange of best practices between European tunnel experts and road safety professionals, which led to the definition of common agreed safety procedures. In the second phase of the project, different groups of experts and observers applied the above common procedures by inspecting five European road sections featuring both open roads and tunnels in Belgium, Albania, Germany, Serbia and Former Yugoslav Republic of Macedonia. This paper shows the feedback of the 5 joint safety operations and how they are being used for a set of - recommendations and guidelines for the application of the RSA and RSI concepts within the tunnel safety operations.

1. Introduction

The general objective of the ECOROADS project is to overcome the barrier established by the formal interpretation of the two Directives 2008/96/EC (on road infrastructure safety management) and 2004/54/EC (on tunnels), which in practice do not allow the same Road Safety Audits/Inspections to be performed inside tunnels, as shown in Figure 1.

The main problem is that, while from the user (driver) point of view a road is a unique linear infrastructure generally in open terrain and sometimes in closed environment (tunnels), the strict application of the two Directives leads to a non-uniform approach to the infrastructure safety management outside and inside tunnels.

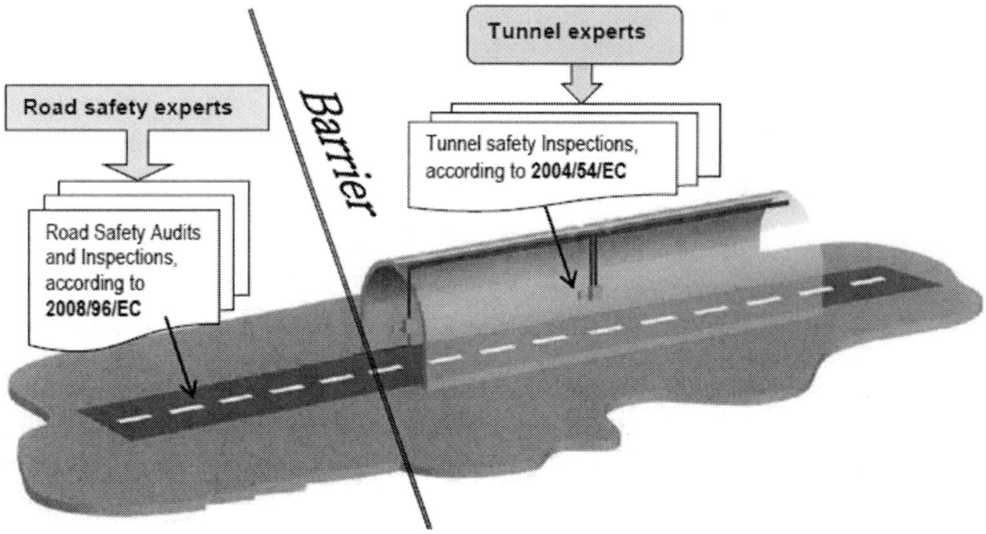

Figure 1. The "barrier" between the two EU Directives.

This project is the follow-up of the initiative related to the European Road Safety Directives and the two workshops held at the European Social and Economic Committee (EESC) by a group of international stakeholders in February and May 2013: a debate that was initiated as a result of the coach crash in Switzerland that caused more than 28 fatalities, including 22 children.

The collision occurred in 2012 with the end wall of an emergency parking facility in the Sierre tunnel, Switzerland, which was opened in 1999 and was rated as "good" in a 2005 European Tunnel Assessment Programme (EuroTAP) test. The end wall was placed at 90 degrees with respect to the direction of the adjacent traffic flow, without any adequate protection from collision.

Figure 2. Left: the lay-by in Sierre Tunnel. Source: [5]. Right: similar situation (source: Hasani A., Albanian Roads Authority).

This feature of tunnel design is typical of many European tunnels, as shown in Figure 2 (90° walls without any protection), where operations such as RSA or RSI according to the prescriptions of the Directive 2008/96/EC, could be beneficial for risk prevention.

Indeed, this Directive does not apply to road tunnels covered by Directive 2004/54/EC (Art.1, point 4 of the Directive 2008/96/CE)

On the other hand, Directive (2004/54/EC) does not deal directly with RSA or RSI inside the tunnels. There is only a general statement about taking "all aspects of the system composed of the infrastructure, operation, users and vehicles" into account in Annex 1. Different interpretation and application by Member States may further amplify the gap between the two Directives.

The ECOROADS consortium submitted to the European Commission a proposal aiming at the deployment of mixed groups of tunnel and road safety experts performing joint safety inspections in both tunnels and open roads, in order to find a common agreed inspection methodology able to enhance safety in roads and tunnels. The project has been financed by the European Commission and started in June 2015

2. Methodology

The overall approach of the ECOROADS is based on the previous successfully project Pilot4Safety [1] and is divided into several phases, including a clear overview of the application of the two Directives in the Member States, a series of workshops with the stakeholders (European tunnel and road managers), and the exchange of best practices between European experts in the two fields:

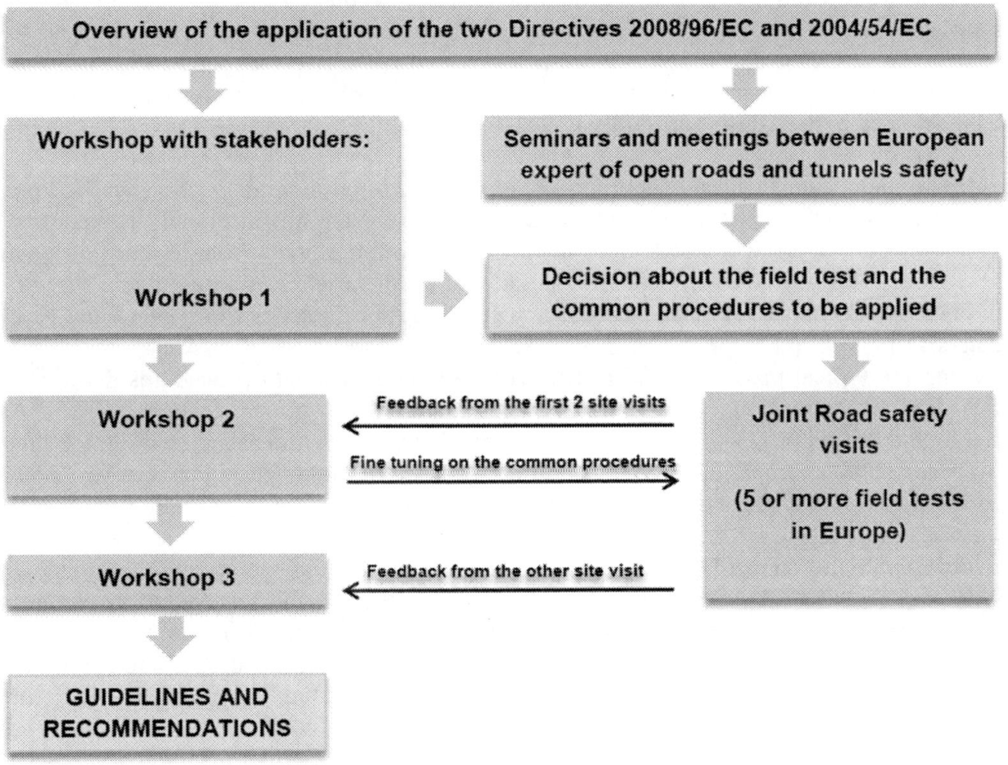

Figure 3. ECOROADS Methodology.

Clear overview of the application of the two Directives in the Member States and the extent of the gap between them already described in the previous section. Such an overview has been conducted on the basis of the results of two previous studies on the effectiveness of the Directive 2008/96/EC[2] and on the implementation and effects of Directive 2004/54/EC [3], plus direct contact with several infrastructure managers.

Workshops with the stakeholders (European tunnel and road managers from at least 10 European countries), organized according to the following Figure 3:

2.1. Joint Audit/Inspection procedures

During the Seminar for exchange of best practices on RSA/ RSI at open roads and tunnels, and during dedicated meetings among the ECOROADS partnership, held between November 2015 and January 2016, a preliminary procedure has been agreed, based on the following considerations.

The Directive 2008/96/EC applies to road sections of the Trans-European Transport Network, whether they are at design stage, under construction or in operation. Its provisions may also be applied to national road transport infrastructure, not being part of the TEN-T but constructed - entirely or partly - using Community financial assistance.

Based on a large literature review, the Pilot4Safety Project Handbook [4] concluded to the following consolidated definitions of RSA and RSI:

- Road Safety Audit describes a systematic and independent examination of a project designed to highlight potential safety issues at the earliest possible stage of planning and construction, to reduce or eliminate these problems and limit the risk for different types of road users.
- Road Safety Inspection is a preventive safety management tool implemented by road authorities/ operators as part of a global Road Safety Management. A RSI is a systematic field survey organised sufficiently frequently on all existing roads or sections of a road to secure adequate safety levels. It is carried out by trained road safety experts to identify hazardous conditions and deficiencies that may lead to serious accidents. RSI results in a formal report on detected road hazards and safety issues.

The Directive on Tunnels (2004/54/EC) makes reference to periodic inspections carried out by the tunnel's Inspection Entity at maximum intervals of 6 years for any given tunnel; it requires a Safety Documentation (Annex II), which describes the processes for approval of the design, for opening of a tunnel, for modifications in the physical and operational characteristics of a tunnel and for performing periodic exercises for tunnel staff and emergency services, and includes the content and results of a Risk Analysis.

Regarding the typical processes of the distinct Road/ Tunnel Safety procedures described above, both RSA/ RSI and tunnel safety inspection (TSI) contain the assignment from the Client/ Assignor (responsible authority/ body/ unit) and an independent approach by the assignee (Auditor/ Inspection Team) to perform the appropriate activities and report back, with interaction between two sides before and after the duration of these activities and reporting.

Therefore, compared to RSA/ RSI, other procedures are foreseen for safety assessment of tunnels that are subjected to the Tunnel Directive.

To this end, the ECOROADS objective was to experiment on the incorporation of the tunnels' safety procedures in an integrated approach for joint safety operations at both tunnels and open roads, with focus on road safety. The ECOROADS approach is purely "operational" and finalized to an integrated practical approach that is going to be substantiated in practical guidelines and recommendations.

On the basis of the feedback from the 1st project Workshop held in September 2015 and the Seminar for exchange of best practices held in November 2015, and considering the experimental approach of the project, the following categories of involvement in the field tests process are foreseen:

- Infrastructure (Road/ Tunnel) Manager(s): the administration/ authority/ manager(s) of the road/ tunnel infrastructure of each of the ECOROADS field tests.
- Host organisation: The organisation/ authority that organises and facilitates the field test.

- Audit/ Inspection Group: the mixed international team of (road/ tunnel) experts and other stakeholders that will take part in a field test. It consists of the Core Audit/ Inspection Team, the "External" observers, the Facilitator , internal observer and other experts.
- Core Audit/ Inspection Team: the mixed international team of experts that are assigned/ authorised to jointly and independently perform an audit/ inspection visit.
- "External" Observers: stakeholders with different competences, representing different authorities accompanying the Core Audit/ Inspection Team in a field test.
- Facilitators: local/national experts ensuring organisation, communication and cooperation between the infrastructure manager(s) and the project.
- ECOROADS "Internal" Observer: A member of the ECOROADS consortium.
- Other "External Experts" and Stakeholders: other local and national interested parties (incl. road user groups) providing complementary information to each Core Audit/ Inspection Team.

The roles and responsibilities of the actors involved in the field tests and their interactions are schematically presented in Figure 4.

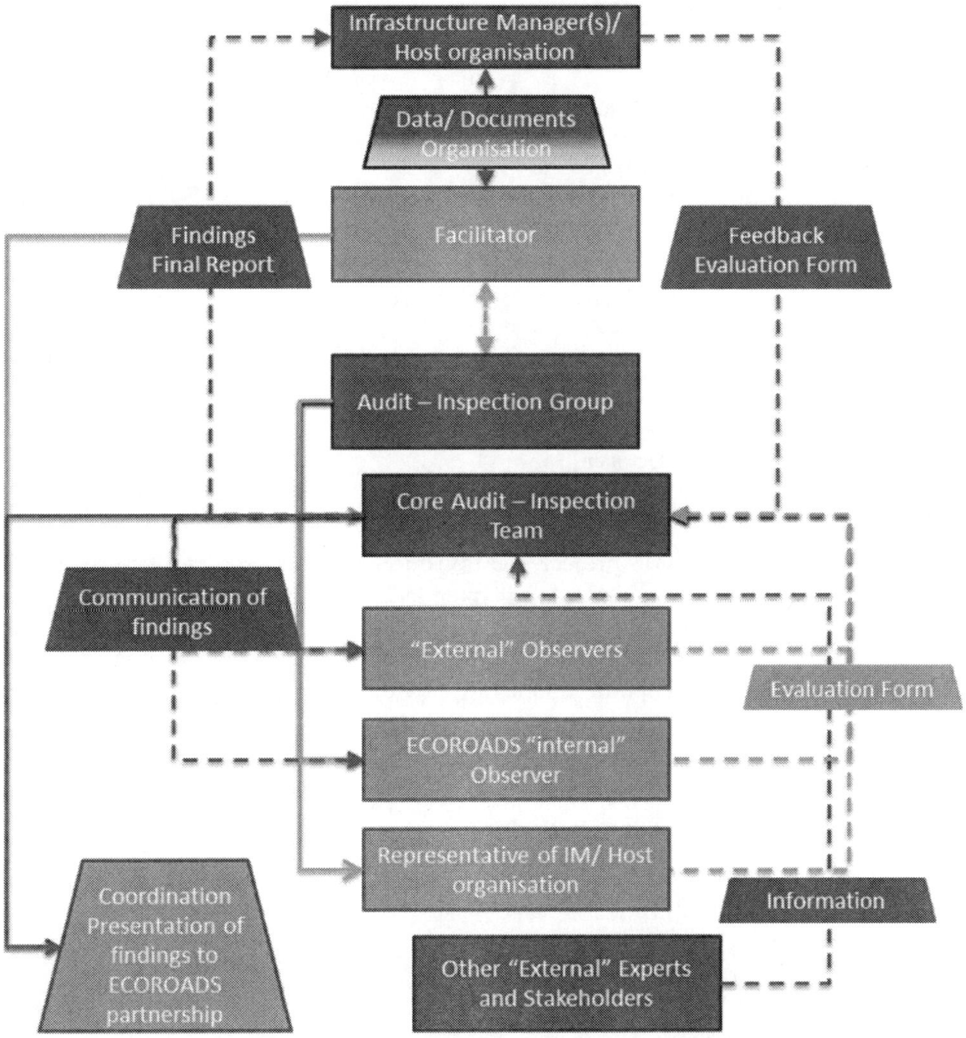

Figure 4. Roles and responsibilities of parties involved in ECOROADS field tests.

IOP Conf. Series: Materials Science and Engineering **236** (2017) 012096 doi:10.1088/1757-899X/236/1/012096

The Deliverable D5.1, available on the website of the ECOROADS project [5], describes the common organisational and technical details for the performance of the ECOROADS joint road safety operations. In the same deliverable it is also explained how these procedures are built on the basis of the Audit and Inspection approach agreed during the Pilot4Safety project.

3. Results

After a call for expression of interest, ECOROADS received 15 applications with a road stretch dossier and a letter of commitment duly signed by the owner of the infrastructure; in August 2015 a specific project committee selected 5 sites and then the operation started as in the following Table1.

Table 1. Overview of the 5 test sites.

TEST SITE, Country	Dates of the joint visits	N. of Experts (core team)	N. of Observers	N. of other Experts	Tunnel type and length	Length of open road inspected
KENNEDY TUNNEL, Belgium	07-08 March 2016	3	3	6	2 tubes, 690 m each	1200 m
KRRABE TUNNEL, Albania	05-06 April 2016	4	4	5	2 tubes, 2230 m and 2500m	1500 m
TUNNEL RENNSTEIG, Germany	17-18 August 2016	3	3	4	2 tubes, 7916 m each	400 m
TUNNEL STRAZEVICA, Serbia	27-28 Sept.r 2016	3	1	12	Single tube 745 m	650 m
TUNNEL DEMIR KAPIJA, Former Yugoslav Republic of Macedonia	18-19 October 2016	4	0	9	Single tube 554 m	400 m

For the scope of the ECOROADS project, the transition area between an open road and a tunnel has been defined: as a minimum requirement, it is intended as the sum of: a) the distance calculated as the distance covered in 10 seconds by a vehicle travelling at the speed limit before the tunnel portal and b) the stopping distance inside the tunnel after the portal, for a vehicle travelling at speed limit, if not identical with design speed. This minimum rule obviously applies on the opposite direction, as shown in Figure 5.

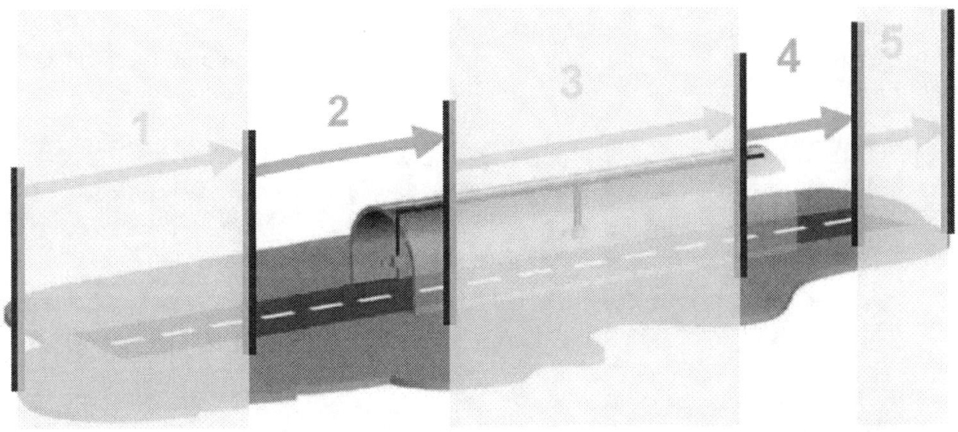

Figure 5. Transition areas 2 and 4 in ECOROADS field tests. Source [5].

4. Conclusions

At the time of preparing this paper, ECOROADS is preparing the final guidelines and recommendations. However, it is possible to resume some important messages the project is delivering:

- Joint safety operations are possible, as demonstrated by their successful completion in the five test sites;
- Joint safety operations are useful: there was an unanimous consent about the joint operations' added value;
- Cost effectiveness of joint safety operations. The complexity of the scheme in the previous Figure 4 was due to organisational and monitoring purposes originated by an external subject. When organized by the infrastructure manager, the scheme is much more simple, as shown by the following Figure 6. Moreover, if the joint operations are organized in parallel with the "current" RSA/RSI (thus having the road safety experts available on site), the cost of having an additional expert is low. It is relatively easy in case of the same Infrastructure Manager (IM), while coordination is needed in case of different ones. In both cases, there is a relevant cost-effectiveness in terms of enhanced safety.

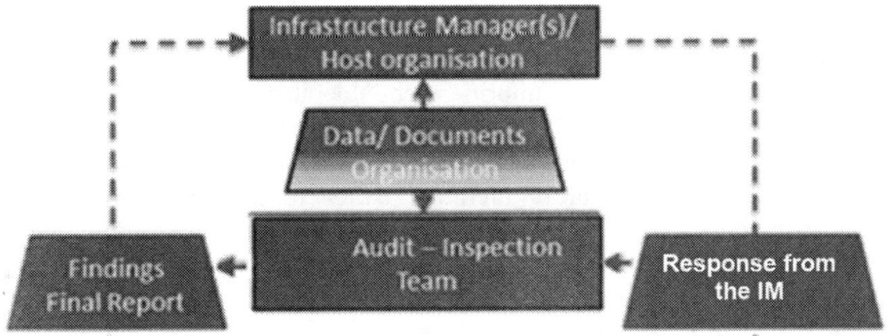

Figure 6. Simplified approach when operated directly by the IM.

- Involving a foreign expert does not necessarily imply more difficulties and adds value, because he/she brings a different approach and a different point of view to the safety team. At least the team leader and the foreign expert should be able to properly communicate.

An excellent cooperation of the Infrastructure Managers (IMs) has been noted: sometime they immediately reacted to the inspection reports by taking immediate countermeasures, like in the following Figure 7.

Figure 7. Installation of guardrail in front of perpendicular wall of lay-by in Krrabe tunnel (source: Hasani A., Albanian Roads Authority).

The following key points and issues are based on the evaluation of the five joint safety operations, exchange of best practices and comments received during and after the 3 workshops with the major stakeholders.

- According to the project results, certain concepts of Directive 2008/96/EC (on road infrastructure safety management) can be applied in the scope of Directive 2004/54/EC (on tunnels) in close cooperation of the managing departments in the two areas.
- Road sections including tunnel sections should be inspected/audited from both tunnel experts and road safety experts.
- Transition areas between tunnels and open roads, as above defined, are of particular interest in terms of their impact on road safety.
- An innovative update of the new safety standards following the technical developments is welcomed. A harmonized approach regarding fire detection, fire-fighting and communication coverage in tunnels should be addressed.
- Member States, as supervision authorities, should ensure the mutual recognition of Road Safety Auditors and Road Safety Inspectors certified by other Member States.
- A coordinated approach to the road safety management of both tunnels and the transition areas is recommended. This will surely facilitate better future integration of road and tunnel infrastructure, taking also into account the costs and benefits of deploying intelligent transport systems and services.
- Since the majority of road fatalities in the EU occur outside the TEN-T, an extension of scope beyond the TEN-T to other roads should be considered.
- Exchange of experts and best practices should be enhanced and facilitated.

5. Final considerations

There are common elements regarding the safety management in the two areas (open roads and tunnels) which could be tackled in an harmonized way through a coordinated communication by the concerned open road and tunnel experts. Coordinated actions can be foreseen by adding/inserting harmonized legal texts in the bodies of the two Directives; any eventual insertion in one Directive should take into adequate consideration the consequences in the other Directive and vice-versa.

The above mentioned coordinated communication between road and tunnel managers (that will in any case maintain their specific roles and responsibilities) should not be demanded to their individual willingness, but somehow made compulsory and periodic, in order to ensure the concrete possibility of conducting joint safety operations.

References

[1] Pilot4Safety - Pilot project for common EU Curriculum for Road Safety experts: training and application on Secondary Roads (http://pilot4safety.fehrl.org)

[2] Study on the effectiveness and on the improvement of the EU legislation framework on Road Infrastructure Safety Management *(Directive 2008/96/EC)* http://ec.europa.eu/transport/facts-fundings/evaluations/doc/2014-12-ex-post-evaluation-study-road-infra-safety-mgmnt.pdf

[3] *Study on the implementation and effects of Directive 2004/54/EC on minimum safety requirements for road tunnels in the trans-European road network* http://ec.europa.eu/transport/facts-fundings/evaluations/doc/tunnel_final_report.pdf

[4] *Safety Prevention Manual for secondary roads: for the international training of road safety auditors and inspectors, complying with the EU Directive 2008/96/CE* www.aipss.it/manual-en.html

[5] www.ecoroadsproject.eu

New research opportunities for roadside safety barriers improvement

Giuseppe Cantisani[1], Paola Di Mascio[2] and Carlo Polidori[3]

[1] "Sapienza" University of Rome - Department of Structural and Geotechnical Engineering, Via Eudossiana 18, Rome 00184 (Italy)
[2] "Sapienza" University of Rome - Department of Civil and Environmental Engineering, Via Eudossiana 18, Rome 00184 (Italy)
[3] AIPSS Italian Association of Road Safety Professionals - via Bergamo 3, Roma 00198 (Italy)

E-mail: giuseppe.cantisani@uniroma1.it

Abstract. Among the major topics regarding the protection of roads, restraint systems still represent a big opportunity in order to increase safety performances. When accidents happen, in fact, the infrastructure can substantially contribute to the reduction of consequences if its marginal spaces are well designed and/or effective restraint systems are installed there. Nevertheless, basic concepts and technology of road safety barriers have not significantly changed for the last two decades. The paper proposes a new approach to the study aimed to define possible enhancements of restraint safety systems performances, by using new materials and defining innovative design principles. In particular, roadside systems can be developed with regard to vehicle-barrier interaction, vehicle-oriented design (included low-mass and extremely low-mass vehicles), traffic suitability, user protection, working width reduction. In addition, thanks to sensors embedded into the barriers, it is also expected to deal with new challenges related to the guidance of automatic vehicles and I2V communication.

1. Introduction

The importance and usefulness of road restraint systems (RRS) are uncontested principles of a scientific and technical approach to the road safety; the general aim is to enhance safety levels by means of these protection systems, because the infrastructure can substantially contribute to the reduction of accidents consequences.

Many products and devices are currently available in order to install longitudinal barriers, crash cushions, transitions, terminals, and so on. The research efforts performed in the last decades, in fact, together with the contribution of industry, have allowed to really improve the technical and physical characteristics of road safety systems [1] [2] [3].

However some performances of road barriers nowadays should be improved, especially for specific topics like: vehicle-barrier interaction, pre-crash information, traffic characteristics and working width reduction. The statistical analyses, in fact, highlight that the accidents in which the barriers are involved have a higher risk of injuries and fatalities. A study leaded in UK on the basis of the STATS 19 DataBase [4] (sample: 1,584,605 accidents on major roads in England, Scotland and Wales, in 1992-2005, involving 3,029,100 vehicles and resulting in 2,233,288 casualties) showed that the severity of crashes involving RRS is higher than the general one. In RRS-involving incidents 2.0% of occurrences are fatal (1.4% for the general incidents); 13.9% serious (12.9% for the total) and 84.0% slight (85.7%).

To deal with this problem, above all it seems necessary to develop new barrier shapes or structural design, in order to moderate crash severity in the case of barriers having high (or very high) containment level. Furthermore, the dynamic interaction of these restraint systems with low mass vehicle and/or very low mass vehicles is another interesting research objective [5] [6] [7], because the deformation of barrier and the controlled redirection of vehicles can significantly reduce the injury risk for users.

On the basis of a review of the technological advancement history and of the analysis of the most critical points in the state of the art, new developments can be outlined, especially by using advanced research tools like simulation models and high-performing materials. The new proposed perspectives deal with these issues and aim to propose new research actions to improve the safety performances of RRS.

2. State of the art analysis

According to a commonly accepted approach, the road safety can be ensured by active and passive measures [8] [9], where the former are devoted to the accidents prevention and the latter are aimed to the protection of users and goods by mitigating the consequences of critical events.

In the field of passive safety factors, the RRS play an important role, since they can significantly contribute to the reduction of risk (especially for run-off-road accidents). As regard of technology of roadside systems, some important developments have been achieved in the last decades, regarding both physical and structural characteristics of the devices.

In particular, for the specific case of longitudinal road barriers, various generations of them have been developed and used, over the time, especially along the UE road networks.

We can remember, for example, about thirty years ago a sort of revolution that changed the conception of roadside safety; in fact, after the previous positive experiences in the U.S., the introduction and the diffusion of the modern concrete barriers came in Europe. New shapes and profiles (in particular: the New Jersey profile [11] [13], see Figure 1) were successfully experimented and new restraint systems were largely installed along primary roads, especially because of their containment performances, using both precast technology and in-situ construction processes.

Figure 1. The New Jersey concrete barriers came in Europe between '80s and '90s of the 20th century, after the previous positive experiences in the U.S.

Later on, in the middle of '90s, the steel industry proposed other high performing devices (three-waves barriers [11] [13], see Figure 2), able to ensure the same containment levels of the concrete barriers. Since then, however, the main technical conception of roadside safety systems have no more significantly changed.

Figure 2. The steel barriers can ensure high containment performances by means of the three-waves profile.

Nowadays, it is important to consider that the vehicle characteristics are not comparable to those of twenty, fifteen or just ten years ago (Figure 3). Technical features like mass, dimensions, shapes, crashworthiness, speed have really changed. Consequently, the typical crashes and their severity are changed too: in particular the protection of vehicle's occupants should be differently addressed, especially because accidents can involve vehicles having various characteristics (trucks or passenger cars, but also motorcycles, low mass vehicles, and so on) [10].

Figure 3. The comparison between road vehicles can immediately show the today's needs for improvement of barrier technology.

In addition, there are new problems related to the automatic guidance technologies and the coexistence between automatic vehicles and traditional ones [14] [15]; in both cases the roadside systems are requested to ensure high containment and low severity characteristics.

In fact, the road barriers can be also used for contributing to the automation technology [16] but, mainly, they have in charge an important role for the roadside safety; these aims require that the marginal spaces of a road infrastructure should be well designed and effective restraint systems should be installed there.

For all the above presented reasons, now it is the right time to improve the features of road edges and performances of roadside safety systems, adapting them to the new needs coming from vehicles and safety issues. A new generation of roadside systems and devices can be explored by means of original and well-addressed research activities.

3. New research opportunities

The control of road exercise and the protection of road users still represent important challenges for the next decades. Big opportunities, in this sense, are provided by possible improvements of road edges, both for passive and active safety issues.

Considering that the road carriageway normally does not allow deep changes in its features, the new problems related to changes in vehicles characteristics should be addressed toward a new conception of road edges and roadside devices and systems. These parts of road infrastructure, in fact, can be developed in order to assist vehicles guidance, to avoid some kinds of accidents and to mitigate their consequences when accidents happen.

The new research opportunities which can be now investigated essentially concern how to obtain barriers having high (or very high) containment level, while maintaining moderate crash severity level. In particular, it is important to control the dynamic interaction of restraint systems with low mass vehicle, in order to obtain lower injury risk for passengers, by means of vehicle post-crash redirection.

To consider how the current standards in concept and design of RRS can be implemented, it is necessary to remind some theoretical fundamentals. The longitudinal road barriers work by means of a complex combination of processes and effects due to their primary properties. In fact, they are able to achieve both containment of errant vehicles and mitigation of crash effects because of their structural resistance and their aptitude to dissipate the kinetic energy transmitted by colliding bodies.

These principles are valid for both concrete and steel barriers, but while the former generally dissipate the energy by means of friction effects on their base, the latter can rely on their internal deformations that allow to absorb a wide part of the crash energy [17] (Figure 4). In addition, the shape and the structural behaviour of the barriers are defined with the aim to redirect the vehicles after the crash, so maintaining a controlled trajectories and avoiding to involve other vehicles in the accident.

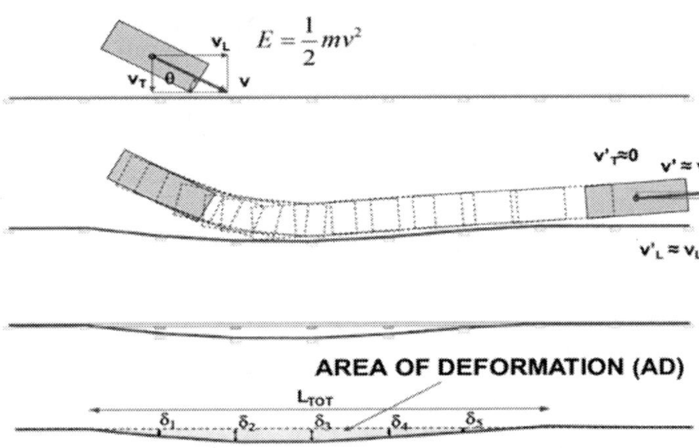

Figure 4. Impact phenomenon evolution for run-off crash collision and energy dissipation by internal and external deformation [17].

In order to pursue the described working principles, the barriers should be designed and realized on the basis of the actual dynamic interaction with the bumping vehicles, considering all possible types of vehicles that can hit the barrier and all crash conditions (i.e.: trajectories, speed and angles, ...). To optimize the barriers' conception and their design, it is essential to consider their detailed shape and the properties of material they are made of.

After the important advancements achieved in the last decades, the design criteria and the construction technology of road safety barriers can now further be improved by means of an advanced use of the properties of different materials, also considering their combination and some integration opportunities.

In addition, it seems important to explore also new shapes and profiles [18], with the aim to obtain a more performing behaviour of the restraint systems under crashes. A comprehensive assessment of the whole technical properties of barriers should also include construction technologies and processes; these can be enhanced, for example, by means of widening of the in-situ activities, in order to obtain a more effective adaptation to the local conditions and characteristics of roads.

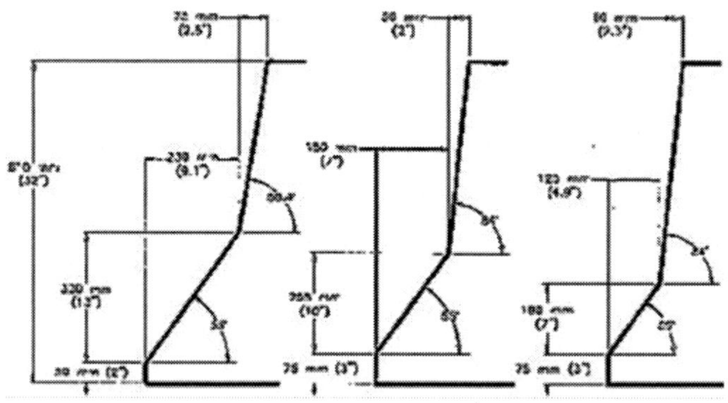

Figure 5. Example of evolution of shapes for concrete safety barriers [11].

It seems now necessary to overpass the traditional division between concrete and steel barriers and/or between high and low containment (low and high severity level) properties. A new generation of road barriers (and/or other restraint systems) should combine different materials, to allow a multi-stage deformation behavior in order to optimize the reaction of barriers in case of crash with both light vehicles and heavy ones.

The research topics, above presented, can be investigated by means of FE models and crash simulations; however, several laboratory tests are needed in order to calibrate the models and to obtain a better knowledge of dynamic actions during impact events.

Another important item should be included in the research on the design and construction processes of road barriers: the integration with sensors and devices needed for the transition toward the guidance of autonomous vehicles. In fact, the predictable developments of the road system and research trends for vehicles, propose a new request in order to provide the road infrastructure of sensors and systems needed for these aims. In this sense, the road edges and, particularly, the restraints systems are the most suitable elements for obtaining the demanded improvements. Special wireless sensors could be embedded in the road barriers, so providing a continuous reference path along the road (Figure 5).

The sensors should preferably be conventional and common devices, in order to ensure easy and low-cost installations together with the interoperability of the system. Nevertheless, it is important to study various possible difficulties or problems regarding the reliability and the effectiveness of the system addressed to the autonomous vehicles guidance (as an example: the shadow areas). The strengths and weaknesses of the integration between safety barriers and wireless sensors has to be specifically evaluated, considering the consistency and redundancy of the system, in case of crashes or other undesired events, that can occur during the road exercise.

Figure 6. Roadside barriers can also provide a I2V infrastructure, but their main task still remain the passive safety performances for both traditional vehicles and innovative ones.

4. Expected Impact

The proposed approach to the research on a new generation of RRS is mainly aimed to consider vehicles characteristics, properties of component materials, safety performances, and I2V communication.

As regard of this last feature, on one hand the future restraint system could enhance the "passive forgiveness" of the roadside respect to light and heavy vehicles: I2V communication warns the vehicle that the upcoming impact will be against a barrier. Therefore, the passive safety devices will activate with the specific countermeasures as earlier time to fire, deployment of side airbags, etc.. On the other hand the barrier will be able to communicate its own plastic-elastics characteristics to the vehicle's on-board intelligence, allowing future generation vehicles to optimize the internal countermeasures, e.g. to set the optimal pre-tensioning of safety belts. The system will be able to communicate the exact location of the impact and its severity to the road operator and the emergency services: this will be extremely useful during the transitional period needed before having a new generation of intelligent vehicles on the roads. In the future, it will also act as a redundant safety system for those new vehicles running on the road network.

More general, the barriers technology implementation will a relevant contribution to the reduction of fatalities and injuries. According to a CEDR publication on forgiving roadside [18], in fact, 45% of fatal accidents are single vehicle, primarily classified as run-off-road accidents, where the vehicle leaves the road and enters the roadside. This indicates that working on a more forgiving roadside can actually have a high safety impact.

After all, a large part of deaths and injuries in run-off-road crashes is related to the imperfect behaviour of the RRS; this is due to a well-known problem: to obtain high containment levels (especially for HGV) the barriers must be very rigid and slightly deformable; but this increases the severity of crashes for passengers cars and low-mass vehicles, together with potential injuries suffered by their occupants. The main research area to provide advancements in the technology of RRS is the pursuit of a better balance between containment, deformation and severity crash level, especially in all those road stretches having very limited space for roadside barriers.

In fact, a new kind of specifically studied restraint system will allow to deal with the problem of RRS installation in sites where there is not enough side space and the design solution leads to ensure the containment of heavy vehicles. This will be a noticeable contribution to the development of "forgiving roadsides", also for some special traffic components.

New researches should also provide proposals for improving technical standards and regulations, both for roadside systems production and installation. Several stakeholders (institutions, professionals and economic operators) are requesting a revision of technical standards and regulations, since restraint systems have a high part in national markets and solutions often differ between countries. Technical standards develop very slowly compared to the situation on the roads: obtaining an updated protocol of installation and use of types of barriers would improve the safety of all roads around Europe. This fact will contribute to obtain a global view of the accidents with the same road conditions and it will be able to compare and analyse accidents between different countries. Moreover, it would be easier to implement safety improvements, protocols and techniques to decrease the accidents in all the interested Countries.

5. Conclusion

After the evaluation of the state of the art and the recall of theoretical fundamentals of roadside safety, it is clear that an improvement of concept and technical characteristics of RRS is now actually needed. The evolution of vehicles, circulation and safety requirements, together with the new challenges due to autonomous vehicles guidance and I2V communication needs, make evident the deficiencies of currently adopted systems respect to the requested performances.

Thus, the paper has highlighted new research opportunities in this field, based on a new approach that overpasses the traditional division between concrete and steel barriers and/or between high and low containment (low and high severity level) properties.

The general objective of a new research is the development, demonstration and validation of a new generation of road barriers and restraint systems, combining different materials, to allow a multi-stage deformation behavior in order to optimize the reaction of barriers in case of crash with both light vehicles and heavy ones. The research can also develop and test innovative uses and functions for roadside devices: in particular, the embedded wireless sensors and other ITS devices, with the aim to provide an advanced infrastructure for autonomous vehicle guidance, pre-crash information (or other on-board safety systems), I2V and V2I communication, as well as the information about conditions of the barrier itself and its maintenance needs.

New barriers should be able to optimize their behavior in case of crash with both heavy and extremely low-mass vehicles and, at the same time, should realize an infrastructure for I2V communication, so promoting an effective advancement of the industrial and commercial standards in the field of RRS technology.

Proposals for improving technical standards and regulations are also requested, both for the production and installation of roadside systems, since a revision of these documents is being waited by various institutions, professionals and economic operators.

Finally, the potential development of new barriers, like the ones whose requested performances have been here described, will contribute to improve the passive safety protection currently offered by road infrastructure. This could be an important strategic target for research, industry and economical system, especially considering that in a large part of European road networks the existent RRS must be implemented or substituted by the next 20 years.

References

[1] Bates L, Soole D and Watson B 2012 The effectiveness of traffic policing in reducing traffic crashes *Policing and security in practice* (Palgrave Macmillan UK) pp 90-109
[2] Hollnagel E 2016 *Barriers and accident prevention* (UK: Taylor & Francis Limited)
[3] Savolainen P T, Mannering F L, Lord D and Quddus M A 2011 The statistical analysis of highway crash-injury severities: a review and assessment of methodological alternatives *Accident Analysis & Prevention* **43**(5) pp 1666-76
[4] UK Government 2006 *STATS 19 Database: Review of Road Accident Statistics* (UK: Crown Copyright)
[5] Reynolds C C, Harris M A, Teschke K, Cripton P A and Winters M 2009 The impact of transportation infrastructure on bicycling injuries and crashes: a review of the literature *Environmental health* **8**(1) p 47

[6] Teschke K et all 2012 Route infrastructure and the risk of injuries to bicyclists: a case-crossover study *American Journal of Public Health* **102**(12) pp 2336-43
[7] Moudon A V and Lee C 2003 Walking and bicycling: an evaluation of environmental audit instruments *American Journal of Health Promotion* **18**(1) pp 21-37
[8] Elvik R, Høye A, Vaa T and Sørensen M 2009 *The handbook of road safety measures* (Emerald Group Publishing Limited)
[9] Jarašūniene A and Jakubauskas G 2007 Improvement of road safety using passive and active intelligent vehicle safety systems *Transport* **22**(4) pp 284-9
[10] Ulfarsson G F and Mannering F L 2004 Differences in male and female injury severities in sport-utility vehicle, minivan, pickup and passenger car accidents *Accident Analysis & Prevention* **36**(2) pp 135-147
[11] McDevitt C F 2000 Basics of concrete barriers *Public Roads* **63**(5) pp 10-14
[12] Ross Jr H E and Sicking D L 1984 Guidelines for placement of longitudinal barriers on slopes *Transportation Research Record 970* pp 3-8
[13] Cota K A 2011 *Roadside Design Guide* (Washington: AASHTO)
[14] Nkoro A B and Vershinin Y A 2014 Current and future trends in applications of Intelligent Transport Systems on cars and infrastructure *Proc. 17th Int. Conf. on Intelligent Transportation Systems - ITSC* (IEEE) pp 514-19
[15] Qian X, Gregoire J, Moutarde F and De La Fortelle A 2014 Priority-based coordination of autonomous and legacy vehicles at intersection *Proc 17th Int. Conf. on Intelligent Transportation Systems - ITSC* (IEEE) pp 1166-71
[16] Coconea L D 2015 *Analysis, simulation and testing of ITS applications based on wireless communication technologies* (Politecnico di Torino)
[17] Bonin G, Cantisani G and Loprencipe G 2005 Analysis of Laboratory Data from Crash Test on Road Safety Barriers *Proc. III Intern. SIIV Conf. on People, Land, Environment and Transport Infrastructures: Reliability and Development (Bari)*
[18] Hou S, Zheng Y, Xie J and Han X 2014 Optimization design of NJ shaped guardrail based on collision safety consideration *International Journal of Computational Methods* **11**(06) 1350083
[19] IRDES ERA NET 2012 *Forgiving roadsides design guide* (Conference of European Directrors of Roads)

IOP Conf. Series: Materials Science and Engineering **236** (2017) 012098 doi:10.1088/1757-899X/236/1/012098

Simulation of crash tests for high impact levels of a new bridge safety barrier

Jiří Drozda[1] and Tomáš Rotter[1]

[1] Czech Technical University in Prague, Faculty of Civil Engineering, Thákurova 7, Prague 6 166 29, Czech Republic

E-mail: jiri.drozda@fsv.cvut.cz, rotter@fsv.cvut.cz

Abstract. The purpose is to show the opportunity of a non-linear dynamic impact simulation and to explain the possibility of using finite element method (FEM) for developing new designs of safety barriers. The main challenge is to determine the means to create and validate the finite element (FE) model. The results of accurate impact simulations can help to reduce necessary costs for developing of a new safety barrier. The introductory part deals with the creation of the FE model, which includes the newly-designed safety barrier and focuses on the application of an experimental modal analysis (EMA). The FE model has been created in ANSYS Workbench and is formed from shell and solid elements. The experimental modal analysis, which was performed on a real pattern, was employed for measuring the modal frequencies and shapes. After performing the EMA, the FE mesh was calibrated after comparing the measured modal frequencies with the calculated ones. The last part describes the process of the numerical non-linear dynamic impact simulation in LS-DYNA. This simulation was validated after comparing the measured ASI index with the calculated ones. The aim of the study is to improve professional public knowledge about dynamic non-linear impact simulations. This should ideally lead to safer, more accurate and profitable designs.

1. Crash barriers

The basic function of the crash barriers is to redirect uncontrollable cars and minimize occupant's injuries. Nowadays the impact severity, energy dissipation and stiffness of the new designs of crash barriers are evaluated with real crash tests. The requirements for these crash tests are prescribed in the European standard EN 1317 [1] and they have to be fulfilled to gain the certificate for using crash barriers in the European Union. The tests must be performed under different conditions, e.g. using passenger car and lorry, with various impact speeds and impact angles. To reduce the development and testing costs of new crash barrier designs, numerical analysis can be used for pre-evaluation of the crash barrier behavior.

2. Model validation

The process of model validation was described in detail in [2]. Basically validation in the context of roadside safety computer simulations is the process of determining the degree to which a roadside safety computer model is an accurate representation of real world crash tests from the perspective of accurately replicating EN 1317 crash test evaluation parameters, the structural performance of the barrier and the response of the vehicle. Validation always involves comparing a computer simulation or numerical experiment to a physical experiment of some type. The question at the root of the validation exercise is whether the simulation replicates the physical experiment and whether it can be used to explore and predict the response of a new or modified roadside hardware in the real world. Validation activities can be performed not only for full-scale crash tests but also for material models, sub-assemblies or components of a full FE model. Ideally, each portion of a large complex model should be validated separately if it is possible to perform a meaningful physical test. This work shows some possibilities to apply Numerical Modal Analysis (NMA) for model validation and mesh calibration. The application of NMA is shown on a prototype design of the bridge crash barrier post.

2.1. Numerical modal analysis

Modal analysis, or a free vibration analysis, is performed to obtain the natural frequencies and mode shapes of a structure. It is a subset of the general equation of motion. In the analysis, the behavior of the structure was assumed to be linear and the response to be harmonic. For solving the damping was not included. ANSYS - Workbench was chosen mainly because the verified FE model was used for the simulation of a crash test. The procedure of verification was based on comparing modal characteristics gained from the FE model and modal characteristics measured on the prototype. The goal of this comparison is to check the appropriate mesh density and to control the geometry 3D model. The first step was to create the 3D geometry, see Figure 1. This shell 3D model was created by using DesignModeler. Parts of the post were divided into subparts, to correspond to the distribution of locations, where excitation forces were applied in the real test. This step is important and necessary for future evaluation and verification of the FE model. It was essential to ensure mesh continuity across subparts of the same plate in order to preserve its real behavior.

The next step was defining material properties and boundary conditions. A mechanical mesh was used with hex dominate elements including mid side nodes. The size of elements was adapted to geometry. Only one boundary condition was defined at the place where the post is bolted to the concrete ledge. The connection was modeled by fixing all degrees of freedom at the edges of bolts holes. The Block Lanczos solver was used to compute the mode shapes.

2.2. Prototype design

A new design of the bridge crash barrier post was created in cooperation with the companies OK-BE and CTU. It consists of two parallel steel plates, which are fillet welded into base plates.

a) b) c) d)

Figure 1. a) Prototype post; b) Certified post; c) Mesh of prototype post; d) Mesh of certified post.

The biggest advantage of this post is that there are no parts, where the dirt can cumulate and the whole construction of the crash barrier can be cleaned by a sweeper-flusher. The serviceability of the bridge crash barrier increases. Another challenge of the designer was to make the post lighter and easy to weld, which makes it more economical. The prototype post was assembled and attached into the massive concrete block by steel bonded anchors. For the comparison between the prototype and the certified post, see Figure 1.

2.3. The measurement
The single point testing was chosen for the experimental modal analysis. It is the most straightforward method of the phase separation techniques. Phase separation techniques rely upon a mathematical assumption that the actual responses are formed from a linear combination of the modes. The forced responses to a known excitation are measured and then the dynamic properties are extracted by means of mathematical curve fitting techniques.

An impact hammer was used to strike the structure at the excitation points. A very short sharp excitation pulse was produced – approximating a Dirac delta function – which has a flat spectrum over a wide frequency range. The amount of energy contained in the impact pulse is small.

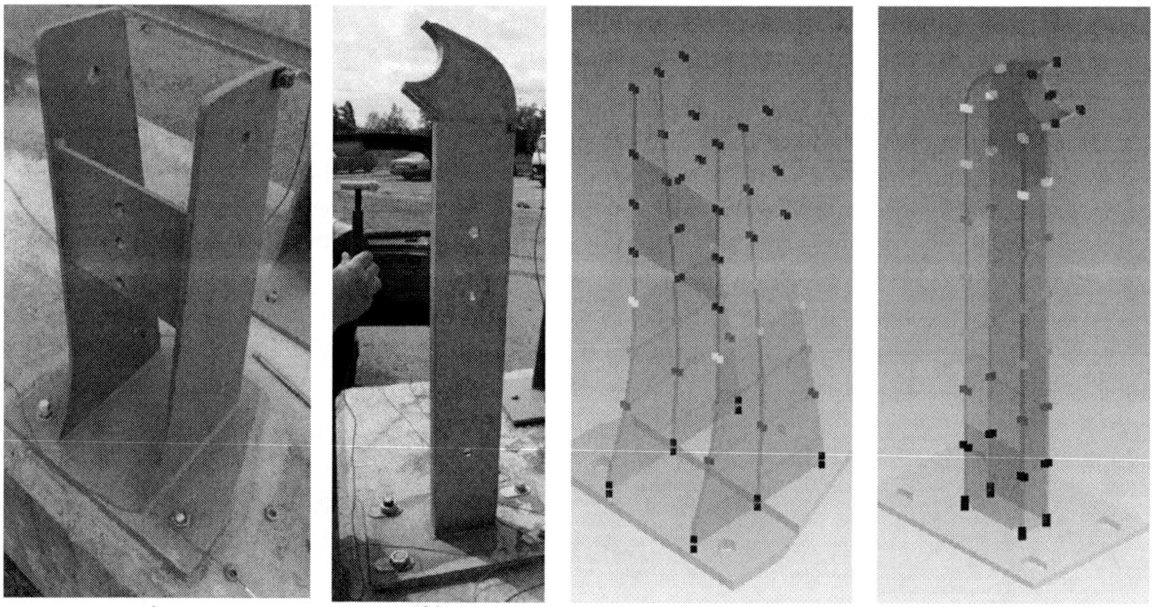

a) b) c) d)

Figure 2. Mounted transducer: a) prototype; b) certified post; Excitation points: c) prototype; d) certified post.

In Figure 2 the mesh of excitation points and the place where the transducer was located are visible. It is important to note, that the reference transducer location must not be in a nodal point of measured natural modes. If the ordinate of a natural mode in a reference point is too small, that natural mode and its corresponding natural frequency cannot be measured. This is one of the reasons for a preliminary NMA, in order to determine the optimal referent transducer placement.

Each excited point was struck 5 times by the impact hammer. Resultant readings were averaged for each point. Both signals, the excitation force and the acceleration, were recorded in the time domain and transformed using Fast Fourier Transform (FFT) to the frequency domain. The Frequency Response Function (FRF) was evaluated from these signals.

2.4. The comparison

The agreement between measured and computed natural mode shapes was evaluated according to MAC values, which were obtained using formula (1) from [3] and [4]:

$$\mathrm{MAC}\left(\{\Phi_X\}_i,\{\Phi_A\}_j\right)=\frac{\left|\{\Phi_X\}_C^T,\{\Phi_A^*\}_M\right|^2}{\left(\{\Phi_X\}_C^T,\{\Phi_X^*\}_C\right)\left(\{\Phi_A\}_M^T,\{\Phi_A^*\}_M\right)} \quad (1)$$

In Table 1 the natural frequencies and the corresponding MAC values, according to Czech technical standards [4] are listed. MAC values near 1.0 refer to perfect match between the compared natural modes, while values near 0.0 refer to perfect orthogonality.

Table 1. Modal assurance criterion (MAC).

	Comp	1	2	3	4	5	6	7	8	9	10
Meas	$f_{(j)}$[Hz]	50.8	154.0	237.9	251.1	258.7	275.0	338.3	413.4	431.2	556.7
1	**52.963**	**0.975**	0.002	0.045	0.001	0.007	0.001	0.072	0.001	0.036	0.001
2	**168.320**	0.037	**0.896**	0.029	0.066	0.012	0.014	0.035	0.000	0.002	0.018
3	**279.583**	0.007	0.006	0.007	0.008	**0.913**	0.035	0.001	0.017	0.09	0.000
4	**299.926**	0.010	0.060	0.014	0.336	0.017	**0.893**	0.025	0.069	0.007	0.008
5	**331.327**	0.037	0.153	**0.527**	0.081	0.003	0.079	**0.511**	0.013	0.008	0.032
6	**447.346**	0.006	0.001	0.005	0.07	0.035	0.251	0.002	**0.834**	0.078	0.053
7	**465.166**	0.046	0.001	0.004	0.005	0.208	0.008	0.054	0.021	**0.959**	0.004
8	**596.518**	0.002	0.018	0.004	0.008	0.006	0.014	0.032	0.05	0.045	**0.733**

Natural mode shapes were obtained from FRFs using polynomial curve fitting method. The comparison of the prototype´s first 4 measured and computed mode shapes is visible in Figure 3.

a) b) c) d)

Figure 3. Natural Mode Shapes: a) 1st mode; b) 2nd mode; c) 3rd mode; d) 4th mode.

3. Finite element simulation of a crash test

The FE simulation of crash tests requires special consideration about modeling parts such as vehicles, vehicle restrain systems and contacts between them. In this part, the primary focus is on the modeling and the performing of dynamic analysis of a bridge safety barrier. Due to large dynamic deformations, it is suitable to employ explicit time integration schemes like LS-DYNA (LSTC, 2002). The FE analysis consists of these main steps: creation of the geometry model and discretization, specification of the material constitutive law and element types for LS-DYNA, performing the calculation and evaluation of results. By the creation of the FE model it is necessary to keep in mind that the calculation time is highly dependent on the size of the mesh, which determines the minimum computational time step, the definition of the material's failure criteria and the selection of an impact contact algorithm [5].

3.1. Vehicles models

Two crash scenarios according to EN 1317 were simulated. The first one, called TB 11, involves a small passenger car (900 kg). The geometrical details of this vehicle are similar to Ford Fiesta for which the FE model Geo Metro Passenger Sedan could be used. This model was developed through the process of reverse engineering at the National Crash Analysis Center (NCAC) of The George Washington University (GWU) [6], (http://www.ncac.gwu.edu) see Figure 4. It contains 193 200 elements. The model includes fully functional capabilities of the suspension and steering subsystems and the whole model was validated by a frontal wall impact test.

Figure 4. Model of passenger vehicle Geo Metro Passenger Sedan.

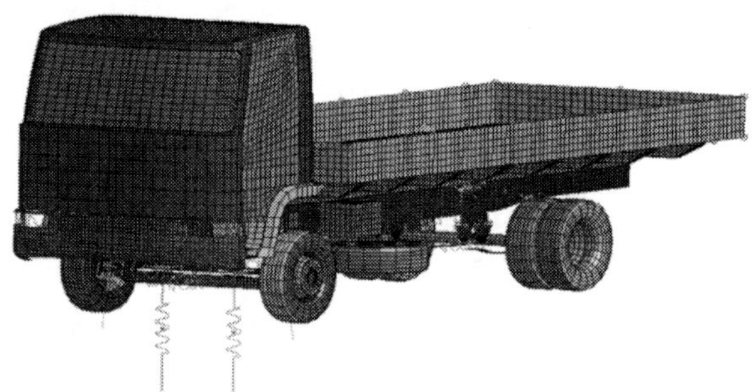

Figure 5. Model of lorry Rigid HGV.

The second simulated impact test is TB42 which covers an impact scenario with a small lorry (10000kg). The model of this vehicle was developed by Politecnico de Milano (POMI). The geometry of this vehicle model is very similar to the geometry of the lorries which are used for full-scale tests in Europe. This model contains real models of tires, frontal and rare axel and also suspension and steering subsystems. The whole model contains 254 000 elements, see Figure. 4.

3.2. Safety barrier

The two FE models of bridge safety barrier used in this study are: the certified type ZSH4 [7] Figure 6, and the newly developed type Figure 7. Each model includes 30 m length of safety barrier. In both models solid elements have been used for the creation of the bridge ledge and shell elements have been used for steel sheets of safety barriers.

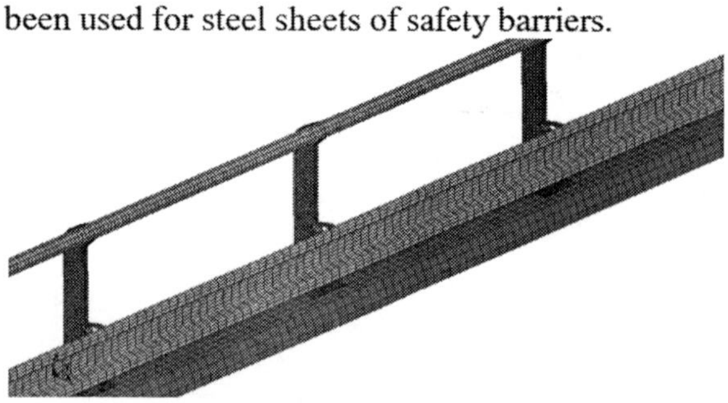

Figure 6. Certified type of vehicle parapet ZSH4.

Figure 7. Newly developed type of vehicle parapet.

The solid elements defined were of type 1, which is a sub-integrated element type with constant stress. With the proper setup of the hourglass stabilization algorithms, this element type has been proved to be very efficient and accurate for crash simulations. The shell elements of steel sheets were considered to be of type 16. This is a fully integrated element type, which delivers accurate results with the proper hourglass control definition. In the case of the newly developed type of bridge safety barrier the accurate modeling of bolts was achieved through beam elements. The beam elements were used due to, that the failure of bolts could be simulated. The connection between beam elements and other steel sheet parts was modeled using a contact algorithm. For this purpose, the bolt nuts and heads were modeled with shell elements, which were included in the contact definition between steel sheets and bolt nut or head.

4. Discussion of results and conclusion

A full-scale crash test was performed for a new design of a vehicle parapet. The guardrail was mounted on the standard concrete ledge with a distance of 1500 mm between posts. The full-scale crash test was performed for the scenario TB1 according to EN 1317. The results of this real crash test were compared with the results of the numerical simulation. Acceleration Severity Index (ASI), a sequence of plots and contact length was used for comparing the results and estimate how good is the representation of the real world crash tests. Firstly, the computational and experimental ASI index histories were compared as shown in Figure 8. The ASI index for the tested system does not exceed the limiting value in any case. The highest value (computational 1.5; measured 1.4), was observed when the guardrail hits the post. Secondly, the 100 km/h and 20 degree impact resulted in an acceptable performance as illustrated in Figure 9. The vehicle was redirected and the contact length was 5.92 m. A similar behaviour was noted in the finite element simulations as shown on the left side of Figure 9.

The usage of computational model provided comparable results to experimental measurements and can thus be used for the computational evaluation of other road safety barriers in order to avoid numerous expensive full-scale crash tests. The tests have also shown that the new safety barrier assures controllable crash energy absorption which in turn increases the safety of vehicle occupants.

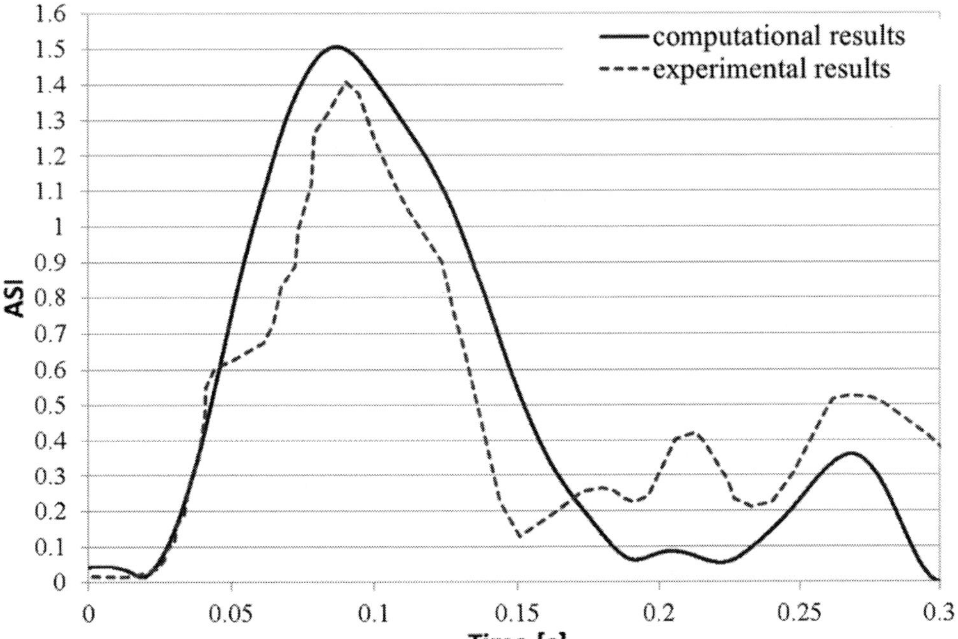

Figure 8. Comparison of ASI.

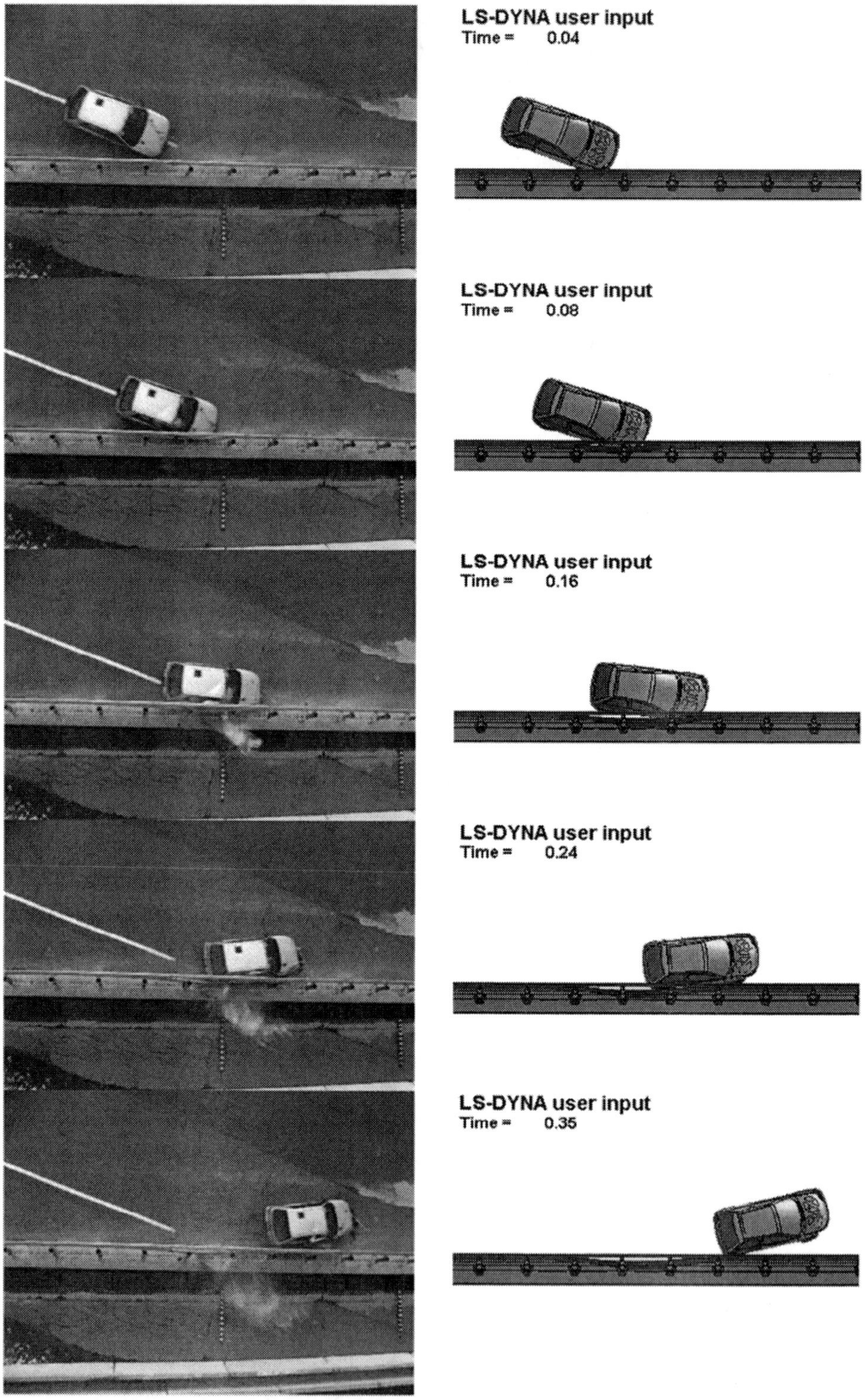

Figure 9. Full-scale test and FE simulation of a 900-kg car.

Acknowledgements
This work was funded and administered by the Faculty of Civil Engineering, Czech Technical University in Prague (CTU). Additional funding was provided by Technology Agency of the Czech Republic (TAČR) within the projects TAČR no. 01031225 and TE01020168 - CESTI.

References

[1] European Committee for Standardization (CEN) 2010 Silniční záchytné systémy - Část 2: Svodidla - Funkční třídy, kritéria přijatelnosti nárazových zkoušek a zkušební metody /Road Restraint Systems – Part 2: Performance classes, impact test acceptance criteria and test methods for safety barriers including vehicle parapets/ (ČSN EN 1317-2)
[2] Ray MH, Plaxico CA and Anghileri M 2010 Procedurs for verification amd validation of computer simulations used for roadside safetz applications *Final Report for NCHRP Projekt 22-24*
[3] Český normalizační onstitut /Czech Standards Institute/ 1984 Dynamické zkoušky stavebních konstrukcí /Dynamic tests of building structures/ *(ČSN 73 2044)*
[4] Český normalizační onstitut /Czech Standards Institute/ 2005 Zatěžovací zkoušky mostů /Load tests of bridges/ *(ČSN 73 6209)*
[5] Ulker MBC and Rahman MS 2008 Traffic barriers under vehicular impact: from computer simulation to design guidelines *Comput-Aid Civ Infrastruct Eng* **23** pp 465-80
[6] Opiela K, Kan S and Marzougui D 1997 *Handbook of Development & Validation of a Finite Element Model for the 1997 Geo Metro Passenger Sedan*
[7] Ministry of Transport of the Czech Republic 2011 Technické podmínky - Ocelové svodidlo ZSH2 *(TP 230)*

IOP Conf. Series: Materials Science and Engineering **236** (2017) 012099 doi:10.1088/1757-899X/236/1/012099

Increasing of visibility on the pedestrian crossing by the additional lighting systems

Richard Baleja[1], Petr Bos[1], Tomas Novak[1], Karel Sokansky[1] and Tomas Hanusek[1]

[1]Faculty of Electrical Engineering and Computer Science, VSB - Technical University of Ostrava, 17. listopadu, 70833 Ostrava, Czech Republic

E-mail: richard.baleja.st@vsb.cz

Abstract. Pedestrian crossings are critical places for road accidents between pedestrians and motor vehicles. For this reason, it is very important to increase attention when the pedestrian crossings are designed and it is necessary to take into account all factors that may contribute to higher safety. Additional lighting systems for pedestrian crossings are one of them and the lighting systems must fulfil the requirements for higher visibility from the point of view of car drivers from both directions. This paper describes the criteria for the suitable additional lighting system on pedestrian crossings. Generally, it means vertical illuminance on the pedestrian crossing from the driver's view, horizontal illuminance on the crossing and horizontal illuminance both in front of and behind the crossing placed on the road and their acceptable ratios. The article also describes the choice of the colours of the light (correlated colour temperature) and its influence on visibility. As a part of the article, there are case designs of additional lighting systems for pedestrian crossings and measurements from realized additional lighting systems by luxmeters and luminance cameras and their evaluation.

1. Introduction

The pedestrian crossings are one of the most dangerous places in the field of transport. The largest number of traffic accidents happens there. To limit the number of accidents, it is necessary to illuminate the crossing enough for pedestrian to be seen on the crossing.

The aim of this article is to get readers acquainted with the criteria which need to be followed when dealing with a design of lighting system for a pedestrian crossing, which would make the pedestrians more visible. In most cases, pedestrian crossings are illuminated insufficiently and they are in conflict with the applicable regulations. At the present time, the lighting systems of pedestrian crossings should meet the requirements of the Ministry of Transport of the Czech Republic - "Technické kvalitativní podmínky staveb a pozemních komunikací, Chapter 15 - Appendix 1 Přisvětlování přechodu". However, a new standard ČSN 36 045 is being prepared. It will deal with the illumination of infrastructure there will be also a detailed description of issues and requirements for the illumination of pedestrian crossing.

The majority of newly implemented pedestrian crossings disregard this regulation. In most of the cases, it is caused by ignorance of the issue and disregarding calculation of the vertical illuminance; in the projects, it is only the horizontal illuminance, which is evaluated in the place of pedestrian crossing. The crossing is obviously highlighted on the road, pedestrians can see which way to go but they may no longer be seen by the approaching driver because proportion of the vertical part of the crossing area can be insufficient for securing the necessary contrast between the pedestrians and the background. The shortage is mainly caused by misplacement of lights or by using lights with inappropriate luminous intensity curve. The lights are placed too close to the crossing transition and therefore it is often impossible to achieve the necessary vertical illuminance.

2. Division of the analysed areas at pedestrian crossings

The pedestrian crossing is divided into several sections; each of these sections has different requirements for additional lighting. Basic space of the pedestrian crossing determines a section which is situated directly on the place where the road is crossed. Unextended complementary space defines a section which is the entrance of the pedestrian crossing on the pavement. Extended complementary space is defined in cases when each driving direction is divided into the left and the right traffic lane by the traffic island.

In the Figure 1, the above-mentioned spaces of the pedestrian crossing are analogously illustrated. Marked areas apply for the left driving direction. For the opposite driving direction (view), the situation is analogous.

A = basic space
B = unextended complementary space
B‘ = extended complementary space

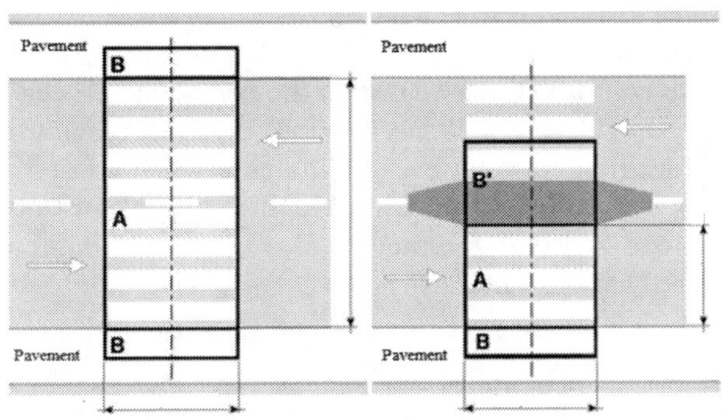

Figure 1. Typical pedestrian crossing (left) and the pedestrian crossing with a median strip (right). [3]

In the regulation "Technické kvalitativní podmínky (TKP15)", these spaces are defined in detail and lighting requirements for additional lighting of these spaces are described there. Logically, the requirements for the basic space illumination are slightly higher than the requirements for the complementary space illumination. It requires higher level of vertical illuminance there which must meet the given requirement for evenness whereas the evenness of illumination in the complementary space does not have to be evaluated. The evenness of illumination needs to be evaluated only in case of the extended complementary space. The choice of the appropriate type of light for the given pedestrian crossing is chosen on the basis of an evaluation of maintained average luminance or illumination of a particular roadway, on which the crossing will be set up. Table 1 depicts the required parameters for the pedestrian crossing illumination. Vertical illumination evaluates them at the height of 1 meter over the roadway from both driving directions. The requirement for the evenness of illumination for the basic and the extended complementary space is 0.4.

Table 1. Average maintained vertical illumination. [3]

Maintained value of the existing lighting		Average maintained vertical illuminance(lx)		
		The lowest		The highest
		Space		
Luminance of roadway ($cd.m^{-2}$)	Horizontal illuminance of roadway (lx)	Basic	Complementary	All spaces
$1.5 \leq \bar{L}$	$50 \leq \bar{E}$	No additional lighting system is set up		
$1.0 \leq \bar{L} < 1.5$	$30 \leq \bar{E} < 50$	75	50	200
$0.75 \leq \bar{L} < 1.0$	$20 \leq \bar{E} < 30$	50	30	150
$0.5 \leq \bar{L} < 0.75$	$10 \leq \bar{E} < 20$	30	20	100
$\bar{L} < 0.5$	$\bar{E} < 10$	15	10	50

It is important to stress the function of extended complementary space. It corresponds with the part of a crossing which is located directly on the roadway but in the opposite direction from the view of approaching driver. Extended complementary space is used as a protection for pedestrians in the middle of the roadway.

3. Making pedestrians more visible on the crossing

There are two ways to make pedestrian more visible while using lights. The first possibility is to achieve the so called negative contrast (Figure 2) when the pedestrian is seen as a dark-coloured figure on the light-coloured background. This effect can be achieved by doing nothing. It means that there is no lighting added on the pedestrian crossing even in the case when constellation of the crossing and illumination of the roadway is suitable. Luminaire which is placed far enough behind the crossing increases luminance of the road and thus the picture of a pedestrian is projected as a dark-coloured silhouette on the light-coloured background.

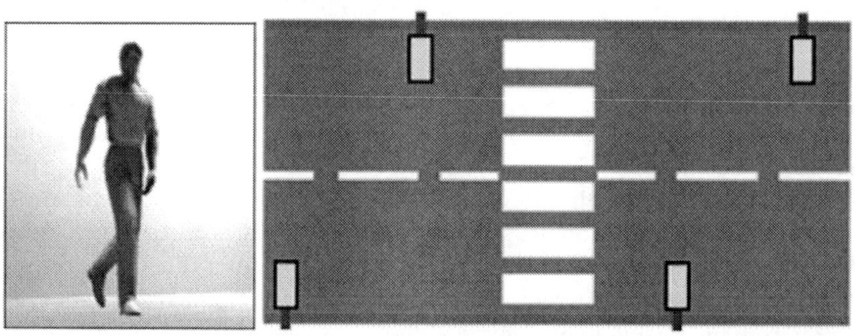

Figure 2. Negative contrast and a way to achieve negative contrast. [4]

The second possibility is the use of positive contrast (Figure 3) when a pedestrian is seen as light-coloured figure on the dark-coloured background. To achieve necessary contrast, it is necessary to add more lighting to the pedestrian crossing. Additional lighting is realized by luminaires which are placed in front of the pedestrian crossing from the view of the approaching vehicles. The luminaires are usually placed 1 to 2 metres from the crossing to achieve the required vertical illumination. This distance may vary depending on the type of luminaires used, that is why it is always necessary to calculate the pedestrian crossing lighting by a program and then, based on the calculation, position of the luminaires needs to be drawn into the plan. For a design of the pedestrian crossing illumination, special luminaires with asymmetrical luminous intensity curve are used. A suitable public lighting placed both in front of and behind the pedestrian crossing must be a part of the quality illuminated crossing. The regulation "TKP15" defines directly these distances from the crossing in both directions which depends on the speed limit for the given roadway.

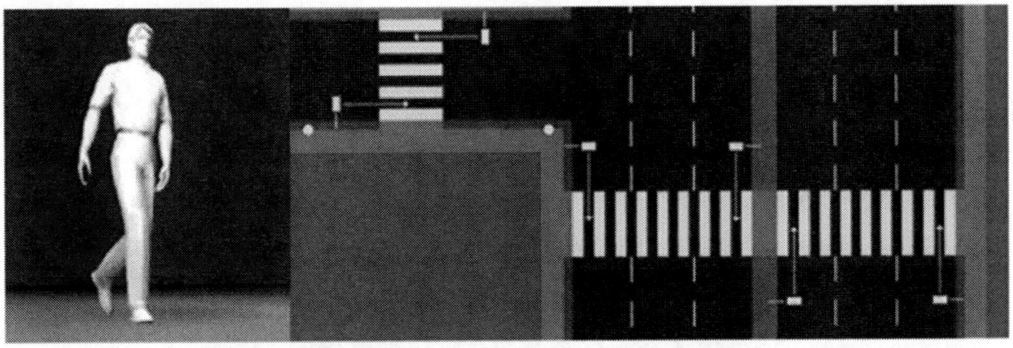

Figure 3. Positive contrast and a way to achieve the positive contrast. [4]

4. Lighting systems on the pedestrian crossings

As already mentioned, it is not easy to project an adequate lighting for a pedestrian crossing. At the beginning of the project, it is important to choose an appropriate place for its implementation. The pedestrian crossing should be placed $A=\pi r^2$ on the well-arranged part of the road, where a pedestrian will not be hidden by parked vehicles of trees and plants. It is necessary to take into consideration the fact that the additional lighting of the crossing can be disturbing for the surrounding residential buildings. For the appropriate project of illumination, it is possible to use nowadays a lot of specialized programs which enable to model the given lighting situation and then determine the type and position of luminaires.

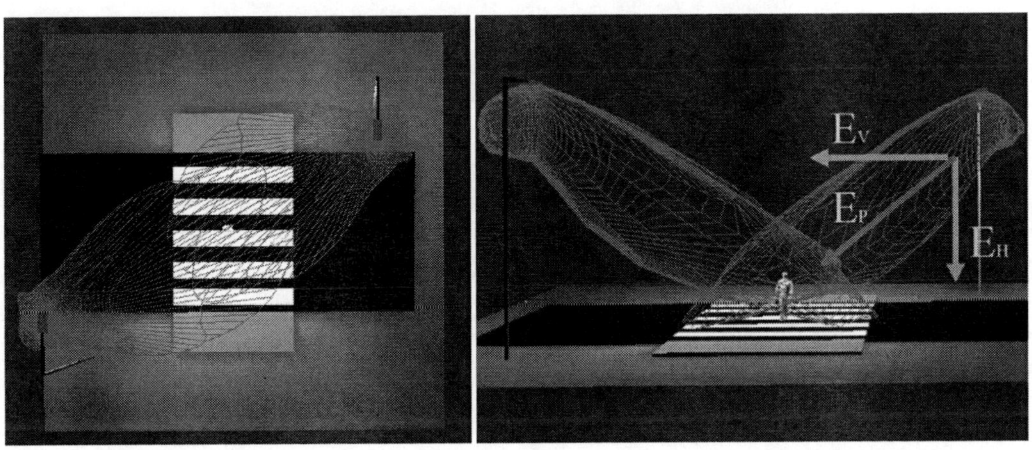

Figure 4. Example of a pedestrian crossing model made in Relux program.

The figure 4 shows a model situation of the pedestrian crossing lighting. On the right side, there are components of illuminance vector of the crossing (E_P), vertical (E_V) and horizontal illumination (E_H).

To demonstrate the lighting design of a pedestrian crossing. The article presents two situations of the crossing illuminance. The first situation, which is represented in the Figure 5, illustrates a luminance analysis of the pedestrian crossing in Petřvald.

Illuminating system of the model pedestrian crossing is realized by two luminaires of Philips LUMA type, with an asymmetric light distribution. The luminaires are placed in front of the crossing (from the view of an approaching driver). By an appropriate combination of the luminaires placement and their light intensity curves, a balanced ratio between vertical and horizontal illuminance was achieved. Thus, not only the pedestrian but also the crossing itself is visible. It is also the public lighting placed both in front of and behind a crossing which forms a quality illuminated pedestrian crossing. As a result of simultaneous light coming from the public lighting there are no significant contrasts and the driver is able to recognize obstacles apart from the crossing in time. LED lamps illuminating the crossing have high correlated colour temperature and the public lighting in front of and behind the crossing is realized by high-pressure sodium lamps. As a result of that lighting colour difference, the pedestrian crossing is recognized from its surroundings not only by the luminance contrasts but also by the colour contrast.

Figure 5. Luminance analysis of the crossing from both directions.

Table 2. Measured values of the picture luminance analysis.

Area	Minimum luminance (cd/m^2)	Maximum luminance (cd/m^2)	Average luminance (cd/m^2)	Area	Minimum luminance (cd/m^2)	Maximum luminance (cd/m^2)	Average luminance (cd/m^2)
Crossing	0.38	13.92	4.06	Crossing	0.22	10.63	2.31
Background	0.06	2.94	0.22	Background	0.05	11.68	0.13
Pedestrian	0.10	17.64	2.7	Pedestrian	0.07	16.03	2.35

On the other hand, there are many crossings which are not laminated in accordance with the regulations. Figure 6 illustrates a crossing with unsuitable lighting system. There are columnar luminaires of height about 1,2 m used but they fail to illuminate pedestrians to the desired extent and direction and in certain cases the can dazzle the approaching driver. From the luminance analysis, it is apparent that the pedestrian on the crossing is illuminated insufficiently and thus there is no contrast which is necessary for the approaching driver to distinguish the pedestrian.

Figure 6. A photograph of a crossing and its luminance analysis. [5]

Table 3. Measured values of the picture luminance analysis

Area	Minimum luminance (cd/m^2)	Maximum luminance (cd/m^2)	Average luminance (cd/m^2)
crossing	1.13	19.31	5.05
background	0.62	13.93	3.03
pedestrian	1.17	16.37	3.59

In this case the calculated contrast between the pedestrian and background of the crossing is only 0.67, this value is too low for the pedestrian to be distinguished on the crossing. The table 3 also shows that the maximum luminance value can cause that a driver may be dazzled and thus will not be able to distinguish the situation happening on the crossing.

Figure 7. In Example of incorrect placement of pedestrian crossing – parked vehicles overshadow the complementary space on the right side. [5]

Nevertheless, neither the most quality pedestrian crossing lighting is not capable to compensate the appropriate clothing of the pedestrian in terms of its reflectivity. Calculated value of the contrast between a pedestrian and a background depends considerably on the reflectivity of the pedestrian's clothing. Dark-coloured clothes show far lower luminance measurements than the clothes made of light-coloured materials.

5. Conclusion

While designing a pedestrian crossing, and dealing with its lighting systems, this issue has to be solved from several points of view. It is necessary to respect not only the legal requirements for the pedestrian crossing itself, but also the requirements for its illuminance. While designing lighting for a pedestrian crossing, it is important to analyse properly the public lighting of the roadway and draw up the project with regard to that lighting. Public lighting (lighting of a roadway) is, in the most cases, a source of background luminance and that is what the contrast value between a pedestrian and a background depends on. Public lighting must ensure the corresponding value of background luminance both in front of and behind the pedestrian crossing and the ratio between a pedestrian luminance and luminance of the background should be in the range of 1 to 10.

Currently, gas-charged lamps are being replaced by LED lamps. In comparison with the gas-charged lamps, the LED lamps provide several advantages. Benefits of LED lamps installation lie especially in significant prolongation of a lighting life expectancy from about 8 000h to 50 000 or even 100 000h. Another significant advantage is the possibility of setting the luminous flux from the luminaire directly in the process of installation (a dimmable ballast) so the balanced luminance contrast between a pedestrian and a background can be set.

Acknowledgements

This article was prepared with the support of the project "Remote Control of Public Lighting Luminaires via the Smart Technology Support" SP2017/97, by institution of VSB-TU Ostrava.

References

[1] Český normalizační institut 2005 Soubor norem pro osvětlování pozemních komunikací (*ČSN CEN/TR 13 201-1 and ČSN EN 13201-2–4*)

[2] Ministry of Transport of the Czech Republic 2006 *Technické kvalitativní podmínky staveb pozemních komunikací, kapitola 15, Osvětlování pozemních komunikací, Dodatek č. 1 – Přisvětlování přechodů*

[3] Bláha Z, Novák T and Sokanský K 2013 Safety and power savings in conflict areas which are illuminated by public lighting *Proc. 14th Int. Scient. Conf. on Electric Power Engineering* (VŠB-TU Ostrava) pp 709-12

[4] Bláha Z, Novák T and Sokanský K 2014 Solution of lighting for pedestrian crossing to increase their visibility *Proc. 15th Int. Scient. Conf. on Electric Power Engineering (Brno)* pp 637-40

[5] Baleja R, Helštýnová B, Sokanský K and Novák T 2015 Measurement of luminance ratios at pedestrian crossings *Proc. 15th Int. Conf. on Environment and Electrical Engineering EEEIC* pp 1306-10

Structural health monitoring system for bridges based on skin-like sensor

Konstantinos Loupos[1], Yannis Damigos[1], Angelos Amditis[1], Reimund Gerhard[2], Dmitry Rychkov[2], Werner Wirges[2], Manuel Schulze[2], Sotiris-Angelos Lenas[3], Christos Chatziandreoglou[3], Christina M. Malliou[3], Vassilis Tsaoussidis[3], Ken Brady[4] and Bernd Frankenstein[5]

[1] Institute of Communication and Computer Systems, 9 Iroon Polytechneiou Street, Athens, 15773, Greece
[2] University of Potsdam, Institute of Physics and Astronomy, Karl-Liebknecht-Strasse 24-25, 14476 Potsdam, Germany
[3] Democritus University of Thrace, Department of Electrical and Computer Engineering, Kimmeria Campus Xanthi, 671 00, Greece
[4] TRL limited, Crowthorne House, Nine Mile Ride, Wokingham, RG40 3GA, UK
[5] Teletronic Rossendorf GmbH, Bautzner Landstraße 45, Radeberg, 01454, Germany

E-mail: kloupos@iccs.gr

Abstract. Structural health monitoring activities are of primal importance for managing transport infrastructure, however most SHM methodologies are based on point-based sensors that have limitations in terms of their spatial positioning requirements, cost of development and measurement range. This paper describes the progress on the SENSKIN EC project whose objective is to develop a dielectric-elastomer and micro-electronics-based sensor, formed from a large highly extensible capacitance sensing membrane supported by advanced micro-electronic circuitry, for monitoring transport infrastructure bridges. Such a sensor could provide spatial measurements of strain in excess of 10%. The actual sensor along with the data acquisition module, the communication module and power electronics are all integrated into a compact unit, the SENSKIN device, which is energy-efficient, requires simple signal processing and it is easy to install over various surface types. In terms of communication, SENSKIN devices interact with each other to form the SENSKIN system; a fully distributed and autonomous wireless sensor network that is able to self-monitor. SENSKIN system utilizes Delay-/Disruption-Tolerant Networking technologies to ensure that the strain measurements will be received by the base station even under extreme conditions where normal communications are disrupted. This paper describes the architecture of the SENSKIN system and the development and testing of the first SENSKIN prototype sensor, the data acquisition system, and the communication system.

1. Introduction

1.1. Challenge and concept

The project concept is based on the currently limited use of sensors being applied to monitor the stability of civil structures and in particular transport infrastructures. Visual inspections tend to overwhelm usual inspections in order to collect information to determine the condition of infrastructures. Inspections nowadays can be considered as rather slow in completion, usually requiring (partial or total) closure of the particular network node or link and/or expose inspectors to dangerous working environments and conditions. Detailed monitoring actions and assessments are (a) particularly slow and expensive, (b) usually based on simplistic and conservative models of structural behaviour, and (c) do not provide a rapid, convenient means of determining structural stability following a major incident or other serious event that could potential limit the structural integrity of the structure[1, 2]. On top of this, quality and reliability of the visually extracted information and measurements really depend on the expertise of the inspector (objectiveness factor) and, to add on this, quite often the inspector is unable to gain access to all parts of a structure. State of the art structural health monitoring (SHM) technologies and integrated systems have proved a major role in the management of the transport infrastructure while currently existing and used SHM technologies and methods rely only on the use of dense networks of point sensors to monitor a structure, which is costly (and sometimes not practicable). All at once, conventional sensors also prove to fail at relatively low strains and their communication system is unreliable in extreme service conditions: thus, they do not provide a fool-proof alarm of an imminent structural collapse.

1.2. The SENSKIN Project

SENSKIN is an EC co-funded project that operates in the framework of EC-FP7-Transport (MG-8.1a-2014 - Smarter design, construction and maintenance). SENSKIN includes a consortium with all the expertise needed in the lifecycle from research to innovation, as well as real-life end-users that can provide solid feedback on the project results. The list of partners of the consortium has been presented below:

Table 1. The SENSKIN Consortium.

Partner Name	Short Name	Country
Institute of Communication and Computer Systems	ICCS	Greece
University of Potsdam, Applied Condensed-Matter Physics Group	UP	Germany
Egnatia Odos A.E.	EOAE	Greece
RISA Sicherheitsanalysen GmbH	RISA	Germany
TECNIC S.p.A.	TECNIC	Italy
Democritus University of Thrace	DUTH	Greece
Mistras Group Hellas A.B.E.E.	MGH	Greece
University of Stuttgart	USTUTT	Greece
TRL Limited, Transport Research Laboratory	TRL	UK
State Enterprise State Road Scientific Research Institute	DNDI	Uktraine
Forum Des Laboratoires Nationaux Europeens De Recherche Routiere	FEHRL	Belgium
Teletronic Rossendorf GmbH	TTRONIC	Germany
Turkish General Directorate of Highways	KGM	Turkey

The consortium includes partners that (a) have already developed a first generation of 'sensing skin' with very encouraging results (University of Potsdam), (b) have already developed relevant sensor near electronics (TTRONIC SpA), (c) have developed LCAs and LCCs for construction works (University of Stuttgart), (d) have developed algorithms for the assessment of deteriorating structural systems in the context of several previous research projects (TECNIC SpA, RISA Sicherheitsanalysen GmbH), (e) have developed and implemented relevant wireless sensor network architectures (Institute of Communication and Computer Systems and Democritus University of Thrace) and (f) have particular experience in performing testing sessions of the project prototypes (TRL Limited) in a lab environment. The work in SENSKIN is totally market driven (there are three major end users in the consortium: Egnatia Motoway (EOAE), Karayollari Genel Mudurlugu (KGM) and Forum de

Laboratoires Nationaux Europeens De Recherche Routiere (FEHRL) (including members that belong to Highway Departments)[1].

SENSKIN aims to: (a) develop a dielectric-elastomer and micro-electronics-based skin-like sensing solution for the structural monitoring of the transport infrastructure that will offer spatial sensing of reversible (repeated) strains in the range of 0.012% to more than 10%, that requires little power to operate, is easy to install on an irregular surface, is low cost compared to existing sensors, allows simple signal processing and includes the ability of self-monitoring and self-reporting. (b) use the new and emerging technology of Delay Tolerant Network to secure that strain measurements acquired through the 'sensing skin' will reach the base station even under extreme environmental conditions and natural disaster events such as, high winds or an earthquake, where some communication networks could become inoperable. (c) develop a Decision-Support-System for proactive condition-based structural intervention under operating loads and intervention after extreme events. It will be based on an accurate structural assessment based on input from the strain sensors in (a) above and will examine the life-cycle economic, social and environmental implications of the feasible rehabilitation options and the resilience of the infrastructure to future changes in traffic demand that these options offer. (d) Implement the above in the case of bridges and test, refine, evaluate and benchmark the monitoring system (integrated a and b) and package (integrated a, b and c) on actual bridges [2].

The practical application of the developed prototypes will significantly enhance structural health monitoring of bridges but could be extended to monitoring of other critical infrastructures like damns, energy related structures, tunnels, concrete pavements (paved or roadway) etc.

2. Elastomer-based sensor for the SENSKIN system

2.1. Soft Capacitive Sensor

The design of the soft skin-like sensor used in the SENSKIN system is shown in Figure 1. The sensor is built around a thin silicone elastomer membrane (1) and can thus detect large reversible strains between 0.01% and 10%.

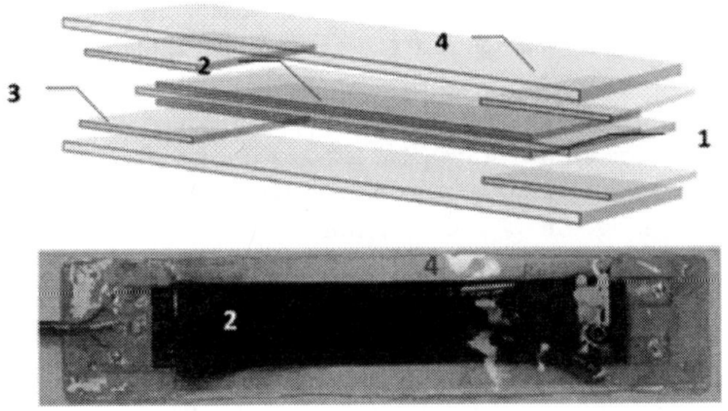

Figure 1. Soft capacitive sensor.

The silicone membrane (1) obtained from Wacker (ELASTOSIL® Film 2030 250/100) is only 100μm thin and coated on both sides with compliant electrodes (2). The top and bottom electrodes are made by spraying a toluene-based solution of the silicone material loaded with carbon nanoparticles (Printex XE-2B) onto the membrane. After solvent evaporation and cross-linking of the silicone, the electrode can be stretched over the whole range of sensor motion with a typical resistance of 8 kΩ across its length (15 cm). Polyester frames (3) are employed to provide electrical terminals and for handling and mounting of the sensor. In a last step, the whole assembly is coated with protective silicone layers (4).

As the active electroded area is 15x3 cm^2, the capacitance without pre-stretching is around 1.5 nF. The resistance of the electrodes changes within 0.02-0.07 Ω per µstrain. The measurement principle of the skin-like capacitive sensor is based on continuously monitoring the capacitance changes caused by sensor deformation [1, 9, 10]. When the sensor is attached to a component of the transport infrastructure, its deformation will cause a corresponding deformation of the sensor and thus also a change in its capacitance (Figure 2).

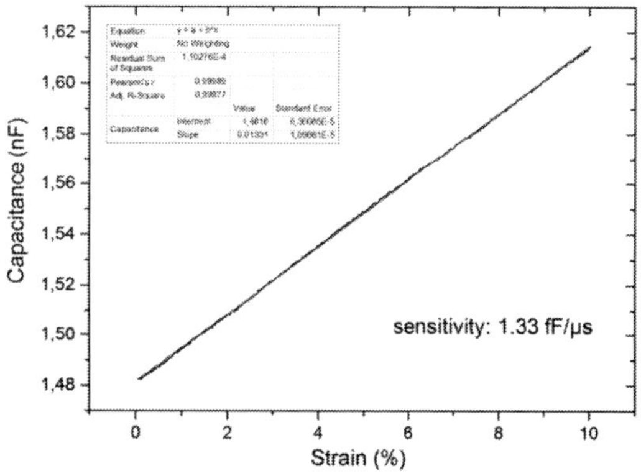

Figure 2. Strain dependence of the capacitance measured directly with an LCR meter.

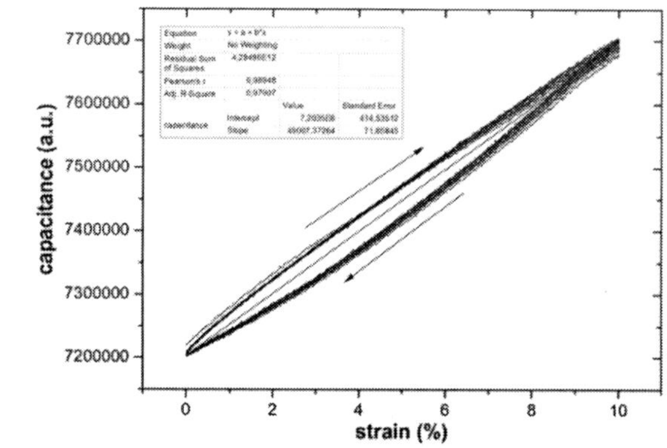

Figure 3. Sensor output as measured via the DAQ (see subsection 2.2 below).

The data shown in Figure 2 were obtained with a linear testing machine (Zwick/Roell Z005) for the controlled deformation of the sensor and with a precision LCR meter (HP 4284A) for measuring the sensor capacitance. The capacitance-strain dependence is linear over the whole range of motion with a typical sensitivity of around 1.3 fF per µstrain. When the sensor output is measured with a SENSKIN Data Acquisition (DAQ) module (see section 1.2 below), the capacitance behaviour becomes more complicated (Figure 3). There is some drift in the measured values at higher strains as well as a hysteresis-like behaviour for the overall curve. The nonlinearity is due to the fact that the PCap system inside the DAQ module actually measures RC times during charging or discharging of the soft capacitor. Even if the capacitance changes linearly (Figure 2), the resistance of the electrodes can change nonlinearly over the range of 10% deformation.

2.2. Data Acquisition (DAQ) from the monitoring sensor

The DAQ is part of a complex sensor NODE situated between the sensing element based on stretchable elastomeric material and the microcontroller for signal pre-processing and communication control. From a measurement point of view, the sensor is a capacitance that must be monitored very precisely. There are three main challenges for the measurement electronics:

a) The sensor being used has a mean capacitance of about 1.5 nF in the prestressed state. However, the sensitivity of the sensor is as low as only 1.33 fF per µstrain. Therefore, a capacitance of approx. 1.5 nF with a noise-limited resolution of ca 40 fF must be determined for the desired resolution of ca. 30 µstrain at a sampling rate of 50 Hz.
b) The compliant electrodes of the stretchable sensors have relatively high internal resistances between 3 and 10 kOhm, and the resistance is also strain-dependent.
c) The strain-dependent capacitance is highly sensitive to temperature. Therefore, the data acquisition system contains a temperature-measuring chip with a resolution of 24 bit for calibration purposes.

For the capacitance measurement, a Charge to Digital Converter (CDC) PicoCAP02 from ACAM electronic is used. It was developed by ACAM Elektronik and adapted by Teletronic to meet the measuring requirements of the project. The operating principle is illustrated in the equivalent circuit:

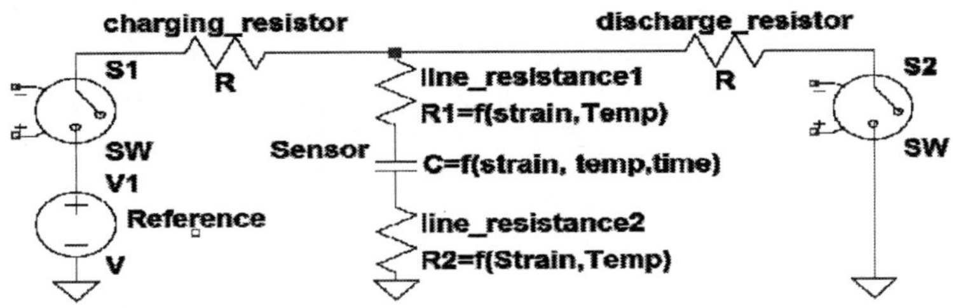

Figure 4. Equivalent Circuit for the PicoCap02 Measuring System.

An internal voltage reference charges the Sensor represented by the Capacitor C across the internal charging resistor R and the compliant-electrode line resistances R1 and R2. The capacitance of the Sensor as well as the two line Resistances depend on both strain and temperature, and the charging process is controlled by the switch S1. The detection process is completed by measuring the discharge time across the discharge resistor controlled by the switch S2. The measuring principle implemented in the application of the PicoCap02 CDC is radiometric, *i.e.* the capacitances are compared to a fixed reference, and the ratio of the discharge times is directly proportional to the ratio of the capacitances, as the discharge time is determined by the respective capacitance and the selected discharge resistor.

Both circuits, the CDC as well as the temperature measurement chip, communicate via the SPI bus with the microcontroller for averaging and data management. Both have internal processors for controlling the measuring process. Measurement of capacitance and temperature may be done in parallel so that an associated temperature is determined for each strain value. In order to compensate for the temperature-dependent internal resistance, as well as for the temperature coefficient of the capacitance, each sensor is calibrated individually. In practical applications, the EMI influence of any nearby 50 Hz power lines is significant because of the length of the sensor and its small capacitance. Up to now, the measuring process is synchronized with the power-line frequency for supressing the interference. In future developments, the influence may be supressed via higher sampling rates for the capacitance and by the use of a 50 Hz notch filter.

3. The SENSKIN Communication system

3.1. Why Delay-/Disruption-Tolerant Networking technology is necessary?

WSNs (Wireless Sensor Networks) deployed in open environments are prone to intermittent connectivity due to power scheduling, channel fading, node failure, and severe packet losses from unpredictable external factors, such as interference. On top of that, the connection of the Gateway to the Internet might also present its own challenges, as it is not uncommon for bridges to be located at places where communication infrastructure to the Internet is only available intermittently (e.g. remote places where only a satellite link is available) or it is not sufficient to cover monitoring needs due to the employment of low-capacity links (e.g. GPRS links).

In this context, SENSKIN communication system utilizes DTN (Delay-/Disruption-Tolerant Networking) protocols [3, 4, 5, 6, 7], which synergistically with mesh networking, address the majority of the aforementioned problems. In particular, DTN communication protocols provide reliable data communication across failure-prone networks by employing various purpose-specific network mechanisms, including store-and-forward data forwarding, multi-path routing and hop-by-hop reliability.

3.2. The SENSKIN Communication System

SENSKIN communication system, depicted in Figure 5, constitutes a fully distributed and autonomous WSN that is able to self-monitor and report itself.

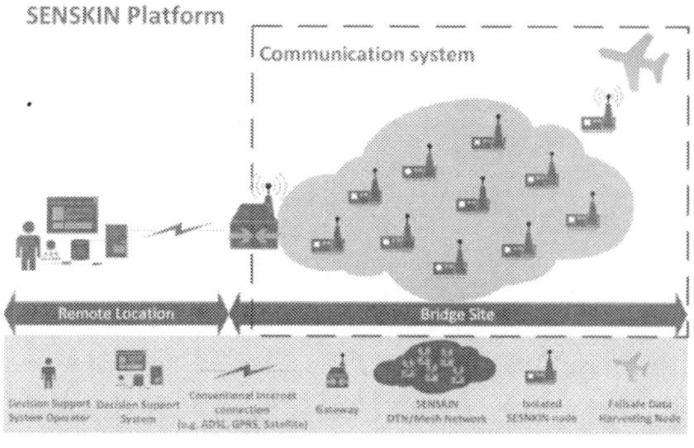

Figure 5. SENSKIN communication system network architecture.

It is composed by three different elements: SENSKIN, Gateway and Failsafe nodes.

3.2.1 SENSKIN nodes. These are essentially SENSKIN devices placed at different locations of a bridge site. Each SENSKIN device is equipped with a communication module. Each communication module implements the network stack presented in Figure 6.

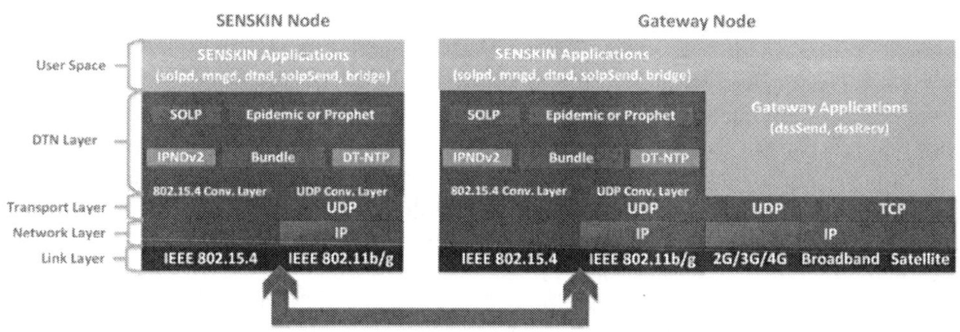

Figure 6. SENSKIN and Gateway nodes network protocol stack.

At link layer, communication modules adopt a dual interface approach, by employing in particular the IEEE 802.11 and 802.15.4 standards on an as-needed basis. This approach helps to increase communication robustness and data delivery reliability, as well as, to balance out the trade-off between performance and energy consumption. Bundle Protocol (BP) runs on top of them [3]. BP constitutes the basic protocol through which a DTN architecture can be established among the nodes of a network. Through appropriate convergence-layer protocols, BP runs directly over IEEE 802.15.4. The same does not stand true for IEEE 802.11, where a UDP/IP stack is employed to allow BP pass its data over IEEE 802.11. Custom-tailored networking solutions for supporting SENSKIN system operation logic, such as the SENSKIN Operation Logic Protocol (SOLP), run at the Bundle layer. Finally, all SENSKIN communication applications used for sending and receiving data (i.e. solpd, solpSend), along with a panic communication mechanism that is implemented as an extension of the node's management application (i.e. mngd), run at the application layer. Panic communication mechanism is triggered under certain network circumstances (e.g. low signal quality, loss of neighbour nodes, extreme weather conditions). In particular, it allows network self-recovery and in extreme cases, (e.g. major catastrophic event like an earthquake) it enables data from potentially isolated

SENSKIN devices to be flushed to external ad-hoc nodes. These ad-hoc nodes essentially act as failsafe data-harvesting elements.

3.2.2. Gateway(s). As depicted in Figure 6, Gateway nodes employ a dual network protocol stack that allows them to serve as a middleware platform for interconnecting the deployed SENSKIN nodes network with a remote administration facility. The first network stack is identical to the one employed by the SENSKIN nodes, while the second one employs mobile telecommunications, broadband or satellite link layer protocols, along with conventional Internet protocols, such as TCP, UDP and IP.

3.2.3. Failsafe node(s). Failsafe nodes are similar devices to Gateways in terms of operational capabilities with their basic difference being that they are mobile.

3.3. PCB Design and Interfaces between the Application MCU and the Communication System

The Application MCU exchanges information with the Communication MCU using a Universal Asynchronous Receiver/Transmitter (UART) serial communication interface. The SENSKIN node provides the functionality to update the Application MCU's firmware remotely. The Application MCU includes a simple, ROM-based bootloader which communicates with the Communication MCU over the serial interface. The Communication MCU uses two digital I/O pins, in addition to the serial communication interface, to take advantage of ROM-based bootloader. One I/O is used to control bootloader's backdoor functionality and one for resetting the Application MCU, in order to force a ROM bootloader entry upon reset of the Application MCU. The hardware interfaces between the Application MCU and the communication system are presented in Figure 7 [8].

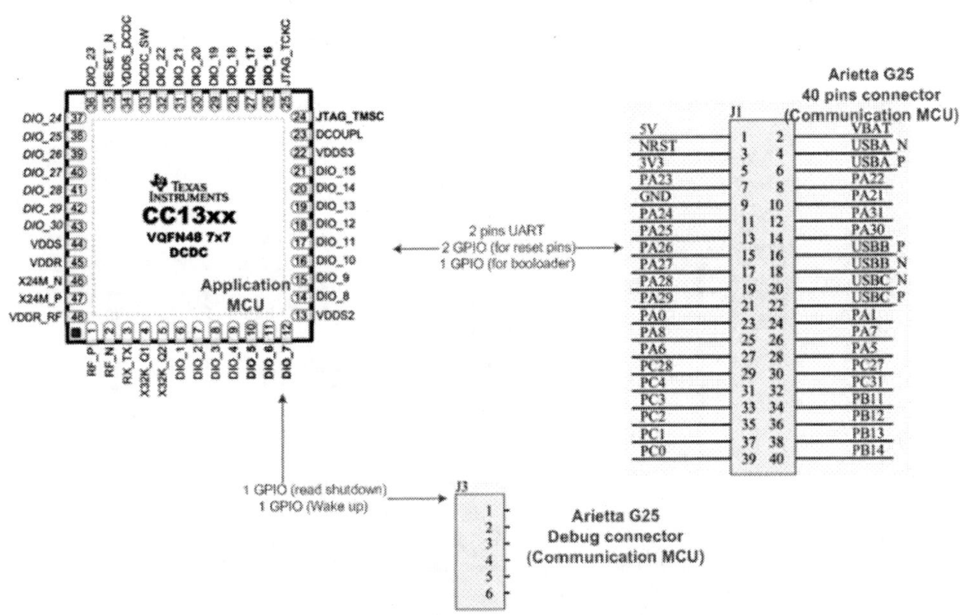

Figure 7. Hardware interfaces between Application MCU and communication system.

The Application MCU handles the power source of the communication module, in order to reduce the power consumption, increase SENSKIN integrated device's lifetime and maintain all the benefits of the communication system without compromising its operation and functionality. If there is no incoming or outgoing information to or from the SENSKIN node, then the Application MCU turns off the power source off the communication module. The application MCU turns on the power source of the communication module incoming or outgoing data needs to be transferred.

The printed circuit board (PCB) of the initial version of the SENSKIN integrated device hosts the Application MCU, the power supply, the necessary physical interfaces for connecting the DAQ and the solar panel, a Ferroelectric RAM (FRAM) for storing measurements and information regarding the operation of the Application MCU and the low power wakeup receiver, which allows the on demand functionality of the SENSKIN node [8].

4. System integration

The SENSKIN integrated device is composed of the following components following an integrated approach:

- Sensor Device: a skin-like sensor capable of measuring strains over a surface area
- Data Acquisition System (DAQ): a capacitance to digital converter unit which will digitize the sensed strains
- Communication Module: the part of the SENSKIN integrated device that is responsible for transmitting and receiving all data related to the operation of the SENSKIN system, as well implementing the communications operational logic
- Processing Module: the computational core of the SENSKIN integrated device
- Low Power Wakeup Receiver: a promising energy-saving approach for on demand wireless network operations
- Power Management, Energy Harvesting and Energy Storage Units: the hybrid power supply of the device
- Packaging: the enclosure of the device

A block diagram of the SENSKIN integrated device is presented in Figure 8.

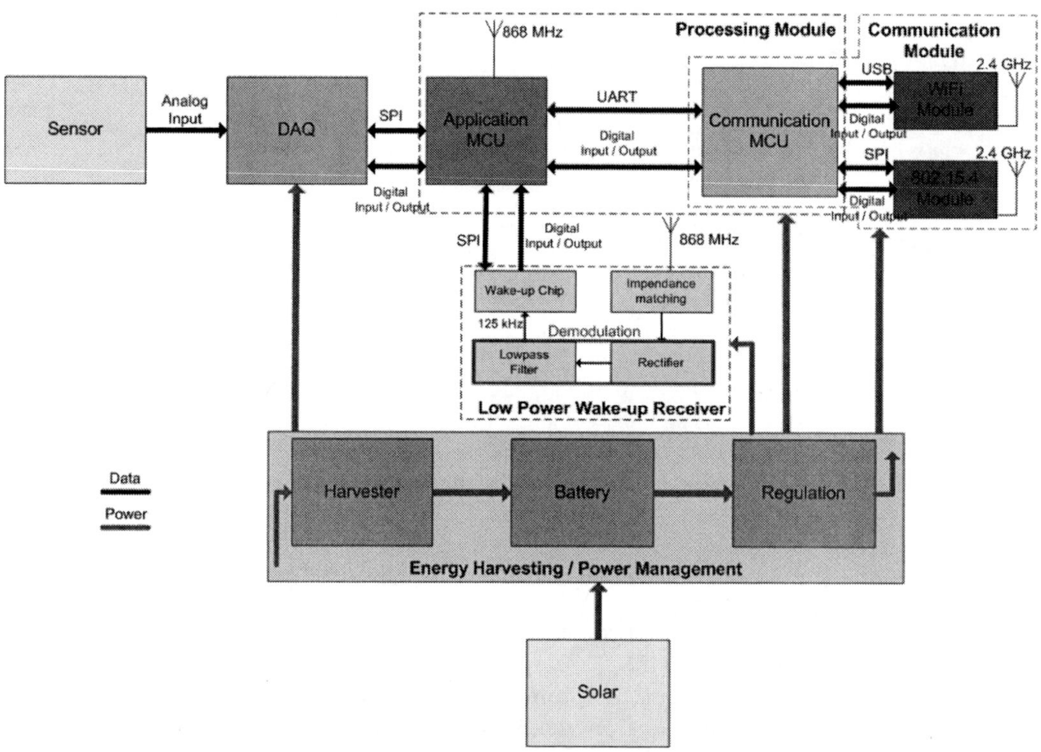

Figure 8. SENSKIN node block diagram.

5. System Experimental Evaluation

The findings of a review showed that it was necessary to (a) design and fabricate purpose-built test rigs for undertaking calibration tests, and (b) that an off-specimen camera-based system was the only practicable means of calibrating the sensor. The calibration tests are undertaken using uniaxial test rigs (in the main) and a flexural rig (see Figure 9). Both types of rig can accommodate steel and concrete test coupons.

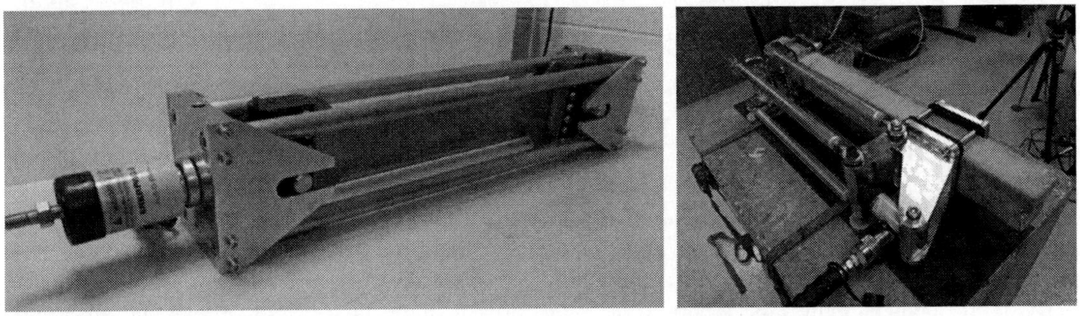

Figure 9. View of uniaxial test rig (left) and flexural test rig (right).

The strains applied to the test coupons, and sensors, are measured using an off-specimen camera-based system obtained from iMETRUM. The variation in the output of two sensors with ambient temperature is investigated using the environmental cabinet at TRL. Prior to undertaking the calibration tests it was necessary to undertake proving tests of the equipment and calculation routines to establish a reliable standard reference procedure.

Tests on the prototype sensor showed that the application of a uniaxial load generated differential movements within the silicone encapsulation. And, as shown in Figure 10, a highly non-uniform strain field was developed around the end-blocks of the sensor.

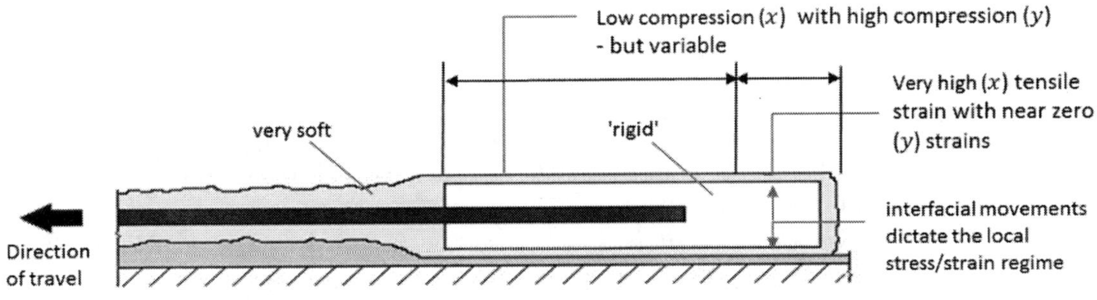

Figure 10. Cross section through the end-block of a sensor

The results of preliminary tests on the prototype sensor showed that the output of a sensor varied non-linearly with the ambient temperature; this was expected. However, more importantly, the temperature sensitivity of the output of the prototype was higher than would be accepted for most structural health monitoring applications.

The findings of the tests show that the prototype needs to be refined to meet a few key target requirements for bridge monitoring: in particular to (a) improve the reliability of the relationship between the applied strain and the output of the sensor, and (b) reduce the temperature sensitivity of that output. The findings also provided suggestions for effecting such refinements.

6. Conclusions

This publication has presented the challenges and system concept of operation as well as the system modules of the SENSKIN monitoring solution that is being developed in the framework of the SENSKIN EC co-funded research project. The project is currently at the first integrated version where the second version of the sensor has been developed followed by the development of the first version of the data acquisition system (DAQ). The first integrated version of the system has been also implemented and tested including the design of the system electronics, interfaces and data considerations. The next steps of the project will be a set of simulated environment testing of the prototype device that will be executed by the partner TRL (TRL Limited) that will provide direct feedback to the technical activities towards the final implementation of the SENSKIN system. In turn the final system prototype will be validated in the actual bridges of Egnatia Motorway (Greece) and the Bosporus Bridge (KGM). SENSKIN is expected to provide high impact to monitoring and management systems increasing infrastructure capacity and optimizing maintenance costs for all transport modes, new construction and maintenance techniques that enhance the performance and reliability of infrastructure, innovative and cost effective approaches to use Green Infrastructure for transport, extension of the life span of ageing transport infrastructure, development and application of effective and efficient materials, technologies and tools to meet cost-effectiveness and sustainability goals, reduction of multi-modal infrastructure construction and maintenance energy intensity and subsequent CO2 pollutants and noise emissions, support the transition towards zero traffic disruption from inspection, construction and maintenance by 2030 and boost the overall performance of the European transport infrastructure and reduce nuisances generated by transport. The first steps towards the system commercialization have already started though a draft exploitation strategy identifying the SENSKIN market opportunity, initial partners' IPRs as well as a concise strategy for the SENSKIN commercialization.

Acknowledgements

The research leading to these results has received funding from the EC H2020 project SENSKIN (Contract N.635844). Authors would like to thank all partners within the SENSKIN consortium.

References

[1] SENSKIN [635844] Annex 1 (part A) Description Of Work
[2] Loupos K et al 2016 Skin-Like Sensor Enabled Bridge Structural Health Monitoring System *8th European Workshop on Structural Health Monitoring – EWSHM (Bilbao)*
[3] Scott K and Burleigh S 2007 Bundle Protocol Specification RFC 5050 (Online) https://tools.ietf.org/html/rfc5050
[4] Morgenroth J 2015 *Event-driven Software-Architecture for Delay-and Disruption-Tolerant Networking* (Technische Universität Braunschweig)
[5] Ellard D et al 2012 DTN IP Neighbor Discovery (IPND) draft-irtf-dtnrg-ipnd-02 (Online) https://tools.ietf.org/html/draft-irtf-dtnrg-ipnd-02
[6] Lindgren A et al 2012 Probabilistic Routing Protocol for Intermittently Connected Networks, RFC 6693 (Online) https://tools.ietf.org/html/rfc6693
[7] Lindgren A, Davies E and Doria A 2013 Epidemic Routing Module for Generic Opportunistic Routing Framework (Online) https://tools.ietf.org/html/draft-lindgren-dtnrg-gorf-epidemic-00.html
[8] SENSKIN [635844] D2.3 SENSKIN Integrated Device Initial Version
[9] Kollosche M, Stoyanov H, Laflamme S and Kofod G 2011 Strongly enhanced sensitivity in elastic capacitive strain sensors *J. Mater. Chem.* **21** pp 8292-94
[10] Carpi E F 2016 *Electromechanically Active Active Polymers: A Concise reference* (Springer International Publishing Switzerland) p 798

Crack identification for rigid pavements using unmanned aerial vehicles

Ahmet Bahaddin Ersoz[1], Onur Pekcan[1] and Turker Teke[1]

[1]Middle East Technical University, Civil Engineering Department, Ankara, 06800, TURKEY

E-mail: opekcan@metu.edu.tr

Abstract. Pavement condition assessment is an essential piece of modern pavement management systems as rehabilitation strategies are planned based upon its outcomes. For proper evaluation of existing pavements, they must be continuously and effectively monitored using practical means. Conventionally, truck-based pavement monitoring systems have been in-use in assessing the remaining life of in-service pavements. Although such systems produce accurate results, their use can be expensive and data processing can be time consuming, which make them infeasible considering the demand for quick pavement evaluation. To overcome such problems, Unmanned Aerial Vehicles (UAVs) can be used as an alternative as they are relatively cheaper and easier-to-use. In this study, we propose a UAV based pavement crack identification system for monitoring rigid pavements' existing conditions. The system consists of recently introduced image processing algorithms used together with conventional machine learning techniques, both of which are used to perform detection of cracks on rigid pavements' surface and their classification. Through image processing, the distinct features of labelled crack bodies are first obtained from the UAV based images and then used for training of a Support Vector Machine (SVM) model. The performance of the developed SVM model was assessed with a field study performed along a rigid pavement exposed to low traffic and serious temperature changes. Available cracks were classified using the UAV based system and obtained results indicate it ensures a good alternative solution for pavement monitoring applications.

1. Introduction

In the last two decades, with the help of developments in high-level processing techniques to extract information from the images and sensing technologies to capture images efficiently and accurately under various lighting conditions, pavement surface monitoring have been brought to advanced levels. Using such superior tools and techniques, the visual examination of pavements becomes now much easier and reliable compared to utilizing conventional methods, which requires massive amount of human work, leading to less accurate analyses and with the potential of producing more biased results.

Technological innovations have helped researchers developing autonomous pavement crack identification systems for flexible pavements. Those developments can generally be grouped in two subfields: (i) automated data collection devices, and (ii) data processing systems. In the first group, data collection systems using either manned or unmanned vehicles are studied for pavement assessment. Fukuhara et al. [1], Klassen and Swindall [2], Wang and Elliot [3], Wang et al. [4,5], Groeger et al. [6], Gavilán et al. [7], Cafiso and Battiato [8], and Sy et al. [9] focused on developing two-dimensional imaging systems based on truck mounted cameras. In addition to those, more recent technologies appeared in the literature. Tsai and Li [10] provided insight into the possible use of emerging three-dimensional laser technologies for crack detection studies. Since UAV platforms were successfully utilized in the natural hazards and structural health monitoring applications, they started to be used in transportation applications. For example, Feng et al. [11] performed a real-time road mapping system using UAVs. Zhang and Elaksher [12] presented a UAV based road condition assessment with the help of 3D surface reconstruction methods. Very recently, Zakeri et al. [13] published a quadcopter-based multi stage system to classify pavement distresses. While the above research studies focus more on the development of pavement data collection equipment, in the second group, advances in image processing and machine learning techniques are mainly emphasized. Huang and Xu [14], and Zhou et al. [15] introduced customized, high-speed image processing algorithms to inspect pavement cracks. Nguyen et al. [16] published a method valid for detecting not only cracks but also joints and bridge defects. Li et al. [17], and Zou et al. [18] proposed new methods called FoSa and CrackTree, to detect cracks from images fully-automatically. Recently, Oliveira and Correia [19] presented a method, called CrackIT, for automatic detection and identification of cracks occurring on flexible pavement's surfaces.

Considering that the majority of the above progresses to develop expert level pavement crack identification systems are limited to flexible pavements, there is still a need for applying these technologies to rigid pavements. With this idea in mind, in this paper, we propose a UAV based pavement crack identification system for monitoring rigid pavements' existing conditions based on the combination of image processing techniques and machine learning. In the following section, the method developed is presented. The next section is dedicated to field experiments for validating the proposed technique. Conclusions are given at the end.

2. Method

A pavement crack identification algorithm is proposed in this part. It is independent of data collection tool, i.e., although the images coming from UAVs are used in this study, the algorithm is also able to work with data coming from different sources such as truck vehicles or manually collected pictures. The method consists of two main steps: namely (i) crack candidate detection, and (ii) crack classification. In the first part, images taken by UAVs are segmented such that crack and non-crack bodies are separated from the background image using image segmentation and enhancement techniques. Afterwards, the geometric properties of these bodies are obtained to be used for training of Support Vector Machine (SVM) model. Finally, crack and non-crack regions are classified using the trained SVM model. The details of these two steps are explained in the following sections.

2.1. Crack Candidate Detection

The algorithm starts with an image resizing operation, during which the size of images are reduced to 512x512 pixels to speed up processing time. Then, grayscale transformation operation is performed. Next, with the aim of extracting crack candidate bodies from pavement surface, a thresholding step is utilized. Since it is hard to define a global threshold for segmentation that is valid for all images, each image is considered individually and a manually chosen threshold value is applied. A binary image is then obtained by applying those thresholding values to the image pixel intensities.

The last step for determining crack candidates is the enhancement of binary images including crack and non-crack regions using median filtering and morphological operations. This can be achieved by replacing median of neighbourhood pixels with the center of 3x3 rectangular filter. After applying median filter, some regions other than crack pixels still remain. For this reason, small regions with areas below a pre-set value are removed from the image. These noise removal processes can cause discontinuities in crack bodies. Finally, a morphological closing operation is performed to connect separated crack bodies and enhance the connected regions. As a result, binary images including only crack and non-crack regions are obtained for further processing in the crack classification step.

2.2. Crack Classification

In this part, crack candidate regions are to be grouped into two classes: namely, cracks and non-cracks. To achieve this, a machine learning based approach is followed as geometric properties of crack and non-crack regions differ from each other. For example, longitudinal cracks are thin and vertically long, on the other hand, the width and height ratios of non-crack regions are different. In, non-crack regions covering rectangle are filled with more pixels compared to crack pixels. Properties like these are used as features to train the machine learning based model. Four different geometric properties are defined for crack identification:

- *Extent:* The ratio of bounding rectangle's area that encloses the region to the area of pixels in the region (Figure 1).
- *Aspect Ratio:* The ratio of the distance between leftmost and rightmost point to the distance between upmost and lowermost points, indicated as (w/h) (Figure 1).
- *Eccentricity:* A value that is calculated by dividing the distance between two focus points of the covering ellipse to the length of its major axis (**Figure 2**).
- *Circularity Ratio:* A value that is computed using the formula given in Equation (1), in which A and P are the area and perimeter of the region, respectively. The value of this measure is 1 for a circular region and π/4 for a square one.

$$R_c = \frac{4\pi A}{P} \tag{1}$$

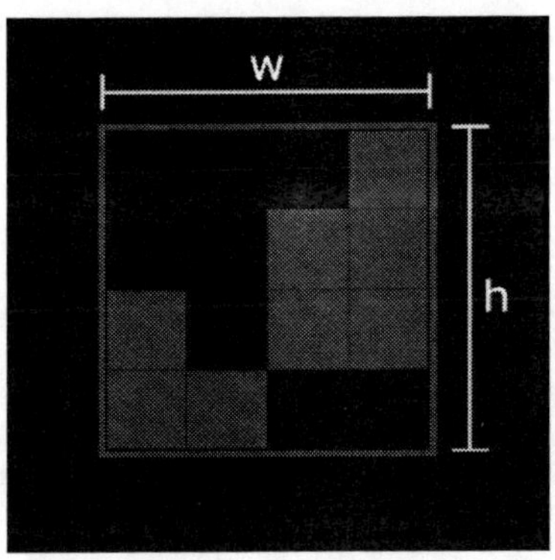

Figure 1. Bounding box and aspect ratio for a given image.

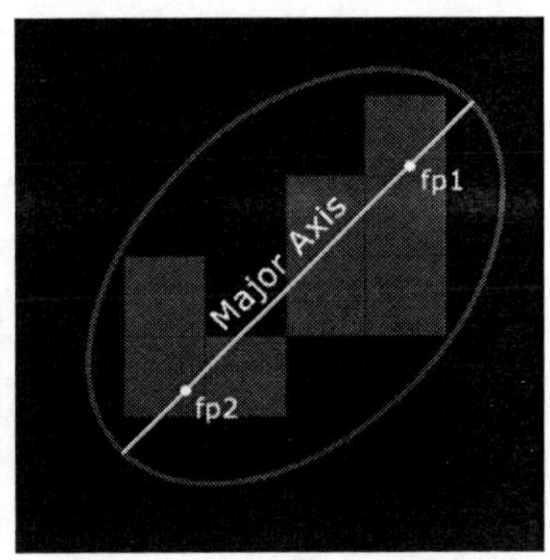

Figure 2. Ellipse covering the region and orientation for a given image.

In the machine learning part, a supervised learning algorithm named Support Vector Machine (SVM) is chosen, since it has performed well in many practical applications. The main idea of SVM is creating decision boundary that separates the classes. The decision boundary, so called, the hyperplane is constructed according to the maximum margin principle, in which the perpendicular distance between data points and the plane are maximized. Briefly, the hyperplane can be formulated by:

$$f(\mathrm{x}) = \mathbf{w}^{\mathrm{T}}\mathbf{x} + w_{N+1} = 0 \tag{2}$$

where $\mathbf{w}$ is the normal vector to the hyperplane and $\mathbf{x}$ is the training data, w_{N+1} is the offset and $\mathbf{w}^{\mathrm{T}}$ is the transpose of $\mathbf{w}$. The aim is finding an optimal hyperplane, which is almost equivalently away from nearest data points. The best position of this hyperplane can be found when the margin is largest, which is defined by:

$$\text{margin} = \frac{|1 - w_{N+1}|}{||\mathbf{w}||} - \frac{|-1 - w_{N+1}|}{||\mathbf{w}||} = \frac{2}{||\mathbf{w}||} \quad (3)$$

This provides an optimal training for a classifier. However, it should be kept in mind that every pixel should stay on its correct side when hyperplane separates them. This defines a set of constraints for training in the following manner, where y_i , equals to 1 for class 1 and -1 for class 2.

$$\min \frac{1}{2} ||\mathbf{w}||^2 \quad \text{such that} \quad y_i(\mathbf{w}^T \mathbf{x}_i + w_{N+1}) - 1 \geq 0 \quad (4)$$

If the above standard quadratic optimization function is solved, the optimal hyperplane orientation can be found. Although this procedure is valid only for linearly separable cases, it can be used for higher order space performing kernel trick transformation. In this study, we selected radial basis function as a kernel. To differentiate crack and non-crack regions, an SVM model is trained using geometric properties mentioned above. While training, the best parameters of SVM are obtained by fivefold cross validation. Finally, the allocated crack candidate regions for testing are classified using the trained model.

3. Experiment and Results

In order to test the performance of the crack candidate detection and classification algorithm, a recent crack data were collected from Middle East Technical University campus rigid pavements. Total of 109 images were taken including crack and non-crack regions using DJI Inspire 1 Quadcopter [20] at different altitudes ranging from 0.5 m to 3 m with a speed of 1 to 3 m/s (Figure 3). Throughout this process, quadcopter's camera was oriented such that it becomes vertical to the surface of the pavement. Pavement images were stored in a micro SD card in the field, then the data processing part was performed in the office.

Figure 3. Data Collection with Drone (METU Campus).

As part of data processing, 157 regions were extracted by applying the crack candidate detection algorithm to a total of 109 pavement photos with 4000 x 3000 pixels resolution. The number of crack candidate regions are greater than the number of photos, because the algorithm is capable of extracting multiple regions from a single image. The results obtained from each step of the algorithm are illustrated in Figure 4. In the field experiments, out of the 157 regions, 124 of them were labelled crack and 33 were non-crack regions. Afterwards, geometric properties of crack/non-crack regions were calculated. 80% of these regions were allocated for training data set and rest of them were used for the testing. Lastly, crack classification step was carried out through testing of the trained SVM model using this data set.

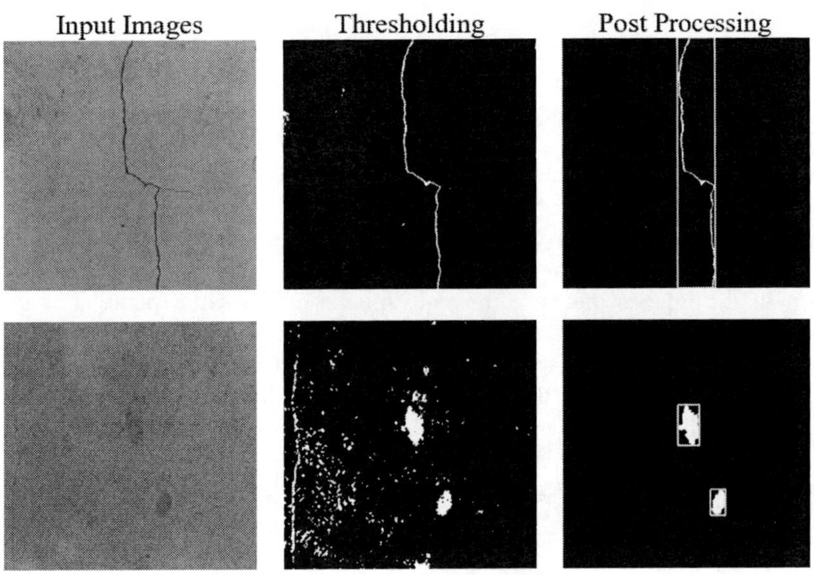

Figure 4. Results of the Algorithm on Crack and Non-crack Regions (Boxes in post processing indicate detected cracks).

Performance estimation of the crack classification algorithm was accomplished by constructing a confusion matrix (Table 1). In this matrix, TP and TN represents the total number of correctly classified and misclassified crack regions, respectively. On the other hand, FP indicates the number of non-crack regions classified as crack and FN denotes the number of crack regions classified as non-crack. Then, most commonly used statistical parameters were calculated using the elements of this matrix.

Table 1. Confusion Matrix.

		PREDICTION	
		Crack	Non-Crack
ACTUAL	Crack	23 (True Positive)	0 (False Negative)
	Non Crack	1 (False Positive)	9 (True Negative)

$$Sensitivity = \frac{TP}{TP + FN} = 100.0\% \tag{5}$$

$$Specificity = \frac{TN}{TN + FP} = 90.0\% \tag{6}$$

$$Accuracy = \frac{TP + TN}{TP + TN + FP + FN} = 97.0\% \tag{7}$$

According to Table 1, sensitivity metric shows that the proposed algorithm is superior to classify crack regions as cracks. It does not misclassify those cracks as non-cracks either. On the other hand, specificity value is not good as sensitivity, because one of the non-crack regions is misclassified as crack region. For this case, a thin non-crack region mimics the properties of longitudinal crack region. Taking everything into account, the total accuracy of 97.0% proves that the algorithm is successful to distinguish crack and non-crack regions.

In the proposed method, the collected images are in RGB color and do not have any depth information. Some of the non-crack regions intersecting with crack ones prevent the detection of the crack region in the images, because they all appear in the same plane. For instance, shadow and concrete joint spaces cover up or intersect with the crack regions and make them difficult to extract crack bodies from the image. Two examples, illustrated in Figure 5 and Figure 6, were included neither in training data set nor the testing one, since the crack and non-crack regions were not distinguished well. Those kind of noises decrease the performance of the crack detection algorithm.

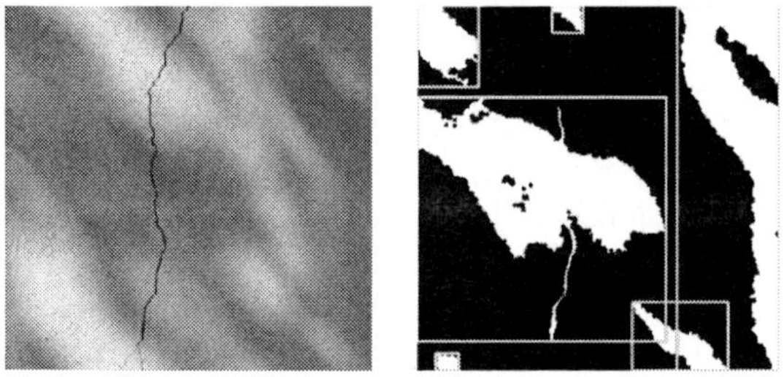

Figure 5. Shadow Included Image.

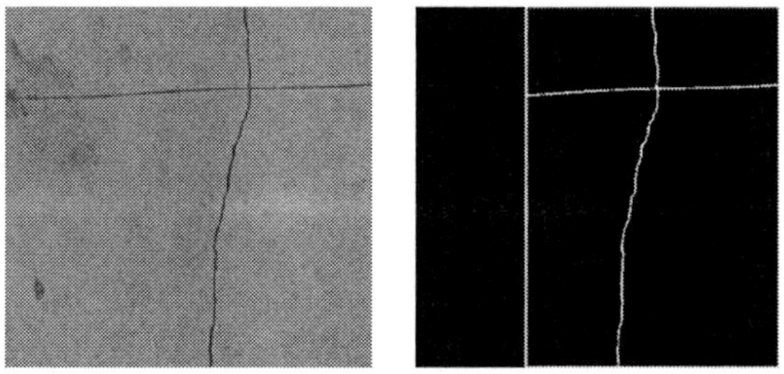

Figure 6. Joint Spacing Included Image.

4. Conclusion

In this paper we propose a pavement crack identification system for monitoring rigid pavements. The system is initialized with images taken by UAVs, which are to be processed using the recently introduced image processing algorithms. Later, distinct features, which are obtained from those images, are used as the inputs of a well-known SVM algorithm for classifying them. The evaluation performance of trained SVM model was measured with a field study performed along a rigid pavement.

The successful results indicate the proposed system ensures a good alternative solution for pavement monitoring applications. Despite some disadvantages such as performance failure in the case of shadowy images or images with low resolution, the main advantage of the system is that, it offers cost-effective solution compared to currently used systems such as truck mounted road monitoring systems as the UAVs are getting cheaper and easily transportable. Meanwhile, the performance of crack detection and classification algorithms should be improved as the performance was tested only with a limited set of data. Within this sense, the future works of this study includes the addition of more features to increase the accuracy of algorithm as well as including additional crack types. After considering safety and flight regulations in the future, the system will have the potential to be used a professional pavement crack identification tool.

References

[1] Fukuhara T, Terada K, Nagao M, Kasahara A and Ichihasi S 1990 Automatic Pavement-Distress-Survey System *J. Transp. Eng.* **116** pp 280–6
[2] Klassen G and Swindall B 1993 Automated Crack Detection System Implementation in ARAN *Proc. Int. Conf. Digital Image Processing: Techniques and Applications in Civil Engineering*
[3] Wang K C and Elliot R P 1999 *Investigation of Image Archiving for Pavement Surface Distress Survey* (Final Report Mack–Blackwellv Transportation Center, Department of Civil Engineering, University of Arkansas, Fayetteville)
[4] Wang K C, Gong W, Li X, Elliott R P and Daleiden J 2002 Data Analysis of Real-Time System for Automated Distress Survey *Transp. Res. Rec.* **1806** pp 101–109
[5] Wang K C P et al 2003 Network Level Crack Survey with the Automated Real-Time Distress Analyzer *Annu. Meet. Transp. Res. Board* pp 1–26
[6] Groeger J L, Stephanos P, Dorsey P and Chapman M 2003 Implementation of Automated Network Level Crack Detection Processes in the State of Maryland *Annu. Meet. Transp. Res. Board*
[7] Gavilán M et al 2011 Adaptive road crack detection system by pavement classification *Sensors* **11** pp 9628–57
[8] Cafiso S, Di Graziano A and Battiato S 2006 Evaluation of Pavement Surface Distress Using Digital Image Collection and Analysis *Seventh Int. Congr. Adv. Civ. Eng (Istanbul, Turkey)*
[9] Sy N T, Avila M, Begot S and Bardet J C 2008 Detection of defects in road surface by a vision system *Proc. Mediterr. Electrotech. Conf. - MELECON* 2, pp 847–51
[10] Tsai Y-C J and L F 2012 Critical Assessment of Detecting Asphalt Pavement Cracks under Different Lighting and Low Intensity Contrast Conditions Using Emerging 3D Laser Technology *J. Transp. Eng.* **138** pp 649–56
[11] Feng W, Yundong W and Qiang Z 2009 UAV borne real-time road mapping system *2009 Jt. Urban Remote Sens. Event* pp 1–7
[12] Zhang C and Elaksher A 2012 An unmanned aerial vehicle-based imaging system for 3D measurement of unpaved road surface distresses *Comput. Civ. Infrastruct. Eng*
[13] Zakeri H, Nejad F M and Fahimifar A 2016 Rahbin: A quadcopter unmanned aerial vehicle based on a systematic image processing approach toward an automated asphalt pavement inspection *Automation in Construction* **72** pp 211-35
[14] Huang Y and Xu B 2006 Automatic inspection of pavement cracking distress *J. Electron. Imaging* **15**(13017)
[15] Zhou J, Huang P S and Chiang F-P 2006 Wavelet-based pavement distress detection and evaluation *Optical Engineering* **45**(027007)
[16] Nguyen T S, Avila M and Begot S 2009 Automatic Detection and Classification of Defect on road Pavement using Anisotropy Measure *Eur. Signal Process. Conf.* pp 617–21
[17] Li Q, Zou Q, Zhang D and Mao Q 2011 FoSA: F* Seed-growing Approach for crack-line detection from pavement images *Image Vis. Comput.* **29** pp 861–72
[18] Zou Q, Cao Y, Li Q, Mao Q and Wang S 2012 CrackTree: Automatic crack detection from pavement images *Pattern Recognit. Lett.* **33** pp 227–38
[19] Oliveira H and Correia P L 2013 Automatic road crack detection and characterization *IEEE Trans. Intell. Transp. Syst.* **14** pp 155–68
[20] DJI Inspire 1 Available at: http://www.dji.com/inspire-1 (Accessed: 10th April 2017)

Using traffic speed deflectometer to measure deflections and evaluate bearing capacity of asphalt road pavements at network level

Ilja Březina[1], Josef Stryk[1] and Jiří Grošek[1]

[1]CDV-Transport Research Centre, Líšeňská 33a, Brno 636 00, Czech Republic

E-mail: ilja.brezina@cdv.cz

Abstract. The paper deals with diagnostics of bearing capacity of asphalt pavements by a Traffic Speed Deflectometer (TSD device), which allows to measure pavement deflections continually at the traffic speed on the basis of dynamic loading induced by moving wheel of a reference axle at the speed of up to 80 km/h. The paper aims to inform of a new method to measure road pavement deflections, describes the principles of measuring pavement deflections by TSD device, and presents results of comparative measurements between FWD (Falling Weight Deflectometer) and TSD devices organized by CDV in Italy and Slovakia. Particular attention was paid to the difference between deflections measured by FWD and TSD devices.

1. Introduction

The diagnostics of asphalt pavements in terms of their bearing capacity is usually performed by the measurements of pavement deflection by a Falling Weight Deflectometer (FWD), which applies loading on the pavement by a stationary impulse. FWD allows to set the loading parameters so that they would be similar to real loading by wheels of moving vehicles. However, the real deflection corresponds the most with the deflection measured during dynamic loading (movement of vehicles).

Traffic Speed Deflectometer (TSD) is the latest device meeting this criterion, which is an alternative to pavement deflection measurement by FWD. The advantage of a TSD is continuous measurement of deflections of the whole diagnosed road segment, which is not limited to measuring of selected points as in the case of an FWD. The deflections measured by a TSD are evaluated in 10-metre intervals.

Contactless measurements at the speed of up to 80 km/h have no restraints to road traffic and the measurement capacity considerably increases in comparison to a FWD. Thanks to this, a TSD device allows fast and safe data recording and is a suitable device for diagnostics at the network level (evaluation of the whole network under the management of a relevant administrator in order to evaluate and identify the weakest road sections).

There are currently 8 of these devices in operation in the world. In Europe there are devices in Denmark (DRD, GE A/S), Italy (ANAS S.p.A.), Poland (IBDiM) and Great Britain (TRL). In order to verify a correlation between FWD and TSD data, comparative measurements are performed for both devices and pilot projects are established for the integration of TSD into the pavement management system.

Traffic Speed Deflectometer, methodology of measurement, analysis and evaluation of measured data are more and more advanced.

2. Traffic Speed Deflectometer

Traffic Speed Deflectometer measures bearing capacity of roads contactlessly at the speed of traffic flow. The road deflection is measured under dynamic loading induced by rolling vehicle wheels, which is most similar to the real deflection induced by loading of heavy vehicles (magnitude, duration).

Based on the recorded deflection values, the user receives complete data on road pavement bearing capacity along the whole measured road section. In contrast to diagnostics with FWD, the issue with road traffic restraints is avoided, safety of road users as well as of measuring vehicle crew during the measurement is higher. Another advantage is high measurement capacity, which allows to measure up to several hundred kilometers of roads per day (approx. 350 – 450 km on average). According to the information from IBDiM organization, daily measurements of more than 800 km are no exceptions. The measurements cannot be performed in rain or on wet road pavements due to the use of laser sensors.

The system consists of a heavy vehicle with a truck and a modified semitrailer (Figure 1). The single axle semitrailer contains a measuring system for measuring road pavement response to loading. The standard loading per axle is 10, or 12 tonnes respectively. Therefore, corresponding weight is installed under the semitrailer container. The main part of the measuring system is a rigid beam with appropriate number of Doppler laser sensors (Figure 2), which scan the road pavement of the right wheel path in the axis (dual tires), and other parts include support and computing systems. The number of laser sensors ranges between 6 and 10.

The crew of the TSD system consists of a vehicle driver and an operator, who controls the measurement and the recording of data during the measurement.

Figure 1. Italian Traffic Speed Deflectometer (author's archive).

Figure 2. The rigid beam with the Doppler laser sensors (author's archive).

3. Principle of TSD deflection measurement

The loading applied to the pavement by a rolling TSD wheel causes deformation of road pavement (deflection basin). Road pavement deflection velocity is measured with the use of Doppler laser sensors, which are fixed to a rigid beam in a corresponding distance from the loading position. Subsequently, the deflection velocity is used for a calculation of the road pavement deflection. The emitted laser beam is reflected by the road pavement and the sensor measures the speed of short-term vertical deformation of the loaded road pavement along the direction of the laser beam.

Since the sensor is not placed exactly perpendicular to the road pavement, the laser beam falls at the angle of approx. 2°. Apart from the required pavement deformation velocity V_V, the Doppler-laser sensor registers vehicle speed V_H. Relatively low angle allows the input of vehicle speed into the calculation and also has only marginal effect on the vertical composition of deformation velocity.

A deflection slope (tangent) [μm/m] is calculated with the use of simultaneously measured vehicle speed and the defined laser beam angle towards the vertical axis. Deflection slope is calculated as the ratio V_V/V_H [5].

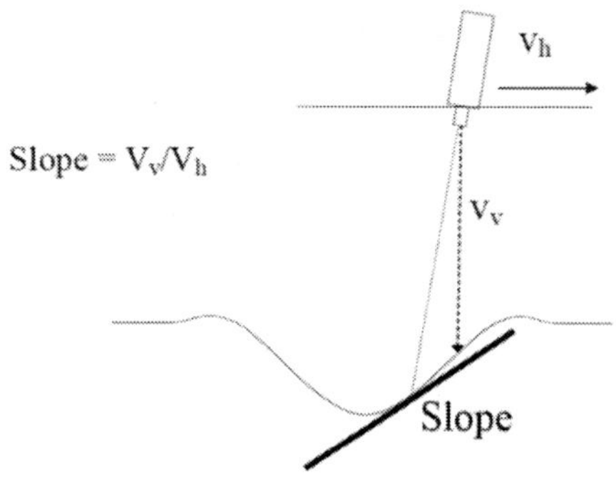

Figure 3. Calculation of slope of deflection bowl. [4]

Figure 4 (left) shows the pavement deflection velocity vectors under a rolling wheel. Together with deflection velocity, the corresponding deflection basin is shown in Figure 4 (right), where deflection slopes (tangents) are displayed.

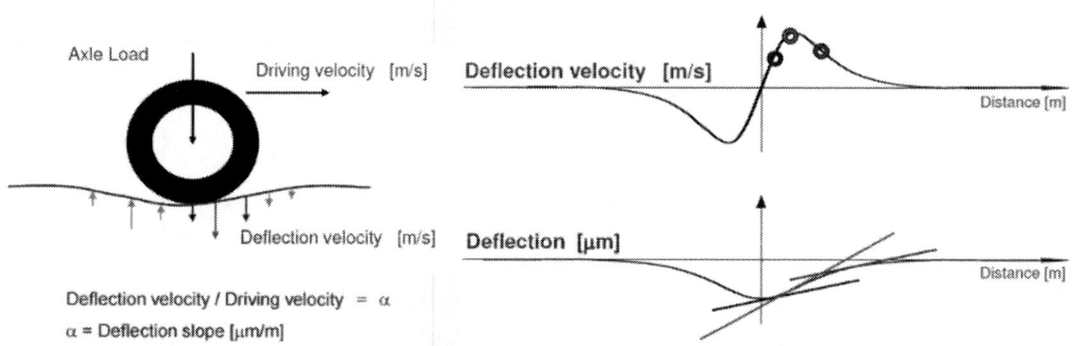

Figure 4. Left: Pavement deflection velocity under rolling load; Right: Pavement deflection velocity and deflection basin with deflection slopes (tangents). [7]

Deflection slope is determined as deflection velocity divided by driving speed. As the slope of deflection is a derivative of the displacement, it is possible to calculate the displacement. This means that it is possible to obtain bearing capacity characteristics such as Structural Curvature Index 300 ($d_0 - d_{300}$) and centre deflection (d_0) based on the measured deflection slopes [5, 7].

$$SCI_{300}=\int_{0}^{300} s(x)dx = d_{300} - d_0 \quad (1)$$

Where s(x) = slope at location x, d(x) = deflection at location x.

The lowest possible step for the export of the measured data by a TSD is 10 metres. This boundary value is given by software used for further processing of measured values, which was developed by Greenwood Engineering A/S. It is possible to evaluate data for a distance lower than 10 metres, but without the values of calculated deflections. When calculating deflections at constant distances, the representative deflection value for each distance is calculated with the use of statistical methods, which are parts of the software.

4. Comparative measurement between TSD and FWD

Comparative measurements between TSD and FWD devices were performed in 2014 and 2015 within the project of the Technological Agency of the Czech Republic No. TE01020168 (CESTI).

In November 2014 near Rome, a comparative measurement was performed with an Italian TSD in the ownership of a research institution ANAS S.p.A. – Centro Sperimentale Stradale and an FWD RODOS 2012 of Czech production, whose owner is a research institute Centrum dopravního výzkumu, v. v. i . The measurements was performed on several road segments with asphalt pavement. One of them was SS1 (Strada Stanale 1 Via Aurelia) with the length of the measured segment of 3 km. The TSD measured road pavement deflections at the speed of 60-80 km/h, depending on traffic flow speed. The measurement step of the FWD device was 20 m. After excluding erroneous data, more than 150 deflection curves were measured. The export of data from the TSD was performed in the same measurement step, i.e. 20 m.

A comparative measurement of an FWD and a TSD, owned by a Polish research institute IBDiM, was performedon 3 road sections with asphalt pavement on D1 (2.5 km), I/18 (1.95 km), II/537 (2.5 km) near Liptovský Mikuláš in Slovakia in June 2015.

Deflections on roads I/18 and II/537 were measured by the TSD at the speed of approx. 70 km/h. In addition, deflections on motorway D1 were measured at different speeds of the TSD (approx. 60 km/h, 70 km/h, 80 km/h). The measurement step of FWD device was 25 m on all measured road segments. After excluding all erroneous data more than 250 deflection curves were measured. The export of data from the TSD was performed in the same measurement step, i.e. 25 metres.

5. TSD and FWD measurement result evaluation

Comparative measurements between FWD and TSD devices which were performed on road segments described in the previous chapter were evaluated in order to find a correlation between deflections from both devices.

Deflection values after corrections for loading force and temperatures were used for the correlation. The deflections measured by FWD and TSD devices were corrected to the comparative force of 60 kN (Italian TSD has loading 12 tonnes per axle), or 50 kN respectively (Polish TSD has loading 10 tonnes per axle). The calculation used in Italy [1, 2], or in Slovakia [6] respectively, were used for the temperature correction. Temperature correction had no considerable effect, since the measurements of FWD and TSD devices were performed in the same temperature conditions.

The degree of correlation between the deflection values measured by a TSD and FWD in Slovakia was determined with the use of linear regression, and the correlation reliability was expressed by reliability coefficient R^2. The results of the determination of deflections correlation in the loading position for linear relationships are shown in Table 1.

Regarding the differences in the method of road pavement loading, the loading time in subgrade and subbase layers is different for TSD and FWD measurements. The loading time of asphalt road layers is relatively comparable. Evaluation of asphalt layers uses the index SCI_{300} (according to [3] a difference of deflections measured in the loading position and at the sensor located 300 mm from the loading position). Therefore, in the next phase the correlation of index SCI_{300}, mentioned in the TSD results and calculated from FWD outputs, was examined. Similarly to the deflection on the loading positin, the relationship between the values of indexes SCI_{300} were examined with the use of linear regression and coefficient of determination R^2. The results for the linear relationship are shown in Table 1.

Table 1. Comparison of TSD – FWD in Slovakia; coefficient of determination R^2.

Parameter	d_0			SCI_{300}		
TSD speed Measured segment	60 km/h	70 km/h	80 km/h	60 km/h	70 km/h	80 km/h
D1	0.36507	0.36262	0.24755	0.05428	0.16924	0.10794
I/18	-	0.03098	-	-	0.09329	-
II/537	-	0.10451	-	-	0.06227	-

The values of coefficients R^2 in Table 1 are relatively low. According to the criteria of determining the relationship of two variables on the basis of the reliability coefficient and regarding the values in the range of 0.1 to 0.25, the relationship is weak. The range from 0.25 to 0.5 characterizes moderate relationship strength, 0.5 to 0.8 shows strong relationship and over 0.8 it is very strong relationship of two variables. Therefore, there was a weak relationship between the values of deflections measured by TSD a FWD devices. Very weak relationship was found for the parameter SCI_{300} in this case. In contrast, the results of measurements from Italy showed similar shape of curves comparing parameters SCI_{300} (Figure 5). When evaluating the results obtained in Slovakia, a problem occurred with the accuracy of localization of both devices. This was probably the reasons for weak relationship of both devices. The localization accuracy during the measurement has the same importance as correct deflection data from the measurement.

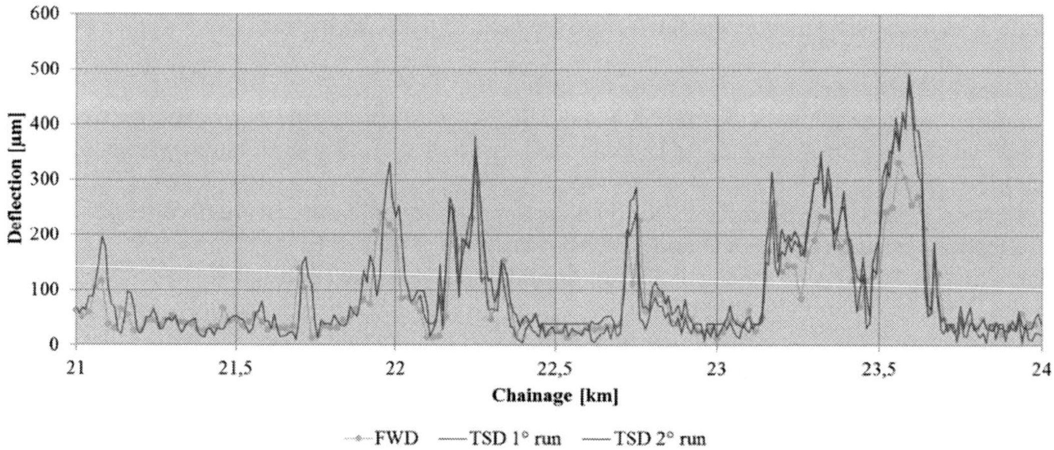

Figure 5. Comparison of index SCI_{300} on road SS1 (Italy).

Regarding the deflection trends, the evaluations of results (SCI300 and D_0) from Italy found the same trends, i.e. the increase in FWD values correlated with the increase in TSD values (Figures 5, 6). This means that if TSD was able to identify differences in bearing capacity of individual road sections. This is important for the usability of TSD for the diagnostics of bearing capacity at the network level.

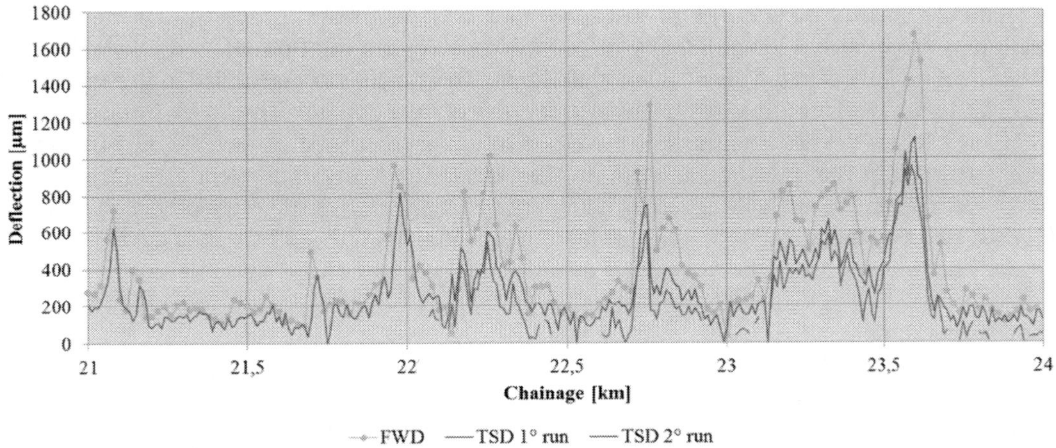

Figure 6. Comparison of index D_0 on road SS1 (Italy).

6. Conclusions

Based on the evaluation of measurement results by TSD and their comparison with the measurement results by FWD, more conclusions can be made in terms of the usability of TSD device for the diagnostics of bearing capacity of asphalt road pavements at the network level.

First of all, TSD is a high performance device, which is able to diagnose a large length of the road network within a short time.

Another advantage is diagnostics at the speed similar to the real road traffic flow speed, while safety protection measures for the measurement vehicle and its crew are unnecessary. In comparison with FWD diagnostics, it is a crucial advantage in terms of traffic safety. Apart from that, the diagnostics outcomes more correspond with the real loading conditions, since they contain dynamic aspects based on the road pavement longitudinal unevenness. Furthermore, the loading time in individual levels (layers) of road structure corresponds with the real loading.

The extent of information from TSD measurements is sufficiently large. It includes all other important input data used for standard evaluation of road pavement bearing capacity. The outcomes not only include the deflection in the position of loading axle and different (standardly used) distances from this position, but also values of different indexes calculated from the mentioned deflections.

Based on the evaluation of performed comparative measurements, the same trend in deflections of TSD and FWD was found. This is important in terms of the usability of TSD device for diagnostics at the network level.

Regarding the evaluation criteria, existence of the relationship between TSD and FWD devices was found, although the relationship differed for individual testing trials. Therefore, we may deduce that it is possible to use the evaluation criteria for road pavement bearing capacity from FWD measurements to derive the evaluation criteria on the basis of TSD measurement results. It is necessary to perform more extensive measurements to collect a larger statistical file of measured data (roads of different classes and constructions, with different thickness of layers and different bearing capacity), and design a system of road pavement bearing capacity evaluation at the network level on the basis of TSD measurements.

The comparative measurements showed the importance of correct localization of measured data. Nowadays, localization is not only performed through the driven distance, but also with the use of GPS coordinates and combination of other systems (IMU, image recording). In addition, it is useful to have reference points on the road with known coordinates, where the localization accuracy can be verified. The measurement localization accuracy at higher speeds is as crucial as the correct deflection data.

A relatively small statistical file of measured data collected from the measurements in Italy and Slovakia may have caused low reliability of results. Therefore, a pilot project with the aim to collect larger amount of TSD data for the evaluation at the network level was performed at the end of 2016 in the Czech Republic. This project was funded by the State Fund for Transport Infrastructure (No. 5006210261). The TSD device measured in total 170 km asphalt road pavements of different classes and constructions in South Moravian region. The selected road sections were also measured by an FWD device. A detailed analysis is currently underway for the results of both methods in order to find differences between the methods and for evaluating potential application of a TSD device in the Czech Republic. The final results of this project was unavailable when this article was written.

Acknowledgements
This article was produced with the financial support of the Technological Agency of the Czech Republic within the project TE01020168 Centre for Effective and Sustainable Transport Infrastructure (CESTI), on the research infrastructure acquired from the Operation Programme Research and Development for Innovations (CZ.1.05/2.1.00/03.0064).

References
[1] Cesolini E 2014 Efficiency and Environmental Sustainability in Based Performance Approach in Building Italian Pavements *3rd Int. Conf on Transportation Infrastructure (Pisa)*
[1] Drusin S 2014 Evaluting the Performance of New and In-Service Pavements in Italy Using High-Speed Non-Destructive Testing *National Pavement Evaluation Conference (Blacksburg, Virginia)*
[2] European commission 2005 *COST 336 Use of Falling Weight Deflectometers in Pavement Evaluation* (Office for Official Publications of the European Communities) p 392
[3] Ferne B, Langdale P, Round N and Fairclough R 2009 Development of the UK Highways Agency Traffic Speed Deflectometer *Eighth Int. Conf. on the Bearing Capacity of Roads, Railways and Airfields (Champaign, Illinois)*
[4] Katicha S W and Flintsch G W 2015 Field Demonstration of the Traffic Speed Deflectometer in New York *Transportation Research Board Meeting*
[5] Ministry of transport ot the Slovac Republic 2009 Meranie a hodnotenie únosnosti asfaltových vozoviek pomocou zariadenia FWD KUAB /Measurement and evaluation of bearing capacity by FWD KUAB device/ *Slovak technical regulation (TP 01/2009)*
[6] Rasmussen S, Aagaard L, Baltzer S and Krarup J 2008 A Comparison of Two Years of Network Level Measurements with the Traffic Speed Deflectometer *6th International Conference Transport Research Arena Europe (Ljubljana, Slovenia)*

Pavement noise measurements in Poland

Ewa Zofka[1], Adam Zofka[1] and Tomasz Mechowski[1]

[1]Road and Bridge Research Institute (IBDiM), ul. Instytutowa 1, 03-302, Warsaw, Poland

E-mail: ezofka@ibdim.edu.pl

Abstract. The objective of this study is to investigate the feasibility of the On-Board Sound Intensity (OBSI) system to measure tire-pavement noise in Poland. In general, sources of noise emitted by the modern vehicles are the propulsion noise, aerodynamic resistance and noise generated at the tire-pavement interface. In order to capture tire-pavement noise, the OBSI system uses a noise intensity probe installed in the close proximity of that interface. In this study, OBSI measurements were performed at different types of pavement surfaces such as stone mastic asphalt (SMA), regular asphalt concrete (HMA) as well as Portland cement concrete (PCC). The influence of several necessary OBSI measurement conditions were recognized as: testing speed, air temperature, tire pressure and tire type. The results of this study demonstrate that the OBSI system is a viable and robust tool that can be used for the quality evaluation of newly built asphalt pavements in Poland. It can be also applied to generate reliable input parameters for the noise propagation models that are used to assess the environmental impact of new and existing highway corridors.

1. Introduction

Noise pollution has long been recognized as an influencing quality of life and well-being. Over past several decades it has been also recognized as an important public health issue. According to a recent report from the World Health Organization (WHO) [1], at least 1 million healthy life years are lost every year in Europe due to health issues associated with the traffic noise alone. The traffic noise can be generated by different sources including propulsion noise, transmission noise, aerodynamic resistance noise or tire pavement interaction [2]. In typical internal combustion engine car (ICE) majority of the noise is generated by its propulsion system up to a certain speed (also called crossover speed). Above this speed tire/pavement contact noise plays a prevailing role in the overall vehicle noise. The crossover speed varies between 16 and 40 km/h for the passenger cars whereas for trucks it varies between 50 and 80 km/h. In modern hybrid-electric vehicles (HEV), electric vehicles (EV) or fuel-cell electric vehicles (FCEV) the propulsion and transmission noise becomes insignificant, so tire/pavement interaction noise becomes dominant regardless of the vehicle speed. Thus there is an urgent need to prepare, verify and validate robust methodology that focuses on tire/pavement noise and can be used by pavement engineers, tire manufactures and traffic safety engineers especially in urban environments.

2. OBSI review

There are several methods to assess noise generated by moving vehicle traffic [3]. These methods can be briefly summarized as follows:

OBSI – On-Board Sound Intensity method (OBSI) – uses two synchronized sound intensity probes mounted on regular vehicle's wheel next to tire/pavement interface [4]. No special sound insulation is needed since intensity probes are directional and insignificantly (if at all) affected by the adjacent environment. Since it is attached to a passenger car it is highly maneuverable and noise assessment can be performed on any road network. This method was developed by General Motors in 1980's but it has been commercially available since 2000's [5]. More details are provided in later paragraphs.

OBSIe – it is an evolution of OBSI system mounted on a electric (EV) or hybrid electric (HEV) vehicle. The main reason is to eliminate the propulsion noise and to conduct assessment of tire/pavement noise at lower speeds, preferably in the urban areas. The tire/pavement interaction noise can be measured from as low speeds as 10 km/h. This method was first time demonstrated by Road and Bridge Research Institute (IBDiM) in Poland in 2015 [6][7].

SPB – Statistical Pass-By method that measures overall noise generated by passing traffic. This method utilizes set of microphones located in a certain distance from the road. Microphones record the overall sound pressure generated by randomly passing vehicles while the observer measures their actual speed. This method records all noise sources from passing vehicles mentioned above as well as amplification phenomena. Different vehicles and different pavement types can be assessed by this method.

CPB – Controlled Pass-By which is similar to SPB method. The difference is that the speed of passing vehicles is strictly controlled as well as the type of vehicle (passenger car, truck or bus) is intentionally selected for evaluation.

CPX – Close Proximity method is another method that focuses on tire/pavement interaction. CPX method typically uses a trailer attached to a leading vehicle (typically a passenger car). Trailer which can be opened or closed-cover comprises a set of microphones mounted around a tested wheel (or wheels). Microphones measure sound pressure that is later processed to obtain sound pressure level (SPL). It is also possible to mount microphones next to a vehicle's wheel however this approach is not common. This method is primarily used in Europe since late 1970's [2].

While both OBSI and CPX are used in the ongoing study in Poland, this paper only discusses the OBSI method. The following paragraphs include the concept of operations and a brief discussion on the effect of surface type and speed on the OBSI noise measurements.

2.1. Concept of operations

The main hardware components of the OBSI system are presented in Figure 1. As mentioned earlier, there are two intensity probes (front and back) located in the close proximity to tire/pavement interface. Each probe consists of two microphones and each microphone measures sound pressure; in order to calculate the sound intensity from each probe, the sound pressure signals from two microphones are combined using so called p-p method [8]. The particle velocity component in the direction of the axis of the probe is obtained by a finite-difference approximation to the pressure gradient in Euler's equation of motion, and the sound pressure is the average of the two pressure signals. The most important limitations of this measurement technique are caused by the finite difference approximation, scattering and diffraction, and instrumentation phase mismatch. More details on computational data processing can be found elsewhere [9]. Overall intensity level from each probe is determined from the energy sum of corresponding A-weighted sound intensities within the one-third octave bands ranging from 400 to 5000 Hz as follows:

$$\text{Overall Sound IL} = 10 \cdot \log_{10}\left(\sum_{i=400}^{5000}\left(10^{L_i/10}\right)\right) \tag{1}$$

The overall ILs from the two probes (measured simultaneously during a single run) are energy averaged to yield an average sound intensity level per test run. The formula for energy averaging is as follows:

$$\text{IL}_{\text{E-avg}} = 10 \cdot \log_{10}\left(\frac{1}{N}\sum_{j=1}^{N}\left(10^{L_j/10}\right)\right) \tag{2}$$

Finally in order to calculate the average of several test runs, overall ILs from the individual valid runs are arithmetically averaged to calculate a single average intensity level value.

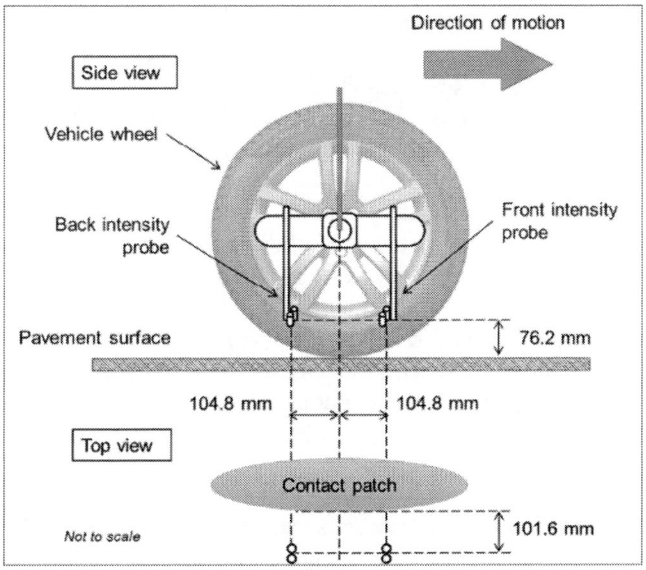

Figure 1. Schematic of OBSI system mounted on a vehicle wheel.

It should be noted that there are two immediate data quality checks that control the validity of each test run: coherence and pressure-intensity index (PI Index). Coherence is defined as a measure of the linear dependency of two signals with a value of zero (0) being no dependency, and a value of one (1) being prefect linear dependence. After AASHTO [10] it was assumed that coherence of sound pressure between the two microphones of each intensity probe should be at least 0.8 for each one-third octave band between 400 and 4000 Hz. On the other hand, PI Index is defined as an arithmetic average of the SPLs of both microphones on each probe minus the calculated sound intensity level (IL) for that probe. The PI index for each one-third octave band is calculated using the sound pressure levels and intensity level corresponding to that one-third octave band. Allowable ranges of PI Index are presented in Figure 4 together with example results of PI Index.

A single testing run is conducted for 5 s which translates to 133 m under constant speed of 96 km/h. For other testing speeds the measurement distance is kept constant which in consequence leads to different testing duration. OBSI measurements can be affected by a number of factors that should be either controlled during testing or accounted for in data processing. Those factors can be attributed to vehicle speed, pavement type, tire type/pressure/temperature, air temperature and pressure, adjacent road furnishing (e.g. barrier, noise walls), vehicle axle load and suspension characteristics among others. The following paragraphs briefly present a summary of relevant literature on the effect of speed and surface type as well as comparison between OBSI and CPX.

2.2. OBSI measurements on different pavement surfaces

There are numerous documented studies showing the OBSI measurements on different pavement surfaces, for example [5][11][12][13]. Extensive study presented in [5] demonstrated OBSI measurements on various pavement sections both in Europe and the USA. 127 combinations of surface, speed, and test tire were evaluated. In terms of pavement surfaces, measurements were taken on dense-graded asphalt (DGA), SMA, single layer porous asphalt (PA), double layer porous asphalt (DLPA) as well as on textured and regular PCC surfaces. At 96 km/h, the loudest pavement (107.6 dBA) was a transversely tined PCC while the quietest at 94.6 dBA was DLPA. While many observations were made in that study it should be noted that measurements were not normalized due to different test temperatures that ranged between 15 and 32 °C. Another study [11] shows OBSI measurements on four sections in the USA: two PCC (burlap drag and longitudinally tined) and two asphalt pavements: open-graded (OGA) and dense-graded (DGA). The results showed similar average overall ILs for PCC section (burlap drag) and OGA section (104 dBA) while both have distinct spectral content. The OGA showed lower level of high-frequency noise, which is indicative of the absorption effect, and high levels of low frequency noise caused by raveling. The quietest pavement was the DGA section (99.7 dBA) while the second PCC (longitudinally tined) had IL of 103.5 dBA. With the intention to reduce noise levels, sections with added rubber component to asphalt wearing surface were assessed in [12]. Specifically, this study focused on the open-graded rubberized asphalt concrete (RAC-O) and gap-graded rubberized asphalt concrete (RAC- G) compared to sections with OGA and DGA. The results indicated that indeed the rubber contributes to noise mitigation but its effectiveness depends on the section age. On average, the reduction is in the order of 1.6 to 2.9 dB. Finally, recent study in Australia [13] presented the OBSI measurements on the OGA, SMA and double layer OGA. The best performing in terms of noise reduction was the OGA section (average of 98 dBA) while the SMA was the loudest with the average IL of 102 dBA. It should be noted that these observations were confirmed not only the OBSI but also by the CPX and SPB measurements.

2.3. Effect of vehicle speed on OBSI measurements

Vehicle speed has indisputable effect on any noise measurements. Typically, regular OBSI measurements are conducted at 96 km/h but other speeds (e.g. 72, 56 or 40 km/h) are also found in the literature. The following bullets provide a brief summary on several studies that accounted for various speeds during the OBSI measurements:

- [5]: testing at 96 km/h and 56 km/h on various surfaces; results at these speeds are highly correlated with average 0.9 dB deviation from line of equality (LOE),
- [14]: testing at several intervals between 56 km/h and 96 km/h allowed to determine a relationship between IL and speed; it depends on the surface and tire type but on average OBSI IL increases 0.18 dBA per 1 km/h,
- [15]: testing at 17 different pavement sections at National Center for Asphalt Technology (NCAT) showed consistent trends between measurements at 72 and 96 km/h,
- [16]: testing at 56, 72 and 96 km/h conducted on the regular asphalt surface and temperatures between 4 and 32 °C demonstrated that average incremental increase in OBSI IL per 1 km/h equals to 0.17 dBA however the observed range was between 0.12 and 0.21 dBA.

2.4. Comparison of OBSI vs. CPX

Table 1 presents several studies that compared the OBSI and CPX methods. In general, both methods show very good qualitative and quantitative agreement in terms of tire/pavement noise assessment.

Table 1. Comparison of CPX and OBSI – representative studies.

Reference	Year	Surface type	Observations / findings
[5]	2007	Various	average offset of 3.3 dB with average difference 0.5 dB (96 km/h)
[17]	2009	Various asphalt	$R^2 = 0.93$, CPX = 1.04*OBSI-6.52 [dBA], highly correlated
[18]	2013	Various asphalt	OBSI-CPX = 3.1 dB at 50 km/h, 2.4 dB at 80 km/h
[13]	2014	SMA, OGA	highly correlated, R^2 above 0.8 (100 km/h)
[19]	2015	Various	$R^2 = 0.93$, CPX = 0.98*OBSI-1.70 [dBA], highly correlated

3. OBSI results from Poland

The following paragraphs demonstrate a subset of the OBSI results collected in Poland. It should be mentioned that all figures depict valid OBSI runs, i.e. runs that passed quality checks. Except for already recognized coherence and PI Index parameters, valid runs under the same conditions should have the range of overall ILs no greater than 1.0 dBA. Further the range of the ILs from the multiple valid test runs within any one-third octave band (between 400 and 5000 Hz) should be less than 2.0 dBA. In the testing presented below, the number of valid runs varied between 2 and 9. The air temperature for all tested runs was approximately 25 °C and all testing was accomplished within one week. The remaining testing conditions were assumed according to AASHTO [10].

3.1. OBSI on different surface types

Figure 2 demonstrates the OBSI measurements obtained for three typical pavement surfaces in Poland: Portland cement concrete (PCC), stone-mastic asphalt (SMA) and regular hot-mix asphalt (HMA). Similar to other studies, there is an apparent effect of surface type. This effect differs depending on the one-third octave bands and it is a function of surface finishing technology (e.g. drag texture, exposed aggregate for PCC pavements) and surface porosity and texture for asphalt pavements. The ongoing analytical research effort focuses on more objective and informative spectra interpretation that accounts for the frequency and connects with passenger annoyance effect inside modern vehicles and sound perception next to the road.

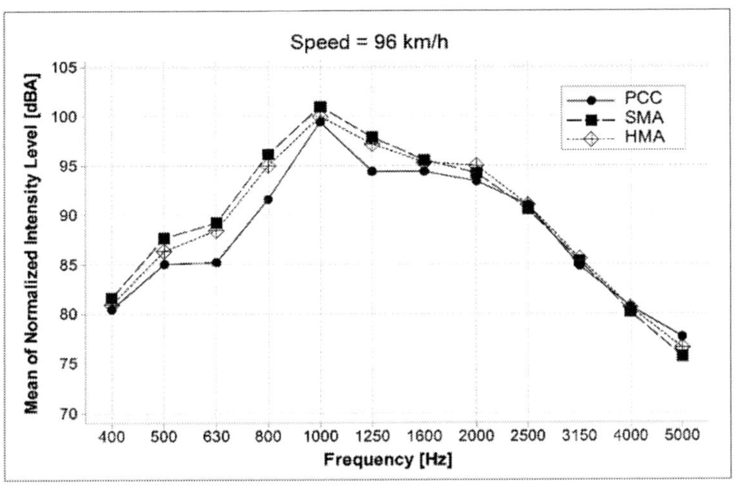

Figure 2. Average intensity spectra for one-third octave bands for different pavement surfaces at 96 km/h.

3.2. OBSI measurements as a function of vehicle speed

Figure 3 presents average values of the overall ILs collected on three surfaces at two different speeds 72 and 96 km/h. One can notice a qualitative agreement in terms of speed dependency for all three surfaces and rather high repeatability that is somewhat dictated by the rigorous data quality checks. It can be also noted that overall ILs clearly differentiate between different speed/pavement combinations and presented PCC pavement sections yields statistically lower sound intensity levels than the corresponding asphalt sections. The average increase in IL per 1 km/h was 0.19 dBA was consisted in the studies reported earlier.

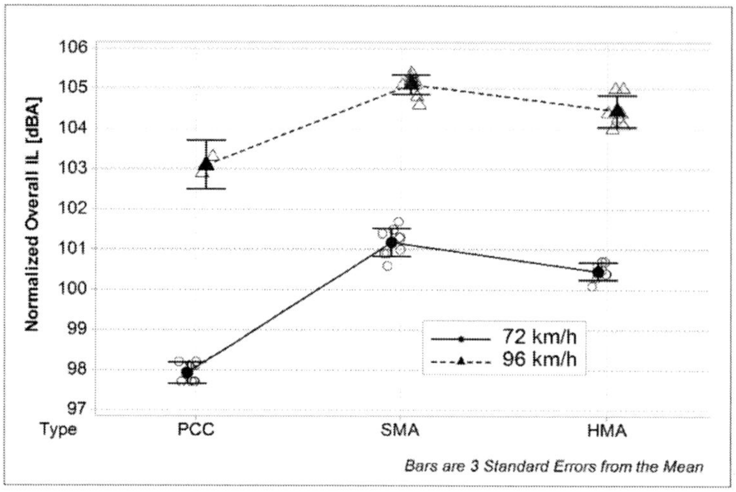

Figure 3. Overall intensity levels (IL) for different speeds and pavement surfaces.

3.3. OBSI data quality checks

Figure 4 depicts the example of data quality check in terms of average PI Index values. Measurements on all three pavement surfaces are located along the center of allowable range defined by the upper and lower bounds [10]. This observation demonstrates a very good accuracy of measurements and phase-matching of microphones.

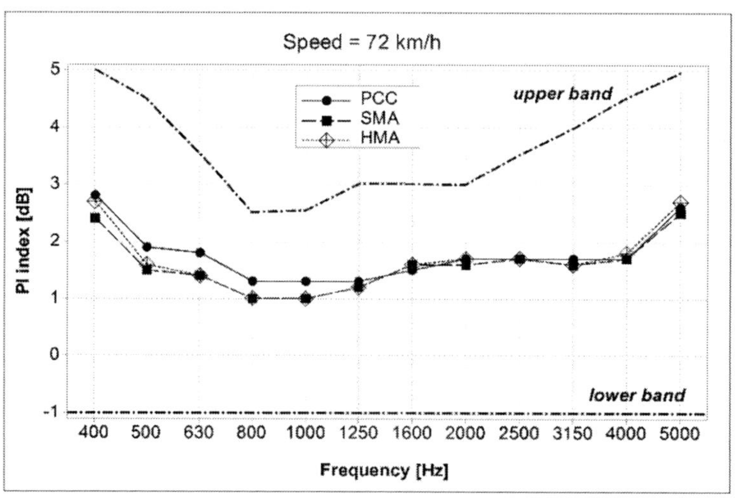

Figure 4. Example of quality check results: PI Index, with allowable upper and lower bands.

4. Discussion

Since OBSI device comprises two intensity probes, one in the front and one behind tire/pavement interface, it is interesting to compare ILs from both probes. Figure 5 presents such a comparison for the measurements collected at 96 km/h on SMA and HMA pavement sections. It can be easily noted that ILs spectra differ for the front and back probes and this trend is consisted for both surfaces. This observation confirms that there are multiple phenomena responsible for the sound generation and they are not symmetrically distributed on both sides of tire/pavement interface. The following phenomena were identified based on the literature [2][8]: radial vibrations of tire tread, air compressing and decompressing during wheel pass, tread tangential motion (referred to as "stick-slip") and breaking of adhesion of tread rubber to the pavement (called "stick-snap"). In addition to these phenomena, there are also amplification mechanisms such as horn amplification, radiation of resonant air (Helmholtz resonance), pipe resonances in channels formed in the tire footprint, tire sidewall vibrations and cavity resonance in tire tube [3].

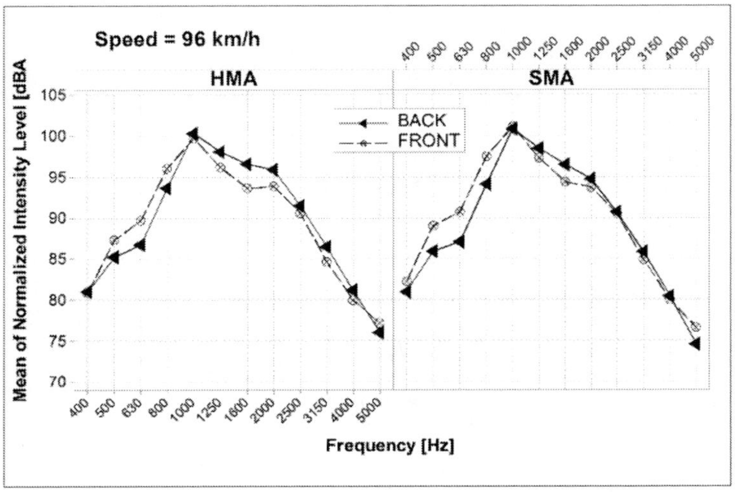

Figure 5. Average Intensity Level (IL) spectra for the back and front probes for two pavement surfaces at 96 km/h.

Basic Analysis of Variance (ANOVA) was performed in order to further assess the significance and contribution of the two main factors presented earlier: pavement surface and measurement speed. The ANOVA results shown in Table 2 confirm earlier observations that both factors are statistically significant and that the speed factor has greater contribution in explaining an observed difference in IL values than pavement type. Finally, performance of the statistical model associated with ANOVA is shown in Figure 6. While it is fairly accurate in predicting IL values for various speeds and pavement surface types, it is just an initial attempt limited only to two factors. More comprehensive analysis will be performed upon completion of the ongoing study in Poland using, for example, Principle Component Analysis (PCA) and Artificial Neural Networks (ANN).

Table 2. Analysis of Variance (ANOVA) table ($R^2 = 98$ %).

Source	DF	Adj SS	Adj MS	F-Value	P-Value
Speed	1	173.657	173.657	1323.28	0.000
Pavement Type	2	53.334	26.667	203.21	0.000
Error	40	5.249	0.131		
Lack-of-Fit	2	2.020	1.010	11.88	0.000
Pure Error	38	3.229	0.085		
Total	43	297.846			

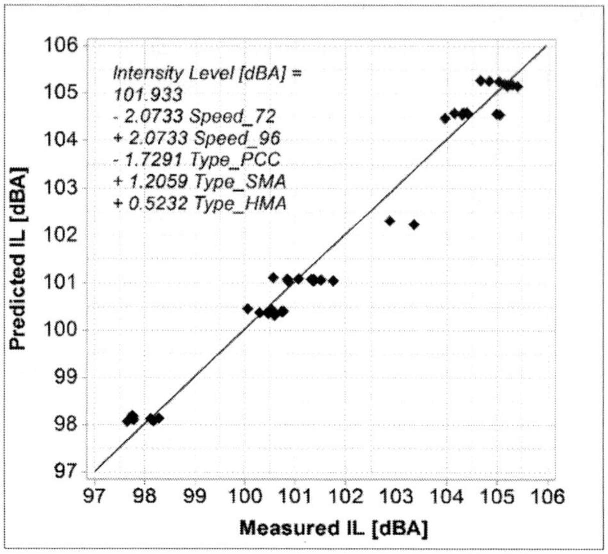

Figure 6. Predicted vs. measured Intensity Levels (IL).

5. Summary

The OBSI method gives an opportunity to evaluate tire/pavement noise using alternative methodology that is predominantly applicable in the urban environments. It is of particular importance due to future developments in vehicle fleet and general societal and economical changes. OBSI method can assist in development and evaluation of various aspects of traffic noise such as:

- tire / pavement noise generation,
- tire construction and thread type,
- sound signals outside the vehicle for VRU (vulnerable road users),
- additional equipment mitigating traffic noise.

In the near future the OBSI methodology will be further verified and validated through extensive testing in Poland. It is planned to conduct OBSI testing under various conditions (speed, temperature, tire pressure, tire hardness, axle weight etc.) on regular and "quiet" pavement surfaces. The detailed analysis will also include individual one-third octave bands and not only the overall OBSI noise level.

Acknowledgements
This paper was created as part of the research project co-funded by the National Centre for Research and Development (NCBiR) and General Directorate for National Roads and Motorways (GDDKiA) and Their Programme for Road Innovations Development – RID.

References

[1] WHO 2014 *Night noise guidelines for Europe* (World Health Organization Regional Office for Europe EU)
[2] Sandberg U and Ejsmont J A 2002 *Tyre/Road Noise Reference Book* (INFORMEX Sweden) p 640
[3] Rasmussen R, Bernhard R, Sandberg U and Mun E 2007 *The Little Book of Quitter Pavements* (FHWA-IF-08-004 USA)
[4] Trinth V 1993 *Measurement of Sound Intensity and Sound Power* (MRL-TR-93-32 DSTO Materials Research Laboratory Australia)
[5] Donavan P R 2007 Comparative Measurements of Tire/Pavement Noise in Europe and the

United States: *Noise Intensity Testing in Europe - NITE Study* (California Department of Transportation Sacramento CA) p 107

[6] Zofka A, Zofka E and Maliszewski M 2016 OBSIe Innovative Noise Measurement Method for Sustainable Urban Development *23rd ITS World Congress (Melbourne)*

[7] Zofka A, Maliszewski M and Zofka E 2016 Innovative method for pavement noise assessment OBSIe *Autostrady* **7** (in Polish)

[8] Jacobsen F and de Bree H-E 2015 A comparison of two different sound intensity measurement principles. Acoustical Society of America **118**(3) pp 1510-17

[9] Vorlander M 2007 *Auralization: fundamentals of acous-tics, modelling, simulation, algorithms and acoustic virtual reality* (Dordrecht Springer Germany)

[10] AASHTO 2016 *Standard Method of Test for Measurement of Tire/Pavement Noise Using the On-Board Sound Intensity (OBSI) Method* (*T 360-16*)

[11] Kohler E 2011 Tire-pavement noise measurement with the on-board sound intensity (OBSI) method *8th International Conference on Managing Pavement Assets ICMPA (Santiago de Chile)*

[12] Lu Q, Harvey J, Kohler E, Rymer B and Motumah L 2011 Comparison of tire-pavement noise characteristics of rubberized and conventional asphalt concrete mixes *Noise Control Engineering Journal* **59**(3) pp 219-25

[13] Buret M, Mcintosh J and Simpson C 2014 Comparative assessment for low-noise pavements by means of the ISO 11819 and the OBSI *Proc. Int. Conf. Internoise (Melbourne Australia)* p 9

[14] Donavan P R 2009 *Measuring Tire-Pavement Noise at the Source Appendix C Results of Test Parameter Evaluation – NCHRP Report*

[15] Donavan R P, Donavan P 2011 Tire Noise Generation and Propagation over Porous and Nonporous Asphalt Pavements *Transportation Research Record: Journal of the Transportation Research Board* **2233** pp 135-144

[16] Mogrovejo D E, Flintsch G W, de Leon Izeppi E D and McGhee K K 2012 Effect of Air Temperature and Vehicle Speed on Tire/Pavement Noise Measured with On-Board Sound Intensity Methodology *Transportation Research Board 93rd An. Meeting (Washington DC)*

[17] Donovan P R and Lodico D M 2009 *Measuring Tire-Pavement Noise at the Source - NCHRP 630 Report* (The Nationam Academies Press) p 37

[18] Oddershede J 2013 *CPX - OBSI Relation in Tyre/Road Noise Measurement Results – Report* (Danish Road Directorate)

[19] Irali F, Kivi A, Tighe S L and Sangiorgi C 2015 Tire-pavement noise and wearing course surface characteristics of experimental Canadian road pavement sections *Canadian Journal of Civil Engineering* **42**(10) pp 818-25

Application of infrared camera to bituminous concrete pavements: measuring vehicle

Michal Janků[1] and Josef Stryk[1]

[1]CDV - Transport Research Centre, Lisenska 33a, 636 00 Brno, Czech Republic

E-mail: michal.janku@cdv.cz

Abstract. Infrared thermography (IR) has been used for decades in certain fields. However, the technological level of advancement of measuring devices has not been sufficient for some applications. Over the recent years, good quality thermal cameras with high resolution and very high thermal sensitivity have started to appear on the market. The development in the field of measuring technologies allowed the use of infrared thermography in new fields and for larger number of users. This article describes the research in progress in Transport Research Centre with a focus on the use of infrared thermography for diagnostics of bituminous road pavements. A measuring vehicle, equipped with a thermal camera, digital camera and GPS sensor, was designed for the diagnostics of pavements. New, highly sensitive, thermal cameras allow to measure very small temperature differences from the moving vehicle. This study shows the potential of a high-speed inspection without lane closures while using IR thermography.

1. Introduction

The quality of transport infrastructure reflects economic and social level of every country. Roads are an inseparable part of the transport infrastructure. Their maintenance in the Czech Republic costs billions of Czech crowns every year. Good technical construction conditions of roads have a direct effect on safety and comfort of driving. Roads are usually designed on the basis of the expected traffic volume in the following years. The time period in the Czech Republic is usually set to 25 years [1]. The expected life span of road wearing courses in relation to traffic volume is given by Annex 4 of Technical Guidelines TP 87 [2]. When evaluating a road with a given traffic volume, its road bearing capacity is expressed by the residual life span as the latest time to perform road pavement reconstruction.

The basic feature of diagnostics is a visual inspection of road pavement, which includes recording and photo documenting of defects while walking or with the use of a video camera fixed to the vehicle. The data can also be collected with the use of laser scanning, which allows us to create a 3D model of road cross section. The evaluation of the data from the visual inspection is performed according to the catalogue of defects [3]. With the use of special multi-function measuring vehicles, it is possible to measure transversal roughness, longitudinal roughness – IRI (International Roughness Index) and macrotexture – MPD (Mean Profile Depth). The measurement of anti-skid properties of road pavements in the Czech Republic is usually performed with the use of a TRT vehicle; the outcome is the longitudinal friction coefficient f_P.

The residual road pavement life span can be estimated on the basis of FWD/HWD measurements. This falling weight device measures deflections under loading in selected points simulating a heavy vehicle overrun. The aim is to find the bearing capacity of road construction layers and its subbase and subgrade [4]. Road construction layers thickness can be determined continually with the use of a georadar – GPR [5]. Both methods can be made even more accurate thanks to the performance of core or drilled or dug probes.

In the USA, IR thermography for detecting the delamination in bridge decks [6] is used. However, the corresponding standard recommends to perform measurements only up to the speeds of 16 km/h. Hiasa *et al.* [7] investigated whether it is possible to check concrete bridge decks using IR thermography at the speed of up to 48 km/h. Solla *et al.* [8] compared IR thermography with GPR when identifying cracks in asphalt pavement. Marchetti *et al.* [9] examined the use of an IR camera for road thermal mapping. In his article [10], Dumoulin *et al.* describes an application of IR thermography to detect non emergent defects in asphalt concrete used for road pavements.

However, the use of IR thermography for pavement condition surveys is not very common in the Czech Republic, therefore, we have focused on this issue in Transport Research Centre and examined the use of a measuring vehicle for the localization of defects on asphalt road pavements. The objective of the work is to evaluate how the use of an infrared camera on a measuring vehicle could improve the determination of road network condition.

2. Description of measuring equipment

2.1. Infrared camera

A FLIR A615 camera was used. This camera is equipped with an uncooled Vanadium Oxide (VoX) detector, which produces thermal images of 640 x 480 pixels. High resolution allows more accuracy and shows more details. In addition, FLIR A615 has a high-speed infrared windowing option. In full resolution the camera is able to record at the frame rate of 50 Hz. At reduced resolution the frequency of imaging can be many times higher. A wide screen 80° lens was used so that the whole traffic lane from relatively short distance could be captured. During the measurement the camera is placed in a protection cover and is fixed to the roof of the measuring vehicle together with an HD camera. The characteristics of the thermal camera are summarized in Table 1.

Table 1. Characteristics of FLIR A615 camera.

Camera Type	**FLIR A615**
Detector type	Uncooled microbolorometer
Thermal sensitivity	<0.05° at 30°C
Accuracy	±2°C or ±2%
IR resolution	640 x 480 pixels
Spectral range	7.5 – 14 μm
Frame Rate	50 Hz (100/200 Hz with windowing)
Field of View	80° × 64.4°
Detector time constant (Electronic Shutter Speed)	8 ms

2.2. Measuring vehicle

During the measurements the thermal camera was fixed to the roof of a vehicle Volkswagen LT35. The dimensions of the vehicle allow to place the camera to a sufficient height, thus the camera view covers the whole traffic lane. The camera was fixed to the roof with the use of roof holders and aluminium profiles. The thermal camera was fixed to the end of an arm by a steel fitting, which allows its turning. The angle of the thermal camera position towards the road plane is recommended to reach at least 45° (Figure 1). In case the angle is too low, undesirable distortion of measured values may occur. On the other hand, higher angle of the thermal camera position is not detrimental during the measurement, although in such case it is necessary to move the camera further ahead of the vehicle so that its front part would not obstruct its field of view.

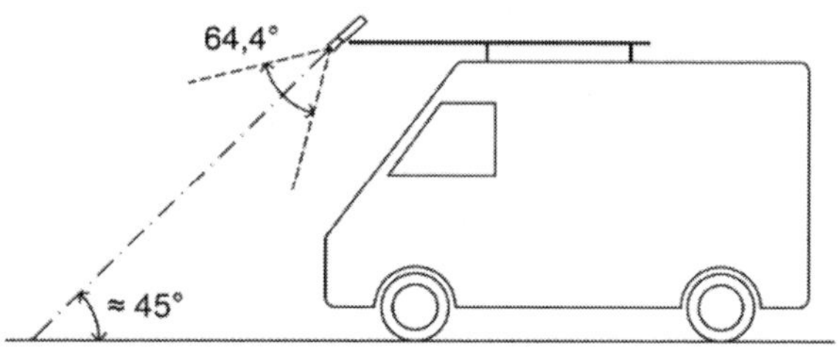

Figure 1. Sketch of measuring vehicle.

Figure 2. Measuring vehicle with both cameras on roof frame.

2.3. Data acquisition and analysis

The cameras are controlled from a laptop in the vehicle cabin. In cooperation with company Workswell, special software was developed for the collection and analysis of data. Thanks to this software it is possible to record synchronized thermographic and digital HD video. In addition, the data are complemented with accurate location by a GPS sensor in the vehicle. Before the beginning of the measurement, frame rate and thermographic video resolution are set through the laptop. The software interface is shown in Figure 3.

Figure 3. Interface of software Workswell CorePlayer; (a) thermographic record; (b) digital video; (c) exact position on map; (d) graph of temperatures over time.

IOP Conf. Series: Materials Science and Engineering **236** (2017) 012104 doi:10.1088/1757-899X/236/1/012104

The software allows simultaneous examination of a thermographic and digital video during the subsequent analysis. The map shows the corresponding location of the vehicle at the moment of measurement. When analysing thermographic videos, it is possible to create graphs of courses of temperatures over time or in selected points or lines. At constant measurement speed the time course of temperatures approximately corresponds with the distribution of temperatures on the road along its whole length. Based on the local minimums or maximums in the graph, it is possible to focus on temperature anomalies that may indicate road pavement defects.

3. Results and discussion

The measurements took place in South Moravian region in the vicinity of the city of Brno. Temperatures were measured on asphalt road pavements in urban as well as rural areas. Different speeds of the vehicle with the camera and different frame rates were tested during the measurements. Unfortunately, the quality of measurement is influenced by the value of the time constant of thermal camera detector (electronic shutter speed). At higher speeds the images were too distorted along the direction of driving. The best results were obtained at the speeds of up to 40 km/h.

As mentioned in the previous chapter, the thermal camera can have frame rate in full resolution of up to 50 Hz. However, this value appeared to be too high in practice. The laptop in the measuring vehicle failed to process such large amount of data from the measurement and the data files were too large. Eventually, the frame rate of 12 Hz was the most used for the measurements. However, even at this frame rate one minute of recording covers approx. 500 MB data. At the measuring vehicle speed of 30 km/h and at the frame rate of 12 Hz the camera produces an image every 0.7 m, which appears to be sufficient for the analysis.

Defects on the road pavement can be recorded thanks to different temperatures on those places. A successful measurement needs a temperature contrast on the road pavement. The asphalt road pavement temperature changes due to ambient temperature changes and solar radiation during the day. Defects may cause changes in the heat transfer in the road pavement. In favourable conditions these changes can be recorded by a thermal camera. The measurements were performed almost over the whole year under different weather conditions. However, the best results were achieved on sunny days, when the weather helped to create a temperature contrast on the road surface.

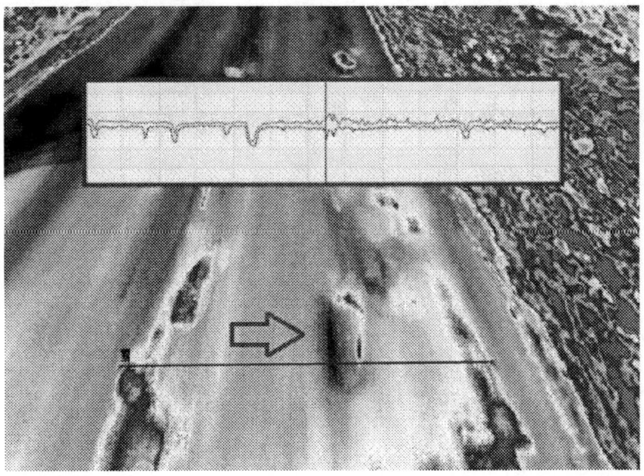

Figure 4. Thermogram of pothole on pavement.

Figure 5. Digital photo of the same pothole on pavement.

Figure 4 shows a thermogram from the measurement on considerably damaged road pavement. Figure 5 shows a digital camera image from the same spot. The measurement was performed by the measuring vehicle at the speed of 30 km/h. In addition, Figure 4 contains a graph of a temperature course. The upper curve corresponds with maximum temperatures and the lower curve with minimum temperatures in a selected measuring line. The arrow points to a pothole in the road pavement. In that spot the measured temperature was lower by approx 3 °C (black color). This drop is apparent on the lower temperature curve. The repairs from darker material heat faster on the sun, therefore, they are warmer. They show yellow-green to blue color on the used color palette in the thermogram. The differences on the curve on the left are caused by cooler road surface at the spots which were shaded by roadside trees.

The arrow in Figure 6 shows a crack in road pavement recorded by the thermal camera. In comparison with Figure 7 from the digital camera, the thermal camera image makes the crack more apparent.

Figure 6. Thermogram of crack on pavement.

Figure 7. Digital photo of the same crack on pavement.

Individual thermograms can be edited and ordered in sequence, which leads to a creation of the temperature map of the measured road pavement. Figure 8 shows an easily distinguishable asphalt filler – irregular red line in the upper part of the image. The filler heats on the sun more than the asphalt road pavement, therefore, it is more apparent thanks to different temperature. Thanks to their white color, road markings reflect heat radiation, thus having lower temperature – it corresponds with blue-white color in the selected color palette in the thermogram.

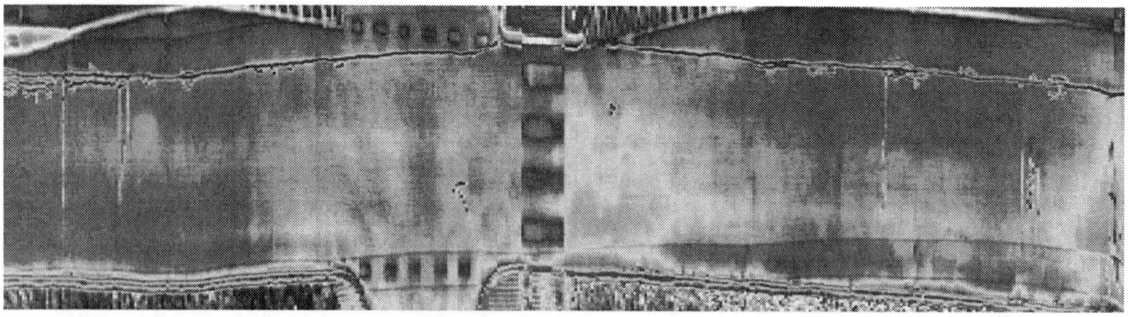

Figure 8. Thermal map of pavement (cross section).

4. Conclusions

The aim of the work was to evaluate the possibility to use an infrared camera on a measuring vehicle to detect defects on asphalt pavements. A fixation of a thermal camera on the measuring vehicle was designed and software for recording and analysis of the measurement data was produced together with a contractor. The research discovered that defects on the road pavement can be successfully detected by a thermal camera. Apart from potholes and cracks, it is possible to localize structural changes such as patching and overlays. There is a weakness concerning the limited speed of the measuring vehicle of up to 40 km/h due to blurred recordings at higher speeds. A solution may be an acquisition of a more expensive thermal camera with a cooled detector which has a shorter integration time.

Even though the best results were reached in sunny weather conditions, an issue occurred. This issue concerns the shadows of roadside trees and buildings, which cause different values of road pavement temperatures. Therefore, subsequent experiments are to test whether it is possible to make the measurements at night, when the road pavement cools down, which has a similar effect on the development of temperature contrast as the heating of the pavement during the day.

Acknowledgements
This article was produced with the financial support of the Ministry of Education, Youth and Sports within the National Sustainability Programme I, project of Transport R&D Centre (LO1610), on the research infrastructure acquired from the Operation Programme Research and Development for Innovations (CZ.1.05/2.1.00/03.0064).

References

[1] Ministry of Transport of the Czech Republic 2004 Technické podmínky - Navrhování vozovek pozemních komunikací (in Czech) /Technical terms - Design of Road Pavement/ *(TP 170)*

[2] Ministry of Transport of the Czech Republic 2010 Technické podmínky – Katalog poruch netuhých vozovek (in Czech) /Technical Terms TP - Catalogue of Defects on Asphalt Pavements) (*TP 82)*

[3] Ministry of Transport of the Czech Republic 2010 Technické podmínyk – Navrhování údržby a opravy netuhých vozovek (in Czech) /Technical Terms - Design of Maintenance and Repairs on Asphalt Pavements/ *(TP 87)*

[4] Stryk J et al 2013 *Měření průhybů a hodnocení únosnosti vozovek rázovým zařízením FWD (in Czech) /Measurement of Deflections and Evaluation of Carrying Capacity of Pavements by Falling Weight Deflectometer – FWD/* (CDV - Transport Research Centre Brno)

[5] Stryk J, Alani A M, Matula R and Pospíšil K 2015 Innovative Inspection Procedures for Effective GPR Surveying of Critical Transport Infrastructures (Pavements, Bridges and Tunnels) *Civil Engineering Applications of Ground Penetrating Radar* (Springer International Publishing) pp 71-95

[6] ASTM International 2013 Standard Test Method for Detecting Delaminations in Bridge Decks Using Infrared Thermography *(ASTM D4788-03)*

[7] Hiasa S, Catbas N, Matsumoto M and Mitani K 2016 Monitoring concrete bridge decks using infrared thermography with high speed vehicles *Structural Monitoring and Maintenance* **3** pp 277-96

[8] Caracelas M S, Lagüela S, Gonzales H and Arias P 2014 Approach to identify cracking in asphalt pavement using GPR and infrared thermographic methods: Preliminary findings *NDT&E Int.* **62** pp 55-65

[9] Marchetti M, Moutton M, Ludwig S, Ibos L, Feuillet V and Dumoulin J 2010 Implementation of an infrared camera for road thermal mapping *10th Int. Conf. on Quant. InfraRed Thermography (Québec City)* ed X Maldaque

[10] Dumoulin J, Ibos L, Marchetti M, Ludwig S and Atef Mazioud A 2009 Active Infrared Thermography applied to detection and characterization of non emergent defects on asphalt pavement *NDTCE'09 (Nantes)* ed Abraham O and Derobert X

Improving linear transport infrastructure efficiency by automated learning and optimised predictive maintenance techniques (INFRALERT)

Noemi Jiménez-Redondo[1], Alvaro Calle-Cordón[1], Ute Kandler[2], Axel Simroth[2], Francisco J Morales[3], Antonio Reyes[3], Johan Odelius[4], Aditya Thaduri[4], Joao Morgado[5] and Emmanuele Duarte[5]

[1]CEMOSA, Benaque 9, Málaga 29004, Spain
[2]Fraunhofer-Institut für Verkehrs- und Infrastruktursysteme IVI, Zeunerstraße 38, 01069 Dresden, Germany
[3]Universidad de Sevilla, C/San Fernando 4, Sevilla 41004, Spain
[4]Lulea Tekniska Universitet, Universitetsomradet Porson, Lulea 971 87, Sweden
[5]Infraestruturas de Portugal, SA, Praça da Portagem, 2809-013, Almada, Portugal

E-mail: noemi.jimenez@cemosa.es

Abstract. The on-going H2020 project INFRALERT aims to increase rail and road infrastructure capacity in the current framework of increased transportation demand by developing and deploying solutions to optimise maintenance interventions planning. It includes two real pilots for road and railways infrastructure. INFRALERT develops an ICT platform (the expert-based Infrastructure Management System, eIMS) which follows a modular approach including several expert-based toolkits. This paper presents the methodologies and preliminary results of the toolkits for i) nowcasting and forecasting of asset condition, ii) alert generation, iii) RAMS & LCC analysis and iv) decision support. The results of these toolkits in a meshed road network in Portugal under the jurisdiction of Infraestruturas de Portugal (IP) are presented showing the capabilities of the approaches.

1. The INFRALERT concept to optimise transport infrastructure maintenance

The condition of transport systems has enormous societal and economic relevance since economic opportunities are likely to arise where transportation infrastructures are able to answer mobility needs and ensure access to markets and resources [1]. Our growing economic require an enlargement of transport capacity, most especially for land transport. However, the economic situation and the lack of land prevents from enhancement of the rail and road network. On the other hand, as our networks are more congested, there are fewer chances to perform maintenance while the extensive use of the infrastructure accelerates its deterioration. In this situation, the only open door to boost land transport capacity is by making a better use of the existing network reducing maintenance interventions and extending the life of existing assets. This is the motivation of the ongoing H2020 project INFRALERT [2] which proves its developments in the road and railways systems.

INFRALERT exploits the similarities of linear infrastructures and develops systems and tools for support Infrastructure Managers (IM) or Maintenance Contractors in maintenance interventions decision making. Figure 1 shows in a dashed box the INFRALERT platform and its interaction with the IM/MC and the different data bases. INFRALERT aims at the development of models and ICT tools to optimise the performance of existing linear land transport infrastructure. It develops an expert-based information system to support and automate infrastructure management from measurement to maintenance. Figure 2 illustrates the concept and the scope of INFRALERT which has been conceived using a modular approach to facilitate its flexibility and applicability. It includes a data management system (i.e. the Data Farm) and a set of toolkits covering Data Analytics modules (asset condition, alert management and the RAMS & LCC) and a decision support tool which receives the results of the Data Analytics modules and optimise maintenance interventions. All these modules are conceived as plug-ins into a common shell, which is the expert-based Infrastructure Management System (eIMS), allowing communication among the different modules, with external data bases and the user. INFRALERT is conceived to be compatible with existing asset management systems.

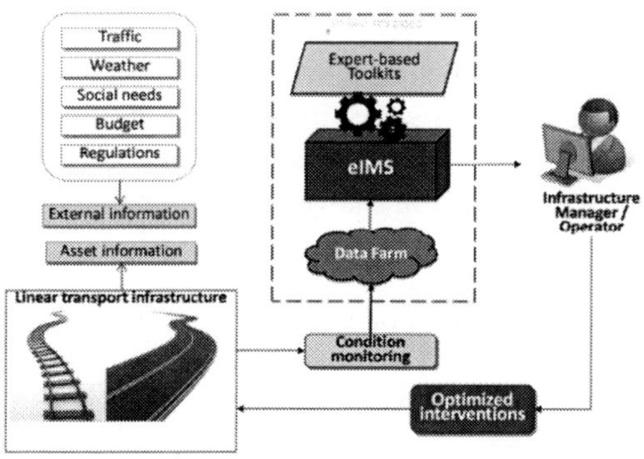

Figure 1. INFRALERT platform.

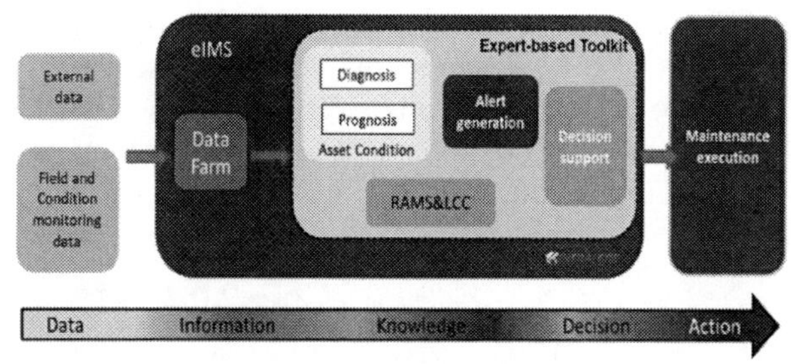

Figure 2. INFRALERT modular concept.

The development of the INFRALERT project will be validated in two real infrastructure systems: a meshed road network in Portugal owned and managed by Infrastruturas de Portugal (IP) and a rail corridor in Sweeden owned by Trafikverket. This paper present the methodologies developed by the different tools in the expert based toolkit (Figure 2) and their preliminary results obtained in part of the road network.

2. Use case: the road pilot

The road pilot in Portugal comprises 539 km of roads in the Coimbra region under IP jurisdiction. It includes a rich variety of road types (principal, national, regional…). The network is classified based on sections of an average length of 6.6km and 87 nodes.

IP Pavement Management System (SGPav) stores information of maintenance activities carried out since 2007 and road condition data such as longitudinal (IRI) and transverse unevenness (Rut Depth), cracked area and pavement macrotexture. IP Maintenance strategy categorises interventions in major or routine maintenance. Major maintenance includes relevant works in terms of cost, length and complexity while routine maintenance includes smaller scale and lower complexity works, such as pavement localised repairs or other activities such as drainage system cleaning, shoulder treatment, minor works performed in bridges and any urgent repairs.

The basis of the data stored in SGPav is related to the section element (start and end node). Inside this database, the section table has a number of selected fields that are relevant in order to summarize the most important information associated with the part of the road it represents. Besides this main table, other relevant tables contain further information based on the field measurements and the pavement historical information with all the road maintenance work performed up to date.

3. Expert-based tookits and test cases

Following the Data Analytics tools (asset condition, alert management and RAMS&LCC) and Decision Support tool are presented together with results from the road pilot in Portugal.

3.1 Nowcasting and Forecasting of Asset Conditions

The main aim of this toolkit is to develop methodologies for linear asset condition assessment and prediction by identifying and segregating hierarchy of conditional information for nowcasting and forecasting. Nowcasting focuses in what is known today while forecasting is the process of exploiting past and present data to make deductions about the future. For nowcasting and forecasting, the asset and condition data is accessed from the Data Farm as shown in Figure 3. Pre-processing methods such as data cleaning and dynamic segmentation is carried out on this data to obtain meaning information.

There are different types of nowcasting and forecasting approaches available based on principle of physical, data-driven, and knowledge-based [3, 4]. Data-driven approaches rely on monitored and historical data that are used to learn the systems behaviour [5]. Regression analysis is a commonly applied approach for modelling of pavement deterioration. Weighted regression approach has been used to better capture the nonlinear behaviour of the degradation [6]. An important consideration in maintenance modelling of linear assets is dynamic segmentation, which allows multiple sets of attributes to be accompanied with any segment of a linear feature. Each time an attribute value changes, it can "dynamically" locate the segment. To achieve good forecasts the input to the models needs to be homogeneous over their respective length. An example for dividing road infrastructure into statistically homogenies segments is the cumulative difference approach which is described in the AASHTO pavement guide [7].

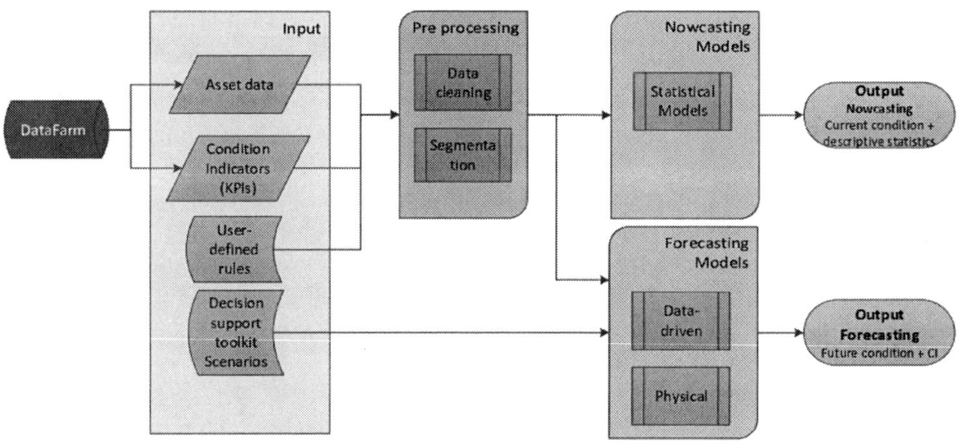

Figure 3. Asset condition toolkit flowchart.

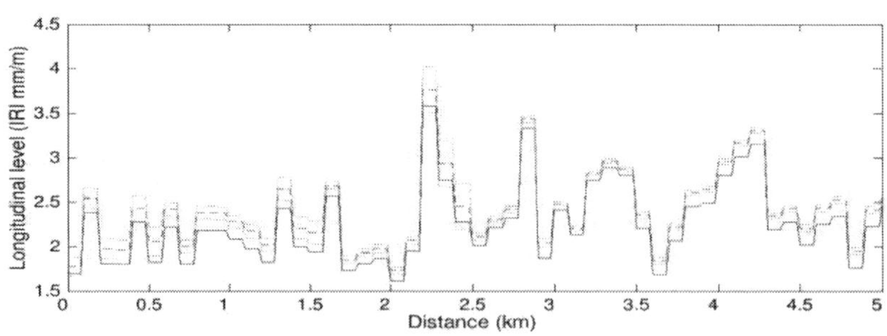

Figure 4. Forecasted output for road case. Current condition (blue solid); forecasted condition after 3 years (red dashed) and associated 95 % confidence levels (dotted)

For the road pilot, forecasting of the longitudinal unevenness (IRI) was performed based on the weighted regression model. The data originates from annual road surveys from 2012-2015 for road section EN101-3 D061. Input to the regression model was the IRI values that first were divided into homogenous segments using the AASHTO approach based on the pavement quality index. The quality index is an aggregated indicator, which is calculated as a function of longitudinal unevenness, transversal unevenness, and cracking. The forecasted output after 3 years with 95% confidence intervals for 100 m sample segments over 5 km is shown in Figure 4. The prediction with uncertainty will be forward to the Alert Management system.

3.2 Alert generation

The aim of the Alert Management toolkit is to prioritise predicted maintenance alerts of linear infrastructure assets, according to the required maintenance interventions based on the forecasted severity of degradation and failure of the assets themselves, and the know-how brought in by the information recorded in the historical maintenance work-orders. Two kinds of alerts are predicted and involved in the proposed methodology shown in the block diagram of Figure 5.

Module AM1 (Alerts based on limits) is responsible for generating alerts from the point of view of those features that overcome their associated limits or reference thresholds, using as inputs the forecasted values of the explanatory features of the asset. As result, the module provides the following outputs (Table 1): i) Alerts indicating that a specific feature exceeds its prescribed threshold and ii) Technical Severity Levels (TSL) of the estimated alerts. The TSL is an objective value used to prioritise the alerts according, for instance, to a distance criterion between the value of the feature and the threshold.

Module AM2 (Alerts based on Work Orders) predicts alerts from the point of view of whether maintenance is or is not required (Yes-No); it also estimates the most probable maintenance interventions to be conducted. To achieve this, the module embodies two different functional submodules. The first one (AM21) is specifically devoted to triggering alerts regarding the need of maintenance and their corresponding level of global technical severity (GTSL) in terms of all forecasted features considered as a whole. Here, the alerts are triggered by the estimator contained in the first block (Alert Estimator) which has been previously trained with the explanatory features (e.g. measurements) and the historical maintenance interventions through a machine learning processing. Submodule AM21 also provides an optional output using a second block (Asset Condition Classifier), which "learns" from the Maintenance Manager (MM) know-how, with the final purpose of predicting a subjective evaluation of the asset condition (from the set of forecasted features) without the intervention of the MM.

Table 1. Results from AM1 module.

ID	CT	IRI	RUT	W CTL1	TSL CT1	W CTL2	TSL CT2	W IRI	TSL IRI	W RUT	TSL RUT
225	**36**	**2.25**	**4.85**	YES	45.76	YES	4.12	No	0.00	No	0.00
243	**4**	**1.75**	**2.71**	No	0.00	No	0.00	No	0.00	No	0.00

The second submodule (AM22) aims at determining the set of k-most probable maintenance interventions that have to be conducted, as well as their corresponding probabilities of occurrence, via a learning procedure based on historical intervention database. As result, the module provides: i) alert triggered: required maintenance; ii) Global Technical Severity Level (GTSL) for the asset; iii) K-Most probable interventions; and iv) probabilities of occurrence of the most probable interventions.

The data used to develop the toolkit are the IRI, Rut and CT corresponding to a set of previous measurement campaigns, from which the empirical statistics were inferred to generate a larger data set. In order to check the accuracy of the different machine learning predicting models used, the last campaign (2014) is chosen as a testing sample, keeping this year out of the training set. Selecting the Decision Tree model (DT) as an example, the results are shown in Figure 6. The upper panel of Figure 6 shows the performance evaluation of the DT model. This evaluation was made by using a confusion matrix, based on counting those test records correctly and incorrectly predicted. This matrix shows the real and the predicted maintenance based on the list of maintenance activities of Table 2. Class T0 is associated to no-alert, and T1 to alert without maintenance required (warning). The lower panel of Figure 6 shows a comparison between the real maintenance carried out and the predicted ones. According to this, results are reliable and the errors are mainly between T4 and T3.1, as suggested by the confusion matrix. This figure represents the most probable maintenance intervention; however, the models offer several possibilities with a score for each one.

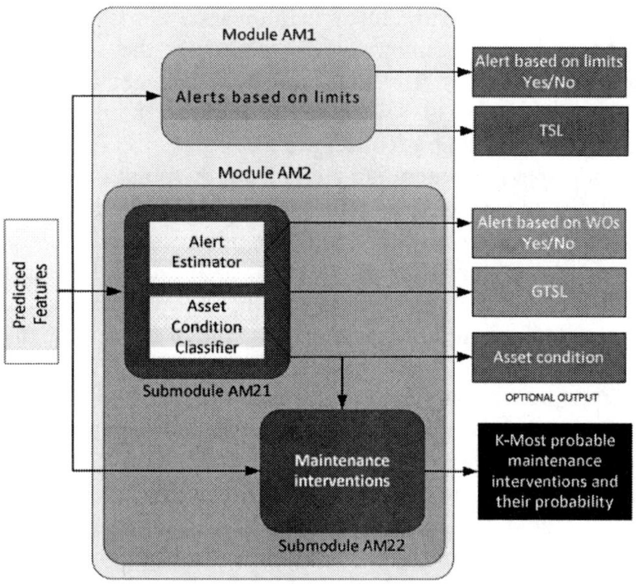

Figure 5. Work-flow diagram of the toolkit.

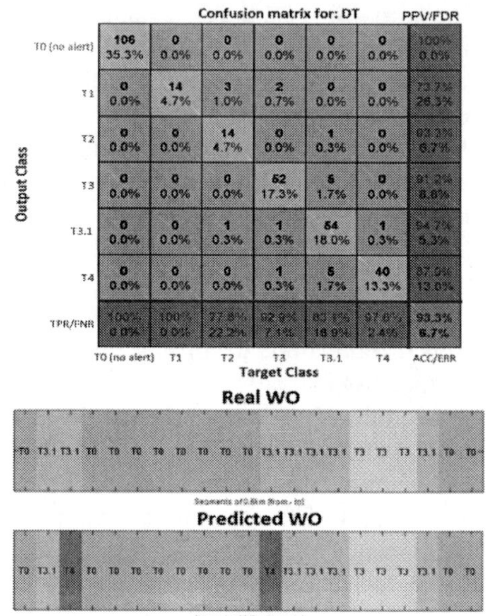

Figure 6. Result example (DT model).

Table 2. Types of maintenance used for alert generation and RAMS/LCC.

M	Alert	Description	M	Alert	Description
T0	No	No request	T3.1	Yes	Profile milling and fill
T1	Yes	Do nothing	T4	Yes	Regulating course
T2	Yes	Surface treatment	PC	Yes	Full-depth patching (corrective)
T3	Yes	Thin asphalt surfacing	PP	Yes	Crack fill (preventive)

3.3 RAMS and LCC analysis

The aim of this toolkit is to perform real-time RAMS&LCC analysis to assess the Reliability, Maintainability, Availability and Safety (RAMS), and the Life-Cycle Cost (LCC) of the infrastructure. RAMS & LCC provides probabilistic information to be considered for decision support.

The application of RAMS analysis [8] in civil engineering, although relatively recent, has a high potential to predict the number and distribution of failures in infrastructures, which in turns provides an estimation of the availability. RAMS enfolds a rich set of parameters [9] and distribution functions characterizing the reliability and maintenance need of the infrastructure. RAMS parameters are provide relevant probabilistic information for the maintenance planning.

LCC analysis takes into account all the costs associated with the lifetime of the system. For complex systems, such as railway and road infrastructures, the cost of maintenance plays a very important role because operation and maintenance comprise a major share of the system's life-cycle and they are the most sensitive to cost uncertainties. These uncertainties are associated to internal risks, which can be quantified through RAMS parameters.

One of the innovations of the INFRALERT's RAMS&LCC toolkit is to make use of the previously calculated RAMS in the assessment of maintenance costs. The integration of stochastic RAMS in the LCC analysis allows obtaining reliable predictions of system maintenance costs and the dependencies of these costs with specific cost drivers through sensitivity analyses. Moreover, using the computed RAMS probability distributions, the associated uncertainties are integrated into the LCC determination. The output provided by these stochastic LCC are finally used for long-term planning in the decision support tool. Figure 7 depicts the general workflow of the RAMS&LCC process.

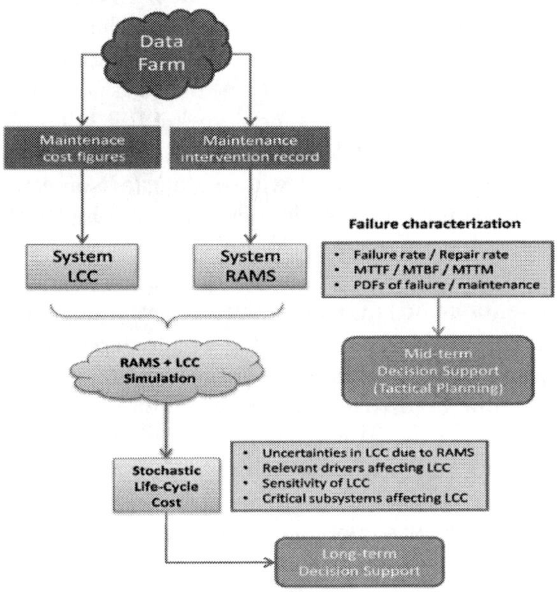

Figure 7. Overview of RAMS&LCC methodology.

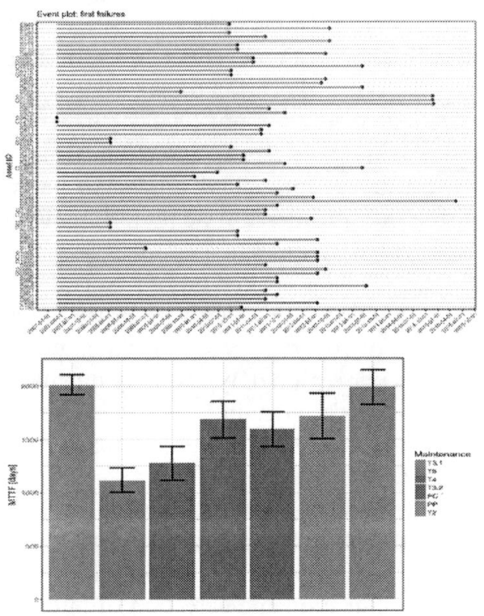

Figure 8. MTTFs by failure mode.

For the road pilot in INFRALERT, different models for RAMS have been applied depending on the characteristics of the road section under study. For sections with a large number of the same intervention, the reliability function is modelled using a Non-Homogeneous Poison Process (NHPP) [10]. Information about Mean Time Between Failures (MTBF) or the degradation trends through the rate of failures [11] can be calculated. However, when studying the road use case there are few sections that allow this kind of study, which forces to apply a time-to-event modelling to obtain Mean Time To First Failure (MTTF). This paper shows the results of the road sections with scarcity of accumulated data.

The upper panel of Figure 8 shows the event plot for the first maintenance interventions performed after January 2007 in the whole road network. Different road sections are displayed in the y-axis, while the x-axis represents the temporal scale from 2007 to 2015. In this figure only time-to-first-event has been plotted. The different colours correspond to the types of maintenance described in Table 2. Weibull distributions are used to model the reliability function for each type of maintenance. From the reliability, it is possible to estimate the probability of these interventions happening in a given time slot, which furthermore can be used for the optimization of tactical planning in the decision support tool. The calculated MTTF with confidence intervals for the different maintenance tasks is shown in the lower panel of Figure 8.

From individual maintenance costs and the calculated mean times, the overall costs per type of maintenance can be calculated using the LCC model. The probability distribution function of MTTFs allows studying the level of uncertainty in the LCC model, which is fed to decision support tool for long-term planning. This simulation is currently under development in INFRALERT.

3.4 Decision support tool

The final objective of the INFRALERT system is to provide Infrastructure Managers/Owners and Maintenance Operators/Contractors with intelligent software tools to support the decision-making process when planning maintenance activities and interventions. This section presents the application of INFRALERT for tactical planning (or mid-term planning) in the road pilot in Portugal (IP owned).

On tactical planning level, IP maintenance department has to allocate major interventions over a 5-year time horizon. The interventions are aggregated as single events over the 500 m-segments of certain road sections in order to avoid multiple traffic interruptions on the same section. The allocation of such intervention events is done on a monthly basis. In detail, the decisions to create a tactical plan include: i) the allocation of a starting month for intervention events (within the next 5 years) and ii) the selection of a minimum level of intervention (corresponding to maintenance types T1, T2, T3, T3.1 and T4 in Table 2) on a section. This selected minimum level determines which segments of the respective section actually have to be maintained, which are the ones whose state (at some point during the considered time period) would cause an alert with an intervention level equal or higher than the selected minimum.

The decision-maker has to consider certain restrictions like: i) minimum quality level for road segments, ii) usage of a given yearly budget for interventions; iii) maximum overall 5-year budget, iv) capacity of supervisory staff (which means that interventions can be allocated to a restricted number of sections per region only), v) only one intervention event per year is allowed in each section.

The tactical planning has to optimise three items in the objective function (which are balanced by weighting factors): i) minimize maintenance interventions costs; ii) maximize quality index (which is a function of the predicted conditions of road segments and predicted failure events coming from RAMS analysis) and iii) maximize availability of the network (which depends on the capacity reduction of sections due to interventions). Maintenance interventions costs are derived from LCC analysis and provided by the toolkit on RAMS & LCC.

The allocation and selection of interventions in the tactical plan is based on the maintenance alerts generated by the Alert Management toolkit, which is based on predicted future conditions coming from the Asset Condition toolkit. Thus, input for tactical planning are no concrete work orders to be scheduled but predicted work orders provided with the corresponding probabilities of occurrence. Due to this uncertainty of the real amount of work to be done the ending time of intervention events at each section will only be known at execution time. Furthermore, the working time needed for interventions cannot be seen as a fixed, deterministic duration, but has to be modelled as a stochastic variable in the model. In particular, the probabilistic information provided by the RAMS analysis toolkit calculations will be used as input in relation to the interventions duration.

These characteristics make the tactical planning a stochastic optimization problem which calls for specific modelling and solution techniques to be applied. A mathematical optimisation model which reflects the uncertainty in the problem description has been developed as foundation for the decision support tool. The handling of such uncertain information in the decision support tool is done using a scenario approach. A set of future scenarios is simulated as follows:

- For each of the condition parameters characterising the road quality (features), a time series of possible values is predicted, using the Asset Condition toolkits for forecasting.
- Based on these predicted conditions the respective intervention alerts are generated by the Alert Management toolkit.
- Additionally, based on RAMS analysis possible failure events are generated, using e.g. results from MTTF, MTBF calculations.

The resulting scenarios represent possible "states of the world" and are used in the optimisation algorithms of the decision support tool to derive decisions, but also to calculate the objective function and restrictions for a selected tactical plan.

4. Conclusions

This paper presents the modular concept of the INFRALERT eIMS for optimizing maintenance interventions in railways and road infrastructures together with the methodologies and preliminary results validated in a meshed road network in Portugal of some of its modules (the items in the expert-based toolkit, Figure 2). In particular, the following methodologies have been explained including the corresponding test results on the road pilot: i) The methodology for nowcasting and forecasting of asset condition which is an input for the following item; ii) the methodology to support and automate the prediction of maintenance intervention alerts (which combines the current and predicted asset condition with operational and historical maintenance data to get information about the needed maintenance tasks) by means of data analytics and machine learning models, which provides forecasted alerts and maintenance alerts to be consider in the maintenance planning; and iii) the methodology to compute probabilistic RAMS parameters which provide relevant information for the maintenance planning. In addition, the following methodologies have been introduced although no test results have been shown: i) the approach to introduce the probabilistic RAMS information for computing of the probabilistic LCC which is a relevant piece of information for the decision support and ii) the formulation and the methodology to solve the tactical planning (mid-term which means up to 5 years in a monthly basis for the road pilot) optimization problem which receives the predictions and computations from the alert management and the RAMS & LCC systems.

Acknowledgements

This research was carried out within the INFRALERT project. This project has received funding from the EU Horizon 2020 research and innovation programme under grant agreement No 636496.

References

[1] European Environment Agency - EEA 2016 *Freight transport demand* (Copenhagen)
[2] Infralert 2016 Linear Infrastructure Efficiency Improvement by Automated Learning and Optimised Predictive Maintenance Techniques *H2020 Programme, European Commission, Research Directorate, Grant agreement No 636496*
[3] Si X S, Wang W, Hu C H and Zhou D H 2011 Remaining useful life estimation - A review on the statistical data driven approaches *European journal of operational research* **213** pp 1-14
[4] Sikorska J Z, Hodkiewicz M and Ma L 2011 Prognostic modelling options for remaining useful life estimation by industry *Mechanical Systems and Signal Processing* **25** pp 1803-36
[5] Sutharssan T, Stoyanov S, Bailey C and Yin C 2015 Prognostic and health management for engineering systems: a review of the data-driven approach and algorithms *Journal of Engineering-JOE*
[6] Lang J M and Dahlgren J M 2001 Prediction model in the Swedish PMS *Fifth International Conference on Managing Pavements (Seatle)* ed Trafikverket (Stockholm)
[7] American Association of State Highway and Transportation Officials - AASHTO 1998 *AASHTO guide for design of pavement structure* (Washington DC)
[8] Smith D J 2011 *Reliability, Maintainability and Risk 8th Edition: Practical Methods for Engineers including Reliability Centred Maintenance and Safety-Related Systems* (Oxford: Butterworth-Heinemann)
[9] International Electrotechnical Commission (IEC) 2016 Mathematical expressions for reliability, availability, maintainability and maintenance support terms *(IEC 61703:2016)*
[10] Meeker W Q and Escobar L A 1998 *Statistical Methods for Reliability Data* (New York: John Wiley & Sons)
[11] Rausand M and Hoyland A 2004 *System reliability theory: models, statistical methods, and applications* (New York: John Wiley & Sons)

IOP Conf. Series: Materials Science and Engineering **236** (2017) 012106 doi:10.1088/1757-899X/236/1/012106

Integration of RAMS in LCC analysis for linear transport infrastructures. A case study for railways.

Álvaro Calle-Cordón[1], Noemi Jiménez-Redondo[1], F J Morales-Gámiz[1], F A García-Villena[1], Amir H S Garmabaki[2] and Johan Odelius[2]

[1]CEMOSA, Benaque 9, Málaga 29004, Spain
[2]Lulea Tekniska Universitet, Universitetsomradet Porson, Lulea 971 87, Sweden

E-mail: alvaro.calle@cemosa.es

Abstract. Life-cycle cost (LCC) analysis is an economic technique used to assess the total costs associated with the lifetime of a system in order to support decision making in long term strategic planning. For complex systems, such as railway and road infrastructures, the cost of maintenance plays an important role in the LCC analysis. Costs associated with maintenance interventions can be more reliably estimated by integrating the probabilistic nature of the failures associated to these interventions in the LCC models. Reliability, Maintainability, Availability and Safety (RAMS) parameters describe the maintenance needs of an asset in a quantitative way by using probabilistic information extracted from registered maintenance activities. Therefore, the integration of RAMS in the LCC analysis allows obtaining reliable predictions of system maintenance costs and the dependencies of these costs with specific cost drivers through sensitivity analyses. This paper presents an innovative approach for a combined RAMS & LCC methodology for railway and road transport infrastructures being developed under the on-going H2020 project INFRALERT. Such RAMS & LCC analysis provides relevant probabilistic information to be used for condition and risk-based planning of maintenance activities as well as for decision support in long term strategic investment planning.

1. Introduction

Railway and road network infrastructures are quite complex systems which components have a very long technical lifetime (40 to 120 years, see e.g. Ref. [1]) and are subject to different failure. As a consequence each decision must consider the usage of the assets in question for at least 40 years into the future. Long term plans, nowadays, are of the order of 10 years, even with the awareness that the remaining live time can be longer. The complexity of these systems is due to the mixture of components of different age and status that have to work together. Replacement of components is also a continuous and on-going process, and changes must be carefully executed.

On the other hand, the pressure for increases in traffic volume leads to a higher utilization of the existing infrastructure and hence to more severe degradation, which therefore requires more maintenance actions on the network. This fact is recognised in the objectives set by the European Commission for 2020. An expected increase of passenger and freight traffic, the reduction of travel time by 25-50% and life-cycle cost (LCC) by 30% and at the same time increasing safety (decreasing fatalities) by a 75% have put strong demands on operational and maintenance optimization (see Refs. [2] and [3]).

In the case of the railway transport, operational quality can be measured by punctuality. In the road case, this concept could be also applicable by considering the planned speed of corridors. Maintenance activities are programmed or scheduled in time slots in which traffic load is low or inexistent, so the maximum priority is to keep the traffic flow in its highest possible values. Maintenance activities must therefore be performed near capacity limits. Time between asset renewal should be long enough to balance maintenance costs and acquisition costs, and components be replaced by deferred or planned maintenance. On top of that, Infrastructure Managers (IMs) must keep infrastructure highly available so that the railway undertakings can deliver a highly quality service at affordable price to the end users.

In general RAMS (Reliability, Availability, Maintainability and Safety) and LCC (Life-Cycle Cost) analyses are used as tools to optimize the performance of the network and make it economically viable. RAMS analysis is used to establish the need of maintenance of the infrastructure by analysing corrective and preventive maintenance data and a central element in many engineering areas, ranging from manufacturing, electrical engineering to the nuclear and space industry. Although they have a high potential for applications in transport infrastructure management, currently there is a lack of standardisation and stated procedures in its implementation [4]. LCC analysis, on the other hand, is the method used to assess the most cost-effective option among competing alternatives considering investment, operation and maintenance, and unplanned interruptions throughout the assets' life cycle [5].

Recently, combined RAMS and LCC analyses have attracted much attention in the railway sector, with a large number of projects devoted to their development and applications [6, 7]. On the contrary, few experiences of implementing this approach are known in the road sector. Moreover, traditional applications of RAMS and LCC in transport infrastructures have followed a deterministic approach [8] in part due to scarce data availability and computer processing capabilities. The predicting deficiencies inherent of such approaches can be overcome using a probabilistic point of view, where RAMS and LCC figures are described statistically.

This paper describes a general approach that combines RAMS and LCC, applicable to linear infrastructures. The methodology is applied to a specific railway use case that is part of one of the demonstrators in the on-going H2020 project INFRALERT [9].

2. Combined RAMS and LCC methodology

The combined RAMS and LCC methodology rests on two basic pillars: *statistical analysis of maintenance interventions* and *information from the accounting system to estimate costs*. Figure 1 depicts the general workflow of the RAMS&LCC process. The idea is to create an optimized maintenance policy to be implemented in the real system in accordance with the needs of the system itself. In order to generate such a policy, access to historical maintenance data is needed. From the maintenance (corrective and preventive) intervention record, the RAMS of the system are calculated. The RAMS are a rich set of parameters extracted from the underlying probability distribution functions of the data. Some relevant RAMS indicators are (see also Ref. [10]): the reliability function, failure/maintenance rates, and mean time to failure/maintenance/repair/restoration. Time to restoration is equivalent to down time and it includes repair and logistic times. The models used and the set of RAMS that are calculated will depend strongly on the level of detail of the available data.

The next step in the calculation consists on building an LCC model. The life cycle of an asset can be subdivided into six phases according to the IEC 60300-3-3 standard [11]. From the ownership point of view, regarding costs, these phases are connected with the LCC of the asset as follows: (i) acquisition costs: concept and definition, design and development, manufacturing and installation, (ii) ownership costs: operation and maintenance, and (iii) termination costs: disposal.

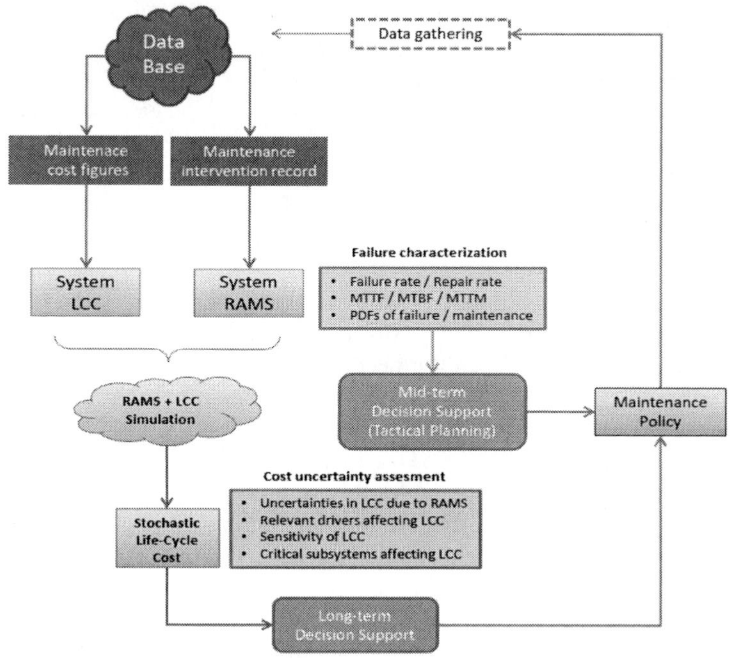

Figure 1. RAMS&LCC workflow.

While acquisition and termination costs are usually fixed or not subject to ownership time variations, ownership costs depend on operation conditions and maintenance policies. The combined RAMS & LCC analysis described here focuses exclusively on the ownership LCC because this phase is the most sensitive to variations and therefore able to be optimised. The first step in the LCC analysis consists in identifying the so-called *cost elements* that considerably influence the total LCC of the system. According to the standard [11], it is recommended to develop a Cost Breakdown Structure (CBS) as a basis to the definition of the cost elements in the LCC analysis. The CBS depends on the system under study, being difficult to define a generic structure for the cost elements in the LCC analysis. Therefore, the CBS has to be tailored to the specific system or subsystem under study. Figure 2 shows a simplified CBS that is suitable for road/railway infrastructures. Once the CBS and the cost drivers have been identified, the next step deals with building a model to quantify the cost elements encompassed in a LCC analysis. That means to find appropriate relations among input parameters and the cost elements.

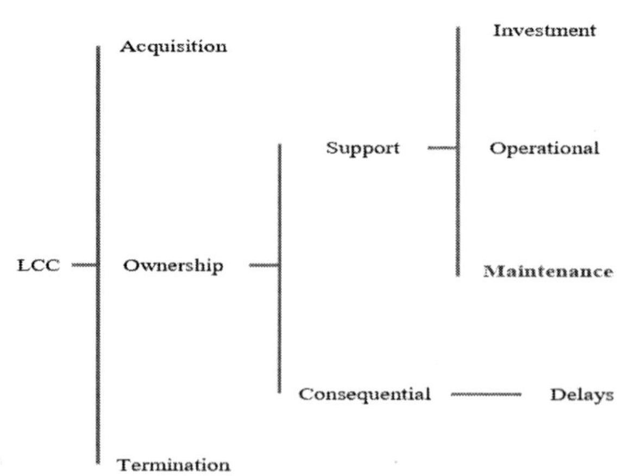

Figure 2. Cost Breakdown Structure.

In this paper only maintenance costs, which depend on the maintenance policy for the different assets and vary among companies, are considered. Maintenance actions can be preventive or corrective. Corrective Maintenance (CM) is assumed to be carried out annually while Preventive Maintenance (PM) can be *annual* and *periodical*. This is so because there may be PM actions carried out off the annual planning with well-defined and fixed costs. The annual costs of maintenance are functions of time while the periodical costs are assumed constant. Moreover, yearly costs are subject to a Net Present Value (NPV) calculation. Annual CM and PM cost can be modelled as follows:

- *Annual Corrective Maintenance Cost*

The costs derived from CM assume failures in the system that lead to replacements or repairs of the components. The replacement/repair cost is therefore of the form:

$$CY_{CM} = \text{Man Hours} + \text{Spare Parts} + \text{Equipment} = \sum_{i=1}^{m}\sum_{j=1}^{n}\lambda_{ij}\left[C_L n_L\left(MRT_{ij} + MLT_{CM}\right) + C_{Pij} + C_{Eij}\right] \quad (1)$$

where λ_{ij} is the failure frequency of action i and unit j, MRT_{ij} is the Mean Repair Time, and MLT_{CM} is the Mean Logistic Time associated to corrective maintenance.

- *Annual Preventive Maintenance Cost*

The cost from PM may include the cost of inspections, condition based and periodical maintenance.

$$CY_{PM} = \text{Man Hours} + \text{Spare Parts} + \text{Equipment} = \sum_{i=1}^{m}\sum_{j=1}^{n} f_{ij}\left[C_L n_L\left(MAT_{ij} + MLT_{PM}\right) + C_{Pij} + C_{Eij}\right] \quad (2)$$

where f_{ij} is the maintenance frequency of action i and unit j, MAT_{ij} is the Mean Action Time, and MLT_{PM} is the Mean Logistic Time associated to preventive maintenance. In both Eq. (1) and (2), C_L is the labour cost, n_L the number of workers, C_P the cost of the component, and C_E the cost of the equipment to carry out the maintenance action.

The maintenance cost figures, together with the result of the RAMS are the inputs of the parametric cost models that will implement the LCC calculation. By means of a simulation, where input parameters take a stochastic character, a stochastic prediction taking into account possible uncertainties is obtained for the LCC of the system.

3. Case study: analysis of railway switches and crossings

In this section we apply the previously described methodology to a use case, specifically to the study of costs associated to failures on switches and crossings (S&C). The data collected for this study has been provided by Trafikverket (TrV) from their failure and maintenance databases and provided for the exclusive use of the INFRALERT consortium for the railway demonstrator. The data belong to a rail corridor in Sweden under the management of TrV, called Iron Ore Line (Malmbanan), in northern Sweden. Some relevant characteristics of the data concerning the case study are: (i) two track sections are considered with a total of 260km and 61 S&C of different types (UIC60 and SJ50), (ii) data contains relevant information about the track section, position, geometric information, type of asset and year of installation, (iii) a total of 6664 records of maintenance Work Orders (WO) from 09/07/2008 to 21/03/2012 (not all of them on S&C).

Switches and crossings (S&C) are one of the most important railway subsystems, causing most train delays due to their frequent maintenance, which usually amounts at least for the 10% of the total maintenance costs [12]. The highly maintenance cost is due to several factors, namely, its complexity and degradation, and the fact that S&C need to be maintained regularly to keep high safety levels. S&C consists of three major parts: switch panel, crossing panel and middle panel (see Figure 3 and Figure 4).

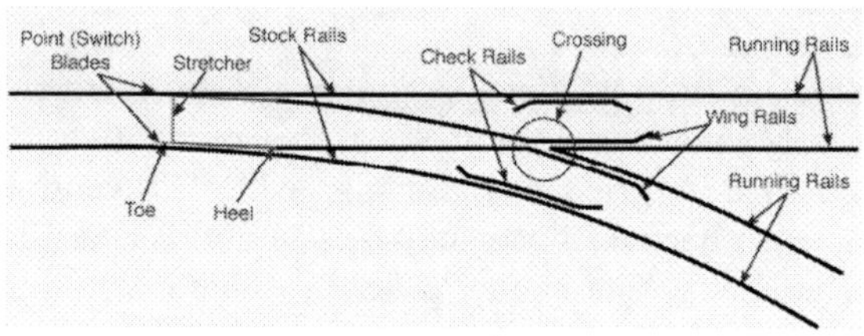

Figure 3: S&C illustration.

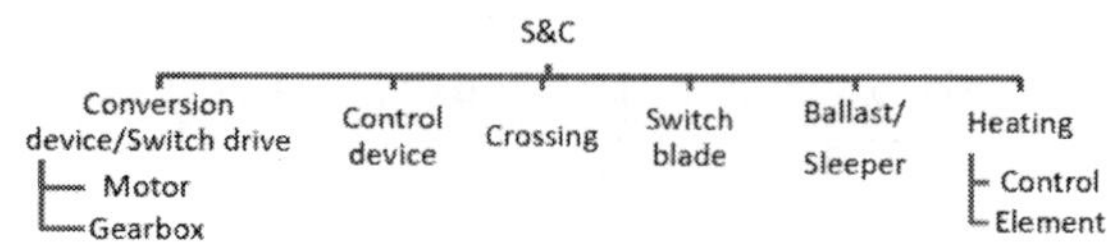

Figure 4. S&C component tree.

The first step in the analysis is the calculation of the RAMS associated to replacement WOs in the switch. It is understood that replacement here means replacement of any of the switch components, i.e. control device, crossing, heating, conversion device, etc. Therefore it is more convenient to say that the switch has been repaired instead of replaced as a whole. The repair brings the system to operation state again, and it is therefore considered that the reliability of the system does not improve substantially after the repair, being the probability of failure the same as before the failure happened. From a statistical point of view the switch is brought to an *as-bad-as-old* state. The event plot shown in Figure 5 provides a clear picture of this repairable characteristic. In this plot, each event, represented by a point in time, corresponds to a CM action associated to a switch restoration. As shown, there are assets (or units) that suffer a larger number of failures than others. This may correspond to failures in different components of the given switch. After the failure occurs, the asset is put back to operating state until another failure happens in the same or a different component.

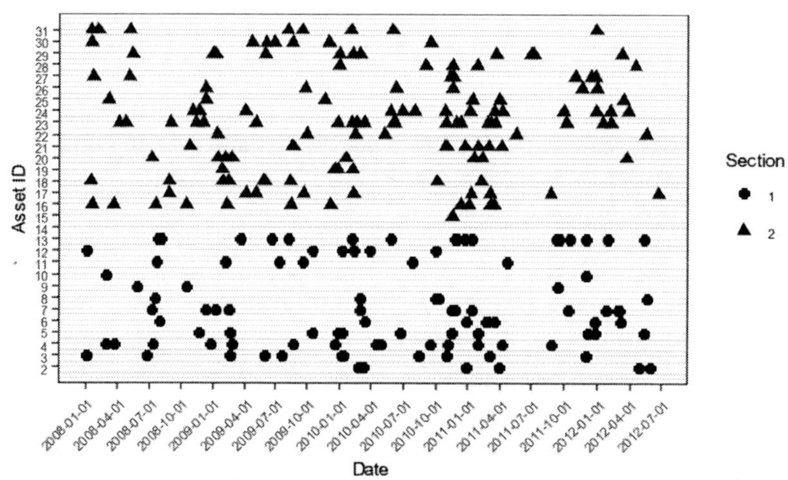

Figure 5. Event plot for replacements in S&Cs components by track section.

These types of processes are described statistically as point processes [13]. To calculate the Time-To-Failure (TTF) and Time-To-Restore (TTR) after failure, each of the different components needs to be analysed independently using probabilistic models. This separation is needed for consistency as it does not take the same time to replace a heating system than a control device for instance and also because their failure modes are different. Moreover, in our study, in order to have statistically significant results all the switches and components have been assumed to be working under identical environmental conditions. Given the available time window, this assumption allows for a larger number of failure events to be analysed. We have 85 events for control device, 45 for conversion device, 25 for crossings and 67 for heating system.

The TTF is determined by transforming the time window between replacements/failures (in days) in Figure 5 to load over the switch in Million Gross Tonnes (MGT) using the known average gross tonnage per year in each track section. In Figure 6 the cumulative distribution of failures of units with at least five replacements is shown as a function of the cumulative load (in MGT). Reliability is measure by the Mean-Time-To-Failure (MTTF) and calculated using a 2-parameters Weibull distribution. The Weibull distribution is widely used in reliability and life data analysis thanks to its versatility and ability to model decreasing, increasing or constant failure rates (the so-called bathtub curve). The averaged MTTF in MGT per component is shown in Table 1 together with the Weibull shape and scale parameters. It is important to notice that expected useful life for this kind of systems is of the order of 20-30 years, and the time range under analysis is only 4 years. This means that the number of registered failures is low and therefore the statistics is not very good. In Figure 7 the calculated distributions of TTF, after removing outliers, are shown in form of boxplots for the different switch components.

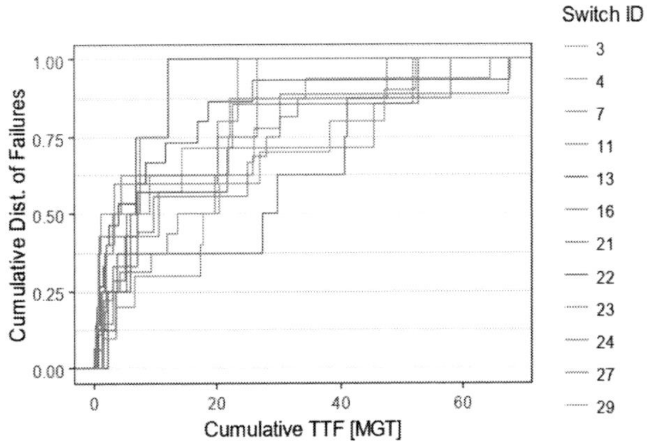

Figure 6. Cumulative distribution of failures for recurrent units.

Table 1. MTTF and MTTR for switch components with significant number of failures. Weibull and lognormal are used as models. Estimated errors are shown in parenthesis and correspond to 65% CI.

Component	Time-To-Failure (MGT)			Time-To-Restore (Hours)		
	Weibull parameters		MTTF	*Log-normal parameters*		MTTR
	Shape	Scale		μ	σ	
Conversion Device	0.87 (14)	29 (4)	31 (6)	0.393 (15)	0.654 (17)	1.8 (1.2)
Control Device	0.88 (10)	13.8 (1.2)	15 (4)	0.29 (1)	0.81 (1)	1.9 (1.3)
Heating	0.71 (9)	19.5 (2.1)	25 (6)	0.282 (15)	0.91 (1)	2.0 (1.5)
Crossing	1.21 (25)	43 (7)	40 (6)	1.42 (12)	0.99 (6)	7 (3)

The database does not provide information about repair times and therefore maintainability is measured by Mean-Time-To-Restore (MTTR). It is possible to present figures for the time spent in the different maintenance activities, as the maintenance crew report on this information. Unfortunately, repair and logistic times cannot be disentangled, and therefore in our model, the MTTR will include the whole restoration time. It is also important to notice that maintenance crew do not report whether the action is partly executed and finished on a later occasion, which may result in non-representative restoration times in some cases. In fact, although most of the restoration times are reasonable figures, a few of them present unusually large times. In order to solve this problem, only maintenance activities with less than 16 man-hours are kept, and unusual long times are considered as outliers and filtered out from the database. The distributions of TTR per switch component are shown in Figure 8 after the outliers have been removed. The calculation of MTTR has been carried out using a log-normal distribution. This assumption is commonly used for restoration times because of the following reason: it is natural to assume that restoration rate increases at least in a first phase. When the restoration has been going on for a rather long time, this indicates serious problems, for example no spare parts available on the site. Then, it is natural to assume that the restoration rate is decreasing after a certain period of time. This process is well described using a log-normal distribution. The averaged MTTR in hours per component is shown in Table 1 together with the log-normal parameters. As can be seen, all components are restored within 8 man-hours, being crossings the ones taking the longest times.

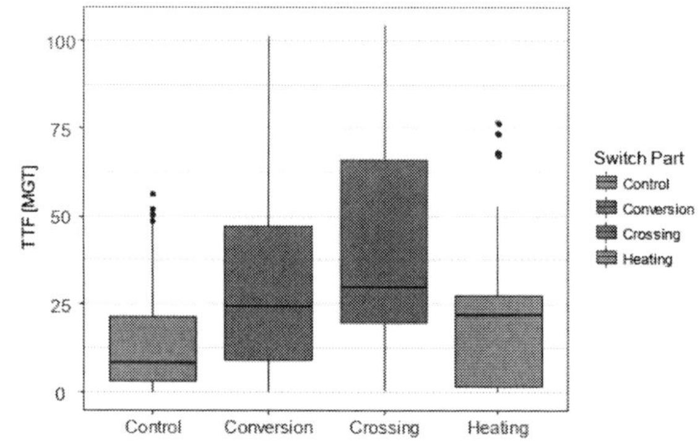

Figure 7. Tukey boxplots for component TTF.

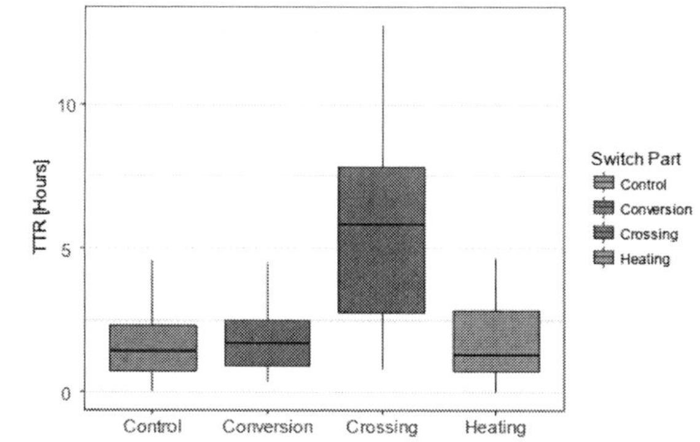

Figure 8. Tukey boxplots for component TTR.

An LCC model for the maintenance of S&C can be built by subdividing the system into a number of subsystems (as shown in the tree of Figure 4). Here we are considering only the *conversion device* (or switch drive), *control device*, *crossing* and *heating system*, because these are the most delicate components and therefore the subject of most of the failures. Within the most frequent maintenance actions for S&C are: (i) corrective or preventive: adjustment, replacement and repair, (ii) preventive: tamping, grinding and inspection.

When building the LCC model, one of the most difficult pieces of information to obtain is the cost per action. As it has been stressed, here we will focus on modelling cost for S&C restoration associated to replacement of components for which RAMS parameters are available. Generally speaking replacement cost is the sum of man-hours plus cost-of-spare-parts plus the cost-of-equipment. Depending on the information available, the model will be more precise on the determination of costs. In our case we will model replacement cost using the following equation:

$$LCC_{replace} = n_s \sum_{i=1}^{K} \sum_{j=1}^{N-1} \frac{1}{(1+r)^j} \frac{M}{MTTF_i} \{ C_{Pi} + MTTR_i (n_{Li} C_L + C_{Ei}) \} \quad (3)$$

Where, n_s is the number of switches; K the number of switch components under analysis; N the life period of the switch in years; M is the Gross Tonnage per year (in MGT); $MTTF_i$ is the Mean-Time-To-Failure for component i (in MGT); C_{Pi} is the cost of component i (in Euro); $MTTR_i$ is the Mean-Time-To-Restore of component i (in hours); n_{Li} is the number of workers needed for replacement of component i; C_L is labour cost (in Euro/hour); and C_{Ei} is the cost of equipment for replacement of component i. For illustration purposes, the following assumptions will be made:

- The average gross tonnage per year is assumed to be $M = 20$ *MGT*.
- The life of the switch will be estimated to be 600 MGT (*N=24 years*).
- The discount rate is taken to be 4% ($r=0.04$).
- Average cost per component (C_P): crossing (500 €/unit), heating device (200 €/unit), control device (1000 €/unit), and conversion device / switch drive (1000 €/unit).
- The average labour cost is $C_L = 25$ *€/hour*.

- The number of workers will be fixed to be $n_L=3$ for the different replacements.
- The equipment cost for replacement will be fixed to $C_E=5$ *€/hour* for the different replacements.

Taking into account the above fixed values and the RAMS previously calculated in Table 1 we obtain the following costs per component for a switch life-cycle of 24 years: conversion device (11.605 €), control device (24.680 €), heating device (8.550 €) and crossing (5.842 €). Using the above set of parameters, the total cost per switch associated to failures that lead to replacements in these components would be around 50.700 €. The differences in cost per component can be easily visualised in the bar-plot of Figure 9.

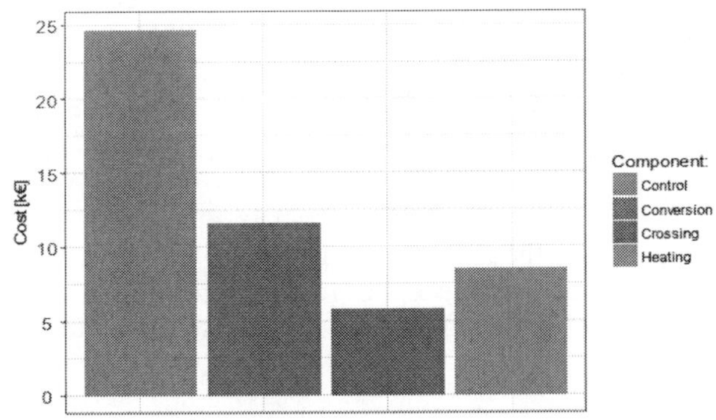

Figure 9. Replacement cost per switch component during 25 years.

Furthermore, a sensitivity analysis has been carried out in order to study how variations in cost parameters affect the total LCC-value. Figure 10 shows the percentage change in the total LCC-value for replacement when different parameters entering the LCC formula are changed a 10%. It can be seen that four parameters (C_P, *M, MTTF and r*) affect the LCC-value significantly. Moreover, replacement equipment cost (C_E), labour cost (either number of workers, n_L, or wage, C_L), switch component costs (C_P), averaged yearly gross tonnes (the load, *M*), and the time to restore failures (MTTR) contribute positively to the LCC-value. On the other hand, the time between failures (MTTF) and the discount rate (*r*) contribute negatively to the total LCC-value. As shown in Figure 10, some values can be considered more critical concerning the preventive maintenance rate and for maintenance optimization purposes.

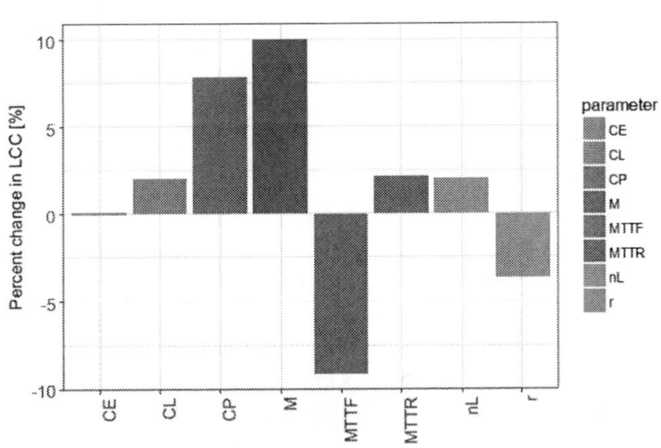

Figure 10. Sensitivity analysis of selected LCC parameters. Parameter changed a 10%.

4. Conclusions

This paper has presented a methodology for the combination of RAMS in the LCC analysis of linear transport infrastructures. The methodology has been demonstrated in a railway use case but it is also suitable for roads and, in fact, it will be applied to a road network within the H2020 INFRALERT project [9]. The paper has focused on the analysis of repair costs of switches and crossings in a railway line that is part of one of the demonstrators of INFRALERT. It has been shown how to use historical maintenance records to statistically characterize system failures in terms of RAMS, and how to use this information, together with individual cost figures, in LCC formulas to obtain cost estimates and cost driver's dependencies. The parameters that influence the LCC results have been identified using sensitivity analysis. This knowledge can be used to apply cost effective long-term maintenance decisions. From the results obtained in this paper it can be concluded that adequate data is crucial to obtain reliable results and that the data must be collected the right way by maintenance operators, since the quality of the reporting and registering is crucial for a good statistical analysis. Nevertheless, this methodology can set the bases for maintenance data collection standards in the railway or road sectors.

Acknowledgements

This research was carried out within the INFRALERT project. This project has received funding from the EU Horizon 2020 research and innovation programme under grant agreement No 636496. The authors also thank Trafikverket for providing the data in the case study analysis.

References

[1] European Committee for Standardization (CEN) 2002 Basis of structural design - Eurocode 0 *(EN-1990)*
[2] Shift2Rail Joint Undertaking (S2R JU) 2016 *Annual work plan*
[3] http://ec.europa.eu/research/transport/index.cfm?pg=transport&lib=rail.
[4] Al-Jibouri S and Ogink G 2009 Proposed model for integrating RAMS method in the design process in construction *Architectural Engineering and Design Management* **5** pp 179–92
[5] Asiedu Y and Gu P 1998 Product life cycle cost analysis: state of the art review *International journal of production research* **36** pp 883-908
[6] Kumar U, Ahmadi A, Verma A K and Varde P 2015 *Current Trends in Reliability, Availability, Maintainability and Safety: An Industry Perspective* (Switzerland: Springer International Publishing)
[7] Zoeteman A 2001 Life cycle cost analysis for managing rail infrastructure *European Journal of Transport and Infrastructure Research* **1** pp 391-413
[8] Smith D J 2011 *Reliability, Maintainability and Risk 8th Edition: Practical Methods for Engineers including Reliability Centred Maintenance and Safety-Related Systems* (Oxford: Butterworth-Heinemann)
[9] Infralert 2016 *INFRALERT: Linear Infrastructure Efficiency Improvement by Automated Learning and Optimised Predictive Maintenance Techniques* (EC H2020 Programme Grant agreement No 636496)
[10] European Committee for Electrotechnical Standardization - CENELEC 1999 Railway applications - The specification and demonstration of Reliability, Availability, Maintainability and Safety Part 1: Basic requirements and generic process *(EN 50126-1)*
[11] International Electrotechnical Commission - IEC 2004 Dependability Management - Part 3-3: Application Guide - Life Cicle Costing *(IEC 60300-3-3)*
[12] Trafikverket 2016 *The Swedish Transport Administration Annual Report* (Borlänge: Swedish Transport Administration) URL http://online4.ineko.se/trafikverket/Product/Detail/44395
[13] Garmabaki A H S, Ahmadi A, Block J, Pham H and Kumar U 2016 A reliability decision framework for multiple repairable units *Reliability Engineering & System Safety* **150** pp78-88

IOP Conf. Series: Materials Science and Engineering **236** (2017) 012107 doi:10.1088/1757-899X/236/1/012107

Historical maintenance relevant information road-map for a self-learning maintenance prediction procedural approach

Francisco J. Morales[1], Antonio Reyes[1], Noelia Cáceres[2], Luis M. Romero[1], Francisco G. Benitez[1], Joao Morgado[3], Emanuel Duarte[3] and Teresa Martins[3]

[1]Transportation Engineering, Faculty of Engineering, University of Seville, Spain
[2]Transportation Research Unit, AICIA, Seville, Spain
[3]Infrastruturas de Portugal, Coimbra, Portugal

E-mail: benitez@us.es

Abstract. A large percentage of transport infrastructures are composed of linear assets, such as roads and rail tracks. The large social and economic relevance of these constructions force the stakeholders to ensure a prolonged health/durability. Even though, inevitable malfunctioning, breaking down, and out-of-service periods arise randomly during the life cycle of the infrastructure. Predictive maintenance techniques tend to diminish the appearance of unpredicted failures and the execution of needed corrective interventions, envisaging the adequate interventions to be conducted before failures show up. This communication presents: i) A procedural approach, to be conducted, in order to collect the relevant information regarding the evolving state condition of the assets involved in all maintenance interventions; this reported and stored information constitutes a rich historical data base to train Machine Learning algorithms in order to generate reliable predictions of the interventions to be carried out in further time scenarios. ii) A schematic flow chart of the automatic learning procedure. iii) Self-learning rules from automatic learning from false positive/negatives. The description, testing, automatic learning approach and the outcomes of a pilot case are presented; finally some conclusions are outlined regarding the methodology proposed for improving the self-learning predictive capability.

1. Introduction

INFRALERT is a European Horizon 2020 project focus on two different pilot cases, one for road and other for rail, whose aim is to develop an expert-based information system to support and automate linear asset infrastructure management from measurement to maintenance. This enfolds the collection, storage and analysis of inspection data, the prediction of interventions to keep the performance of transport networks in optimal conditions, and the optimal planning of maintenance interventions. The results also facilitate the assessment of new construction strategic decisions.

For this purpose, one of the goals of the project focusses on the development of an intelligent alert management system [1], with the purpose to analyse asset conditions and operational information, to provide alerts whenever the infrastructure reaches, or is close to reaching, a critical level either in the present time or in a forthcoming scenario. It combines the current and predicted asset condition with operational and historical maintenance data, to get information about the needed maintenance tasks to avoid later severe degradation, safety mismatching and/or comfort conditions. By means of data analytics and machine learning methodologies, the system generates a prioritised listing (ranked on severity level) corresponding to the alerts generated by all assets of a linear transport infrastructure.

The concept of maintenance alerts is linked to any asset (e.g. part, element, component, subsystem, system) when its functional condition is in danger or will be jeopardised in a future scenario. An alert is generated when the condition of an infrastructure asset crosses a threshold limit value defined by a standard in a specific forecasted scenario and/or recorded know-how information from previous maintenance interventions. The evolving condition, soundness/unsoundness, of an asset in forecasted scenarios depends on the technical characteristics of the asset itself, the set of other assets implied in the infrastructure, the evolution of the stresses and environmental loads the infrastructure is undergoing, and many other characteristics/attributes denoted under the generic term "feature".

Service loadings and their frequency acting on an asset are the main reason for a decrease in the reliability and occurrence of failures, which obligate to carry out maintenance interventions. The historical evolution of the asset features and the historical recording of maintenance operations are valuable sources of information to predict further maintenance interventions.

The existing knowledge on historical maintenance works, and their corresponding asset condition encountering during the repair, can be used in a non-structured manner to gather a repository of knowledge on severity of failures and the according maintenance interventions carried out, besides the resources demanded, to bring the infrastructure back into service.

A structured database of historical maintenance interventions, founded on quantifying the information available on the said repository of knowledge, paves the way for using data analytic techniques in order to infer tools for decision making, be either fully or semi-automatic.

There follow the procedures and codes to ascertain the reliability of positive alert levels from false positives/negatives based on existing recorded knowledge of previous maintenance interventions and activities. The use of machine learning, data analytics, data inference and statistic techniques, allows addressing this goal compiling and processing the prior available pieces of information regarding maintenance interventions and operations.

2. Alert Management toolkit

The aim of this toolkit is to prioritise assets of the infrastructure, according to the required maintenance interventions based on the forecasted severity of degradation and failure of the assets themselves, and the know-how brought in by the information recorded in the historical maintenance interventions.

There are two types of alerts. The first ones are triggered by the deviation of the predicted condition of an asset from the Standards, the other ones are inferred by correlating the recorded information from previous maintenance interventions. Figure 1 depicts a block diagram reflecting the automated learning algorithms associated to the prediction of the two alert types.

Module AM1 (Alerts based on limits) is responsible for generating alerts from the point of view of those features that overcome their associated limits or reference thresholds, using as inputs the forecasted features of the asset. In particular, the goal of this module is to compare the value of each forecasted feature with the corresponding limit to determine/quantify the asset state condition. As results, the module provides the following pieces of information/outcome:

- Alerts indicating that a specific feature exceeds the prescribed threshold limit.
- The technical severity levels (TSL) of those alerts based on limits. The TSL is an objective value used to prioritise the alerts according, for instance, to a distance criterion between the value of the feature and the threshold.

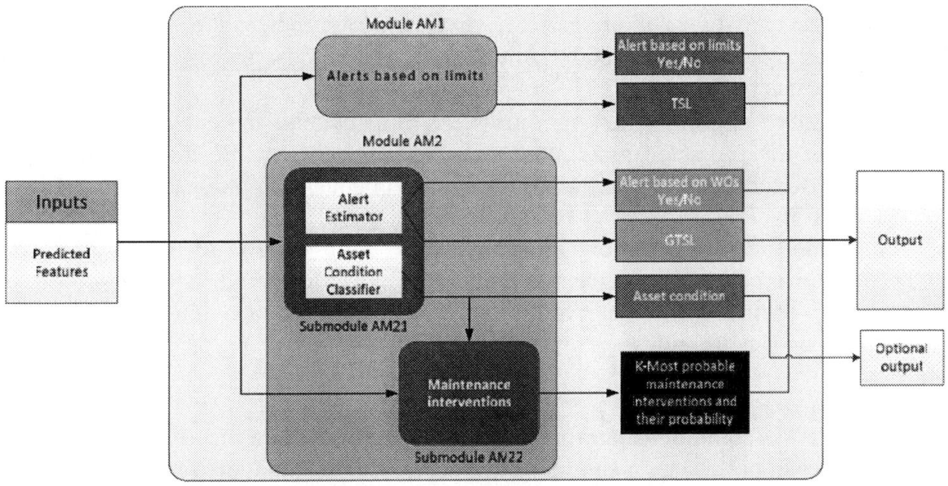

Figure 1. Block diagram of the general work-flow of the alert predictor.

Module AM2 (Alerts based on Work Orders) predicts alerts based on whether maintenance is or is not required (YES-NO), according to the forecasted features of the asset; it also estimates the most probable maintenance interventions to be conducted. To achieve this, the module embodies two different functional submodules. The first one (AM21) is specifically devoted to triggering alerts regarding the need of maintenance and their corresponding level of global technical severity (GTSL) in terms of all forecasted features considered as a whole; here, the alerts are not based on overcoming specific limits but they are triggered by the estimator contained in the first block (Alert Estimator) which has been previously trained through a machine learning processing with the measured features and the historical maintenance interventions. These outputs (alerts and GTSL) are used by the Maintenance Manager (MM) for designing maintenance plans. Submodule AM21 also provides an optional output, to be used by the MM, of the asset condition associated to the forecasted features both at individual feature level and at the overall performance level of the asset (regarding the simultaneous contribution of all feature effects as a whole). This last task is performed by the second block (Asset Condition Classifier), which has been also trained with historical intervention data to "learn" from the MM know-how, with the final purpose of predicting a subjective evaluation of the asset condition (from the set of forecasted features) without the intervention of the MM.

The second submodule (AM22) aims at determining the set of k-most probable maintenance interventions that have to be conducted, as well as their corresponding probabilities of occurrence, via a learning procedure based on historical intervention database. As results, the module provides the following outputs:

- Alert triggered: required maintenance (ALERT).
- Global Technical Severity Level (GTSL) for the asset.
- K-Most probable interventions: a listing of maintenance intervention types ordered according to the estimated probability of occurrence, regarding to the triggered alert.
- Probability of occurrence of the most probable interventions.

As this last module is based on machine learning techniques, it is necessary to define the self-learning procedure. Next section focusses on this point and the methodology used discerning from "false positives" and "false negatives" predictions, according to the following rules:

- A false positive arises when an estimate (from models) indicates a given condition has been reached, when it has not; it is commonly regarded as a "false alarm". Therefore, an erroneous positive case has been assumed.
- A false negative appears when an estimate (from models) indicates that no alert has been detected (i.e. the predicted asset state condition is right), while it was later detected by a corrective maintenance intervention; therefore, erroneously no failure/fault was forecasted.

Before getting into the various cases which may arise, a description of the way-of-proceeding to follow by the maintenance team is suggested when an alert is triggered, whether it be corrective or predictive; the case of preventive intervention is not taken into account herein as it follows a predefined plan laid on specific rules.

3. Triggered-alert procedure

When a triggered alert is communicated to the maintenance team, this may have either a "corrective" or "predictive" cause. In both cases a summary of the steps to be conducted by the MM follows and it is shown in Figire 2:

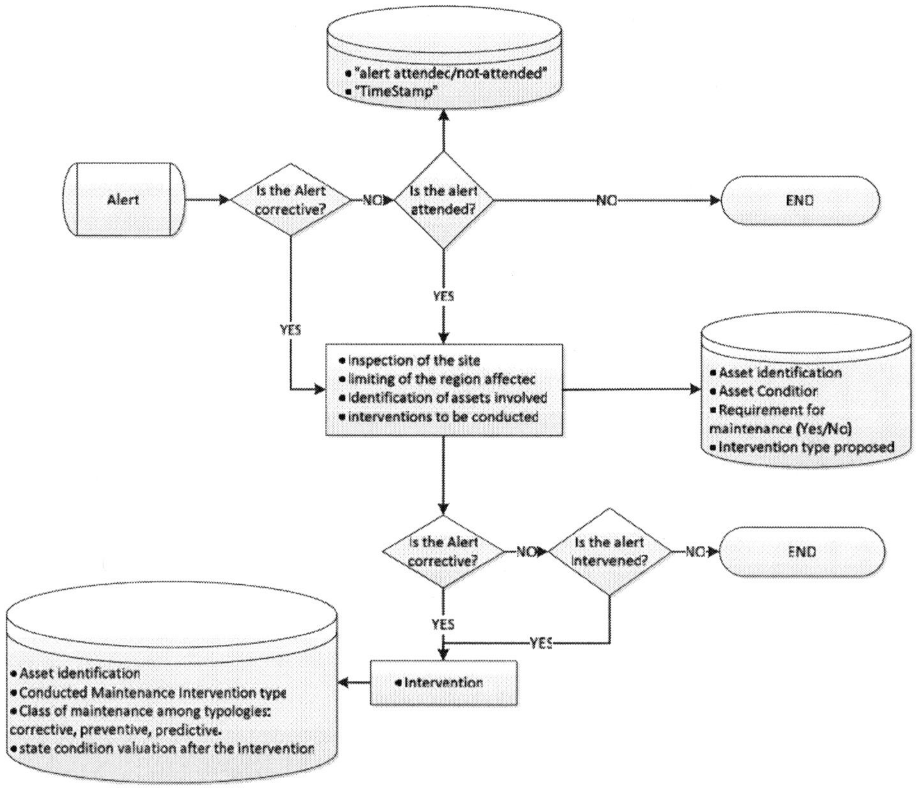

Figure 2. Triggered-alert procedure diagram flow.

I. A decision has to be made regarding whether the triggered alert is attended or not, according to the information available with respect to the severity of the alert and the internalities and externalities affected; therefore, two cases are presented:
   - If the alert is not attended, the procedure stops (it does not continue) after recording this decision. In case of a corrective alert this decision may imply safety action to guarantee the integrity of the infrastructure, service and other affected personal/material values.

- If the alert is attended, the procedure continues to step II after recording this decision, and the following information:
  - ❖ Information recorded in the Data Base: "alert attended/not-attended", the "Timestamp" to reflect the time the decision regarding the alert is made at. Other time logs relative to the alert (e.g. the time scenario the predicted alert is estimated at).

II. In case the alert is attended/inspected, a second level of ordered actions should take place:

- Inspection of the site where the alert takes place.
- Identification of the region affected by the alert.
- Identification of all assets involved.
  - ❖ From those actions, the following information is captured:
  - o General qualitative assessment of the full set of assets involved:
    - ▪ Qualitative (lexicographic) or quantitative (numeric, if possible) state condition valuation.
  - o Assessment of each asset involved:
    - ▪ Qualitative (lexicographic) or quantitative (numeric, if possible) state condition valuation.
  - ❖ Other pieces of information recorded and stored in the Data Base are:
    - ▪ Asset identification (Asset-ID).
    - ▪ Asset Condition:
      - Subjective evaluations associated to any individual feature (A_i).
      - Subjective evaluations associated to a combination of features (C_i).
      - A global subjective evaluation associated to the asset condition as a whole (G).
    - ▪ Requirement for maintenance (RfM): Yes/No.
- (Optional) Identification of the overall cause/origin of the alert.
  - ❖ Information recorded and to be stored in the Data Base:
    - ▪ General qualitative (lexicographic) or quantitative (numeric, if possible) assessment of the general cause and origin.
- (Optional) Identification and assessment of each unit cause/origin of the alert:
  - ❖ Information recorded to be stored in the Data Base:
    - ▪ Unit cause/origin.
    - ▪ Qualitative (lexicographic) or quantitative (numeric, if possible) cause valuation.
- Identification of the interventions to be conducted for eliminating/diminishing the alert:
  - ❖ Information recorded in the Data Base:
    - ▪ Asset implied.
    - ▪ Intervention type proposed (Mproposed).

III. In case the alert is intervened (a maintenance intervention is carried out to re-establish the performance of the transport network), a third level of ordered actions should take place:

- Identification of the real interventions carried out which eliminated/diminished the alert. The information captured and recorded in the Data Base for each involved asset is:
    - ▪ Asset identification (Asset-ID).
    - ▪ Conducted Maintenance Intervention type (Mconducted).

- Class of maintenance (CM) regarding the typology: corrective, preventive, predictive.
- Qualitative (lexicographic) or quantitative (numeric, if possible) state condition valuation just after the intervention: asset's state condition nowcast (MaintDescrip).

Besides, all recorded data have associated a timestamp field to identify the date and time when the action is carried out (when the alert is regarded as either "not-attended", "attended" or "intervened").

4. Predicted false cases considered

The false cases may arise when wrong estimates are due to: i) wrong forecasting of a feature, ii) wrong prediction for requesting maintenance (Yes/No), iii) wrong prediction for maintenance type. A description of the cases follows.

4.1. Wrong forecasting of a feature

Two possible cases, false positive and false negative, are analysed in separated manner:

a) False positive. This case takes place when the alert is triggered by the expected value of a single feature, a prior combined single feature, or a multiplicity of features, due to a wrong forecasting of the features' values. The Triggered-Alert Procedure, activated by the MM, ends by assessing all assets implied in the alert, and by recording information of the state condition of those assets. The new information will enrich the data base corresponding to the state condition of the assets involved. The machine learning (ML) procedure will learn from the enriched data base information, improving the success rate according to the quality of the captured information in further prediction runs.

b) False negative. This case arises when the expected value of the feature (single, combined or a multiplicity of features) is below the value-to-be and also falls below the TSL (Technical Severity Level) threshold [1]. In this case the Triggered-Alert Procedure is not activated and no alert is triggered. Eventually a corrective warning may brought up and the Triggered-Alert Procedure activated, which will record the pieces of information identified in section 3.

4.2. Wrong prediction for requesting maintenance

Similar to the previous paragraph, two possible cases, false positive and false negative, may arise:

a) False positive. This takes place when Submodule AM21 [2] predicts the need of maintenance, and an alert is triggered. This case follows a similar pattern than case a) of section 4.1. The ML procedure will be improved according to the data base enrichment.

b) False negative. This takes place when Submodule AM21 does not detect the need of maintenance and no alert is triggered. This case follows a similar pattern than case b) of section 4.2.

4.3. Wrong prediction for maintenance type

In this case, the requesting for maintenance is supposed to be correctly estimated as positive, otherwise the procedure goes back to section 4.2. Only the general "false type" case prediction is possible:

a) False type. The alert has been triggered due to a requesting for maintenance (Module AM21), and an estimated maintenance intervention type is provided. The Triggered-Alert Procedure is activated by the MM, and the assets implied are assessed by the maintenance intervention team. The team detects that the maintenance type estimated does not corresponds to the predicted type; it means a wrong maintenance type prediction by module AM22 [2] has taken place. The Triggered-Alert Procedure is continued, acting according the unit actions specified in the procedure. The information captured by this procedure (reflecting all pieces presented in section 3) is recorded and stored in the Data Base, which enriches the data base information. Later estimations by the ML algorithms will be benefited by the new enriched Data Base.

5. Pilot case

The pilot case is a road network in Coimbra, region in the centre of Portugal, managed by Infraestruturas de Portugal (Figure 3). The available information consists in two different data bases. The first one corresponds to the results of different measurement campaigns in which representative features of the state condition of the road are obtained, and the second one compiles the historical maintenance interventions carried out in the net from years 1933 to 2014. By correlating both data bases the inputs and targets for the machine learning models are inferred. However, the data recorded has previously filtered to extract the relevant information to be used by the system before triggering the whole maintenance predictive methodology.

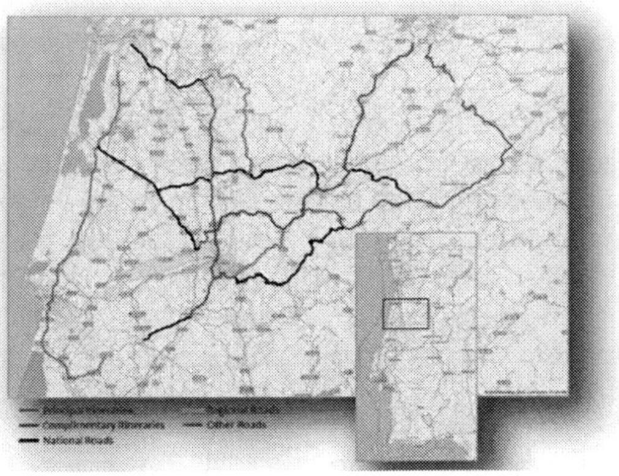

Figure 3. Road pilot case demonstration in Portugal.

With the information available for the pilot case, the models have been implemented by means of Matlab code. In particular, these are based on Decision Trees (DT), k-Nearest Neighbours (KNN), Support Vector Machines (SVM) and Artificial Neural Networks (ANN). In order to check the accuracy of those models, the last year data-set (2014 campaign) is used as a testing sample, keeping this year out of the training set. The performance evaluation of the models was made by using a confusion matrix, based on counting those test records correctly and incorrectly predicted. In those matrixes appear the real (M) and the predicted maintenance type based on Table 1. Class T0 is associated to no-alert, and the rest of classes belong to different maintenance alerts. Figure 4 shows the confusion matrix for the DT model where the columns correspond to the known class (target class specified by the work-order) and the rows correspond to the predictions made by the model (output class). Thus, the diagonal elements show the number of correct classifications made for each class, and the off-diagonal elements show the errors made by the model predictions. Each cell also contains the same information as a percentage of the total test set size. In the last row and column, these plots also contain the performance measures derived from the confusion matrix; those are the accuracy (ACC), recall or true positive rate (TPR) and precision or positive predictive value (PPV), showed in green; as well as the respective complements: error rate (ERR), false negative rate (FNR) and false discovery rate (FDR), showed in red.

Table 1. Description of maintenance types.

	Alert	Description
T0	No	No maintenance requested
T1	Yes	Do nothing
T2	Yes	Microsurfacing, Surface dressing
T3	Yes	Thin Hot-Mix Asphalt overlay (thickness ≤ 5 cm)
T3.1	Yes	Surface milling with Thin Hot-Mix Asphalt overlay (thickness ≥ 5 cm)
T4	Yes	Thick Hot-Mix Asphalt overlay (thickness > 5 cm) combined or not with milling

Once the models provided a good level of prediction; the self-learning process described in this paper was checked. A simulator, based on the available real data, was carried out in order to increase gradually the input data for the models. The learning curves of the example model (DT) are shown in Figure 5. The two bold lines represent the average value of the train and test sets; the rest of lines are generated with the 20th and 80th percentiles of train and test sets and provide an insight of the variance of the predictions (thin lines). As the training set sizes get larger, these curves converge toward a threshold representing the amount of irreducible error in the data.

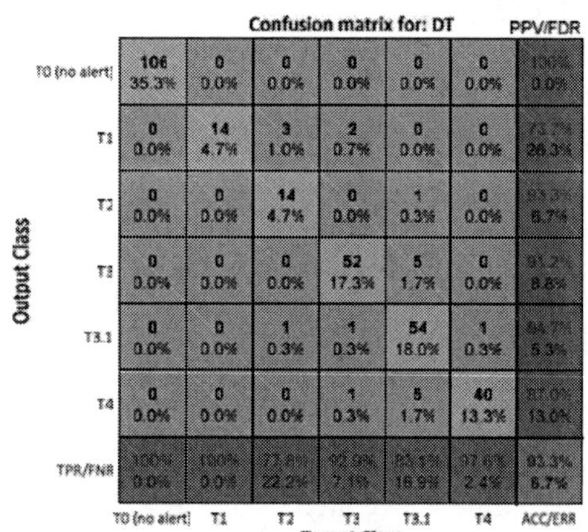

Figure 4. Confusion Matrix (DT model).

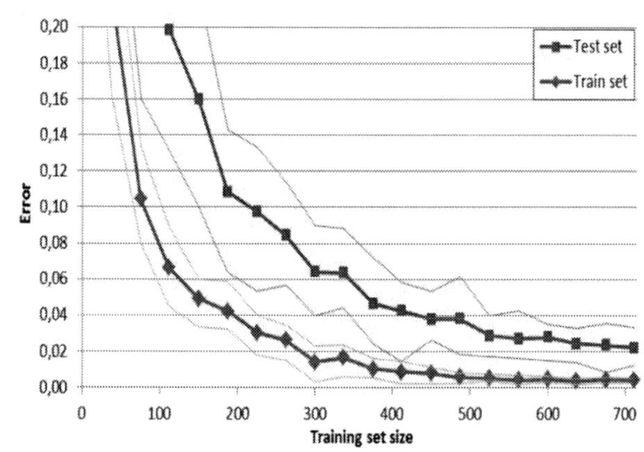

Figure 5. Learning curve (DT model).

6. Conclusions

This communication presents a procedure to record the adequate information in order to be used by automatic learning systems to predict maintenance interventions in future scenarios. The methodology presents evidences, through a pilot case, that with new relevant information it is possible to reduce the error in the prediction of the maintenance intervention to be carried out in the linear assets of a road network. A diagram flow is presented in order to have a general view of the procedure to follow in order to learn from false positive/negative predicted cases and the way they are considered are analysed.

Acknowledgements
The research leading to these results has received funding from European Union's Horizon 2020 Research and Innovation Programme (grant agreement n° 636496). Some of the authors express their gratitude to the Spanish Ministry of Economy and Competitiveness for the partial subsidy granted under the national R&D program (TRA2015-65503) and the Torres Quevedo Programme (PTQ-13-

06428). The content reflects only the authors' view and it is stated that the Union is not liable for any use that may be made of the information contained therein.

References

[1] INFRALERT 2016 *Linear Infrastructure Efficiency Improvement by Automated Learning and Optimised Predictive Maintenance Techniques* (H2020 Programme European Commission Research Directorate Grant agreement No 636496) http://infralert.eu

[2] INFRALERT 2017 *Deliverable D4.3. Methodologies and procedures for inferring three-dimensional alert-severity-intervention pattern space. Supervised and unsupervised approaches*. (H2020 Programme European Commission Research Directorate Grant agreement No 636496) http://infralert.eu

Utilization of building information modeling in infrastructure's design and construction

Josef Zak[1] and Helen Macadam[2]

[1]Czech Technical University in Prague, Faculty of Civil Engineering, Thakurova 7, Prague, Czech Republic
[2] Skanska UK, Maple Cross, Denham Way, Rickmansworth, Hertfordshire, United Kingdom

E-mail: josef.zak@fsv.cvut.cz

Abstract. Building Information Modeling (BIM) is a concept that has gained its place in the design, construction and maintenance of buildings in Czech Republic during recent years. This paper deals with description of usage, applications and potential benefits and disadvantages connected with implementation of BIM principles in the preparation and construction of infrastructure projects. Part of the paper describes the status of BIM implementation in Czech Republic, and there is a review of several virtual design and construction practices in Czech Republic. Examples of best practice are presented from current infrastructure projects. The paper further summarizes experiences with new technologies gained from the application of BIM related workflows. The focus is on the BIM model utilization for the machine control systems on site, quality assurance, quality management and construction management.

1. Introduction

BIM is an acronym that currently resonates in a number of conferences, seminars, roundtables and special events in the construction industry. In English, this acronym is known as Building Information Modeling, however in North America the same principles are called VDC (Virtual Design and Construction). On the basis of this term and differences in Czech technical language, the expression "Information Modeling" has become recognized as the Czech language alternative together with the use of English abbreviation "BIM". Therefore in this article both fixed expressions (BIM and Information Modeling) are used.

Building Information Modeling is the process involving the creation, use and management of construction data throughout the asset's entire life cycle [1]. Very closely related to Information Modeling are two other terms nowadays better known from mechanical engineering: Digitization and automation. This phrase means the same for construction as in other industries : With the increasing performance of computers and more specialized software, the use of information technologies in construction is ever expanding.

Processes commonly using paper-based documentation are changing due to the use of electronic data. Information Modeling is not about database-driven document management solutions with file extensions pdf, doc, xls, DGN, DWG, and their management and distribution; the information model contains the data and therefore the preparation of these documents is in many cases redundant. This process is called digitization of project documentation.

When digitization and automation took place in the automotive industry it increased production efficiency by 75%. A parallel can be drawn in construction: The introduction of Information Modeling in the life cycle of construction increases automation and prefabrication in the construction industry. Quantifying the savings in transportation construction by individual authors ranges between 8-35% [2]–[5].

2. Current state of BIM implementation in Czech Republic

2.1. Public procurement

One of the bases for the drafting of the new Czech Law on public procurement was Directive of the European Parliament and of the EP Council 2014/24 / EU, which recommends using the principles of Information Modeling. According to the Law on Public Procurement, § 103, paragraph 3 "contracting authority may specify in tender conditions mandatory requirement to use special electronic formats, including the tools of information modeling." The law came into force on 1st October 2016, allowing governmental authorities to require the use of the information model for the construction and preparation works.

The Czech State Fund for Transportation Infrastructure (SFDI) issued a press release at the end of 2016, expressing the intention to gradually introduce BIM into design and construction. Press releases followed Government Resolution dated 2nd November 2016 no. 958, on the importance of BIM for civil engineering practice and suggest next steps for its introduction. In acknowledgement of the need to tackle the topic, the SFDI established an expert working group dedicated to the topic of Information Modeling and its implementation in the transportation sector.

SFDI declared that in cooperation with Road and Motorway Directorate of Czech Republic and Railway Infrastructure Administration, they already plan to use building information modeling in several pilot projects in 2017. The goal is to broaden the use of digital methods in projects financed by SFDI. Detailed description of steps made by SFDI and other agencies can be found in [5].

The expert working group started work on two basic parallel tasks:

- Preparation of the strategy for the introduction of BIM in construction of transport infrastructure,
- Support for the preparation of pilot projects and their implementation.

Road and Motorway Directorate of the Czech Republic and Railway Infrastructure Administration will begin their selected pilot projects in 2017.

2.2. Associations

The Czech BIM council (CZ BIM) elected a new Board of Directors on 15th November 2016. Following this, the expert working group for Information Modeling in transportation was established. Currently, the following industry partners are actively represented in the Expert working group: Skanska, Metrostav, Hochtief, Sudop Prague, Mott MacDonald, Metroprojekt, CCE Prague and BIM Project.

The objectives of this working group are:

- Share news, experience and knowledge in the field (BIM);
- Prepare other materials relating to the topic as a team of co-authors;
- Comment on material originating outside the expert working group;
- Lay the foundations for cooperation in projects where the use of BIM is desired;
- Define a common approach to the topic of BIM.

IOP Conf. Series: Materials Science and Engineering **236** (2017) 012108 doi:10.1088/1757-899X/236/1/012108

An internal working group dedicated to the topic of Information Modeling is also based at the think-tank Association for Infrastructure Development (ARI). The first internal ARI members meeting was held on 13th January 2017.

2.3. Standards

Czech office for Standards, Metrology and Testing have established '152 TNK': Technical Normalization Committee (TNC) for BIM. The committee has been named as Information Organization of construction and building information modeling.

Comments and revision of following standards are currently ongoing: ISO 12006-2 Building construction – Organization of information about construction works – Part 2: Framework for classification and ISO 16757-1 - Data structures for electronic product catalogues for building services – Part 1: Concepts, architecture and model.

3. Examples of BIM in transportation in the Czech Republic and Slovakia

One of the contractors who sees significant advantages in the use of BIM and is not waiting for the introduction of an investor-led standardized process is Skanska. Examples of Skanska projects which use BIM are: D4 Skalka - Crossroads II / 118, modernization of railway stations Česká Lípa, D1 Přerov - Lipník nad Bečvou, R1 Nitra - Mlynárce and D1 Budimir-Bidovce. The following figures include several photos of information models from these projects. Data is used to detect inaccuracies in project documentation and resolve potential issues before they happen on site; information on the construction works is taken from the information model in certain cases instead of project documentation. One very important use for this data has been for earthworks and geodetic works.

Figure 1. Modernization of the railway station Česká Lípa – BIM model.

Figure 2. Excavator with GNSS machine control system (left), UAV before take-off (right).

Figure 3. Wired (left) and object (right) model of D4 Skalka – Crossroads II / 118.

4. BIM in other countries

BIM is commonly used in the Netherlands, Norway, Finland, Sweden and Singapore in civil engineering [6]–[8]. A new country that plans to introduce the obligation to use BIM for procurement from 2017, is France. In Germany this is planned to be the case from 2020. Unlike Europe, North American countries know BIM under the term Virtual Design and Construction; It is an inseparable part of the preparation of project documentation, construction and management of large highway projects.

4.1. BIM in UK

One of the examples which may be mentioned is the use of BIM in the United Kingdom (UK), where it is compulsory to use BIM principles for publicly funded civil projects.

BIM in the UK is governed by British Standard BS 1192:2007 Collaborative production of architectural engineering and construction information – Code of practice: It defines the process for using a Common Data Environment and code conventions for document status and numbering. It is supported by the suite of Publicly Available Specifications PAS 1192 documents, which give practical guidance for information management during the delivery of projects using Building Information Modeling. It outlines the information flow through-out the asset life cycle, management documents that should be produced, and project delivery team roles and responsibilities.

IOP Conf. Series: Materials Science and Engineering **236** (2017) 012108 doi:10.1088/1757-899X/236/1/012108

In 2011 the British Government published a strategy for reducing the cost of publicly funded assets, which required the use of BIM level 2 (as defined by the Bew-Richards BIM Maturity model below) for all centrally-procured government construction projects by 2016.

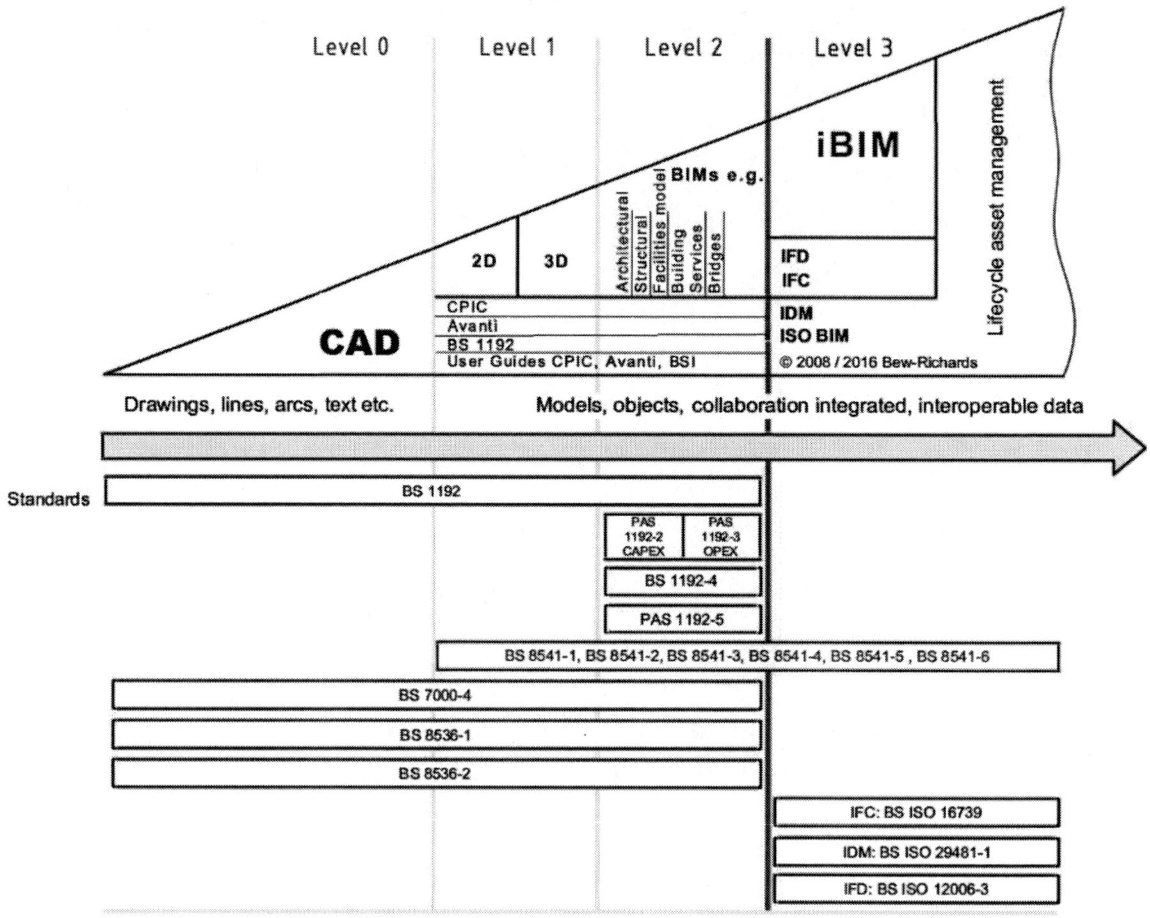

Figure 4. BIM Maturity Model [9].

In practice this means that all project data and asset information is electronic, with the design in the form of one 3D coordinated model. Although this is only a requirement for government projects, the use of these BIM processes is being widely adopted across other projects as the industry recognizes the benefits. Some examples of how Skanska are advancing the use of this information during the delivery phase are: 4D planning by adding a time dimension to the 3D model, and 5D modeling by using the 3D model to produce a bill of quantities and a calculation of embodied carbon, during the tender phase. The use of 3D information has also led to the use of survey techniques such as the use of 3D laser scanning and the production of Digital Terrain Models for earthworks.

Mobile technology has also become a familiar sight on Skanska sites, with smart phones and tablets being linked to the Common Data Environment. This access to the coordinated model, drawings and other project documentation improves the quality of work, reducing rework and time spent looking for information. The use of this mobile technology has also grown organically, with the people in possession of the devices finding ever more innovative uses for them. The devices are currently also used to produce paperless quality documentation, carry out site inspections, mark-up drawings and keep site progress records. The move towards paperless documentation has created efficiencies in administrative processes, and also created data that can be used to drive continuous improvement during the construction process, for example using data from health and safety inspections and quality checks to analyze trends and address them accordingly.

5. Conclusion

The transportation sector will be implementing pilot projects using Building Information Modeling principles in the near future. There are many questions associated with broader BIM implementation to be solved on governmental level. Thus it is planned that even during the preparation of pilot projects, the expert working group will continue to prepare the standardization and methodologies for processes associated with the digitization of the construction industry.

Future steps to be taken have been summarized by Ministry of Industry and Trade and SFDI in their strategies defining mandate for BIM implementation. Both of the strategies are currently in agencies internal document review process and should be issued in august 2017.

The reported use of Information Modeling in the Czech Republic and Slovakia is an example of collaboration between design firms, construction contractor, and surveying offices for achieving savings during construction, whilst also improving quality and safety. The contractor was a leader of the whole process in the presented examples, which defined the terms of cooperation, the technical content of BIM models and methods for its use. Experience of these pilot projects will be very valuable in the further development of this theme.

Acknowledgment:
This work was supported by Competence Centers program of Technology Agency of the Czech Republic (TA CR), project no. TE01020168.

References:

[1] Eastman C, Fischer D, Lafue G, Lividini J, Stoker D and Yessios C 1974 *An Outline of the Building Description System* (Carnegie-Mellon University, Pittsburgh, Institute of Physical Planning)
[2] *National BIM Report 2016* (RIBA Enterprises) https://www.thenbs.com/knowledge/national-bim-report-2016
[3] Köhler N 2008 *Virtuellt byggande ska sänka NCCs byggkostnader /Virtual construction will reduce NCC's construction cost/* (Byggindustrin)
[4] Eastman C, Teicholz P, Sacks R and Liston K, *Wiley: BIM Handbook: A Guide to Building Information Modeling for Owners, Managers, Designers, Engineers and Contractors, 2nd Edition - Chuck Eastman, Paul Teicholz, Rafael Sacks, et al.* 2011.
[5] Hořelica Z, Mertlová O, Vykydal I and Žák J 2017 Využití digitálních metod a zavedení informačního modelování staveb jako nástroj k dosažení vyšší efektivnosti staveb financovaných Státním fondem dopravní infrastruktury *Silnice a železnice* **1**
[6] Matějka P, Růžička J, Žák J, Hájek P, Tomek A, Kaiser J and Veselka J 2016 The implementation of building information modeling into educational programs at CTU in Prague *Int. Conf. Central Europe towards Sustainable Building 2016 Innovations for Sustainable Future (Prague)*
[7] Zak J and Brouwers J 2015 Informační modelování v infrastrukturních stavbách *Proc. 7th Conf. on Projektování pozemních komunikací (Prague)*
[8] Zak J and Macadam H 2017 Informační modelování staveb (BIM) a digitalizace v dopravním stavitelství *Silnice a železnice* **1**
[9] Bew M and Richards M 2008 *Bew-Richards BIM Maturity Model* http://bim-level2.org/globalassets/pdfs/bim-maturity-model.pdf.

Sustainability of transport structures – some aspects of the nonlinear reliability assessment

Radomír Pukl[1], Tereza Sajdlová[1], Alfred Strauss[2], David Lehký[3] and Drahomír Novák[3]

[1]Červenka Consulting s.r.o., Na Hřebenkách 55, 150 00 Prague 5, Czech Republic
[2]University of Natural Resources and Life Sciences, Peter Jordan Strasse 82, 1190 Vienna, Austria
[3]Faculty of Civil Engineering, Brno University of Technology, Veveří 95, 602 00 Brno, Czech Republic

E-mail: radomir.pukl@cervenka.cz

Abstract. Efficient techniques for both nonlinear numerical analysis of concrete structures and advanced stochastic simulation methods have been combined in order to offer an advanced tool for assessment of realistic behaviour, failure and safety assessment of transport structures. The utilized approach is based on randomization of the non-linear finite element analysis of the structural models. Degradation aspects such as carbonation of concrete can be accounted in order predict durability of the investigated structure and its sustainability. Results can serve as a rational basis for the performance and sustainability assessment based on advanced nonlinear computer analysis of the structures of transport infrastructure such as bridges or tunnels. In the stochastic simulation the input material parameters obtained from material tests including their randomness and uncertainty are represented as random variables or fields. Appropriate identification of material parameters is crucial for the virtual failure modelling of structures and structural elements. Inverse analysis using artificial neural networks and virtual stochastic simulations approach is applied to determine the fracture mechanical parameters of the structural material and its numerical model. Structural response, reliability and sustainability have been investigated on different types of transport structures made from various materials using the above mentioned methodology and tools.

1. Introduction

Non-linear analysis is nowadays used by engineers for design of new and assessment of existing structures. This development is supported by the new ***fib*** Model Code 2010 [1] where rational safety assessment approach is presented, which reflects new developments in safety formats based on probabilistic methods. In the Chapter 4 on *Principles of structural design* the probabilistic safety format is introduced as a general and rational basis of safety evaluation. In addition to the partial factor safety concept (which remains as the main safety format for most practical cases) a *global resistance format* is recommended for nonlinear analysis. This concept can be applied in design practice and can offer advanced and rational solutions to modern structural technologies [2].

The fully probabilistic approach represents the most accurate method for the safety assessment of civil engineering structures. The accuracy of this approach is much higher if the non-linear structural analysis is used as a limit state function. The numerical simulation resembles a real testing of structures by considering a representative group of samples, which can be statistically analyzed for the assessment of safety.

The probabilistic analysis is performed with software SARA, which integrates program tools ATENA and FReET. The variability of basic properties is described by distribution functions and its parameters (mean, standard deviation, etc.). Probabilistic analysis of the resistance is performed by numerical method such as Latin hypercube sampling method. Resulting array of resistance values is approximated by a distribution function of global resistance, and describes the random properties of the resistance. Finally, for a required reliability index β, or probability of failure P_f, a value of the design resistance R_d shall be calculated.

The fully probabilistic analysis is the ultimate tool for the safety assessment. It is superior to simplified methods because it provides information on the variability of resistance. However, it is computationally demanding and requires good information about random properties of input variables. Therefore, it should be applied in special cases, where consequences of failure substantiate the increased effort.

2. Nonlinear simulation of engineering structures

The nonlinear finite element simulation is recently a well-established approach for analysis of concrete structures. Behavior of the structure under service as well as ultimate conditions can be virtually simulated using computer methods quite realistically. The nonlinear fracture analysis accounting tensile capacity of material enables to exploit reserves, which are usually neglected or diminished in codes or in linear analysis.

The ATENA software [3] was developed for realistic simulation of reinforced concrete structures. It is based on the finite element method with non-linear material models, and utilized for analysis of beams and girders, plates and shells, composite structures, strengthenings, structural details, fastenings, fibre reinforced concrete structures, timber, stonebrick and masonry structures [4], [5], [6], [7], [8], [9]. Considerable part of the analyzed civil engineering constructions belong to the structures of transport infrastructure, such as bridges, tunnels, railway sleepers or load-carrying rail plates etc.

The ATENA software consists of calculating core ensuring the non-linear numerical analysis, and a user-friendly graphical interface for an efficient communication between end-user and program core. The numerical core covers the finite element technology, non-linear material models and non-linear solution. Since concrete is a complex material with strongly nonlinear response, special constitutive models for the finite element analysis of structures made of concrete and similar materials are employed. The non-linear material models are based on the orthotropic damage theory and special concrete-related theory of plasticity. As one of the main features the non-linear fracture mechanics is employed for concrete cracking in tension. Based on the fracture energy approach the tensile cracks are modeled as smeared material damage which enables utilization of the continuum mechanics even for the damaged material. For the shape of stress-crack opening curve an exponential law derived by Hordijk [10] is used in case of normal concrete. Objectivity of the solution (independency on the finite element mesh) is ensured using crack band method [11], [12]. The material law for concrete exhibits softening after reaching the tensile strength. The behavior of concrete in compression is defined by special theory of plasticity (three-parameter model [13], [14]), with non-associated plastic flow rule and softening. This material model for concrete can successfully reproduce also other important effect, such as volume change under plastic compression or compressive confinement. The native graphical user-interface in ATENA Engineering supports all the specifics of reinforced concrete, e.g. input of discrete reinforcing bars, or evaluation of crack patterns in the damaged structural model.

ATENA Engineering is perfectly suitable for static analysis of concrete structures, obtaining their load-displacement response and resistance, crack pattern including crack widths and identification of the failure mode. It can be used for structure optimization, assessment of retrofitting or reinforcement detailing.

In order to extend ATENA potential and features, the recent development combines the calculating core with AtenaStudio runtime and post-processing environment, and a powerful third-party program

IOP Conf. Series: Materials Science and Engineering **236** (2017) 012109 doi:10.1088/1757-899X/236/1/012109

GiD for pre-processing. The resulting product ATENA Science covers broad range of structural and material behavior in time. It enables to model geometrically complicated shapes and it is suitable for analysis of complex structural problems, such as dynamic implicit analysis, dynamic eigenvalue analysis, static stress analysis, creep analysis, transport of heat and fluids or fire analysis. The coupling of the above effects can be often achieved through simultaneous solution of various constitutive models. Thus, dynamic analysis can capture non-linear material response due to cracking, etc. In the eigenvalue analysis vibration frequencies reflect the stiffness changes due to material damage. In creep analysis the cracking of concrete and redistribution of stress due to plastic deformations is reflected. In fire analysis material response is strongly dependent on changing temperature fields.

However, the advanced non-linear analysis is performed in the ATENA software using purely deterministic way, i.e. it is assumed, that all the input parameters (material properties, geometry of the structure, etc.) are fixed, well known values. This is in contradiction to reality, where these values are usually uncertain or random, evtl. variable in space and time. Therefore, stochastic analysis would be much more suitable for obtaining really realistic results.

3. Probabilistic assessment of structural reliability

The probabilistic statistical, sensitivity and reliability analysis of engineering problems is based on the efficient reliability techniques with emphasize on small/sample simulation techniques, in order to be suitable for time demanding nonlinear finite element modeling. The main purpose of this kind of analysis is to account for randomness and epistemic uncertainties of material input parameters and other structural properties in the nonlinear structural analysis.

The probabilistic software FReET [15] allows simulations of uncertainties of the analyzed problem basically at random variables level (typically in civil/mechanical engineering – material properties, loading, geometrical imperfections). The attention is given to those techniques that are developed for analyses of computationally intensive problems; nonlinear FEM analysis is a typical example. Stratified simulation technique Latin hypercube sampling (LHS) is used in order to keep the number of required simulations at an acceptable level. This technique can be used for both random variables and random fields levels. Statistical correlation is imposed by the stochastic optimization technique called simulated annealing. Sensitivity analysis of the input parameters to resulting values is based on nonparametric rank-order correlation coefficients.

The procedure can be briefly outlined: Random input parameters are generated according to their PDF using LHS sampling. Samples are reordered by the Simulated Annealing approach in order to match the required correlation matrix as closely as possible. Generated realizations of random parameters are used as inputs for the analyzed function (computational model). The solution is performed many times and the results (structural response) are saved. At the end of the whole simulation process the resulting set of structural responses is statistically evaluated. The results are: estimates of the mean value, variance, coefficient of skewness and kurtosis, and the empirical cumulative probability density function estimated by an empirical histogram of structural response.

This basic statistical assessment is visualized through the "Histograms" window. It is followed by reliability analysis based on several approximation techniques: (i) the basic estimate of reliability by the Cornell safety index, (ii) the curve fitting approach applied to the computed empirical histogram of response variables and (iii) the simple estimate of probability of failure based on the ratio of failed trials to the total number of simulations.

State-of-the-art probabilistic algorithms are implemented to compute the probabilistic response and reliability. The main features of the FReET software are:

1) Stochastic modeling (inputs)

- Direct connectivity to the nonlinear analysis input data
- Friendly Graphical User Environment (GUE)

- 30 probability distribution functions (PDF), mostly 2-parametric, some 3-parametric, two 4-parametric (Beta PDF and normal PDF with Weibullian left tail)
- Unified description of random variables optionally by statistical moments or parameters or a combination of moments and parameters
- PDF calculator
- Statistical correlation (also weighting option)
- Categories and comparative values for PDFs
- Basic random variables visualization, including statistical correlation in both Cartesian and parallel coordinates

2) Probabilistic techniques (solution)

- Crude Monte Carlo simulation
- Latin Hypercube Sampling (3 alternatives)
- First Order Reliability Method (FORM)
- Curve fitting
- Simulated Annealing
- Bayesian updating

3) Response/limit state function (evaluation)

- Numerical form directly connected to the results of nonlinear FE analysis
- Multiple response functions are assessed in same simulation run

4. Identification of material parameters by inverse analysis

The key parameters for the nonlinear fracture mechanics modeling are fracture energy [16], tensile and compressive strength, and modulus of elasticity of the structural material, which are generally not well known, but crucial for a successful computer simulation of the structural response. For identification of fracture mechanical parameters of concrete from experimental results of three-point bending tests on notched-beam specimens an inverse method based on artificial neural networks has been introduced by Lehký et al. [17], [18].

Determination of parameters values is performed by inverse analysis using artificial neural network based method. The background of the inverse analysis is finite element method model which is used for numerical simulation of three-point bending fracture test; the model was created in ATENA software. Subject of identification are the basic three fracture mechanical parameters of the concrete model: modulus of elasticity, tensile strength, fracture energy. These material model parameters are considered as random variables described by probability distribution. The rectangular distribution was chosen as the lower and upper limits represent the bounded range of physical existence. The variables are then simulated randomly based on the small-sample simulation Latin Hypercube Sampling. Multiple calculation of deterministic computational model using random realizations of material model parameters is performed, and statistical set of the virtual structural response is obtained. Random realizations and the corresponding responses from the computational model serve as the basis for the training of the neural network [19]. After the training the neural network is ready to solve the main task: To provide the material parameters for which the numerical simulation will result in the best agreement with the provided experimental data. This task is performed by means of the simulation of the neural network using measured response as an input. It results in a set of identified optimal material input parameters. The last step of the procedure is results verification – calculation of computational model using identified parameters and comparison with the measured data. To obtain statistical characteristics of material parameters inverse analysis is performed for each specimen (L–D diagram) individually. The set of identified values is obtained as the result of individual identification and can be assessed statistically as it is usually done for experiments.

5. Sustainability of reinforced concrete engineering structures

Sustainability of civil engineering structures regards reliability and durability issues, both of which are among the decisive structural performance characteristics [20]. Assessment, together with inspection, maintenance and/or replacement plans, should guarantee the correct performance of a structure (e.g. a bridge) during its intended or residual service life. This time t_S is defined as the assumed period of time, obtained after an assessment process, for which the structure will continue to serve its purposes. When t_S elapses, a new assessment is requested. The intended service life, agreed by the client and justified on the basis of socio-economic criteria, is the design life t_D. This is decided on a case by case basis. An assessment process must always define the reliability (or safety) level of the particular limit state, the residual service life inherent to the final results and the conclusions, decisions and optimization to be carried out for the bridge under consideration

The program FReET-D (FReET Deterioration/Degradation Module [21]) was developed in order to support the life-cycle, sustainability and performance-based assessment and design of reinforced concrete structures. It is an extension to the probabilistic software FReET, and contains a deterioration module for statistical, sensitivity and reliability assessment of degradation effects of reinforced and pre-stressed concrete structures. Number of recognized models for concrete carbonation and deterioration as well as reinforcement corrosion are included. This deterioration module extends abilities of the basic program FReET to degradation effects assessment. It provides modeling of degradation phenomena in concrete structures, assessment of service life and assessment of reliability measure.

The (stochastic or deterministic) results can be directly used as inputs in ATENA or SARA programs in order to account for material deterioration and degradation and to perform life-cycle analysis or performance-based design or assessment of concrete engineering structures.

6. Nonlinear reliability assessment of transport structures

Probabilistic analysis is a general tool for safety assessment of civil engineering structures, in particular of concrete structures. In the fully probabilistic non-linear approach the structural resistance R_d is calculated by means of the probabilistic non-linear analysis. The classical statistical and reliability approach is to model material parameters as random variables with prescribed distribution function. The stochastic response requires repeated analyses of the structure with these random input parameters, which reflects randomness and uncertainties in the input values [22].

In this approach the resistance function $r(r)$ is represented by nonlinear structural analysis and loading function $s(s)$ is represented by action model. Safety can be evaluated by the reliability index β, or alternatively by failure probability P_f taking into account all uncertainties due to random variation of material properties, dimensions, loading, and other.

Probabilistic analysis based on the non-linear numerical simulation includes following steps:

1) Numerical model based on non-linear finite element analysis. This model describes the resistance function $r(r)$ and can perform deterministic analysis of resistance for a given set of input variables.
2) Randomization of input variables (material properties, dimensions, boundary conditions, etc.). This can also include some effects of actions, which are not in the action function $s(s)$ (for example pre-stressing, dead load etc.). Random properties are defined by random distribution type and its parameters (mean standard deviation, etc.). They describe the uncertainties due to statistical variation of resistance properties.
3) Probabilistic analysis of resistance and action. This can be performed by stratified method of Monte Carlo-type of sampling, such as LHS sampling method. Results of this analysis provide

random parameters of resistance and actions, such as mean, standard deviation, etc. and the type of distribution function for resistance.
4) Evaluation of safety using reliability index β or probability of failure.

Probabilistic analysis can be also used for determination of design value of resistance function $r(r)$ expressed as R_d. Such analysis involves the steps 1) to 3) above and R_d is determined for required reliability β or failure probability P_f.

In order to make the application of the probabilistic non-linear analysis user-friendly, special software tool has been developed by the authors and their co-workers The resulting software SARA (Structural Analysis and Reliability Assessment) integrates the above mentioned software tools ATENA and FReET. It is equipped with a user friendly shell SARA Studio, which leads the user interactively through the modelling and randomization process of the solved problem as described above. All the above mentioned features (or selected) of the involved programs including deterioration/degradation phenomena are utilized in the reliability analysis and sustainability and performance-based assessment of concrete structures [23], [24], [25], [26], [27], [28].

7. Examples of sustainability and nonlinear reliability assessment of transport structures

7.1. Bridge Kristienberg, Stockholm

The newly constructed intersection of highway No. E4 is the main traffic node for Kungsholmen Island in Stockholm, Sweden. A reinforced concrete bridge with two-span frame structure is a part of the intersection. Total bridge length is 26 m, bridge deck has width of 7 m. There are two lateral abutments and one intermediate support. The abutments have a significant inclination with respect to road axis and they have a different shape and size. Bridge has been subjected to several loading tests under the loads as two or four heavy trucks (weight of one truck is between 24 and 29 tons). Reliability assessment of this bridge has been performed and presented in [29].

Figure 1. Bridge Kristienberg.

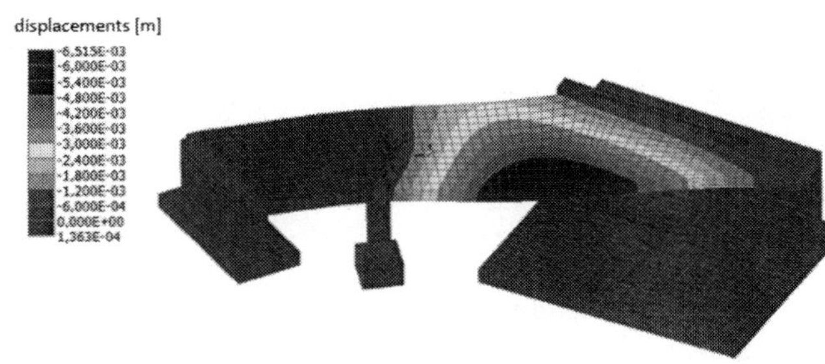

Figure 2. Simulated vertical displacement (colour scale) and cracks (short black lines) in the structural model loaded by four heavy trucks.

The numerical model of the bridge was created in ATENA 3D. Parameters of concrete are based on laboratory experiments, e.g. compressive strength 41.2 MPa. Steel reinforcement (class B500B) is modelled according to the project documentation, shear reinforcement is modelled as smeared reinforcement. For the model consisting of 11351 finite elements both deterministic and probabilistic analyses were performed. For the probabilistic approach, basic material parameters of concrete and reinforcement were randomized, total number of 20 samples-simulations has been calculated. Displacements were compared between experimental and numerical results.

The simulations were in very good accordance with the experimental results. Moreover, it was possible to predict and asses crack width in critical parts of the bridge, i.e. above the central pillar in this case (see black lines in Figure 2).

7.2. Rohrbach arch bridge, Mattersburg

Simulation of an old masonry railway bridge was presented in [30]. The bridge consists of five arch elements and was built between 1845 and 1847 in Rohrbach bei Mattersburg in Austria. To investigate the sustainability and load bearing capacity of the bridge, model of one typical arch was build in laboratory and tested [31]. The experimental arch has a clear span of 2.96 m, a rise of 1.00 m and for technical reasons a thickness of 0.60 m. In order to replicate the abutment portion of the modelled arch and the associated stiffness springing, these were attached to the steel abutments.

The experiments were reproduced by ATENA 2D model. The individual bricks and the mortar were modelled by a simplified micro modelling. The mortar joints were modelled mostly with so called "Interface Material". The monitoring points in the ATENA 2D model were placed at the same positions as the measuring points of the laboratory arch.

To adjust the material parameters as close as possible to the laboratory arch, probabilistic simulation was performed and parameters of bricks and mortar were dispersed. The initial material parameters were taken from the PhD thesis [31]. By comparing load-displacement diagrams optimal material parameters were identified.

According to the laboratory test, the arch in the model was loaded to four different damage levels. Model shown in Figure 4 is at damage level 1 where bricks on the left side of the arch are already destroyed and model does not account them in the analysis.

Figure 3. Rohrbach arch masonry bridge.

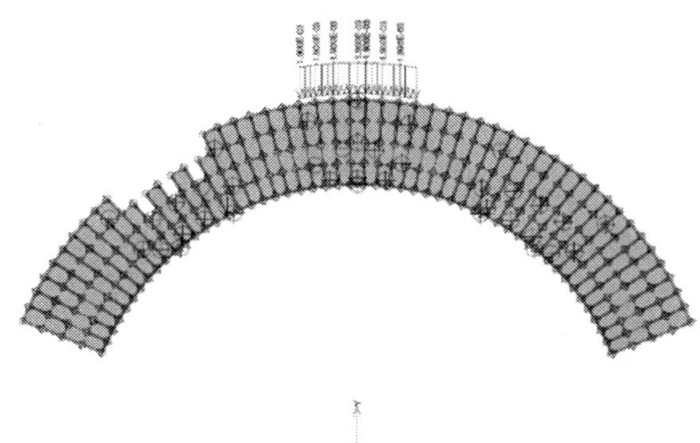

Figure 4. ATENA 2D model at damage level 1.

Comparison of numerical and experimental results on the small scale model documented that the computer model can reproduce real structural behaviour. The material parameters for all structural components are of crucial importance. The loading history and damage level of the structure should be accounted as well. If such preconditions are fulfilled, behaviour and response of the real (full scale) bridge under various actions and its damage under limit conditions can be successfully reproduced by the numerical modelling. Corresponding tests are hardly possible at the real bridge or full scale laboratory specimen.

8. Conclusions

- Assessment of performance and sustainability of civil engineering structures based on randomization of advanced nonlinear computer analysis is presented.
- Structural response, reliability and sustainability have been investigated on different types of transport structures made from various materials using the above mentioned methodology and tools.
- Appropriate identification of material and structural parameters of the numerical model is crucial for the virtual response and failure modelling of structures and structural elements.

- The probabilistic analysis could predict and assess material damage (crack width) in critical parts of the structure.
- Two examples of the numerical simulation for different types of bridge structures are shown in order to document feasibility of the presented approach.
- Very good agreement of simulation with the measured data and experimental results has been obtained.
- The examples demonstrate that the presented techniques can be applied for complex transport structures.
- The results from the nonlinear probabilistic analysis can serve as a rational basis for design, assessment and maintenance of concrete structures.

Acknowledgements
Research reported in this paper was partially performed within the Czech Science Foundation (GA CR) project 16-04132S "Epistemic uncertainty of crack models in reinforced concrete structures" and supported by Competence Centres program of Technology Agency of the Czech Republic (TA CR), project Centre for Effective and Sustainable Transport Infrastructure (No. TE01020168).

References

[1] FIB 2013 *FIB Model Code for Concrete Structures 2010* (Berlin: Wilhelm Ernst & Sohn) p 434
[2] Červenka V 2013 Reliability-based non-linear analysis according to fib Model Code 2010 *Structural Concrete* **14** pp 19-28
[3] Červenka V, Červenka J and Pukl R 2002 ATENA – A tool for engineering analysis of fracture in concrete *Sadhana - Academy Proceedings in Engineering Sciences* **27**(4) pp 485-92
[4] Tej P, Vacek V, Kolísko J and Čech J 2014 Computer Nonlinear Analysis of the Formation and Development of Cracks in a Reinforced Concrete Slab Loaded by a Planar Uniform Load *Key Engineering Materials* **606** pp 229-32
[5] Pukl R, Červenka V, Červenka J and Novák D 2012 Computer Simulation of Concrete Bridges *Proc. IABMAS 2012 (Stresa, Italy)* (London: CRC Press Taylor & Francis) pp 3680-4
[6] Červenka J, Proske D, Kurmann D and Červenka V 2012 Pushover analysis of nuclear power plant structures *Proc. FIB Symposium (Stockholm)* pp 245-48
[7] Tej P, Vacek V, Kolísko J and Čech J 2014 Computer Nonlinear Analysis of the Formation and Development of Cracks in a Travertine Stone Pavement Exposed to Bending Stress Caused by a Single Load *Key Engineering Materials* **606** pp 225-28
[8] Červenka V, Červenka J and Šístek M 2011 Verification of global safety assisted by numerical simulation *Proc. FIB Symposium (Prague)*
[9] Červenka V, Doležel J and Novák D 2010 Shear Failure of Large Lightly Reinforced Concrete Beams: PART II – Assessment of Global Safety of Resistance *Proc. 3rd FIB International Congress (Washington DC)*
[10] Hordijk D A 1991 *Local Approach to Fatigue of Concrete* (Delft University of Technology)
[11] Bažant Z P and Oh B H 1983 Crack Band Theory for Fracture of Concrete *Materials and Structures* **16** pp 155-77
[12] Červenka V, Pukl R, Ožbold J and Eligehausen R 1995 Mesh Sensitivity Effects in Smeared Finite Element Analysis of Concrete Fracture *Proc. FRAMCOS2 (Zurich) Vol. II* pp 1387-96
[13] Menétrey P and Willam K J 1995 Triaxial failure criterion for concrete and its generalization *ACI Structural Journal* **92** (3) pp 311-8
[14] Červenka J and Papanikolaou V K 2008 Three Dimensional Combined Fracture-Plastic Material Model for Concrete *Int. Journal of Plasticity* **24** (12) pp 2192-2220
[15] Novák D, Vořechovský M, Rusina R 2003 Small-sample probabilistic assessment – software FREET *Proc 9th Int. Conf. on Applications of Statistics and Probability in Civil Engineering – ICASP 9* (Rotterdam: Millpress) pp 91-6
[16] Bažant Z P and Planas J 1998 *Fracture and Size Effect in Concrete and other Quasibrittle*

Materials (Boca Raton: CRC Press)

[17] Lehký D, Keršner Z and Novák D 2014 FraMePID-3PB software for material parameter identification using fracture tests and inverse analysis *Advances in Engineering Software* **72** pp 147–54

[18] Novák D and Lehký D 2006 ANN Inverse Analysis Based on Stochastic Small-Sample Training Set Simulation *Engineering Application of Artificial Intelligence* **19** pp 731-40

[19] Cichocki A and Unbehauen R 1993 *Neural networks for optimization and signal processing* (Stuttgart: John Wiley & Sons Ltd. & B.G. Teubner)

[20] Teplý B, Chromá M, Rovnaník P and Novák D 2012 Role of modeling in probabilistic durability assessment of concrete structures *Proc. Life-Cycle and Sustainability of Civil Infrastructure Systems (IALCCE)* ed A Strauss, D M Frangopol and K Bergmeister (CRC Press, Taylor & Francis Group)

[21] Novák D, Vořechovský M and Teplý B 2014 FReET: Software for the statistical and reliability analysis of engineering problems and FReET-D: Degradation module *Advances in Engineering Software* **72** pp 179-92

[22] Bergmeister K, Novák D, Pukl R and Červenka V 2009 Structural assessment and reliability analysis for existing engineering structures, theoretical background *Structure and Infrastructure Engineering* **5** (4) pp 267-75

[23] Strauss A, Hoffmann S, Wendner R and Bergmeister K 2009 Structural assessment and reliability analysis of existing engineering structures, applications on real structures *Structure and Infrastructure Engineering* **5** (4) pp 277-86

[24] Novák D, Pukl R, Teplý B, Lehký D, Šomodíková M and Doležel J 2013 Reliability and durability of concrete structures under complex conditions *Proc. CONSEC 2013 (Nanjing)*

[25] Strauss A, Bergmeister K, Hoffmann S, Pukl R and Novák D 2008 Advanced life-cycle analysis of existing concrete bridges *Journal of Materials in Civil Engineering ASCE* **20** pp 9-19

[26] Novák D, Teplý B, Pukl R and Strauss A 2012 Reliability assessment of concrete bridges *Proc. Int. Conf. on Bridge Maintenance, Safety, Management, Resiliance and Sustainability - IABMAS (Stresa)* (London: CRC Press Taylor & Francis) p 160

[27] Novák D and Pukl R 2012 Simulation of random behavior of engineering structures: From parameters identification to reliability assessment *Proc. Life-Cycle and Sustainability of Civil Infrastructure Systems - IALCCE (Vienna)* (London: CRC Press Taylor & Francis) p 446

[28] Novák D, Fekete L and Pukl R 2012 Statistical analysis of crack width by virtual modeling of reinforced concrete beams *Proc. Numerical Modeling Strategies for Sustainable Concrete Structures - SSCS (Aix-en-Provence)* pp 75-6

[29] Pukl R, Sajdlová T, Řoutil L, Novák D and Šeda P 2016 Case study - Nonlinear reliability analysis of a concrete bridge *Proc. Maintenance, Monitoring, Safety, Risk and Resilience of Bridges and Bridge Networks (Foz do Iguaçu)* (Taylor & Francis Group) pp 2503-07

[30] Mendlig K, Proske D, Strauss A, Krawtschuk A, Zeman O 2016 Structural behavior of damaged arch bridges *Proc. Maintenance, Monitoring, Safety, Risk and Resilience of Bridges and Bridge Networks - IABMAS (Foz do Iguaçu)* (Taylor & Francis Group) pp 334-342

[31] Krawtschuk A 2014 *Optimierung von Monitoringkonzepten für die Erhaltungsplanung von Bogenbrücken* (University of Natural Resources and Life Sciences Vienna)

IOP Conf. Series: Materials Science and Engineering **236** (2017) 012110 doi:10.1088/1757-899X/236/1/012110

Utilization of BIM for automation of quantity takeoffs and cost estimation in transport infrastructure construction projects in the Czech Republic

Stanislav Vitásek[1] and Petr Matějka[1]

[1]CTU in Prague, Faculty of Civil Engineering, Department of Management and Economics in Civil Engineering, Thakurova 7/2077, Prague 166 29, Czech Republic

E-mail: petr.matejka@fsv.cvut.cz

Abstract. The article deals with problematic parts of automated processing of quantity takeoff (QTO) from data generated in BIM model. It focuses on models of road constructions, and uses volumes and dimensions of excavation work to create an estimate of construction costs. The article uses a case study and explorative methods to discuss possibilities and problems of data transfer from a model to a price system of construction production when such transfer is used for price estimates of construction works. Current QTOs and price tenders are made with 2D documents. This process is becoming obsolete because more modern tools can be used. The BIM phenomenon enables partial automation in processing volumes and dimensions of construction units and matching the data to units in a given price scheme. Therefore price of construction can be estimated and structured without lengthy and often imprecise manual calculations. The use of BIM for QTO is highly dependent on local market budgeting systems, therefore proper push/pull strategy is required. It also requires proper requirements specification, compatible pricing database and software.

1. Introduction

Transport infrastructure construction projects are one of widely debated topics in public discussion. There has been substantial discontentment with both quality and costs of new construction works or repairs of current roads and motorways. The costs of such construction projects or future maintenance costs can be influenced primarily in pre-investment and investment phase [1]. That is a good reason to pay detailed attention to these phases of project life cycle. It is in the best interest of any investor to know the costs of construction project in as much detail as needed, in order to make economically sound decisions concerning the development of construction work. When we compare construction to other branches of industry, it is rather specific because every construction work is a unique product. Therefore, it is difficult to estimate construction costs very precisely from design. Even with the latest tools, it is next to impossible to reach a perfect match of real and estimated costs of construction. Real costs are obvious only after the project is finished. But there are methods of bringing the real and estimated costs closer together. Building Information Modeling (BIM) is one of them. This paper deals with current options of using a BIM model for quantity takeoffs in the Czech Republic (CR) and connecting this model with public price databases of Directorate of Highway and Motorway in CR (DHM).

Current system of estimating price in Czech construction companies is highly labour-intensive and there is much room for error. Making quantity takeoffs from 2D documentation, which is still very common, is obsolete when compared to modern tools. The reasons for using inadequate methods are many, and they are not specific only to the CR – it can be inadequately prepared legislation or incompatibility of model and price system.

2. Estimated costs of transportation infrastructure in the CR

The method of estimating costs of transportation infrastructure projects in the CR depends on the phase of investment process, i.e. the level of detail given in entry data for estimation. There are four different phases of project documentation, all being parts of the investment and legislative procedure which leads up to a building permit.

In the first phase (project beginning), the estimated cost of construction is stated in the investment plan (IP), it is established according to Database of cost estimate (DCE). DCE establishes so-called base cost of realizing individual construction objects (tunnels, groundwork, etc.). Base cost always responds to criteria of Czech National Standards, and can be modified by so-called expert modification, which only serves for conforming the norms to situation in construction zone. The process of estimating total cost of investment plan is listed here:

- major construction objects are estimated according to database of base costs,
- other construction objects are estimated according to percentual rates,
- base costs are modified, if needed,
- risk assessment for all construction objects is carried out,
- total price estimate for investment plan is finalized, possible conversion to current price system. [2]

When the investor authorizes the investment plan, project documentation is prepared for Zoning (ZO) and Building permit (BP). Certified public institution issues these documents. The estimated price during this phase follows Proceedings of aggregated items for valuation of construction (PAI). PAI delimits major categories of construction object, construction costs of each category represent 80 % and more out of the total investment cost. These are motorways, roads, bridges, tunnels, walls, sewage, water conduit, gas conduit, tanks, noise protection, fencing (including subcategories of new construction and reconstruction for motorways, roads and bridges). The process (see Figure 1) of estimating total costs during ZO and BP is similar to the process of estimating the investment plan, only the databases differ in each phase.

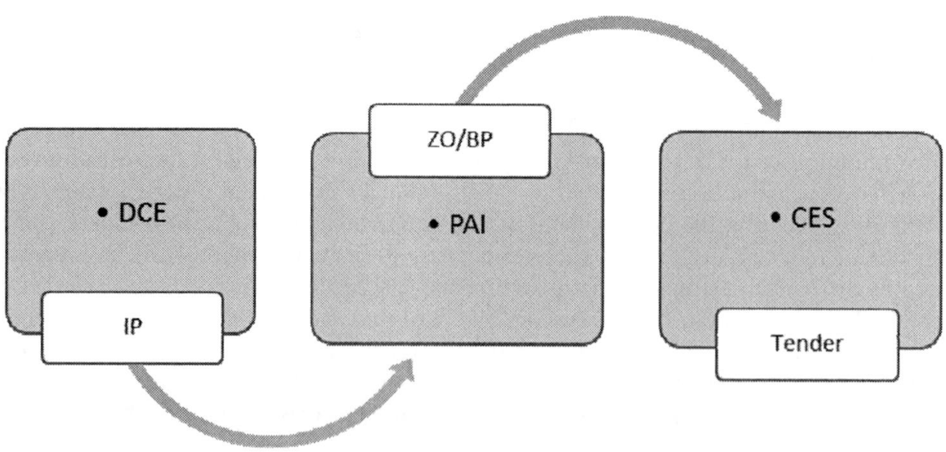

Figure 1. Estimate data and different databases used for estimate in phases of investment process (source: authors).

When a certified institution issues a permit for realization of construction work, drafts of competition are created, in order to choose general contractor. Then the most detailed estimate follows, one that is based on Classification of Engineering Structures (CES). CES specifies the standard of transport constructions in great detail, and also products needed to reach standard quality of construction. CES enables the estimator to compile a quantity takeoff according to the character of construction, and the contractor's demands regarding estimation process. Quantity takeoff defines all contractor's demands – estimate of required amounts of work, supplies and services. It gives full information about types and amounts of supplies and services, therefore it is used as input for subcontractors' offers. The takeoff states manner of estimation of construction works, as defined by a specific list of prices and tariffs. CES is a summary of all individual construction objects paired with proper code, description, unit of measure, and unit price. Unit price within CES is called „expert price". The total price estimated with use of CES is considered the most precise estimate possible. Figure 1 summarizes the whole process – from estimate of investment plan making use of DCE to the final estimate of contracting documentation based on CES [3].

3. Quantity takeoff and BIM

Quantity takeoff is a list of individual construction units as taken from project documentation. It states the required amounts of work, constructions, supplies and services rendered in units of measure. It is expected to be well arranged, structured and revisable [4]. In case of transport infrastructure construction projects, a supplier is contracted by Directorate of Highway and Motorway in CR (DHM) or Ministry of Transport. This engineering company creates quantity takeoff and creates all project documentation for all phases of the investment process.

Estimator creates quantity takeoff and estimation of construction work (from IP to drafts of competition). Estimator is a specialized profession in civil engineering, expert who reads individual quantities of construction work as defined in project documentation and classifies them using a given database (DCE, PAI, CES) [5].

A large part of BIM model is made up of databases, which allow extracting information. Quantity takeoff may be one way of extracting information from a BIM model. It can also be varied in level of detail. The creator of information is also the creator of BIM model but the level of detail within the model can vary and ought to be specified in contract. Standard model requirements specify creating a model in one of the phases of project documentation, mostly there is no need to link the information in the model with a specific price estimate system. If BIM model is created with all the additional data to enable price estimate, it results in extrawork, and the price of such detailed model rises. However, most models already contain large amounts of data needed for price estimates. Unfortunately, these models are often incomplete, created by different contractors, and construction estimator is not supplied with the entire model, only its exported output (e.g. IFC or NWD). This leads to unfavourable loss of information. Construction estimator should specify requirements for additional information, so that model data can be used efficiently, especially when it concerns measurements of construction objects in BIM model. Such specification may be a part of model requirements stated in construction contract.

When BIM model is efficiently linked to cost estimate system via partially automated quantity takeoff, the estimator's role in the process is significantly simplified. Estimator ceases to be the creator of quantity takeoff, rather he becomes quantity takeoff reviewer. Estimator may use BIM model for reference and additional calculations – such that cannot be automated or are not part of model.

4. BIM model analysis with regard to quantity takeoffs

Due to current situation in transport infrastructure in the CR, there have not been many BIM-based models so far. Investment in BIM model is not supported, the contractors need to use their own resources to pay for BIM model. For the purpose of the research, models from Skanska AB were used. These models were originally used for different purposes than quantity takeoffs, namely for visualising and coordination. Skanska has offered several models for analysis (see Figure 2).

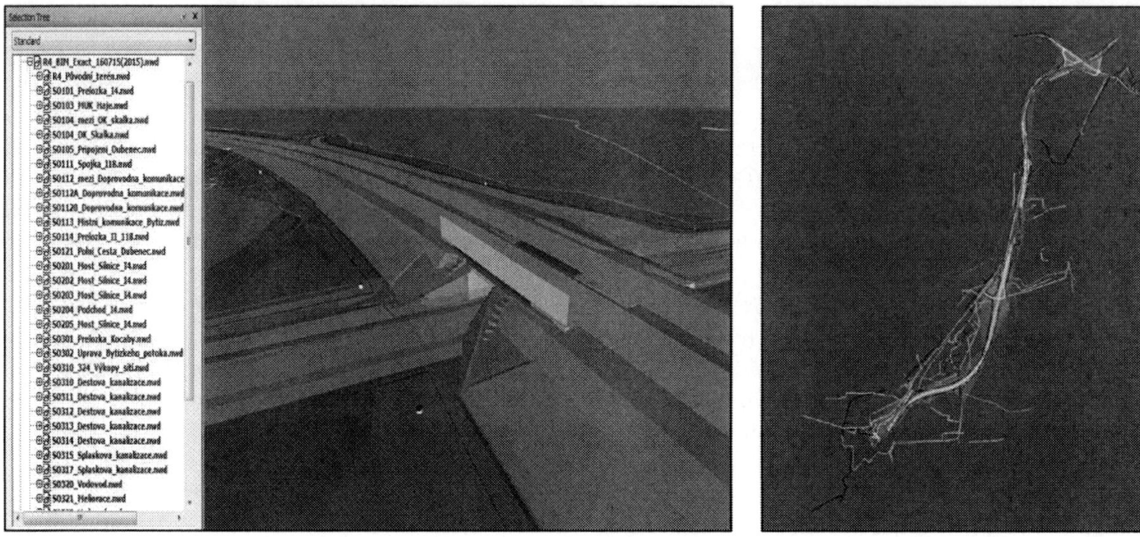

Figure 2. Layout drawing of motorway D4 in BIM (source: Skanska).

Models have been analyzed from two viewpoints. Analysis of model objects has been done to verify that the model contains necessary data for measurements or required information as such. The second aspect of analysis was the possibility to match individual objects and specific work or one unit of CES. Autodesk Naviswork Manage (further referred to as Navisworks) was used to create CES quantification workbook structure, as shown in Figure 3.

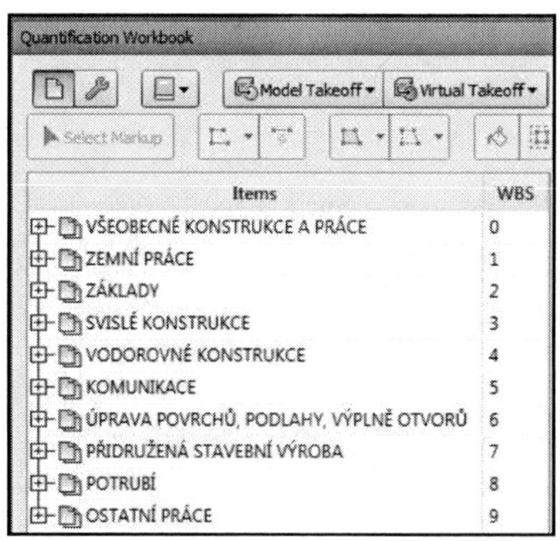

Figure 3. CES quantification workbook structure adapted to BIM software (source: authors using Navisworks).

Objects in BIM model have been classified. Analysis, which followed, showed that exported models, currently created for use in the CR, do not contain necessary information, not even for partial automation of quantity takeoff. The major reason for this situation is that models are passed via IFC or NWD format for use in Navisworks. Original model information is lost in the process because it is not considered necessary for coordination. Moreover, the medium does not enable recovery of data. The second important reason for data losses is high degree of aggregation of objects contained in a model. Road as an object is not differentiated in its partial levels.

Current state of models allows for price estimate only in the beginning phases of investment process. DCE and PAI allow us to generate quantity takeoff, making use of integrated software tools. Unfortunately, full automation is currently impossible.

5. Case studies

For the purpose of the research, the consistency of the theoretical findings and practice has been examined on two case studies.

5.1. D4 (Skalka) motorway

The first case study focuses on the completion of a nearly five kilometre section of D4 motorway (Skalka) in the Central Bohemian Region. It is the first transport infrastructure project in Czech Republic, where the contractor uses the BIM model during construction. Specifically, it is the construction of multilevel intersection of D4 motorway with road II/118 nearby Háje municipality and the construction of a new supporting II. class road. Table 1 shows basic information about the construction.

Table 1. Basic information about completion of D4 motorway (source: [7]).

Beginning of construction	04/2014	Opening	09/2017
Length of road	4 788 m	Time of construction	2 years
Term of guarantee	120 months	Winning bid	417 485 287 CZK

5.2. D1 (Přerov – Lipník nad Bečvou) motorway

The second case study focuses on the construction of 14 kilometres long section of D1 motorway between towns Lipník nad Bečvou and Přerov in Olomouc Region. There will be 24 bridge structures, 3 multilevel intersections and over 7 noise barriers or support walls. The construction of the new section of D1 includes, in addition to the main route, the relocation of I.-III. class roads and paths in total length of almost four kilometers [8]. Table 2 shows basic information about the construction.

Table 2. Basic information about construction of new D1 motorway section (source: [9]).

Beginning of construction	08/2015	Opening	08/2018
Length of road	14 300 m	Time of construction	2 years
Term of guarantee	120 months	Winning bid	2 717 043 775 CZK

5.3. Case studies conclusion

The main use of BIM model in both cases is mainly for guidance of GNSS guided construction machinery for excavation and laying works. Unfortunately, information models were not used to generate quantity takeoffs of individual construction works [6, 8].

Winning bid for this stretch of highway is roughly fifty per cent lower than original estimate made according to CES. Price tendering in the range of 50 to 70 % of the presumed price, determined with the use of CES, represents the current situation in the Czech infrastructure construction projects. The possible sources of difference between corporate and CES prices from CES technical perspective may be (especially, but not exclusively):

- different profit rates (CES calculates with higher profit rates),
- achievable rebates for the purchase of building materials (corporate prices are lower),
- lower machines costs, when compared to CES (machines are used even when already depreciated).

Other possible market and economical sources are not mentioned. It is important to state that such sources might even be marginal when compared to those mentioned. To identify these other sources, further and more thorough analysis is necessary. Such analysis is not part of this paper.
Further difference between unit prices in cost estimate systems is caused by quantity discounts in case of construction material purchases made by large companies. Building machinery costs in price estimate systems also count on depreciation costs. When the machinery is already depreciated, machinery costs are lowered to mere operating costs. The difference is also aggravated by use of different type of machinery (lower or higher output, universal machinery, etc.). This also influences evaluation of economic effectiveness of projects – when acquisition costs in a project are high, the probability of positive evaluation of project becomes substantially lower.

Practical application of these findings made use of BIM in order to establish estimated price for construction mentioned in chapters 5.1 and 5.2. Quantity takeoff, which was created with the use of the model, was then utilized to establish costs based on DCE and PAI. The output is presented in Figure 4.

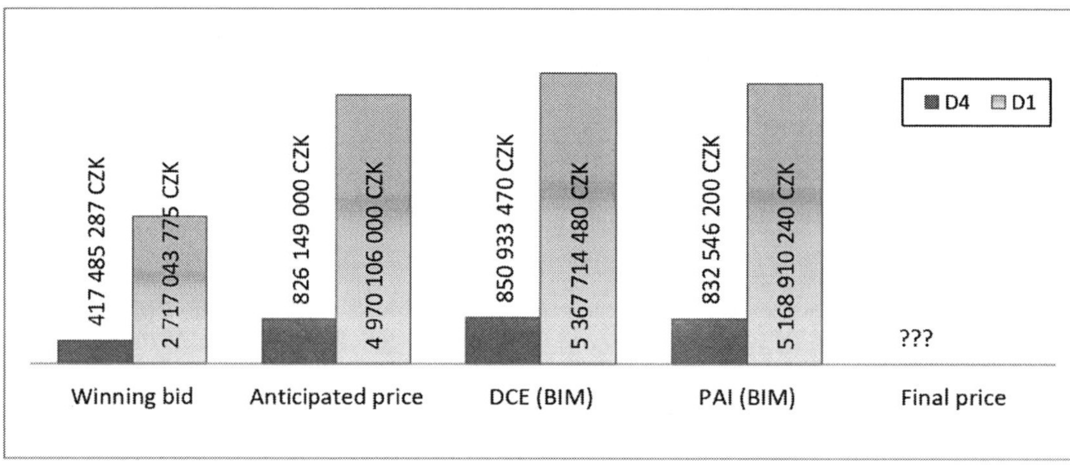

Figure 4. Comparison of price estimates for stretch of highway D4 and D1 (source: authors with use of [6, 7]).

The difference between estimated price as stated by DHM and price obtained by matching cost estimate database with BIM model is roughly under 10 %. There are several possible factors that may have caused the difference, such as the interference of expert commission in total price estimate – for one thing due to potential risks, for another to conform some of the prices to specifics of construction zone. However, it is more interesting to compare DCE and PAI estimates. The substantial difference is likely to have originated in lower level of aggregation in units of DCE, which results in a more precise estimate of construction costs.

6. Conclusion

As the text of conducted studies obviously states, the issue of estimating construction costs in transport constructions is very broad and interesting from its economic-technical aspect. New solutions, which support precision in calculating expected costs, are fully in accordance with public interest, especially in such large and financially demanding constructions as demonstrated [10]. BIM tools and their proper usage are a potential solution to one of the most topical dilemmas of cost estimates in construction projects.

Analysis of BIM models provided by private subject unfortunately pointed out inadequacy of marketed models for automated evaluation, according to price systems. Models lack measurements of objects and objects themselves are highly aggregated. Case studies of a stretches of D4 and D1 motorways connects BIM model to cost estimates in systems DCE and PAI. Implementing appropriate tools was rather unobjectionable, supportive BIM tools made it possible to establish price estimates of D4 and D1 motorway stretch construction. The estimates were compared to expected price published by Directorate of Highways and Motorways, as created with use of CES database. The difference between any pair of prices was not higher than roughly 10 %.

Fully automated generation of quantity takeoffs with price estimates based on BIM models (from DCE o CES) needs to go through these phases:

- specify requirements for designers (method of proceeding, specification of commissioner's requirements),
- implement price databases used by DHM in a BIM-compatible database,
- software extension to convert generated data into price databases.

For the successful adoption of BIM in cost estimation in transport infrastructure projects, proper supportive environment is necessary. This is especially:

- specification and standardization of requirements for quantity takeoffs for both client and contractor in typical project stages in standardized digital format, according to defined Common Data Environment (CDE) for transport infrastructure projects,
- creation and development of freely accessible manual for methods of quantity takeoffs and estimation using BIM in specified CDE in accord with Czech long-term BIM implementation strategy, Czech legislation and best practice,
- consistent testing and improvement of best practice, preferably government-driven,
- implementation of innovative methods into education (both tertiary and life-long).

Since the use of BIM for quantity takeoffs is highly dependent on market and pricing system, market push (especially from software developers and pricing systems) and pull strategy is very important.

One of the possibilities, which would allow easier implementation of price databases into BIM model structures, would be general revision of estimating costs of construction production in the Czech Republic. It would be suitable to apply foreign ways of classification and estimation of construction costs, along with utilization of best practices in calculation of quantity takeoffs. CES, which is used in the Czech Republic, classifies units according to type of work, regardless of kind of construction. Foreign classifications, on the other hand, sort construction works according to kind of construction. Another subpart of a more complex solution may also be the necessity to recalculate unit prices in classification systems so that the differences between price estimate and winning bid are not so marked.

Acknowledgements
The article has been created with the support of the Technology Agency of the Czech Republic (TACR), project CESTI, number TE01020168.

References
[1] Hromada E, Schneiderová Heralová R and Johnston H J 2014 Cost Structure of the Highway Projects in the Czech Republic *Procedia Engineering* **123** pp 496-503
[2] Ředitelství silnic a dálnic 2016 Technické předpisy (online) https://www.rsd.cz/wps/portal/web/technicke-predpisy/soupisy-a-ceny-praci
[3] Ministerstvo dopravy ČR 2012 *Směrnice č. V-2/2012 upravující postupy Ministerstva dopravy, investorských organizací a Státního fondu dopravní infrastruktury v průběhu přípravy a realizace investičních a neinvestičních akcí dopravní infrastruktury, financovaných bez účasti státního rozpočtu*

[4] Schneiderová Heralová R 2013 *Oceňování v rámci výstavbového projektu: (propočty, položkové rozpočty)* (České vysoké učení technické v Praze, Fakulta stavební)
[5] Tomek T and Vitásek S 2016 Improvement of economic effectiveness of road highway projects *Procedia Engineering* **164** (Amsterdam: Elsevier)
[6] Skanska a.s. 2016 D4 Skalka – křižovatka II/118 http://www.skanska.cz/cz/o-nas/bim/probihajici-projekty/ZST-Ceska-Lipa/
[7] České dálnice 216 Dálnice D4 http://www.ceskedalnice.cz/dalnice/d4/
[8] Skanska a.s. 2017 D1 Přerov – Lipník nad Bečvou http://www.skanska.cz/cz/O-nas/BIM/Probihajici-projekty/Dalnice-D1/
[9] České dálnice 2017 Dálnice D1 http://www.ceskedalnice.cz/dalnicni-sit/ve-stavbe/
[10] Matějka P 2014 The Importance of a Transport Infrastructure Construction for the Implementation of BIM *Int. Conf on People, Buildings and Environment (Brno)* pp 277-287

User costs as one of main advantages of precast concrete application in highway construction

Radan Tomek[1]

[1]Czech Technical University in Prague, Faculty of Civil Engineering, Thákurova 7, 166 29 Praha 6 – Dejvice, Czech Republic

E-mail: radan.tomek@fsv.cvut.cz

Abstract. Road user cost primarily refer to the monetized components of road (re)construction impacts, such as the user delay costs, vehicle operating costs, crash costs and emission costs. Objective of this paper is to analyze and appraise the advantages and benefits of the innovative prefabrication approach in contrast to traditional cast-in-place construction method. The goal is to reduce these additional costs borne by motorists and the community at-large as a result of road construction activity to their minimum through application of the prefabrication. Assessing two basic possible approaches to highway infrastructure construction - casting the road pavements and structures either in place or precast off the site - it can be concluded that the initial capital investment costs do not vary much. Substantial differences can be recognized when comparing their life-cycle costs and an extent to which their construction process affects the public, environment and the local economy. Prefabrication of any structure component off-site offers major construction time and user cost savings in comparison with the traditional cast-in-place methods of construction. Precast prestressed road pavements' technology and precasting bridges' parts and elements offers dramatic increase in durability, while it also substantially decreases construction time and resulting user costs.

1. Introduction

European highway infrastructure is aging at higher pace than expected as it is used by increasing volumes of road traffic. As an example to illustrate the above stated - volume of traffic on the backbone Czech highway "D1" approximately doubled between years 2010 and 2016. Even though an increase in traffic volumes on the rest of European highways is not that high, it is a steady reality in every country, only with a few years' pause as the result of world financial recession between 2008 and 2012, according to Eurostat [1]. Economic recovery since this hiatus in a long-term traffic volume growth can be tracked by another Eurostat's data on total number of vehicles, i.e. total car fleet in almost all of the EU member states which has grown over the last five years again [2].

1.1. Alternative means of transport

Changing the ways public uses personal means of transport is a very complex issue. So far it has been partly successful regarding long distance travel as a result of high capital investments into the fast train railroads. Otherwise lowering the numbers of passenger cars on highways for the short and mid haul was not achieved either. Regarding heavy transport, efforts to relieve overloaded highway network by shifting its volumes to the railroad so far mostly fail. This is mainly due to high demands on flexibility and speed of supply required by the manufacturing and retail industry and also due to various complications that transport companies must face while switching from road to track and back.

1.2. Growing need of road maintenance

European highway network is aging faster than expected during its planning and construction. Such an unexpectedly increasing usage and consequent hard wear of the whole road infrastructure requires big volumes of repair and reconstruction works. Logically, the needs of its maintenance grow proportionally to its increasing usage. At the same time, this maintenance must be performed on a continuous basis while serving steadily growing traffic flow with minimal disturbances. Road network users demand that this refurbishment and new construction is done faster and with limited road closures, traffic congestion, delays and detour complications. This applies for all the European highway networks in general, but for the high-traffic areas the most.

1.3. Proposing the problem solution

The current situation of steadily increasing demand for highway construction capacities can be without any doubt defined as extraordinary. Analyzing existing capacities of the construction industry in Europe, as of March 2017, they still have not got over 80% of their pre-recession volumes [3]. To answer the demand for investment and maintenance construction services, current capacities are evidently not sufficient. There is a solution for such an extraordinary capacitive demands in the analogy to the residential housing problem in the sixties and seventies: application of the prefabricated construction systems using the technology of precast prestressed concrete elements.

2. Precast prestressed concrete

Public highway administrations are challenged to find ways to build new and more durable routes and to restore those already existing quickly and with minimal impact to its users and with the repairs being as durable as possible. Precast concrete road pavement and bridge construction systems - cast off-site (in a plant or next to the site) and installed rapidly off the rush hours - have that potential to transform highway network construction and maintenance. Not only do they cut limitations to the traffic and enhance safety during construction, they also bring in the higher quality product meaning the increase in its durability.

To summarize above stated, precast prestressed concrete technology brings unmatchable durability for the newly constructed road pavements and bridges (for both – substructure and superstructure; tunnels are not addressed in this paper due to their relatively rare occurrence and technical specificity) and along with this feature also the lowest level of disruption to the traffic during its implementation.

2.1. Technology and its advantages

Precast concrete technology is not new as it had been used by the ancient Romans already. However it considerably developed since then from the composition of the concrete mix, through adding steel to increase tensile strength to new ways of casting and curing.

Main advantages of precast concrete over the cast-in-place (i.e. in target placement of the pavement or object) are generally as following. Precast pavement panels or bridges' elements can be cast and cured in a controlled environment at a precast plant, providing greater control over consistency of the concrete mix, procedures of vibration and proper curing. Precast objects reduce or eliminate curling, strength, and air-entrainment problems that are common with conventional concrete paving [4]. Importantly, precast objects are also being both prestressed during their plant production and post-tensioned during their installation, which is not possible for the cast-in-place approach.

Prestressing in the plant significantly improves performance (most importantly the tensile strength) by inducing compressive stresses in the panels. This effectively prevents cracking and strengthens them for better onsite manipulation preventing them from bending and torquing. This durability not only slashes maintenance costs over the life of the roadway but significantly cuts user costs through the reduction of the repair cycles.

As proved by testing project of the Texas Department of Transportation from 2002 [4] and suggested by El-Reedy [5] post-tensioned panels provide the same design life as thicker conventional concrete pavements using thinner slabs; therefore, a 200-millimeter post-tensioned pavement would have the same life as that of a 355-millimeter conventional pavement. Post-tensioning also increases durability by minimizing or even eliminating cracking, and ties the individual panels together, promoting load transfer between the panels. Additionally, post-tensioning not only reduces the required pavement thickness, but also greatly increases durability, lessening or even preventing cracking in the pavement [6]. This increases the life of the pavement and contributes to a reduction in maintenance costs.

3. Application

As mentioned before, there is a vast and successful experience with the prefabrication in Europe. In the Czech Republic, resp. Czechoslovakia, it is the same, mainly in the housing construction. Even though it is in decline recently, the expertise and skilled labor force is still available. This provides a solid basis for its new way of mass application.

3.1. Precast Concrete Pavement Systems

In precast concrete pavement system construction, adjacent panels are assembled sequentially and tied together onsite through either post-tensioning or cast-in-load transfer systems. Precast concrete pavement systems can be used for single-lane replacements, multiple-lane replacement (an additional lane may be needed to accommodate materials and equipment) or full-width road construction [7]. Using the precast concrete parts for the road's pavement speeds up the construction time two to three times against the traditional cast-in-place approach, if well organized. Such a pavement can be used, i.e. driven on the very next day after its installation as no curing of the newly laid road surface is further necessary.

3.2. Accelerated Bridge Construction

Precasting bridges' parts and elements offsite is also very beneficial as bridges are generally among the most expensive objects constructed and also serve as a natural bottleneck for the traffic flows and so speeding up the construction process is very beneficial. The precast technology is applicable and needed for both – replacement or rehabilitation of existing bridges and construction of the new ones. It is the main factor in the new approach to process these two construction needs called Accelerated Bridge Construction (ABC). ABC concentrates on the innovative planning, subordinates design to the technology, uses resources and techniques to accelerate the bridge construction and maintenance without compromising the safety anyhow. It is mainly about the new approach to the construction management system. However, our main concern lies in the possibilities and the role that the precast technology plays in the ABC concept. Similarly, with the precast pavements, it is mainly about cutting the construction time, i.e. the limitations to the existing traffic, and about increasing the durability due to higher quality of the factory-made concrete. Based on that, it does make sense to precast most of the bridge elements - a pile, pier column, pier cap, beam, deck and barrier/railing as well, i.e. all but column's footing, which is usually cast-in-place. Regarding the construction cost – the result is dependent on the quantities as when being high, they definitely turn the economics in favor of the prefabrication, unification and standardization. Regarding user costs, which might easily exceed those of construction if limiting the high traffic areas for a long time, precast is a clear winner (see the Case Study below).

3.3. Case study

As an example of the precast pavement technology and organizational approach it is this project organized by the Virginia Department of Transportation in fall 2009 illustrating its potential very well. The objective was to rehabilitate the 40-year-old exit ramp from highway I-66 to U.S. no.50 in

Virginia, USA, with the precast concrete pavement systems. It was a $5 million project to replace distressed pavement slabs in a high-traffic area outside Washington D.C., i.e. a heavy-traffic area.

Conventional repair with cast-in-place concrete would have required about 100 days with traffic congestion from lane closures, but the precast slab approach allowed closure of one lane at a time for about 35 nights of work and made all lanes available for rush-hour traffic. Every night's cycle included the removal of the existing concrete, preparation and grading of the roadway sub-base, and preparation and placement of the new pavement slabs [7].

3.4. Work zones safety

Reducing time of work zones' existence in the moving traffic also dramatically increases safety of everyone involved. It is mainly workers, who are fully exposed to the risk of collision with the consequences that are mostly fatal for them. Nevertxheless on average, 85% of deaths in work zones were to drivers and passengers in cars according to U.S Federal Highway Administration [8]. To illustrate the occurrence of these accidents, based on the same source, there were approx. seven hundred fatal accidents in the work zones in the USA in 2014 and slightly above 200 fatal plus non-fatal injuries in the UK [9].

4. User costs

Road user cost (RUC) is defined as the additional costs borne by motorists and the community at-large as a result of road construction activity. User cost primarily refer to the monetized components of road construction impacts, such as the user delay costs, vehicle operating costs, crash costs and emission costs. Above these, there are other off-site components such as noise, business and local community impacts which are hard to monetize since the factors that influence their computation are often site-specific and no generalized method or tool is yet able to determine them accurately [10]. Nevertheless, it is the software tool HDM-4 which is in Europe most widely used to assess the mentioned costs with relatively satisfying precision [11].

Qin and Cutler [13] developed the detailed methodology on the road user costs' calculation aiming to set incentive or disincentive compensation for contractors, quantify project-specific liquidated damages, select the ideal sequencing of a project, and forecast the long-term effects new construction will have on the traveling public. Gathering and maintaining local and up-to-date RUC data is also extremely important to any public highway administration, as underestimation of road user costs results in increased costs to the traveling public, while overestimation results in overpaying on incentives for early completion of construction projects. Obtaining proper RUC also helps to justify the deployment of new technologies that can accelerate the construction process for pavement and bridges [13]. Precast technology for pavements and accelerated bridge construction approaches comply with that perfectly.

But limiting the time of the road construction time does not only reduce the probability of work zone crashes, it also brings substantial cost savings to the road network users. Any road or even lane closure time brings substantial costs to the general public and business sector regarding loss of labor hours, lowered comfort, gasoline wastage and resulting decrease in overall society's production effectivity [12].

As mentioned in the Case Study, the difference in closure times with the precast and cast-in-place approach can be as big as days and weeks versus months and years. Possible size in all above mentioned types of savings and benefits is evident. Problem is that many European governments, respectively their highway network administrating bodies generally do not attribute that much importance to the user costs inflicted by the traffic limitations. This is most probably caused by their relative "invisibility" in comparison to the construction costs. It is out of question that high user costs have serious national economic impacts and deserve a lot more attention.

5. Conclusions

As indicated above, selecting the method to use for the highway (re)construction should be significantly influenced by the value of the corresponding user costs.

Traffic volumes continue to rise in the European highway infrastructure and its users increase their demands regarding its comfort and quality. Because the capacities of the construction industry are limited, utilization of rapid construction methods is becoming to be a necessity. Prefabrication, respectively precasting seems to offer all that is needed in the current situation:

- Higher pace of construction works to answer the need to cut the time of traffic flow limitation.
- Higher and stable quality due to better controlled production process in the dedicated facility and the benefits of pre- and post-tensioning.
- Long-term durability thanks to the higher quality material used.
- Less frequent maintenance of the critical parts of the network – highways' pavements and bridges.
- Higher economical effectiveness due to mass production of all standardized elements (mainly pavement panels and bridge elements) and achieved economy of scale.

In comparison with the very slow and labor intensive cast-in-place method, precast prevails in the current conditions and requirements of current highway network. Nevertheless, it has its opponents too – many designers and contractors. It is unpopular because of lower possibility to create every project unique and to achieve higher possible mark-up. Public demand for the quality and comfortable highway travel will hopefully move the state highway administrations to overcome their influence and let them exploit the potential that the precast prestressed method certainly has.

Acknowledgements
This research was conducted within the Centre for Effective and Sustainable Transport Infrastructure (CESTI), the project of the Competence Centres Program of Technology Agency of the Czech Republic (TA CR) number TE01020168.

References

[1] Statistical Office of the European Union – Eurostat 2016 *Road freight transport statistics* http://ec.europa.eu/eurostat/statistics-explained/index.php/Road_freight_transport_statistics
[2] Statistical Office of the European Union – Eurostat 2015 *Passenger cars in the EU* http://ec.europa.eu/eurostat/statistics-explained/index.php/Passenger_cars_in_the_EU
[3] Statistical Office of the European Union – Eurostat 2017 *Construction production (volume) index overview 03/2017* http://ec.europa.eu/eurostat/statistics-explained/index.php/Construction_production_(volume)_index_overview
[4] Merritt D K, McCullough B F and Burns N H 2002 Texas Tests Precast for Speed and Usability *Public Roads Magazine* **66**(1)
[5] El-Reedy M A 2009 *Advanced materials and techniques for reinforced concrete structures* (CRC Press, Taylor & Francis Group)
[6] Tayabji S, Ye D and Buch N 2013 *Precast Concrete Pavement Technology - The Second Strategic Highway Research Program Report S2-RO5-RR-1* (Transportation Research Board Washington D.C.) p 165
[7] Portland Cement Association 2010 Precast Concrete Pavement Systems Save Time, Cut Congestion on Highway Repair *Roads & Bridges magazine* **5** (Scranton Gillette Communications) https://www.roadsbridges.com/precast-concrete-pavement-systems-save-time-cut-congestion-highway-repair
[8] U.S. Department of Transportation - Federal Highway Administration 2015 *Facts and Statistics – Work Zone Safety* (Work Zone Management Program) https://ops.fhwa.dot.gov/wz/resources/facts_stats/safety.htm
[9] The Road Worker Safety Forum http://www.rowsaf.org.uk/
[10] Mallela J and Sadasivam S 2011 Work Zone Road User Costs - Concepts and Applications Chapter 2 *Work Zone Mobility and Safety Program* (Federal Highway Administration – U.S. Department of Transportation) https://ops.fhwa.dot.gov/wz/resources/publications/

IOP Conf. Series: Materials Science and Engineering **236** (2017) 012111 doi:10.1088/1757-899X/236/1/012111

fhwahop12005/sec2.htm

[11] Tomek R 2017 Improving Effectiveness of Public Spending on Transport Infrastructure *Periodica Polytechnica Architecture* **1**

[12] Federal Highway Administration – U.S. Department of Transportation 2017 *Traffic Congestion and Reliability: Trends and Advanced Strategies for Congestion Mitigation* https://ops.fhwa.dot.gov/congestion_report/chapter2.htm

[13] Qin X and Cutler Ch E 2013 *Review of Road User Costs and Methods* (South Dakota State University Department of Civil & Environmental Engineering, Brookings) p 124 http://www.mountain-plains.org/pubs/pdf/MPC13-254.pdf

IOP Conf. Series: Materials Science and Engineering **236** (2017) 012112 doi:10.1088/1757-899X/236/1/012112

Risk variables in evaluation of transport projects

Petr Vařbuchta[1], Hana Kovářová[1], Vít Hromádka[1] and Eva Vítková[1]

[1]Brno University of Technology, Faculty of Civil Engineering, Veveří 331/95, 602 00 Brno, Czech Republic

E-mail: petr.varbuchta@seznam.cz

Abstract. Depending on the constantly increasing demands on assessment of investment projects, especially assessment of large-scale projects in transport and important European projects with wide impacts, there is constantly increasing focus on risk management, whether to find mitigations, creating corrective measures or their implementation in assessment, especially in the context of Cost-Benefit analysis. To project assessment is often used implementation of certain risk variables, which can generate negative impacts of project outputs in framework of assess. Especially in case of transportation infrastructure projects is taken much emphasis on the influence of risk variables. However, currently in case of assessment of transportation projects is in Czech Republic used a few risk variables, which occur in the most projects. This leads to certain limitation in framework of impact assessment of risk variables. This papers aims to specify a new risk variables and process of applying them to already executed project assessment. Based on changes generated by new risk variables will be evaluated differences between original and adapted assessment.

1. Introduction

Transportation has always been very important part of the economy. Its importance is rooted in our society since ancient times. The trade starts on local level, gradually regional, national and transnational level, and couldn´t be so profitable and feasible without possibilities that transportation allows us. Equally transportation is important for people themselves. Only a few of us can imagine living in isolated place, without larger possibility to travel to work, to the family, or entertainment. Only just transportation, modern means of transportation and gradually development of transportation infrastructure allows this for all of us.

Investment directed to the transportation infrastructure reaching enormous amounts of financial sources. For this great emphasis the comprehensive evaluation of these investments in relation to expected effects in the context of benefits is important.

We actually know and use predominantly four types of transportation: by vehicle (road), by plane (air), by ship (water) and by train (railway). Road transportation has its own transportation network, which includes roads, highways, speedways etc. Air transportation uses point infrastructure of airports all around the world. Water transportation uses natural and artificial water surfaces, which are rivers, canals, dams, seas etc. And at last railway transportation uses its network of different types of rails, which includes local railways, major corridors etc.

For assessment of investment in transportation and for road transportation a lot of assessment methodologies exist. Some of them are the same, but some methodologies for roads have its specifics and differences. Actually there are tendencies to improve the methodologies of assessment with respect to business as usual, real opportunities of construction projects and their risk. This paper is aimed just to road transportation infrastructure and precisely at its risk and risk variables, which are used for investment projects assessment.

2. Literature review

As already mentioned above, building of transportation infrastructure is necessary in framework of sustainable development. Transportation infrastructure is intermediary of economic improvement. But it´s important to come out of sophisticated analysis, which can evaluate every benefits and costs of the transportation project. We must realize that only well infrastructure equipped regions have the sufficient potential for economic growth. Compared to that, isolated regions will never dispose by attractiveness for entering of new economical subjects to the local market. It´s simple, because these activities wouldn´t be profitable.

In assessment of infrastructure investment there are many default aspects and methodologies, but in business as usual there is analysis of costs, benefits and great consideration is taken to the risks. All this is included in the most used analytical tool, invented for assessment of large-scale projects with enormous impacts: Cost-Benefit Analysis (CBA). This analysis is used in assessment of every important transportation projects. Methodology itself coming out of document called Guide to Cost-Benefit Analysis of Investment Projects and actually from its version for programing period 2014 – 2020. This document was published by European Commission [1]. Some authors study and adjust this methodology with aim to its use in response to feasibility studies and building of new roads and corridors as del Giudice and collective [2]. Or aim their research to perspective in transportation of specific state as MacKie in paper called Cost-Benefit Analysis in Transport: A UK perspective [3].

J Korytárová and P Papežíková deal detailed with this problematics in case of extensive projects using Cost-Benefit Analysis in their paper called Assessment of Large-Scale Project Based on CBA. [4]

In the Czech Republic is effort to the nearest connecting to Guide to Cost-Benefit Analysis of Investment Projects [1] in assessment of transportation infrastructure. This is conditioned by regulation of European Commission. Actually the document called Methodology for assessment of Economical Effectiveness [5], related to assessment of transportation projects, exists. This document is published by Road and Motorway Directorate of the Czech Republic (ŘSD ČR). This institution has in its jurisdiction the most important road corridors in the Czech Republic. This methodology is temporary document, before implementation of final methodology. This final methodology should be released in 2017.

But our attention should be focused mainly on risks of projects, which is able to affect project itself by higher costs, or its infeasibility in reality. This all is coming out of imminent risks. Andrei Soeanu and collective pursued more attention to problematic of risks in paper Transportation risk analysis using probabilistic model checking. [5]

J Korytárová together with B Pospíšilová describes risks in framework of Cost-Benefit Analysis in paper called Evaluation of Investment Risks in CBA with Monte Carlo Method [6]. This describes the relation between default cash flow and affected cash flow by risk. In this case they also handle with its impacts to principle of 3E (Economy, Effectiveness and Efficiency). This paper output is connected to decision making bounded to risk analysis and its managing.

Many of authors aim its research to specific branch of risks connected with environment. This risk impacts are described in paper aimed to environmental impact assessment of transportation projects inscribed by E Igondova, K Pavlickova and O Majzlans called The ecological impact assessment of a proposed road development (the Slovak approach) [8]. This paper describes main ecological risks of transportation infrastructure projects.

Risks are usually similar in assessment of projects in transportation infrastructure. The most emphasis is given to analysis of risks bound to increasing of total costs and extension of construction time. This paper is based on this problematics and its aim is to specify new and detailed important risks.

3. Methodology

In the assessment of investment projects of transportation infrastructure is used the methodology of Directorate of Roads and Motorways of the Czech Republic, consequential from Guide to Cost-Benefit Analysis of Investment projects, as stated above. In one of the most important chapter, risk analysis, ongoing sensitivity analysis, quantitative and qualitative analysis are defined.

Risks for detailed analysing and their choice are misleading. In feasibility studies it is not often explained, how the choice of risks was carried out. We can assume that the choice is based on subjectivity and business as usual. The most often risks are related to extension of construction time or increase of total costs of projects. This is proved by searches of realised project assessments.

Especially for projects in transportation we should take into the consideration the environmental risk, because transportation construction can severely disrupt ecosystem, the integrity of landscape, bio corridors etc. The aim of this paper is to refer on threats connected with environment and their impact on results of project assessment and project realisation itself.

Exactly in case of project assessment the environmental impacts are implemented in external costs. This paper applies methodology Monte Carlo connected to one part of total costs. Total costs are except external costs consisted from investment and operation costs, costs of travel time and costs from car accidents.

For thus chosen risk variable the appropriate setting of Monte Carlo Methodology (distribution, deviation etc.) is chosen. Following simulation shows relations of economic indicators.

All of this will be done on the case study, which is developed for one real project and its most effective variant (based on conducted feasibility study and verified variants).

4. Results and discussion

For this paper the most important part of Cost-Benefit Analysis is the economic analysis. Inputs for economic analysis are costs/benefits from basic and design variant of the technical-economic study, used as case study, of project "I/13 Děčín – Nový Bor." [9] This paper works with summary of discounted financial cash flow.

In original case study as risk variables for the analysis were used change in construction costs and change of total benefits (and costs) of project. In detail, new risk variables are sought among external costs, otherwise the external costs are new risk variable. External costs consist of several parts: valuation of air impacts, valuation of noise impacts and valuation of CO_2 pollution. In the case study several design variants were solved:

- option S1: transfer of the road Folknáře junction and bypass of village called Markvartice,
- option S2a: transfer of the road Děčín – Manušice,
- option S2b: transfer of the road Děčín – Markvartice – Volfartice – Manušice,
- sub-option S3a: transfer of the road Folknáře junction and Děčín – Manušice,
- sub-option S3b: transfer of the road Folknáře junction + bypass of Markvartice + transfer of the road Děčín – Markvartice – Volfartice – Manušice (without transfer in village Huntířov).

Considering to construction and assessment was selected variant S2a (transfer of road Děčín – Manušice), which includes the carrying out of transportation infrastructure through Central Bohemian Uplands. Central Bohemian Uplands is nature reserve, and it is the reason, why it could be really risky option. There are threatens of disruption of local ecosystem, there could be impact on bio corridors and biotopes, which weren´t included in external costs.

With regard on new risk variables we take into account a possibility of change in external costs/benefits. This change means the possible impact on environment. We can consider this paper like an appropriate adjustment for assessment of transportation infrastructure, especially in places with preserved nature and ecosystem.

4.1. Analysis setting

Analysis method Monte Carlo is used for evaluation of changes in the assessment. For this method triangular distribution of external costs in basic and design variant of project is set. Triangular distribution is selected as best for this type of analysis. Limits of this distribution are set to anticipated minimum -10% and anticipated maximum +30% against likeliest value (based on the basic and design external costs in summary of discounted cash flows in the case study).

Table 1. Input values for the triangular distribution and assumption [own elaboration].

	Basic variant			Design variant		
(mil. CZK)	Lower limit	Likeliest	Upper limit	Lower limit	Likeliest	Upper limit
Value	9 456.208	10 506.898	13 658.97	8 613.769	9 570.855	12 442.11

Triangular distribution represents random variables in basic and design variant of the project. The correlation coefficient is set for these random variables to the value 0.5, which is anticipated as suitable value for this economical issue.

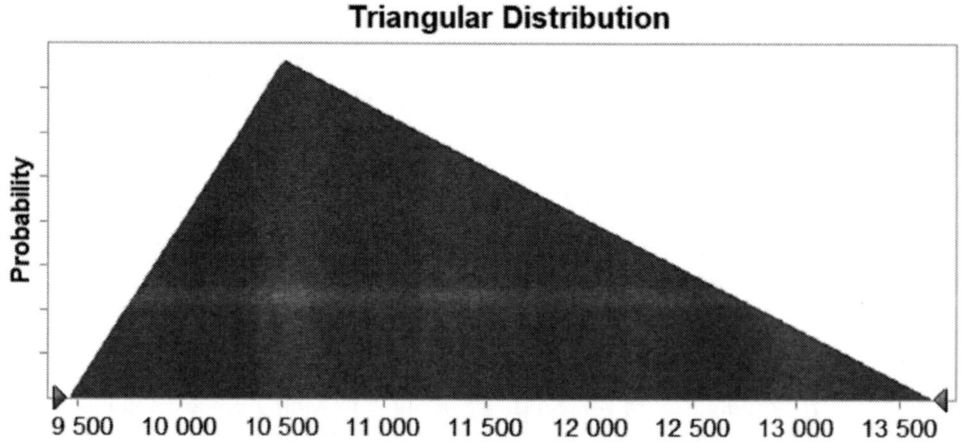

Figure 1. This is a figure which represents a triangular distribution set for assumption of external costs in basic variant of a project assessment.

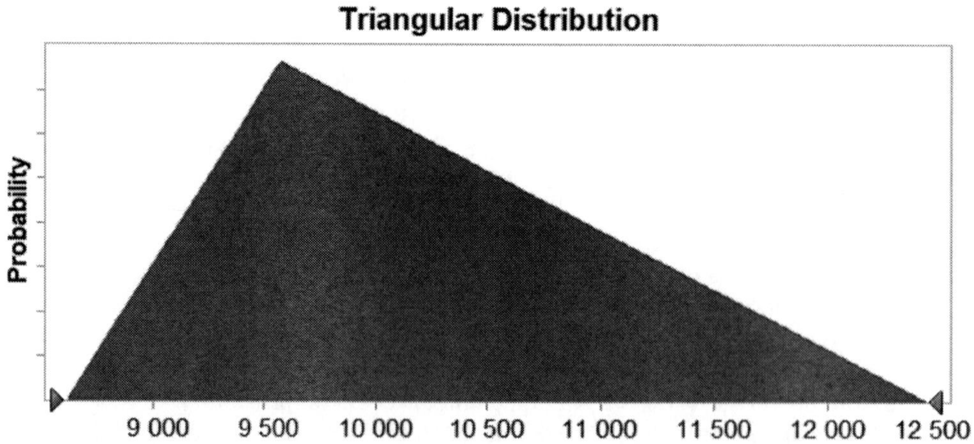

Figure 2. This is a figure which represents a triangular distribution set for assumption of external costs in designed variant of a project assessment.

IOP Conf. Series: Materials Science and Engineering **236** (2017) 012112 doi:10.1088/1757-899X/236/1/012112

For the Monte Carlo analysis are set 1 000 000 trials, with confidence level 95.00% and there are set two assumptions (external costs), and correlation between them. Then there are three forecasts, which are columns representing Benefit Cost Ratio, Net Present Value of project and Net Present Value of separated External costs.

4.2. Analysis outputs
Output values of the Monte Carlo analysis are impacts of risk variables (external costs) to potentially raise/decrease against default values in case study of investment project.
The following paragraphs and figures of the output of analysis show us the impact of changes of external costs/benefits caused by triangular distribution on default external costs.

4.2.1. Benefit Cost Ratio (BCR). Results of indicator BCR show, how can be the default values of this indicator affected by the Monte Carlo analysis. The results aren't alarming, but there is a 10% probability, that the project may not be approved from the perspective of indicator BCR. Minimal and maximal value of the indicator BCR reach sufficient reserve against potential environmental risk.

Table 2. Impact on the economic indicator Benefit Cost Ratio [own elaboration].

(mil. CZK)	Base case	Mean	Median	Standard deviation	Coefficient of variation	Minimum	Maximum
Result	1.574	1.591	1.587	0.239	0.1502	0.645	2.561

4.2.2. Net Present Value for separated external costs/benefits. Another monitored economic indicator is Net Present Value. As risk variables are contemplated only external costs/benefits and achieved values are marked risk variables of investment projects in transportation infrastructure. In this case there is evident impact. Default value of case study (936.04 mil. CZK) is considerably impacted in both sides, positive and negative. Minimum value achieved with analysis Monte Carlo and with assumption on basic and design variant, reached -2.361.11 mil. CZK and maximum value reached +4.442.88 mil. CZK. Negative values are achieved in 20 % of cases and decrease compared to default value is achieved in 50 % of cases.

Table 3. Impact on the separated economic indicator Net Present Value (External costs/benefits) [own elaboration].

(mil. CZK)	Base case	Mean	Median	Standard deviation	Coefficient of variation	Minimum	Maximum
Result	936.04	997.94	981.23	849.00	0.8508	-2 361.11	4 442.88

Value of Net Present Value of external costs/benefits, is seriously impacted by the Monte Carlo analysis and this impact is reflected in total Net Present Value, which is verified in next paragraphs.

4.2.3. Net Present Value of whole project assessment. Analysis Monte Carlo and its results have proper impacts to total Net Present Value. But the most important are changes of external costs/benefits. These changes have their impacts to deviation from default value of total Net Present Value of case study (2 038.04 mil. CZK). Minimum value of Net Present Value from analysis Monte Carlo decrease to value -1.259.11 mil. CZK and maximal increase to +5.544.88 mil. CZK. Net Present Value reaches negative in approximately 10 % of cases and decrease compared to default value is achieved in 50 % of cases.

IOP Conf. Series: Materials Science and Engineering **236** (2017) 012112 doi:10.1088/1757-899X/236/1/012112

Table 4. Impact on the economic indicator Net Present Value [own elaboration].

(mil. CZK)	Base case	Mean	Median	Standard deviation	Coefficient of variation	Minimum	Maximum
Result	2 038.04	2 099.94	2 083.23	849.00	0.4043	-1 259.11	5 544.88

Next graph shows total Net Present Value and process of distribution function, which response to normal distribution. Graph also shows distribution of reached values, lower and higher than default value of Net Present Value by analysis Monte Carlo.

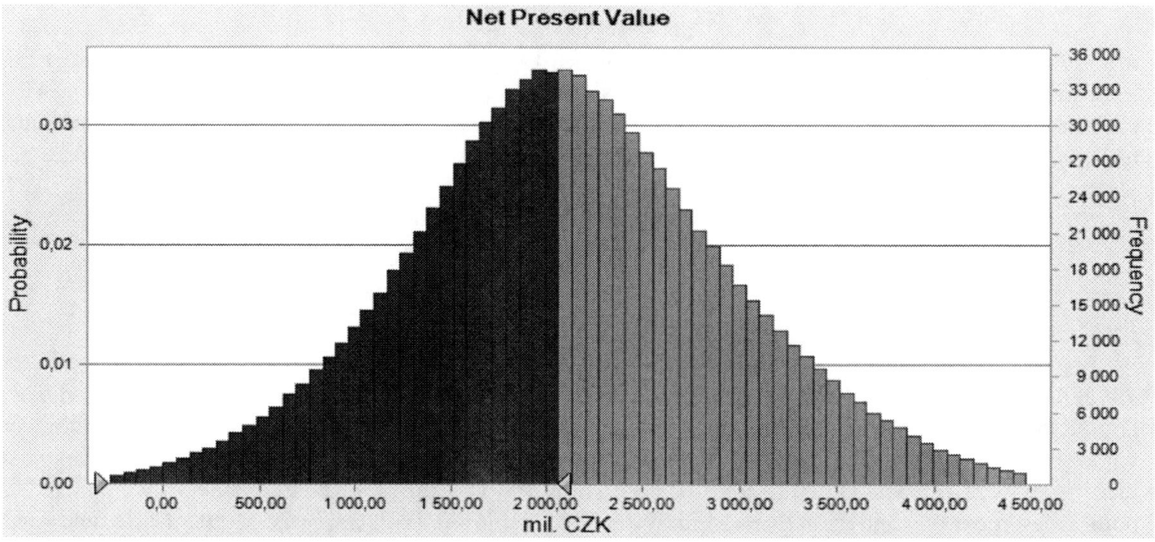

Figure 3. Normal distribution of Net Present Value, impacted by changes of external costs/benefits.

4.3. Discussion

There is a sufficient information based on the Monte Carlo analysis and from impacts of implementing of triangular distribution on default external costs/benefits in basic and design variant of the case study.

Impact of the risk variables (external costs/benefits) is considerable and this problematic is bearer of many other unknowns. These risk variables may not be directly exhaustively parts used in external costs/benefits, but there could be others, that aren't taken into account in external costs and benefits. Indicator of total Net Present Value and its evaluation shows strong impact of changes in external costs/benefits. This is approved by differences between analysis of outputs and default value of Net Present Value from case study. [9]

The threat of whole project assessment and realization of transportation investment project is inconsiderable. Threat arises from risk variables hidden in external costs/benefits. It is alarming, that the main parts of external costs/benefits are only air pollution impact, noise impact and CO_2 impact. Especially in cases as that one represented in case study, is necessary to focus in detailed assessment of environmental risk variables and its impacts.

Ecosystem damage and disruption of corridors in construction process or in process of usage in sensitive locations are examples of neglected external impacts. This is connected with possibility of damage of important biotopes. After construction process is nature able to revitalize any of environmental damages, but not all of them and not at all if people will be unscrupulous.

The most important question is connected with future assessment of this problematics, which can contain more detailed Monte Carlo analysis.

Detailed Monte Carlo analysis could be able to specify risk variables in detailed perspective of external costs/benefits. Future assessment could contain evaluation of biotopes, which are neglected in default project assessment in case study. For example: change of quality of biotopes and inspiration for their evaluation could be found in methodology of evaluation of biotopes by AOPK Czech Republic [10].

5. Conclusion

Finding of new risk variables in assessment of transportation investment projects was aim of this paper. These variables were predicted and identified in one part of financial cash flow of economic assessment. This part is external costs/benefits. In this part is hidden a lot of risk variables bound with ecosystem and ecological impacts. Assessment has to be more focused on impacts to landscape and nature. Permanently marginalization of these impacts could bring important problems to project realisation and natural incalculable consequences.

It´s necessary to be more focused on risk variables bound to environment and on solving of its reasons and impacts (financial, spiritual or material). This paper fulfilled its purpose and pointed to issue which need to be developed and not neglected. Assessment methodologies of investment projects are still developed and it´s important to edit them with regard to environmental risk variables.

Acknowledgement

This paper has been worked out under the project of the specific research at Brno University of Technology no. FAST-J-17-4636 Evaluation of investment plans in transport infrastructure.

References

[1] Sartori D, Catalano G, Pancotti C, Sirtori E, Vignetti S and Del Bo C 2014 Guide to Cost-Benefit Analysis of Investment Projects – Economic Appraisal Tool for Cohesion Policy 2014-2020 (European Commission - Directorate-General for Regional and Urban policy) http://ec.europa.eu/regional_policy/sources/docgener/studies/pdf/cba_guide.pdf p 346

[2] del Giudice V, Passeri A, Torrieri F and da Paola P 2014 Risk Analysis with Feasibility studies: An Application to Cost-Benefit Analysis for the Construction of a New Road *Applied Mechanics and Materials* **651-653** pp 1249-54

[3] MacKie P 2010 Cost-benefit analysis in transport: A UK perspective *OECD/ITF Round Table of on Improving the Practice of Cost Benefit Analysis in Transport*

[4] Korytárová J and Papežíková P 2015 Assessment of Large-Scale Project BASED on CBA *Procedia Computer Science* **64** pp 736-43

[5] Road and Motorway Directorate of the Czech Republic (ŘSD ČR) 2016 Methodology for assessment of Economical Effectivenes https://www.rsd.cz/wps/portal/web/technicke-predpisy/HDM-4

[6] Soeanu A, Debbabi M, Alhadidi D, Makkawi M, Allouche M, Bélenger M and Léchevin N 2015 Transport risk analysis using probabilistic model checking *Expert Systems With Applications* **42** pp 4410-21

[7] Korytárová J and Pospíšilová B 2015 Evaluation of Investment Risks in CBA with Monte Carlo Method *Acta Universitatis Agriculturae et Silviculturae Mendelianae Brunensis* **63** pp 245-51

[8] Igondova E, Pavlickova K and Majzlan O 2016 The ecological impact assessment of a proposed road development (the Slovak approach) *Environmental Impact Assessment Review* **59** pp 43-53

[9] AF-CityPlan s.r.o. 2014 *I/13 Děčín – Nový Bor: Technicko-ekonomická studie, optimalizace vedení trasy /Technical-Economic study, optimisation of routing/*

[10] Seják J, Cudlín P, Dejmal I, Petříček V and Černý K 2010 *Metodika oceňování biotopů Agentury ochrany přírody a krajiny ČR /Methodology of evaluation of biotopes according to the Agency for Nature and Landscape Protection/* http://users.prf.jcu.cz/kucert00/HABIT/METODIKA_OCENOVANI_BIOTOPU.pdf

IOP Conf. Series: Materials Science and Engineering **236** (2017) 012113 doi:10.1088/1757-899X/236/1/012113

Research of cost aspects of cement pavements construction

Artem Bezuglyi[1], Sergii Illiash[2] and Oleksandr Tymoshchuk[3]

[1]Deputy of director of DerzhdorNDI SE, Peremohy Avenue 57, Office 1402, Kyiv, Ukraine, 03113
[2]Head of Department of DerzhdorNDI SE, Peremohy Avenue 57, Office 1407, Kyiv, Ukraine, 03113
[3]Researcher of DerzhdorNDI SE, Peremohy Avenue 57, Office 1607б, Kyiv, Ukraine, 03113

E-mail: vtdr@ukr.net

Abstract. The tendency to increasing traffic volume on public roads and to increased axle loads of vehicles makes the road scientists to develop scientifically justified methods for preserving the existing and developing the new transport network of Ukraine. One of the options for solving such issues is the construction of roads with rigid (cement concrete) pavement. However, any solution must be justified considering technical and economic components. This paper presents the results of the research of cost aspects of cement pavements construction.

1. General instructions

The dynamic growth of traffic volume on public roads and increased axle loads of vehicles that has been observed in Ukraine over the past decade requires continuous development of the road network and upgrading the road pavement designs. Unavailability of highways to traffic flow demands results in speed reduction, increased wear of motor vehicles and growth of repair costs, causes increased fuel consumption and reduced traffic safety level, as well as an increased number of traffic accidents.

Modern road design and construction technologies are based on two alternative competing pavement types - asphalt and cement pavements.

Significant reduction in service life of asphalt pavements which is caused by the dynamic growth of traffic volume on major trunk roads stimulates a constant search for an alternative to the construction of highways with asphalt pavement. One of the alternatives to solve such issue is the construction of roads with rigid (cement concrete) pavement.

In order to determine the feasibility of using a particular type of pavement for the construction of the highway a feasibility comparison study of different design options was fulfilled.

1.1. Calculation of pavement design

Regardless of the type, road pavement is to be calculated considering the composition of traffic flow, predicted traffic volume by the end of its service life, as well as geological and climatic conditions.

Non-rigid type of pavement (asphalt concrete) is designed considering the reliability of pavement structure.

Design and calculation of non-rigid pavement type is fulfilled in accordance with the Departmental Construction Norms VBN V.2.3-218-186 [1].

IOP Conf. Series: Materials Science and Engineering **236** (2017) 012113 doi:10.1088/1757-899X/236/1/012113

Design of rigid pavement type (cement concrete) is fulfilled in accordance with VBN V.2.3-218-008 [2]. The calculation includes the definition of design and standard loads, the size of basic structural elements (thickness and length of slabs, thickness of foundation layers, reinforcement plates and layers, distance between expansion joints) for different types of pavement, road categories, for any traffic loads, soil and climatic conditions.

1.2. Methodology of pavement designs comparing assessment

Defining the effectiveness of different pavement design options is carried out by comparing labor and material costs incurred throughout the comparative period *(T)*.

Pavement design options under consideration are distinguished by the design type and by their operating maintenance methods during the period T (the distinctive features of operating maintenance of rigid and non-rigid pavement types are taken into account).

Evaluation of the effectiveness of placement and operation of road pavement designs is fulfilled per 1 running km based on the parameters and operation conditions of a certain road category.

As a criterion for assessing the comparative effectiveness of placement and operation of various road pavement types the minimum value of integrated costs B_{gen} is used which includes:

- The cost of new construction of road pavement
- Costs for capital and current repairs of road pavement
- Costs for operational maintenance of road pavement

Calculation of reduced costs in the i-th year starting with the next year following the beginning of the new pavement construction is carried out considering the risk-free discount rate E (in relative units).

For comparison of the costs for the new construction, repair and operational maintenance of pavement designs of various types the following formula is used:

$$B_{gen} = B_{cons} + B_{Mj} \times (1+E)^t + \sum_{i=1}^{n} \left[(B_{Mid} \times (1+E)^t) \times n \right] + \sum_{i=1}^{m} \left[(B_{Mid+Op} \times (1+E)^t) \times m \right] + \sum B_{User} \times (1+E)^t \quad (1)$$

B_{cons} – the cost of new construction of 1 km of pavement;
B_{Mj} – the costs for major repairs of pavement;
B_{Mid} – the costs for current mid-life repairs of pavement;
n – the number of current mid-life repairs performed during the comparative period T;
B_{Mid+Op} – the costs for current mid-life repairs and operating maintenance of pavement;
B_{User} – the losses of road users when performing repair works.
m – the number of years of operating maintenance of pavement of the highway when using the selected option of works fulfillment;
E – the risk-free discount rate in relative units;
$(1 + E)$ – the discount rate costs.

Table 1. Initial data.

№	Initial data	Road pavement design type	
		Non-rigid	Rigid
1	Beginning of works fulfillment, year	2014	2014
2	Road category	II	II
3	Initial traffic volume on the section, vehicles/day	5000	5000
4	Road-climatic zone	У-II	У-II
5	Interrepair time for roads (year):		
	- current mid-life repair after construction (second current repair)	6 (17)	9 (18)
	- capital repair	12	21
6	Growth of traffic volume, %	4	4
7	Discount rate, %	5	5

Considering the initial traffic volume, the alternative options of non-rigid and rigid pavements design types were calculated.

Non-rigid pavement design type (Figure 1) is adopted on the basis of the requirements of VBN V.2.3-218-186 [1]. Rigid pavement design type (Figure 2) is adopted on the basis of the requirements of VBN V.2.3-218-008 [2].

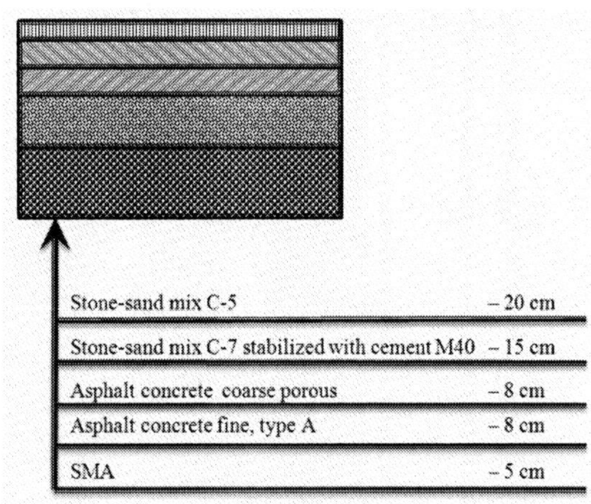

Figure 1. Non-rigid pavement design type.

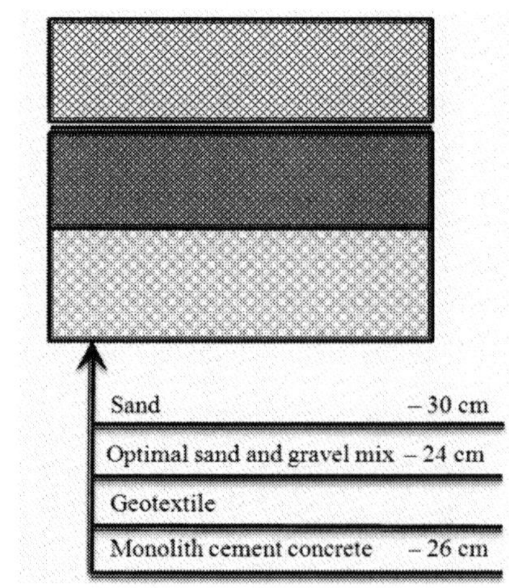

Figure 2. Rigid pavement design type.

The costs for the construction and maintenance of subgrade, placing and arrangement of road equipment elements during the feasibility comparison study were not taken into account because they are the same for both options.

Process sequence of works performance, the associated machines and mechanisms were adopted to meet the requirements of existing standards and regulations [3] and flow charts.

1.3. Calculation of costs for capital and current repairs of pavement design

1.3.1. Non-rigid pavement design type. According to Table 2 VBN G.1-218-050 [4], the first current mid-life repair after the new construction is carried out after 6 years of the highway operation.

According to Table 1 VBN G.1-218-050 [4], capital repair of pavement design is performed after 12 years of the highway operation.

The next current mid-life repair is performed after 5 years of operation after capital repair, according to Table 2 VBN G.1-218-050 [4], that is the 17th year of the highway operation.

*1.3.2. Rigid pavement design type.*Process sequence of works performance is adopted to meet the requirements of the Guidelines MR V.3.2-218-03449261-674 [5].

According to Table 2 VBN G.1-218-050 [4], the first current mid-life repair after the new building is to be performed during the 9th year of the highway operation.

The next current mid-life repair is to be performed in 7 years, according to Table 2 VBN G.1-218-050 [4] after the first current mid-life repair- that is the 16th year of the highway operation.

Major repair of pavement is not taken into account, as a comparative term is 20 years.

Based on the initial data, consolidated construction costs for the construction and functioning of different road pavement design types are defined. The calculation results for non-rigid and rigid pavement types are shown in Tables 2 and 3.

Table 2. Total cost of the construction and maintenance of 1 km of non-rigid pavement design type.

Year	Discount rate	Costs. ths UAH				Total for the year	Accumulation of costs
		B_{cons}	B_{Mj}	B_{Mid}	B_{Mid+Op}		
0	1.0	9325.495	-	-	-	9325.495	9325.495
1	1.05	-	-	-	19.04	19.04	9344.535
2	1.103	-	-	-	21.001	21.001	9365.536
3	1.158	-	-	-	46.787	46.787	9412.323
4	1.216	-	-	-	49.13	49.13	9461.453
5	1.276	-	-	-	51.555	51.555	9513.008
6	1.34	-	-	1221.744	25.513	1247.257	10760.265
7	1.407	-	-	-	26.789	26.789	10787.054
8	1.477	-	-	-	59.676	59.676	10846.73
9	1.551	-	-	-	62.666	62.666	10909.396
10	1.629	-	-	-	65.817	65.817	10975.213
11	1.71	-	-	-	69.09	69.09	11044.303
12	1.796	-	4474.087	-	34.195	4508.282	15552.585
13	1.886	-	-	-	35.909	35.909	15588.494
14	1.98	-	-	-	79.999	79.999	15668.493
15	2.079	-	-	-	83.999	83.999	15752.492
16	2.183	-	-	-	88.2	88.2	15840.692
17	2.292	-	-	2089.729	43.64	2133.369	17974.061
18	2.407	-	-	-	45.828	45.828	18019.889
19	2.527	-	-	-	102.099	102.099	18121.988
20	2.653	-	-	-	107.19	107.19	18229.178
	Total	**9325.495**	**4474.087**	**3311.473**	**1118.123**	**18229.18**	-

Table 3. Total cost of the construction and maintenance of 1 km of rigid pavement design type.

Year	Discount rate	Costs. ths UAH				Total for the year	Accumulation of costs
		B_{cons}	B_{Mj}	B_{Mid}	B_{Mid+Op}		
0	1.0	9188.441	-	-		9188.441	9188.441
1	1.05	-	-	-	17.033	17.033	9205.474
2	1.103	-	-	-	17.893	17.893	9223.367
3	1.158	-	-	-	22.048	22.048	9245.415
4	1.216	-	-	-	23.152	23.152	9268.567
5	1.276	-	-	-	24.295	24.295	9292.862
6	1.34	-	-	-	60.632	60.632	9353.494
7	1.407	-	-	-	63.663	63.663	9417.157
8	1.477	-	-	-	66.831	66.831	9483.988
9	1.551	-	-	125	29.53	154.53	9638.518
10	1.629	-	-	-	31.015	31.015	9669.533
11	1.71	-	-	-	32.558	32.558	9702.091
12	1.796	-	-	-	34.195	34.195	9736.286
13	1.886	-	-	-	35.909	35.909	9772.195
14	1.98	-	-	-	37.698	37.698	9809.893
15	2.079	-	-	-	94.07	94.07	9903.963
16	2.183	-	-	-	98.775	98.775	10002.738
17	2.292	-	-	-	103.707	103.707	10106.445
18	2.407	-	-	2329.913	45.828	2375.741	12482.186
19	2.527	-	-	-	48.113	48.113	12530.299
20	2.653	-	-	-	50.512	50.512	12580.811
	Total	**9188.44**	-	**2454.913**	**937.457**	**12580.8**	-

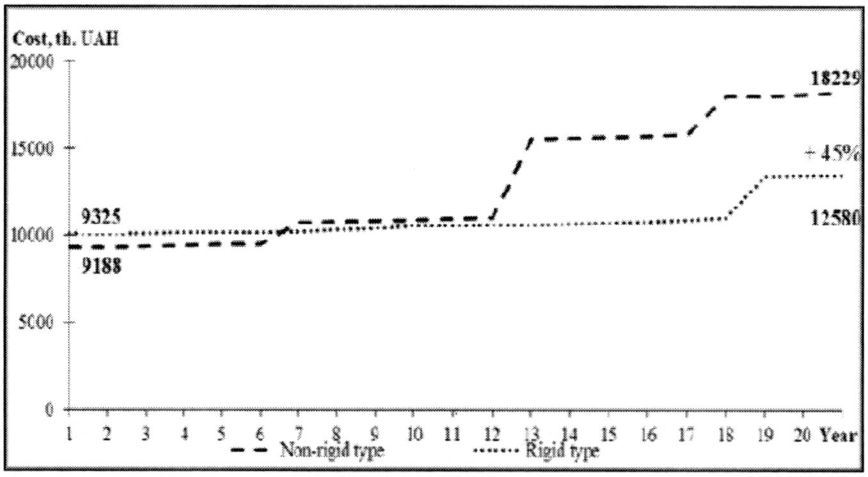

Figure 3. Comparison of new construction and maintenance costs of 1 km of pavement of rigid and non-rigid design types.

Graph in the Figure 3 shows that considering the average prices of materials as of the third quarter 2016, the construction of cement pavements is economically sounder because savings at the end of the calculation period will make 45%.

Starting with the second quarter of 2012, DerzhdorNDI SE monitors changes in prices of basic road construction materials. Dynamics of price of asphalt and concrete mixes during the period from 2012 through the third quarter of 2016 are shown in Figures 4 and 5.

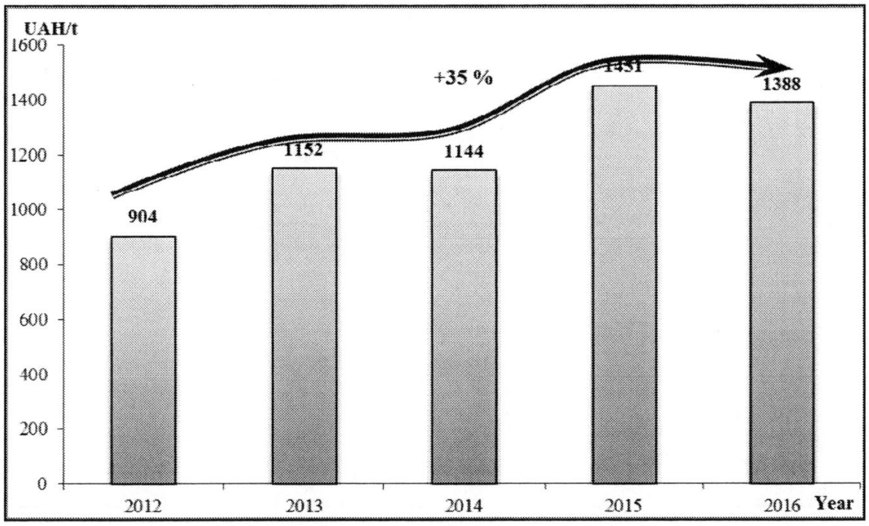

Figure 4. Dynamics of prices of asphalt mix.

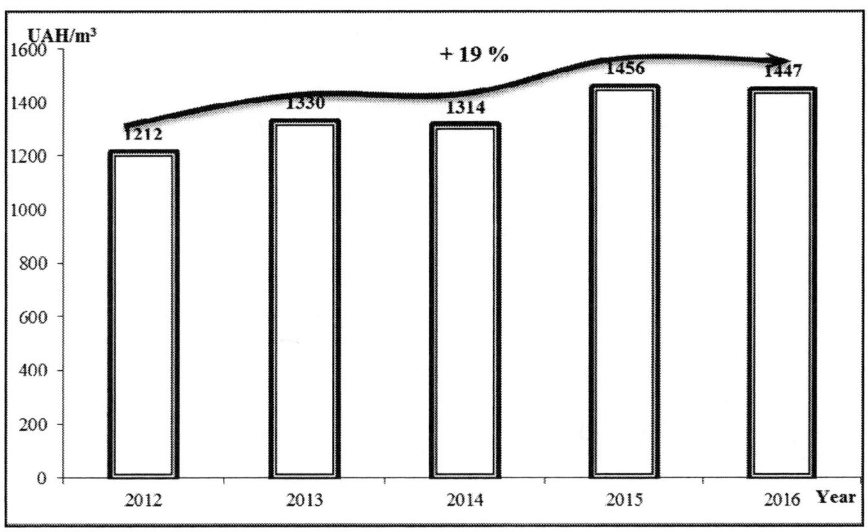

Figure 5. Dynamics of prices of concrete mix.

To compare the effects of the material component price on the final cost of construction and maintenance of pavement designs, 2 scenarios have been modeled that are shown in Figures 6 and 7.

IOP Conf. Series: Materials Science and Engineering **236** (2017) 012113 doi:10.1088/1757-899X/236/1/012113

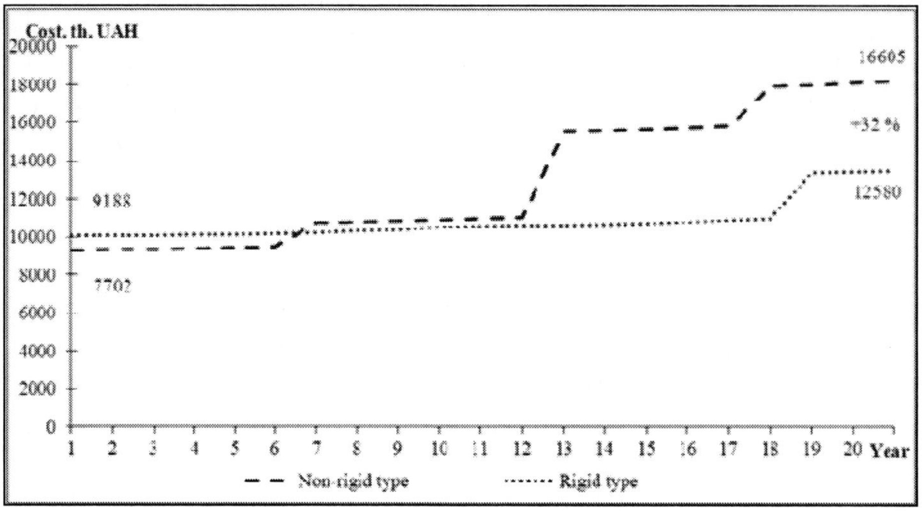

Figure 6. Cost of construction and maintenance of 1 km of various pavement design types (cost of asphalt mix is reduced by 35%).

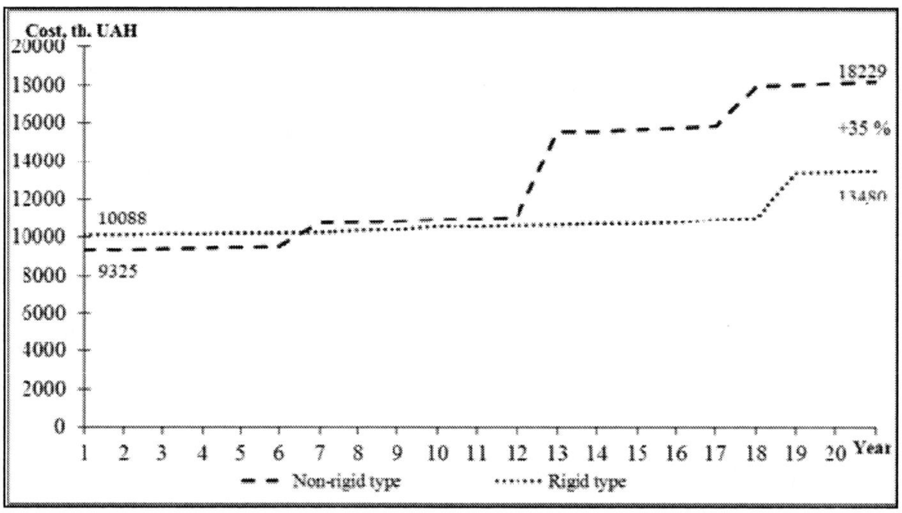

Figure 7. Cost of construction and maintenance of 1 km of various pavement design types (cost of concrete mix is increased by 30%).

With the reduced cost of asphalt mix by 35% (Figure 6), the situation changed dramatically. The efficiency of the construction and operation of cement pavements starts with the 12th year only. But at the end of the calculation period savings will make about 30%.

Increased cost of concrete mix by 30% results also in the growth of cement pavements construction costs compared to asphalt ones (Figure 7). But ultimately, there is saving due to lower repair and maintenance costs of cement pavements.

IOP Conf. Series: Materials Science and Engineering **236** (2017) 012113 doi:10.1088/1757-899X/236/1/012113

Conclusions

Thus, the following conclusions can be drawn on the basis of the conducted analysis and the results of calculation:

- The construction cost of any type of pavement design is too dependent on the cost of construction materials and the share of local materials used
- In the long term (10 - 20 years) the construction of cement pavements is more expedient. It should also be taken into consideration that cement pavements can bear heavier loads than asphalt pavements which provides additional social and economic efficiency
- Taking into account the fact that the price of bitumen is more tied to the exchange rate, the use of cement is the best solution
- In each specific case, when choosing the type of pavement design, the pre-design feasibility and comparison study should be fulfilled which will allow proper assessment of the efficiency of the opted pavement design type use.

References

[1] Kyiv National Transport University 2004 Споруди транспорту. Дорожній одяг нежорсткого типу (in Ukrainian) /Transport facilities. Non-rigid pavement design type./ *(VBN V.2.3-218-186-2004)*

[2] Kyiv National Transport University 2010 Проектування і будівництво жорстких та з жорсткими прошарками дорожніх одягів (in Ukrainian) /Transport facilities. Design and construction of rigid pavements and of pavements with rigid layers./ *(VBN V.2.3-218-008-97)*

[3] Kyiv SE - State Road Research Institute named after M.P. Shulgin 2016 Автомобільні дороги. Частина I. Проектування. Частина II. Будівництво (in Ukrainian) /Transport facilities. Highways. Part I. Design. Part II. Construction/ *(DBN V.2.3-4:2015)*

[4] Kyiv SE - State Road Research Institute named after M.P. Shulgin 2002 Міжремонтні строки експлуатації дорожніх одягів та покриттів на автомобільних дорогах загального користування (in Ukrainian) /Interrepair terms of road pavements and surfaces operation on public roads and of concrete pavements on the highways./ *(VBN G.1-218-050-2001)*

[5] Kyiv Ukrainian State Production and Technological Organization Ukrdortekhnologia 2007 Методичні рекомендації з ремонту цементобетонних покриттів автомобільних доріг (in Ukrainian) /Guidelines for feasibility comparison of pavement structures/ *(MR V.3.2-218-03449261-674:2007)*

[6] Kyiv SE - State Road Research Institute named after M.P. Shulgin 2012 Порядок визначення вартості капітального ремонту автомобільних доріг загального користування (державного та місцевого значення) (in Ukrainian) /The procedure for determining the cost of major repairs of public roads (of state and local importance]/ *(SOU 42.1-37641918-050:2012)*

[7] Kyiv SE - State Road Research Institute named after M.P. Shulgin 2017 Правила визначення вартості робіт з експлуатаційного утримання автомобільних доріг загального користування, мостів та інших транспортних споруд (in Ukrainian) /Rules of costs determining of operating maintenance works of public roads, bridges and other transport facilities/ *(SOU 42.1-37641918-085:2012)*

[8] State Road Agency of Ukraine 2014 Методичні рекомендації з техніко-економічного порівняння конструкцій дорожнього одягу різного типу (in Ukrainian) /Guidelines for feasibility comparison of pavement structures of various types/ *(MR V.2.3-03450778-847:2014*)

Institute of Physics
Dirac House, Temple Back
Bristol BS1 6BE UK

ISSN: 1757-8981
ISBN 978-1-5108-4810-8

9 781510 848108

CURRAN ASSOCIATES INC.
proceedings.com